Lecture Notes in Computer Science 686

Edited by G. Goos and J. Hartmanis

J. Mira J. Cabestany A. Prieto (Eds.)

New Trends in Neural Computation

International Workshop on Artificial Neural Networks, IWANN '93
Sitges, Spain, June 9-11, 1993
Proceedings

Springer-Verlag
Berlin Heidelberg New York
London Paris Tokyo
Hong Kong Barcelona
Budapest

Series Editors

Gerhard Goos
Universität Karlsruhe
Postfach 69 80
Vincenz-Priessnitz-Straße 1
W-7500 Karlsruhe, FRG

Juris Hartmanis
Cornell University
Department of Computer Science
4130 Upson Hall
Ithaca, NY 14853, USA

Volume Editors

José Mira
Dept. Informática y Automática, Universidad Nacional de Educacion a Distancia
Senda del Rey s/n, E-28040 Madrid, Spain

Joan Cabestany
Dept. de Ingeniería Eléctrica, Universidad Politécnica de Catalunya
Campus Nord. Edifici: C4, Gran Capitán s/n, E-08034 Barcelona, Spain

Alberto Prieto
Dept. de Electrónica y Tecnología de Computadores, Facultad de Ciencias
Universidad de Granada, E-18071 Granada, Spain

CR Subject Classification (1991): B.3.2, B.7.1, C.1.3, C.3, C.5, F.1.1-2, F.2.2, H.3.m, I.2.6-10, I.4.5, J.2, J.7

ISBN 3-540-56798-4 Springer-Verlag Berlin Heidelberg New York
ISBN 0-387-56798-4 Springer-Verlag New York Berlin Heidelberg

Printed in Germany

Typesetting: Camera ready by author
Printing and binding: Druckhaus Beltz, Hemsbach/Bergstr.
45/3140-543210 - Printed on acid-free paper

Preface

Neural computation arises from the capacity of nervous tissue to process information and accumulate knowledge in an intelligent manner. Perception, learning, associative memory, self-organization, fault tolerance and self-repair, planning, reasoning and creativity are basic properties of biological systems computed by the neural tissue. By way of contrast, computational machines with von Neumann architecture and conventional external programming, including knowledge based systems, have encountered enormous difficulties in duplicating these functionalities.

In an effort to escape this impasse, the scientific community has turned its attention to the anatomy and physiology of neural networks and the structural and organizational principles at the root of living systems. This has given rise to the development of the field of Artificial Neural Networks (ANN), where computation is distributed over a great number of local processing elements with a high degree of connectivity and in which external programming is replaced with supervised and non-supervised learning.

The papers presented here are carefully reviewed versions of the talks delivered at the International Workshop on Artificial Neural Networks (IWANN '93) organized by Universities of Cataluña (Politécnica, Autónoma and of Barcelona) and the Spanish Open University at Madrid (UNED) and held in Sitges (Barcelona), Spain, from 9 to 11 June, 1993. More than 160 papers were submitted, of which 111 were accepted for oral presentation and are included in these proceedings. Extended papers originated from invited talks related to the main topics considered are also included as introductions to the corresponding sections.

This workshop has been organized in cooperation with the Spanish RIG of the IEEE Neural Networks Society, and the IFIP WG 10.6, and has been sponsored by the Spanish CICYT, the Catalan CIRIT, and the organizing universities.

Collaboration of the Spanish chapter of the IEEE Computer Society, the UR&RI Communication chapter of IEEE, and the AEIA (Spanish Association for Computing and Automation) has been obtained.

We would like to thank all the authors as well as all the members of the International Program Committee for their labour in the production, evaluation and refinement of the papers. Furthermore, the editors would like to thank Springer-Verlag, in particular Alfred Hofmann, for excellent cooperation.

The papers published in this volume present the current state in neural computation and are organized in seven sections:

- Biological perspectives,
- Mathematical models,
- Learning,
- Self-organizing networks,
- Neural software,

  - Hardware implementation,
  - Applications:
    - Signal processing and pattern recognition,
    - Communications,
    - Artificial vision,
    - Control and robotics,
    - Other applications).

We begin with biological perspectives, including studies of the anatomical and physiological roots of neural computation. The biophysical level is enhanced and some claims on more realistic models of natural computation are included.

Thus far we have examined biology. Now we turn our attention to the world of mathematical models and organizational principles. A strong theoretical perspective is needed to seek organizational knowledge that will enable us to reproduce through synthesis some of the properties observable in living beings. Self-organization, continuous learning, and genetic algorithms are the topics more frequently addressed.

Learning is the key to neural computation. If we say that learning (self-programming) should substitute external programming, we must develop algorithms and methods of local learning comparable in clarity, completeness, and efficiency to those in conventional computation. It is true that local training requires more complex connections and redundant computations, but it simplifies the global design, includes intrinsic parallelism, and goes closer to biology. As long as the learning algorithms are executed in a general purpose computer separated from the network we are far from biology.

The next step in the proceedings is related to the development of neural software (languages, tools, simulations and benchmarks) and hardware implementations. Programming environments are usually classified as application-oriented, algorithm-oriented, and general programming systems. The simulation of neural networks in conventional computers can only be considered as a first step in the training and evaluation of models, architectures, and algorithms on the pathway towards intrinsically parallel hardware implementations.

The implementation of neural networks depends directly on which neural model and learning algorithm we seek to implement. In other words, it is necessary to distinguish between (a) what we want to implement and (b) how we do it. Once we have agreed on which computational model and what degree of autonomy we want to implement, the next step is how to do it. In all the cases the implementation can be analog, digital, or hybrid and is within a concrete technology (electronic or optical). There are two options, which we could call the simple model and the complex model.

In the first case, it is accepted that there is little autonomy. If the second option (complex model) is selected, we are forced to think in terms of neurocomputers, specifically designed for the implementation of neural networks with local computation, structural and functional parameters adjustment, and several modes of functioning (initialization, training/learning and use). Between this level (nothing in the host, all in the network) and the software simulations in conventional computers previously mentioned (nothing in the network, all in the host) there should be an ample range of intermediate situations (specific and general purpose neurocomputers) so that the closer we come to "all in the network", the closer we will be to the biological computation from which we drew inspiration.

The last part in the proceedings is related to applications. The basic question here is: what type of applications possess the computational requirements for the solution of which it would be advisable to use neural networks? Not all functions are capable of being distributed.

The majority of application tasks in neural computation can be formulated as multilayer classification functions in which a set of input configurations X = { Xm } associates itself to a set of classes Y = { Yn} after supervised or unsupervised learning. Signal and image processing and pattern recognition are the known examples in this line. Artificial vision, adaptive control, systems identification, and sensory-motor control loops are also adequate tasks to be solved using neural nets.

The most serious computational problem in the field of artificial neural nets (ANN) is the lack of theory, with direct and inverse constructive theorems. Given a specific computational family, which would be the map of individual functions and learning algorithms such that – when they operate linked by the data – they synthesize the global function)? Conversely, given a net of thousand of individual processors with local learning, which would be the global computation that emerges as a results of the cooperative integration of these local computations? It is clear that an enormous amount of work still remains to be done in neural computation, and this is a challenge for all of us.

Madrid, April 1993

J. Mira
J. Cabestany
A. Prieto

Contents

1. Biological Perspectives

2. Mathematical Models

3. Learning

4. Self Organizing Networks and Vector Quantizer

5. Neural Software

6. Hardware Implementation

7. Applications

7.1. Pattern Recognition and Signal Processing

7.2. Communication

7.3. Artificial Vision

7.4. Control and Robotic

7.5. Other Applications

7.4. Control and Robotic

7.5. Other Applications

BIOPHYSICS OF NEURAL COMPUTATION

K. N. Leibovic

State University of New York at Buffalo

Department of Biophysical Sciences

and

Center for Cognitive Science

120 Cary Hall

Buffalo, NY 14214-3005

Abstract:

This paper discusses neural computation in vision. Optimal design links structure-function relationships. This is evident at the cellular level with rod photoreceptor structure subserving detection in the presence of noise; and it is evident in the architecture of neural networks in which parallel computation is carried out in converging and diverging lines between different levels of the nervous system. Such an architecture makes possible some interesting schemes for information processing, including the computation of explicit parameters, resolution and reliability.

Keywords:

Biophysics, optimality, cells and networks, parallel processing, convergence-divergence.

Introduction:

Knowledge is an interconnected web, and so biophysical models of the brain serve as patterns for computational implementations while advances in communications and computers lead to theories, by analogy, of mental processes. In this paper I shall discuss some characteristics of neural computation in relation to vision. I have chosen vision because it is our primary sensory modality and it provides us with a richness of mental images expressed in memory and language. We know from our own experience that memory is largely visual - some people are said to have a photographic memory - and our language is full of visual metonyms for intelligent operations such as "expressing a view", "I see what you mean" and many others (Leibovic 1990).

A principle which seems to guide biological evolution is the optimization of structure and function. Therefore I shall take optimality as a theme of this presentation. I shall consider photoreceptor cells as an example of the biological components and I shall consider the architecture of neural networks for information processing. We can see at the cellular level how the components are designed for optimal detection in the presence of noise, while in neural networks we find a design that is particularly well-adapted for multivariable, parallel computations.

The Optimal Design of Photoreceptors:

The rods and cones of the retina absorb light and convert the photon energy into a neural response. These cells are more or less cylindrical and they present their circular cross sections to the pattern of light incident on the retina.

Design Considerations for Rod and Cone Diameters:

Figure 1 illustrates the arrangement of photoreceptors in the retina. It is clear that there are requirements for photon capture and image resolution, both of which are related to the aperture and cross section of the cells. A large aperture is desirable in dim illumination, but this would compromise resolution in daylight. In our case evolution has solved the problem without compromise by devoting the cones, especially in the fovea, to daylight vision, and the rods to vision in dim illumination. The diameter and spacing of our foveal cones are at the limit of optical diffraction, while our rods are capable of counting sinle photons.

In some regions of our retina (e.g., at 18° from the fovea) the rods are as slender and tightly packed as the cones in the fovea, while farther from the fovea cones are quite fat and tapered towards the tip. These variations in shape reflect different requirements subserving a foveal design, such as movement sensitivity in the periphery versus form perception in the center. Different species, in different ecological niches also show different designs. This is illustrated for rods in Figure 2: rods, like cones, can be fat or thin, long or short.

While high resolution requires a small cell diameter a large rod diameter can be advantageous not only in dim illumination, and in movement detection, but also in a noisy photic environment where the absorption of more than one photon is needed to reduce false alarms.

Considerations such as these determine photoreceptor diameter. There is an extensive literature devoted to this topic (see reviews by Snyder 1978, Pugh 1988). On the other hand, photoreceptor length has hardly been considered. I shall therefore devote the rest of this section to photoreceptor length, focussing especially on rods.

Significance of Length in Rod Function:

Responding with great sensitivity to one or a few photons, noise is an important factor in rod function. The photon energy is amplified by as much as 10^5 times in producing the neural response. This is mediated by a biochemical cycle and the longer the rod for a given diameter, the more thermal noise is produced. This argues for a short length. On the other hand, the efficiency of absorbing a photon depends on the number of absorbing layers, and this argues for a long rod. Noise and absorption efficiency are the primary variables in rod function. How can one achieve a compromise between them and optimize the design?

I have developed an optimization criterion to address this question (Leibovic 1990, Leibovic and Moreno-Diaz 1992). It is given by:

$$R = [1-\exp(-\alpha s)][1-(D/T)(s/L)]A-N(s/L) \quad (I)$$

where R = useful response signal, α = concentration of the light absorbing molecules (rhodopsin) x mean absorption cross-section, s = rod (outer segment) length considered as a variable, L = actual rod (outer segment) length, D = duration of the single photon response, T = interval between spontaneous thermal activations of rhodopsin (false alarms), A = RMS peak response to a photon, N = RMS continuous noise. D, T, A and N refer to a rod (outer segment) of length L.

When R is maximized with respect to s we find the optimal length.

The meaning of (I) is as follows: The rod response A is weighted by the photon absorption probability (first bracket on the right of (I)) and the probability that the response is not a false alarm (second bracket on right side of (I)). From this we subtract a weighted continuous noise term. Thus, we maximize with respect to length the absorption probability and minimize the false alarm probability against a baseline of a noise relative to the signal.

This criterion is designed especially for the responses to one or two photons, i.e., for those rods which are efficient single photon detectors. At higher light-intensities, photon absorption probabilities and false alarms are of decreasing importance, and the continuous noise becomes a small fraction of the response amplitude.

It should be noted that rod diameter is not explicitly included in (I). It is assumed that is already determined by considerations such as discussed earlier. Nevertheless, the diameter is implicit in (I),since D and T as well as N depend on rod diameter.

We have compared rod (outer segment) lengths, using this criterion, in a number of species, based on our own and published data from other sources. Our calculated optimal lengths are in very good agreement with the actual lengths in species in which rods respond sensitively to single photons. This is detailed in Table 1.

Table 1: Comparisons of Actual and Optimal Lengths

Species	Bufo (red rods)	Bufo (green rods)	Rana	Rabbit	Macaca
Actual Length (μm)	60-70	33	58	20	25
Optimal Length (μm)	65	31	60	22	28

On the other hand, our criterion does not apply to rods which are not designed as reliable, sensitive single photon detectors. This is the case, for example, in the mudpuppy, a relatively primitive North American salamander living in muddy waters. By nature, its photic environment is rather noisy and its rod responses to single photons are small although more intense stimuli evoke large responses. When false alarm probabilities are relatively high as here, single photon absorptions should be given little weight and then our criterion is not applicable.

Convergence and Divergence in Neural Nets:

I now turn from the optimal design of a cellular unit to operations in neural networks.

Computer scientists are working intensively on systems for parallel computation. Nervous systems, in addition to serial processing, use massive parallel computing as was already recognized by the fathers of the digital computer (von Neumann 1958).

Signal Flow Architecture:

What is the brain architecture for parallel computation? Neural tissue is organized into a series of layers, with signals transmitted from one layer to the next. This is serial computation. But in addition, there is processing of signals in parallel channels. There are channels for vision, hearing and so on and channels within the same modality. For example, motion, color and form are processed in segregated streams in the visual pathway. But, within each channel, and in integrating the information from different channels the architecture takes the form of convergence and divergence between different levels. Neighborhoods of cellular processors (receptive fields) at one level converge onto target cells at subsequent levels and the signals from one cell at one level diverge to many cells (responsive fields) at a subsequent level. At first sight this looks like a diffusion into vague haziness, and yet we are capable of accuracy and high resolution in our motor-sensory activities.

Receptive Field Tuning:

In man-made applications we may have narrowly tuned detectors which are optimized for a specific variable with values in a narrow range. In the nervous system, on the other hand, target cells and their associated receptive fields respond to several variables and they are broadly tuned For example, a target cell may respond broadly to light intensity, movement in a preferred direction, speed and position within its receptive field.

What are the advantages of separage channels and of broadly tuned, multivariable dotoctors?

The Need for Channels and Broad Tuning:

Consider the photoreceptor array on the retina. It can only convey a pointwise map of stimulation. To compute velocity, edges, shading and so forth requires receptive fields, and it makes sense to segregate the variables to which the receptive fields are tuned, into channels as long as they require serial processing. We can gain some insight into the need for broadly tuned channels by considering the example of color.

It is estimated from color matching experiments that we can distinguish some 10^6 different colors (see Boynton 1990). A dictionary of color names published by the US National Bureau of Standards includes about 7500 entries (Kelly and Judd 1976). Suppose that color was signaled by narrowly tuned cones. Then even for the lower figure of 7500 colors we would need that number of different cone types and for a 400 nm visible spectrum, each cone would have to be tuned to a 0.5 Å wavelength width. 7500 cones spaced 2 μm apart, as they are in the human fovea, would occupy an area of approximately 173 x 173 μm^2 or some 3% of the fovea. This would pose problems with the spatial resolution of colors. But the situation is even worse than this would suggest: From the uncertainty principle, a 0.5 Å tuning would have a position uncertainty of about 800 μm, which would cover more than 60% of the fovea. Clearly, this would be a very crude design.

How does the biological system handle this problem? I am not asking this in the general context of color vision, which is very complicated. My purpose is to illustrate a principle.

As we know there are three types of cone in our retina, the red (R), green (G) and blue (B) cones. Each of these responds to a broad range of wavelengths with overlapping tuning curves, as shown in Figure 3. Instead of 7500 different cones we identify color with 3! The actual neural computations for color perception remain elusive. But we can show that the information is available for unique wavelength identification. This is illustrated in Figure 4 for the R and G cones. The responses of these cells depend on light intensity and wavelength entering the eye. The horizontal and vertical axes are marked in units of wavelength. The curves based on the horizontal axis denote the hypothetical responses of an R cone for three different light intensities as a function of wavelength, and similarly the curves based on the vertical axis denote the responses of a G cone. For any wavelength, such as λ, there is a unique pair of (R,G) responses for a given intensity. Thus, when the R and G responses are known, so are intensity and wavelength. Wavelength mixtures give rise to some ambiguity, which is reduced by the presence of a third, the B, cone type. This ambiguity is a psycho-physical fact and is known as metamerism: physically different stimuli appear the same.

Similarly to the encoding of color and intensity of a spot of light we can determine its position on the retina.

Again, consider a broadly tuned, multivariable cell, specifically a target cell with an ON center and OFF surround receptive field, responding to position and intensity of a spot of light as illustrated in Figure 5. Similarly as for color, if we have two overlapping receptive fields as in Figure 5, the simultaneous responses of the corresponding target cells are unique for pairs of values of position and intensity (Leibovic 1969). We can have information on color, position and intensity by combining Figures 4 and 5 into a 3-dimensional picture. We know there are target cells with receptive fields as illustrated in Figure 5, whose center input comes from either R or G cones and whose surround input comes from G or R cones respectively (cf. e.g., Dow 1990), i.e., their responses are R-G in one case and G-R in the other. Thus, the information contained in Figures 4 and 5 is also contained in these responses.

The above examples illustrate some of the advantages of broad tuning and how overlapping, multivariable representations can contain accurate information in parallel streams. I shall now make this more precise in the context of convergence and divergence.

Fidelity of Information:

Recall that convergence-divergence implies an overlapping of receptive fields as illustrated in Figure 6. Each level in the nervous system has a two dimensional topology with converging and diverging lines between the levels. Consider a linear system for the purpose of illustration:

$$y_{ij} = \Sigma \Sigma K_{ij}^{rs} x_{rs} \qquad \text{(ii)}$$

y_{ij} are the target cell responses at a given level, x_{rs} the responses of the cells within the receptive field of y_{ij} and K^{rs} is a kernel, specifying how the outputs from x_{rs} are scaled. The suffices i, j, r, s index the positions of the respective x and y.

Due to the overlap of receptive fields (ii) is a set of simultaneous equations which has a unique solution for the x_{rs} when the y_{ij} are known. The conditions for this to hold are contained in two lemmas and a theorem (Leibovic 1988):

Lemma 1: The maximum number of distinct m x m "receptive fields" overlapping on a pixel is m^2.

Lemma 2: The maximum number of distinct "receptive fields" of size m x m on a "retina" of size n x n is $\{n-(m-1)\}^2$; $m < n$.

Theorem: A unique solution exists for the pixel grey levels in a n x n "retina" containing m x m receptive fields with maximum overlap, when the contributions of {2n-(m-1)} (m-1) of the boundary pixels is negligible and set to zero.

These results are independent of the size of the receptive fields and have the following consequences:

1. We can have receptive fields of any size without losing information, provided there is sufficient overlapping of receptive fields. The larger (smaller) they are the more (less) they will need to overlap.

2. There is not necessarily a saving of units at the level of the target cells, but we have acquired computing power which we do not have with point to point transmission. The price for this is the greater number of lines in the network as compared to pointwise transmission.

As regards receptive field size, in general, small fields, because of the smaller number of convergent lines, are more economical than large fields in both the "hardware" as well as the computations. They are thus more suitable for computing detail. Large fields are required for computing global features, and for this there is no need for high resolution and extensive overlap. In computational and machine vision, e.g., for scene analysis, there are advantages to performing computations over a range of scales, like using large and small receptive fields. The analysis I have given provides a rationale for such a scheme.

Reliability with Convergence - Divergence:

Finally, I want to briefly address the problem of reliability with convergence-divergence.

Suppose we want to determine speed and direction of a moving object. We could do this by having two receptive fields overlapping the area of interest on the retina and either (1) one target cell computing speed and the other direction, each with the required accuracy, or (2) both target cells broadly tuned and each responding to both direction and speed. In both cases different directions and speeds evoke different responses. In the first case, if one of the target cells was to fail we would completely lose one variable. In the second case the output of the remaining target cell would still encode directed motion, though this would be hazy: we would not know whether we were dealing with a fast moving object in a direction producing a weak response or slow movement in a direction with a vigorous response. If we wanted to insure against failure by duplicating the units, we would need to duplicate both units in case (1), but only one unit in case (2). For, as we saw in the earlier example of Figure 5, just two overlapping receptive fields are sufficient to specify two variables.

We see from this the superiority of multivariable broad tuning over single variable, narrow tuning.

If we wanted to have an error detection scheme, we could proceed as follows.

We can rewrite (ii) by forming vectors out of x_{rs}, y_{ij}, where $1 \leq r,s,i,j \leq n$,

$$\underline{y} = A\underline{x}$$

There can be errors due to malfunctioning in a y_k or in the connections a_{ij}. But assume that the probability of more than one error is negligible, and consider $\underline{y}$ first.

Construct

$$y_{n+1} \equiv \sum_{i=1}^{n}\sum_{j=1}^{n} a_{ij} x_j$$

by duplicating the lines a_{ij} and feeding them into the target cell y_{n+1}; and let S be the sum formed from the outputs y_i, $i=1,2,\ldots n$. This is sketched in Figure 7. Suppose there is an error in y_k, $k \leq n$, but all a_{ij} and all other y_i are correct. Then if $y_{n+1}^{(k)}$ and $S^{(k)}$ are the sums obtained by omitting $\Sigma a_{kj} x_j$ and y_k respectively, from y_{n+1} and S, it follows that

$$S^{(k)} - y_{n+1}^{(k)} = 0 \text{ if } y_t,\ t \neq k \text{ are all correct}$$

$$\neq 0 \text{ if some } y_t,\ t \neq k, \text{ is wrong}$$

The calculations $S^{(k)} - y^{(k)}$ can be performed serially for all k, and it can be implemented by serially inhibiting y_k by adding a set of inhibitory lines from $S - y_{n+1}$ to the a_{kj} in Figure 7. Such inhibition could be activated whenever $S - y_{n+1} \neq 0$. A similar scheme can be devised for error detection in a_{ij}. Again it is assumed that only one a_{st} can be in error, but not its duplicate nor any other a_{ij}.

A switching network would be required: for error detection in y_k it would serially inhibit the subsets (a_{kj}) connecting the receptive field (x_j) to y_k and the duplicate subset to y_{n+1}, while for error detection in a_{ij}, the serial inhibition would work on one a_{st} and its duplicate at a time.

The expected number of steps in locating the error can be reduced by eliminating subsets of $\underline{y}$ instead of one y_k at a time from the sums of S and y_{n+1}. Thus, if we divide $y_1, y_2, \ldots y_n$ into p subsets of q elements each, so that $pq = n$, then the expected number of steps to locate a faulty subset is 1/2 p and then the expected number of steps to locate the faulty y_k within the subset is 1/2 q, giving altogether 1/2(p+q) steps. This compares with 1/2 n = 1/2pq if we eliminate one y_j at a time. It can easily be shown that to minimize the number of steps for locating the fault, $p \approx (n)^{1/2}$, i.e., the set $(y_1, y_2, \ldots y_n)$ should be divided into approximately $(n)^{1/2}$ equal subsets.

<u>Summary and Conclusion:</u>

Biological organisms live in a world in which many and diverse stimuli have to be processed and appropriate actions must be performed. This requires good detection and simultaneous, parallel computation.

I have considered the retinal rods as an example of cells which are exquisitely sensitive and I have shown how their structure might have evolved to optimize the detection of photons in the presence of noise.

I have described the architecture of converging and diverging signal transmission in the nervous system, and while we do not know at present the actual neural computations, we can investigate the general properties of such a design. This includes multivariable, broadly tuned and overlapping representations being processed in channels in parallel. I have discussed some of the advantages of such a system and the processes it employs. From the viewpoint of optimality, one can be confident that the architecture is well suited to the biological function.

<u>References</u>:

Boynton, RM (1990): Human color perception. In Science of Vision (K.N. Leibovic, ed.) Springer Verlag, New York.

Kelly, KL, Judd, DB (1976): The ISCC-NBS method of designating colors and a dictionary of color names. NBS Special Publication 440, US Dept. of Commerce, US Superintendent of Documents, Washington, DC.

Leibovic, KN (1988): Parallel processing in nervous systems with converging and diverging transmission. In "Biomathematics and Related Computational Problems", (L.M. Ricciardi, ed.), Kluwer Academic Publishers, Dordrecht.

Leibovic, KN (1990): Some conjectures on the design of a rod outer segment. Biol. Cybern. 63, 359-361.

Leibovic, KN (1990): Overview in "Science of Vision" (K.N. Leibovic, ed.) Springer Verlag, New York.

Leibovic, KN, Moreno-Diaz, R. Jr. (1992): Rod outer segments are designed for optimum photon etection. Biol. Cybern. 66, 301-306.

Pugh, EN (1988): Vision: physics and retinal physiology, in "Stephen's Handbook of Experimental Psychology, Vol. 1, 2nd ed., Wiley, New York.

Snyder, AW (1979): Physics of Vision in compound eyes, in "Handbook of Sensory Physiology" Vol. II/6A, Springer Verlag, Berlin.

von Neumann, J (1958): The Computer and the Brain. Yale U. Press, New Haven.

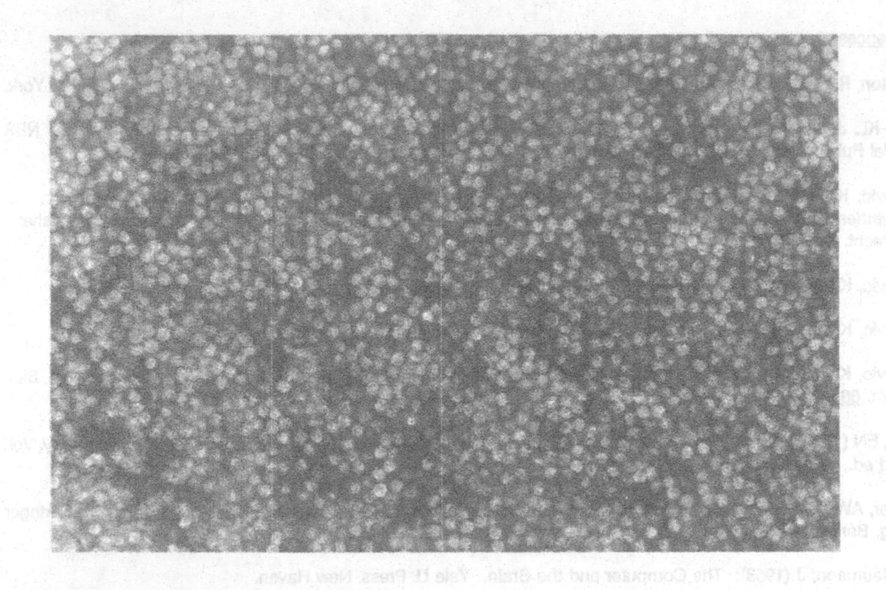

Figure 1: A photoreceptor mosaic seen end on. This one is from the retina of Bufo marinus and shows only rods, but it is typical of the appearance of retinae also from other species.

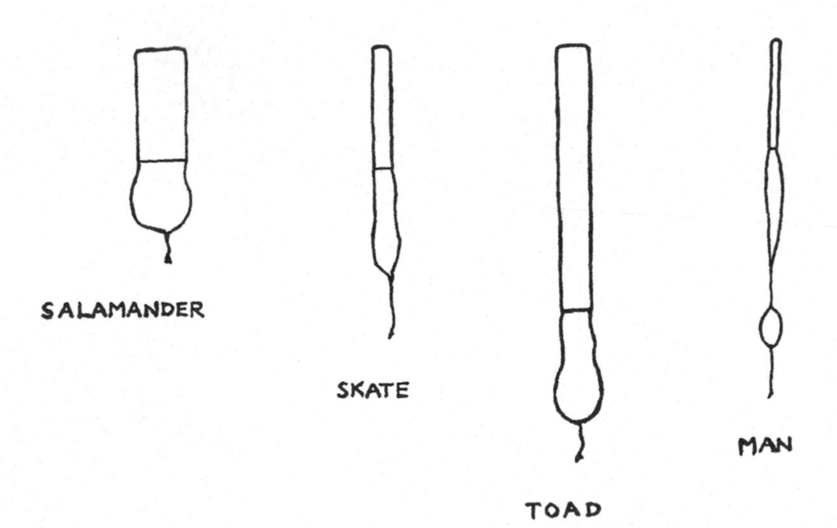

Figure 2: Illustrations of rods from four species. The cylindrical portion at the top is the photon absorbing outer segment. Below this, the inner segment contains the nucleus, mitochondria and otehr organelles. The fiber at the bottom ends in a synaptic bouton.

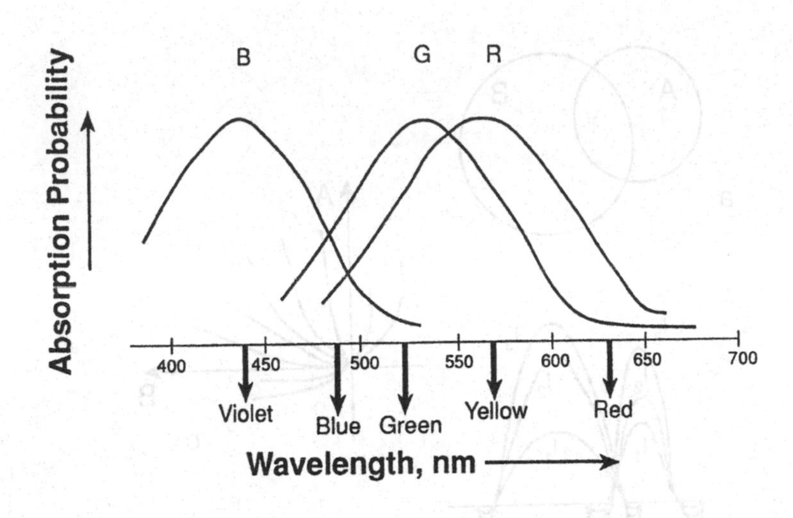

Figure 3: The absorption probabilities of the blue (B), green (G) and red (R) cones as functions of wavelength. The midpoint positions of five colors are marked on the wavelength scales.

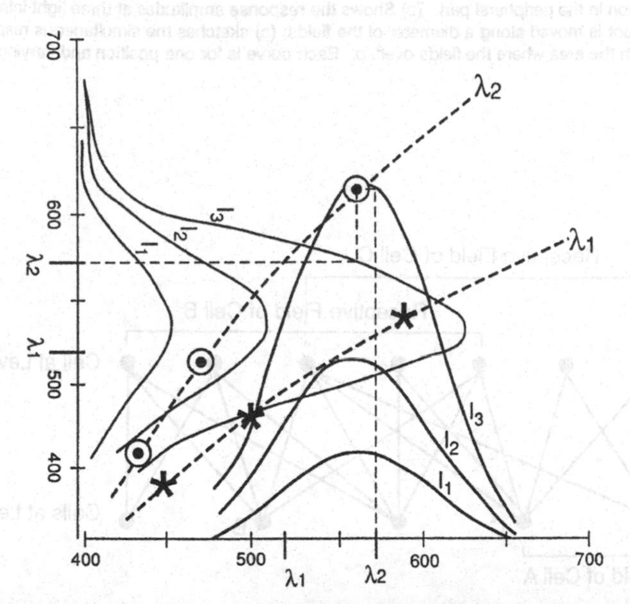

Figure 4: Demonstration that to each wavelength and intensity of light entering the eye there corresponds a pair of responses of the R, G cones. For the R cone curves based on the horizontal axis, the x-axis measures wavelength and the y-axis response amplitude. Similarly, for the G cone curves based on the vertical axis, the y axis measures wavelength and the x-axis response amplitude. The curves marked I_1, I_2, I_3 are hypothetical responses at three light intensities. Then the pairs of responses at wavelength λ_1 and varying intensity lie on the curve marked λ_1. Similarly for λ_2. The dotted lines illustrate the construction of the curves marked λ_1 and λ_2.

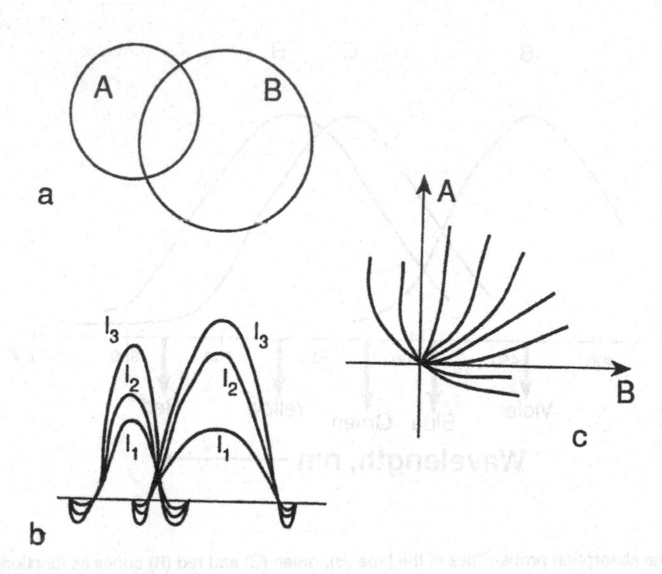

Figure 5: (a) Two overlapping receptive fields A and B respond with excitation to a spot of light in the central part of the field and with inhibition in the peripheral part. (b) Shows the response amplitudes at three light intensities I_1, I_2, I_3, for A and B as the spot is moved along a diameter of the fields; (c) sketches the simultaneous responses of A and B to a spot of light in the area where the fields overlap. Each curve is for one position and varying intensity of the spot of light.

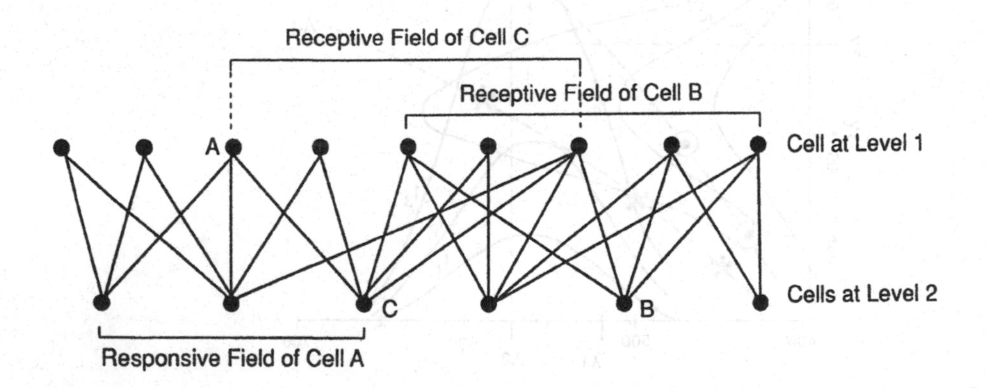

Figure 6: Illustration of Overlap with Convergence and Divergence.

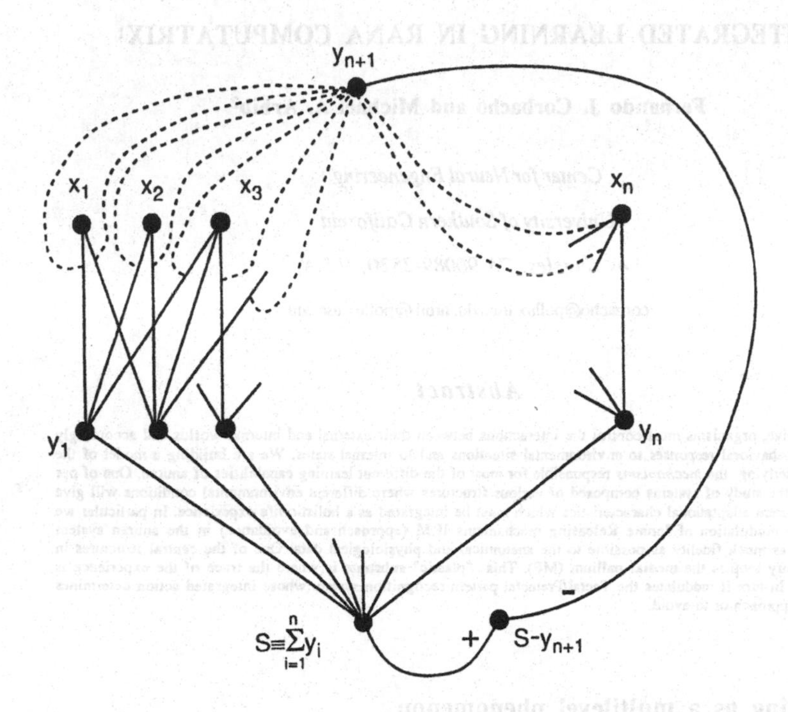

Figure 7: A scheme for error detection. Cells x_i produce inputs to cells y_i ($i=1,2,...n$). The connections between x_i and y_i give off collaterals (duplicates) which form the input to y_{n+1}. S sums the y_i, $i \neq n+1$, and then the difference is formed $S-y_{n+1}$. (See text.)

INTEGRATED LEARNING IN RANA COMPUTATRIX[1]

Fernando J. Corbacho and Michael A. Arbib[2]

Center for Neural Engeneering

University of Southern California

Los Angeles, CA 90089-2520, U.S.A.

corbacho@pollux.usc.edu, arbib@pollux.usc.edu

Abstract

To survive, organisms must control the interactions between their external and internal worlds and accordingly adapt their behavioral responses to environmental situations and to internal states. We are building a model of the circuitry underlying the *mechanisms* responsible for most of the different learning capabilities of anuran. One of our goals being the study of systems composed of various structures where different environmental conditions will give place to different adaptational characteristics which must be integrated as a holistic life experience. In particular we simulate the modulation of Innate Releasing mechanisms IRM (approach and avoidance) in the anuran system maintaining as much fidelity as possible to the anatomical and physiological data. One of the central structures in the modulatory loop is the medial pallium (MP). This "plastic" substrate is where the trace of the experience is "imprinted". In turn it modulates the Tectal/Pretectal pattern recognition system whose integrated action determines whether to approach or to avoid.

1. Learning as a multilevel phenomenon

Learning and memory can be analyzed at different levels of biological organization (Dudai, 1990). The highest level is that of the behaving organism input-output relationships (e.g. prey approach, bee avoidance). Next the neuronal system level: how the system functions in an integrated manner, circuit level role of specific brain structures and neuronal pathways in learning. Next is the cellular level analysis of the function of individual neurons and their connections in the previous structures. And finally, there is the molecular level: the study of the biochemical and biophysical events associated with learning and memory (i.e. plasticity mechanisms). This research project is therefore structured in the same fashion. At the top level we investigate the different learning capabilities in terms of behaviors (e.g. habituation, conditioning). Focusing down we study the integrated circuitry and modulatory loops, then reach the neuron level and study the physiological properties of neurons in different structures (e.g. T5.2 tectal cells physiology, functionality, and behavioral responses); and finally go down to a level where plasticity mechanisms can be accounted for.

1.1. Categories of Learning

Behavioral observations clearly show that learning occurs in very many species under very different circumstances (Dudai, 1990). To introduce some order into a very rich and heterogeneous body of observations, it is useful to classify learning into a few categories. Non-associative, in which the organism habituates or is sensitized to a stimulus. Where habituation is a decrease in the strength of a behavioral response that occurs when an initially novel stimulus is presented repeatedly. Sensitization is an augmentation of a response to a stimulus, following presentation of the same or another stimulus. The sensitizing stimulus is usually strong or noxious (Dudai, 1990).
Associative learning, in which the organism learns the causal relations among events in the world -either among stimuli, or among stimuli and actions (Thompson, 1986). The two main categories of associative learning are classical conditioning and instrumental conditioning. In classical conditioning experiments the conditioned stimulus (CS) always precedes the

---

[1] Rana Computatrix (Arbib, 1989) is an evolving computer model of anuran visuomotor coordination.

[2] M.A. Arbib and F. J. Corbacho are supported in part by grant no. 1RO1 NS 24926 from the National Institutes of Health (M.A. Arbib, principal investigator).

unconditioned stimulus (US). The strength of the association being directly proportional to the proximity of both events (within a range). In the instrumental conditioning paradigm the probability or intensity of a spontaneous behavioral response is altered by a reinforcing stimulus. Here the experimenter manipulates the relationship between a spontaneous response and a reinforcer, so that delivery of the latter depends on the occurrence of the former.
Animals thus learn the structure of the world by different processes which result in the generation and modification of internal representations. One of the goals of this research is the investigation of the mechanisms that allow the agent to form and integrated/holistic life experience out of temporal/spatial partial experiences.

2. Learning behaviors in Rana Computatrix

Anurans represent an excellent testbed for the study of different forms of learning and their possible interactions. The *merging* of structure and function is more feasible in systems with this level of complexity. One of the main tasks of the visual system is to classify signals from the environment into innate and learned categories of functional significance. Frog has two main types of antagonistic behaviors, prey catching and predator avoiding. Species examples of more general approach and avoidance behaviors. In an individual toad, responses to visual objects can be influenced by various modes of learning or by changes in motivation or attention. Learning in this model is exhibited as the capacity of the system to choose for a specific situation the proper motor pattern rather than as the mere storage of information. Learning in this sense can be regarded as the dynamic changes of the system that allow the animal to shift from one to another motor schema or to assign different perceptual schemas to the same motor schema.

2.1. Some anuran Adaptive behaviors

• Toads decrease the number of orienting responses toward the moving stimulus after repeated presentation of the same prey dummy in their visual fields. In particular the response intensity decreases exponentially. Wang (1991) has modeled the hierarchical dishabituation of visual discrimination in toads.
• Predator avoidance habituation is produced when a predator stimulus is repetitively presented without any negative experience for the frog. Predator habituation lasts for a shorter period than that of prey-orienting habituation (Ewert & Rehn, 1969; Ewert & Traud, 1979; Ewert 1984).
• In the condition where toads were fed mealworms out of the experimenter's hand for a certain time period, animals learned to associate the experimenters hand (CS) with food (US), and finally responded with catching behavior to the moving hand alone (Brzoska & Shneider, 1978; Burghagen, 1979). This association was generalized to other large objects as well.
• More recently in a similar paradigm, Finkenstasdt and Ewert (1988b) quantitatively studied conditioning of a large moving square (predator dummy) with mealworms. After a few weeks, the large square became effective to trigger the conditioned response (CR), and they found that, after training, the maximum response activity toward square, antiworm-l and worm-like objects is all shifted towards larger size.
• Recent studies of association between a prey dummy (US) and olfactory cues (CS) confirmed the same phenomenon (Merkel-Harf & Ewert, 1991). After successful associations, with presence of the odor, toads respond with prey-catching behavior to virtually any moving object.
• Toads and frogs after they swallowed honeybees and received a pain in their stomach can learn to avoid these painful insects, and their acquired preference can be retained for different intervals of time, ranging from hours, days and weeks to several months (Cott, 1936; Brower & Brower, 1962). Nevertheless this conditioning is stimulus specific. The frog will snap at other worm-like stimuli while retaining the avoidance behavior for bees.

3. Anatomy of the Learning Circuits

All animals -anuran in particular- are provided with a set of Innate Releasing Mechanisms (IRM) which provide the animal with a "toolkit" for survival (Duday 1990). The "gross" structure of every frog is provided by its "genetic memory". This basic architecture gives rise to elementary behaviors so that frogs do not have to learn everything from scratch when they are born. Nevertheless animals are also provided with adaptational capabilities to be able to cope with an "unpredictable" changing environment. The modulatory loops mediated by higher brain centers will provide this adaptational capabilities therefore increasing the animal chances of survival and procreation. They allow for the adaptation and modulation of the IRM to account for the individual experiences under different environmental conditions. In our model of Rana Computatrix we provide with some prewired structure underlying the IRM[3] . All this structure is prewired in the

[3] The evolutionary and developmental stages are therefore manifested into the given prewired structure in our model.

sense that every frog will have the same structure regardless of its own experience. This structure allows the naive frog to Approach prey-like stimuli and Avoid predator-like stimuli without any previous experience.

3.1. Loops and Modulation

The model of habituation (Wang and Arbib, 1991) has tested several hypotheses about the functioning of the modulatory loop as well as the memory role of the MP in the loop. Wang & Arbib have built on Ewert's analysis to suggest that there are two distinct neural loops for modulating innate releasing behaviors, namely:

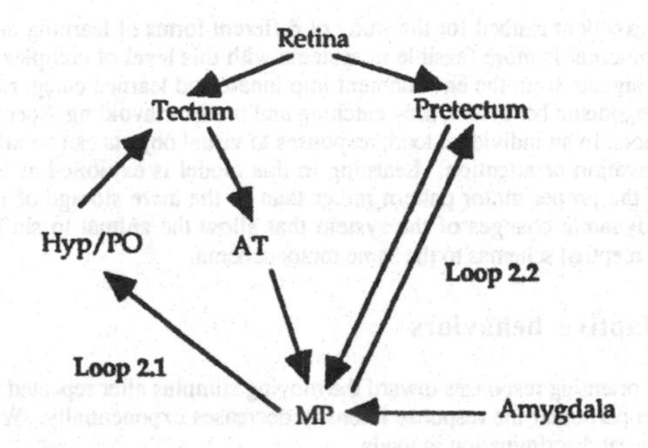

Figure 1: Modulatory loops; the architecture of Learning

Loop (2.1) is: retina → tectum → Anterior Thalamus (AT) → Medial Pallium (MP) → Preoptic/Hypothalamus (PO/HYP) → tectum. Loop 2.1 modulates prey-related behaviors centered in the tectum and learning in this loop is stimulus-specific (Specialization)
Loop (2.2) is: retina → Pretectum (TP) → MP → TP. Loop(2.2) modulates predator-related behaviors centered in TP and learning in this loop is stimulus non-specific (generalization).
Both conditioning and habituation in toads exhibit stimulus-specific properties with prey-catching behavior (Ewert & Kehl, 1978). In both also the effects of learning generalize along stimulus configurations for the predator stimuli (Ingle, 1976; Merkel-Harff & Ewert, 1991). We may think that this property of Generalization in loop (2.2) involving pretectum vs. Specialization in loop (2.1) involving tectum is a result of the discrimination capabilities of tectum versus pretectum. The integration of retina and tectum by the AT is able to produce a finer discrimination among different prey-like stimuli whereas TP is not able to produce such a fine discrimination among different predator-like stimuli.

4. Structures and Neurons in the learning circuits

We build on our current models of pattern recognition in the anuran system. The visual input is reflected as a 2D representation of the environment with gray levels on the receptors of the retina.
Retina: "extracts" information about area of objects, location, shape, edges[4], velocity of moving objects, and so on.
Tectum: is an integrative, and triggering (for approach behaviors) center. It is the place of convergence of several sensorial modalities as well as target of pretectal and forebrain modulatory connections. In particular our model consists of tectal cells T5.2 and T3. T5.2 cells (Ewert, 1984) are monocularity driven and have relatively small excitatory receptive fields of about 25°. As every T5 cell they are distributed in all parts of the visual field according to the retino-tectal map (Ewert, 1984; Fig. 9B). They exhibit selective sensitivity to wormlike objects moving in the direction of their main axis and strongly reduce their activity with progressive stimulus extension in that dimension. T3 tectal cells are monocularly driven neurons with ERFs about 20-30° in diameter and located in nasal visual field positions. They are optimally activated when an object larger than 3° moves toward the frog's eye (Ewert, 1984).
Pretectum mainly "detects" stationary and big moving objects. It inhibits the tectum to "control" approach behaviors and also plays an important role in triggering avoidance behaviors. Pretectal TH3 neurons receive visual input from the

---

[4] For a detailed current description of our retina model see Teeters (et.al.).

contralateral eye, have ERF's of about 30° diameter and are best activated by moving visual stimuli that fill the ERF. TH3 neurons have been traditionally associated with predator-like detection systems.
Anterior Thalamus integrates information from retina (R2) and tectum (SP) as to provide the finer visual discrimination of different worm-like stimuli. It functionally provides a more "integrated" representation of this type of visual input.

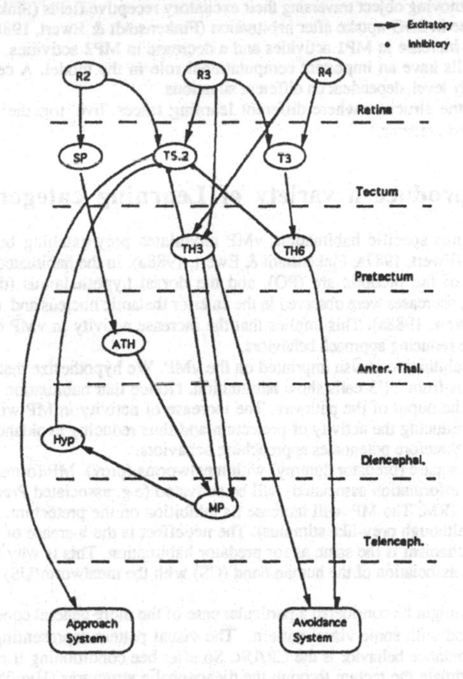

Figure 2: Circuit and neurons.

Hypothalamus: receives input from MP and inhibits tectal T5.2 cells.

Medial Pallium

Learning processes are assumed to take place in the medial pallium, a structure in the telencephalon, that receives direct projections from AT. This is based on both lesions experiments, physiological and 2DG studies. MP can modulate/override the IRM after learning experiences between different perceptual schemas and motor schemas. It does this by two mechanisms: "controlling" the inhibition to the Pretectum (which in turn controls the inhibition of the Tectum) and by directly "modulating" the tectum through its diencephalic projection. We propose that MP indirect projections via the preoptic area and hypothalamus modulate the state of the T5.2 cells by scaling prey catching behavior. This is consistent with the lesion results that after bilateral lesions of MP toads still exhibit accurate prey-catching response but they show no hint of habituation (Finkenstädt & Ewert, 1988a). It has also been shown that bilateral transection of medial MP destroys the aftereffects in previously habituated (retrieval phase affected) toads (Ewert, 1965; 1970). In addition, both the effects of conditioning and the associative learning ability in naive toads are abolished due to MP -lesion (Finkenstadt & Ewert, 1988a; Finkenstadt, 1989b). Anatomically the medial pallium has been thought to be the homologue of the mammalian hippocampus ("primordium hippocampi" by Herrick, 1933; Finkenstädt, 1989). Regional distribution in glucose utilization of various visual areas has been studied by the 14C-2DG method (Finkenstädt et al., 1988; Finkenstädt et al., 1989). Of particular interest is the comparison of 2DG-uptake across brain structures between naive and trained toads. In habituated toads, compared to naive toads, vMP, a certain portion of the PO region, and dHYP showed a statistically significant increase (Finkenstadt & Ewert, 1988a). They found, however, no habituation-related changes in the TP area. It has been shown that, when toads are trained to associate mealworms with predator sign stimuli, the trained toads in comparison with naive toads display a strong increase in glucose utilization in vMP (Finkenstadt & Ewert, 1988; Merkel-

Harff & Ewert, 1991). Preliminary physiological studies indicate that visual telencephalic neurons exhibit long-term adaptation, in comparison with those obtained in the tectum and the pretectal region (Ewert, 1984). A recent recording investigation identified three types of visual sensitive neurons exhibiting *spontaneous* firing activities. MP1 neurons strongly increase, MP3 neurons decrease, and MP2 neurons do not alter their discharge rates in response to 30 minutes of repetitive stimulation with a visual moving object traversing their excitatory receptive fields (Finkenstadt, 1989a). The fact that MP shows a significant increase in 2DG uptake after habituation (Finkenstädt & Ewert, 1988a) may be a result of the net outcome of a relatively greater increase in MP1 activities and a decrease in MP2 activities. Spontaneous activities as shown in MP1, MP2, and MP3 cells have an important computational role in the model. A cell of this type can either increase or decrease its stable activity level, dependent on different situations.
Hypothetical conclusion: MP is the structure where different learning traces 'live' together. Where habituation and conditioning coexist for both prey and predator.

5. Mechanisms to produce a variety of Learning categories

In the case of prey stimulus specific habituation, vMP modulates prey-catching behavior via the PO/HYP pathway, inhibiting tectal activities (Ewert, 1987a; Finkenstadt & Ewert, 1988a). In the habituated toads, the caudal ventral medial pallium (vMP), a portion of the preoptic are (PO), and the dorsal hypothalamus (dHYP) show statistically significant increases in 2DG-uptake; decreases were observed in the anterior thalamic nucleus and the medioventral layers of the optic tectum (Finkenstädt & Ewert, 1988a). This implies that the increase activity in vMP causes a reduction on the activity on T5.2 tectal cells therefore reducing approach behaviors.
The trace for Predator avoidance habituation is also imprinted on the vMP. We hypothesize that a group of MP synapses in the pathway receiving projections from TH3 cells show habituation. Notice that habituation in an inhibitory pathway may cause increase of activity on the ouput of the pathway. The increase of activity in MP will augment the inhibition from MP to pretectum. Therefore reducing the activity of pretectum and thus reducing avoidance behaviors. This in turn reduces the inhibition to tectum and therefore potentiates approaching behaviors.
On Conditioning of a large moving square (predator dummy) with mealworms (prey) MP forms a trace of the CS so that when similar input is presented the information associated will be activated (e.g. associated Predator-Hand with the food that it provides) thus overriding the IRM. The MP will increase its inhibition on the pretectum. Contrary to the IRM the Tectum will not be inhibited now (although prey-like stimulus). The net effect is the increase of MP activity projecting to pretectum and from that on the mechanism is the same as for predator habituation. This is why both paradigms show the same generalization properties. The association of the human hand (CS) with the mealworm(US) is a particular case of the above mentioned.
Bee avoidance conditioned learning might be considered a particular case of the more general conditioning paradigm where an unpalatable stimulus is associated with some visual pattern. The visual pattern representing the bee is the CS[5], the "pain" signal is the US and the avoidance behavior is the CR/UR. So after bee conditioning if a bee is in the visual field of the animal then the MP will modulate the tectum through the diencephalic structures (Hyp/PO) inhibiting the tectum. This will reduce the activity of tectal T5.2 cells and therefore the animal will not snap. Notice the similarity with prey habituation. That is why both are stimulus specific and reduce approaching behaviors.

5.1. Role of Modulatory Loops through Tectum/Pretectum

If there is no "stored" experience associated with a particular prey-like perceptual schema the loop going from MP back to the tectum will "reinforce" the activity of the tectum when that shema is activated (warm-up effect). This loop is therefore providing build-up of excitation to tectum. We consider T5.2 as a high threshold cell. To reach the region of high firing frequency the neuron either needs a facilitation effect from within the tectal column itself or a "push" by the modulatory loop from the MP. On the other hand, if the animal did have any previous learned experience associated with this perceptual schema then the MP may override the IRM. The hypothesis is that of a control mechanisms based on inhibitory "chains": The MP by increasing or decreasing its inhibition to the Pretectum modulates it. In turn the Pretectum will control the Tectum by the same mechanism. Also the MP can modulate the Tectum through the indirect inhibition from the Diencephalon. The tectum is thus the common locus for two loops and the place for interactions between these two loops.

---

[5] Assuming that anuran discriminate bees based on visual features and not odor or sound. This assumption is consistent with our worm discrimination model in the anterior thalamus.

5.2. Interactions among learning substrates

The interface of the two learning loops provides the basis for integrating a group of evolving models of *Rana computatrix* (Arbib 1991). Loop (2.1) modulates prey-related behaviors whereas loop(2.2) modulates predator-related behaviors. Both conditioning and habituation in toads exhibit stimulus-specific properties with prey-catching behavior (Ewert & Kehl, 1978). In both also the effects of learning generalize along stimulus configurations for the predator stimuli (Ingle, 1976; Merkel-Harff & Ewert, 1991). For example toads will avoid honeybees through conditioning but not all prey-like stimuli such as mealworms. Also both Habituation and Conditioning with predators show generalization from one predator to another.

Ingle (1976) observed that the releasing value of prey-catching behavior with prey stimuli increases following habituation of a predator stimulus, i.e., habituation with a large object (predator) facilitates the response to a small prey. The interaction among both loops is due to the inhibitory connections from pretectum to tectum. So predator-avoidance habituation will cause a shift in tectal activity towards bigger prey-like stimuli since the inhibition from pretectum to tectum would be reduced and therefore will "produce" a less selective T5.2 cell. So we find that predator habituation produces an interference effect where prey selection becomes less selective. In Conditioning of a predator dummy with a prey a similar phenomenon takes place. In particular it has been found that after training the maximum response activity toward square, antiworm-l and worm-like objects is all shifted towards larger size. Recent studies of association between a prey dummy (US) and olfactory cues (CS) confirmed the same features (Merkel-Harf & Ewert, 1991).

Other possible interactions are: compound conditioning where two different CSs are paired with the US, with the result that each will generate a CR; and 2nd order conditioning effects (Tesauro, 1986) since MP cells may receive convergent input from at least three pathways (e.g. T5.2, TH3, olfactory, ...). For the same reason blocking phenomena -where CS1 may block the formation of the CS2-US association- may also be produced in this structure.

5.3. Refining the MP model.

The model has to account for the temporal specificity characteristics of conditioning: duration of the US and the CS, their rate, the interval between them, and the order of their presentation. As well as data on retention phenomena (after extinction of the association), disassociation by CS without Us, 2nd order conditioning, blocking, and so on. For a review see Tesauro (1986). MP must be able to associate a perceptual schema (e.g. bee schema) with another perceptual or motor schema. In the new model sets of parallel fibers carrying the US converge with the fibers that carry the CS patterns. For instance to have predator/prey conditioning we hypothesize that both T5.2 projections as well as TH3 terminals will converge in some area of the MP. It is in these synapses that the trace of the conditioning is formed. At a biophysical level habituation occurs due to a decrease of the "weight" of a synapse in the CS pathway. For conditioning a different synapse must be implicated to avoid interference.

Visual habituation in toads shows Locus specificity whereas conditioning shows some spatial generalization. Wang (1991) achieved locus specificity by including a retinotopic array of MP columns as the elementary unit in the MP. In our new model the parallel fibers run through all the columns thus providing with spatial generalization for conditioning.

Prediction 1: if we were to "cut" the parallel fibers carrying the US in the middle of the column (e.g. unit 25) then conditioning will occur only in the retinotopic portion which receives input from parallel fibers. Nevertheless habituation will be unaffected.

Wang's habituation model (1991) shows Hierarchical stimulus specificity. He argued that the hierarchy was first produced in the AT and then "used" in the MP. It can also be argued that the AT may provide another hierarchy for another class of conditioning stimuli and therefor produce a conditioning hierarchy. Where by conditioning a stimulus higher in the hierarchy, stimuli lower in the hierarchy will also be conditioned, but not the other way around.

6. Plasticity Mechanisms

As an example let us review the neural mechanisms used to model prey habituation (Wang, 1991b). Current neurobiological studies suggest that short term habituation operates on presynaptic terminals as a result of reduced neurotransmitter release (Kandel, 1976; Thompson, 1986; Dudai 1991). Long-term habituation, however, may be accompanied by structural changes. For instance, the frequency of active zones in presynaptic terminals and the average size of each active zone can be modified by long-term habituation training (Bailey & Chen, 1983; Bailey & Kandel, 1985). Mathematically, the decrease of synaptic efficacy is mostly modeled by build-up of inhibition V (see among others Lara & Arbib, 1985; Gluck & Thompson, 1987)

$$\tau \frac{dV(t)}{dt} = \alpha (V_0 - V(t)) + S(t)$$

where V_e is the normal, initial value of V; S(t) is the activity transmitted through the synapse, symbolizing training; τ governs the rate of habituation, and α regulates the rate of recovery. This simple differential equation models the exponential curve of habituation and spontaneous recovery. For predator habituation we use similar dynamics only changing the parameters of the equations.

The simplest circuit that can produce conditioning uses one neuron for each sensory stimulus and one neuron for the conditioned response, with synaptic connections from each sensory neuron to the response neuron. The plasticity rule must contain the key ingredients of timing sensitivity, and a mechanism of generating sign changes in the synaptic modification. For biologically plausible local learning rules of the generalized Hebb form, the timing sensitivity comes from using a "stimulus trace" of the presynaptic activity, and the sign change comes from using two postsynaptic terms of opposite sign. Kandel et. al. (1992) review some basic associative plasticity rules: a) Hebb's rule requires both presynaptic and postsynaptic coincident neural activity. b) Kandel's requires presynaptic and modulatory neuron activity. It does not require activity of the postsynaptic cell to strengthen the connection. c) A combination of both.

Selected References

Arbib, M. A. (1989). *The metaphorical brain 2: neural networks and beyond.* New York: Wiley Interscience.

Arbib, M. A. (1991). Neural Mechanisms of visuomotor coordination: the evolution of Rana computatrix. In M.A. Arbib, J.-P. Ewert (Eds.), *Visual structures and integrated functions.* Research notes in neural computing. Berlin: Springer-Verlag.

Bailey, C.H., & Kandel, E.R. (1985). Molecular approaches to the study of short-term and long-term memory. In C.W. Coen (ed.), *Functions of the Brain* (pp. 948-129). Oxford: Clarendon Press.

Brower, J. V. Z. & Brower, L. P. (1962). Experimental studies of mimicry 6. The reaction of toads (Bufo terrestris to honeybees (Apis millifera) and their dronefly mimics (Eristalis vinetorum). *American Naturalist*, 96, 297-307.

Cott, H.B. (1936). The effectiveness of protective adaptations in the hive-bee illustrated by experiments on the feeding reactions, habit formation and memory of the commonctoad (Bufo bufo bufo). *Proceedings of the Zoological Society London*, 1, 111-133.

Dudai, J. (1990). *The Neurobiology of Memory*, Oxford University Press.

Ewert, J.-P. (1984). Tectal mechanisms that underlie prey-catching and avoidance behaviors in toads. In: H. Vanegas (ed.), *Comparative neurology of the optic tectum* (pp. 246-416). New York: Plenum.

Ewert, J.-P. & Rehn, B. (1969). Quantitative Analyse der Reiz-Reaktionsbeziehungen bei visuellem Auslösen des Fluchtverhaltens der Wechselkröte (Bufo viridis Laur.). *Behaviour*, 35, 212-234.

Ewert, J. -P., & Traud, R. (1979): Releasing stimuli for antipredator behaviour in the common toad *Bufo bufo* (L.). *Behaviour*, 68, 170-180.

Finkenstädt, T., & Ewert, J.-P. (1988b). Efects of visual associative conditioning on behavior and cerebral metabolic activity in toads. *Naturwissenschaften*, 75, 85-97.

Gluck, M.A., & Thompson, R.F. (1987). Modeling the neural substrates of associative learning and memory: A computational approach. *Psychological Review*, 94, 1-16.

Ingle, D. (1976). Spatial vision in anurans. In K.V. Fite (Ed.), *The Amphibian visual system: A multidisciplinary approach* (pp. 119-141). New York: Academic Press.

Kandel E. R., & Hawkins R. D. (1992). The Biological Basis of Learning and Individuality. *Scientific American*, September 1992, 79-86.

Merkel-Harff, C., & Ewert, J.-P. (1991): Learning-related modulation of toad's responses to prey by neural loops involving the forebrain. Im M.A. Arbib, & J.-P. Ewert (Eds.), *Visual structures and integrated functions* (in press). Research notes in neural computing. Berlin: Springer-Verlag.

Sutton, R. S., & Barto, A.G. (1981). Toward a modern theory adaptive networks: expectation and prediction. *Psych. Rev.* 88, 135-170.

Teeters, J.L., Arbib, M.A., Corbacho. F., Lee. H.B. (1992). Quantitative modeling of Responses of Anuran Retina: Stimulus shape and size dependency. Submitted to *Vision Research.*

Tesauro G. (1986). Simple Neural Models of Classical Conditioning. *Biological Cybernetics*, 55, 187-200.

Thompson, R. F. (1986). The neurobiology of learning and memory. *Science*, 233, 941-947.

Wang, D.L., & Arbib M. A. (1991a). How does the toad's visual system discriminate different worm-like stimuli? *Biological Cybernetics*, 64, 251-261.

Wang. D.L. (1991b). Neural Networks for Temporal order learning and stimulus specific habituations. Technical Report 91-06. Center for Neural Engineering. University of Southern California.

A MODEL FOR CENTERING VISUAL STIMULI THROUGH ADAPTIVE VALUE LEARNING

Murciano, A.; Zamora, J.; Reviriego, M.;

Dpto. Matemática Aplicada (Biomatemática). Facultad CC. Biológicas. Univ. Complutense. Madrid.

ABSTRACT

Learning through adaptive value leads to environment-dependent behaviors. This paper introduces a model for stimuli centering within a visual field following the principles of this kind of learning. The neurons responsible for the eye movement execute a mapping of the visual field by means of the adaptive value of each movement. This value is the result of the interaction of biologically inspired layers, chosen in an evolutionarily and without any planning or supervision, throughout the learning process. To check the stabilization and learning abilities of the model three measurements have been used: an energy function representing weight variations; a measurement of the increment of the distances through the chosen trajectories; and, a discriminant linear analysis of the model behavior.

INTRODUCTION

One of the main goals when developing models that respond as biological organisms is the acquisition of learning mechanisms that allow our models to react in a environment-dependent way. These mechanisms are the basis of an opportunity for an organism to become adapted to the environment and to evolve as a species. It is also necessary that its whole learning-set be versatile enough to be directed to one function or another, depending on the environment in which it is performed. Thus, we have to search for learning models that allow a neural network to be able to perform a function, if it has grown in an environment in which that function involves an advantage, in any sense, to the development of this network [1]. This advantage represents the adaptive value of that function in that environment.

Models capable of representing the adaptive value as a result of the non-linear interaction of a multivariate set, show high efficiency when performing on-line processes in unknown or changing environments like the real world [2].

When speaking about primate vision, it is evident that the image-centering process within the visual field involves a great adaptive value. Retinal central areas are linked to small high-resolution visual fields whereas peripheral areas are linked to wider low resolution fields [3]. This distribution produces a major representation of retinian central area to the detriment of peripheral one.

In our model, the adaptive value of each eye movement represents the stimuli approximation to the fovea. We use this adaptive value to perform network training.

MODEL DESCRIPTION

The model consists of 4 layers of neurons (Fig.1):

i) an input layer of n x n neurons whose activation values conform to a square matrix $I=(I_{ij})_{1\leq i,j\leq n}$;

ii) an adaptive value computing layer with serially connected neurons and simulating antagonistic neurotransmitter release;

iii) a motor layer made up of 4 sub-layers, in each case firing in one of the four directions of a two-dimensional space; and
iv) a motor-activation layer firing to activate motor neurons if all of them have not been activated.

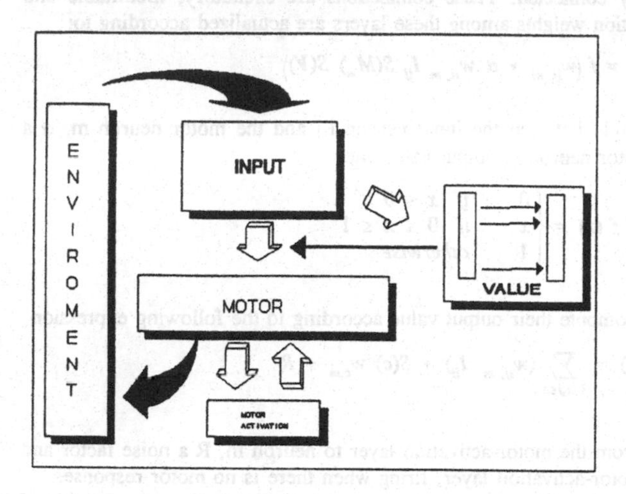

Figure 1: Network scheme: Input layer receives and processes stimuli from environment. Modifiable and excitatory connections from input layer produce an activation pattern in motor layer. A competition cycle among motor cells decides the eye movement. This movement produces a change in the relative position of the stimuli to the eye. Input layer also makes connections to value layer. Cell activities in value layer increase when the stimuli are approximated to the fovea (see text). These activity changes lead to modify connection weights from input cells to those motor neurons responsible for last movement. Motor activation ensures a motor layer response.

The input layer carries out a topological mapping of the visual field having an identity activation and output functions. These neurons have excitatory connections to value and motor layers. Fixed weights have been used for the connections to the adaptive value layer increasing them inversely to the Manhattan distance from each input cell to the center of the visual field. Such an ascription of value is based on a biological fact. The connection pattern among the visual receptors and the ganglionar cells (bipolar cells mediated) vary depending on the proximity between any area and the central point of the retina [4].

$$W_{ij,v} = 1 - \frac{\left|i-\frac{n+1}{2}\right| + \left|j-\frac{n+1}{2}\right|}{n}$$

where (n+1/2,n+1/2) are the central coordinates of the visual field, and $w_{ij,v}$ the connection weight from the input neuron i,j to the value cell v.
The value layer is divided into two strata. The first one receives connections from the input layer at a time t and sends them to the second stratum in t+1. In this manner, these two strata represent the value at a time t and the immediately previous one. Because the output connection weights in both strata are equal, fixed, and have opposite signs, the conjoint effect represents the increment in adaptive value, as a result of the movement chosen by the motor layer:

$$S(V_1) = \sum_{1 \le i,\, j \le n} w_{ij,v}\, I_{ij}$$

$$S(V_2(t+1)) = S(V_1(t))$$

$$S(V) = S(V_1) - S(V_2) - \theta$$

where $S(V_1)$ and $S(V_2)$ are the output values of both strata in value layer and S(V) the conjoint output value. Θ is an adjustable parameter, avoiding movements without value increment. Thus, the value increment is not drawn from comparisons with any optimum previously fed, but it is inherent to the network architecture (input-value connections structure) described above. Input and motor layers are fully connected. These connections are excitatory, modifiable and randomly initialized. The connection weights among these layers are actualized according to:

$$w_{ij,m} = f\left(w_{ij,m} + \alpha\, w_{ij,m}\, I_{ij}\, S(M_m)\, S(V)\right)$$

where $w_{ij,m}$ is the connection weight between the input neuron i,j and the motor neuron m, α a learning coefficient, $S(M_m)$ is motor neuron m output value and

$$f(x) = \begin{cases} 0 & \text{if } x < 0 \\ x & \text{if } 0 \le x \le 1 \\ 1 & \text{otherwise} \end{cases}$$

The neurons in the motor layer compute their output value according to the following expression

$$S(M_m) = \sum_{1 \le i,j \le n} (w_{ij,m}\, I_{ij}) + S(c)\, w_{c,m} + R$$

where $w_{c,m}$ is a random weight from the motor-activation layer to neuron m, R a noise factor and S(c) the output value from the motor-activation layer, firing when there is no motor response. A competition cycle among motor neurons innervating antagonistic muscles leads to the deletion of the output from motor cells that did not win, and to the movement selection.

RESULTS

In the earliest phases of learning, the implemented model (input 19 x 19) has a weight distribution from input to motor layer as shown in figure 2. At the same phase, for each step, we measured the change in distance between the presented object and the central point of the visual field. This is graphically represented in figure 4. In both cases, the network has a highly disordered behavior. After 50,000 movements of the visual field, with a noise activity of motor neurons at each 8,000, the motor neurons correctly map the input area (figure 3). Thus, they have a high efficiency in centering the object (figure 5).

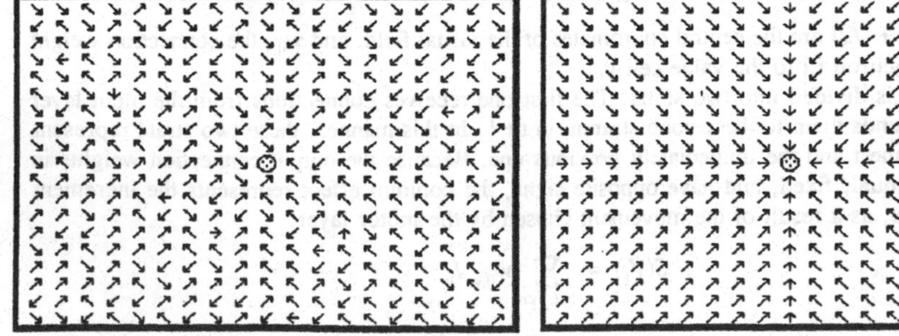

Figure 2: Winner weights at early phases of learning. Figure 3. Winner weights after 50000 steps

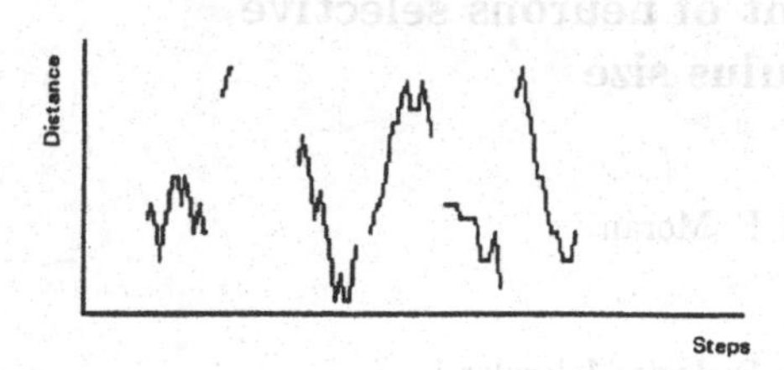

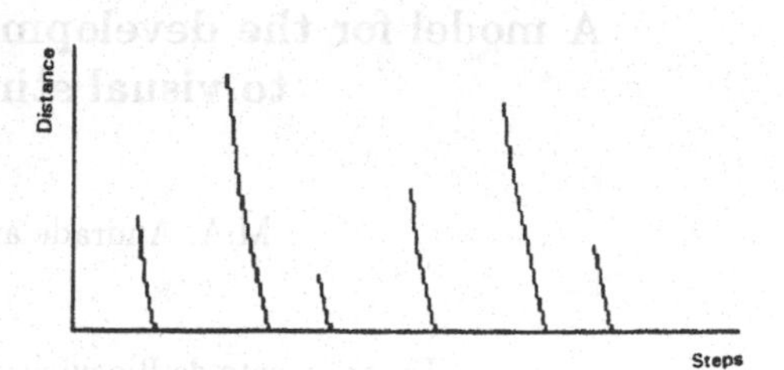

Figures 4 and 5: Distances vs steps. Each 15 steps a new stimulus is presented then representing and its distance from the center is represented.

An energy function was defined from the stability of the weights as learning evaluation measurement: $E = \sum_{\substack{1 \le i,j \le n \\ m \in M}} | \Delta w_{ij,\,m} |$ where M is the motor neurons set.

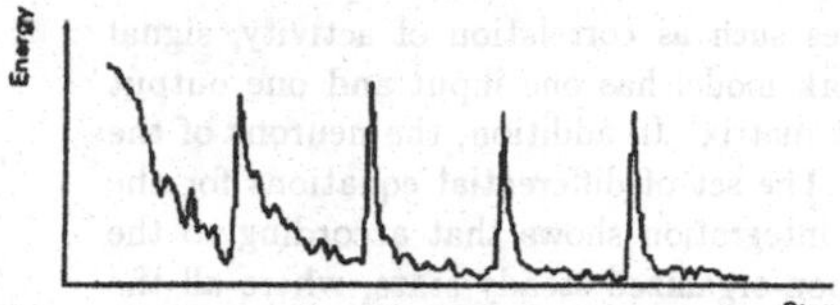

Figure 6: Energy along learning.

The results (figure 6) show that this function decrease throughout the learning process, giving faster responses to the periodic noise disturbances. Finally, a linear discriminant analysis was performed as a learning measurement, to classify each input neuron in a given quadrant, using its weights to the motor neurons as independent variables. The correctly classified percentage reached 100%.

In conclusion, the model introduced for centering stimuli shows a high efficiency in performing this task without feeding it any information about the stimuli type or situation, planning anything about preferred trajectories, or supervision during the learning process. Other models, including value layer with changeable weights and nested structures, will allow these systems to perform more complex tasks, and to select a preferred adaptive value depending on the environment.

REFERENCES

[1] Changeux, J.P.; Heidmann, T.; Patte, P., 1984. Learning by selection. In *The Biology of Learning*. Marler, P. & Terrace, H.S. (eds) Berlin. Springer-Verlag. pp. 115-133.

[2] Reeke, G.N.; Sporns, O.; Edelman, G.E., 1990. Synthetic Neural Modeling: The "Darwin" Series of Recognition Automata. *Proc. IEEE*. 78(9):1498-1530.

[3] Hubel, D.H.; Wiesel, T.N., 1977. Functional architecture of macaque monkey visual cortex. *Proc. Roy. Soc. Lon.* 198:1-59.

[4] Schwartz, E.L., 1977. Spatial mapping in the primate sensory projection: Analytic structures and relevance to perception. *Biol. Cyber.* 25:181-194.

A model for the development of neurons selective to visual stimulus size

M.A. Andrade and F. Morán

Departamento de Bioquímica y Biología Molecular I.
Facultad de Ciencias Químicas.
Universidad Complutense de Madrid.
Madrid. Spain, E-28040.

Abstract

In this work, a neural network model for the development of variable sized receptive fields is presented. The system self-organizes under simple rules such as correlation of activity, signal diffusion, and competitive synaptic growth. The network model has one input and one output layer. They are fully connected by an excitatory weight matrix. In addition, the neurons of the output layer are interconnected by inhibitory weights. The set of differential equations for the time evolution of the system is calculated. Numerical integration shows that according to the set of network parameters the system reaches either a non-organized steady state, where all the connections have the same value, or any of two organized states, one of them having connections that represent mexican hat shaped receptive fields of variable size.

1 Introduction

In an early stage of the development of the mammal embryo, the nervous fibres coming from the ganglionar cells grow from retina to brain establishing connections into the visual cortex. After this gross primary connection, a self-organizing process, dependent on neural activity, takes place and the connections are pruned giving functional characteristics to the visual system. Through this mechanism, several superimposed mappings appear in the visual cortex. Groups of neurons organized in areas (columns, blobs, etc.) selective to specific characteristics of the visual stimulus, such as orientation of a bar, colour, position in the visual field, velocity and size appear. Additionally, alternated regions of the visual cortex become dominated either by signals coming from one eye or by the other. [See for example references 3, 4, 6, 7, 8, 11, 15 and 16].

The receptive field is a characteristic organization of the visual system. A receptive field of a neuron is the compact region of the visual space that affects the activity of that neuron. A typical example is the *mexican-hat* shaped receptive field. This field has a circular symmetry. Visual stimulation of the central area produces an excitatory response in the target neuron, whereas stimulation in the surroundings produces inhibition.

Current models based on neurophysiological knowledge are used to explain visual cortex organization using neural network architectures and activity dependent rules. They show that simple rules can explain how could complex structures in the visual cortex appear without the necessity of a detailed genomic information [e.g., references 9, 10, 12, 13, 17, 18, 20, 21, etc.].

Accordingly, this paper presents a self-organizing network to demonstrate that *diffusion of synaptic activity*, *competitive synaptic growth* and *synaptic evolution* driven by either *activity correlation* or *activity anticorrelation* (Hebbian [5, 14] or anti-Hebbian learning rules, respectively), explain the development of variable sized mexican hat shaped receptive fields, and consequently of neurons selective to stimulus size in a development stage of a mammal embryo prior to visual experience.

2 The model

The network is composed of one input (a) and one output layer (b) of neurons, fully connected by excitatory connections whose weights are $W_{ij} > 0$ $(i, j = 1, \ldots, n)$ and $j = 1, \ldots, m$. Intra-layer inhibitory connections are present in layer b, whose weights are $Q_{ij} > 0$, where $i, j = 1, \ldots, m$.

The following equation describes the time evolution of the weights:

$$\dot{W}_{ij} = \alpha + \beta W_{ij}(t)[F^W_{ij}(t) - \gamma W^2_{ij}(t)] \quad (1)$$

$$\dot{Q}_{jk} = \alpha + \beta Q_{jk}(t)[F^Q_{jk}(t) - \gamma Q^2_{jk}(t)] \quad (2)$$

The first term, α, is a positive constant that accounts for the autonomous generation of new synaptic connections. The second term, $\beta W_{ij}(t)F^W_{ij}(t)$ or $\beta Q_{jk}(t)F^Q_{jk}(t)$, describes the increase (or decrease) of a particular synapse dependent on the *global* state of the network synapses. Parameter β controls the change rate. F^W_{ij} and F^Q_{jk} are the growth factors of the synaptic weights W_{ij} and Q_{jk}, respectively. As will be shown below, these factors depend on the activity of the neurons. The third part of the equations, $\beta W_{ij}(t)\gamma W^2_{ij}$ or $\beta Q_{jk}(t)\gamma Q^2_{jk}$, is a decaying term. It reflects an *individual* restriction over the growth of each synaptic connection. γ indicates the relative importance of *individual* restrictions and the *global* effect of the network.

Growth factors are calculated according to a Hebbian rule, i.e., the evolution of each factor depends on the temporal correlation between the activity values of the neurons connected by the weight whose growth factor is being calculated:

$$F^W_{ij}(t) = < A^a_i(t), A^b_j(t) >_t \quad (3)$$

$$F^Q_{jk}(t) = < A^b_j(t), A^b_k(t) >_t \quad (4)$$

where $A^a_i(t)$ represents the activity of neuron N^a_i (for $i = 1, \ldots, m$) and $A^b_i(t)$ the activity of neuron N^b_i (for $i = 1, \ldots, n$). These A values can be defined in terms of weight values and constants.

Layer a is considered the only source of activity in the system. In the modelled stage of development the system does not receive any coherent visual input from the environment. Therefore, this activity is assumed to be spontaneous and uncorrelated in both space and time. Hence, a random activity vector $\vec{f}^a(t)$, whose components are uncorrelated, is taken as the source of activity. This activity can diffuse within layer a, so it spreads from every neuron to their neighbouring neurons. The signal diffusion from neuron N^a_i to neuron N^a_j is regulated by $D^a_{ij} = G^a(|i-j|)$ terms. G^a is a function of the distance between the neurons and it is assumed to be gaussian:

$$G^a(x) = \delta_a \frac{1}{s_a} \exp(-(x/s_a)^2/2) \quad (5)$$

So, the output activity of neuron N^a_i is calculated as follows:

$$A^a_i(t) = \sum_j^n D^a_{ji} f^a_j(t) \quad (6)$$

The output activity from layer a is transmitted to layer b by the W excitatory weights. Then, the input of a neuron N^b_j is calculated through the following expression:

$$\vec{I}_j(t) = A^a_i(t) W_{ij}(t) \quad (7)$$

In layer b, signal diffusion and inhibition processes occur, and the resulting output activity of neuron N^b_k is:

$$A^b_k(t) = \sum_i^n \sum_j^m D^b_{jk}[1 - Q_{jk}(t)] \quad (8)$$

where $D^b_{ij} = G^b(|i-j|)$, being G^b a gaussian function of the distance between two neurons:

$$G^b(x) = \delta_b \frac{1}{s_b} \exp(-(x/s_b)^2/2) \tag{9}$$

Now, the activity of each layer b neuron can be described as a function of the input random activity:

$$A_o^b(t) = \sum_i^n \sum_j^n \sum_k^m f_i^a(t) D_{ij}^a W_{jk}(t)[D_{ko}^b - \sum_l^m D_{kl}^b Q_{lo}(t)] \tag{10}$$

Assuming $< f_i^a(t), f_j^a(t) >_t = \delta_{ij}$ for $i, j = 1, \ldots, n$ the equations (3) and (4) are reduced to:

$$F_{ij}^W(t) = \sum_k^n D_{ik}^a E_{kj}(t) \tag{11}$$

$$F_{ij}^Q(t) = \sum_k^n E_{ki}(t) E_{kj}(t) \tag{12}$$

where

$$E_{ij}(t) = \sum_k^n D_{ik}^a \sum_l^m W_{kl}(t) \left(D_{lj}^b - \sum_o^m D_{lo}^b Q_{oj}(t) \right) \tag{13}$$

$E_{ij}(t)$ representing the effect that the stimulation of the N_i^a neuron produces on the N_j^b neuron. This effect can be either positive or negative, i.e., excitatory or inhibitory.

Finally, the time evolution of the weights is described as follows:

$$\dot{W}_{ij} = \beta W_{ij}(t) \left[\sum_k^n D_{ki}^a E_{kj}(t) - \gamma W_{ij}^2(t) \right] + \alpha \tag{14}$$

$$\dot{Q}_{ij} = \beta Q_{ij}(t) \left[\sum_k^n E_{ki}(t) E_{kj}(t) - \gamma Q_{ij}^2(t) \right] + \alpha \tag{15}$$

Either W or Q weights are positive. The excitatory or inhibitory effect is implemented through the sign of the equation terms (see equations (6) and (7)). Notice that the weight values can not cross the zero value, since their derivatives at zero values are $\alpha > 0$.

3 Results

Numerical integration of the differential equations (14) and (15) has been performed and graphically displayed using the Neural Simulation Language[19] (NSL) in a SUN SparcStation 2. The steady state solutions of the network weights are multiple, but they have similar form. There is a strong dependence on the initial conditions. Depending on them, the system will fall into one or another solution. To impose the retinotopical one, a slight retinotopic bias is imposed in the random initial conditions.

The resulting steady states belong to one of the following classes:

1. *Uniform steady state.* Both the excitatory connections and the inhibitory connections reach the same value. This state is obtained when the inhibitory weights are removed.
2. *Uniform diagonal.* Every neuron of output layer, b, has a receptive field on imput layer, a, these receptive fields being equal to each other.
3. *Non-uniform diagonal.* Receptive fields with different size appear.

Results obtained from the simulation of the system using a particular parameter set are shown in figures 1 and 2. In this example, two one-dimensional layers of ten neurons each one have been used. Figures 1a-1c show the matrix of excitatory connections (left) and inhibitory connections (right) at sucessive steps of the run.

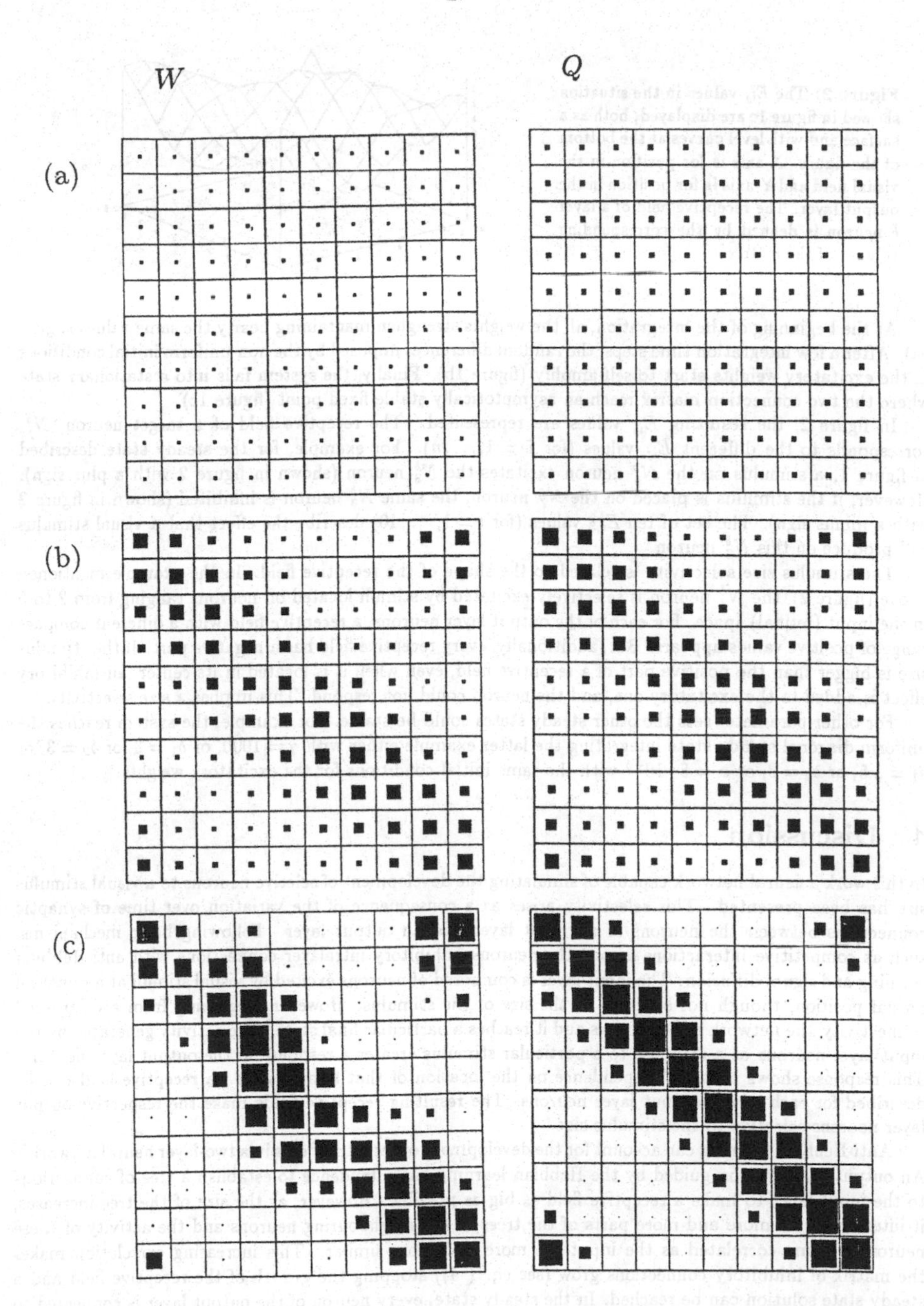

Figure 1. $\mathcal{W}$ **and** $\mathcal{Q}$ **matrix at different times of a concrete running. The size of the square is proportional to the value of the weight in the [0.0,0.3] range. (a)** $t = 103$, **(b)** $t = 153$, **(c)** $t = 500$. **The parameter values used were:** $\beta = 2 \cdot 10^{-3}$, $\delta_1 = \delta_2 = 4$, $s_1 = s_2 = 1$, $\alpha = 10^{-4}$, $\gamma = 500$ **and the** $\mathcal{W}$ **initial values were set randomly in the** $[0, 0.001]$ **range.**

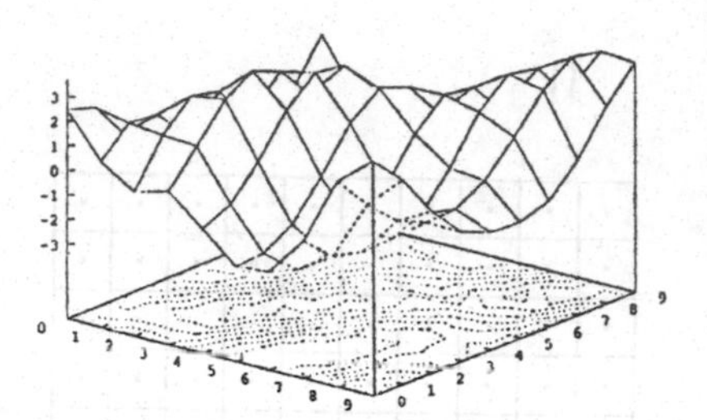

Figure 2. The E_{ij} values in the situation showed in figure 1c are displayed, both as a surface and with level curves at the bottom of the figure. X-axis is for position in the visual field and Y-axis is for position in the output layer. The receptive field of a layer b neuron is defined by the corresponding column of the matrix.

At the beginning of the integration, all the weights strenghen mantaining nearly the same values (figure 1a). After a few integration time steps, the random differences imposed by the non-uniform initial conditions in the excitatory weights start to self-amplify (figure 1b). Finally, the system falls into a stationary state where the two connection matrix reach an asymptotically stable fixed point (figure 1c).

In figure 2, the resulting E_{ij} values are represented. The receptive field of a target neuron, N_j^b, corresponds to the different E_{ij} values (for $i = 1, \ldots, n$). For example, for the steady state described in figure 1, a stimulus on the N_4^a neuron excitates the N_3^b neuron (shown in figure 2 with a plus sign). However, if the stimulus is placed on the N_7^a neuron, the same N_3^b neuron is inhibited (shown in figure 2 with a minus sign). The set of ten $E_{i,3}$ values (for $i = 1, \ldots, 10$) describe the effect that a visual stimulus will produce on this N_3^b neuron.

The stimulus size selectivity is related to the shape of the receptive field. In the example mentioned above (figure 2), the N_3^b neuron is selectively excitated by stimuli located on neurons ranging from 2 to 5 in the input (retinal) space. For each of the output layer neurons, a receptive field with a different compact range of positive values appears. But, additionally, every receptive field has a negative part. If the stimulus size is bigger than the positive part of a receptive field, even when it is located in its center, an inhibitory effect is added to the excitatory one, and the neuron could not respond. This implies a size selectivity.

For other parameter sets the other steady states could be stable. For example, the system reaches the uniform diagonal steady state integrating the latter example either with $\gamma = 1000$, or $\delta_1 = 3$, or $\delta_2 = 3$, or $s_1 = 1.5$, or $s_2 = 2$, or $\alpha = 5 \cdot 10^{-5}$ with the same initial conditions for the excitatory weights.

4 Discussion

In this work a neural network capable of simulating the development of selctive neurons to a visual stimulus size has been presented. This selectivity arises as a consequence of the variation over time of synaptic connections between the neurons of an input layer and an output layer. Following basic mechanisms, such as competitive interactions among the neurons, inhibitory intralayer connections with anti-Hebbian learning and signal diffusion. The input layer is composed of neurons excited by visual stimuli at a punctual retinal position, though not selective to the size of the stimulus. However, beginning from a disordered connectivity, the network self-organizes and it reaches a particular final state. The activity generated by the input layer neurons as a response to a particular stimulus creates a response in the output layer neurons. This response shows a special dependence on the location of that stimulus, i.e., a receptive field can be described for each of the output layer neurons. The resulting receptive fields make the respective output layer neurons selective to the stimulus size.

Anti-Hebbian learning can account for the development of receptive fields in a two-layer neural network[1,2] An output layer neuron guided by the Hebbian learning has a tendence to establish a tree of connections to the input layer to make a receptive field as big as possible. However, as the size of the tree increases, it intersects with more and more parts of the trees of the neighbouring neurons and the activity of these neurons become correlated as the input are more and more similar. This increasing correlation makes the matrix of inhibitory connections grow (see eq. (4)) stopping the growth of the receptive field and a steady state solution can be reached. In the steady state, every neuron of the output layer is connected to neighbouring neurons with inhibitory conections that module the input from the input layer.

This mechanism accounts for the development *mexican-hat* shaped receptive fields: an input close to the centre of one neuronal tree gives strong activation to that neuron (though the inhibitory connections subtract some activity). However, as the inputs are far and far away from the centre, the direct activation is less due to the tree structure. In addition, the inhibition from other neighbouring cells is enhanced as the input is closer to the center of the tree of a neighbour cell strongly connected to the first one by an inhibitory weight. At larger distances from the tree center there are no connections. The result is an inhibitory ring surrounding an excitatory region: an mexican hat shaped receptive field.

This study suggests, to investigate how the development of the receptive fields of the ganglionar cell of the mammal retina is affected by either avoiding the normal inhibitory synapses performance or the spontaneous activity in rods and cones.

References

1. Carlson, A. 1990. Anti-Hebbian learning in a non-linear neural network. *Biol. Cybern.* **64**, 171-176
2. Földiak, P. 1990. Forming sparse representations by local anti-Hebbian learning. *Biol. Cybern.* **64**, 165-170
3. Frégnac, Y. and M. Imbert. 1984. Development of neuronal selectivity in primary visual cortex of cat. *Physiol. Rev.* **64**, 325-434
4. Gizzi, M.S., E. Katz, R.A. Schumer, J.A. Movshon. 1990. Selectivity for orientation and directon of motion of single neurons in cat striate and extrastriate visual cortex. *J. Neurophysiology.* **63**, 1529-1543
5. Hebb, D.O. 1949. *Organization of Behavior* John Wiley & Sons, New York
6. Hubel, D.H. and T.N. Wiesel. 1963. Receptive fields of cells in striate cortex of very young visually inexperienced kittens. *J. Neurophysiol.* **26**, 994-1002
7. Kuffler, S.W., J.G. Nicholls, A.R. Martin. 1984. *From Neuron to Brain.* Sinauer Associates Inc. Publishers.
8. Le Vay, S., M.P. Stryker and C.J. Shatz. 1978. Ocular dominance columns and their development in layer IV of the cat's visual cortex: a quantitative study. *J. Comp. Neurol.* **179**, 223-224
9. Miller, K.D., J.B. Keller and M.P. Stryker. 1989. Ocular dominance columnar development: analysis and simulation. *Science.* **245**, 605-615
10. Miyashita, M. and S. Tanaka. A mathematical model for the self-organizaton of orientation columns in visual cortex. *NeuroReport.* **3**, 69-72
11. Orban, G.A. 1984. *Neuronal Operations in the Visual Cortex.* Springer-Verlag.
12. Singer, W. 1987. Activity-dependent self-organization of synaptic connections as a substrate of learning. In: *The Neural and Molecular Bases of Learning*, eds. J.-P. Changeaux y M. Konishi, pp 239-262. Dahlem Konferenzen. Chichester: John Wiley & Sons Ltd.
13. Spillman, L. and J.S. Werner. 1990. *Visual Perception. The Neurophysiological Foundations.* Academic Press, Inc.
14. Stent, G.S. 1973. Physiological mechanism for Hebb's postulate of learning. *Proc. Natl. Acad. Sci. USA* **70**, 997-1001
15. Stryker, M.P. 1986. The role of neural activity in rearranging connections in the central visual system. In *The Biology of Change in Otolaryngology*, Elsevier Science Publishers B V, R.W. Ruben *et al* eds., pp 211-224
16. Tootell, R.B., M.S. Silverman and R.L. de Valois. 1981. Spatial frequency columns in primary visual cortex. *Science.* **214**, 813-815
17. von der Malsburg, C. 1987. Synaptic plasticity as basis of brain self-organization. In: *The Neural and Molecular Bases of Learning*, eds. J.-P. Changeaux and M. Konishi, pp 411-431. Dahlem Konferenzen. Chichester: John Wiley & Sons Ltd.
18. von der Malsburg, C., and W. Singer. 1988. Principles of cortical network organization, in *Neurobiology of Neocortex*, eds. P Rakic y W Singer, pp 69-99
19. Weitzenfeld, A. 1992. *A Unified Computational Model for Schemas and Neural Networks in Concurrent Object-Oriented Programming.* Technical Report 92-03. Center for Neural Engineering. University of Southern California.
20. Wörgotter, F. and G. Holt. 1991. Spatiotemporal mechanisms in receptive fields of visual cortical simple cells: a model. *J. Neurophysiology.* **65**, 494-510
21. Yuille, A.L., M.D. Kammen, and D.S. Cohen. 1989. Quadrature and the development of orientation selective cortical cells by Hebb rules. *Biol. Cybern.* **61**, 183-194

AN INVARIANT REPRESENTATION MECHANISM AFTER PRESYNAPTIC INHIBITION

Roberto Moreno-Díaz Jr. and Olga Bolívar Toledo

Dept. de Informática y Sistemas
Universidad de Las Palmas de Gran Canaria
Campus de Tafira
35017 Las Palmas, España

Abstract

The search for mechanisms that allow invariant neuronal representations has been always of great interest since the achievement of invariances still is, at different levels, an open question in the theory of neuronal function. In this paper we present the formulation of a theoretical mechanism for visual representation which is invariant against general changes of light intensity. The model, which is an example of the interaction among Cybernetics, Systems Theory and Artificial Perception Systems, takes into account the different kinds of operations that so far have been described in different cells of the visual pathway and the biologic architecture that is necessary to perform them together with the necessary connections among processors. Structural basis of this schema in natural systems are found in Lateral geniculate Body. Finally, we include an example on Image Processing and a discussion on the conditions in which the model can be of practical use.

Preprocessing in artificial and natural systems.

One of the main goals in Artificial Image Processing is to get invariances against changes in the illumination conditions [1]. This is part of the so-called preprocessing of a given image and the image resulting is sent to a higher-level stage to be analyzed. On the other hand, some preprocessing characteristics (e.g. fast and retarded signals to compute constrast and dimming, brightening...) have been described to be present in retinal computations [2,3]. In both cases, artificial and natural, a non-linear operation is assumed to be present and, in the artificial world, several procedures have already been proposed to obtain invariances as part of a preprocessing system [4].

In natural systems, however, we have to cop with a given computing structure whose function is not always clear and that presents two key characteristics to take necessarily into account: parallelism and versatility. In the model, whose mathematical formalism is presented below, parallelism is a need for the system to work properly. From Systems Theory, we borrow the convolution-like type of operation performed by the neural system, in any case widely used to simulate the functioning of parallel systems. Thus, it is assumed that the processors (cells) are arranged in layers; that some kind of computation is done by every cell in such layer over the receptive field (RF), and that the output of the layer is a transformation of the original data. Also, there will be a plexiform (in the synaptic sense) layer where an inhibition between cells may take place. The assumptions on the properties of the neuron's computation are expressed in the following section.

Formalism and implications in Neural Nets.

The mathematical basis of our model is an extension of a Theorem stated by Muñoz-Blanco in his PhD Thesis [5]. That Theorem showed that given two purely spatial transformations T_1 and T_2 with certain properties on the spatial region they act, the ratio of both produces an spatial invariance against traslations and homotecies. The strict aplication of said Theorem in Neural Nets and Retinal Theory needed a double number of bipolar cells than photoreceptors in the fovea. The following generalization does not require a specific number of cells or processors provided they are arranged in a layer. We will formalize it using the proposed terminology by Mira and Moreno-Díaz [6].

According to it, let be I(x,y,t) the volume of intensities that constitutes our Input Space (that is, the incident image) and O(x,y,t) the Output Space of our system which can be again an image (in the artificial processing) or a pattern of responses (in the natural counterpart). The Relational Structure that links both Spaces will be defined in two parts: let be $T_1(I(x,y,t))$ and $T_2(I(x,y,t))$ two linear transformations whose kernels are $W_1(x,y,x',y',t)$ and $W_2(x,y,x',y',t)$:

$$T_i(I(x,y,t)) = \int_c I(x,y,t) W_i(x,y,x',y',t) dx'\,dy'$$

with i=1,2

such that the kernels can be expressed as:

$$W_i(x,y,x',y',t) = E(x,y,x',y')\Theta(t) \quad \{1\}$$

The Theorem of invariances can be written, then, as follows:

Theorem: Given two transformations T_1 and T_2 of the type described by Eq. {1} such that:

$$\int_c W_i(x,y,x',y',t)\,dx'\,dy' = 0 \quad \{2\}$$

for i=1,2, where c is the domain for every (x,y), that is, the receptive field, then the non-linear transformation

$$T = T_1/T_2$$

is invariant against homotecies and traslations of intensity of the form kI(x,y,t)+R(t), where k is a constant and R(t) is an arbitrary function of time t.

Proof: Let us begin with T_1,

$$T_1(kI(x,y,t)+R(t)) = \int_c kI(x,y,t)E(x,y,x',y')\Theta(t)dx'\,dy' +$$

$$+\int_c R(t)E(x,y,x',y')\Theta(t)\,dx'\,dy' =$$

$$= k\int_c I(x,y,t)E(x,y,x',y')\Theta(t)\,dx'\,dy' +$$

$$+R(t)\Theta(t)\int_c E(x,y,x',y')\,dx'\,dy' =$$

$$= kT_1(I(x,y,t)).$$

For T_2, the analysis is identical, and then for T we have:

$$T(kI(x,y,t)+R(t)) = \frac{T_1(kI(x,y,t)+R(t))}{T_2(kI(x,y,t)+R(t))} = \frac{kT_1(I(x,y,t))}{kT_2(I(x,y,t))} =$$

$$= \frac{T_1(I(x,y,t))}{T_2(I(x,y,t))} = T(I(x,y,t)) \quad \text{q.e.d.}$$

In our model $O(x,y,t) = T(I(x,y,t))$.

It is easy to see that a basic requisite in this description is expressed by {2}, which means, in Neural Nets language, that the receptive field structure of the cells computing T_i have to have antagonistic structures like center-surround or vertical (horizontal) bars canceling each other. These kinds of arrangements of the receptive fields have been found almost everywhere in Nervous Systems by neurophysiological experiments. On the other hand, it is also necessary that the units computing T_i have the same, or almost the same, receptive field localization.

Also, the structure of T makes it invariant against temporal global traslations of light intensity. Local traslations would destroy the invariant representation, as is easy to prove substituting R(t) by R(x,y) in the above analysis. Given some R(x,y), in order to maintain the invariance of T we would need to formulate very restrictive conditions both on R and the spatial factor of W_i.

The fact that T is calculated as a ratio, which is the key non-linearity of the construct (a divisive process), does not affect biological realism of the model since this kind of operations are comparable to presynaptic inhibitions, and as such they are known since more than twenty years ago [7].

Physiological Basis in the geniculo-cortical pathway.

The computational structure of the net computing the previously explained calculation is shown in Figure 1. The first layer of computing units calculate T_1 and T_2 and both results interact

The output of information from retina through the optic nerve follows three channels toward higher brain areas. Thus, the geniculo-cortical path, the tectopulvinar and the pretectum pathways have been described. The first of them, the geniculo-cortical, seems to be specialized in pattern analysis [8] and is the biggest of them, the one receiving more fibres from ganglion cells. The first stop of the signals on his way is at the Lateral Geniculate Nucleus, where a topographic representation of the whole retina can be found. The cells in LGN are arranged in six perfectly defined layers, and attending to the size of the cells, this layers could be divided again into two groups: the manocellular layer and the parvocellular layer [9]. The magnocellular layer is formed by big cells image illumination characteristics and the parvocellular includes smaller cells involved in color coding.

Our model can be applied on the magnocellular layer. The transformations T_1 and T_2 are calculated by the ganglion cells and the result reaches the LGN via the optic nerve. The cells at the magnocellular layer of LGN would be the processors to compute T using presynaptic inhibition. The simplicity of the figure mimics the simplicity found in the physiology of LGN, where each cell receives only a few input lines from retina including inhibitory effects [10]. Also, it has been found that stimulation of the retina with a spot of light produces the same kind of responses in both, ganglion and LGN cells which is consistent with our model. Then, one of the outputs of LGN cells would be an invariant representation of the light pattern already coded by ganglion cells. The signals from LGN to cortex are perfectly segregated: there is a zone in layer IV that receives input only from the parvocellular layer of LGN, other only from the magnocellular and still a third one receiving from both. This segregation is maintained in Area 18. Using this mechanism of invariances calculation by a zone of the LGN, the visual cortex has, among others, a fixed representation of outside world independent from sudden global changes of light intensity which is useful when discriminating movement, one of the largely known tasks of cortical cells connecting to the magnocellular layer of LGN [11,12,13].

An example on Image Processing.

As an example of the explained method, we took a 512x512 pixel black and white image with 256 grey levels (Picture 1). The image was divided into receptive fields of size 3x3 to convolve it using a Laplacian-type mask with maximum pixel overlapping between neighbours. The resulting processed image is again 512x512 in size. On every field, two operations were performed, the ones whose kernels can be seen in one dimension in Figure 2. For both kernels (convolution matrices), the sum of coeficients is zero. These operations correspond to T_1 and T_2. From them, T was calculated as a quotient and the result is given in Picture 2.

Then, the original image was displaced with k=3 and R(t)=70, according to the mathematical formalism used here (Picture 3), and then re-scaled to fit pixel intensities into the limits of representation (0-255). Both T_1 and T_2 were again computed. The result of T on the original and displaced images is shown in Pictures 2 and 4. The subtraction of both is a black picture , showing that both are the same. Thus, we get an invariant representation of the original picture no matter the general level of illumination.

The practical use of this technique presents, besides obvious advantages, some inconvinients. In natural environments, where the variation of light does not follow such restrictive law as expressed in the theoretical part of this paper, the invariance can be affected. Then, the local computation should be accompanied by a global thresholding to get a pseudo-invariance. This, however, would cancel the validity of the biological basis for the model, since no known global computation takes place in the retina or in the Lateral Geniculate Body.

Conclusions.

Mechanisms for obtaining invariant representations of features in the first steps of information processing are important in Nervous Systems for they provide the basis of reliable pattern and movement discrimination. In order to do this, one can expect that the visual cortex work, at least to some extent in the first steps, with low-level invariat versions of the image impinging on the retina. One of the roles of the cells in LGN, from our point of view, is to get this kind of operation performed for the cortex. Then, the Lateral Geniculate Nucleus would be the first place where a generalization of the input data is available to higher centers of the brain.

The mechanism presented in this paper, that only needs divisive inhibition, uses a very simple mathematical formulation and is also valid to explain neuronal adaptation, for instance in fovea. In this case, the Input Space corresponds to the output of the receptors, the Computational Layer I corresponds to the bipolar layer and the presynaptic inhibition would take place in the Inner Plexiform Layer.

On the artificial side, the presented parallel mechanism is suitable of being used in artificial perceptual systems since it is completely based on tools widely used in Systems Theory. Some problems, however, may arise in the practical use of this method, concerning the eventual saturation of intensity in some images as well as a redefinition of the level 0 of light intensity to avoid illegal operations.

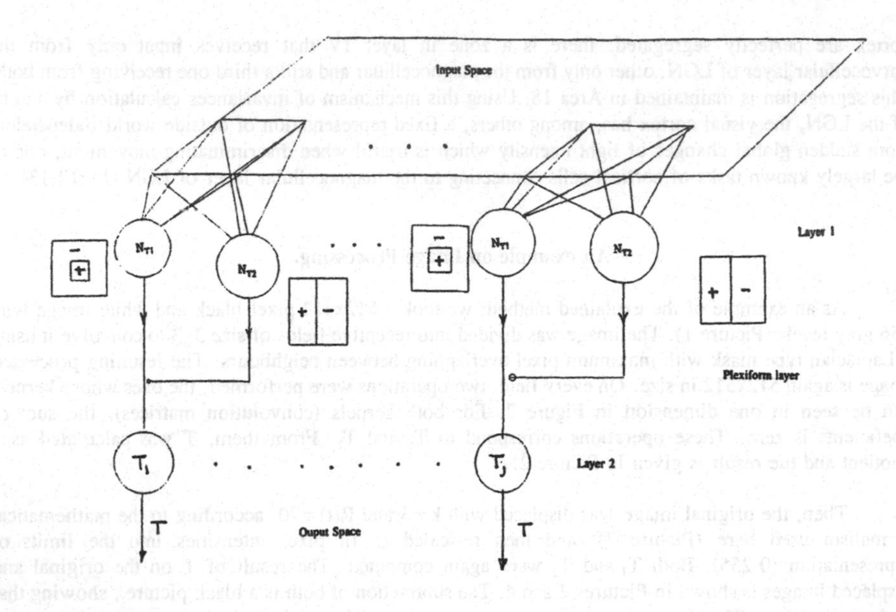

Figure 1: Sketch of a two dimensional Net to compute the invariant transformation T as explained in text. The boxes to the right and left of the cells show the structure of their kernels.

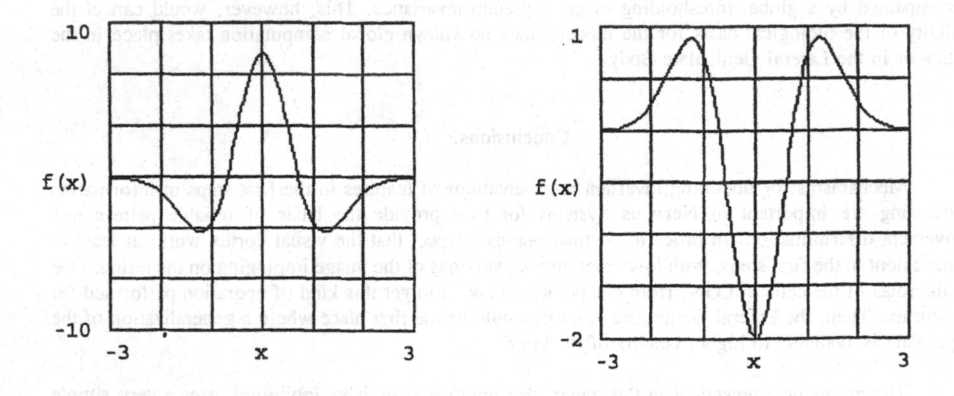

Figure 2: The two kernels used in the computer example. The needed bidimensional extension for image processing is obtained as a rotation in both cases.

Picture 1 Picture 2

Picture 3 Picture 4

Figure 3. An axample on image processing. Picture 1, 2, 3 and 4. See description in the text.

References.

[1] Ballard DH, Brown CM. "Computer Vision", Prentice Hall, New Jersey, USA, 1982.
[2] Moreno-Díaz R, Rubio E, Nuñez A. "A layered model for visual processing in avian retina", Biol. Cybernetics V38, pp85-79, 1980.
[3] Fernández Escartín V, Moreno-Díaz R. "A spatio-temporal model of cat's retinal cells", Biol. Cybernetics V30, pp15-22 (1978).
[4] Rosenfeld A, Kak C, "Digital Picture Processing", Academic Press, New York, USA 1982.
[5] Muñoz-Blanco JA, "Jerarquización de estructuras de nivel bajo y medio para Reconocimiento Visual", Tesis Doctoral, Universidad de Las Palmas, 1987.
[6] Mira J, Moreno-Díaz R. "Un marco teórico para interpretar la función neuronal a altos niveles" in "Biocibernética", Mira y Moreno-Díaz Eds. Siglo XXI, Madrid, 1984.
[7] Lettvin JY, Quarterly Progress Report No. 73, Research Lab. of Electronics, M.I.T., Cambridge, Mass, 1965. pp199-208.
[8] Kolb B, Whishaw I, "Human Neuropsychology", Third Ed. Freeman and Co., New York, 1990.
[9] Truex RC, Carpenter MB "Human Neuroanatomy", Williams and Wilkins, Baltimore, 1969.
[10] Stone J, Freeman Jr RB, "Neurophysiology of form discrimination" in Handbook of Sensory Physiology, Central Visual Information, VA pp154-198, Springer Verlag, New York 1973.
[11] Dow MB, "Nested maps in Macaque Monkey Visual Cortex" in Science of Vision, KN Leibovic Ed., Springer Verlag, New York, 1990.
[12] Livingstone MS, Hubel DH "Psychophysical evidence for separate channels for the perception of form, color, movement and depth", J. of Neurosci. V7, pp3416-3468, 1987.
[13] Hubel DH, Wiesel TN "Receptive fields, binocular interactions ans functional architecture in cat's visual system", J. of physiol. V160, pp106-154, 1962.

THE PANCREATIC B-CELL AS A VOLTAGE-CONTROLLED OSCILLATOR.

J.V. Sánchez-Andrés and B. Soria.

Dept. de Fisiología. Instituto de Neurociencias. Univ. Alicante. Aptdo. 374, 03080 Alicante. Spain. Phone: (346) 565 98 11; Fax: (346) 565 85 39; E-mail: andres@EALIUN11

ABSTRACT

We recorded the intracellular activity of pancreatic B-cells. The B-cells are the biological elements of the glucose control system. The action of glucose on the pancreatic B-cells take place in a two step-fashion: 1) concentrations lower than 6 mM glucose induces a progressive depolarization, 2) for higher concentrations the cells start to oscillate. This oscillation develops between two levels of potential that remain stable independently of the glucose concentration. The effect of increased glucose concentrations is transduced in terms of modulation of frequency of the oscillations rather than in a sustained depolarization. We stress the analogies between this control system and a voltage-controlled oscillator fed by a comparator.

1. INTRODUCTION.

The absorption of glucose by the majority of the tissues in the organisms is controlled by the hormone insulin. These tissues use glucose or other metabolites as a function of the insulin levels in blood. The more remarkable exception is the nervous tissue: glucose penetrates inside the neurons with independence of insulin. Additionally, neurons are unable to use other substrates apart of glucose as metabolic substrates. The abnormalities in these parameters produce the diabetes when chronically installed, and hypo- and hyper-glycemic commas when installed acutely. The deleterious consequences of the last syndromes over the brain come from the deviations of the glucose levels far away of the physiological levels (for a review see ref. 5)

Insulin is produced in the endocrine pancreas, consisting of several thousands groups of cells called islands of Langerhans distributed inside the exocrine pancreas (Fig 1A). Each island contains 2-3000 cells, 80% out of them are insulin-secretors (B-cells) (Fig 1B). Pancreatic B-cells are equipped with a series of specific membrane channels which have now been well-defined in microelectrode and patch-clamp studies. So far, a Na^+ channel, two types of voltage-activated Ca^{2+} channels, referred to as fast and slowly inactivating, three types of K^+ channels, referred to as ATP-sensitive, Ca^{2+} activated and delayed rectifier, and a voltage independent but glucose sensitive Ca^{2+} channel have been identified in normal B-cells (14). These channels provide the basis for the excitable behavior of the B-cells. On the other hand, the same type of channels have been described in neurons, underlying the same functions. Under this scope the pancreatic B-cells turns out to be a potentially interesting cell model for neuronal simulations as far as: a) they are endowed with sets of channels, very like regular neurons, but b) they lack the geometrical complexity provided by the dendritic and axonal branching. The glucose sensitivity is coupled to the electrical pattern and to the insulin secretion process through a chain of metabolic a biophysics steps. Briefly, when a resting pancreatic B-cell is challenged with glucose (or other metabolic susbstrate), the sugar interacts with a tissue-specific glucose transporter, enters the cell and is rapidly phosphorilated. The ATP generated by glucose metabolism blocks the K^+ channel controlling resting membrane potential. As these channels close, the cells depolarizes, causing the opening of voltage-activated Ca^{2+} channels, which, in turn, lead to Ca^{2+} entry. The increased intracellular calcium concentration triggers insulin release (9, 15), which stimulates glucose absorption by tissues, then, reducing blood glucose concentration. The whole process constitutes a loop of negative feed-back, permitting the control of the blood glucose concentration (Fig. 3).

Islands of Langerhans constitute an unique example of electrically and metabolically coupled network. As a system they show several prominent properties: a) B-cells into an island constitute an heterogeneous population coupled through gap-junctions, b) they are able to respond with an oscillatory pattern in given experimental conditions (1), and c) the electrical pattern is synchronic for every cell into an islet (4).

Although there are evidences supporting that the whole islet works as a functional unit or "syncithium", studies with isolated B-cells have shown a high degree of heterogeneity with respect to insulin biosynthesis (10), glucose-induced intracellular calcium ($[Ca^{2+}]_i$) changes (12) and insulin release (11). We have recently shown that, at intermediate glucose concentrations (7-16 mM), $[Ca^{2+}]_i$ measured in single islets undergo oscillations which are due to glucose-induced bursting of electrical activity (13, 9) indicating that the whole islet is simultaneously active. Despite their heterogeneous responses to glucose the whole population of B-cells has the ionic mechanisms that lead to a rise of $[Ca^{2+}]_i$ when ATP-regulated K^+ channels are blocked (14). Under the framework of a coupled tissue, cell-to-cell differences have attenuated functional consequences (13). B-cells grouped in island of Langerhans constitute an unique example of paraneural network in which electrical and metabolic coupling through gap junctions permits to build homogeneous responses from an heterogeneous population of cells. On the other hand, the islands of Langerhans have a rich inervation that provides the possibility of a precise regulation from the nervous system (7, 8).

The aim of this paper is to show the parallelism between the functionality of an island of Langerhans and a voltage-controlled oscillator (VCO).

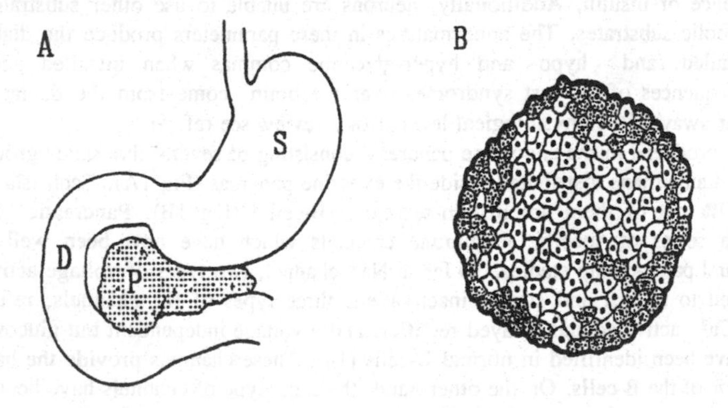

Fig 1. A. Schematic representation of the pancreas position in the abdomen. S: stomach; D: duodenum; P: exocrine pancreas. Dots represent the islands of Langerhans. B. Diagram representing an island of Langerhans. White cells: B-cells.

2. MATERIALS AND METHODS.

Intracellular electrical activity of B-cells from mouse islets of Langerhans was recorded with a bridge amplifier. Recordings were made with thick wall borosilicate microelectrodes pulled with a PE2 Narishige microelectrode puller. Microelectrodes were filled with 3 M potassium citrate, resistance around 100 MΩ. The modified Krebs solution used had the following composition (mM): 120 NaCl, 25 $NaHCO_3$, 5 KCl, 2.6 $CaCl_2$ and 1 $MgCl_2$ and was equilibrated with a gas mixture containing 95% O_2 and 5% CO_2 at 37°C. Islands of Langerhans were microdissected by hand. Data recorded were directly monitored on a Tektronix oscilloscope and stored with a magnetic tape recorder for further analysis.

3. RESULTS.

3.1. Effect of glucose on the steady membrane potential.

In the absence of glucose the membrane potential of the pancreatic B-cells remains stable around a value of -60 mV. Glucose addition induces an steady concentration-dependent depolarization (Fig. 2, top). If the glucose concentration in the bath arrives to a level of about 10 mM, the cell is depolarized to -50 mV and starts to oscillate. Further increases in the glucose concentration will drive the cell into an oscillatory pattern (see 3.2). The maximal response to glucose is obtained for concentrations bigger than 20 mM, consisting of a continuous depolarization at the level of -35 mV (Fig. 2, bottom).

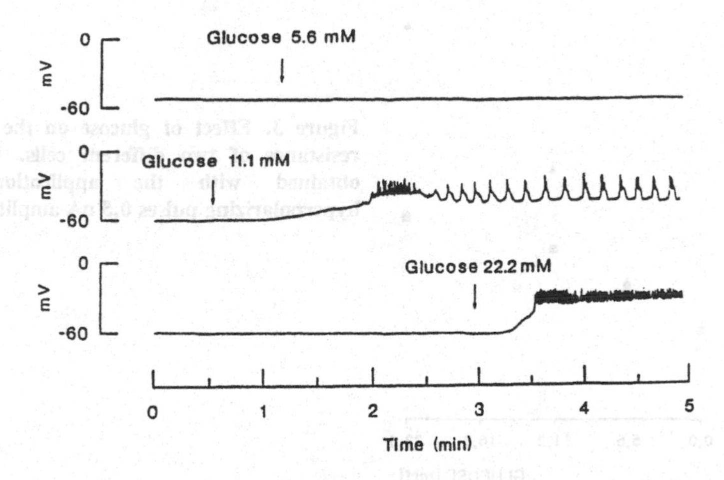

Figure 2. Effect of the addition of three glucose concentrations over the membrane potential of a pancreatic B-cell. Before the arrows, no glucose. After the arrows, glucose as indicated in the figure. Records were obtained from the same cell and are representative of the typical B-cell pattern.

3.2. Effect of glucose on the oscillatory pattern.

The installation of the oscillatory pattern takes place typically in two phases: <u>1st. phase</u>. After the appropriate glucose concentration is added the cell steadily depolarizes up to a threshold level. Then the membrane depolarizes abruptly up to a potential identical with the top of the further oscillations. The cell stays at this level for a relative prolonged period (average 30 s) before getting into oscillations (Fig. 2, center). The duration of this phase depends on the previous concentration of glucose, lasting longer as lower was that glucose concentration (data not shown). <u>2nd. phase</u>. After the sustained depolarization the cell starts to oscillate with a frequency of 4-8 oscillations/min. for a glucose concentration in the bath of 11.1 mM. The frequency of the oscillation is stable for every cell and is the same in the other cells of the same island (data not shown), but it can be different in cells from other islets. The oscillations develop between two states of potential that constitute the top and the bottom of the oscillations. A complete oscillation is classified in two phases: active phase, when the membrane is depolarized, and silent phase, when the membrane is hyperpolarized. A further increase of the glucose concentration does not modify the absolute value of both states. Conversely, this increase of the glucose concentration modifies the ratio between the duration of the active and the silent phases of the oscillation. Then, a glucose concentration increase is followed by an increase in the duration of the active phase, versus a reduction of the silent phase. As described in 3.1, when the glucose concentration is high enough the silent phase is canceled, becoming the cell continuously at the level of potential of the active phase. Additionally, the action of glucose induces a linear increase in the input resistance of the pancreatic B-cells (Fig. 3)..

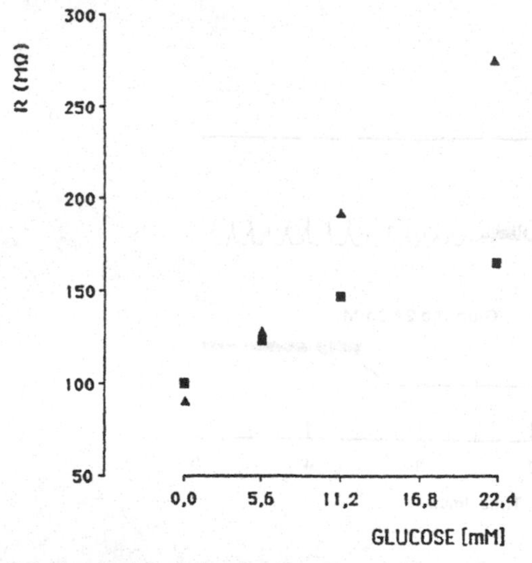

Figure 3. Effect of glucose on the input resistance of two different cells. Values obtained with the application of hyperpolarizing pulses 0.5 nA amplitude.

3.3. Analogies of the glucose effects with a voltage-controlled oscillator.

The basic principle of the pancreatic B-cell work consists of in monitoring the actual glucose level in order to compare it with a reference level physiologically pre-established. If the recorded concentration is higher than the reference level a series of mechanisms are going to be activated. The most remarkable is the activation of the oscillatory behavior of the cells. The final result is the insulin secretion which will activate the glucose metabolization in the peripheral tissues. This last step will reduce the glucose concentration to the reference level. The whole process can be understand as a voltage-controlled oscillator fed by a comparator (Fig. 4). The comparator is biologically implemented through the capability of the B-cells to act as glucose sensors. The change in the glucose concentration induces a change in the voltage level associated to changes in the input resistance and capacitance of the cells. Simultaneously the coupling between cells is modified. This whole mechanism is able to switch the system to an oscillatory pattern and to modulate its frequency, depending on the voltage change generated in the comparator step.

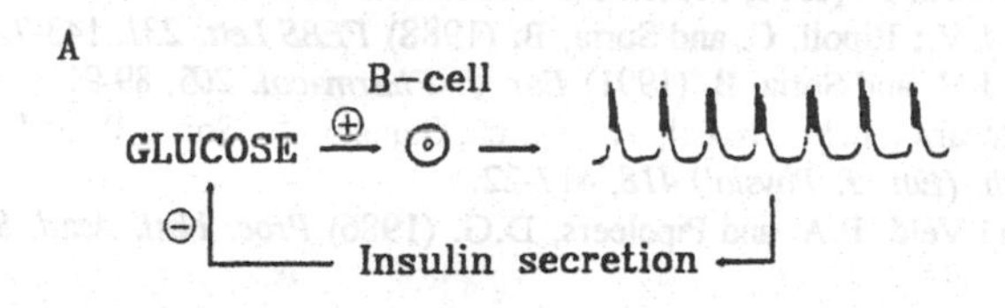

Figure 4. Physiological diagram of the control system for the insulin-glucose system.

4. DISCUSSION.

The role of glucose is complex: a. initial depolarization; b. reset the system from working as an RC circuit to behave as an oscillator; and c. modulation of the frequency of the oscillations. These actions are the result of the B-cell capability of working as a glucose sensor. These capabilities are mediated through the glucose metabolism, which affects the activity of several ionic channels. Particularly, it has been demonstrated the reduction in the conductance of an ATP-sensitive K^+ channel, which is used to depolarize the cell into the level where bursting is possible. Additionally, it has been hypothesized that the regulation of the rate of activation and inactivation of two different Ca^{2+} channels can be used to induce bursting activity (2). The good knowledge about the biophysical constituents of the B-cell has permitted to develop computational models to simulate the B-cell activity (6). These models are based on the single-cell level. Unfortunately, the isolated B-cells are not able to oscillate. To obtain oscillations are required at least small clusters of cells. This strongly suggests the convenience of incorporating more than one cell into the models.

The oscillatory nature of the system suggests the existence of a positive feed-back. this feed-back can be provided by the neighbor cells. This hypothesis rests on the next experimental evidences: a. the cells into an islet are electrically coupled, b. the degree of coupling is regulated by glucose (3) and, c. all the cells in every islet oscillate simultaneously (9). The lack of feed-back would be the element that impairs the isolated cells to oscillate. Under this scope the understanding of the oscillatory behavior of the B-cells would require the appropriate study of its effects on the balance of the coupling resistance and capacitance between cells. Whether these features turn out to be compatible with future experimental work remains to be seen.

ACKNOWLEDGMENTS. Supported by Spanish DGICYT PL87-0789 and PM92-0115, and FISss 90E-1262-6F. We thank Mr. A. Pérez Vergara for technical support and valuable advice.

5. REFERENCES.

1. Atwater, I.; Dawson, C.M.; Scott, A.; Eddlestone, G. and Rojas, E. (1980) *J. Horm Metab. Res. Suppl. 10*, 100-7.
2. Cook, D.L.; Satin, L.S. and Hopkins, W.F. (1991) *TINS 14*, 411-4.

3. Eddlestone, G.T.; Goncalves, A.A.; Bangham, J.A. and Rojas, E. (1984) *J. Membr. Biol. 77*, 1-14.
4. J.C. and Meissner, H.P. (1984) *Experientia 40*, 1043-52.
5. Kahn, S.E. and Porte Jr. D. In: *Diabetes Mellitus*, Rifkin and Porte eds., 4°. ed, Elsevier, NY, 1990, pp 436-56.
6. Keizer, J. and Smolen, P. (1991) *Proc. Natl. Acad. Sci. USA 88*, 3897-901.
7. Sánchez-Andrés, J.V.; Ripoll, C. and Soria, B. (1988) *FEBS Lett. 231*, 143-7.
8. Sánchez-Andrés, J.V. and Soria, B. (1991) *Eur. J. Pharmacol. 205*, 89-91.
9. Santos, R.M.; Rosario, L.M.; Nadal, A.; García Sancho, J.; Soria, B. and Valdeolmillos, M. (1991) *Pflugers Arch. (Eur. J. Physiol) 418*, 417-22.
10. Schuit, F.C.; In t Veld, P.A. and Pipeleers, D.G. (1986) *Proc. Natl. Acad. Sci. USA 85*, 3865-9.
11. Solomon, D. and Meda, P. (1986) *Exp. Cell Res. 162*, 507-20.
12. Valdeolmillos, M.; Nadal, A.; Contreras, D. and Soria, B. (1992) *J. Physiol. 455*, 173-86.
13. Valdeolmillos, M.; Nadal, A.; Soria, B. and García-Sancho, J. (1993) *Diabetes. In press.*
14. Soria, B.; Chanson, M.; Giordano, E.; Bosco, D. and Meda, P. (1991) *Diabetes 40*, 1069-78.
15. Wollheim, C.B. and Pralong, W.F. (1990) *Biochem. Soc. Trans. 18*, 111-4.

APPROXIMATION OF THE SOLUTION OF THE DENDRITIC CABLE EQUATION BY A SMALL SERIES OF COUPLED DIFFERENTIAL EQUATIONS

Jaap Hoekstra
Delft University of Technology, Dept. Electrical Engineering,
P.O. Box 5031, 2600GA Delft, The Netherlands
e-mail: jaap@neuron.et.tudelft.nl

Abstract

In most literature on artificial neural network models the node (neuron) is represented as a point, learning rules are based on activity of other nodes and the activity of the node itself. Models incorporating local learning rules allows increase and decrease of weights on bases of activities of other nodes if the synapses are placed close together on the same dendrite. These models model the biological dendrite by solving an equivalent electrical circuit, consisting of several compartments. Current models solve the electrical circuit model by the numerical calculation of a recurrence relation in which the current of a compartment is expressed in the value of the current in a next compartment. In this way it is always necessary to model the whole dendrite, even if we are only interested in local effects. In this paper an alternative mathematical description is proposed on bases of which local activities can be calculated by only simulating a part of the dendrite model.

1 Introduction

In artificial neural network research the dendritic membrane is modeled by the cable equation, see for example [1]. In this model the axial current in the dendritic membrane is assumed to flow along an unbranched cylinder of uniform cross-section [3]. The axial current is then one-dimensional, and the current density changes *only* if current enters or leaves through the cell membrane. The model is based on two assumptions concerning the impedances. These are that the intracellular medium acts as an ohmic resistance and that the surface membrane acts as a leaky capacitance.

In general the transport of potential pulses in the membrane cannot be solved analytically in the model. Numerical solution can be obtained by using compartmental modeling, in which the membrane is modeled by a cable of equivalent electrical circuits. Synaptical inputs on the dendritic membrane can be translated into (weighted) currents at compartments, so the model can be used to describe local learning, and can direct the development of local learning rules.

2 Mathematical Description (1)

The mathematical description of compartmental models of dendrites contains a system of N coupled first order differential equations, if we consider a discrete model of the dendrite with N compartments. In the model, only information transport in dendrites towards the cell body will be considered, and the boundary conditions give rise to a initial value problem. The system of coupled first order differential equations can be approximated by a set of finite difference equations.

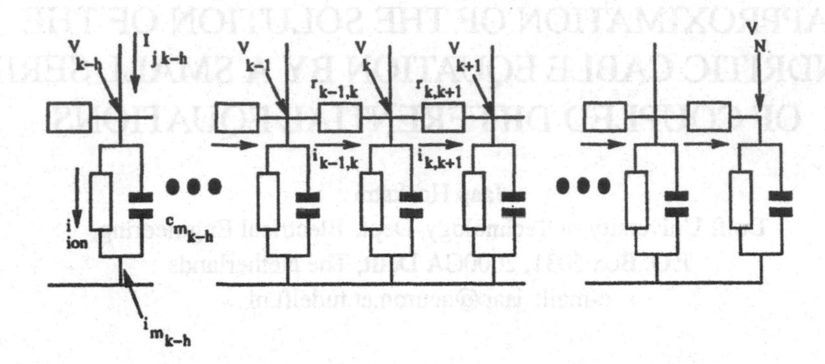

Figure 1: Equivalent electrical circuit for dendritic membrane

The equations are derived from Kirchhoff's current law, which states that in each compartment, k, the net current through the membrane, i_{m_k}, must equal the sum of the axial current that enters that compartment and the radial current that may enter as a consequence of a synaptic input at that compartment, minus the axial current that leaves it. For the kth compartment the membrane current is:

$$i_{m_k} = i_{k-1,k} + I_{j_k} - i_{k,k+1} \tag{1}$$

where, see fig. 1, $i_{k-1,k}$ is the axial current that flows from compartment $k-1$ to k, where $i_{k,k+1}$ is the current that flows from compartment k to $k+1$, and where I_{j_k} is the current flow caused by a synaptic input from neuron j. If there is no synapse attached to the kth compartment than I_{jk} is zero.

In the model the membrane impedance is represented by a parallel resistance and capacitance circuit. The membrane current will then be given by the sum of two components. The ionic current, i_{ion_k}, which is carried by the flow of ions across the membrane, and the capacitive current flow which results in a change in the amount of charge separated by the membrane. For the kth compartment the membrane current can be expressed as:

$$i_{m_k} = c_{m_k} \frac{dV_k}{dt} + i_{ion_k} \tag{2}$$

where V_k is the membrane potential measured with respect to the resting potential.

Equations 1 and 2 relate the currents, the capacitance and voltage of a compartment. To obtain an expression which includes the axial resistance within a compartment, the axial current is described as the voltage gradient between directly connected compartments divided by the resistance between these compartments:

$$c_{m_k} \frac{dV_k}{dt} = \frac{V_{k-1} - V_k}{r_{k-1,k}} + I_{j_k} - \frac{V_k - V_{k+1}}{r_{k,k+1}} - \frac{V_k}{r_{m_k}} \tag{3}$$

where i_{ion_k} is expressed as $\frac{V_k}{r_{m_k}}$ in which r_{m_k} the radial membrane resistance is.

This system of differential equations is discussed in standard literature on compartmental modeling [8, 7, 6, 4, 5, 9]. A disadvantage of this system is that the influence of individual compartmental parameters of neighboring compartments, such as the compartmental capacitances and radial membrane resistance, on potentials introduced in the cable are not formulated explicitly, nor is the time-derivative of the potential in neighboring compartments formulated explicitly. Insight in these parameters and the time-derivative of the potential in neighboring compartments is essential in the understanding of spatial-temporal learning.

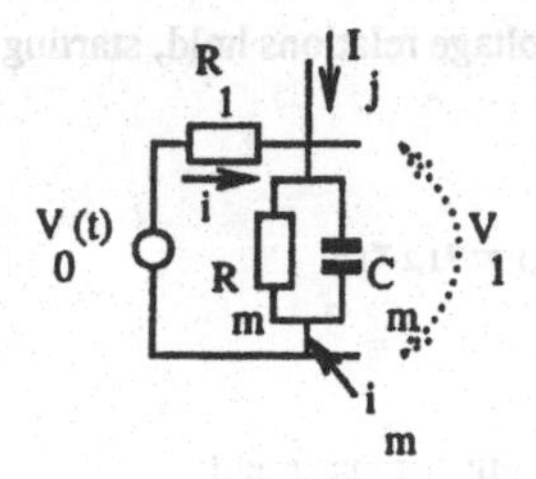

Figure 2: Simple membrane circuit

3 Mathematical Description (2)

In this section an equation for the voltage of a compartment is derived in which the individual capacitances and resistances of neighboring compartments explicitly appear. First the voltage in a simple dendritic membrane circuit is considered. Then the relations for the dendrite model consisting of N compartments are derived. Hereafter the equivalence of this equation and equation (3) is shown.

Fig. 2 shows a simple membrane circuit. The membrane has a axial resistance R_1 from the place where a synaptic input causes a voltage, V_0, towards the cell body. The the membrane impedance is represented by a parallel resistance and capacitance circuit, respectively R_m and C_m. As the voltage is applied a certain current i will flow in the circuit. The current through the membrane impedance equals the current through the axial resistance. For the potential across the membrane V_1 holds:

$$V_1 = V_0 - iR_1 \tag{4}$$

while for the currents the following equations hold:

$$i_m = i + I_j \tag{5}$$

$$i_m = C_m \frac{dV_1}{dt} + \frac{V_1}{R_m} \tag{6}$$

Combining 4, 5, and 6 we end up with the description of the system in terms of a differential equation:

$$(1 + \frac{R_1}{R_m})V_1 + R_1 C_m \frac{dV_1}{dt} = V_0 + R_1 I_j \tag{7}$$

Only for special forms of the pulse $V_0(t)$ can the differential equation be solved analytically. In general a numerical solving method will be used.

To obtain a solution for a finite cable of N compartments, see fig. 1, we start from the end of the cable and sum the currents through the different axial resistances:

$$i_{N-1,N} = i_{m_N} - I_{j_N} = c_{m_N} \frac{dV_N}{dt} + \frac{V_N}{r_{m_N}} - I_{j_N} \tag{8}$$

$$i_{N-2,N-1} = i_{N-1,N} + i_{m_{N-1}} - I_{j_{N-1}} = c_{m_N} \frac{dV_N}{dt} + c_{m_{N-1}} \frac{dV_{N-1}}{dt} + \frac{V_N}{r_{m_N}} + \frac{V_{N-1}}{r_{m_{N-1}}} - I_{j_N} - I_{j_{N-1}} \tag{9}$$

•

•

Resulting in the current $I_{k-1,k}$ towards compartment k:

$$I_{k-1,k} = \sum_{l=k}^{N} (c_{m_l} \frac{dV_l}{dt} + \frac{V_l}{r_{m_l}} - I_{j_l}) \tag{10}$$

At the same time the following voltage relations hold, starting from the beginning of the cable:

$$V_1 = V_0 - i_{0,1}\, r_{0,1} \tag{11}$$

$$V_2 = V_1 - i_{1,2}\, r_{1,2} = V_0 - i_{0,1}\, r_{0,1} - i_{1,2}\, r_{1,2} \tag{12}$$

•

•

Resulting in the voltage V_k in the kth compartment:

$$V_k = V_0 - \sum_{f=1}^{k} r_{f-1,f}\, i_{f-1,f} \tag{13}$$

Combining 10 and 13 we obtain in case of a potential caused by a synaptic pulse on h compartments from k:

$$V_k = V_{k-h} - \sum_{f=k-h+1}^{k} r_{f-1,f} \sum_{l=f}^{N} \left(c_{m_l}\frac{dV_l}{dt} + \frac{V_l}{r_{m_l}} - I_{jl}\right) \tag{14}$$

An equation in which all resistors and capacitances as well as the time-derivatives explicitly appear.

4 Equivalence of both descriptions

This section shows the equivalence of both descriptions.

Consider a part of the cable as shown in fig. 3

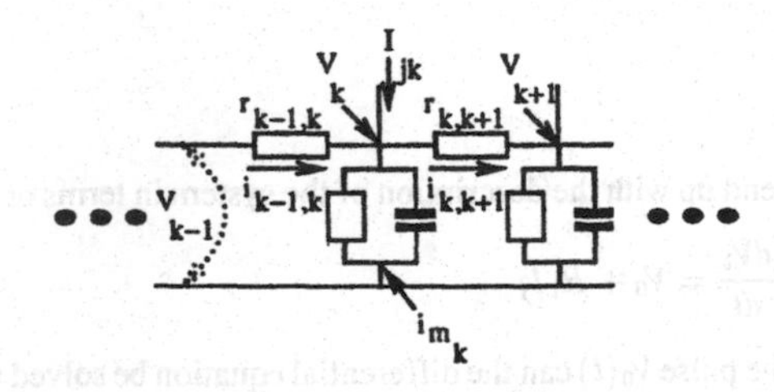

Figure 3: Part of the cable

Using the equations for V_k and V_{k+1} in case that $h = 1$:

$$V_k = V_{k-1} - r_{k-1,k} \sum_{l=k}^{N} \left(c_{m_l}\frac{dV_l}{dt} + \frac{V_l}{r_{m_l}} - I_{jl}\right) \tag{15}$$

and

$$V_{k+1} = V_k - r_{k,k+1} \sum_{l=k+1}^{N} \left(c_{m_l}\frac{dV_l}{dt} + \frac{V_l}{r_{m_l}} - I_{jl}\right) \tag{16}$$

We obtain

$$V_k = V_{k-1} - r_{k-1,k}\left(c_{m_k}\frac{dV_k}{dt} + \frac{V_k}{r_{m_k}} - I_{jk}\right) - r_{k-1,k} \sum_{l=k+1}^{N} \left(c_{m_l}\frac{dV_l}{dt} + \frac{V_l}{r_{m_l}} - I_{jl}\right) \tag{17}$$

Rewriting this equation gives:

$$c_{m_k}\frac{dV_k}{dt} = \frac{V_{k-1}-V_k}{r_{k-1,k}} + I_{j_k} - \frac{V_k}{r_{m_k}} - \sum_{l=k+1}^{N}\left(c_{m_l}\frac{dV_l}{dt} + \frac{V_l}{r_{m_l}} - I_{j_l}\right) \tag{18}$$

Which equals:

$$c_{m_k}\frac{dV_k}{dt} = \frac{V_{k-1}-V_k}{r_{k-1,k}} + I_{j_k} - \frac{V_k - V_{k+1}}{r_{k,k+1}} - \frac{V_l}{r_{m_l}} \qquad \square \tag{19}$$

5 A cut-off approach for realistic models

Basic to a local learning scheme is that a voltage pulse induced at a specific compartment only affects other compartments within a limited range. Compartmental models of dendrites based on biological plausible data also show such a limited range [8, 2].

In relation to the above described electrical model it means that the influence of a voltage pulse strongly decreases after a limited number of compartments. And on the other hand that the effect of neighboring compartments on a voltage pulse is only limited to a small number of them.

In relation to the above described mathematical model (2) it means that the formula for the compartmental voltage can be approached by taken not the complete summations but by using only summations over a limited set of compartments. This can be expressed by a the use of cut-off parameters k_{max} and h_{max} :

$$For: h \leq h_{max}: \tag{20}$$

$$V_k = V_{k-h} - \sum_{f=k-h+1}^{k} r_{f-1,f} \sum_{l=f}^{k_{max}}\left(c_{m_l}\frac{dV_l}{dt} + \frac{V_l}{r_{m_l}} - I_{j_l}\right) \tag{21}$$

To get an idea of the values for h_{max} and k_{max} we have performed some simulations.

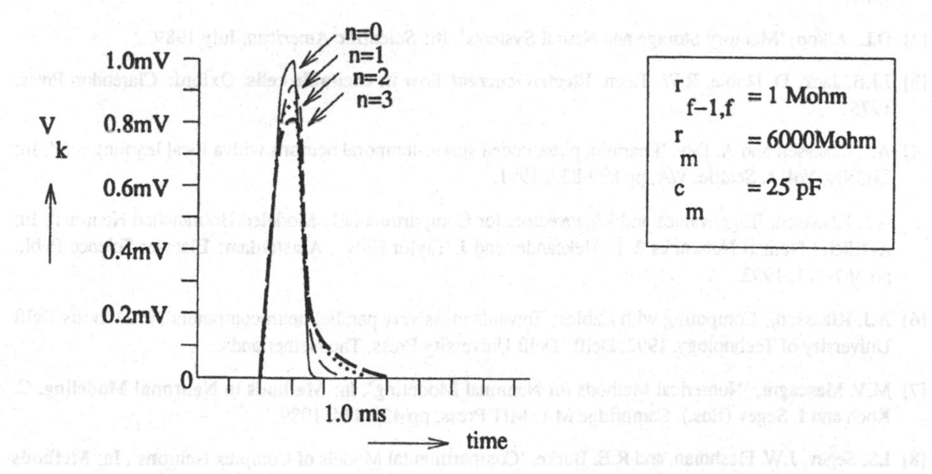

Figure 4: Simulation of the potential in a compartment, caused by a voltage pulse in a previous compartment, as a function of the number of identical compartments following it.

Fig 4 illustrates the effect of the loading of the capacitances in forthcoming compartments. The effects decreases with a increasing number of compartments. Only the decrease of the potential of the

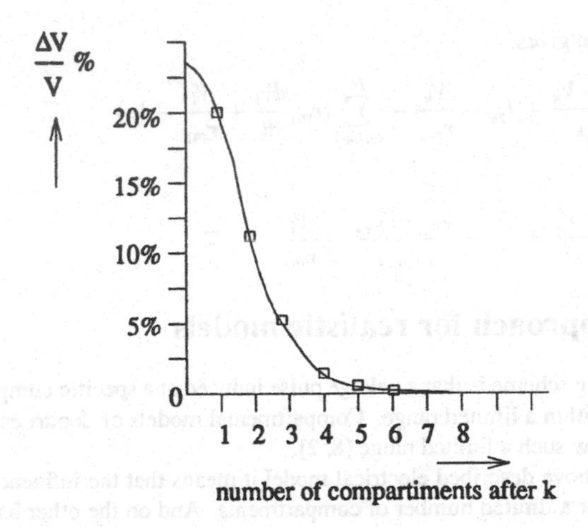

Figure 5: the relative effect of the numbers of compartments.

compartment up to a load of three compartments is shown. To show the effect of more compartments, the relative decrement is shown as a function of the number of following compartments in fig 5. It shows that a reasonable value for both k_{max} and h_{max} is 5.

The author is thankful to Mohammed Maouli (Delft University of Technology) for doing the simulations.

References

[1] **Methods in Neuronal Modeling, C. Koch and I. Segev (Eds.), Cambridge MA: MIT Press, pp 63-97, 1989.**

[2] **D.L. Alkon, 'Memory Storage and Neural Systems', In: *Scientific American*, July 1989.**

[3] **J.J.B. Jack, D. Noble, R.W. Tsien, Electric-current flow in excitable cells, Oxford: Clarendon Press, 1975.**

[4] **A.J. Klaassen and A. Dev, 'Learning pulse coded spatio-temporal neurons with a local learning rule', In: IJCNN, Vol. 1, Seattle, WA, pp 829-837, 1991.**

[5] **A.J. Klaassen, 'Eigenvalues and Eigenvectors for Compartmentally Modeled Unbranched Neurons', In: Artificial Neural Networks 2, I. Aleksander and J. Taylor (Eds.), Amsterdam: Elsevier Science Publ., pp 367-371, 1992.**

[6] **A.J. Klaassen, 'Computing with Cables: Towards massively parallel neurocomputers', PhD-thesis Delft University of Technology, 1992, Delft: Delft University Press, The Netherlands.**

[7] **M.V. Mascagni, 'Numerical Methods for Neuronal Modeling', In: Methods in Neuronal Modeling, C. Koch and I. Segev (Eds.), Cambridge MA: MIT Press, pp 439-484, 1989.**

[8] **I.S. Segev, J.W. Fleshman, and R.E. Burke, 'Compartmental Models of Complex Neurons', In: Methods in Neuronal Modeling, C. Koch and I. Segev (Eds.), Cambridge MA: MIT Press, pp 63-97, 1989.**

[9] **M.A. Wilson, U.S. Bhalla, J.D. Uhley, J.M. Bower, 'GENESIS: A System for Simulating Neural Networks', In: Advances in Neural Information Processing Systems 1, D.S. Touretzky (Ed.), Palo Alto: Morgan Kaufmann Publ., pp 485-492, 1989.**

A NEURAL NETWORK MODEL INSPIRED IN GLOBAL APPRECIATIONS ABOUT THE THALAMIC RETICULAR NUCLEUS AND CEREBRAL CORTEX CONNECTIVITY

Javier Ropero Peláez

ICAI Electronic Engineer.
C/Cantalejos 2 Madrid 28035 Spain.

ABSTRACT

This paper introduces a neural network architecture inspired in general rules of connectivity inside the thalamic reticular nucleus (TRN) and between it and the cerebral cortex. Applying several statistical theoremes it shows how this structure is able to learn, discovering distinctive features from the input information and projecting new informations over distinctive ancient ones so that the size of the storage is minimized. It doesn't need any type of error backpropagation mechanism because the input data is statistically orthogonalized.

1. RULES OF INTERCONNECTIVITY

Most of the information that arrives to the neocortex is relayed through the TRN. Many neurons in the spinal cord synapse with the relay neurons of the TRN.These relay neurons send its information to two principal sites: the cortex middle layer and the interneurons, inside the thalamus. Speaking in general terms the information that arrives to the middle layer is spread over the star shaped neurons, and from these it goes back to the interneurons inside the TRN through the pyramidal shaped ones. The interneurons are inhibitory and projects over other interneurons and over the relay neurons inside the TRN so that the interneurons inhibit each other (in a competitive way) and modulate the activation of the relay ones.

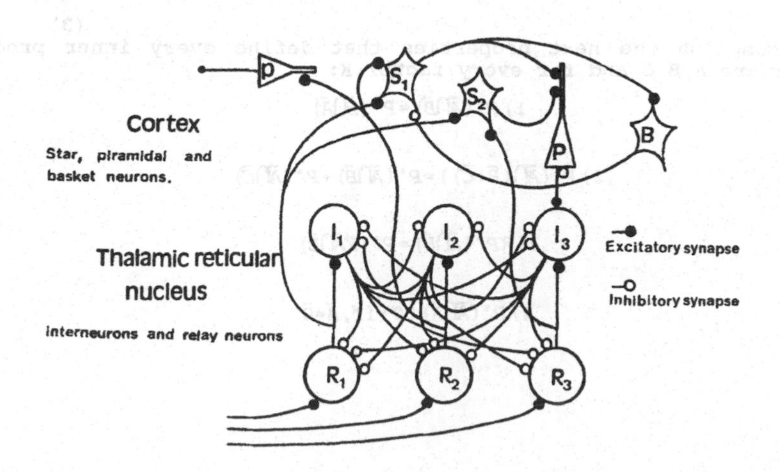

We've tryed to reflect this architecture in the previous, very simplified, figure. Many other types of interaction exist in the neocortex like the inhibitory effect of basket shaped neurons in the cortex and the association effect of small pyramidal neurons in the upper layer. However we believe that the type of interaction explained above is one of the most studied and documented (Crick F. & Asanuma., 1.989;Jahnsen,1984; Jones,1.985).

2. STATISTICAL ORTHOGONALIZATION OF THE INPUT DATA

The orthogonalization of input information is a necessary condition in order to achieve a hebbian learning in pattern associator models (McClelland,1988; Silva,1991). Because our model has a statistical background we´ll define a statistical orthogonalization of input data. Let´s suppose that an statistical event could be defined like a linear combination of other incompatible events that, joint, form the complete space, like this:

$$\vec{A}=0,3\vec{a}_1+0,5\vec{a}_2+0,2\vec{a}_3$$

$$\vec{a}_1\cup\vec{a}_2\cup\vec{a}_3=E \tag{1}$$

We can demonstrate that this kind of vectors constitute a linear space, with the addition between components with the same index and the multiplication defined in the usual way. These vectors form a linear or euclidean space because the whole of them are closed for the addition and multiplication and they accomplish the conmutative, associative and distributive laws for the addition and the multiplication and there exist the zero, identical and opposite elements. Let´s define a P^* function like this:

$$P^*(K\vec{A})=KP^*(\vec{A}) \tag{2}$$

P is the habitual probability function and "A" an event. Once we´ve got a linear space and a P^* function we are able to define the inner product:

$$P^*(\vec{A}\cap\vec{B}) \tag{3}$$

that accomplish the next properties that define every inner product wichever are A,B,C and for every factor K:

$$1)\; P^*(\vec{A}\cap\vec{B})=P^*(\vec{B}\cap\vec{A})$$

$$2)\; P^*(\vec{A}\cap(\vec{B}+\vec{C}))=P^*(\vec{A}\cap\vec{B})+P^*(\vec{A}\cap\vec{C})$$

$$3)\; KP^*(\vec{A}\cap\vec{B})=P^*(K\vec{A}\cap\vec{B})$$

$$4)\; P^*(\vec{A}\cap\vec{A})>0;\, if,\, \vec{A}\neq\vec{0}$$

If..

$$0<K\leq 1$$

(5)

then P and P_* have the same mathematical properties and we can put P instead P_* in the above expressions.

Therefore, knowing that in a generic linear space a vector could be expressed like a weighted sum of its components..

$$\vec{X}=\sum_{i=1}^{n}\frac{(\vec{X},\vec{e}_i)}{(\vec{e}_i,\vec{e}_i)}\vec{e}_i$$

(6)

(the parenthesis means an inner product)

..we can substitute these inner products by our, previously defined, statistical inner product, so that we could obtain a vectorial event expressed like a weighted sum of incompatible events.

$$\vec{A}=\sum_{i=1}^{n}\frac{P(\vec{A}\cap\vec{a}_i)}{P(\vec{a}_i\cap\vec{a}_i)}\vec{a}_i=\sum_{i=1}^{n}P(\vec{A}/\vec{a}_i)\,\vec{a}_i$$

$$\bigcup\vec{a}_i=\vec{E}$$

(7)

And, as the probability of a sum of incompatible events is the sum of the probabilities:

$$P(\vec{A})=P(\sum_{i=1}^{n}P(\vec{A}/\vec{a}_i)\,\vec{a}_i)=\sum_{i=1}^{n}P(\vec{A}/\vec{a}_i)\,P(\vec{a}_i)$$

(8)

..which is the total probability theoreme. The above expression helps us to understand the meaning of the factors that multiply an event in our euclidean space. Every factor is a conditioned probability: it shows how many times "A" occurs if we´ve got the previous ocurrence of "a_i". We´ve said before that our model orthogonalizes the afferent information. In our context two events are orthogonal or incompatible if their inner product is zero:

$$P(\vec{A}\cap\vec{B})=0$$

(9)

We can obtain a base of orthogonal events y_i from other non-orthogonal ones by substituting our inner product in the Gram-Schmidt formula:

$$\vec{y}_1=\vec{a}_1;\vec{y}_{r+1}=\vec{a}_{r+1}-\sum_{i=1}^{r}\frac{P(\vec{a}_{r+1}\cap\vec{y}_i)}{P(\vec{y}_i\cap\vec{y}_i)}\vec{y}_i$$

(10)

Applying to this expression the property:

$$P(A-B)=P(A)-P(b)\,;\,if,\,B\subseteq A \tag{11}$$

we obtain:

$$P(\vec{y}_1)=P(\vec{a}_1)\,;\,P(\vec{y}_{r+1})=P(\vec{a}_{r+1})-\sum_{i=1}^{r}\frac{P(\vec{a}_{r+1}\cap\vec{y}_i)}{P(\vec{y}_i\cap\vec{y}_i)}P(\vec{y}_i) \tag{12}$$

or :

$$P(\vec{y}_1)=P(\vec{a}_1)\,;\,P(\vec{y}_{r+1})=P(\vec{a}_{r+1})-\sum_{i=1}^{r}P(\vec{a}_{r+1}/\vec{y}_i)\,P(\vec{y}_i) \tag{13}$$

The weights in the synaptic connection of our model are defined as conditional probabilities. In this way $P(Y_1/Y_2)$ indicates how many times the interneuron 1 has been depolarized if the interneuron 2 has previously shot. The interneuron 1 probability of receiving a shot from interneuron 2 through the synapse that joints both neurons is $P(Y_1/Y_2)*P(Y_2)$ where $P(Y_2)$ is the shot probability of the interneuron 2. The shot probability in a neuron is the sum of the shot probability of its synapses. If there are inhibitory synapses their influences are substracted. In the particular case of interneurons all of them inhibit each other in a competitive way. Therefore the event of a shot in a particular interneuron tends to be orthogonal to the occurrence of the same event in the remaining interneurons. Suposse that we excite the relay neurons R_1 and R_2 with two stimulus "a" and "b" with shot probabilities P(a) and P(b).

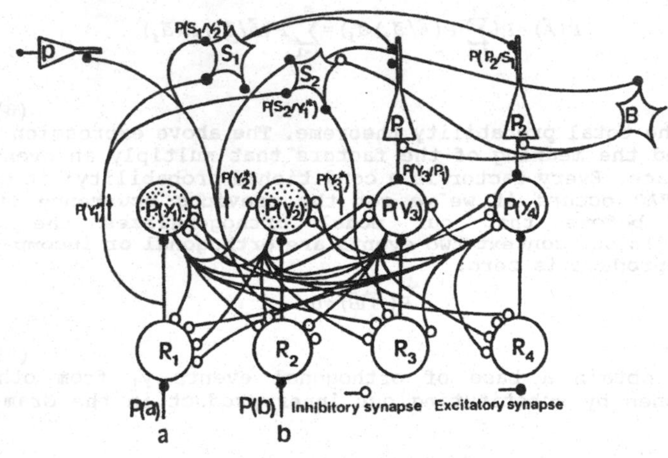

According to the previous explanations and if we suppose that the only interneurons activated are 1 and 2 (dotted in the picture), we can calculate the shot probability $P(y_1^*)$ and $P(y_2^*)$ of the relay neurons:

$$P(y_1^*)=P(a)-P(a/y_1)\,P(y_1)-P(a/y_2)\,P(y_2) \tag{14}$$

$$P(y_2^*)=P(b)-P(b/y_1)P(y_1)-P(b/y_2)P(y_2)$$

If these expressions are compared with equation 13 it could be seen that there are two interneurons whose shots are orthogonal events with probabilities P(y_1) and P(y_2). They also show that y_1^* and y_2^* are orthogonal to y_1 and y_2 but not between them because:

$$P(y_1^* \cap y_2^*) \neq 0$$

(15)

The non-orthogonal convergence of the shots "y_1^*" and "y_2^*" over the star shaped neurons S_1 and S_2 will produce the shot of these star neurons that will fire a pyramidal one that, in the same way, will draw a new interneuron to be active. With the activation of the new interneuron "y_3" and, after adjusting the conditional probabilities, the new shot probabilities of the relay neurons R_1 and R_2 will be:

$$P(y_1^*)=P(a)-P(a/y_1)P(y_1)-P(a/y_2)P(y_2)-P(a/y_3)P(y_3)$$

$$P(y_2^*)=P(b)-P(b/y_1)P(y_1)-P(b/y_2)P(y_2)-P(b/y_3)P(y_3)$$

(16)

That in short is:

$$P(\vec{Y}^*)=P(A\vec{B})-P(A\vec{B}/\vec{y}_1)P(\vec{y}_1)-P(A\vec{B}/\vec{y}_2)P(\vec{y}_2)-P(A\vec{B}/\vec{y}_3)P(\vec{y}_3)$$

or:

$$P(\vec{Y}^*)=P(A\vec{B})-P(A\vec{B}/\vec{Y})P(\vec{Y})$$

(17)

Althoug P(y_1^*) and P(y_2^*) are not necessay cero y_1^* and y_2^* tend to be orthogonal to each other. Consequently the probability of the compound vector Y^* tends towards cero and therefore we can obtain the expression of the total probability from the last equation.

$$P(A\vec{B}) \approx \sum_{i=1}^{3} P(A\vec{B}/\vec{y}_i)P(\vec{y}_i)$$

or:

$$P(A\vec{B}) \approx P(A\vec{B}/\vec{Y})/P(\vec{Y})$$

(18)

If we add all the interneuronal influences over the relay neurons we obtain the above expressions that provides a "mirror" of the input data. While learning, the paths in the different loops "relay-star-pyramidal-interneurons-relay" (RSPIR loops) are reinforced by increasing their synaptical connections defined by conditional probabilities. The basket neurons with its inhibitory effect make the star neurons to "fight" among them in a competitive way so that the knowledge would be spread over different paths. The recovering of knowledge could be understand in a pattern completion manner. If we trained the network to associate the inputs "a" and "b" the interneuronal "mirror effect" with its inner representation will recover "b" if we only apply "a".
If we apply another input say, c and d, and the previous interneurons

keep the inertia of their last pacemaker rithms, at least another group of interneurons Y´ will play its role as a new orthogonal axe:

$$P(\vec{Y}^{**})=P(\vec{CD})-P(\vec{CD}/\vec{Y})P(\vec{Y})-P(\vec{CD}/\vec{Y}')P(\vec{Y}') \tag{19}$$

Every orthogonal axe represents a new distinctive feature that is filtered through a moving mask, the thalamus, (Taylor, 1.992) from the input information. Every new feature is projected over these axes so the size of the information storage in the brain is minimized.

3. EXPERIMENTAL RESULTS

The previous model has been implemented in a Lotus worksheet that is refreshed continously using a "macro". It has thirty columnar structures with a relay, an interneuron, four star, one pyramidal and one basket neuron in each column. Several input patterns have been applied while controlling the interneuronal "mirror effect". The next conclusions must be remarked:

a/The model is able to learn non-orthogonal patterns.

b/We can eliminate or modify connections at will discovering the characteristic effects of different kind of neurons. For example without the reinforcement of the RSPIR loops the model only learns the last pattern. Without basket neurons the learned patterns are mixed when we want to recover the information of a unique one. This is due to the reinforcement of quite the same RSPIR loops with different patterns without a competitive mechanism among star neurons.

c/If a random input pattern is applied and the model is allowed to continue its inner dynamic we can see how the previous patterns appear in the "interneuronal mirror" single or mixed varying its appearance gradually from one pattern to the others or to a combination of them.

d/The star neurons dynamic of activation is oscillatory. A random pattern produce slower and bigger amplitude oscillations than coherent patterns, emulating respectively the alfa and beta brain rithms.

e/Interneurons seem to detect the border line between discontinuous zones of patterns.

REFERENCES:

Crick F. , Asanuma C. 1.989 Certain aspects of the anatomy and physiology of the cerebral cortex. In Parallel Distributed Processing. Explorations in the microestructure of cognition. Vol.2. Psychological and Biological Models. McClelland J.L, Rumelhart D.E., MIT Press.

Jahnsen, H. and Llinás, R. 1984, Electrophysiological properties of guinea-pig thalamic neurones: an in vitro study. J. Physiol. (Lond.), 349:205-226.

Jones, E.G. ,1.985 The Thalamus, Plenum Press New York, pag. 193.

McClelland, James L. & Rumelhart David E.,1.988, Explorations in Paralłallel Distributed Processing. MIT Press.

Silva, Fernando M. & Almeida Luis B. Speeding-Up Backpropagation by Data Ortonormalization, 1.991 Artificial Neural Networks, Elsevier Science Publishers.

Taylor, J.G., 1.992. Temporal Processing in Brain Activity. In Neural Networks Dynamics. Springer-Verlag, pag.276

TOWARDS MORE REALISTIC SELF CONTAINED MODELS OF NEURONS: HIGH-ORDER, RECURRENCE AND LOCAL LEARNING

J. Mira, A.E. Delgado, J.R. Alvarez, A.P. de Madrid and M. Santos

Dpto. de Informática y Automática
Facultad de Ciencias, UNED
C/ Senda del Rey s/n. MADRID, SPAIN
Phone: 34-1-3987155; Fax: 3986679
e-mail: jose.mira@human.uned.es

ABSTRACT: ***The anatomy and physiology of biological neurons is revisited looking at a minimum set of computational requirements to be included in new and more complex models of self-contained local computation ANN. Some of these functionalities are then integrated and the corresponding model is evaluated. Properties included are: (1) locality and autonomy in all the computations including the learning algorithms, (2) a layered architecture with high-order recurrent neurons, (3) self and external programming via input spaces and (4) fault tolerance after physical lesion, or even elimination of one or more neurons.***

Given the case that the improvements included in this proposal are of some value, it is nonetheless clear to us that the more genuine properties of biological computation are still lost. If we were to look for implementation, we would say that a minimal model of a neuron should include at least the computational capacity of a microprocessor with self-programming and fault tolerance facilities as addenda.

KEYWORDS: Self-contained models, recurrent neurons, local learning, input programming, fault tolerance.

1. INTRODUCTION AND PROBLEM STATEMENT

By neural computation we mean the distributed calculations carried out by tridimensional modular architectures organized in layers and made up of a great number of processing elements with a high degree of connectivity and in which programming is replaced with supervised or non-supervised learning. The most frequently used model of local computation consists of the weighted sum followed by a non-linear and derivable decision function. Digital delay and feedback (recurrent nets) are also included, although with less frequency.

This analog model of local computation is based on what was known about the neural function around 1940 [1],[2]. Even then, W.S. McCulloch recognized (personal communication of R. Moreno-Díaz) that in order to duplicate the functionalities of a biological neuron (and this only at a logical level), hundreds of formal neurons would be required. Since 1940 more than enough anatomical electrophysiological and inmunocytochemical evidence [3],[4],[5],[6],[7],[8] has been produced to encourage the search for new models of local computation which come closer to biology than does a mere linear adder followed by threshold. This is independent of the usefulness of the latter in a multitude of applied domains [9], [10], [11], [12],[13].

It is important to point out that we are seeking an approximation of biology not at the biophysical level [37], but at the computational one in the sense given to it by David Marr [14]. In other words, we are looking for new computational models of the neuron which incorporate some of the known functionalities, these being additional to those already integrated and independent of the implementation level. Examples include autonomy, multiplicative inhibition (pre-synaptic and axon-axonic), absolute and relative refractory periods, specific functionality in neurotransmitters, the conditionals of excitation and/or inhibition, reconfiguration after lesion, etc.

Our initial idea is based on the profound feeling that the neuron is an autonomous and computationally very complex system of acquisition, integration and selective re-distribution of information. In order to duplicate it, it is imperative to keep in mind the electronic complexity of a special purpose microprocessor along with the organizational and architecture capacities required to sustain parallel processing, local self-programming and fault tolerance.

In this work we will begin with a partial summary of the computational properties associated with the anatomy, physiology and organization of biological neurons. We will subsequently select an initial subset of these properties and propose the model by which they are integrated. In the final section we will partially evaluate the advantages and disadvantages of the proposed model for two classic examples, the XOR problem and the prediction of temporal series in nonlinear systems.

2. ANATOMO-PHYSIOLOGICAL AND ORGANIZATIONAL REQUIREMENTS

Examining the majority of nervous systems we are left unclear about what should be considered the elementary autonomous processor (EAP). The neuron is evidently just this, however at the subcellular level there exists sufficient wealth, complexity and diversity for both the electric and the chemical synaptic contacts to each be considered a different type of EAP possessing its own characteristics. This is also the case with each area of membrane and with the axon, in such a way that a neuron with over 60.000 dendro-dendritic, dendro-somatic, axon-axonic and axon-somatic synaptic contacts should be considered a system which integrates subcellular microcomputation with the following anatomical, physiological and organizational properties. These properties ought then to be considered as essential requirements to a neuron model.

2.1. Anatomical requirements

A.1 Modular and layered computation [2], [8], [15],[16].
A.2 Superimposed hierarchical and heterarchical organizations (columns of ocular dominance, barrels, etc.) [7], [8],[16],[17],[18].
A.3 Tridimensional dendritic and axonic fields with overlapping and with computationally significant shapes and sizes (symmetry, directionality, centro-peripheric organizations,...) [17],[18].
A.4 Convergent and divergent processes [15],[16],[17].
A.5 Dendro-dendritic and axon-axonic local circuits (anatomical support for nonlinear codifiers in the input and output spaces [17],[20],[21].
A.6 Recurrent and non-recurrent lateral interaction [3], [7],[22],[23].
A.7 Local and distal feedback with ascending and descending collaterals [15],[16],[17],[7],[8].
A.8 Specialization in form-function relations [3],[17],[19],[21].
A.9 Specificity and plasticity of connections (growth, contacts and selective deaths) [3],[17],[19],[21].

2.2. Physiological requirements

P.1 Functional multiplicity of the same anatomical units [24], [25].
P.2 Specific functional modes with partial programming through the input spaces (i.e. oscillatory and tonically activated modes,...) [26].
P.3 Multiple inputs with analog delay and attenuation in the transmission, a function of the coordinates of the dendritic field where they make synaptic contact [27].
P.4 Multiplicative pre-synaptic and axon-axonic inhibition (products, modulations and multiplexings) [5],[20].
P.5 Algebraic sum (weighted excitation and inhibition) [1],[2],[17].
P.6 Absolute facilitation and inhibition (synergies of the "IF ... THEN ..." type) [17],[20],[26].
P.7 Spatio-temporal filtering in membranes (tunning and resonance) [17],[20].
P.8 Specific functionalities influenced by neurotransmitters [5],[7],[17].
P.9 Analog delay on axon, dendrites and soma [1],[2],[17],[20],[27].
P.10 Digital delay in synapses [1],[2],[5],[17],[20],[27].
P.11 Time and activity dependent threshold functions (absolute and relative refractory periods) [1],[17].
P.12 A great diversity of response firing patterns [3],[4],[5],[7],[8],[17],[20],[26],[28].

2.3. Organizational requirements

O.1 Autonomous modules with local learning.
O.2 Plastic and exuberant connectivity. What is used is reinforced and what is not used is not stabilized (incremental architectures), in such a way that learning operates by reinforcing connections as much as by eliminating those which do not surpass a certain value of threshold activity.
O.3 Reconfigurability and fault tolerance.
O.4 Non-empty initial anatomy and physiology. Neural learning does not start at zero; there is an enormous amount of initial knowledge in the topology of the net.
O.5 Input programming via external sensory information.
O.6 Independence and heterogeneity in learning processes. Each neuron modifies itself in the manner and at the time suggested by its stimulus history.

3. EXTENSIONS OF THE ANALOG MODEL

In the most common analog models, the neural function is interpreted in terms of a weighted sum followed by threshold and delay. This means that it would be difficult to even partially include the properties described in A.1, A.3, A.6, and P.5. There are thus many functionalities which remain unintegrated. In this work we propose the integration of A.5, A.8, P.10, O.1 and O.5 and the partial consideration of A.9, P.1, P.2 and P.4. We are interested in putting special emphasis on P.4 and O.1, that is to say, on the polynomic extension of the input and output spaces and on the autonomy of the processors. In addition to biological considerations there are arguments of flexibility, fault tolerance and ease of implementation which would also advise local learning. First we will discuss the static situation (dendritic and axonic fields) and will later introduce the dynamic components (local learning and modes of functioning).

3.1. High-order dendritic and axonic fields

In figure 1 we illustrate the computational frame used to embody multilayer and recurrent neural modeling [29],[30],[31] and obtained as a generalization of the concept of layered computation [2], [32]. A neural net is structured in layers of similar elementary autonomous processors (EAP). Each (EAP)j of the i-esime layer samples data from receptive fields R_i^j (dendritic) and $R^*{}_i^j$ (axonic) in the corresponding input and output spaces (FIFO like) to provide a new value into its output space performing some general type of local computation. Both spaces are of representation, where each "axis" represents a property (values and semantics) pertinent to the description of the world external to this layer. Please note that input space coincides with a physical space only when we consider layers situated at the end of receptors. Analogously, the output space is only a physical one close to the effector side of the overall network. For central layers we have purely representational spaces.

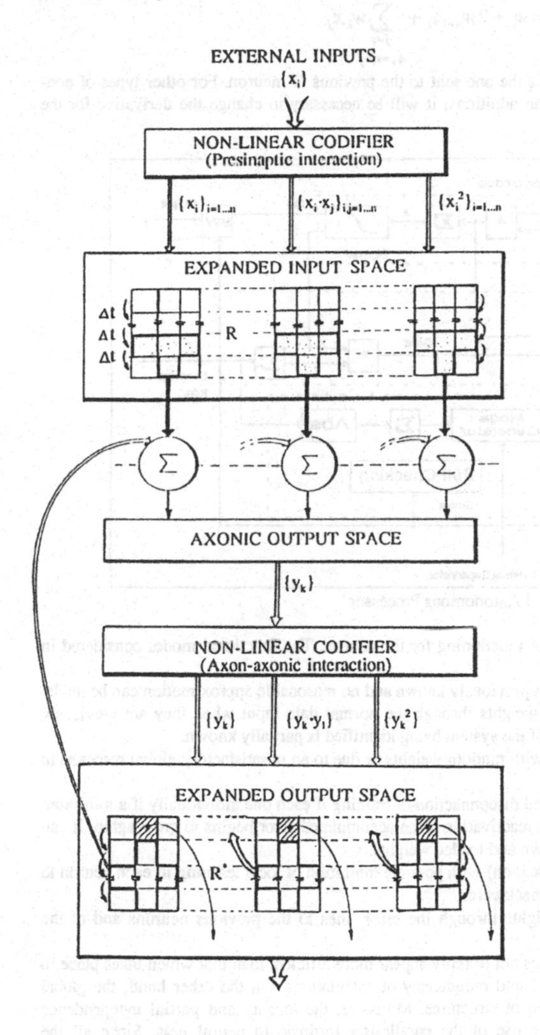

Fig. 1.- Computational frame to embody recurrent neural modelling with expanded input and output spaces

To include the interaction between afferents and axon-axonic contacts (A.5), it is necessary to include non-linear codifiers for each EAP on all the layers. We thus obtain expanded spaces in which the input variables to each layer are not only the direct inputs of the linear model $\{x_i\}$ but also the expanded ones which result from the pre-synaptic non-linear interaction that generates the terms $\{x_i \cdot x_j\}$ and $\{x_i^2\}$ for the quadratic case [21],[33],[34],[35]. In the same manner the axon-axonic multiplicative interaction allows us to extend the output spaces with terms of the $\{y_i \cdot y_j\}$ and $\{y_i^2\}$ type. As there is delay in propagation and overlapping in the dendritic and axonic fields, we have generalized recurrent quadratic nets in which each EAP has accessible a field of data which includes its inputs at that instant and those preceding it, as well as its responses and those of its neighboring cells at previous instants.

These non-linear extensions of the input and output spaces introduce redundancy which is later pruned in the learning process when it eliminates the redundant connections which have not been activated beyond a certain threshold value (O.2 and O.6). This generates non-linear representations which are complete and which have low redundancy for each training set.

In the same way that functionality of dendritic and axonic fields has been extended we could also expand the decision function (passing from the sigmoid to a conditional) as well as the local computation (passing from the algebraic sum to the module-2 sum in Galois fields for logical nets, or to the operators of fuzzy logic, max., min. and comp.). Its development lies beyond the scope of this paper [29],[30],[31],[36].

3.2. Local learning

As long as the learning algorithms are executed in a host separated from the net we are far from biology. For the model to be self-contained, the EAP must be provided with the necessary additional architecture to (a) estimate its contribution to the global error (b) consequently modify its computational parameters. The modification depends on the learning algorithm although it is possible to find a single structure capable of accommodating the most usual learning mechanisms (i.e. backpropagation [13] and associative).

Figure 2 shows the diagram of EAPs with local learning capacity. There is one module which calculates the output according to the diagram in figure 1, a second which actualizes the parameters, and a third module in charge of control, with special emphasis on the different modes of functioning and the additional module to expand the inputs.

These non-linear extensions [35] introduce modifications in the local synthesis of the learning algorithms. In the input codifier, it is necessary to include the contribution of the new non-linear terms in the calculation of error. In this way during local backpropagation the partial derivative with respect to an input will have crossed terms from the other inputs. If we call Φ(x) the expanded vector for an input x, we have

$$\Phi(\mathrm{x}) = \left[x_1, x_2, \ldots, x_n, (x_1)^2, \ldots, (x_n)^2, x_1 x_2, \ldots, x_1 x_n, \ldots, x_{n-1} x_n\right]$$

$$\Phi(\mathrm{x}) = \left[\phi_1(\mathrm{x}), \phi_2(\mathrm{x}), \phi_3(\mathrm{x}), \ldots, \phi_N(\mathrm{x})\right]$$

and assuming an associative addition function with N weights, the partial derivative of the sum with respect to a primitive variable will be of the following form:

$$\frac{\partial}{\partial x_i} \sum_{j=1}^{N} w_j \phi_j(\mathrm{x}) = w_i + 2 w_{n+i} x_i + \sum_{\substack{j \neq i \\ \phi_k = x_i x_j}} w_k x_j$$

This value multiplied by the contribution of the neuron is the one sent to the previous "i" neuron. For other types of non-linear codification or for associative operations other than additions, it will be necessary to change the derivative for the operation corresponding to the functions used.

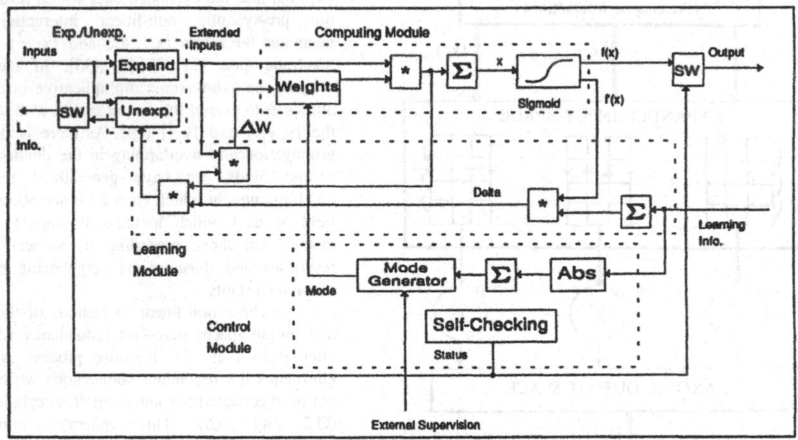

Fig. 2.- Elemental Autonomous Processor

Control block decides which is the local mode of functioning for the neuron. The functional modes considered in this model are:

I.1 Initialization with random weights when they are not previously known and no reasonable approximation can be made.

I.2 Initialization with the option of loading the initial weights through the normal data input when they are previously known from similar trainings or when the dynamic of the system being identified is partially known.

E.1 Forced global training at the beginning if you start with random weights or due to an unsatisfactory global response in the system.

E.2 Local training with error detection in each neuron and disconnection of training if each one individually if a minimum error is achieved with the possibility of independent reactivation if the accumulated error begins to grow again. It can also be directly used if it has been initialized by known and loaded weights.

C.1 Normal calculation mode (without parameter modification) with possible inhibition of local learning in each neuron to increase the speed of response and thus avoid a step backwards.

C.2 Stop or debugging mode with output of actual weights through the error lines to the previous neurons and of the threshold through the normal activation output.

At the level of implementation, local learning does not initially appear more efficient than that which takes place in the host, as it requires greater complexity of connections and redundancy of calculations. On the other hand, the global design is simplified due to the recurrence and repetition of structures. Moreover, the locality and partial independence enable the construction of neurocomputers which make use of the parallelism intrinsic to neural nets. Since all the computation is local (including learning), it is easier to mix up different types of neurons which are compatible in connections and functioning modes but which have different calculations and learning procedures. Finally, the possibility of considering fault tolerance, the residual function after the network lesion and the functional reorganization modifying the synaptic efficiency of the surviving EAPs is a further advantage of locality. The inclusion of an additional module to detect "sameness" or "newness" in the medium enables the autonomous EAPs to constantly adapt themselves to the modifications of the medium in which they operate.

Control block has inputs for external supervision, so it is possible for an external supervisor to put the neuron in any of the modes (initialization, training, normal use, ...). But, once the neuron has been properly initialized, it can locally decide whether it must be trained or not. To do so, it receives the learning information δ_i from that neurons connected to its output in the following layer. There exist several different decision criteria, but one that has proved simplicity and good performance is

$$\sum_i |\delta_i| > \theta_{learning}$$

If this threshold value is reached, learning is activated, and weights are modified according to a learning mechanism (in figure 2 this mechanism is backpropagation), so the network does not need to be disconnected from the system for being trained in a host.

Another feature in the control block is a module to check if the neuron is working well or not. If neuron fails, it can be disconnected form the rest of the network. To physically separate the neuron outputs, there are two switch blocks (SW) activated by the status signal. If a neuron fails, it is disconnected but the rest of the network keeps working (if necessary, neurons can decide to go into the local training mode).

4. EVALUATION

In addition to examining the biological motivations from which the model originated, we are also interested in demonstrating its value in the field of applications. For this reason we have done a comparative study. Results from simulation of two kinds of nets are shown in figs 3 and 4. One net has only 1 neuron of cuadratically expanded inputs and the other has 2 neurons with linear inputs, the second neuron has the other one output as input besides the network inputs. The two nets have the same learning algorithm respect to their weights and threshold and they have the sigmoid function as output transformation. Learning rate is 0.6, without momentum term and random initialization around 0.3 for weights. These networks have been trained in two examples: the XOR logical function and the non-linear temporal series $x_n = \cos(3.05\, x_{n-1}\, x_{n-2})$ over 1000 examples.

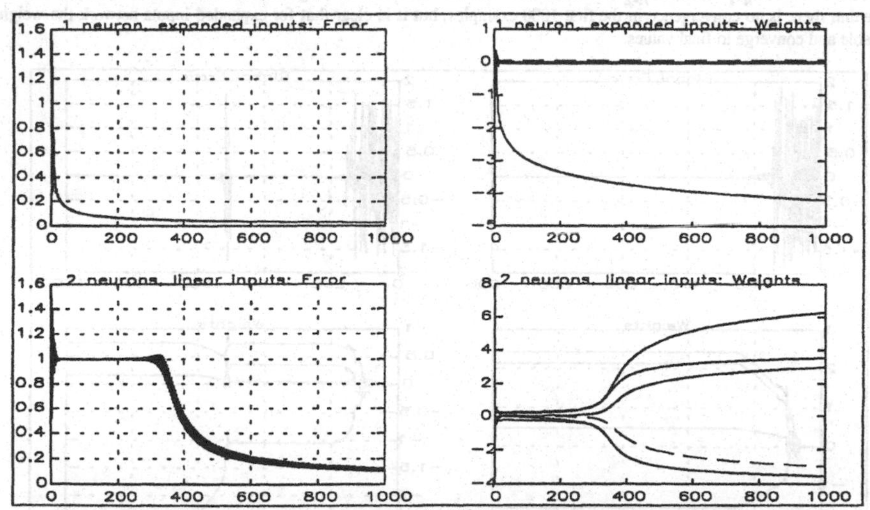

Fig. 3.- Comparisons of error and weight traces for XOR example on expanded and linear inputs nets

The resulting weights after 1000 training examples of XOR were for the expanded inputs neuron: $0.0006\,x + 0.0001\,y - 0.0111\,x^2 - 4.1913\,x\,y + 0.0589\,y^2$ and threshold - 0.0478. The more important weight is the cross product because of the non linearity of XOR. The rest of weights are negligible and can be pruned. The weights for the network with linear inputs were for the first neuron: $-3.5378\,x + 3.5378\,y$ and threshold - 3.5378, for the second neuron were $6.2596\,z + 2.9753\,x - 2.9753\,y$ and threshold 2.9753. These weights and thresholds are proportional to those theoretical in this case. It is quite clear that the expanded inputs neuron response is better and faster than the linear inputs

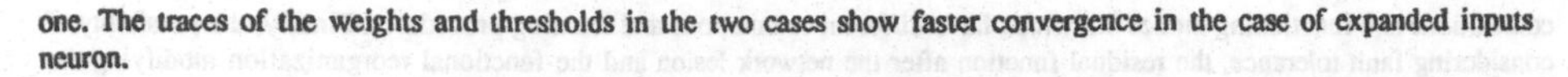

one. The traces of the weights and thresholds in the two cases show faster convergence in the case of expanded inputs neuron.

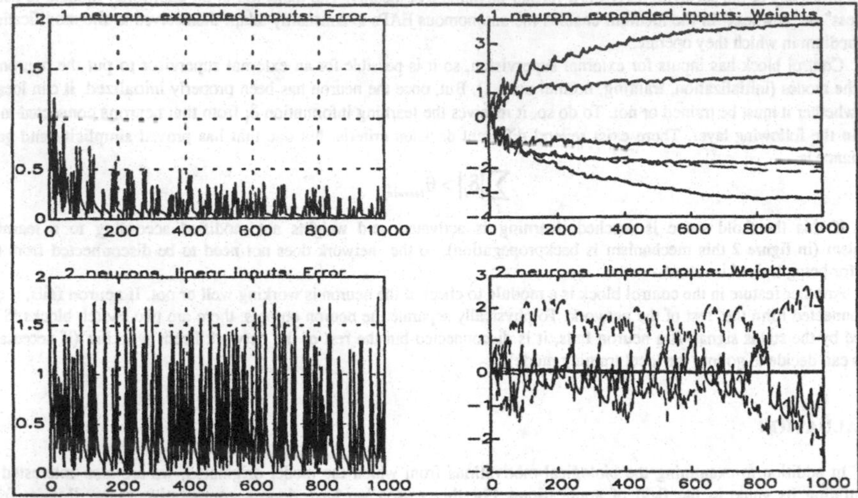

Fig. 4.- Comparisons of error and weight traces for non-linear temporal series

The weights for non-linear temporal series after 1000 training examples were for the expanded inputs neuron: 1.6718 x_{n-1} - 1.2834 x_{n-2} - 3.2702 $(x_{n-1})^2$ - 1.8565 x_{n-1} x_{n-2} - 2.1780 $(x_{n-2})^2$ and threshold 3.7577. In this case all the weights are important because the example is highly non-linear. The weights for linear inputs network were for the first neuron: - 0.0234 x_{n-1} + 0.0138 x_{n-2} and threshold 0.0105, for the second neuron were: - 0.0153 z + 2.4548 x_{n-1} - 2.2190 x_{n-2} and threshold - 0.3746. The traces of weights and threshold of linear inputs network shows that there is no convergence in the first 1000 examples, but it is clear that for expanded inputs network the weights are stable and converge to final values.

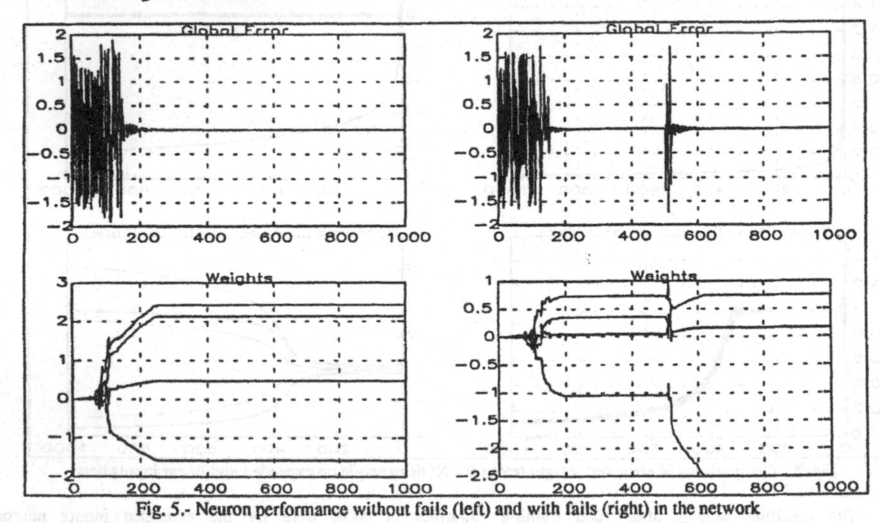

Fig. 5.- Neuron performance without fails (left) and with fails (right) in the network

The reliability and fault tolerance of the neural tissue is well recognized. Our purpose therefore is to attempt to create a neural model which can cope with functional stability after physical damage in the net. In figure 5 it is shown how a neuron behaves when another neuron in the network fails. The normal behaviour is displayed in the left : global error tends to 0 and, when this happens, learning is disabled so weights are no longer modified. If at t=500 another neuron fails

and it is self-disconnected, global error becomes bigger and the rest of the neurons can locally enable learning, so the network finds a new set of weights to compute the target functions with one less module. This example has been done for the XOR function, with 2 neurons in the hidden layer; one of these neurons fails. Data are displayed for the other one.

5. PRELIMINARY CONCLUSIONS

In this paper we have presented a summary of anatomo-physiological and organizational requirements for a model of neuron. Some of these requirements have been integrated (polynomic extensions of input and output spaces, recurrence and self-contained computation of learning algorithms, modes of functioning and fault-tolerance). Nevertheless, even if the improvements included in this proposal were of some practical value, it will be clear to us that the more genuine properties of the nervous tissue (A.8, A.9; P.1, P.8, P.11, P.12; O.2, O.3, O.4 and O.5) are still lost. Autonomy, acceptance of a large complexity in the structure and local function of the EAPs (always very distant from simple analog sum and threshold), self-organization and reprogramming after lesion are possible clues to be followed in the attempt to duplicate the computational capacity of nervous tissue.

ACKNOWLEDGMENTS

We acknowledge the economical support of the Spanish CICYT under project TIC-92/0136.

REFERENCES

[1] McCULLOCH W.S. and PITTS (1943): "A logical calculus of the ideas immanent in nervous activity". Bulletin of Mathematical Biophysics, Vol. 5, pp. 115-133. Chicago Univ. Press.
[2] McCULLOCH W.S. (1965): "Embodiments of Mind". The MIT Press. Mass.
[3] KANDELL E.R. (1967): "Cellular Studies of Learning". In: "The Neurosciences: A Study Program". Ed. by G.C. Quarton, T. Melnechukand, F.O. Schmitt, pp. 666-689. Rockfeller University Press. New York.
[4] KANDELL E.R. (1983): "From Metapsychology to Molecular Biology: Explorations into the Nature of Anxiety". The American Journal of Psychiatry, 140:10. pp 1277-1293.
[5] SCHMITT F.O. and WORDEN F.G. (1979): "The Neuroscience Fourth Study Program". The MIT Press. Mass.
[6] HEBB D.O. (1949): "The Organization of Behavior". J. Wiley. N. York.
[7] DE FELIPE J. et al (1990): "A Microcolumnar Structure of Monkey Cerebral Cortex Revealed by Inmunocitochemical Studies of Double Bouquet Cell Axons". Neuroscience, vol 37, No 3, pp. 655-673.
[8] DE FELIPE J. and FARIÑAS I. (1992): "The Pyramidal Neuron of the Cerebral Cortex: Morphological and Chemical Characteristics of Synaptic Inputs". Progress in Neurobiology, vol 39 pp. 563-607.
[9] KOHONEN T. (1988): "The 'Neural' Phonetic Typewriter". Computer, Vol. 2, No. 3, pp. 11-22.
[10] KOHONEN T. et al, eds. (1991): "Artificial Neural Networks", Vol. 1 and 2. North-Holland.
[11] PRIETO A. ed. (1991): "Artificial Neural Networks". Lecture Notes in Computer Science, 540. Springer-Verlag.
[12] HOPFIELD J.J. and TANK D.W. (1986): "Computing with neural circuits: a model". Science, Vol. 233, pp. 625-633.
[13] RUMELHART D.E. et al. (1986): "Learning internal representations by error propagation". In D.E. Rumelhart and J.L. McClelland (Eds.): "Parallel Distributed Processing: Explorations in the Microstucture of Cognition", Vol. 1: Foundations. MIT Press, Mass.
[14] MARR D. (1982): "Vision". W.H. Freeman and Company. New York.
[15] CAJAL S.R. (1952): "Histologie du Systéme Nerveux de l'Homme et des Vertébrés". Vol. 1 y 2. Maloine. Parfs.
[16] MOUNTCASTLE V.B. (1979): "An Organizing Principle for Cerebral Function: the Unit Module and the Distributed System", in F.O. Schmitt and F.G. Worden (Eds.): The Neuroscience Fourth Study Program, pp. 21-42. The MIT Press, Cambridge. Mass.
[17] BULLOCK et al. (1977): "Introduction to Nervous Systems". W.H. Freeman and Company. San Francisco.
[18] HUBEL D.H. and WIESEL T.N. (1962): "Receptive fields, binocular interaction and functional architecture in the cats visual cortex". J. Physiol. 166, pp. 106-154. London.
[19] BRAITENBERG V. (1977): "On the texture of brains: An introduction to neuroanatomy for the cybernetically minded". Springer-Verlag. New York.
[20] KOCH C. and POGGIO T. (1985): "Biophysics of Computation: Neurons, Synapsis and Membranes" in New Insights Into Synaptic Function", Edelman, Gall and Cowan, eds. J. Wiley and Son.
[21] LETTVIN J.Y. (1962): "Form-function Relations in Neurons". Q. P. R. No 66, pp. 333. Research Lab. of Elect. MIT
[22] HARTLINE H.K. et al (1965): "Inhibitory Interactions in the Retina and its Significance in Vision". In Nervous Inhibition. E. Florey ed. pp. 241-284. Pergamon. Oxford.
[23] RATLIFF F. (1965): "Mach bands". Holden Day, San Francisco.
[24] LASHLEY K.S. (1950): "In search of the engram". In: "Society for Experimental Biology". Symposium No. 4: Physiological Mechanims in Animal Behavior. Cambridge Univ. Press.

[25] LURIA A.R. (1974 a): "El cerebro en acción". Ed. Fontanella. Barcelona.
[26] STERIADE, M. and LLINÁS, R.R. (1988): "The Functional States of the Thalamus and the Associated Neuronal Interplay". Physiological Reviews, Vol. 68, nº. 3, July 1988. pp.649-742.
[27] RALL, W. (1977): "Core Conductor Theory and Cable Properties of Neurons". Chapter, 3 of the Handbook of Physiology. The Nervous System I, pp. 39-97. R. Kandel ed.
[28] PERKEL D.H. and BULLOCK T.H. (1968): "Neural Coding. A Report Based on an NRP Work Session". Neurociences Research Program Bulletin, Vol. 6, No. 3. Brookline, Mass.
[29] MIRA J. et al. (1983): "A theoretical proposal to embody co-operative decision in the nervous system". International Conference on World Problems and Systems Learning, pp. 23-27. Detroit. Michigan.
[30] MIRA J. and DELGADO A.E. (1991a): "Always Trying to Write an Equation for the Brain". In Lecture Notes in Computer Science, 540. Artificial Neural Networks. A. Prieto ed. pp. 93-100. Springer-Verlag. Berlin.
[31] MIRA J. and DELGADO A.E. (1991b): Linear and Algorithmic Formulation of Co-operative Computation in Neural Nets. In Lecture Notes in Computer Science, 585: Computer Aided Systems Theory. EUROCAST'91. F. Pichler and R. Moreno Díaz Eds. pp. 2-20. Springer-Verlag. Berlin.
[32] MORENO-DIAZ R. and RUBIO E. (1979): "A Theoretical Model for Layered Visual Processing". Int. J. Bio-Med. Comp. 10, pp. 134-143.
[33] FONSECA, J.S. da and McCULLOCH, W.S. (1967): "Synthesis and Linearization of Non-linear Feedback Shift Registers - Basis of a Model of Memory" Q.P.R. nº. 86. MIT, pp. 355.
[34] FUKUNAGA K. (1964): "A Theory of Non-linear Autonomous Sequential Nets Using Z-Transformes". IEEE Trans., 1964, vol. EC-13, pp. 310-313.
[35] RÖCKMANN D. and MORAGA, C. (1991): "Using Quadratic Perceptrons to Reduce Interconnection density in multilayer neural networks". In Lecture Notes in Computater Science, nº 540. Artificial Neural Networks. A. Prieto, ed. pp. 86-91. Springer-Verlag. Berlin, 1991.
[36] MIRA J. (1992): "Computación Neuronal". UNED, Madrid.
[37] SEGEV I. (1992): "Single Neurone Models: Oversimple, Complex and Reduced". Trend in Neuroscience. Vol. 15, nº 11 pp. 414-421.

McCULLOCH'S NEURONS REVISITED

Robert J. Scott

Information Systems Department
The University of Maryland Baltimore County
Baltimore, MD 21228-5398 USA
rscott@icarus.ifsm.umbc.edu

Abstract The 1943 McCulloch and Pitts seminal paper on artificial neurons (ANs) provided the stimulus for early neural network research. However, McCulloch's work in the 50's has been mainly overlooked. The importance of this later work is shown in this paper not only because McCulloch's neuron design was significantly more powerful than either the Perceptron or the Adeline, but also because of its ability to satisfy the ***XOR*** function 12 years before the publication of Minsky and Papert's book, ***Perceptrons***. The similarities between McCulloch's work in the 50's and 60's and more current AN research are also identified here. Some rudimentary experiments with AN design using a delta learning rule are also included. It is hoped that by recasting McCulloch's and the author's early work in terms of current technology, further research will be stimulated in designing efficient artificial neural systems that provide the stability and reliability of biological systems. These were the goals of McCulloch's original work.

1 Introduction

We are indebted to McCulloch and Pitts for giving us a simple model of neural activity that stimulated research not only in neural networks but early digital computer design as well. Most are familiar with their widely cited 1943 seminal paper [1], but few appear to be aware of the subsequent contributions of McCulloch and his research team at M.I.T. My purpose here is to review some key aspects of this work and to suggest the relevancy of this research to that which is currently pursued in neural networks.

Every dedicated neural network researcher knows of Rosenblatt's Perceptron [2] and of the controversial book, ***Perceptrons*** [3], by Minsky and Papert that was critical of Rosenblatt's neuron design for failing to handle a specific logical function of two variables, the Exclusive Or (***XOR***). The deleterious effect of this disclosure has been cited by many authors as a major reason for neural network research faltering in the 60's [4-6].

One of McCulloch's goals in the 1943 paper was to lay the foundation for a ***"Principia Mathematica"*** for neural networks [7, p.209]. Specifically, he wished to show that with his neural model he could produce the logical functions ***AND, OR,*** and ***NOT*** from which any other logical function can be expressed [8]. In other words, any logical function can be represented in a canonical form that is composed only of the three logical functions just mentioned. It follows that by using only these three neuron types which McCulloch designed it is possible to build a neural network that will execute that function. This means that the elusive ***XOR*** could be realized by a network of McCulloch neurons connected to perform: X ***XOR*** Y = (X ***AND*** (***NOT*** Y)) ***OR*** ((***NOT*** X) ***AND*** Y). Still, Minsky's concern prevails since each individual neuron, whether McCulloch's or Rosenblatt's produces only a linear discriminant which in two dimensions defines a dividing line that categorizes all points (XY), into class 1 or class 0. The points satisfying the ***XOR*** as Minsky and Papert proclaimed cannot be separated by a single linear discriminant or a single perceptron type neuron (PN).

Before continuing with McCulloch's research in the 50's, let me relate what is to come to the Perceptron (PN) with which we are familiar. Restricting our attention to two dimensions, Figure 1 shows a neuron with fixed weights on the X and Y binary inputs. Using a step transfer function, we are able to alter the logical function that the neuron performs by simply reducing the bias weight or threshold. As the linear discriminant sweeps across the four points defined by the two binary inputs, the neuron progresses through the five logical functions from contradiction (a, no firing) to tautology (e, firing regardless of input state). Figure 2a shows a McCulloch neuron (MCN) that performs identically to the neuron in Figure 1. The strange shape of the neuron's body or soma gives the MCN a degree of biological similarity.[1] Rather than using conventional weights, the MCN uses multiple afferent connections. For example, while the single fiber from the X input in figure 1's PN has a weight of +2, the MCN in figure 2a provides the same stimulus from the X input with two fibers, but with fixed weights of +1 each. The fiber that teminates directly on the cell's body near the top with a "doughnut" is inhibitory in that it contributes a weight of −1 when its source is active (1 state) A threshold is provided which permits the neuron to fire (1 output) whenever the sum of the input stimuli equals or exceeds the threshold. Up to this point, except for representional differences, PNs or MCNs can produce logically equivalent functions. The ***truth table*** [10] in figure 2b is useful to show the neuron's total excitation under each of the four input states. These excitation values are the larghest thresholds at which an input state will cause the neuron to fire. The only other representational difference in the MCN is the compact notation shown in figure 2c. Much like a Venn diagram, this X-shaped symbol completely describes the order of logical change in the neuron's output as the threshold is reduced by unit increments beginning with the smallest threshold that causes the neuron to fire. McCulloch was only interested in what he called ***non-degenerate*** neurons, those in which the logical function changed for each unit decriment in threshold. The pattern shown in figure 2d is that of the MCN in figure 2a which is a non-degenerate form.[2]

2 The Exclusive OR

In 1957, McCulloch [7] showed that all but two of the sixteen logical functions of two variables could be constructed with conventional neurons employing suitable synaptic weights , a summing function, a step function, and a threshold which set the conditions under which the binary values 0 and 1 applied to the inputs X and Y would cause the neuron to fire or output a 1 otherwise a 0. The two missing logical functions which could not be satisfied by a single neuron were the ***XOR*** and its complement ***NOT(XOR)***. McCulloch [7, p.210] states, "The missing pair can be constructed by using three neurons, but this requires three times the number of neurons and an extra synaptic delay. Nature has to compute efficiently in true time, i.e., a minimum number of neurons and minimal delay." Figure 3 shows sketches from [4] of the pair of missing neurons. While these are degenerate types, they do realize both the ***XOR*** function and its complement. The addition of fibers that terminate not on the cell's body but on opposing inputs' excitatory or inhibatory fibers provide the "power" necessary to realize the ***XOR*** function (figure 3a) which Minsky and Papert found missing in the PN. A circle at the fiber's end signifies an inhibitation of the effect of any fiber on which it terminates. I call such action "unique relative inhibition" [12]. . Figure 3b shows a MCN with both direct and relative inhibitory fibers.

---

[1]This similarity, however, proved to be a nuisance to Minsky [9, p.33] in his reference to McCulloch's 1943 paper because the schematic has the form of an arrowhead which one might believe to be the direction of signal flow, when in reality the flow is in the opposite direction (top to bottom).

[2]Although not a topic of this paper, the non-degenerate forms of threshold patterns allowed McCulloch to design networks that were logically stable under changes in the overall network threshold [11].

Regardless of the fact that the Perceptron had been implemented in hardware and the MCN had not, it is hard to understand why Minsky and Papert in their criticism of the PN's failing to realize the *XOR* function did not acknowledge the existence of an MCN type that could accomplish the task. The following section describes how the addition of relative inhibitory fibers to the model can provide the power to overcome the limitations of a linear discriminant.

3 Analysis

Both the PN in figure 1 and the MCN in figure 2 realize a linear discriminant (line) that can be represented algebraically as: $(2*X) - (1*Y) - T = 0$. Regardless of the value of the threshold, T, this will always be the equation of a straight line. In figure 3a, however, we find a more interesting representation. Here, since the X and Y inputs can only take on the values 0 or 1, the equation representing this realization must account for not only the individual fibers connected directly to the cell's body each contributing either +1 or −1 (excitatory or inhibatory) when active, but also for the relative inhibitory action of a fiber when active on the contribution of the fiber to which it is terminated when it is also active. For example, if the input pair (X,Y) assume the point (1,0), then the total stimulus of the neuron is the contribution from X, which is 1. The point (0,1) provides the same contributrion but this time from Y. When both X and Y are active (1,1), each active relative inhibitory fiber eliminates the contribution from the direct fiber to which it is terminated. The total effect of this neural "wiring" then can be characterized by the following equation: $(3*X) + (2*Y) - (4*X*Y) - T = 0$. It is the second-ordered product term in this equation that provides the power to overcome the limitations of a linear discriminant and solve the *XOR* problem when the threshold, T=2. Figure 4 shows a mapping of the points in the positive orthant over the range (0,1) for the MCN in figure 3a with T=2. Since the second-order product term here is a function of two-valued (0,1) inputs it is equivalent in this restricted domain to the *AND* function.

4 Related research

The utility of using non-linear methods for solving classification problems is not a new idea. Nilsson [13] introduced the *quadric machine* which generated its powerful mapping capability by introducing new non-linear variables made of product terms from the inputs. This mapping into a Φ space essentially warped a sample space that was not linearly classifiable into a space that could be linearly separable. Non-linear separability using higher ordered discriminants was considered in Duda and Hart's often-referenced book [14]. A number of researchers [15,16] have also used the power of product terms to increase neural net efficiency. The concept of the *Functional-Link Net* [17] was introduced in 1988. More recently, Pao [18] expanded this concept in his book with solutions to the XOR and Parity3 problems. This net incorporates two higher-ordered representations of input variables, a functional-expansion model that uses only functions of a single input variable, such as sin(x) or x^2 or the outerproduct model that combines different inputs into product terms. Both of these models The outerproduct model could produce the $(2*X*Y)$ needed in the figure 2's MCN.

Goodman et al [19] use pair-wise product terms in the "conjunctive rules" layer to build a rule-based neural network classifier. In the context of input variables having only zero or one values, product is synonymous with *AND*.

5 Example

The Parity 3 problem was used by Pao [17] to demonstrate the power of functional links in solving the 3-dimensional form of the *XOR*. Without a hidden layer, he was able to solve this problem by using three second-order terms and a third-order term in addition to the three input variables. These seven terms were summed after solving for the appropriate weights. Pao calls this a "flat functional-link net".

This same technique can be used to design MCNs that can realize any logical function of n binary variables. If product terms are used as in [17] and the weights after learning are integers, they can be used to design MCNs. First, the Parity 2 (*XOR*) problem was attempted using an approach that yielded only integer weights after training: the delta learning rule, $w_{ij}'=w_{ij}+\varepsilon_i*x_{ij}$ (learning coefficient is 1 with no momentum term), zero initial weights, and a step transfer function (sum $\geq 0 \rightarrow 1$, else sum $\rightarrow 0$). Convergence occurred after training on 40 input points. The weights (all integers) yielded the equation: $X+Y-(3*X*Y)-1=0$. The Parity 3 problem using the same paradigm converged in 245 training points, with the weights (again all integers) produced the equation: $X+Y+Z-(3*X*Y)-(3*X*Z)-(3*Y*Z)+(9*X*Y*Z)-1=0$.

Proposed Research

Scott [11] has defined the MCN structure in terms of linear inequalities in a way that allows linear programming to optimally design an MCN by minimizing the number of input fibers needed to define a particular logical function. It appears that designing MCN's of arbitrary complexity can be accomplished by extending the Parity 3 example. Further research is suggested to determine the utility of using complex MCN neurons in the design of fault tolerant neural nets that may eventually be implemented in hardware.

References

[1] W.S. McCulloch and W. Pitts, "A Logical Calculus of the Ideas Immanent in Neural Activity," *Bulletin of Mathematical Biophysics*, Vol. 5, pp. 115-133, 1943.

[2] F. Rosenblatt, "The Perceptron: a Probabalistic Model for Information Storage and Organization in the Brain," *Psychological Review*, 65, pp. 386-408, 1958.

[3] M. Minsky and S. Papert, *Perceptrons*, MIT Press, Cambridge, MA, 1969.

[4] R. Eberhart and R. Dobbins, Ed., *Neural Network PC Tools*, Academic Press, San Diego, CA, 1990.

[5] M. Caudill and C. Butler, *Naturally Intelligent Systems*, MIT Press, Cambridge, MA, 1990.

[6] W. Altman, *Apprentices of Wonder*, Bantam Books, New York, 1989.

[7] W. McCulloch, "The Stability of Biological Systems," *Homeostatic Mechanisms*, Brookhaven Symposia in Biology: No. 10, Uptown, NY, 1957.

[8] J.E. Whitesitt, *Boolean Algebra and its Applications*, p. 33, Addison-Wesley Publishing Co., Reading, MA, 1961.

[9] M. Minsky, *Computation: Finite and Infinite Machines*, Prentice-Hall, Inc., Englewood Cliffs, NJ, 1967.

[10] N. Scott, Analog and Digital Computer Technology, McGraw-Hill Co., New York, 1960.

[11] W. McCulloch, "What is a Number That a Man May Know it and a Man, That he May Know a Number?," in W. McCulloch, *Embodiments of Mind*, MIT Press, Cambridge, MA, 1988.

[12] R. Scott, "Construction of a Neuron Model," *IRE Transactions on Bio-Medical Electronics*, Vol. BME 8, No.3, July, 1961.

[13] N. Nilsson, *The Mathematical Foundations of Learning Machines*, Morgan Kaufman Publishers, San Mateo, CA, 1990.

[14] R. Duda and P. Hart, *Pattern Classification and Scene Analysis*, Wiley, New York, 1973.

[15] D. Rumelhart, G. Hinton, and R. Williams, "Learning Internal Representations by Error Propagation," in D. Rumelhart and J. McClelland, Eds., *Parallel Distributed Processing: Explorations in the Microstructures of Cognition*, Vol.1, MIT Press, Cambridge, MA, 1986.

[16] T. Sejnowski, "Higher Ordered Boltzman Machines," *American Institute of Physics Conference Proceedings, No. 151, Neural Networks for Computing*, pp. 398-403, Snowbird, UT, 1986.

[17] M. Klassen and Y. Pao, "Characteristics of the Functional-Link Net: A Higher Order Delta Rule Net," *IEEE Proceedings of the 2nd Annual International Conference on Neural Networks*, San Diego, CA, June, 1988.

[18] Y. Pao, Adaptive Pattern Recognition and Neural Networks, Addison-Wesley Publishing Co., Reading, MA, 1989.

[19] R. Goodman, C. Higgins, J. Miller, and P. Smyth, "Rule-Based Neural Networks for Classification and Probability Estimation," *Neural Computation*, pp. 781-804, Vol. 4, No. 6, Nov, 1992.

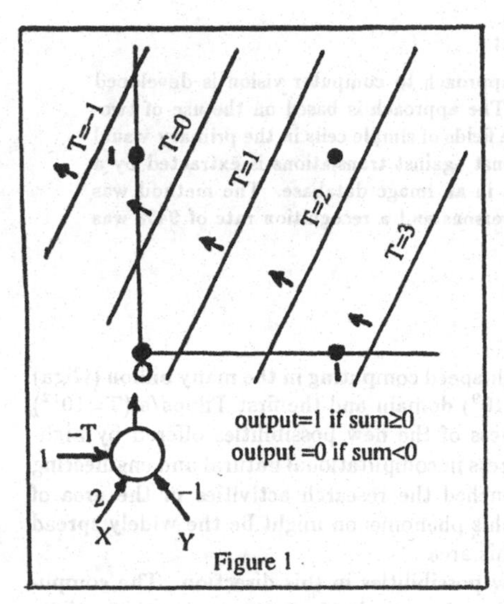

Figure 1

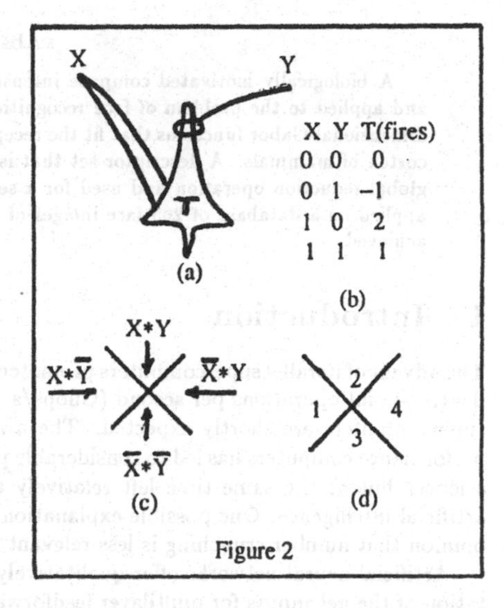

Figure 2

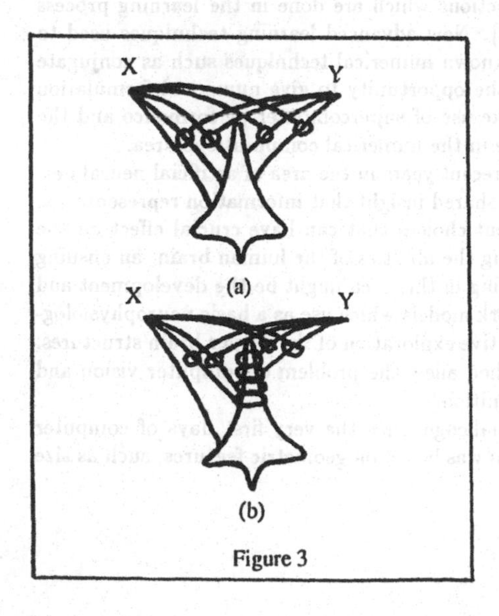

Figure 3

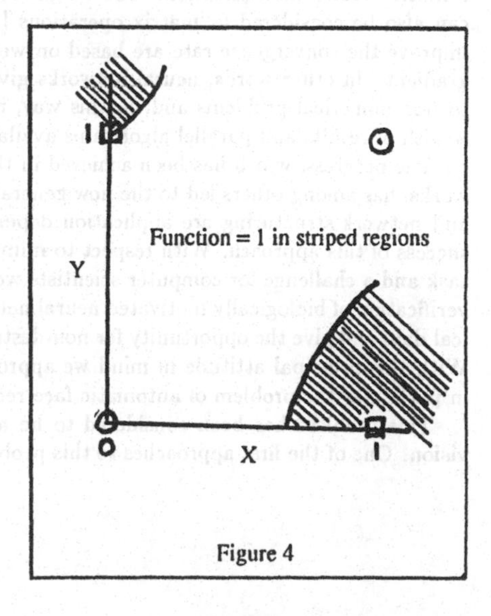

Figure 4

Biologically Motivated Approach to Face Recognition

N. Petkov, P. Kruizinga, T. Lourens

Department of Mathematics and Computer Science
Rijksuniversiteit Groningen
P.O. Box 800, 9700 AV Groningen
The Netherlands

Abstract

A biologically motivated compute intensive approach to computer vision is developed and applied to the problem of face recognition. The approach is based on the use of two-dimensional Gabor functions that fit the receptive fields of simple cells in the primary visual cortex of mammals. A descriptor set that is robust against translations is extracted by a global reduction operation and used for a search in an image database. The method was applied on a database of 205 face images of 30 persons and a recognition rate of 94% was achieved.

1 Introduction

The advent of parallel supercomputers promoted high-speed computing in the many billion (Giga) floating-point operations per second (Gflops/s, G=10^9) domain and the first Tflops/s (T=10^{12}) supercomputers are shortly expected. The awareness of the new possibilities offered by high-performance computers has led to considerable progress in computational natural and engineering sciences but at the same time left relatively untouched the research activities in the area of artificial intelligence. One possible explanation of this phenomenon might be the widely spread opinion that number crunching is less relevant in this area.

Artificial neural networks offer qualitatively new possibilities in this direction. The computation of the net inputs for multilayer feedforward neural networks, for instance, is substantially a matrix-vector multiplication. The weight corrections which are done in the learning process can also be considered as matrix operations [1-2]. New advanced learning techniques used to improve the convergence rate are based on well-known numerical techniques such as conjugate gradient. In other words, neural networks give the opportunity to give numerical formulation to non-numerical problems and, in this way, make use of supercomputer performance and the wealth of results and parallel algorithms available in the numerical computations area.

The progress, which has been achieved in the recent years in the area of artificial neural networks, has among others led to the now generally shared insight that information representation and network structuring are application dependent choices that can have crucial effect on the success of this approach. With respect to mimicing the abilities of the human brain, an ensuing task and a challenge for computer scientists working in this area might be the development and verification of biologically motivated neural network models which use as a basis neurophysiological data and give the opportunity for non-destructive exploration of the deeper brain structures. With this principal attitude in mind we approached anew the problem of computer vision and in particular the problem of automatic face recognition.

This problem has been considered to be a challenge since the very first days of computer vision. One of the first approaches to this problem was based on geometric features, such as size

and relative positions of eyes, mouth, nose and chin [3-6]. Another basic technique is template matching which has reached a considerable level of sophistication [7-9,30]. Further approaches to face recognition use graph matching [10], Karhunen-Loewe expansion [11,12], algebraic moments [13], isodensity lines[14], etc. Connectionists approaches to the problem are described in [15-18,30-32]. We refer the reader to [19] for a comprehensive discussion of various aspects of face recognition and to [20] for a collection of recent works in this area.

Our approach is biologically motivated and based on the use of Gabor functions which have been shown to fit well the receptive fields of the majority of simple cells in the primary visual cortex of mammals. The data obtained by projecting a two-dimensional signal (image) onto a set of Gabor functions can be interpreted as the activities of individual cells in the primary visual cortex (area V1 of the human brain). This data is then reduced to obtain a representation in a lower dimension space and use it for database storage and searching. We use an extended set of Gabor functions: 8 orientations and 8 scales give rise to a set of 64 Gabor functions and one copy of this set is centered on each point of the visual field. The compute intensiveness of the approach is due to the large resulting number of Gabor functions onto which an input image has to be projected.

The paper is organized as follows: In Section 2 we introduce the reader to two-dimensional Gabor functions and their relation to the mammalian visual system and propose a simple model for descriptor extraction. Section 3 presents our experimental setup and results on face recognition. Section 4 summarizes the approach and the results and outlines planned future work.

2 Gabor functions for computer vision

Our approach is biologically motivated in that it mimics the image transforms which take place in the mammalian visual cortex. It is well known from neurophysiological research that a large amount of neurons in the primary visual cortex react strongly to short oriented lines [21,22].

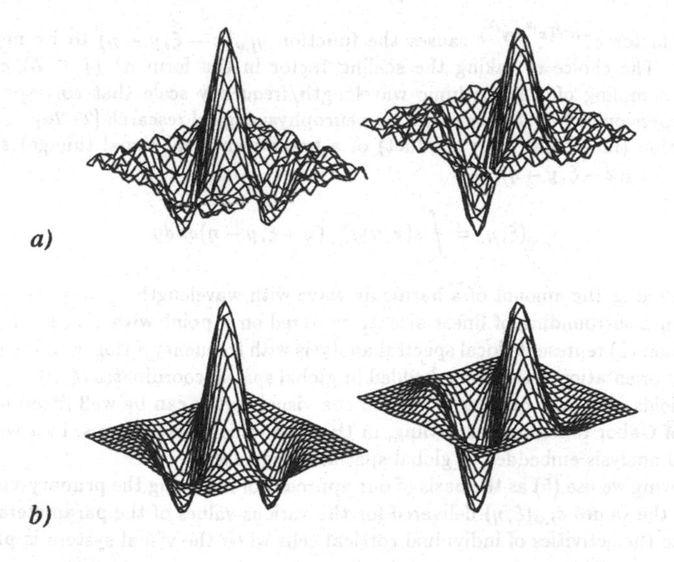

Figure 1: Receptive fields of simple cells in the primary visual cortex (a) and the respective Gabor functions which approximate them (b) (adapted from [26]).

A more precise study shows that the receptive fields of the overwhelming majority of the simple cells in the primary cortex can be well fit by two-dimensional Gabor functions [23-26]. Figure 1 shows the receptive fields of a pair of such cells with the real and imaginary part of the corresponding Gabor function.

The basic two-dimensional Gabor function has the following form:

$$g(x,y) = \frac{1}{\pi} e^{-(x^2+y^2)+i\pi x} \tag{1}$$

By means of translations parameterized by a pair (ξ, η), delations parameterized by an integer j and rotations parameterized by an angle φ, one gets the following family of two-dimensional Gabor functions (ξ and η have the same domain as x and y, respectively):

$$g_{j,\varphi}(x-\xi, y-\eta) = \frac{1}{\pi}\alpha^{2j} e^{-\alpha^{2j}(x'^2+y'^2)+i\pi\alpha^j x'} \qquad (j \in \mathbf{Z}, \varphi \in [0,\pi)) \tag{2}$$

$$x' = (x-\xi)\cos\varphi + (y-\eta)\sin\varphi$$

$$y' = -(x-\xi)\sin\varphi + (y-\eta)\cos\varphi$$

Fig. 1b shows the real and imaginary parts of one such function. By comparing Fig. 1a and Fig. 1b one can see that both the real and imaginary part of a Gabor function are represented by cells of the visual cortex. According to neurophysiological data, such cells are even adjacent [27] (however, one has to fairly admit that only 12 such pairs have been found in [27]).

The oscillations of $g_{j,\varphi}(x-\xi, y-\eta)$ are due to the harmonic wave factor $e^{i\pi\alpha^j x'}$ with wavelength

$$\lambda_j = \frac{2}{\alpha^j} \tag{3}$$

and spatial frequency

$$f_j = \pi\alpha^j. \tag{4}$$

The Gaussian factor $e^{-\alpha^{2j}(x'^2+y'^2)}$ causes the function $g_{j,\varphi}(x-\xi, y-\eta)$ to be negligible for $|x-\xi| > \lambda_j$. The choice of taking the scaling factor in the form α^j ($j \in \mathbf{Z}$) corresponds to equidistant sampling of a logarithmic wavelength/frequency scale that corresponds to the logarithmic dispersion of frequencies found by neurophysiological research [23-26].

The projection (functional inner product) of a two-dimensional signal (image) $s(x,y)$ on a Gabor function $g_{j,\varphi}(x-\xi, y-\eta)$

$$\tilde{s}_{j,\varphi}(\xi,\eta) = \int s(x,y) g^*_{j,\varphi}(x-\xi, y-\eta)\,dx\,dy \tag{5}$$

may be considered as the amount of a harmonic wave with wavelength $\lambda_j = \frac{2}{\alpha^j}$ and wavevector orientation φ in a surrounding of linear size λ_j centered on a point with coordinates (ξ, η). In this way, equation (5) represents local spectral analysis with frequency j (logarithmic magnitude) and wavevector orientation φ that is embedded in global spatial coordinates (ξ, η). The fact that the receptive fields of a large number of cells in the visual cortex can be well fitted by members of the family of Gabor functions is startling, in that the visual cortex seems, in a way, to make a local spectral analysis embedded in global spatial coordinates.

In the following we use (5) as the basis of our approach of mimicing the primary visual cortex, assuming that the values $\tilde{s}_{j,\varphi}(\xi,\eta)$ delivered for the various values of the parameters j, φ, ξ and η correspond to the activities of individual cortical cells when the visual system is presented an image $s(x,y)$.

Note that for fixed j and φ and variable ξ and η the above quantities present a function $\tilde{s}_{j,\varphi}(\xi,\eta)$ which can be computed as convolution of the signal $s(x,y)$ with a function $g_{j,\varphi}(x-\xi, y-\eta)$. We use this fact for the efficient computation of these quantities using fast Fourier transform (FFT). The coefficient $\frac{1}{\pi}\alpha^{2j}$ in front of the exponent in (2) is a normalization factor that is chosen in such a way that

$$|\int e^{i\pi\alpha^j x'} g^*_{j,\varphi}(x-\xi, y-\eta)dxdy| = 1, \tag{6}$$

i.e. for an input signal $s(x,y) = e^{i\pi\alpha^j x'}$ with magnitude 1, which is the harmonic wave factor in $g_{j,\varphi}(x-\xi, y-\eta)$, the above normalization delivers a quantity of magnitude one, $|\tilde{s}_{j,\varphi}(\xi,\eta)| = 1$.

Note that the functions $\tilde{s}_{j,\varphi}(\xi,\eta)$ comprise more data than the original image $s(x,y)$. This is in contrast to traditional approaches to computer vision where the amount of data is reduced at each stage of a hierarchical image analysis process. At present, neither we nor neurobiologists seem to know how this data expansion is used to effectively recognize an object. What one is certain about is that this data expansion is actually carried out in the brain as confirmed by the fact that the visual information is transferred from the retina to the primary visual cortex via 10^6 fibers of the optic nerve but in the primary cortex it is encoded by $10^8 - 10^9$ simple cells (100-1000 expansion at cortical level) [28]. What one may wish to do on a computer is *simulate this expansion, make hypotheses about the further processing stages and verify them by applying the model to a set of test images.* As we shall see below, even a very naive model delivers startling results.

Let us consider the following quantities:

$$S_{j,\varphi} = \int | \tilde{s}_{j,\varphi}(\xi,\eta) | d\xi d\eta, \quad j \in \mathbb{Z}, \varphi \in [0,\pi). \tag{7}$$

Each of them presents the cumulative activity of all cells with the same wavevector orientation φ and spatial frequency center j independently of their positions (ξ,η) in the visual field. The naive premise is that cells doing similar things (in this case cells with identical receptive fields but responsible for different areas of the visual field) might contribute in a similar way to quantities computed at higher stages. Each of the quantities (7) might, for instance, correspond to the activity of a corresponding higher abstraction level cell that receives activating stimuli from all lower level cells with the same wavevector orientation φ and spatial frequency center j. We have to admit that we are not aware of any neurobiological evidence that would confirm this hypothesis. Computing the quantities $S_{j,\varphi}$ according to (7) might make sense for one reason: they are not sensitive to the particular position of an object in the visual field, property which we refer to as *translational invariance*. More precisely, if $s(x,y)$ and $s'(x,y)$ are two images such that

$$s'(x,y) = s(x-t_1, y-t_2) \tag{8}$$

i.e. $s'(x,y)$ is produced by shifting $s(x,y)$ by a constant vector (t_1,t_2), one can easily show that for the respective quantities $S_{j,\varphi}$ and $S'_{j,\varphi}$ holds

$$S'_{j,\varphi} = S_{j,\varphi}. \tag{9}$$

Let us now represent two images $s(x,y)$ and $w(x,y)$ by the respective sets of quantities $S_{j,\varphi}$ and $W_{j,\varphi}(j \in \mathbb{Z}, \varphi \in [0,\pi))$ according to (7) and define the *dissimilarity* of the two images as follows

$$D_{s,w} = \sum_{j,\varphi} | S_{j,\varphi} - W_{j,\varphi} | . \tag{10}$$

The above defined dissimilarity is a non-negative quantity. It is zero for two identical images and for any two images which differ only by a translation as defined by (8). The relations (5), (7) and (10) are the basis of our approach to automatic face recognition. The quantities (7), to be referred to in the following as the *descriptors*, are computed for all images in a database and for each new input image. The descriptor set of an input image is then used for a best match search in the database to find the prestored image for which the dissimilarity (10) is minimal.

3 Implementation and Results

A database of face images has been built and this database is still being extended. The results reported below refer to the time when the database comprised 205 different face images of 30 persons. Several images were taken of each person, with the exact number varying from 5 to 9. The individual images of each person exhibit differences in facial expressions and/or orientations. Similarly, the individual images of one person show small deviations in size (a tolerance of approximately 5-10%) due to the fact that the distance between a subject and the camera was not controlled to keep it exactly constant. (From person to person, there are size deviations of up to 20%). Illumination was strived to be constant from session to session but no special effort was given to achieve exactly the same illumination conditions. Deviations in the illumination were due to changes in the position of the lamps and by sun light coming through the windows. The face pictures are stored as graylevel images with spatial resolution of 500×400 pixels and 8-bit quantization (256 gray levels).

Discretization is necessary for the practical computation of of the quantities $\tilde{s}_{j,\varphi}(\xi, \eta)$ according to (5) and we use the following one:

$$x, \xi = 1, 2, \ldots, 500 \tag{11}$$

$$y, \eta = 1, 2, \ldots, 400 \tag{12}$$

$$\varphi = \varphi_k = k\frac{\pi}{8}, \qquad k = 0, 1, \ldots, 7 \tag{13}$$

$$j = -1, -2, \ldots, -8 \tag{14}$$

The basic scaling factor α was taken as follows:

$$\alpha = \sqrt{2}. \tag{15}$$

This choice of α and the range of the parameter j allow for covering a wavelength domain that ranges from 2 to 32 pixels with logarithmic dispersion of the wavelength averages of the basic Gabor functions (see (3)).

The convolutions (5) were computed by applying FFT. In spite of the efficiency of FFT, the convolution computation is quite intensive and comprises more than 99% of the used computing time. On a 17 Mflops/s workstation, half a minute is required to compute the convolution of an image with one of the Gabor functions . For the set of 64 Gabor functions used this amounts to half an hour exclusive (non-timesharing) use of the workstation and for a database of 200 images this means 100 hours computing time. Timesharing and system failures effectively lead to a time of several weeks for one run of the model. This explains the relatively rough angle discretization (eight orientations) and the limited amount of (eight) basic spatial frequencies that were used.

After computing a convolved image $\tilde{s}_{j,\varphi}(\xi, \eta)$ for a given input image $s(x, y)$ and a given Gabor function $g_{j,\varphi}(x, y)$, it is reduced to a single number $S_{j,\varphi}$ according to (7). In this way 64 numbers (descriptors) are computed for each input image, one number for each of the 64 basic Gabor functions, and only this information is used to represent the image for database searching.

To obtain statistics on the recognition rate, we applied the above approach to all images in the database, considering each image in turn as an input image and the rest as prestored images. For each image the first four nearest neighbours were determined but only the first match was used to determine whether the search was succesful (delivering an image of the same person) or not (delivering an image of another person). For 192 out of 205 images the search was successful as illustrated by Fig. 2. The model failed in 13 cases as illustrated by Fig. 3. This gives a recognition rate of approximately 94%.

Figure 2: Examples of successful matches: the leftmost image in each row is a test image for which best match search is done in the whole image database; the images right to it are the first four matches.

Figure 3: Examples of failure of the model: the best match (the second image in each row) is an image of a different person.

4 Summary and Future Plans

In this paper we have shown how a biologically motivated model can be used for automatic face recognition. The biological relevance refers to the use of Gabor functions that fit the receptive fields of the overwhelming number of simple cells in the primary visual cortex of mammals. In the rest of its part, the approach is an attempt to guess what might be happening in the further form analysis structures of the visual cortex. In this case, we have no neurophysiological and neurobiological data to build on and, therefore, we rely only on general principles such as achieving robustness for image translations. Besides this uncertainty in the biological relevance of the final processing stages, our model comprises a certain simplification of the earlier stages: only 8 orientations and 8 basic frequencies are used and local information is completely lost. In spite of these shortcomings, we achieve a recognition rate of 94% on a database of 205 face images of 30 persons, a result which is startling with respect to previous work in this area and the short time (a few months) we have been dealing with the problem.

Our future work on this problem will focus on:

(i) *Increasing the number of Gabor functions used*: A larger number of Gabor functions is needed to improve the sensitivity of the model and to explore the possibilities for rotation and scaling compensation [29]. We intend to proceed with 32 orientations and 32 wavelengths which amount to 1024 Gabor functions.

(ii) *Improving the model in its higher stages*: The current model is an oversimplification of the processes taking place in the higher form-analysis structures of the visual cortex. We intend to improve the model by introducing local sensitivity by decomposing an input image into parts and applying the model to each individual part. This, however, will also be heuristics and we see a reasonable solution to the problem only in getting new neurobiological and neurophysiological data that can confirm or reject our hypotheses and give us hints on the ways to go.

(iii) *Parallel supercomputer implementation*: The computational intensiveness of the approach has become inhibitive for further investigations of the model. We estimate that one year computing time will be needed to apply 1024 Gabor funcions on a database of 1000 images, a delay which is unacceptable with respect to the fact that we are interested in experimenting with the model and have every day a new idea how it may be changed. The use of a supercomputer is inevitable and we already port our programs to a Connection Machine CM-5 which will be installed at our university in the near future. (Currently, the Connection Machines CM-2 and CM-5 of the University of Wuppertal are used for code transfer).

After the results presented above had been obtained and the above part of this paper had been prepared we received a copy of a paper to be published which presents the work done on face recognition by the group of C. von der Malsburg at the University of Bochum. They use Gabor functions for feature extraction and labeled graph matching. Their approach includes greater local sensitivity than as intended in point (ii) of our future plans mentioned above. Although they report lower recognition rate, it has been obtained on a different database of face images and at present we are not able to say how our respective results compare. Our future work may be extended by applying our model to other databases of face images and comparing the recognition rates obtained by different methods (e.g. [30-32]).

References

[1] N. Petkov: "Systolic simulation of multilayer, feedforward neural networks", *Proc. Int. Conf. on Parallel Processing in Neural Systems and Computers*, Düsseldorf, 1990, ed. by R.

Eckmiller, G. Hartmann and G. Hauske (Amsterdam: North-Holland, 1990) pp. 303-306.

[2] N. Petkov: *Systolic Parallel Processing* (Amsterdam: North-Holland, Elsevier Sci. Publ., 1992).

[3] W.W. Bledsoe: "Man-machine facial recognition", Technical Report PRI:22, Panoramic Research Inc., (Paolo Alto, CA, 1966).

[4] A.J. Goldstein, L.D. Harmon, and A.B. Lesk: "Identification of human faces", In *Proc. IEEE*, Vol. 59 (1971) pp. 748.

[5] T. Kanade: "Picture processing by computer complex and recognition of human faces", Technical Report, Kyoto University, Dept. of Information Science, 1973.

[6] Y. Kaya and K. Kobayashi: "A basic study on human face recognition", in S. Watanabe (ed.) *Frontiers of Pattern Recognition* (1972) pp. 265.

[7] D.J. Burr: "Elastic matching of line drawings", *IEEE Transactions on Pattern Analysis and Machine Intelligence*, Vol. 3 (1981) No. 6, pp. 708-713.

[8] J. Buhmann, J. Lange, and C. von der Malsburg: "Distortion invariant object recognition by matching hierarchically labeled graphs", *Proceedings of IJCNN'89* (1989) pp. 151-159.

[9] A.L. Yuille: "Deformable templates for face recognition", *Journal of Cognitive Neuroscience*, Vol.3 (1991) No.1, pp. 59-70.

[10] B.S. Manjunath, R. Chellappa, and C. von der Malsburg: "A feature based approach to face recognition", *Proc. 1992 IEEE Computer Society Conference on Computer Vision and Pattern Recognition* Champaign, Illinois, June 1992, pp. 373-378

[11] M. Turk and A. Pentland: "Eigenfaces for recognition", Technical Report 154, MIT Media Lab Vision and Modelling Group, 1990.

[12] M. Turk and A. Pentland: "Face recognition using eigenfaces", *Proc. IEEE Computer Society Conference on Computer Vision and Pattern Recognition*, Maui, Hawaii, June 1991, pp. 586-591.

[13] Zi-Quan Hong: "Algebraic feature extraction of image for recognition" *Pattern Recognition* Vol. 24 (1991) No.3, pp. 211-219.

[14] O. Nakamura, S. Mathur, and T. Minami: "Identification of human faces based on isodensity maps", *Pattern Recognition*, Vol. (1991) No.3, pp.263-272.

[15] M. Lades, J.C. Vorbrüggen, J. Buhmann, J. Lange, C. von der Malsburg, R.P. Würtz, and W. Konen: "Distortion invariant object recognition in the dynamic link architecture", 1991 (preprint).

[16] T. Kohonen: *Self-Organization and Associative Memory*, (New York: Springer Verlag, 1989).

[17] A. Fuchs and H. Haken: "Pattern recognition and associative memory as dynamical processes in a synergetic system II". *Biological Cybernetics*, Vol. 60 (1988), pp. 107-109.

[18] G. Cottrell and M. Fleming: "Face recognition using unsupervised feature extraction", *Proceedings of the International Neural Network Conference*, 1990.

[19] V. Bruce and M. Burton: "Computer recognition of faces". in *Handbook of Research on Face Processing*, A.W. Young and H.D. Ellis (eds.), (Amsterdam: Elsevier Sci.Publ., 1989) pp. 487-506.

[20] A.W. Young, and H.D. Ellis (eds.): *Handbook of Research on Face Processing*, (Amsterdam: Elsevier Sci. Publ., 1989).

[21] D. Hubel and T. Wiesel: "Receptive fields, binocular interaction, and functional architecture in the cat's visual cortex", *J. Physiol.(London)*,1962, vol. 160, pp. 106-154.

[22] D. Hubel and T. Wiesel: "Sequence regularity and geometry of orientation columns in the monkey striate cortex", *J. Comput.Neurol.*, Vol. 158 (1974) pp. 267-293.

[23] J.P. Jones and L.A. Palmer: "An evaluation of the two-dimensional Gabor filter model of simple receptive fields in cat striate cortex", *Journal of Neurophysiology*, Vol.58 (1987) pp. 1233-1258.

[24] J. Daugman: "Uncertainty relation for resolution in space, spatial frequency, and orientation optimized by two-dimensional visual cortical filters", *J. Opt.Soc.Amer.*, Vol.2 (1985) No. 7, pp. 1160-1169.

[25] J. Daugman: "Two-dimensional spectral analysis of cortical receptive field profiles", *Vis.Res.*, Vol. 20 (1980) pp. 847-856.

[26] J.G. Daugman: "Complete discrete 2-D Gabor transforms by neural networks for image analysis and compression", *IEEE Trans. on Acoustics, Speech and Signal Processing*, Vol.36 (1988) No. 7, pp. 1169-1179.

[27] D.A. Pollen and S.F. Ronner: "Phase relationships between adjacent simple cells in the visual cortex", *Science*, Vol. 212 (1981) pp. 1409-1411.

[28] M. Connoly and D. van Essen: "The representation of the visual field in parvocellular and magnocellular layers in the lateral geniculate nucleus in the macaque monkey", *J. Comput. Neurol.*, Vol.226 (1984) pp. 544-564.

[29] N. Petkov, T. Lourens and P.Kruizinga: "Computationally intensive approach to face recognition", Comp. Sc. Notes, CS9207, Department of Computer Science, University of Groningen, December 1992.

[30] M. Lades, J.C. Vorbrüggen, J. Buhmann, J. Lange, C. von der Malsburg, R.P. Würtz, W. Konen: "Distortion invariant object recognition in the dynamic link architecture", to appear in IEEE Trans. on Comp.

[31] H. Boattour, F. Fogelman Soulié and E. Viennet: "Solving the human face recognition task using neural nets", *Proceedings of the ICANN-92, Brighton, September 1992*, pp.1595-1598.

[32] E. Viennet and F. Fogelman Soulié: "Scene segmentation using multiresolution analysis and MLP", *Proceedings of the ICANN-92, Brighton, September 1992*, pp.1599-1602.

LEARNING BY REINFORCEMENT: A PSYCHOBIOLOGICAL MODEL

Francisco J. Vico[1], F. Sandoval[1] and J. Almaraz[2]

[1]**E.T.S.I. de Telecomunicación, Dpto. Tecnología Electrónica**
[2]**Facultad de Psicología, Dpto. de Psicobiología**
Universidad de Málaga
Plaza El Ejido s/n. 29013 Málaga. Spain

ABSTRACT

The current connectionist models of the learning by reinforcement paradigm make use of the delta rule to back-propagate the error. The work we present proposes a biologically inspired learning by reinforcement method. It uses only biological concepts to learn the desired outputs, as chemical substances and homeostatic regulation. On the other hand, the formulae that rule their dynamics are made with respect to the constraints imposed by the observed phenomena in behaviorist experiments of operant and classical conditioning. The authors propose that an input-output map can be expressed as a combination of these phenomena, in the sense that the task of teaching a function is to make the network to adapt itself to verify a set of phenomena between its inputs and outputs.

I. INTRODUCTION

There are different learning strategies depending on the amount of information used to teach a network. The *supervised learning* feeds the network with patterns consisting on the correct input-output pairs that the network must map. The weights are adjusted according to the differences between the actual and desired output.

If the information given to the network for its learning is just an estimation of the correctness of the outputs, the learning method is said to be *by reinforcement*. The recent actions are good or bad depending respectively on the reward or punishment that the network receives, and these actions are generated by the network in a deterministic or random way. As the networks implementing this learning method do a trial-and-error search of the correct outputs, the learning time is supposed to be longer that those using supervised learning. On the other hand, the supervised methods have the disadvantage that the teacher is not always able to specify the exact outputs of the network for each input pattern; as long as the problem to solve is not a classification task this job becomes a very hard question.

Finally, the networks that don't use any information about the desired output implement a *non-supervised learning* method. Usually, this paradigm classifies the input patterns according to a distance criterion.

The learning by reinforcement method we present in this paper is highly inspired by the fields of neurobiology and behaviorist psychology. The knowledge coming from this fields was used as constraints that the model has to verify in order to obtain the learning curves observed at different levels in living organisms, from general behaviour to biochemical reactions in the synapse.

The main connectionist implementations of learning by reinforcement models backpropagate the error to adjust the weights, such as AHCON that uses the adaptive heuristic critic learning architecture (Barto et al., 1990), and QCON using Q-learning (Watkins, 1989). The proposed method uses homeostatic regulation to change and select the synaptic weights in order to improve the outputs fittness.

II. THE PSYCHOLOGICAL INTERPRETATION OF LEARNING BY REINFORCEMENT

Learning by reinforcement occurs when living organisms adopt new behavioral patterns due to the reinforcing consequences that these conducts gave in the past.

E. L. Thorndike asserted the *law of effect* (Thorndike, 1911), that was the first psychological theory of learning by reinforcement. This law states that, from the different responses that an animal yields in a given situation, those that are followed by a satisfaction will have a stronger connection with that situation, in the sense that when this situation arises again their probabilities of occurrence will be larger. Those that carry dissatisfaction to the animal will have a smaller probabilities of occurrence in this situation.

The correct output (e.g. to move a bar) gets more consolidated because its connection (its association) with the situation (the current environment) is reinforced due to satisfaction of the responses (food). These connections get weaker if these effects were unpleasant for the animal. This law proposes a symmetric effect on the association of the reward and the punishment.

Skinner used the law of effect to develop the *operant conditioning* theory (Skinner, 1938, 1974), whose varieties (positive reinforcement, negative reinforcement, positive punishment and negative punishment) are the four different kinds of learning that support most of the behaviors observed in animals and human beings.

The work of Skinner involved a behavioral technology that is applied to clinic, educative, labour problems, etc. Presently, the study of the operant conditioning gave place to theories (Dickinson, 1980; Mackintosh, 1983) that are far away from the descriptive viewpoint of Skinner and closer to the early theories of Thorndike, whose first connectionist approach is near the current connectionist perspective in which this work is included.

III. DESCRIPTION OF THE MODEL

This learning method was first used at a neuronal level to verify associative and non-associative phenomena (Vico et al., 1992). The aim of this experiment was to obtain a learning rule capable of implementing what the authors consider that are the *building blocks* in the learning process. This is to state that all the knowledge that can be stored in a neural network is organized in the synaptic weights according to simple rules, operating on previously stored information, where the sensations are the elemental items that support the knowledge.

III.a. Synaptic variables and their dynamics

The model proposed in this paper adds some constraints to work at the network level.

At the synaptic level the learning rule operates varying three internal variables:

S : recent activation rate
E : synaptic efficacy
N : available neurotransmitter

These variables are hypothetical chemical substances whose concentration acts as a memory of useful information to implement the learning process.

The formulae that gobern the dynamics of this variables for an excitatory synapse are:

if *reward* is present in the network,

$$E_{t+1}=E_t+kE\cdot Reinforcement\cdot(S_t/TotalS)\cdot(Esat-E_t)/Esat \quad (1)$$

if *punishment* is present in the network,

$$E_{t+1}=E_t+kE\cdot Reinforcement\cdot(S_t/TotalS)\cdot E_t/Esat \quad (2)$$

where *kE* is a constant that specifies the slope of the synaptic efficacy evolution, *Reinforcement* represents the amount of reward or punishment (positive or negative sign, respectively) present in the network due to the recent outputs, *TotalS* is the total amount of substance *S* in the synapsis of the network, and *Esat* is the concentration at which substance *E* reaches the saturation in the synapse.

(1) and (2) state that the weight in a synapse varies (incrementing or decrementing) depending on: the reinforcement in the network, the relative activity of the synapse with respect to all synapsis, and the actual value of the weight. For an inhibitory synapse (1) and (2) are applied in the opposite way.

To compute the value of *S* in a neuron (all the synapsis of a neuron have the same activity level) the following rules are applied:

if the neuron is *excited*,

$$S_{t+1}=S_t+kS\cdot(Ssat-S_t) \quad (3)$$

if the neuron is *resting*,

$$S_{t+1}=S_t-kS\cdot S_t \quad (4)$$

where *the constant kS* gives the slope of the recent activation rate evolution, and *Ssat* is the saturation concentration of *S* in a synapse.

With this dynamics, the current value of *S* enhances the changes in the activity level of the synapse in order to implement the short term memory (STM) with the amount of available neurotransmitter in the pre-synaptic site (variable N).

III.b. Neural network functioning

The following steps are performed to evaluate the activation state in the network:

- to compute the membrane potential (MP) as the summation of the post-synaptic potentials (PSP's) of all the synapsis connecting the neuron, being the PSP:

$$PSP=Hm\cdot S\cdot N\cdot E \quad (5)$$

where *Hm* represents the homeostatic state of the network, and the presence of *S* cause STM to alter the functioning of the network

- to activate a neuron if its MP exceeds a threshold value or let it in resting state if not

- to update the internal variables S and N of every synapse

- to evaluate the network state according to the incoming reinforcement

- learning (the variations in the substance E are the learning process)

This algorithm ensures that no inputs are lost due to the evaluation order of the neurons, because the whole network is evaluated concurrently.

We have developped a neural networks simulator called *BioNet*. It implements this algorithm on a PC. *BioNet* works with two text files, one to specify the architecture and simulation parameters of the network, and another to determine the incoming stimuli to the input neurons and the representation specifications. As the simulations are doing according to the biological constraints, the neurons' output has two possible values: activation or resting. Neurons don't work with frequencies.

III.c. Biological considerations

As said before, the homeostatic level is taken into account for the evaluation of the network. The main reason is that the network needs some stimulation to search for new outputs, in the sense that if the current output is being punished this punishment makes the homeostatic level go further from the optimal value, and this enhances the effect of the excitatory synapsis and reduces the effect of the inhibitory synapsis. This results on a higher general excitation of the network, that is the observed behaviour in the living beings.

On the other hand, a correct output implies a reward and makes the homeostatic level to be closer to the optimal level with the logical reduction in the network excitation; this is: the network gets more relaxed.

IV. SIMULATION AND RESULTS

We verified that simple tasks could be expressed as a combination of classical conditioning phenomena; in this way, the rewarded phenomena that the network carries out are stored in the weights.

For instance, the XOR (exclusive or) function consists of two conditioning and two conditioned inhibitions. Representing the inputs of the XOR function as the CS1 and CS2 conditioned stimuli, to teach the input-output relation (1,0) --> 1 we must reward the activation of the output neuron when the CS1 is present. We should proceed in the same way for the relation (0,1) --> 1 and CS2; doing this, we have conditioned the stimuli CS1 and CS2 to the response of the network. Now, the relation (1,1) --> 0 implies two simultaneous conditioned inhibitions between CS1 and CS2.

To implement a more difficult function we tried with the multiplexor function whose boolean table is presented in *table 1*. The function of a multiplexor is to set the output (*O*) equal to the value of input *I1* if *Ct* is 1 and equal to the value of input *I2* if *Ct* is 0. Teaching the network this function we verify the presence of the main classical conditioning phenomena: conditioning, extinction, conditioned inhibition, blocking and overshadowing. The training of this function on *BioNet* is shown in *figure 1*. After initializing the network with random and small weights the conditioning is learned by rewarding the first activations and the conditioned inhibition by punishing the simultaneous activation of two stimuli. Phenomena as blocking and extinction can make the learning time longer, but they are necessary to complete the set of implementable functions.

The simulation results in *figure 1* show how the reinforcement drives the network to give the right outputs, rewarding the desired activations and punishing the non-desired activity.

It is our belief that very few modifications will be necessary on this method to describe any other function in terms of classical conditioning phenomena.

Table I. The multiplexor function.

I1	I2	Ct	O
0	0	0	0
0	1	0	1
1	0	0	0
1	1	0	1
0	0	1	0
0	1	1	0
1	0	1	1
1	1	1	1

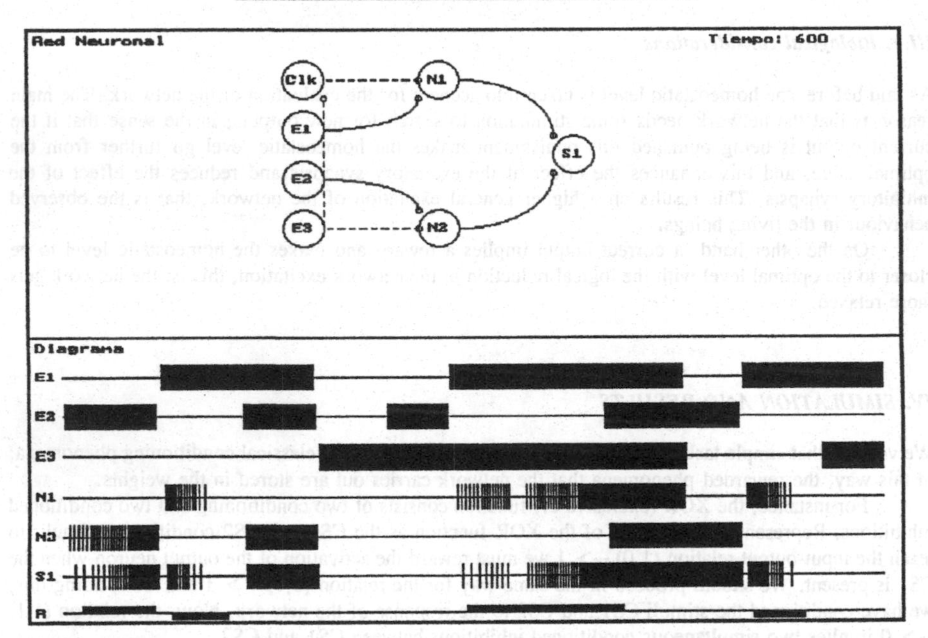

Figure 1. Training the multiplexor function. The inputs of the table, I1, I2 and Ct, are represented by the neurons E1, E2 and E3; Clk is a constant input (a source of stimuli or clock) not controlled externally and the output neuron S1 gives the output O. The excitatory synapsis are represented with solid line while the inhibitory are in dashed line. At the bottom of the figure, R represents the reinforcement: reward above the middle line and punishment below it.

V. CONCLUSIONS

The most outstanding aspects of the proposed model are:

- coincidence with the experimental results in predicting the operant conditioning phenomena

- agreement with the chemical bases involved in cellular functioning

- mathematical simplicity of the model

- it works on a biological model of neuron, so its predictions can be applied to the field of neurobiology and psychology

Although, at present, the relation between the knowledge engineering and the artificial neural networks is more based in the search for the same goals than in the methods they use, it is important to note the way in which the neural networks discipline is taking part in the cognitive field. Giving connection patterns and learning rules that make possible simulations more in agreement with the experimental results, and whose functioning will give a better understanding of the superior functions of the human brain. In this sense, we can consider that the behavioral experiments will be an inspiration source until the artificial neural networks theory is more established.

VI. FUTURE WORK

The actual line of work is oriented to the automatic generation of minimal neuronal networks to support a set of given experiments. The final goal of this project is to obtain automatically networks to implement functions that can be expressed as a combination of classical conditioning phenomena. On a different way this learning model will be used to train neural networks obtained by a genetic algorithm.

ACKNOWLEDGMENTS

This work has been partially supported by the Spanish Comisión Interministerial de Ciencia y Tecnología (CICYT), Project No. TIC91-0965.

REFERENCES

Barto, A.G., Sutton, R.S., & Watkins, C.J.C.H. (1990). Learning and sequential decision making. In: M. Gabriel & J.W. Moore (Eds.), *Learning and computational neuroscience.* MIT Press.

Dickinson, A. (1980). *Contemporary animal learning theory.* Cambridge University Press.

Mackintosh, N. (1983). *Conditioning and associative learning.* Oxford University Press.

Skinner, B. F. (1938). *The behavior of organism.* Appleton: New York.

Skinner, B. F. (1974). *About behaviorism.* Knopf: New York.

Thorndike, E. L. (1911). *Animal intelligence.* McMillan: New York

Vico, F.J., Sandoval, F. & Almaraz, J. (1992). A learning by reinforcement model to predict associative and non associative phenomena. In: *Proceedings of the Fifth International Symposium on Knowledge Engineering*, November, Seville (Spain), pp. 190-194.

Watkins, C.J.C.H. (1989). *Learning from delayed rewards.* Ph.D. Thesis, King's College, Cambridge.

A NEURAL STATE MACHINE FOR ICONIC LANGUAGE REPRESENTATION

Igor Aleksander[1]
Helen Morton[2]

[1]Neural Systems Engineering Group
Department of Electrical and Electronic Engineering Department
Imperial College of Science Technology and Medicine
London SW7 2BT UK

[2]Department of Human Sciences
Brunel University
Uxbridge UB8 3PH UK

Abstract

Three key points are made in this paper. The first is that "iconic" states arise naturally in all recursive neural nets which are trained on world states. The second point is that this leads to a new paradigm of cognitive representation. Language is seen as a vehicle for retrieving iconic representations in a recursive system - this is dubbed the "iconic hypothesis". The third point is that while a similar idea has been presented as a "symbol grounding problem" (Harnad, 1990) the "iconic hypothesis" goes further in suggesting that a recursive neural system can operate in both a symbolic fashion and use grounded internal states. To illustrate these points we introduce an architectural concept (the Neural State Machine Model - NSMM) which allows a clear formalisation of the concept of iconic representations. Examples of the application of this concept to representation of visuo-linguistic data are given.

1. Why iconic states?

In a totally reduced view of the interaction of a learning machine with a "world", such a machine can receive world-state information, it can act on the world and it can receive data from the world that controls learning. Kelly (1955) believed that a person learns in order to build up an internal system of constructs that enables that person to predict and control the world. We adopt this view and see the learning machine as a classical state machine whose states are its constructs. Learning must then be a process of developing state structure that is related to taking appropriate actions in the world. There are only two ways that states can acquire an assignment. The first is what in connectionism is called unsupervised learning. That is, the representation of the world is done by arbitrary assignments for the machine states. The state structure will then be a model of world behaviour, and the learning control effort will be largely devoted to a determination of *when* a machine state has to be defined. The hope is that the machine will find internal representations that are more distinct than the world states, while remaining prototypical of them. The difficulty with this is that should the machine ever be required to describe some of its own constructs they will need to be internal decodings of the arbitrarily assigned states.

We believe that this does not remove the need for some states having arbitrary assignments (as we shall see) but the only way that the machine can have an "understanding" of the world is through the following alternative way of assigning values to states. It is suggested that the states of the world are best represented through a many-to-some relationship between the state variables of the world and the state variables of the state machine. As the perception of the states of the world can only occur through some senses our definition of an *iconic state* is 'a state assigned to the state variables of the state machine in such a way that it represents the sensory pattern that is generated by a world state'.

The first key point we make in this paper is that this assignment comes naturally from the limited number of ways that a recursive network can be taught to retrieve the assigned state. To illustrate this we show, in fig. 1, a weightless neural system (see Aleksander, 1992) which is trained as follows. The triangles are generalized neurons that receive input from the external world and from other neurons. The feedback variables form the state of the machine. Training proceeds by a world state being sensed on some inner sensory surface which feeds the input *and* the synapses that control the firing of the neurons. This produces a transition to an iconic state through an input of the projection of the world state on a sensory surface. The state can be made reentrant as shown. Note that the many-to some mapping happens between the world state and the input and that the mapping to the learning control terminals need not be the same as that for the input terminal.

In summary, this method creates reentrant states that represent the world states in a way that can be accessed from perceptual input. The generalisation of the neurons makes these states into attractors giving them the properties of distinctiveness and being prototypes.

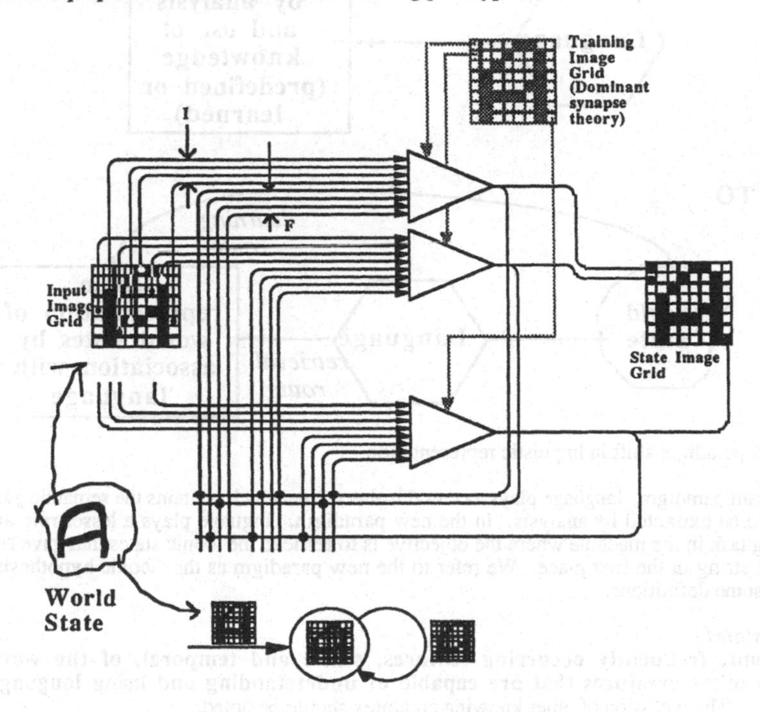

Fig. 1. A General Neural Unit

2. A paradigm shift in language understanding - the iconic hypothesis.

The target for much modelling work in cognitive science is to provide a good theory for language understanding. This is rooted in a definitional paradigm of what human language understanding is. To summarize a vast corpus of literature on the subject, language understanding is defined as an analytical process. This takes differing forms, but what they have in common is a process which decomposes language into standard classes. Well known is the work of Schank and Abelson (1977) where the analysis takes the form of breaking down linguistic utterances into semantic primitives which through evoking scripts fill in the gaps in stories. This is a classical example from the symbolic processing camp. But connectionists too have accepted the same paradigm. Rumelhart and Kawamoto (1986) have used feedforward nets to analyse utterances to 'case roles'. Famous are the sentences

"The boy broke the window with a rock"
and "The boy broke the window with a curtain".

Knowledge stored in the net drives the analysis which results in binding "with a curtain" to "the window" as a modifier and "with a rock". Ellman (1990) too works within the analysis paradigm - he finds that the formation of arbitrary states in a neural state machine causes words in sentences to cluster in terms of their syntactic and semantic categories. So, within the notion of language understanding being an analytical process, the aim of symbolicists and linguists alike is to build 'machines' (algorithms) that take linguistic strings as input and provides analytic labels as output.

This leads to the **second key point** that we make in this paper. The use of iconic states as a way of representing linguistic utterances may be illustrated as shown in fig 3.

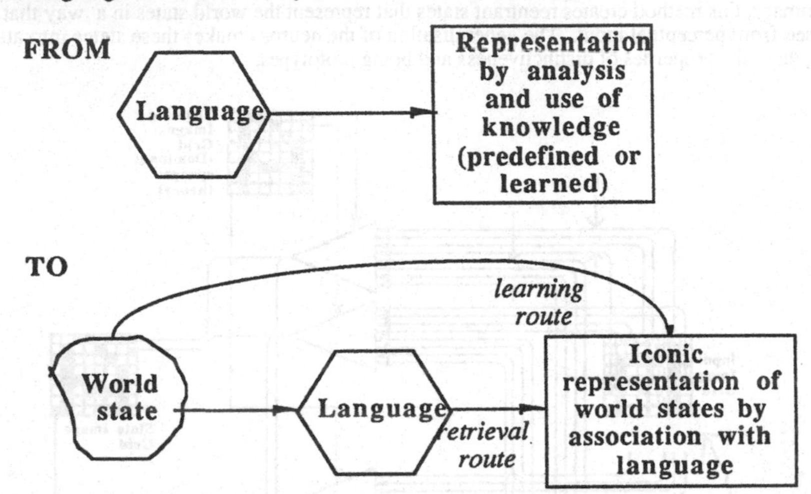

Fig. 2. A paradigm shift in linguistic representation

In the extant paradigm, language plays a central isolated role which contains the semantic parameters that are to be extracted by analysis. In the new paradigm, language plays a lesser role and has a triggering task in the machine where the objective is to retrieve the iconic states that gave rise to the linguistic string in the first place. We refer to the new paradigm as the "iconic hypothesis". This requires some definitions.

World states

Significant, frequently occurring features, static and temporal, of the world that contains other creatures that are capable of understanding and using language.

The inclusion of other knowing creatures should be noted.

Acquisition of iconic representation

The creation, in the state structure of a learning state machine, of states that have a many-to-some relationship to the states of the world and a state structure that parallels that of the world.

The implication here is that the state machine is capable not only of mirroring world states that are static but also the temporal changes in such states as they might occur in the world.

Language

A string-like symbolic encoding of world states and state structures, optimized by evolution and skillfully used by organisms that form part of the sensed world.

Note that this definition allows for both the world state and its linguistic representation to be present at the same time so that the learning organism can use a process of neural association to label the representation of world states with linguistic symbols.

Understanding

The retrieval of iconic representations of world states in response to a linguistic stimulus.

This is the crux of the paradigm shift. In a sense it demotes the role of language from its central position as an entity that needs to be analysed to that of an intermediary, a medium which is driven by world states.

As far as AI and cognitive science are concerned this, at first sight might appear to be the denial of a great deal of symbolic modelling. However, it does resolve a problem that is central to representation of the symbolic kind: the symbol grounding problem.

3. The symbol grounding problem.

In this paper we have shown that the iconic state arises naturally in neural systems where it is required that the inner state structure should form a non-arbitrary model of the world in which the neural net is required to operate. Steven Harnad (1990) has also evoked the concept of iconic states to resolve the problem of intentionality as raised by John Searle (1980) in his celebrated critique of language understanding in AI. Harnad characterizes the symbolic paradigm as containing <u>physical tokens</u> and <u>strings of physical tokens</u> which are <u>manipulated by explicit rules</u> that are syntactic and based on the <u>shape</u> of tokens and strings. This is a closed system the very closure of which excludes the possibility of making the tokens and their strings relate in a meaningful way to events in the world. Harnad equates this closure to the usefulness of a Chinese to Chinese dictionary to someone who does not know Chinese. No amount of manipulation will unlock the true meaning of symbols unless at least some of them are linked to objects in the world. This is the *symbol grounding problem.*

Harnad defines an iconic state as: "internal analog transforms of projections of distal objects on our sensory surfaces". (Indeed this definition may be found in other discussions on mental states, e.g. Shepard and Cooper, 1982). He goes on to argue that recursive neural networks can form such representations but are not capable of the full range of symbolic manipulations required by a symbolic system. So he concludes that the symbol grounding problem is resolved through *hybrid* models that contain *both* connectionist and rule-based systems. The **third key point** we make in this paper is that a connectionist system *is* capable in itself both of the symbol grounding and the symbolic processing necessary in language understanding. We briefly illustrate this in the rest of this paper.

4. The neural state machine model (NSMM).

Fig. 3 shows a new development of the original neural state machine first described in Aleksander (1992).

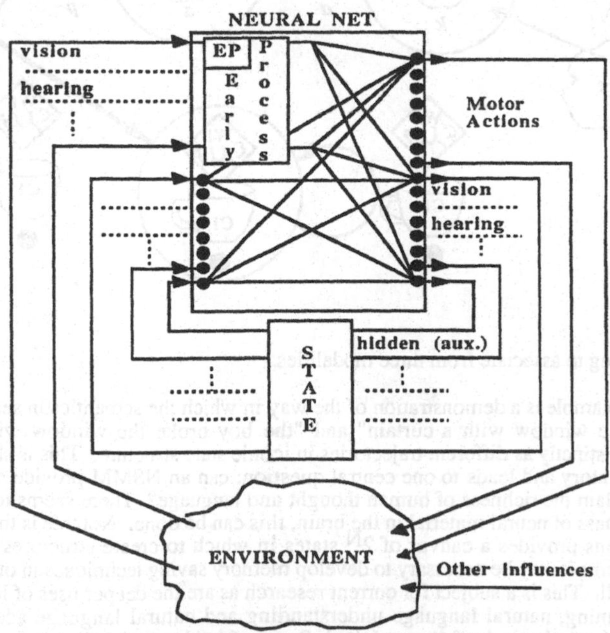

Fig. 3. A neural state machine model (NSMM) for cognitive studies.

Each of the right-hand dots in the "neural net" represent a neuron. They all receive inputs from the input interface and any architectural bias is determined by a connection matrix from these inputs to the neurons. One group is devoted to providing motor actions on the environment, while another group forms the internal state in which the iconic representations are to be represented. This is partitioned into a number of sensory modalities (all five if needs be) and makes provision for auxiliary state variables on which concepts such as ordinality and duration can be represented.

One of the features of this model, one that differentiates it from the model published previously, is the presence of an early processing block (EP). This is necessary for realism, as it is unlikely that iconic states occur in the brain at the raw sensory level. It is assumed that the early processing stages are pre-fashioned. That is they do not learn but perform fixed tasks such as edge extraction or auditory frequency encoding. One thing not shown in fig. 3 is the way in which training data is distributed to the neurons. On the whole, the method shown in in fig. 1 is applied. This does not indicate how neurons in the motor region are trained. This depends on what is being learned and is best illustrated by the two examples which follow.

5. Examples.

As a first example it is worth noting that the training shown in fig. 1 enables objects to be associated with their phonemic input and lexical symbol. Assuming that all input is processed by the early processor and say that *k-a-p* is the phonetic input which is heard at the same time as the image of a cup, Ci, as well as the written word Wi. The learned state sequence would be <Ci,Cw,*k*> followed by <Ci,Wi,*a*> and then followed by <Ci,Wi,*p*> finally followed by <Ci,Wi,• > which becomes reentrant. The symbol • is used to indicate that the phonetic input has stopped. Learning to say *k-a-p* is a matter of associating heard sounds with those uttered by the organism itself through the motor action neurons. This can occur independently of context. This is shown in fig. 4.

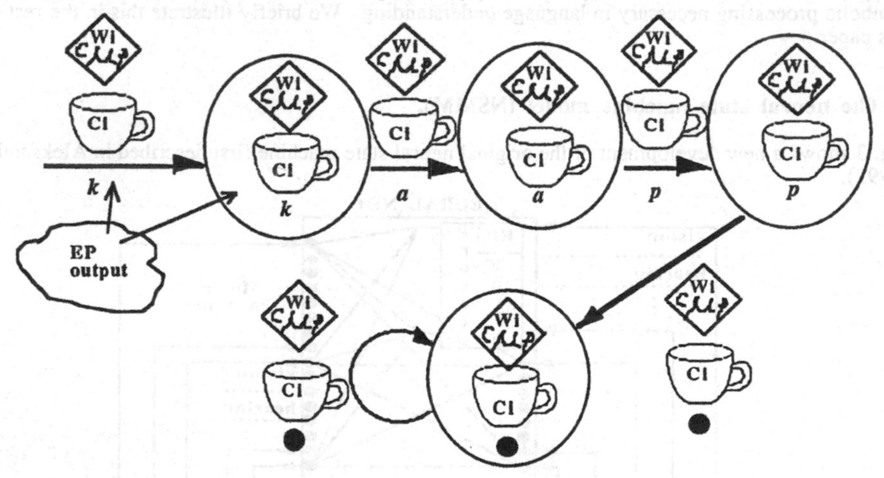

Fig. 4 Learning to associate from three modalities.

The second example is a demonstration of the way in which the semantics in sentences such as "the boy broke the window with a curtain" and "the boy broke the window with a rock" may be represented distinctly as different trajectories in iconic state structure. This is shown in fig. 5. This is self-explanatory and leads to one central question: can an NSMM provide a sufficient capacity model to explain the richness of human thought and language? There seems to be little doubt that with the richness of neural material in the brain, this can be done. Notable is the fact that a system with N neurons provides a canvas of 2^N states in which to create structures of iconic states. In artificial systems it may be necessary to develop memory saving techniques in order to build systems that work well. This is a subject for current research as are the deeper uses of iconic representation in robot planning, natural language understanding and natural language acquisition. The new paradigm still remains to be fully exploited. Some of its deeper implications are in the process of being published (Aleksander and Morton, 1993).

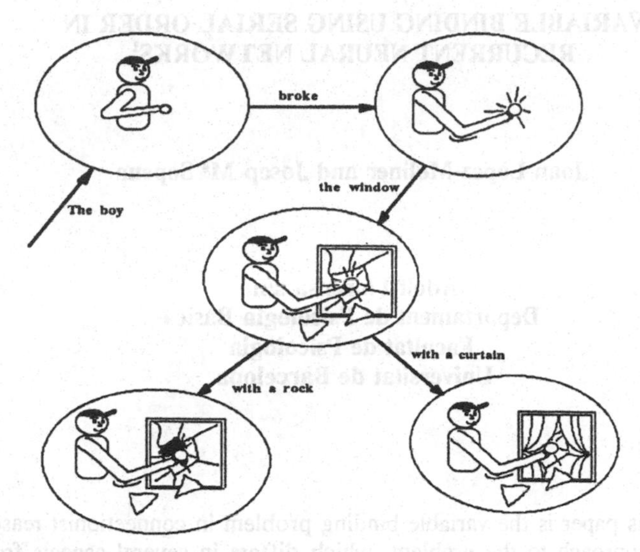

Fig. 5 Representation of sentences with different case roles.

6. Conclusion

Recursive nets are usually thought of as being autoassociators. In the first key point that we have made in this paper we have suggested a stronger role where such systems have a properly defined input and action output as one would expect of general state machines. A 'natural' form of training is to create states in the machine that have a relationship between their state variables and objects in the world as such objects appear on the sensory surfaces of the machine. The second key point is that this enables a definition of the concept of understanding of language as a transferral of world events to previously experienced iconic states that relate to these events. This is a paradigm shift in language understanding, but one which creates models which (as our third key point) are capable of representing grounded symbolic operations in an integrated fashion, rather than a hybrid way as has been suggested by others. Two examples, one of lexical knowledge acquisition and the other of case role resolution, suggest that the opportunities for exploring the iconic state hypothesis are plentiful.

References

Aleksander, I. An automata-theoretic assessment of the cognitive debate. In: Aleksander and Taylor(Eds), Artificial Neural Networks II, 625-630. Amsterdam: Elsevier.(1992)

Aleksander, I. and Morton, H.B. *Neurons and symbols: the stuff that mind is made of.* London: Chapman and Hall (in press), (1993).

Ellman, J. L.: Finding Structure in Time, *Cognitive Science*, **14**, 179-211. (1990).

Fodor, J.A. & Pylyshyn, Z.W. Connectionism and cognitive architecture: A critical analysis. *Cognition*, **28**, 3-71 (1988).

Kelly, G. *A theory of personal constructs*. New York: Norton. (1955).

Rumelhart, D.E. and Kawamoto, A.H. Mechanisms of sentence processing: assigning roles to the constituents of sentences. In Rumelhart, D.C.. & McClelland, J.L. (eds) *Parallel Distributed Processing*, Cambridge: MIT Press (1986).

Schank, R. C. & Abelson, R. *Scripts, plans , goals and understanding*. Hillsdale, N.J.: Lawrence Erlbaum. (1977).

Searle, J. Minds brains and programs. *The behavioural and brain sciences*, **3**, 417-457. (1980).

Shepard, R.N. and Cooper, L.A. *Mental images and their transformations*. Cambridge, Mass - MIT Press (1982)

VARIABLE BINDING USING SERIAL ORDER IN RECURRENT NEURAL NETWORKS[1].

Joan Lòpez-Moliner and Josep Mª Sopena

Adolf Florensà s/n,
Departament de Psicologia Bàsica
Facultat de Psicologia
Universitat de Barcelona.

ABSTRACT.
The scope of this paper is the variable binding problem in connectionist reasoning. We present a new approach to the problem, which differs in several aspects from former solutions. Mainly, an inductive reasoning stance, which seems more suitable for neural networks, is taken into account. The proposed solution deals with some of the limitations that are often present in earlier models, such as cross-talk, hardware limitations and static knowledge. A modified Elman architecture (Sopena, 1991) is used to store and retrieve items sequentially, so that temporal order allows us to bind different predicates which are represented by distributed patterns. This process makes it possible to represent conditional rules with a good degree of generalization. Distributed representations and temporal order offer a simpler treatment of variable binding than localist and spatial models do.

1. Introduction.

Neural Networks must deal with the variable binding problem in order to represent structure and show high-level cognitive skills. Most of the models are presented thus far, either do not deal with the binding problem or they do so in such a way that some important limitations result. Sun (1992b) treats the problem in order to get a reasoning based on logic rules where aspects like completeness and soundness become essential. We think that by assuming this perspective, connectionist implementations of reasoning undergo the problems that logic does. In the same way, some of connectionist reasoning models represent symbols and binding between them in a static way and not in one emerging from the dynamics of the network. Our goal starts from a more inductive view

[1] **This work was supported by Grants DGICYT PB89-0276 from the Ministerio de Educacion y Ciencia and FI92/64 from the government of the Generalitat of Catalonia.**

of reasoning, instead of a deductive one that can be better seen as a finite set of rules manipulating atomic symbols. The binding between predicates is learned. The representation of this binding takes the form of a distributed pattern of activity as a result of the dynamics of the network. First, the binding of the different arguments with the corresponding predicate takes place either at the input or output layers and at the same time step, so that no sequential processing is needed. Second, the binding between two predicates (and their corresponding arguments) is made by using serial order in a Recurrent Neural Network (RNN).

2. The model.

Backpropagation in a RNN permits the treatment of sequences, which means processing items through time. The representation of a concrete input pattern depends not only on the item itself, but also on its serial order. The distributed representation of an item is a function of the item itself and of the previous items of the sequence. As we will show later, this capacity allows us to bind items through time. Most of the works have offered spatial solutions to the variable binding problem (Barnden, 1988; Smolensky, 1990 and Sun, 1992a); Shastri & Ajjanagadde (1990) use a temporal approach in a localist implementation rather than a distributed one.

Architecture.

A modified Elman architecture (Sopena, 1991) is used to implement the reasoning system. Our network version differs from the original architecture (Elman, 1990) in the use of linear units in the hidden layer so that they have a no saturated activation function in order to improve short term memory. Figures 1a and 1b show the two basic architectures. In Elman networks, hidden units compute sums of context activations corresponding to hidden representations of former input patterns. The computation for each hidden unit at time t+1 is given by:

$$h_i(t+1)=F\left(\sum_{j=1}^{ninputs} input_j(t+1)\ w_{ij}+\sum_{k=1}^{ncontext} context_k(t+1)\ w_{ik}\right) \quad (1)$$

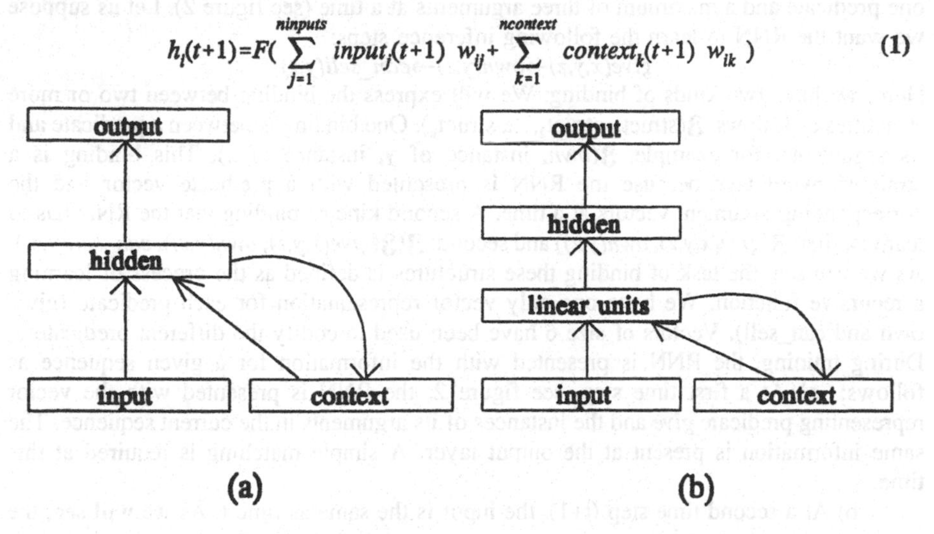

Figure 1. (a) Elman architecture; (b) Recurrent architecture with linear units (Sopena, 1991).

where $context_k(t+1)$ is an identical copy of hidden activations at time t:

$$context_k(t+1)=h_i(t), \quad k=i; \tag{2}$$

F is the sigmoid function, since this function is saturated it loses the capacity to store more than three or four items. If we have the identity function the computation of a hidden unit is given by:

$$h_i(t+1)= \sum_{j=1}^{ninputs} input_j(t+1)\ w_{ij} + \sum_{k=1}^{ncontext} h_k(t)\ w_{ik} \tag{3}$$

If we substitute (3) in (2), further analysis evidenciates that the hidden units compute a weighted sum, every input at time t-n is weighted by $(w_{ij})^n$. Then the sum returns different values for each different input sequence. The units of the first hidden layer (linear) are partially connected. We feed each linear unit with blocks of only 5 context units. Also, each block of 5 linear unit is fed with one input unit (there is an independent RNN for each input unit). The information is integrated in the second hidden layer. Sopena (1991) and Sopena (in preparation) for more details.

Codification and training.

This sequential approach to binding permits us a straightforward codification of items. Both arguments and predicates are codified at the input layer by binary valued vectors. A vector of size 10 allows us to codify 2^{10} different instances of one argument, so that no recruitment of units is needed in order to codify all possible instances.

Figure 2 shows how the information is entered into the RNN and the corresponding output. For each predicate, the number of arguments has been limited to three, so that we can have up to ternary predicates. Both input and output layers allow us to represent only one predicate and a maximum of three arguments at a time (see figure 2). Let us suppose we want the RNN to learn the following inference steps:

give(x,y,z)→own(y,z)→can_sell(y,z)

Here, we have two kinds of binding. We will express the binding between two or more structures as follows: $\mathfrak{B}(struct_1, struct_2, ..., struct_n)$. One binding is between a predicate and its arguments, for example, $\mathfrak{B}$(*own*, instance of *y*, instance of *z*). This binding is a straightforward task because the RNN is presented with a predicate vector and the corresponding argument vectors at a time. A second kind of binding that the RNN has to learn is, first: $\mathfrak{B}$(*give(x,y,z)*, *own(y,z)*) and second: $\mathfrak{B}$($\mathfrak{B}$(*give(x,y,z)*, *own(y,z)*), *can_sell(y,z)*). As we can see, the task of binding these structures is defined as the process of learning a recursive function. We have one only vector representation for each predicate (give, own and can_sell). Vectors of size 6 have been used to codify the different predicates. During training, the RNN is presented with the information for a given sequence as follows: a) At a first time step, see figure 2, the RNN is presented with the vector representing predicate *give* and the instances of its arguments in the current sequence. The same information is present at the output layer. A simple matching is required at this time.

b) At a second time step (t+1), the input is the same as time t. As we will see, the

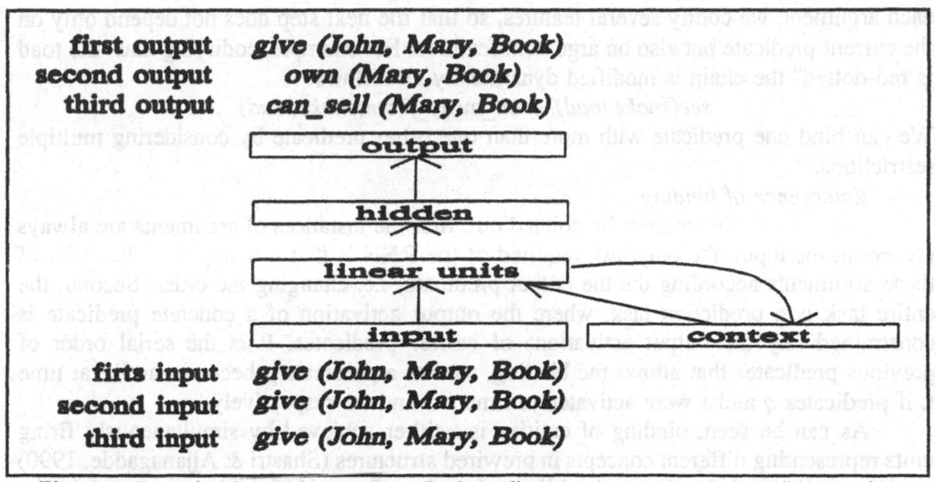

Figure 2. Example of an entire sequence. Both predicates and arguments are codified by using binary-valued vectors.

input is always the same, so that the first predicate of the sequence plays the role of a plan generator (Jordan, 1986). The output required is the predicate *own* and its arguments. Then the output is a function not only of the input but also of the previous internal state, which generated the output of the first predicate.

c) Finally, at a third time step, the output required is the predicate *can_sell*, note that the input is not variable. This third predicate is a function of the current input, and the two previous internal states, which generated the first and second predicate respectively.

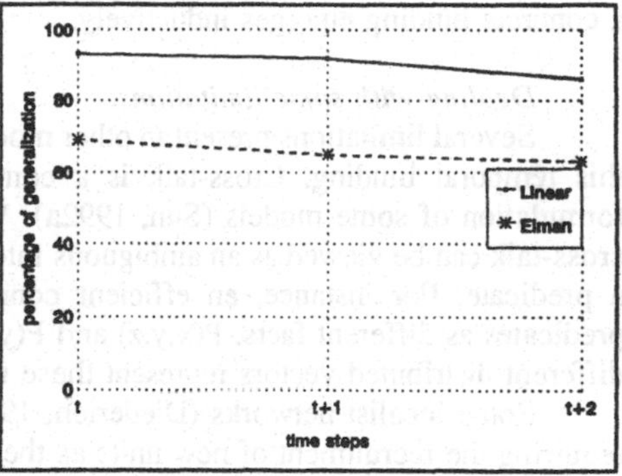

Figure 3. Generalization for linear and traditional Elman architectures for each time of the sequence.

Figure 3 compares the degree of generalization achieved by both our linear Elman architecture and the traditional one. All training sequences consisted of chains of three predicates. At each time step we have the degree of binding generalization for new instances.

So far, each inference step is a function only of the antecedent predicate, but inferences can be more complicated. Let us suppose we have the following inference chain:

see(snake,toad)→catch(snake,toad)→eat(snake,toad)

The RNN can learn variations of this sequence depending on the input constraints. For each argument, we codify several features, so that the next step does not depend only on the current predicate but also on argument features. For example, codifying that "the toad is red-dotted" the chain is modified dynamically as follows:

see(snake,toad)→run_away_from(snake,toad)

We can bind one predicate with more than one other predicate by considering multiple restrictions.

Emergence of binding.

At least two things must be pointed out, first the instances of arguments are always present at the input, the only task required of the RNN is that it manipulate the order of these arguments according the the output predicate, i.e. changing the order. Second, the entire task is a prediction task, where the output activation of a concrete predicate is constrained by the output activations of earlier predicates. It is the serial order of previous predicates that allows the binding, so that a predicate *p* becomes active at time t, if predicates *q* and *s* were activated at time t-2 and t-1 respectively.

As can be seen, binding of entities is neither achieved by simultaneously firing units representing different concepts in prewired structures (Shastri & Ajjanagadde, 1990) nor by static and localist assemblies of units (Sun, 1992a). In our model, the representation of a symbol predicate or a rule is not given beforehand, so that the representation of the binding itself is an outcome of the learning process as well.

As a result of training, the distributed representation of hidden units accounts for the recurrent function of binding as defined above. A binding of two structures has been being approximated through learning, so that a distributed representation accounting for a concrete binding emerges inductively.

Dealing with some limitations.

Several limitations present in other models can be straightforwardly addressed from this temporal binding. Cross-talk is a common and undesirable consequence of the formulation of some models (Sun, 1992a). When dealing with connectionist reasoning, cross-talk can be viewed as an ambiguous interpretation of a concrete pattern representing a predicate. For instance, an efficient connectionist system must treat the following predicates as different facts: P(x,y,z) and P(y,z,x). As we pointed out before, in our case different distributed vectors represent these facts unambiguously in the hidden units.

Some localist networks (Diederich, 1988; Sun, 1992a) have hardware limitations, requiring the recruitment of new units as the number of items to be processed increases. In our case, the use of distributed representations permits the representation of a large amount of information in a relatively small number of units. Binding representation takes place in a distributed manner through hidden units. Although generalization is a difficult problem for localist models, the use of distributed representation gives the net a good performance in testing generalization.

Some approaches to variable binding have to increase the size of a vector representing the binding of two items. For example, the tensor product (Smolensky, 1990) of two vectors **v** and **w** of dimension n and m respectively, is a nm-dimensional vector. Circular convolution operation used by Plate (1991) avoids this problem. Our sequential treatment of input patterns and the use of distributed representations impose no limitations of this nature and the same hidden layer can accomodate different size of data.

3. Conclusions.

Although the biological plausibility of the model has not been considered in depth, there are some experimental data (Damasio, 1989) that point to the presence of recurrent feedback interactions playing an important role in binding entities. We think that tretement of sequences as trajectories through a representational state space, offer new alternatives in modeling and formalizing reasoning processes. From this point of view, an independent consideration of the binding problem may not be required, since the network dinamically implements the binding mechanism as a trajectory that starts at the input predicate. The presence of multiple constraints, codified at the input layer, allows us to get several bindings for a concrete predicate. This inductive procedure makes connectionist models more suitable for dinamically achieving representations of useful rules (not necessarily in a logical sense). These rule representations should be able to account for new data with similar regularities, which means generalization. The improvement of memory capacity to deal with more complex rules, and a deeper treatment of feature codification to address analogical and similarity based reasoning processes in an integrated way are points current interest to us.

References.

Barnden, J. (1989). The right of free association: relative-position encoding for connectionist data structures. *Proceedings of the 11th Cognitive Science Society Conference*, 396-403. Hillsdale, NJ.: Lawrence Erlbaum Associates.

Damasio, A.R. (1989). The Brain Binds Entities and Events by Multiregional Activation from Convergence Zones. *Neural Computation, 1*, 123-132.

Diederich, J. (1988). Knowledge Intensive Recruitment Learning. Technical Report TR-88-010. Computer Science Institute, Berkeley, California.

Elman, J.L. (1988). Finding structure in time. *Cognitive Science, 14*, 179-211.

Jordan, M. (1986). Serial Order: A Parallel Distributed Approach. ICS Report 8604. Institute for Cognitive Science. University of California, San Diego.

MacLennan, B. (1991). Continuous Symbol Systems. The Logic of Connectionism. Technical Report CS-91-145.

Plate, T. (1991). Holographic Reduced Representations. Technical Report CRG-TR-91-1. University of Toronto.

Shastri, L. & Ajjanagadde, V. (1990). From Simple Association to Systematic Reasoning. Technical Report MS-CIS-90-95. University of Pennsylvania. (To be appeared in BBS).

Smolensky, P. (1990). Tensor Product Variable Binding and the Representation of Symbolic Structures in Connectionist Systems. *Artificial Intelligence, 46*, 159-216.

Sopena, J.M. (1991). ERSP: A Distributed Connectionist Parser that Uses Embedded Sequences to Represent Structure. Technical Report UB-PB-1-91. Universitat de Barcelona.

Sopena, J.M. (in preparation). Linear Recurrent Networks.

Sun, R. (1992a). A Connectionist Model for Commonsense Reasoning Incorporating Rules and Similarities. Technical Report, Honeywell SSDC.

Sun, R. (1992b). On Variable Binding in Connectionist Networks. *Connection Science, 4*, 93-124.

REGION OF INFLUENCE (ROI) NETWORKS. MODEL AND IMPLEMENTATION.

F.Castillo (1), J.Cabestany(2), J.M.Moreno*(2)

Universidad Politecnica de Catalunya UPC.Departament d'Enginyeria Electronica.

(1) E.U.P. Vilanova i la Geltru. Victor Balaguer, s/n 08800 VILANOVA Spain
(2) E.T.S.E. Telecomunicacion. PO Box 30.002 08080 BARCELONA Spain

Abstract:

Two different approaches in constructing Neural Network (NN) classifiers are discussed - discriminant-based networks and Region of Influence networks. A general model for ROI networks is presented, and the different functionalities of this structure are discussed: classification, vector quantization and associative memory.

Also, an architecture for this model's implementation is presented, and the hardware realization of each layer is reviewed in detail.

1. Introduction

Two broad type of Neural Network (NN) classifiers presently exist. The first, based on Multi-layered Perceptrons, tries to discriminate between classes using some function (usually a linear function which produces hyperplane discriminants), while the second (herewith refered to as Region Of Influence - ROI networks) attempt at approximating regions in which a class is predominant (as in the RCE [1] and GAL [2] networks). For this second type of NNs, patterns are stored sequentially using some learning rule during training, while in recall phase, patterns are presented to the network and compared to all stored patterns. Classification is then performed by comparing the distances obtained from the presented pattern and each stored pattern, and selecting the class associated to some pattern which produces a minimum distance.

ROI-type of networks are not limited to classification applications but as will be shown, may also be used for other purposes. In this article we shall present a general model for ROI-type networks and shall attempt to implement such a structure in hardware, layer by layer.

2. The model and its functionalities.

In this section, a general model for ROI networks is presented. The model proposed contains three different layers, each layer having a specific task. The functionality of the model is based on the so-called "competitive" neural networks. When an input pattern is presented, each class neuron compares this pattern to a set of stored patterns

---

* co-author J.M.Moreno is an FI scholar under the Generalitat de Catalunya's Education Dept.

characteristic of its class. Figure 1 shows the model:

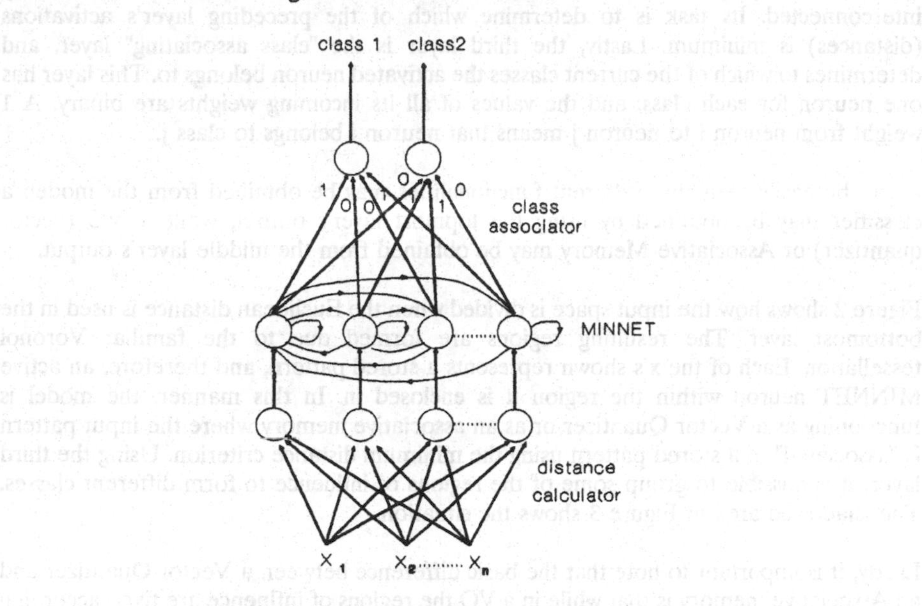

Figure 1. Region of Influence Network model.

The bottommost layer is the "distance calculating layer", and is in charge of evaluating the distance between the input pattern X, and each and every pattern stored in the network, X^j. Figure 1 depicts the n components of pattern X: $x^1, x^2, ..., x^n$. The M stored patterns are not shown in this figure, but each one's components $x^j_1, x^j_2, ... x^j_n$ may be thought of as the first layer's weights. The number of neurons in this first layer must be equal to the number of patterns stored in the network.

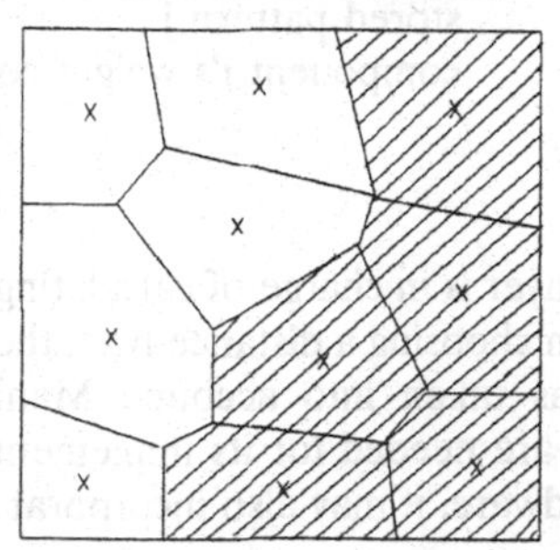

Figure 2. Sample two classes formed in the input space.

The second layer is a MINNET layer. It also contains M neurons, and is fully interconnected. Its task is to determine which of the preceding layer's activations (distances) is minimum. Lastly, the third layer is the "class associating" layer, and determines to which of the current classes the activated neuron belongs to. This layer has one neuron for each class, and the values of all its incoming weights are binary. A 1 weight from neuron i to neuron j means that neuron i belongs to class j.

It can be easily seen that different functionalities may be obtained from the model: a classifier may be obtained by using the topmost layer's output, while a VQ (vector quantizer) or Associative Memory may be obtained from the middle layer's output.

Figure 2 shows how the input space is divided when the Euclidean distance is used in the bottomost layer. The resulting regions are formed due to the familiar Voronoi tessellation. Each of the x's shown represents a stored pattern, and therefore, an active MINNET neuron within the region it is enclosed in. In this manner, the model is functioning as a Vector Quantizer or as an associative memory where the input pattern is "associated" to a stored pattern using the minimum distance criterion. Using the third layer, it is possible to group some of the regions of influence to form different classes. The shadowed area in Figure 3 shows the situation.

Lastly, it is important to note that the basic difference between a Vector Quantizer and an Associative memory is that while in a VQ the regions of influence are fixed according to some learning scheme, the regions in an Associative memory vary according to the search criterion (e.g. making one component more significant than another changes the regions formed). To take this into account, a weighted distance may be used, as an example, take the Manhattan distance:

$$d^j(X,X^j) = a_1|x_1-x_1^j| + a_2|x_2-x_2^j| + a_3|x_3-x_3^j| + \ldots \qquad (1)$$

where: $d^j(X, X^j)$ - distance between X and stored pattern X^j

X - input pattern

X^j - stored pattern j

a_i - component i's weight coefficient

3. Parts and implementation.

In the above model, the first layer is in charge of calculating distance between the actual pattern and the stored ones. In choosing a distance type, the complexity of the metric for hardware implementation was taken into account. Manhattan distance was chosen because the amount of hardware needed for its implementation is minimal, and with a slight modification on the hardware, it may also incorporate weighted components in its distance measure.

Figure 3 shows the elements needed to implement this layer using Manhattan distance. The leftmost shift register representation is shown just to ilustrate the method, and the actual implementation is show on the right. On the left, each shift register stores an integer which is represented as a series of 1s starting from the LSB, so that the value 6 for example would be represented as a series of six 1s from the LSB (as is shown on the upper register). At each cycle the registers are shifted left and the MSBs compared, if unequal, then a distance counter is incremented. After all bit positions have been evaluated, the value in the counter shall correspond to the Manhattan distance value between the two register contents. However, this implementation is inefficient in storing values, so that the two counters scheme on the right is actually used. In such a scheme each counter initially contains its corresponding integer value. On each cycle the counter is decremented and their zero flag compared (indicating whether a zero value has been reached), if unequal, then the distance counter is incremented. Take note that both structures on figure 3 are in fact equivalent.

If the weighted Manhattan distance is to be used, an AND gate must be added as in figure 4, where the other input to the AND gate is a mask value which is inputed sequentially in synchronization with the count-down action. This mask implements the coefficients of equation (1) and is chosen according to the rule:

> If a_i represents field i's weight coefficient and if $a_i \leq 1$, then in field i's mask every $(1/a_i)^{nth}$ mask bit must be set to 1, and the remaining bits to zero

or in other words, 1s are distributed uniformly within the mask with probability a_i. It can be shown [3] that the maximum error obtained is equal to 1 - a_i, which is quite acceptable considering that no adders and multipliers have been used in performing the calculations.

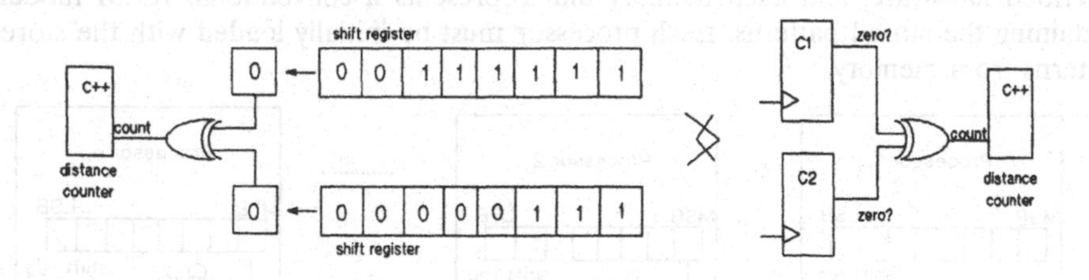

Figure 3. Methods used in calculating the Manhattan distance.

Our approach for MINNET layer implementation is presented in Figure 5. Supposing that all processors have calculated in parallel the distance for different stored patterns and that upon initiating the MINNET calculation process, processor j already has the distance $d^j(X,X^j)$ computed and available locally. A control processor shall function as the arbitrator, and shall determine which among the processors contains the minimum value. The process is as follows. First, each processor's not(MSB) is ORed, and if any MSB is equal to zero then all other registers are then set (all bits to 1). The process is again repeated after shifting left, so that after all bit positions have been evaluated, only one register contains a non-FFh value.

The advantages of this scheme are that the number of stored patterns to be processed is not limited by the number of processors available, as the processors may reprocess another batch of patterns. Take note that while the MINNET calculation is being performed, the CP is updated with the minimum value, which may in turn be used for comparing to the next batch of values. Other advantages of the scheme are that it is simple, modular, and easily expandable.

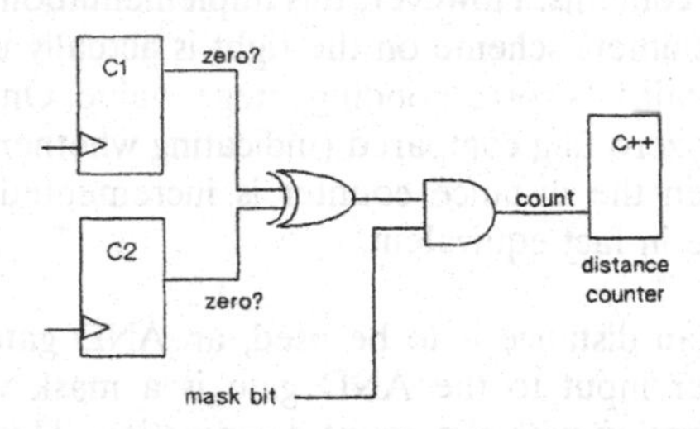

Figure 4. Hardware for the weighted Manhattan distance

The last layer, the "class associator", is the easiest to implement, and is really a pointer (register) which stores the class label to which the pattern has been attached to.

4. General architecture.

Figure 6 illustrates the proposed global architecture. Each processor P_i contains the described hardware, and each memory unit represents a conventional RAM module containing the stored patterns. Each processor must be initially loaded with the stored patterns from memory.

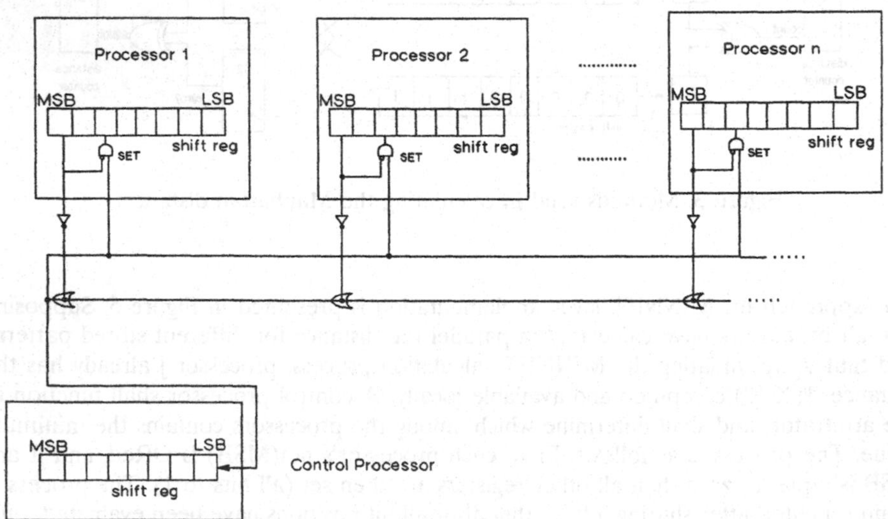

Figure 5. Digital MINNET implementation.

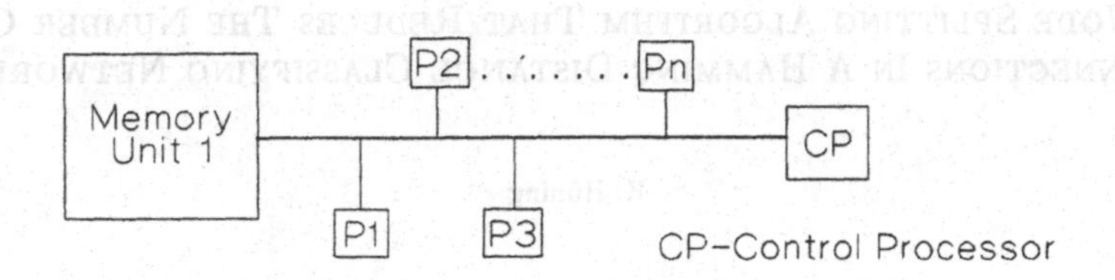

Figure 6. Inter-Processor Architecture.

Upon finalizing the process, the minimum is obtained and stored in the control processor. The process may then be repeated for another set of stored patterns. Obviously, a bottleneck effect seems to appear due to the common bus configuration, yet this is not critical as pipelining techniques can be used (e.g.dispatch the next set of patterns to be processed while the processors are still calulating). An alternative is to use more of these structures in parallel and simply obtain the overall minimum from all the CPs.

5. Conclusions.

A general model for a particular type of NNs has been presented along with the general lines for its digital implementation. The different functionalities of the model have been discussed, and a general architecture proposed. Possible applications of such an architecture include classification, high-speed Vector Quantizers and Associative memories. Finally, it is necessary to point out that on-line learning is possible for RCE [1] and GAL [2] algorithms (see [3]).

6. Bibliography

1. Reilly D.L., Cooper L.N., Elbaum C. A neural model for category learning. Biological Cybernetics, 45, 35-41 1982

2. Alpaydin A.I. Neural models of incremental supervised and unsupervised learning. PhD Thesis Lausanne EPFL 1990

3. Castillo F. Digital VLSI Architectures for Neural Networks. PhD Thesis Universidad Politécnica de Catalunya 1992

A Node Splitting Algorithm That Reduces The Number Of Connections In A Hamming Distance Classifying Network

H. Hüning

Aachen University of Technology, Communication Networks,
Kopernikusstrasse 16, W-5100 Aachen, Germany, e-mail: harry@dfv.rwth-aachen.de

Abstract

This paper describes an algorithm that builds up a network to classify binary patterns by Hamming distance using less connections than Lippmann's Hamming Net. The underlying principle is to have nodes with a variable number of connections taking their inputs from the overlapping parts of several patterns. Combining the outputs of these nodes in an uncompletely connected second layer gives the distances between a test input and all training patterns. The network structure is grown by a node splitting algorithm, which is a one-shot learning method and leaves the neurons with small sets of input connections. A comparison of different approaches to minimize the total number of connections is presented. Finally, the limitations of pattern classification by Hamming or vector distances are discussed.

1 Introduction

Many problems can be considered as classification problems, where some property of the patterns to be classified gives rise to a categorization into one of several classes. A classifier performs a mapping of input data to a number of outputs that usually are mutually exclusive (fuzzy outputs are not considered here). In particular neural network models have successfully been applied to classification problems where the desired mapping is represented by a set of *training* patterns carrying the labels of given classes. During the training of a neural net some parameters such as connections, neuron functions, and network structure may be modified in order to achieve the correct mapping for each training or *exemplar* pattern. After training the net can *test* new patterns that are generated in the same way as the training patterns. By the property of generalization the net is supposed to associate any new pattern with the class of patterns it most likely belongs to.

The Hamming distance is defined as the number of bits in which two binary vectors or patterns differ [1]. A Hamming distance classifier finds the closest exemplar pattern to a novel binary pattern by comparing the 'bit distances' of the pattern to a set of exemplars. The Hamming distance is usually applied as the criterion for an optimal correction of codewords that may have some bit errors with symmetric bit-error probability. Assume that a code is used for transmission where all codewords have a minimum Hamming distance of $d_{min} = 2n + 1$ bits to each other. Then the receiver of the codewords can correct n bit errors by finding the nearest valid codeword.

In contrast to such decoding problems, pattern classification problems typically have no exact prototype of a codeword or exemplar pattern for each class, but the classes are represented by several patterns each. For example, digitized images from a camera may have to be discriminated as different scenes. The binary patterns may also be composed of non-image data which should be coded with much redundancy (see [7]) or equal significance of every bit. Such patterns can have several thousand pixels and it suggests itself that not every tiny difference is supposed to indicate a different classification.

When a certain data processing task can be achieved by a network architecture, a parallel hardware implementation can be very efficient. R. P. Lippmann has proposed an artificial neural network architecture that applies Hamming distance to find nearest exemplar patterns, the *Hamming net* [2]. It has been taken up by V. G. Dobson [3]. In Lippmann's Hamming net there are nodes that compute the overlap to one examplar pattern each. On top of these nodes he uses a lateral inhibition layer 'MAXNET' to find the maximum overlap. Dobson [3] shows that a "minimum-mismatch" selection has advantages over the maximum overlap selection with respect to analog circuitry. Lippmann claims that the Hamming net is equivalent or better than a Hopfield net with respect to the total number of connections. Other neural network models have also been interpreted as reacting to pattern overlap (WISARD [4]) or do find a closest match by Hamming distance (Sparse distributed memory [5] and G-RAM [6]).

The current work has developed from an application of a growing weightless neural network model [7], which makes elaborate searches for statistically relevant n-tuple RAM-neuron connections and produces a relatively small and sparsely connected net. Its generalization performance on a smaller problem has been comparable to a Hamming distance classifier. The logical (RAM-neuron) net and the Hamming distance classifier are two extremes of unequal and equal significance of the input components.

In this context, a logical network has been created in a straightforward manner with connections to only those input pixels which are common to all the patterns of one class [8]. However, in general an overlapping part of *all* patterns in a class does not necessarily exist. An approach to draw advantage of the pattern overlap in a more general sense has lead to the network described in this paper.

2 A Hamming Distance Node

In Lippmann's Hamming net there are nodes with weights and thresholds set in a way to compute the overlap of input and exemplar patterns. Here a rather logical view is taken that allows a clearer graphical representation. The knowledge of an exemplar pattern is stored in the logical inversion of particular inputs to a summing node without threshold. Figure 1a shows three sample patterns:

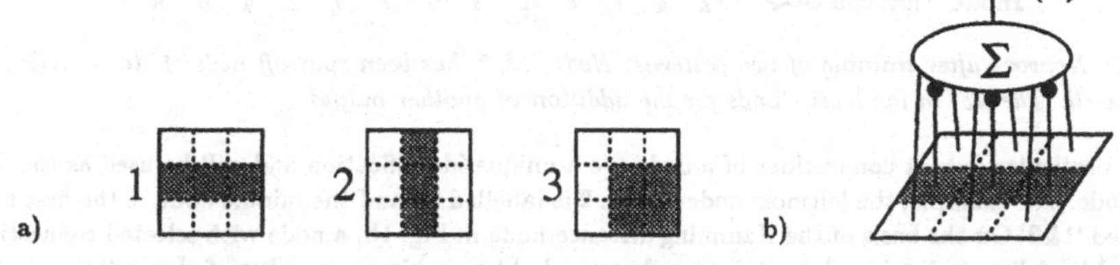

Figure 1: *a) Three exemplar patterns where shaded areas denote value 1 and white is 0. b) Hamming distance summing node that computes the distance to pattern 1. Inverted inputs are marked with a black dot which is a symbol for the logical NOT function.*

The summing node in Fig. 1b computes the Hamming distance $d(1)$ of any binary input to exemplar pattern 1, where the pixels have either value 0 or 1. Using inverted inputs where the exemplar pattern has value 1 makes every bit that does not match the exemplar add 1 to the distance. For example, the Hamming distance between patterns 1 and 2 is $d = 4$ bits. The Hamming distance can be expressed as:

$$d = \sum_i |p_i - e_i| \tag{1}$$

where p_i denotes the components of a binary pattern vector, e_i carries an exemplar pattern, and i runs through the components to compare. Formula (1) is a special case of the vector distance of nth order d_n with $n = 1$ (Minkowski metric, see [9]).

$$d_n = \sqrt[n]{\sum_i |p_i - e_i|^n} \tag{2}$$

For example, $n = 2$ in (2) gives Euclidean distance. The relation of Hamming distance to Euclidean distance by the square root has been pointed out in [10]. For binary values the comparison of $d = d_1$ is equivalent to the comparison of d_n, in so far as finding the nearest exemplar vector is concerned. Thus the distance function (2) determines what patterns are more similar than others for the classifier presented here. An interesting property of functions (1) or (2) is the linearity in the summation of absolute differences. The distance of the ith pattern component contributes linearly to the whole pattern distance. Therefore the calculation of distances can be shared for the overlaps of many exemplar patterns and summed up by smaller nodes.

3 Node Splitting Algorithm

The node splitting algorithm takes advantage of the *overlap* of training patterns, where overlap refers to equal pixel values of two patterns, both 0 or 1. The algorithm can detect the overlapping parts of any subset of the training set by keeping records of the input indices where patterns overlap and labelling them with a list of corresponding patterns. The overlap records are equivalent to one network layer of Hamming distance nodes and their labels give the connections to a second layer of summing nodes.

The training of this classifier network is described in terms of pattern or exemplar training, because every pattern is actually stored in the connections and could be retrieved by backtracing the links that are leading to its response node in the 2. layer. For the very first pattern a completely connected node is created just as in Fig. 1b. For the second pattern such a node is created as well, but the overlapping part with the first one is split off into a shared node whose response in used for d(1) and d(2) as illustrated in Fig. 2, where d(k) means distance to exemplar k.

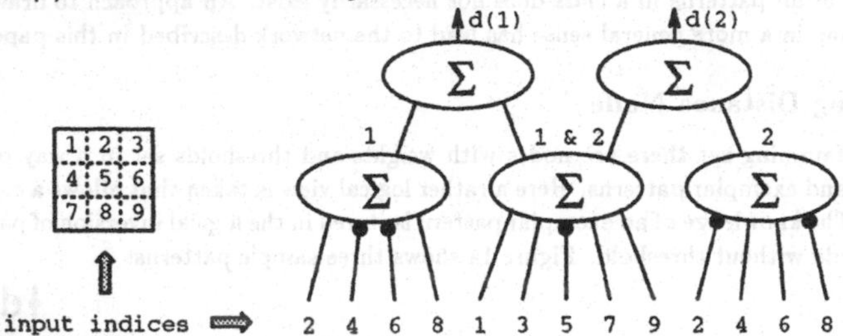

Figure 2: ***Network after training of two patterns. Node '1&2' has been split off node '1' to contribute to d(2) as well. The '&' in the label stands for the addition of another output.***

The particular output connections of a node are a unique identification and will be used as the *label* of the node. For example, the leftmost node in Fig. 2 is labelled '1' and the middle node in the first layer is labelled '1&2' On the basis of the Hamming distance node in Fig. 1b, a node with selected connections is defined by 3 lists (a list is a data structure that can hold an arbitrary number of elements):

- **a list of inputs from pixels with value 0 in the exemplar**
- **a list of inverted inputs from pixels with value 1 in the exemplar**
- **a list of corresponding patterns as output connections or *label***

The splitting of nodes refers to their input lists, as the inputs of the shared node '1&2' are not maintained in the parent nodes '1' and '2'. The output connections (label) of an overlap node are merely a combination of the outputs of its parent nodes.

Before the effect of training the third pattern is explained, the node splitting algorithm is expressed in terms of two operators where a, b, and o are nodes that consist of the three lists described above:

- **the node overlap operator '&': $o = a\&b$**
  generates a new node o with the overlapping inputs of a and b and the outputs of both
- **the node subtraction operator '−': $a = a - o$**
  eliminates those inputs in a that exist in o, and leaves all outputs unchanged

All three components of the nodes are affected by the node splitting algorithm that is now defined for the training of another pattern to an empty or existing network:

1. **create a node p taking the pattern as exemplar**
2. **let a take the previously existing nodes in turns and carry out the following steps:**
   2a) $o = a\&p$ 2b) $a = a - o$ 2c) $p = p - o$
3. **delete nodes that have no input connections left**

An illustration of the creation of nodes and modification of their connections is given in the example in Fig. 3 where all patterns from Fig. 1a are trained. Symbols for each type of input are placed in an input frame to show what pixels each node is connected to. The second layer connections are given by

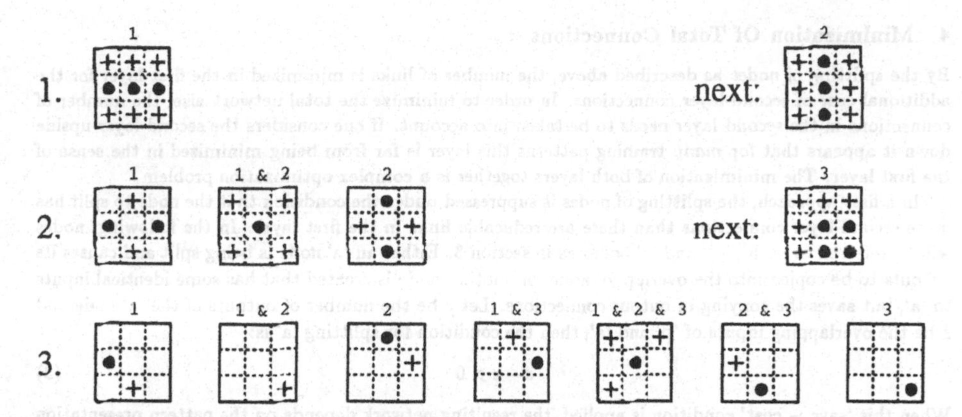

Figure 3: ***Training steps after 1, 2, and 3 patterns. The node inputs are entered as symbols into a pattern frame for each node. Inverted inputs are marked with • and direct inputs with +. Labels on top of the frames indicate the node outputs.***

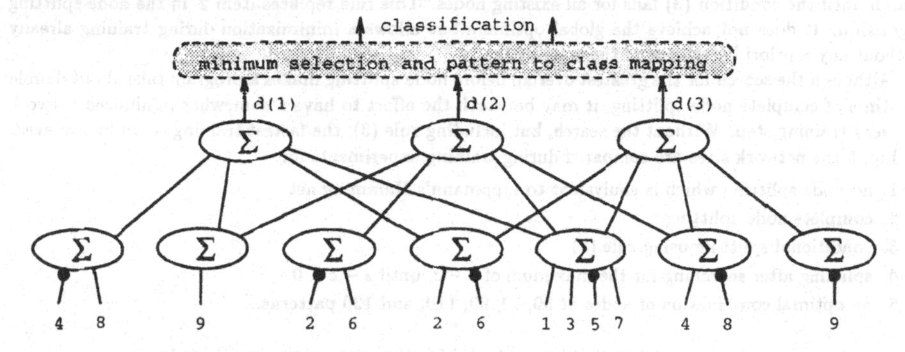

Figure 4: ***Network after training of three patterns. In this example no empty node occurs. On top of the second layer the best match is searched for and mapped to a classification.***

the node labels. The three nodes in the second row are equivalent to the network in Fig. 2. A network that corresponds to the 7 nodes in the bottom line of Fig. 3 is depicted in Fig. 4.

Classification is achieved by finding the minimum Hamming distance in the second-layer output and combining several of these outputs to classes by the logical OR function. The search for the minimum in the second layer is performed by serial comparison in software but can also be implemented in hardware as decrementing fastest-take-all devices [3]. The program that is called NOSDIC (Node Splitting Distance Classifier) can also output the name of the closest exemplar pattern as well as the Hamming distance as useful side-information.

The network seems to grow rapidly. The theoretical number of first-layer nodes after training of x patterns is $2^x - 1$, and this number is reached in the case of the small example: $2^3 - 1 = 7$. On the other hand the number of nodes is limited by the total number of different events in the disjointed inputs which is two times the number of input pixels. One may conclude that in large training sets not all kinds of different overlap exist. In practical networks a lot of the theoretical nodes are empty, so they do not consume any memory and processing time.

The NOSDIC program that implements the node splitting algorithm makes extensive use of dynamic memory allocation in particular for the lists of connections in the nodes. Because of frequent problems with loss of deallocated memory in blocks of varying size, it is advisable to reduce any deallocations during training to list elements of uniform size.

4 Minimization Of Total Connections

By the splitting of nodes as described above, the number of links is minimized in the first layer for the additional cost of second-layer connections. In order to minimize the total network size, the number of connections in the second layer needs to be taken into account. If one considers the second layer upside down it appears that for *many* training patterns this layer is far from being minimized in the sense of the first layer. The minimization of both layers together is a complex optimization problem.

In a first approach, the splitting of nodes is suppressed under the condition that the node to split has more second-layer connections than there are reducable links in the first layer. In the following, nodes will be referred to as 'a', 'p', and 'o' nodes as in section 3. Either an 'a' node is being split and causes its outputs to be copied into the overlap 'o' node, or another node is created that has some identical inputs to 'a', but saves the copying of output connections. Let c be the number of outputs of the 'a' node and s be the overlapping inputs of 'p' and 'a', then the condition for splitting 'a' is:

$$s - c > 0 \tag{3}$$

When this 'save − cost' condition is applied, the resulting network depends on the pattern presentation order during training, because the next node with partial overlap to 'a' and 'p' could cause splitting of either one with different results. Thus there will be many nodes with partially identical inputs. It makes sense to search for the particular 'a' node with greatest $s - c$ before node splitting and repeating this search until the condition (3) fails for all existing nodes. This rule replaces item 2. in the node splitting algorithm. It does not achieve the global optimum but allows a minimization during training already without any a-priori knowledge of the training set.

Although the search for the greatest overlap before node splitting makes a program take about double the time of complete node splitting, it may be worth the effort to have a somewhat minimized network at every training step. Without the search, but including rule (3), the fastest training could be achieved. In Fig. 5 the network sizes are compared during training experiments for:

1. no node splitting which is equivalent to Lippmann's Hamming net
2. complete node splitting
3. conditional splitting using rule (3)
4. splitting after searching for the maximum of $s - c$, until $s - c < 0$
5. an optimal combination of nodes at 50, 70, 90, 110, and 130 patterns.

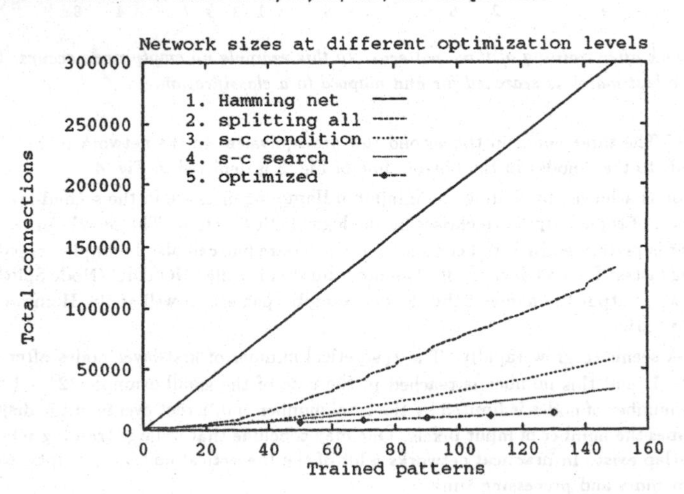

Figure 5: *Comparison of network sizes that result from different degrees of optimization.*

In the fifth experiment an elaborate optimization method has been applied for comparison. Starting with one Hamming distance node for each pattern, always those two nodes have been split and combined in turns that provide the greatest reduction of total connections, until the possible reduction becomes zero. However, this minimization method takes very long, because its cost increases quadratically with the number of nodes. For example, with 110 training patterns up to 131841 possible combinations of any two from 514 nodes had to be compared.

The data used in these experiments represents signal levels from wireless transmission coded by the position of a bar that is 6 bits high in 40 bit columns, next to a small part of circumstantial information [8]. 15 classes have been used, each represented by 10 patterns of 2000 bits.

5 Discussion and Conclusion

The comparison in figure 5 shows that the NOSDIC network can reduce the number of connections considerably in comparison to Lippmann's Hamming Net. The proposed training method is of the one-shot learning type which is much faster than iterative training methods. Of course, the function performed by this network is identical to a nearest neighbour look-up table. But the rather classical concept of Hamming distance has been applied several times in neural models [5, 6] apart from the Hamming Net. In the current approach very simple model neurons are equipped with such inputs that exhibit strong statistical dependence, equality in the extreme, which suggests using a distance function for generalization. The Hamming distance gives an exact measure of the quality of classification decisions, and allows the rejection of too close patterns with contradictory classification during training.

Also the general limitations of Hamming or vector distance classification must be considered. Every input value is equally significant to the resulting decisions, which might not be desired. In particular the classifier is not a vision system and cannot recognize objects with tolerance of shift, scale, and background variations. The NOSDIC network can be applied to binary codes with redundancy or patterns where characteristic changes do occur at fixed positions, as in scene observation for example.

Possible extensions of the NOSDIC network are weighting of the component distances in formula (1) or inclusion of a-priori probabilities of a Bayesian classifier [3], but the node splitting algorithm is only worthwhile if enough identical calculations can be carried out for different exemplar patterns. By the modification of the connectivity during training the cost of a complete connectivity like in other models could be avoided. The development of networks with an economical amount of connections may lead to systems that can deal with greater complexity. The prominent advantages of the NOSDIC network are the very fast training algorithm and the automatic adaptation of its structure and size according to a given problem.

References

[1] S. Lin, D. J. Costello *Error Control Coding,* Prentice Hall 1983.

[2] R. P. Lippmann *An Introduction to Computing with Neural Nets,* IEEE Acoustics, Speech, and Signal Processing Magazine, April 1987, pp. 4 - 22.

[3] V. G. Dobson, J. M. Salinas *Decrementing Hamming and Bayesian Neural Networks: Analog Implementations and Relative Performance,* in A. Prieto (ed.) *Artificial Neural Networks,* Int. Workshop IWANN '91 Proceedings, LNCS 540 Springer-Verlag 1991.

[4] I. Aleksander and H. Morton *An Introduction to Neural Computing,* Chapman & Hall 1990.

[5] P. Kanerva *Associative-memory models of the cerebellum,* in I. Aleksander, J. Taylor (editors) *Artificial Neural Networks, 2,* ICANN '92 Proceedings, North-Holland 1992.

[6] I. Aleksander *Connectionism or Weightless Neurocomputing?* In T. Kohonen, K. Mäkisara, O. Simula, J. Kangas, editors, *Artificial Neural Networks,* Proc. ICANN '91, North-Holland 1991.

[7] H. Hüning *Cognition and Neural Network Modelling,* in I. Aleksander, J. Taylor (editors) *Artificial Neural Networks, 2,* ICANN '92 Proceedings, North-Holland 1992.

[8] G. Ratayczak *Verarbeitung der im GSM-Mobilfunksystem anfallenden Funkmeßdaten,* Diploma Thesis at Aachen University of Technology, Communication networks, Prof. Dr.-Ing. B. Walke, Kopernikusstr. 16, W-5100 Aachen, Germany, 1992.

[9] M. Zeidenberg *Neural Networks in Artificial Intelligence,* Ellis Horwood 1990.

[10] J. A. Freeman, D. M. Skapura *Neural Networks,* Addison Wesley 1991.

A HIGH ORDER NEURAL MODEL

LOPEZ ALIGUÉ, FRANCISCO J.; ACEVEDO SOTOCA, Mª ISABEL ; JARAMILLO MORAN, MIGUEL A.

Departamento de Electrónica e Ingª Electromecánica, Universidad de Extremadura, Avda. de Elvas, s/n. 06071 - Badajoz, Spain

ABSTRACT

The interaction between afferent nerves had always been regarded as a phenomenon which is produced outside the neuron. This work presents an extension of the classic concept of interaction between inputs, including the possibility of higher-order effects at the level of neuronal activity function. The mathematics is formulated with a view to its algorithmic simulation being implemented in a multiprocessor system by means of an adequate programming language running as a multi-elemental processor parallel computer. The system that is finally presented is a higher-ordered neural network with nonsupervised learning implemented in a multilayer structure.

INTRODUCTION

Baldi and Xu (1988) first proposed the concept of generalized neural network, which was subsequently enlarged by Xu and Tsai (1990), demonstrating its effectiveness when used in networks configured as associative memories. In this generalized network, the state $y_i \in \{0, 1\}$ of each neuron is taken to be defined by its outputs at each instant "t", and it is supposed that the set of outputs $\{y_j\} = \{y_1, y_2, \dots, y_N\}$ act on the i-th neuron in the form:

$$y_i(t+1) = \sigma\Big[\frac{1}{(N-1)!}\sum_{i_1=1}^{N}\cdots\sum_{i_{N-1}=1}^{N} w^{i}_{i_1 \dots i_{N-1}}\, y_{i_1}(t)\cdots y_{i_{N-1}}(t) +$$

$$\frac{1}{(N-2)!}\sum_{i_1=1}^{N}\cdots\sum_{i_{N-2}=1}^{N} w^{i}_{i_1 \dots i_{N-2}}\, y_{i_1}(t)\cdots y_{i_{N-2}}(t) + \dots$$

$$+\frac{1}{2}\sum_{i_1=1}^{N} w^{i}_{i_1}\, y_{i_1}(t) + \theta_i\Big] \qquad (1)$$

Here the $w^{i}_{i_1 \dots i_p}$ represent coupling factors between the inputs $y_{i_p}(t)$ that reach the i-th neuron, so that $w^{i}_{i_1}$ is the weight associated with the union of the i_1-th neuron's output to the i-th neuron in question. Thus, while in the normally accepted model the neuron's reponse depends on the weighted sum of the inputs from other neurons, in Xu and Tsai's case, the neuronal activity is a function not only of the usual first-order term, but also of other higher-order terms obtained from the various products between the inputs, and where we denominate by order the number of elements intervening in each of these products. In its original expression, each term was divided by the number of possible permutations of the elements of each product in order to eliminate repetitions. The parameter θ_i represents a threshold below which the neuron will not fire. The function $\sigma()$ is the sigmoid function which can adopt any form which limits its value to:

$$\sigma(x) \in \{0,1\} \ , \ \forall \ x \in \{-\infty, +\infty\}$$

The fundamental motive for developing the inclusion of higher-order terms is to simulate the possible interactions between the input signals arriving at a neuron, and that had not been taken into account in earlier models. We shall in this way be developing a more evolved neuronal model, thanks to which the neural networks that have been constructed from first-order neurons up to now, will be able to present far richer patterns of behaviour (learning and recognition).

THE MODEL

Starting from the aforementioned idea of making the neuron's activity depend not only on the weighted sum of the activities converging on it but also on the interactions established between them represented in the form of higher-order terms, we shall present a model which improves a series of fundamental characteristics, particularly to allow its later algorithmic simulation. The first thing for this is to define the neuron as the basic element of a layer able to receive inputs both from outside the layer and from other neurons in the same layer. In any case, the behaviour of the neuron will not distinguish the origin of the excitation signal, so that its functioning, with respect to inputs, will be identical independently of the source of the signal. One can therefore define the network's activity vector as

$$Y(t) = (y_1(t), y_2(t), ..., y_N(t)) \quad , \quad Y(t) \in R^N \tag{2}$$

and the vector of inputs to the i-th neuron as

$$X_i(t) = (y_{i_1}(t), y_{i_2}(t), ..., y_{i_M}(t), z_{i_1}(t), z_{i_2}(t), ..., z_{i_L}(t)) \tag{3}$$

$$X_i(t) \in R^P \ , \ P = M + L$$

$$y_{i_m} \in \{0,1\} \ , \quad z_{i_j} \in \{0,1\}$$

Here now they have all been taken with positive values defined in the interval {0,1}, since we consider that no signal reaching an entry has by itself an excitory or inhibitory character, but rather gets this character from the synaptic weight of the connexion.

The $y_{i_m}(t)$ represent the inputs coming from the lateral interaction with neurons of the same layer, and the $z_{i_l}(t)$ represent the inputs coming from outside the layer, either from a preceding layer or from the outputs of appropriate transducers, which means that in this model the feedback is a natural consequence of the whole structure of the formulation and does not need any special considerations. In any case, from the point of view of inputs to neuron, their origin is irrelevant. Moreover, the behaviour is indistinguishable with inputs coming from the lateral interaction or feedback (spatial interaction).

Accordingly, the neuronal activity (1) will now be controlled by the expression

$$y_i(t+1) = \sigma\Big[\sum_{i_1=1}^{P} \dots \sum_{i_P=1}^{P} w^i_{i_1 \dots i_P} x_{i_1}(t) \dots x_{i_P}(t) + \sum_{i_1=1}^{P} \dots \sum_{i_{P-1}=1}^{P} w^i_{i_1 \dots i_{P-1}} x_{i_1}(t) \dots x_{i_{P-1}}(t) + \dots + \sum_{i_1=1}^{P} w^i_{i_1} x_{i_1}(t) + \theta_i \Big] \quad (4)$$

where now the dividing terms employed by Tsai and Xu to renormalize the value of the new "synaptic weights" have disappeared. This is clearly one of the most significant features of the model, at the same time as being that which presents the greatest difficulty of mathematical definition. An added difficulty is that deriving from the fact that the model we are presenting is aimed at being implemented algorithmically in a very complex network: in our case, this is a network formed by an input and an output layer and various hidden layers, each consisting of an array of, at least, 10^3 by 10^3 neuronal elements and with a process of non-supervised adaptation.

To resolve it, we shall make use of the fact that the inputs can be grouped into two subsets: that of the excitors and that of the inhibitors. We thus must consider the contribution from each of these blocks of inputs separately, defining the higher-order excitory interaction and its corresponding inhibitory counterpart.

As a result, the functions (1) and (4) which describe the behaviour of the neuron becomes

$$\begin{aligned} y_j(t+1) = \sigma \Big\{ \Big[& \sum_{i_1=1}^{P} \dots \sum_{i_P=1}^{P} w^j_{i_1 \dots i_P} x_{i_1}(t) \dots x_{i_P}(t) \\ & + \sum_{i_1=1}^{P} \dots \sum_{i_{P-1}=1}^{P} w^j_{i_1 \dots i_{P-1}} x_{i_1}(t) \dots x_{i_{P-1}}(t) + \dots + \sum_{i_1=1}^{P} w^j_{i_1} x_{i_1}(t) \Big] \\ & - \Big[\sum_{k_1=1}^{Q} \dots \sum_{k_Q=1}^{Q} w^j_{k_1 \dots k_Q} x_{k_1}(t) \dots x_{k_Q}(t) + \\ & + \sum_{k_1=1}^{Q} \dots \sum_{k_{Q-1}=1}^{Q} w^j_{k_1 \dots k_{Q-1}} x_{k_1}(t) \dots x_{k_{Q-1}}(t) + \dots + \sum_{k_1=1}^{Q} w^i_{i_1} x_{i_1}(t)) \Big] \\ & + \theta_j \Big\} \\ = \sigma \{ & E^+ - E^- + q_j \} \end{aligned} \quad (5)$$

which represents the combination of the affected inputs with positive weights (excitors) E^+ and that of the inputs with negative weights (inhibitors) E^-. In the equation, P and Q are the order of each term, which do not have to be equal in the two cases. Also, there is no a priori requirement for the excitory and inhibitory inputs to belong to disjoint subsets, but rather a given input will be able to belong to both groups, although the weights that will be associated with it in each case will, in principle, be different.

The weights affecting each of the terms of the summation are defined as

$$w_i^j \in \{-1, +1\}$$

$$w_{i_1 \ldots i_p}^j = w_{\pi(i_1 \ldots i_p)}^j \; , \quad \forall \; i_1, \ldots, i_p \tag{6.1}$$

where $\pi(i_1 \ldots i_p)$ is a unique permutation (at any order) of $i_1, \ldots, i_p$. Its value is calculated from the first-order weights, in the form

$$w_{i_1 \ldots i_p}^j = w_{i_1}^j w_{i_2}^j \ldots w_{i_p}^j \in \{+1, -1\} \tag{6.2}$$

for which one must take into account that repetitions have been avoided, so that the set of the weights is

$$W_i^j = (w_{i_1}^j, w_{i_2}^j, \ldots, w_{i_N}^j, w_{i_1 i_2}^j, w_{i_1 i_3}^j, \ldots, w_{i_N i_N}^j, \ldots, w_{i_1 \ldots i_N}^j) \tag{7}$$

Given that, as a general rule, the first-order weights are less than unity, the value of the weight of a determined order will be less than that of a lower order. The resulting activity will thus decline as the order being considered rises:

$$w_{i_1 \ldots i_{p1}}^j < w_{i_1 \ldots i_{p2}}^j \quad \forall \quad p1 > p2$$

In this way, one avoids having to predetermine the values of the weights for each order of the interaction, and neither is it necessary to consider the damping factors of the different summations in the neuron function, thus obviating their storage in memory, making the implementation of the algorithm less exigent in memory requirements and hence in computation time.

Another difficulty eliminated by this method is that of the adaptation of the weights. According to this model, the first-order adaptation automatically carries with it the adaptation of all the other orders.

With regards to the activity function of the neuron, one can obviously choose any of those that one has been able to use up to the present, although throughout the work we use the traditional sigmoid definition

$$\sigma(x) = \frac{1}{1+e^{-Ax}}$$

THE ADAPTIVE MECHANISM

The dynamic behaviour of the synapse showed in (7) is one of the most important aspects of the definition of the activity of a neuron. Indeed, the definition of a new model of neuronal function is always made in light of the outlook for its utilization in a network.

As has been noted previously, the present goal is to improve the learning and pattern recognition capacity, so that one requires a model of adaptation that responds to the following premises:

a) The weight increases, without changing sign, when the associated input grows in absolute value, decreases in the contrary case, and stays unchanged if the input signal does the same.

b) The speed of adaptation during growth does not have to coincide with that of decline. As a general rule, one takes it in a form such that growth (learning) is always greater than decline (forgetting).

c) A synapse that is little used will have a low weight. It will also have a small tendency to vary, which will increase as the activity increases. Likewise, as the weight approaches its maximum value, the variation will again have to diminish.

These premises provide a model that is simple but powerful at describing synaptic evolution. One supposes, also, that this evolution is independent of the origin of the input to the neuron, i.e., its form is the same for inputs from outside the layer as for those from the lateral interaction. Its transformation to mathematical expressions can be made in the following form:

$$w_j^i(t+1) = \text{sign}(w_j^i)\, \sigma_{1/2}(\bar{x}_j(t+1)) \qquad (8)$$

In this expression, an intermediate variable $\bar{x}_j(t+1)$ has been defined that represents the "average value" of the inputs to the neurons, and in which resides the memory of the network. This intermediate variable will evolve according to the expression

$$\bar{x}_j(t+1) = \bar{x}_j(t) + \frac{x_j(t+1) - \bar{x}_j(t)}{\Delta} \qquad (9)$$

in which the term Δ controls the speed of the evolution, a large value slowing it down. a small value speeding it up. Hence, point b) of the above premises is fulfilled. Point a) is also clearly satisfied by the above expression.

The above rule of evolution presents a clearly linear behaviour, contradicting premice c). To get over this inconvenience, one introduces the function $\sigma_{1/2}(\,)$ which responds to the expression

$$\sigma_{1/2}(x) = \frac{1}{1+e^{-4(x-1/2)}} \qquad (10)$$

The synaptic weight $w_j^i(t + 1)$ from (7), is thereby provided with the alinearity needed to comply with premise c) commented on above. It is also given a fixed sign which determines its excitory or inhibitory character.

As can readily be verified, the rule of adaptation of the weights is a convergent expression, since $\overline{x}_j(t)$ evolves towards x_j as the time increases. Also, one supposes that the speed of evolution of the synapse is much less than that of the neuronal activities, since one defines the dynamics of the latter by means of a lag-free transfer equation. In this way, and as the synapses' behaviour is clearly convergent, the dynamic behaviour of the network will be dominated by that of the transfer equation.

ACKNOWLEDGEMENT

The authors wish to express their gratitude to the C.I.C.Y.T. for its financial support through the grant TIC - 0671 / 89.

REFERENCES

Baldi, P.: Neural networks, orientations of the hypercube, and algebraic threshold functions. *IEEE Transactions on Information Theory,* Vol. 34 No. 3, pp. 523 - 530 . 1988.

Grossberg, D. : Nonlinear neural networks : Principles, mechanisms, and architec tures. *Neural Networks,* Vol. 1. pp. 17 - 61. 1988.

Simpson, P. : Higher-ordered and intraconnected bidirectional associative memories. *IEEE trans. on Systems, Man, Cybernetics.* Vol. 20, No. 3, pp. 637 - 653, 1990.

Xu, X. and Tsai, T.: Constructing associative memories using neural networks. *Neural Networks.* Vol. 3, pp. 301 - 309. 1990.

Higher-Order Networks for the Optimization of Block Designs *

Pau Bofill
Departament d'Arquitectura de Computadors—UPC
Campus Nord. Mòdul D-4. c/ Gran Capità s/n. 08071 Barcelona
email: pau@ac.upc.es
Carme Torras
Institut de Cibernètica—CSIC/UPC
c/ Diagonal 647. 08028 Barcelona
email: torras@ic.upc.es

Abstract

The existence of a block design with a given set of parameters is a well-known problem from combinatorial theory. Although there exist some constructive methods, the generation of block designs in the general case is an NP-complete problem.

In this paper we use optimizing neural networks as an heuristic approach to the generation of block designs. First, a cost function is defined and mapped onto a network which has connections of arity four. This network is then used for the generation of some designs that are known to exist. For some designs the results are good, but for some others the system fails to find an optimal solution in a reasonable time. The problem is shown to be a good example to test the performance of optimizing neural networks.

1 Introduction

A *Balanced Incomplete Block Design* (BIBD) [Hall, 86, Street & Street, 87] is an arrangement of *elements* of a set in several subsets called *blocks* in such a way that every element belongs to the same number of blocks, every block has the same number of elements, and every pair of elements cooccur in the same number of blocks. The *incidence matrix* A of a BIBD with v elements and b blocks is an $v \times b$-binary matrix with exactly r ones per row and exactly k ones per column, such that the dot product (or correlation) of every pair of rows is exactly λ. The constants v,b,r,k,λ are called the *parameters* of the design, and the resulting configuration is called an (v,b,r,k,λ)-BIBD.

The parameters of a block design are said to be *admissible* when their "physical" meaning is consistent, that is, when the following two conditions are met:

$$vr = bk, \tag{1}$$

$$r(k-1) = \lambda(v-1). \tag{2}$$

Equation 1 says that the total number of ones of the incidence matrix must be the same when summed up by rows or by columns. Equation 2 counts the number of ones cooccurring with those in a given row by considering first that the r columns that have a one in that row have $k-1$ ones in the remaining rows and, then, that each of these $v-1$ remaining rows cooccur λ times with the original row. Since we have 2 equations, only 3 of the 5 parameters are independent.

*This work has been partially supported by the project CICYT TIC-91-0423

The admissibility of the parameters of a design is a necessary but not sufficient condition for the *existence* of a BIBD. For a given parameter set, the number of existing non-isomorphic BIBDs is not known in the general case. In [Mathon & Rosa, 90] there is a listing of all admissible parameter sets with $r \leq 41$ and $k \leq v/2$, together with the number of known non-isomorphic designs. In some cases it has been proved that no designs exist, in other cases we only have a lower bound, and sometimes the existence of a particular design is still an open problem. Although there exist construction methods for some particular parameter sets, the algorithmic generation of BIBDs in the general case is an NP-complete problem.

Like most combinatorial configurations, BIBDs have their origins and main applications in the field of experimental design (arranging experiments in such a way that the statistical analysis of their results is well defined). Although Fisher is considered the precursor of experimental design, and provided some important contributions [Fisher, 40], BIBDs were first introduced by Yates in [Yates, 35].

BIBDs have another important application in the field of code theory [Hall, 86, Hill, 86]: a BIBD, its complementary, a row with all ones, a row with all zeros, and an extra column (half ones and half zeros) make up a *perfect code*.

In this paper, we approach the generation of BIBDs using optimizing neural networks. Like other heuristics, the interest of combinatorial optimization techniques (neural networks included) is that they yield pseudooptimal solutions in polinomial time. In our case, since only optimal solutions are of interest (actual BIBDs) and they are easy to recognize, rather than attempting to "solve" the problem, we will "challenge" our networks with a difficult, NP-complete work. Global optima versus local optima will be our performance ratio, and we will find out that, sometimes, this ratio goes to zero, even for relatively small designs.

The organization of the paper is as follows. In Section 2 the generation of BIBDs is formulated as an optimization problem. In Section 3, the proposed cost function is mapped onto the energy of an optimizing neural [Hopfield, 82, Ackley et. al., 84, Aarts & Korst,87] with connections of arity 4 [Sejnowski, 86, Torras & Bofill, 89]. Section 4 describes the relaxation strategy, together with the experimental results. And Section 5 is devoted to conclusions and further work.

2 A cost function for the generation of block designs

Let us define a cost function

$$F([x_{pq}]) = \alpha \sum_{i=1}^{v} (\sum_{j=1}^{b} x_{ij} - r)^2 + \sum_{i=1}^{v} \sum_{k>i}^{v} (\sum_{j=1}^{b} x_{ij} x_{kj})^2 \tag{3}$$

over the set $\mathcal{A}$ of all possible $v \times b$-binary matrices or *configurations* of the kind $A \equiv [x_{pq}]$, where x_{pq} represents the incidence of element p in block q. In this section we show that, if there exists a BIBD with independent parameters v, b, r, and incidence matrix A^*, a lower bound for α can be found that ensures that $F(A^*)$ is a *global minimum* of F over $\mathcal{A}$.

Proposition 2.1 *If $\alpha > (k-1)(2\lambda - 1)$ then*

i) *If there exists a (v,b,r,k,λ)-BIBD with incidence matrix A^*, then A^* is a global minimum of F over the set $\mathcal{A}$ of all $v \times b$-configurations and $F(A^*) = \frac{v(v-1)}{2}\lambda^2$.*

ii) *If there exists a $v \times b$-configuration A^* such that $F(A^*) = \frac{v(v-1)}{2}\lambda^2$, then there exists a (v,b,r,k,λ)-BIBD with incidence matrix A^*.*

Proof. In a BIBD, by definition, all rows have r ones and all pairs of rows have correlation λ, and thus from equation 3, $F(\text{BIBD}) = \frac{v(v-1)}{2}\lambda^2$. Part i) of the proposition states that this value is a lower bound, and part ii) states that only BIBDs reach this value.

The proof is based on the quadracity of both terms in the right hand side of equation 3. The second term, while minimizing the sum of the correlations, will favour those configurations with the same correlation value for every pair of rows. This term, by itself, would lead to $x_{ij} = 0, \forall(i,j)$. The first term is then intended to force all rows to have exactly r ones. The worst case are those configurations that are close to a BIBD but have a 1 less. The cost increment from an hypothetic BIBD to one of this configurations is

$$\Delta F = \alpha + (k-1)(2\lambda - 1)$$

and forcing this cost to be positive leads to the selected value for α. The fact that quadratic functions grow faster when their slope is bigger ensures that for all remaining configurations (and therefore, for all configurations that do not correspond to BIBDs) the positive contribution from uncompleted rows will always be bigger than the negative contribution from lower correlations. □

3 Mapping onto a higher-order optimizing network

The interactions between the x_{pq} variables in the cost function of equation 3 can be made explicit by expanding this function into the following terms:

$$\begin{aligned} F([x_{pq}]) &= -\alpha(2r-1)(\textstyle\sum_{i=1}^{v}\sum_{j=1}^{b} x_{ij})+ \\ &+ 2\alpha(\textstyle\sum_{i=1}^{v}\sum_{j=1}^{b}\sum_{l>j} x_{ij}x_{il})+ \\ &+ \textstyle\sum_{j=1}^{b}\sum_{i=1}^{v}\sum_{k>i}^{v} x_{ij}x_{kj}+ \\ &+ 2(\textstyle\sum_{i=1}^{v}\sum_{k>i}^{v}\sum_{j=1}^{b}\sum_{l>k}^{b} x_{ij}x_{il}x_{kj}x_{kl})+ \\ &+ \alpha v r^2 \end{aligned} \quad (4)$$

Next, we define a network with $v \times b$ units arranged in a bidimensional array in direct correspondance with the $[x_{pq}]$ matrix, and we define the following sets of symmetric connections:

- Bias connections, with weight I: $\{(x_{ij}, x_{ij})\}$
- Horizontal connections, with weight H: $\{(x_{ij}, x_{il}) | l \neq j\}$
- Vertical connections, with weight V: $\{(x_{ij}, x_{kj}) | k \neq i\}$
- Fourth order connections or *quadruples*, with weight Q: $\{(x_{ij}, x_{il}, x_{kj}, x_{kl}) | (k \neq i) and (l \neq j)\}$

The *Energy* E of this network is defined as the sum of the weights of the *active* connections and can be written as:

$$\begin{aligned} E([x_{pq}]) &= I\textstyle\sum_{i=1}^{v}\sum_{j=1}^{b} x_{ij}+ \\ &+ H\textstyle\sum_{i=1}^{v}\sum_{j=1}^{b}\sum_{l>j}^{b} x_{ij}x_{il}+ \\ &+ V\textstyle\sum_{j=1}^{b}\sum_{i=1}^{v}\sum_{k>i}^{v} x_{ij}x_{kj}+ \\ &+ Q\textstyle\sum_{i=1}^{v}\sum_{k>i}^{v}\sum_{j=1}^{b}\sum_{l>j}^{b} x_{ij}x_{il}x_{kj}x_{kl}. \end{aligned} \quad (5)$$

The mapping between E and F is then straightforward if the weights are set to

$$\begin{aligned} I &= -\alpha(2r-1) \\ H &= 2\alpha \\ V &= 1 \\ Q &= 2, \end{aligned} \quad (6)$$

leading to

$$F = E + \alpha v r^2. \quad (7)$$

Minimizing E is thus equivalent to minimizing F. In the next section we present some examples of the generation of BIBDs by minimizing the energy of the corresponding network.

(v,b,r,k,λ)	$v \times b$	n	*global/local*	*tics*
$(7,7,3,3,1)$	49	1	$\frac{319}{3022} = 10.5\%$	15341
$(6,10,5,3,2)$	60	1	$\frac{282}{3097} = 9.1\%$	19352
$(7,14,6,3,2)$	98	4	$\frac{95}{2612} = 3.6\%$	30737
$(9,12,4,3,1)$	108	1	$\frac{4}{2755} = 0.1\%$	33901
$(11,11,5,5,2)$	121	1	0	38215
$(10,15,6,4,2)$	150	3	0	48052
$(13,13,4,4,1)$	169	1	0	53567

Table 1: Experiments and results (see text).

4 Experiments and results

The simplest relaxation strategy for energy minimization uses local search with a "downhill" decision rule [Hopfield, 82]. The system is initialized at a random configuration A_0 and randomly selected units are updated whenever this leads to a decrease in the global energy of the system (downhill movement), until all unit updates lead uphill. If we happen to "fall" in a local minimum that does not correspond to a BIBD (i.e., it is not a global minimum), the system is "restarted" to a new, random configuration. The ratio of global minima versus local minima is a measure of the success of the search procedure for each particular instance of the problem (that is, for each set of admissible parameters (v,b,r,k,λ)).

The appeal of the neural network formulation is that it enhances the *locality* in the computation of the energy increment with respect to an arbitrary unit x_{ij}, in terms of the connections arriving to this unit:

$$\begin{aligned} \Delta E_{ij} = E|_{x_{ij}=1} - E|_{x_{ij}=0} \quad &= I + \\ &+ H \sum_{l=1, l\neq j}^{b} x_{il} + \\ &+ V \sum_{k=1, k\neq i}^{v} x_{kj} + \\ &+ Q \sum_{k=1, k\neq i}^{v} \sum_{l=1, l\neq j}^{b} x_{il} x_{kj} x_{kl}. \end{aligned} \tag{8}$$

The unit x_{ij} is set to 1 whenever $\Delta E_{ij} < 0$, and to 0 otherwise.

Table 4 shows the experimental results for several parameter sets corresponding to designs of small to medium size. (v,b,r,k,λ) are the parameters of the design, $v \times b$ is the total number of units and n stands for the number of known non-isomorphic BIBDs with those parameters, according to [Mathon & Rosa, 90]. The ratio of *global* versus *local* optima shows the performance of the system in each case after 10,000 iterations (an iteration is the update of each unit of the network once), and *tics* shows the computing cost expressed in system clock "tics". The performance clearly decreases with the number of units, and no solutions are found for networks with more than 100 units. The computing cost grows slightly faster than the size of the network.

The above results show that the landscape is too difficult for a simple downhill search, and more sophisticated relaxation strategies are called for. At the present moment we are working with simulated annealing [Ackley et al., 84] and mean field theory [Peterson & Anderson, 87], and preliminary results show an important improvement with respect to table 1. (From 35.6% in the $(7,7,3,3,1)$ case to 0.3% in the $(10,15,6,4,2)$ case). Further results will be presented at the conference.

5 Conclusions

In this paper we have proposed a cost function for the generation of Balanced Incomplete Block Designs (BIBDs) from a combinatorial optimization approach. An extension of the Hopfield network architecture has been used, with connections of arity 4, and a simple downhill relaxation strategy. From the point of view of BIBD generation, the results are good for networks with less than 100 units, but fail for bigger networks. The performance of other relaxation strategies (simalated annealing and mean field theory) is currently being tested, and preliminary results show an important improvement,

BIBD generation has proved to be a good testbed for the study of optimizing neural networks. The problem is a especially difficult one, that really "challenges" the performance of any system, and has the advantage, on the other side, that optimal solutions are easy to recognize. From a methodological point of view, the explicitation of the local interactions between the x_{ij} variables leads to a straightforward mapping of the cost function onto higher-order neural networks.

Further work involves the comparison between the different relaxation strategies and the definition of other cost functions [Bofill & Torras, 93a]. At the same time, the combinatorial optimization approach to BIBD generation suggests the use of pseudooptimal solutions for the design of experiments, especially when the BIBD does not exist, or when the parameters are not admissible [Bofill & Torras, 93b].

References

[Aarts & Korst,87] Aarts,E.H.L, and Korst,J.H.M., "Boltzmann machines and their applications", *Proc. PARLE*. Lecture Notes in Computer Science, Vol 258, p 34-50, 1987

[Ackley et al., 84] Ackley D.H., Hinton G.E. and Sejnowsky T.J., "A Learning Algorithm for Boltzmann Machines", *Cognitive Science*, Vol 9, 147, 1985.

[Bofill & Torras, 93a] Bofill P. and Torras C., "Combinatorial Optimization for Block Designs, and Vice Versa", working paper.

[Bofill & Torras, 93b] Bofill P. and Torras C., "Maximally Balanced Designs", working paper.

[Fisher, 40] Fisher R.A., "An examination of the different possible solutions of a problem in incomplete blocks", *Ann. Eugen.*, Vol 10, p 52-57, 1940.

[Hall, 86] Hall M., *Combinatorial Theory*, Ed. John Wiley & Sons, Second Edition 1986.

[Hill, 86] Hill R., *A First Course in Coding Theory*, Oxford University Press, 1986.

[Hopfield, 82] Hopfield J.J., "Neural Networks and Physical Systems with Emergent Collective Computational Abilities", *Proc. Nat. Academ. Sciences USA*, Vol 79, 2554, 1982.

[Mathon & Rosa, 90] Mathon R. and Rosa A., "Tables of parameters of BIBD with r≤41 including existence, enumeration and resolvability results: an update", *Ars Combinatoria*, Vol 30, December, Winnipeg, Canada, 1990.

[Peterson & Anderson, 87] Peterson C. and Anderson J.R., "A Mean Field Theory Learning Algorithm for Neural Networks", *Complex Systems*, Vol 1, 995-1019, 1987.

[Sejnowski, 86] Sejnowski T.J., "Higher-Order Boltzmann Machines", *Proc AIP*, Snowbird 1986.

[Street & Street, 87] Street A. P. and Street D. J., *Combinatorics of Experimental Design*, Oxford Science Publications, Claredon, Oxford 1987.

[Torras & Bofill, 89] Torras C., & Bofill, P., "A neural solution to finding optimal multibus interconnection networks", *Proc. IX Conf. de la Sociedad Chilena de Ciencias de la Computaci n*, Vol 1, p 446- 454, July, 1989.

[Yates, 35] Yates F., "Complex Experiments (with discussion)", *Journ. of the Royal Statist. Soc.*, Suppl. 2, 181-247, 1935.

NEURAL BAYESIAN CLASSIFIER

Christian JUTTEN [(1)], Pierre COMON [(2)]
(1) INPG-TIRF 46, avenue Félix Viallet F-38031 Grenoble Cedex
(2) Thomson-Sintra BP 157 F-06903 Sophia-Antipolis Cedex

ABSTRACT.

Numerous applications of Artificial Neural Networks (ANN) are devoted to classification,which is a key part of the framework of Pattern Recognition [6] and Detection Theory [11] for many years. It is thus not surprising that close connections have been recently pointed out between ANN and these areas. However, some theoretical results as well as practical tricks still remain quite unknown.

In this paper, the advantages of Bayesian detection are recalled, emphasizing possibilities of parallel implementation, and the practical difficulties that prevent to reach the expected optimal performances are highlighted.

1. BAYESIAN DETECTION THEORY.

Assuming M exclusive hypothesis H_i, and suppose we want to choose which hypothesis H_j is the most likely to be true, according a given observation r belonging to R^n. To do this, we define a risk $\mathcal{R}$:

$$(1) \quad \mathcal{R} = \sum_{i=1}^{M} \sum_{j=1}^{M} C_{ij} P_j \int \cdots \int_{r \in D_i} p(r/H_j)\, dr,$$

where P_i is the prior probability of H_i, C_{ij} is the cost associated to the decision : (say H_i while H_j is true), and D_i is the region (unknown for the moment) where decision H_i is made. The integral on the region D_i is thus the probability to choose H_i while H_j is true (this is one of the M^2 probabilities related to the M^2 possible situations, M^2 - M over them corresponding to errors).

It is quite natural to assign a smaller cost C_{jj} to correct classification than to any of the misclassification costs C_{ij}. Under this assumption, it can be seen that the integrand I_k is positive:

$$(2) \quad I_k = \sum_{j \neq k} P_j \left(C_{kj} - C_{jj}\right) p(r/H_j).$$

Then it can be shown [6] [11] that the risk (1) achieves its minimum if and only if the domains D_i are defined by:

$$(3) \quad D_i = \{ r \,/\, I_i(r) = \min_j I_j(r) \}.$$

Relation (3) defines the optimal classifier, in the sense of the minimisation of a combination of all probabilities of error. However, the optimality holds only if priors P_j and densities $p(r/H_j)$ are given. In practice, they are of course generally unknown, and must be estimated from a finite set of examples.

2. INFLUENCE OF COSTS AND PRIORS

To be more precise, we illustrate the influence of these parameters on a simple example in binary case. We assume we want to detect the presence (H_1) or the absence (H_0) of a target. Such a situation is the simplest case in Radar or Sonar problem. Four different situations can occur :

1. say H_0 while H_0 is true,
2. say H_1 while H_0 is true,
3. say H_0 while H_1 is true,
4. say H_1 while H_1 is true.

It is clear that situations 2 and 3 correspond to decision errors. Situation 2 means we make a "false alarm", for instance, we say there is a target, while it is only a bird ! Situation 3 means we make a "missing" : there is a target but it has not been detected. It is clear that these two errors have not the same seriousness. Generally, we then associate a larger cost for the "missing" than for the "false alarm". Therefore, it is clear from relation (4) that cost values strongly change the decision.

We now show than priors also influence the optimal decision. For sake of simplicity, we always consider the simple binary example, and we choose the following costs :
$\forall$ i, $C_{ii} = 0$,
$\forall$ i, j, $C_{ij} = 1$.

Then, the set of M - 1 inequalities derived from (3) reduces to (4). We chose H_1 if :

(4) $P_0 \left(C_{10} - C_{00}\right) p(r/H_0) > P_1 \left(C_{01} - C_{11}\right) p(r/H_1)$,

and H_0 otherwise. Therefore, with the above value of costs, the decision (4) can be still simplier written :

$$(5) \quad P_1\, p(r/H_1) \underset{H_0}{\overset{H_1}{\gtrless}} P_0\, p(r/H_0).$$

This relation means that the best decision is achieved by comparing the conditional pdfs of the observations, weighted by the priors.

Then, assuming the observation r is a scalar, we can very simply show the effect of variation of priors. Figure 1 shows the optimal bayesian decision if $P_0 = P_1 = 0.5$, $P_0 = 0.3$ and $P_1 = 0.7$ and $P_0 = 0.1$ and $P_1 = 0.9$. We assume the two conditional pdfs are gaussian with the same variance $\sigma^2 = 1$ and means equal to 0 and 2 respectively.

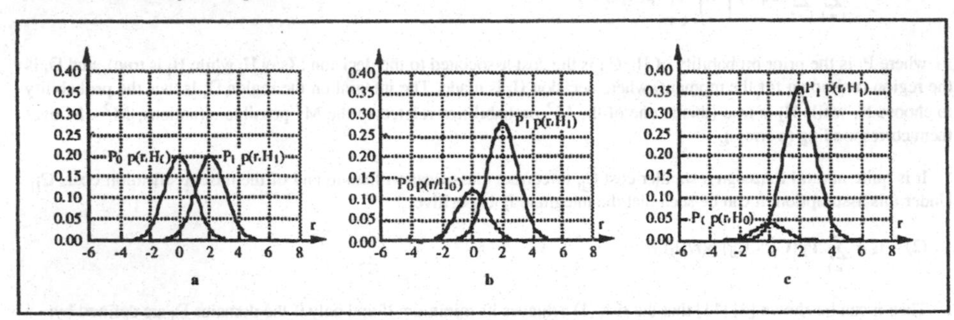

Figure 1. *Influence of priors on decision.*
The optimal bayesian decision corresponds to the value of r where the 2 curves are equal. If we do not know the prior probabilities, by assuming that $P_0 = P_1 = 0.5$, we chose H_1 if $r > 1$, and H_0 otherwise, and we remark this threshold is different of the optimal one (see also Table 1).

If priors are unknown, we may assume equiprobability : $P_0 = P_1 = 0.5$, but thus, we are no more able to do optimal decision. In fact, it is easy to compute the error probability implied by this bad choice of priors. If actual priors are $P_0 = 0.3$ and $P_1 = 0.7$ (or $P_0 = 0.1$ and $P_1 = 0.9$), optimal receiver gives an error probability equal to 0.138 (or 0.069) while the receiver assuming priors $P_0 = P_1 = 0.5$ gives an error probability equal to 0.158.

The interest of the optimal approach is clear. However, it needs a perfect knowledge of priors and especially of conditional probability densities $p(r/H_i)$. Therefore, two relevant questions are :
- How is it possible to estimate probability density functions (pdf) ?
- Is it possible to build a network which implement both pdf estimation and minimize a Bayesian risk ?

The answer to the first question is well known in statistics. A few methods exist to estimate pdf, the simplest one is the histogram method. A more interesting method is Kernel method [8]. This method is especially interesting because it seems very simple to implement with neural networks. In fact, Kernel estimation of pdf is a special case of approximation of functions with Radial Basis Functions [1, 9]. The difference lies in the two following properties : a pdf is a positive function and its integral is equal to one.

Priors	$P_0 = P_1 = 0.5$	$P_0 = 0.3, P_1 = 0.7$	$P_0 = 0.1, P_1 = 0.9$
Optimal error probability	0.158	0.138	0.0692

Table 1. *Optimal decision thresholds and optimal error probabilities for various priors.*
If the decision is done assuming priors are equal, the error probability is always equal to 0.158, which is only optimal for $P_0 = P_1 = 0.5$.

3. KERNEL ESTIMATION

Kernel estimators are commonly used for pdf estimation, the most well known being the simplest one, the Parzen windows method [6]. Assume we want to estimate the pdf $f_x(u)$ from N examples x(n) belonging to R^p. We define an even function, called kernel, K(x), such that :

$$(6) \quad \int_{R^p} K(x)\, dx = 1.$$

The pdf $f_x(u)$ can be estimated from a set of N examples $x^{(n)}$:

$$(7) \quad \hat{f}_x(u) = \frac{1}{N} \sum_{n=1}^{N} \frac{1}{h^p} K\left(\frac{u - x^{(n)}}{h}\right),$$

where p denotes the dimension of vectors x(n). The calculation of bias and variance can be derived explicitly under reasonable additional assumptions. In order for this calculation to be of practical interest, for instance to find the best Kernel K(u) or the best width factor h in some Mean Square Error (MSE) sense, it is necessary to resort to a series expansion of density $f_x(u)$, which is then required to be twice differentiable. One of the key papers on the subject is due to Cacoullos [2]. There are also a number of other results that have been obtained since Parzen, see [4] for complementary information. Note that in (7), the width factor h can be either constant (depending only on N), or may vary with n. In the latter case, the estimator is ofter referred to as the "variable kernel" estimator, because the width of the bump of K(u) is varying form one example to another.

Denote Rect(x/2) the function that is equal to 1 if $-1 \leq x \leq +1$, and 0 otherwise. In dimension 1, several standard kernels are worth mentioning :

- uniform : $\frac{1}{2}\,\mathrm{Rect}\left(\frac{x}{2}\right)$,
- triangle : $(1 - |x|)\,\mathrm{Rect}\left(\frac{x}{2}\right)$,
- Epanechnikov : $\frac{3}{4}(1 - x^2)\,\mathrm{Rect}\left(\frac{x}{2}\right)$,
- gaussian : $\frac{1}{\sqrt{2\pi}} \exp\left(-\frac{1}{2}x^2\right)$.

Note that in principle, kernels do not need to be positive, but this is often an assumption that is added for practical purposes, and is consistent in case of small values of N. The Gaussian kernel has shown to be very satisfactory in dimension 1, so that it is probably the most widely used.

In higher dimensions, the Gaussian kernel is not recommended, especially for dimensions larger than 5. In fact, it is well known that in dimension 1, 90% of the samples are likely to occur within the interval [-1.6, 1.6]. But when the dimension gets larger, the integral of the density in the hypersphere of radius 1.6 **tends to zero!** In other words, the Gaussian kernel will not allow an example $x^{(n)}$ to increase the estimated density (7) locally, but on the contrary will increase the density rather in the tails of the kernel. This "empty space" phenomenon, as named by Scott and Thompson in 1983, imposes to use kernels that decrease faster than Gaussian in the tails.

In [5], kernel estimators with Gaussian kernels have been utilized to build a Bayesian classifier in dimension 10. Experiments run on real data showed that performances were not as good (though of same order) as a simple MultiLayer Perceptron (MLP). This disappointing result was merely due to the use of a kernel with too heavy tails. Experiments performed with a kernel of the form [7] :

(8) $K(u) = B \exp\left[-\left(A \|u\|^2\right)^g\right]$,

with g = 1.5, have led recently to better results, that are not reported here for reasons of space. The practical evaluation of performances is also a key issue. In order to avoid both overfitting and loss of information, the best way to proceed seems to be the Averaged Leave-One-Out approach, which turns out to be nothing else but the well-known Jacknife approach to confusion estimation.

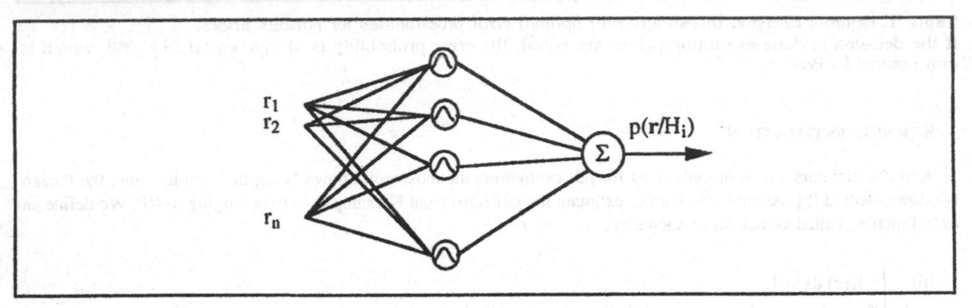

Figure 2. *Pdf estimation with Radial Basis Function (Kernel) neurons.*

4. ASYMPTOTIC THEOREMS

In Neural networks, and in MLP in particular, an input-output relation is attempted to be coded with the help of fixed non-linear functions and a number of free parameters. If these parameters are sought in such a way that a quadratic error of the form

$$(10)\quad \varepsilon(N) = \frac{1}{N} \sum_{n=1}^{N} \| y^{(n)} - \phi(W, x^{(n)}) \|^2,$$

is minimized, it will be referred to Minimum Quadratic Error (MEQ) classifiers. In definition (10) above, $y^{(n)}$ are the desired outputs associated to example $x^{(n)}$, W is the connection matrix and ϕ is a fixed non-linear function (activation function). For a 3-layers MLP for instance, this input-output relation applies. As pointed out in [5], there are at least four theorems stating asymptotic performances of such classifiers. We give these theorems below, but refer to [4] for proofs and related references. These theorems state somewhat more formally the assertion often admitted that: "the outputs of the MLP tend to Bayesian probabilities as the size of the learning set is growing to infinity".

Let $A_0(N) = \{ (x^{(n)}, y^{(n)}), 1 \le n \le N \}$ be a learning base, having N_k examples in each class ω_k, $1 \le k \le K$. Let $\phi(W, .)$ be a mapping parametrized by W that associates any vector x of size p to a K dimensional output vector $y = \phi(W, x)$ from which the decision will be made. When N is finite, a classical method to enlarge the data base consist in creating new prototypes by adding independent samples $z^{(n,r)}$ of a zero mean noise with a pdf $p_z(u)$. Thus, we get R N-size data bases $A_r(N) = \{ (x^{(n)} + z^{(n,r)}, y^{(n)}), 1 \le n \le N \}$, $1 \le r \le R$, and we denote $A(N, R) = \cup_{i=1}^{R} A_r(N)$.

Moreover, the learning can be done using two ways of defining the desired outputs :
(11) if $x^{(n)} \in \omega_j$, then $y_i^{(n)} = \delta_{ij}$,
(12) if $x^{(n)} \in \omega_j$, then $y_i^{(n)} = C_{ij}$.

Theorem 1
Let $A_0(N)$ be the data base, and let $\phi(W, .)$ be the function computed by minimizing the error (10) with respect to W. If N_k tends toward infinity for each k, $1 \le k \le K$, the function $\phi(W, .)$, using the output **coding (11)**, tends toward the best MQE estimate (in the class of functions of the form ϕ) of the Bayesian receiver with **equal error costs** ($C_{ij} = \delta_{ij}$).

Theorem 2
Let $A_0(N)$ be the data base, and let $\phi(W, .)$ be the function computed by minimizing the error (10) with respect to W. If N_k tends toward infinity for each k, $1 \le k \le K$, the function $\phi(W, .)$, using the output **coding (12)**, tends toward the best MQE estimate (in the class of functions of the form ϕ) of the **general Bayesian receiver.**

Theorem 3

Let $A_0(N)$ be a data base, and let $A(N, P)$ be its expanded version. If R tends toward infinity, the function $\phi(W, .)$, using the output coding (11) and computed to minimize the following error (13) tends toward the best MQE estimate (in the class of functions of the form ϕ) of the Bayesian receiver with equal error costs ($C_{ij} = \delta_{ij}$), in which pdf are kernel estimation using the kernel $p_z(u)$. The error (13) is :

$$(13)\quad \varepsilon(N, R) = \frac{1}{N}\frac{1}{R}\sum_{n=1}^{N}\sum_{r=1}^{R} \| y^{(n)} - \phi\left(W, x^{(n)} + z^{(n, r)}\right) \|^2.$$

Theorem 4

Let $A_0(N)$ be a data base, and let $A(N, P)$ be its expanded version. If R tends toward infinity, the function $\phi(W, .)$, using the output coding (12) and computed to minimize the error (13) tends toward the best MQE estimate (in the class of functions of the form ϕ) of the **general Bayesian receiver**, in which pdf are kernel estimation using the kernel $p_z(u)$.

But these properties can hold true only when the absolute minimum of error $\varepsilon(N)$ is indeed reached, which cannot be guarranteed in an algorithm such as the gradient backpropagation. A direct implementation of the Bayesian solution, in the form of kernel estimator, is thus preferred.

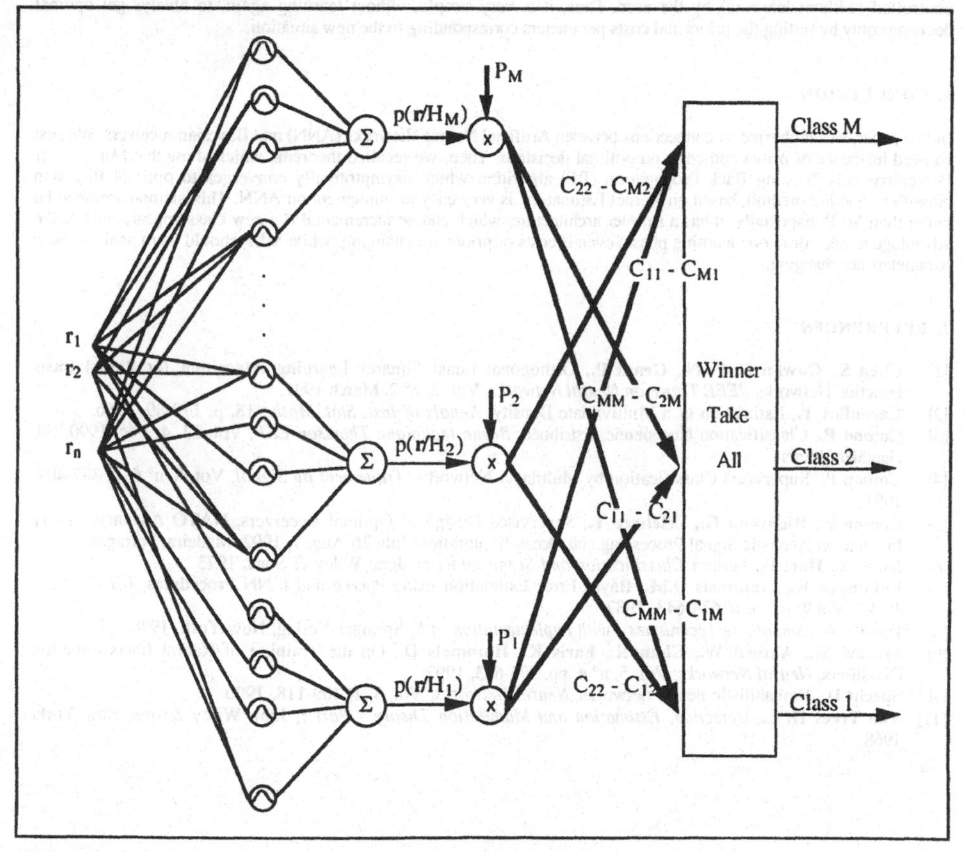

Figure 3. *Neural implementation of a complete bayesian classifier.*
The first layer provides estimations of conditional pdf with Radial Basis Function (kernel method). These pdf are multiplied by priors P_l. Then, integrands - I_k are computed by sum of $P_l\ p(r/H_l)$ weighted by the difference of costs $C_{jj} - C_{kj}$. The optimal decision is done by computing the winner of all integrands - I_k.

5. NEURAL IMPLEMENTATION

It has been noticed independently in 1990 by Specht [10] and Comon [3] that there exist obvious parallel implementations of the Bayesian classifier. See for instance [4] and references therein. Here we show it is very simple to implement a general Bayesian receiver using a first layer Radial Basis Function neurons followed by a Winner-Take-All layer.

Figure 2 shows the neural implementation of formula (7). By using M neural networks as the one proposed in Figure 2, we can compute all the M conditional pdf $p(r/H_i)$. Now the question is the following : Is it possible to use the outputs of these networks to compute the optimal Bayesian decision ? That is, is it possible to built a neural net which solves the set of inequalities (3) ?

In fact, each term I_k can be viewed as a sum over j of probability densities $P_j\, p(r/H_j)$ weighted by the difference of cost $C_{kj} - C_{jj}$. These operations are very simple to do with neural nets. And the set of M - 1 inequalities (equivalent to (3)) will be solved by searching the smallest I_k. In fact, we can use a Winner-Take-All network by computing the opposites $- I_k$. Finally, we propose the network of Figure 3.

In this scheme, the learning acts on the first layer and provides kernel estimation of conditional pdfs. The priors and the connections between this first layer and the second one (in bold in Figure 3), corresponding to the error costs, are controlled (without learning) by the user. Thus, it is very simple, wihout learning again, to always get optimal decisions only by setting the priors and costs parameters corresponding to the new situation.

6. CONCLUSION

In this paper, we emphasize on connections between Artificial Neural Networks (ANN) and Bayesian receivers. We first showed influence of priors and costs on optimal decisions. Then, we recalled theorems which show that Multi Layer Perceptron (MLP) using Back Propagation (BP) algorithm which asymptotically converges to optimal Bayesian classifier. Another method, based on Kernel Estimators, is very easy to implement on ANN. This method seems to be better than MLP. Especially, it has a simpler architecture, which can be incremental if a new class appear, and has the advantage to need only one learning phase, even if costs or priors are changing, while MLP should learn again if these parameters are changing.

7. REFERENCES

[1] **Chen S., Cowan C. F. N., Grant P.**, Orthogonal Least Squares Learning Algorithm for Radial Basis Function Networks, *IEEE Trans. on Neural Networks*, Vol. 2, n° 2, March 1991.

[2] **Cacoullos T.**, Estimation of a Multivariate Density, *Annals of Inst. Stat. Math.*, 18, p. 178-89, 1966.

[3] **Comon P.**, Classification bayésienne distribuée, *Revue technique Thomson-CSF*, Vol. 22, 4, Déc. 1990, Ed. Gauthier-Villars

[4] **Comon P.**, Supervised Classification by Multilayer Networks, *Traitement du Signal*, Vol. 8, n° 6, p. 387-407, 1991.

[5] **Comon P., Bienvenu G., Lefebvre T.**, Supervised Design of Optimal Receivers, NATO Advanced Study Institute on Acoustic Signal Processing and Ocean Exploration, July 26-Aug. 7, 1992, Madeira, Portugal.

[6] **Duda R., Hart P.**, *Pattern Classification and Scene Analysis*, John Wiley & Sons, 1973

[7] **Fukunaga K., Hummels D.M.**, Bayes Error Estimation using Parzen and k-NN Procedures, *IEEE Trans. PAMI*, Vol.9, n.° 5, p. 634-643, 1987.

[8] **Härdle W.**, *Smoothing Techniques : with implementation in S*, Springer-Verlag, New-York, 1990

[9] **Musavi M., Ahmed W., Chan K., Faris K., Hummels D.**, On the Training of Radial Basis Function Classifiers, *Neural Networks*, Vol. 5, n° 4, pp. 595-603, 1992

[10] **Specht D.**, Probabilistic neural networks, *Neural Networks*, Vol. 3, 1, 109-118, 1990

[11] **Van Trees H. L.**, *Detection, Estimation and Modulation Theory - Part I*, John Wiley &sons, New York, 1968

CONSTRUCTIVE METHODS FOR A NEW CLASSIFIER BASED ON A RADIAL-BASIS-FUNCTION NEURAL NETWORK ACCELERATED BY A TREE

Philippe GENTRIC, Heini C.A.M. WITHAGEN
Laboratoires d'Electronique Philips
22, av. Descartes
94453 LIMEIL-BREVANNES
FRANCE

Abstract

We present a new constructive algorithm for building Radial-Basis-Function (RBF) network classifiers and a tree based associated algorithm for fast processing of the network. This method, named Constructive Tree Radial-Basis-Function (CTRBF), allows to build and train a RBF network in one pass over the training data set. The training can be in supervised or unsupervised mode. Furthermore, the algorithm is not restricted to fixed input size problems. Several construction and pruning strategies are discussed. We tested and compared this algorithm with classical RBF and multilayer perceptrons on a real world problem: on-line handwritten character recognition. While instantaneous incremental learning is the major property of the architecture, the tree associated to the RBF network gives impressive speed improvement with minimal performance losses. Speed-up factors of 20 over classical RBF have been obtained.

1 Introduction

Classification is a very general task that may be used in a great number of fields. The desired properties for a classifier are: high generalization rate, high processing speed, low memory requirement and the ability to give a likelihood measurement of some sort. Another important requirement is a short training time. Also, some tasks require an additional capability: very fast incremental learning (for example, on-line user adaptation in handwriting recognition modules for small computer).

Many of the most popular neural algorithms are only able to take into account a new information through a full re-training. If we take for example a multilayer perceptron trained by error back-propagation, even with special acceleration techniques [1] such a learning would require several passes over the training set which is simply not feasible if one is to stick to the requirements of real-world "cheap" applications.

The method we present, named Constructive Tree Radial-Basis-Function (CTRBF), is a method that allows to build very rapidly a structure made of a Radial-Basis-Function (RBF) network and a tree that will be much faster than a classical RBF in resolution mode with minimal performance loss. The network is able to provide a very reliable likelihood estimate and can be easily adapted to changes (incremental learning). Furthermore, *all* the information stored in the network and the tree can be very precisely accounted for (we know exactly and declaratively the purpose of every component of the net) and consequently, the network and/or the tree can be pruned very efficiently when a smaller system is required, with control over the subsequent loss of performance.

In this paper, we will describe the tree and network structure as well as the algorithms needed to build and optimize this structure. Finally we will present a real-world application: on-line handwritten character recognition.

2 Radial-Basis-Function network

Radial-Basis-Function Networks are known to be capable of universal approximation [2] and the output of a RBF network can be related to Bayesian properties.

A RBF net has 2 layers. The hidden layer is (usually) fully connected to the input units $X = (X_i)$ of size N_{input}. A hidden unit j has an input vector of synaptic weights $Win^j = (Win^j_i)$ and is evaluated using a metric, for example in the following formula the Euclidean metric, and a nonlinear function f(x) such as exp(-x) or 1/(x+a). σ_j is an adjustable parameter. As this process has a radial symmetry of center Win^j, the output of a hidden unit j will increase when the input pattern vector X comes "closer" (according to the metric) to the synaptic weights vector Win^j:

$$OUT_j = f(||Win^j - X||) = f(\sum_{i=0}^{N_{input}-1} (X_i - Win_i^j)^2/\sigma_j)$$

The next layer gathers the activities of the hidden units with the purpose of taking a decision on the class of the input prototype. For classification tasks with C classes, this output layer will have C output units. The classification result is obtained from this layer on a "winner-take-all" (WTA) basis: the class of the input pattern will be given by the most active output unit. This layer is usually made of linear units because the computation of a monotonic non-linear function adds useless computations in a winner-take-all context and/or leads to useless confusions (loss of information) in case of hard limiting function or limited precision outputs.

Assuming N_{hidden} hidden units, the activity of output unit k is:

$$OUT_k = \sum_{j=0}^{N_{hidden}-1} (Wout_j^k * f(||Win^j - X||))$$

and the resulting class $w \mid OUT_w = max_{(among\ k)}(OUT_k)$.

One of the most interesting properties of RBF networks is that they provide intrinsically a very reliable rejection of "completely unknown" patterns. Indeed, in a classical multilayer perceptron there is no guaranty that a prototype very far from any previously presented data will not produce a positive output.

Note also that the synaptic vector Win^j *stores a location in the problem space*, in other words, it stores an input pattern. It is then straightforward to imagine how (incremental) learning can be performed by creating a new hidden unit whose input synaptic weight vector will store the new training pattern. The synaptic weight from this new hidden unit to the output unit corresponding to the class of the new pattern is set to a positive value (say 1). One way to build a RBF network for a given classification problem is thus to create one hidden unit per training prototype. In real-world problems though, the number of prototypes can get very large. For this reason we developed CTRBF, a technique that allows to reduce the number of stored patterns accessed for each classification. Note that pruning techniques can also be used in order to reduce the number of hidden units (see below).

Most of the RBF literature [3] [4] is devoted to the use of optimization techniques in order to compute an optimized set of synaptic weights in the RBF network. Our experience is that for real world problems these methods demand a great amount of computations and in the end produce systems that do not perform much better than CTRBF.

3 Tree accelerated RBF

The idea of associating a tree to a RBF network is based upon the simple remark that all of the hidden units of the RBF network are not active at the same time. More precisely, only the hidden units storing a pattern that is "close" to the input pattern are active. Consequently, when a pattern is presented to the input, instead of computing the full network, we will search which hidden units store patterns that are close to this input pattern. Then, just like a classical RBF, we will evaluate these units and propagate their activity to the output. Of course, the aim is to find these active units with as little distance computations as possible. The tree is not used for classification but for speed.

3.1 Tree structure

The tree structure is binary and asymmetric, as is illustrated in figure 1. It can be described as follow: to each node is associated a stored pattern and a distance threshold named "radius" equivalent to σ_j. Each node may lead to a "son" node and a "(younger) brother" node, (the reciprocal pointer is described as "father" and "elder brother"). If the "son" node exists, its radius is smaller than its father radius and the radius of a node is *bigger* than the distance between this node and any node up the tree. If a younger brother node exists, its radius is equal to its elder brother radius (hence the name "brother"). If a node has no "son" then it is a "leaf" and each leaf points to one hidden unit of the RBF net. If a node has no younger brother then it is a terminal node. The set of the brothers up to a terminal form a "phratry". The radius of the root node should be bigger than the biggest possible distance between two patterns.

3.2 Tree evaluation

When an input pattern is presented we start the evaluation of the tree at its root. We compute the distance between the input pattern and the pattern stored by the current node. If the distance is bigger than the node radius, we go to the younger brother, otherwise, we go to the son. When the evaluation of the tree leads to a leaf,

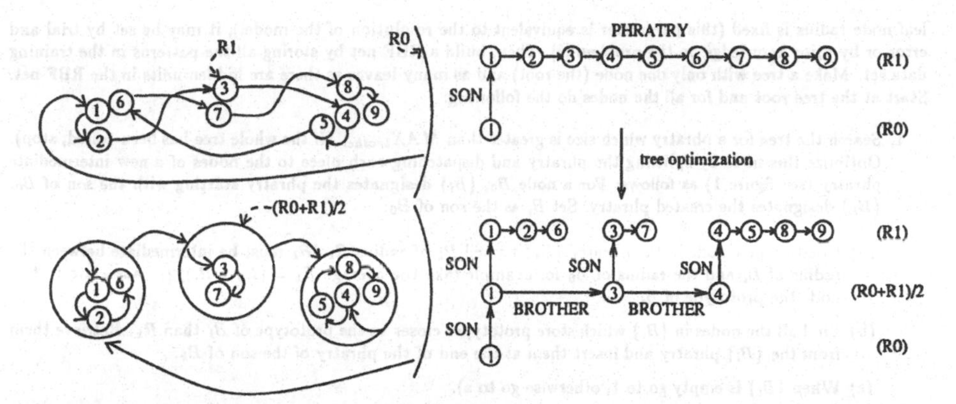

Figure 1: *Tree structure: an example of very simple tree. On the right, a representation of what the corresponding topology could be with 2 dimensional input data and a Euclidean metric. On the left an example of optimization: on the top the initial structure, underneath the result after optimization*

the corresponding hidden unit of the RBF network is evaluated. When the evaluation of the tree leads to a terminal node, the search is ended (see below for exception: extended search). In order to know how much acceleration the tree gives, we compute the speed-up factor α : the total number of hidden units (or leaves) divided by the average number of distance computations performed per evaluation (tree + network distance computations); hence we have $\alpha = 1$ for a classical RBF network and $\alpha < 1$ if the tree adds computations instead of preventing some.

3.2.1 Extending the search

We may extend the search in order to get a bigger number of selected hidden units and a better classification. First, the tree may be seen as a variable-K-NN: when a leaf is reached all leaves along the phratry are evaluated. In the following "CTRBF" refers to this extended search mode. Also, the search may be extended in the following fashion: when a terminal node is reached, instead of ending the search we may backtrack down the tree and at each node test the younger brother. Note that the tree is not completely explored because the search backtracks again if the test of the younger brother is negative. This mode of extended search (that we call "backtracking") is most useful when used in conjunction with a bigger radius than the stored radius [5]. The factor by which the radii are multiplied is called the *external search factor*. Experimentally, the results show that extending the search using these methods is efficient in terms of performance while allowing to keep a good speed-up factor α.

3.2.2 Rejection

Rejection is provided in three different ways. First, the tree search may end while no leaf was selected: this rejection mode is typical of a pattern that is "very different" from what was in the training data set. Secondly, the maximum output unit (winner of the WTA) activity may be below a given threshold : this is a likelihood rejection. Thirdly, the difference of activity between the winner and the second best is less than a second prefixed threshold: this is a confusion rejection.

4 Tree Construction algorithm

There are several ways to build the tree. Whether the class of a pattern is used or not, the construction is supervised or unsupervised.

4.1 Unsupervised learning by dichotomy

This method has the advantage of making few hypothesis on the structure of the data. We set the following constraints: the maximum size of a phratry $MAX_{brother}$ is fixed (this parameter will influence the speed-up factor α). The root node radius is fixed (the biggest possible distance between two prototypes is a good estimation). The leaf node radius is fixed (this parameter is equivalent to the resolution of the model), it may be set by trial and error or by prior knowledge on the problem [6]. Then, build a RBF net by storing all the patterns in the training data set. Make a tree with only one node (the root) and as many leaves as there are hidden units in the RBF net. Start at the tree root and for all the nodes do the following:

1. Search the tree for a phratry which size is greater than $MAX_{brother}$ (If the whole tree has been tested, stop). Optimize this node by splitting the phratry and dispatching each piece to the nodes of a new intermediate phratry (see figure 1) as follows: For a node B_0, $\{B_i\}$ designates the phratry starting with the son of B_0. $\{B_k\}$ designates the created phratry. Set B_i as the son of B_0:
   (a) Create a new node in the phratry $\{B_k\}$ named B_k of radius R_k (R_k must be intermediate between the radius of B_i and the radius of B_0 for example take the average $R_k = (R_0 + R_i)/2$). Associate to this node the prototype of B_i
   (b) Find all the nodes in $\{B_i\}$ which store prototypes closer to the prototype of B_k than R_k. Retrieve them from the $\{B_i\}$ phratry and insert them at the end of the phratry of the son of B_k.
   (c) When $\{B_i\}$ is empty go to 1, otherwise go to a).

4.2 Supervised learning by distribution

We start with the same constraints as before:

1. Get at random a prototype from the training data set (stop when the set is empty). Use it as input pattern for an evaluation.
2. If the classification decision is correct, discard the pattern.
3. Otherwise, create a RBF hidden unit and a leaf storing this pattern.
4. Find through the tree the first node which stores a prototype that is closer to this new pattern than its radius (in fact, this search was done during evaluation). Insert the new leaf at that location in the tree.
5. If no node can be found, create (recursively up the tree) nodes storing this new pattern with had hoc radius(ii).
6. (optional) When a phratry becomes bigger than $MAX_{brother}$ apply the optimization strategy presented above.
7. Go to 1.

This tree and network construction mode is the mode used in case of incremental learning. Also, this is a supervised algorithm but removing the test in step 2 provides an unsupervised strategy.

4.3 Tree pruning

Branch pruning may be needed because, during construction, nodes that have no brother but only a son may be created: these node are useless and should be pruned (the connection from their father to their son is made directly). Also, after a network pruning (see below), some node may loose all their upper nodes, of course they should also be pruned.

4.4 Network pruning versus efficient construction

The most simple pruning methods are based upon the "leave-one-out" strategy: if the global system performs just as well without a given unit then this unit is not needed and must be pruned. However, this method leads to N^3 complexity (if N is the training data set size) because every pattern in the data set (N) must be evaluated (*N) for each hidden unit tested for pruning (*N). You may benefit from a kinetic factor: the network gets smaller as the pruning proceeds, thus less computation is needed. This is dual to the "efficient construction" obtained by distribution. Another strategy is the "self-leave-one-out" strategy. When a hidden unit is tested for pruning, the temptation is to perform the test with as few patterns as possible: ultimately, the unit is tested only with its own stored pattern (N^2 complexity but poor performance). With CTRBF one can choose to test one hidden unit using only the tree-designated closest patterns. The overall complexity is then reduced to $MAX_{brother} * N^2/\alpha$ because only the "close by" patterns (less than $MAX_{brother}$) must be evaluated ($*N/\alpha$) for each hidden unit tested for pruning (*N). The CTRBF architecture allows to choose any of these strategies, depending on the amount of data, time and processing power available.

4.5 Metric and *a priori* information

Very often we have *a priori* information about the problem to be solved. The possibility to incorporate such information into the classification task is known to be fundamental for the final performance. For example, in multi-layer perceptrons the *a priori* information can be built into the system using local connections, constrained weights and back-propagation of special quantities depending on known properties in the data such as translation invariance [7]. In RBF networks this kind of information can also be incorporated by changing the metric used (Euclidean in the original RBF). For example an image metric may be computed allowing local translations. Or a dynamic programming technique may be used to compute a metric between two character strings, which also allows to compute a metric between patterns that are not described with the same amount of information (in case of character strings, one can compare the word "wagon" and the word "wgon") etc..

5 Application to on-line handwritten character recognition

The handwritten characters considered have been acquired using an electronic paper interface, and pre-processed into a pattern vector having 481 components. A detailed report on this application will be published elsewhere [8]. Here we present two problems, the first problem is the "bars" problem. There are 4 kinds of bars: vertical, horizontal, slanted to the left, slanted to the right. We used 544 prototypes for learning and 544 others for testing. The second problem is a upper case character recognition problem. We used 3698 prototypes for learning and 3697 for testing. For both problems we compare the results obtained with various parameter values against the performance of multi-layer perceptrons. For the "bars" problem, the results are summarized in table 1. For the upper case characters problem, the results are summarized in table 3. One can see that the performances are still good even for α as big as 20 (note that the node count does not include leaves).

In table 2 we compare the rejection ability. Here we tested the system learned with the "bars" data (4 output classes) with the upper case letter test set: all the test patterns should be rejected. One can clearly see that in this domain CTRBF is far superior to multi-layer perceptron.

The evaluation times of CTRBF and perceptrons are comparable. For a CTRBF network learning and evaluation takes about the same time. On the contrary, the learning time of the perceptron is much bigger. For example the learning time of the perceptron with 50 hidden units in the second problem is far superior to all the learning and evaluation times of the tests on the CTRBF for the same problem. It is also important to remark that in character recognition the number of classes can get up to 100, now the size of the perceptron grows at least proportionally to the number of classes while with CTRBF the size only depends on the number of hidden units (that should depend on the complexity of the problem). Of course, the instantaneous incremental learning is not possible with a perceptron.

6 Conclusion

CTRBF has been software-implemented in a user-adaptable on-line handwritten character recognition and runs at 3 characters per second on a 386 PC. CTRBF allows to explore constructive neural network techniques in supervised and unsupervised mode. The tree structure provides a clustering of the training data. The fact that all the information in the system can be accounted for, the intrinsic robust rejection capability, the short training time and the instantaneous incremental learning coupled with a high processing speed are the assets of this new architecture. Speed-up factors of 20 over a classical RBF can be obtained on a real world application. Furthermore, the parameters of the constructive algorithm allow to tailor the system to the application requirements. Many questions are still to be answered, such as: how can we systematically build a metric taking into account *a priori* information ? Is CTRBF suitable for efficient hardware implementation ? Nevertheless, we already know that CTRBF is a very versatile new architecture for pattern classification.

References

[1] S. Makram-Ebeid, and al., A rationalized error backpropagation algorithm, *Proc. INNS*, Washington, DC, June 18-22,(1989)

[2] J.Park, I.W. Sandberg, Universal Approximation Using Radial-Basis-Function Networks, *Neural Computation* 3,246-257 (1991)

[3] E.Hartmann, J.D.Keeler, Predicting the Future: Advantage of Semilocal Units, *Neural Computation* 3,566-578 (1991)

[4] J. Moody and C. Darken, Learning with localized receptive fields, *Proceedings of the 1988 Connectionist Models Summer School*, ed. D. Touretzky, Morgan Kaufmann Publishing, San Mateo, CA, pp. 133-143, (1988).

[5] A. Saha and J.D. Keeler, Algorithms for better representation and faster learning in radial basis function networks, *Neural Information Processing Systems*, ed. D. Touretzky, Morgan Kaufmann, San Mateo, CA, 482-489, (1990).

[6] M.T. Musavi and al. On the training of radial basis function classifiers *Neural Networks* 5, 595-603 (1992).

[7] P. Simard, Y. Le Cun, J. Denker, B. Victorri, An efficient algorithm for learning invariances in adaptive classifiers. To be published in IAPR.

[8] P. Gentric, On-line handwriting recognition for small computer. Sixth International Conference on Handwriting and Drawing, Paris July 5-6-7, (1993).

network used	nodes	hidden units	error rate	rejection rate	$\alpha(speed)$
CTRBF	149	544	1.1 %	9.2 %	33
idem + backtracking	149	544	2.0 %	1.5 %	10
idem + external search factor 1.2	149	544	0.9 %	1.1 %	6.2
idem + external search factor 2	149	544	1.1 %	0.6 %	2.7
idem + external search factor 10	149	544	1.5 %	0.6 %	0.9
normal RBF	0	544	1.5 %	0.6 %	1
1 layer perceptron (no hidden neurons)		0	1.1 %	1.1 %	
2 layer perceptron (10 hidden neurons)		10	1.8 %	1.2 %	

Table 1: Compared performances on a 4 classes problems: the bars. One can see that a speed-up factor from 5 to 10 does not affect the performance. Also the perceptrons do not perform much better although (especially the two-layer perceptron) they are much slower to learn.

network used	error rate	rejection rate	$\alpha(speed)$
CTRBF + backtracking + external search factor 1.2	4.4 %	95.6 %	33
2 layer perceptron (10 hidden neurons)	45 %	55 %	

Table 2: Rejection capability: with the 4 bars data for learning and upper-case data (3697 prototypes) for testing. On this problem the error rate cannot go much below 6% because there are 240 'I' among which many that may be confused with a vertical bar. This is a clear confirmation of the superiority of classifiers based on distance-to-pattern over classifiers based on distance-to-linear-separator

network used	nodes	hidden units	error rate	rejection rate	$\alpha(speed)$
normal RBF	0	3698	4.3 %	0.6 %	1
CTRBF	1099	3698	7.9 %	25.5 %	200
idem + backtracking	1099	3698	12.5 %	4.1 %	42
idem + external search factor 1.2	1099	3698	5.4 %	0.8 %	21
idem + external search factor 2	1099	3698	4.3 %	0.5 %	5
idem + external search factor 10	1099	3698	4.3 %	0.6 %	1
esf 1.2 + pruning	675	742	5.6 %	3.8 %	35
esf 1.2 + learning by distribution	137	601	5.1 %	4.5 %	38
1 layer perceptron (no hidden neurons)		0	10.4 %	12.3 %	
2 layer perceptron (10 hidden neurons)		10	16 %	20 %	
2 layer perceptron (50 hidden neurons)		50	7.8 %	8.6 %	

Table 3: Compared performances on a 26 classes problem: the upper case letters. One can see that the performance/speed compromise is easy to find with CTRBF ($MAX_{brother}$ is set to 30).

Practical realization of a Radial Basis Function Network for handwritten digit recognition

Bernard Lemarié

La Poste
Service de Recherche Technique de la Poste
10, Rue de l'Île Mabon
F-44038 Nantes Cedex, France
Tel.: +33.40.69.97.91
Fax: +33.40.89.60.00
E-mail: lemarie@srtp.srt-poste.fr

Abstract : We present a practical realization of a Radial Basis Function Network for handwritten digits recognition task . Inspired from regularization theory and Parzen windows non parametric estimator, Radial Basis Function networks are tested for a classification task. Reduction of the number of hidden nodes which is an important and necessary step to obtain a computationally tractable network is made using an original technique. A comparison is made with the k-nearest neighbour and Parzen windows methods. Results appear better for the network at a much lower computational cost.

1. Introduction

In the field of neural networks Radial Basis Function models have been introduced in the context of the regularization theory[1]. Regularization models allow to assure the existence and also to obtain a global minimum of a function interpolation problem under some constraints of regularity. Therefore, when viewing layered Neural Networks as universal approximators, RBF models offer a theoretical framework.

For a classification task, the function to be approximated by the neural network is the a posteriori probability and we recall that, in a statistical approach, RBF networks are also near of Parzen windows methods. This should help to introduce more statistical notions as the consistency. For a problem of classification statistical techniques have been investigated for many years and offer today good references. Most used is certainly the method of k-nearest neighbour. Thus we consider the comparison of this method and also of the Parzen windows method with the network.In previous experiments RBF-Networks have yet been compared with the k-nearest neighbour method and also with more classical Neural Network models, like the Multi Layer Perceptron Network [4]. Such a comparison remains very dependant on experiment conditions like the database for learning and testing and also on the precise realization of the model under investigation. We chose to realize most of parameter learning steps by a unique backpropagation algorithm and to limit the number of free parameters so as to strengthen the comparison and the understanding of the network dynamics. Another important point, may be the most relevant, is the method of reduction of centres which might considerably influence the performance of the recognition system.

On the other part, recent papers focus the attention on the theoretical comparison between neural networks and statistical methods, considered in the context of non parametric estimation and especially point out the «bias Variance/Dilemma» [7]. Our purpose in this paper is to illustrate the fact that neural networks, if they actually do not avoid nor solve this difficult theoretical problem could nevertheless help in practice to build a effective estimator in the sense of bias/variance.

We use here a database of 28,000 handwritten digits from french postal code and introduce the network as a classifier after a process of morphological extraction of characteristics. We argue that for the present comparison with statistical methods, morphological or topological aspects of neural networks do not have to be considered and thus classification of characteristic vectors is convenient.

The organization of the paper is as follows. Design of the network on the basis of regularization theory and Parzen windows model is sketched in section 2. Section 3 presents the method of reduction of the number of centres.In section 4, the morphological method, the database and the resulting network are briefly described. Section 5 reports the results obtained with the proposed method after learning and its comparison with k-Nearest Neighbour and Parzen windows methods. The last section discusses some of the remaining issues and extensions of the method.

2. Regularization network and Parzen windows estimator.

RBF networks are issued from regularization theory. In this model, one considers that «world is smooth»: Given $S = \{(x_i, y_i) \in R^n \times R, i = 1, 2, ..., N\}$ a set of points we try to build a function from $R^n \rightarrow R$ which minimizes the sum of Mean Square Error on the given points and a regularity constraint on the unknown function:

$$H(f) = \sum_{i=1}^{N} (y_i - f(x_i))^2 + \lambda \|P(f)\| \qquad (1)$$

The regularity constraint is represented by a differential operator on the unknown function f. As described in [1] for some form of regularity operator we can obtain and build the solution of (1). In particular for:

$$\|P(f)\| = \sum_{m=0}^{\infty} a_m \left(\int_{R^n} \sum_{i_1 i_2 \ldots i_m} \left(\frac{\partial^m}{\partial i_1 \partial i_2 \ldots \partial i_m} f(x) \right)^2 dx \right) \qquad a_m = \frac{\sigma^{2m}}{m! 2^m} \qquad (2)$$

the solution of (1) is given by:

$$f(x) = \left(\sum_{i=1}^{N} C_i \times \left(exp\left(-\frac{\|x - x_i\|^2}{2\sigma^2} \right) \right) \right) \qquad \forall x, x \in R^n \qquad (3)$$

$$with \qquad C_i = (y_i - f(x_i)) / \lambda$$

C_i can be found by solving the N equations system obtained with (3) applied to each (x_i, y_i) . As pointed out in [1][2], the formula (3) can easily be viewed as the output of a neural network with one hidden layer of N neurons. The formalism is also easily extended to the approximation of functions from R^n to R^c and then can be applied to the estimation of a posteriori density of probability in a problem of classification.

Thus under a condition of regularity we can obtain a global minimum and solve the learning task. In practice however some problems appear.The first difficulty is to select the values of the regularization and kernel width parameters σ and λ . The second difficulty is due to the computational cost of the method because we have to compute the kernel activity for each element of the learning set. Also it's interesting to consider the regularization method under the aspect of neural networks and to try to apply a learning method for some parameters on a network. For a neural network a mean square error criteria is equivalent to (1) except the regularization coefficient.We would like to suppress this one but doing this with a complete network the trivial solution of (1) is $\sigma = 0$ (a Dirac function) and thus without any interest. However on a reduced network this solution disappears and it becomes relevant to minimize the mean square error.In this sense reduction of centres appear to counterbalance the omission of the regularization parameter avoiding thus a too important overfitting on the learning set.

Several solutions have been proposed and tested for practical realizations [3-6]. In many cases, after a reduction of centres algorithm the output coefficients are solved with a pseudo matrix inversion algorithm.In other cases finding the ray of each centre is included in the algorithm of minimization of the mean square error. This last solution appears to be more generic and should be extended to other parameters like centres position and covariance matrix.

We have to note that the formula (3) is closed to the expression of Parzen windows non parametric estimator of a priory density with gaussian kernel. Given a n elements data set in a d-dimension space, one writes:

$$\hat{p}_n(x) = \left(\frac{1}{nh_n^d} \right) \sum_{i=1}^{n} exp\left(-\frac{\|x - x_i\|^2}{2h_n^2} \right) \qquad (4)$$

Let's recall that this estimator is consistent under the following conditions:

$$\lim_{n \rightarrow inf} h_n = 0 \qquad and \qquad \lim_{n \rightarrow inf} h_n^d \times n = \infty \qquad (5)$$

For a classification task we then estimate for each class i of n_i elements the a priory probabilities and get the a posteriori probabilities by Bayes' rule:

$$\hat{p}(Ci|x) = \left(\frac{K}{h_{n_i}^d} \right) \sum_{j=1}^{n_i} exp\left(-\frac{\|x - x_{ij}\|^2}{2h_{n_i}^2} \right) \qquad (6)$$

Here again such a formula can be interpreted as the dynamic of a neural network with one hidden layer. This relation have been pointed out in [10], but without any subsequent learning. Considering only the network at age zero, we can therefore affirm that if the learning set grows neural network first and second order moments will converge if the conditions (5) are respected. On the other part dealing with risk of local minimum, results of learning are very sensitive to the initial state of the networks. In this sense taking the Parzen estimator as the initial state of the network would ensure a convenient performance. Here again, the reduction of the references and the choice of the kernel width are the main difficulties. As for the regularization reduction is required so as to decrease computational cost and also to avoid the trivial solution. Again, we propose the reduction as the first step following by the learning step of kernels width.

Finally, Radial Basis Function networks offer a new model of neural network architecture with theoretical justification and proof of existence[2]. The parallelism with the Parzen windows estimator gives us an initial state of the network.At the last step, using a learning algorithm like back propagation should help us to improve the performance of such a network with respect to the bayesian probabilities. However, this network presents too much hidden nodes to be tractable for any learning database and we have to try to reduce this number. Moreover, this reduction turns out to be required to avoid overfitting.

3. Method of reduction.

Clustering methods like K-means are the most used tools of reduction of data sets and have been applied to RBF-Networks[5]: Neighbour points in the data set are grouped into clusters which become the hidden neurons of the network. In this method however, one have to choose the number of final clusters and clustering algorithm object is mainly to adapt the position of the initial centres so as to minimize a specified criterion. On the other part, for neural networks the number of hidden neurons is a fundamental parameter of the functional capacity of the networks to approximate with more or less consistency the unknown function. This subject has been extensively discussed during the last years with for example the introduction of the famous notion of VC-dimension, see for example [12][13] for most recent applications of this concept to the field of neural networks. So, we can regret the absence of an automatic selection of number of clusters but this is certainly not an easy task and we will look for a reduction method which at least would help us to interpret the choice of the number of clusters.

Next we note that the cluster positioning processing before the learning by the neural network could be included in the network learning with this time the criterion of the network instead of the clustering method's one. And at last as pointed out very recently in [11] classical clustering does not take into account the classification problem in the sense that clustering is done on the whole data set. Thus, a reduction method introducing differentiation between classes would be welcomed.

An interesting reduction method for RBF networks has been proposed in [6] and applied in [6][4] which consists in randomly selecting a given number of elements in the learning sets. Apart from the choice of the number of centres, this method appears to be interesting because of its parallelism with the Parzen windows method and could be considered as a reduction of the Parzen windows estimator for each class followed by a learning step with all the learning set elements. However, by this method, less frequent points of the learning set do not produce centres and therefore as pointed in [4], estimation remains bad in the low density regions.

Finally we have tried to concentrate on a reduction method according to the following guidelines Number of elements, classification aspects and relation with consistency.

The first step is to reduce each Parzen estimator of the a priory density. For that few solutions exist, but we can hope that grouping most frequent elements of a given class should not affect too much the consistency of the non parametric estimator. the formula (6) becomes:

$$\hat{p}_n(x) = \left(\frac{1}{nh_n^d}\right)\sum_{i=1}^{n} exp\left(-\frac{\|x - C(x_i)\|^2}{2h_n^2}\right)$$

where $C(x_i)$, $i = 1, 2, n$ is the cluster assigned to xi and finally for our classification problem:

$$\hat{p}(Ci|x) = \left(\frac{K}{h_{n_i}^d}\right)\sum_{j=1}^{nc_i} n_{ij} \times exp\left(-\frac{\|x - c_{ij}\|^2}{2h_{n_i}^2}\right)$$

where nc_i is the number of clusters found for the class i and $Cij, j = 1, 2, nci$ the j-th centre of the class i containing n_{ij} points. This formula gives us the initial state of the network where h_{n_i} is chosen so as to respect the condition (5).

for each class cluster list is built from a chosen ray r. Taking randomly one element x we pick up all the other elements within the same class which are included in the hypersphere centred at x and with a ray of r. If r is too large, not only Parzen estimation will be false but classification task won't be realized. So we try to choose the clustering ray for each class in a classification objective. For that, we plot for each pair of classes the average number of points of the two classes which distance is below r:

$$N(i, j, r) = \frac{1}{N_i \times N_j} \times \sum_{x \in C_i} \sum_{y \in C_j} H(x, y, r)$$

where H is the Heavyside function:

$$H(x, y, r) = 1 \quad if \quad d(x, y) \le r \quad and \quad H(x, y, r) = 0 \quad otherwise$$

Figures 1 and 2 show these curves for classes 0 and 9:

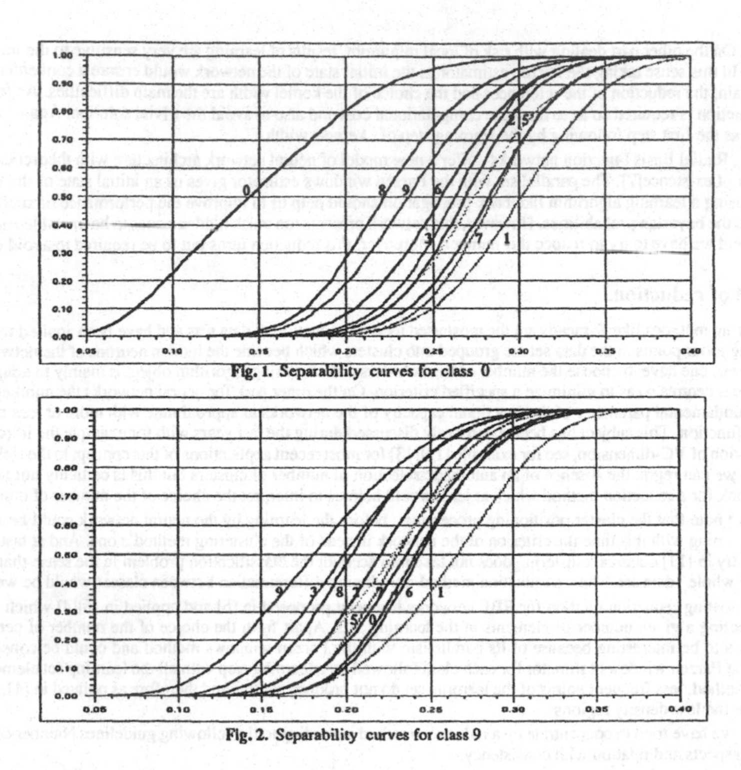

Fig. 1. Separability curves for class 0

Fig. 2. Separability curves for class 9

The curves are used to select the ray of the clustering for each class. For example for class 0 we take the ray 0.175 while for class 9 we take the ray 0.13.

4. Application: Building Network from learning set.

We now present the application of the RBF Networks to the recognition of handwritten digits issued from the segmentation of french postal codes. A base of 8783 examples is extracted for learning. A test base is also used for testing and finally an evaluation base is required for final comparison with other methods.

Three particularities of the base must be pointed. First the base has been labelled by a human operator and in some cases (1%) label is wrong. Also, some characters are not perfectly segmented but are well labelled. And finally classes are not equiprobably represented as illustrated by the following chart:

	0	1	2	3	4	5	6	7	8	9	total
learn	2391	1313	456	876	824	790	640	673	376	444	8783
test	570	338	126	230	218	189	158	178	88	102	2197
eval.	1240	799	658	862	772	879	553	728	379	524	7398
total	4094	2179	1196	1853	1667	1733	1240	1491	807	1025	17289

Fig. 3. Description of the database

The morphological preprocessing extracts a vector of length 143 from the bitmap image of the character. The method is inspired from «characteristic loci method» [8][9]: For each white point we consider the axis issued from the point and oriented in one of the eight main cardinal directions. We characterise the axis and therefore the point to cross the character (black pixel).

Curves of distance estimation as described in the preceding section have been plotted for each class and allow to get

the following clustering:

	0	1	2	3	4	5	6	7	8	9	total
ray	0.175	0.2	0.16	0.15	0.17	0.16	0.16	0.18	0.17	0.13	xxxxx
ncentre	45	31	22	55	41	28	29	28	18	59	356

Fig. 4. clustering on the earning set.

Ncentre is the number of centres within each class. We remark that although class 0 represent 30% of the learning set only 13% of the centres are in the class 0. On the other hand class 9 represents only 5% of the learning set but 17% of the centres. This comes in first place from the smaller ray selected for the class 9 than for the class 0. But we must also note that elements in class 0 are well grouped while in class 9 dispersion is more important and for example the first centre of class 0 contains 1681 elements.

Next, the network adjusted the ray of each centre and the output weights via a stochastic learning backpropagation algorithm.

5. Results.

Both statistical methods use all of the 8783 learning set elements as references.

	learn	test	validation
nel	8783	2197	7394
Parzen 0.1	84.87/15.13	84.93/15.07	83.30/16.7
Parzen 0.05	96.29/3.71	93.31/6.69	92.36/7.64
1-NN	100/0.00	96.13 /3.87	94.96/5.04
3-NN	97.51/1.75	96.13/2.81	95.00/3.97
5-NN	96.85/2.33	96.22/3.03	94.95/4.01
RBF-Network	98.14/1.86	97.04/2.96	96.29/3.71

Fig.5.. Comparative performances of statistical methods and RNF-Network.

For each case the first value represents the lecture rate TL the second value the substitution rate TS according to the formulas:

$$TL = \frac{nelements - nsubstitution - nreject}{nelements} \qquad TS = \frac{nsubstitution}{nelements - nreject}$$

Those results first show that practical use of Parzen windows estimator remains arduous because performance is very sensitive to the choice of the kernel width. As a crude interpretation the width of 0.05 gives us an overfitted estimator with bad results on test and evaluation sets, while the width of 0.1 turns out to be a low performance estimator but without important degradation of performances on test and evaluation sets. This illustrates the bias/variance dilemma and we have to notice that the choice of the kernel width has been studied for a long time so as to try to reduce the problem [14][15]. The direct suggestion here is that the kernel width could be learnt using the test set to valid the choice. Such a proposal is not far from neural network's learning mode and we believe it's better to include these parameters within a set of parameters like we do in the neural network formalism for after concentrating the investigation onto the selection of the most convenient parameters to be learnt.

With the k-nearest neighbour method results are better and quite comparable to the RBF-Networks ones. Here again the choice of the number of neighbours to be considered is very important. Taking into account only the first neighbour, we observe overfitting of the system. Using the test set to choose this parameter, best results are obtained with the 3-nearest neighbour but remain below the network's one on the valid set.

Finally the RBF-Network presents the best performance. The network shows an interesting balance between performance and overfitting i.e. between bias and variance of the estimator. Let's recall that backpropagation learning reduces mainly bias and that the measure of performance on the test set is generally used to prevent overfitting. With the RBF network the limitation of overfitting is reinforced by the structure of the network and by the reduction of centres. In fact at a local level presence of several examples of the same class near a centre forbids the reduction of this centre width leading thus to a local equilibrium closed to the variance of the set of the initial points included in the hypersphere around the centre and in counterpart presence of near elements of a different class leads to a reduction of the width. For centres with low probabilities i.e. centres with few near examples of the same class the variance should be low. But because this centre has a low probability, gradient updating is also rare and the initial width does not decrease too much.Thus for those points overfitting is reduced. We can illustrate this fact by an other experiment: We used the same reduction method but for each cluster we chose the middle point for the centre and the variance for the initial kernel width. Centres with only one point have a zero variance and we took a non zero value say 0.005. Learning was faster in the sense of comparison with the test set but for the few probability centres the width of the gaussian kernel remained very small. Therefore, those centres weren't able to generalize as much as in the previous experiment.

As an important aspect of comparison we have also to consider the computational cost of the two best methods in recognition mode. The most expensive operation in the processing is the computation of the distance between centres or references and the pattern to recognize embedded in the 143 dimension space. k-NN method requires 8783 computations of distance while RBF-Networks process only 356 ones. We get thus a cost reduction by a factor of 20.

6. Conclusion

The application of Radial Basis Function presented in this paper gives fairly good results.The comparison with statistical non parametric estimators reveals that in a practical situations the RBF network will be well suited because it gives better performance at a much lower computational cost. Moreover this model should raise in performance by adding more free parameters in the network as reviewed in [1]: Moving centres, multidimensional gaussian functions or equivalently customised distance. We think that the interest of this latter is reinforced by considering the dimensional aspect of the data, because in spite of a high dimensional embedding space, data appear to present low intrinsic dimensionality in the sense discussed in [16]. Working with multidimensional gaussian functions should allow to point out those specific directions and for that it's should be sufficient in a first step to work with a diagonal covariance matrix. But finally we must recall the necessity to control the liberation of parameters first in order to avoid bad convergence with back propagation and secondly to limit the variance problem. In this sense regularization networks if they do not solve the theoretical bias variance/dilemma present at least a fruitful framework for non parametric estimation with practical applications. This mainly comes from the neural architecture which allows the integration of most parameters in an unique formalism.

Another required improvement of the method is the automation of the selection of the centres nodes: For this a criterion is needed for separability of classes against ray of clustering. A fast but still empirical method would consist for example in taking into account the closest class on the separability curves and selecting the ray with a chosen coefficient [11]. Another approach is to try to integrate directly the selection of the centres in the learning step. This point is currently under investigation.

7. References

[1] Poggio T., Girosi F., Networks for approximation and learning.Proceedings of the IEEE, Vol 78, No. 9, 1990..

[2] Girosi F., Poggio T., Networks and the best approximation property. Biological Cybernetic 63, 169-176, 1990.

[3] Richard M. D., Lippman R. P., Neural Networks Classifiers estimate a posteriori Probabilities. Neural Computation, 4,461-483, 1991.

[4] Lee Y., Handwritten recognition using K Nearest-Neighbour, Radial Basis Function and Backpropagation Neural Networks. Neural Computation ,3,440-449, 1991.

[5] Moody J., Darken C. J. , Fast learning in Networks of locally tuned processing units, Neural Computation, 1, 281-294, 1989.

[6] Ng, Lipmann R.P. , A comparative study of the practical characteristics of neural networks and conventional pattern clasifiers, in Neural Information Processing Systems 3, 1991, D.S. Touretzky, ed. Morgan Kaufmann, San Mateo, Ca.

[7] Geman S, Bienenstock E.,Boursat R., Neural Networks and the bias variance dilemma, Neural Computation, 4, 1-58,1992.

[8] Gluksman H. A., Classification of Mixed Font alphabetics by characteristics loci., 1st annual IEEE Computer Conference, 138-141,1967.

[9] Gaillat G., Berthod M., Panorama des techniques d'extraction de traits caractéristiques en lecture, optique des caractères , Revue Technique THOMSON-CSF, Vol 11, No 4, 1979.

[10] Specht D. F., Probabilistic Neural Networks, Neural Networks, vol.3, 109-118, 1990.

[11] Musavi M. T., Ahmed W.,Chand K. H., Faris K. B., Hummels D. M., On the Training of Radial Basis Function Classifiers, Neural Networks , Vol 5,595-605, 1992.

[12] Hausler D., Decision Theoretic Generalization of the PAC Model for Neural Net and Other Learning Applications, Technical Report, UCSC-CRL-91-02, 1991.

[13] White H, Conectionist Non Parametric Regression: Multilayer feedforward Networks can learn Arbitrary Mappings, Neural Networks, Vol.3, 535-549,1990.

[14] Devroye L.,Automatic Pattern Recognition: A study of the Probability of Error, IEEE Transactions on Pattern Analysis and Machine Intelligence,Vol. 10, No.4, 1988.

[15] Chiu S.T., Bandwith Selection for Kernel Density Estimation, The Annals of Statistics, vol. 19, no. 4,1883-1905,1991.

[16] Somorjai R.L., Ali M.K.,„an efficient algorithm for estimating dimensionalities,Can. J. Chem. 66,979,1988.

Design of Fully and Partially Connected Random Neural Networks for Pattern Completion

Christine HUBERT

Ecole des Hautes Etudes en Informatique (EHEI), Université René Descartes,
45 rue des Saints-Pères, 75006 Paris, France
chris@ehei.ehei.fr

Abstract

In previous works [1,2,3], the behavior of a fully-connected single-layer Random Neural Network (RN) [4,5] has been illustrated in a problem of pattern completion. We applied the gradient-descent learning algorithm which has been introduced by Gelenbe [6,7] for recurrent RN networks. The recall of any training pattern from a corrupted version consists in a progressive retrieval process with adaptive threshold. We have reduced the influence of the pattern geometry on the performance by modifying the computation of the network state. The experimental results are now compared to thoses obtained with Hopfield's network. As the learning times in such a model become rapidly prohibitive, we look into the use of a single-layer network with local interactions between neurons. The connectivity influence on the convergence of the learning algorithm and on the recognition rates is particularly examined.

1. INTRODUCTION

A new neural network model called the Random Neural Network (RN) has been introduced in [4,5] by Gelenbe. Recently, a training algorithm which uses the gradient descent method, has been proposed for the recurrent RN model [6,7]. The ability of this model using the latter learning algorithm to act as an associative memory, has been shown up in previous works [1,2,3]. We considered a single-layer network of fully-interconnected neurons. However, the geometry of the typed patterns to be stored had a certain effect on the learning and recognition rates. The second disadvantage of the model used was the long learning times. In this paper, we show that a simple modification of the computation of the network state improve the performance and especially the learning and recall times. In order to reduce again these computation times, we investigate the use of a partially connected RN network.

The paper is organized as follows. In Section 2, we first recall the main properties of the basic Random Neural Network. Then, we present the design of fully and partially connected RN for pattern completion. Section 3 and 4 respectively describe the supervised learning procedure and the progressive retrieval process of binary patterns. In Section 5, the learning algorithm convergence and the success rates in pattern completion are examined for both networks.

2. PATTERN COMPLETION WITH THE RANDOM NEURAL NETWORK

We only recall here the main theoretical aspects of the RN model. Then, we present its application to pattern completion using either fully interconnected or locally connected neurons.

2.1. The basic Random Neural Network model

The Random Neural Network [4,5] consists of a set of N neurons which exchange positive (excitatory) and negative (inhibitory) signals. The signals which arrive at neuron i, may come from the outside of the network or from other neurons. The exogenous signal arrivals follow

Poisson process of rate $\Lambda(i)$ for positive signals and $\lambda(i)$ for negative signals. Each neuron i is represented by its potential $k_i(t)$ at time t. The arrival of a positive signal increments $k_i(t)$ by one, while the arrival of a negative signal decrements $k_i(t)$ by one if it isn't already zero. When $k_i(t) > 0$, the neuron i sends out signals which may either depart from the network with probability $d(i)$ or head for neuron j as positive signals with probability $p^+(i,j)$ or as negative signals with probability $p^-(i,j)$. The intervals between successive signal emissions by neuron i follow an exponential distribution with the mean $1/r(i)$.

It has been shown in [4] that the steady-state probability that the neuron i is excited, is given by

$$q_i = \lambda^+(i) \,/\, [r(i) + \lambda^-(i)] \tag{1}$$

with $\lambda^+(i) = \sum_{j=1}^{N} q_j\, r(j) p^+(j,i) + \Lambda(i) \qquad \lambda^-(i) = \sum_{j=1}^{N} q_j\, r(j) p^-(j,i) + \lambda(i).$ (2)

Let $w^+(i,j) = r(i)p^+(i,j)$ and $w^-(i,j) = r(i)p^-(i,j)$ be the synaptic weights for positive and negative signals. Hence, the firing rate of the neuron i is :

$$r(i) = \sum_{j=1}^{N} [w^+(i,j) + w^-(i,j)] \tag{3}$$

2.2. Pattern completion

Pattern completion operation may be performed by an associative single-layer RN network. Such a network associates a given information to itself and is able to correctly recall it from a corrupted version. The network input may consists in binary images which are represented by binary vectors $X_k = (x_{1k},...,x_{Nk})$ where N is the number of neurons. Each component x_{ik} is associated to neuron i and converted into arrival rates of positive and negative exogenous signals as follows :

$$x_{ik} = 1 \rightarrow (\Lambda_k(i), \lambda_k(i)) = (\Lambda, 0) \qquad x_{ik} = 0 \rightarrow (\Lambda_k(i), \lambda_k(i)) = (0, \lambda \tag{4}$$

where Λ and λ provide the network stability.

The network is recurrent and the probability that emitting signals depart from the network is $d(i) = 0$. The network state is represented by the vector of firing probabilities $Q_k = (q_{ik},...,q_{Nk})$. For each input pattern, the corresponding binary output vector $Y_k = (y_{1k},...,y_{Nk})$ is determined from Q_k by applying the progressive retrieval procedure described in Section 4.

During the learning phase, synaptic weights are adjusted to allow the network to properly associate the training images to themselves.

The steady-state probabilities q_{ik} that each neuron i is excited, are determined from the non-linear equations (1) and (2). In previous works [1,2,3] we iteratively solve these equations after initializing every probability q_{ik} to 0.5. We have succeeded in having better performance results and much shorter computation times by initializing q_{ik} to the input value x_{ik} and by performing only one iteration for the determination of q_{ik}.

We may design a fully or partially connected associative network. Concerning the second network, each neuron may be connected to its nearest neighbors only. Thus, the state of each neuron depends only of the state of its neighbors and the computations are much faster.

3. THE SUPERVISED LEARNING PROCEDURE

Let $S = \{X_1,...,X_M\}$ be the set of M vectors to be associated to themselves. Using Gelenbe's learning algorithm [6,7], we adjust synaptic weights so as to reduce the global error $E = \sum_{k=1}^{M} E_k$ where $E_k = 1/2 \sum_{i=1}^{N} (q_{ik} - y_{ik})^2$ with the desired output $y_{ik} = x_{ik}$. The different steps of the learning phase are :

- initialize the weight matrices $W^+_0 = [w^+_0(i,j)]$ and $W^-_0 = [w^-_0(i,j)]$,
- compute the initial firing rates $r_0(i)$ for $i = 1,...,N$ using (3),
- determine the arrival rates of exogenous signals Λ and λ,
- update weight matrices so as to reduce the global error E.

3.1. Weight matrix initialization

The choice of initial weights may influence the convergence of the learning process which uses gradient descent [8]. Simulations have lead us to initialize W^{+}_{0} referring to Hebb's law [9] :

$$w^{+}_{0}(i,i) = 0\,,\quad \text{if } i \neq j \quad w^{+}_{0}(i,j) = \begin{cases} w_{ij} = \sum_{k=1}^{M} (2\,x_{ik} - 1)(2\,x_{jk} - 1) & \text{if } w_{ij} > 0 \\ 0 & \text{otherwise} \end{cases} \tag{5}$$

where x_{ik} and $x_{jk} \in \{0,1\}$. On the other hand, W^{-}_{0} was initialized with small positive random values uniformly distributed between 0 and $V_{max} = 0.2$.

3.2. Arrival rates of exogenous signals

Considering the vector encoding (4), the network hyper-stability condition [10] becomes for each vector X_k :

$$\text{if } x_{ik} = 1,\ \ \Lambda < r_0(i) - \sum_{i=1}^{N} w^{+}_{0}(j,i) \qquad \text{if } x_{ik} = 0,\ \ \lambda > \sum_{i=1}^{N} w^{+}_{0}(j,i) - r_0(i)$$

Let $\Lambda_k = min_{i,x_{ik}=1}[r_0(i) - \sum_{i=1}^{N} w^{+}_{0}(j,i)]$ and $\lambda_k = max_{i,x_{ik}=0}[\sum_{i=1}^{N} w^{+}_{0}(j,i) - r_0(i)]$. The network is damped for the M training vectors if $\Lambda \leq min_k[\Lambda_k]$ and $\lambda \geq max_k[\lambda_k]$. Here, we take $\Lambda = min_k[\Lambda_k]$ and $\lambda = 0$.

3.3. Weight matrix update

The learning algorithm uses the gradient descent method and updates the weights for each training vector, according to the following general rule [6,7] :

$$w_p(u,v) = w_{p-1}(u,v) - \eta \sum_{i=1}^{N} (q_{ik} - x_{ik})[\partial q_{ik} / \partial w(u,v)]_{W^{+}_{p-1},\,W^{-}_{p-1}} \tag{6}$$

The learning rate η is initialized to 0.5 and when oscillations occur, we decrease it down to 0.01 :

$$\text{if } 0.5 \geq \eta > 0.1, \text{ then } \eta = \eta - 0.05 \qquad \text{if } 0.1 \geq \eta > 0.01 \text{ then } \eta = \eta - 0.01$$

We choose to present the training vectors in a random order and more often when they are badly learned.

Let NB_CYC be the number of presentation epochs of the M training images and f_{max} be the maximal number of presentations per epoch of each example. The number of operations that the learning algorithm requires is proportional to $NB\_CYC\ M\ f_{max}\ N^3$. Thus, the algorithm is $O(N^3)$.

4. THE PROGRESSIVE RETRIEVAL PROCESS

Once the network has been trained, it must perform the completion of noisy versions of the stored vectors. Let $X = (x_1,\ldots,x_N)$ be any binary input vector. To retrieve the corresponding output vector $Y = (y_1,\ldots,y_N)$, we first compute the vector of firing probabilities $Q = (q_1,\ldots,q_N)$ and treat the neurons whose state is considered certain, that is, whose $q_i \notin Z = [1-b,b]$ with for instance $b = 0.6$. We thus obtain the output vector $Y^{(1)} = (y^{(1)}_1,\ldots,y^{(1)}_N)$ with

$$y^{(1)}_i = g_Z(q_i) = \begin{cases} 1 & \text{if } q_i \geq b \\ 0 & \text{if } q_i \leq 1 - b \\ x_i & \text{otherwise} \end{cases} \tag{7}$$

where g_Z is the thresholding function by intervals. If the number NB_Z of neurons whose $q_i \in Z$ is zero, we consider that the network has stabilized to an attractor state and then the output vector is $Y = Y^{(1)}$. Otherwise, Y is obtained after applying the thresholding function f_α where α is the selected threshold, as follows :

$$y_i = f_\alpha(q_i) = \begin{cases} 1 & \text{if } q_i > \alpha \\ 0 & \text{otherwise} \end{cases} \tag{8}$$

Each value $q_i \in Z$ and also the lower bound $1\text{-}b$ are potential thresholds. For each potential value of α, we present the vector $X^{(2)}(\alpha) = f_\alpha(Q)$ to the network. Then, we compute the new vector of firing probabilities $Q^{(2)}(\alpha)$ and the output vector $Y^{(2)}(\alpha) = g_Z(Q^{(2)}(\alpha))$. We keep the cases where $NB\_Z = 0$ and $Y^{(2)}(\alpha) = X^{(2)}(\alpha)$. If these two conditions are never satisfied, the initial input X is considered too much different of any training vector and α is set to 0.5. If several thresholds are candidate, we choose the one which minimizes $E(\alpha) = 1/2 \sum^{N}_{i=1} (q^{(2)}_i(\alpha) - y^{(2)}_i(\alpha))^2$.

The complexity of the retrieval process is $O(N^2)$ if the thresholding function by intervals g_Z is used only. In the worst case where $NB\_Z = N/3$, it becomes $O(N^3)$.

5. LEARNING ALGORITHM CONVERGENCE AND RECOGNITION RATES

The fully and partially connected Random Neural Network models we have described above, have been experimentally evaluated in the completion of typed images. We have considered different learning sets S_M = *{image 1,..., image M-1, image M}* composed of *M* images chosen from among thoses of Figure 1. Each image was represented by 6x6 bits. Thus, we used single-layer networks of $N = 36$ neurons.

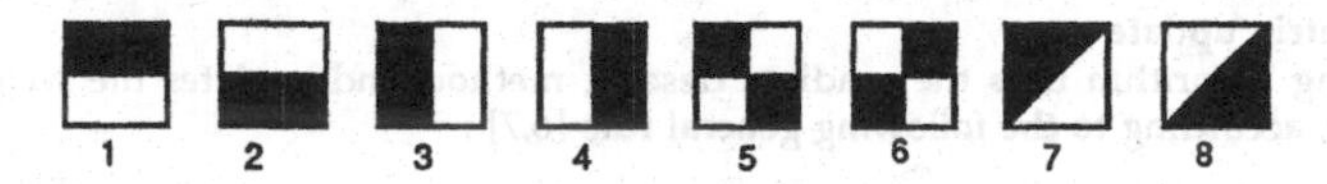

Figure 1 : Typed training images

In the partially connected network, each neuron was connected to its Nv = 8 nearest neighbors - when they exist - in the directions N, NE, E, SE, S, SW, W, NW.

For both types of network, we examine the learning algorithm convergence and the success rates in pattern completion (recognition rates).

5.1. Learning algorithm convergence

For the learning procedure, we have used the following set of parameters :

$\varepsilon 1 = 10^{-3}$: limit on the global error E,
$\varepsilon 2 = 10^{-2}$: minimal value of the learning rate η,
$f_{max} = 5$: maximal number of presentations per epoch of each training vector,
$STOP = 10$: maximal number of iterations in a local minimum.

For each type of network and for each learning set, Table 1 shows the initial and final global errors, and the number of matrix updates.

Table 1 : Learning algorithm convergence

SET	FULLY CONNECTED RN						PARTIALLY CONNECTED RN		
	Case 1			Case 2			Case 1		
	Init Error	Final Error E	No of updates	Init Error	Final Error E	No of updates	Init Error	Final Error E	No of updates
S1	1.36	0.0007	94	0.10	0.0009	15	0.20	0.0007	43
S2	2.74	0.0007	469	0.05	0.0009	45	0.12	0.0003	113
S3	6.90	0.6351	2745	1.34	0.2399	617	0.39	0.1698	644
S4	5.48	0.0006	1458	0.15	0.0009	83	0.13	0.0009	82
S5	6.94	0.0007	1153	0.12	0.0009	113	0.12	0.0009	219
S6	8.19	0.0009	2402	0.10	0.0009	130	0.08	0.0009	74
S7	11.06	1.1510	9333	1.83	1.1218	6678	0.97	0.8256	2601
S8	13.86	2.5569	1880	2.64	0.8386	7246	1.24	0.8360	3233

We compare the two methods to compute the firing probabilities q_i :
- case 1 : initialization to 0.5 - iterative computation,
- case 2 : initialization of q_i to the input value x_i - single iteration.

At first, we can notice that the second way of computing q_i (case 2) really decreases the number of weight updates to reach the desired global error. Equally, the learning convergence is improved in some difficult cases where the geometry pattern is involved (sets S_3 , S_7 and S_8). With S_3, it's hard to reduce the global error because of the symetry between the images *1* and 2. With S_7 and S_8 , the learning of the diagonals is problematic.

Compared to the fully connected network, the partially connected one needs a similar number of matrix updates except for the sets S_7 and S_8 whose learning becomes easier. On the other hand, learning times are much shorter.

5.2. Recognition rate

To evaluate the exact success rates in pattern completion, we have generated 30 randomly corrupted patterns for each training image and for a given distortion rate. For the fully connected network, the results obtained with the two different computation methods of q_i , are presented on Figure 2 and 3.

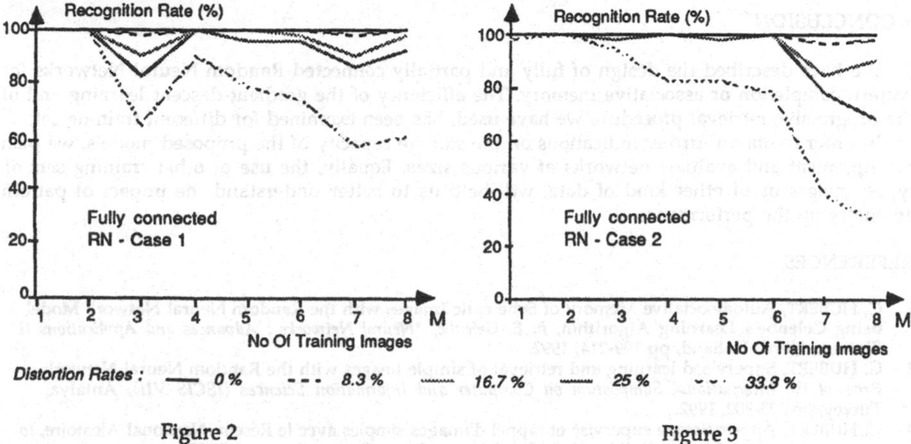

Figure 2 Figure 3

The new computation method we have introduced (case 2) attenuates the influence of pattern geometry on the performance, especially in case S_3. Indeed, apart from the cases S_7 and S_8 where learning remains unsatisfactory, the recognition rates are greater than 98% if the distortion rate is less than 25%. Compared to Hopfield's network [11] whose performance results are given on Figure 4, the RN network is much less sensible to symmetry and diagonal effects.

The recognition rates obtained with the partially connected network are shown on Figure 5. As expected, this model is not so robust to pattern distortion but the performance remains almost the same whatever the learning set is.

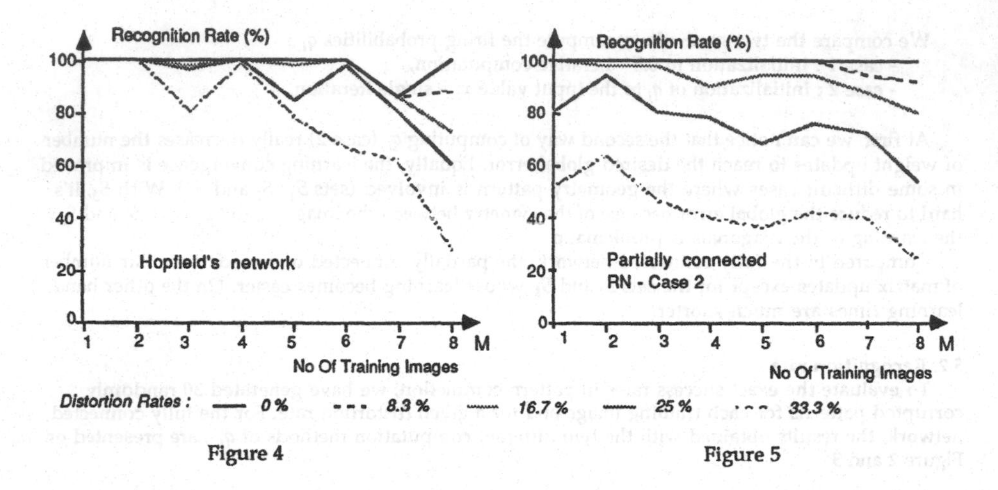

Figure 4 Figure 5

6. CONCLUSION

We have described the design of fully and partially connected Random Neural Networks for pattern completion or associative memory. The efficiency of the gradient-descent learning and of the progressive retrieval procedure we have used, has been examined for different training set.

In order to obtain further indications on the storage capacity of the proposed models, we plan to implement and evaluate networks of various sizes. Equally, the use of other training sets of typed images or of other kind of data, will help us to better understand the impact of pattern geometry on the performance.

REFERENCES

1. C. HUBERT, Autoassociative Memory of Schematic Images with the Random Neural Network Model using Gelenbe's Learning Algorithm, *in E. Gelenbe, "Neural Networks : Advances and Applications II"*, Elsevier, North-Holland, pp 199-214, 1992.
2. C. HUBERT, Supervised learning and retrieval of simple images with the Random Neural Network, *Proc. of the International Symposium on Computer and Information Sciences (ISCIS VII)*, Antalya, Turkey, pp 295-302, 1992.
3. C. HUBERT, Apprentissage supervisé et rappel d'images simples avec le Réseau Neuronal Aleatoire, *to appear in Comptes Rendus de l'Académie des Sciences*, Paris, France, January 1993.
4. E. GELENBE, Random Neural Networks with Negative and Positive Signals and Product Form Solution, *Neural Computation*, Vol. 1, No. 4, pp 502-510, 1989.
5. E. GELENBE, Theory of the Random Neural Network Model, *in E. Gelenbe, "Neural networks: advances and applications"*, Elsevier, North-Holland, 1991.
6. E. GELENBE, Learning in the Recurrent Random Neural Network Model, *in E. Gelenbe, "Neural Networks: Advances and Applications II"*, Elsevier, North-Holland, 1992.
7. E. GELENBE, G-nets and Learning Recurrent Random Networks, *Proc. of the International Conference on Artificial Neural Networks (ICANN-92)*, Brighton, UK, pp 943-946, 1992.
8. Y. LE CUN, Modèles connexionnistes de l'apprentissage, *Ph.D Thesis*, Paris 6 University, France, 1987.
9. D.O. HEBB, The organization of behavior, *John Wiley and Sons*, New York, 1949.
10. E. GELENBE, Stability of the random neural network model, *Neural Computation*, Vol. 2, No. 2, pp 239-247, 1990.
11. D.E. RUMELHART, G.E. HINTON and R.J. WILLIAMS, Learning internal representations by error propagation *in D.E. Rumelhart and J.L. McClelland "Parallel Distributed Processing"*, Vol. 1, Bradford Books and MIT Press, Cambridge, Massachussets, pp 318-362, 1986.
12. J.J. HOPFIELD, Neurons with graded response have collective computational properties like those of two-state neurons, *Proc. of the National Academy of Sciences 81*, USA, pp 3088-3092, 1984.

REPRESENTATION AND RECOGNITION OF REGULAR GRAMMARS BY MEANS OF SECOND-ORDER RECURRENT NEURAL NETWORKS

R. Alquézar[1] and A. Sanfeliu

Institut de Cibernètica (UPC - CSIC)
Diagonal 647, 2a, Barcelona (Spain)

Recently, some models of neural networks, recurrent neural networks, have been used in conjunction with their associated neural learning schemes to infer regular grammars from a set of sample strings. The representation of the inferred automata is hidden in the weights and connections of the net, this being a common feature in emergent subsymbolic representations. In order to relate the symbolic and connectionist approaches to the tasks of grammatical inference and recognition, we address and solve a basic problem, which is, how to build a neural net recognizer for a given regular language specified by a deterministic finite-state automaton. A second-order recurrent network model is employed, which allows to formulate the problem as one of solving a linear system of equations. These equations directly represent the automaton transitions in terms of static linear approximations of the network running equations, and can be viewed as constraints to be satisfied by the network weights. A description is given both for the weight computation step and the string recognition procedure.

1 Introduction

Neural networks are being extensively studied in a large spectrum of fields. In recent years, some models of neural networks, recurrent neural networks, have been used in conjunction with their corresponding neural learning schemes to infer regular grammars from sets of positive sample strings ([6],[7]), and in some cases from negative examples as well ([2],[8]). The reported approaches can be grouped into three basic models: the simple recurrent network (SRN) ([1],[6]), the fully-connected first order recurrent network trainable by the RTRL algorithm (FORN) ([7],[9]), and second order recurrent networks (SORN) ([2],[8]).

In a recent paper [5], we explained how to map a deterministic finite-state automaton (FSA) into a fully-connected first-order recurrent neural network so that the latter may be used as a recognizer for the associated regular language. The proposed method included as well the representation of stochastic regular grammars (i.e. their equivalent stochastic automata). The key of the above mapping was an algebraic model in which automata transitions were viewed as linear equations to be satisfied by the network weights.

In the present work, we develop a similar model for the representation of a deterministic FSA in a class of second-order recurrent neural networks (SORN) that has been suggested for grammatical inference by Giles et al. [2]. Although the approach is analogous to that described in [5], the architecture of the second-order recurrent network imposes some relevant differences in both the weight computation and string recognition procedures, which will be commented later.

[1]The author is supported through a grant from the Government of Catalonia.

2 Regular Grammars and Finite State Automata (FSA)

Some definitions and known results from automata and formal languages theory are recalled next. A grammar G is a four-tuple (V_N, V_T, P, S), where $V_N = \{A_1, ..., A_n\}$ is a set of nonterminal symbols, $V_T = \{a_1, ..., a_l\}$ is a set of terminal symbols, P is a set of production rules, and $S = A_1$ is the start symbol. For every grammar G, there exists a language $L(G)$, a set of strings of the terminal symbols (i.e. $L(G) \subseteq V_T^*$), that the grammar generates or recognizes. A grammar G is regular iff, each production rule P_r in P, for $r = 1, ..., |P|$, has one of the two following structures:

$$A_j \longrightarrow a_k A_i \qquad A_j \longrightarrow a_k$$

As we will see later, a second-order recurrent neural net can be built to recognize the language generated by any regular grammar. Actually, we will show how to build a neural net to recognize the language accepted by a given deterministic finite-state automaton (FSA), but it is well known that any regular grammar is equivalent to a deterministic FSA [3]. It is readily proved that for every regular grammar $G = (V_N, V_T, P, S)$ there exists a (non-deterministic) FSA $A = (\Sigma, Q, \delta, q_0, F)$ that accepts the language generated by the grammar, and viceversa, where $\Sigma = V_T$ is a finite set of input symbols (alphabet), $Q = V_N \cup F$ is a finite set of states, $q_0 \epsilon Q$ is the initial state ($q_0 = S$), $F \subseteq Q$ is the set of final states, and $\delta : (Q \times \Sigma) \rightarrow 2^Q$ is a state transition (partial) function, such that for each production rule of type $A_j \longrightarrow a_k A_i$ or $A_j \longrightarrow a_k$ a transition $\delta(A_j, a_k) = A_i$ or $\delta(A_j, a_k) = q_f$ ($q_f \epsilon F$) is assigned respectively. It can be demonstrated that, given a non-deterministic FSA, there exists a deterministic FSA, where the transition function is restricted to $\delta : (Q \times \Sigma) \rightarrow Q$, that accepts the same language. So, for every regular grammar G, there exists a deterministic FSA A_D that accepts the language generated by G.

3 Second-Order Recurrent Neural Networks (SORN)

Recently, Giles et al. [2] and Watrous & Kuhn [8] have used second-order recurrent networks and complete gradient learning schemes to induce a set of simple regular languages over the alphabet {0,1} from both positive and negative examples. The networks were trained to accept or reject a whole string after presentation of all of its symbols (including a special "end" symbol). We have chosen for discussion here the Giles' network because is simpler and uniform (each neuron just receives second-order inputs), but the approach presented in next sections may be modified easily to work with Watrous' model.

The Giles' architecture [2] (see Fig.1) consists of N recurrent hidden neurons labeled S_j, L nonrecurrent input neurons labeled I_k, and N^2xL real-valued weights labeled w_{ijk}. The values of the hidden neurons are referred collectively as a state vector S in the finite N-dimensional space $[0,1]^N$. The recurrent network accepts a time-ordered sequence of inputs and evolves with dynamics defined by the following equations

$$S_i(t+1) = f(\sigma_i) \qquad \sigma_i = \sum_{j,k} w_{ijk}\, S_j(t)\, I_k(t) \qquad \forall i \epsilon \{1, .., N\} \tag{1}$$

where f is a sigmoid function, such that $f(x) \epsilon (0,1)$.

The network is built with one input neuron for each terminal symbol in the alphabet of the relevant language, plus one input neuron for the "end-of-string" symbol (say \$). A special "response" neuron S_0 is selected, whose value at the end of an input string (after feeding the \$ symbol) is above $1-\epsilon$ if the string is accepted, or below ϵ if rejected, where ϵ is a tolerance term. The training data consists of a series of example-response pairs, where the example is a sequence of symbols that are presented as input one at each discrete time step t (following a unary encoding), and the response is "1" for positive examples or "0" for negative examples, and it is used as training signal for neuron S_0.

A procedure for extracting a FSA from the neural network, that can be executed during or after training, is described in [2] based on the hypothesis that, during training, the network partitions its state space into well-separated regions or clusters, which represent corresponding states in some FSA. Giles et al. claimed that once the FSA is extracted and minimized it might be used for recognition instead of the network, since, in some cases, the extracted FSA outperformed the neural nets from which they were built.

Conversely, if the FSA is already known and we just want the network to act as a recognizer, then an explicit coding of the FSA states may be represented in the state units. Either a unary or a binary representation can be followed, so the number N of required units is $|Q|$ or $\lceil log_2|Q|\rceil$ respectively, where $|Q|$ is the number of FSA states. We will see in the next sections that, for a given deterministic FSA, the weights of the SORN that recognizes the same language can be obtained analytically, and in a straightforward way, if a unary state encoding is employed; if a binary or a higher-base coding is used, further non-linear terms may be required to obtain a solution. The method to be explained allows to go in a reverse symbolic-connectionist direction through the tasks of grammatical inference and recognition, since a grammar could be inferred from a set of examples through some symbolic algorithm [4] and then, after transformation to a deterministic FSA, could be "implemented" in a SORN acting as recognizer.

4 FSA Representation in Second-Order Recurrent Networks

When a SORN is running at a discrete time t, a terminal symbol a_k is fed to the net by means of the input unit I_k, the feedback of recurrent units represent in some way the sequence of previous symbols (or the FSA state q_j reached after it), and the computed output of the N network neurons, given by equation (1), stand for the incoming state q_i (i.e. the destination of the current transition). Thus, the set of N network equations are implementing the transition $\delta(q_j, a_k) = q_i$ that occurs at time t.

Since the studied SORN architectures do not include any symbol output unit (in contrast to FORN models) in which to learn or represent the legal (grammatical) successor symbols, the way to detect an illegal transition is by using a no-exit "garbage" state q_G where all illegal transitions must lead. This implies that, the network must implement a fully-defined transition function δ_E, which coincides with δ in the domain of pairs $(q\epsilon Q, a\epsilon\Sigma)$ where δ is defined, and is extended by the mappings $\delta_E(q', a') = q_G$ for the rest of pairs (q', a'). Moreover, $\delta_E(q_G, a) = q_G$ $\forall a\epsilon\Sigma$. Furthermore, for any final state $q_f \epsilon F$, an "end-of-string" transition must be included of the type $\delta_E(q_f, \$) = q_0$, where q_0 is the start state, and for any $q \notin F$, $\delta_E(q, \$) = q_G$. The former extensions allow to recognize or not a given string (see Section 6).

Hence, if we enumerate all the state transitions $\delta_E(q\epsilon Q, a\epsilon\Sigma)$ with an index $m\epsilon\{1..M\}$, the following finite set of equations should be satisfied by the network weights w_{ijk} at any arbitrary time t:

$$S_{mi}(t+1) = f(\sum_{j,k} w_{ijk} S_{m'j}(t) I_{mk}(t)\ )\ \ \forall m\epsilon\{1,..,M\}\ \ \forall i\epsilon\{1,..,N\} \tag{2}$$

where m' refers to a predecessor transition of m.

The system of equations (2) describes a complete FSA. It is evident that for a non-deterministic FSA, we will have two or more equations with the same right part (representing the outgoing state and current symbol) and different left part (representing the incoming state) for each non-deterministic transition. Clearly, since the function f is single-valued, the derived system of equations will not be solvable. Therefore, in the following, we restrict our discussion to the class of deterministic finite-state automata.

We realize that, if the FSA is known, the values S_{mi}, $S_{m'j}$ and I_{mk}, that appear in the equations corresponding to any transition $m \equiv \delta(q_j, a_k) = q_i$, are all known, since they are determined by the coding of the incoming state q_i, the outgoing state q_j, and the transition symbol a_k, respectively. Hence, the only variables which are unknown are the weights of the neural network.

In order to employ the above algebraic model for weight computation, it is preferable to transform it in a linear system of equations. To that end, it is necessary to convert the dynamic equations into static ones and to use the inverse of the nonlinear function f. The first step is straightforward. It has been pointed out that the set of equations (2) must be fulfilled independently of the time instant t, so the variables t

and $t-1$ can be ruled out from (2). However, in order to launch the sequence of automaton transitions, the network must be initialized to reproduce the start state code on its unit activation values.

Because $S_{mi} = f(\sigma_{mi})$ is known, a good approximation of the required linear output σ_{mi} can be obtained by accurately computing $f^{-1}(S_{mi})$. This is possible since f is single-valued and strictly monotonous, so $f^{-1}(S)$ does exist and is unique for all S in the open interval (0..1). When the ideal values 0 and 1 are needed as output S_{mi}, then σ_{mi} is chosen to be equal to $-C$ and C respectively, where C is an enough large number such that $f(C) \simeq 1$ with a desired precision.

Hence, the nonlinear system of equations (2) can be converted in the linear system:

$$f^{-1}(S_{mi}) = \sum_{j,k} w_{ijk} S_{m'j} I_{mk} \quad \forall m \epsilon \{1,..,M\} \; \forall i \epsilon \{1,..,N\} \tag{3}$$

which can be described in the following matrix representation:

$$\begin{pmatrix} S_{1'1}I_{11} & .. & S_{1'1}I_{1L} & .. & S_{1'N}I_{11} & .. & S_{1'N}I_{1L} \\ \vdots & & \vdots & & \vdots & & \vdots \\ S_{m'1}I_{m1} & .. & S_{m'1}I_{mL} & .. & S_{m'N}I_{m1} & .. & S_{m'N}I_{mL} \\ \vdots & & \vdots & & \vdots & & \vdots \\ S_{M'1}I_{M1} & .. & S_{M'1}I_{ML} & .. & S_{M'N}I_{M1} & .. & S_{M'N}I_{ML} \end{pmatrix} \begin{pmatrix} w_{11} & .. & w_{1N} \\ \vdots & \vdots & \vdots \\ w_{P1} & .. & w_{PN} \end{pmatrix} = \begin{pmatrix} f^{-1}(S_{11}) & .. & f^{-1}(S_{1N}) \\ \vdots & & \vdots \\ f^{-1}(S_{m1}) & .. & f^{-1}(S_{mN}) \\ \vdots & & \vdots \\ f^{-1}(S_{M1}) & .. & f^{-1}(S_{MN}) \end{pmatrix}$$

or

$$AW = B \tag{4}$$

where $A(M \times P)$ is the array of the neuron inputs, $W(P \times N)$ is the array of weigths and $B(M \times N)$ is the array of the neuron linear outputs. Moreover, L is the number of input neurons, which is defined as $L = |\Sigma|$ (i.e the number of alphabet symbols including the special symbol \$), $|N|$ is the number of state units, $P = N \times L$ is the number of inputs that arrive to each neuron, and $M = |Q| \times |\Sigma|$, where $|Q|$ includes the garbage state. It's important to note that, for a unary state encoding ($N = |Q|$), A is an identity diagonal matrix.

As an example, consider the simple regular grammar known as Tomita's 4th grammar, that have been employed in previous studies [2][8], whose associated deterministic FSA is shown in Figure 2. The coefficient arrays A and B that are obtained from the FSA by using a unary state encoding are displayed in Figure 3.

5 Recognizer SORN construction procedure

The building steps of the SORN equivalent to a given deterministic FSA are based on the algebraic model which has been described. Firstly, the values M, P and N, which specify the dimensions of the A, W and B arrays are determined. Secondly, the a_{mp} and b_{mn} elements of the A and B arrays respectively are fixed from the FSA information and the state encoding scheme.

If a unary encoding is used for the states, we have seen that the A array is always an identity matrix $I(M \times M)$, so the weights w_{mn} of the array W can be assigned directly to the corresponding coefficients b_{mn} of the B array. However, if a higher base is employed, the linear system (4) should be solved to obtain the weights. Due to the state encoding, $P = N \times |\Sigma|$ is always less than $M = |Q| \times |\Sigma|$, so (4) has no solution. Therefore, the original system must be extended with new elements (new columns in A and new rows in W) until the rank of the extended A array be equal to M, the number of equations. When this condition is satisfied, a solution for the network weights can be reached by applying the Gauss-Jordan method for solving linear systems.

The new elements in the extended A array are treated as external inputs to the network (with associated weights for each neuron) and can be evaluated by using a family of predefined nonlinear functions g, of the form

$$g(a,q,r) = a^{u_r} q^{v_r} \tag{5}$$

indexed by $r \geq 2$, where $u_r \geq 1$ and $K_v \geq v_r \geq 1$ are positive integers such that $u_r = ((r-1)\ div\ K_v) + 1$ and $v_r = ((r-1)\ mod\ K_v) + 1$, being $K_v > 1$ an arbitrary constant. The arguments a and q are the ordinal numbers (beginning at 1) of the symbol a_k and the outgoing state q_j in any given transition, which can be computed on-line during network running. The above g functions may be normalized if desired to the interval (0, 1] by dividing their output by the greatest possible value which is $|\Sigma|^{u_r}|Q|^{v_r}$ for each one of them, thus generating a new set of functions $g'(a,q,r)$. Therefore, we have an ordered set of functions g (or g') from which we can take any finite number R required to solve the extended linear system of equations

$$f^{-1}(S_{mi}) = \sum_{j,k} w_{ijk} S_{m'j} I_{mk} + \sum_{r\epsilon\{2,..,R+1\}} w_{ir}\, g(a,q,r) \quad \forall m \epsilon \{1,..,M\} \; \forall i \epsilon \{1,..,N\} \tag{6}$$

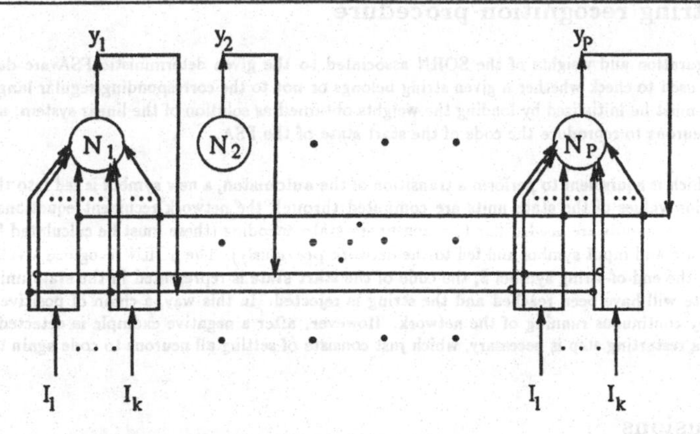

Fig. 1 ***Second-Order Recurrent Network (SORN) model***

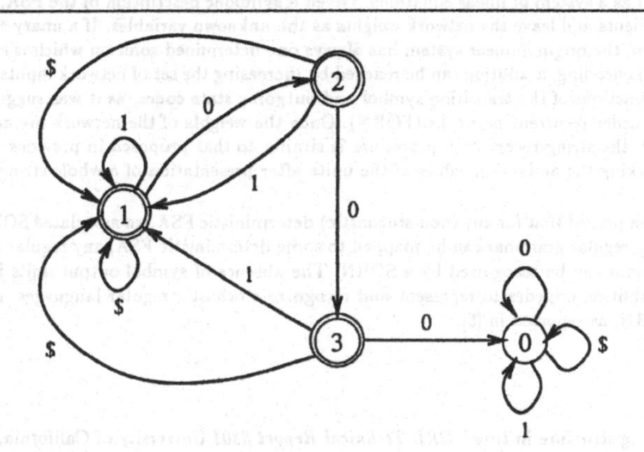

Fig. 2 ***Tomita's 4th grammar***

$$
\begin{pmatrix}
S_0 0 & S_0 1 & S_0\$ & S_1 0 & S_1 1 & S_1\$ & S_2 0 & S_2 1 & S_2\$ & S_3 0 & S_3 1 & S_3\$ \\
1 & 0 & 0 & 0 & 0 & 0 & 0 & 0 & 0 & 0 & 0 & 0 \\
0 & 1 & 0 & 0 & 0 & 0 & 0 & 0 & 0 & 0 & 0 & 0 \\
0 & 0 & 1 & 0 & 0 & 0 & 0 & 0 & 0 & 0 & 0 & 0 \\
0 & 0 & 0 & 1 & 0 & 0 & 0 & 0 & 0 & 0 & 0 & 0 \\
0 & 0 & 0 & 0 & 1 & 0 & 0 & 0 & 0 & 0 & 0 & 0 \\
0 & 0 & 0 & 0 & 0 & 1 & 0 & 0 & 0 & 0 & 0 & 0 \\
0 & 0 & 0 & 0 & 0 & 0 & 1 & 0 & 0 & 0 & 0 & 0 \\
0 & 0 & 0 & 0 & 0 & 0 & 0 & 1 & 0 & 0 & 0 & 0 \\
0 & 0 & 0 & 0 & 0 & 0 & 0 & 0 & 1 & 0 & 0 & 0 \\
0 & 0 & 0 & 0 & 0 & 0 & 0 & 0 & 0 & 1 & 0 & 0 \\
0 & 0 & 0 & 0 & 0 & 0 & 0 & 0 & 0 & 0 & 1 & 0 \\
0 & 0 & 0 & 0 & 0 & 0 & 0 & 0 & 0 & 0 & 0 & 1
\end{pmatrix}
W =
\begin{pmatrix}
S_0 & S_1 & S_2 & S_3 \\
C & -C & -C & -C \\
C & -C & -C & -C \\
C & -C & -C & -C \\
-C & -C & C & -C \\
-C & C & -C & -C \\
-C & C & -C & -C \\
-C & -C & -C & C \\
-C & C & -C & -C \\
-C & C & -C & -C \\
C & -C & -C & -C \\
-C & C & -C & -C \\
-C & C & -C & -C
\end{pmatrix}
$$

Fig. 3 ***Representation in the algebraic model of Tomita's 4th grammar of Figure 2.***

6 The string recognition procedure

Once the configuration and weights of the SORN associated to the given deterministic FSA are determined, the network can be used to check whether a given string belongs or not to the corresponding regular language. To that end, the SORN must be initialized by loading the weights obtained as solution of the linear system, and setting the outputs of all neurons to reproduce the code of the start state of the FSA.

At each step, which is equivalent to perform a transition of the automaton, a new symbol is fed into the input units, and the activation values of the state units are computed through the network recurrent equations (1), or (6) if additional non-linear inputs are needed due to a non-unary state encoding (these must be calculated from the codes of the current state and input symbol and fed to the network previously). The SORN recognize a valid string when after processing the end-of-string symbol $, the code of the start state is represented in the state units. Otherwise, the garbage state will have been reached and the string is rejected. In this way, a chain of positive examples can be recognized by continuous running of the network. However, after a negative example is detected (through the garbage state), a restarting step is necessary, which just consists of setting all neurons to code again the start state.

7 Conclusions

We have shown that after some transformations, the representation of a deterministic FSA in a second-order recurrent network can be viewed as a system of linear equations. Given a symbolic description of the FSA, it is possible to fill up the equation coefficients and leave the network weights as the unknown variables. If a unary encoding is followed for state representation, the original linear system has always one determined solution which is easily obtained. For binary (or higher base) encoding, a solution can be reached by increasing the set of network inputs with a computable number of nonlinear functions of the transition symbol and outgoing state codes, as it was suggested in [5] for FSA representation in first-order recurrent networks (FORN). Once the weights of the network are analitically obtained by the former method, the string recognition procedure is similar to that proposed in previous approaches ([2],[8]) and it is based on checking the activation values of the units after presentation of a whole string.

In summary, it has been proved that for any (non-stochastic) deterministic FSA, an associated SORN-type recognizer can be built. Since any regular grammar can be mapped to some deterministic FSA, any regular language generated by some regular grammar can be recognized by a SORN. The absence of symbol output units in SORN, where to place transition probablities, impedes to represent and recognize stochastic regular languages, in contrast to what can be done with FORN, as reported in [5].

References

[1] J.L. Elman "Finding structure in time" *CRL Technical Report 8801* University of California, San Diego, Center of Research in Language, 1988.

[2] C.L. Giles et al. "Learning and extracting finite state automata with second-order recurrent neural networks" *Neural Computation*, vol.4, pp.393-405, 1992.

[3] J.E. Hopfcroft and J.D. Ullman "*Introduction to Automata Theory, Languages and Computation*", p.68, Addison-Wesley, Reading MA, 1979.

[4] L. Miclet "Grammatical Inference" *Chapter 9 "Syntatic and Structural Pattern recognition: Theory and Applications, H. Bunke and A. Sanfeliu editors*, World Scientific, 1990.

[5] A. Sanfeliu and R. Alquezar. "Understanding neural networks for grammatical inference and recognition" *IAPR Int. Workshop on Structural and Syntactic Pattern Recognition*, Bern, August 26-28, 1992.

[6] D. Servan-Schreiber, A. Cleeremans and J.L. McClelland "Graded state machines: the representation of temporal contingencies in simple recurrent networks" *Machine Learning*, vol.7, pp. 161-193, 1991.

[7] A.W. Smith and D. Zipser "Learning sequential structure with the real-time recurrent learning algorithm" *Int. Journal of Neural Systems*, vol.1 n.2, pp.125-131, 1989.

[8] R.L. Watrous and G.M. Kuhn "Induction of finite state languages using second-order recurrent networks" *Neural Computation*, vol.4, pp.406-414, 1992.

[9] R.J. Williams and D. Zipser "A learning algorithm for continually running fully recurrent neural networks" *Neural Computation*, vol.1, pp.270-280, 1989.

CONNECTIONIST MODELS FOR SYLLABIC RECOGNITION IN THE TIME DOMAIN

J. Santos* and R. P. Otero**

*Dept. Enxeñeria Industrial, Universidade da Coruña
E-15405 Ferrol (A Coruña), Spain

**Dept. Computación, Facultade de Informática
Universidade da Coruña
E-15071 A Coruña, Spain

ABSTRACT

In this work we study the possibility of identifying syllables in the temporal domain of the speech signal using connectionist models. We set out a distributed neural network for the separated recognition of vowels and consonants. The network will act as a speech to text translator performing the recognition at the minimum level neccesary for its use as input to a semantic system. We have studied the features that the network shows when the input is a temporal-based or a frecuency-based signal.

1. INTRODUCTION

In this work we present a connectionist model for speech recognition that will be integrated with a system for semantic analysis of Natural Language. The Neural Network will act as a speech to text translator performing the recognition at the minimum level neccesary for its use as input to the semantic system. In particular, the identification of all the phonemes produced, in a first stage, by a single speaker, but in different speech situations, is required in order to deal with real speech. From the point of view of the whole system, there is no need for a very accurate recognition at the network level because the semantic system can solve ambiguities and mistakes. We must take into account that 'holes' -missing phonemes- and mistakes are likely to be present in real speech thus a phoneme oriented recognition alone cannot solve the problem.

We choose a connectionist approach as interface to speech for three main reasons. Firstly, the well known adequacy of Neural Networks for pattern recognition without needing to consider a study of the features of the pattern at a deep level. Secondly, and more importantly is the use of the learning properties of the connectionist models to endow the whole system with the ability to adapt and deal with many speakers through a learning process that can be supervised by the semantic system. We could even include the identification of the speaker. Finally, the last reason is to check the connectionist model as a knowledge acquisition system. Our aim is to translate the knowledge from the network representation into a symbolic representation that is more useful for system integration. The interpretation of the states of the hidden units [Hinton 86] together with the weights is the first step for carrying out this traslation.

Few neural networks have been developed for the identification of all the phonemes in a language. It is well known that the network quality strongly depends on the number of patterns it must deal with.

We present here a neural network of distributed architecture in which we have studied the features it shows when the input was a temporal-based signal or a frecuency-based signal. Some preprocessing of the speech signal was carried out in order to help the network in the recognition task.

2. THE ARCHITECTURE OF THE NETWORK

After attempting to handle the many-phoneme recognition problem using a single network and, going to an inadecuate low level of recognition, we have moved on to a distributed architecture. A first 'vowel' network is trained to detect the five vowel patterns in the Spanish language. The output of this first network activates just one of the second five networks, one for each vowel. These second layer networks are trained to detect the syllables in which a given vowel appears. The whole distributed network acts as a syllabic recognition system simultaneously detecting the consonants and the vowel forming the syllable. The syllables including two or more vowels will be considered separatly for each vowel (semisyllable).

All the subnetworks present a similar architecture, a single hidden layer and backpropagation learning [Rumelhart et al. 86]. These networks consists of 64 input units and 50 units in the hidden layer. The vowel network has five output units and the syllabic networks have 17, one for each Spanish syllable of two phonemes with initial consonant.

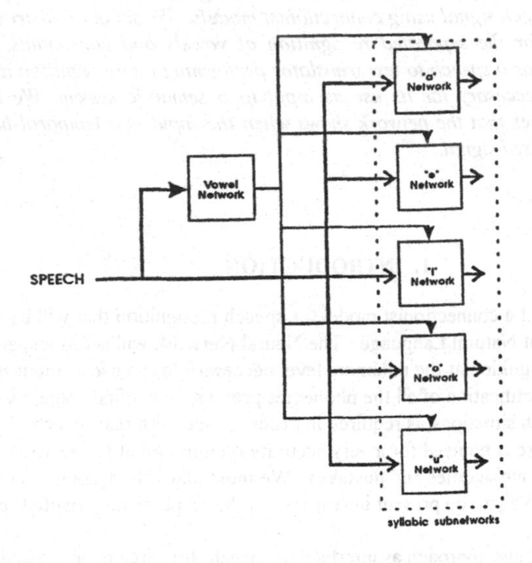

Fig. 1 Distributed network to detect the vowel and the consonants in a syllable.

3. THE INPUT TO THE NETWORK

We start from a speech signal sampled at a low rate, about 8 KHz, such as those that can be obtained using low cost machines. We have considered different ways of presenting the signal to the network. Of special importance was the decision of chosing between time-based or frecuency-based signals.

Inspired in other methods for speech recognition, in particular, Dynamic Time Warping [Bellman 72] that pay particular attention to the temporal alignment of the signal, we have studied the speech signal and we have carried out a light preprocessing on it. The aim is to unload the network of signal dissimilarities that are not directly related to speech itself but to the way the signal is placed in the input units.

First we identify the microphonemes present in a syllable signal and we present each one of them to the network separately. Due to the different frecuencies of the five vowels these microphonemes are of different lengths. A linear transformation is carried out for each microphoneme in order to traslate the five lengths in an intermediate one we have established. Then the signal is placed in the input units of the network, spreading it out over the whole input set. On performing this transformation we have lost the information of the particular microphoneme length, retaining only a normalized representation of its internal features. This loss of information may be considered minimal not affecting the quality of the network because the length of a microphoneme only depends on the vowel of the syllable and not on the consonants. It is well known that the vowels can be easyly identified. The benefits of presenting a normalized microphoneme to the network are very important. The network becomes independent from the 'base frecuency' of the speaker that may be very variable. More relevant is that the need for translating to the frecuency domain can be avoided because a network trained on time-based normalized microphonemes obtains similar results.

The advantages of microphoneme normalization can also appear in the frecuency domain. In figure 2 we show two FFTs over the same signal. In the first one we have considered the 'classical' method [Kohonen 88, Waibel 89], that is, obtained the FFT at time intervals of fixed length, about 10 msecs -greater than the longest microphoneme- and we have represented all the FFTs for the whole syllabic signal in the same graph to show the similarities between them. In the second part of the figure the individual FFTs have been obtained for the normalized microphonemes.

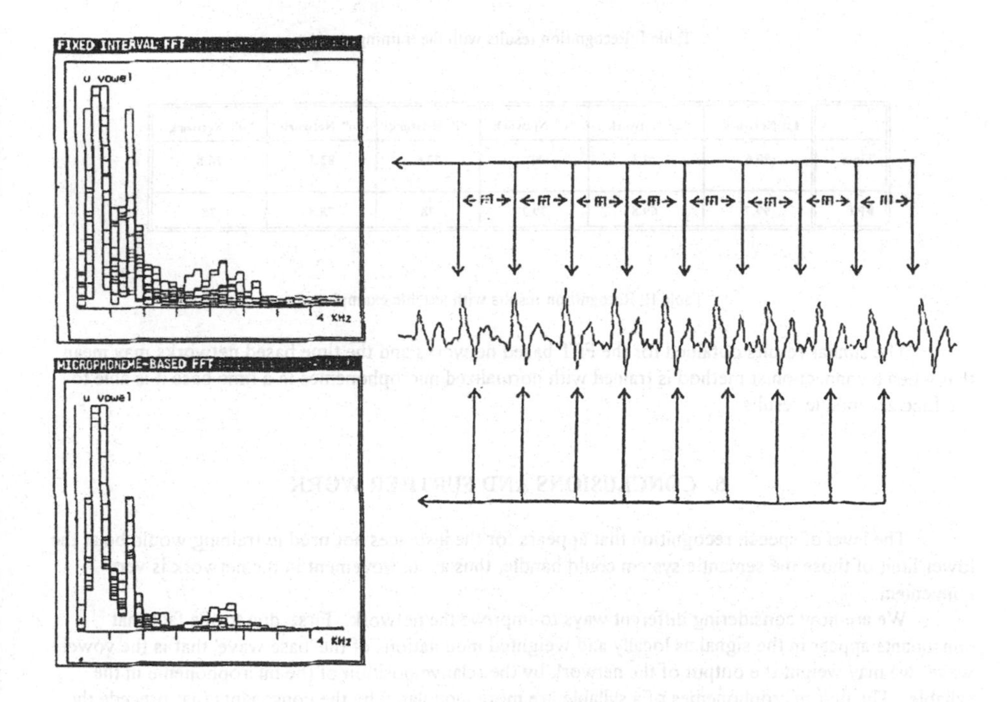

Fig. 2 FFT based on fixed intervals and on normalized microphonemes.

4. NETWORK TRAINING AND RESULTS

From a set of five instances for each syllable of a single speaker, we presented the network for training with the first six microphonemes of the syllable. The vowel network is trained on all the syllables, but for the second level syllabic network only the one corresponding to the known vowel is trained.

We have constructed two networks with the same training set and architecture but with different input signal representations. One network received the normalized microphonemes in a temporal base and the other network received the FFT of the normalized microphonemes. Table I shows the recognition results using this training set for both networks and table II shows the average network recognition on speech instances not used in training. The relative low level on the recognition of the network can be explained by the large number of patterns the syllabic networks must consider and the very similar features that the microphonemes of some different syllables present. Another reason may be the different instances of syllables we have used in training that have been taken from different intonation samples, in order to increase the performance of the network in real speech recognition.

	1st Network	"a" Network	"e" Network	"i" Network	"o" Network	"u" Network
Time	100	89	94	87	80	83
FFT	100	86.5	90.8	85.4	80.6	80

Table I. Recognition results with the training set.

	1st Network	"a" Network	"e" Network	"i" Network	"o" Network	"u" Network
Time	99.6	70.2	80	77.6	82.2	76.8
FFT	99.7	69.8	79.2	78	78.8	78

Table II. Recognition results with testable examples.

The similar results obtained for the FFT based networks and the time based networks may mean that when a connectionist method is trained with normalized microphonemes in a time base it is able to produce acceptable results.

5. CONCLUSIONS AND FURTHER WORK

The level of speech recognition that appears for the instances not used in training would be at the lower limit of those the semantic system could handle, thus an improvement in the network is very convenient.

We are now considering different ways to improve the network. First, due to the fact that consonants appear in the signal as locally and weighted modulations of the 'base wave' that is the vowel wave, we may weight the output of the network by the relative position of the microphoneme in the syllable. The first microphonemes of a syllable are more modulated by the consonants that precede the vowel and the microphonemes of the end of the syllable are modulated by its final consonants, whareas the intermediate microphonemes are almost 'vowel waves'. We are now considering this weigthed output in the training process of the network. A related improvement may be made on syllables of some unvoiced consonants which do not appear as modulations of the vowel wave but as separate waves before or after the microphoneme zone. These consonants are better recognized in non syllabic oriented systems and perhaps their recognition must be improved using a separate network with input from the boundaries of the micophoneme zone.

Increasing the sampling frecuency can also improve the system, because microphoneme detection as a local maxima detection algorithm depends on the sampling rate near the maxima. For example, the e vowel signal has two very close and similar local maxima at the begining of their microphonemes.

One of our aims in developing a connectionist system for speech interfaces was to find a method for knowledge translation from the connectionist representation to a symbolic one. Figure 3 shows a representation of the states of the units and network weights for a network with a few syllables.

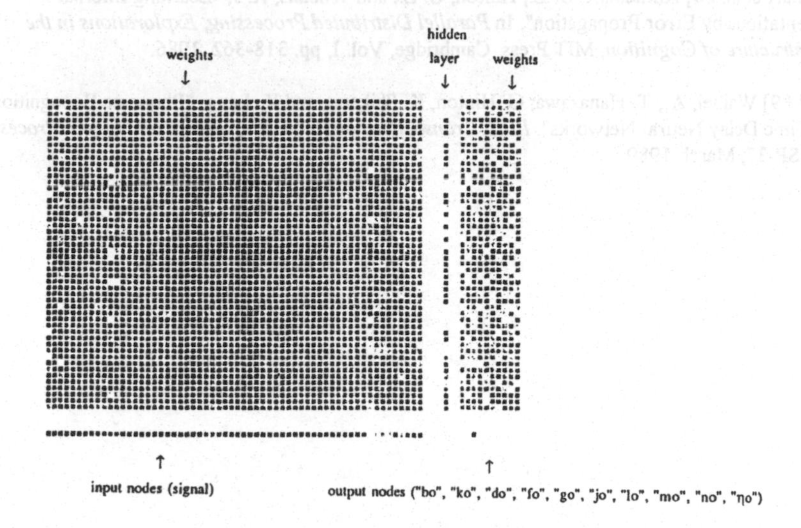

Fig. 3 "O" network schematized with grey levels.

We have not seen any way to interpret the knowledge representation of the network from a symbolic point of view. For system integration considerations we have moved on to another approach consisting in representing the general connectionist mechanisms in the Generalized Magnitudes knowledge representation scheme [Otero 91, Otero and Mira 88] on which the semantic analysis system is based.

ACKNOWLEDGEMENTS

We acknowledge the support of the Government of Galicia under project XUGA10502A92 and the support of the University of A Coruña under a project directed by the authors.

REFERENCES

[Bellman 72] Bellman, R. "*Dynamic Programming*". Princenton University Press, 1972.

[Hinton 86] Hinton, G. E., Learning Distributed Representations of Concepts, in: *Proceedings Eighth Annual Conference of the Cognitive Science Society*, Amherst, MA, 1986.

[Kohonen 88] T. Kohonen, "*Self-Organization and Associative Memory*", Series in Information Sciences, Vol. 8, Springer-Verlag, Berlin-Heildelberg-New York-Tokyo, 1984; 2nd ed. 1988.

[Otero and Mira 88] Otero, R.P. and Mira, J., MEDTOOL: A teachable Medical Expert System Development tool, in: *Proceedings Third International Symposium on Knowledge Engineering*, Madrid, Spain, 1988 191-200.

[Otero 91] Otero, R.P., *MEDTOOL: Una Herramienta para el desarrollo de Sistemas Expertos*, Tesis Doctoral, Univ. de Santiago, Dpto. Electronica y Computacion, Santiago de Compostela, 1991.

[Rumelhart et al. 86] Rumelhart, D. E., Hinton, G. E., and Willians, R. J. "Learning Internal Representations by Error Propagation". In *Parallel Distributed Processing, Explorations in the Microestructure of Cognition*, MIT Press, Cambridge, Vol. I, pp. 318-362, 1986.

[Waibel 89] Waibel, A., T. Hanazawa, G. Hinton, K. Shikano, and K. Lang. "Phoneme Recognition Using Time Delay Neural Networks". *IEEE Transactions on Acoustics, Speech and Signal Procesing*, Vol ASSP-37, March 1989.

SPARSELY INTERCONNECTED ARTIFICIAL NEURAL NETWORKS FOR ASSOCIATIVE MEMORIES *

Derong Liu Anthony N. Michel

Department of Electrical Engineering
University of Notre Dame
Notre Dame, IN 46556, U. S. A.

Abstract

We develop in the present paper a design procedure for neural networks with *sparse* coefficient matrices. Our results guarantee that the synthesized neural networks have *predetermined* sparse interconnection structures and store *any* set of desired memory patterns as reachable memory vectors. We show that a sufficient condition for the existence of a sparse neural network design is self feedback for every neuron in the network. Our design procedure for neural networks with sparse interconnecting structure can take into account various problems encountered in VLSI realizations of such networks. For example, our procedure can be used to design neural networks with few or without any line-crossings resulting from the network interconnections. Several specific examples are included to demonstrate the applicability of the methodology advanced herein.

1 Introduction

In the present paper, we consider neural networks described by equations of the form

$$\begin{cases} \frac{dx}{dt} = -Ax + T\text{sat}(x) + I \\ y = \text{sat}(x) \end{cases} \quad (1)$$

where $x \in R^n$, $y \in D^n \triangleq \{x \in R^n: -1 \le x_i \le 1, i = 1, \cdots, n\}$, $A = \text{diag}[a_1, \cdots, a_n]$ with $a_i > 0$ for $i = 1, \cdots, n$, $T = [T_{ij}] \in R^{n \times n}$, $I = (I_1, \cdots, I_n)^T \in R^n$, $\text{sat}(x) = [\text{sat}(x_1), \cdots, \text{sat}(x_n)]^T$, and $\text{sat}(x_i) = 1$ if $x_i > 1$, $\text{sat}(x_i) = x_i$ if $-1 \le x_i \le 1$, and $\text{sat}(x_i) = -1$ if $x_i < -1$. We assume that the initial states of (1) satisfy $|x_i(0)| \le 1$ for $i = 1, \cdots, n$. System (1) is a variant of the analog Hopfield model with a piecewise linear activation function sat(·). In the analog Hopfield model [4], one requires that $T_{ii} = 0$ for $i = 1, \cdots, n$. We do not make this assumption for (1).

In the present paper, we concern ourselves primarily with the implementation of *associative memories* by means of artificial neural networks (modeled by (1)). One of the major difficulties encountered in VLSI implementations of artificial neural networks is the realization of extremely large numbers of interconnections in the networks. To reduce the number of connections is of great interest from a practical point of view. Most of the existing synthesis procedures for associative memories [2], [3], [5]–[10], [12] were developed for fully interconnected neural networks and none of them result in neural networks with *prespecified* partial or sparse interconnection structure. Synthesis procedures for neural networks with arbitrarily (prespecified) sparse interconnection structure, or equivalently, with sparse coefficient matrices constitute a major addition to the development of neural network theory, and such procedures will have potentially many practical applications, especially in the areas of associative memories and pattern recognition (we will define the exact meaning of sparse coefficient matrix later).

In Section 2, we will develop a synthesis procedure for neural networks with sparse coefficient matrices in which the interconnection structure is predetermined. In Section 3, we consider a specific example to demonstrate the applicability of our synthesis procedure. Special emphasis is placed on networks with different sparse interconnection structures. We conclude with several pertinent remarks in Section 4.

*This work was supported by the National Science Foundation under grant ECS 91-07728.

2 Synthesis Procedure for Sparsely Interconnected Neural Networks

The synthesis techniques developed in [6], [9] will result in neural networks with *symmetric* and *non-sparse* coefficient matrix T. Implementations of artificial neural networks with *perfectly* symmetric interconnections are not practical. Furthermore, it has been argued by some workers [3], [11] that symmetric interconnections in artificial neural networks are not necessarily always desirable. Moreover, artificial neural networks with even a moderate number of interconnections will give rise to large numbers of line-crossings, and thus pose formidable obstacles in VLSI implementations. For these reasons, it is desirable to establish a synthesis procedure which will result in an interconnecting structure which does not require symmetry and which does not demand large numbers of connections.

By adapting the synthesis techniques (the eigenstructure method) of [6] and [9], we develop in the following a design procedure for neural networks which will result in few line-crossings or no line-crossings at all in the interconnections, and which does not require that the interconnection matrix be symmetric. *Cellular neural networks* [1] are special cases of such *sparsely interconnected artificial neural networks*.

We let $C(\alpha) = \{x \in R^n: x_i\alpha_i > 1, i = 1, \cdots, n\}$ for each $\alpha \in B^n \triangleq \{x \in R^n: x_i = 1 \text{ or } -1, i = 1, \cdots, n\}$. We give next a result which serves as the basis for our synthesis procedure.

Lemma 1 Suppose $\alpha \in B^n$. If $\beta = A^{-1}(T\alpha + I) \in C(\alpha)$, then β is an asymptotically stable equilibrium point of (1).

Proof: Since $\alpha_i = \pm 1$ for all i, we see that $\text{sat}(x) = \alpha$ for all $x \in C(\alpha)$. For $x \in C(\alpha)$, the first equation of (1) can be written as

$$\dot{x} = -Ax + T\alpha + I. \tag{2}$$

System (2) has a unique equilibrium at $x_e = A^{-1}(T\alpha + I)$, and $x_e = \beta \in C(\alpha)$ by assumption. Clearly, this equilibrium is also asymptotically stable, since system (2) has n negative real eigenvalues. ∎

If $\xi \in R^n$ is an asymptotically stable equilibrium point of (1), then $y_\xi = \text{sat}(\xi)$ is said to be a *memory vector* of (1). A memory vector y_ξ of (1) is said to be *reachable*, if there exists a neighborhood V of y_ξ, such that for any $x(0) \in V \cap D^n \neq \phi$, the output vector $y(t)$ of (1) tends to y_ξ asymptotically as $t \to \infty$.

A matrix $S = [S_{ij}] \in R^{n \times n}$ is said to be an *index matrix*, if it satisfies $S_{ij} = 1$ or 0. The restriction of matrix $W = [W_{ij}] \in R^{n \times n}$ to an index matrix S, denoted by $W|S$, is defined by $W|S = [h_{ij}]$, where $h_{ij} = W_{ij}$ if $S_{ij} = 1$ and $h_{ij} = 0$ otherwise. We will say that (1) is a *neural network with a sparse coefficient matrix* if $T = T|S$ for some given index matrix S.

We are now in a position to address the following synthesis problem.

Sparse Design Problem: Given an $n \times n$ index matrix $S = [S_{ij}]$ with $S_{ii} \neq 0$ for $i = 1, \cdots, n$, and m vectors $\alpha^1, \cdots, \alpha^m$ in B^n, choose $\{A, T, I\}$ with $T = T|S$ in such a manner that $\alpha^1, \cdots, \alpha^m$ are reachable memory vectors of system (1). ∎

A solution for the above sparse design problem is as follows.

Sparse Design Procedure 2.1: Suppose we are given an $n \times n$ index matrix $S = [S_{ij}]$ with $S_{ii} \neq 0$ for $i = 1, \cdots, n$, and m vectors $\alpha^1, \cdots, \alpha^m$ in B^n which are to be stored as reachable memory vectors for system (1). We proceed as follows:

1) Choose matrix A as the identity matrix.

2) Choose a real number $t > 1$ and m vectors $\beta^1, \cdots, \beta^m$, such that $\beta^i = t\alpha^i$.

3) Compute the $n \times (m-1)$ matrices $Y = [y^1, \cdots, y^{m-1}] = [\alpha^1 - \alpha^m, \cdots, \alpha^{m-1} - \alpha^m]$ and $Z = [z^1, \cdots, z^{m-1}] = [\beta^1 - \beta^m, \cdots, \beta^{m-1} - \beta^m]$. We let $y^i = (y_1^i, \cdots, y_n^i)^T$ and $z^i = (z_1^i, \cdots, z_n^i)^T$.

4) Denote the ith row of the index matrix S by $S_i = (S_{i1}, \cdots, S_{in})$. For each $i = 1, \cdots, n$, construct two sets M_i and N_i, such that $M_i \cup N_i = \{1, \cdots, n\}$, $M_i \cap N_i = \phi$, and $S_{ij} = 1$ if $j \in M_i$, $S_{ij} = 0$ if $j \in N_i$. Let $M_i = \{\sigma_i(1), \cdots, \sigma_i(m_i)\}$, where $m_i = \sum_{j=1}^n S_{ij}$ and $\sigma_i(k) < \sigma_i(l)$ if $1 \leq k < l \leq m_i$. (Note that m_i is the number of nonzero elements in the ith row of matrix S.)

5) For $i = 1, \cdots, n$ and $l = 1, \cdots, m-1$, let $y_{Ii}^l = (y_{\sigma(1)}^l, \cdots, y_{\sigma(m_i)}^l)^T$.

6) For $i = 1, \cdots, n$, compute the $m_i \times (m-1)$ matrices $Y_i = [y_{Ii}^1, \cdots, y_{Ii}^{m-1}]$ and the $1 \times (m-1)$ vectors $Z_i = [z_i^1, \cdots, z_i^{m-1}]$.

7) For $i = 1, \cdots, n$, perform singular value decompositions of Y_i, and obtain

$$Y_i = [U_{i1} \vdots U_{i2}] \begin{bmatrix} D_i & \vdots & 0 \\ \cdots & \vdots & \cdots \\ 0 & \vdots & 0 \end{bmatrix} \begin{bmatrix} V_{i1}^T \\ \cdots \\ V_{i2}^T \end{bmatrix},$$

where $D_i \in R^{p_i \times p_i}$ is a diagonal matrix with the nonzero singular values of Y_i on its diagonal and $p_i = \text{rank}(Y_i)$.

8) Compute for $i = 1, \cdots, n$, $G_i = [G_{i1}, \cdots, G_{im_i}] = Z_i V_{i1} D_i^{-1} U_{i1}^T + W_i U_{i2}^T$, where W_i is an arbitrary $1 \times (m_i - p_i)$ real vector.

9) The matrix $T = [T_{ij}]$ is computed as $T_{ij} = 0$ if $S_{ij} = 0$, and $T_{ij} = G_{ik}$ if $S_{ij} \neq 0$ and if $j = \sigma_i(k)$.

10) The bias vector $I = (I_1, \cdots, I_n)^T$ is computed by $I_i = \beta_i^m - T_i \alpha^m$ for $i = 1, \cdots, n$, where T_i is the ith row of T.

Then, $\alpha^1, \cdots, \alpha^m$ will be stored as memory vectors for system (1) with A, T, and I determined as above. The states β^i corresponding to α^i, $i = 1, \cdots, m$, will become the asymptotically stable equilibrium points of the synthesized system. ∎

Our next result addresses the existence of a solution for the sparse design problem and the validity of the above design procedure.

Theorem **1) Solutions for the sparse design problem always exist if $S_{ii} = 1$ for $i = 1, \cdots, n$.**

2) The Sparse Design Procedure 2.1 guarantees that $T = T|S$.

3) The Sparse Design Procedure 2.1 guarantees that $\alpha^1, \cdots, \alpha^m$ are stored as reachable memory vectors of system (1).

4) The Sparse Design Procedure 2.1 can be applied to any set of desired memory patterns in B^n.

Proof: **In order for the synthesized system to be a solution of the sparse design problem, we need $G_i Y_i = Z_i$ in steps 7 and 8 of the sparse design procedure. Thus, G_i in step 8 is a solution for the sparse design procedure** ***if and only if***

$$\text{rank}[Y_i] = \text{rank} \begin{bmatrix} Y_i \\ \cdots \\ Z_i \end{bmatrix}.$$

This condition is satisfied if $S_{ii} = 1$, $i = 1, \cdots, n$, since under these conditions, Z_i becomes a row vector which is one of the rows in Y_i multiplied by t. This proves part 1 of the theorem.

Part 2 is clear from step 9.

To prove part 3, we first check the equilibrium conditions for $\alpha^1, \cdots, \alpha^m$, in which case we require that $T\alpha^l + I = A\beta^l = \beta^l$, for $l = 1, \cdots, m$, where $\beta^l = t\alpha^l$ and $t > 1$ ($\beta^l \in C(\alpha^l)$). Using the notation given in the design procedure, we write for $l = 1, \cdots, m-1$, $y_{|i}^l = Y_i e_l$, where $e_l \in R^{m-1}$ is a column vector with all elements zero except the lth element which is 1. According to the design procedure and the properties of singular value decomposition, we compute, for $i = 1, \cdots, n$, $U_{i2}^T y_{|i}^l = 0$ and

$$\begin{aligned} T_i\alpha^l + I_i &= T_i y_i^l + T_i\alpha^m + I_i = G_i y_{|i}^l + T_i\alpha^m + \beta_i^m - T_i\alpha^m \\ &= Z_i V_{i1} D_i^{-1} U_{i1}^T Y_i e_l + \beta_i^m = Z_i e_l + \beta_i^m = z_i^l + \beta_i^m = \beta_i^l - \beta_i^m + \beta_i^m = \beta_i^l. \end{aligned}$$

Hence, $T\alpha^l + I = \beta^l$ for $l = 1, \cdots, m-1$. For $l = m$, $T\alpha^m + I = \beta^m$ is clear from step 10. By Lemma 1, we see that $\alpha^1, \cdots, \alpha^m$ are stored as memory vectors for (1). The states β^i corresponding to α^i, $i = 1, \cdots, m$ will become the asymptotically stable equilibrium points of the synthesized system. By choosing $t > 1$ sufficiently small in step 2, we can guarantee that $\alpha^1, \cdots, \alpha^m$ become reachable memory vectors.

Part 4 follows from the design procedure since we do not have any restrictions on m as long as $\alpha^1, \cdots, \alpha^m$ are in B^n. ∎

Remark 1 **For given positive integers (n, M, N, r), with $n = MN$, choose the index matrix $S = Q = [S_{ij}] \in R^{n \times n}$, where $Q = [Q_{ij,kl}] \in R^{MN \times MN}$ with $Q_{ij,kl} = 1$ if $(k,l) \in N_r(i,j)$ and $Q_{ij,kl} = 0$ otherwise, and $N_r(i,j) \triangleq \{(k,l)\colon \max\{|k-i|, |l-j|\} \leq r, 1 \leq k \leq M, 1 \leq l \leq N\}$. When defined in this way, we see that in system (1), if $T = T|S$, then (1) is equivalent to a nonsymmetric cellular neural network [1].** ∎

3 An Example

To demonstrate the applicability of the present results, we consider the following example.

Example We now present several problems which to the best of our knowledge cannot be addressed by other synthesis procedures for associative memories. In all cases, we consider a neural network with 16 neurons ($n = 16$) and in all cases our objective is to store the four patterns shown in Fig. 1 as reachable memories. As indicated in this figure, sixteen boxes are used to represent each pattern (in R^{16}), with each box corresponding to a vector component which is allowed to assume values between -1 and 1. For purpose of visualization, -1 will represent white, 1 will represent black, and the intermediate values will correspond to appropriate grey levels, as shown in Fig. 2.

The four cases which we consider below, were synthesized by the Sparse Design Procedure 2.1. These cases involve different prespecified constraints on the interconnecting structure of each network.

Case I: Cellular neural network. We designed a cellular neural network (cf. Remark 1) with $r = 1$, $M = 4$, and $N = 4$. (Due to space limitations, we will not display the interconnecting matrix T for the present case, as well as for the three subsequent cases. For more specifics, see [7].) The performance of this network is illustrated by means of a typical simulation run of equation (1), shown in Fig. 3. In this figure, the desired memory pattern is depicted in the lower right corner. The initial state, shown in the upper left corner, is generated by adding to the desired pattern zero-mean Gaussian noise with a standard deviation SD=1. The iteration of the simulation evolves from left to right in each row and from the top row to the bottom row. The desired pattern is recovered in 14 steps with a step size $h = 0.2$. All simulations for the present paper were performed on a Sun SPARC Station using MATLAB.

Case II: Reduction of line-crossings. We arranged the 16 neurons in a 4×4 array and we considered only horizontal and vertical interconnections. For this case, the index matrix $S = Q = [S_{ij}] \in R^{16\times16}$ and $Q = [Q_{ij,kl}] \in R^{(4\times4)\times(4\times4)}$ assumes the form

$$Q_{ij,kl} = 1 \text{ if } i = k \text{ or } j = l, \text{ and } Q_{ij,kl} = 0 \text{ otherwise} . \tag{3}$$

A typical simulation run for the present case is depicted in Fig. 4. In this figure, the noisy pattern is generated by adding to the desired pattern uniformly distributed noise defined on $[-0.5, 0.5]$. Convergence occurred in 7 steps with $h = 0.2$. We emphasize that by choosing the index matrix as in (3), we were able to reduce significantly the number of line-crossings, which is of great concern in VLSI implementations of artificial neural networks.

Case III: Two rows of S identical. In this case, we chose an index matrix $S = [S_{ij}] \in R^{16\times16}$ of the form $S_{ij} = 1$, if $i = 1$ or $i = 16$ or $j = 1$ or $j = 16$ or $|i - j| \leq 1$, and $S_{ij} = 0$, otherwise. This requires that the T matrix has zero elements everywhere except in its first and last rows, its first and last columns, and in its tridiagonal elements. Rows 2 to 15 are designed by using the Sparse Design Procedure 2.1, step by step. Since the first row S_1 and the last row S_{16} of S are identical, we can design the rows T_1 and T_{16} of T simultaneously. To see this, we take in step 5 of the design procedure $y^l_{I1} = [y^l_{\sigma(1)}, \cdots, y^l_{\sigma(m_1)}]^T$ and $y^l_{I16} = [y^l_{\sigma(1)}, \cdots, y^l_{\sigma(m_{16})}]^T$ for $l = 1, \cdots, m-1$. Clearly, $m_1 = m_{16} = n = 16$ and $y^l_{I1} = y^l_{I16}$ for $l = 1, \cdots, m-1$, since $S_1 = S_{16}$. In step 6, we take $Y_1 = [y^1_{I1}, \cdots, y^{m-1}_{I1}]$ and the $2 \times (m-1)$ vector

$$Z_1 = \begin{bmatrix} z^1_1 & \cdots & z^{m-1}_1 \\ z^1_{16} & & z^{m-1}_{16} \end{bmatrix}.$$

In step 7, we perform a singular value decomposition of Y_1 and obtain U_{11}, U_{12}, D_1, and V_{11}. In step 8, we compute $G_1 = Z_1 V_{11} D_1^{-1} U_{11}^T + W_1 U_{12}^T$, where W_1 is an arbitrary $2\times(m_1 - p_1)$ real matrix and $p_1 = \text{rank}(Y_1)$. In step 9, we determine T_1 from the first row of G_1 and T_{16} from the second row of G_1. A typical simulation run for this network is shown in Fig. 5. In this case, the noisy pattern was generated by adding Gaussian noise $N(0, 0.5)$ to the desired pattern. Convergence occurred in 24 steps with $h = 0.2$.

Case IV: Quinquediagonal matrix S resulting in an interconnecting structure without line-crossings. We chose $S = [S_{ij}] \in R^{16\times16}$ as

$$S_{ij} = 1 \text{ if } |i - j| \leq 2 \text{ and } S_{ij} = 0 \text{ otherwise} . \tag{4}$$

This will result in a quinquediagonal matrix S, enabling us to arrange the $n = 16$ neurons in the configuration shown in Fig. 6. Note that in this figure there are no line-crossings. Furthermore, note that this configuration can be generalized to arbitrary n. A typical simulation run for the present case is depicted in Fig. 7. In this figure, the noisy pattern was generated by adding Gaussian noise $N(0, 0.7)$ to the desired pattern. Convergence occurred in 13 steps with $h = 0.2$. ∎

We note that for the above example, many other interesting design cases can be addressed in a *systematic* manner, including a neural network (1) with lower or upper triangular matrix T, combinations of Cases I, II, III, and IV given above, and so forth.

4 Conclusions

By adapting the synthesis procedures of [6] and [9], we developed in Section 2 a design procedure for sparsely interconnected neural networks (Sparse Design Procedure 2.1). This procedure results in neural networks which satisfy a *prespecified* interconnecting structure. In Section 3, we demonstrated the applicability and the versatility of the results developed herein by means of a specific example.

The significance of the results presented in this paper is that we can synthesize by the present method neural networks which have a *prespecified interconnecting structure* and which will *guarantee to store any desired set of memory patterns in* B^n as reachable memories provided that the interconnecting structure includes self feedback for all neurons.

References

[1] L. O. Chua, L. Yang, "Cellular Neural Networks: Theory," *IEEE Transactions on Circuits and Systems*, Vol. 35, pp.1257–1272, Oct. 1988

[2] S. R. Das, "On the Synthesis of Nonlinear Continuous Neural Networks," *IEEE Transactions on Systems, Man, and Cybernetics*, Vol. 21, pp.413–418, March/Apr. 1991

[3] J. A. Farrell, A. N. Michel, "A Synthesis Procedure for Hopfield's Continuous-Time Associative Memory," *IEEE Transactions on Circuits and Systems*, Vol. 37, pp.877–884, July 1990

[4] J. J. Hopfield, "Neurons with graded response have collective computational properties like those of two-state neurons," *Proc. Nat. Acad. Sci. USA*, Vol. 81, pp.3088–3092, May 1984

[5] J.-H. Li, A. N. Michel, W. Porod, "Analysis and Synthesis of a Class of Neural Networks: Variable Structure Systems with Infinite Gain," *IEEE Transactions on Circuits and Systems*, Vol. 36, pp.713–731, May 1989

[6] J.-H. Li, A. N. Michel, W. Porod, "Analysis and Synthesis of a Class of Neural Networks: Linear Systems Operating on a Closed Hypercube," *IEEE Transactions on Circuits and Systems*, Vol. 36, pp.1405–1422, Nov. 1989

[7] Derong Liu, A. N. Michel, "Sparsely Interconnected Neural Networks for Associative Memories with Applications to Cellular Neural Networks," Submitted to *IEEE Transactions on Circuits and Systems*

[8] A. N. Michel, J. A. Farrell, "Associative Memories via Artificial Neural Networks," *IEEE Control Systems Magazine*, Vol. 10, pp.6–17, Apr. 1990

[9] A. N. Michel, J. Si, G. Yen, "Analysis and Synthesis of a Class of Discrete-Time Neural Networks Described on Hypercubes," *IEEE Transactions on Neural Networks*, Vol. 2, pp.32–46, Jan. 1991

[10] L. Personnaz, I. Guyon, G. Dreyfus, "Collective Computational Properties of Neural Networks: New Learning Mechanisms," *Physical Review A*, Vol. 34, pp.4217–4228, Nov. 1986

[11] F. M. A. Salam, Y. Wang, M.-R. Choi, "On the Analysis of Dynamic Feedback Neural Nets," *IEEE Transactions on Circuits and Systems*, Vol. 38, pp.196–201, Feb. 1991

[12] G. Yen, A. N. Michel, "A Learing and Forgetting Algorithm in Associative Memories: The Eigenstructure Method," *IEEE Transactions on Circuits and Systems-II: Analog and Digital Signal Processing*, Vol. 39, pp.212–225, Apr. 1992

Figure 1. The four desired memory patterns used in the Example

−1 0 1

Figure 2 Grey levels

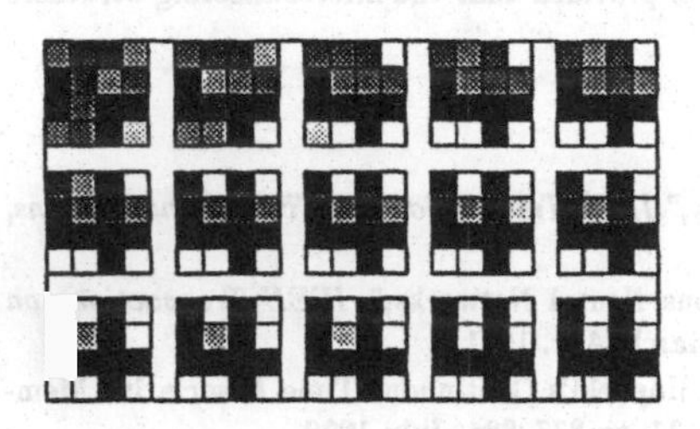

Fig. 3 A typical evolution of pattern No. 1 of Fig. 1

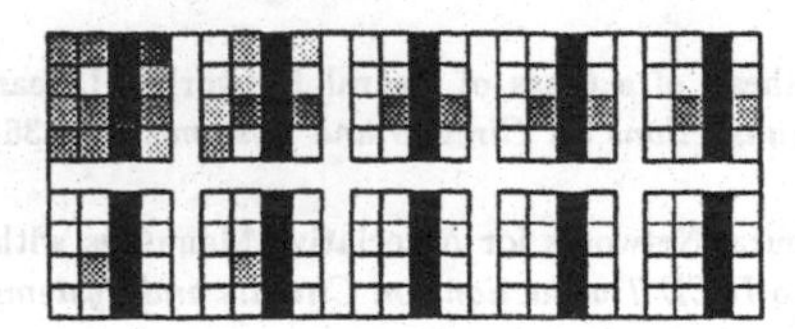

Fig. 4 A typical evolution of pattern No. 2 of Fig. 1

Figure 5. A typical evolution of pattern No. 3 of Figure 1

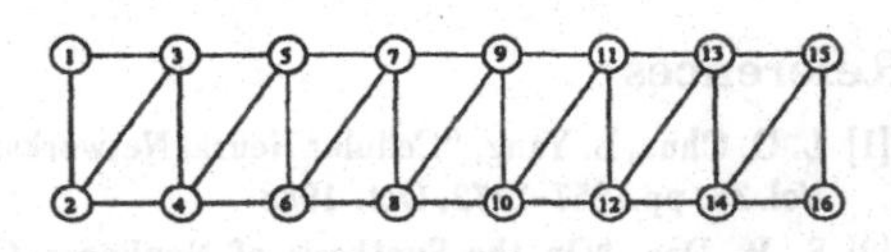

Fig. 6. A possible structure for a neural network without line-crossings in the interconnecting structure

Fig. 7 A typical evolution of pattern No. 4 of Fig. 1

DYNAMIC ANALYSIS OF NETWORKS OF NEURAL OSCILLATORS

A.Arenas and C.J. Pérez Vicente
Departament de Física Fonamental, Facultat de Física
Universitat de Barcelona,
Diagonal 647, 08028 Barcelona, Spain.

Abstract

We present here a dynamic formalism which allow to compute analitically the stationary properties of networks of neural oscillators. This technique, derived originally to study situations away from equilibrium, is an alternative to standard methods developed to analyze the behaviour of attractor neural networks.

1 Introduction

The analysis of the dynamical properties of attractor neural networks (ANN) are the focus of important works in the last years. Not only because it can provide information about the short and long time behaviour of networks characterized by symmetric and asymmetric couplings but also because they are the only way to understand the nature of some collective phenomena, such as mutual synchronization in the temporal activity of large assemblies of neurons [1], which are responsible of interesting effects related to the processing of information observed in real experiments performed in the visual cortex of monkeys [2].

The conventional models of ANN characterize the activity of the neurons through binary values, corresponding to the active and non-active state of each neuron [3]. However, in order to reproduce synchronization between members of a population it is convenient to introduce new variables which could provide information about the degree of coherence in the temporal response of active neurons. A possible way to do this, is by associating a phase to each element of the system and consequently to model neurons as oscillators. One of the most common models of phase oscillators is the so called Kuramoto's model [4], whose dynamics is governed by

$$\frac{d\theta_i}{dt} = \omega_i + \gamma_i(t) + \sum_{j=1}^{N} K_{ij} \sin(\theta_j - \theta_i) \tag{1}$$

where K_{ij} is the coupling matrix, θ_i the phase of the i-th oscillator, ω_i is a random frequency for each oscillator that obeys a certain distribution $g(\omega)$, N the size of the population and $\gamma_i(t)$ independent white noise random processes with zero mean and correlation

$$< \gamma_i(t)\gamma_j(t') >= 2D\delta_{ij}\delta(t-t'), \quad D \geq 0 \tag{2}$$

An important point in our discussion is the specific form proposed for the couplings since it is the bridge that allows to make the analogy between models of phase oscillators and ANN. After a suitable choice of K_{ij} it is possible to wonder about the ability of the system to work as an associative memory [5].

Let us consider a population of N neurons, active at high rate during a given period of time that can carry information about their phase. As usual in ANN we want to store p sets of random patterns (phases) $\{\xi\}$ and a simple way to do this task is to assume that the synaptic efficacies (couplings) are given by

$$K_{ij} = \frac{k}{N} \sum_{\mu=1}^{p} \cos(\xi_i^\mu - \xi_j^\mu) \tag{3}$$

where k is the intensity of the coupling. This form preserves the basic idea of the Hebb´s rule but now adapted to the symmetry of our problem.

Our goal is to determine the stationary properties of the model described by equations (1) and (3) through a mean-field formalism widely used in the analysis of large populations of coupled oscillators [6], but new in the treatment of the features of ANN. We will show that this technique is an excellent alternative to conventional methods of analysis of associative memories, such as the replica method.

Notice that when the distribution of frequencies vanishes ($g(\omega) = \delta(\omega)$) our neurons are no longer oscillators. In this case our system becomes a Q-state clock model of neural network in the limit $Q \to \infty$, which has been extensively studied by Cook [7] in the replica symmetry approximation. We will show that with our method it is possible to reproduce the results of [7] in a simpler way, emphasizing the relevant influence of $g(\omega)$ on the long time properties of the system.

Description of problem and results

In this preliminar study we have only considered the behaviour of the system in the low loading limit, i.e., in the case where the capacity α=p/N, defined as the ratio between the number of patterns and the number of units of the system, goes to zero. To analyze our model it is convenient to introduce the following order parameters

$$q^{\mu}_{\pm} e^{i\phi^{\mu}_{\pm}} = \frac{1}{N} \sum_j e^{i(\theta_j \pm \xi^{\mu}_j)} \tag{4}$$

$\phi^{\mu}_{\pm}$ play the role of a mean phase, q^{μ}_{-} measures the correlation between the state of the system and the pattern ξ^{μ}, and q^{μ}_{+} is another correlation not relevant in our study. Then the evolution equation for the phase oscillators is

$$\frac{d\theta_i}{dt} = \omega_i + \frac{k}{2} \sum_{\mu=1}^{p} [q^{\mu}_{-} \sin(\phi^{\mu}_{-} - \theta_i + \xi^{\mu}_i) + q^{\mu}_{+} \sin(\phi^{\mu}_{+} - \theta_i - \xi^{\mu}_i)] + \gamma_i(t) \tag{5}$$

In the thermodynamic limit $N \to \infty$, it is possible to derive a non-linear Fokker-Planck equation for the one oscillator probability density $\rho(\theta, t, \omega, \xi)$

$$\frac{\partial \rho}{\partial t} + \frac{\partial}{\partial \theta} [\mathcal{V} \rho] - D \frac{\partial^2 \rho}{\partial \theta^2} = 0 \tag{6}$$

where $\mathcal{V}$ is the drift velocity term

$$\mathcal{V} = \left[\omega + \frac{k}{2} \sum_{\mu=1}^{p} [q^{\mu}_{+} \sin(\phi^{\mu}_{+} - \theta - \xi^{\mu}) + q^{\mu}_{-} \sin(\phi^{\mu}_{-} - \theta + \xi^{\mu})] \right] \tag{7}$$

If $g(\omega)$ and $f_{\mu}(\xi^{\mu})$ are the frequency and pattern distribution, respectively, the order parameters (4) become

$$q_{\pm}^{\mu}e^{i\phi_{\pm}^{\mu}} = \int \dots \int d\omega d\theta e^{i(\theta \pm \xi^{\mu})} \rho(\theta, \omega, \xi^1, \dots, \xi^p) g(\omega) \prod_{\mu=1}^{p} f_{\mu}(\xi^{\mu}) d\xi^{\mu} \quad (8)$$

Since we are interested in the long time behaviour of the system we have solved the equation (6) for the stationary case

$$\rho(\theta, \omega, \xi^{\mu}) = \frac{F(\theta) \int_0^{2\pi} d\eta H(\theta, \eta)}{\mathcal{Z}} \quad (9)$$

where

$$F(\theta) = exp\left[\frac{k}{2D}\sum_{\mu=1}^{p}[q_{+}^{\mu}\cos(\phi_{+}^{\mu} - \theta - \xi^{\mu}) + q_{-}^{\mu}\cos(\phi_{-}^{\mu} - \theta + \xi^{\mu})]\right] \quad (10)$$

$$H(\theta, \eta) = exp\left[-\frac{\omega\eta}{D} - \frac{k}{2D}\sum_{\mu=1}^{p}[q_{+}^{\mu}\cos(\phi_{+}^{\mu} - \theta - \eta - \xi^{\mu}) + q_{-}^{\mu}\cos(\phi_{-}^{\mu} - \theta - \eta + \xi^{\mu})]\right] \quad (11)$$

and

$$\mathcal{Z} = \int_0^{2\pi} d\theta F(\theta) \int_0^{2\pi} H(\theta, \eta) d\eta \quad (12)$$

Equations (9)-(12) describe the behaviour of the system in the most general case. However since we are interested in the limit of $\alpha \to 0$ we can assume that if the initial state of the system has a macroscopic correlation with a pattern μ, then only the order parameter $q_{-}^{\mu} \equiv q$ will be relevant, what simplifies notably the nature of the problem. The situation with $\alpha \neq 0$ will be considered elsewhere. Notice that q

$$q = < \cos(\phi - \theta - \xi) > \quad (13)$$

plays the role of the overlap in classical models of ANN, i.e., the projection of the state of the system over the memories except for a mean phase ϕ that goes to zero when the distribution of natural frequencies $g(\omega)$ is even and has zero mean.

To calculate q we can proceed in two different manners, either by solving directly equation (8), what is complex because it means to solve an integral equation impliying to get values of q through numerical integration, or by identifying $\mathcal{Z}$ as a generating functional of the order parameters. This method is more elegant and give algebraic expressions easier to deal with. Let us rewrite $\mathcal{Z}$ as

$$\mathcal{Z} = \int_0^{2\pi} d\theta F(\theta,\sigma) \int_0^{2\pi} H(\theta,\eta) d\eta \tag{14}$$

where

$$F(\theta,\sigma) = exp\,[\sigma \cos(\phi - \theta + \xi)] \tag{15}$$

then it is straightforward to see that

$$q = << \frac{\partial}{\partial \sigma} \ln \mathcal{Z} \,|_{\sigma = \frac{kq}{2D}} >> \tag{16}$$

where $<< .. >>$ is an average over ω and ξ. To carry out the calculation we apply the following identity

$$e^{\frac{kq}{2D}\cos(\phi-\theta-\eta+\xi)} = I_0(\frac{kq}{2D}) + 2\sum_{n=1}^{\infty}(-1)^n \cos n(\phi - \theta - \eta + \xi) I_n(\frac{kq}{2D}) \tag{17}$$

Integrating (12), averaging over ξ and evaluating the partial derivate (16) we obtain a self-consistent equation for the q parameter

$$q = < \frac{\frac{D}{\omega} I_0(\beta q) I_1(\beta q) + \sum_1^{\infty} \frac{(-1)^n}{(\omega/D)^2+n^2} I_n(\beta q)(I_{n-1}(\beta q) + I_{n+1}(\beta q))(\frac{\omega}{D})}{\frac{D}{\omega} I_0^2(\beta q) + 2\sum_1^{\infty} \frac{(-1)^n}{(\omega/D)^2+n^2} I_n^2(\beta q)(\frac{\omega}{D})} >_\omega \tag{18}$$

where I_n are the modified Bessel functions of first kind of order n, $\beta = \frac{k}{2D}$ and $<>_\omega$ means an average over the distribution of frequencies. Taking into account the symmetry properties of the modified Bessel functions for n integer ($I_n = I_{-n}$), we can summarize this formula in

$$q = < \frac{\sum_{-\infty}^{\infty} \frac{(-1)^n}{\omega^2+D^2n^2} I_n(\beta q) I_{n-1}(\beta q)}{\sum_{-\infty}^{\infty} \frac{(-1)^n}{\omega^2+D^2n^2} I_n^2(\beta q)} >_\omega \tag{19}$$

In practice the numerical computation of this algebraic expression is not difficult because the maximum contribution to the infinity sum comes from the modified Bessel functions of lower orders.

It is interesting to compare our results with those given by Cook in [7] in the limit of $Q \to \infty$. We observe from (18) that when $\omega \to 0$ (absence of frequencies), the overlap is

$$q = \frac{I_1(\beta q)}{I_0(\beta q)} \tag{20}$$

which is exactly the same expresion reported by Cook. However our result is more general because we have included the effect of a distribution of frequencies. Additionally it is not difficult to deal with more complex situations (e.g. random fields). This shows the power of the formalism developed in this paper.

References

[1] S.H. Strogatz and R.E. Mirollo, J. Stat. Phys. **63**, 613 (1991).

[2] R.Eckhorn, R.Bauer, W.Jordan, M.Brosch, W.Kruse, M.Munk and H.J.Reitboeck,Biol. Cyber **60**,121 (1988).

[3] D.Amit, *Modeling Brain Function*, Cambridge University Press 1989.

[4] Y. Kuramoto, *Chemical oscillations, waves and turbulence.* Springer, Berlin 1984.

[5] C.J.Perez-Vicente, A.Arenas and L.L.Bonilla, submitted to Phys. Rev. Lett.

[6] L.L. Bonilla, C.J. Perez-Vicente and J.M. Rubi, J. Stat. Phys. (in press).

[7] J. Cook, J. Phys. A: Math. Gen. **22**, 2057 (1989).

OPTIMISED ATTRACTOR NEURAL NETWORKS WITH EXTERNAL INPUTS

Anthony N. Burkitt
Computer Sciences Laboratory
Research School of Physical Sciences
Australian National University
GPO Box 4, Canberra, A.C.T. 2601, Australia
Email: `tony@nimbus.anu.edu.au`

Abstract. Attractor neural networks resemble the brain in many key aspects, such as their high connectivity, feedback, non-local storage of information and tolerance to damage. The models are also amenable to calculation, using mean field theory, and computer simulation. These methods have enabled properties such as the capacity of the network, the quality of pattern retrieval, and robustness to damage, to be accurately determined. In this paper a biologically motivated input method for external stimuli is studied. A straightforward signal-to-noise calculation gives an indication of the properties of the network. Calculations using mean field theory and including the external stimuli are carried out. A threshold is introduced, the value of which is chosen to optimize the performance of the network, and sparsely coded patterns are considered. The network is shown to have enhanced capacity, improved quality of retrieval, and increased robustness to the random elimination of neurons.

1. Introduction

The wide interest in attractor neural networks (ANN) as a model for the cortex follows from the many features that they share in common, such as their high degree of connectivity, feedback, non-local storage of information, and tolerance to damage. The large number of neurons in the human brain (of order 10^{10}) and their high connectivity (of order 10^4–10^5 connections per neuron) suggest that a statistical approach which captures the essential details of the network, without all the detail and complexity of individual neurons, may be a fruitful way to proceed [1].

Recent modifications have sought to make these models even more biologically plausible by showing that various constraints and approximations of the model can be relaxed (e.g., the symmetry of the synaptic matrix). Such models can accommodate, for example, low firing rates and neural specificity (Dale's law) quite naturally. In this paper a further biologically motivated modification is examined, namely looking at a more realistic model of the input side of such networks. The usual assumption about external stimuli is that they provide an initial stimulus that forces the attractor neural network into an initial state. Once this impinging stimulus has prepared the network in an initial state, it disappears and the network is left to evolve according to its own dynamics, a comprehensive review of which is given in [2].

More recently a number of authors have investigated alternative scenarios involving the effect of external stimuli. Amit, Parisi and Nicolis [3] consider an impinging stimulus that introduces strong PSPs that force the ANN into an initial state. This external stimulus is then weakened, but persists for an extended period of time. They consider the effect of the external stimuli being different from the memorized patterns according to three different types of noise distributions. They further found that rather strong persistent stimuli did not damage the retrieval properties of unsaturated networks. Even for saturated networks, moderate amplitudes of the external stimuli would ensure substantial error correction. Similar conclusions were reached independently by Engel *et al.* [4].

The approach considered here is that the external potentials remain fixed and constant during the retrieval time of the network. The calculations are carried out using mean field theory on an optimized network for sparsely coded patterns [5]. Sparse coding [6] describes networks that differ from the original proposal of Hopfield [1] in having mean spatial levels of activity less than 50%. Vicente and Amit [5] have shown that such a network with an optimally chosen neuronal potential (but in which they did not consider external stimuli) has improved retrieval properties that are close to the bound found by Gardner [7].

In this paper the effect of external stimuli upon this network is investigated for the case in which all external stimuli have the same strength. Section 2 gives a signal-to-noise ratio analysis that provides a rough guide to the behaviour of the network. The *capacity* of the network is then examined in more detail in Section 3 using the zero temperature mean field equations to order to see how the maximum number of patterns that the network can store increases with the introduction of the external stimuli (the capacity α is the ratio of the number of memorized patterns, p, to the number of neurons, N). Secondly the *quality of retrieval* of the memorized patterns is studied as a function of both the storage capacity of the network and the strength of the external stimulus. The quality of retrieval is defined by the overlap of the attractor with the memorized pattern. Thirdly we are also interested in the *robustness* of the results to external potentials that are incomplete patterns, in order to determine the amount of degradation of the external potential that the network can tolerate and still perform satisfactorily. All three measures of network performance show substantial enhancement when external potentials are included.

Following the notation of [5], the synaptic matrix is given by:

$$J_{ij} = \frac{1}{N}\sum_{\mu}^{p}(\xi_i^\mu - a)(\xi_j^\mu - a), \quad i \neq j; \quad J_{ii} = 0 \tag{1}$$

where the ξ^μ, $\mu = 1, \ldots, p$ are the random N-bit memorized patterns and the dynamics is given by:

$$S_i(t+1) = \mathrm{sgn}(h_i), \quad h_i = \sum_j J_{ij}(S_j - b) - U + h(\xi_i^1 - a) \tag{2}$$

(here the pattern ξ^1 is presented as an external stimulus with strength h) where U is determined by optimizing the performance of the network, a is the bias and b is a parameter that can take values in [+1,-1]. In [5] it is shown that the network performs best when $b = a$.

2. Signal-to-noise Ratio

The signal-to-noise ratio gives a useful preliminary guide to the performance of the network, and shows how the behaviour of the network differs from that of [5]. Consider what happens when the network state coincides with pattern 1:

$$\begin{aligned} h_i &= \frac{1}{N}(\xi_i^1 - a)\sum_{j,j\neq i}^{N}(\xi_j^1 - a)(\xi_j^1 - b) - U + h(\xi_i^1 - a) \\ &+ \frac{1}{N}\sum_{j,j\neq i}^{N}\sum_{\mu=2}^{p}(\xi_i^\mu - a)(\xi_j^\mu - a)(\xi_j^1 - b) \end{aligned} \tag{3}$$

The signal term is:

$$\begin{aligned} S &= \frac{1}{N}(\xi_i^1 - a)\sum_{j,j\neq i}^{N}(\xi_j^1 - a)(\xi_j^1 - b) - U + h(\xi_i^1 - a) \\ &\approx (1 - a^2 + h)\xi_i^1 - a(1 - a^2) - ah - U \end{aligned} \tag{4}$$

which is optimized when $U = -a(1-a^2) - ah$. The noise term is exactly the same as that given in [5], which is minimal when $a = b$, in which case:

$$\langle R^2 \rangle \approx \alpha(1-a^2)^3$$

where $\alpha = p/N$, and the signal-to-noise ratio is therefore:

$$\frac{|S|}{R} = \frac{|S|}{\sqrt{\langle R^2 \rangle}} \approx [\alpha(1-a^2)]^{-1/2} + h[\alpha(1-a^2)^3]^{-1/2} \tag{5}$$

The first term is simply that obtained in [5] while the second term, which is due to the external stimulus, increases the signal to noise ratio dramatically for biases near ± 1.

3. Mean-field Equations and their Solutions

3.1. $T = 0$ Mean-field Equations

In the limit $\beta \to \infty$ the mean field equations reduce to [5]:

$$\begin{aligned}
m &= \frac{(1-a^2)}{2}[\mathrm{erf}(\phi_1) + \mathrm{erf}(\phi_2)] \\
x &= b - \frac{(1+a)}{2}\mathrm{erf}(\phi_1) + \frac{(1-a)}{2}\mathrm{erf}(\phi_2) \\
C &= \frac{1}{\sqrt{2\pi\alpha r}}[(1+a)\exp(-\phi_1^2) + (1-a)\exp(-\phi_2^2)] \\
r &= \frac{(1-a^2)^2(1+2bx-b^2)}{[1-(1-a^2)C]^2}
\end{aligned} \tag{6}$$

where:

$$\begin{aligned}
\phi_1 &= \frac{(m+h)(1-a) - U}{\sqrt{2\alpha r}} - \frac{ab(1-a^2)^2 C}{\sqrt{2\alpha r}[1-(1-a^2)C]} \\
\phi_2 &= \frac{(m+h)(1+a) + U}{\sqrt{2\alpha r}} + \frac{ab(1-a^2)^2 C}{\sqrt{2\alpha r}[1-(1-a^2)C]}
\end{aligned}$$

3.2. Capacity of the Network

We present here numerical solutions to the above mean-field equations at $T = 0$. The lower curve in Figure 1 shows how the capacity (α) of the network decreases over the range of the bias (b is put equal to the bias a throughout) when $U = 0$. Below the curve the retrieval states are dynamically stable attractors, whereas above this critical value α_c the network is in an $m = 0$ spin-glass phase. The overlap (m) displays a discontinuity as α is increased above α_c.

The increasing curve directly above this lower one shows the effect of introducing U, as was demonstrated in [5]. The value of U is chosen to give the largest capacity possible, and thus optimize the network. The two upper curves correspond to $h = 0.1$ and $h = 0.2$ respectively, also with U chosen optimally. The values of α_c at $a = 0$ are simply those also given in [8], and the shape of the curves is that expected by the signal-to-noise estimate. The combined effect of introducing both U and h can clearly be seen to substantially increase the capacity of the network. These two curves have not been continued to higher values of the bias since the transition that characterises the saturation of the network memory becomes continuous, and there is therefore no sharp transition.

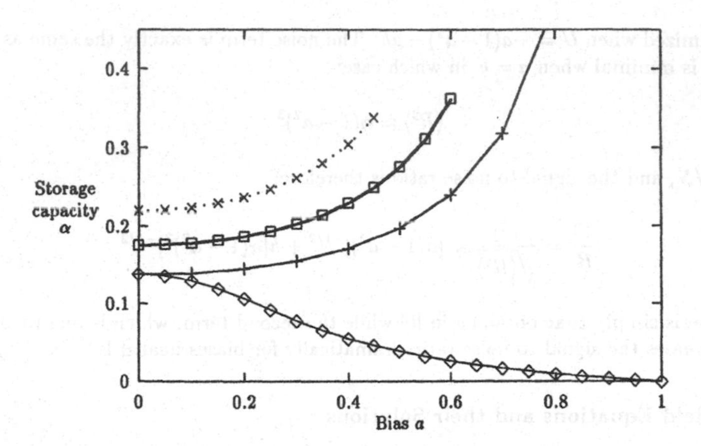

Figure 1: Storage capacity (α) against the bias (a). The upper three lines show (from top) $h = 0.2$, $h = 0.1$ and $h = 0.0$ with optimally chosen values of U. The lower line is for $h = 0.0, U = 0.0$. The symbols correspond to numerical computation.

3.3. Quality of Retrieval

In order to more fully characterise the properties of the network it is necessary to examine more closely the quality of retrieval of the patterns stored by the network. The overlap below the critical storage capacity has been found to remain very high right up to α_c. Moreover, as the bias is increased there is not only an increase in storage capacity, but also for fixed storage capacity the quality of retrieval improves further. The discontinuity of the overlap at the critical capacity α_c can be seen in Figure 2, where the overlap at a bias of $a = 0.2$ is plotted as a function of the capacity (in this figure U has been chosen to have its optimal value at α_c). The $h = 0.0$ results show a discontinuity from a state of nearly perfect memory recall ($m/m_0 = 0.967$) to a state of almost zero overlap at a critical capacity of $\alpha = 0.145$, with $U = -0.24$. It can be seen that increasing the strength of the external potential results in larger values of α_c while the discontinuity becomes smaller. At larger values of the bias the discontinuity of the overlap disappears as U is increased, and it is therefore no longer possible to give a value for α_c or an optimum value of U.

3.4. Robustness

The next question about the behaviour of the network concerns how it handles incomplete and noisy external potentials. Amit *et al.* [3] look at three different types of noise, namely Gaussian noise, discrete noise and hidden units type of noise. The case that will be examined more closely here, in which the external potential is incomplete, corresponds to the hidden units noise distribution where the external potential is of the form:

$$h_i^{ext} = h(\xi_i^1 - a)\eta_i \tag{7}$$

in which η is +1 with probability P and 0 with probability $1 - P$. This is called a hidden unit noise distribution since a fraction $1 - P$ of neurons are hidden from the external potential and

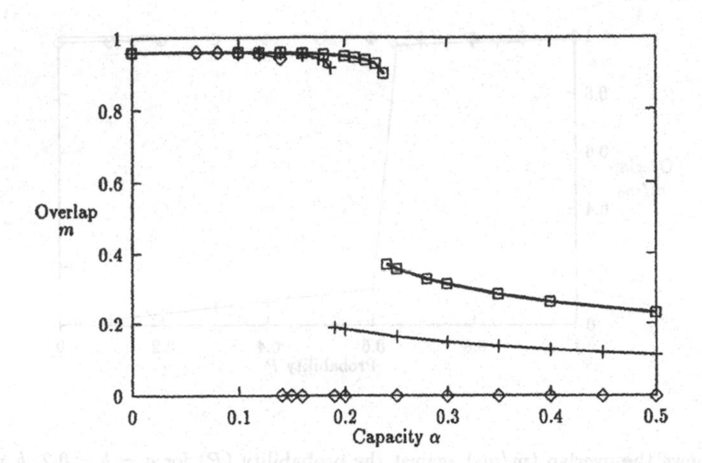

Figure 2: Typical plot showing how the overlap (m) depends upon the capacity (α) for $h = 0.0$ (lower curve, diamonds), $h = 0.1$ (middle curve, crosses) and $h = 0.2$ (upper curve, squares) with $a = b = 0.2$ and U chosen in each case optimally.

receive inputs only from other neurons in the network. The overlap of the external stimulus with the pattern is simply $m_0(P)/m_0 = P$.

Typical behaviour of the network when hidden unit noise is introduced is illustrated in Figure 3. At low capacities the noise makes essentially very little difference to the retrieval properties of the network, and indeed at small values of the external potential the network behaves very similarly to that described in [5] without external potentials, namely there is only a very small degradation of the network performance. At higher capacities there is a discontinuous jump in the overlap as the noise increases, as can be seen in the lower curve in Figure 3, which represents the results for a capacity approximately 10% below the critical capacity. Here there is very little degradation of the performance of the network until a critical value of approximately $P_c \approx 0.65$ is reached and the overlap falls to a value near zero. Similar behaviour is observed at both other biases and larger values of h.

4. Discussion

The calculations presented here show that a more biologically realistic method of introducing the external stimuli into an optimized network *increases* both the capacity (α) of the network and the quality of retrieval of the stored patterns. Calculations have been carried out at various strengths of external potential to see where memory saturation occurs and to investigate how well the memorized patterns are capable of being retrieved, both at the saturation threshold and below this limit. These results are also robust to the degradation of the external potential, showing only a very slight reduction in the quality of retrieval when a substantial proportion of the pattern is missing from the external potential. Computer simulations are also being carried out to investigate the behaviour of the network both for finite temperature and in the spin-glass state, where mean-field theory does not capture the behaviour of the network.

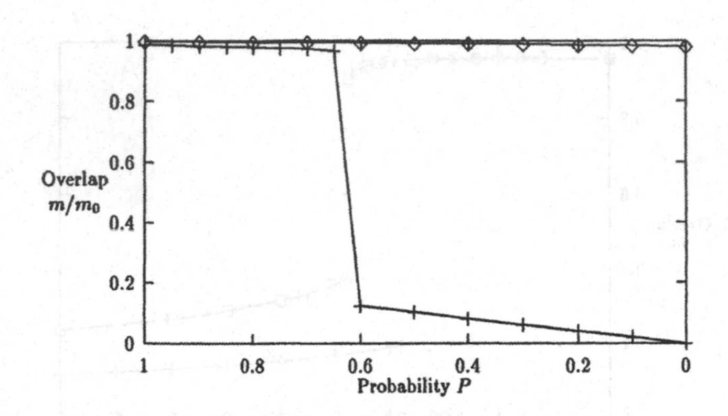

Figure 3: Shows the overlap (m/m_0) against the probability (P) for $a = b = 0.2$, $h = 0.1$ and two values of the capacity: $\alpha = 0.138$ (upper curve) and $\alpha = 0.168$ (lower curve), both with $U = -0.28$.

References

[1] Hopfield J. J. (1982), "Neural networks and physical systems with emergent collective computational abilities," *Proc. Natl. Sci.*, **79**, pp. 2554–2558.

[2] Amit D. J. (1989), *Modeling Brain Function: The world of attractor neural networks*, Cambridge: Cambridge University Press.

[3] Amit D. J., Parisi G. & Nicolis S. (1990), "Neural potentials as stimuli for attractor neural networks," *Network*, **1**, pp. 75–88.

[4] Engel A., Englisch H. & Schütte A. (1989), "Improved retrieval in neural networks with external fields," *Europhys. Lett.*, **8**, pp. 393–397.

[5] Vicente C. J. P. & Amit D. J. (1989), "Optimised network for sparsely coded patterns," *J. Phys.*, **A22**, pp. 559–569.

[6] Amit D. J., Gutfreund H. & Sompolinsky H. (1987), "Information storage in neural networks with low levels of activity," *Phys. Rev.*, **A35**, pp. 2293–2303.

[7] Gardner E. (1988), "The phase space of interactions in neural network models," *J. Phys.*, **A21**, pp. 257–270.

[8] Amit D. J., Gutfreund H. & Sompolinsky H. (1987), "Statistical mechanics of neural networks near saturation," *Ann. Phys., NY*, **173**, pp. 30–67.

NON-ORTHOGONAL BASES AND METRIC TENSORS: AN APPLICATION TO ARTIFICIAL NEURAL NETWORKS

Konrad Weigl and Marc Berthod*

INRIA - B.P. 93 - 06902 Sophia Antipolis Cedex
France

Abstract:
We consider neural networks such as Radial Basis Function networks as projection operators from function space onto a submanifold. We then interpret learning in such a network as the rotation and shifting of that submanifold in such a way that the projection of a function to be approximated unto the submanifold is as close as possible to the function itself. That rotation and shifting, executed through modification of the parameters of the basis functions of the network, is computed with the help of *metric tensors*, a geometric object of differential geometry.
The resulting network displays graceful degradation, and adapts dynamically to changes in the environment.

Keywords: Tensor Theory, Projection Operators, Metric Tensors, Radial Basis Functions

Function Approximation via Non-orthogonal Bases

Let F be a function to be approximated, from $\Re^n$ to $\Re$, thus a vector in a function space H^n. We want to approximate that function by another function $A(\gamma)$ which belongs to a submanifold of H^n, and which is the projection of F unto that submanifold. Let $B(\gamma)$ be the base of elementary functions g_μ of H^n spanning that submanifold, and let $B(\gamma)$ be parametrized by a parameter vector γ, $\gamma \in \Re^p, p \leq n, \Re^p$ thus a parameter space. $A(\gamma)$ will be the best approximation $A(\gamma_{opt})$ to F if the distance between them in H^n is minimal:

$$D(F, A(\gamma_{opt})) = min(D(F, A(\gamma))) \quad (1)$$

where D can be a distance such as the classical L^2, or another norm, as will be discussed below.

*email: weigl@sophia.inria.fr berthod@sophia.inria.fr
Correspondence to: Konrad Weigl

Note that that is exactly the task which a neural network such as a Radial Basis Function network [5] has to solve via learning: F is given by examples only, and within the class of functions which a network such as RBF can represent, we look for the one closest to F. Such a single-layer neural network can be seen as an implementation of the algorithm above: Each node is assigned an elementary function with which to process the input; these functions span a submanifold of function space, and every function of that submanifold can be learned perfectly by the network, others only to a specific degree of approximation. *Learning* in such a network is the minimization of D; it is done by rotating and shifting the submanifold spanned by $B(\gamma)$ in such a way that it is as close as possible to F through a suitable modification of the parameters γ (Ref. fig. 1): Expressing D as an energy functional of γ, and differentiating it w.r.t. γ, we obtain a set of differential equations for the modification of γ:

$$\frac{d\gamma}{dt} = -\frac{\delta D}{\delta\gamma} \tag{2}$$

Here we implement gradient descent directly. Obviously, conjugate gradient, or other methods, are equally suitable. By our experience, however, gradient descent implemented with Runge-Kutta seemed to be the most robust method. If D, expressed as an energy, is not convex, other suitable methods such as Graduated Non-Convexity, Simulated resp. Mean-Field Annealing or other methods would have to be used.

In order to compute $\frac{\delta D}{\delta\gamma}$ above, we need to compute $A(\gamma)$ as a function of γ. We have:

$$A(\gamma) \;=\; \textstyle\sum_\mu A^\mu g_\mu \;=\; A^\mu g_\mu \tag{3}$$

where A^μ, scalars, are the expansion coefficients of F in the base $B(\gamma)$. The equation shows application of the Einstein notation: We sum over any index appearing twice, e.g. here over the μ-index. Since we already know the dependence of g_μ on γ, we need to compute only the dependence of the A^μ on γ. In order to do this, we use the metric tensor of the base:
The so-called *metric tensor* of a base $B(\gamma)$ of functions g_μ is a square matrix defined by the computation of its components $g_{\mu\nu}$:

$$g_{\mu\nu} = \langle g_\mu, g_\nu \rangle \tag{4}$$

where the $\langle , \rangle$ operator denotes the scalar product. Commutativity of $\langle , \rangle$ $\Rightarrow$ symmetry of $g_{\mu\nu}$. Inversion results in the *contravariant metric tensor* $g^{\mu\nu}$, allowing us to compute the normalprojection $A(\gamma)$ of any function F unto the subspace spanned by $B(\gamma)$:

$$A(\gamma) \;=\; \textstyle\sum_\mu \sum_\nu A_\mu g^{\mu\nu} g_\nu \;=\; A_\mu g^{\mu\nu} g_\nu \;=\; A_\mu g^\mu \;=\; A^\nu g_\nu \tag{5}$$

with $A_\mu = \langle g_\mu, F \rangle$ the so-called *covariant components* and $A^\mu = g^{\mu\nu} A_\nu$ the *contravariant components*. The base $g^\nu = g^{\mu\nu} g_\mu$ is called the *contravariant* or *biorthogonal* basis.

Inverting a matrix generated via a scalar product is similar to an approach used in the class of so-called Least-Squares Problems in the field of probability theory [3], for the approximation of given data samples, where similar relationships are derived by analytical, non-geometrical means.

Let now D be e.g.the distance using the L^2-norm; for the minimization, we use the distance squared, as an energy (Ref. fig. 1):

$$D(F, A(\gamma)) = (F - A(\gamma))^2 = (F - A^{\nu} g_{\nu})^2 \qquad (6)$$

We then minimize D via gradient descent: Initially starting with a vector γ_0, we compute:

$$\gamma_{n+1} = \gamma_n - \left. \frac{dD(F, A(\gamma))}{d\gamma} \right|_{\gamma_n} \epsilon \qquad (7)$$

where ϵ is a suitably chosen stepsize, via Runge-Kutta in the present case. Inserting for $\frac{dD(F,A(\gamma))}{d\gamma}$:

$$\frac{dD(F,A(\gamma))}{d\gamma} = \frac{d((F - A^{\nu} g_{\nu})^2)}{d\gamma} = -2(F - A^{\nu} g_{\nu})(g_{\nu} g^{\nu\mu} \frac{dA_{\mu}}{d\gamma} + g_{\nu} A_{\mu} \frac{dg^{\nu\mu}}{d\gamma} + A^{\nu} \frac{dg_{\nu}}{d\gamma}) \qquad (8)$$

Once the algorithm has converged, we have found the best approximation $A(\gamma_{opt})$ to F possible, for given distance metric, number and kind of elementary functions, assuming D is convex.

Fig. 1 shows an example in three dimensions: Modifying γ moves g_0 and g_1, which moves the plane spanned by them, diminishing the distance between F and A_{γ} until it is minimal, giving us $A(\gamma_{opt})$.

Fig. 2 shows the evolution of the system for a simple NON(XOR)-function, computed with five gaussfilters; γ, consisting of the means and variances of the filters, is initialized at random, but ensuring linear independence.

Once the system has converged, we store the set of coefficients A^{μ} and the parameter vector γ; given as input the values (x,y), the output will be $A(x,y) = A^{\nu} g_{\nu}(x,y)$ or $A(x,y) = A_{\nu} g^{\nu}(x,y)$.

2 Other Metrics for Functional Spaces

Obviously, we are not restricted to linear distance metrics in our system, nor to linear domain spaces; as an example for another metric, we can take the symmetrical version of the Kullback-information, or cross-entropy, as a metric for a space of probability functions [2], which transforms our system to a parameter estimator for given histogram distributions, for example:

$$D = \int_x F(x) log(\frac{F(x)}{A^{\mu} g_{\mu}(x)}) dx + \int_x A^{\mu} g_{\mu}(x) log(\frac{A^{\mu} g_{\mu}(x)}{F(x)}) dx \qquad (9)$$

where D is the distance to be minimized, x is the domain of the function, F(x) again the function to be approximated, and $A^{\mu} g_{\mu}(x)$ the approximated function. For details, refer to [6].

3 Further Features of Dynamical Non-orthogonal Bases

We show in [6] that the system breaks down gracefully when neurons/elementary

proof, that the system will as well integrate further neurons/elementary functions, and reconverge to the new optimal subspace, with the new optimal approximation. This allows for an efficient approximation to a function, adding elementary functions until the desired quality of approximation is reached, or until the distance converges to a minimum.
Furthermore, if the function F changes in time, e.g. is a histogram of samples from the evolution of a non-stationary random process, the system is able to follow, assuming a change which is slow w.r.t. the intrinsic dynamics of the network.

4 Conclusion and Outlook

The learning algorithm presented above is in its prototype phase: Learning is rather slow, and scales badly with the size of the network; the curse of dimensionality, the classical problem with locally weighted neural networks, strikes here again. This is why we consider implementing the algorithm on a massively parallel Connection Machine.
We think that the use of tools from differential geometry in the field of neural networks, as suggested and initiated by Pellionisz [4], with another approach by Amari [1], and of which the present paper is a further simple example, is only at its beginning, and should be fruitful for many aspects of the field.

References

[1] Amari, S-I, Information Geometry of Boltzmann Machines, IEEE Trans. on Neural Networks, Vol. 3, No. 2, March 1992, pg. 260-271

[2] Caianiello, E.R., Quantum and Other Physics as Systems Theory, La Rivista del Nuovo Cimento, della Societa Italiana di Fisica, 1992, Editrice Compositon, Bologna, Italy

[3] Lawson, L.L., and Hanson, R.J., Solving Least Squares Problems, Prentice-Hall, Inc., Englewood Cliffs, New Jersey, 1974

[4] Pellionisz, A., Discovery of Neural Geometry by Neurobiology and its Utilization in Neurocompute Theory and Development, in Proc. ICANN Helsinki, Elsevier Pub. North Holland, pg. 485-493, (1991)

[5] Poggio, T., and Girosi, F., Networks for Approximation and Learning, in Proc. IEEE, Vol. 78. No. 9, Sept. 1990, pg. 1488

[6] Weigl, K., and Berthod, M., Metric Tensors and Dynamical Non-Orthogonal Bases: An Application to Function Approximation, in Proc. WOPPLOT 1992, Workshop on Parallel Processing: Logic, Organization and Technology, Springer Lecture Notes in Computer Sciences, to be published.

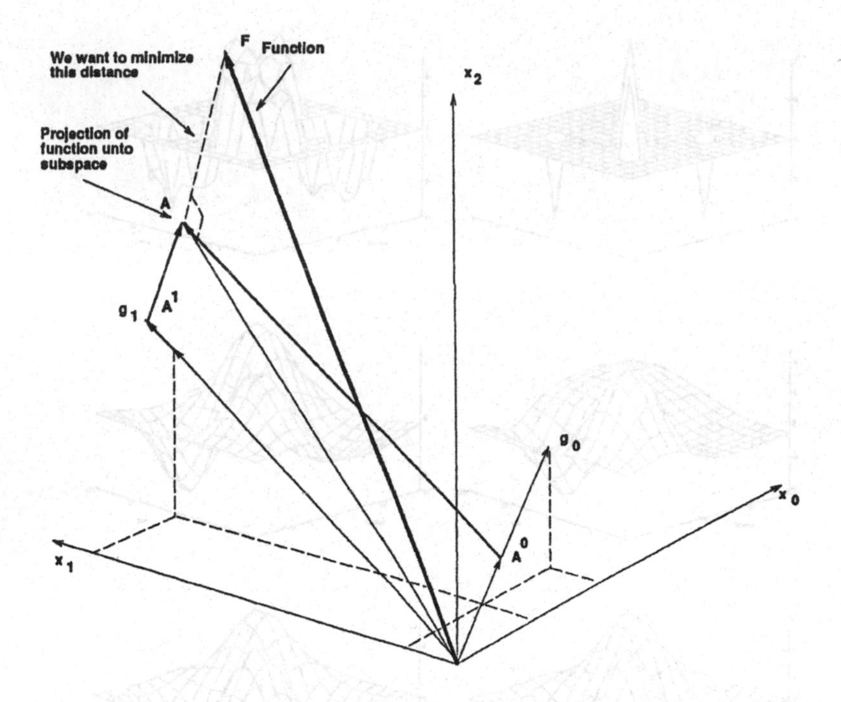

Fig. 1: An example in three dimensions

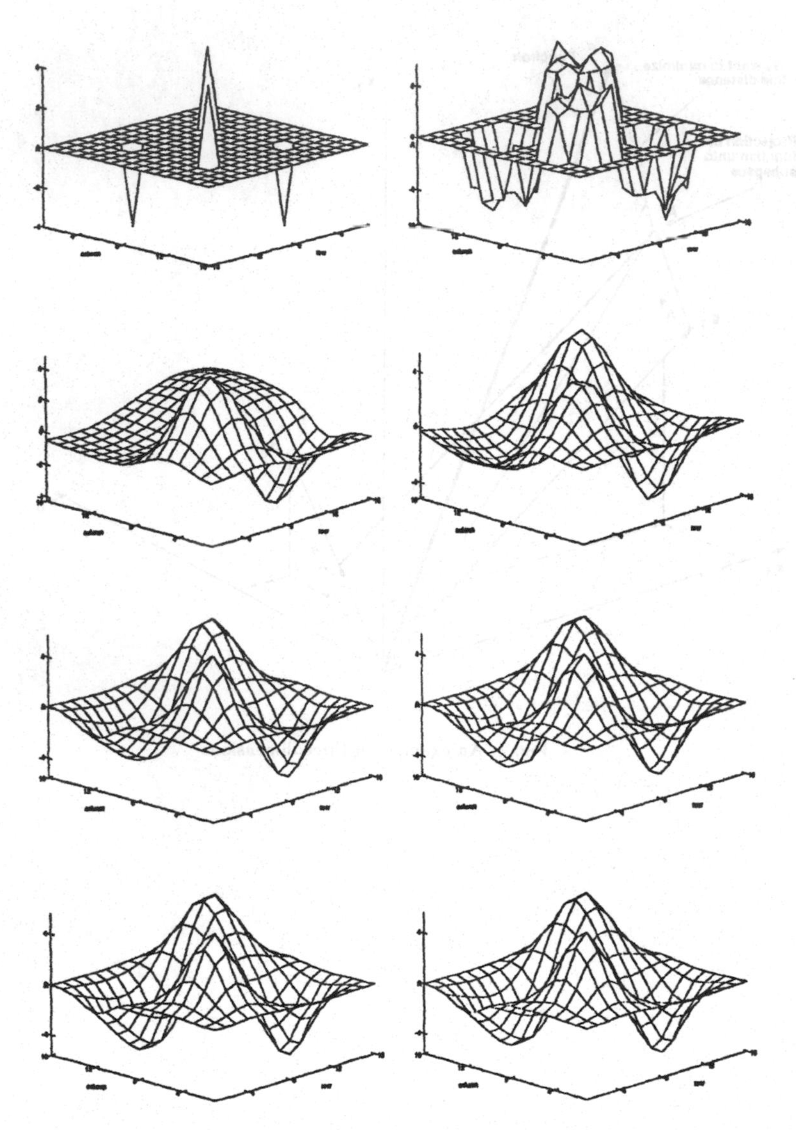

Fig. 2: Evolution of the system towards an optimal approximation; five gauss filters
Top left image shows the original NON(XOR)-function
We add noise to it to ensure good generalization
The remaining images show the evolution of the system, top left linewise to bottom right

GENETIC SYNTHESIS OF DISCRETE-TIME RECURRENT NEURAL NETWORK

F.J. Marín[(1)] and F. Sandoval[(2)]

(1) Dpto. Arquitectura y Tecnología de Computadores y Electrónica.
(2) Dpto. Tecnología Electrónica.
Universidad de Málaga, Plaza El Ejido s/n, 29013 Málaga (SPAIN).

ABSTRACT

In this paper, we proposed a different genetic model for optimizing both network architecture and connection weights of Discrete-Time Recurrent Neural Networks in evolutionary process. Empirical studies show that our model can efficiently generate the appropiate network size and topology for small applications. We have used two experiments: a parity function and a finite state machine for detection of sequences.

1. INTRODUCTION.

In the choice of a neural network for a certain application, the designer must resolve the question of drawing up its internal structure, that is, the number of layers, the number of neurons in each layer and the connections themselves. Unfortunately, there is no acceptable metodology for this decision; at the moment it's resolved by tedious task of trial-and-error, which make the search space be excessively large, taking an enormous CPU time.

There are some attempts for optimising the architecture of the network by constructive/destructive algorithms [3,5,8] which begin with a random initial solution and by small adjustments add or eliminating neurons one after the other during the learning process. These algorithms depend strongly on the initial solution and present certain facilities for falling in local minima.

The genetic algorithms are search evolutive algorithms based in process of natural selection [9]. They behave as an efficient tool to face problems which present a complex and noisy surface with multiple local minima and large search spaces.

The application of the genetic algorithm in the design of neural networks to solve a determinated task can be addressed from three fields of application: first, in the search of the optimal set of connection weights for a fixed architecture and learning rule [15]; second, in the search of the optimal architecture given the learning rule [7]; and third, in the search of the optimal learning rule once the number of neurons and the connectivity among themselves have been selected [4].

There isn't much research in the design of neural networks where several or all the previous fields have been combined. Most of the efforts are intendet towards an only field and for feed-forward networks, and using the back-propagation learning algorithm [13].

In this paper we study the genetic synthesis of discrete-time recurrent neural network, including both the search of the architecture as well as the connection weigths. Some results that show the design capabilites are presented and we conclude with a section of further tasks.

2. DISCRETE-TIME RECURRENT NEURAL NETWORK

The recurrent neural networks (RNN) were independently introduced by Almeida [1] and Pineda [11]. These networks are characterized as dynamic networks, where the equations of the neurons are time dependent function. Therefore, they are specially appropiate in those areas where a dynamic control system is necessary. We can underline: robotic control, speech recognition, sequence elements prediction, etc.

The topology of a RNN consists of a single layer of neurons fully interconnected, including self-feedback and where the weights can not be symmetric ($w_{i,j} \neq w_{j,i}$). The inputs and outputs of the network are chosen as subsets of the neuron set of the network.

The dynamic of the network is governed by differential equations:

$$\tau_i \frac{\partial y_i}{\partial t} = -y_i(t) + f_i\left(\sum_j W_{j,i} y_j(t) + I_i(t)\right) \qquad (1)$$

where y_i is the state of ith neuron, $w_{j,i}$ is the weight of the connection from jth neuron to ith neuron, I_i is an external input for ith neuron, f_i is the sigmoid activation function and τ is a time constant.

The states of the neuron change with time and the network can model spatio-temporal patterns. A solution of the differential equations describes a trajectory in the state space.

A specific implementation of RNN in discrete time was carried out by Williams and Zisper [16]. In this case

$$y_i(k+1) = f_i\left(\sum_j W_{j,i} y_j(k)\right) \qquad (2)$$

$$\text{where} \quad y_i(k) = \begin{cases} y_i(k) & i=1,2,\ldots,N \\ I_{i-N}(k) & i=N+1,\ldots,N+M \end{cases}$$

being N the total number of neurons and M the size of the input vector I.

In a recurrent network the weight matrix $\mathbf{W} = \{ W_{i,j} \}$ have all or several main diagonal elements different from zero. If the weight matrix is lower triangular with zeros in the diagonal, it is a feed-foward network. In addition, if the matrix has symmetric weights and the diagonal weights are zero, we have a Hopfield network with weight learning.

The learning algorithm most frequently used for recurrent network is a generalization of the back-propagation obtained by linearization and transposition [2,12]. It's a lineal optimizing technique based on the gradient descendet method to minimize a criterion function defined as the total squared error between an output $y_j(k)$ and a teacher $d_j(k)$ of the jth output neuron:

$$F(W) = \sum_k \sum_j (d_j(k) - y_j(k))^2 \qquad (3)$$

The main drawbacks that present these networks, in addition to their complexity, are the difficulties for being trained and optimized. This is a natural field of application of the genetic algorithms.

3. NETWORK REPRESENTATION SCHEME

Our objective is to test the capability of the genetic algoritms in the design of the architecture and connection weights for recurrent neural networks, considering that our scheme must be able to perfom a feed-forward for combinatorial problems (the weight matrix W is lower triangular with zeros in the main diagonal) and a feed-back network for sequential problems (W can have elements different from zero in any row or column). Thus, our scheme is fully open to design topologies with or without feed-back links. The choice of the paradigm to be used to resolve a specific problem is completely transparent for the designer. His only task is to adapt the fitness function to reward or punish the kind of connectivity selected.

The network architecture is represented by a binary string and is composed of only one layer which includes the input and output neurons. Before the codification of the structure, the subsets of neurons that implement the input and output of the network are defined according to the problem to be resolved. The representation scheme must include the number of neurons,the allowable feed-forward and feed-back interconnection degree, and the connection weight values.

Figure 1 shows the representation scheme of the network. Each individual is composed of two well defined areas: (1) the area of connectivity definition (ACD), which includes the following fields: input neurons (I), with a maximum of four (two bits); output neurons (O), two bits too; total number of neurons in the net (N), being $N > I + O$; feed-forward and feed-back connectivity degree (β), both represented with two bits and being the possible values 0.25, 0.5, 0.75 and 1, indicating the last one total connectivity. (2) the area of weight space (AWS) where each weight is codified with four bits, taking values in the range between -1 and 0.75; this election is completely arbitrary, and codification can be done with any other amount of bits.

The genetic algorithm starts generating a random population of individuals which codify a specific network. This is, each individual of the population represents a possible architecture where the weights are obtained after training. The individuals (genotypes) are evaluated to obtain a quantitative measurement of their fitting to the problem solution. This evaluation is carried out by the fitness function which is coincident with the criterion function (3) plus a term that punishes both a larger number of neurons as an operator connectivity degree.

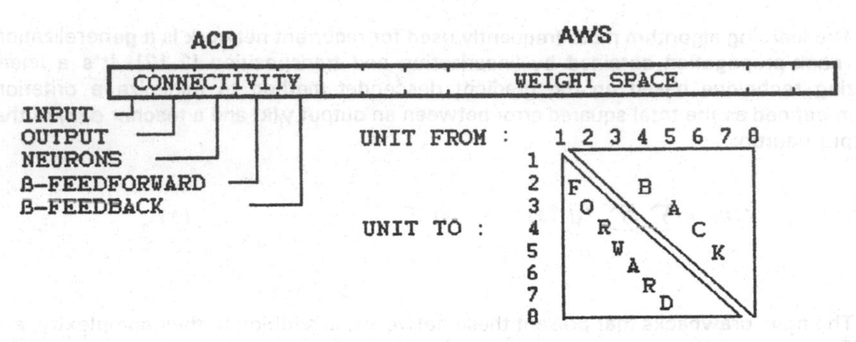

Figure 1. Individual representation of a network topology.

Individuals with lowest fitness are those with the highest probability to be selected for reproduction. The crossover and mutation genetic operator are the charged of leading the search for the most optimal network, removing or establishing connections and/or neurons. The crossover operator works independently in each area of the string.

4. EXPERIMENTS AND RESULTS

All the experiments have been carried out with the GAucsd1.4 software tool, developed by Schraudolph [14], written in C under the UNIX operating system.

Several experiments have been carried through, both with feed-forward as with feed-back networks. Two of them are outlined to be shown in this section: the three bits parity combinatorial problem, and the sequential detection problem, specifically to detect the 110 sequence.

Figure 2 shows the network topology and connection weight that resolve the three bits parity problem. The selected parameters for this experiments are the following: each weight is codified with four bits,taking values between -1 and 0.75; total feed-forward connectivity is allowed, but feed-back connectivity is removed. The population dimension is fixed with 100 individuals; the two points crossover operator is used with probability 0.6, and the mutation operator with probability 0.01. The most optimal solution was achieved as an average about the 70th generation.

Figure 3 shows the network topology and connection weight obtained to resolve the 110 sequence detection problem. The network consists of four neurons, being the first neuron the input and the last one the output; the value of the bits arriving to the input of the network are previously weighed up. The parameters selected for this experiments are equal to that described above, but now total feed-back connectivity is allowed. With 10,000 random training patterns the network presents an error rate less than 12% for the same training pattern. If the number of training pattern increases, the error rate decreases; with some 50,000 training patterns the network has completely learned. In all the experiments the number of evaluations was 10,000, i.e. dimension of the population times the number of generations.

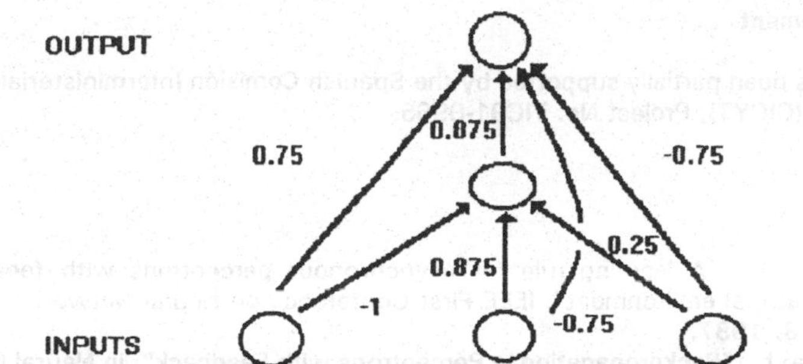

Figure 2. Network topology and weights for the parity problem.

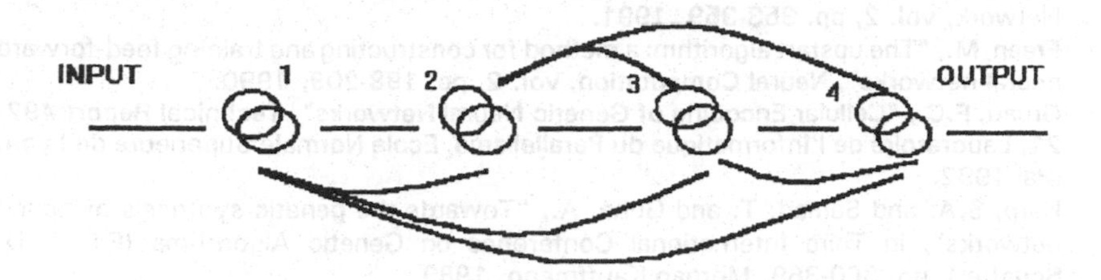

Figure 3. Recurrent Network topology for 110 sequence detection.

5. CONCLUSIONS AND FURTHER WORKS

The combination of genetic algorithms and neural networks is an efficient tool for resolution for small problems. We have shown that it is possible to optimize both the architecture, the connection weights and the connectivity in feed-forward and feed-back networks in a simultaneous way if the problem dimension is small. On larger networks, the size of the genotype rises enormously and the amount of time required to find a topology that learns quick and accurately is quite significant.

Therefore, it is necessary to use other methods of representation more compacts taking their function into consideration(to ensure that the codes are minimal in size) and with a higher degree of scalability. We are investigating methods of codification with attribute grammar on line with the works of Kitano [10] y Gruau [6]. Their results suggest that the application of genetic algorithm to synthesis of neural networks is a promising area of researchers, because it is possible some interactive regularities similar to systolic implementation.

We will also investigate the possibilities of incorporating advanced operators such as invertion, dominance, niche and speciacion in order to test the genetic algorithms capacities.

Finally, parallelization of genetic algorithm will be an important research topic.

Acknowledgement

This work has been partially supported by the Spanish Comisión Interministerial de Ciencia y Tecnologia (CICYT), Project No. TIC91-0965.

REFERENCES

[1] Almeida, L., "A learning rule for asynchronous perceptrons with feedback in a combinatorial environment", IEEE First Conference on Neural Networks, vol. 2, pp. 609-618, 1987.

[2] Almeida, L., "Backpropagation in Perceptrons with Feedback", in Neural Computers, (Eds. Roff Eckmiller and C.v.d. Malsburg), pp.199-208, Springer-Verlag, 1989.

[3] Fahlman, S.E. and Lebiere, C., "The cascade-correlation learning architecture", in Advances in Neural Information Processing Systems (Ed. D. S. Touretzky), vol. 2, pp. 524-532, Morgan Kauffmann, 1990.

[4] Fontanari, J.F. and Meir, R., "Evolving a learning algorithm for the binary perceptron", Network, vol. 2, pp. 353-359, 1991.

[5] Frean, M., "The upstart algorithm: a method for constructing and training feed-forward neural networks", Neural Computation, vol. 2, pp. 198-209, 1990.

[6] Gruau, F.C., "Cellular Encoding of Genetic Neural Networks", Technical Report #92-21, Laboratoire de l'Informatique du Parallelisme, Ecole Normale Superieure de Lyon, Mai 1992.

[7] Harp, S.A. and Samad, T. and Guha, A., "Towards the genetic synthesis of neural networks", in Third International Conference on Genetic Algorithms (Ed. J. D. Schaffer), pp. 360-369, Morgan Kauffmann, 1989.

[8] Hirose, Y. and Yamashita, K. and Hijiya, S., "Back-propagation algorithm which varies the number of hidden units", Neural Networks, vol. 4, pp. 61-66, 1991.

[9] Holland, J.H. "Adaptation in Natural and Artificial Systems", University of Michigan press, Ann Arbor, 1975.

[10] Kitano, H., "Designing neural networks using genetic algorithms with graph generation system", Complex Systems, vol. 4, pp. 461-476, 1990.

[11] Pineda, F.J., "Generalization of Back-propagation to Recurrent Neural Networks", Physical Review Letters, vol. 59, pp. 2229-2232, American Physical Society, 1987.

[12] Pineda, F.J., "Dynamics and Architecture for Neural Computation", Journal of Complexity, vol. 4, pp. 216-245, Academic Press, 1988.

[13] Rumelhart, D.E., Hinton, G.E. and Williams, R.J., "Learning internal representations by error propagation", Parallel Distributed Proccessing, vol. 1, pp. 310-362. MIT Press, 1986.

[14] Schraudolph, N.N., Computer Science & Engeniering Department, University of California, San Diego, La Jolla, CA 92093-0114.

[15] Whitley, D., Starkweather, T. and Bogart, C., "Genetic algorithms and neural networks: optimizing connections and connectivity", Parallel Computing, vol. 14, pp. 347-361, 1990.

[16] Williams, R.J. and Zisper, D., "A learning algorithm for continually running fully recurrent neural networks", Neural Computation, vol. 1, pp. 270-280, 1989.

OPTIMIZATION OF A COMPETITIVE LEARNING NEURAL NETWORK BY GENETIC ALGORITHMS

Merelo, J.J.[1]; Patón,M. [1]; Cañas,A.[1]; Prieto,A.[1]; Morán,F.[2]

[1] Dpto. de Electrónica y Tecnología de Computadores.
Facultad de Ciencias. Universidad de Granada. 18071 Granada. Spain.

[2] Dpto. de Bioquímica.
Facultad de Químicas. Universidad Complutense. 28040 Madrid. Spain.

In this paper we present the use of a genetic algorithm (GA) for the optimization, in clustering tasks, of a new kind of fast-learning neural network. The network uses a combination of supervised and un-supervised learning that makes it suitable for automatic tuning -by means of the GA- of the learning parameters and initial weights in order to obtain the highest recognition score. Simulation results are presented showing as, for relatively simple clustering tasks, the GA finds in a few generations the parameters of the network that lead to a classification accuracy close to 100%.

1. INTRODUCTION

Clustering algorithms are widely used for classification tasks (Makhoul et al., 1985). A simple task issued to them is to correctly classify 2 dimensional points as belonging to one out of several Gaussian distributions, as seen in Figure 1, called clusters or classes. The complexity of this problem increases with the number of clusters to be classified.

Clustering algorithms have two different phases: training and recognition. During the training phase, initial random dictionary vectors are given, and the clustering algorithm tries to place them in the middle of each Gaussian distribution, or cluster. Once training has been performed, the classification or exploitation phase follows: a point is classified as belonging to a given class when that class's dictionary vector is the closest. Several methods, both neural (Kohonen's self-organizing maps, competitive learning, multilayer perceptrons) (Kohonen, 1991; Rumelhart, et al. 1991) and non-neural (k-means) (Makhoul et al., 1985) try to achieve this with varying degrees of success.

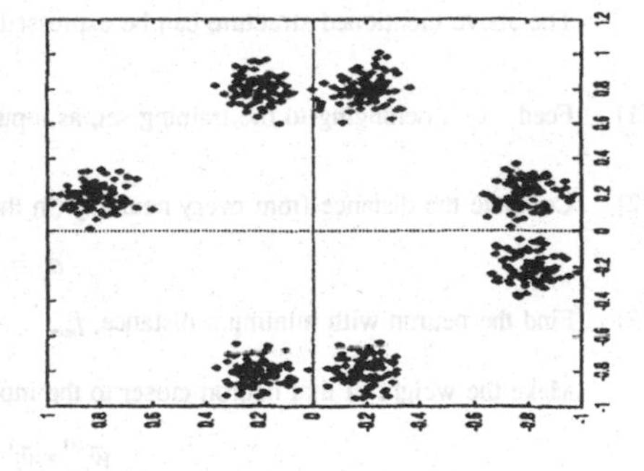

Fig 1. Example of distribution in clusters of bidimensional Cartesian points used for the training and testing of the algorithm. The points of a given cluster are distributed according to a Gaussian function.

Any one of the above algorithms usually has several parameters - that must be set heuristically - and initial dictionary vectors (or weights, in the case of neural algorithms) - also set heuristically or randomly. Subtle changes in parameters or initialization can lead to vast improvements, or, conversely, to a severe fall in speed or performance, or both. A good set of parameters can be set by trying them out sequentially in different combinations, but the sheer quantity of such combinations makes it impossible to find an "optimum".

Any optimization technique can be used to find an optimum, such as a gradient descent techniques (like backpropagation, although this is more difficult to apply to the training parameters themselves), simulated annealing, or, in our case, a genetic algorithm.

This paper presents a multilayer neural network for classification tasks together with a genetic algorithm that optimizes its parameters and structure.

2. NEURAL NETWORK FOR CLASSIFICATION

The multilayer neural network here proposed has got the structure shown in Fig.2. This network uses simple neurons as processing elements that add their weighted inputs, and pass the sum through a function to give an output. However, this function varies for the different layers.

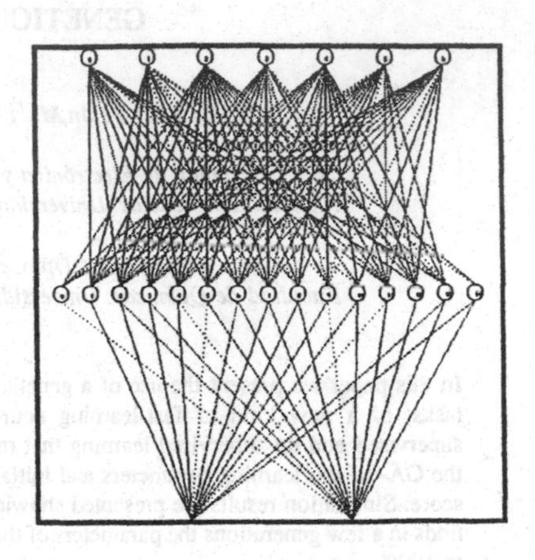

Figure 2: Neural network structure: 2 inputs (cartesian coordinates), hidden layer genetically optimized, and output layer (4 clusters)

The first layer of neurons receives the input vectors (two-component vectors in the example of figure 2) and is trained following a competitive learning algorithm. Each neuron applies a non-linear function to its output, which is passed to the second layer. The first layer itself could work well as classifier (as neural network with non-supervised learning) if, before recognition, a hand labelling is carried out, thus assigning each neuron to a class. However, for optimization, an automatic assignation of classes and a straight evaluation of the performances is needed both during and after training. In order to do this, another supervised learning layer has been added as second (or output) layer, which has as many neurons as classes, and uses a pseudo-hebbian algorithm for training; the unit whose input is maximum is activated (set to 1).

The above mentioned structure can be expressed by the following algorithm:

[1] Feed $\vec{x}_s$, belonging to the training set, as input vector.

[2] Compute the distance from every neuron j (in the first layer) to the input vector:

$$d_j = \|\vec{w}_j - \vec{x}_s\| \tag{1}$$

[3] Find the neuron with minimum distance, j_{min}

[4] Make the weight of that neuron closer to the input vector

$$\vec{w}_{j_{min}}^{\,t+1} = \vec{w}_{j_{min}}^{\,t} - \alpha(\vec{w}_{j_{min}}^{\,t} - \vec{x}_s) \tag{2}$$

where α is the gain parameter for the first layer.

[5] Apply the non-linear function to all the distances

$$h_j = -\log(d_j) \tag{3}$$

This function is larger for smaller distances, 0 for distance equal to 1, and less than 0 for distances greater than 1. The resulting h_j are the inputs to the output layer.

[6] Compute output, o_k activate the neuron with the highest output (set its output value to 1), deactivate all the other neurons (set their values to -1).

[7] Modify weights, v_{jk}, between the neuron j of the first layer and the neuron k of the output layer

$$v_{jk}^{t+1} = v_{jk}^{t} + \beta\gamma h_j o_k \quad (4)$$

where $\gamma = 1$ if the input is correctly classified, and $\gamma = -1$ if it is incorrectly classified; β is the gain parameter for the output layer, o_k is the output of neuron k, and h_j the value obtained in step 5.

[8] Back to step [1].

The neural network described has been successfully used for several classification criteria. As has been seen in the algorithm, there are several parameters that must be user-selected:

- The number of neurons in the first layer.
- The gain parameter for the first layer, α.
- The gain parameter for the output layer, β.

Besides, the initial weights must be either fixed or else randomly generated. All these parameters and initial values will be genetically optimized.

3. GENETIC CODING

Genetic algorithm theory states that all parameters of the solutions to the problem (in this case, all the parameters of the neural network) must be coded into a gene. By a process of selection of the fittest (in our case, the best classifying neural networks) only the best solutions are chosen for reproduction, to give new solutions, hopefully better than those of the previous generation. After each generation, new solutions are generated by means of reproduction between solutions and their associated mutation and mixing, and are only selected if they are good enough for the classification task.

Whenever the parameters of a problem are genetically coded, the *building block theory* applies (Belew, 1991; Paredis, 1991). Traits of the parents that should remain together in the offspring must be close in the coded gene. Two different genes have been devised, namely the *weights gene*, and the *somatic gene*. The first one codes the initial network weights, whereas the second codes the network parameters (learning parameters and maximum number of units in the first layer).

The weight genes have a length equal to

8×(*no. output neurons*)×(*no. inputs*)×(*max. no. input neurons*)

where 8 corresponds to the number of bits in a byte (each weight is represented by a byte). If the actual number of neurons is less than the maximum number of neurons, part of the information is not used in the building of the neural network. This adds some randomness to the evolutionary process, as the offspring are using information for which their parents have not been selected (if their number of neurons in the first layer is greater than that of their parent), and, therefore, may be a mechanism to inherit certain traits that are not actually used.

In the *weights* gene, the hidden-layer neuron has been considered as the main building block. In this way, all weights belonging to the same hidden-layer neuron are clumped together in the gene, first the first-layer weights, and then the output-layer weights. Only one byte is given to every weight that, at birth, is scaled to the interval [-1, 1].

In the *somatic gene*, one byte is assigned to α and another to β. The value of this byte plus 1 is divided by 256 to obtain values $0 < \alpha, \beta \leq 1$. The number of neurons in the first layer fills only 3 or 4 bits out of a byte, depending on the maximum number of neurons allowed during training. This adds up to 3 bytes.

4. GENETIC ALGORITHM

A *steady state* algorithm (Goldberg, 1989) has been chosen because it obtains solutions faster than other kinds of selection procedures, and because we are seeking only one of all the possible solutions -if it is good enough- despite the possible problem of premature convergence. *Rank-based* selection (Goldberg, 1989) has also been used, the number of hits reached during the last half of the training being considered for fitness evaluation. There are two ranking criteria: the first one is raw fitness (in our case accuracy in recognition i.e., the percentage of correct classifications); and the second, in cases of equal fitness, whereby the network with the lowest number of neurons is considered the fittest since it will obviously be faster at training and recognition.

Every generation, the last *g* individuals of the population (i.e., those with the lower fitnesses) are chosen for oblivion, and the first *g* for reproduction. In the examples of this work we use *g* equal to 20. The fittest then mate randomly with the rest of the population, and their offspring replaces the worst individuals. Whenever two individuals mate, their two genes always undergo crossover and mutation, and give only 1 offspring. Mutation rate is 0.01 for the weights gene and 0.1 for the somatic gene.

5. SIMULATION RESULTS

Several experiments have been carried out in order to test the efficiency of the combination of the genetic and neural algorithms. The efficiency of inheriting the weights was also tested. Different training runs were carried out with several combinations between a number of clusters (from 4 to 7) and a maximum number of neurons in the first layer.

Figs. 3 and 4 summarize the results obtained during training runs. Figure 3 (a) and (b) show the results for 4 clusters, while Figure 3 (c) and (d) show the results for 7 clusters. Bot figures represent the evolution of the fitness, taking the mean for all the population. Comparing the results with and without inheriting initial weights, it can be seen that the mean population fitness is always greater in the case of inherited initial weights. In spite of the maximum number of neurons, in 20 generations the population reaches a maximum of mean fitness that is maintained throughout the following generations. The parameters and weights obtained for the best individual of the last generation are a good choice for classification tasks.

The difference between networks with a different maximum number of neurons is only revealed when the weights are not inherited, as can be seen in Figures 3 (a) and (c)

One of the main interests in this study was to obtain a selected neural network with very good recognition properties, that can be used as a neural classifier. This is obviously related with the best fitness; the larger the best fitness, the greater the possibility of obtaining an optimal individual. In Fig.4, the best individual is represented for the case of classification of 7 clusters. The best individual in the case of inherited initial weights always has higher fitness than in the case of random initial weights.

Fig. 5 shows the connectivity for the best individual in one of the cases shown in Fig.4, for inherited (a) and random (b) initial weights. While for a small number of clusters, the number of neurons in the first layer is close to the number of clusters, for more complex problems, the system tends to use a larger number of neurons. In the final generation, the values of learning constants is consistent within all the training runs. The value for α is close to 0.004 in most cases, and the learning parameter β always ranges between 0.5 and 1.

6. CONCLUSSION

In this paper we have introduced a neural network algorithm with simplified learning rules, and the use of a GA for its optimization in clustering tasks. The population is evolved by means of the GA, which provides inheritance of neural network structure and learning parameters. Two different cases have been studied, moreover, where the initial weights values for the network are either inherited or not. The main result obtained here is that, in very few generations, the genetic algorithm actually found a neural network which approaches 100% accuracy in classification (only 1 misclassification out of 200, or 350).

The GA evaluates fitness as a combination of accuracy on the classification problem and the number of neurons in the first layer. The inclusion of an evolutive strategy (based in a GA), that includes inheritance of the initial weights of the network, is proved to work more efficiently than with random weights initialization. The weights for the starting generation are randomly generated, and these weights are the ones involved in the GA algorithm. It must be pointed out that inheritance involves the initial values and not the final values modified by the individual's learning. With this strategy we have tried to avoid Lamarckism (i.e., make the individual inherit the already trained weights of their parents) and give to the system more significance as a possible model for natural evolution (Merelo et al, 1992).

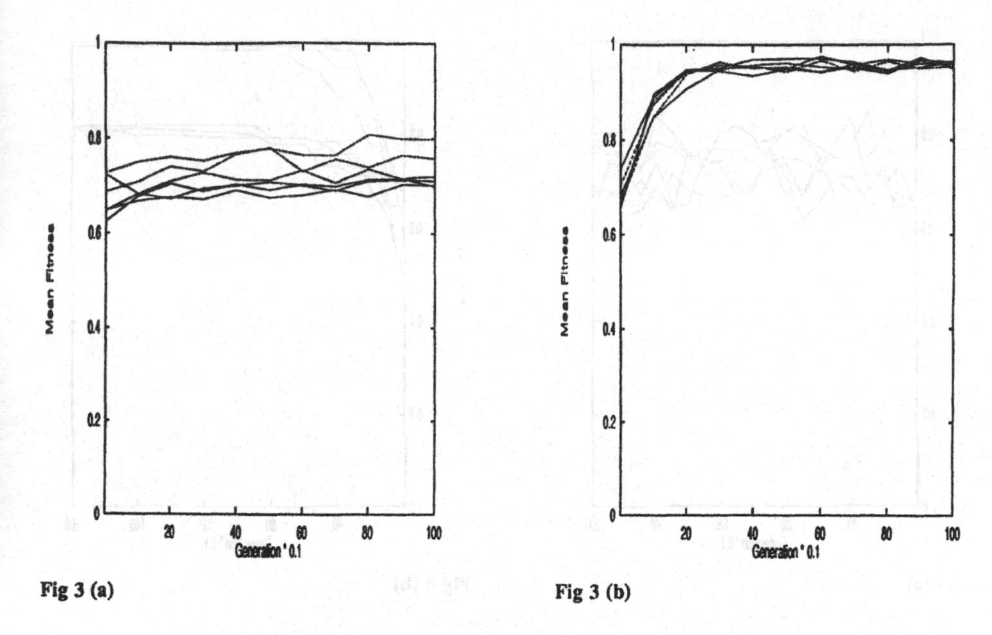

Fig 3 (a)

Fig 3 (b)

Figure 3: Evolution of the mean population fitness during the training test process for a 4-cluster problem. The lines represent 3 different simulations with different initial weights: (a) inheritance of initial weigh values; (b) random initial weights. The maximum number of neurons of the first layer is 8 (solid lines) or 16 (dashed lines). (c) and (d) are the same than (a) and (b), for 7 clusters classification.

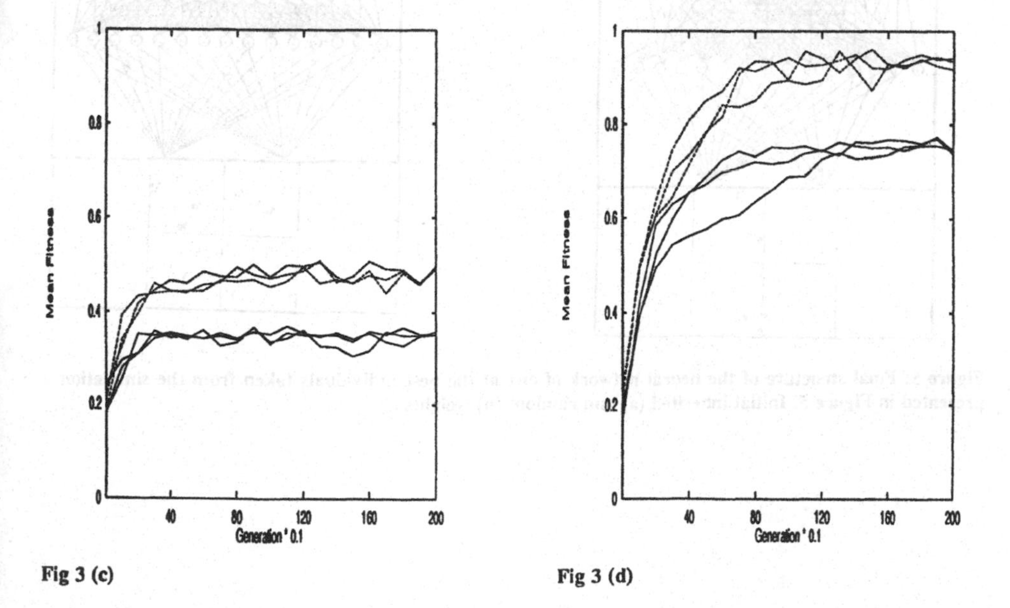

Fig 3 (c)

Fig 3 (d)

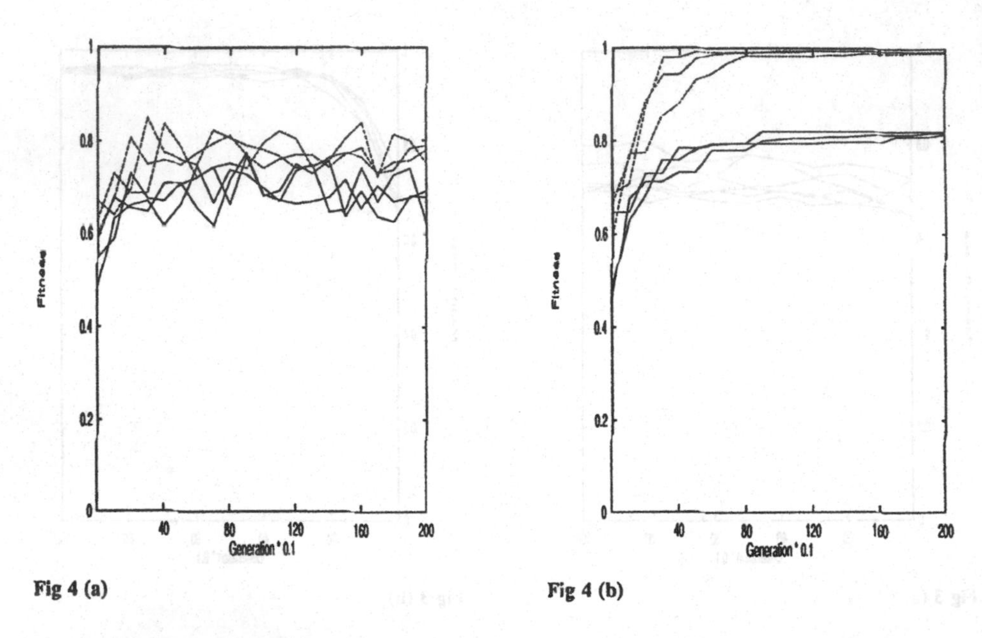

Fig 4 (a) Fig 4 (b)

Figure 4. Evolution of the fitness for the best neural network for each of the simulations presented in figure 3 (c) and (d), with initial inherited (a) and random (b) values.

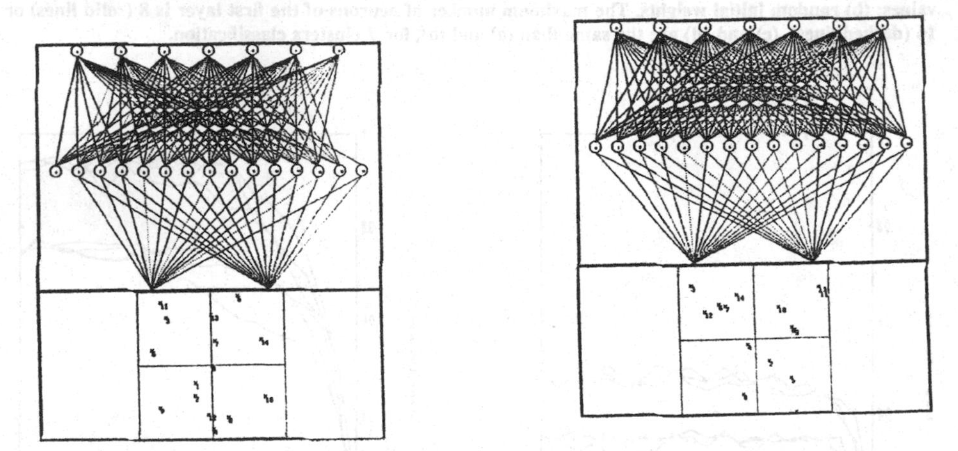

Figure 5: Final structure of the neural network of one of the best individuals taken from the simulation presented in Figure 5. Initial inherited (a) and random (b) weights.

With the system presented here, we have obtained the simultaneous optimization of initial weights, learning constants and the internal network structure (number of neurons). This algorithm looks for the minimum number of neurons necessary for the clustering task. When the number of clusters is small (4 or 5), the number of units in the first layer is close to this number of clusters. In more complex cases, the number of neurons is larger than the number of clusters but lower than the maximum allowed for this network. This result is due to the inclusion of number of neurons as an additional fitness criterion.

The value of the learning parameter α is related to the range of the input vectors (in the examples shown here the coordinates range from -1 to 1). This is a logical result because α regulates the grade of approximation of the weight vector to the input vector during each training step.

The algorithm has proved to be fast and reliable, despite the large number of iterations required for a solution. For example, in the case of 7 clusters and a maximum of 16 neurons in the first layer, the number of connections is up to 144 the number of individuals always being 100. The training set involves a number of points that is 100 times the number of clusters. This training set must be presented to each individual, followed by the modification of the weights, which represents about 10^7 steps for each generation. In this case a fitness close to 100% is reached in 100 generations, and running time laps close to half an hour in a SUN SPARCstation 1+ computer.

Several other neural algorithms have been genetically optimized (Whitley et al., 1991; Ackley et al., 1991), but this one, for the simple problem used, seems to be faster and to offer better recognition results. Besides, the genetic and neural algorithms have been successfully tested in an Artificial Life environment, as has been published in Merelo et al. (1992).

The experimental results of more complex classification problems are currently being studied, using the above presented neural network and optimizing its parameters and initial weights by means of the genetic algorithm.

Acknowledgements

This work was supported in part by grants PB89-0108 from DGICYT (Spain) and TIC 324.90 from PRONTIC (Spain).

References

Ackley, D. and Littman, M: Interactions Between Learning and Evolution, in Artificial Life II, Sante Fe Institute in the Sciences of Complexity, vol. X, pp 487-509. Edited by C. G. Langton, C. Taylor, J. D. Farmer, and S. Rasmussen. Redwood City, CA: Addison Wesley, 1991.

Belew, R.K., McInerney, J., and Schraudolf, N.N.: Evolving Networks: Using the Genetic Algorithm with Connectionist Learning, in Artificial Life II, Sante Fe Institute in the Sciences of Complexity, vol. X. pp 511-547. Edited by C. G. Langton, C. Taylor, J. D. Farmer, and S. Rasmussen. Redwood City, CA: Addison Wesley, 1991.

Goldberg, D.: Genetic Algorithms in Search, Optimization and Machine Learning, Reading, MA: Addison Wesley, 1989.

Kohonen, T.: The Self-Organizing Map, Proceedings of the IEEE, Vol 78, 9, 1464-1480, 1990.

Makhoul, J., Roucos, S., and Gish, H.: Vector Quantization in Speech Coding, Proceedings of the IEEE 73 (1985), 1551-1587.

Merelo, J.J., Patón, M., Cosano, J., Cillero, M., Bermúdez, A., Molina, J.A., and Morán, F.: Combination of Hebbian Neural Networks and Genetic Algorithms in Artificial Life Organisms, communication in Artificial Life III Conference, Santa Fe, NM, 1992.

Paredis, J.: The Evolution of Behaviour: Some Experiments, Procs. of the 1st Int. Conf. on Simulation of Adaptive

Behaviour, Wilson, Meyer, Eds. 1991, pp 419-426.

Rumelhart, D.E., McClelland, J.L.: Parallel Distributed Processing: Explorations in the Microstructure of Cognition. Volume 1: Foundations, Cambridge, MA: The MIT Press, 1986.

Whitley, D.; Dominic, S.; Das, R.; Genetic Reinforcement Learning with Multilayer Neural Networks, Procs, ICGA 91, pp 562-569.

Adaptive Models in Neural Networks

Panos A. Ligomenides
Cybernetics Research Lab., Electrical Eng. Dept.,
University of Maryland, College Park, MD 20742

Abstract

Artificial neural networks (ANNs) are principally attractive for their high degree of parallelism, for their associative memory properties, and for their ability to swiftly compute "near-optimal" solutions to highly constrained optimization problems. In this paper we examine the essential adaptive models that have been proposed for ANNs.

1. Introduction

We are interested to discover the "Computation and Learning - C&L" principles used by the brain to compute cognitive or sensorimotor functions and to adapt itself to its needs in a varying environment. We seek to understand the principles used for "computation" of visual or auditory recognition, of motor commands or of decision making, and for "learning", both for their own sake and for finding the means to build artificial systems having similar capabilities. Until we have some more specific answers from empirical studies of the brain itself, we approach the C&L problem by studying simplified formal models.

Somewhere between those psychologists and AI researchers who believe that brain-like processing (at least for high level cognitive functions) is best expressed in the language of abstract symbol manipulation, and those neurophysiologists who explain brain functioning in terms of biochemical details, are the students of *connectionist systems, parallel distrubuted processing networks* [7] and *artificial neural systems.*

The underlying feature of these systems is that they consist of highly interconnected networks of relatively simple processing units, where the computational properties of the system are a result of the *collective* dynamics of the network. The term "neuron-like", or even "neuron" is used for the relatively simple individual processing unit, although they are not constrained to match the details of the biological neuron but in the most superficial way.

2. Adaptive Neuron

The commonly used "neuron" model is,

$$q_i = f_{NL}(q_j \cdot w_{ij} \mid j \neq i)$$

where q are time varying scalars representing the state of a neuron, j represents a subset of other neurons and/or external inputs, f is a some simple nonlinear function (deterministic or stochastic) of the w-weighted analog summation of the inputs. We, often, avoid making explicit the time dependence by synchronized action. The states and weights may be assumed to be discrete or continuous.

As discussed in the next section, a simple rule for altering the connectivities between neurons is essentially a mathematical representation of Hebb's postulate of pre- and postsynaptic activity correlation. The adaptive element in this kind of model, learns to increase its response rate *in anticipation* of increased stimulation (Sutton-Barto model). This view of classical conditioning provides a simple explanation of stimulus substitution theory. The adaptive element extracts *predictive relationships* among its inputs, thus extending the Hebian element's ability to extract simultaneous associations. A dinstinct time, with respect to a conditioned stimulus, is provided for reinforcement, which replaces the need for distinct reinforcing signals.

Klopf [2] proposed a broader theoretical framework for adaptive behavior, which places the Hebian and the Sutton-Barto models within its bounds. Concerning the general class of reinforcement models of neural learning, (the other class is that of association models), Klopf's model of the "heterostat" is a generalized one. He was motivated by the similarity of adaptive behavior he perceived between neural and social systems. Both systems possess the ability to learn from experience and to have "plasticity" (i.e. to recover from damage to individual units). The primary questions are the nature and the location of the adaptive mechanism. Just as people in a social system are "hedonistic" in the sense of being pleasure-maximizers and pain-minimizers, so are neurons, in which the states of pleasure and pain are identified as states of depolarization and hyperpolarization respectively.

The view of goal-seeking systems made-up of non-goal seeking components, is now replaced by the view of goal-seeking systems with goal-seeking components. Non-goal-seeking neuronal nature would make "intelligence" an emergent phenomenon of the brain. This view, according to Klopf, prevents brain models from accounting for local feedback mechanisms that are crucial to the emergence of intelligence. Neurons which are goal-seeking in their own right would allow for a richer operational environment for the system.

In his argument, Klopf goes even further to suggest that the assumption of essential homeostasis in systems with non-goal-seeking components is misleading. The amount of homeostatic behavior in organisms decreases as the intelligence of the organism increases. Humans in particular are known to display nonhomeostatic behavior, which strives to maximize the difference between the amount of reward and punishment obtained. This maximal condition is refered to as *heterostasis.* At conditions far-from-dymanic-equilibrium, goal-seeking adaptive units try to establish a consensus of systemic activity that will support each component's needs and will allow "intelligence" to emerge as a outgrowth. This model certainly finds justification in social and in corporate business environments.

3. Learning Models

Artificial neural networks are designed to explore the behavioral (i.e. the computational) possibilities of physical systems with modifiable structures [3-6]. The structures of ANNs are represented by the distribution of pair-wise connectivity between neurons, i.e. by the topology of their the synaptic strengths. During "learning", they are being altered on the basis of some pre- and post-synaptic correlation rule.

The pairwise connection between neurons is characterized by its nature (whether it is excitatory or inhibitory) and by the degree of influence that the source-unit has on the incident unit. The latter charactewristic is represented by the synaptic weight associated with the connection. As the neural network learns something in response to environmental inputs, the weights are modified to "memorize" the learned experience and to reflect it in future dynamic behavior of the network. The definition of learning laws has been the focus of much of the research in the field. We will attempt here a quick overview of the most prominent learning strategies developed so far.

4.1 Hebbian learning.

By far the most common learning strategy, with its many modifications, is the elegent Hebb's rule [1], which suggests that when a neuron j repeatedly and persistently participates in firing neuron i, then j's efficiency (synaptic efficacy) in firing i is increased. A widely accepted mathematical approximation to the Hebbian postulate [8], is:

$$w_{ij}(t+1) = w_{ij}(t) + c\,x_j(t)\,y_i(t)$$

where $x_j(t)$ is one of the inputs to neuron i, whose output is $y_i(t)$, and c is a positive constant determining the rate of learning.

By the Hebbian rule, the network can learn to associate a stimulus pattern x with another pattern y. Further, presentation of a portion of x can cause the network to generate the complete version of y. This is made possible by the distributed memorization of the properties of input pattern x, which is embedded in the connectivity matrix of the network, and which associates input properties with output activities. When a portion of these properties are activated by the input pattern, simultaneous activation of all units that share these properties is caused, resulting in a combinatorial explosion of sorts. To control this situation, several neural models include provisions for lateral inhibition, with mutual inhibitory connections between incompatible concepts, as for instance in the case of competitive learning models.

This important function of pattern association by content addressability is most effectively implemented in neural networks. Although one might implement this function on a digital computer, e.g. by cycling through all possible pattern-pairs in search of least disimilarity (minimal Hamming distance), this programmed process is easily proven to be very inefficient and error prone.

4.2 Reinforced signals.

A generalized rule, whose special case is the Hebbian rule, dictates that the synaptic weight increase or decrease in proportion to a reinforcement signal r :

$$w_{ij}(t+1) = w_{ij}(t) + c r_{ij}(t)$$

Besides the Hebbian rule and its modifications, another case fitting this generalized rule of learning is the Widrow-Hoff rule, where the reinforcement signal is:

$$r_{ij}(t) = [d_i(t) - y_i(t)] x_j(t)$$

where

$$y_i(t) = \sum_{j=1}^{n} w_{ij}(t) x_j(t)$$

is the actual output and $d_i(t)$ is the desired output. Although Hebbian synapses would perform perfect recall only when the stimulus patterns x form an orthogonal set, the Widrow-Hoff rule allows perfect recall even if the stimulus pattern set is only linearly independent. Because of its orthogonalizing capabilities, the Widrow-Hoff learning is often refered to as orthogonal learning.

4.3 Binary reinforcement learning with a critic.

Reinforcement learning includes the case of supervised learning (with a "critic"), where a binary reinforcement signal is used as the only feedback from the environment, signaling whether each output is "right" or "wrong". This kind of supervised learning, applicable to either recurrent or non-recurrent ANN topologies, is only evaluative, not instructive, since the teacher, - a critic in this case -, provides single-bit feedback information merely as to whether the output is right or wrong. The single-bit reinforcement signal gives no clue as to what the right output should be, and therefore it provides no gradient information in terms of a "cost" function. This learning paradigm must provide some mechanism of random search, so that the space of possible outputs may be explored to determine a correct output state. Stochastically behaving neurons are used to achieve random search of the output space.

The reinforcement signal is produced by the environment (the critic) by some defined (possibly probabilistic) procedure based on input-output mappings evaluation in a dynamical environment of input stimulations.

4.4 Weakness of instantaneous conditioning.

The generalized connectionistic Hebbian rule can implement any application of simultaneous, or spatial, correlation, which need not imply a locationalistic view of memory. However, temporal relations of *classical conditioning*, as known from empirical psychological data, are not accomodated by the Hebbian rule even with the use of delays and other modifications.

4.5 The Sutton-Barto model.

Consider a threshold element that fires whenever one of its input pathways shows elevated activity. If another input pathway also happens to have elevated activity when this firing occurs, then it will also, according to the Hebbian rule, have its synaptic weight strengthened. After sufficient repetitions, an elevated activity along the second pathway alone will be enough to cause firing. In this fashion, Hebbian conditioning is able to account for stimulus substitution when the conditioned and unconditioned stimuli are presented simultaneously.

Sutton and Barto [8] saw problems with this approach and suggested that the Hebbian rule provides a much simplified explanation of the stimulus substitution theory. They have taken the view that classical conditioning involves temporal relationships and an interplay between expectations and stimulus patterns, in ways that are too complex to incorporate into a simple correlation rule like Hebb's. The predictive aspect of classical conditioning is largely overlooked by the conventional Hebbian rule. Empirical data suggests that adaptive elements learn to increase their response *in anticipation* of increased stimulation, thus producing a conditioned response before the occurence of the unconditioned stimulus. Further, the suggestion is made that the computational power needed for anticipative conditioning may not reside in a single neuron element.

To formulate a mathematical model for the Sutton-Barto learning mechanism, we need to consider a separate stimulus trace ξ_j in addition to the input stimulus signals x_j. If we indicate by $x_j(t)=1$ the occurence of a conditioned stimulus at time t, then we say that a prolonged trace of nonzero values of ξ_j is initiated for some period after t. This is done by letting $\xi_j(t)$ be a weighted average of the values of x_j for some period preceding t. In a similar manner, we require an output trace $\psi_i(t)$, which is a weighted average of the values of the variable y_i over some period preceding t. Then, we have

$$\xi_j(t+1)=\alpha\xi_j(t)+x_j(t)$$

$$\psi_i(t+1)=\beta\psi_i(t)+(1-\beta)y_i(t)$$

where $\alpha \geq 0, \quad \beta < 1$

The weight modification rule is given by

$$w_{ij}(t+1)=w_{ij}(t)+c[y_i(t)-\psi_i(t)]\xi_j(t)$$

where the constant c determines the rate of learning.

According to the Sutton-Barto model, if activity on an input pathway causes an immediate change in the neuron output y_i, it will also cause the connection from that pathway to be "tagged" by the stimulus trace ξ_j as being eligible for modification for the duration of the trace ξ_j. A connection is actually modified only if it is eligible and the current value of y_i differs from the value of the associated trace ψ_i. The effectiveness of the reinforcement for the conditioning process depends on the difference $y_i(t)-\psi_i(t)$. The characteristic of this learning strategy is that while a set of conditions may make synapses eligible for modification of their efficacies, actual modifications will occur due to other influences during the period of eligibility. Thus, eligibility is separated from electrical activity, so that prolonged presynaptic activation would not necessarily evoke synaptic modifiability. The mechanism that causes eligibility, such as a transient increase in the concentration of some chemical in biological neurons, may not participate directly in the electrical signaling of the neuron.

The Hebbian conjunctive correlations between input and output signals is replaced in the Sutton-Barto model by correlations between ξ-traces of input stimuli and changes of output. With reference to reinforcement signal, we have

$$r_i(t) = [y_i(t) - \psi_i(t)]\xi_j(t)$$

Since $y_i(t)$ can be effected by activity on any input pathway, any input signal can bring about changes in the efficacies of other pathways. This property allows the adaptive neuron to extract *predictive relationships* among its inputs in the same way that a Hebbian neuron extracts simultaneous associations. Providing a distinct time for reinforcement (with reference to a conditioned stimulus), eliminates the need for a distinct channel for reinforcing signals.

4.6 Learning as an Optimization Problem

The learning algorithm may be formulated as an optimization problem by specifying a performance criterion, and then by using the simple but powerful technique of stochastic hill-climbing along the gradient. Importantly, such a procedure is locally implementable. Learning is guided by a "teacher" or by a "critic" using a finite set of "exemplars". The nature of the feedback provided by the external trainer necessitates different weight adjustment procedures, such as the various versions of the Back Propagation algorithm, or the reinforcement methods outlined before.

In supervised learning by a teacher, a criterion (cost) function to be minimized may be the expected value of the mean square error,

$$J = E\left\{\frac{1}{2}\sum_k (y_k - d_k)^2 \mid W\right\}$$

where d_k is the desired response, conditioned on the selection of W. We wish to find W* that minimizes J for input patterns p which are drawn from a given distribution of possible input patterns. The value J_p for input p is an unbiased estimate of J. Because of the linearity of the expectation operator E, $\nabla_W J_p$ is an unbiased estimate of $\nabla_W J$. Thus we will minimize J by adjusting W along the sample gradients of J, using

$$\Delta W = -\alpha \nabla_W J_p$$

where α is a positive rate factor. Rumelhart et al [7] developed the Back Propagation algorithm as a straightforward computational technique for computing $\nabla_W J_p$, as long as the transfer function f of the neurons in the network are differentiable.

References

[1] Hebb,D.O, (1949), *The Organization of Behavior,* New York: Wiley

[2] Klopf,A.H., (1982), *The Hedonistic Neuron: A Theory of Memory, Learning and Intelligence,* New York: Hemisphere Publ. Corp.

[3] Ligomenides,P.A., (1993), "Cooperative Computers With Extended Computational Capabilities", *Proceedings of European Symp. on ANNs, ESANN'93,* Brussels, Belgium, April 7-9

[4] Ligomenides,P.A., (1992), "Computation and Learning Paradigms for Biologically Inspired Intelligent Computing", Proc. 2nd Int'l Conf. on Fuzzy Logic and Neural Networks, Iizuka'92, Iizuka, Japan, July 17-22, 1992

[5] Ligomenides,P.A., (1991), "Cooperative Computing and Neural Networks", in *Lecture Notes in Computer Science: Artificial Neural* Networks, A.Prieto (Ed), vol. 540, Springer-Verlag

[6] Ligomenides,P.A., (1991), "Computation and Uncertainty in Regulated Synergetic Machines", in *Lecture Notes in Computer Science: Uncertainty in Knowledge Bases,* B.Bouchon-Meunier, R.R.Yager, L.A.Zadeh, (Eds), vol. 521, Springer-Verlag

[7] Rumelhart,D., et al, (1986), "A general Framework for Parallel Distributed Processing", Ch.2, in *Parallel Distributed Processing, Explorations in the Microstructure of Cognition,* vol.1, Cambridge, Mass, The MIT Press

[8] Sutton,R. and A.Barto, (1981), "Toward a Modern Theory of Adaptive Networks: Expectation and Prediction", *Psychological Review 88,* no2:135-170

SELF-ORGANIZING GRAMMAR INDUCTION USING A NEURAL NETWORK MODEL

Christian Mannes*
Department of Cognitive and Neural Systems, Boston University
111 Cummington Str., Boston MA 02215

Abstract

This paper presents a self-organizing, real-time, hierarchical neural network model of sequential processing, and shows how it can be used to induce recognition codes corresponding to word categories and elementary grammatical structures. The model, first introduced in Mannes (1992), learns to recognize, store, and recall sequences of unitized patterns in a stable manner, either using short-term memory alone, or using long-term memory weights. Memory capacity is only limited by the number of nodes provided. Sequences are mapped to unitized patterns, making the model suitable for hierarchical operation. By using multiple modules arranged in a hierarchy and a simple mapping between output of lower levels and the input of higher levels, the induction of codes representing word category and simple phrase structures is an emergent property of the model. Simulation results are reported to illustrate this behavior.

Introduction

Serial order plays a key role in many aspects of behavior, most notably speech, language, and motor control. Modeling these phenomena, therefore, requires an explanation of how organisms deal with time-varying patterns whose order and timing matters. Sequential patterns in language occur on several levels and time scales, from the succession of acoustic features to sequences of words. Based on a model of spatiotemporal pattern recognition, recall, and timing, first described in Mannes (1992), this paper describes how a hierarchy of spatio-temporal pattern recognition modules can account for the induction of grammatical categories from raw sequences of patterns.

The model, based on Grossberg (1978), unifies treatment of order and timing information processing in a way that provides possible explanations of STM storage and recall, LTM learning, timed recall, completion of learned sequences, distinction between sub- and supersequences, and hierarchical chunking. Stable LTM memory codes for sequences of patterns, and their timing, are learned by a mechanism similar to ART (Carpenter and Grossberg, 1987) which maps temporal sequences of patterns into a single, static pattern. Stable memory codes can be learned in a single pass, and the capacity of the system is only limited by the number of nodes. Since sequences are chunked into single unitized patterns, sequences of sequences can be learned and performed, enabling the hierarchical model to learn arbitrarily long and complex sequences.

Several aspects of the model, in particular timing and performance, will not be discussed in this paper. Instead, I will focus on issues that arise in hierarchical serial learning, and show how the model can induce simple grammatical categories. While experiments have confirmed that the model is able to learn arbitrarily long sequences by building chunks of chunks on several hierarchical levels, it is doubtful that intelligent beings learn complex sequences in this way: Combinatorial explosion would make representation of sequences of sequences infeasible with any finite memory capacity. Thus, we need to ask how to make higher-level memory codes more meaningful.

In our language, possible successions of phonetic units are not only constrained by the physical limitations of our vocal apparatus, but also what seem to be rules governing what can follow what. These rules are in effect from the phonetic to the semantic level. Given that humans can easily understand

*This research was partially supported by NSF grant # IRI-9024877. The author would like to thank Daniel Bullock, Gail Carpenter, Michael Cohen, and Stephen Grossberg, for their valuable advice and support.

fluent speech while isolated speech segments are poorly recognized, it seems that human listeners make extensive use of contextual information to disambiguate speech—they are using grammatical rules.

The Model

The unit of learning—at least according to current theories of brain function—is the spatial pattern (Grossberg, 1978). Therefore, any model of serial learning must find a way to convert a temporal stream of input into spatial patterns. Typical approaches to this problem involve the use of feedback to associate an accumulation of past patterns with the next one (Jordan, 1986; Port 1990 for a review). A different approach is STM working memories (Grossberg, 1978; Bradski, Carpenter, and Grossberg, 1992) which employ short-term memory mechanisms to convert a temporal list of items into a spatial pattern that represents the order of items by relative activity. Our model uses a similar mechanism, called an STM primacy gradient (Grossberg, 1978), which codes precedence, or earlier occurence in a list by higher activity in a field of neurons (Figure 1). Using short-term memory mechanisms has the distinct advantage that storage and recall of patterns is possible without any learning, which is a necessary feature in order to explain how we can remember and dial a telephone number for a short time—and forget it after a while without any trace in long-term memory.

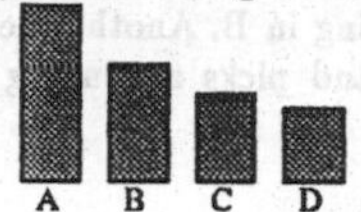

Figure 1: Representation of the sequence ABCD as an STM primacy gradient. Higher neural activity, shown as higher bars, codes precedence or earlier occurence in the sequence.

Representing sequences by gradients is advantageous because they transform a dynamical, spatio-temporal pattern into a static pattern which is then amenable to spatial pattern learning mechanisms. Given that the gradients are normalized appropriately, it is possible to classify the gradient using competitive learning (v.d. Malsburg, 1973, Grossberg 1976). However, competitive learning is inherently unstable, and requires additional mechanisms to stabilize learned memory codes. This was accomplished in Adaptive Resonance Theory (ART, Carpenter and Grossberg, 1987) by simultaneously learning an expected pattern and matching it with the input pattern in order to ensure that the pattern in question is not only the best match, but also a good match. Our model employs a similar strategy. However, the nature of spatio-temporal patterns and the requirement that the model also be able to perform sequences dictated a specialized design, shown in Figure 2.

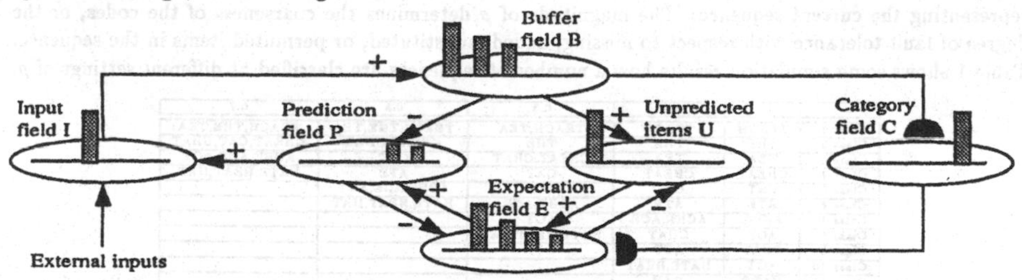

Figure 2. The architecture of the model. Ovals denote fields (layers), bars denote activity of nodes within these fields. Arrows stand for topographical, fixed-weight connections and are labelled + for excitatory and − for inhibitory. Half circles indicate full, plastic connections. See text for details.

A single module of the model consists of six fields, or layers, labelled I, B, U, P, E, and C. The input layer I registers incoming unitized patterns, which are thought to be the outcome of an underlying clustering algorithm, e.g. ART (Carpenter and Grossberg, 1987) or Kohonen's (1988) "Neural Phonetic Typewriter." This field projects via fixed-weight, topographical connections (one-to-one correspondence of nodes) to a buffer field B, whose STM interactions store the sequence of inputs I_i as a primacy gradient. The activities of the i-th node of B are given by

$$\frac{d}{dt}B_i = I_i(-B_i + A - B_i D \sum_j g(B_j)) + (1 - I_i)(B_i - B_i||B||) \quad (1)$$

where $||B|| = \sqrt{\sum_j B_j^2}$ and $g(x) = 1$ if $x > \theta$ and 0 otherwise.

B connects via fixed-weight, topographical connections to two fields, called U and P, which in turn are connected to the expectation field, called E. U and P represent the vector difference between the accumulating sequence in B and the expectation of what the sequence will be in E as given below:

$$\frac{d}{dt}U_i = \mu(-U_i + [B_i - E_i]^+) \qquad \frac{d}{dt}P_i = \mu(-P_i + [E_i - B_i]^+) \tag{2}$$

where $[x]^+ = \max(x, 0)$. By this equation, P represents all items that are in E, but not in B—the *predicted* future part of the sequence. Conversly, U represents all items which are in B, but not in E—the *unpredicted* past items. Furthermore, U serves to copy the pattern from B into E, such that an expectation can be learned by adapting the weights w_{ij}:

$$\frac{d}{dt}E_i = \nu LT_i - \nu E_i \sum_j T_j g(B_j) + U_i - P_i \tag{3}$$

where $T_i = \sum_j C_j w_{ji}$ and $L = \min(\sum_i B_i, 1)$. The dynamics of E ensure that both new patterns from B and the top-down expectation T_i can be represented, and that the total activation level of items that are both in B and E is matched.

As a new sequence is presented, the gradient pattern begins accumulating in B. Another field, called C for category field, receives input from B via modifiable weights z_{ij}, and picks a winning node by competition:

$$C_i = \begin{cases} 1 & \text{if } \sum_j B_j z_{ji} = max_k\{\sum_j B_j z_{jk}\} \\ 0 & \text{otherwise} \end{cases} \tag{4}$$

The weights z_{ij} connecting B and C are adapted such that the current winner in C will be more likely to get activated in the presence of the current pattern in B. At the same time, the winning node in C learns to reproduce the pattern in E, the expectation, via the pathways w_{ij} from C to E.

$$\frac{d}{dt}z_{ij} = \epsilon C_j[B_i - z_{ij}] \qquad \frac{d}{dt}w_{ji} = \epsilon C_j[E_i - w_{ji}] \tag{5}$$

Therefore, when the winning C-node gets activated after learning, it can read out an expectation in E.

Taken together, the combined activity in U and P is a measure of discrepancy between input and expectation. If this discrepancy, or error r, exceeds a parameter called ρ, the categorization field C is reset, and a new winner is chosen. This constitutes a memory search for the best matching memory code representing the current sequence. The magnitude of ρ determines the coarseness of the codes, or the degree of fault tolerance with respect to missing, added, substituted, or permuted items in the sequence. Table 1 shows some simulation results how a number of sequences are classified at different settings of ρ.

$\rho =$	0.1	0.2	0.3	0.5	0.7
Class 1	TEACH	TEACH	TEACH,TEA	TEACH,THE,TEA	TEACH,THE,TEA
Class 2	THE	THE	THE	CHEAT,CHAT	CHEAT,CAT,CHAT
Class 3	TEA	TEA	CHEAT,CHAT	CAT	ACHE,ACE,ACT
Class 4	CHEAT	CHEAT	CAT	ATE	HATE,HEAT,HAT
Class 5	CAT	CAT	ATE	ACHE,ACE,ACT	
Class 6	ATE	ATE	ACHE,ACE	HATE,HEAT,HAT	
Class 7	ACHE	ACHE,ACE	ACT		
Class 8	ACE	CHAT	HATE,HEAT,HAT		
Class 9	CHAT	ACT			
Class 10	ACT	HATE,HEAT			
Class 11	HATE	HAT			
Class 12	HEAT				
Class 13	HAT				

Table 1. Categorization results for different settings of the tolerance parameter ρ: In each case, the sequences TEACH, THE, TEA, CHEAT, CAT, ATE, ACHE, ACE, CHAT, ACT, HATE, HEAT, HAT were presented to the network in this order, where letters are labels, or indices, for nodes in the input field. Note that all sequences are composed of the elements of TEACH, and exhibit a very high degree of overlapping elements. A low tolerance setting forces the system to categorize every different sequence into a different class, while at high tolerance settings sequences are grouped according to similarity.

The fact that P contains the expected future part of a sequence at any time can be used for prediction, performance, completion, and disambiguation. By using feedback connections from P to the input field I, the expectation can shape the perception of future items. The input field is assumed to be a winner-take-all field, receiving inputs both from other modules, and from P.

$$I_i = \begin{cases} 1 & \text{if } \sum_k J_k m_{ki} + \alpha(r)P_i = \max_j\{\sum_k J_k m_{kj} + \alpha(r)P_j\} \\ 0 & \text{otherwise} \end{cases} \quad (6)$$

where J_i are external inputs, $\alpha(r)$ is a function of the error r, in the simplest case $\alpha(r) = \alpha_0$ if $r < \rho_s$ and 0 otherwise. Thus, the expectation can modify which node in the input field will be active when the error is low, effectively enabling the model to disambiguate and complete input sequences with missing or ambiguous items. This fact is critically important in the construction of hierarchies and for grammar induction, as is the learning law for the adaptive filter m_{ij} gating the inputs J_i.

Hierarchical Learning

The model described in the last section maps temporal sequences of unitized patterns into a single unitized pattern. This makes it straightforward to stack together multiple modules in a hierarchy, such that the C-nodes of one level are the inputs to the next level. Without further modification, such a system is capable of learning sequences of sequences, which makes it possible to store and recall arbitrarily long sequences. If we take the chunks of one level to be words, the chunks at the next level could be phrases or sentences.

The example of words and sentences clarifies the problem with this setup: The phrase-level chunks would have to have a node for every possible combination of words, which would exhaust any realistic memory capacity very soon. For example, the phrases *this is a cat* and *this is a dog* would be different sequences, and there would have to be a master chunk at the highest level for every sequence stored in the system. This is clearly a very undesirable property. What should higher level chunks represent, then?

Research on speech errors (Fowler, 1985) provides helpful hints in this respect: Slips of tongue seem to indicate that errors occur and are tolerated on several distinct levels: For instance, on a phonetic level (*pace fainted* instead of *face painted*), on a level of words (*the room to my door* instead of *the door to my room*), but rarely in between. Elements of speech which are confused most often are similar on the phonetic level, and of the same grammatical class on the word level.

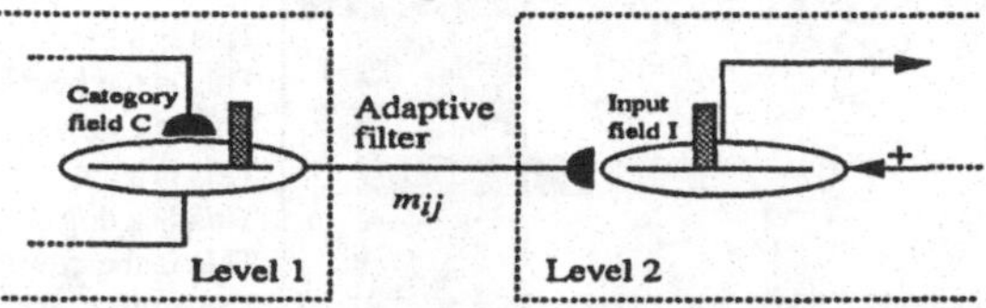

Figure 3: Connection of multiple modules in a hierarchy. The C-layer of level 1 is connected to the input layer I of the next level via an adaptive filter m_{ij}.

Given that error tolerance in our model is based on similarity, we have to think what similarity means at higher levels of the hierarchy: the words *cat* and *dog* are similar in the sense that they are both nouns. However, they do not share *any* phonetic features, yet, they are *instances* of the same grammatical *class*. Since the members of a class are not necessarily similar to each other, competitive learning will not in general put them into the same category. Yet, at some level, they should be treated as one.

This dilemma can be solved by using different neural sites for classes and instances, and learning their membership relation. The connection between modules that makes this possible is sketched in Figure 3. If we interpret C-nodes as instances of a class, and I-nodes as class representations, higher level sequences would not merely represent sequences of sequences, but sequences of equivalence classes. For example, the sequences *this is a dog, this is a cat,* and *this is a cow* could all be represented by *this is a* <noun> , thus separating structural information from item information.

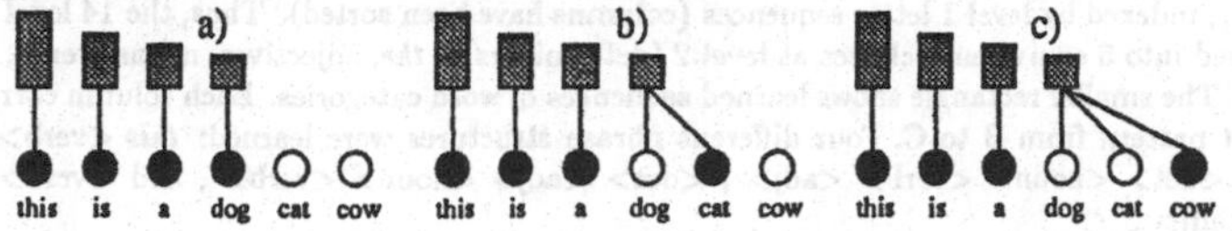

Figure 4. An example how grammatical class membership relations are learned. See text for details. Filled circles represent active C-nodes at level 1, lines strong weights m_{ij}, and bars show the expectation pattern at level 2.

To see how these equivalence classes can be learned, consider Figure 4. Suppose that words have been categorized into unitized patterns by a module 1, and are the input to another module 2, and that the

word sequences *this is a dog* and *this is a cat* are presented to the network. First, the sequence *this is a dog* will be learned by the level 2 module (Figure 4a). As *this is a* is presented again, this same chunk will be accessed, and the prediction will be *dog*. This prediction then may bias the input field sufficiently such that when *cat* is presented, the prediction can cause the node corresponding to *dog* to win (Figure 4b), effectively replacing the input with the expectation. The learning law governing the weights between the the C-nodes of level 1 and the input field of level 2 then adapts those pathways such that this substitution is more likely to occur in the future:

$$\frac{d}{dt}m_{ij} = \eta J_i[I_j - Fm_{ij}] \tag{7}$$

The weights m_{ij} are initialized as $m_{ii}(0) = 1$ and $m_{ij}(0) = m_0$ if $i \neq j, 0 < m_0 < 1$. When *this is a cow* is learned, the same substitution is carried out, with the effect that *dog, cat,* and *cow* all develop strong weights to the node that originally stood for *dog*, which now stands for the equivalence class with the members *dog, cat,* and *cow*.

By Equation(6), substitutioning input items with expected items, and hence learning word categories, is only possibly close to the end of the sequence, when the discrepancy between input sequence and expectation is low. This is in agreement with research on mother-child conversations in the early stages of linguistic development, which has shown that mothers tend to put 89% of novel words in utterance-final position, sometimes even if this produces an ungrammatical sentence (Aslin, 1993).

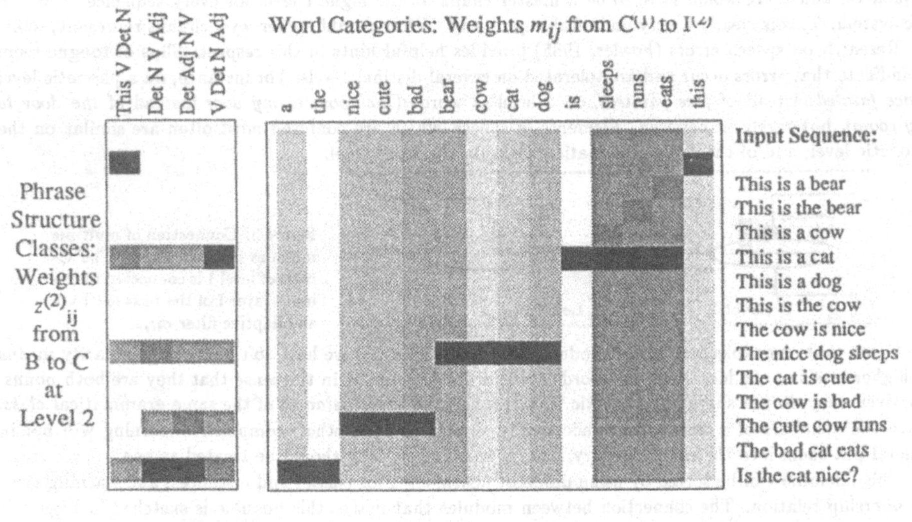

Figure 5. Typical simulation result for inter-module weights and level 2 classes after learning. Darker rectangles correspond to stronger weights (black:1, white: 0). The sequence of letters given on the right was presented to the network, where letters index input nodes. All words (letter sequences separated by blanks) were classified into different categories at level 1. The larger square represents the inter-module weights m_{ij}, indexed by level 1 letter sequences (columns have been sorted). Thus, the 14 level 1 nodes were grouped into 5 equivalence classes at level 2 (determiners *a*, *the*, adjectives, nouns, verbs, and the word *this*). The smaller rectangle shows learned sequences of word categories. Each column corresponds to a weight pattern from B to C. Four different phrase structures were learned: *this* <verb> <det> <noun> , <det> <noun> <verb> <adj> , <det> <adj> <noun> <verb> , and <verb> <det> <noun> <adj> .

Simulation Results

To test the hypothesis that this simple scheme actually leads to induction of codes corresponding to grammatical classes, I set up a system of two modules, connected by the modifiable pathways m_{ij}. All

variables were initialized to zero, except the weights as follows: $z_{ij}(0) = \frac{1}{\sqrt{n}}$ where n is the number of nodes in B, and $m_{ii}(0) = 1, m_{ij}(0) = 0.8$ if $i \neq j$. Then, a sequences of letters like *this is a cat* were presented to the network, where letters are labels, or indices of input nodes. For example, an input "b" means that $J_2 = 1, J_i = 0, \forall i \neq 2$. Blanks denote segmentation signals, which cause the buffer field B to be cleared. The learning rate was set to fast learning ($\epsilon = 1$), so that asymptotic performance was reached in a single trial. Figure 5 shows a typical result: All words were classified into different categories at level 1, then remapped into equivalence classes at the input field of level 2 via adaptation of m_{ij}. At level 2, sequences of classes rather than sequences of sequences were learned, corresponding to *phrase structures*. After learning, sequences containing familiar elements are represented as word codes, word classes, and the syntactical structure of the sequence.

Several other simulations of the same sort have confirmed this ability to induce grammatical classes. An intuitive classification of words and phrase structures, however, depends a great deal on the learning history. The grouping of words into syntactical equivalence classes depends on putting words in the final position of a familiar context, as also seems to be true of human language learning (Aslin, 1993). If novel words are presented in unfamiliar contexts only, or only at the beginning of "utterances", their word classes will not be learned, and the network can only learn sequences of sequences, with little generalization.

Discussion

This paper has shown how a model of hierarchical sequence recognition can be used to induce syntactical word classes and simple phrase structure patterns, using only the information implicit in the serial order of the input sequence. This scheme affords significant compression of temporal patterns by exploiting the regularities of grammatical sequences. The self-organizing approach to grammar induction is attractive in view of some of the alternative engineering solutions to language processing, such as Hidden Markov Models or traditional parsing, where grammar has to be hand-coded by the designer. The system also shares the advantages of ART-based systems, in that codes are stable, yet learning is never shut off, such that new categories can be added at any time without the danger of forgetting old categories. Fault tolerance with respect to perturbations of the input is in effect at every level of the system, which lets the system tolerate misclassifications from the acoustic to the syntactic level. Furthermore, the instance-class mappings can be used to disambiguate input based on contextual information embedded in the syntactical structure of sequences. However, the present model is not perfect. Good results depend on the learning history, and segmentation signals have to be provided as breaks between sequences at every level, which is probably unrealistic for speech input. Future research will focus on the segmentation issue, and on ways to make the system less dependent on presentation order.

References

Aslin R.N. (1993) *Models of word segmentation in fluent maternal speech to infants*, Lecture given at the International Conference "From Signals to Syntax", Feb 20, Brown University, Providence R.I.

Bradski G., Carpenter G.A., Grossberg S. (1992) Working memory networks for learning temporal order, with application to three-dimensional visual object recognition. *Neural Computation*, 4:270–286.

Carpenter G.A., Grossberg S. (1987) A massively parallel architecture for a self-organizing neural pattern recognition machine. *Computer Vision, Graphics, and Image Processing*, 37:54–115.

Fowler C.A. (1985) Current perspectives on language and speech production: a critical overview. In: Daniloff R.G. (Ed.) *Speech science*. San Diego: College Hill Press, pp. 195-278.

Grossberg S. (1976) Adaptive Pattern Classification and Universal Recoding I: Parallel Development and Coding of Neural Feature Detectors. *Biological Cybernetics*, 23:121–134.

Grossberg S. (1978). A theory of human memory: Self-organization and performance of sensory-motor codes, maps, and plans. In *Progress in Theoretical Biology* 5, New York: Academic Press, 233-374.

Jordan M.I. (1986) Serial Order: A Parallel Distributed Processing Approach. *TR-8604*, UCSD.

Kohonen T. (1988) The "Neural" Phonetic Typewriter. *Computer*, March 1988, 11-22.

V.d.Malsburg, C. (1973) Self-organization of orientation sensitive cells in the striate cortex. *Kybernetik*, 14:85-100.

Mannes C. (1992) A neural network model of spatio-temporal pattern recognition, recall, and timing. *Proceedings of the International Joint Conference on Neural Networks, Baltimore, 1992*, IV, 109-114.

Port R.F. (1990) Representation and Recognition of Temporal Patterns. *Connection Science*, 2:151-176.

THE ROLE OF FORGETTING IN EFFICIENT LEARNING STRATEGIES FOR SELF-ORGANISING DISCRIMINATOR-BASED SYSTEMS.

G. Tambouratzis & T.J. Stonham

Electrical Engineering Department,
Howell Building, Brunel University,
Uxbridge, UB8 3PH,
Middlesex, U.K.

An unsupervised-learning algorithm which endows a logical neural network model with the ability to separate patterns belonging to different pattern classes while at the same time creating a topology-preserving mapping of the input space is examined in this paper. Emphasis is placed in the storage efficiency of the algorithm and its ability to maximise the number of pattern classes that can be stored in the network. To achieve that objective, the introduction of a decay mechanism in the learning rule is proposed. The development of this mechanism has been based on a detailed analysis of the self-organisation process. Experimental results indicate that this forgetting mechanism maximises the storage capacity of the network, considerably improving the system performance.

1. Introduction.

The training phase is of paramount importance to neural networks, as it determines to a large extent their satisfactory behaviour during operation. Training algorithms can be distinguished in supervised and unsupervised ones. In supervised learning, the network is provided for each training pattern with the expected response. The network response is then compared to the expected one and the network weights are modified to reduce the difference in the two responses. Hence, a teacher is required to provide the correct response for each training pattern. On the contrary, in unsupervised learning the desired response is not specified for any training pattern. The neural network is presented with unlabelled training patterns which are clustered into a number of classes, based on the correlations between these patterns. Thus, unsupervised learning eliminates or at least minimises the amount of supervision required. Unsupervised learning is biologically plausible, as it is impossible to provide every neuron in the brain with a teacher. Learning must proceed mostly without supervision, the paradigm of supervised learning reserved only for high-level learned system behaviour.

Logic neural networks [1] are based on the structure of Random Access Memories. This enables them to overcome the limitations in hardware implementation and training speed which restrict the usefulness of analogue neural networks. At the same time, they possess a very high and flexible functionality, which enables them to be used in a wide variety of applications.

The subject of this paper is a logic neural network, trained using an unsupervised learning algorithm. Hence, it combines the ease of implementation and short response time of logic networks with the minimal supervision requirements of unsupervised-learning systems. The clustering capabilities of this model are investigated, with emphasis being placed on maximising its class-separation performance for a given number of nodes.

2. System Description.

The discriminator network [1] belongs to the family of logic neural networks. It consists of a number of logic functions, each of which samples n pixels from an input image. Each function consists of 2^n memory locations which correspond to all possible combinations of the binary pixels sampled by the function. The memory locations have a capacity of a single bit where information concerning the occurrence of the tuple combination is stored. For example, a stored value of "1" signifies that the combination has occurred at least once during training, while a "0" indicates that it has never occurred.

In order to perform unsupervised learning tasks, the discriminator structure has been modified [5]. The main change involves extending the memory locations of the discriminator functions so as to store a k-bit number (where k>1 and typically equal to 8). This number represents the relative frequency of occurrence of each tuple and is constantly updated during training by applying the adaptive algorithm described in the following section. The neural network comprises a number of discriminators which are laterally interconnected in order to form a one-dimensional (or higher-dimensional) structure as in Kohonen's [4] self-organising maps. During training the interconnections serve to define a neighbourhood of discriminators whose size is reduced as training continues. All discriminators compete for each input pattern, the winning discriminator together with its neighbours being adapted towards that pattern by applying the adaptation rule.

3. Adaptation rule.

The adaptation rule [5] differs significantly from Kohonen's learning rule, being designed from the outset for logical neural networks. The discriminators are initialised by inserting units in randomly-selected addresses of each function and during training the total number of units in each logic function is kept constant. When an input pattern is presented to the network, the discriminator which is best-matched to this input (according to criteria described later) together with its nearest neighbours - as defined by the neighbourhood function - are adapted towards the current input. The adaptation process involves locating for each function the memory address a_j which is designated by the input pattern. The memory content of that address is increased by one to record the new occurrence of that tuple. At the same time, the content of address a_j', where a_j' has the maximum Hamming distance from a_j and a non-zero content, is decreased by 1. Address a_j' is decreased as it represents the most dissimilar tuple value to the current input and is consequently the least likely to correspond to patterns belonging to the same class as the input. Removing this item of knowledge is therefore desirable and endows the system with noise-reduction capabilities. The update of the frequencies of occurrence m in a discriminator comprising d functions is described by the following formula:

$$\text{for } 0 \le j < d: \quad \begin{aligned} m_j(a_j)(t) &= m_j(a_j)(t-1) + 1 \\ m_j(a_j')(t) &= m_j(a_j')(t-1) - 1 \end{aligned} \qquad (1)$$

An example of the algorithm operation is shown in figure 1. It is worth noting that the update operation described by (1) ensures that the sum of contents over all memory locations of each function remains constant during the learning phase, addressing the problem of overgeneralisation and subsequent loss of discrimination capability in discriminator networks which may result from the presentation of an excessive number of training patterns. Instead, the proposed system will operate correctly for training phases of unlimited length.

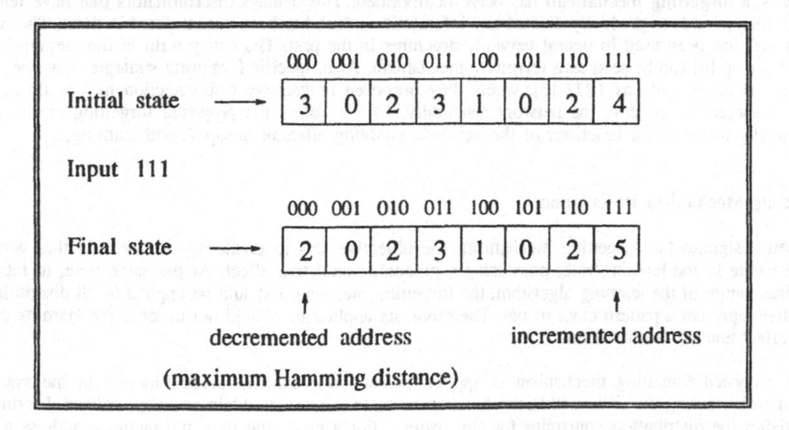

Figure 1 - An example of the adaptation rule application.

The **distribution constraint** [6], has been introduced in the selection of the discriminator to be assigned to the current input, in order to ensure the separation of classes which are closely positioned in the pattern space. The constraint examines the number of functions that contribute significantly to the discriminator's response ("active functions"). A function is considered to be active if the content of its currently-designated address does not differ by more than a given amount from the maximum address content of this function. All discriminators generate responses to the input pattern, and the highest-responding discriminator which satisfies the distribution constraint is selected. The constraint ensures that the discriminator assigned to the input is well-matched to it and enables the system to separate classes which present similarities yet remain considerably different, by utilising the knowledge accumulated in the discriminators. It assists the formation of stable pattern categories and prevents the erasure of useful information, fulfilling a role similar to Adaptive Resonance Theory's top-down expectation [2]. The stability of the class formation process has been demonstrated by applying the proposed network to pattern completion tasks [7].

This learning algorithm has been shown to successfully separate classes of patterns while forming a topology-preserving mapping on the neural network [6]. As is the case in most self-organising neural networks, in order to achieve this separation a number of redundant discriminators is required. As reported in [8], the required amount of redundant discriminators has been estimated to be equal to 50% of the number of classes. In experiments performed using a system with such a level of redundant nodes, the separation of all classes was achieved in 90% of the simulations [8]. However, in some cases the system failed to separate all classes, due mainly to a very unfavourable initialisation, when many classes were assigned to a small proportion of the network. In order to create a reliable and robust learning algorithm, it is essential to eliminate the probability of such a failure. To achieve this aim, a more detailed inspection of the system learning process was performed.

During training and while the system has not settled, especially when the neighbourhood activation function is relatively wide, patterns are temporarily assigned to different discriminators. Due to the use of the neighbourhood in the learning phase, a number of discriminators are adapted towards each input sample. Thus, each discriminator will move from its initial, relatively unbiased state (due to the random initialisation) to a more specialised state, generating a higher response to inputs it has been adapted to. As the system gradually settles, each pattern class becomes constantly assigned to a single discriminator. However, it is possible that two similar classes C_1 and C_2 are assigned to the same discriminator D_x. As the training progresses, discriminator D_x becomes focused to these two classes, the distribution constraint signalling the conflict. Ideally, an unused discriminator will then be located and the weakest of the pattern classes (suppose C_2) shall gradually be displaced to the unused discriminator. In the cases where the system failed to separate the pattern classes [8], the distribution constraint was constantly signalling the conflict but no discriminators that satisfied the distribution constraint were found. This was caused by the relatively small amount of "transient" training at the early stages of the learning process which sufficed to render unusable discriminators that had not been assigned to any pattern class. Thus, these could not be used to resolve distribution constraint violations in future stages of the self-organising procedure.

In order to erase this out-of-date information and enable the discriminator to be used to represent a pattern class, a **forgetting mechanism** has been incorporated. This enables discriminators that have remained unused for long periods to gradually discard any information which has been accumulated in them. The concept of "forgetting" has been used in neural network structures in the past. The decay term in the learning rule of the Kohonen map [4] can be seen as a forgetting mechanism. Also, specific forgetting strategies have been used by Fairhurst et al. as early as 1972 [3], where they proposed progressive disconnection in a feedback logic system as a means of reducing the network sensitivity. In this paper, the proposed forgetting mechanism is applied directly to the nodal functions of the network, enabling efficient unsupervised learning.

3. Forgetting Mechanism Description.

When designing the forgetting mechanism, the objective was to create an on-line rule that would be similar in nature to the learning rule, possessing a gradual, cumulative effect. At the same time, to retain the unsupervised nature of the learning algorithm, the forgetting mechanism should be applied to all discriminators, whether they represent a pattern class or not. Therefore, its application should not unsettle the learning process but only affect unused discriminators.

The proposed forgetting mechanism is applied in each activation to every function in the system. It consists of redistributing the units within the function so as to eventually obtain an unspecialised discriminator which satisfies the distribution constraint for any pattern. For a given function, the memory address a_m with the highest content is located. The content of a_m is reduced by 1, while to compensate the content of a randomly-selected address a_r (where a_m is different to a_r) is increased by 1. This is formulated as follows:

$$\text{for } 0 \le j < d: \quad m_j(a_m)(t) = m_j(a_m)(t-1) - 1 \qquad (2)$$
$$m_j(a_r)(t) = m_j(a_r)(t-1) + 1$$

An example of the application of rule (2) is shown in figure 2. It is notable that (2) is very similar in nature to (1). The incorporation of this mechanism on the system is straightforward as the main requirement is the need to locate address a_m with the highest content for each function. However, since the maximum-content address needs to be known in order to implement the distribution constraint, the implementation complexity is not considerably increased. Notably, by embedding rule (2) in the training algorithm, the system discards information in two different levels. At a local level, discriminators adapted to the current input have information contradicting that input (in the maximum Hamming distance) removed by rule (1). At a more global level, rule (2) is applied to all discriminators to remove any redundant information.

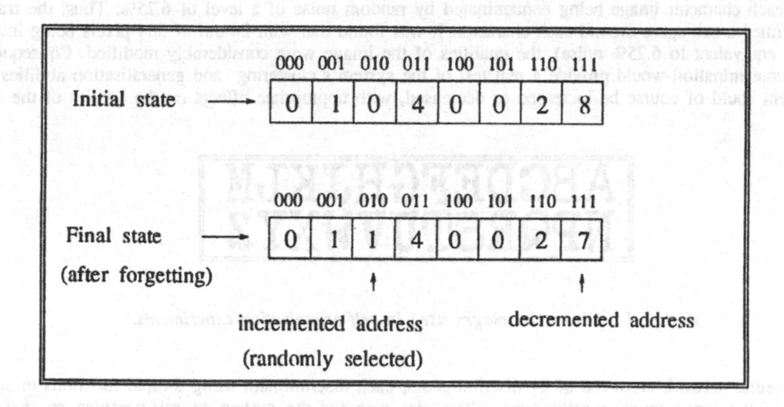

Figure 2 - Application of forgetting rule (2).

Preliminary simulations using the forgetting mechanism indicated that the system performance was considerably improved in comparison to the basic system. However, in some cases the system still failed to separate all classes. Interestingly, in these cases, after a large number of iterations the distribution constraint stopped signalling the clustering of two distinct classes on the same discriminator, effectively ensuring that the system would fail to separate these classes. Investigation of the system evolution showed that at the end the discriminator response to the two classes was almost identical. This can be explained by the fact that the forgetting mechanism always reduces the content of the highest-frequency address in the function (in our case, the address corresponding to C_1). Thus the difference between the two classes is gradually reduced in subsequent activations of the forgetting mechanism provided that patterns from both classes are presented with approximately the same frequency. As the gap between the responses to the two classes is reduced, the distribution constraint will signal more weakly up to the point where it fails to signal at all. Hence, if the conflict is not resolved by displacing one of the two classes to a vacant discriminator before the response difference falls below a critical value, the system will fail to separate the classes.

This problem can be solved if the difference in response for classes C_1 and C_2 is retained. This can be achieved by reducing for each function the two addresses a_{m1} and a_{m2} with the highest contents in each application of the forgetting mechanism. Similarly, two randomly selected addresses a_{r1} and a_{r2} (excluding a_{m1} and a_{m2}) are increased by one. Thus, (2) is replaced by:

$$\text{for } 0 \le j < d: \quad m_j(a_{m1})(t) = m_j(a_{m1})(t-1) - 1 \qquad (3)$$
$$m_j(a_{m2})(t) = m_j(a_{m2})(t-1) - 1$$
$$m_j(a_{r1})(t) = m_j(a_{r1})(t-1) + 1$$
$$m_j(a_{r2})(t) = m_j(a_{r2})(t-1) + 1$$

A final note concerns the frequency of application of the forgetting mechanism. This should be set at such a level that it does not unsettle the learning process by affecting discriminators to which a single pattern class has been assigned. Since the forgetting mechanism is applied to all discriminators, its frequency of application should be set at such a level that the discriminators can recover any erased information by sampling new examples of their corresponding pattern class.

4. Experimental Results.

A data-clustering task was simulated in order to evaluate the forgetting mechanism's abilities to maximise the storage capacity of a neural network (i.e. to maximise the number of classes formed for a given population of available nodes). The data set comprised images of 24 machine-printed capital letters, as shown in figure 3. The dataset was selected as it consisted of a large number of readily-classifiable (by humans) pattern classes. These 16x24-sized binary images were presented to the network in succession for a number of training iterations, each character image being contaminated by random noise of a level of 6.25%. Thus, the training patterns defined a sub-space around each character. It was found that with 24 out of 384 pixels being inverted (this being equivalent to 6.25% noise), the qualities of the image were considerably modified. Consequently, this noise contamination would provide a real test of the system's clustering and generalisation abilities. The noise content could of course be increased or decreased, with appropriate effects on the quality of the data.

ABCDEFGHJKLM
NPQRSTUVWXYZ

Figure 3 - Character images used in self-organisation experiments.

The neural network consisted of 24 discriminators, each discriminator using 8-tuple functions to sample all pixels of the input retina exactly once. The aim was for the system to self-organise so that each discriminator would learn to recognise a single character class. It is worth noting that this task is exceptionally hard, as the system needs to utilise 100% of the available discriminators. As reported in [6], a similar simulation had resulted in the system using only 21 out of 24 available discriminators. The system was initialised by inserting a total of 256 units in randomly generated addresses of each function. The neighbourhood size used was relatively small, its width initially being equal to 5 (for iterations 0 to 600). It was afterwards reduced to 3 (iterations 601 to 800) and finally set to 1 for the remaining training iterations. The forgetting mechanism was activated after the pattern-to-discriminator assignments had been stabilised (in step 840). In order to ensure the stability of the system, the forgetting mechanism described by (2) was applied once in every 5 training iterations. As rule (3) has effectively a forgetting rate twice as high as (2), the corresponding forgetting mechanism was applied once every 10 iterations, to obtain an equivalent forgetting rate for both cases.

The basic system (without forgetting) was compared to system variants enhanced by the two forgetting mechanisms. All three system configurations were simulated with 10 different settings, the settings differing in the system initialisation and the noise-contamination of the input patterns. Three attributes were used to evaluate the system performance. The most important was the number of classes created by the system, or equivalently the number of discriminators to which at least one class had been assigned, which indicates the discrimination efficiency of the system. The second one was the settling step, which indicates how many presentations of patterns from each class were required before the system converged to the final assignment of classes corresponding to each of the discriminators. The third one was the **similarity measure** [8] (calculated by comparing pixelwise the pattern classes assigned to neighbouring discriminators), which reflects how reliably the inter-class relationships existing in the pattern space were preserved in the network lattice. The results obtained by averaging over the 10 settings for each system configuration are summarised in table 1.

According to table 1, the basic system had a rather low utilisation factor (less than 85%), while in the worst case it only used 19 out of 24 available discriminators. However, it is worth noting that in one of the 10 simulations, the system failed to settle after 12000 training iterations, as it proved unable to assign one of the letter classes to a single discriminator. Instead, it kept shifting that letter class from discriminator to discriminator, without managing to permanently assign it to any of them. This was clearly a sign that the system storage capacity had been exceeded, preventing it from settling to even a sub-optimal solution.

System Configuration	Classes Formed	Settling Step	Similarity Measure
no forgetting	21.3	1135.9	6053.9
forgetting rule (2)	23.8	2041.3	5926.0
forgetting rule (3)	24.0	3331.8	5917.1

Table 1 - Comparative Performance of System Configurations.

When the forgetting rules described by (2) and (3) were added, the system performance rose considerably. The optimum performance in terms of storage efficiency and class separation was obtained using rule (3), when all classes were constantly separated. This improvement in storage efficiency was obtained at the expense of learning speed, as the system required a much larger number of training iterations in order to fully settle. This is expected, as in the simulations performed, the forgetting mechanism was only activated after the basic system was expected to have settled. This was done in order to compare the performance of the standard system to the one enhanced by incorporating a decaying term. It is possible that the forgetting rule could be activated at an earlier stage (although this has not been investigated yet). Finally, the topology preservation characteristics of the system were affected by the forgetting term. This can be explained by the fact that the forgetting mechanism enables the displacement of pattern classes from areas of the network that are heavily populated towards previously unused areas, thus affecting the topological mapping. However, it is worth mentioning that the average similarity measure even for rule (3) was comfortably better than a very high proportion of all possible mappings for the given dataset (the similarity measure value exceeding 99.5% of all possible mappings was determined to be equal to 5709 [8]). Hence, the improvements in utilisation efficiency are achieved without seriously compromising the topology-preserving properties of the system.

5. Conclusion.

As demonstrated by the experimental results, the forgetting rules proposed enable the system to maximise the network's storage capacity. This has been demonstrated for a very hard case where the system possessed no redundancy, the numbers of network nodes and pattern classes being equal. By adding the decay rule during training, cases where the system fails to settle are avoided, and the resulting discriminator utilisation is higher than with the standard system. It should be noted that the topology-preserving capabilities of the system are affected, though the most important topology relationships are still replicated in the network lattice. In most cases, it would be worth providing the system with more discriminators than classes, in order to combine the advantages of topology-preservation due to the neighbourhood training with the ability to discard redundant knowledge, offered by the forgetting mechanism.

References.

1. Aleksander, I. & Morton, H. (1990) *An Introduction to Neural Computing*, Chapman and Hall.
2. Carpenter, G.A. & Grossberg, S. (1988) The ART of Adaptive Pattern Recognition By a Self-Organising Neural Network. *IEEE Computer*, March 1988, pp. 77-88.
3. Fairhurst, M.C. & Aleksander, I. (1972) Dynamics of the Perception of Patterns in Random Learning Nets. *Proceedings of the Conference on Machine Perception of Patterns and Pictures*, Teddington, England, April 1972, pp. 311-316.
4. Kohonen, T. (1989) *Self-Organisation and Associative Memory* (3rd edition). Springer-Verlag.
5. Tambouratzis, G. & Stonham, T.J. (1992) A Logical Neural Network that Adapts to Changes in the Pattern Environment. *11th IAPR Conference Proceedings*, The Hague, Netherlands, August 1992, Vol. 2, pp. 46-49.
6. Tambouratzis, G. & Stonham, T.J. (1992) Implementing Hard Self-Organisation Tasks Using Logical Neural Networks. In *Artificial Neural Networks-II*, Aleksander, I. & Taylor, J. (eds.), Vol. 1, pp. 643-646, North-Holland.
7. Tambouratzis, G. & Stonham, T.J. (1993) A Self-Organising Logic Neural Network For Pattern Completion Tasks. In *Neural Networks: Techniques and Applications*, Lisboa, P.J.G. & Taylor, M.J. (eds.), Ellis-Horwood (in print).
8. Tambouratzis, G. & Stonham, T.J. (1993) Evaluating The Topology-Preservation Capabilities Of A Self-Organising Logical Neural Network. *Pattern Recognition Letters*, (in print).

SIMULATION OF STOCHASTIC REGULAR GRAMMARS THROUGH SIMPLE RECURRENT NETWORKS†

M.A. Castaño[††] , F. Casacuberta , E. Vidal

Dpto. Sistemas Informáticos y Computación.
Universidad Politécnica de Valencia
Camino de Vera s/n 46071 Valencia (Spain)

Abstract

Formal grammars have been successfully simulated through Artificial Neural Networks. This fact has established a new approach to the problem of Grammatical Inference. First, [Pollack,91], [Giles,92] and [Watrous,92] trained network architectures from positive samples or positive and negative samples generated by regular grammars to accept or reject new strings. On the other hand, [Servan,88] and [Smith,89] used nets in which strings were fed character by character, so that the possible successors for each character were predicted. Later, [Servan,91] suggested that these networks could also predict the generation probabilities of each character in the strings generated by Stochastic Regular Grammars. Our present work shows empirical evidence supporting this suggestion.

1. Introduction

The history of Finite-State Automatas (or, equivalently, Regular Grammars (RGs)) and Neural Networks (NNs) began when Minsky proved that "Every finite-state machine is equivalent to and can be simulated by some neural network" [Minsky,67]. Later, [Servan,88] and [Smith,89] trained first order *Recurrent Neural Networks* (RNNs) that recognized regular languages, using the methods shown in [Elman,88] and in [Williams,89], respectively. Both studies aimed at predicting the character (or characters) that may follow the current input symbol generated by the target grammar.

The problem of obtaining NNs that simulate RGs has also been approached through a second-order RNN trained to accept or reject strings instead of predicting the next symbol. So, [Pollack,91] established a new inference method called "Induction by Phase Transition", where the connectionist architecture was a cascaded recurrent network to which the back-propagation technique of adjustment [Rumelhart,86] could be applied. In [Giles,92] and [Watrous,92] the simulation of these NNs was carried out by using both positive and negative data; in the first one, by means of a second order form of the real time recurrent net [Williams,89], and in the second one, through a method of accumulating the weight dependencies backward in time.

On the other hand, the capacity of these connectionist models to capture the implicit statistical regularities that are exhibited by the input data has been used to estimate the *a posteriori probabilities* required by Hidden Markov Models [Boulard,92]. Now then, if the network has a feedback from hidden layers to the input field and the training set is large enough, the estimation can be more general and these architectures should be useful for estimating the probabilities with which a particular character appears as the successor of a given sequence generated by a grammar, as is suggested in [Servan,91]. The work we present below follows the idea presented in [Servan,88] and adds evidence to the truth of the previous statement. More specifically, for the three main experiments we have carried out, a Simple RNN was able to predict the next symbol(s) of strings randomly generated by a *Stochastic Deterministic Regular Grammar* (SDRG); furthermore, our results show that such network was also able to predict the generation probabilities of each character quite accurately.

---

† Work supported in part by the Spanish CICYT, under grant TIC-1026/92-CO2.
†† Supported by a Spanish MEC postgraduate grant.

2. Processing strings through Simple Recurrent Networks

2.1. Network architecture

The difficulty of simulating deterministic regular grammars through NNs arises from the fact that a substring can be followed by different characters and that the length of every string is variable. These two drawbacks lead to modifying the basic Multilayer Perceptron proposed in [Rumelhart,86] in order to represent time and to make the net remember past events which have already been presented (Dynamic Networks). With regard to overcoming these problems, some techniques have been proposed in the literature, such as [Jordan,86], [Elman,88], [Pineda,88] or [Williams,89], among others.

The network architecture used in this paper was a *Simple Recurrent Network* (SRN) introduced by [Elman,88] and shown in Figure 1. The net has only a hidden layer which is fed with the input units and the state of the hidden units in the previous step. This SRN was already employed in [Servan,91] for predicting successive elements in a sequence. Although, in principle, less powerful than general RNNs trained with the Real Time Recurrent Learning algorithm [Smith,89], this kind of SRNs seams to be powerful enough to learn a wide range of RGs with much lower computational requirements.

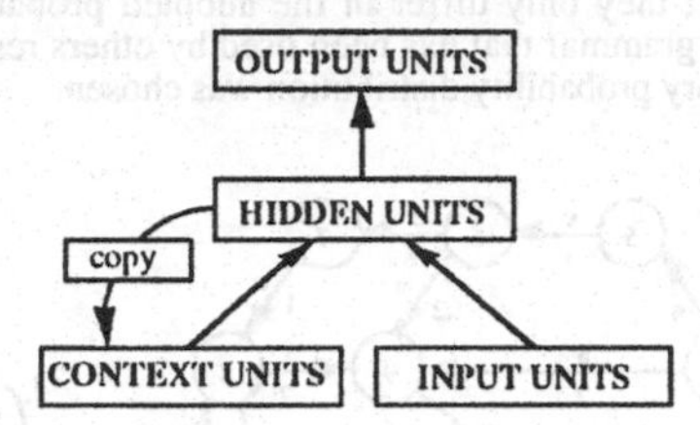

Figure 1. Elman's Simple Recurrent Network.

2.2 Training procedure

The *basic algorithm* for training this SRN was an adequate modification of the well-known Error-Back-Propagation algorithm [Rumelhart,86], in which strings were presented character by character to the net so that the next character(s) of the string may be predicted; that is, for each input symbol (character of the string) the target output consisted of the letter that followed this input symbol in such string. The output character was then applied to the network's input units and the next letter was used as the target, and so on.

In both the input and output layers, a *local representation* of the alphabet was used in such a way that every input (and output) unit was dedicated to the representation of one of the possible characters of the alphabet. Two labels that symbolized the beginning and end of a string were included. At the beginning of each string, the activations of the context units were initialized to zero. Later, for each character of the string, the forward step, the back-propagation of the error and the updating of the context layer were carried out. The optimization criterion for back-propagation was the average squared error.

The *complete training algorithm* repeatedly processed the basic algorithm by using an adopted fixed number of random training samples and evaluated the net on a validation set, until some established criterion was verified.

2.3. Correct estimation criterion

Let $p_1p_2...p_n$ be the probability of generating a given string $x_1x_2...x_n$ of length n by the stochastic regular grammar to be simulated, where each p_i corresponds to the probability of the grammar rule used to generate the i-th symbol; and let $q_1q_2...q_n$ be the *predicted probability* of the same string with the inferred network , where q_i corresponds to the output activation associated to the i-th character of the string. This *string* was considered as *correctly estimated* if the following condition was verified:

$$\left| \ln \left[\frac{(p_1 p_2 ... p_n)^{1/n}}{(q_1 q_2 ... q_n)^{1/n}} \right] \right| < Threshold. \tag{1}$$

Note that the expression between brackets corresponds to the *normalized likelihood quotient*. As our aim is to achieve $p_i = q_i$ $i=1,...,n$, the leftmost term should be very close to zero for every sample in the validation set.

3. Experimentation

3.1. Stochastic regular grammars to be simulated

The simulation of Stochastic Deterministic Regular Grammars using the above proposed SRN has been evaluated through its application to three different SDRGs G_1, G_2 and G_3, shown in Figure 2. The first two SDRGs generate those strings that begin with the substring "*ba*", include any number of "*c*"s except two and four and end with "*ae*"; they only differ in the adopted probability distributions. The third grammar is the, so-called, Reber grammar that has been used by others researchers in RNN [Servan,88] [Williams,89], in which an arbitrary probability distribution was chosen.

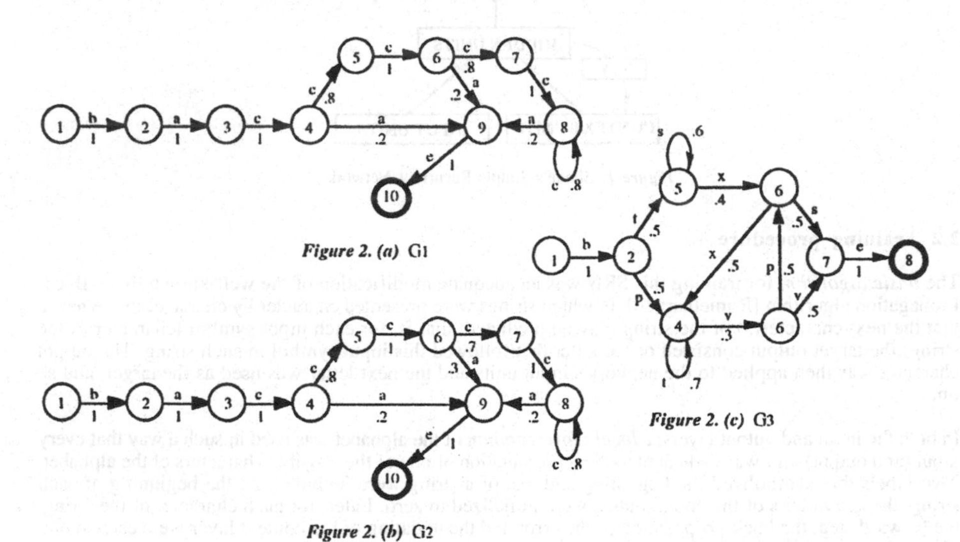

Figure 2. (a) G1

Figure 2. (b) G2

Figure 2. (c) G3

Figure 2. Deterministic Regular Grammars to be simulated.

3.2. Training and validation corpora

In all of these three experiments, the training set consisted of 30.000 strings generated by the grammar with an average length of 10 characters. These strings ranged from 5 to 59 symbols for G_1, from 5 to 54 for G_2 and from 5 to 57 for G_3. Each experiment was repeatedly evaluated on 10.000 new samples, corresponding to the validation set.

3.3. Features of the networks

The three SRNs used during the experiments consist of a two-layer Elman architecture with 20 hidden units; 4 binary input units and 4 outputs were adopted in the first two experiments and seven binary input and output units, in the last one, corresponding to the local representation of the associated alphabet, respectively. In every network, the activations of the context units were reset to 0.5 at the beginning of every string. On the other hand, the weights were initially set to random values between -0.5 and +0.5 and the sigmoid function in the range 0 to 1 was chosen as the non-linear activation function. Preliminary explorations aimed at 0.01 and 0.5 as adequate values for the learning rate and momentum, respectively, to be used for training G_1, 0.03 and 0.9 for G_2 and 0.09 and 0.5 for G_3.

3.4. Results

The probabilities of the SDRGs, G_1, G_2 and G_3, were induced by training three networks according to the features and learning method indicated in sections 3.3 and 2.2, respectively. Each net was estimated three times (trials) with the weights initialized to different random values. The validation of each net was carried out after each block of 5.000 samples randomly drawn from the training set (30.000 strings) and the inference stopped when 200.000 (repeated) training strings were presented to the net.

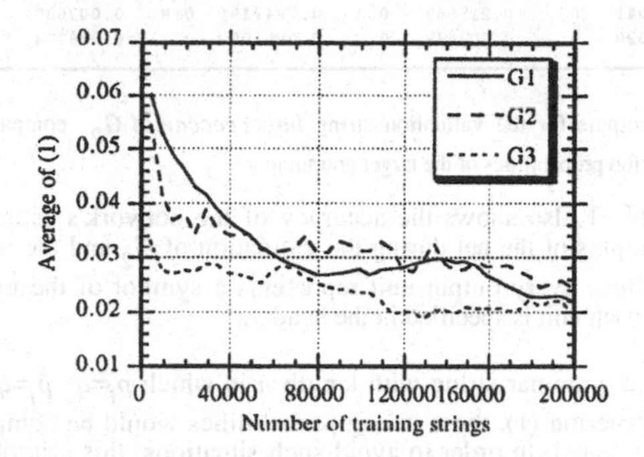

Figure 3. Evolution of the average absolute logarithm of the normalized likelihood quotient during validations for G_1, G_2 and G_3.

Figure 3 shows the evolution, for one of the three trials in each of the three experiments, of successive validations with regard to the average normalized likelihood quotient expressed in terms of absolute logarithms (leftmost term in (1)). These curves were smoothed using a window with length 10. The other runs which are not reported here produced almost identical results. These results show that these averages were very close to zero, which means that the network's outputs really approached the probabilities of the grammar.

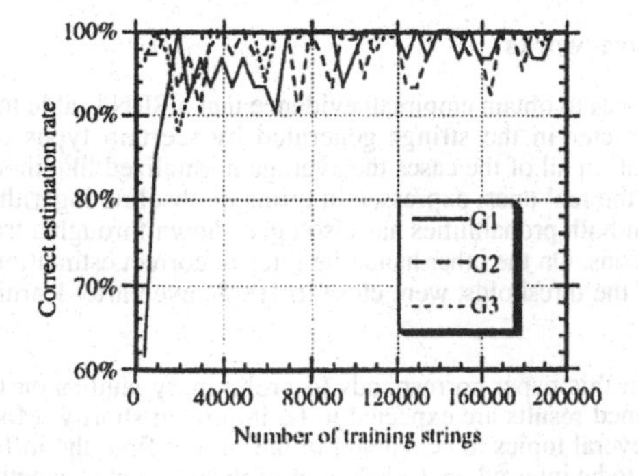

Figure 4. Evolution of the number of strings correctly estimated during validations, averaged over the three trials of G_1, G_2 and G_3.

The evolution of the correct string estimation rate during the validation is given in Figure 4. Each curve was averaged over the three trials associated to each experiment, assuming 0.08 as the value of the threshold used in the first two grammars and 0.15 in the last one. These results show that very good rates were achieved, even after learning only 5.000 or 10.000 samples.

	b		a		c		e	
	NET OUTPUT	GRAMMAR PROBAB.	NET OUTPUT	GRAMMAR PROBAB.	NET OUTPUT	GRAMMAR PROBAB.	NET OUTPUT	GRAMMAR PROBAB.
b	0.000225	0	0.997043	1	0.002046	0	0.001172	0
a	0.000011	0	0.000009	0	0.999285	1	0.001340	0
c	0.002724	0	0.194824	0.2	0.805240	0.8	0.002863	0
c	0.001143	0	0.010434	0	0.988384	1	0.001727	0
c	0.002109	0	0.288695	0.3	0.709830	0.7	0.002932	0
c	0.001559	0	0.075073	0	0.926458	1	0.002063	0
c	0.002019	0	0.246459	0.2	0.753762	0.8	0.002752	0
c	0.001747	0	0.136199	0.2	0.864369	0.8	0.002361	0
c	0.001941	0	0.215669	0.2	0.784719	0.8	0.002639	0
a	0.000028	0	0.003648	0	0.000109	0	0.996774	1

Table 1. Net outputs for the validation string *bacccccccae* of G_2 compared to the generation probabilities of the target grammar.

The trace depicted in Table 1 also shows the accuracy of the network's estimation of the grammar probabilities, where the outputs of the net during the validation of G_2 and the real probabilities of this grammar are compared. Since every output unit represents a symbol of the grammar's alphabet, the character corresponding to each unit is specified in the header.

Now then, let us suppose a grammar string with length n in which $p_i=q_j$, $p_j=q_i$ and $p_k=q_k$ $i \neq j$ $k \neq i,j$ $i,j,k=1,\ldots,n$. Attending to criterion (1), these wrong probabilities would be compensated and the string considered as correctly estimated. In order to avoid such situations, this criterion was changed to the following expression:

$$\frac{1}{n}\sum_{i=1}^{n}|\ln p_i - \ln q_i| < Threshold. \qquad (2)$$

The nine preceding runs were then carried out again assuming the new criterion. The results corresponding to the successive values of the leftmost term in (2) during the validations as well as the rates of strings correctly estimated were just slightly worse than those previously obtained.

4. Conclusions and future works

The main goal of this paper was to obtain empirical evidence that a SRN is able to estimate the generation probabilities of every character in the strings generated by (certain types of) SDRG. Three main experiments were carried out. In all of the cases the average normalized likelihood quotient between the estimated probabilities and the real ones, expressed in terms of absolute logarithms, gradually tended to zero. The closeness between both probabilities has also been shown through a trace showing the outputs of the net during the validations. On the other hand, the rates of correct estimation during the validations, using the adopted values of the thresholds, were close to 100%, even after learning only a few thousand samples.

Since the work presented in this paper corresponds to preliminary studies on this issue, the inference method as well as the obtained results are expected to be improved shortly. Moreover, the experiments carried out so far opened several topics to be studied in the future: first, the influence of the probability distribution of the grammar to be inferred on the behavior of the net, that is, whether or not any Stochastic Regular Grammar can be induced by a SRN; second, extract the whole structure of the grammar through the inferred SRN using simpler techniques than those used so far in the literature; and finally, to use this approach in Language Modeling for Automatic Speech Recognition.

References

[Boulard,92] *Continuous Speech Recognition: From Hidden Markov Models to Neural Networks.* H. Boulard. EUSIPCO'92, vol.1, pp. 63-70. 1992.

[Elman,88] *Finding Structure in Time.* Elman,J.L. Technical Report 8801. Center for Research in Language. University of California. La Jolla. 1988.

[Giles,92] *Learning and Extracting Finite State Automata with Second-Order Recurrent Neural Networks.* Giles,C.L et al. Neural Computation, no. 4, pp. 393-405. 1992.

[Jordan,88] *Serial order: A parallel distributed processing approach.* Jordan,M.I. Technical Report No. 8604. Institute of Cognitive Science. University of California. San Diego. 1988.

[Minsky,67] Minsky,M.L. Computation: Finite and Infinite Machines, Chap. 3.5. Ed. Prentice-Hall, Englewood Cliffs, New York. 1967.

[Pineda,88] *Generalization of BackPropagation to Recurrent and Higher Order Neural Networks.* Pineda,F.J. Neural Information Processing Systems. Ed. D.Z. Anderson. American Institute of Physics. New York. 1988.

[Pollack,91] *The Induction of Dynamical Recognizers.* Pollack,J.B. Machine Learning, no. 7, pp. 227-252. 1991.

[Rumelhart,86] *Learning sequential structure in simple recurrent networks.* Rumelhart,D.E. Hinton,G and Williams,R. Parallel distributed processing: Experiments in the microstructure of cognition, vol. 1. Ed. Rumelhart,D.E. McClelland,J.L. and the PDP Research Group. MIT Press. Cambridge. 1986.

[Servan,88] *Encoding sequential structure in simple recurrent networks.* Servan-Schreiber,D. Cleeremans,A. and McClelland,J.L. Technical Report CMU-CS-183. School of Computer Science. Carnegie Mellon University. Pittisburg, PA. 1988.

[Servan,91] *Graded State Machines: The Representation of Temporal Contingencies in Simple Recurrent Networks.* Servan-Schreiber,D. Cleeremans,A. and McClelland,J.L. Machine Learning, no. 7, pp. 161-193. 1991.

[Smith,89] *Learning Sequential Structure with the Real-Time Recurrent Learning Algorithm.* Smith,A.W. and Zipser,D. International Journal of Neural Systems, vol. 1, no. 2, pp. 125-131. 1989.

[Watrous,92] *Induction of Finite-State Languages Using Second-Order Recurrent Networks.* Watrous,R.L. and Kuhn,G.M. Neural Computation, no. 4, pp. 406-414. 1992.

[Williams,89] *Experimental Analysis of the Real-time Recurrent Learning Algorithm.* Williams,R.J. and Zipser,D. Connection Science, vol. 1, no.1, pp. 87-111. 1989.

Local Stochastic Competition and Vector Quantization

M. Graña, A. D'Anjou, F.X. Albizuri, F.J. Torrealdea, M.C. Hernandez
Dept. CCIA Univ. Pais Vasco/EHU[1]
Aptdo 649, 20080 San Sebastián, España
e-mail: ccpgrrom@si.ehu.es

Abstract: Stochastic relaxation techniques and competitive neural networks have been applied to Vector Quantization (VQ). We provide a short review of the relevant approaches and define a Local Stochastic Competition rule, relating it to the Soft Competition algorithm.

1 Introduction

Vector Quantization (VQ) is a technique taht can be used to map analog waveforms or discrete vector sources into a sequence of digital data for storage or transmission over a channel. A vector quantizer is a mapping of input vectors to one of a finite collection of predetermined codevectors. The set of all codevectors is called the codebook. In designing a vector quantizer, the goal is to construct a codebook for which the expected distortion, introduced by approximating an input vector by a codevector, is minimized. The VQ design problem is a nonconvex optimization problem. Recent approaches to VQ design include the application of competitive neural netwoks, and variations that involve a probabilistic interpretation of competition among units, such as the Soft Competiton [13]. In this paper, we propose a rather general Local Stochastic Competition scheme whose detailed analysis is on the way. In section 2 a formal definition of VQ will be given, in section 3 the most relevant approaches to its solution will be revised, in section 4 Local Stochastic Competition will be presented, and finally section 5 gives some conclussions and directions of future work.

2 Vector Quantization [3,4,9]

Given an space E, the Vector Quantization of this space can be formally defined through a couple of mappings: the coding and decoding maps.

$$C: E \rightarrow N \qquad D: N \rightarrow E$$

Coding and decoding are based on the existence of a partition of the space, and a set Y of representatives (codebook), one for each region.

$$E = \bigcup_{i=1}^{N} R_i \qquad \bigcap_{i=1}^{N} R_i = \emptyset \qquad Y = \{y_1, \ldots, y_N\}$$

---

[1] This work is being supported by a research grant from the Excma. Diputación de Guipuzcoa

The coding of a point (vector) $x \in E$ is its region index, and the decoding of a region index is its representative point (codevector).

$$C(x) = i \qquad s.t.\, x \in R_i$$

$$D(i) = y_i$$

Usually, the partition of the input space is not explicit, and so the coding is computed looking for the nearest-neighbor codevector, according to some distance d conveniently defined.

$$C(x) = i \qquad s.t. \quad d(x, y_i) = \min\{d(x, y_k) \;\; k = 1..N\}$$

Most of the literature assumes that $E=R^n$ and d is the squared euclidean metric. VQ design implies the search for the best codebook: the one that minimizes the expected codification-decodification error (distortion) δ:

$$\langle\delta\rangle = \int_E d(x, D(C(x)))P(x)dx$$

Given a sample $\{x_1,...,x_M\}$ of P(x), the probability distribution defined over E, and a codebook Y, the estimate of the distortion is

$$\delta(Y) = \frac{1}{M}\sum_{i=1}^{M} d(x_i, D(C(x_i)))$$

The next section revises some of the approaches to VQ learning developed so far. Two basic searching strategies can be defined depending on the search space chosen. The first goes on over the possible partitions of the sample $\{x_1,...,x_M\}$, computing the codevectors as the region centroids. The second updates the codebook and assumes that the space partition is always defined by the nearest-neighbor codevector rule.

3 Approaches to VQ

Classical approach

This approach is closely related to the Isodata procedure in Statistical Pattern Recognition [2], and was proposed in [9], and is sometimes referred as the generalized Lloyd algorithm (GLA). Starting from a random codebook, the algorithm computes the partition given the nearest neighbor rule, recomputes the codebook as the centroids of each region and iterates until a stopping criterion is met. Let m be the iteration number, then $R_i^{(m)}$ denotes the i region at the m iteration, $y_i^{(m)}$ its codevector and δ_m the estimate of the distortion. An iteration step is computed as follows:

for each x_i in the sample

find j s.t. $d(x_i, y_j) = \min\{d(x_i, y_k) \;\; k = 1..N\}$

Put x_i in $R_j^{(m)}$ and Accumulate $d(x_i, y_j)$ to δ_m

Recompute codebook $\quad y_k^{(m+1)} = \frac{1}{\left|R_i^{(m)}\right|}\sum_{R_i^{(m)}} x_i$

And the stopping criterion is usually of the form:

$$\frac{\delta_{m-1}-\delta_m}{\delta_m}<\varepsilon$$

The GLA is a local search procedure and, as such, it converges to (good) local minimum of the distortion. To transform GLA into a global search procedure, the sample is perturbed randomly at each iteration, or the GLA is embedded into other global search schemas [14].

Neural network approaches:

Competitive neural networks have been (proposed to be) applied to VQ. Neural network learning algorithms are introduced as on-line algorithms that produce the adaptation of the codebook as the sample points are presented. The weights of the neural network units are the codevectors, no explicit representation of the space partition is used. The Self-Organizing Map (SOM) [7] has been found of straightforward application to VQ [1,10,11,13], its formulation is as follows:

$$y_i(n)=y_i(n-1)+\alpha_i(n)\vartheta_i(i^*,n)\big(x(n)-y_i(n-1)\big)$$

where i^* is the winning unit

$\alpha_i(n)$ is the learning rate or step size

ϑ_i is the neighborhood function

The SOM has been shown [13] to converge to optimal VQ in the special case of codebook of cardinality 1, but a generalization of this result doesn't seem to be trivial.

A key problem in competitive (unsupervised) learning is the monopolization of the space by a subset of the codevectors. The neighborhood function of the SOM is an approach to solve this problem. Another approach, that has also been applied to VQ in [1], is Frecuency Sensitive Competitive Learning (FCSL). In FCSL, fair competition is encouraged by self-inhibition of above-average winning units, more formally:

$$y_i(n)=y_i(n-1)+\alpha_i(n)z_i(n)\big(x(n)-y_i(n-1)\big)$$

where $z_i(n)=\begin{cases}1 & d(x(n),y_i(n-1))c_i(n)=\min_k\{d(x(n),y_k(n-1))c_k(n)\}\\ 0 & \text{otherwise}\end{cases}$

$\alpha_i(n)$ is the learning rate or step size

$c_i(n)$ is the number of times that the i unit has win

Stochastic Relaxation approaches

Stochastic Relaxation procedures are introduced to overcome the local character of GLA. The Simulated Annealing (SA) algorithm [5,6,8] is a well known stochastic optimization procedure. In the proposed application of SA to VQ [13,14], an state (feasible solution) is defined as a partition of the sample: a vector $s=(s_1,\ldots,s_M)$ where $s_i=k$ if x_i is attached to the region R_k coded by the codevector y_k. Codevectors are computed as the centroids (averages) of each region, so each partition perturbation is followed by a codebook updating. Transitions are performed by randomly choosing a vector x_m whose assignement is given by s_m, and randomly choosing a new assignement $s_m=k$ according to the probability distribution:

$$P_m(k) = P\{s_m = k|s, Y\} = \frac{e^{-\beta d(x_m - y_k)}}{\sum_{j=1}^{N} e^{-\beta d(x_m - y_j)}}$$

It has been shown in [13] that $P_m(k)=\exp(-\beta\Delta_{m,k}\delta(Y))$, where $\Delta_{m,k}\delta(Y)$ is the variation in the distortion due to the reassignement of x_m. The described procedure fits into the pattern of SA, with the advantage of its ability to avoid rejected trials by computing the complete transition distribution and sampling it. Therefore the states (partitions) obtained with this procedure will follow (at termal equilibria) the Boltzmann-Gibbs distribution at each temperature ($1/\beta$), with the distortion playing the role of the energy. Proper temperature schedules must give near global minima of the distortion. Combined with GLA (performing a sequence of random transitions followed by a GLA iteration at each temperature) has been succesfully applied to image coding [14], with an appreciable reduction in computation time. Other stochastic relaxation approaches [14] involve the addition of noise to the sample or the codevectors, and are not relevant for the purposes of this paper.

Soft-Competition

Soft Competition as proposed in [13] is a mixture of the above SA approach and the SOM. Formally it is stated as follows:

$$y_i(n) = y_i(n-1) + \alpha_i(n) P_n(i)\big(x(n) - y_i(n-1)\big)$$

where $\alpha_i(n)$ is the step size

$$P_n(i) = \frac{e^{-\beta(n)d(x(n)-y_i(n-1))}}{\sum_{j=1}^{N} e^{-\beta(n)d(x(n)-y_j(n-1))}}$$

The transition probabilities of the former SA approach play the role of neighborhood functions, scaling the codevector adaptation. Each codevector has its own iteration counter n, which is updated by addition of the winning probability. All the codevectors are updated simultaneously at the presentation of a sample point x(n), in a sort of deterministic difussion process.

In this case, the transition probabilities do not fill in the SA requisites, so it can not be expected that the states (codebooks) follow the Boltzmann-Gibbs distribution at thermal equilibrium. Soft Competition can not be considered as a deterministic SA. However good experimental behavior has been reported, although the approach involves a lot of numerical subtleties concerning the step size and temperature schedules.

4 Local Stochastic Competition

In this section we define a Local Stochastic Competition rule over a set of units. This rule simulates probabilistically the competition among the units avoiding any global computation. This approach is closely related to the SA and Soft Competition approaches described above. We consider a layer of units whose weights are the codevectors in the VQ problem. Let us define the probability that a codevector decides to be the winner upon presentation of the x(n) input independently of the remaining codevectors as:

$$p_n(i) = e^{-\beta(n)d(x(n)-y_i(n-1))}$$

The Local Stochastic Competition (LSC) consists of the independent sampling by each unit of the Bernouilli distributions defined by these probabilities. This will result in one, some or none of the units declaring themselves as the winners. Note that when $\beta(n)\rightarrow\infty$ (low temperatures) x(n) must be very close to y to produce the firing of the unit. Thus the sequence $\{\beta(n)\}$ must not go to infinity if we want a limited number of units to clasify the inputs. On the other hand when $\beta(n)\rightarrow 0$ (high temperatures) any input will produce the firing. Thus to get a probabilistic approximate of the competition process, a critical value of β must be found. This value is related to the number of codevectors, and could be different from unit to unit. The work reported in [12] gives some directions in the search for the determination of β, although the problem statement found there may seem unrelated to ours. An Stochastic Learning rule for a set of units under LSC would consist in the independent updating of those units that declare themselves as winners.

$$P\left[y_i(n) = y_i(n-1) + \alpha_i(n)\left(x(n) - y_i(n-1)\right)\right] = p_n(i)$$

$$P\left[y_i(n) = y_i(n-1)\right] = 1 - p_n(i)$$

The learning process must start with low values of $\beta(n)$ (high temperatures), increasing them up to its critical value. It must produce the codevector values and the critical value of $\beta(n)$ that produces the best classification.
Let us define A_i as the event "unit i declares itself the winner upon the presentation of the x(n) input", and A as the event "at least one of the units declares itself the winner upon the presentation of the x(n) input". It is easy to see that the former Soft-Competition winning probabilities $P_n(i)$ are the conditional probabilities of A_i upon A.

$$P[A_i|A] = \frac{p_n(i)}{\sum_{j=1}^{N} p_n(j)} = P_n(i)$$

Note that under this interpretation, the Soft-Competition rule appears as a deterministic approximation of the Local Stochastic Competition, under the assumption that always at least one unit would be updated by the stochastic rule.
The event "none of the units declares itself as the winner" is the complementary of A and can have nonzero probability (1-P[A]). The empirical determination of the critical value of $\beta(n)$ follows from the observation that P[A], computed as the addition of the $P[A_i]$, can be greater than 1, an improper value for a probability. This results from the overlapping of the regions probabilistically attached to each codevector, due to the low value of $\beta(n)$. Thus $P[A]\leq 1$ could be used as a criterion to determine the critical value for $\beta(n)$. Alternatively P[not A]>0 could be used as an easier to compute criterion.

4.1 Preliminary experiments

We have performed some experiments trying to evaluate the feasibility of the LSC. Those experiments consisted in the codification/decodification of an image using LSC and the usual nearest neighbor (competitive) rule. The codebook used was obtained applying the Isodata procedure. Figure 1(a,b,c) shows the result of applying the common competitive rule to the image using codebooks obtained after 0, 20 an 100 iterations of the Isodata algorithm. Figure 1 (d,e,f) shows the results of applying LSC using the same codebooks.

Note that, as the codebook improves, the differences between the LSC and the become less appreciable.

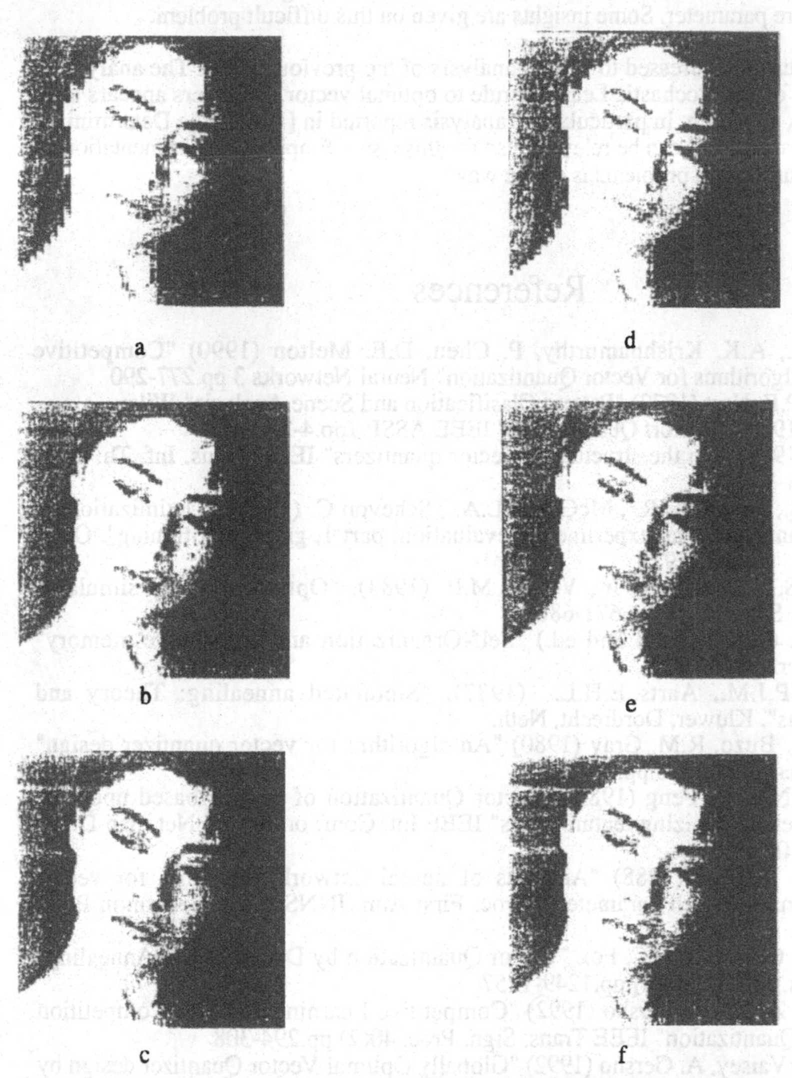

Figure 1. Codification using NN (a,b,c) and LSC (d,e,f)

5 Conclussions and further work

A Local Stochastic Competition rule has been presented which simulates probabilistically the competition among a set of units. This rule avoids any global computation. It uses an exponential probability of the unit declaring itself as the winner. To minimize the probability of more than one unit declaring itself as the winner, the temperature parameter must be adjusted. The Soft Competition can be interpreted as a deterministic approximation to the Local Stochastic Competition. An Stochastic Learnig rule, associated with the Local Stochastic Competition set of units, must determine the units weights and the critical value of the temperature parameter. Some insights are given on this difficult problem.

Further work must be addressed to formal analysis of the previous ideas. The analysis of the convergence of the Stochastic Learning rule to optimal vector quantizers appears to be related to the SA approach, in particular the analysis reported in [12] for the Deterministic Annealing approach to VQ can be relevant also for this issue. Empirical experimentation on some Vector Quantization problems is on the way.

References

[1] Ahalt S.C., A.K. Krishnamurthy, P. Chen, D.E. Melton (1990) "Competitive Learning Algorithms for Vector Quantization" Neural Networks 3 pp.277-290

[2] Duda R.O., P.E. Hart (1973) "Pattern Clasification and Scene Analysis" Wiley

[3] Gray R.M. (1984) "Vectort Quantization" IEEE ASSP 1pp.4-29

[4] Gersho A. (1982) "On the structure of vector quantizers" IEEE Trans. Inf. Th. 28(2) pp.157-166

[5] Johnson D.S., Aragon C.R. , McGeoch L.A. , Schevon C. (1989). "Optimization by simulated annealing: an experimental evaluation; part 1, graph partitioning". Oper. Res. 37(6), pp.865-892.

[6] Kirpatrick S., Gelatt C.D. Jr., Vecchi M.P. (1983). "Optimization by simulated annealing". Science 20, pp.671-680.

[7] Kohonen T. (1984) (1988 2nd ed.) "Self-Organization and associative memory" Springer Verlag

[8] Laarhoven P.J.M., Aarts E.H.L. (1987). "Simulated annealing: Theory and Applications". Kluwer, Dordrecht, Neth.

[9] Linde Y., A. Buzo, R.M. Gray (1980) "An algorithm for vector quantizer design" IEEE TRans. Comm. 28 pp.84-95

[10] Nasrabadi NM, Y. Feng (1988) "Vector Quantization of images based upon the Kohonen Self-Organizing feature maps" IEEE Int. Conf. on Neural Net. San Diego pp.1101-1108

[11] Naylor J., K.P. Li (1988) "Analysis of neural network algorithm for vector quantization of speech parameters" Proc. First Ann. INNS Meet. Pergamon Press p.310-315

[12] K. Rose, E. Gurewitz, G.C. Fox "Vector Quantization by Deterministic Annealing" IEEE Trans. Inf. Th. 38(4) pp.1249-1257

[13] Yair E., K. Zeger, A. Gersho (1992) "Competitive Learning and Soft Competition for Vector Quantization" IEEE Trans. Sign. Proc. 40(2) pp.294-308

[14] Zeger K., J. Vaisey, A. Gersho (1992) "Globally Optimal Vector Quantizer design by stochastic relaxation" IEEE Trans. Sign. Proc. 40(2) pp.310-322

MHC - AN EVOLUTIVE CONNECTIONIST MODEL FOR HYBRID TRAINING

José M. Ramírez
Paradigma C.A.
Apartado 67079, Caracas 1061
Venezuela
Phone: +58-2-2836942, Fax: +58-2-2832689
email: jramire@conicit.ve

Abstract

We present a Connectionist architecture called Modified Hyperspherical Classifier (MHC) based on: 1) the work of Cooper [5][6] and Batchelor [2][3][4] about Hyperspherical Classifiers, 2) the RCE paradigm as described by Scofield et al. [10] and 3) some new considerations derived from the search of an efficient model to perform heterogeneous pattern processing using Hybrid training algorithms depending on the nature of the problem, the availability of the correct output during the training process and the operational state of the network. We use the term Hybrid Training to define the use of a supervised or an unsupervised strategy to train the same network; this definition differs from the presented by Hertz et al. [7] as Hybrid Learning, which refers to different learning strategies for each layer.

HC, RCE and MDC

Like the nearest-neighbor classifier, the HC is based upon the storage of patterns that represent points in a space. The association of an unknown pattern with a known category is made by the distance function (Cartesian, Hamming, etc.). The main difference with the nearest-neighbor model is that each point has a region of influence that is defined by a sphere with the pattern's location point as the center. Each stored pattern with its region of influence defines a decision region associated with the category or class of the stored point.

RCE networks can be viewed as special cases of HC in that the stored patterns are stationary in the space and the regions of influence can only shrink and can not expand, this approach is also known as DSND (Disjoint Spheres/ No Drift) because the training process try to disjoint the spheres of different categories to avoid confusion in the classification but the stored point, the center of each sphere, remains stationary.

The MDC model uses the N-dimensional feature space as RCE and HC, but its functionality is based on the definition of at least one prototypical point for each category to be considered, this definition is made in the initialization stage. The classification is performed based on a distance metric between the prototypes and the pattern being processed. details of MDC can be found in [10],[12] and [13].

Modified Hyperspherical Classifier

The proposed architecture (MHC) has the following properties:

1. High storage density based on the N-dimensional feature space model.

2. The stored points (hidden layer units) act like discriminant functions.

3. The connections between the input and hidden layer are weighted. The weights represent the relative importance of each feature (input unit) to classification . This importance can be set "a priori" or by an analysis (e.g. covariance) of the input patterns.

4. The hidden units contain the location of the stored points associated and the size of the region of influence (radius of the sphere).

5. The connections between the hidden layer and the output layer are weighted. the weights are the result of a probabilistic density function applied to the region of influence of the unit. If an unit H1 shares part of the space with an unit H2 of another category, the weight of the connections of H1 and H2 with the corresponding output units will be <1 and will depend on the probability that a point in the shared region belongs to H1's or H2's category. The density function can be easily implemented as a count of the correct classifications performed over patterns that "fires" both neurons H1 and H2.

6. The spheres can shrink and expand during the training process to form the category areas in the feature space. Several spheres can be summarized into one or an sphere can be moved to a different location.

7. All the adjustments concerning the classification of the pattern being processed are made before processing the next pattern. The convergence of the net is always reached in just 2 epochs.

8. The initial radius of the spheres are not initially set to a fixed default value, but to the maximum value that can be assigned without include a point of a different category.

Figure 1. illustrates a MHC network

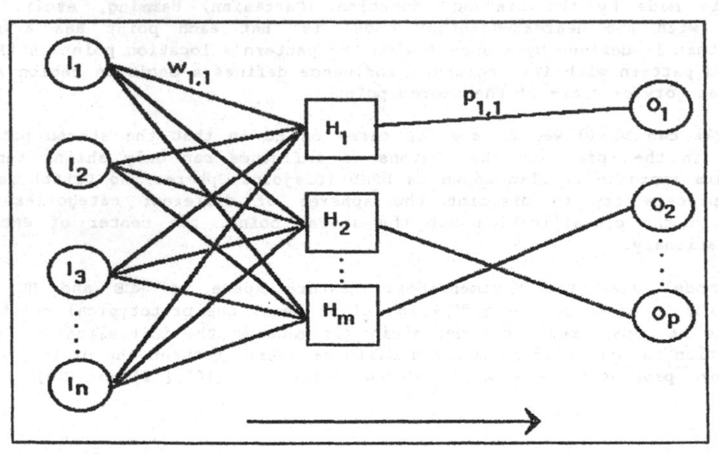

Figure 1.

Each input unit corresponds to a feature and each output unit corresponds to a pattern category. The hidden units are the stored points (spheres).

The key characteristics of MHC are: Feedforward, High storage density, Reduced connectivity, Partially distributed, Dynamic category learning and hybrid training.

Operation and Training of the MHC

When a pattern is presented to the network the activation of the hidden units will be given by:

$$H_j = G((SW_{ij}*(X_i-P_i)^2)^{1/2}) - T_j)$$

$$G(x) = \begin{cases} 0 \text{ if } x > 0 \\ 1 \text{ if } x <= 0 \end{cases}$$

where:

H_j **is the activation of the jth hidden unit**

X_i **is the ith feature of the input pattern**

P_i **is the ith component of the stored point**

W_{ij} **is the connection between the ith input unit and the jth hidden unit**

T_j **is the threshold or radius of the sphere (the region of influence)**

The activation of the output units will be given by:

$$O_j = H_i * P_{ij}$$

where:

P_{ij} **is the weight of the connection between the hidden unit and the associated output unit. Corresponding to the probabilistic density function.**

The function G(x) associates each hidden unit with a region in the feature space. The location of the region is stored in the hidden unit and the size of the region is determined by the threshold also stored in the hidden unit. Any input pattern falling within the region of influence of a given unit will cause the unit to transfer output to the corresponding output unit. A pattern will cause the firing of all the units that share the region where the pattern is located, if the hidden units project theirs output to more than one output unit (categories) the result will be an "ambiguous" classification; if only one output unit fires the answer of the network will be "known", the other possible output is "unknown", this results when no output unit fires.

The training process of MHC networks involves the execution of several actions: a) Creation of hidden units (spheres) and/or output units (categories). b) Adjustment of the spheres, traduced in shrinking, expansion, movement or jointing and c) Adjustment of the hidden to output connections using a probabilistic density function.

The conditions to perform the above actions are different in each training strategy. The control over their execution resides in a preprocessing module that determines if

the pattern is prepared (or intended) for unsupervised or supervised training, based in the availability of target responses and/or the nature of the patterns.

Hybrid training is necessary since there are problems where the solution is poorly known or just a subset of the possible categories are needed; in this case the network can be trained in a supervised way using the patterns which solution is known or just with patterns representing the desired categories. This process initializes the network and then a unsupervised training can take place until the entire feature space is covered. If the solution of a problem is totally unknown (e.g. exploratory data analysis) the network can be trained in an unsupervised way, using all the patterns available. The network will form clusters that can be analyzed for the experts in the domain and the categories can be defined to proceed with supervised training over the same network or a new one.

In the supervised strategy, if the output of the network is "unknown", a new hidden unit is created using the location of the current pattern and this unit is associated to the output unit of the correct category (target), if the category is not in the network (there is no output unit for this category) a new output unit is created, this process is known as dynamic category learning. the weight of the connection between the hidden unit and the output unit will be 1 and the size of the region of influence (threshold) will be the maximum value that does not include a point of another category. In an "ambiguous" situation, the spheres that are not associated to the correct category are shrunk so that they no longer cover the current pattern. If that size of the spheres reached the minimum value and remains the overlapping with spheres of another category, the probabilistic density function is used to adjust the connections between the overlapping units and the associated output units.

In the unsupervised strategy, if the output of the network is "unknown" a new category is created (hidden unit and output unit) using the same procedure described above, the category identifier is generated by the network. If more that one output unit fires, the categories involved are ranked according to the computed degree of belonging (as in MDC). The initial size of the spheres in the unsupervised strategy is settled to a default maximum value given by the user, obviously the maximum possible size can not be used as in the supervised strategy.

Results

Problem 1. Analysis and classification of seismic data used for petroleum exploration. This is a typical example of exploratory data analysis; a huge amount of data is presented and the model must created clusters identifying "categories" or subsets. For this unsupervised problem we used three training sets and one testing set:

Set	Patterns	Categories (estimated)
UTrain-1	13,100	12
UTrain-2	27,000	12
UTrain-3	11,500	8
UTest-1	5,000	12

Table 1.

The train sets were obtained from different petroleum fields. The encoding method generated 15 features, according to a classification scheme given by the experts in petroleum exploration. The results are shown in the following tables. The sets were presented to the model in the order shown in the tables:

RCE

Set	Elapsed time (minutes)	Categories created	Prototypes created	Epochs
UTrain-1	15.05	21	217	4
UTrain-2	10.15	1	13	4
UTrain-3	7.25	0	22	3
Totals	32.45	22	252	
Set	**Elapsed time (minutes)**	**Categories found**	**Ambiguous responses**	**Unidentified**
UTest-1	4.00	15	10%	5%

Table 2.

MHC

Set	Elapsed time (minutes)	Categories created	Prototypes created	Epochs
UTrain-1	17.00	10	160	2
UTrain-2	13.40	3	7	2
UTrain-3	5.25	0	3	2
Totals	35.65	13	170	
Set	**Elapsed time (minutes)**	**Categories found**	**Ambiguous responses**	**Unidentified**
UTest-1	2.35	13	3%	2%

Table 3.

Problem 2. Classification of petroleum wells according to the interpretation of pressure charts. An analysis of pressure charts can lead to a classification of the wells; This classification is used as an input in the diagnostic/optimization process followed by petroleum engineers. For this unsupervised problem we used three training sets and one testing set:

Set	Patterns	Categories
STrain-1	25	5
STrain-2	50	8
STrain-3	15	5
STest-1	30	7

Table 4.

The train sets were obtained from digitized images of the pressure charts. The encoding method generated 28 features, according to a topological study made by J. Ortiz [8]. The results are shown in the following tables. The sets were presented to the model in the order shown in the tables:

RCE

Set	Elapsed time (minutes)	Prototypes created	Epochs
STrain-1	2.15	15	4
STrain-2	4.10	12	4
STrain-3	1.45	0	3
Totals	7.70	27	
Set	**Elapsed time (minutes)**	**Ambiguous responses**	**Unidentified**
STest-1	2.05	15%	6%

Table 5.

MHC

Set	Elapsed time (minutes)	Prototypes created	Epochs
STrain-1	3.20	12	2
STrain-2	5.30	5	2
STrain-3	2.00	0	2
Totals	10.50	17	
Set	**Elapsed time (minutes)**	**Ambiguous responses**	**Unidentified**
STest-1	1.50	5%	0%

Table 6.

Conclusions

1. The MHC adapt to a wide range of applications, since there's no restriction derived from the size of the problem (storage requirement) or nature (unsupervised or supervised).

2. Compared with RCE, the elapsed time for MHC is higher, but the convergence is reduced to 2 epochs, no matter what training strategy is used (see tables 1 to 6).

3. The design of the network is simplified, due to the evolving nature of the model. The network designs itself.

4. The MHC is able to learn to separate very complex decision surface (see tables 1,2 and 3).

5. The hardware implementation of MHC is easy, compared to parametric feedforward networks (e.g., backpropagation).

Acknowledgments

Thanks to my colleague Mauricio Paletta who wrote most of the C++ code for RCE and also the base code for MDC.

References

[1] B. Batchelor, B. Wilkins. "Adaptive Discriminant Functions". Pattern Recognition. 1.968 IEE Conf. Publication, 42, pp 168-178.

[2] **B. Batchelor.** "Learning machines for pattern recognition". Ph.D. Dissertation, Southampton.

[3] **B. Batchelor.** "Practical approach to pattern classification". 1.974. New York. Plenum Press.

[4] **P. Cooper.** "The Hypersphere in pattern recognition". 1.962. Information and Control, 5, pp. 324-346.

[5] **P. Cooper.** "A note on an adaptive hypersphere decision boundary". 1.966. IEEE Transactions on Electronic Computers, pp. 948-959.

[6] **J. Hertz et al.** "Introduction to the theory of Neural Computation", Addison-Wesley, 1.991.

[7] **N. Nilsson.** "Mathematical Foundations of Learning Machines". 1.990. Morgan-Kaufmann.

[8] **J. Ortiz.** "An Artificial Intelligence aproach to model-based gaslift troubleshoting". M.S. Thesis. Texas A&M Univ. Texas. USA. 1.990

[9] **J. Ramírez.** "Use of Connectionist Classifiers to detect regularities in data". 1.991. Intevep S.A. tech. report.

[10] **J. Ramírez, I. Torres.** "Conclusions of the use of connectionist models as feature detectors", Intevep S.A., 1.991. tech. report.

FAST-CONVERGENCE LEARNING ALGORITHMS FOR MULTI-LEVEL AND BINARY NEURONS AND SOLUTION OF SOME IMAGE PROCESSING PROBLEMS

Naum N. Aizenberg*, Igor N. Aizenberg**

* University of Uzhgorod, Professor of the Department of Cybernetics
Geroev Stalingrada 28, kv. 4, Uzhgorod, 294015, UKRAINE.
Tel (+7 03122) 23908; Fax (+7 03122) 36120

** Joint Venture PGD, Research Department Chief, Dr.Sc.;
Engelsa 27, kv. 32, Uzhgorod, 294015, UKRAINE.
Tel. (+7 03122) 63269; Fax (+7 03122) 36120

This work is supported by Maltian company DKL Ltd -198 Old Bakery Street, Valetta, Malta. Fax (356) 221893.

Abstract

In this paper we consider fast-convergence learning algorithms for multi-valued and universal binary neurons. These neurons are suggested to be used for the design of neural networks based on Cellular Neural Networks (CNN) —in the sense of connections between neurons. On the basis of such networks we offer a solution to some problems of image processing. For instance, a highly efficient method for contours distinguishing, obtained by the learning algorithm described in this paper is presented.

I. INTRODUCTION

Intensity of applying neural networks is closely connected with the design of both efficient learning algorithms and connections between neurons. CNN, introduced in [1] and being extensively developed recently [2, 3], became a brilliant alternative to conventional computers for image processing and recognition. Despite the fact that a lot of complicated tasks are solved on the basis of CNN [1-3 and others], we still assume that in a great number of papers the range of processed signals is "unfairly" restricted, since mostly binary (bipolar) signals are considered. Another restriction, not only in CNN but in most neural networks schemes, is to develop and apply only Perceptron (or Perceptron-like) learning. This implies that neural elements taught by such algorithms can perform only threshold functions. But the number of such functions is very small in comparison with the number of all Boolean (or multi-valued) functions of n variables (for instance, from 65536 Boolean functions of 4 variables only nearly 2000 are of the threshold type). To break this restrictions and extend both the neural networks functionality and the range of problems that are solved on neural networks basis, we recently suggested using neural elements with complex-value weights —universal neural element [4] which performs arbitrary (not only threshold) Boolean functions of n variables and multi-valued neural element [5] which performs multi-valued functions. In [5] a learning algorithm for multi-valued neurons was also considered and proposed, using CNN based on such neurons, as associative memory for grey-scale images storing. Now, developing the ideas considered in [5], we suggest the fast convergence learning algorithm for the universal neural element considered in [4] and apply such algorithm for image processing on CNN based on such neurons.

II. MULTI-VALUED AND UNIVERSAL NEURONS AS BASIC NETWORK ELEMENTS.

Suppose we have a CNN of dimension $N \times M$. We will apply *multi-valued* and *universal neural elements* as basic neurons of this network. Each of these elements performs the following transformation:

$$Y_{ij}(t+1) = F\,[\,W_0 + \textstyle\sum_m W^{ij}_m X^{ij}_m(t)], \tag{1}$$

where Y_{ij} is a neuron state, W^{ij}_m (compare it with the control operator B in the original CNN [1]) is the connection weight corresponding to the m-th input of the ij-th neuron and X^{ij}_m is the input signal value on the m-th input of the ij-th neuron. $F(\bullet)$ represents the output function of the i-th neuron which will be defined further on.

If (1) describes multi-valued neurons [5], input signal X and output signal Y for each neuron are located in the range $0,\ldots,k$-1, i.e. each neuron at each particular moment performs some function of k-valued logic determined by the weights W (i.e., for byte images k=256). In [5], it was proposed to codify input and output signals of multi-valued neuron by the k-th power roots of a unit:

$$\text{if} \quad j = 0, 1, \ldots, k-1, \text{ then } \quad r_j = \exp(\mathrm{i}\ 2\pi\ j/k), \tag{2}$$

where i is the imaginary unit. According to (2), $r_0 = 1$, $r_1 = \varepsilon$, the primitive k-th power root of a unit, $r_2=\varepsilon^2,\ldots,\ r_{k-1}=\varepsilon^{k-1}$.

If (1) describes a universal binary neuron [4], the input signal X and the output signal Y for each neuron are coded by a binary alphabet {1,-1}. The output function F of both the multi-valued and the universal binary neurons is defined similarly in the next way. Let arg(Z) be the argument of a complex number Z ($0 < \arg(Z) < 2\pi$). Then the output function for a multi-valued neuron is:

$$\mathrm{SIGN}(Z) = \exp(\mathrm{i}\ 2\pi\ j/k) = \varepsilon^j, \text{ if } 2\pi j/k < \arg(Z) < 2\pi(j+1)/k \tag{3}$$

and for a universal binary neuron it is

$$P(z)= \exp(i\ 2\pi\ j/m) = (-1)^j, \text{ if } 2\pi j/m < \arg(Z) < 2\pi(j+1)/m, \quad (4)$$

where m is a natural number depending on n (the number of neuron inputs). It must be noted that, in order to ensure the universal functionality of neurons, estimates for m are: low - $2n$ and up one equal to number of sign changes in the xor function of n variables (i.e., for n=9, what corresponds to a CNN with 3*3 nearest neighbours, the number of changes equals 320 and this guarantees the universal functionality of neurons). If m=2, then (4) is transformed into (3) in the Boolean case (k=2), but then neurons can perform only threshold functions. If m=4, then any n-neuron described by (4) is multi-functional and can perform a considerably larger number of functions in comparison with the number of threshold functions of n variables. Naturally, since Y and X are complex numbers, then the weights W are complex too.

III. LEARNING ALGORITHMS

In [5], we proposed a learning algorithm for the multi-valued neuron described by (3). This algorithm is reduced to the iterative evaluation of the sequence of weighting vectors W

$$W_{m+1} = W_m + \omega_{Bm}\, \varepsilon^q\, \bar{X}\ , \tag{5}$$

where W_m and W_{m+1} are the current and next weighting vectors, w is the correction coefficient, X is the vector of neuron input signals, and ε^q is the value of the neuron output signal. The initial value W_0 for vector W can be assigned arbitrarily and (5) is applied to figuring out vector W (if it exists) satisfying

$$\text{CSIGN}(X, \bar{W}) = \varepsilon^q \tag{6}$$

for all values of input signals X (or for all learning subsets). The correction coefficient ω_{Bm} in (5) is introduced to make the value of expression

$$\text{CSIGN}(X, \bar{W}_{m+1}) = \text{CSIGN}\ [\ (X, \bar{W}_m) + (n+1)\ \omega_{Bm}\, \varepsilon^q]$$

closer to ε^q at each successive step.

Let us consider a learning algorithm for a universal binary neuron. First of all, a learning algorithm for the multi-functional case will be considered. Let m=4 in (4). Then it becomes

$$\begin{cases} P(Z) = 1, \text{ if } \quad 0 < \arg(Z) < \pi/2 \quad \text{or} \quad \pi < \arg(Z) < 3\pi/2 \\ P(Z) = -1, \text{ if } \quad \pi/2 < \arg(Z) < \pi \quad \text{or} \quad 3\pi/2 < \arg(Z) < 2\pi \end{cases} \tag{7}$$

The learning procedure in this case is the iterative procedure based on (5). q corresponds in this case to the right bound of the complex-plane sector defined by (7). The weighted sum

$$Z=(W,X)=W_0 + W_1X_1 + ... + W_nX_n$$

must be obtained in order to under the current learning subset (or current value of input vector X).

The complex plane is separated by (7) into four equal sectors. If Z gets into an "incorrect" sector, weights must be corrected by (5) to get Z into one of the neighbour sectors (on the left or on the right, depending on which is closer to the current value of Z). This determines the value of coefficient ωB_m in (5) (i is the imaginary unit):

$$\omega B_m = \begin{cases} 1, & \text{if } Z \text{ must be "moved" to the right sector} \\ i, & \text{if } Z \text{ must be "moved" too the left sector} \end{cases} \tag{8}$$

Many experiments showed that the convergence of such learning process is pretty fast (only a few iterations are necessary) provided that the neuron output function is P-realizable [4]. In the case of universal binary neurons, learning is reduced to multi-valued neuron learning (iterative procedure (5), or look at [5] for a more precise presentation). Let $Y = (y_1 ,..., y_s)$ be a vector of neuron output function values ($y_i \in \{1,-1\}$, $0 < s < 2^n$). Let us recode the binary vector Y to vector $\tilde{Y}$ with k-valued components, where $4 < k < \tilde{m}$ ($\tilde{m}$ being equal to number of sign changes in the xor function of n variables). This recoding is given by: $\tilde{y}_0 = \varepsilon^0 = 1$; if $y_1 = y_0$, then $\tilde{y}_1 = \varepsilon^0$, else $\tilde{y}_1 = \varepsilon$, and further (let $\tilde{y}_{i-1} = \varepsilon^p$):

$$\tilde{y}_i = \begin{cases} \tilde{y}_{i-1} = p & \text{if} \quad y_i = y_{i-1} \\ \varepsilon(p+1) \bmod k & \text{if} \quad y_i \neq y_{i-1} \end{cases} \tag{10}$$

After that recoding, one has to apply the learning procedure for multi-valued neurons [5] defined by (5) and (6) in order to train the neuron to perform the output function $\tilde{Y}$, or, what is the same, the output function Y. It is evident that the weight vector W evaluated by such procedure for the multi-valued function $\widetilde{Y}$ obtained from the binary function Y by (10) is the same as for function Y.

IV. APPLICATION TO IMAGE PROCESSING

Lots of applications of CNN are closely connected to image processing and recognition [1-3]. One of the main advantages of the universal neuron and its learning algorithm is the possibility of training each neuron (and the neural network, respectively) to solve various image processing problems in the same way, namely by one of the learning algorithms described above. Now we will consider image processing learning on the example of contours distinguishing. This problem will be solved on the base of the learning algorithm for multi-functional neurons defined above by (5), (7) and (8). Let us consider an N*M dimensional CNN where each neuron is connected only with 8 neurons from the nearest neighborhood (3*3 subnet). In this case, each neuron performs a Boolean function of 9 variables. First of all, we have to define the function that will be implemented by each neuron. That function may be easily defined by the following formula:

$$Y\begin{pmatrix} x1 & x2 & x3 \\ x4 & x5 & x6 \\ x7 & x8 & x9 \end{pmatrix} = \begin{cases} -1 & \text{if } x5 = -1 \text{ or } x1 = \ldots = x5 = \ldots = x9 = 1 \\ 1 & \text{if } x5 = 1 \text{ and if only one of } x1, \ldots, x4, x6, \ldots, x9 \neq 1 \end{cases}$$

In other words, a contour is detected if and only if it passes through the central point of the 3*3 window (the central neuron of the 3*3 subnet). Afterwards, the generation of the output function learning process can start. Just 51 iterations (less than 2 min. on a 286 IBM PC) are necessary for solving this problem. The connection weights (the template) obtained by this procedure are :

W = (6, 17);

(-3.23, 3.64)	(-7.83,-0.02)	(-2.50,-0.13)
(-6.94, 3.34)	(35.06, 0.66)	(-6.5 , 2.13)
(-1.83, 0.02)	(-7.23,-3.64)	(-1.00, 1.00)

(11)

It is evident that the problem of contours distinguishing on any binary pattern is solved by template (11). In Fig. 1, one can see an example of the solution of this problem by software simulation. But it is more interesting the problem of contours distinguishing not on binary but on grey-scale images. In that approach, the problem can be solved as following. A byte grey-scale image is separated into 8 bit planes and then contours have to be detected on each bit plane on the network by template (11). Then the 8 contours planes obtained have to unite into a byte image which will represent the grey-scale contours of the input grey-scale image. This approach gives much better results in comparison with differential operators, i.e. the Laplace operator [6]. One may compare results of contours distinguishing on the image "Moscow Kremlin" (Fig.2) in the approach presented here (Fig.3) with those obtained by using the Laplace operator (Fig. 4).

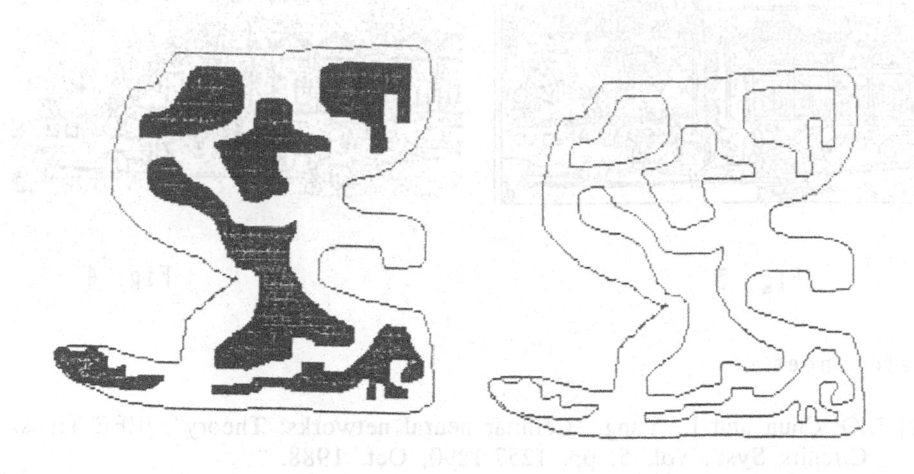

Fig. 1

Fig. 2

Fig. 3

Fig. 4

References

[1] L.O. Chua and L. Yang, "Cellular neural networks: Theory", IEEE Trans. Circuits Syst., vol. 5, pp. 1257-1290, Oct. 1988.

[2] Proceedings of the 1990 IEEE International Workshop on CNN and their applications (CNNA-90). Budapest, 1990, IEEE Catalog No. 90TH0312-9.

[3] Proceedings of the Second International Workshop on Cellular Neural Networks and their Applications (CNNA-92), Munich 1992. IEEE Catalog No. 92TH0498-6.

[4] N. N. Aizenberg and I. N. Aizenberg, "Model of the Neural Networks basic elements (cells) with universal functionality and various hardware implementations", Proc. of the 2nd International Conf. on Microelectronics for Neural Networks, Munich, Kirill & Methody Verlag, 1991, pp. 77-82.

[5] N. N. Aizenberg and I. N. Aizenberg, "CNN based on multi-valued neuron as a model of associative memory for grey-scale images", Proc. of the 2nd International Workshop CNNA-92, Munich, 1992, IEEE Catalog No. 92TH0498-6, pp. 36-41.

[6] W. K. Pratt "Digital Image Processing" John Wiley & Sons Publisher House, N.Y. 1978.

Invariant Object Recognition using Fahlman and Lebiere's Learning Algorithm

Kazuki ITO Masanori HAMAMOTO Joarder KAMRUZZAMAN Yukio KUMAGAI

Department of Computer Science and Systems Engineering
Muroran Institute of Technology
27-1 Mizumoto-cho, Muroran-shi, 050 JAPAN

Abstract A new neural network system for object recognition is proposed which is invariant to translation, scaling and rotation. The system consists of two parts. The first is a preprocessor which obtains projection from the input image such that, for any rotation and scaling of standard image, the projection results are reduced to cyclically shifted ones, and then adopts the Rapid Transform [9] which makes the projected images cyclic shift invariant. The second part is a neural net classifier which receives the outputs of preprocessing part as the input signals. The most attractive feature of this system is that, by using only a simple shift invariant transformation (Rapid Transform) in conjunction with the projection of the input image plane, invariancy is achieved and the system is reasonably small. Experiments with six geometrical objects with different degree of scaling and rotation show that the proposed system performs excellent when the neural net classifier is trained by Fahlman and Lebiere's learning algorithm [8].

1. Introduction

One of the most fundamental problems in the area of pattern recognition is associated with the scaling, translation and rotation of the patterns, and design of an invariant system with less complexity and possible minimum size is of extremely importance. For invariant recognition of planary object (binary image), many of the existing algorithms simplify the problem by reducing it to a shape or contour recognition. Classical approaches are Fourier descriptors [1], moment invariants [2], stochastic models [3] etc. Methods usually adopted for classification by these approaches are nearest neighbour, Bayesian decision theory.

Several invariant systems have been reported using neural networks. In [4], Widrow's ADALINE is proposed to build an invariant object recognition system which needs large number of slabs, each slab being invariant to a specific degree of translation, rotation, or scaling. This makes the resultant network very large and complex since the network must accommodate all the possible degrees of translation, rotation and scales. Neocognitron developed by Fukushima [5] is not so effective for rotation-invariant. Higher order neural networks [6] are also proposed to achieve invariance but use of higher order increases the number of connections astronomically and makes its implementation for large scale image planes extremely difficult. Recently, in [7] a projection based object recognition system has been proposed whose preprocessing part consists of projections from the input image plane and two transformations (Mellin and Rapid Transforms). The classification network used here is the nearest neighbour classifier.

In this paper, we present an invariant object recognition system using neural network. The system is simple and of reasonable size. It consists of two parts, one preprocessing part and the other a multilayer neural net classifier. The preprocessing obtains projection from the input image in such a way that for any rotated or scale input images, the projected results are reduced to cyclically shifted versions of those of standard images, and then employs the Rapid Transform (RT) [9] which makes the projected results cyclic shift invariant. Thus using only a simple shift invariant transform (RT), we can achieve rotation, scaling and translation invariant system. The classifier is a multilayer feedforward network. Two types of networks, one trained by Backpropagation (BP) algorithm [10] and the other by cascade correlation learning algorithm proposed by Fahlman and Lebiere (FL algorithm) [8] were considered. Experiments with six geometrical objects show the excellent performance of the proposed system for various degrees of translation, rotation and scaling; especially when the neural network classifier is trained by FL algorithm.

2. Description of the system

The proposed system as shown in Fig. 1 mainly consists of two parts; preprocessing and classification network and is described below.

2.1 Preprocessing

Preprocessing is done in two stages. The first stage takes the projection from the input image plane and the second stage performs Rapid transformation on the projection results.

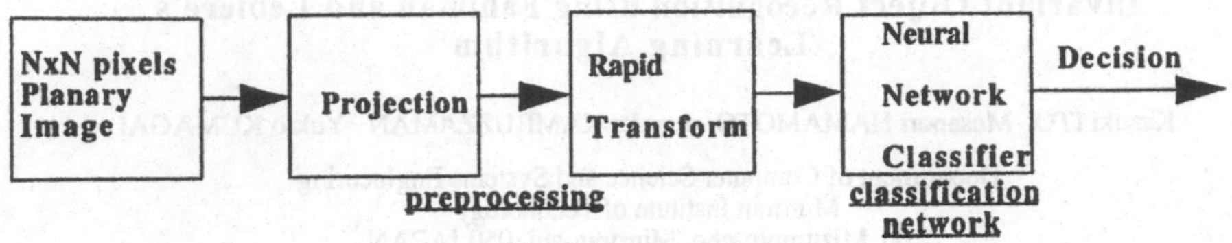

Fig. 1 System diagram

2.1.1 Projection

The projection block, at first, computes the center of gravity $(x^{(\alpha)}_g, y^{(\alpha)}_h)$ of the object $f^{(\alpha)}(x,y),(\alpha=1, 2,...,v)$ by averaging the values of pixels of object $f^{(\alpha)}(x,y)$ which take on the values of 1 or 0, depending upon a black or white pixel, respectively. Computational method is given below:

$$x^{(\alpha)}_g = \sum_{i=1}^{N}\sum_{j=1}^{N} x_i\, f^{(\alpha)}(x_i,y_j) \Big/ \sum_{i=1}^{N}\sum_{j=1}^{N} f^{(\alpha)}(x_i,y_j),$$

$$y^{(\alpha)}_h = \sum_{i=1}^{N}\sum_{j=1}^{N} y_j\, f^{(\alpha)}(x_i,y_j) \Big/ \sum_{i=1}^{N}\sum_{j=1}^{N} f^{(\alpha)}(x_i,y_j),$$

where N is the number of pixels on the x and y axes respectively. Then, the pixel $f^{(\alpha)}(x_i,y_j)$ of object $f^{(\alpha)}(x,y)$ is projected onto the r-θ plane spanned by two mutually orthogonal axes, r and θ, associated with the half non-negative real number. The concrete computational method of this projection is as follows:

$$r^{(\alpha)}_{(ij)} = \sqrt{\left(x_i - x^{(\alpha)}_g\right)^2 + \left(y_j - y^{(\alpha)}_h\right)^2} \cdot f^{(\alpha)}(x_i, y_j),$$

$$\theta^{(\alpha)}_{(ij)} = \left(\tan^{-1}\frac{y_j - y^{(\alpha)}_h}{x_i - x^{(\alpha)}_g}\right)\cdot f^{(\alpha)}(x_i, y_j),$$

where, in the above expressions, multiplication of $f^{(\alpha)}(x_i,y_j)$ in the right hand side implies that the projection is actually carried out only with the black pixels of object.

That is, $r^{(\alpha)}_{(ij)}$ is computed as the Euclidian distance between the pixel $f^{(\alpha)}(x_i,y_j)$ and the center of gravity $(x^{(\alpha)}_g, y^{(\alpha)}_h)$, and $\theta^{(\alpha)}_{(ij)}$ is computed as the angle measured counterclockwise between the pixel $f^{(\alpha)}(x_i,y_j)$ and the straightline $y=x^{(\alpha)}_g$ through the center of gravity. And, for all i and j, the pairs $(r^{(\alpha)}_{(ij)}, \theta^{(\alpha)}_{(ij)})$ are described on the r-θ plane according to the increasing order of $\theta^{(\alpha)}_{(ij)}$ and, for simplicity, this projection(the relationship as a function $r^{(\alpha)}_{(ij)}$ of $\theta^{(\alpha)}_{(ij)}$) will be denoted by $r^{(\alpha)}_{(ij)}(\theta^{(\alpha)}_{(ij)})$ or $r^{(\alpha)}(\theta^{(\alpha)}_{(ij)})$.

Now, the results of projection $r^{(\alpha)}(\theta^{(\alpha)}_{(ij)})$ give a cyclically shifted result for any rotation of the standard object since any rotation of the object by γ degree will cause the projection $r^{(\alpha)}(\theta^{(\alpha)}_{(ij)})$ to be shifted by γ degree along the θ-axis. If the projected results are normalized by their maximum value, we obtain projection results which are irrespective to scale transformation. Thus to make the system rotation and scale invariant, we need a cyclic shift invariant transform. Here, the projection results go through the Rapid Transform in the next stage and this makes the output of this transformation invariant to translation, scaling and rotation.

In order to use the Rapid Transform which performs cyclic shift invariant transform, the projection results of the object image are quantized into the averaged data $S^{(0)}_k(\alpha)$, $(k=0,1,..,N-1; \alpha=1,2,...,v)$, which number 2^M, $(M=\log_2 N)$, as follows:

$$S^{(0)}_k(\alpha) = \frac{1}{n_k} \sum_{\theta^{(\alpha)}_{(ij)} \in [k\cdot\Delta\theta,\,(k+1)\cdot\Delta\theta]} r^{(\alpha)}_{(ij)}(\theta^{(\alpha)}_{(ij)}), \qquad (k=0,1,2,...,N-1; \alpha=1,2,...,v),$$

where $\Delta\theta = 2\pi/2^M$ is the quantized width of θ-axis, and n_k is the number of $r^{(\alpha)}_{(ij)}(\theta^{(\alpha)}_{(ij)})$ included in k-th quantized interval $[k\Delta\theta, (k+1)\Delta\theta]$ of θ-axis.

2.1.2 Rapid Transform

The Rapid Transform (RT) was developed by Reitboeck and Brody [9]. This transform has the property of cyclic shift and reflection invariance of the input data sequence and is attractive for its computational simplicity. The RT is not an orthogonal transform, and has no inverse transform. The computation of RT is as follows:

$$S^{(R)}_{2u} = \left| S^{(R-1)}_{u} + S^{(R-1)}_{u+\frac{N}{2}} \right| ,$$
$$S^{(R)}_{2u+1} = \left| S^{(R-1)}_{u} - S^{(R-1)}_{u+\frac{N}{2}} \right| , \quad (u = 0, 1, \cdots, N/2-1) ,$$

where N is the number of data, and R is transformation step (R=1,2,...,M(=$\log_2 N$)), and $\{S^{(0)}_k : k=0,1, \ldots, N-1\}$ are needed to be substituted with the projection results $\{S^{(0)}_k(\alpha) : k=0,1, \ldots, N-1\}$ obtained in subsection 2.1.1 for every α (α=1,2,...,v) and $S^{(R)}_u$ is the intermediate data of the R-th transformation step.

Since this transformed result $\{S^{(M)}_k : k=0,1, \ldots, N-1\}$ is cyclic shift invariant, the output of the preprocessing is invariant to scaling, translation and rotation. Compared to the other invariant systems, this preprocessing is very simple, computationally less expensive and of reasonable size. The interesting thing that should be mentioned here is that, by incorporating only a simple shift invariant transform, an invariant system can be designed.

2.2 Neural network classifier

The next step is the classification of objects for which a neural network classifier is used. The classifier is a multilayer feedforward network trained with the outputs of the preprocessor obtained from the input image plane of the object as the input signals and the corresponding desired output assigned to that object as the target output signals. In our experiment, we used locally represented target vectors for classification. Two types of neural networks were considered as classifier. One was a 3-layer network trained by standard Backpropagation learning algorithm [10]. The other was a similar network trained by Fahlman and Lebiere's learning algorithm which has been reported to be much more robust than Backpropagation network [11]-[14]. Both the systems (i.e., one trained with BP algorithm and the other with FL algorithm) were tested on the objects transformed by translation, scaling and rotation, and performances of the systems were compared. Construction of a 3-layer FL network similar in architecture to a BP network is briefly described below.

Learning by FL algorithm is divided in two stages, namely, learning of output units and learning of hidden units. In this paper, to construct a layered network, all the weights between input and output units are first set to zero and from then on, all the weights except the biases of the output units are kept frozen. To reduce the error, the algorithm adds a hidden unit. Fig. 2 shows the network architecture on this stage. In Fig. 2, the unit whose output is not connected to the active network is called candidate hidden unit. At this stage, the candidate hidden unit's input weights are adjusted to maximize S, i.e., the sum of the absolute values of covariances C_j between its output value and the observed errors at the output units by the hill climbing method. When S stops improving, this unit is connected to the output units and its incoming weights are frozen. Equations in this stage;

$$S = \sum_j |C_j| , \quad C_j = \sum_p (V_p - \overline{V})(E_{jp} - \overline{E_j}) , \quad E_{jp} = T_{jp} - O_{jp} , \quad V_p = f(N_p) ,$$
$$\overline{V} = \sum_p V_p / M , \quad \overline{E_j} = \sum_p E_{jp} / M , \quad N_p = \sum_i W_i I_{ip} \text{ (including bias)} ,$$

where T_{jp}, O_{jp} and E_{jp} are the target output, actual output and error respectively at output unit 'j' for training pattern 'p'. N_p, f and V_p are the net input to the candidate hidden unit, its activation function and output respectively, M being the total number of training patterns. I_{ip} is the input that the candidate hidden unit receives from input unit 'i' corresponding to pattern 'p' and W_i is the weight between them. In order to maximize S, computing $\partial S/\partial W_i$ in a manner similar to the derivation of back propagation rule, weight change for W_i is given by

$$\Delta W_i = \eta \sum_j \sum_p \sigma_j (E_{jp} - \overline{E_j}) f'(N_p) I_{ip} , \text{ where } \sigma_j = 1 \ (C_j > 0) , \sigma_j = -1 \ (C_j < 0) ,$$

and η is the learning rate.

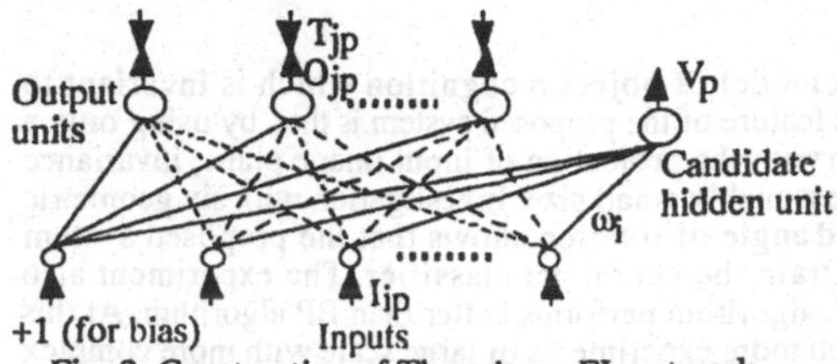

Fig. 2. Construction of Three layer FL network

Next the algorithm begins output stage learning and modifies the weights from hidden units (the newly added hidden unit and pre-existing hidden units, if any) to the output units along with the biases of the output units by delta rule. After the sum-squared error at the output units for the whole training set reaches an asymptote and if the designer is not satisfied with the network behavior, the algorithm generates a new hidden unit. This process is repeated until the designer is satisfied with the network.

3. Experimental Results and Discussions

To evaluate the performance of the proposed system, experiments were done with 6 geometrical objects, namely, square, triangle, pentagon, ellipse, circle and cross. The exemplar (standard) objects are of 64 × 64 black and white pixels as shown in Fig. 3. As described in the previous section, each input image undergoes through the preprocessing step consisting of projection and Rapid Transform and produces an output of 32 data sequence which become the input signals of the classifier network. Thus the classifier network has 32 input and 6 output units. When FL learning algorithm was used for training the classifier, it finally generated 3 hidden units creating a 32-3-6 FL network. Similarly, a 32-3-6 BP network was also trained as classifier in order to make the comparison of using both networks on the same cost performance basis. The performance of the overall system also depends on how well a classifier performs and FL algorithm is used with the expectation of achieving better overall performance as this network is found to be more robust than BP network [11]-[14]. Experiments were conducted in three parts as described below.

Experiment 1 (Scaling)

Test patterns were formed by scaling the exemplar patterns by factor 0.5 (50% reduction) and 1.5 (150% enlargement). Test patterns with scaling factor 1.5 are shown in Fig. 4.

Experiment 2 (Rotation)

The system was tested on patterns rotated by 10°, 15°, 20°, 30°, 40°, 45°, 50°, 60°, 75° angle. Test patterns by rotating the exemplar patterns by 30° are shown in Fig. 5.

Experiment 3 (Scaling and Rotation)

Exemplar patterns were first scaled by factor 0.5 and 1.5 as mentioned in Experiment 1 and then rotated by 10°~75° in step of 5° as done in Experiment 2. The test patterns are thus both scaled and rotated. Test patterns for 150% enlargement and 30° rotation are shown in Fig. 6.

Figs. 7(a), 7(b), 7(c) and 7(d) show the output of the projection block in system diagram for exemplar, scaled, rotated, scaled and rotated patterns, respectively. This shows that the projection results are scale invariant, and shifts nearly γ degree when the exemplar object is rotated by γ degree, and since projection is made considering the center of gravity of the pattern, it is also translation invariant.

Recognition ability of the system was investigated under two different recognition criteria as follows. Upon presenting a test pattern to the system, the maximum and the second maximum value at output layer were detected and recognized by :

Criterion 1 : If the maximum value was greater than 0.9 and the second maximum value was lower than 0.1, the test pattern was recognized to belong to the category represented by the output unit of maximum value. This was a rather severe condition.

Criterion 2 : The test pattern was recognized to belong to the category represented by the output unit of maximum value. This was a rather mild condition.

Figs. 8(a), 8(b), and 8(c), 8(d) show the percentage of correct recognition by the system for scaling (Experiment 1), rotation (Experiment 2), and scaling and rotation (Experiment 3), respectively. Results show that, under both recognition criteria, when FL network is used as classifier the system recognizes all of the test patterns correctly but the performance degrades with BP network, especially when the input object is scaled and rotated. The better performance of FL classifier is due to the fact that, in FL network, hidden unit outputs in response to the exemplar input object yields well saturated values whereas in BP network most of hidden unit outputs are non-saturated intermediated values. This fact is also observed in other studies [11]-[14]. Since the preprocessing is the same in both the cases, excellent performance of the system with FL classifier indicates the powerful and effective preprocessing adopted in the proposed system. With a robust classifier network like FL network the proposed system is an effective model for invariant object recognition.

4. Conclusion

In this paper, we proposed a neural network model of object recognition which is invariant to translation, scaling and rotation. The most attractive feature of the proposed system is that, by using only a simple shift invariant transformation in conjunction with the projection of input image plane, invariance recognition can be achieved and the system is of reasonably small size. Investigation with six geometric objects transformed by different scaling factor and angle of rotation shows that the proposed system performs excellent when FL algorithm is used to train the neural net classifier. The experiment also supports the previous observation [11]-[14] that FL algorithm performs better than BP algorithm. At this stage experiments were done with simple objects, and more experiments of large scale with more complex objects like English characters and alpha-numeric digits are currently in progress.

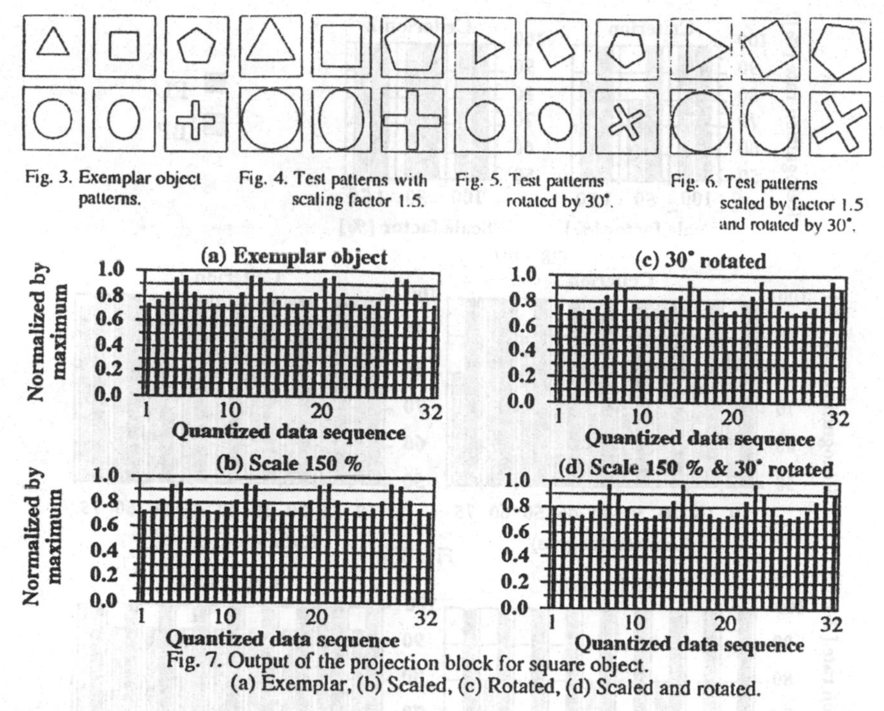

Fig. 3. Exemplar object patterns.

Fig. 4. Test patterns with scaling factor 1.5.

Fig. 5. Test patterns rotated by 30°.

Fig. 6. Test patterns scaled by factor 1.5 and rotated by 30°.

Fig. 7. Output of the projection block for square object.
(a) Exemplar, (b) Scaled, (c) Rotated, (d) Scaled and rotated.

References

[1] C. T. Zahn and R. Z. Roskies, "Fourier descriptions for plane closed curves," IEEE Trans. Computers, vol. C-21, pp. 269-281, 1972.

[2] S. A. Dudani, K. J. Breeding and R. B. McGhee, "Aircraft identification by moment invariants," IEEE Trans. Computers, vol. C-26, pp. 39-45, 1977.

[3] R. L. Kashyap and R. Chellappa, "Stochastic models for close boundary analysis : representation and reconstruction," IEEE Trans. Information Theory, vol. IT-27, no. 5, pp. 627-637, 1981.

[4] B. Widrow, R. G. Winter and R. A. Baxter, "Layered neural nets for pattern recognition," IEEE Trans. Acoustics, Speech, and Signal Processing, vol. ASSP-36, no.7, pp. 1109-1118, 1988.

[5] K. Fukushima, S. Miyake and T. Ito, "Neocognitron : a neural network model for a mechanism of visual pattern recognition," IEEE Trans. Syst., Man, and Cyber., vol. SMC-13, pp. 826-834, 1983.

[6] L. Spirkovska and M. B. Reid, "Connectivity strategies for higher-order neural networks," in Proc. International Joint Conference on Neural Networks, IJCNN'90, San Diego, vol. 1, pp. 21-26, 1990.

[7] S. D.You and G. E. Ford, "Object recognition based on projection," in Proc. International Joint Conference on Neural Networks, Baltimore, vol. 4, pp. 31-36, 1992.

[8] S. E. Fahlman and C. Lebiere, "The Cascade-correlation learning architecture," in D. S. Touretzky (ed.), Advances in Neural Information Processing Systems, vol. 2, pp. 524-532, Morgan Kaufmann, 1990.

[9] H. Reitboeck and T. P. Brody, "A transformation with Invariance under cyclic permutation for applications in pattern recognition", Information and Control, vol. 15, pp. 130-154, 1969.

[10] D. E. Rumelhart, J. L. McClelland and the PDP Research Group, "Parallel Distributed Processing," vol. 1, M.I.T. Press, 1986.

[11] M. Hamamoto, J. Kamruzzaman and Y. Kumagai, "Generalization ability of artificial neural network using Fahlman and Lebiere's learning algorithm," in Proc. of IEEE/INNS Int. Joint Conf. on Neural Networks, IJCNN'92, Baltimore, vol. I, pp. 613-618, 1992.

[12] M. Hamamoto, J. Kamruzzaman and Y. Kumagai, "Network synthesis and generalization properties of artificial neural network using Fahlman and Lebiere's learning algorithm", to be published in Proc. of IEEE 35th midwest symp. on circuits and systems, MWSCAS'92, Washington DC, 1992.

[13] M. Hamamoto, J. Kamruzzaman, Y. Kumagai and H. Hikita, "Generalization ability of feedforward neural network trained by Fahlman and Lebiere's learning algorithm," IEICE Tran. Fundamentals of Electronics, Communications & Computer Science, Japan, vol. E75-A, no. 11, pp. 1597-1601, 1992.

[14] M. Hamamoto, J. Kamruzzaman, Y. Kumagai and H. Hikita, "Incremental learning and generalization ability of artificial neural network using Fahlman and Lebiere's learning algorithm," IEICE Tran. Fundamentals of Electronics, Communications & Computer Science, Japan, vol. E76-A, no. 2, pp. 242-247, 1993.

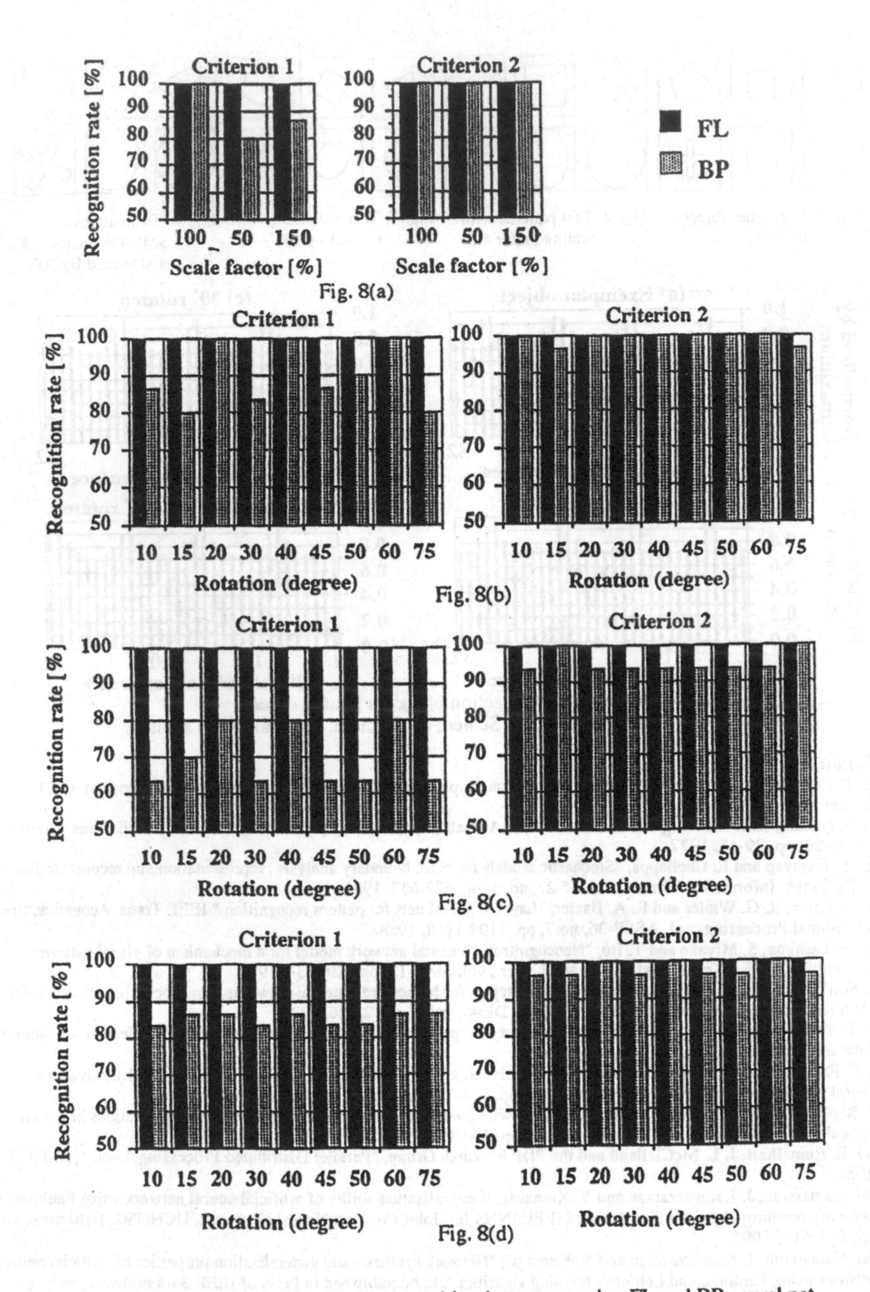

Fig. 8. Percentage of correct recognition by system using FL and BP neural net classifiers under two differnt recognition criteria.
(a) with test patterns scaled by factor 0.5 and 1.5.
(b) with test patterns rotated by angle 10°~75° in step of 5°.
(c) with test patterns scaled by factor 0.5 and rotated by angle 10°~75°.
(d) with test patterns scaled by factor 1.5 and rotated by angle 10°~75°.

Realization of Surjective Correspondence in Artificial Neural Network Trained by Fahlman and Lebiere's Learning Algorithm

Masanori HAMAMOTO Kazuki ITO Joarder KAMRUZZAMAN Yukio KUMAGAI

Department of Computer Science & Systems Engineering
Muroran Institute of Technology
27-1 Mizumoto-cho, Muroran-shi, 050 JAPAN

Abstract

In realizing surjective correspondence as an incremental learning by feedforward neural network system, it is desirable that the network designer be able to make use of hidden outputs in the already-trained network realizing injective correspondence in order to be adapted to a changeable environment that may demand learning of newly added patterns into the same category. To design a system that performs an extended task without destroying the hidden outputs gained by the previously trained network, some new hidden units are incorporated to acquire additional information required to realize the newly defined task. Fahlman and Lebiere's (FL) learning algorithm is particularly suitable for this purpose since this algorithm can gradually add the required number of new hidden units. Previous studies show that FL network generalizes far better than Backpropagation (BP) network [10], [11]. And it has also been reported that an extended FL network which realizes an incremental learning with increased category have generalization ability superior to BP network [12], [13]. In this paper, we describe a realization of surjective correspondence as an incremental learning by FL algorithm. Investigation shows that FL network trained surjective correspondence has better generalization ability than BP network due to the attainment of well-saturated hidden outputs in FL network.

1. INTRODUCTION

The usual practice in multilayer neural networks is to retrain a new network when the environment has to be changed. In the previous study [12], [13] we showed that one alternate way to do this is to retain the information acquired by the already-trained network which performs a specific pattern classification and add new hidden units to realize the newly defined task. Previous study dealt with injective correspondence in which one training pattern belongs to one category and the extended task was a kind of incremental learning needing more categories to be learned with the newly added training set. In this paper, we deal with surjective correspondence in which multiple training patterns belong to same category and a different kind of incremental learning in which a network is already trained for injective correspondence, and the newly defined extended task is to train more patterns to belong to the existing categories. Here, we show that this can be done without modifying the weights between the input and hidden layer of the previously trained network. Thus the previously trained network will retain the feature extraction capability already learned and the newly added hidden units will gain the necessary additional capability to realize the task.

To perform above mentioned task, standard Backpropagation learning algorithm [1] is not suitable since in this case the required number of newly added hidden units has to be pre-specified and a priori specification of the number of hidden units sufficient to realize the task is rather difficult, and trial and error is the only solution. Thus we used a learning algorithm that can gradually add new hidden units one by one and stop creating new hidden units when the desired task is realized.

Several algorithms that gradually build network topology have been reported [2]-[9]. Gallant [2] proposed perceptron based pocket algorithm in which networks are constructed by adding cells during learning according to several algorithms. Mezard and Nadal [3] proposed tiling algorithm in which new ancillary units are added to the layer, if the output of the master unit of that layer is not identical to the desired output and continues adding until this layer gives a faithful internal representation. Frean [4] proposed upstart algorithm in which daughter units are generated by the parent unit in order to correct its mistakes according to its response. Fujita [5] proposed construction of feedforward network by optimizing each hidden unit's function as it is included in the network and possibly can generate a network with minimum hidden units. Hanson [6] proposed meiosis network in which the number of hidden units increases according to a splitting policy governed by stochastic delta rule. Fahlman and Lebiere [7] proposed cascade-correlation learning algorithm in which a network is constructed by adding hidden units one by one based on maximizing the correlation between the residual error at the outputs and the output of candidate hidden unit. Littmann and Ritter [8], [9] proposed CASQEF, CASER and CASLLM network using modified cascade-correlation algorithm.

However, the algorithms proposed in [2]-[4] have been mainly analyzed for a single output unit and use of these algorithms for multiple output units might need an excessively large network. Even though the algorithm proposed in [5] seems to be able to build a network of optimum or near-optimum size when the output units are linear, it does not perform well for nonlinear output units. For a given task, simulation results in [6] show wide variation in network size with different trials even though small or no variation is desired. The approach in [8] and [9] needs more sweeps, more cascaded hidden units and additional powers of their activity values than cascade-correlation algorithm to achieve similar results for XOR problem.

The basic idea of cascade-correlation learning algorithm proposed by Fahlman and Lebiere [7] is that, to realize any input-output relationship, a network is constructed by adding hidden units one by one based on maximizing the correlation between the residual error at the outputs and the output of candidate hidden unit. This algorithm learns very quickly [7], and is capable of constructing near-optimum network automatically [7], builds a network with high generalization ability [10], [11] and can be applied to an incremental learning with increased category [12], [13].

In this paper, we trained networks which perform a specific job (injective or surjective correspondence) both by FL algorithm and BP algorithm of similar architecture. We then applied FL algorithm to realize extended task (surjective correspondence) as an incremental learning without increasing category by making use of weights between the input and hidden layers in already-trained network realizing injective correspondence. And we made a comparison between the generalization properties of those networks.

2. DESCRIPTION OF THE ALGORITHM

Learning by Fahlman and Lebiere's algorithm [7] is divided in two stages, namely, learning of output units and learning of hidden units. This algorithm begins with a two layer network of input and output units and then continue adding hidden units until the designer is satisfied with the network. In this paper, our approach is to generate a three-layer network similar in structure to a BP network by this algorithm which is constructed in the following way [10]-[13]. To construct a layered network, all the weights between input and output units are first set to zero and from then on, all the weights except the biases of the output units are kept frozen. Fig. 1 shows network construction as a hidden unit is added during learning. In Fig. 1, the unit whose output is not connected to the active network is called candidate hidden unit. At this stage, the candidate hidden unit's input weights are adjusted to maximize S, i.e., the sum of the absolute values of covariances C_j between its output value and the observed errors at the output unit j by gradient ascent.

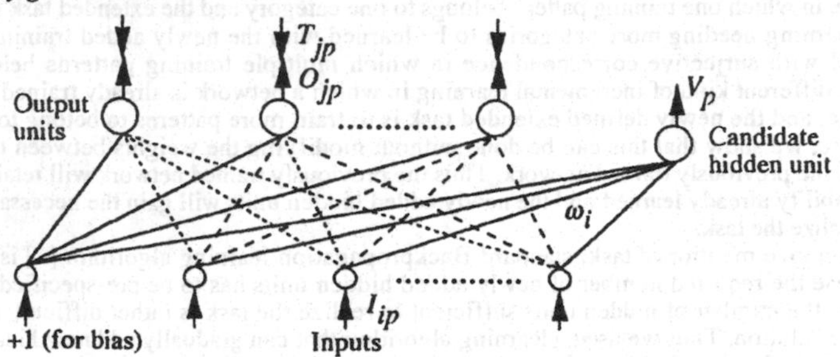

Fig. 1. Construction of network architecture used by FL algorithm.
The dashed-line indicates weight of zero value.

Equations on learning of hidden unit :

$$S = \sum_j |C_j| \ , \quad C_j = \sum_p (V_p - \overline{V})(E_{jp} - \overline{E_j}) \ , \quad E_{jp} = T_{jp} - O_{jp} \ , \quad V_p = f(N_p) \ ,$$

$$\overline{V} = \sum_p V_p / M \ , \quad \overline{E_j} = \sum_p E_{jp} / M \ , \quad N_p = \sum_i \omega_i I_{ip} \ \text{(including bias)} \ ,$$

where T_{jp}, O_{jp} and E_{jp} are the target output, actual output and error respectively at output unit j for training pattern p. N_p, f and V_p are the net input to the candidate hidden unit, its activation function and output respectively, M being the total number of training patterns. I_{ip} is the input that the candidate hidden unit receives from input unit i corresponding to pattern p and ω_i is the weight between them.

In order to maximize S, computing $\partial S/\partial \omega_i$ in a manner similar to the derivation of Backpropagation rule [1], weight change for ω_i is given by

$$\Delta\omega_i = \eta \sum_j \sum_p \sigma_j (E_{jp} - \overline{E_j}) f'(N_p) I_{ip} \ ,$$

where $\sigma_j = 1$ $(C_j > 0)$, $\sigma_j = -1$ $(C_j < 0)$, and η is the learning rate.

When S stops improving, the candidate hidden unit is connected to the output units and its incoming weights are frozen. Next the algorithm begins output stage learning to reduce the network error and modifies the weight ω_{hj} between hidden unit h (the newly added hidden unit and pre-existing hidden units, if any) and the output unit j along with the biases of the output units by delta rule. After the network error E, i.e., sum-squared error at the output layer for the whole training set, reaches an asymptote and if the designer is not satisfied with the network behavior, the algorithm generates a new hidden unit.

Equations on learning of output units :

$$E=\frac{1}{2}\sum_j \sum_p (T_{jp}-O_{jp})^2 \ , \quad O_{jp}=f(N_{jp}) \ , \quad N_{jp}=\sum_h \omega_{hj} O_{hp} \ \text{(including bias)} \ ,$$

where N_{jp} and O_{hp} are the net input to the output unit j and the output at the hidden unit h respectively for the training pattern p. Weight change for ω_{hj} is given by

$$\Delta\omega_{hj}=\eta \sum_p (T_{jp}-O_{jp}) f'(N_{jp}) O_{hp} \ .$$

This process, which has two stages as stated above, is repeated until the designer is satisfied with the network. A three-layer feedforward network is constructed by this process.

3. SIMULATION RESULTS AND DISCUSSIONS

3. 1 Realization of injective and surjective correspondence

One of the essential properties of neural network for real world application is its ability to generalize any pattern not included in the training set. Previous works in [10] and [11] show that generalization ability of a network trained by FL algorithm is significantly better than that of a similar network trained by BP algorithm. And also for performing incremental learning with increased category, FL network generalizes far better than BP network [12], [13]. In the present work which deals with the realization of a different type of incremental learning without increasing category, we explored the possibility of realization of surjective correspondence both by FL and BP algorithms, and made a performance comparison between the networks.

The experiment we performed was a capital letter recognition task (A~Z) and small letter recognition task (a~z). Each character was 8x8 black and white pixels, and the component of each input vector consisted of +1 or -1, +1 representing the black pixel. Target vectors were binary vectors [0, 1] and were locally represented, i.e., only one bit representing a specific category was '1' and the rest were '0' ("A" and "a" belong to same category and so on). Activation function used for the output and hidden units was sigmoidal function within the range (0, 1).

Initially two different networks were trained independently, one for the capital letter recognition (injective correspondence) and the other for the capital and small letter recognition (surjective correspondence). To realize injective correspondence (26 categories, 26 patterns), FL algorithm began with a minimal network of 64 inputs, 26 output units and finally generated 6 hidden units, i.e., a 64-6-26 network was constructed (we call it "injective FL network"). To realize surjective correspondence (26 categories, 52 patterns), FL algorithm finally generated 6 hidden units (we call it "surjective FL network"). Similarly, three-layer networks with 6 hidden units were trained by BP algorithm for both tasks mentioned above (we call those networks "injective BP network" and "surjective BP network", respectively). The generalization abilities of the resultant networks trained by FL and BP algorithms were investigated for both the tasks (injective and surjective correspondence) as follows.

In order to form test patterns, each training pattern was corrupted by noise. This noise changes white pixel (-1) into black (+1), or black pixel into white. All the training patterns were corrupted by changing 1, 2, 3, and 4 pixels. All of the possible test patterns were formed by injecting noise upto 4 bits (in total 17551636 test patterns for capital letter recognition task, 35076944 test patterns for capital and small letter recognition task). Test pattern database also includes such patterns that the Hamming Distance (H.D.) between the test pattern and its closest training pattern is one less than the H.D. between the test pattern and its second closest training pattern; these test patterns are the most difficult ones to be recognized. Recognition ability was investigated under two different recognition criteria as stated below.

Criterion 1 : Upon presenting a test pattern to the network, maximum value and the second maximum value at the output layer were detected. If the maximum value was greater than 0.8 and the second maximum value lower than 0.2, the test pattern was recognized to belong to the category represented by the output unit of maximum value. This was a rather severe criterion.

Criterion 2 : Upon presenting a test pattern to the network, maximum value at the output layer was detected and the test pattern was recognized to belong to the category represented by the output unit of maximum value. This was a rather mild criterion.

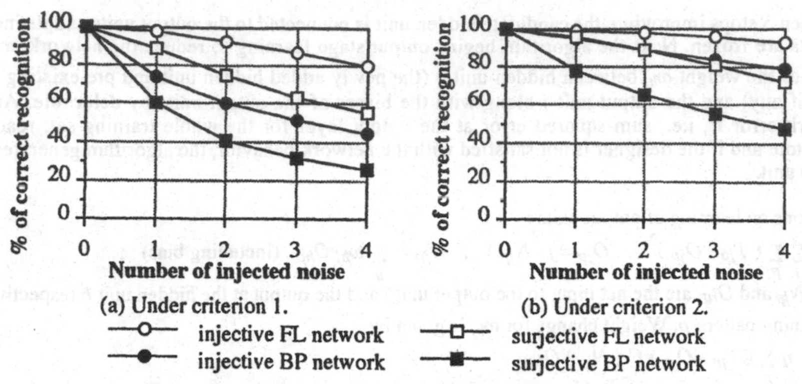

Fig. 2. Percentage of correct recognition by each network.

Fig. 2 shows the percentage of correct recognition by each network under each recognition criterion. From the simulation results, it is obvious that generalization ability of FL network is always far better than that of BP network. Moreover, recognition ability of BP network decreases more than that of FL network as more corrupted test patterns are presented to the network. BP network has totally degraded performance while FL network still performs well. Surjective FL network which is trained with more number of training patterns than injective BP network maintains better generalization ability than injective BP network under criterion 1.

The reason of better performance in FL network than that of BP network can be explained by making a comparative study of the behavior of hidden units in both the networks. It can be reasonably assumed that if the hidden units attain saturation for almost all of the training patterns, noise will be well absorbed by the hidden units. Then hidden layer can effectively filter out the noise and the network is expected to have good generalization ability.

Fig. 3 shows frequency distribution of the net input to the hidden units for all the training patterns after the completion of learning in surjective FL and BP networks. In FL network, hidden units have the tendency to converge to the extreme value and the hidden units attain saturation for almost all the training patterns. Here, we considered a hidden unit to be saturated if its net input was greater than +2.5 or less than -2.5. In contrast, in BP network almost all the training patterns except a few use intermediate values of hidden units for internal representation and this greatly reduces the generalization ability. Similar trends in the saturation of hidden outputs in response to the training patterns are observed in the injective FL and BP networks. In both injective and surjective networks, it is the saturation of hidden outputs in FL network that makes this network perform far better than BP network.

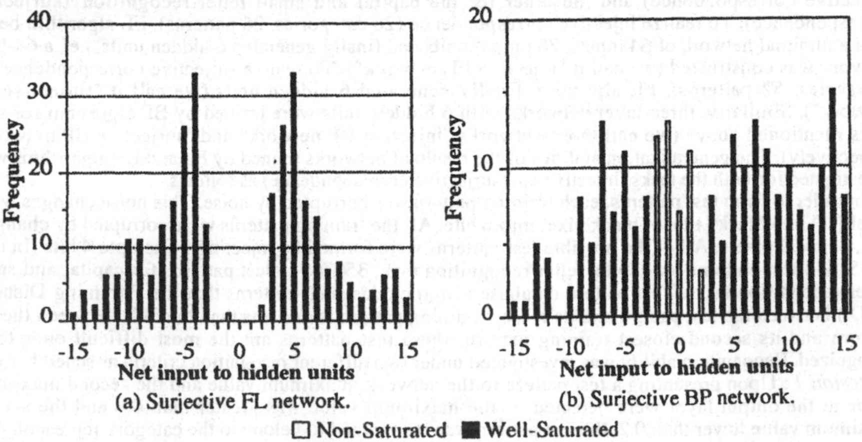

Fig. 3. Frequency distribution of the net input to hidden units for all the training patterns

3. 2 Realization of surjective correspondence by incremental learning without increased category

Given that a network is already trained for injective correspondence and now if a task of realizing surjective correspondence is to perform, one way is to train a new network as described above. In this case, starting with anew network destroys the previous information gained by already-trained network. In this paper, our approach is to do this by keeping the weights between input and hidden layers in the previously trained network fixed and then extending this network by adding new hidden units. Fig. 4 illustrates the network construction for realization of surjective correspondence as an incremental learning without increasing category. For this task, FL algorithm is suitable one since this algorithm gradually adds on new hidden units without being pre-specified by the designer.

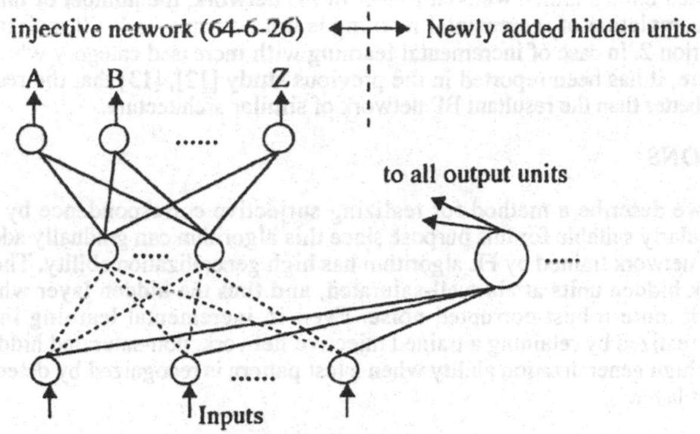

Fig. 4. Network construction for realization of surjective correspondence as an incremental learning without increasing category. The dashed-line indicates fixed weight.

To perform this task, FL algorithm began with a 64-6-26 injective FL network and finally generated 6 hidden units. Thus the resultant 64-12-26 FL network (we call it "extended surjective FL network") is constructed without destroying the weights between input and hidden layers in injective FL network.

The same task may be able to be performed by using BP algorithm. However, in this case the number of newly added hidden units must be pre-specified. To make a performance comparison with FL algorithm, we tried to construct a BP network having similar architecture to the extended surjective FL network described above. But, to perform this task, a 64-12-26 BP network which includes 6 hidden units in injective BP network and 6 newly added hidden units did not converge in our experiment even though many trials were performed. Thus, BP algorithm seems not to be suitable for incremental learning mentioned here. In this situation, new network has to be trained as it done in subsection 3. 1.

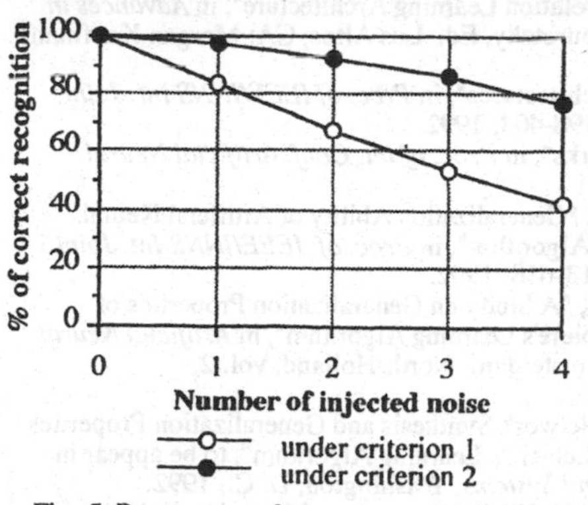

Fig. 5. Percentage of correct recognition by the extended surjective FL network.

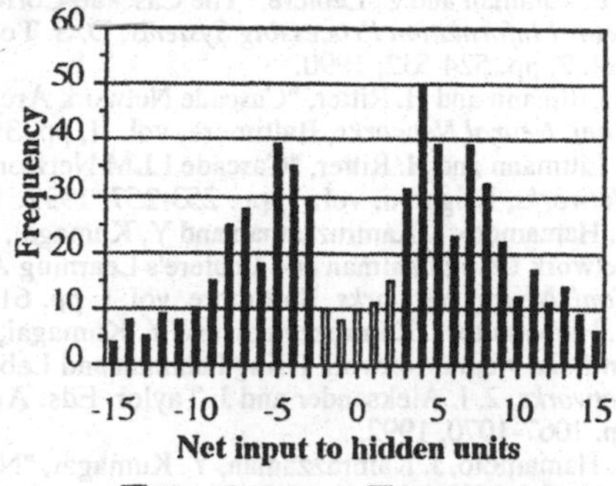

Fig. 6. Frequency distribution of the net input to the hidden units in extended surjective FL network.

Fig. 5 shows the percentage of correct recognition by the extended surjective FL network under each criterion. This network is trained by the same training pattern set and tested on the same pattern set as the surjective network. The generalization ability of the extended surjective FL network is better than that of a surjective FL network under criterion 2 and that of a newly trained surjective BP network under both the criteria (Fig. 2).

Fig. 6 shows the frequency distribution of the net input to the hidden units in response to all the training patterns in extended surjective FL network. Since the hidden outputs in previously trained network is fixed, when the training set a~z is presented to the resultant network, its portion which was previously trained for A~Z will not necessary produce saturated hidden outputs. Thus, the resultant network produces some non-saturated hidden outputs even though in the previously trained network almost all the hidden units attained well-saturated. In FL network, the number of non-saturated hidden outputs after the completion of incremental learning is not so many and still maintains generalization ability under criterion 2. In case of incremental learning with increased category which is different from the one treated here, it has been reported in the previous study [12], [13] that the resultant FL network generalizes much better than the resultant BP network of similar architecture.

4. CONCLUSIONS

In this paper, we describe a method for realizing surjective correspondence by FL algorithm. FL algorithm is particularly suitable for this purpose since this algorithm can gradually add necessary hidden units. Moreover, a network trained by FL algorithm has high generalization ability. The underlying fact is that in FL network hidden units attain well-saturated, and thus the hidden layer which acts as a filter makes FL network more robust corrupted noise. Even in incremental learning in which surjective correspondence is realized by retaining a trained injective network, non-saturated hidden outputs are few and still maintains high generalization ability when a test pattern is recognized by detecting the maximum output at the output layer.

REFERENCES

[1] D. E. Rumelhart, G. E. Hinton and R. J. Williams, "Learning Internal Representations by Error Propagation", in *Parallel Distributed Processing*. D. E. Rumelhart and J. L. McClelend, Eds. Cambridge, MA: MIT Press, vol. 1, pp. 318-362, 1986.

[2] S. I. Gallant, "Perceptron-Based Learning Algorithms", *IEEE Trans. Neural Networks*, vol. 1, no. 2, pp. 179-191, 1990.

[3] M. Mezard and J. -P. Nadal, "Learning in Feedforward Neural Networks : the Tiling Algorithm", *J. Phys.*, A: Math. Gen. 22, pp. 2191-2203, 1989.

[4] M. Frean, "The Upstart Algorithm : A Method for Constructing and Training Feedforward Neural Networks", *Neural Computation*, vol. 2, pp. 198-209, 1990.

[5] O. Fujita, "Optimization of Hidden Unit's Function for Feed-forward Neural Networks", *IEICE Tech. Rep.*, NC90-75, pp. 43-48, 1991 (in Japanese).

[6] S. J. Hanson, "Meiosis Networks", in *Advances in Neural Information Processing Systems*, D. S. Touretzky, Ed. Los Altos, CA: Morgan Kaufmann, vol. 2, pp. 533-541, 1990.

[7] S. E. Fahlman and C. Lebiere, "The Cascade-Correlation Learning Architecture", in *Advances in Neural Information Processing Systems*, D. S. Touretzky, Ed. Los Altos, CA: Morgan Kaufmann, vol. 2, pp. 524-532, 1990.

[8] E. Littmann and H. Ritter, "Cascade Network Architectures", in *Proc. of IEEE/INNS Int. Joint Conf. Neural Networks*, Baltimore, vol. II, pp. 398-404, 1992.

[9] E. Littmann and H. Ritter, "Cascade LLM Networks", in *Proc. of Int. Conf. Artificial Neural Networks*, Brighton, vol. 1, pp. 253-257, 1992.

[10] M. Hamamoto, J. Kamruzzaman and Y. Kumagai, "Generalization Ability of Artificial Neural Network Using Fahlman and Lebiere's Learning Algorithm", in *Proc. of IEEE/INNS Int. Joint Conf. Neural Networks*, Baltimore, vol. I, pp. 613-618, 1992.

[11] M. Hamamoto, J. Kamruzzaman and Y. Kumagai, "A Study on Generalization Properties of Artificial Neural Network Using Fahlman and Lebiere's Learning Algorithm", in *Artificial Neural Networks*, 2, I. Aleksander and J. Taylor, Eds. Amsterdam: North-Holland, vol. 2, pp. 1067-1070, 1992.

[12] M. Hamamoto, J. Kamruzzaman, Y. Kumagai, "Network Synthesis and Generalization Properties of Artificial Neural Network Using Fahlman and Lebiere's Learning Algorithm", to be appear in *Proc. of 35th Midwest Symposium on Circuits and Systems*, Washington, D. C., 1992.

[13] M. Hamamoto, J. Kamruzzaman, Y. Kumagai and H. Hikita, "Incremental Learning and Generalization Ability of Artificial Neural Network Trained by Fahlman and Lebiere's Learning Algorithm", to be published in *IEICE Trans. Fundamentals of Electronics, Communications and Computer Sciences*.

BIMODAL DISTRIBUTION REMOVAL

P. Slade & T.D. Gedeon

School of Computer Science and Engineering,
The University of New South Wales, Australia

Abstract:

A number of methods for cleaning up noisy training sets to improve generalisation have been proposed recently. Most of these methods perform well on artificially noisy data, but less well on real world data where it is difficult to distinguish between noisy data points from valid but rare data points.

We propose here a statistically based method which performs well on real world data and also provides a natural stopping criterion to terminate training.

Background:

Some researchers (White, 1989, Geman et al, 1992) have compared neural networks to non-parametric estimators. As such, the limitations of neural networks can be explained by a well understood problem in non-parametric statistics, namely the 'bias and variance' dilemma. Basically, the dilemma is that to obtain a good approximation of an input-output relationship using some form of *estimator*, constraints must be placed on the structure of the estimator and hence introduce bias, or a very large number of examples of the relationship must be used to construct the estimator.

Non-parametric statistics is concerned with model free estimation. When employed for classification both parametric and non-parametric statistics seek to construct decision boundaries between the various classes using a collection of training samples. Non-parametric methods differ from parametric methods in that there is no particular structure assigned to the decision boundaries, *a priori*.

The obvious advantage of parametric techniques is efficiency. By setting the structure of the decision boundary before estimation begins then fewer data points (or training examples) are required. This is because there are (hopefully) a small number of parameters in the parametric model that require estimation. Non-parametric or model free estimation potentially requires the estimation of an infinite number of parameters and hence needs a much larger number of training examples. However, the efficiency of parametric methods comes at a cost. If the actual form of the decision boundary departs substantially from the assumed form, then parametric methods can result only, in the "best" approximation for the decision boundary from within the adopted class of decision boundaries. Non-parametric methods place no restriction on the class in which the decision boundary used in estimation must reside.

Informally, *consistency* is the asymptotic convergence of the estimator to the object of estimation. In this context asymptotic refers to the sample size or the number of patterns in the training set approaching infinity. Most non-parametric algorithms are consistent for any regression function $E[y|x]^2$. Indeed it has been shown (White, 1989, and Gallant and White, 1988) that feed forward networks are consistent, under appropriate conditions relating to the architecture of the network. Consistent in the sense that the weights in the network will, in the limit of training set size approaching infinity, converge to the optimal weights $\mathbf{w}^*$. Although this is an encouraging property, non-parametric methods can be very slow to converge, and this has indeed been observed in the training times for neural networks.

Non-parametric estimators are guaranteed to perform optimally in the limit. In the context of neural networks, they are only guaranteed to outperform other parametric estimators when the size of the training set approaches infinity. For a finite sample, non-parametric estimators can be very sensitive to the actual realisations of (x,y) contained in the sample. This sensitivity results in an estimator that is high in what is known as *variance*. The only way to control this variance is to introduce some *a priori* structure into the estimator, that is to use parametric methods. This approach also has its pitfalls. In complex classification problems, it is difficult to know the structure to impose on the estimator. As mentioned above, this can result in estimators that converge to an incorrect solution. This creates models that are high in what is known as *bias*. The performance function used in back-propagation can be readily decomposed into a bias and a variance term (Geman et al, 1992).

It is this dilemma between bias and variance that can explain the limitations of non-parametric learning. Low bias and low variance requires large numbers of training examples. In situations where it is not possible to obtain sufficiently large numbers of training examples it is necessary to allow some bias into the neural network training procedure.

A number of methods have been used to introduce bias including:

- Pruning - by the removal of hidden units, the class of functions the network can produce is restricted (eg Sietsma and Dow, 1991, Gedeon and Harris, 1991).
- Dynamic node addition - network training starts with few units and thus high bias (eg Ash, 1989, White, 1989, Harris & Gedeon, 1991).
- Extra terms in performance function - act as smoothing terms decreasing variance. The cost is an increase in bias as details of the object of estimation *E*[y|x] are blurred and lost.
- Cross validation - used to halt training, the network is restricted from building some decision boundaries.
- Outlier removal - reduce the initial variance in the training set and thus improve the variance/bias trade-off (Geman et at, 1992).

In the next section we will examine a number of outlier detection methods before introducing our own.

Outlier detection methods:

In terms of the statistical framework we use, there are several ways that a noisy training set can occur. Either, the input pattern **x** does not obey the environmental probability law ν, or the target pattern **y** does not obey the conditional probability law γ. Both these cases result in an atypical or irregular mapping between input and output. In its quest to identify γ, network training cannot help but be adversely affected by the presence of these errors.

During the back error propagation step of the back-propagation algorithm, each weight is changed by an amount which is a function of the discrepancy between actual network output and desired output. When presented to the network, the erroneous patterns in the training set produce a high disparity between desired and actual output. This produces large weight changes as the network tries to minimise the error on these patterns. As most networks are trained until the mean square error over the training set is below some threshold, these erroneous patterns will prolong training. This growth in training time greatly increases the chance of the network overfitting the training set. So these inaccurate patterns seem to have two possible effects on training:

- they force the network to slow its learning of the majority of patterns in order to learn those few erroneous patterns, and
- they cause training time to escalate, thereby increasing the effect of overfitting.

A simple way of lessening the effect of these incorrect patterns is to remove them altogether. This approach, though rudimentary in theory, is complicated in practice. Joines and White (1992) identify a number of approaches, all of which have some problems:

Absolute Criterion Method

This method attempts to minimise the absolute value of the error as opposed to normal back-propagation which attempts to minimise the mean square error. The outliers in the training set will have larger errors relative to the rest of the training set. This method does not propagate these large errors by design, consequently the changes in weight are smaller. This method may allow backpropagation to find simple models. The obvious problem with the Absolute Criterion Model is of stability or oscillation. Being parabolic, very small weight changes are needed to minimise the error function for normal back propagation, when the error is near zero.

This method propagates the same error throughout training, resulting in relatively large weight changes, which prevent the network from reaching a stable minimum. It is possible to revert to normal backpropagation error function in the region of small error, to alleviate the problem of oscillation, since the normal backpropagation function is very smooth around the origin.

Least Median Squares (LMS)

This method seeks to minimise the median of the residual errors. The procedure involves calculating the mean square error for each pattern in the training set then back-propagating the median of those errors.

This method is similar to batch learning in normal back-propagation and shows the same slowness in convergence.

Least Trimmed Squares (LTS)

This method is designed to speed up convergence of normal back-propagation training. The basic premise is to minimise the mean square error over only a percentage of the training set. At every 5th epoch all the patterns in the training set are sorted in ascending order based on their mean square error. The patterns connected with the lowest mean square errors are used to train the network for the next 5 epochs. The process is then repeated and a new subset of the full training set is selected as training patterns. This method may result in better generalisation because the outliers in the training set will never be used to train the network, because they produce large mean square errors. As a result, the weights will never adjusted to fit these outliers.

Joines and White (1992) tested the above method on a clean data set with noise specifically introduced into the training set (but not the test set). The result was excellent, with the mean square error over the training set increasing and decreasing over the test set - the reverse of the usual case. There are some major problems. Firstly in the real world our test set would also contain noise. Secondly the details of the method requires knowing (or assuming) *a priori* how many outliers are present in the training set.

Bimodal distribution removal:

This section introduces our method for outlier detection called *Bimodal Distribution Removal* (BDR), which addresses all the weaknesses of the other methods indicated in the previous section.

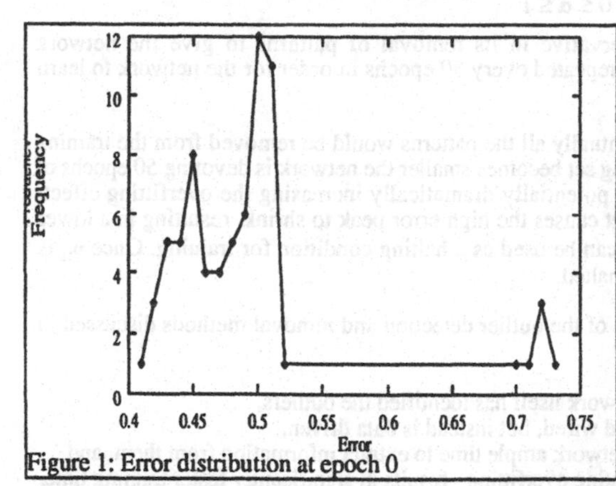

Figure 1: Error distribution at epoch 0

In order to investigate the behaviour of outliers in the training set during training, frequency distributions of the errors for all patterns in the training set were produced every 50 epochs of training. shows that very early in training (0-100 epochs) the distribution of the pattern errors is approximately normal with a large variance. Very quickly however, the network dramatically reduces the errors for a majority of the training set. However, as evident from Figure 2 there remains patterns with relatively high error. This creates an almost bimodal error distribution, with the low error peak containing patterns the network has learnt well, and the high error peak containing the outliers. Figure 2 also reveals that the distribution is only approximately bimodal. There exist patterns with errors which are between the two peaks. These patterns are being slowly learnt by the network and shall hereafter be referred to as *slow coaches.*

From the two peaks in the error distribution it is clear that the network can identify outliers itself. The network is learning $E[y|x]$, the expected value of y given x. The patterns that appear in the high error peak are outliers in the sense that they are *not* what the network 'expects' y to be given x.

It would be difficult and time consuming to identify a bimodal error distribution during training. Fortunately, a measure of the variance will achieve the same effect. As mentioned above, the variance

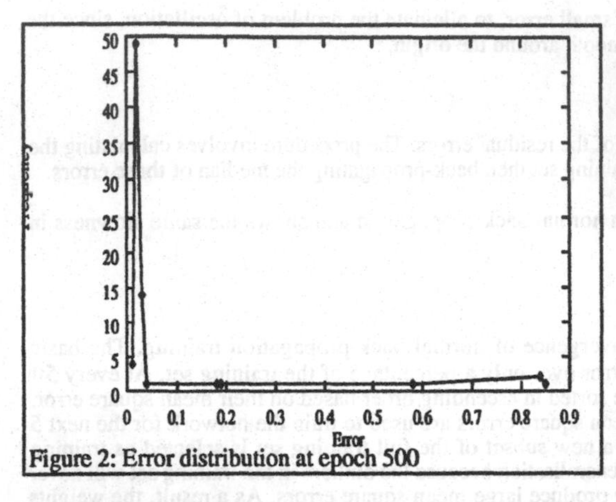

Figure 2: Error distribution at epoch 500

υ_{ts}, of the error distribution is quite large very early in training. As the network begins to learn the majority of the patterns, υ_{ts} drops sharply (see Figure 3). This is due to the lower error peak dominating the distribution. Low variance ($\upsilon_{ts} \approx 0.1$ and below) indicates that the two error peaks have formed, so the removal of outliers can begin. All this usually occurs very quickly, normally within 200-500 epochs.The next step is to decide which patterns to remove. Patterns should not be removed too quickly, as those patterns with midrange errors could eventually be learnt by the network. In choosing which patterns to remove the first step is to calculate the mean of the errors for all the patterns in the training set $\overline{\delta}_{ts}$. Due to the dominance of the low error peak (Figure 2) $\overline{\delta}_{ts}$ will be very low, but greater than nearly all errors in the low error peak (due to the presence of the high error peak). Those patterns in the low error peak are not outlier candidates. In order to isolate potential outliers, all those patterns with error greater than $\overline{\delta}_{ts}$ are taken from the training set.

This subset will contain the patterns from the high error peak (outliers) and the slow coaches between the two peaks. The dominance of the outliers in the subset will skew the distribution towards the outliers. The mean $\overline{\delta}_{ss}$ and standard deviation σ_{ss} of this skewed distribution is calculated. $\overline{\delta}_{ss}$ will be heavily influenced by the outliers and hence will be relatively high. From these two statistics is possible to decides which patterns to permanently remove from the training set. Those patterns with

$$\text{error} \geq \overline{\delta}_{ss} + \alpha\sigma_{ss} \qquad \text{where } 0 \leq \alpha \leq 1$$

are removed. BDR is intentionally conservative in its removal of patterns to give the network opportunity to learn the slow coaches. It is repeated every 50 epochs in order for the network to learn the features of each new training set.

Should BDR be continued indefinitely, eventually all the patterns would be removed from the training set. This is undesirable because as the training set becomes smaller the network is devoting 50 epochs of training to a reduced set of examples, thus potentially dramatically increasing the overfitting effect. Removal of the outliers from the training set causes the high error peak to shrink, resulting in a lower $\overline{\delta}_{ts}$ and a very much lower υ_{ts}. It is υ_{ts} that can be used as a halting condition for training. Once υ_{ts} is below a constant (typically 0.01) training is halted.

BDR attempts to address all the weaknesses of the outlier detection and removal methods discussed in the previous section in that:

- pattern removal does not start until the network itself has identified the outliers,
- the number of patterns removed is not hard wired, but instead is data driven,
- patterns are removed slowly, to give the network ample time to extract information from them, and
- a halting criterion naturally evolves preventing overfitting - results in significantly faster training time.

The remainder of this paper describes an experiment to gauge the effectiveness of BDR and some of the outlier detection methods presented in the previous section. The methods will be compared to normal back propagation, using a real world data set. The effectiveness of each method will be measured by the method's ability to control both bias and variance.

Experiment:

The data used in this experiment comprises student information, assessment and final mark for a sample of students from a first year Computer Science subject at The University of New South Wales.

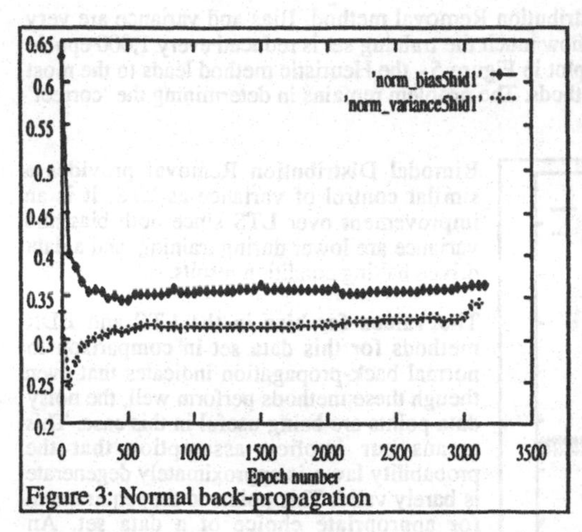

Figure 3: Normal back-propagation

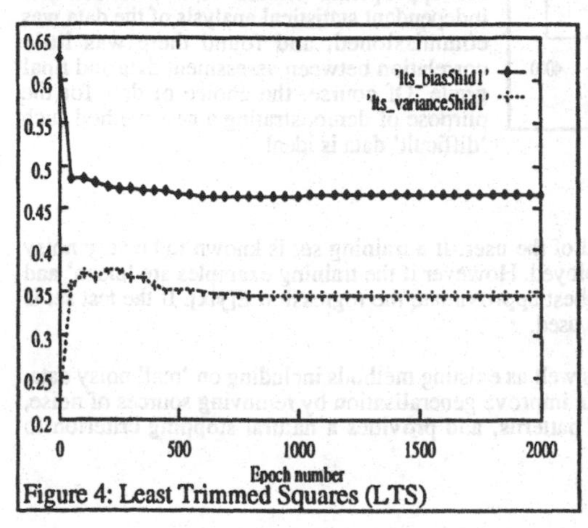

Figure 4: Least Trimmed Squares (LTS)

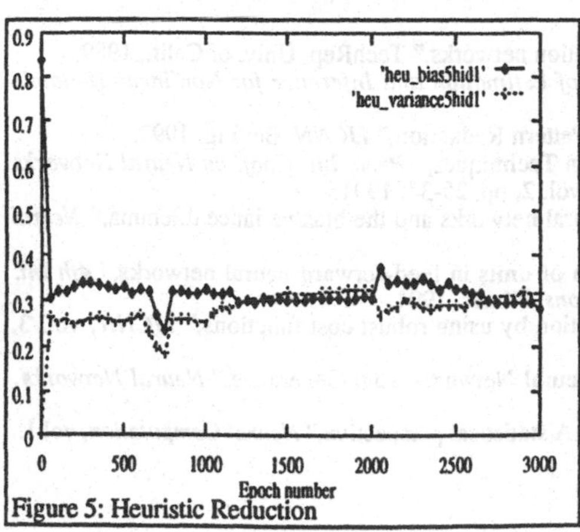

Figure 5: Heuristic Reduction

The sample contains the marks for 150 students. The exam mark is excluded to introduce noise, the task of the network is to predict the final grade (HD, D, CR, P) based on the 40% of the mark which comes from assessment prior to the exam. For further details see Gedeon and Bowden (1992). A student's assessment may not reflect their grade at one extreme by copying assignments and so on to get a good mark during the year but not understand the material and hence do very badly in the final exam, or at the other extreme expend very little effort during the year and achieve low course assessment marks but study very hard for the final exam and do well. An analysis of the data indicates that these two cases are relatively rare, of the order of 10%, and can be classed as noise.

Normal backpropagation is usually classified as a high variance / low bias estimator. As such, normal backpropagation will control variance if the network does not begin overfitting the training set. Overfitting can happen if as in this experiment, the training set is noisy.

The network structure is 14-5-4, being input, hidden and output units. The experiment was repeated with a 14-10-4 network with no significant difference in the results.

Fifty patterns were set aside as a test set and were never used in training. The remaining 100 patterns were used to create 50 sets of 70 pattern training sets at random. Fifty networks for each of the Absolute Criterion, LMS, LTS, Heuristic Reduction (described in Gedeon and Bowden, 1992), and Bimodal Distribution Removal were trained, as well as normal back-propagation. The integrated bias and variance were then calculated. The results for the latter 4 cases are shown. The Absolute Criterion and LMS methods performed less well than the LTS method and are not shown.

Least Trimmed Squares provides the necessary control of variance at the expense of higher bias. This control of variance becomes significant if training is continued for a long period of time (the overfitting effect increases). The LTS method to control variance is hard wired and requires *a priori* knowledge of the amount of noise in the training set.

Heuristic Pattern Removal produces an almost contradictory result. The asymptotic nature of neural networks indicates that network performance becomes optimal as the size of the training set approaches infinity. Yet, measurements of bias and variance for training on a half size training set show the Heuristic method performs as well as the Bimodal Distribution Removal method. Bias and variance are very sensitive to the complexity of the data and by how much the training set is reduced every 1,000 epoch. This can be seen by the slope of the variance plot in Figure 5 - the Heuristic method leads to the most uncontrolled increase in variance of all the methods. The problem remains in determining the 'correct' time to halt training.

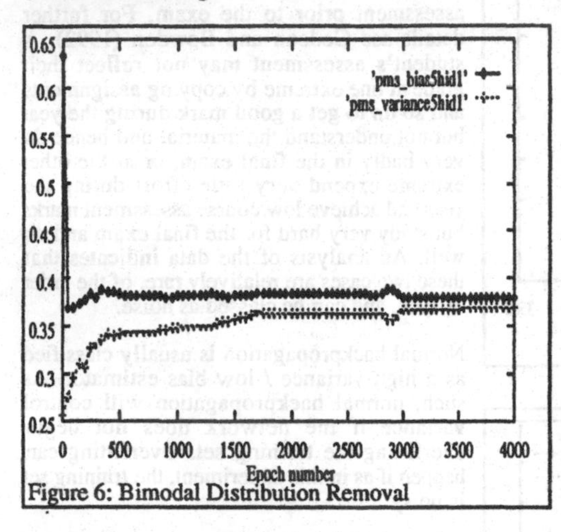

Figure 6: Bimodal Distribution Removal

Bimodal Distribution Removal provides a similar control of variance as LTS. It is an improvement over LTS since both bias and variance are lower during training, and a data driven halting condition results.

The values for bias in the LTS and BDR methods for this data set in comparison to normal back-propagation indicates that even theugh these methods perform well, the noisy data points are being useful in this case. This means our implicit assumption that the probability law γ is approximately degenerate is barely valid. This points to the requirement for appropriate choice of a data set. An independant statistical analysis of the data was commissioned, and found there was little correlation between assessment data and final grade. Of course, the choice of data for the purpose of demonstrating a new method such 'difficult' data is ideal.

Conclusion:

The choice of training method depends on goal of the user. If a training set is known to be very noisy then an outlier detection method should be employed. However if the training examples are 'clean' and large in number, normal backpropagation will best approximate the regression $E[y|x]$. If the test set is known to be clean, then the LTS method can be used.

The BDR method has been shown to perform as well as existing methods including on 'real' noisy data, with none of the disadvantages. The method can improve generalisation by removing sources of noise, speed up training by reducing the number of patterns, and provides a natural stopping criterion to terminate training.

References:

Ash, T, "Dynamic node creation in back-propagation networks," TechRep, Univ. of Calif., 1989.

Gallant, AR and White, H, *A Unified Theory of Estimation and Inference for Nonlinear Dynamic Models*, Basil Blackwell, Oxford, 1988.

Gedeon, TD and T.G. Bowden, TG, "Heuristic Pattern Reduction," *IJCNN*, Beijing, 1992.

Gedeon, TD and Harris, D, "Network Reduction Techniques," *Proc. Int. Conf. on Neural Networks Methodologies and Applications*, San Diego, vol. 2, pp. 25-34, 1991.

Geman, S, Bienenstock, E and Doursat, R, "Neural networks and the bias/variance dilemma," *Neural Computation*, vol. 4, pp. 1-58, 1992.

Harris, D and Gedeon, TD, "Adaptive insertion of units in feed-forward neural networks," *4th Int. Conf. on Neural Networks and their Applications*, Nîmes, 1991.

Joines, M and White, M, "Improving generalisation by using robust cost functions," *IJCNN*, vol. 3, pp. 911-918, Baltimore, 1992.

Sietsma, J and Dow, RF, "Creating Artificial Neural Networks That Generalize," *Neural Networks*, vol. 4, pp. 67-79, 1991.

White, H, "Learning in artificial neural networks: A statistical perspective," *Neural Computation*, vol 1., pp. 425-464, 1989.

A SIMPLIFIED ARTMAP ARCHITECTURE FOR REAL-TIME LEARNING

Alex Guazzelli*
Dante Barone*
E. C. de B. Carvalho Filho[†]

Universidade Federal do Rio Grande do Sul*
CPGCC – Instituto de Informática
Porto Alegre – Brazil
e-mail: alex@inf.ufrgs.br
Universidade Federal de Pernambuco[†]
Departamento de Informática
Recife – Brazil
e-mail: ecdbcf@di.ufpe.br

Abstract

This paper presents a simplified version of the ARTMAP system, which is based on the Adaptive Resonance Theory (ART). The simplified architecture is designed from ART 1 with Quasi-supervision and ARTMAP, being classified as a Predictive ART architecture.

1 Introduction

This article introduces a simplified version of the ARTMAP system, which is based on the Adaptive Resonance Theory proposed by Grossberg [GRO76].

The Adaptive Resonance Theory deals with a system involving self-stabilizing input patterns into recognition categories while maintaining a balance between the properties of *Plasticity* (discrimination) and *Stability* (generalization).

The simplified system is strongly influenced by ART 1 with Quasi-supervision [GUA91] and ARTMAP (designed from ART 1) [CAR88c]. It is a simplification of the ARTMAP system without loss of computational performance. The differences between the ARTMAP system and its simplified version are described in the sequel.

The ART 1 architecture, which is embedded into the simplified ARTMAP system, was proposed by Carpenter and Grossberg in 1987 [CAR87a]. It was designed to treat only binary input data, while other ART systems like ART 2 [CAR87b], ART 3 [CAR88a] and Fuzzy ART [CAR88b] deal with analog patterns.

The simplified ARTMAP system is a supervised neural network. By means of this, it is used to predict input data according to a distributed learned map obtained from an input vector I and a target vector T. For this reason, it can be viewed as a Predictive ART architecture.

2 The Simplified ARTMAP System

The simplified system (Figure 1) involves two different processes. The *bottom-up process*, often refered, in ART 1, as an adaptive filtering or contrast-enhancement process, which provides the pattern $Y = (Y_1, ..., Y_N)$ (the F_2 layer has N neurons), and the *top-down process* which performs template matching and serves to stabilize the learning, producing the match patterns X_a^* in the F_{1_a} layer and X_b^* in the F_{1_b} layer.

The input pattern $I = (I_1, ..., I_{M_a})$ is clamped at F_{0_a}, activating the F_{1_a} layer and producing the pattern $X_a = (X_{a_1}, ..., X_{a_{M_a}})$, identical to I (the F_{0_a} and F_{1_a} layers have M_a neurons). The pattern X_a

learns the input pattern I through the adaptation of the bottom-up (w_{ij}) and top-down weights (z_{ij} and u_{jk}).

2. when both equations 1 and 2 are false: the winner F_2 neuron is reset until another input pattern is presented, the input pattern I and the target pattern T are restored in F_{1_a} and F_{1_b} respectively, and the search for a better resonant neuron in F_2 ensues. If no categories are found to match the input pattern, a node at F_2 that has not yet been assigned to a recognition category will be assigned to the new category.

3. when equation 1 is true and equation 2 is false: a process called match-tracking takes place [CAR88c]. This process increments the vigilance parameter ρ_a until a minimum amount necessary to turn equation 1 to false, thus reporting the system to situation two. After learning the input vector, the vigilance parameter ρ_a is set, again, to its baseline value.

4. when equation 1 is false and equation 2 is true: the same actions described in situation two are performed.

Although the ARTMAP system and the simplified one are both designed from ART 1, they are very different if we think about their architecture. For example, ARTMAP uses two ART 1 modules, an inter-ART associative memory and an internal control to learn the given distributed map between I and T. On the other hand, the simplified system has only one ART 1 module and an additional layer to do the same work. Despite of this, the simplified version conjointly maximizes generalization and minimizes predictive error, retaining the good aspects observed in ARTMAP.

During the recall phase, the F_2 element activated by the input vector I outputs a signal through the $F_2 \to F_{1_b}$ connections, which is compared to T.

3 Fast Learning

The top-down $F_2 \to F_{1_b}$ connections weights u_{jk}, like the top-down $F_2 \to F_{1_a}$ connections weights z_{ji} in ART 1, receive one as initial value.

$$z_{ji} = u_{jk} = 1 \tag{3}$$

Otherwise, the bottom-up $F_{1_a} \to F_2$ connections weights w_{ij} are initialized according to the following equation:

$$w_{ij} = \alpha_j \tag{4}$$

where $\alpha_1 \geq \alpha_2 \geq ... \geq \alpha_N$ and

$$0 \leq \alpha_j = z_{ij}(0) \leq \frac{1}{\beta + |I|} \tag{5}$$

β and α receive small values, and $\beta > 0$.

Learning is really performed changing the connections weights (top-down and bottom-up) after a stable category, or a resonant F_2 neuron, has been chosen.

The top-down weight z_{ji} connecting a neuron v_j of F_2 to a neuron v_i of F_{1_a} is given by

$$z_{ji} = \begin{cases} 1 & \text{if } v_i = v_j = 1 \\ 0 & \text{if } v_i = 0 \text{ e } v_j = 1 \\ z_{ji} & \text{if } v_j = 0 \end{cases} \tag{6}$$

and the top-down weight u_{jk} connecting a neuron v_j of F_2 to a neuron v_k of F_{1_b} is given by

$$u_{jk} = \begin{cases} 1 & \text{if } v_k = v_j = 1 \\ 0 & \text{if } v_k = 0 \text{ e } v_j = 1 \\ u_{jk} & \text{if } v_j = 0 \end{cases} \tag{7}$$

and, finally, the bottom-up w_{ij} connecting v_i of F_{1_a} to a neuron v_j of F_2 is given by

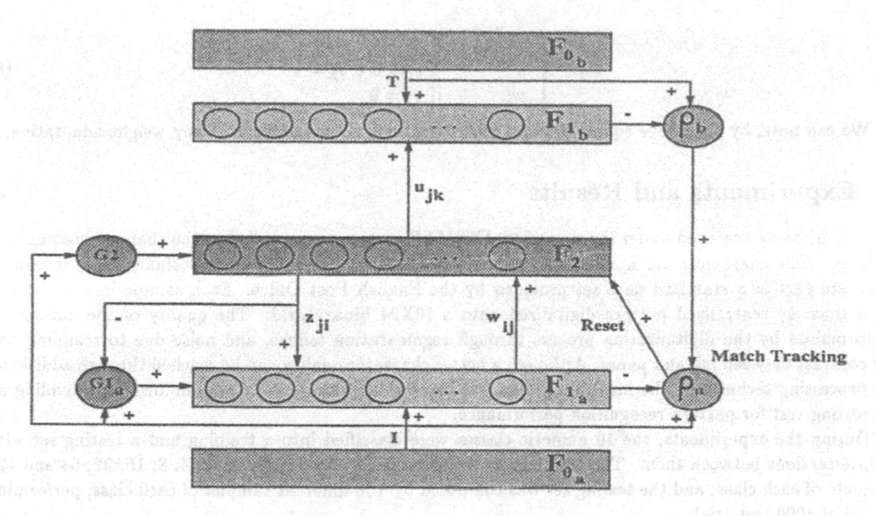

Figure 1: Super ART

is multiplied by the bottom-up weights (LTM) $w_{ij} = (w_{11}, ..., w_{M_aN})$ at each F_2 node. This produces the contrast-enhanced pattern Y at the F_2 layer. The pattern Y is the result of a *Winner-Take-All Process*, where only one neuron from F_2 can be the winner, or resonant neuron. To this neuron, it is given the chance to produce an output value equal to one.

On the other hand, the target vector $T = (T_1, ..., T_{M_b})$ is clamped at F_{0_b}, activating F_{1_b} layer and producing the pattern $X_b = (X_{b_1}, ..., X_{b_{M_b}})$, identical to T (the F_{0_b} and F_{1_b} layers have M_b neurons). But, here, there is no bottom-up output signal toward the F_2 layer.

At this point the bottom-up process finishes and the top-down process starts. The pattern Y activates the top-down LTMs $z_{ji} = (z_{11}, ..., z_{NM_a})$ and $u_{jk} = (u_{11}, ..., u_{NM_b})$, which are binary valued weights, producing the patterns $V_a = (V_{a_1}, ..., V_{a_{M_a}})$ and $V_b = (V_{b_1}, ..., V_{b_{M_b}})$ in F_{1_a} and F_{1_b}, respectively. The pattern V_a, in an ART 1 system, is known as the *Top-Down Expectation* or critical feature.

Whereas the top-down pattern V_a and the input pattern I activate the F_{1_a} layer, where the matching process takes place (this is simply an AND function), producing the match pattern X_a^*. The top-down pattern V_b and the target vector T activate the F_{1_b} layer, resulting in the match pattern X_b^*.

The match pattern X_a^* indicates the similarity of the input pattern I and the prototype pattern V_a. A rule of similarity is applied to this pattern matching, deciding whether the resonant category must be activated. This rule obeys the following equation:

$$\frac{|X_a^*|}{|I|} \geq \rho_a \tag{1}$$

where $|X_a^*|$ is the number of ones in X_a^*, $|I|$ is the number of ones in I, and ρ_a is a given *Vigilance Parameter* which is set by the user, $0 \leq \rho_a \leq 1$.

On the other hand, the match pattern X_b^* is responsible for determining if the winner F_2 node can represent the input vector I. To do this, a matching process between T and X_b^* is performed, which obeys the following equation:

$$\frac{|X_b^*|}{|T|} = \rho_b \tag{2}$$

where $|X_b^*|$ is the number of ones in X_b^*, $|T|$ is the number of ones in T, and ρ_b is equal to one.

So, as we can see, there are a number of different possibilities at this point in the process. Among them, we can identify four situations and the consequences they hold in the following way:

1. when both equations 1 and 2 are true: the recognition category is deemed to be stable, the network

$$w_{ij} = \begin{cases} \frac{X_a^*}{(\beta+|X_a^*|)} & \text{if } v_i = vj = 1 \\ 0 & \text{if } v_i = 0 \text{ e } v_j = 1 \\ w_{ij} & \text{if } v_j = 0 \end{cases} \tag{8}$$

We can note, by the above equations, that when F_2 is not active, there isn't any weight adaptation.

4 Experiments and Results

The experiments executed under the simplified ARTMAP system were applied to the character recognition problem. The characters are handwritten numeric post codes, extracted from ordinary mail envelopes. They are part of a standard data set provided by the English Post Office. Each sample is a scaled and approximately centralized pattern digitalized onto a 16X24 binary grid. The quality of the data set is compromised by the digitalization process through segmentation failure, and noise due to scanning and low contrast between ink and paper. Although a better character quality can be reached through additional pre-processing techniques, the most important feature of this data set is its non-uniformity, providing an interesting test for pattern recognition performance.

During the experiments, the 10 numeric classes were classified into a training and a testing set with no intersections between them. The training set was composed respectively by 2, 4, 8, 16, 32, 64 and 128 elements of each class, and the testing set was composed by 100 different samples of each class, performing a total of 1000 test trials.

The target vectors used in the experiments described below used local representation. Thus, each numeric class is represented by a vector composed by only one 1 and nine 0s, starting by $< 1, 0, 0, 0, 0, 0, 0, 0, 0, 0 >$ representing the class of number 0, and finishing by $< 0, 0, 0, 0, 0, 0, 0, 0, 0, 1 >$ representing the class of number 9.

The experiments performed with the simplified system incorporates off-line learning, where the vigilance parameter assumed different baseline values (0.0, 0.7 and 0.9).

In order to compare results, the experiments were also taken under an ART 1 system with three different vigilance values (0.0, 0.7 and 0.9), and in a Multi-layer Perceptron (MLP) using the standard Backpropagation algorithm [RUM86] with the following topology: 384 input neurons, 30 hidden neurons and 10 output neurons, where the learning rate was set to 0.2, and the momentum to 0.95. If the maximum absolute error for all output nodes was less than 0.1 for all patterns, then the MLP network was said to have converged. The threshold function was chosen to be sigmoid.

Figure 2, 3 and 4 show the recognition rate per cent obtained through the variation of the training set length under the analysis of the three systems. In all figures, the curve showing the MLP performance rates is the same.

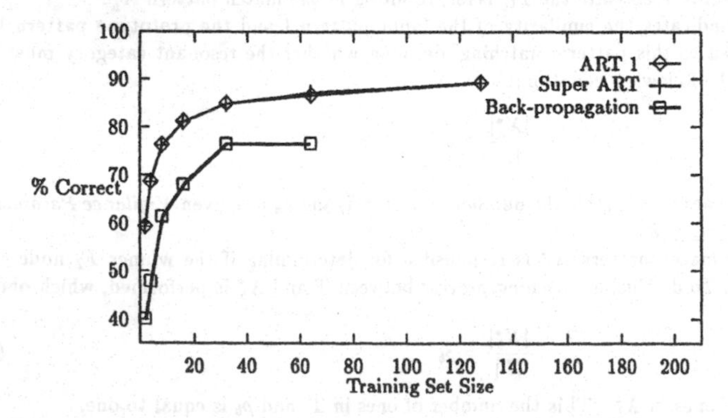

Figure 2: ART 1 (vigilance = 0.9), simplified ARTMAP system (baseline value = 0.9) and Backpropagation Performances After Training on Input Sets Ranging in Size From 2 to 128 Exemplars.

In figure 2 the simplified ARTMAP system (baseline value = 0.9) is a little bit better than ART 1 (vigilance = 0.9). It can be seen that MLP doesn't present a good performance through this pattern data base. It was tried to simulate the network with the training set length equal to 128, but the network didn't converge in an accepted number of iterations (1500,000).

Figure 3 shows that the differences between the recognition rates obtained with the simplified ARTMAP system (baseline value = 0.7) and ART 1 (vigilance = 0.7) become larger. At this time, the simplified system retained its predictive success rate, while ART 1, classified more training examples in less recognition categories, turning the prototypes more general and less specialized (coarse classifications). For this reason, during the recall phase, there were misclassification of more noise characters. They have more in common with other category classes than with the categories represented by their class.

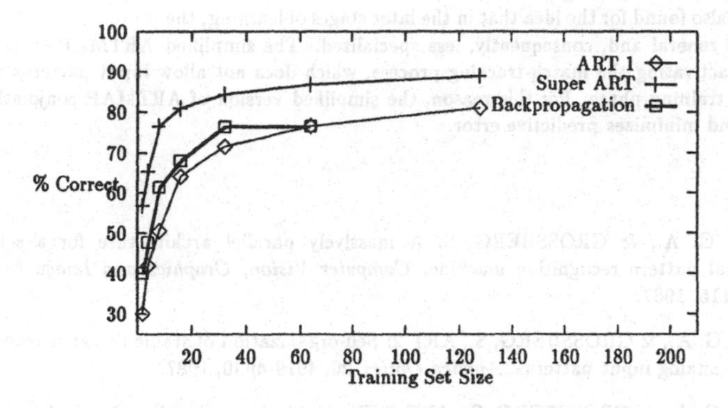

Figure 3: ART 1 (vigilance = 0.7), simplified ARTMAP system (baseline value = 0.7) and Backpropagation Performances After Training on Input Sets Ranging in Size From 2 to 128 Exemplars.

When the vigilance parameter is set to 0.0, the process started in figure 3 becomes stronger, as shown in figure 4. At this time, the more the learning proceeds, the more generalized simplified ARTMAP's prototype categories become. The simplified system creates the necessary categories to classify the training samples in non-mixed input classes. On the other hand, ART 1 has no way to classify its inputs in more than 3 different categories, turning impossible the correct classification of the input patterns during the recall phase. In this situation, the MLP network presents better results than both ART systems.

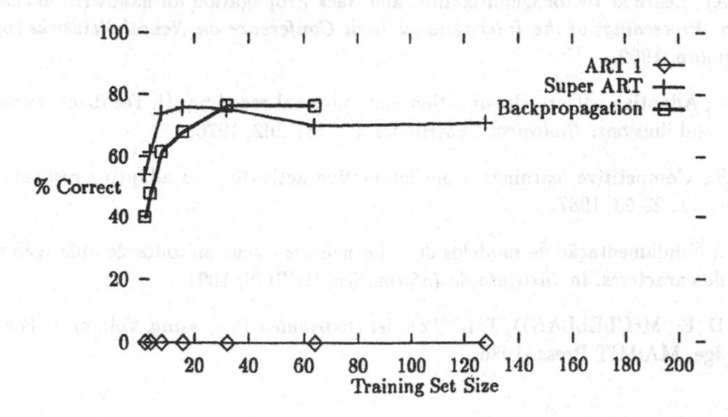

Figure 4: ART 1 (vigilance = 0.0), simplified ARTMAP system (baseline value = 0.0) and Backpropagation Performances After Training on Input Sets Ranging in Size From 2 to 128 Exemplars.

5 Conclusions

This paper has described a simplified version of the ARTMAP architecture, a supervised learning system designed from ART 1. The simplified ARTMAP system retains the computational advantages observed on an ARTMAP system. A similar simplification is reported in [CAR92].

Implementations of the simplified ARTMAP system, ART 1 and MLP applied to the handwritten characters recognition problem were developed. Recognition performance rates have been calculated and compared among the three systems. They have shown that in an ART 1 system, it is very easy to misclassify the input patterns, disrupting the category prototypes and, consequently, compromising the recall phase.

The MLP system showed a bad performance considering the data base used, and when presented to a larger training set the system didn't converge.

Additional support was also found for the idea that in the later stages of learning, the prototypes in ART categories tend to be more general and, consequently, less specialized. The simplified ARTMAP system attenuates this process by activating the match-tracking process, which does not allow input patterns to be misclassified during the training phase. For this reason, the simplified version of ARTMAP conjointly maximizes generalization and minimizes predictive error.

References

[CAR87a] CARPENTER, G. A., & GROSSBERG, S. A massively parallel architecture for a self-organizing neural pattern recognition machine. *Computer Vision, Graphics and Image Processing*, 37, 54-115, 1987.

[CAR87b] CARPENTER, G. A., & GROSSBERG, S., ART 2: Self-organization of stable category recognition codes for analog input patterns. *Applied Optics*, 26, 4919-4930, 1987.

[CAR88a] CARPENTER, G. A., & GROSSBERG, S., ART 3: Hierarchical search using chemical transmitters in self-organizing pattern recognition architectures. *Neural Networks*, 3, 129-152, 1990.

[CAR88b] CARPENTER, G. A., GROSSBERG, S., & ROSEN, D. B., Fuzzy ART: Fast stable learning and categorization of analog patterns by adaptive resonance system. *Neural Networks*, 4, 759-771, 1991.

[CAR88c] CARPENTER, G. A., GROSSBERG, S., & REYNOLDS, J. H., ARTMAP: Supervised real-time learning classification of nonstationary data by a self-organizing neural network. *Neural Networks*, 4, 565-588, 1991.

[CAR92] CARPENTER, G. A., GROSSBERG, S., & IIZUKA, K., Comparative performance measures of Fuzzy ARTMAP, Learned Vector Quantization, and Back Propagation for handwritten character recognition. *Proceedings of the International Joint Conference on Neural Networks* (pp. I 794-799). Baltimore, 1992.

[GRO76] GROSSBERG, S., Adaptive pattern classification and universal recoding, II. Feedback, expectation, olfaction, and illusions. *Biological Cybernetics*, 23, 187-202, 1976.

[GRO87] GROSSBERG, S., Competitive learning: from interactive activation to adaptive resonance. *Cognitive Science*, 11, 23-63, 1987.

[GUA91] GUAZZELLI, A., Fundamentação de modelos de redes neurais e seus métodos de aplicação no reconhecimento de caracteres. In *Instituto de Informática*, UFRGS, 1991.

[RUM86] RUMELHART, D. E., McCLELLAND, J. L., *Parallel Distributed Processing*. Volume 1: Foundations, Cambridge, MA:MIT Press, 1986.

B-LEARNING: A REINFORCEMENT LEARNING ALGORITHM, COMPARISON WITH DYNAMIC PROGRAMMING

Thibault LANGLOIS†‡ Stéphane CANU‡*

March 12, 1993

†INESC, rua Alves Redol, 9
Apartado 10105
1017 Lisboa Codex PORTUGAL
Tel. (351)1 31 00 206 Fax: (351)1 52 58 43
E-mail: tl@neura.inesc.pt

‡Université de Technologie de Compiègne,
U.R.A. 817 du CNRS BP649
F-60206 Compiègne Cedex FRANCE
E-mail: scanu@hds.univ-compiegne.fr

Abstract

In this paper we present a Reinforcement Learning method — B-Learning — for the control of a water production plant. A comparison between B-Learning and Dynamic Programming is provided from both theoretical and performance points of view. It is shown that Reinforcement-based neural control can lead to results comparable in quality to Dynamic Programming-based though less computationnally expensive.

1 Introduction

Reinforcement Learning is an unsupervised method mainly used in the control area. It is based on a binary signal $r(t)$[1] called **external reinforcement** provided by the environment. This signal indicates whether the system is in an allowed state or not. Two functions have to be modeled for the construction of the controller. An **evaluation function** which produces an estimate of a long-term evaluation of the system state and a **control function** which provides the actions. Barto et al. [3] has introduced this learning method (Adaptive Heuristic Critic) for the control of an inverted pendulum. Both functions were modeled with a single neuron and the state space was discretized. Anderson [1] extended the method and used MLP to model the functions. The two algorithms were used by Jouse [5] for the control of a nuclear reactor and Hoskins [4] for the control of a CSTR.

Watkins [13] proposed a slightly different method — Q-Learning — with only one Q-function producing an estimate of the long-term evaluation of a couple (state, action). At each time-step, the action leading to the best evaluation is chosen.

Several methods have been proposed in order to improve the learning speed. Sutton [12] introduced an architecture called Dyna: A model of the system is used to achieve an off-line learning for a given number of time-steps and increase the learning speed. Lin [9] proposed a method called "experience replay". Training experiences are stored and a kind of batch learning is performed. The more efficient use of data yields increases of the learning speed.

*The authors are grateful to Patrick Lesueur who performed the experiments using Dynamic Programming.

[1] $r(t)$ is equal to 0 if the system is in an allowed state and equal to -1 if not.

Notations

$x(t)$: system state at time t	$\underline{X}$: the set of all system states
$a(.), \underline{A}$: an action, the set of all actions	$e(t)$: the long-term evaluation of state $x(t)$
$\widehat{e}(t)$ an estimate of $e(t)$ (see 1.1)	$\gamma \in [0,1]$ a discount parameter usually $\gamma = 0.9$
E : the evaluation function (see 1.1)	C : the control function (see 1.1)
$r(t)$: the external reinforcement	J : a cost function

1.1 Adaptive Heuristic Critic

AHC algorithms are based on the following architecture:

- An evaluation function, modeled with a MLP, gives an estimate ($\widehat{e}(t)$) of long-term evaluation ($e(t)$) of the current state. The evaluation is defined as:

$$e(t) = \sum_{T=t}^{\infty} \gamma^{T-t} r(T)$$

The function is:

$$E : \begin{cases} \underline{X} \rightarrow \mathbb{R} \\ x(t) \mapsto \widehat{e}(t) \end{cases}$$

The method used to train the network is the Temporal Difference (TD) algorithm [11]. This prediction method uses the forecast value at the next step as the target for the current step. A special form of Temporal Difference is obtained with the cost function:

$$J_E = \frac{1}{2} ||\widehat{e}_t(x(t)) - [r(t+1) + \gamma \widehat{e}_t(x(t+1))]||^2 \quad (1)$$

- A control function which maps states to actions:

$$C : \begin{cases} \underline{X} \rightarrow \underline{A} \\ x(t) \mapsto a(t) \end{cases}$$

This function is also trained using TD algorithm. The cost function is (1). If there are n possible actions at each time step, $a(t)$ represents a vector with long term evaluations estimated for each action. The action corresponding to the best evaluation is chosen at each time step. This function can be build using a single MLP or with a MLP for each possible action. Lin [10] reports better results in the later case.

1.2 Q-Learning

Only one function is necessary for Q-Learning [13, 12, 10]. This function provides estimates of the long-term evaluation of state-action pairs:

$$Q : \begin{cases} \underline{X} \times \underline{A} \rightarrow \mathbb{R} \\ x(t), a(t) \mapsto Q(x,a) \end{cases}$$

At each time-step, an action is chosen according to the maximum of this function. A special form of Temporal Difference is used for training. The cost function used is:

$$J_Q = \frac{1}{2} ||Q(x,a) - r(t+1) - \gamma \max_a Q(x(t+1), a)||^2$$

2 B-Learning

B-Learning is a Reinforcement Learning algorithm we have described in [7]. It is based on an architecture similar to AHC i.e. with an evaluation function — described above — and a policy function. The purpose of our policy function is to provide estimates of *benefits* associated with each action. The *benefit* is defined as the derivative of the long-term evaluation with respect to time:

1. $x(t) \leftarrow$ intial state
2. $\widehat{e}(t) \leftarrow \mathrm{E}(x(t))$
3. $\widehat{B}(t) \leftarrow \mathrm{B}(x(t)); a(t) \leftarrow \max_i(\widehat{B}_i(t))$
4. $x(t+1) \leftarrow \mathrm{S}(x(t), a(t))$
5. $\widehat{e}(t+1) \leftarrow \mathrm{E}(x(t+1))$
6. $\mathrm{BP}(\mathrm{B}, \widehat{B}(t) - \widehat{e}(t+1) + \widehat{e}(t))$
7. $\mathrm{BP}(\mathrm{E}, -r(t) - \gamma\widehat{e}(t+1) + \widehat{e}(t))$
8. $t \leftarrow t+1$; goto 2.

Table 1: *B-Learning algorithm. $S(.,.)$ is the system and $\mathrm{BP}(N, e)$ is the back propagation algorithm applied to the network N with error e.*

$$B(t) \triangleq \frac{\partial e(t)}{\partial t}$$

As we use a fixed time-step, we write:

$$B(t) \triangleq \frac{e(t+1) - e(t)}{\Delta t} = e(t+1) - e(t)$$

At each time-step there are n possible actions. The function $B(.)$ is defined:

$$B : \begin{cases} X \mapsto \mathbb{R}^n \\ x(t) \rightarrow B(t) \end{cases}$$

The output of this function is a vector:

$$B(t) = \begin{bmatrix} e_1(t+1) - e(t) \\ \vdots \\ e_i(t+1) - e(t) \\ \vdots \\ e_n(t+1) - e(t) \end{bmatrix}$$

$e_i(t+1)$ is the evaluation of the state reached with action a_i. Each component is the expected benefit for each action. At each time-step an action is chosen according to the maximum of this function. These values are estimated using the evaluation function and the cost function used for training is:

$$J_{Bi}(t) = \frac{1}{2}||\widehat{B}_i(t) - [\widehat{e}_i(t+1) - \widehat{e}(t)]||^2 \quad (2)$$

The algorithm is described in table 1. Unlike Q-Learning, the purpose of B-Learning is to build an evaluation function not dependent on particular actions. This function provides a coarse mapping of state-space and memorizes "good" states and "bad" states. Then the B-function, uses this information to achieve a more accurate mapping from state-space to the expected benefit gained when applying each possible action. As the B-function is modeled with only one neural network, it is possible to add constraints to the output layer in order to work in a space of probabilities. In this case, the B-function would compute the probability of best benefit, given action a_i.

3 Dynamic Programming and Reinforcement Learning

3.1 Dynamic Programming

Dynamic Programming is a method used to solve optimization problems given a set of constraints. This method is based on the optimality principle stated by Bellman:

"An optimal solution must be composed of partial solutions which are themselves optimal"

It requires a cost function and an accurate model of the system. The cost function must be decomposable. For dynamic processes, a time decomposition is done. At each time-step, all possible actions are applied on the model and a cost function is derived. The tree-searching procedure is achieved in a breadth-first fashion. The optimality principle is used to reduce the exploration field to relevant sub-trees. If the same state is reached with different paths, then only the minimal-cost path is kept. Despite this reduction method, in real applications, there are so many states that a combinatorial explosion cannot be avoided. It is computationnally expensive and several methods exist to decrease the number of branches to be explored. For example, at each step, it is possible to develop only low-cost sub-trees. It is also possible to use Dynamic Programming in conjunction with Branch and Bound methods. In this case, an estimate of the optimal cost is derived and all branches whose cost is superior to this estimate are not developed. As the model is used to evaluate system states along the tree, the method is sensitive to its accuracy.

All these improvements are made at the cost of a lower precision and the modified DP procedures lead to **sub-optimal** solutions.

3.2 Reinforcement Learning

Reinforcement Learning (and B-Learning) is unsupervised. The single information available to the learning system is the external reinforcement signal. A function giving an estimate of the long-term evaluation of the system's state with respect to the current policy is built ($\widehat{e}(.)$). It is a noteworthy fact that the function provides information about *future states*, it is an evaluation of states from current time t to the end of the sequence. Therefore it is not necessary — like in the DP procedure — to use the model to evaluate all future states. Only a one-step forward evaluation is necessary. In the B-Learning procedure, the best action is chosen on the basis of the difference between two long-term evaluations. This mapping between states and expected benefits is modeled by the B-function. The Reinforcement Learning procedure (and B-Learning procedure) requires the training of two functionals, evaluation and control function (B-function). We must point out, that the convergence of the latter depends on the convergence of the former. Therefore it is important to tune the respective learning rates in order to observe the convergence of the whole. Moreover, these functions provide estimates $\widehat{e}(x(t))$ and $\widehat{B}(x(t))$, and consequently the algorithm will lead to **sub-optimal** solutions.

Theses methods are compared from an industrial application point of view in section 4.4.

4 Control of a water production plant

4.1 Description of the plant

Our intent is to control a water-production plant. Water arrives in the plant by the mean of five fixed-flow pumps[2] . After a cleaning process, it is dumped into a buffer-tank. The task is to maintain the level of water in the tank within a specified range, by actioning the pumps at the entrance of the plant.

The upper (h_{max}) and lower (h_{min}) limits of the tanks are 2m and 4.50m and the control interval is a quarter of hour. The state of our system is the height of water in the tank, $h(t)$. The system's equation of motion is:

$$h(t) = h(t-1) + F_i(t) - F_o(t) \tag{3}$$

F_i is the input flow (the command), F_o the output flow (the water demand). The system is simulated using real data obtained from the plant. Data is a measurement of the time-dependent output-flow function. We used 15 days of data for the training set.

[2] The pumps flows are 2160 m^3/h, 3450 m^3/h, 4750 m^3/h, 5560 m^3/h, 6400 m^3/h.

This rather simple problem becomes more complex with the following constraints:

- The output flow which empties the tank is unknown. The variations of the output flow are very significant and larger than the range of the entrance flow.
- In order to maintain a good water quality it is important to keep the input flow as constant as possible.
- The cleaning process causes an half-an-hour delay.

The problem of control of time-delay systems has been addressed in [8]. In this paper we assume the system to be time-delay free.

4.2 The Architecture

The evaluator is modeled with a MLP. A representation of the system state is used for the input of the MLP. This representation is the distance between the actual height $h(t)$ and the limits h_{min} and h_{max}. The external reinforcement is given by:

if $h_{min} < h(t) < h_{max}$ $r(t) = 0$ else $r(t) = -1$.

The cost function (1) — i.e. the TD method — is used to adapt MLP's weights in order to maximize the time before failure. The controller is also modeled with a MLP. The same inputs are used and five outputs match the expected benefit of using each of the five flows. We used the following cost function:

$$J_c(t,i) = \begin{cases} \frac{1}{2}\|\widehat{B}_i(t) + 1\|^2 & \text{if the systems gets out of the limits} \\ \frac{1}{2}\|\widehat{B}_i(t) + \widehat{e}(t,i) - \widehat{e}(t+1,i)\|^2 & \text{else} \end{cases} \quad (4)$$

This cost function is slightly different from (2), since we add the first case in order to punish actions which cause failure. The second case increases or decreases the expected benefit according to the reinforcement variation.

One of our constraints is to minimize the variations of the input flow. The B-Learning provides flexibility in the choice of the control rule. At each time-step we derive a rich information: We know the **best command** to apply to the system, as well as the **amount of risk** we take if another command is chosen. The flexibility allows us to chose the actions according to the following heuristic in order to meet the constraint: *"if the best action causes a change then use the second best action if it is not too bad"* By "not too bad" we mean that the difference between the expected benefit for the best action and the expected benefit for the second best action must be smaller than a threshold value ϵ.

4.3 The Dynamic Programming approach

In order to make a comparison, we used a Dynamic Programming approach on the same problem, with the same set of data. The problem is solved using a finite horizon equal to one day and the time step used was one hour. In order to reduce the computation time, the number of sub-trees was limited the 500 minimal cost subtrees. Moreover, in order to simulate the whole tree was simulated by randomizing three actions at each node. An A^{*}[3] algorithm was used to limit the number of sub-trees to be explored. The whole method is depicted in [6]. The controller was tested on 15 days of data.

4.4 Results

Recall that the aim of the controller is to maintain the water level between the limits and minimize the number of input flow changes. As both methods succeeded in reaching the first objective, the comparison is based on the number of changes:

[3] A^* is an admissible search algorithm

day#	1	2	3	4	5	6	7	8	9	10	11	12	13	14	15	*Total*
B-Learning	5	7	4	6	7	5	6	6	6	4	4	6	5	5	4	80
DP	4	5	3	11	6	4	3	10	6	7	4	3	4	4	3	77

One can see that the B-Learning method performs nearly as well as the DP procedure. The advantage of B-Learning over the Dynamic Programming-based method is that during the learning phase, an identification of the evaluation function is done. There is no learning phase for the DP-based method and this function has to be estimated at each time-step. Moreover, the control function identified by B-Learning is far more cheaper to compute than the DP-based procedure.

5 Conclusion

The similarities between Reinforcement Learning and Dynamic Programming have been stated by Sutton in [12] and Barto et al.[2]. In this paper a variant of Reinforcement Learning (B-Learning) is proposed and we compare the principles involved in Dynamic Programming and Reinforcement Learning methods. We point out the fact that both algorithms lead to a sub-optimal solution if used for realistic problems. The Reinforcement Learning (and B-Learning) methods can be viewed as a kind of pre-programming of Dynamic Programming methods. Indeed, the evaluation of a sub-tree calculated through tree exploration in DP methods is provided by the evaluation function in Reinforcement Learning. Reinforcement Learning methods require a training phase but are very fast in utilization phase whereas Dynamic Programming methods are relatively easy to develop but are very computationnally expensive.

References

[1] Charles W. Anderson. Learning to control an inverted pendulum using neural networks. *IEEE Control Magazine*, pages 31–37, april 1989.

[2] Andrew G. Barto, Steven J. Bradtke, and Satinder P. Singh. Real-time learning and control using asynchronous dynamic programming. Technical Report 91-57, University of Massachusetts, Dept of Computer Science, Amherst MA 01003, August 1991.

[3] Andrew G. Barto, Richard S. Sutton, and Charles W. Anderson. Neuronlike adaptive elements that can solve difficult learning problems. *IEEE Transactions on Systems, Man and Cybernetics*, SMC-13(5):834–846, september october 1983.

[4] J. C. Hoskins and D. M. Himelblau. Process control via incremental neural networks and reinforcement learning. In *Chicago Meeting of American Institute of Chemical Engineers*, Chicago Illinois, November 1990.

[5] Wayne C. Jouse and John G. Williams. The control of nuclear reactor start-up using drive reinforcement theory. In Cihan H. Dagli, Soundar R. T. Kumara, and Yung C. Shin, editors, *Intelligent Engineering Systems Through Artificial Neural Networks*, pages 537–544, St. Louis, Missouri, USA, November 1991.

[6] R. Kora, P. Lesueur, and P. Villon. An adaptive optimal control algorithm for water treatment plants. In *to be published*, 1992.

[7] Thibault Langlois and Stéphane Canu. B-learning: a reinforcement learning variant for the control of a plant. In *Intelligent Engineering Systems Through Artificial Neural Networks (ANNIE'92)*. ASME Press, 1992.

[8] Thibault Langlois and Stéphane Canu. Control of time-delay systems using reinforcement learning. In *Artificial Neural Networks, 2*. Elsevier Science Publishers, 1992.

[9] Long-Ji Lin. Programming robots using reinforcement learning and teaching. In *NinthNational Conference on Artificial Intelligence*, pages 781–786, 1991.

[10] Long-Ji Lin. Self-improving reactive agents based on reinforcement learning, planning and teaching. *Machine Learning*, 8(3–4):923–321, 1992.

[11] Richard S. Sutton. Learning to predict by the method of temporal differences. *Machine learning*, 3:9–44, 1988.

[12] Richard S. Sutton. Integrated modeling control based on reinforcement learning and dynamic programming. In Richard P. Lippman, John E. Moody, and David S. Touretzky, editors, *Advances in Neural Information Processing Systems*, volume 3, pages 471–478. Morgan Kaufmann, 1990.

[13] Christopher J. C. H. Watkins. *Learning with Delayed Rewards*. PhD thesis, Cambridge University Psychology Department, 1989.

INCREASED COMPLEXITY TRAINING

Ian Cloete, Jacques Ludik
ian@cs.sun.ac.za, jludik@cs.sun.ac.za

Computer Science Department, University of Stellenbosch, Stellenbosch 7600, South Africa

Abstract

The training strategy used in connectionist learning has not received much attention in the literature. We suggest a new strategy for backpropagation learning, *increased complexity training*, and show experimentally that it leads to faster convergence compared to both the conventional training strategy using a fixed set, and to combined subset training. Increased complexity training combined with an incremental increase in the success ratio required on the training set produced even quicker convergence.

1 Introduction

In backpropagation training the choice of parameters plays an important role in determining the success of learning a particular problem. Equally important is the training strategy for successful learning. We introduce a new training method, *increased complexity training* (*ict*) and show that the number of epochs to successful learning can be reduced drastically by first learning easy problems, and gradually increasing the complexity of the problem to be learned. A further reduction in training time is accomplished by using an *incremental success ratio*, thus reducing the overall training time compared to training on a fixed set by 50% for a counting experiment.

The paper is organised as follows: In section 2 we present the methodology for increased complexity training. A problem for experimental testing is explained in section 3, and the results are compared with those obtained by *combined subset training* [Cottrell 91] and by conventional training on a fixed set. In section 4 the results are discussed and improved training time is obtained using an *incremental success ratio* for each subsequent training set. We conclude with suggestions for further research.

2 Increased Complexity Training

Various methods have been investigated to speed up learning using the backpropagation algorithm [Rumelhart 86]. These include adaptive learning rate adjustment [Silva 90], use of a momentum term in weight updates [Rumelhart 86], and second order derivatives [Fahlman 89]. To a large extent, however, the training strategy has been neglected. Usually the experimenter selects a fixed set of training patterns and a disjoint set of test patterns, where the assumption is that these sets are representative of the problem to be learned. Then the connectionist network is trained using the fixed set of training patterns until a success criterion is met. This may require repeated trials with varying initial parameters, since the network may get stuck in a local minimum. When the training set is large, this problem is more pronounced. After training the generalization performance of the network is determined on the test set.

Cottrell and Tsung (1991) suggested an improved training strategy, called *combined subset training* (*cst*) which works as follows: Select randomly a manageable subset of training patterns to train the network

Time	Inputs	Outputs
t0	0	0000
t1	1	0001
t2	1	0010
t3	0	0000
t4	1	0001
t5	1	0010
t6	1	0011

Table 1: Example input and output patterns for counting

initially. When the network has learned this set fairly well, select randomly another subset of equal size and add it to the first. Train the network further with the combined set. Repeat this procedure until the whole training set is included, or until the network is able to generalize to the rest of the original training set.

We conjecture that training time can similarly be improved by using *increased complexity training* (*ict*). The idea is simply that it should be quicker to learn the easy part of a problem first, and then gradually increase the complexity of the problem to be learned by giving "more difficult" training patterns in addition to those that the network have already learned. That means that the fixed training set be split into subsets of "increasing complexity", and having learned the initial subset, the next more complex subset is added to the first for subsequent training. This process is repeated until the whole training set is learned. So rather than selecting patterns randomly, patterns are selected "intelligently" by learning easier tasks first. This is reminiscent of human learning – for instance, first learn to count to small numbers, before proceeding to larger numbers.

3 Experimental Design

In this section we outline experiments to test and compare the proposed method with both *cst* and training on a fixed set. We elected to use an Elman recurrent network [Elman 90] which sequentially learns to count to 15. We chose to experiment with a recurrent network known to have a slow learning convergence [Hoekstra 92, p.28]. Recurrent connectionist topologies offer the attractive facility over time-delay neural networks [Waibel 89] that the exponential growth in the number of training patterns (2^N) for a window of size N can be avoided by processing input serially [Hoekstra 92, p.28].

This experiment is similar to the counting experiment reported by [Hoekstra 92]. Assume that pulses have to be counted sequentially up to 15. The absence of a pulse is represented by a zero, and resets the counter. The network receives one input at a time, and outputs the binary representation of the counter, i.e. four bits. An example of two such temporal patterns and the corresponding outputs are given in Table 1.

A test set (128 temporal patterns) was generated which contains all counting sequences randomly varying in length. The same test set and initial parameters for the network (i.e. weights, fixed learning rate, zero momentum term) were used in all experiments. Weights were updated after every input. The optimal learning rate was determined on a fixed training set and then used for all experiments. Whenever the error for a particular pattern was below 0.02 the pattern was deemed to have been successfully learned, thus determining the success ratio on the training set. Training was continued until either a 100% success ratio was reached, or the training performance did not improve.

For increased complexity training four training sets, numbered one to four of increasing complexity, were constructed as follows: The first contained 10 temporal patterns in random sequential order to learn to count from 0 to 1. This set has only two different training patterns which causes the low order output digit to vary. The second set includes the first, but in addition it contains the temporal patterns to count from 2 to 3, causing both the low order digits of the output to vary. Again the temporal patterns are permuted and each different pattern occurs approximately 5 times. Similarly the third set contains patterns for counting to 7, while the fourth contains those for counting to 15. In this way the maximum length of input sequences increases with every training set, while an additional output digit is involved as well.

For combined subset training four random training sets of identical size as for increased complexity training were generated. Each subsequent set includes the previous temporal patterns, all in permuted order. Figure 1 shows the error curves for *ict* and *cst*.

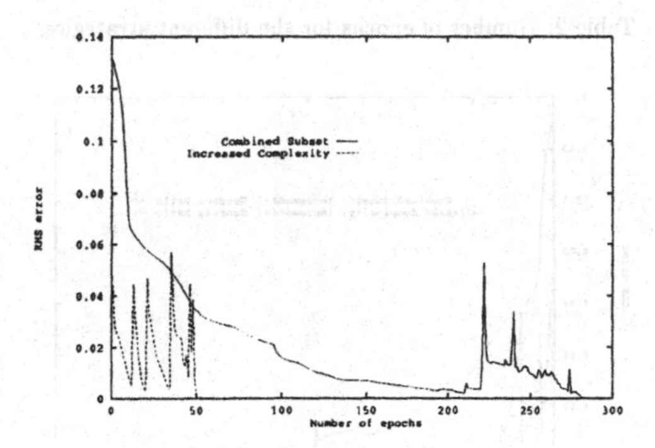

Figure 1: Error curve for *ict* and *cst*

For comparison, the last set for *ict* was also used as the fixed set on which the network was trained in a non-incremental fashion. Figure 2 shows the corresponding error curve.

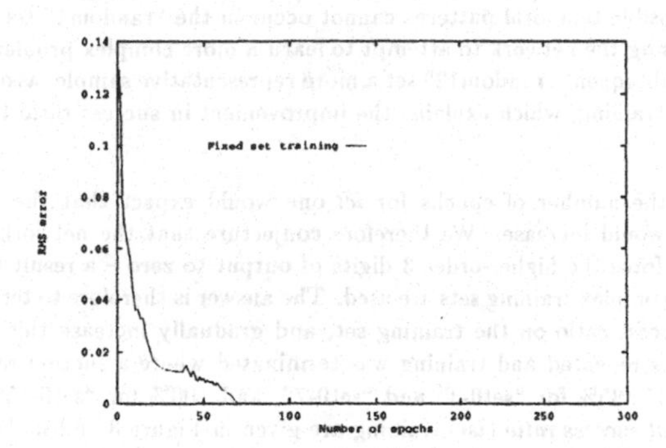

Figure 2: Error curve for the *fixed* set

The results are summarised in Table 2. The maximum success ratio obtained on the training set is given in brackets.

4 Discussion

The jumps on the error curves in Figure 1 roughly correspond to the incremental modification of the training sets, i.e. the addition of new temporal patterns. Comparing *ict* to training on the fixed set shows a 25% improvement (53 vs. 71 epochs), while *ict* compared to *cst* shows a dramatic improvement. A possible reason for the weak performance of *cst* in this case may be due to the small size of the randomly chosen training sets. The number of temporal patterns per subsequent set was kept identical for fair comparison,

# of patterns		Fixed	*icl*		*cst*
10	set0-1		12 (100%)	random1	95 (75%)
20	set0-3		8 (100%)	random12	115 (98%)
40	set0-7		14 (100%)	random123	3 (98%)
80	set0-15	71 (100%)	19 (100%)	random1234	69 (100%)
Total epochs		71	53		282
Success ratio	test set	100%	100%		100%

Table 2: Number of epochs for the different strategies.

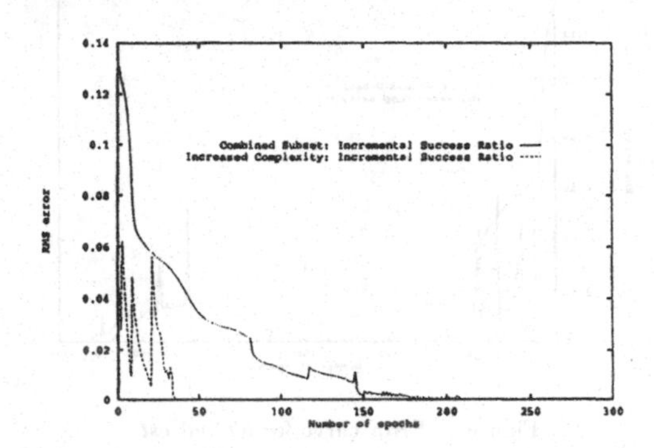

Figure 3: Error curve for *ict* (*isr*) and *cst* (*isr*)

but obviously all the possible temporal patterns cannot occur in the "random1" set. The pattern lengths, however, are larger, causing the network to attempt to learn a more complex problem than that presented in the "set0-1". In the subsequent "random12" set a more representative sample, assuming uniform pattern distribution, is used for training, which explains the improvement in success ratio to 98% on the training set.

For the comparison of the number of epochs for *ict* one would expect that the number of epochs for each more complex set would increase. We therefore conjecture that the network is erroneously trying to adapt the weights to force the higher-order 3 digits of output to zero – a result that would have to be "unlearned" when more complex training sets are used. The answer is therefore to terminate training earlier by requiring a lower success ratio on the training set, and gradually increase this termination criterion. Thus the experiment was repeated and training was terminated where a success ratio of greater than or equal to 70% for "set0-1", 85% for "set0-3" and "set0-7", and 100% for "set0-15" were obtained. The results of the *incremental success ratio* (*isr*) training are given in Figure 3 and in Table 3. In this case *ict* improves training time by 50% (35 epochs vs. 71) compared to training on a fixed set, and by 34% (35 epochs vs. 53) over its previous value when *isr* is used.

For *cst* comparison with its previous result (not using *isr*) shows a 26% improvement, but again the total number of epochs required is much more than that using *ict*. It is interesting to note that the number of epochs (68) for set "random1234" is comparable to that obtained by training on the fixed set (71), indicating that *cst* is not suitable for this problem.

From the results in Table 3, therefore, it seems as if a doubling in complexity using an incremental success ratio and increased complexity training only produces a linear increase in the number of epochs.

# of patterns		Fixed	icl (isr)		cst (isr)
10	set0-1		2 (70%)	random1	81 (70%)
20	set0-3		6 (93%)	random12	35 (93%)
40	set0-7		12 (90%)	random123	24 (90%)
80	set0-15	71 (100%)	15 (100%)	random1234	68 (100%)
Total epochs		71	35		208
Success ratio	test set	100%	100%		100%

Table 3: Number of epochs using *isr*.

5 Conclusions

Increased complexity training, suggested as a new training method for backpropagation networks, produced a shorter learning time compared to *cst* and training on a fixed set. This incremental method can be used to overcome the problems of training with large pattern sets when the problem can suitably be decomposed into training sets of increasing complexity. The training method also showed the same generalization capability as the other methods on a disjoint representative training set.

Further research is needed to determine the optimal termination point for incremental success ratio training in relation to the increase in complexity for each subsequent training set. It also remains to compare the methodologies on other problem domains. We conjecture that increased complexity training may also produce improved training time using connectionist training methods other than backpropagation.

References

[Cottrell 91] Cottrell, G.W., and Tsung, F.S., "Learning Simple Arithmetic Procedures", *High-Level Connectionist Models*, eds. J.A. Barnden, J.B. Pollack, in the series Advances in Connectionist and Neural Computation Theory, Vol. 1, pp.305-321, 1991.

[Elman 90] Elman, J.L., "Finding Structure in Time", *Cognitive Science*, Vol. 14, pp. 179-211, 1990.

[Fahlman 89] Fahlman, S.E., "Faster Learning Variations on Back-propagation: An Empirical Study", *Proceedings of the 1988 Connectionist Models Summer School*, Morgan Kaufmann Publishers, pp. 38-50, 1989.

[Hoekstra 92] Hoekstra, J., "Is Counting with Artificial Neural Networks a Problem?", Proceedings of the 1st IFIP Working Group-10.6 Workshop, INPG, Grenoble, France, pp. 27-30, 2-3 March 1992.

[Rumelhart 86] Rumelhart, D.E., Hinton G.E., Williams R.J., "Learning Internal Representations by Error Propagation", *Parallel Distributed Processing: Vol. 1, Foundations*, eds. Rumelhart, D.E., McClelland, J.L., Cambridge, MA: MIT Press, 1986.

[Silva 90] Silva, F.M., and Almeida, L.B., "Speeding up Backpropagation", *Advanced Neural Computers*, ed. R. Eckmiller, Elsevier Science Publishers B.V., North Holland, pp. 151-158, 1990.

[Waibel 89] Waibel, A., "Consonant Recognition by Modular Construction of Large Phonemic Time-Delay Neural Networks", *Advances in Neural Information Processing Systems 1*, ed. D.S. Touretzky, Morgan Kaufmann Publishers, pp. 215-223, 1989.

OPTIMIZED LEARNING FOR IMPROVING THE EVOLUTION OF PIECEWISE LINEAR SEPARATION INCREMENTAL ALGORITHMS

J.M. Moreno[1*], F. Castillo[2], J. Cabestany[1]

[1]Universitat Politécnica de Catalunya, Departament d'Enginyeria Electrónica, E.T.S.I. Telecomunicació de Barcelona, c/Jordi Girona Salgado s/n, P.O. Box 30002, 08080 Barcelona, SPAIN

[2]Universitat Politécnica de Catalunya, Departament d'Enginyeria Electrónica, E.U.P. Vilanova i la Geltrú, c/Victor Balaguer s/n, 08800 Vilanova i la Geltrú, SPAIN

Abstract : In this paper we address the problems which may appear when using the classical Perceptron or Pocket algorithms in order to train the units generated by Piecewise Linear Separation (PLS) incremental algorithms. These problems are due to the type of optimal solutions found by such training algorithms. Some of these solutions force a useless separation of input data , resulting in that the new units added to the network by the incremental algorithm are again faced with the same problem. The final network would then be composed of a large number of redundant units, each of them trying to solve exactly the same problem and arriving at exactly the same solution. We review some modifications proposed for improving the training algorithms, which are mainly based on the evaluation of entropy-like functions calculated for the input distributions. Furthermore, an alternative solution is proposed which has the advantage of the low computational cost associated to it. This method compares well, as simulation results show, with the methods based on Information Theory concepts.

1. Introduction

In the last years a considerable effort in the field of Artificial Neural Network's Theory has been devoted to the study and development of the so called Evolutive or Incremental Neural Networks. One possible definition of this kind of Neural Networks may be as follows : They are parallel processing systems whose functional structure, i.e., number of units and network topology, is determined during the network's training phase. Hence, contrary to the usual approach of finding a priori the right structure for a particular problem (usually by trial and error methods), training in fact determines the number of units and/or layers which are needed.

As was pointed out in [1], there are several types of incremental algorithms, and in this paper we will consider the Piecewise Linear Separation (PLS) incremental algorithms. This class of algorithms, used mainly for classification purposes, attempts at separating differently-classed input patterns using combinations of linear functions as their discriminant functions.

This global separation task is obtained by combining individual separations produced by perceptron-like units, which are able to generate linear decision regions in their training data set. In this way the incremental algorithm begins by training a small network (usually one unit), and if the linear decision region generated by this network is not capable of solving the whole classification problem, new units are added in order to obtain better solutions. The process ends when the resulting combination of linear decision regions performs the input-output mapping required by the classification problem reliably.

The subproblem assigned to each unit consists of finding a discriminant for only part of the input space. When a good solution is found, then the next unit is given the next subproblem - finding a discriminant for the remaining or part of the remaining input space.

In this paper, we shall review some classical methods used in training the individual units generated by the incremental algorithms. Next, we shall analyze the problems encountered as the network evolves (grows). Finally, we shall present new criterions which improve the performance in training individual units. These modified unit training algorithms will then be used on some of the most representative PLS algorithms, and the simulation results for

*Holder of an FI research Grant under the Generalitat de Catalunya's Educ. Dept.

different problems shall be presented.

2. Training of individual units

The units generated by the PLS incremental algorithms are usually trained by means of the perceptron learning algorithm [2]. When the training process is finished the unit owns a set of connection weights which optimally performs the input-output mapping presented as a problem to the unit. In other words, this set of weights defines a hyperplane in the n-dimensional input space which optimally performs the separation of the input data patterns.

Another training algorithm commonly used in the training of units generated by the incremental algorithms is the Pocket algorithm [3]. This consists on running the perceptron algorithm and keeping in memory (in one's pocket) the set of weights which yields the largest number of correct classifications. This algorithm also guarantees that an optimal set of weights is found after a finite number of iterations.

However, since these algorithms find a weight set which maximizes the number of patterns correctly classified by the unit, the separation performed by this unit may not be useful for constructing a network as this separation may not reduce the complexity of the problem.

An example of such an optimal, in the sense of maximum classification rate stated above,but not useful, solution is depicted in figure 1, which represents the problem of classifying input patterns belonging to two non-linearly separable classes.

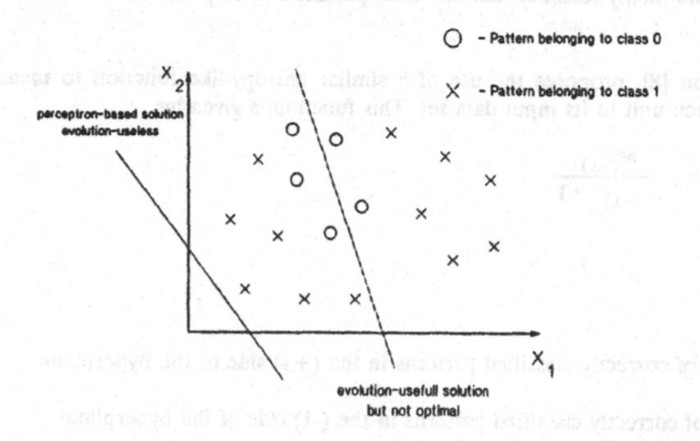

Figure 1 - Optimal and useful solutions

In this figure, the solid line represents a possible optimal solution found by perceptron-like learning algorithms. As can be seen, such a solution yields the best correct classification rate for the unit, but in fact represents no separation in the input data space. Therefore, the next unit generated by the incremental algorithm will have to solve the same problem as the previous one, arriving at the same solution. In this way, the incremental algorithm will always generate redundant units only, each of them trying to solve the same problem and arriving at the same solution.

However, a solution like the one represented by a dotted line in this figure would avoid the previous problem, as it would be able to perform a separation of the input space. In such a case, lower complexity problems are generated for the new units added to the network, construction of which ends after a finite number of steps.

Some preprocessing techniques have been proposed [4], [5], in order to avoid the problems indicated above. These methods consist mainly of increasing the dimensionality of the input space by projecting the n-dimensional input vector onto a n+1-dimensional input space. There are several types of such projection techniques (e.g., stereographic projection, gnomonic projection, ortographic projection, etc.), though the improvements obtained by such methods are highly problem-dependent.

In the next section, we present a review of some of the criterions used to modify the unit training algorithms so as to obtain suboptimal yet useful solutions.

3. Enhanced unit training

As was pointed out in the previous section, some modifications must be introduced in the unit training algorithms in order to avoid redundant solutions for the final network. As was also indicated, it is important in PLS incremental algorithms that each unit performs a "good" separation of the patterns in its training data set, so that the complexity of the problem is reduced at each step of the incremental algorithm.

Bearing in mind the previous considerations, one criterion has been proposed in [6], [7], for improving the quality of the separation performed by each individual unit in its training data set. This method consists on running the standard Pocket algorithm, but considering that the function to be maximized is the following entropy-like function, as opposed to directly maximizing the number of input patterns correctly classified by the unit.

$$I = \log C_p^q - \log C_{p'}^{q'} - \log C_{p-p'}^{q-q'}$$

C is a combinatorial term, and p, p', q, q' are, respectively, the total number of patterns in the set under consideration, the number of patterns belonging to the first class, the number of patterns being classified as belonging to the first class and the number of patterns of the first class which are correctly classified.

The function I can be considered as the difference in mixing entropy before and after the partition has been performed. In this way, we measure the "goodness" of the partition performed by the unit in terms of the value given by I for the separation, and finally consider that the "best" partition corresponds to the set of weights which maximizes I.

Another criterion [8], proposes the use of a similar entropy-like function to measure the quality of the partition produced by each unit in its input data set. This function is given by:

$$J = - \frac{w_{(+1)}}{m_{(+1)} + 1} - \frac{w_{(-1)}}{m_{(-1)} + 1}$$

where :

$w_{(+1)}$: number of correctly classified patterns in the (+1) side of the hyperplane

$w_{(-1)}$: number of correctly classified patterns in the (-1) side of the hyperplane

$m_{(+1)}$: number of misclassified patterns in the (+1) side of the hyperplane

$m_{(-1)}$: number of misclassified patterns in the (-1) side of the hyperplane

As in the previous method, the best partition will be obtained by running the Pocket algorithm for a maximum entropy function, which also gives information about the quality of the partition performed by the unit in its input data set.

As will be explained in the next section through simulation results, both methods allow the individual units to obtain satisfactory partitions of the input data space, so that a reasonable reduction in the complexity of the problem is observed through the evolution of the incremental algorithm. This is demonstrated by the small number of units required to solve the whole problem, as opposed to the endless number of redundant units obtained when using the Pocket algorithm without separation quality criterions.

The method we propose in this paper is not based on the evaluation of entropy-like functions, whose main drawback is the computational cost, when specialized hardware architectures are considered, associated to the operations performed in obtaining the separation quality indicator. Our method consists of running the Pocket algorithm on the input training set associated to each unit, but we don't require that the unit performs the best possible classification on the input data set, but rather, require it to classify correctly patterns of only one side of the separating hyperplane.

Therefore, the proposed method for improving the Pocket algorithm for PLS incremental algorithms could be explained as follows : Let run the perceptron algorithm on the training data set, and, after a certain number of iterations, prove the actual weight vector with the training data patterns. If the input space separation given by this weight vector causes that patterns belonging to only one class be correctly placed in one side of the hyperplane, and the number of correctly classified patterns is greater than the number given by the previous weight vector, then store the actual weight vector as the new best weight vector and continue the weight adaptation process. Otherwise, continue with the perceptron algorithm. Figure 2 depicts the principle on which this method is based.

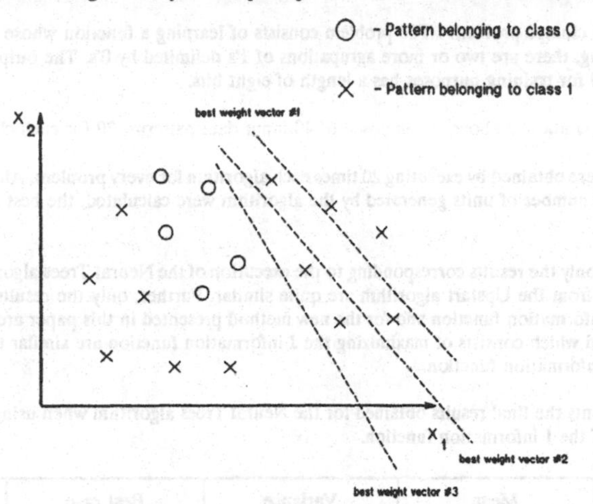

Figure 2 - Principle of the proposed method

In this figure, the dotted lines represent the successive hyperplanes associated to the best weight vector sets found by the training algorithm. The numbers associated to the weight vectors indicate the order in which these vector sets are obtained.

As can be seen, this method allows for a step by step simplification of the original problem by eliminating from the training data set patterns belonging to one class. In this way, the problems presented to the new units generated by the incremental algorithm will be reduced versions of the previous ones, and so the complexity of the classification problem is reduced as the incremental algorithm evolves.

However, as will be indicated in the next section, when the input classes to be separated are intertwined, the efficiency of this method, measured in number of units required to solve the problem, is smaller than the one yielded by the methods based on the optimization of distribution entropy functions.

In the next section we shall present the simulation results obtained by applying the previous methods to two of the most representative PLS incremental algorithms: the Neural Trees algorithm [6], and the Upstart algorithm [9].

4. Simulation results

For comparison purposes, we have selected two of the most representative PLS incremental algorithms: the Neural Trees algorithm and the Upstart algorithm. Both of them begin with a single perceptron-like unit, and then add units to the network until the classification problem is solved. In the Upstart algorithm, the new units are added in order to correct the errors produced by the previous ones, which are trained with the Pocket algorithm modified by one of the criterions mentioned in the previous section. On the other hand, the Neural Trees algorithm tries to obtain a faithful internal representation of the input data patterns; for this purpose, each unit obtains, by means of a training phase using the modified Pocket algorithm, a separation of its input data space, and, in the case that this separation doesn't match the classification task, two new units are added to the network. The input patterns assigned to each of these new units are, respectively, the patterns situated at each side of the separating hyperplane generated by the previous unit.

The problems we have selected to probe the properties of the incremental algorithms are :

* Separation of input data patterns belonging to two bidimensional, concentric distributed classes.

* Separation of input data patterns belonging to two bidimensional, normal distributed classes.

* Separation of input data patterns belonging to two bidimensional reflected spirals.

* Random boolean function of eight inputs

* Two or more clumps problem : The problem consists of learning a function whose output is 1 when, in a certain bit string, there are two or more agrupations of 1's delimited by 0's. The output will be 0 otherwise. The string used for training purposes has a length of eight bits.

Each problem mentioned above is composed of 40 input data patterns, 20 for each class.

These results were obtained by executing 20 times each algorithm for every problem. After execution, the mean and the variance of the number of units generated by the algorithm were calculated; the best and the worst case are also indicated.

For simplicity, only the results corresponding to the execution of the Neural Trees algorithm will be presented as the results obtained from the Upstart algorithm are quite similar. Further, only the results corresponding to the maximization of the I information function and for the new method presented in this paper are indicated. The results obtained for the method which consists of maximizing the J information function are similar to the obtained for the maximization of the I information function.

Table 1 represents the final results obtained for the Neural Trees algorithm when using the Pocket algorithm for the maximization of the I information function.

Problem	Mean	Variance	Best case	Worst case
Concentric classif.	6.59	0.1	5	7
Gaussian classif.	6.32	0.11	5	7
Spiral classif.	16.41	0.91	13	23
Random boolean function	4.41	0.07	3	5
Two or more clumps	5.05	0.09	3	7

Table 1 - Neural Trees maximizing the I Information function

As was expected, the worst results are obtained for the problem which consists of separating two reflected spirals, due to the high number of linear decision regions required to solve the problem.

Table 2 shows the results obtained for the Neural Trees algorithm when the Pocket algorithm is modified so as to have correctly classified patterns belonging to only one class in one side of the separating hyperplane.

Problem	Mean	Variance	Best case	Worst case
Concentric classif.	28.68	2.75	19	39
Gaussian classif.	10.77	0.78	5	19
Spiral classif.	46.82	18.85	39	128
Random boolean function	6.5	0.24	5	11
Two or more clumps	9.14	0.35	7	15

Table 2 - Neural Trees with the proposed criterion

As can be seen the results compare well (even better results are obtained for the Upstart algorithm) with the results obtained when using the Pocket algorithm modified for maximizing the entropy functions. As commented in the previous section, the worst results are obtained for the spiral classification problem, since this method tends to generate a large number of units when the classes to be separated are intertwined.

5. Conclusions

In this paper we have investigated the problems which may appear when the PLS incremental algorithms use the classical perceptron or Pocket learning algorithms for unit training. As a result, the need for suboptimal but evolution-useful solutions have been pointed out.

Some of these suboptimal solutions can be found by modifying the unit training algorithms so as to maximize entropy-like functions. These functions determine the quality of the separation performed by each unit in its input data set, and hence its maximization implies that the unit will obtain such a separation that the complexity of the problem assigned to the new units generated by the incremental algorithm will be considerably reduced. The main drawback of these methods lies however on the computational cost associated with the evaluation of the entropy functions used as a criterion for determining the quality of the partition.

Finally, a new method has been presented for avoiding the problems generated by the unit training algorithms. This method consists of running the Pocket algorithm on the input data set associated to each unit, but trying to have on one side of the separating hyperplane only patterns belonging to onc class, which must also be correctly classified. As can be deduced from the simulation results, this method compares well with the methods based on the evaluation of entropy functions, exception made for the problems in which the classes to be separated are complexly intertwined. In addition, this method has the advantage of the low computational cost associated to the evaluation of the criterion for obtaining the best weight set for each unit.

We are currently investigating another criterions for enhancing the unit training algorithms used by PLS incremental algorithms. Furthermore, we are considering simplifications of the methods based on the evaluation of entropy functions, so as to obtain hardware realizable structures.

Acknowledgment :

This work has been partially funded by ESPRIT III Project ELENA-Nerves 2 (no. 6891).

6. References

[1] F. Castillo, "Incremental Neural Networks-A Survey", Technical Report, INPG Grenoble, 1991.

[2] D.E. Rumelhart, J.L. McClelland,"Parallel Distributed Processing: Explorations in the Microstructure of Cognition", MIT Press, 1986.

[3] S.I. Gallant, "Optimal Linear Discriminants", Proc. of the 8th. Intl. Conf. on Pattern Recognition, Vol. 2, pps. 849-854, Paris, 1986.

[4] D. Martinez, M. Chan, D. Estève, "Construction of Layered Quadratic Perceptrons", Proc. of Neuronimes'92, pps. 655-665, Nimes, 1992.

[5] J. Saffery, C. Thornton, "Using Stereographic Projection as a Preprocessing Technique for Upstart", Proc. of IJCNN 91, pps. 441-446, Seattle, 1991.

[6] J.A. Sirat, J.P. Nadal, "Neural Trees: A New Tool for Classification", Technical Report, Laboratoires d'Electronique Philips, 1990.

[7] J.P. Nadal, G. Toulouse, "Information Storage in Sparsely-Coded Memory Nets", Network, Vol. 1, pps. 61-74, 1990.

[8] S. Knerr, L. Personnaz, G. Dreyfus, "A new Approach to the Design of Neural Network Classifiers and its Application to the Automatic Recognition of Handwritten Digits", Proc. of IJCNN 91, pps. 91-96, Seattle, 1991.

[9] M. Frean, "The Upstart Algorithm: A Method for Constructing and Training Feedforward Neural Networks", Neural Computation 2, pps. 198-209, 1990.

A Method of Pruning Layered Feed-Forward Neural Networks

Marcello Pelillo and Anna Maria Fanelli

Dipartimento di Informatica
Università di Bari
Via G. Amendola, 173 - 70126 Bari, Italy

Abstract — *The problem of reducing the size of a trained multilayer artificial neural network is addressed, and a method of removing hidden units is developed. The method is based on the idea of eliminating units and adjusting remaining weights in such a way that the network performance does not worsen over the entire training set. The pruning problem is formulated in terms of a system of linear equations, and a very efficient conjugate-gradient algorithm is used for solving it, in the least squares sense. The algorithm also provides a sub-optimal criterion for choosing the units to be removed, which is proved to work well in practice. Preliminary results over a simulated pattern recognition task are reported, which demonstrate the effectiveness of the proposed approach.*

1. Introduction

In the last few years, artificial neural networks have been widely used to solve difficult problems in many different domains, especially because of the development of powerful learning algorithms [1], [2]. However, one of the major open questions that arises in applying neural networks to a particular task, consists of choosing the optimal architecture for the problem at hand. In the case of layered feed-forward neural networks, this means determining the optimal number of hidden layers and the number of units for each layer. It is widely recognized that too many hidden units usually result in a poor generalization ability [2], [3]; on the contrary, too small a network could not be able to satisfactorily learn the desired input-output mapping [1], [4].

Several techniques have been proposed for finding the optimal network architecture (see e.g. [2] for a detailed description). One possible approach involves training an over-dimensioned network and then pruning off unimportant units and/or connections. As pointed out by Karnin [5], this presents several advantages. Firstly, we are guaranteed that learning takes place with the desired degree of accuracy; moreover, larger networks may result in faster learning rate and, finally, the pruning method is independent of the particular learning algorithm used. Sietsma and Dow [6], [7], also advocated the same approach: they suggested some rules for detecting redundant units and removing them, after the learning process.

In this paper we propose a formal method of pruning previously trained feed-forward neural networks, which can be regarded as a generalization of Sietsma and Dow's rules. Our idea consists of removing hidden units and distributing the remaining weights in such a way that the overall input-output behavior of the network over the entire training set remains unchanged. This leads to a formulation of the pruning problem in terms of a system of linear equations that we solve by means of an efficient projection algorithm.

Some experiments over a simulated pattern classification task were conducted, which prove the effectiveness of the proposed approach.

2. The Pruning Method

In this paper we consider a layered feed-forward network with L layers (we do count the input units as a layer) and assume that the network has been previously trained over a sample of P training patterns. As a conventional shorthand we shall denote unit i of layer l by u_{il}. The number of units in layer l will be denoted by n_l, w_{ijl} will represent the weight between units u_{il} and $u_{j,l+1}$, and y_{il} will denote the output of unit u_{il}. Each non-input unit u_{il} receives from the preceding layer a net input ξ_{il} given by

$$\xi_{il} = \sum_{j=1}^{n_{l-1}} w_{ji,l-1} y_{j,l-1} + \theta_{il}$$

where θ_{il} is a bias value for u_{il}; the unit will produce its output according to some nonlinear activation function f:

$$y_{jl} = f(\xi_{il}).$$

Commonly, the activation function f is the "logistic" function $y=(1+\exp\{-\xi_{il}\})^{-1}$ [8] but our pruning algorithm does not require any particular form for f. In the following, $y_{il}^{(p)}$ will denote the actual output of unit u_{il} following presentation of pattern p.

Now, suppose that hidden unit u_{kl} $(2 \leq l \leq L-1)$ is to be removed. One reasonable strategy is to remove the unit and adjust the remaining weights of the same layer so that the net inputs to the succeeding layer remain approximately unchanged. This amounts to requiring that the following relations hold:

$$\sum_{j=1}^{n_l} w_{jil} y_{jl}^{(p)} + \theta_{i,l+1} = \sum_{\substack{j=1 \\ j \neq k}}^{n_l} (w_{jil} + \delta_{ji}) y_{jl}^{(p)} + (\theta_{i,l+1} + \gamma_i) \tag{1}$$

for all $i=1\ldots n_{l+1}$ and $p=1\ldots P$. Here the δ_{ji}'s and γ_i's are appropriate adjusting factors for weights w_{jil}, and biases $\theta_{i,l+1}$, respectively. Simple algebraic manipulations yield

$$\sum_{\substack{j=1 \\ j \neq k}}^{n_l} y_{jl}^{(p)} \delta_{ji} + \gamma_i = w_{kil} y_{kl}^{(p)} \tag{2}$$

which is a sparse system of Pn_{l+1} linear equations in the unknowns $\{\delta_{ji}\}$ and $\{\gamma_i\}$.

Studying the consistency of system (2) can help derive rules for locating redundant units and removing them. As an example, it turns out that Sietsma and Dow's stage-one pruning rules [6], [7] correspond to very special consistency conditions for system (2). However, more general rules could be derived.

Here we shall pursue a different approach. Instead of analyzing system consistency and deriving rules for detecting and removing redundant units, we use an iterative method for solving the system, in the least squares sense; as a consequence, the choice of the particular method used will result in a criterion for choosing the units to be removed, which is proved to work well in practical experience.

Linear least squares problems are as follows (see e.g. [9], [10]):

$$\text{minimize} \| A\bar{x} - \bar{b} \|_2$$

where A is an $m \times n$ rectangular matrix and $\bar{b}$ is an m-dimensional column vector. Björck and Elfving [11] developed an efficient algorithm for solving sparse linear least squares problems, referred to as *CGPCNE*; it is a preconditioned conjugate-gradient method that begins with an initial point $\bar{x}_0 \in \text{range}(A^{\mathrm{T}})$ and iteratively produces a sequence of points $\{\bar{x}_k\}$ so as to decrease residuals

$$r_k = \| A\bar{x}_k - \bar{b} \|_2^2.$$

The algorithm is terminated when some stopping condition is met, commonly when

$$\| \bar{x}_k - \bar{x}_{k+1} \|_2 < \epsilon.$$

As *CGPCNE* algorithm is a residual reducing method it makes sense to choose the unit to be eliminated in such a way that the initial residual

$$r_0 = \| A\bar{x}_0 - \bar{b} \|_2^2$$

is minimum among all hidden units of the layer under consideration. As $\bar{x}_0$ is usually chosen to be the null vector, this corresponds to choosing the unit for which $\|\bar{b}\|$ is minimum. Notice that if we denote by $\bar{y}_{kl}$ the P-dimensional column vector consisting of the output of u_{kl} upon presentation of training patterns, the vector $\bar{b}$ is simply obtained by concatenating the n_{l+1} vectors $w_{kil}\bar{y}_{kl}$, for $i=1\ldots n_{l+1}$:

$$\bar{b} = (w_{k1l}\bar{y}_{kl}^{\mathrm{T}}, w_{k2l}\bar{y}_{kl}^{\mathrm{T}}, \cdots, w_{kn_{l+1}l}\bar{y}_{kl}^{\mathrm{T}})^{\mathrm{T}}.$$

Clearly our criterion of choosing the units to be removed is far from being optimal, and indeed it seems a little strange because we decide to remove units by simply looking at their weights and outputs. However, as we shall see in the next section, the proposed approach resulted to work very well in practice.

Summarizing, the proposed pruning algorithm proceeds as follows. Starting from the first hidden layer the "minimum-norm" unit is located and the corresponding system is solved (this can be thought of as a "relearning" phase). The performance of the reduced network is tested over the training set: if it is sufficiently close to the performance of the initial trained network, then the process is repeated over the same layer; otherwise, the last reduced network is rejected, and the algorithm continues over the next layer.

3. Results

In order to assess the effectiveness of the proposed approach we conducted some experiments over a simulated pattern classification task suggested by Niles *et al.* [12] and used as a test problem by many authors [13], [14]. This is a two-class problem with a two-dimensional continuous feature space, and known probability distributions. The class-conditional densities are shown in Fig. 1 and are:

$$p_1(x,y) = N(x,0,\sigma^2)N(y,\mu_y,4\sigma^2)$$

for class 1 and

$$p_2(x,y) = \frac{N(x,\mu_x,\sigma^2) + N(x,-\mu_x,\sigma^2)}{2} N(y,-\mu_y,\sigma^2)$$

for class 2, where

$$N(t,\mu,\sigma^2) = (2\pi\sigma^2)^{-1/2}\exp\{-(t-\mu)^2/2\sigma^2\}$$

is a Gaussian with mean μ and variance σ^2. In the experiments reported here the values μ_x=2.30σ, μ_y=2.106σ, and σ=0.2 were used, which correspond to "Mixture 2" data used both in [12] and [14].

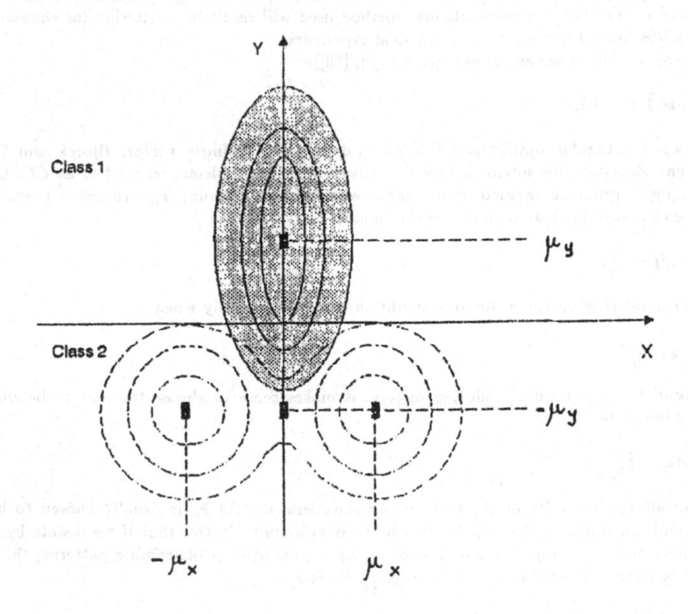

Fig. 1. Gaussian densities used in the experiments.

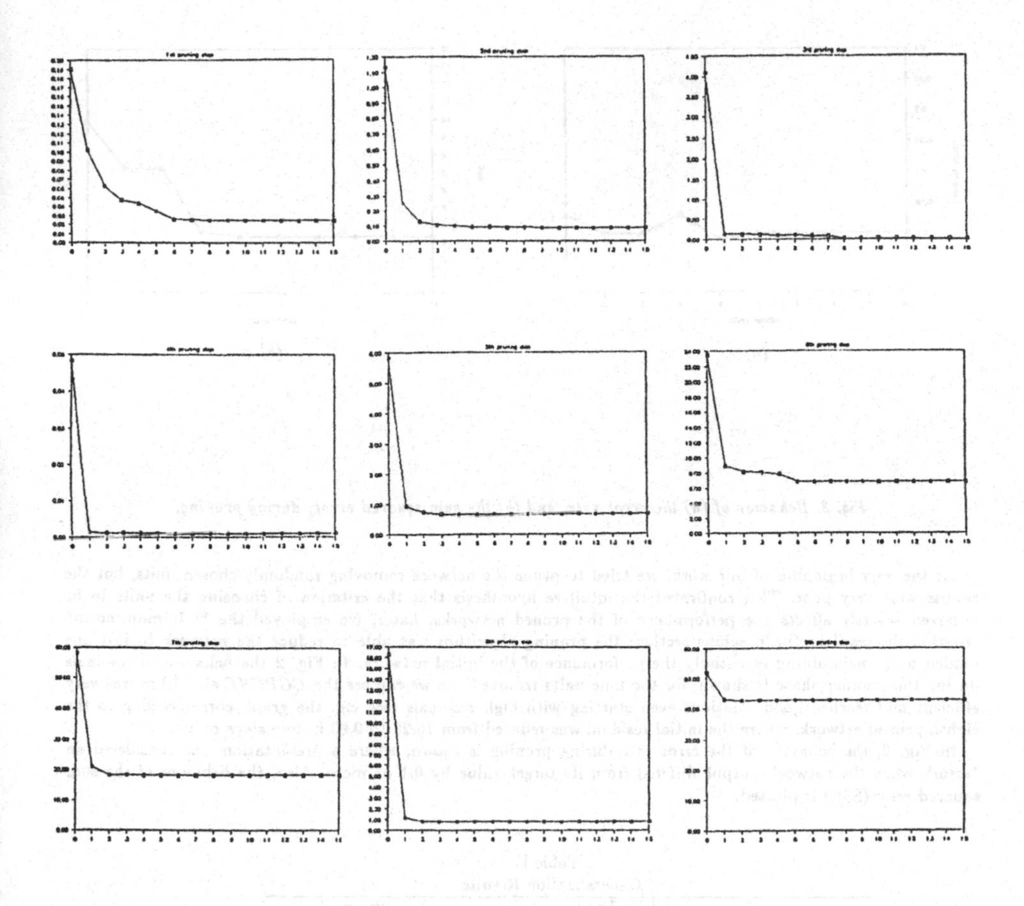

Fig. 2. Behavior of residuals during the pruning phase.
The x-axis represents the iteration number, the y-axis represents the residual.

We generated two separate data sets of 100 exemplars, one for training and one for testing. Each data set contained the same number of exemplars from the two classes, i.e. uniform a-priori distribution was supposed. A three-layer neural network consisting of 2 input units, 10 hidden units, and one output unit (2-10-1) was trained using the back-propagation algorithm [8] with learning rate η=1, and momentum term α=0.7. The learning was stopped after 1000 presentations of the training set.

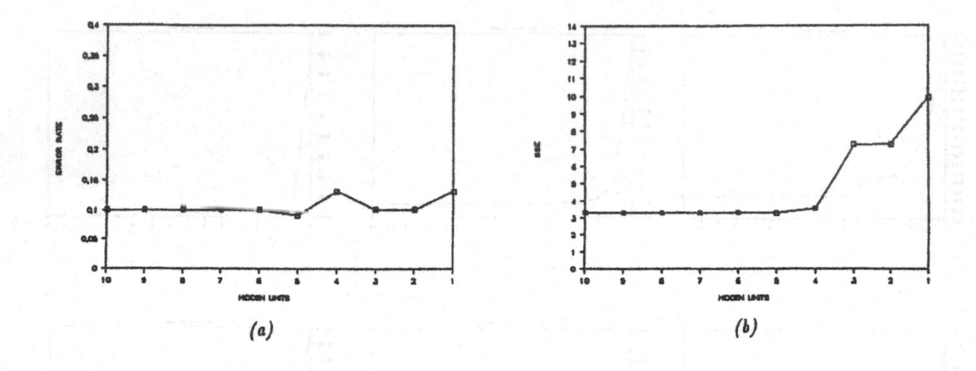

Fig. 3. Behavior of (a) the error rate, and (b) the sum squared error, during pruning.

At the very beginning of our work, we tried to prune the network removing randomly-chosen units, but the results were very poor. This confirmed the intuitive hypothesis that the criterion of choosing the units to be removed severely affects the performance of the pruned networks. Later, we employed the "minimum-norm" criterion discussed in the previous section: the pruning algorithm was able to reduce the network to just one hidden unit, maintaining essentially the performance of the initial network. In Fig. 2 the behavior of residuals during the pruning phase is shown, for the nine units removed. As we can see the *CGPCNE* algorithm was very efficient and reached good solutions even starting with high residuals (see e.g. the graph corresponding to the eighth pruned network, where the initial residual was reduced from 16.28 to 0.83 in two steps only).

In Fig. 3, the behavior of the error rate during pruning is shown, where a presentation was considered an "error" when the network output differed from its target value by 0.1 or more. Also, the behavior of the sum squared error (SSE) is plotted.

Table I
Generalization Results

	Original 2-10-1 network	Pruned 2-1-1 network
Error rate	15%	11%

Finally, Table I presents the results obtained over the test set, using both the initial 2-10-1 network and the reduced 2-1-1 network. As we can see, the pruned network reached better generalization performance than the original back-propagation trained one.

4. Conclusions

In this paper a formal method of reducing the size of trained feed-forward artificial neural networks has been developed. The key idea of our work consists of removing hidden units and adjusting the remaining weights in such a way that the overall input-output behavior of the network over the training set remains unchanged. This allows to formulate the pruning problem as a system of linear equations, that we solve with an efficient conjugate-gradient projection algorithm. The method also provides a criterion of choosing the units to be removed which leads to finding small nets that are well suited to the problem at hand. Preliminary results over a simulated pattern classification task has proved the effectiveness of our approach, and the resulting pruned network showed better generalization ability than the BP-trained one. Future work is in progress aiming at testing the performance of the proposed method over different problems, and comparing our approach with other ones.

Acknowledgments

The authors wish to thank Fabio Abbattista for the many stimulating discussions and his contribution in preparing the paper.

References

[1] R. P. Lippmann, "An introduction to computing with neural nets," *IEEE ASSP Mag.*, pp. 4-22, Apr. 1987.

[2] J. Hertz, A. Krogh, and R. G. Palmer, *Introduction to the Theory of Neural Computation.* Redwood City, CA: Addison-Wesley, 1991.

[3] Y. Chauvin, "Generalization performance of overtrained back-propagation networks," in L. B. Almeida and C. J. Wellekens (eds.), *Neural Networks - Proc. EURASIP Workshop 1990.* Berlin: Springer-Verlag, 1990, pp. 46-55.

[4] K. Hornik, M. Stinchcombe, and H. White, "Multilayer feedforward networks are universal approximators," *Neural Networks*, vol. 2, pp. 359-366, 1989.

[5] E. D. Karnin, "A simple procedure for pruning back-propagation trained neural networks," *IEEE Trans. Neural Networks*, vol. 1, no. 2, pp. 239-242, 1990.

[6] J. Sietsma and R. J. F. Dow, "Neural net pruning - Why and how," in *Proc. ICNN*-88, San Diego, 1988, vol. 1, pp. 325-333.

[7] J. Sietsma and R. J. F. Dow, "Creating artificial neural networks that generalize," *Neural Networks*, vol. 4, pp. 67-79, 1991.

[8] D. E. Rumelhart, G. E. Hinton, and R. J. Williams, "Learning internal representations by error propagation," in D. E. Rumelhart and J. L. McClelland (eds.), *Parallel Distributed Processing*, Vol. 1, Cambridge, MA: MIT Press, 1986, pp. 318-362.

[9] G. H. Golub and C. F. Van Loan, *Matrix Computations.* Baltimore, MD: Johns Hopkins, 1989.

[10] A. Björck, "Methods for sparse linear least squares problems," in J. R. Bunch and D. J. Rose (eds.), *Sparse Matrix Computations.* New York: Academic, 1976, pp. 177-199.

[11] A. Björck and T. Elfving, "Accelerated projection methods for computing pseudoinverse solutions of systems of linear equations," *BIT*, vol. 19, pp. 145-163, 1979.

[12] L. Niles, H. Silverman, G. Tajchman, and M. Bush, "How limited training data can allow a neural network to outperform an optimal statistical classifier," in *Proc. ICASSP-89*, Glasgow, 1989, vol. 1, pp. 17-20.

[13] P. Burrascano, "Learning vector quantization for the probabilistic neural network," *IEEE Trans. Neural Networks*, vol. 2, no. 4, pp. 458-461, 1991.

[14] M. J. J. Holt, "Comparison of generalization in multi-layer perceptrons with the log-likelihood and least-squares cost functions," in *Proc. 11th ICPR*, The Hague, 1992, vol. 2, pp. 17-20.

TESTS OF DIFFERENT REGULARIZATION TERMS IN SMALL NETWORKS

Crespo, J.L.

Mora, E.

Applied Mathematics and Computer Science Department
University of Cantabria
Avda. Los Castros, s/n
39005 Santander
Spain

Abstract

Several regularization terms, some of them widely applied to neural networks, such as weight decay and weight elimination, and some others new, are tested when applied to networks with a small number of connections handling continuous variables. These networks are found when using additive algorithms that work by adding processors. First the different methods and their rationale is presented. Then, results are shown, first for curve fitting problems. Since the network constructive algorithm is being used for system modeling, results are also shown for a toy problem that includes recurrency buildup, in order to test the influence of the regularization terms in this process. The results show that this terms can be of help in order to detect unnecessary connections. No clear winner has been found among the presented terms in these tests.

Background

An automatic network construction method, described elsewhere [Crespo, 1992], has been developed for system modeling purposes. This method includes overfitting control as a way of limiting the network size, and builds up an hybrid network with feedback that is easy to interpret in a useful way. It works by adding processors and fully connecting each one before trying the new network. This may add unnecessary connections, so, in order to overcome this problem, regularization techniques have been tried.

The idea is to control the weights value in order to avoid considering variables that may be non relevant for a particular task, or a particular processor; this control would suppress non needed dependencies added during the network setup process.

One of the possibilities of the algorithm is adding feedback, and eventually it should select the right lagged terms to be considered in both independent variables and output. The regularization terms influence in this selection process have also been investigated.

Options

In order to test their efficiency, several methods are presented below. Each one adds a particular term to the objective function being minimized. The most usual methods and the corresponding terms are:

Weight decay [Hinton, 1986]:

$$\lambda \sum w^2$$

where w represents the weights.

Weight elimination [Weigend et al, 1991]:

$$\lambda \sum \frac{w^2}{1+w^2}$$

This terms can be seen [Weigend et al, 1991] as a certain assumption about the a priori distribution of the weights. In particular, weight decay assumes a normal distribution centered in 0 and with variance $1/2\lambda$. Weight elimination is assuming a combination of a uniform and a normal distributions, both centered in 0. Doing something similar to what Movellan did when studying the error distribution we can propose different a priori distributions, hence other terms, such as:

Absolute value:

$$\lambda \sum |w|$$

From a bayesian viewpoint the absolute value term conveys the assumption that the a priori weight distribution is a double exponential.

Even without assuming a distribution we can add just penalty terms to the error function. For instance we may be interested in a term that is quadratic for small weights and becomes linear as their value increases, that is, in-between absolute value and weight decay. This may be called *intermediate* term.

Intermediate:

$$\lambda \sum w \tanh(w)$$

We can also be interested in a term that makes a more clear distinction among small and big weights by considering that big weights add no cost but small non-zero ones do. Such a term could be obtained by truncating weight decay.

Truncated weight decay:

$$\lambda \sum w^2 \text{ if } w > \alpha \text{ and } 0 \text{ otherwise}$$

The intermediate, though it hasn't been derived from any a priori distribution function, this should be in between normal and double exponential, similar to Huber's. The truncated weight decay term would come from some kind of combination between a truncated distribution for small weights and a uniform distribution for large ones.

When gradient descent is applied to the modified cost function, a negative term appears for each weight. These are shown in fig. 1 as a function of weight value for all different methods. Truncated weight decay is not represented, since it would be exactly like weight decay for small weight values and zero above certain cut point.

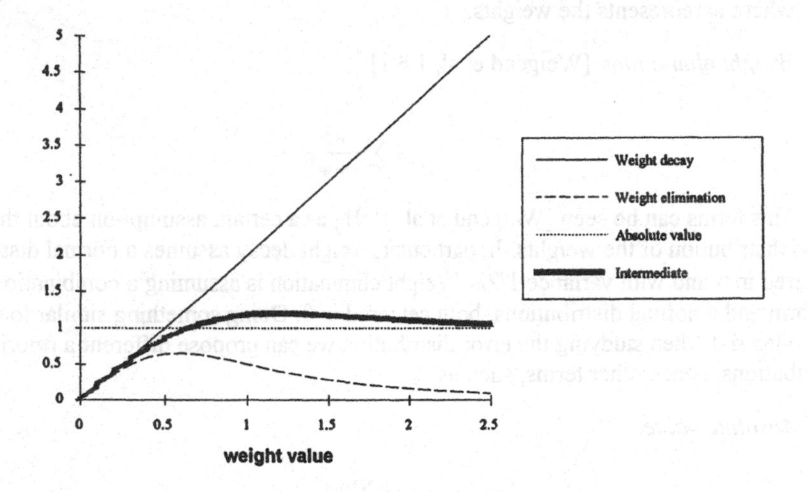

Fig. 1. Terms that affect each weight for the different methods tested here, expressed as times λ.

Test results

In order to test their effect on distinguishing non relevant variables, all methods have been tested with a combination of two straight lines with a null coefficient in each one, so the correct result would have been to assign a zero weight to the connections corresponding to those input elements representing the non relevant variables. These lines are:

$$2x \text{ for } 0 < x < 0.5$$

$$1 \text{ for } 0.5 < x < 1$$

The sample size was 600 points from a uniform distribution.

The results obtained so far are presented in tables 1 and 2. The constructive algorithm built 4 connections for the cases presented in table 1 and 6 for the ones in table 2. Ideally two of these connections should be discarded.

Table 1. Results achieved with different methods in a problem with a noise standard deviation of 0.028.

Term	λ	residual error	coefficient 1	coefficient 2
Weight decay	0.01	0.034	0.17	0.010
Weight decay truncated at 0.01	0.01	0.028	0.0027	< 0.001
Weight decay truncated at 0.1	0.01	0.028	0.010	< 0.001
Absolute value	0.01	0.029	0.027	0.0048
Weight elimination	0.01	0.029	0.033	0.0046
Intermediate	0.01	0.031	0.13	< 0.001
None	NO	0.028	0.0046	0.0015
Weight decay	0.0001	0.028	0.0025	< 0.001
Weight decay truncated at 0.01	0.0001	0.028	0.0028	< 0.001
Absolute value	0.0001	0.028	0.010	0.0012
Weight elimination	0.0001	0.028	0.020	< 0.001

Next, in order to test the effects on recurrency selection, a toy problem is setup following the equation:

$$y_n = 0'5(y_{n-1} + x_{n-2})$$

The sample includes 200 points starting x=0, y=0 and keeping x between 0 and 1. No noise is added.

Table 2. Results achieved by different methods when applied to a sample a noise standard deviation of 0.12.

Term	λ	residual error	coefficient 1	coefficient 2
Weight decay	0.01	0.12	0.055	>0.2
Weight decay truncated at 0.01	0.01	0.12	0.039	0.16
Weight decay truncated at 0.1	0.01	0.12	0.044	0.31
Absolute value	0.01	0.12	0.072	0.048
Weight elimination	0.01	0.12	0.034	0.090
Intermediate	0.01	0.12	0.053	>0.2
None	NO	0.12	0.019	0.017
Weight decay	0.0001	0.12	0.021	0.015
Weight decay truncated at 0.01	0.0001	0.12	0.039	0.16
Absolute value	0.0001	0.12	0.0089	0.0034
Weight elimination	0.0001	0.12	0.019	0.019

The results are shown in table 3 for different methods used along the whole construction process.

In order to perform a more difficult test, another equation is chosen, including a larger dead time. This is:

$$y_n = 0.5(y_{n-1} + x_{n-6})$$

The results achieved with all methods show a residual error equal to the noise standard deviation, but none is similar to the proposed equation. This is due to autorregresive relationships found along the step by step procedure.

Table 3. Results achieved with different methods, using them along the whole construction process.

Term	residual error	final models
None	0.0051	o $0.59x_{n-2}+0.045x_{n}-0.12x_{n-1}+0.48y_{n-1}$ o $0.19x_{n-2}-0.19x_{n}+0.44x_{n-1}+0.55y_{n-1}$
Weight decay	0.0013	o $0.54x_{n-2}+0.024x_{n}-0.060x_{n-1}+0.49y_{n-1}$
Weight decay truncated at 0.01	0.027	o $1.19x_{n-1}-0.59x_{n}-0.028y_{n-1}+0.42y_{n-2}$ o $1.26x_{n-1}-0.63x_{n}-0.093y_{n-1}+0.46y_{n-2}$ o $1.15x_{n-1}-0.62x_{n}-0.002y_{n-1}+0.45y_{n-2}+0.060$
Weight elimination	0.0013	o $0.55x_{n-2}+0.026x_{n}-0.063x_{n-1}+0.49y_{n-1}$
Absolute value	0.0016	o $0.55x_{n-2}+0.026x_{n}-0.066x_{n-1}+0.49y_{n-1}$

Further work is necessary, including some way of avoiding autorregresive terms, what may be related to sample setup.

Conclusions

For a constructive method such as the one being used here, regularization methods do not show very glossy results. This is due to the fact that there are not many extra connections.

This methods do not totally cancel weights, but reduce the value of the non necessary ones. Certainly they are a clever choice, but user interpretation is still necessary.

Considering the results achieved so far, there is no clear winner for all cases. As it was already known the regularization coefficient should be kept very small. Truncated terms have not showed advantages in the test cases.

When it comes to large lags, none of the approaches used so far has been successful at eliminating autorregresive effects. This means that the network is not capable yet of finding the right lag by itself. The user analysis is again necessary.

References

Crespo, J.L. "Procesamiento paralelo y distribuido aplicado a la simulación de sistemas". Ph.D. Thesis, Universidad de Cantabria, Santander (1992)

Hinton, G.E. ."Learning Distributed Representations of Concepts". Proceedings of the Eighth Annual Conference of the Cognitive Science Society (Amherst 1986), 1-12. Hillsdale: Erlbaum.(1986)

Movellan, J.R. "Error functions to improve noise resistance and generalization in backpropagation networks". (1989)

Weigend, A.S.; Rumelhart, D.E. y Huberman, B.A. . "Generalization by Weight-Elimination with Application to Forecasting". Advances in Neural Information Processing 3. ed.: R.P. Lippmann, J. Moody and D.S. Touretzky. Morgan Kaufmann, San Mateo, CA.(1991)

ON THE DISTRIBUTION OF FEATURE SPACE IN SELF-ORGANISING MAPPING AND CONVERGENCE ACCELERATING BY A KALMAN ALGORITHM

Hujun Yin and *Nigel M Allinson*

Image Engineering Laboratory
Department of Electronics, University of York
York, YO1 5DD, U.K.

In this paper we present a detailed investigation of the statistical and convergence properties of the Kohonen's Self-Organising Mapping (SOM) in any dimension. Using an extend Central Limit Theorem, we prove that the feature space in SOM learning is an approximation to multiple Gausssian distributed stochastic processes, which will eventually converge in the mean-square sense to the density centres of the input probabilistic sub-spaces. We also demonstrate that combining the SOM with a Kalman filter can smooth and accelerate the learning and convergence of the SOM. In our applications, we show that such a modified SOM achieves a much better performance, namely a lower distortion than the original algorithm, especially in early training stages, and at low extra computational cost. This modification will be particular useful when the available training set is small.

1 INTRODUCTION

For many years, artificial neural networks have been used to model alternate information processing systems based on biological neural structures. They may provide not only results with improved performance compared with traditional methods, but also deeper understanding of human cognitive abilities. Among the various neural network architectures and learning algorithms, Kohonen's Self-Organising Mapping (SOM) (Kohonen 1982) possesses some special properties, e.g. unsupervised learning, simple computation, a cortex-like array of the neurons with topologically preserved mapping from input to output space. It is biologically inspired. Numerous researchers have already examined on the SOM algorithm and many of its problems, however, there are still many aspects need to be exploited. Even the most general theory about this algorithm is far from complete or is lacking in vigorous mathematical explanation, as Kohonen (1991) and other researchers have remarked (Lo and Bavarian 1991; Erwin et al. 1992; Bauer and Pawelzik 1992).

As the reasons for the self-ordering and the convergence phenomena are very subtle, the ordering and convergence of the SOM have been proved by Kohonen (1984) only for the simplest case, i.e. one-dimensional array of neurons with a one-step neighbourhood function. Erwin et al. (1992a, 1992b) extended the proof of this Kohonen chain's ordering and convergence from the one-step neighbourhood function to any monotonically decreasing function centred on the winning neuron, even though they showed that non-convex neighbourhood functions may cause the existence of metastable states. Lo and Bavarian (1991) also analysed the effects of stepped and Gaussian type neighbourhood functions on the ordering of the SOM. They gave a comparison of both through simulations of two-dimensional arrays of the neurons. However in high dimensional cases or for mappings from a high dimensional input to a low dimensional output, the ordering and convergence are very difficult to describe or examine. Allinson (1990) gave some applications of dimensional reduction from some regular patterns. More recently, Bauer and Pawelzik (1992) proposed the use of a topographic product to measure the neighbourhood preservation or violation in the SOM.

In this paper we will not further the discussion on the effects of the neighbourhood functions even though they are very important; we will assume that they are adequate for the ordering of the SOM. We will analyse the learning dynamics of the SOM from probability theory since we treat the feature space as stochastic processes. In Section 2, we briefly review the SOM algorithm. In Section 3, the diminishing effect, as training progresses, of initial states on the feature space will be shown. In Section 4, we will extend the Central Limit Theorem to a particular case, from which we will prove that the distribution of the feature space will tend to approximate Gaussian functions, and that the feature space will converge in the mean-square (m.s.) sense to the probabilistic density centres of sub-spaces of the training set. Ritter and Schulten (1988) have derived a Fokker-Planck equation to describe the transitional properties of the distribution of the feature space from the viewpoint of the SOM's Markovian properties. However their results are obtained only in the vicinity of equilibrium where the "gain coefficient" and fluctuations around the stationary states are already very small. They also demonstrated that the deviations of the feature space are Gaussian distributed.

Section 5 will show that the fluctuations or variances of the feature space during the initial period of training can be greatly reduced when the SOM algorithm is modified by a Kalman filter. We demonstrate that the proposed algorithm, based on simulations, will reduce the distortion to very near the optimal level after a greatly reduced number of iterations.

2 KOHONEN SOM ALGORITHM

The SOM algorithm attempts to use a set of neurons to form a topology conserving discrete map of input space. Let $X \in R^N$ represent the input space, where N is the dimension of the input space. Let Y represent the neuron set which is arranged in a M-dimensional space, so Y is a $C_1 x C_2 x ... C_M$ array, where $\{C_i, i=1,2,...M\}$ represent the number of the neurons along each dimensional side of the neuron space, and $C=C_1+C_2+...C_M$ is the total number of neurons. Every neuron or cell, $c \in Y$, is connected, in parallel, to all dimensional components or variables of input sample, $x \in X$, $x=(x_1,x_2,...x_N)^T$. The connection strengths or weights are:

$$W_c(n) = (W_{c1}(n), W_{c2}(n), \dots W_{cN}(n))^T, \qquad \forall c \in Y \qquad (2.1)$$

Where n is the discrete time and $n \geq 0$.

The initial weights are randomly set., At every training step, a randomly selected input sample, x(n), from the input space X is presented to the network. Every neuron compares its weights with the input x(n), and the best match neuron, v(n), can be found through:

$$\left\| x(n) - W_{v(n)} \right\| = \min \{ \left\| x(n) - W_{c(n)} \right\| \}, \qquad \forall c \in Y \qquad (2.2)$$

Then the weights are updated according to the following rules:

$$W_c(n+1) = W_c(n) + \alpha(n) h(c,v,n) \, (x(n) - W_c(n)), \qquad \forall c \in Y \qquad (2.3)$$

Where $h(c,v,n)$ is termed the neighbourhood function. There are many types of neighbourhood functions, generally a stepped function is used, that is:

$$h(c,v,n) = \begin{cases} 1, & \text{if } c \in N_v(n) \\ 0, & \text{if } c \notin N_v(n) \end{cases} \qquad (2.4)$$

Where $N_v(n)$ is the neighbourhood set around the winner, $v(n)$, at time n. $N_v(n)$ should be very wide in the beginning of the training, and should shrink monotonically with time until the winner is the only member of the neighbourhood set. A topology preserved mapping can then be guaranteed. The coefficients $\{\alpha(n), n \geq 0\}$ are scalar-valued adaptation gains, which also monotonically decrease and should satisfy:

$$0 < \alpha(n) < 1, \qquad \lim_{n\to\infty} \Sigma\, \alpha(n) \to \infty, \qquad \lim_{n\to\infty} \Sigma\, \alpha^2(n) \to \infty \qquad (2.5)$$

In most applications, only scalar-valued $\{\alpha(n)\}$ and $\{h(c,v,n)\}$ terms are used, that implies that all dimensional weight components are independent, thus each neuron's ith weight, $W_{ci}(n)$, is only affected by the ith dimensional component, $x_i(n)$, of the input x(n). Then (2.3) can be rewritten as:

$$W_{ci}(n+1)=\prod_{k=0}^{n}(1-\alpha(k)h(c,v,k))\,W_{ci}(0)+\sum_{k=0}^{n}\prod_{l=k+1,k<n}^{n}(1-\alpha(l)h(c,v,l))\alpha(k)h(c,v,k)\,x_i(k),$$

$$\forall\, i\in(1,2,\ldots N);\ \forall\, c\in Y \tag{2.6}$$

3 EFFECT OF INITIAL STATES

To examine the first term of (2.6), we can write:

$$b_{ci}(n)=\prod_{k=0}^{n}(1-\alpha(k)h(c,v,k)) = \prod_{k=0,c\in N(k)}^{n}(1-\alpha(k)) \tag{3.1}$$

Only if the neuron, c, is in the neighbourhood set, $N_v(n)$, at time n, will its weights be modified and corresponding term will appear in (3.1). Let $\{D(m) = \textit{interval of c not in } N_v(n),\ n\geq 1,\ m\geq 0\}$ represent the frequency of neuron c not firing. Then $D_{max}=\max\{D(m)\}$ will be a finite number, otherwise c will never fire, Hence:

$$b_{ci}(n)=\prod_{k_m=0,stepD(m)}^{n}(1-\alpha(k_m)) \tag{3.2}$$

Taking natural logarithms of both sides, gives:

$$\ln b_{ci}(n) = \sum_{k_m=0,stepD(m)}^{n}\ln(1-\alpha(k_m)) \leq \sum_{k_m=0,stepD(m)}^{n}(-\alpha(k_m)) = \frac{1}{D_{max}}\sum_{k_m=0,stepD(m)}^{n}D_{max}\alpha(k_m))$$

$$\leq -\frac{1}{D_{max}}\sum_{k_m=0,stepD(m)}^{n}\overbrace{(\alpha(k_m)+\alpha(k_m)+\ldots\alpha(k_m))}^{D_{max}}\leq -\frac{1}{D_{max}}\sum_{k=0}^{n}\alpha(k) \tag{3.3}$$

And because $\sum_{k=0}^{n}\alpha(k)\xrightarrow{n\to\infty}\infty$. Hence:

$$b_{ci}(n)\leq \exp\left(-\frac{1}{D_{max}}\sum_{k=0}^{n}\alpha(k)\right)\xrightarrow{n\to\infty}0 \tag{3.4}$$

So the effect of initial states will tend to zero if the initial states are finite. This is why the initial states of the SOM can be randomly selected.

4 GAUSSIAN APPROXIMATION OF SOM FEATURE SPACE DISTRIBUTION

As the input sequence $\{x(n),\ n\geq 0\}$ of the SOM are treated as independent random variables (r.v.s.) with probability density function $f(x)$, $x\in X$, then from (2.6) every weight is a weighted sum of $\{x(n)\}$. Each neuron receives input from a subspace, $X_c(n)$, of the input space X. At the beginning of the training phase, each subspace is almost the same as X. As the training progresses and shrinking rate of the neighbourhood size is properly selected, the order, or topologically preserved map, will be formed in a finite time, and input subsets $\{X_c(n),\ c\in Y,\ n\geq 0\}$ will eventually be separated with:

$$\bigcup_{c\in Y}X_c(n)\xrightarrow{n\to\infty}X,\ \text{and}\ X_c(n)\bigcap X_{c'}(n)\xrightarrow{n\to\infty}\phi,\qquad c\neq c',\ \forall\, c, c'\in Y \tag{4.1}$$

As time tends to infinity, $\{X_c(n),\ c\in Y\}$ tend to $\{X_c,\ c\in Y\}$ called the final input subspaces

The Central Limit Theorem is concerned with the stochastic properties of the sum of independent r.v.s. The difference in the present case, is that every variance of weighted r.v.(random variable) in (2.6) will tend to zero, rather than to be a finite number, and in the following we will show that the sum of them will also tend to zero (otherwise the SOM will not converge). We cannot apply directly any existing version of the Central Limit Theorem to our analysis. Herein we introduce an extended form of Central Limit Theorem, which has been proved by authors(1992).

Theorem: If $\{X_n, n\geq 0\}$ are independent r.v.s. with density functions $f(X_n)$, finite means of $\{m_n, n\geq 0\}$, finite variances of $\{\sigma_n^2, n\geq 0\}$, and finite higher moments, i.e. for any $\delta>0$,

$$\mu_n^{(2+\delta)} = \int_{X_n} X_n^{2+\delta} f(X_n)dX_n < \infty \tag{4.2}$$

$\{a_k(n), k=0,1,...n, n\geq 0\}$ is a coefficient set of time varying real numbers, which satisfy:

$$\text{i. } 0<a_k(n)<1; \quad \text{ii. } \sum_{k=0}^{n} a_k(n) \xrightarrow{n\to\infty} 1; \quad \text{iii. } \sum_{k=0}^{n} a_k^2(n) \xrightarrow{n\to\infty} 0 \tag{4.3}$$

The weighted sum $\{\sum_{k=0}^{n} a_k(n)X_n, n\geq 0\}$ will tend to a Gaussian distributed process with means of $\{m(n)=$

$\{\sum_{k=0}^{n} a_k(n)m_k, n\geq 0\}$ and variances of $\{\sigma^2(n)=\sum_{k=0}^{n} a_k^2(n)\sigma_k^2, n\geq 0\}$, and with $n\to\infty, m(n)\to E\{m_n\}, \sigma^2(n)\to 0$. Furthermore if $X_n \to X$, then such a weighted sum will converge in the m.s. sense to m, the mean of X.

Returning to the second term of (2.6), a weighted sum of independent r.v.s. and here the coefficient set $\{a_k(n), k=0,1,...n; n\geq 0\}$ are:

$$a_k(n) = [\prod_{l=k+1,k<n}^{n}(1-\alpha(l)h(c,v,l))]\alpha(k)h(c,v,k) \tag{4.4}$$

Next we shall prove that this set will satisfy the three conditions of (4.3). The first condition $0<a_k(n)<1$ holds because of (2.4) and (2.5), then the second condition holds because:

$$\sum_{k=0}^{n} a_k(n) = \sum_{k=0}^{n}[\prod_{l=k+1,k<n}^{n}(1-\alpha(l)h(c,v,l))]\alpha(k)h(c,v,k) = (1-(1-\alpha(n)h(c,v,n)))$$
$$+[1-\alpha(n)h(c,v,n)](1-(1-\alpha(n-1)h(c,v,n-1)))$$
$$+...[1-\alpha(n)h(c,v,n)][1-\alpha(n-1)h(c,v,n-1)]...[1-\alpha(1)h(c,v,1)](1-(1-\alpha(0)h(c,v,0)))$$
$$= 1-\prod_{k=0}^{n}[1-\alpha(k)h(c,v,k)] \tag{4.5}$$

From Section 3, the second term of the above will tends to zero. The last condition, namely:

$$\sum_{k=0}^{n} a_k^2(n) = \sum_{k=0}^{n}[\prod_{l=k+1,k<n}^{n}(1-\alpha(l)h(c,v,l))^2]\alpha^2(k)h^2(c,v,k) \tag{4.6}$$

Since $\sum\alpha^2(k)$ converge, so for any arbitrary small value ε, there exists a number of N, for $\sum_{N}^{\infty}\alpha^2(k)<\varepsilon$, and because : $0<[1-\alpha(l)h(c,v,l)]<1$, then:

$$\lim_{n\to\infty}\sum_{k=N}^{n} a_k^2(n) = \sum_{k=N}^{\infty}[\prod_{l=k+1}^{\infty}(1-\alpha(l)h(c,v,l))^2]\alpha^2(k)h^2(c,v,l) < \sum_{k=N}^{\infty}\alpha^2(k) < \varepsilon \tag{4.7}$$

For a finite N, since $\sum_{k=0}^{\infty}\alpha(k)$ diverge, $\sum_{k=N}^{\infty}\alpha(k)$ will also diverge, and from Section 3, $\prod_{l=N+1}^{\infty}(1-\alpha(l)h(c,v,l))$

will also tend to zero, and since $\sum_{k=0}^{N}\alpha^2(k)<\theta$, (a constant). Thus:

$$\lim_{n\to\infty}\sum_{k=0}^{N}a_k^2(n)=\sum_{k=0}^{N}[\prod_{l=k+1}^{\infty}(1-\alpha(l)h(c,v,l))^2]\alpha^2(k)h^2(c,v,k)<\theta\prod_{l=N+1}^{\infty}(1-\alpha(l)h(c,v,l))\to 0 \quad (4.8)$$

We can conclude from (4.7) and (4.8) that the last condition also holds.Therefore, the feature space of the SOM are approximate mutiple Gaussian distributed stochastic processes, and will converge in the m.s. sense to the means of the final input subspaces,

5 KALMAN FILTER MODIFIED SOM

We have shown that for the SOM algorithm the feature space are multiple time-varying or non-stationary stochastic processes. They are also asymptotic processes, since they will eventually converge to finite states. This means that these processes begin with large fluctuations or variances, and gradually the training will cause them decrease. In this Section, we combine a Kalman filter with the SOM to moderate the effects of large variances or noise in the training process. There have been previous examples of applications of classic optimal filter theory in neural networks (e.g., Cho and Don 1991; Ruck et al. 1992). A Kalman filter is a linear estimation method which can produce an optimal estimation of model states when applied to Gaussian distributed processes. It is also a recursive algorithm which is based on the predictions of system states from the latest states and linear measurements such that the expected sum of the squared errors between actual and estimated states are minimised.

When applying a Kalman filter to multilayer perceptrons (i.e. a non-linear system), the model of the system should be linearized. Cho and Don (1991) and Ruck et al.(1992) used an *extended Kalman filter* as an alternate training algorithm for multilayer perceptrons. Ruck et al.(1992) also compared it with the *back propagation* algorithm and concluded that *back propagation* is a degenerate form of the *extended Kalman filter* under the assumptions of: 1) $[\mathbf{HPH}^T+\mathbf{R}]^{-1}=a\mathbf{I}$, and 2) $\mathbf{P}=p\mathbf{I}$. In general, these conditions are not satisfied. However in the SOM, where the system model is linear, the Kalman algorithm must be applied in different way. We use a Kalman algorithm as a *post-filter* to the SOM. The updated weights of the SOM are considered as the measurements of the Kalman filter. The system models are asymptotic, thus we can use some asymptotic functions to describe them, such as an exponential function. The model states are the true weights of the network. Since every weight of each neuron is independent as we analysed in Section 2, the **F**, **H**, **K**, **P** matrices are all diagonal, so there will no need for matrix computation. Furthermore, it is possible to consider every weight having the same speed of convergence, so we can use the simplest computational form of the algorithm.

For a one-dimensional input to output example. the state transltion **F**, can be modelled as:

$$\mathbf{F}(k)=\frac{\partial \mathbf{S}(k)}{\partial k}=e^{\frac{g}{k(k+1)}}\,\mathbf{I} \quad (5.1)$$

Where g is the underlying converging constant, which is depend on the number of neurons. Its value is not very critical and can be easily chosen experimentally.

The measurement matrix **H** is identical to **I** since we treat the output of the SOM as the measurement of the filter. The noise in every neuron's state is considered to be the same scalar-valued, and to be decreasing with time. So is the measurement noise. Thus:

$$\mathbf{U}(k)=e^{-k/u}\,\mathbf{I} \quad (5.2)$$

$$\mathbf{V}(k)=e^{-k/v}\,\mathbf{I} \quad (5.3)$$

Where u, v are decay constants of state noise and measurement noise respectively.

Figure 5.1 shows the comparison in performance (one simulation sample, and the averaged of over 100 independent simulations, in a 16-neuron network) of this modified, or co-operated, algorithm with the original SOM algorithm. The data used are uniform distributed over the input interval. Both adaptation gains and the neighbourhood size shrinking function are inversely proportional to time. The figure shows that the modified algorithm can greatly reduce the distortion especially in the early iterations. It almost reaches the final distortion level (optimal level) after only few hundreds iterations. It is not difficult to understand why the original SOM can, in time, reach the final distortion level since it is an *optimization approach* algorithm.

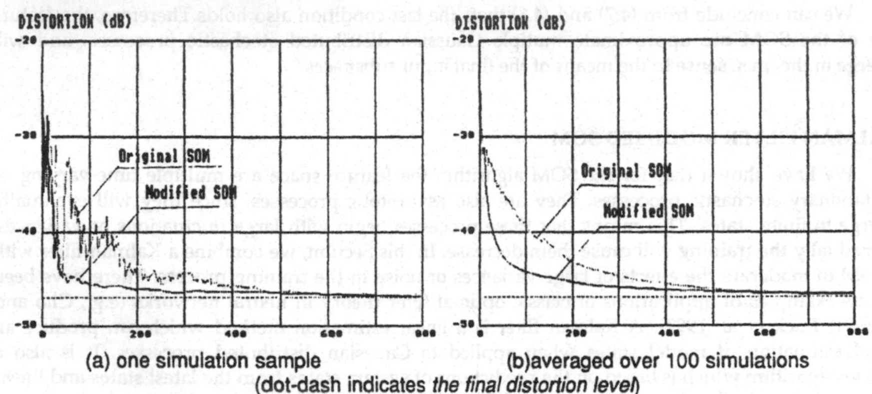

(a) one simulation sample (b)averaged over 100 simulations
(dot-dash indicates *the final distortion level*)
Fig 5.1 Comparison of original SOM and Kalman Filter modified SOM

6 CONCLUSION

We have analysed the stochastic properties of the feature space of the SOM. From the proof of its Gaussian distribution approximation we have also formally proved the convergence of the SOM algorithm. The clear understanding of the learning dynamics of the SOM will give some inspiration on improvement or better implement of the SOMalgorithms in actual applications.

REFERENCES

Allinson, N. M. 1990. Self-organising maps and their applications. *Theory and Applications of Neural networks*. J. G. Taylor and C. L. T. Mannion eds. London, Springer-Verlay. 101-118.

Bauer, H.-U., and Pawelzik, K. R. 1992. Quantifying the neighborhood preservation of self-organizing feature maps. *IEEE Trans*. Neural Networks 3(4), 570-579.

Cho, C.-M., and Don, H.-S. 1991. A parallel Kalman algorithm for fast learning of multilayer neural networks. *Proc. IJCNN-91 (Singapore)*, 2044-2049.

Erwin, E., Obermayer, K., and Schulten, K. 1992a. Self-organizing maps: ordering, convergence properties and energy functions. *Biol. Cybern*. 67, 47-55.

___ 1992b. Self-organizing maps: stationary states, metastability and convergence rate. *Biol. Cybern*. 67, 35-45.

Kohonen, T. 1982. Self-organized formation of topologically correct feature maps. *Biol. Cybern*. 43, 59-69.

___ 1984. *Self-Organization and Associative Memory*. Springer-Verlay.

___ 1991. Self-Organizing Maps: Optimization Approaches. *Artificial Neural Networks*. Elsevier. 981-990.

Lo, Z. P., and Bavarian, B. 1991. On the rate of convergence in topology preserving neural network. *Biol. Cybern*. 65, 55-63.

Ritter, H. and Schulten, K. 1988. Convergence properties of Kohonen's topology conserving maps: Fluctuations, stability, and dimension selection. *Biol. Cybern*. 60. 59-71.

Ruck, D. W., Rogers, S. K., Kabrisky, M., Maybeck, P. S., and Oxley, M. E. 1992. Comparative analysis of backpropagation and the extended Kalman filter for training multilayer perceptrons. *IEEE Trans. PAMI*-14(6), 686-691.

Yin, H. and Allinson, N. M. 1992. Stochastic Analysis and Treatment of the Kohonen's Self-Organising Map.

A LEARNING ALGORITHM TO OBTAIN SELF-ORGANIZING MAPS USING FIXED NEIGHBOURHOOD KOHONEN NETWORKS

Martin-Smith,P; Pelayo,F.J.; Diaz,A; Ortega,J.; Prieto,A.

Departamento de Electrónica y Tecnología de Computadores
Universidad de Granada, Spain.

Abstract: ***In this paper, a learning algorithm that leads to an efficient self-organization in a Kohonen Neural Network (KNN) with fixed neighbourhood is presented. This algorithm may be faster than the originally proposed for KNNs, produces in general better covering of the input stimulus space, and can be more easily implemented in hardware due to the fixed neighbourhood it manages.***

1. INTRODUCTION

Due to their ability to learn autonomously by extracting statistical features (without supervision) from the input pattern population, neural networks with competitive learning are very interesting in the general context of artificial neural network applications. From a biological point of view, they constitute a more plausible approach than those networks that require the desired output be presented for each input stimulus (supervised learning).

One of the networks with competitive learning with a wider use is that proposed by Teuvo Kohonen [KOH82], which has as a relevant characteristic its ability to obtain self-organized maps. This feature has been widely exploited to solve a great variety of problems [KOH88,KOH90]. However, from the perspective of its hardware implementation, a KNN presents a great difficulty due principally to the neighbourhood treatment, which is essential in the learning phase in order to obtain a suitable self-organization. The lack in the literature of hardware proposals for implementing KNNs that really self-organize themselves, is mainly due to the connectivity required to carry out in an effective way the neighbourhood treatment. In a KNN, the use of variable size neighbourhoods may even imply the interconnection of all the neurons in the network.

The learning algorithm we propose allows a KNN with a fixed neighbourhood to learn, producing maps with a good global organization. This algorithm is advantageous from both the software and hardware implementation points of view with respect to that originally proposed by Kohonen. In the former, because it consumes less learning time and, for its hardware implementation, the advantage is due to the great reduction in connectivity requirements.

The rest of the paper is organized as follows: Section 2 deals with the paradigm of self-organizing maps, Section 3 describes the proposed learning algorithm, Section 4 is devoted to present experimental simulation results with the new algorithm and, in Section 5, some conclusions are given.

2. SELF-ORGANIZING MAPS

A KNN is a particular case of competitive learning neural networks. Basically, given a set of neurons, they can learn competitively if they have common input connections and they learn stimuli patterns selectively, in such a way that this selectivity depends only on the specialization that neurons themselves develop during the training phase. The simplest scheme of competitive learning consists in the iterative execution of the following steps [RUM86,PRI90]:

1.- *present a stimuli vector,*
2.- *compute the winning neuron,*
3.- *change its synaptic weight vector to shift it, in a small quantity, towards the applied input vector.*

In the Kohonen network the winning neuron is, by definition, the neuron that has the weight vector $W_g(t)$ closest (usually using euclidean distance $d(X,W)$) to the input stimulus $X(t)$. Moreover, the weight modification not only affects the winning neuron but also those neurons belonging to a certain neighbourhood $N_g(t)$, this neighbourhood being variable in time. Concretely, in each iteration, the winning unit U_g is that verifying:

$$d(X(t),W_g(t)) = \min_i\{d(X(t),W_i(t)\} \quad (1)$$

The adaptation of weights affects the neighbourhood $N_g(t)$ as follows:

$$\begin{aligned} W_i(t+1) &= W_i(t)+\alpha(t)\cdot(X(t)-W_i(t)) \quad &for\ all\ \ i\in N_g(t) \\ W_i(t+1) &= W_i(t) \quad &for\ all\ \ i\notin N_g(t) \end{aligned} \quad (2)$$

where $\alpha(t)$ is an empirical and monotone decreasing function, which forces the convergency of weights in the learning process.
Subindeces in expression (2) are labels of neurons. Usually, these labels are defined as elements from a set $LAB(n)\subset N^n$, i.e. a subset of indeces from an n-dimensional matrix.

An alternative notation is to introduce a scalar "kernel" function h, and to adapt the weights as [KOH82,MÜL90]:

$$\begin{aligned} &W_i(t+1) = W_i(t)+\epsilon(t)\cdot h(i,g,t)\cdot(X-W_i(t)) \quad for\ all\ i\in LAB(n) \\ &where\ h(i,g,t) = e^{\frac{-\|i-g\|^2}{2\cdot\Delta^2(t)}} \end{aligned} \quad (3)$$

In the following, the learning algorithm that uses this adaptation law will be referred as KKA.

We have tested several simple schemes of competitive learning which include some kind of interaction between the neurons of a pre-defined neighbourhood, leading in some cases to the obtention of self-organizing maps. The self-organization paradigm usually emerges when the following operations are added to a basic competitive learning scheme:

1.- *define a neighbourhood criterium among the neural elements by means of its labels,*
2.- *establish a dependence or correlation, in the learning process, among the elements belonging to the winner's neighbourhood.*

There are several ways to establish an interaction among the neural elements. For example, we have found a simple learning law that leads to the obtention of self-organized maps:
If the winning neuron is U_g, then the weight vector $W_{\delta(g)}$ of the neuron $U_{\delta(g)}$ is modified as:

$$W_{\delta(g)} = W_{\delta(g)}+\alpha(t)\cdot(X(t)-W_{\delta(g)}) \quad (4)$$

where $\delta(g)$ is randomly selected from a neighbourhood N_g.

This law may be summarized in the following two points:

1.- *Only one neuron learns for each iteration of the learning process.*
2.- *The neuron that learns, i.e. whose weights are modified, is randomly selected from the winning neuron's neighbourhood.*

3. THE HKFN LEARNING ALGORITHM.

The learning algorithm we propose is a variant of the learning process used by Kohonen. This variant (in the following termed HKFN: Hierarchical learning algorithm for Kohonen networks with Fixed Neighbourhood) presents several advantages from the point of view of its implementation, either by hardware or by software. In this way, while the software implementation of the HKFN algorithm requires less time in the learning process than the algorithm KKA of expression (3), a HKFN hardware implementation would relax the requirements dealing with the connections between neighbour elements which is one of the most important drawback in the attempt to obtain a full (i.e. network topology plus learning laws) hardware implementation of a Kohonen neural network.

In the Kohonen network the situation is very complex: the need of a neighbourhood with variable size makes necessary even to have links between every possible pair of elements in the network. These neighbourhoods with variable size are required to produce an adequate global organization. In the HKFN algorithm, this global organization is obtained by means of a sucesive organization of maps using Kohonen networks with fixed neighbourhood and a growing number of neurons.

Description of the algorithm:

The HKFN algorithm is executed in *NS* steps, each step representing a hierarchical level *lev* (within the learning process) for which only a network *NET(lev)* learns. The topology of the network *NET(lev)* depends on the level and on the neighbourhood pattern as follows: starting with *lev=1* and *NET(1)* defined by the concrete kind (or pattern) of neighbourhood, then *NET(lev)* is obtained replacing each element U_j of *NET(lev-1)* by a neighbourhood of neurons following the used pattern, giving as a result a partition in classes $C_j(lev)$ in the set of neurons belonging to *NET(lev)*.

In our experiments, we have used, as shown in Fig.1, a neighbourhood pattern with n_e=9 elements, and *NS*=3 steps. This implies the learning of networks *NET(lev)* with *lev*=1,2,3 and $NU(lev)=n_e^{lev}$ elements, i.e. 9 elements for *NET(1)*, 81 for *NET(2)* and 729 for *NET(3)* as is illustrated in Fig.2.

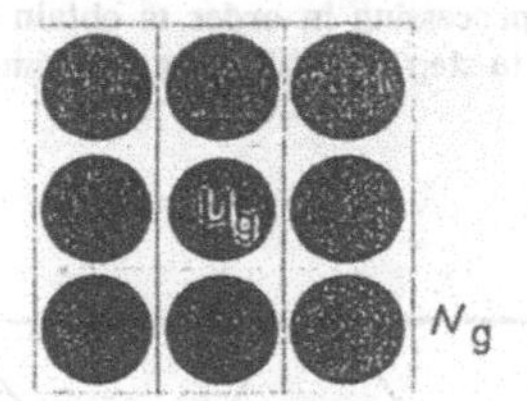

FIGURE 1: Bidimensional neighbourhood (N_g) used in the experiments.

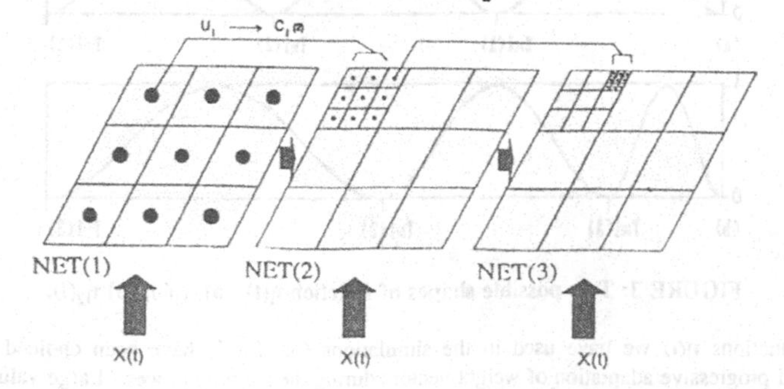

FIGURE 2: Hierarchies of networks NET(lev) with the neighbourhood pattern of Fig.1

For adapting the weights, the following learning law will be used:

$$
\begin{aligned}
W_i(t+1) &= W_i(t)+\beta\cdot\eta(t)\cdot(X(t)-W_i(t)) && \text{for all } i\in N_g\cap LAB(n) \\
W_{g^*}(t+1) &= W_{g^*}(t)+\beta_{lev}\cdot\eta(t)\cdot(X(t)-W_{g^*}(t)) && \text{for all } i=i^*\in N_g\cap\overline{LAB(n)} \\
W_i(t+1) &= W_i(t) && \text{if } i\notin N_g
\end{aligned}
\quad (5)
$$

where g* is the element label on the border of LAB(n) which is nearest to each i*.

The parameter β_{lev} in expression (5) has been introduced to reinforce the learning of the elements in the boundary of the neuron matrix *LAB(n)* towards the periphery of the input stimuli space, thus obtaining a better covering of the area defined by these stimuli.

The function $\eta(t)$ can have several shapes, whenever it have minimum values that will be used to indicate the end of the learning phase for each network NET(lev). In the simulations we have carried out, the functions $\eta(t)$ showed in Fig.3 have been used. Different intervals between two minimum values of $\eta(t)$ will be referred by a number of cycle *nc*.

The HKFN algorithm may be summarized in the following steps:

1) set *lev=1; nc=1; t=0.*
2) *t=t+1*; use the *X(t)* stimulus as input and compute the winning unit U_g in *NET(lev)* as expression (1) indicates.
3) apply the learning law of expression (5).
4) if $\eta(t)$ is a minimum and *lev<NS*, then:
   - 4.1) set *lev=lev+1; nc=nc+1*
   - 4.2) initialize the weights of *NET(lev)* by means of a weight transfer from each unit U_j of *NET(lev-1)* to all the units in the class $C_j(lev)$ of *NET(lev)*.
   - 4.3) go to step 2)
5) if $\eta(t)$ is a minimum and *lev=NS*, a result for *NET(NS)* is generated. Optionally, it is possible to continue the processing in order to obtain more precise results by keeping *lev=NS* ; *nc=nc+1* and go to step 2) until a given maximun number of cycle N_{cf}.
6) go to step 2)

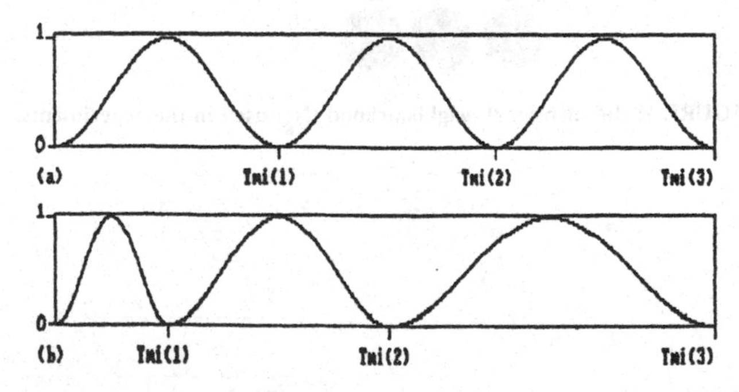

FIGURE 3: Two possible shapes of function $\eta(t)$. (a) $\eta_1(t)$, (b) $\eta_2(t)$.

The functions $\eta(t)$ we have used in the simulations (see Fig.3) have been choiced to produce an appropriate and progressive adaptation of weight vectors during the learning process. Large values for $\eta(t)$ lead to big and fast displacements of weights, which tend to follow the applied input stimuli. On the contrary, small values for $\eta(t)$ produce smaller and slower displacements of weights, which tend to a statistical average over a wider set of input stimuli, thus obtaining a more accurate adjustment of weight vectors.

In this way, it is interesting $\eta(t)$ to have small values at the beginning and the end of the learning process of each *NET(lev)*. At the beginning, i.e. after the initialization of step 4.2, to obtain a correct initial differentiation of the state of elements belonging to the same class, and at the end to perform an accurate tuning of the self-organized map.

Although the function $\eta_2(t)$ is more complex than $\eta_1(t)$, the former makes more efficient the learning process since it assigns less iterations to the networks with less neurons.

4. EXPERIMENTAL RESULTS

In this section simulation results that illustrate the working of the proposed learning algorithm are presented. To limit the extent of the paper only the results of a reduced number of experiments are included. The stimuli patterns for these experiments are randomly generated within the dark areas of Fig.4, which are included in the square area $[0,1]^2$. Without loss of generality of the algorithm, a dimension $n=2$ for the neuron matrix *LAB(n)*, and the neighbourhood pattern of Fig.1, have been used in the simulations. Figures 5.a to 5.l show the maps obtained with the HKFN algorithm using the function $\eta_2(t)$ (see Fig.3) in the learning phase. These results present a organization degree similar to that obtained with the KKA algorithm.

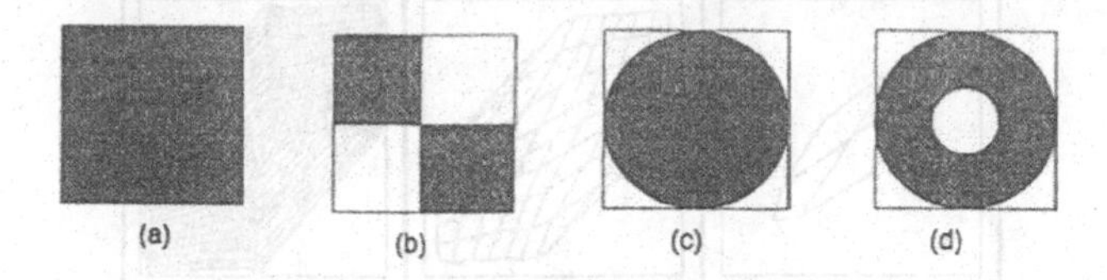

FIGURE 4: Areas where the input simulus patterns are generated for the learning tests.

In order to give a approximate measure of the covering of the input stimulus space that the HKFN produces, as a function of the number of iterations, a fast test has been designed which computes the following quantity for a square matrix *LAB(2)* and a circular stimuli pattern (see Fig.4.c):

$$Dm(t) = \frac{|W_{00}(t) - W_{ZZ}(t)| + |W_{0Z}(t) - W_{Z0}(t)|}{4} \tag{6}$$

where W_{00}, W_{ZZ}, W_{0Z}, and W_{Z0} are the weight vectors of the neurons in the four corners of the square matrix.

Fig.6 shows the values of function $F(t)=-log(0.5-Dm(t))$ when the algorithms HKFN and KKA are used to obtain the maps in figures 5.j to 5.l and 5.m to 5.p respectively. The number of iterations (t) would have a linear correspondence (approximately) to the processing times if dedicated hardware (that take advantage of the parallelism of the algorithms) were used. In a software simulation, the number of iterations has not a linear correspondence with the actual computing time for the algorithm HKFN. More accurate measures of the computing times (in terms of the total number of iterations) for the examples of figures 5.a to 5.p and 5.m to 5.p using the algorithms HKFN and KKA respectively, are:

$$T_{HKFN}(Nit) = (393 \cdot P_\alpha + 9 \cdot P_\beta) \cdot Nit = 813 \cdot Nit \ \mu sec. \tag{7}$$

$$T_{KKA}(Nit) = 729 \cdot (P_\alpha + P_\beta) \cdot Nit = 5103 \cdot Nit \ \mu sec. \tag{8}$$

where the time parameters P_α and P_β depend on the kind of computer and compiler. For a workstation Sun S/10 and its standard C-compiler the above parameters have the following experimental values:

Algorithm	P_α	P_β
HKFN:	2 μsec.	3 μsec.
KKA:	2 μsec.	5 μsec

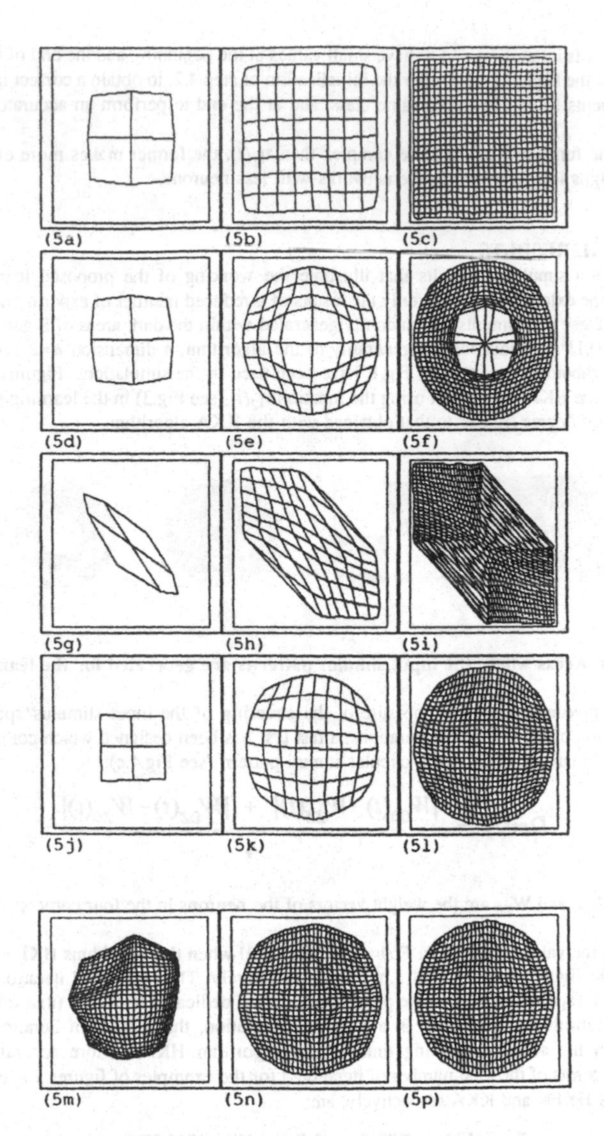

FIGURE 5: Self-organized maps obtained with the algorithms HKFN (figs. (a) to (l)) and KKA (figs. (m) to (p)) from test patterns in Fig.4:
(a) to (c): maps obtained with NET(1), NET(2), and NET(3) after 50,000, 150,000, and 300,000 iterations respectively. The parameters used in the simulation are: Nit=300,000; β=0.005; β_1=4; β_2=16; β_3=8.
(d) to (f): maps for inputs from Fig.4.d using the above simulation parameters.
(g) to (i): referred to Fig.4.a after 30,000, 90,000, and 180,000 iterations for the respective hierarchical levels. Nit=180,000; β=0.015; β_1=4; β_2=17; β_3=8.
(j) to (l): maps for inputs from Fig.4.c after 10,000, 30,000, and 60,000 iterations respectively. Nit=60,000; β=0.015; β_1=1; β_2=20; β_3=10.
(m) to (p): Results obtained with the algorithm KKA for inputs from Fig.4.c using a network with the same size than NET(3) in the above results (i.e. a matrix of 27x27 elements). The maps (m), (n), and (p) are obtained after 10,000, 30000, and 60,000 iterations respectively. The parameters used in the learning law of expression (3) are: $\varepsilon(0)=1$; $\Delta(0)=13$; $\varepsilon(t+1)=0.9999.\varepsilon(t)$; $\Delta(t+1)=0.9999.\Delta(t)$

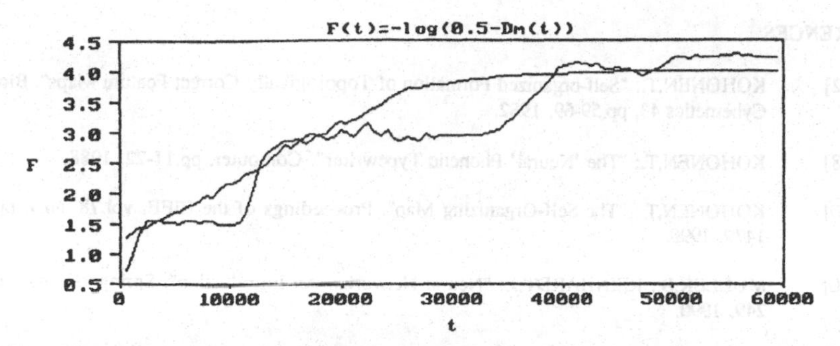

FIGURE 6: Values of F(t) as a measure of the covering obtained with the algorithms HKFN and KKA, using the simulation parameters that produce the maps in figures 5.j-5.l and 5.m-5.p respectively. The "staircase" trace corresponds to the proposed algorithm.

In order to give a more clear comparative idea of the actual computing times to obtain organization with the HKFN and KKA algorithms, expressions (7) and (8) have been evaluated on the results of Fig.5. From less to more execution time, maps in Fig.5 are obtained with the following order:

5.l, 5.m, 5.i, 5.n, 5.c, 5.f, and 5.p

5.- CONCLUSIONS

In this paper a learning algorithm to obtain self-organized maps using networks with fixed neighbourhood has been presented. As it has been proved by simulations, the proposed algorithm (HKFN) is able to produce a good global organization, similar to that obtained with the Kohonen algorithm, but it would be more easily hardware implemented due to the limited neighbourhood it manages.

It is difficult to establish a comparison of performance between algorithms to obtain self-organized maps. This comparison must be based on the required learning times and on the organization characteristics of the resulting maps. However, does not exist suitable general criteria to measure the quality of such organization. We have used some test problems (see Figs. 4 and 5) to show as the proposed algorithm lead to well organized maps and a good coverage of the input stimulus space. The results presented in section 4, although they must be considered orientative, show the following general features for the HKFN algorithm:

1.- The algorithm is able to use efficiently Kohonen networks with fixed neighbourhood. Without the hierarchic processing imposed by the algorithm, a fixed neighbourhood network could also produce self-organized maps but with poor performance in terms of learning speed and the global organization of the resulting maps.

2.- The maps obtained with the HKFN, having a global organization degree similar to that obtained with conventional Kohonen algorithms as the KKA, exhibit a better covering of the area defined by the stimuli vectors. This is due to the introduction of a parameter that modifies the learning law for the boundary neurons (see expression (5) and Fig.6)

3.- In a software implementation of algorithms HKFN and KKA, the former seems to be faster (see expressions (7) and (8) and the learning time order to obtain the maps in section 4).

4.- If the algorithms were hardware implemented, the HKFN introduces the simplification of using fixed neighbourhood networks, thus reducing the connectivity complexity among the neurons forming each network. This feature is not only interesting when the neural network has to be mapped on a reduced number of processors, but also if a full hardware implementation (one processor per neuron) is required. In this case it would be useful to implement the network of the last hierarchic level, whose initial weights will be pre-determined by simulation but, after the initial training, the network could be used in autonomous systems for applications that require continuous adaptability in field conditions.

REFERENCES

[KOH82] KOHONEN,T.: "Self-organized Formation of Topologically Correct Feature Maps". Biological Cybernetics 43, pp.59-69. 1982.

[KOH88] KOHONEN,T.: "The 'Neural' Phonetic Typewriter". Computer. pp.11-22. 1988.

[KOH90] KOHONEN,T.: "The Self-Organizing Map". Proceedings of the IEEE. vol.78. no.9. pp.1464-1479. 1990.

[MÜL90] MÜLLER,B.; REINHARDT,J.: "Neural Networks. An Introduction". Springer-Verlag. pp.245-249. 1990.

[PRI90] PRIETO,A.; MARTIN-SMITH,P.; MERELO,J.J.; PELAYO,F.J.; ORTEGA,J.; FERNANDEZ,F.J.; PINO,B: "Simulation and Hardware Implementation of Competitive Learning Neural Networks". in Statistical Mechanics of Neural Networks, L.Garrido (Ed.). Springer-Verlag. pp.189-204. 1990.

[RUM86] RUMELHART,D.E.; ZIPSER,D: "Feature Discovery by Competitive Learning". in Parallel Distributed Processing, Rumelhart & McClelland (Eds.) MIT Press, Vol.I, pp.151-193. 1986.

ANALYSING A CONTINGENCY TABLE WITH KOHONEN MAPS : A FACTORIAL CORRESPONDENCE ANALYSIS

Marie COTTRELL, Patrick LETREMY, Elisabeth ROY

SAMOS - Université Paris 1 - 90, rue de Tolbiac

F-75634 PARIS CEDEX 13

Abstract :

The Kohonen self-organizing algorithm is a powerful tool to achieve a categorization of vectorial stochastic data into classes. Many researchers use it to get a preliminary reduction of the data complexity in numerous application fields. They address some problems which are usually solved by means of statistical methods like Classification, or Principal Component Analysis. In this paper, we propose to extend this approach to another data analysis method : the simultaneous analysis of two qualitative variables which are crossed in a contingency table.

I – INTRODUCTION

One of the most important tasks to achieve in Statistics is to analyze complex and numerous data in order to simplify them, to extract relevant and useful information and to represent them in a simpler form. Many techniques have been developed : Principal Components Analysis, Hierarchical Clustering, Factorial Correspondence Analysis ... Each one has its own application field, but all of them have a common characteristic : to provide a low-dimensional graphical representation of the data. Most of them are linear (projection on a linear sub-space, diagonalization of variance-covariance matrix, ...), but, although these classical methods are well-known, their computational aspects are sometimes heavy, and anyway not easily parallelizable. In some particular geometrical disposition of the data, they may be irrelevant.

The connectionist models can be an interesting alternative to address this kind of tasks, since they have shown their capability to extract features, to classify patterns or to compress the information. Their characteristics (adaptation, parallelism, high speed, easy programmation, possible implementation in analog VLSI,...), make them very attractive to try providing new solutions to traditional problems.

Among the connectionist models, the Kohonen self-organization algorithm ([KOH-a], [KOH-b]) is widely used to approximate a probability density function with a relatively small number of points. Each input vector in $\mathbf{R}^n$ can be represented by a weight vector (or *codebook vector*), in such a way that

input vectors close together are associated with the same codebook vector or with close codebook vectors. *The Kohonen algorithm is said to preserve the neighborhoods and to preserve the topology of the input vectors.*

Various non-supervised classification methods (different versions of the LVQ -Learning Vectorial Quantization) are based on this property, and have been successfully applied for phoneme classification and recognition, signal pattern categorization, ... See for example [KOH-b].

II – PRINCIPAL COMPONENT ANALYSIS WITH KOHONEN ALGORITHM (KPCA)

Recently, some researchers ([BLA], [VAR]) used the Kohonen self organization algorithm to get a two-dimensional representation of numerous multi-dimensional data. In the referenced papers, the authors consider socio-economic data, but the method can be applied to a very large class of data.

The data are given in a $(N \times p)$-matrix, where the rows represent the N individuals, described by p quantitative components. The p columns correspond to the variables measured on each individual j (country, person, point, ...). Thus the data consist of a N-sample of p-components vectors $(x_1, x_2, \ldots, x_N)$ in $\mathbb{R}^p$. The most classical method used to treat this kind of data is the PCA (Principal Component Analysis). One looks for a new basis $(v_1, v_2, \ldots, v_p)$ in $\mathbb{R}^p$, so as to get the following : for every $k \leq p$, find, in the least square sense, the best possible projection of the N points on the subspace $\mathcal{L}(v_1, v_2, \ldots, v_k)$ spanned by the vectors $(v_1, v_2, \ldots, v_k)$. Then the best two-dimensional representation of the data is the projection of the plane spanned by (v_1, v_2).

To get a similar plane representation using the Kohonen self-organization, Blayo [BLA] and Varfis et al. [VAR] choose a $(n \times n)$ grid. Around each unit i, a topological neighborhood $\mathcal{V}(i)$ is defined, and each unit i has a weight vector W_i in $\mathbb{R}^p$. During the training phase, the weights are updated according to the Kohonen adaptation rule ([KOH-a], which makes the weight vectors of the "winning" unit i_c and its neighbors in $\mathcal{V}(i_c)$ moving towards the current input $X(t)$.

After the learning phase, each data vector x_j is represented by one unit of the network. The graphical representation is built as a table $(n \times n)$, : each cell i contains the indices of the inputs related to itself : the input $j \in$ the cell i means that i is the "winning" unit for the input x_j. Because of the topology preserving property, the representation respects the neighborhood relations. In fact, after a very small number of iterations, the rough features of the data are visible in the self-organization map.

The main difference between this algorithm and the general Kohonen algorithm that is studied in [BOU], [COT], [KOH-a] for instance, is that the inputs are picked at random in a finite set (the rows of the matrix data). In this case, it was pointed out ([KOH-b], [RIT]), that the Kohonen algorithm is a ***stochastic gradient descent method*** minimizing the potential function V :

$$V = \sum_{i=1}^{n \times n} \sum_{\substack{x_j \in \mathcal{C}(i') \\ i' \in \mathcal{V}(i)}} \| x_j - w_i \|^2$$

where $x_j \in \mathcal{C}(i)$ means that i is the "winner" unit when x_j is presented as input.

This potential is clearly an intra-class variance extended to the neighbor codebook vectors. It is important to note that such a potential does not exist in the general continuous Kohonen algorithm, as it has been proved by Erwin et al [ERW]. It has also to be noted that this definition supposes that the neighborhood size is constant, so as $\mathcal{V}(i)$ is well-defined, independently of the time t. So the V value gives an indicator of the convergence.

The algorithm that we have just described, (after [BLA], [KOH-a], [KOH-b], [VAR]), when it is used to give this kind of representation, will be referred as KPCA (Kohonen Principal Component Analysis) algorithm in the following.

III – CORRESPONDENCE ANALYSIS

We are interested in extending the above approach to *simultaneously analyze two qualitative variables which are crossed in a contingency table*. The classical statistical method to perform it is the Correspondence Analysis, introduced by Benzecri [BEN]. It yields a low-dimensional graphical representation of the association between rows and columns of a contingency table.

In this case, the data are displayed in a $(p \times q)$-table. The first qualitative variable has p levels and corresponds to the rows. The second one has q levels and corresponds to the columns. The entry n_{ij} $(i = 1,\ldots, p,\ j = 1,\ldots,q)$, is the number of observations categorized by the row i and the column j. From the contingency table, the matrix of relative frequencies is computed, with entry $f_{ij} = \dfrac{n_{ij}}{\sum_{ij} n_{ij}}$.

To correctly normalize the rows and the columns, the row i is represented by the *row profile* $r(i) = (f_{ij}/\sum_j f_{ij},\ j \in \{1,\ldots,q\})$, and the column j by the *column profile* $c(j) = (f_{ij}/\sum_i f_{ij},\ i \in \{1,\ldots,p\})$. The elements of each profile sum to one. Let R (resp. C) be the $(p \times q)$ – matrix whose rows are the row profiles (resp. the $(p \times q)$ – matrix whose columns are the column profiles). The classical Correspondence Analysis is a weighted Principal Component Analysis on the p rows of R, considered as points in $\mathbb{R}^q$, and on the q columns of C, considered as points in $\mathbb{R}^p$, which provides a simultaneous representation of the $p + q$ points giving some ideas about the relations between the variables of each group.

In this contribution, we propose *two Kohonen-algorithm-based methods.*

The first one is similar to a raw Principal Component Analysis on the row profiles taken together with the typical profile of the column levels, or, conversely, on the column profiles taken together with the typical profile of the row levels. It is compared with the so-called *barycentric correspondence analysis.*

The second one consists in a Kohonen map algorithm, where the inputs are the rows and the columns coupled in some way. We compare it with the so-called *simultaneous correspondence analysis.*

IV – BARYCENTRIC METHOD

Let us present the technique for the row profiles in **R**. The same computations can be done for the column profiles, by transposing the matrix **C**.

If a row i is very strongly associated with some column j, the entry (i,j) of this row will dominate, and the row will be close to a row (0, 0,..., 0, 1, 0,..., 0), with all elements 0, except the jth which is taken equal to 1. Thus we can add q more rows to the $(p \times q)$ – matrix **R**, so that the data matrix used for the KPCA analysis is the $((p+q) \times q)$ matrix $\begin{bmatrix} \mathbf{R} \\ \mathbf{I_q} \end{bmatrix}$, where $\mathbf{I_q}$ is the identity matrix. The dimension of the input space and of the weights is q.

The KPCA algorithm provides a two-dimensional representation of the p rows and of the q columns, which emphasizes the neighborhoods between the rows, between the columns, between the rows and the columns.

V – SIMULTANEOUS METHOD

This second method consists in performing the KPCA algorithm applied to a new data matrix **M** which is built in the following way : for each row $r(i)$ in **R**, the column $c(j(i))$ which maximizes the entry $\mathbf{R}_{ij}$ is determined, (j is the more probable column given i) and for each column $c(j)$ in **C**, the row $r(i(j))$ which maximizes the entry $\mathbf{C}_{ij}$ is computed. Then the new data matrix **M** is the $(p+q) \times (q+p)$ – matrix whose first p rows are the $(q+p)$ – vectors $(r(i), c(j(i)))$ and q last rows are the $(q+p)$ – vectors $(r(i(j)),\ c(j))$. The input vectors are picked at random among the rows of the matrix **M**. The weights of each unit of the Kohonen network are parted into (W_r, W_c), with $W_r \in \mathbb{R}^q$, and $W_c \in \mathbb{R}^p$ and are updated according to the usual rule. It is important to note that, as in classical correspondence analysis, we use the χ^2 – distance to compare the current input and the weight vectors in each linear space $\mathbb{R}^q$ and $\mathbb{R}^p$.

This "coupled" KPCA algorithm also provides a two-dimensional representation of the proximity between the variables of both kinds.

VI – EXAMPLES

In the following we present two examples and the results that we obtained

1) – with the barycentric and simultaneous Kohonen methods

2) – with the classical barycentric and simultaneous Correspondence Analysis method introduced by Benzecri.

A – MARRIAGES ACCORDING TO THE SOCIO-PROFESSIONAL CATEGORY

Répartition suivant la catégorie socioprofessionnelle des deux conjoints

(DONNEES SOCIALES 1990)

Femme / Homme	*FAGR*	*FART*	*FCAD*	*FINT*	*FEMP*	*FOUV*
HAGR	77.1	2.1	0.9	5.4	11.1	3.4
HART	1.2	42.8	4.2	12.7	33.0	6.2
HCAD	0.3	3.2	30.5	35.9	28.3	1.8
HINT	0.3	2.4	5.1	34.1	48.8	9.2
HEMP	0.3	2.2	2.4	16.9	68.2	10.1
HOUV	0.7	1.9	0.7	9.0	58.3	29.4

FINT : Profession intermédiaire	FAGR : Agricultrice	FCAD : Cadre
FART : Artisan-commerçante	HAGR : Agriculteur	FOUV : Ouvrière
HINT : Profession intermédiaire	FEMP : Employée	HCAD : Cadre
HART : Artisan-commerçant	HEMP : Employé	HOUV : Ouvrier

BARYCENTRIC KOHONEN METHOD AND SIMULTANEOUS KOHONEN METHOD

▫ BARYCENTRIC METHOD with a 4x4 grid and 200 steps ▫

HOUV	*FOUV*	*FEMP*	HEMP
		HINT	
	FCAD HCAD	*FINT*	
FART HART			*FAGR* HAGR

▫ SIMULTANEOUS METHOD with a 4x4 grid and 200 steps ▫

FAGR HAGR		HART	*FART*
			FCAD HCAD *FINT*
FOUV HOUV	*FEMP* HEMP	HINT	

BARYCENTRIC CORRESPONDENCE ANALYSIS AND SIMULTANEOUS CORRESPONDENCE ANALYSIS

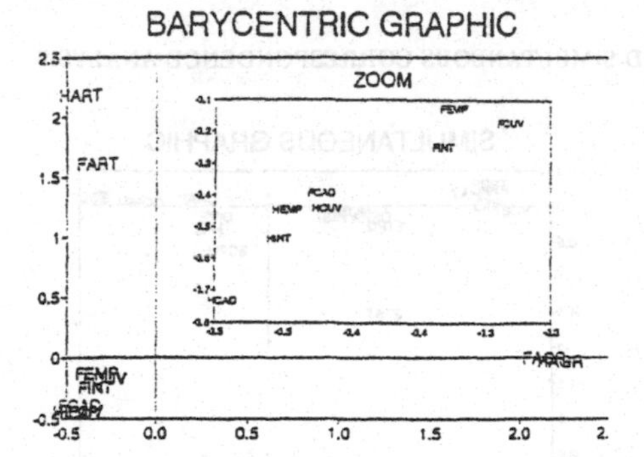

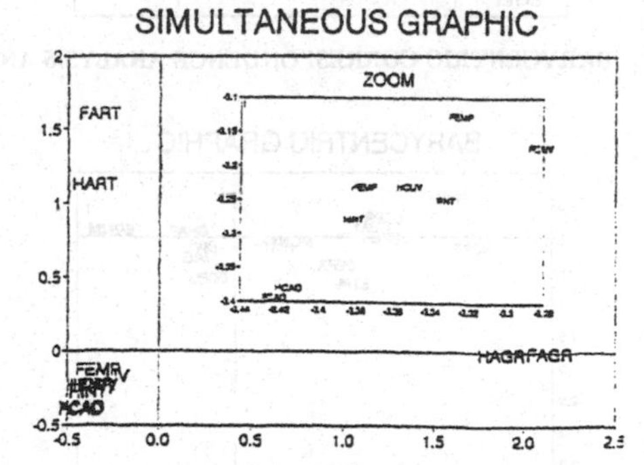

The two representations that we obtained with the Kohonen methods show the strong association between the professions of the husband and of the wife. The conclusions are the same as those we can deduce from the correspondence analysis.

B – HISTORIC MONUMENTS CLASSIFIED BY CATEGORY AND OWNER

Monuments historiques classés par catégorie et type de propriétaire en 1990
(Source MCC/DP/DEP)

	COMM	*PRIV*	*ETAT*	*DEPA*	*ETPU*	*NDET*
PREH	244	790	115	9	12	144
HIST	246	166	46	23	11	31
CHAT	289	964	82	58	40	2
MILI	351	76	59	7	2	0
CATH	0	0	87	0	0	0
EGLI	4298	74	16	5	4	2
CHAP	481	119	13	7	8	4
MONA	243	233	44	37	18	0
ECPU	339	47	92	19	41	2
ECPR	224	909	46	7	18	4
DIVE	967	242	109	40	10	9

ETPU : Etablissement public	NDET : Non détermine	PRIV : Privé
PREH : Antiquit. préhistoriques	DEPA : Departement	ETAT : Etat
HIST : Antiquités historiques	COMM : Commune	EGLI : Eglises
MILI : Architecture militaire	MONA : Monastères	DIVE : Divers
ECPR : Edifices civils privés	CATH : Cathédrales	CHAT : Chateaux
ECPU : Edifices civils publics	CHAP : Chapelles	

BARYCENTRIC KOHONEN METHOD AND SIMULTANEOUS KOHONEN METHOD

▫ BARYCENTRIC METHOD with a 5x5 grid and 200 steps ▫

PRIV ECPR CHAT				*ETPU*
PREH		*DEPA*		
MONA	HIST			*ETAT* CATH
CHAP	MILI DIVE			
COMM EGLI		ECPU		*NDET*

▫ SIMULTANEOUS METHOD with a 5x5 grid and 200 steps ▫

NDET PREH				EGLI
			CHAP	*COMM*
PRIV CHAT ECPR		HIST	MILI	DIVE
		MONA		ECPU
ETAT CATH			*DEPA*	*ETPU*

BARYCENTRIC CORRESPONDENCE ANALYSIS AND SIMULTANEOUS CORRESPONDENCE ANALYSIS

BARYCENTRIC GRAPHIC

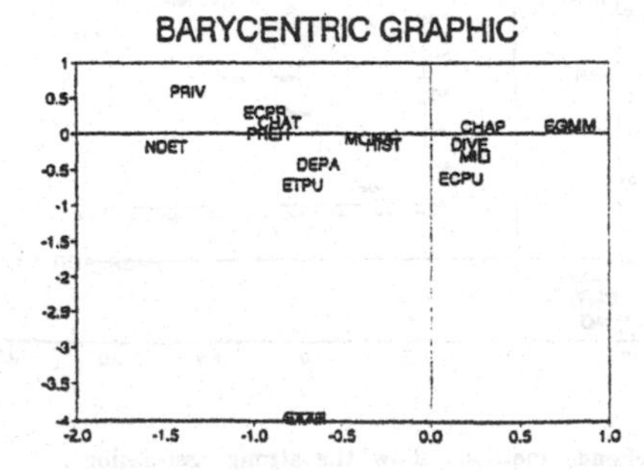

SIMULTANEOUS GRAPHIC

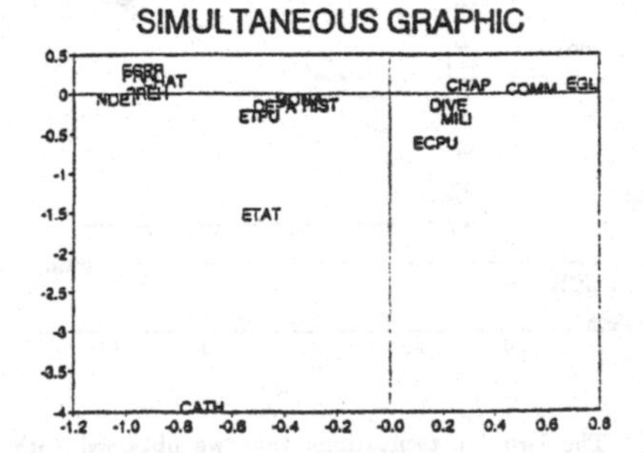

The four representations show the same associations : The cathedrals mainly belong to the State, the churches to the communes, the castles to private owners, and so on.

VII – CONCLUSION

We observe that the two Kohonen methods provide very easily and very quickly a good bidimensional representation of the links between the variables. These methods need few computations, no matrix diagonalization and use only elementar operations. On the other hand, the potential function can decrease until a local minimum and it is difficult to characterize the fitness of the representation. We are now working to improve the methods to overcome these problems, and to apply them to very numerous data. Another direction of research consists of the extension of these methods to multiple correspondence analysis.

ACKNOWLEDGMENT

The authors would like to thank F.Blayo and J.Ambuhl for stimulating discussions and P.Vintras for his preliminary contribution.

REFERENCES

[BEN] BENZECRI J.P. :*L'analyse des Données, T. 2, L'Analyse des Correspondances*, Dunod, Paris, 1973

[BLA] BLAYO F., DEMARTINES P. : *Data analysis : How to compare Kohonen neural networks to other techniques ?* In Proceedings of IWANN 91, Prieto ed., Lectures Notes in Computer Science, Springer-Verlag, 469-476, 1991

[BOU] BOUTON C., PAGES G. : *Self-Organization and convergence of the one-dimensional Kohonen algorithm with non uniformly distributed stimuli.* To appear in Stochastic Processes and their Applications, 1993

[COT] COTTRELL M., FORT J.C. : *Etude d'un algorithme d'auto-organisation.* Annales de l'Institut Henri Poincaré, Vol. 23, #1, 1-20, 1987

[ERW] ERWIN E., OBERMAYER K., SCHULTEN K. : *Convergence properties of self-organizing maps.* In Artificial Neural Networks, Kohonen T. et al eds, Vol. I, North-Holland, 409-414, 1991

[HER] HERTZ J., KROGH A., PALMER R.G. : *Introduction to the Theory of Neural Computation.* Lecture Notes Vol.1, Santa Fe Institute, Addison-Wesley, 1991

[KOH][a] KOHONEN T. : *Self-Organization and Associative Memory.* 3rd Edit. Springer-Verlag, 1989

[KOH][b] KOHONEN T. : *The Self-Organizing Map.* Proc. of the IEEE, Vol. 78, #9, 1464-1480, 1990

[LEB] LEBART L., MORINEAU A., WARWICK K.M. : *Multivariate Descriptive Statistical Analysis : Correspondence Analysis and Related Techniques for Large Matrices.* John Wiley, 1984

[RIT] RITTER H., MARTINETZ T., SCHULTEN K. : *Neural Computation and Self-Organizing Maps : An Introduction.* Addison-Wesley, Reading, 1992

[VAR] VARFIS A., VERSINO C. : *Clustering of socio-economic data with Kohonen maps.* Proc. 3'rd Int. Workshop on Parallel Applications in Statistics and Economics(PASE), Prague, Dec. 1992

Dynamics of Self–Organized Feature Mapping

R. Der, Th. Villmann
Universität Leipzig, Institut für Informatik
Augustusplatz 10-11
D–O–7010 Leipzig

November 26, 1992

Abstract

The dynamics of the feature maps created by Kohonen's algorithm is studied by analyzing the spectral density of synaptic fluctuations both analytically and by means of computer simulations. We consider unsupervised learning as a stochastic process and investigate the usefulness of the Fokker–Planck approach for the case of a topological mismatch between input and output space. A breakdown of the Fokker–Planck description is observed if the mismatch exceeds a critical value.

1 Introduction

Kohonen's self organized feature mapping [2] plays an important role in many real world applications of neural nets. In control problems like the one faced in the LADY–project a Kohonen neural net is usefully applied as a partitioner (or structured vector quantizer) partitioning a high–dimensional input space into a restricted set of discrete categories of low topological dimensionality. This is useful for the formulation of the control strategies. Despite its popularity and many efforts by several authors, cf. Ref. [1], the synaptic dynamics created by the Kohonen algorithm is not fully understood. In particular, this concerns the critical effects arising if there is a dimensional mismatch between the space of the stimuli and that of the neurons.

A systematic approach to these effects has been developed by Ritter et al. based on a Fokker–Planck description of the stochastic time evolution of the synaptic weights driven by Kohonen's algorithm [1]. We have continued this work by concentrating our efforts on displaying a complete record of the space time characteristics of the evolution of the map. The appropriate tool of such an investigation is the wave length and frequency depending spectral density of synaptic fluctuations.

2 Learning as a stochastic process

Let us consider a set of neurons situated at sites $\vec{r}_i, \vec{r}_i \in \mathcal{R}^d, i = 1, \ldots, N$ in some pre–specified setting which includes a metrics in neuron space. Each neuron is connected with the input units by synaptic connectivities $w_{i,j}$ where $i, j = 1, \ldots, N$. The inputs to the network are given by random stimuli $v \in \mathcal{R}^d$. Kohonen's learning rule is given by

$$\Delta \vec{w}_i(t) = -\epsilon h_{i,s}(\vec{w}_i - \vec{v}) \tag{1}$$

where $\vec{w}_i$ is the synaptic vector of neuron i, s denotes the winner neuron and $h_{i,s}$ is the neighborhood function decaying on length scale σ. $\vec{v}$ is chosen randomly according to the distribution of input stimuli $P(\vec{v})$.

Eq. 1 generates the Markov stochastic process $\mathbf{w}(t)$ of the matrix of synaptic weights. Under specific conditions the probability density $U(\mathbf{w},t)$ is governed by the Fokker–Planck equation

$$(\partial_t + L(\mathbf{w}))\, U(\mathbf{w},t) = 0 \tag{2}$$

the Fokker–Planck operator L being given explicitly in Ref. [1]. As discussed in more detail there, eq. 2 describes the dynamics of the map deployment sufficiently well in the asymptotic time region if there is no dimensional mismatch. The arisal of a dimensional conflict between input and neuron space ($d \neq n$) is signalled by the behaviour of the eigenvalues of L. In fact, if the dimensional mismatch approaches a critical value certain modes corresponding to specific spatial patterns begin to destabilize which is seen in characteristic foldings of the map.

Our main interest consists in studying these critical dynamic effects in more detail. As an appropriate tool for displaying these structures we use the spectral density of the synaptic fluctuations, i.e. consider

$$S(k,\omega) = \left\langle \|\vec{u}(k,\omega)\|^2 \right\rangle \tag{3}$$

where $\vec{u}(k,\omega)$ is the Fourier transform of $\vec{u}(\vec{r}_i,t) = \vec{w}_i(t) - \langle \vec{w}_i \rangle$ with respect to both the time and the sites $\vec{r}_i, i = 1,\ldots,N$ of the neurons, the brackets denoting an ensemble average.

For the derivation of an explicit expressions for $S(k,\omega)$ we use that S can be obtained as the Fourier transform of the space time correlation function of the synaptic fluctuations, i.e.

$$S(\vec{k},\omega) = \int_{-\infty}^{\infty} dt e^{i\omega t} \left\langle \vec{u}(\vec{k},t)\vec{u}(-\vec{k},0) \right\rangle \tag{4}$$

where $\vec{u}(\vec{k},t) = \sum_{i=1}^{N} \vec{u}(\vec{r}_i,t)$. The correlation function is evaluated by means of the two point probability density $p_2(\mathbf{w},t;\mathbf{w}',t')$ using

$$(\partial_t + L(\mathbf{w}))\, p_2(\mathbf{w},t;\mathbf{w}',t') = 0 \tag{5}$$

and $p_2(\mathbf{w},t;\mathbf{w}',t) = \delta(\mathbf{w}-\mathbf{w}')$. After some algebra one obtains by means of the explicit expressions for the Fokker–Planck operator derived in Ref. [1] the following

$$K(\vec{k},t) = \left\langle \vec{u}(\vec{k},t)\vec{u}(-\vec{k},0) \right\rangle = \left\langle \left[e^{-\mathbf{B}(\vec{k})|t|} \vec{u}(\vec{k},0) \right] \cdot \vec{u}(-\vec{k},0) \right\rangle \tag{6}$$

where the matrix $\mathbf{B}(\vec{k})$ is the kernel of the drift term of the Fokker–Planck operator.

In the case of the mapping of a two dimensional rectangle onto a one dimensional chain of neurons, the matrix $\mathbf{B}$ is two dimensional. If the height s of the rectangle is sufficiently small ($s \ll \sigma$). i.e. if the input space is pseudo one–dimensional the chain is mapped onto the line which cuts the rectangle symmetrically into two halves. For $s > s_c$ this symmetric mapping is broken spontaneously and the chain begins to fold into the input space. This is seen best by studying the transversal synaptic fluctuations, i.e. we consider $S_\perp(k,\omega)$

$$S_\perp(k,\omega) = \left\langle \|u_\perp(k,\omega)\|^2 \right\rangle = \left\langle \|u_\perp(k)\|^2 \right\rangle \frac{2\lambda}{\lambda^2 + k^2} \tag{7}$$

where $\lambda(k) = \sqrt{2\pi}\sigma \left(1 - \frac{2}{3}s^2 (1-\cos k) \exp\left(\frac{-k^2\sigma^2}{2}\right) / N\right)$is the lowest eigenvalue of the matrix $\mathbf{B}$ and $\left\langle \|u_\perp(k)\|^2 \right\rangle$ has been obtained before, cf. Ref. [1].

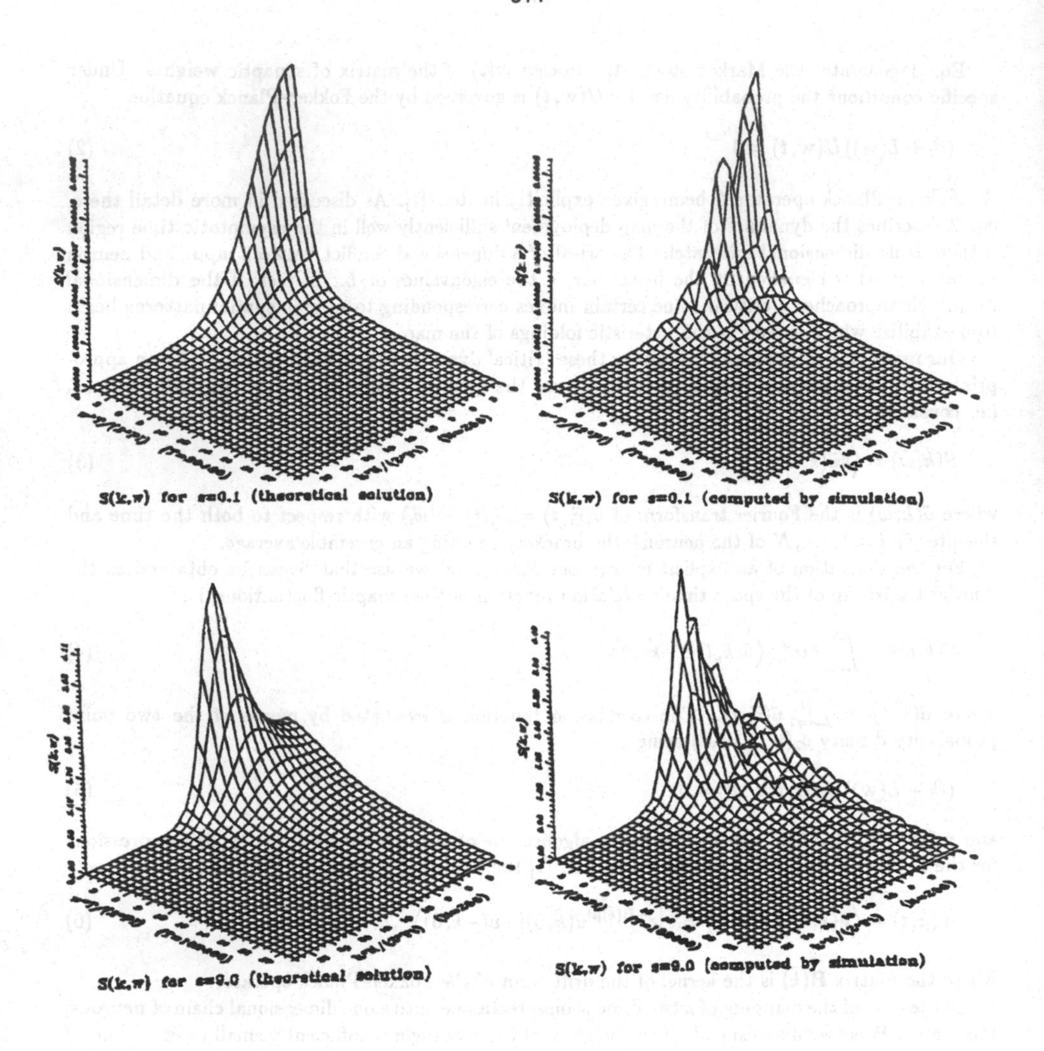

S(k,w) for s=0.1 (theoretical solution)
S(k,w) for s=0.1 (computed by simulation)
S(k,w) for s=9.0 (theoretical solution)
S(k,w) for s=9.0 (computed by simulation)

Figure 1: *The dynamic spectral density of the transversal fluctuations $S_\perp(k,\omega)$ as evaluated from eq. 7 (left) and from computer simulations based on eq. 1 (right). The mapping of a rectangle of height s on a chain of neurons is considered. The width σ of the neighborhood function is $\sigma = 5$.*

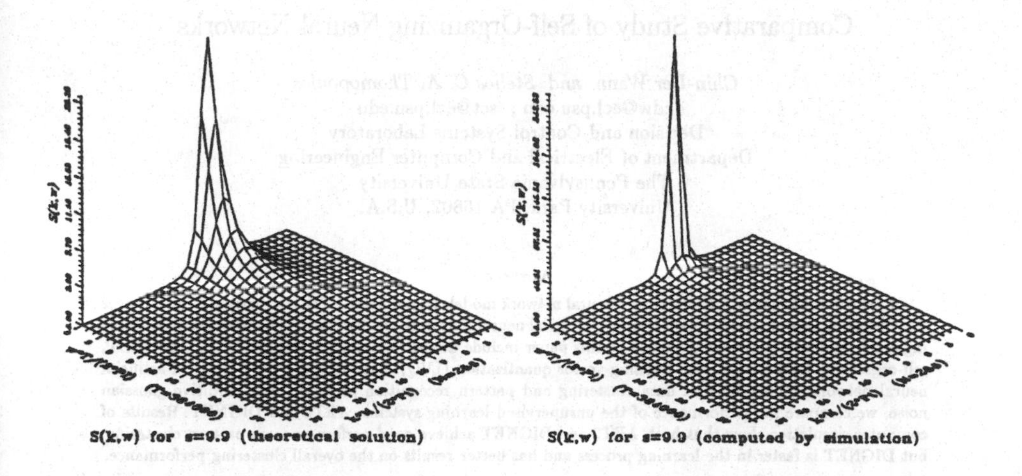

S(k,w) for s=9.9 (theoretical solution)
S(k,w) for s=9.9 (computed by simulation)

Figure 2: *Same as Figure 1 for the near critical case of* $s = 9.9$.

3 Results

We have evaluated the dynamic spectral density of the transversal fluctuations $S_{\perp}(k,\omega)$ both from eq. 7 and from computer simulations based on eq. 1. In the figures we display the landscape of $S_{\perp}(k,\omega)$ over k and ω. This landscape is a complete record of the length and time scales involved in the evolution of the map. In particular we observe that fluctuations on length scales $l_c \cong 1/k_c$,where $k_c = \sqrt{2}/\sigma$ [1], become very much enhanced as soon as s approaches its critical value s_c. These may be considered as precursor states to the destabilizing mode with $k = k_c$ at $s = s_c$. Essentially this result has been already obtained in Ref. [1], which studied the static spectral density, obtained from summing our $S_{\perp}(k,\omega)$ over ω. The important new information contained in the ωdependence of $S_{\perp}(k,\omega)$ is a corresponding record of the relevant *time* scales involved in the evolution of the map. This is of immediate interest for designing cooling strategies for the learning parameters ϵ, σ.

The Fokker–Planck theory is seen from the figures for $s = 0.1$ and $s = 9.0$ to be in nice agreement with the simulations. However, for $s = 9.9$ which is still well below the critical value $s_c = 2.02\sigma = 10.1$ the peak at $k \approx k_c$ is higher by a factor of ten for the simulation result. This is an effect due to the nonlinearities and in fact invalidates any results obtained from the Fokker–Planck equation for that case. An analysis in terms of the time dependent Ginzburg–Landau theory which takes care of these effects is under way.

References

[1] H. Ritter, T. Martinetz, K. Schulten: *Neural Computation and Self-organizing Maps.* Reading (Mass.): Addison–Wesley, 1992.

[2] T. Kohonen: *Self-Organization and Associative Memory.* Springer Series in Information Science 8, Berlin, Heidelberg: Springer–Verlag, 1984.

Comparative Study of Self-Organizing Neural Networks

Chin-Der Wann and *Stelios C. A. Thomopoulos*
cdw@ecl.psu.edu ; sct@ecl.psu.edu
Decision and Control Systems Laboratory
Department of Electrical and Computer Engineering
The Pennsylvania State University
University Park, PA 16802, U.S.A.

Abstract

A benchmark study of self-organizing neural network models is conducted. The comparison of advantages and disadvantages of unsupervised learning artificial neural networks are discussed. The unsupervised learning artificial neural networks discussed in this paper include adaptive resonance theory (ART2), DIGNET, self-organizing feature map, and learning vector quantization (LVQ). For the benchmark study of artificial neural network applications on data clustering and pattern recognition problems with additive gaussian noise, we compare the performance of the unsupervised learning systems, ART2 and DIGNET. Results of computer simulation show that both ART2 and DIGNET achieve good performance on pattern clustering, but DIGNET is faster in the learning process and has better results on the overall clustering performance.

1 Introduction

The artificial neural network approaches apply parallel distributed processing models to achieve high performance via highly interconnected simple computational elements. Among the artificial neural network approaches, models are categorized to supervised learning and unsupervised learning. In the supervised learning category, several networks such as Adaline, perceptron, and back-propagation algorithm have been used in the fields of pattern recognition, control systems, and signal processing[1, 2]. They require target patterns or teachers during the classification and learning procedures. In the real world, there are many situations where supervised learning models are not able to handle the received data:

- when target signals or teachers are not available;
- when we are able to obtain samples, but unable to label them;
- when the characteristics of the class-specific pattern generating systems are changing over time.

In these cases, unsupervised learning methods must be used to achieve the clustering and classification tasks.

In the unsupervised learning category, many algorithms have been proposed and applied to pattern classification and recognition problems. Some of the unsupervised learning models are adaptive resonance theory (ART)[3, 4, 5], DIGNET[6], Kohonen's self-organizing feature map[7], and learning vector quantization (LVQ)[8]. In this paper, we focus on the comparison between ART2[4] and DIGNET, which are truly self-organizing networks. The basic theories behind these two models are briefly discussed in the next section. A comparison between them is presented in the following section. The performance of the two networks is illustrated and compared through computer simulations in two benchmark problems.

2 General Background

2.1 Adaptive Resonance Theory (ART2)

Adaptive resonance architectures are neural networks that self-organize stable pattern recognition codes in real-time in response to arbitrary sequences of analog and binary input patterns[4, 5]. The concepts of competitive learning and interactive activation are applied on the architecture, in a manner that leads to a stable learning algorithm.

One of the ART2 architectures[4, Fig. 10] has three levels in its model. The F_0 level, which can be treated as the pre-processing level, normalizes the input patterns and performs contrast enhancement and noise suppression on the components of input pattern vectors. The middle level F_1 is where the input pattern and stored exemplar match. The top level F_2 performs a competition among the stored exemplars. Only the weights associated with the winning node in the top level are updated during the learning cycle. An *orienting subsystem* resets the winning node if the measure of similarity between input pattern and the related stored exemplar is less than a pre-specified threshold called *vigilance*. The competition and learning rules of the ART2 architecture in [4, Fig. 10] are summarized as follows.

Let T_j be the summed $F_1 \rightarrow F_2$ input to the jth F_2 node:

$$T_j = \sum_{i=1}^{M} p_i z_{ij}, \quad j = 1, \ldots, N. \tag{1}$$

If the Jth F_2 node receives the largest total input, F_2 makes a choice, i.e.,

$$T_J = \max\{T_j : j = 1, \ldots, N\}. \tag{2}$$

The update equations of the top-down and bottom-up LTM (long term memory) traces for ART2 are given as

$$\tfrac{d}{dt} z_{ji} = g(y_j)(p_i - z_{ji}), \quad \text{top-down } (F_2 \rightarrow F_1), \tag{3}$$

$$\tfrac{d}{dt} z_{ij} = g(y_j)(p_i - z_{ij}), \quad \text{bottom-up } (F_1 \rightarrow F_2), \tag{4}$$

where $g(y_j)$ equals d if the jth node has not been reset on the current trial; otherwise, $g(y_j)$ equals 0. ART2 models calculate the degree of match between an STM (short term memory) pattern at F_1 and an active LTM pattern in the *orienting subsystem*. The degree of match r_i is given by

$$r_i = \frac{u_i^0 + cp_i}{\|\mathbf{u}^0\| + \|c\mathbf{p}\|}, \tag{5}$$

where $c > 0$. The *orienting subsystem* is assumed to reset F_2 whenever an input pattern is active and $\|\mathbf{r}\| < \rho$.

2.2 DIGNET

DIGNET is a self-organizing network based on competitive learning[6]. The stored exemplar called *attraction wells* are generated or eliminated according to the results of competition in the pattern space.[6] The DIGNET model uses the measure of inner-product to compare the similarity between the normalized input pattern vector and the exemplar vectors. The competition and learning rules of DIGNET are summarized as below.

1. From a given signal-to-noise ratio (SNR), determine the threshold of each attraction well.

2. Pre-process and normalize the input pattern $\mathbf{x}_n$.

3. Compute the similarity between the normalized input pattern and each stored exemplar. Each exemplar is defined as the center of an attraction well, denoted by $\mathbf{e}_{n-1}$.

4. Determine the update status c_n of each exemplar from the results of threshold functions and MAXNET, which select an exemplar to which the input pattern has the maximum similarity.

5. Update the exemplar by using the following rules:

$$\mathbf{e}_n = \frac{c_n}{d_n}\mathbf{x}_n + \frac{d_{n-1}}{d_n}\mathbf{e}_{n-1}, \text{ with initial conditions } \mathbf{e}_0 = 0, \tag{6}$$

   where d_{n-1} is the depth of the well after the presentation of last input pattern. The depth of a well is updated according to $d_n = d_{n-1} + c_n$ with initial conditions $d_0 = 0$.

In order to avoid excessive and spurious wells, a *stage age* is defined in the algorithm. The depth of each well is periodically examined at the end of each *stage age*. If the depth of a well does not exceed a certain threshold (age) at the end of a *stage age*, the well is eliminated; otherwise, it survives this stage age.

As it has been pointed out above, the threshold of each *attraction well* can be determined from *a priori* information on the SNR. Given an SNR (in dB), the threshold is determined by

$$\text{threshold} = \frac{1}{\sqrt{1+10^{-\frac{SNR}{10}}}}, \tag{7}$$

and the width (in radian) of each well is $\theta = \arccos(\text{threshold})$.

3 Comparison of the Models

3.1 Real-time Consideration

Both ART2 and DIGNET architectures apply highly interconnected network structures to process the classification of input patterns. For ART2, the classification process is fast, but the slow convergence during the learning cycles in "slow learning" mode is a drawback. When "fast learning" is used, the learning speed is much improved; however, some other problems still remain. This will be discussed in the following subsection. On the other hand, DIGNET has very fast learning procedures. From the simulations, we have seen that DIGNET exhibits considerably faster classification and learning on input patterns, and needs fewer epochs than ART2 to reach a stable performance. The results are shown in Fig. 1. (Detailed description of the simulations is given in Section 4.)

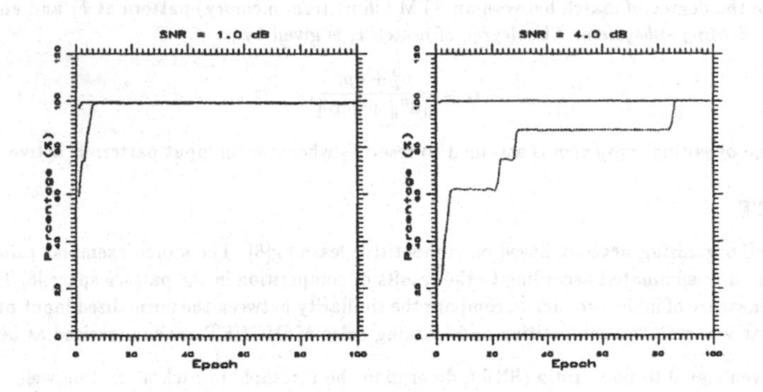

Figure 1: The percentage of successful clustering of DIGNET and ART2 versus epochs. Solid lines: DIGNET. Dotted lines: ART2. (Every input pattern is presented to the systems once in an epoch. The threshold value used in DIGNET is 0.707, and the vigilance used in ART2 is 0.924.)

3.2 Architecture Differences

Generally speaking, both ART2 and DIGNET are based on the process of competitive learning. In the applications, *a priori* information of the number of clusters is not required for these two models, unlike the K-means algorithms[9], in which the number of clusters must be specified beforehand. A comparison of processing procedures in DIGNET and ART2 is illustrated in Fig. 2. The choices of system parameters in the artificial neural networks affect the results of classification. In ART2, the parameter *vigilance* determines the criteria of classification of each stored exemplar, and the performance is sensitive to the choice of *vigilance*. Unfortunately,

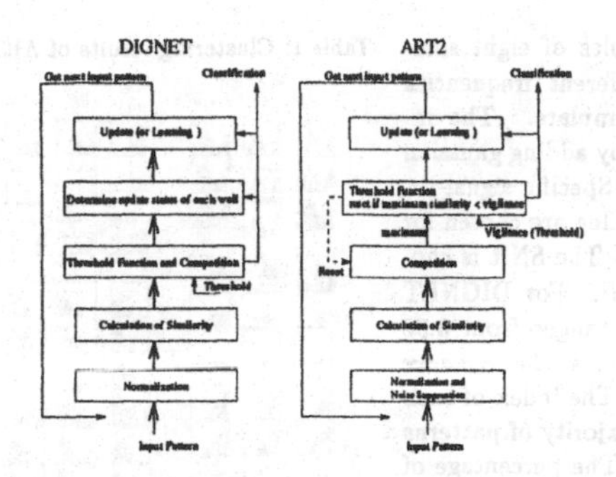

Figure 2: Comparison of processing functions in DIGNET and ART2

there is no direct and effective method to determine the value of *vigilance*. In DIGNET, a threshold value also determines the attraction region of each well, but it can be determined from a pre-specified signal-to-noise ratio (Eq. (7)) that relates to the desirable maximum tolerable noise level[6]. Therefore, the parameters of DIGNET can be chosen efficiently and directly.

In DIGNET, the "depth" of a well has been defined and is applied to the weighted learning rules. This variable is similar to the monotonically decreasing scalar gain factor, $\alpha(t)$ in self-organizing feature map and LVQ[7, 8]. But, unlike the well depth, the factor function $\alpha(t)$ must be chosen externally.

The ART2 architectures use some fixed variables (i.e., the a, b, c and d in [4, Fig. 10]) to determine the weighting of the input pattern and stored exemplar during the competition and learning process. These fixed variables may not be appropriate in some cases. Also, these variables must be very carefully chosen since the performance of ART2 models are sensitive to changes of these variables. Another condition that a "reset" can be initiated during the learning cycle is also not suitable. The learning cycle of LTM traces (or the stored exemplar) in ART2 stops when the "convergence" is achieved. This process may take a long period of time unless a checking condition is set. Another problem with ART2 is that learning of a new pattern may overwrite the previously stored information since the value of the exemplar will approach to the value of this input pattern.

3.3 Measures of Similarities

The DIGNET model uses inner-product of the normalized input pattern vector and the exemplar vectors as the measure of similarity[1]. The ART2 architectures use the concept of inner-product in the competition stage, but they apply the form of Eq. (5) in the reset stage to evaluate the similarity between input patterns and the stored exemplar. So, the measures of similarities in ART2 models are more complex.

Since DIGNET and ART2 use the normalized vectors in the processes, similar input pattern vectors which have different magnitudes may be considered as the same within the systems. Therefore, when the magnitudes of pattern vectors are treated as important features, some preprocessing on the features may be required.

4 Simulations and Discussions

We use an input data set which consists of the 50 patterns originally used in the ART2 and ART2-A[10] simulations. The DIGNET simulations give results essentially similar to the results of "fast-learn" ART2[4, Fig. 3] and ART2-A[10, Fig. 2] with comparable choices of parameters. Table 1 shows the summary of clustering results.

[1]Other measures of similarities are possible depending on the problems as described in [6].

In other simulations, samples of eight sinusoidal waveforms with different frequencies are used as noise-free exemplars. The input patterns are generated by adding gaussian noises to those exemplars. Specific signal-to-noise ratio and threshold value are chosen for each individual simulation. The SNR is chosen from $-10dB$ to $+15dB$. For DIGNET simulations, the ***threshold*** ranges from 0.38 to 0.98. For ART2 simulations, the ***vigilance*** ranges from 0.837 to 0.99. The index of each cluster is labeled by the majority of patterns classified into the cluster. The percentage of successful clustering (PSC) is determined by the total number of patterns which have the same indexes as the clusters they are with. The percentage of successful clustering during the simulations is illustrated in Fig. 3, as function of the SNR. The number of clusters generated are shown in Fig. 4.

Table 1: Clustering results of ART2, ART2-A and DIGNET

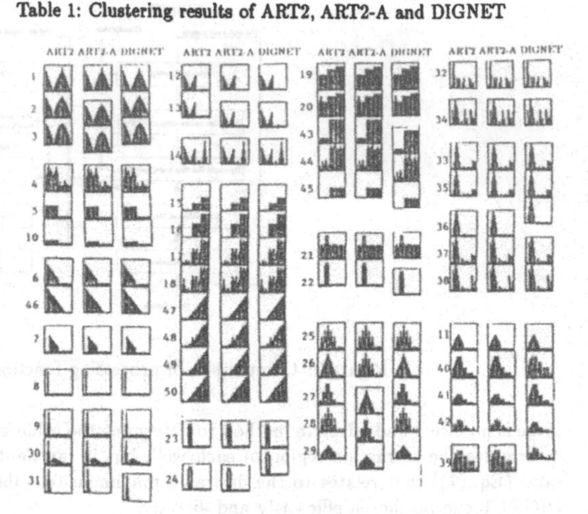

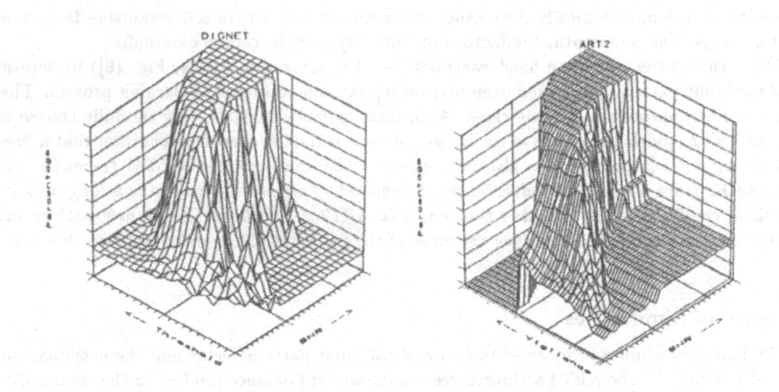

Figure 3: Percentage of successful clustering of DIGNET and ART2 after 30 epochs. The scale of percentage is from 0 to 100. The values of ART2 systems in the region with low SNR and high *vigilance* are set to zero since the numbers of clusters exceed the number of F_2 nodes used in the simulations.

If low threshold (or vigilance) values are used, both of DIGNET and ART2 create fewer clusters since some of the exemplars though from different classes are treated alike. With properly chosen thresholds, good performance is obtained. The 100% successful clustering regions are shown in Fig. 3 for both DIGNET and ART2. For most of the 100% successful clustering, DIGNET generates the correct number of clusters, i.e., eight; whereas ART2 generates nine clusters instead, as shown in Fig. 4. In the cases of low SNR, spurious clusters are created due to the high noise level.

If the threshold value or *vigilance* increases, more clusters will be created by the systems. However, DIGNET keeps the number lower than 30, while the ART2 model creates more spurious clusters.

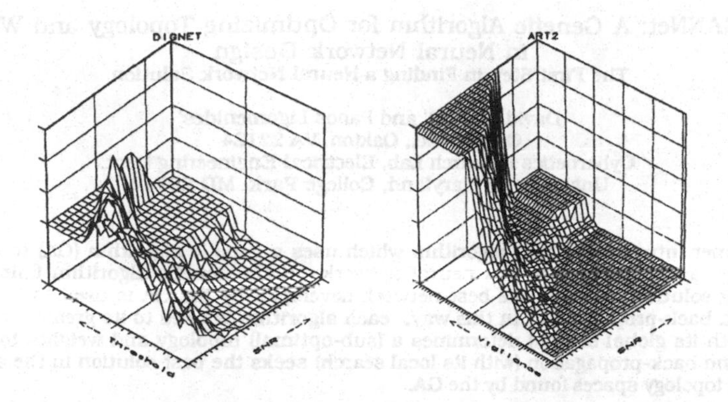

Figure 4: Numbers of clusters created by DIGNET and ART2 after 30 epochs. The shown scale varies from 0 to 50 clusters. The region with low SNR and high *vigilance* in ART2 plot has been truncated.

5 Conclusions

Both ART2 and DIGNET architectures apply highly interconnected network structures and competitive learning algorithms to perform clustering of input patterns. Although both of them have good performance, we have seen that DIGNET is faster in the learning procedures. One advantage of DIGNET over ART2 is that the threshold values can be determined from a pre-specified signal-to-noise ratio. Also, the architectures and the measures of similarities in DIGNET models are less complex than those in ART2. Results of computer simulations show that DIGNET model overperforms the ART2 systems. DIGNET is faster in the learning process and has more accurate results on the overall clustering performance.

References

[1] R. P. Lippmann, "An introduction to computing with neural nets," *IEEE ASSP Magazine*, pp. 4–22, April 1987.

[2] D. E. Rumelhart and J. L. Mcclelland, *Parallel Distributed Processing: Explorations in the Microstructure of Cognition*, vol. 1. The MIT Press, 1986.

[3] G. A. Carpenter and S. Grossberg, "A massively parallel architecture for a self-organizing neural pattern recognition machine," *Computer Vision, Graphics, and Image Processing*, vol. 37, pp. 54–115, 1987.

[4] G. A. Carpenter and S. Grossberg, "ART 2: Self-organization of stable category recognition codes for analog input patterns," *Applied Optics*, vol. 26, pp. 4919–4930, December 1987.

[5] G. A. Carpenter and S. Grossberg, "The ART of adaptive pattern recognition by a self-organizing neural network," *Computer Magazine*, pp. 77–88, March 1988.

[6] S. C. A. Thomopoulos and D. K. Bougoulias, "DIGNET: A self-organizing neural network for automatic pattern recognition, classification, and data fusion," in *SPIE's Proceedings on Sensor Fusion*, (Boston, MA), pp. 478–495, November 1991.

[7] T. Kohonen, *Self-Organization and Associative Memory*. Springer-Verlag, second edition ed., 1988.

[8] T. Kohonen, "The self-organizing map," *Proceedings of the IEEE*, vol. 78, pp. 1464–1480, September 1990. Special issue on neural networks, I: theory & modeling.

[9] J. T. Tou and R. C. Gonzalez, *Pattern Recognition Principles*. Addison-Wesley, 1974.

[10] G. A. Carpenter, S. Grossberg, and D. B. Rosen, "ART 2-A: An adaptive resonance algorithm for rapid category learning and recognition," *Neural Networks*, vol. 4, pp. 493–504, 1991.

GANNet: A Genetic Algorithm for Optimizing Topology and Weights in Neural Network Design
The First Step in Finding a Neural Network Solution

David White@' and Panos Ligomenides'
@ MRJ, Inc., Oakton, VA 22124
' Cybernetics Research Lab, Electrical Engineering Dept.,
University of Maryland, College Park, MD 20742

This paper introduces a new algorithm which uses a genetic algorithm (GA) to determine the topology and link weights of a neural network. If the genetic algorithm fails to find a satisfactory solution network, the best network developed by the GA is used to try to find a solution via back-propagation. In this way, each algorithm is used to its greatest advantage: the GA (with its global search) determines a (sub-optimal) topology and weights to solve the problem, and back-propagation (with its local search) seeks the best solution in the area of the weight and topology spaces found by the GA.

The intent is to develop an algorithm which can be used as a first attempt to solve unknown problems. If a solution is not found immediately by the algorithm, at least an appreciable amount of information about the solution can be gleaned from its results, and a lot of the initial guesswork that currently exists in finding a neural network solution to a problem can be eliminated.

Some of the features of the GANNet algorithm are:

1. The networks produced are feed-forward layered networks, which means that there exist many techniques for further refinement and analysis of the resulting networks.
2. The network nodes can have fan-in and fan-out limits, permitting relatively simple hardware implementation using modular units.
3. Both Gaussian and sigmoid node transfer functions are available.
4. No initial knowledge about a solution network is needed, other than the obvious (the number of inputs and outputs).

Introduction

When applying neural networks to a new problem, determining the architecture of the network is a difficult task. Generally, some attempt is made to determine a near optimal number of hidden units, and training is attempted. If training fails (which is usually the case), little insight is gained about the desired architecture, because training could have failed for any of the following reasons:

1. The initial conditions guaranteed failure, even though the architecture was acceptable.
2. The network has too few hidden neurons, and therefore can not learn the input - output mapping desired.
3. The network has too many hidden neurons, slowing training and creating a more complex error surface.

Therefore, this process must be repeated until an acceptable architecture is found. Many attempts have been made to algorithmically alter the network architecture during training [1][2][3][4][5][6][7][8][9][10], and these methods show an improvement over the static architecture methods. However, some initial guess of the number of hidden units must often still be made, and the architecture modifications are usually limited to adding hidden nodes ,and the modifications are made on the basis of local information.

The Promise of Genetic Algorithms

Genetic algorithms (GAs) offer a general and global optimization procedure. Since the GA is a global search method, it will be less susceptible to the failings of a local search procedure such as back-propagation. Also, the GA will not have the stability problems back-propagation does (i.e., if the learning rate is too high, the link weights will approach infinity). And with a change in perspective, the GA can be used to develop the architecture of the network as well as the weight set.

Some attempts have been made to use GAs to find the architecture of a neural network [11][12], and to find the link weights for a fixed architecture network [13]. At least two attempts have been made to use a GA to find both the architecture and link weights [14][15]. Unfortunately, in [14] the architectures produced are not layered, and in [15], the fitness function consists of training each network each generation.

The GANNet Algorithm

Figure 1 presents a flowchart of the training procedure. First, the normal steps of a GA are performed:

1. Generate a population of "individuals" to be evaluated and operated upon.
2. Select candidates to be used to generate new individuals.
3. Generate a new individual from the selected "parents" by performing a "crossover", mixing of the "genes" of each.
4. "Mutate" the new individuals to allow the search to explore new areas of the search space.
5. Replace an individual in the old population. The algorithm provides either worst-only or whole population (with one or more reserved individuals) replacement.

The new individuals are evaluated, and steps 2. to 5. are repeated until the desired performance is acheived or the individuals converge to a sub-optimal network (once the individuals in the population become too alike, little further progress can be made). If the performance of the best network is acceptable, nothing more need be done. If it is not, the best network is used in an enhanced back-propagation training algorithm.

The GA used has some differences with the standard GA. The standard GA has one population, and although one population can be used in this implementation as well, normally a number of subpopulations will be used instead (i.e., this is a distributed GA a la [16] and [17]). This allows the use of small subpopulations, while retaining the global nature of the GA. After a user-specified number of generations (presently set at 5 times the number of individuals per population), a user-specified number of the best individuals from each subpopulation are copied into other subpopulations.

Also, individuals are selected for reproduction according to their rank in the population, not their raw fitness (as in [18]). This provides a steady selection pressure throughout the trial, avoiding both the complete dominance of the population by a superior (but not completely acceptable) individual early in the trial, and the lack of selective pressure towards the end of the trial.

Each individual is a complete neural network with the prescribed number of inputs and outputs, and the "allele" is a hidden or output node with its associated input links. Therefore, because of the need to maintain the prescribed number of outputs (and to reject any one-node layers), some justification of the results of the crossover operation must be performed. The crossover variations currently implemented are one-point, two-point, uniform, and binomial. Also, it should be noted that both sigmoid and Gaussian node transfer functions are used, with the distribution of node transfer functions in the initial population being 80% sigmoid and 20% Gaussian.

The use of nodes as alleles is motivated by the following observations:

1. GAs are relatively indifferent to the complexity of the allele. They seem to work well with both binary and floating point representations [19][20]. Therefore, a GA may work equally well with a very complex allele, such as a neuron.
2. Unfortunately, GAs *are* sensitive to the length of the chromosome string. Thus, the shorter this string can be made, the better. Using link weights as the alleles creates far longer chromosome strings than using the nodes as alleles.
3. The link weight is not really the "unit of information" in the neural network; a link weight by itself is meaningless. Only in the larger context of its exact position in the network does it have meaning. However, a neuron and its associated input links does have meaning: it is a "feature detector", i.e., it responds to some feature in the input set. The task of the GA, then, is to find a proper set of feature detectors for the problem at hand.
4. In addition to its hardware-implementation considerations, the fan-in limit permits the differentiation of the hidden nodes, since the connection pattern of each node will likely be unique. The current fan-in limit is set to 4 inputs pre node (plus a bias input) as suggested in [21].
5. The "architecture space" can be searched at the same time the "weight space" is searched, without modifying the GA theory at all.

Results

GANNet always produces improvement from its starting population, i.e., the best individual at population convergence is always much better than the best individual at the start. Unfortunately, it suffers from the problem of "premature convergence" along with the rest of the GA universe. There are known methods of dealing with this problem (as in [14] and [22]), and these methods can be employed here as well.

The first case to study is the exclusive-or. This is a small, well-known problem, which (although not the best test case for general purposes) at least proves the operation of the algorithm. (It was also used to debug the code.) A set of ten trials was performed, with a successful network being produced in all ten. The results are presented below. In each trial, the following parameters were used:

Subpopulation size:	10
Number of subpopulations:	20
Number of individuals reserved in each subpopulation:	2
Probability of crossover:	0.9
Probability of mutation:	0.6
Minimum initial number of nodes per layer:	2
Maximum initial number of nodes per layer:	10
Minimum initial number of layers:	2
Maximum initial number of layers:	5
Fan-in limit:	10
Fan-out limit:	4
Maximum generations limit:	1000
Number of generations between swapping individuals between subpopulations:	50
Number of individuals to swap:	2
Backprop learning rate:	0.9
Backprop momentum term:	0.6

The termination condition was determined by using the mean sum-squared error. Since this is a binary problem, on a scale of 0.0 to 1.0, we need only have a "0" be below 0.5, and a "1" be above 0.5. As an approximation, a mean sum-squared error of 0.2 was used to verify network operation. However, to maximize the probability that all vectors meet the final requirement, the network error goal during training was set to be a mean sum-squared error of 0.02. All ten networks generated during this test succeeded in meeting this goal

Trial	Number of Generations	Number of Epochs	Number of Layers	Number of Hidden Nodes
1	68	0	2	3
2	56	0	2	3
3	46	0	1	0
4	40	0	2	5
5	62	0	1	0
6	49	0	2	2
7	45	0	2	5
8	88	0	2	3
9	54	0	2	4
10	60	0	2	3
Avg	56.8	0	1.8	2.8

The results show that the algorithm finds a solution. Since the algorithm was set to find a full solution via the GA, no back-propagation epochs were needed to find an acceptable solution network. In addition, the networks are usually near the optimal size, and twice out of ten trials the optimum size was obtained. However, any of the networks would work. Since the initial population of networks vary from two to five layers, this implies that for the problem selected, deeper networks are unnecessary, and get dropped from the population.

Conclusion

GANNet shows promise of providing a way of attacking new problems, providing valuable information on network topology, if not an outright solution, on the first attempt. All the techniques and progress in layered network analysis and design can be incorporated into this algorithm, as well as most of the advances in genetic algorithm design. Therefore, the number of advances available to this algorithm is greater than any of the alternatives.

Some enhancements would make this tool even more automatic. The maximum and minimum number of nodes per layer in the initial population should be determined automatically, taking into account the possible need for more than one hidden layer that the fan-in and fan-out limits will force. The range of nodes per layer in the initial population is currently set a priori, as is the range of node layers in the initial population. Also, the number of subpopulations, and the number of individuals per population, should also be determined automatically.

As mentioned above, implementing ways to stave off population convergence would greatly help the performance of this algorithm, as well as one of the more advanced generative back-propagation algorithms (for refinement of the best individual). Currently, the techniques presented in [4] and [23] are used.

Bibliography

[1] M. Mezard and J.P. Nadal, "Learning in Feedforward Layered Networks: the Tiling Algorithm", Journal of Physics A: Math. Gen., Vol. 22, pp. 2191 - 2203.

[2] S.I. Gallant, "Perceptron-Based Learning Algorithms", *IEEE Transactions on Neural Networks*, Vol. I, No. 2, pp. 179 - 191. June 1990.

[3] M. Frean, "The Upstart Algorithm: A Method for Constructing and Trainging Feed-Forward Neural Networks", *Neural Computation*, Vol. 2, No. 2, pp. 198 - 209. MIT Press.

[4] Y. Hiroshe, K. Yamashita, and H. Shimpei, "Back-Propagation Algorithm Which Varies the Number of Hidden Units", *Neural Networks*, Vol. 4, No. 1, pp. 225 - 229. Pergamon Press.

[5] Wilson Wen, Huan Lin, and Andrew Jennings, "Self-Generating Neural Networks", *Proceedings of the International Joint Conference on Neural Networks*, Vol. IV, pp. 779 - 784. Baltimore, June, 1992.

[6] Juyang Weng, Narendra Ahuja, and Thomas Huang, "Cresceptron: A Self-Organizing Neural Network Which Grows Adaptively", *Proceedings of the International Joint Conference on Neural Networks*, Vol. I, pp. 577 - 581. Baltimore, June, 1992.

[7] Sheng-De Wang and Ching-Hsao Hsu, "A Self Growing Learning Algorithm for Determining the Appropriate Number of Hidden Units", *Proceedings of the International Joint Conference on Neural Networks*, Vol. II, pp. 1098 - 1104. Singapore, November, 1991.

[8] Paul Baffes and John Zelle, "Growing Layers of Perceptrons: Introducing the Extentron Algorithm", *Proceedings of the International Joint Conference on Neural Networks*, Vol. II, pp. 392 - 397. ????????????

[9] Kazuyuki Murase, Yutaka Matsunaga, and Yoshiaki Nakade, "A Back-Propagation Algorithm which Automatically Determines the Number of Association Units", *Proceedings of the International Joint Conference on Neural Networks*, Vol. I, pp. 783 - 788. Singapore, November, 1991.

[10] Dit-Yang Yeung, "Automatic Determination of Network Size for Supervised Learning", *Proceedings of the International Joint Conference on Neural Networks*, Vol. I, pp. 158 - 164. Singapore, November, 1991.

[11] Nigel Dodd, "Optimisation of Network Structure using Genetic Techniques", *Proceedings of the International Joint Conference on Neural Networks*, Vol. III, pp. 965 - 970. San Diego, June, 1990.

[12] Geoffrey Miller, Peter Todd, and Shailesh Hegde, "Designing Neural Networks using Genetic Algorithms", *Proceedings of the Third International Conference on Genetic Algorithms*, pp. 379 - 384. Morgan Kaufmann, 1989.

[13] D. Whitley, T. Starkweather, and C. Bogart, "Genetic Algorithms and Neural Networks: Optimizing Connections and Connectivity", *Parallel Computing*, Vol. 14, pp. 347 - 361.

[14] John Koza and James Rice, "Genetic Generation of Both the Weights and Architecture for a Neural Network", *Proceedings of the International Joint Conference on Neural Networks*, Vol. II, pp. 397 - 404. ??, July 1991.

[15] Steven Harp, Tariq Samad, and Aloke Guha, "Towards the Genetic Synthesis of Neural Networks", *Proceedings of the Third International Conference on Genetic Algorithms* , pp. 360 - 369. Morgan Kaufmann, 1989.

[16] Reiko Tanese, "Distributed Genetic Algorithms", *Proceedings of the Third International Conference on Genetic Algorithms*, pp. 434 - 439. Morgan Kaufmann, 1989.

[17] Darrell Whitley and Timothy Starkweather, "Optimizing Small Neural Networks Using a Distributed Genetic Algorithm", *Proceedings of the International Joint Conference on Neural Networks*, Jan. 1990, Vol. I, pp. 206 - 209. Washington, D.C., Jan. 1990.

[18] D. Whitley, "The GENITOR Algorithm and Selection Pressure: Why Rank-Based Allocation of Reproductive Trials is Best", *Proceedings of the Third International Conference on Genetic Algorithms*, pp. 116 - 121. Morgan Kaufmann.

[19] David Goldberg, *Genetic Algorithms in Search, Optimization & Machine Learning*, Addison-Wesley, 1989.

[20] Xiaofeng Qi and Francesco Palmieri, "Analyses of the Genetic Algorithms in the Continuous Space", *Proceedings of the International Joint Conference on Neural Networks*, Vol. IV, pp. 560 - 565. Baltimore, June, 1992.

[21] L. Akers and M. Walker, "A Limited-Interconnect Synthetic Neural IC", *Proceedings of the IEEE International Conference on Neural Networks*, Vol. II , pp. 151 - 158.

[22] Larry Eshelman and J. David Schaffer, "Preventing Premature Convergence in Genetic Algorithms by Preventing Incest", *Proceedings of the Fourth International Conference on Genetic Algorithms*, pp. 115 - 122. Morgan Kaufman, 1989.

[23] Masafumi Hagiwara, "Novel Back Propagation Algorithm for Reduction of Hidden Units and Acceleration of Convergence using Artificial Selection", *Proceedings of the*

International Joint Conference on Neural Networks, **Vol. I, pp. 625 - 630. San Diego, June, 1990.**

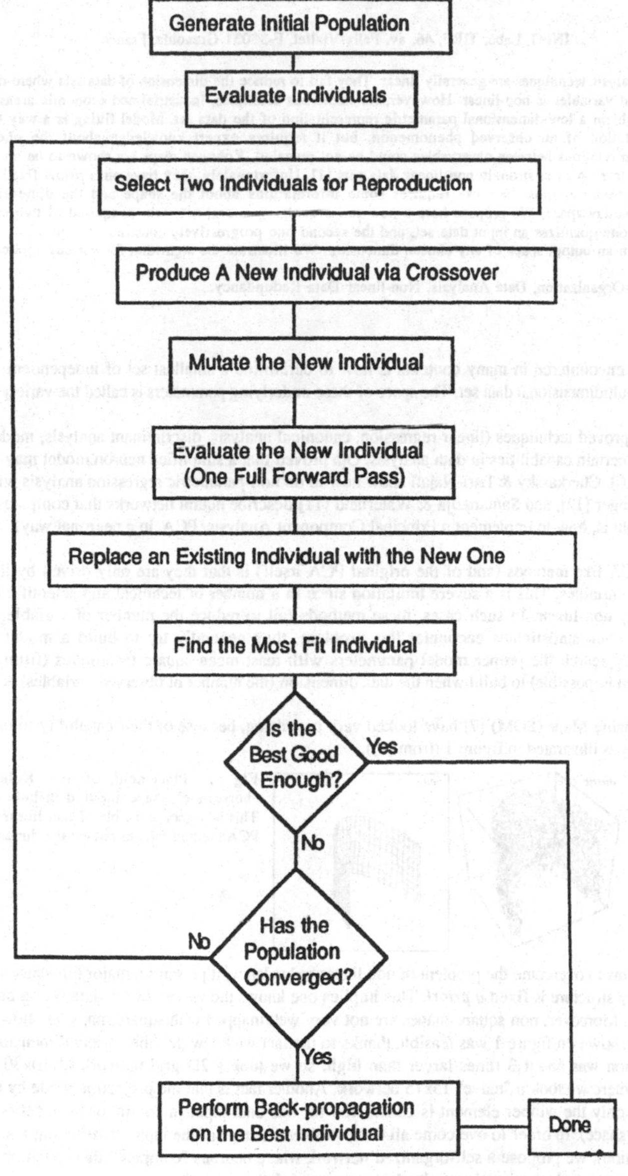

Figure 1.

Vector Quantization and Projection neural network

P. Demartines & J. Hérault

INPG, Labo. TIRF, 46, av. Félix-Viallet, F-38031 Grenoble, France

Classical data analysis techniques are generally linear. They fail to reduce the dimension of data sets where dependence between observed variables is non-linear. However, for numerous scientific, industrial and economic areas, it should be desirable to obtain a low-dimensional parametric representation of the data set. Model fitting is a way to obtain a usable representation of an observed phenomenon, but it requires expert knowledge about the phenomenon. Moreover, hidden relations between observables could be not revealed. Kohonen maps are shown to be an alternative techniques, able to map even strongly non-linear data sets [1]. Unfortunately, they have an *a priori* fixed shape and neighbourhood structure, thus their use requires some informations about the shape and the dimension of the underlying parameters space. We propose here a new self-organizing neural network, composed of two connections layers. The first one quantizes an input data set, and the second one progressively constructs the projected shape and neighbourhood on an output space of any chosen dimension. We illustrate the algorithm for various applications.

Keywords: Self-Organization, Data Analysis, Non-linear Data Redundancy.

1. Introduction

A problem frequently encountered in many contexts is how to determine a smallest set of independent variables able to describe a redundant multidimensional data set. The space of these underlying parameters is called the variety of the data set.

In margin of many improved techniques (linear regression, canonical analysis, discriminant analysis, model fitting), neural networks have proven certain capabilities in data analysis. Oja proved that a simplified neuron model may act as a principal component analyzer [10]. Cherkassky & Lari-Najafi show how to do non-parametric regression analysis with self-organized neural network [2]. Sanger [12], and Samardzija & Waterland [11] describe neural networks that compute eigenvectors and eigenvalues of data (that is, how to implement a Principal Component Analysis, PCA, in a neuronal way).

The problem of all PCA-like methods (and of the original PCA itself) is that they are only driven by linear dependence between the observed variables. This is a severe limitation since in a number of technical and scientific applications, this dependence is strongly non-linear. In such cases, these methods fail to reduce the number of variables without loosing information. In fact, when statisticians encounter this problem, they generally try to build a model of the observed phenomenon, then they search the proper model parameters with least mean square techniques (fitting). The model is generally not easy (even impossible) to build when the data dimension (the number of observed variables) is high.

Kohonen's Self-Organizing Maps (SOM) [7] have looked very promising, because of their capability to *map* the input data set shape. This property is illustrated in figure 1 (from [1]).

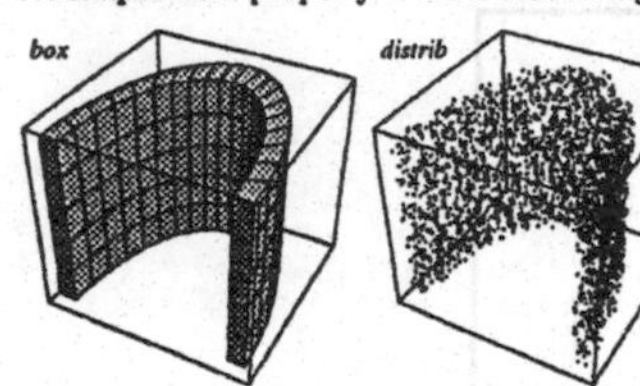

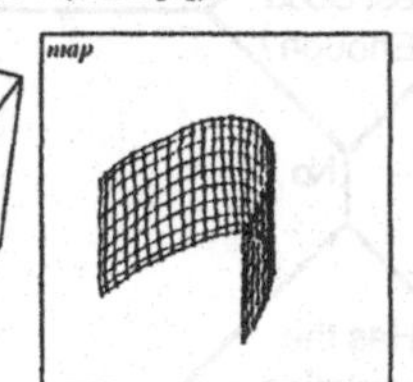

map

Fig. 1. Placement of the Kohonen grid on a "horseshoe" shape input distribution in a 3D space. This is a good example of non-linear distribution where PCA method fails to reduce the dimension.

Although the SOM seems to overcome the problem of non linear redundancy, it presents a major drawback being the fact that the projection geometry structure is fixed *a priori*. This implies one knows the variety of the data set, in order to choose the correct grid dimension. Moreover, non square shapes are not very well mapped with square maps, as shown in figure 2. For example, the mapping shown on figure 1 was feasible thanks to the fact we knew that the variety dimension was 2, and that the unfolded distribution was about 3 times larger than high, so we took a 2D grid network of 10x30 neurons. On the contrary, in figure 2, where we took a "naive" 15x15 network. Another fact is that the projection made by the original SOM is discrete, as long as only the winner element is considered for one data input (a continuous projection is therefore not available in the output space). In order to overcome all these problems (shape of the input distribution, non-linear projection and continuous projection), we propose a self-organized network where neurons "compute" their position in the projection space in an adaptive manner, and where their activity drives an interpolation algorithm.

2. The VQP network

Considering the function of a Kohonen map, one should focus on two main features:

1) The Vector Quantization property, for which the particular structure of the SOM algorithm is interesting because it reduces the number of dead units, that is, the number of units out of the input distribution [6].

2) **The topologically correct projection, where neighbour neurons have neighbour synaptic weights, that is, small distances in the output space (the space of the neuron grid) reflect small distances in the input space.**

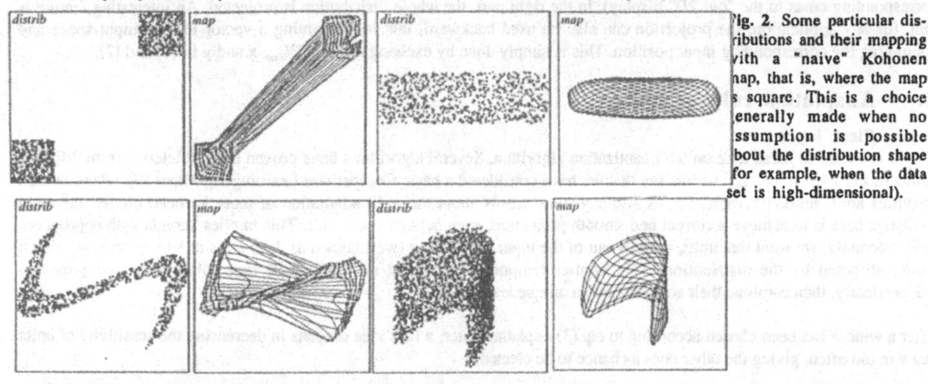

ig. 2. Some particular distributions and their mapping vith a "naive" Kohonen hap, that is, where the map s square. This is a choice enerally made when no ssumption is possible bout the distribution shape for example, when the data set is high-dimensional).

The new self-organized neural network proposed here, Vector Quantization and Projection network, presents these two main features, but does not impose any particular *a priori* defined shape, as a rectangular or hexagonal map. This network is composed of two connection layers as shown in figure 3:

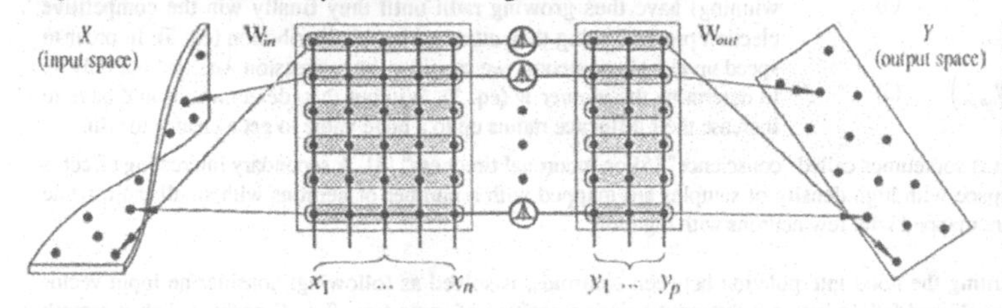

Fig. 3. Structure of the VQP network. The first layer realizes a vector quantization of the input distribution. The second layer reproduces the configuration of the first one in a self-organized way.

The aim of the network is to project the *n*-dimensional input vector x $[x_1, ..., x_n]^T$, to a *p*-dimensional space, the result being the output vector y $[y_1, ..., y_p]^T$, preserving as much as possible the topological information. The definition of output vector y is not supervised; the only constraint attached to it is to try to mimic, at least locally, the topology of the input space. The dimension *p* is the only information fixed (by the user) for y. The first layer realizes a *vector quantization* of the input space, while the second one performs a *projection* towards the output space. As well as in Kohonen maps, where neurons may be considered to point in the input space with their own weight vector, here, each neuron *i* points to $W_{in,i}$ in the input space, and also to a corresponding position $W_{out,i}$ in the output space. When an input vector x is presented, each neuron *i* computes an activity a_i reflecting the proximity of $W_{in,i}$ to the vector x through a radial basis function (RBF). Such a function is for example the Gaussian function of the euclidean distance:

$$a_i = \exp\left(-\frac{1}{\lambda_i^2}\|\mathbf{x} - \mathbf{W}_{in,i}\|^2\right) \qquad (1)$$

where $\|x - W_{in,i}\|$ is the euclidean distance between the input vector *x* and the input weight $W_{in,i}$ of neuron *i*, while λ_i is the influence radius of the neuron *i*.

Then, the projection is computed as the sum of all output vectors, weighted by the neuron activities (this is an interpolation within the output vectors with Gaussian kernels):

$$\mathbf{y} = \frac{\sum_{i=1}^{N} a_i \mathbf{W}_{out,i}}{\sum_{i=1}^{N} a_i} \qquad (2)$$

where *N* is the number of neurons. The adapted values through self-learning are the input matrix W_{in}, the influence radius λ_i of each neuron *i*, and the output matrix W_{out}.

An example of a projection (after adaptation) is given in figure 4:

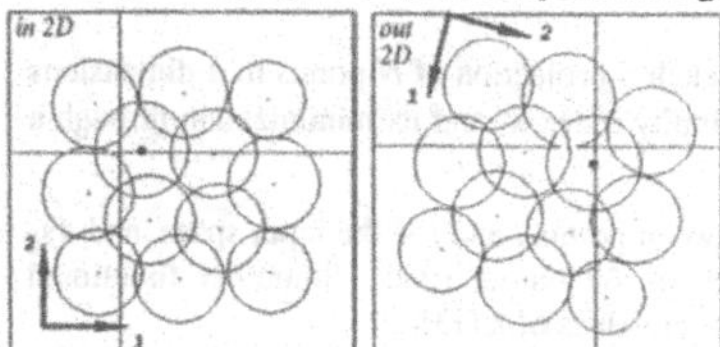

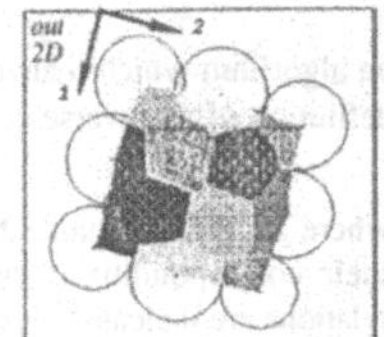

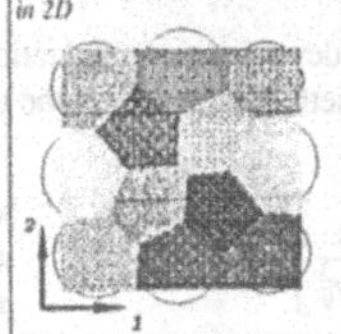

Fig. 4. The network is able to project input vectors into output space. This is shown here after organization on a 2D square distribution for one vector (left side), and for the whole distribution (right side).

The network was trained with a square uniform data set in 2 dimensions, the output space being also in 2 dimensions. In the left part, we show a particular vector presented on input (the cross in the "*in 2D*" display) and its projection (the corresponding cross in the "*out 2D*" display). In the right part, the whole distribution is projected. An interesting feature is that, for any application, the projection can also be used backward, that is, presenting a vector in the output space and computing the corresponding input position. This is simply done by exchanging $\mathbf{W}_{in}$, $\mathbf{W}_{out}$, x and y in (1) and (2).

3. Adaptation rules

3.1. First layer

We do not want to focus here on the quantization algorithm. Several algorithms have proven their efficiency from different points of view [6, 9, 14]. Let us just say that we have considered a basic Competitive Learning algorithm [9], where neuron activities are Gaussian kernels. Let us also say some words about the radii adaptation of these kernels. Firstly, the main challenge here is to achieve a correct and smooth projection, even between centroids. This implies kernels with appropriate radii. Secondly, we want that units, initially out of the input distribution (well known as dead units in VQ theory) are, after a while, attracted by the distribution. This implies temporarily growing radii. We will first solve these two problems independently, then combine their solutions into a unique learning law.

After a winner has been chosen according to eq. (3) explained later, a first idea consists in decreasing the sensitivity of units that win too often, giving the other ones a chance to be elected.

$$w \mid \lambda'_i a_i \le \lambda'_w a_w \forall i \qquad (3)$$

$$\lambda'_i \leftarrow \begin{cases} \lambda'_i + \frac{\Delta\lambda}{N} - \Delta\lambda & i=w \\ \lambda'_i + \frac{\Delta\lambda}{N} & i \ne w \end{cases} \qquad (4)$$

$$\mathbf{W}_{in,w} \leftarrow \mathbf{W}_{in,w} + \alpha\left(\mathbf{x} - \mathbf{W}_{in,w}\right) \qquad (5)$$

This is done through eq. (4), at each iteration, by reducing the influence radius of winning unit by a small value $\Delta\lambda$, while all others are increased by $\Delta\lambda/N$ (thus the mean of all radii is left constant). Dead units (never winning) have thus growing radii until they finally win the competitive election process, being then attracted by the distribution (eq. 5). In order to speed up this phenomenon, we consider the expression $\lambda_i a_i$, and not only a_i, to determine the winner w (eq. 3). Without this, dead units would have to increase their influence radius up to a huge value to get a chance to win.

This idea is similar to what is sometimes called "conscience" [5] or "neuronal tiredness" [4]. A secondary interesting effect is that regions of the input space with high density of samples are mapped with a number of neurons with small radii, while regions with low density are mapped with few neurons with big radii.

The other problem, concerning the good interpolation between centroids, is solved as following: consider an input vector being presented to the network, and falling between the input pointing position of some (say, 3 or 4) neurons. Let us remark that almost only these neurons are responsible for the y position resulting of eq. (2), because other neurons are too far, thus their activity is negligible. Then, a good empirical constraint to obtain smooth interpolation is to impose the sum of these maximal activities to be roughly 1 in average. This has been found a good approximation of the optimal value (that is, the one which gives the minimal distortion) in one dimension.

$$\lambda''_i \leftarrow \lambda''_i\left[1 + a_i \varepsilon\left(1 - \sum_{i=1}^{N} a_i\right)\right] \qquad \text{(4 bis)}$$

Since other neurons activity is near to 0, we may in fact consider the sum of all activities and drive its average value to be approximately 1.

When the sum is too high, the influence regions of neurons pointing near the presented input are supposed to overlap too much. Then, the radii of these units are reduced in function of their contribution in the sum. On the contrary, when the sum is too low, the radii are increased. The weighting by a_i in (4 bis) avoids not concerned kernels to have their radius changed.

We may now consider λ'_i and λ''_i as a unique value λ_i to simplify the algorithm, as long as the two described effect (dead units suppression and good interpolation) are not antagonistic. To summarize, the first layer adaptation is described by two equations (one for the centroids position, and the other one for their radius), whose dynamics is controlled by three parameters, $\Delta\lambda$, ε, and α. In our simulations, we have taken $\alpha = 10^{-1}$, $\Delta\lambda = 10^{-3}$, and $\varepsilon = 10^{-3}$.

3.2. Second layer

The second layer aims at reflecting the topology of the first one. When 2 units are neighbours in the input space, they should be also neighbours in the output space. On the contrary, if two units are far in the input space, they should be far in the output space.

Shepard and Carroll [13] describe an optimization algorithm which realizes such a projection of N points in n dimensions onto p dimensions. The method is based on the definition of an inverse continuity index κ, and its minimization through a gradient descent algorithm.

$$\kappa = \sum_i \sum_{j \ne i} \frac{dx_{ij}^2}{dy_{ij}^4} \bigg/ \left[\sum_i \sum_{j \ne i} \frac{1}{dy_{ij}^2}\right]^2 \qquad (6)$$

where dx_{ij} is the distance between points i and j in the input space, and dy_{ij} their corresponding distance in the output space. Smoother functional relations are indicated by smaller values of κ [13].

From a computational point of view, the minimisation of κ is very heavy. The main problem of this method, for a neuronal implementation, is the globality of κ, whose computing involves all inter-neurons distances. This becomes a real problem when the number of representative points is large (i.e. $N > 100$). For this reason, we decided to provide a simpler and more local algorithm: the ideal expected state where distances are conserved is reached simply by maximizing the weight distances correlation between the input and the output spaces. That is, considering the winner at time t ($w(t)$), and the winner at time t-1 ($w(t$-1)), we compute in (7) their distance in the input space (dx), respectively in output space (dy). Then, the output weights of both neurons $w(t)$ and $w(t$-1) are adapted to obtain a better matching between the distances dx and dy (8 and 9).

$$\Delta x = W_{in,w(t)} - W_{in,w(t-1)} \quad , dx = \|\Delta x\|, \qquad \Delta y = W_{out,w(t)} - W_{out,w(t-1)} \quad , dy = \|\Delta y\| \tag{7}$$

$$\beta = \frac{dx-dy}{dx+dy} \tag{8}$$

$$W_{out,w(t)} \leftarrow W_{out,w(t)} + \frac{\beta}{2}\Delta y \qquad W_{out,w(t-1)} \leftarrow W_{out,w(t-1)} - \frac{\beta}{2}\Delta y \tag{9}$$

In (8), β represents a distance adaptation factor. That is, the output distance should be multiplied by β to become more correct: if $dx < dy$, then $\beta < 1$, and if $dx > dy$, then $\beta > 1$. In (9), $W_{out,w(t)}$ and $W_{out,w(t-1)}$ are moved away or brought closer together in function of β. These both cases are illustrated on figure 5. This very simple law unfolds the projected space and drives our projector network toward a state where distances are approximately preserved. Let us remark that, due to the form of (9), the center of gravity of output weights remains obviously constant. On the other hand, the orientation is undefined.

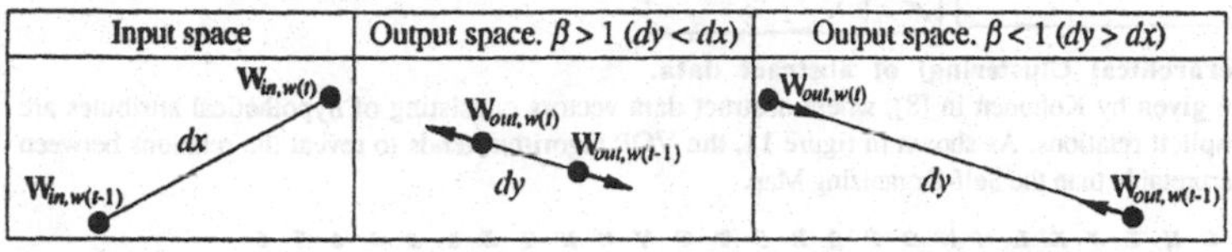

Fig. 5. Adaptation of output weights for winner and last winner in both cases $\beta > 1$ and $\beta < 1$.

In practice, when the dependence in input space is strongly not linear, it is not possible to obtain a correct projection for both short and large distances. In order to unfold the data cloud, it is desirable to give more importance for short distances than for large ones. Then, the projection will be at least locally correct. This is obtained by the adjunction in (8) of a weighting factor monotonically decreasing with the output distance:

$$\beta = \frac{dx-dy}{dx+dy}\left(\frac{1}{1+\left(\frac{dy}{dy_0}\right)^p}\right) \tag{10}$$

The exponent p controls how quickly the weighting factor has to decrease; in our simulations, p was fixed to 5, in order to ensure a good separation between short and long distances in output space. The value dy_0 is a parameter that give an order of value for dy.

In the complete algorithm, the two layers are adapted independently but at the same time. That means the two described phases (quantization, then projection) are in fact realized simultaneously in one. It has been observed that once the network has found a correct projection for a given distribution, it converges more easily to another one than from a random state.

The algorithm is illustrated by various examples in section 4. Because it is impossible, in high dimensions, to visually assess the quality of the topological correspondence between W_{in} and W_{out}, we compute a graphical representation between the input and output space interpoint distances introduced in [13], and defined for Kohonen maps under the name of "δ_i, δ_w relation" in [3]. In this last representation, some couples of units are randomly taken. These couples are represented on a 2D plot by points whose x-coordinates are the output space distances, and the y-coordinates the input space distances. A strictly linear relation between the two distances (i.e. all point are on a line starting from <0,0>) reflects a perfect projection. As shown in [3] for Kohonen maps, folded maps give δ_i, δ_w relation where the input distance no longer grows after a certain output distance. In this paper, the distances are called dx resp. dy, instead of δ_w and δ_i, thus the relation is called here "dy-dx relation" (dy, output distances, are on the horizontal axis).

4. Examples

4.1. 2D → 2D examples

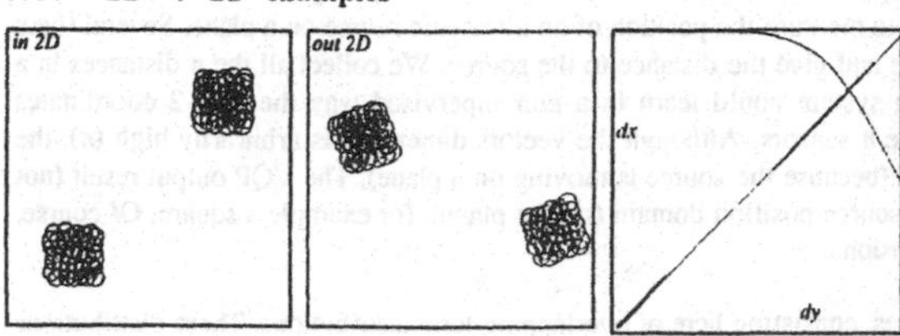

Fig. 6. Here, we show resulting states (after approximately 5000 iterations), given a uniform distribution in two boxes. The dy-dx diagram shows the good matching between output and input distances. As in all following figures, we show on the same graphic the value of dy_0 (the vertical line), and the weighting factor of eq. (10).

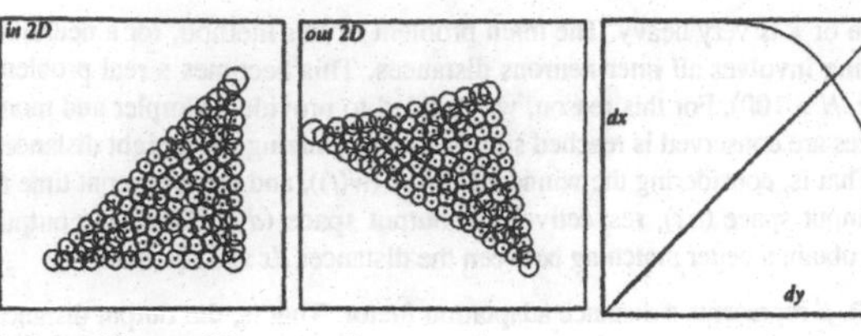

Fig. 7. Same representation as for fig. 6, but with a different input distribution (uniform distribution through a triangular mask).

4.2. 3D → 2D example

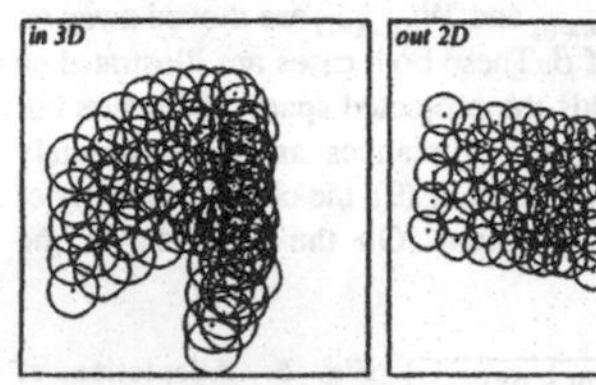

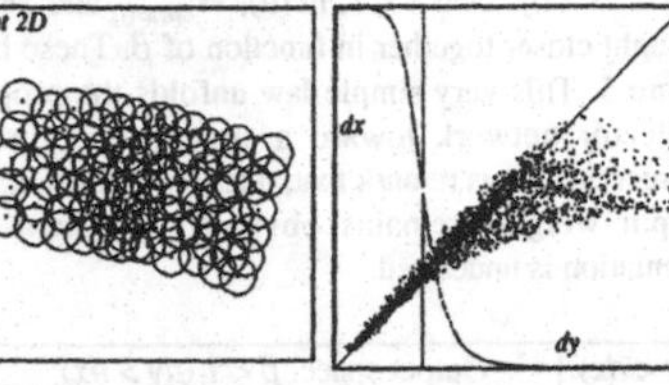

Fig. 8. This is the result obtained on the "horseshoe" shape of figure 1. The network automatically unfolds the distribution, revealing its rectangular shape. The *dy-dx* representation show that the projection is strongly non-linear (*dy* distances after the vertical line are not correlated with corresponding *dx* distances, therefore the dots are not on the line *dy=dx*).

4.3. Taxonomy (Hierarchical Clustering) of abstract data.

We take here the example given by Kohonen in [8], where abstract data vectors consisting of hypothetical attributes are analyzed to reveal their implicit relations. As shown in figure 11, the VQP algorithm tends to reveal the relations between data and is more easily interpretable than the Self-Organizing Map.

	A	B	C	D	E	F	G	H	I	J	K	L	M	N	O	P	Q	R	S	T	U	V	W	X	Y	Z	1	2	3	4	5	6
x_1	1	2	3	4	5	3	3	3	3	3	3	3	3	3	3	3	3	3	3	3	3	3	3	3	3	3	3	3	3	3	3	3
x_2	0	0	0	0	0	1	2	3	4	5	3	3	3	3	3	3	3	3	3	3	3	3	3	3	3	3	3	3	3	3	3	3
x_3	0	0	0	0	0	0	0	0	0	0	1	2	3	4	5	6	7	8	3	3	3	3	6	6	6	6	6	6	6	6	6	6
x_4	0	0	0	0	0	0	0	0	0	0	0	0	0	0	0	0	0	0	1	2	3	4	1	2	3	4	2	2	2	2	2	2
x_5	0	0	0	0	0	0	0	0	0	0	0	0	0	0	0	0	0	0	0	0	0	0	0	0	0	0	1	2	3	4	5	6

Table 1. (Redrawn from Kohonen [8]). Input data matrix consisting of 32 vectors, each one collecting 5 hypothetical attributes. They are labelled from "A" to "6" for later identification.

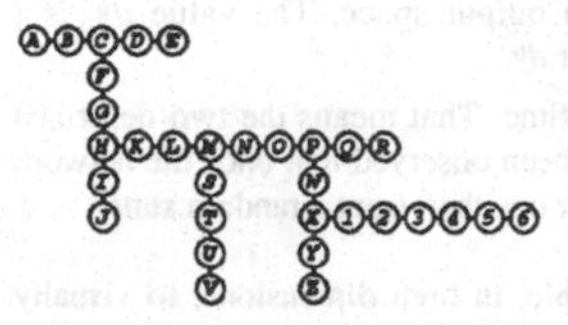

Fig. 9. (Redrawn from Kohonen [8]). Minimal spanning tree (where the most closely similar pairs of items are linked by hand) corresponding to Table 1.

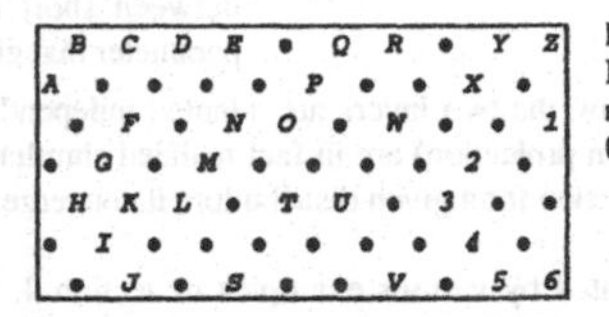

Fig. 10. (Redrawn from Kohonen [8]). Self-organized map of the data matrix of Table 1.

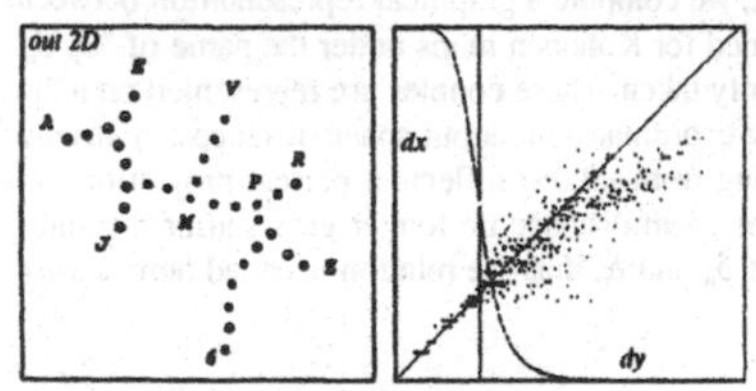

Fig. 11. The projection obtained by the VQP algorithm of the data matrix of Table 1. The output space has been chooosed bidimensional. The *dy-dx* representation show that the projection is strongly non-linear (*dy* distances after the vertical line are not correlated with corresponding *dx* distances, therefore the dots are not on the line *dy=dx*). The VQP algorithm reduces directly the hierarchical clustering in a much more explicit way than the SOM (fig. 10).

4.4 15D → 2D example

We have also used the VQP algorithm to project a particular 15-dimensional distribution onto a plane. The distribution is generated by simulating a system whose purpose is to measure the position of an ultrasonic source on a plane. Several (here $n = 15$) sensors are randomly disposed on the plane and give the distance to the source. We collect all the n distances in a vector, constituting the input distribution. Such a system could learn in a non-supervised way the n to 2 coordinates transform, with no need to know the position of the n sensors. Although the vectors dimension is arbitrarily high (n), the *degree of freedom* of the system is naturally only 2 (because the source is moving on a plane). The VQP output result (not shown here) in two dimensions is the shape of the source position domain (on the plane), for example a square. Of course, such a system could also be implemented in 3 dimensions.

4.4 4D → 2D example

In this example, we show how to separate two classes, consisting here of overlapping toric distributions. These distributions are in 3 dimensions, but we append a class label (0 or 1) magnified by 10 times radius of the torus, in order to achieve the correct separation, giving in fact a 4D input distribution.

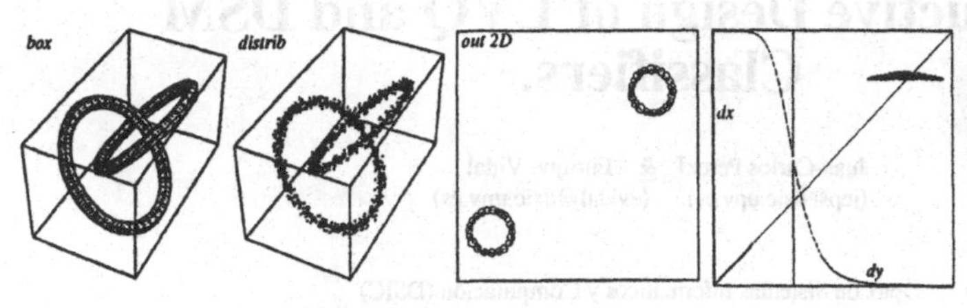

Fig. 11. The two classes are well separated by the algorithm. The *dy-dx* relation is composed of two clouds: one at left of the vertical line ($dy < dy_0$) perfectly linear, corresponding to couples of points of the same class. The other cloud (on the right part of the graphic) is due to couples of points of different classes. For these couples, the projection is strongly non-linear.

5. Conclusion

We have described a new algorithm for data analysis, consisting of a vector quantization of a given distribution (that can be arbitrarily high-dimensional), and its topologically correct projection onto another space of lower dimension. It has be shown some possible applications of this algorithm. We believe that much more applications could be envisaged in various areas:

- Monitoring and control of industrial processes
- Robotics
- Signal (especially speech) processing
- Function approximation

For several applications (especially in robotics), it is interesting to be able to obtain the reverse projection of the data analyzed (that is, to project from the output space to the input one). This can be simply obtained by inverting the respective roles of input and output layer, once the network has been organized.

Further work has to be done to ameliorate the quantization process and to increase the projection fidelity of data itself (not only the centroids). Another hard point is the time wasting for very high dimensional and numerous data vectors (like millions of points in several hundreds dimensions). Evolutive implementation should be envisaged to overcome this problem.

9. References

[1] Blayo F., Demartines P.: *Data analysis: How to compare Kohonen neural networks to other techniques ?* Artificial Neural Networks, International Workshop IWANN'91. A. Prieto Ed. Lecture Notes in Computer Science, Vol. 540, pp. 469-476. Springer-Verlag, 1991.

[2] Cherkassky V., Lari-Najafi H.: *Constrained Topological Mapping for Nonparametric Regression Analysis.* Neural Networks, Vol. 4, pp. 27-40, 1991.

[3] Demartines P.: *Organization measures and representations of the Kohonen maps.* Proc. of the First IFIP Working Group-10.6 Workshop. J. Hérault Ed. Grenoble, 1992.

[4] Demartines P., Blayo F.: *Kohonen Self-Organizing Maps: Is the Normalization Necessary ?* Complex Systems, Vol. 6, No. 2, pp. 105-123, 1992.

[5] DeSieno D.: *Adding a Conscience to Competitive Learning.* Neural Networks, Vol. 1, pp. 117-124, San Diego, 1988.

[6] Hertz J., Krogh A., Palmer R. G.: *Introduction to the Theory of Neural Computation.* Santa Fe Institute Lecture Notes Volume I, Addison-Wesley Publishing Company, 1991.

[7] Kohonen T.: *Self-Organization and Associative Memory (3rd ed.).* Springer-Verlag, Berlin, 1989.

[8] Kohonen T.: *The Self-Organizing Map.* Proc. of the IEEE, Vol. 78, No. 9, pp. 1464-1480, 1990.

[9] Linde Y., Buzo A., Gray R. M.: *An algorithm for vector quantizer design.* IEEE Trans. Commun., Vol. COM-28, pp. 84-95, 1980.

[10] Oja E.: *A Simplified Neuron Model as a Principal Component Analyzer.* IEEE International Conference on Neural Networks, Vol. 15, pp. 267-273, 1982.

[11] Samardzija N., Waterland R. L.: *A neural network for computing eigenvectors and eigenvalues.* Biological Cybernetics 65, pp. 211-214, 1991.

[12] Sanger T. D.: *Optimal Unsupervised Learning in a Single-Layer Linear Feedforward Neural Network.* Neural Networks, Vol. 2, pp. 459-473, 1989.

[13] Shepard R. N., Carroll J. D.: *Parametric Representation of Nonlinear Data Structures.* Proc. of an International Symposium on Multivariate Analysis, pp. 561-592. P. R. Krishnaiah Ed. Academic Press, New-York and London, 1965.

[14] Zeger K., Vaisey J., Gersho A.: *Globally Optimal Vector Quantizer Design by Stochastic Relaxation.* IEEE Transactions on Signal Processing, Vol. 40, No. 2, pp. 310-322, 1992.

Constructive Design of LVQ and DSM Classifiers.

Juan-Carlos Perez[†] & Enrique Vidal
(jcp@dsic.upv.es) (evidal@dsic.upv.es)

Dpto. de Sistemas Informáticos y Computación (DSIC)
Universidad Politécnica de Valencia. 46071 Valencia (Spain)

Abstract

LVQ and DSM are adaptive learning techniques that represent simple and effective approaches to classifier design. They rely on the Nearest Neighbor classification model and can be regarded as adaptive condensing methods, since a reduced number of prototypes is adapted to represent the whole set of samples. If the specified number of prototypes is too small, the classifier will perform poorly. Conversely, if too many prototypes are allocated, some of them will not be useful and can even cause unwanted generalization effects. Two basic strategies can be used to automatically estimate the number prototypes needed: *constructive* or evolutive, in which we start with a small number of prototypes and more are added as necessary, and *destructive* or substractive where a large initial model is prunned until its performance falls below a certain level. In this work, an evolutive method is presented along with a number of experiments showing its usefulness.

1. Introduction.

LVQ (Learning Vector Quantization) methods [Kohonen,88] [Kohonen,90] are simple and effective adaptive learning techniques. They rely on the Nearest Neighbor classification model and are strongly related to the condensing methods [Hart,68], where only a reduced number of prototypes are kept from the whole set of samples. This condensed set of prototypes is then used to classify unknown samples using the Nearest neighbor rule. Originally, the LVQ methods were conceived as an extension to the Self-Organizing Map used to determine the labels of the cells in the map. The vectors associated to the cells are used as initial values for the prototypes (codebook vectors) of the LVQ model.

Given a training set X of N d-dimensional vectors (samples), each labelled with a class identifier, and a set M of P prototypes (also class-labelled d-dimensional vectors), the LVQ methods adaptively modify these prototypes so as they represent as faithfully as possible the class probability distributions in the training set. This modification of prototypes consists of applying a "punishment" when a prototype is near a sample of a different class and a "reward" when it is near a sample of its own class.

There are several variants of this basic idea. The original LVQ1 method updates the prototypes, for every sample $x_i \in X$, $1 \leq i \leq N$, of class b, using the following rule where $m_c(t) \in M$, $1 \leq c \leq P$ is the closest prototype to x_i in the Euclidean metric:

$$m_c(t+1) = m_c(t) + \alpha(t)(x_i - m_c(t)) \quad \text{[reward]} \qquad (1)$$

if $m_c(t)$ belongs to class b, and:

$$m_c(t+1) = m_c(t) - \alpha(t)(x_i - m_c(t)) \quad \text{[punishment]} \qquad (2)$$

if $m_c(t)$ belongs to another class. The training set can be presented repeatedly and, in this case, i cyclically indexes the samples a certain number of times. The learning rate $\alpha(t)$ is a scalar gain $(0 < \alpha(t) < 1)$ monotonically decreasing with t.

This means that the prototypes of each class tend to come near the samples of its same class and far from the samples of the other classes. The prototype density around the Bayesian decision surfaces is therefore reduced and the prototypes tend to reflect the probability density of the training set. Of course, once the model is trained, the classification of new samples (test set) is performed using the nearest neighbor rule (with the Euclidean distance as metric) on the set of prototypes.

If the initialization of the set of prototypes is performed using the values of the cells of the Self-Organizing Map or another clustering method, the initial value of α should be small (less than 0.05). Another option is to determine the initial prototypes randomly. In this case, α should start at a higher value. In any case, the proportion of prototypes from each class should match the *a priori* probabilities of the classes.

In order to follow more closely the philosophy of the Bayes theory, a variant of the previous method is introduced by the same author [Kohonen,88]. The LVQ2 technique, as it is called, defines a "window" between each pair of prototypes of different classes. When the sample x_i, of class b, is considered, *both* the nearest and next-to-nearest prototypes, m_c and $m_{c'}$ respectively, are modified ($m_{c'}$ is rewarded (1) and m_c is punished (2)) only if $c \neq b$, $c' = b$ and the sample lies in a window between them. A point of the space is defined to lay in the window if its distance to m_c is "approximately" equal to its

† The first author is supported by a grant from the spanish *Ministerio De Educación y Ciencia*.

distance to $m_{c'}$. Formally, this condition is stated as: x_i is in the window if $min(d(x_i,m_c)/d(x_i,m_{c'}), d(x_i,m_{c'})/d(x_i,m_c))>s$, where $d(\cdot)$ is the Euclidean distance and $s \in [0,1]$. If s is near one, the window is very "narrow" and it gets "wider" as s decreases. If w is the relative width of the window (the ratio between the distance from m_c to $m_{c'}$ and the width of the window in its narrowest point), then $s=(1-w)/(1+w)$. In [Kohonen,90], a 20% window ($s=0.66$) is cited as a reference value and it is also noted that this algorithm *does not converge* to the best solution, but it "drifts" to a worse configuration if the learning is applied too long (a maximum of 10^4 learning steps with an initial $\alpha=0.02$ are the values mentioned).

Given the problems encountered in the previous method (although good results have been also obtained), a third version of LVQ has been devised [Kohonen,90]. The LVQ3 algorithm differs from the previous version in two points: the first is that m_c and $m_{c'}$ are now the two closest prototypes (not necessarily in this order, in the previous case, m_c was the nearest and $m_{c'}$ the next-to-nearest prototype) and the second is that both prototypes are partially rewarded if they belong to the class of the sample. This means that a coefficient $\varepsilon<1$ multiplies $\alpha(t)$ in the corresponding reward equation (1). Typical values for ε are 0.1 to 0.5, being larger for large windows. This algorithm seems to *converge* to a good solution. Details on the theoretical motivations and basis of these algorithms can be found in [Kohonen,88] and [Kohonen,90].

The DSM (Decision Surface Mapping) classification method was recently introduced by Geva & Sitte [Geva,91] as another member of the LVQ family. In this case, when the sample and the nearest prototype, m_c, belong to different classes ($c\neq b$, i.e. the sample is not well classified) reward and punishment are always applied, as opposed to LVQ2 where no action is taken if $m_{c'}$, the next-to-nearest prototype, is not of the same class as the sample. In DSM $m_{c'}$ is defined differently, as the ***nearest prototype that belongs to the same class as the sample*** ($c'=b$), and reward (eq. (1)) is applied to it when a misclassification occurs. Punishment (eq. (2)) is applied, as always, to the nearest prototype m_c.

The DSM technique is not intended to produce, as all the LVQ variants do, a set of prototypes that reflect the probability distribution of the classes. Instead, modeling the decision boundaries between the classes is the target of the method. The experiments in [Geva,91] show a substantial improvement in the classification accuracy of DSM over all the LVQ variants on many problems. Unfortunately, when a significant overlap between classes is found in the training set, the decision boundaries are not well defined and the performance of the method can drop substantially.

2. Incremental Prototype Generation.

All the adaptive learning methods reported in the previous section work by adjusting the position of a *given* number of prototypes to obtain an accurate and efficient classifier. How many of these prototypes are necessary is a question difficult to answer in most cases. If the number of prototypes specified is less than the optimal, the classifier will perform poorly. Conversely, if too many prototypes are allocated, some of them will not be useful and can even cause unwanted generalization effects. It is usually accepted that, for the same training set performance, a smaller model (in this case, with less prototypes) provides better generalization. Therefore, finding the minimum classifier is an important issue.

Two basic strategies can be used to automatically estimate the amount of resources (prototypes, parameters, weights, etc.) needed to perform a concrete classification task: *constructive* (or evolutive) and *destructive* (or substractive). In the first case, the model starts with a small number of resources and more are added as needed. In the second, an initially (too) large model is prunned until its performance falls below a certain level.

The constructive approach has been traditionally considered more natural and computationally efficient. Substractive techniques are generally slower because they start with a large model. No theoretical evidence seems to exist in favor of either of these approaches and, in fact, the work carried out in dynamically built neural network models [Thodberg,91] [Weigend,91] [Fahlman,90] suggests similar features in both cases..

In [Poirier,91], a modified version of LVQ2 called DVQ (Dynamic Vector Quantization) is presented. In that work, a constructive strategy is employed: at the initialization, each class is assigned one prototype (the mean vector of the class) and more prototypes are added as necessary. The rule employed to generate these new prototypes is applied for every new sample as follows: if the closest prototype to the sample is in a different class and the closest reference in the class of the sample is farther than a given value, then the sample is taken as a new prototype, else the LVQ rule is used.

The method presented here also starts with one prototype for each class and proceeds by adding a new prototype whenever it is needed. However, the decision to add a new prototype is not taken for every sample, but a number of complete presentations of the training set is performed before. This means that the process is more exhaustive and a new prototype is only added if it is completely necessary. The results confirm that the number of prototypes obtained in this way is much smaller than with DVQ. Clearly, the reduction in the time and space resources used in the final classifier often justifies a small increase in the training effort.

As has already been said, the initial location of the prototypes is very important for a good performance of LVQ methods. Therefore, an adequate initialization has to be provided. In the proposed algorithm, the initial prototype for each class is carefully chosen as the most centered sample in the class. Since in a vector space the nearest vector to the mean is also the most centered vector, this process can be performed efficiently (in a time linear with the number of samples). The most centered sample has been preferred over the mean vector to avoid problems with, for example, "concentric" (non-overlapping) class distributions where an outside class is distributed around an inside class. In this case, the mean vector of the outside class lays pathologically in the center of the inside class.

With this initial distribution of prototypes, one per class, a complete learning process using either LVQ method or DSM is performed. At this stage, the prototypes tend to reflect as faithfully as possible the class probability distributions (for LVQ) or the decision surface of the classes (in DSM). If the required performance has been achieved, the process can be considered complete. On the other hand, if the training set error is still too high, new prototypes have to be added to the set. In fact, *each class is independently considered* and a new prototype is added to it if the number of misclassified samples in the class is greater than a pre-specified limit.

As in the initial stage, the location of the new prototypes is again a critical issue. In the proposed method, the new prototype for each class is the sample, among those classified incorrectly in the last stage, which is further away from any prototype of the class. For class *b*, the new prototype *q* can be found using the following equation:

$$q = \underset{\substack{x \in X \\ class(x)=b}}{arg\,max} \left(\min_{\substack{m \in M \\ class(m)=b}} d(x,m) \right) \quad (3)$$

where $d(\cdot)$ is the Euclidean distance.

Every sample used to generate a new prototype is conveniently marked so as to be not considered again as a candidate to generate new prototypes. This is aimed at avoiding that an outlier far from the final placement of the correctly trained prototypes can repeatedly bias the generation of the new ones.

It is possible that a certain prototype becomes unused when it is "pushed away" from samples of its own class (for example, by a larger number of nearby samples from other classes) or when another prototype of the same class assumes its function (for example, coming from a better location). In these cases, the prototype can be easily identified given that it is no longer the nearest prototype to any sample of its class. In the method presented here, this condition is checked after every run of the basic learning phase.

Given that deletion of prototypes is allowed, it can happen that all the samples are marked as "already used to generate a new prototype" and that new prototypes are still needed. In this case, all the marks can be deleted and any sample can be used again to generate new prototypes. This only happens when the maximum error specified is very strict and the class overlap is important. In fact, this can lead to a non-convergent process if prototypes are continuously created and deleted. To avoid this problem, when the number of deleted prototypes gets too high (for example, when it is larger than the number of valid prototypes), prototype deletion is not allowed any longer. In the case of DSM, punishment is also supressed to avoid the effect of the prototypes being "pushed away" from the samples of their own class. Of course, this mechanism importantly changes the nature of the basic DSM procedure and should be taken as an extreme action, useful when this procedure is not powerful enough to solve the problem under the constraints specified. A summary of the whole constructive scheme is algorithmically presented hereafter:

1- Assign one initial prototype to each class. In this work, the most centered sample in each class has been used as the initial prototype. Label this sample as "already used to generate a prototype".

2- Repeat until all the available prototypes have been used or the training-set error is sufficiently small:

2.1- If it is the first iteration or at least one new prototype has been added in the last step, then

2.1.1- Perform again a complete learning using LVQx or DSM.

2.1.2- Delete any prototype which is not directly responsible of the good classification of at least one sample. If the number of deleted prototypes is larger than the current number of valid prototypes, avoid applying punishment from now on.

2.2- Compute the training-set error of each class. Add a prototype to every class that exceeds the maximum error. Among the samples of the class which are not well classified by the present model, the sample which is further away from any prototype of the class (eq. 3) and has still not been used to generate a prototype, is chosen as the new prototype. Label the sample as "already used to generate a prototype". (If all the samples have been used, delete all the labels to allow the samples to be selected again)

3- Delete any prototype which is not directly responsible of the good classification of at least one sample.

3. Experiments.

Three experiments on synthetic data and one on an isolated-words speech database are presented. The basic techniques used were LVQ1 and DSM. Although it is likely that the more recent versions of LVQ (LVQ2 & 3) produce better results for some of these problems, the results obtained with LVQ1 are probably fairly representative.

3.1. Experiments on synthetic data.

In each experiment with synthetic data, two sets, each containing 2000 samples, were used; one as the training set and the other one as the test set. In the conventional DSM and LVQ experiments, three random initializations have been performed and its results averaged in each case. To test the constructive scheme (as it does not depend on a random initialization), each average comes from three experiments using a different number of training-set presentations (50, 75 and 100) in each basic LVQ or DSM learning phase. The number of presentations used in the conventional (non-constructive) LVQ and DSM tests is always 100. The initial value of the learning parameter was $\alpha(0)=0.01$ in all the cases. The maximum number of prototypes allowed in the experiments using the constructive procedure was 25 per class. When this number of prototypes per class was not enough to meet the error bound specified, that fact has been represented as ">25" in the tables.

The first problem studied was defined by Hart [Hart,68] and has been used in [Geva,91] to test the performance of the DSM technique. It is a nonlinearly separable problem with no overlap between classes. As has been already mentioned before, the results given in [Geva,91] show a dramatic performance improvement of DSM over LVQ1. However, it is suggested in that work that, for problems with an important overlap between classes (fuzzy boundaries), DSM exhibits instabilities and the performance of LVQ can be superior. This has been also reported in [Ferri,92] and the experiments shown below seem to confirm this effect also for the constructive scheme presented here.

Table 1: Hart's non-linearly separable problem using the *constructive method*. The average number of prototypes per class needed and the mean training-set/test-set errors obtained are represented for different values of the maximum per-class resubstitution error allowed.

Max. training set error	30%	20%	10%	5%
LVQ1	1 (20.4%/19.5%)	2.5 (18%/17.4%)	5 (3.8%/3.5%)	6 (3.4%/3.43%)
DSM	1 (23.1%/22.5%)	2 (16%/16%)	6.5 (7.3%/7.3%)	9.6 (4.4.%/4.3%)

Table 2: Hart's non-linearly separable problem using the *conventional method*. The training-set and test-set errors obtained are represented for different numbers of prototypes per class. A total of 100 presentations of the training set were used in each experiment.

Prtypes/class	1	2	4	8	16	32	64
LVQ1	20.5%/19.6%	19.8% / 19%	17.1%/17.1%	6.7% / 6.9%	3.9% / 4.4%	3% / 3.3%	1.9% / 2.1%
DSM	25.8%/25.2%	22.5%/22.2%	14.7%/14.6%	6.6% / 6.8%	2.5% / 2.9%	0.8% / 1.2%	0.8%/1.2%

In Table 1, the average number of prototypes per class needed to achieve different training-set errors are shown. It should be noted the effect of the construction method on the performance of the system: the results with 5 and 6 prototypes for LVQ *using the evolutive scheme* is only achieved using 16 and even 32 prototypes *with the conventional LVQ* (Table 2). A similar effect is found using DSM. The performance of DSM is similar to LVQ in the conventional as well as in the new case (given the small number of repetitions of the experiment, the figures are worse in some cases and better in others). It is likely that, using better tuned values for $\alpha(0)$ and for the number of training-set presentations, the results of DSM would outperform the ones of LVQ for this zero-overlapping prolem, as in [Geva,91]. However, the interest of the experiments shown resides on the comparison between the classical and the constructive methods in the same conditions for LVQ1 and for DSM.

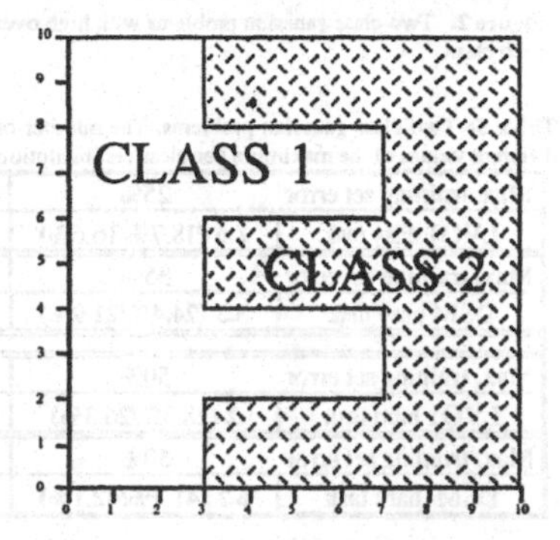

Figure 1. Two-class classification problem introduced by Hart. Although the problem is not linearly separable, the class boundaries are well defined (there is no overlap between classes).

The second problem is introduced in [Ferri,92] and consists of two interlaced spirals ($x_1(t)=10t\,sin(t);\ y_1(t)=10cos(t)$ and $x_2(t)=-y_1(t);\ y_2(t)=-x_1(t)$, for $t\in[0,10]$) with additive bivariate gaussian noise ($\mu_1(t)=(x_1(t),y_1(t))$, $\mu_2(t)=(x_2(t),y_2(t))$ and diagonal covariance matrices with $\sigma_{11}^2(t)=\sigma_{22}^2(t)=t+3.5$ for both classes). The two-spirals problem, without added noise, was introduced in [Lang,88] and is already a classical classification task in the Neural Networks literature.

Table 3: Interlaced spirals with additive noise using the *constructive method*. The number of prototypes needed and the training-set/test-set error obtained are represented for different values of the maximum per-class resubstitution error allowed.

Max. training set error	50%	40%	30%	20%
LVQ1	1 (36.8%/38.8%)	2 (36.7%/38.2%)	5 (25.1%/26.7%)	7.8 (17.8%/19.5%)
DSM	4.5 (41.1%/43.4%)	6.1 (34.9%/36.7%)	12.1 (26.5%/28.7%)	>25

Table 4: Interlaced spirals with additive noise using the *conventional method*. The training-set and test-set errors obtained are represented for different numbers of prototypes per class.

Prtypes/class	1	2	4	8	16	32	64
LVQ1	36.9%/38.8%	37.6%/39.9%	33.7%/36%	17.8%/20.2%	13.3%/15.4%	12.5%/14.6%	12.5%/16.3%
DSM	50.5%/51.5%	48.6%/48.8%	43.8%/44.1%	35.5% / 35.8%	25.8% / 27.6%	18.5% / 21.8%	15% / 19.6%

In Table 3, the mean number of prototypes per class needed for this problem are represented. In this case, the performance of DSM is clearly worse than the one of LVQ in all cases given the important overlap between classes. The advantage of the initialization scheme used in the constructive method is here less aparent because of the "spherical" shape of the class distributions. Therefore, the results of both methods (constructive (Table3) and conventional (Table 4)) are comparable.

The third classification problem, introduced by Kohonen [Kohonen,88], consists of two different two-class tasks. In the first one ("easy task"), each class is a two bi-variate normal distribution: class 1 with $\mu_1=(0,0)$ and diagonal covariance matrix with $\sigma^2_{11}=\sigma^2_{22}=1$; class 2 with $\mu_2=(2.32,0)$ and diagonal covariance matrix with $\sigma^2_{11}=\sigma^2_{22}=4$ (see Figure 2). The second task ("hard task") is similar to the first one, but in this case the mean of class 2 is also zero ($\mu_2=(0,0)$) which means that the smaller gaussian lays completely inside the larger one. Both tasks represent classification problems with overlapping classes and, in fact, the second one can be considered an extreme case.

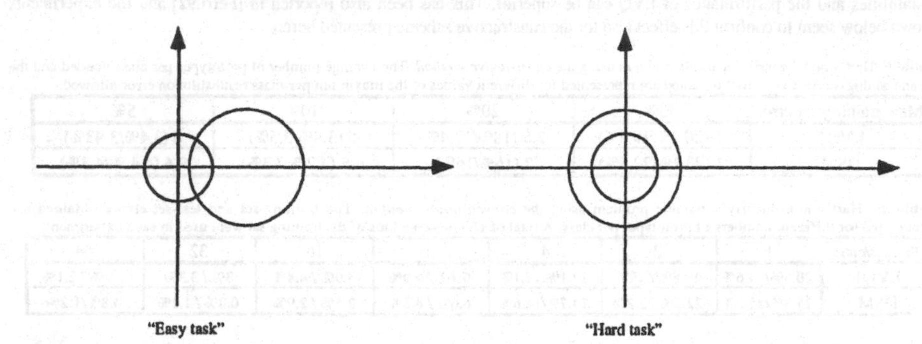

Figure 2. Two-class gaussian problems with high overlap between classes. The circles indicate areas with the same probability mass for each class.

Table 5: Two-class gaussian problems. The number of prototypes needed and the training-set/test-set error obtained are represented for different values of the maximum per-class resubstitution error allowed.

Max. training set error	25%	23%	21%	19%
LVQ1 easy task	2.6 (18.7%/16.6%)	2.5 (18.7%/16.5%)	3.2 (18.8%/16.2%)	4 (18.1%/16%)
Max. training set error	35%	30%	28%	25%
DSM easy task	3.3 (24.4%/21.9%)	5 (25.4%/23.5%)	5.1 (25.4%/23.9%)	>25
Max. training set error	50%	40%	30%	25%
LVQ1 hard task	2 (28.3%/26.3%)	2 (28.3%/26.3%)	4.5 (26.5%/26%)	>25
Max. training set error	50%	40%	30%	25%
DSM hard task	6.2 (41.7%/42.1%)	9.3 (36.6%/36.9%)	>25	>25

Table 6: Two-class gaussian problems. The training-set and test-set error obtained are represented for different numbers of prototypes.

Prtypes/class	1	2	4	8	16	32
LVQ1 easy	22%/20.9%	20.3%/18.8%	20%/18.2%	21.7%/21.1%	19.6%/18.2%	20.5%/21%
DSM easy	44.4%/45%	40.4%/40.8%	40%/40.1%	37.1%/35.5%	33.7%/32.7%	28.5%/30.4%
LVQ1 hard	47.6%/47.1%	34.3%/34.8%	27.1%/27.6%	26.6%/27.3%	28%/30%	27.8%/31.6%
DSM hard	54.8%/54.2%	49.1%/48.9%	46.3%/45.9%	41.2%/40.4%	39.3%/39.4%	38.5%/40.4%

In Tables 5 and 6, similar conclusions as in the previous case can be extracted. These tasks involve an important overlap between clases and the superiority of LVQ1 over DSM is clear here. The performance benefit of the constructive method is also apparent.

In these experiments, the fact that less prototypes produce better generalization is evidenced. In the hard task, it is even apparent the fact that the test-set error increases when the number of prototypes increases from 16 to 32 (for LVQ as well as for DSM). Of course, the training-set error always decreases when more prototypes are allowed. The fact that some error values are even better for the test-set than for the training-set is not surprising given the stochastic nature of the data which is generated following exactly the same probability distributions.

3.2. Experiments on real data.

In order to test the method on real data, a difficult classification problem on isolated word recognition has been used: the Spanish E-Set. It is a nine-class problem where the classes are the most confusable subset from the Spanish alphabet: */efe/, /ele/, /eλe/, /eme/, /ene/, /eɲe/, /ere/, /er̄e/, /ese/*. The corpus consisted of 900 utterances (10 repetitions of each word uttered in a laboratory environment by 10 different speakers, 5 male and 5 female). The signal was sampled at 8533 Hz and

transformed into sequences of *11*-dimensional parameter vectors with 10 Cepstral coefficients plus the Energy at a rate of 133.3 vectors per second. To obtain a fixed-size vector for every sample (word), Trace Segmentation has been used. This technique accomplishes a uniform sampling on the trajectory of an utterance in the parametric space of representation (in this case, Cepstral coefficients). A detailed description of this procedure and of the E-Set corpus can be found in [Castro,91]. The dimension of the vectors obtained is 220 (20 frames × [10 cepstrals+1 energy]).

Five experiments have been performed, each with 8 training speakers and 2 test speakers (one male and one female). The results shown in Table 7 are the mean of these speaker-independent experiments. The results obtained on the same data with other classification techniques such as K-Nearest Neighbors and Back-Propagation are also shown.

Table 7: Results of several classification techniques on the Spanish E-Set speech corpus.

DSM (conventional)	20.3% (Set to 2 prototypes per class. $\alpha(0)=0.2$. Average of 3 random initializations)							
DSM (constructive)	18.2% (Result: 1 prototype per class. $\alpha(0)=0.2$. Average of 3 experiments with 100,200 and 300 presentations, respectively, of the training set)							
K-NN	K =	1	4	8	16	32	36	40
(Euclidean distance)		30.9%	28.7%	27.2%	24.4%	21.6%	20.1%	22.2%
Back Propagation	18.0% (Two-layer perceptron with 220-20-9 units, see [Castro,91] for details)							

The best results are obtained for Back-propagation and the technique proposed here. It is noticeable that, in the last case, *no parameter tuning at all* was attempted; using other values for $\alpha(0)$ and a different number of presentations could lead to better results. Also, the LVQ1 technique could be used instead of DSM. The Back-propagation result, on the other hand, is the best one among a number of experiments with different learning rates and network topologies [Castro,91].

Aside from this, the most remarkable fact that can be learned from this experiment is that the training set can be classified at 100% accuracy using only *one prototype per class*. This is surprising if we consider the difficulty of the problem. The conclusion that can be drawn is that there is no overlap between classes (in the training set) but the number of samples used is clearly insufficient to cope with the complexity of the task (the test set is not adequately represented).

4. Conclusion.

A constructive scheme for the automatic design of LVQ and DSM classifiers have been presented. The method is intended to produce models with a small number of prototypes. The inclusion of a new prototype is allowed only when, after a complete LVQ or DSM training process, the maximum specified error is not met. The generalization obtained with a classifier is often better if the number of prototypes is kept to a minimum. In embedded applications, where the quality of the classifier is more important than the learning time, a minimal classifier is important.

The scheme proposed can be implemented as a layer on top of a conventional LVQ or DSM classifier with only minor modifications. The experiments presented show that the constructive design leads to classifiers that, for the same number of prototypes, offer better performance than randomly-initialized models. Also, the results suggest that this approach may be useful for many kinds of problems without requiring excessive parameter tuning.

5. References.

Castro, M.J. ; Casacuberta, F. ; Puchol, C. (1991) "Trace Segmentation with Artificial Neural Networks", *Universidad Politécnica de Valencia.* Technical Report DSIC-II/13/91.

Duda, R.D. ; Hart P.E. (1973) "Pattern Classification and Scene Analysis", New York: *Willey.*

Fahlman, S.E. ; Lebiere, C. (1990). "The Cascade Correlation Learning Architecture", *Tech. Report CMU-CS-90-100.* Carnegie Mellon university.

Ferri, F. ; Vidal, E. (1992). "Small Sample Size Effects in the Use of Editing Techniques", *International Conference on Pattern Recognition (ICPR 92).* The Hague. pp. 607-610.

Geva, S. ; Sitte, J. (1991). "Adaptive Nearest Neighbor Pattern Classification", *IEEE Trans on Neural Networks, Vol.2, No.2,* pp.318-322.

Hart, P.E. (1968). "The Condensed Nearest Neighbor Rule", *IEEE Trans. on Information Theory, Vol.125,* pp.515-516.

Kohonen, T. (1988). "Statistical Pattern Recognition with Neural Networks: Benchmarking Studies.", *IJCNN-88 San Diego.* Vol. I, pp. 61-68.

Kohonen, T. (1990). "The Self-Organizing Map", *Proc. of the IEEE* Vol. 78, No. 9. pp. 1464-1480.

Lang, K.J. ; Witbrock, M.J. (1988). "Learning to Tell Two Spirals Apart", *Proceedings of the 1988 Conectionist Models Summer School,* Morgan Kaufmann.

Poirier, F. (1991). "Improving the Training and Testing Speed and the ability of generalization in Learning Vector Quantization: DVQ", *ICASSP-91,* Toronto. pp. 649-653.

Thodberg, H.H. (1991). "Improving Generalization of Neural Networks through prunning", *International Journal of Neural Systems.* Vol. 1, pp. 317-326.

Weigend, A.S. ; Rumelhart, D.E. ; Huberman, B.A. (1991). "Generalization by Weight Elimination with Application to Forecasting", *NIPS-3,* pp. 877-882.

LINEAR VECTOR CLASSIFICATION: AN IMPROVEMENT ON LVQ ALGORITHMS TO CREATE CLASSES OF PATTERNS

Michel Verleysen, Philippe Thissen, Jean-Didier Legat

Université Catholique de Louvain
Microelectronics Laboratory
3, pl. du Levant
B-1348 Louvain-la-Neuve
Belgium

Learning Vector Quantization algorithms (LVQ1 and LVQ2), proposed by Kohonen, are widely used for the quantization and the classification of vectors into clusters. These algorithms quantize each class of vectors in the space into a defined number of 'prototypes'. Despite an efficient quantization of the stimuli space, these algorithms are not well adapted to classification tasks where the distribution of prototypes inside a single class is not important, provided that the boundaries between classes are adequately approximated through the prototypes. We propose here an adaptation of the LVQ1 algorithm where the resulting prototypes will approximate the boundaries between classes; by this way, stimuli located as well near the border as in the center of a class will be correctly classified, even if they are not adequately quantified in the sense of 'Vector Quantization'.

1. Introduction

Learning Vector Quantization algorithms [1], are used in data processing to quantize and to classify a K-dimensional input space represented by N vectors (or 'stimuli'), into a fixed number P of prototypes, approximating the probability density of the N stimuli, and dissociating the different classes to which the stimuli (and the prototypes) belong. LVQ are iterative algorithms where all stimuli are sequentially (or randomly) presented; the basic principle is to raise the number of prototypes of the right class in regions where the stimuli density is important, and to move away prototypes leading to misclassification. The difference between the LVQ1 and LVQ2 algorithms resides in the criterion to satisfy in order to move prototypes, and in the way of how the final position of the boundaries between classes are determined.

2. Learning Vector Quantization

2.1. LVQ1

The LVQ1 algorithm can be described as follows.

Before the first iteration, P K-dimensional prototypes p(i) (i=1...P) are randomly initialized. If a priori limits of the set of input vectors in the space are known, the prototypes will be chosen inside these limits, in order to accelerate the convergence of the algorithm. One possibility is to initialize the prototypes to any P of the N input vectors. At each iteration, a K-dimensional input vector x(j) (j=1...N) is compared to all prototypes. We then select the prototype p(a) for which the standard Euclidean distance between p(a) and x(j) is minimum

$$\mathrm{dist}\big(p(a),x(j)\big) \le \mathrm{dist}\big(p(k),x(j)\big) \quad \forall k \in \{1 \ldots P\} \setminus \{a\} \tag{1}$$

($1 \le a \le P$). If vectors x(j) and p(a) belong to the same class, p(a) is moved in the direction of x(j) following equation (2).

$$p(a) = p(a) + \alpha\,(x(j) - p(a)) \quad (2)$$

where α is the adaptation factor (0<α<1). If the two vectors belong to different classes, p(a) is moved in the opposite direction of x(j):

$$p(a) = p(a) - \alpha\,(x(j) - p(a)) \quad (3)$$

The adaptation factor α must decrease with time to obtain a good convergence of the algorithm [2]. Usually, the same value of the adaptation factor α is kept for a whole 'epoch' (an epoch consists in presenting once the whole set of input patterns), and is decreased before the next one. Figures 1a) and 1b) respectively show the distribution of prototypes for an uniform two-class distribution (prototypes are marked '+' and 'x' for the two classes, and input vectors are marked '.'), and for a single class with 2-nd order decreasing (in the two directions) probability density of stimuli.

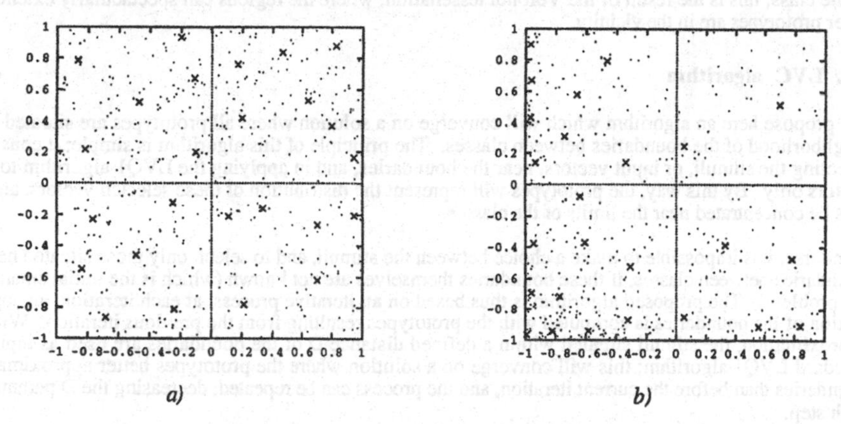

fig.1 examples of LVQ convergence on 2-D vectors (100 iterations)

2.2. LVQ2

LVQ2 is an extension of the LVQ1 rule, where several conditions must be verified before moving prototypes [3]. In this rule, the two prototypes p(a) and p(b) closest to input vector x(j) are selected according to equation (1.1). The three conditions to satisfy in LVQ2 algorithm are then:

- **stimulus x(j) and nearest prototype p(a) must belong to different classes**
- **stimulus x(j) and second nearest prototype p(b) must belong to the same class**
- **stimulus x(j) must be inside a predefined symmetrical window centered on the midpoint of segment p(a)-p(b)**

If these conditions are satisfied, prototypes p(a) and p(b) are moved according to:

$$p(a) = p(a) - \alpha\,(x(j) - p(a)) \quad (4)$$

$$p(b) = p(b) + \alpha\,(x(j) - p(b)) \quad (5)$$

The three mentioned conditions for moving the prototypes guarantee a better approximation of the Bayes decision boundary between prototypes p(a) and p(b); a justification of this can be found in [1]. This algorithm suffers from a huge amount of computations necessary to reach a quasi-equilibrium point.

3. Learning Vector Classification (LVC)

The above described LVQ algorithms are efficient to quantize the space into a defined number of regions. The influence regions defined by the prototypes are the result of a "Voronoï tessellation": the influence region of one defined prototype is the set of all points in the input space which are closer from this prototype than from all others. If the probability density of input vectors is constant, the influence regions will have similar volumes. On the other hand, if input vectors are not equiprobable, the volumes of the influence regions will be modulated in order to have the same probability to be selected at each presentation of a stimulus.

Nevertheless, if the distribution of prototypes in the whole input space is adequate for vector quantization, it would be more efficient for vector classification to move all prototypes near the boundaries between classes instead to have most of them inside the classes. Indeed, if the vectors near the boundaries between classes are correctly classified, the influence regions of the prototypes near these boundaries will automatically extend inside the classes, leading to a correct classification of vectors further from the limits of the class; this is the result of the Voronoï tessellation, where the regions can spectacularly extend if no other prototypes are in the vicinity.

3.1. LVC algorithm

We propose here an algorithm which will converge on a solution where all prototypes are situated in the neighborhood of the boundaries between classes. The principle of this algorithm is simple: it consists in selecting the stimuli, or input vectors, near the boundaries, and in applying the LVQ1 algorithm to these vectors only. By this way, the prototypes will represent the distribution of these selected vectors, and will thus be concentrated near the limits of the classes.

Of course, it is impossible to make a choice between the stimuli, and to select only those situated near the boundaries between classes, if these boundaries themselves are not known (which is the initial situation of the problem). The proposed algorithm is thus based on an iterative process: at each iteration, an approximation of the boundaries is computed with the prototypes resulting from the previous iteration. With this approximation, the stimuli situated within a defined distance D of the boundaries are used as input of a standard LVQ1 algorithm; this will converge on a solution where the prototypes better approximate the boundaries than before the current iteration, and the process can be repeated, decreasing the D parameter at each step.

To reduce the amount of computations, a stimulus x will be deemed to be within a distance D of the boundaries between classes if the Euclidean distance between x and the closest prototype from a different class is less than D. The region where stimuli are considered is thus approximated by the union of all hyperspheres centered on the prototypes and with radii equal to D, instead of being an hypersurface situated at a distance D of the boundary determined by the Voronoï tessellation. Since only an approximation of the limits between classes is needed at each iteration, this way to estimate the boundaries is valid and will not decrease the performances of the algorithm.

When decreasing parameter D between two iterations, it may happen than some prototypes fall outside the region where input patterns are considered, i.e. the above-mentioned estimated volume (union of hyperspheres centered on prototypes from other classes). In this case, only the prototypes located inside the region are kept for the next iteration. To keep the same number of prototypes for each iteration, the remaining ones are again randomly chosen in the list of input vectors, as at the beginning of the LVQ algorithm, but of course inside the region where stimuli will be considered at the next iteration. All prototypes could also be randomly chosen inside this region at each iteration; keeping the results of the previous step however turned out to be efficient for a reduced amount of computations.

More precisely, with the same notations as in section 2, the algorithm can be described as follows:

1. Initialize P prototypes to any of the N input vectors (stimuli). For example, choose

$$p(i) = x(i) \qquad \langle 1 \leq i \leq P \rangle \tag{6}$$

2. Choose initial distance $D=D_{init}$ (input vectors are only considered within a distance D of any prototype from another class). Since it is important to consider all prototypes during the first iteration, distance D_{init} can be chosen as the maximum distance between patterns in the input distribution.

If Nsteps is the number of epochs (see section 2.1), the following is repeated Nsteps times:

3. Execute LVQ1 algorithm on the stimuli inside the considered region.
4. Decrease parameter D linearly:

$$D = D - \frac{D_{final} - D_{init}}{Nsteps} \tag{7}$$

where D_{final} is the value of D after completion of the algorithm (D_{final} depends on the precision required on the boundaries).

5. Eliminate prototypes falling outside the region determined by the actual set of prototypes and the new distance D.
6. Replace missing prototypes by stimuli falling inside the region.

Step 3, the LVQ1 algorithm executed at each iteration, is in the reality not critical; keeping the prototypes found at the previous iteration, as explained above, allows in the reality to execute only one epoch of the LVQ1 algorithm (one presentation of the set of input patterns) at each cycle. By this way, the execution of the LVQ1 algorithm and the diminution of the regions where stimuli are considered are simultaneously performed, and the total number of cycles is similar to what would be necessary in a standard LVQ1 process; verifying if the stimuli and/or prototypes fall inside the regions determined by the estimated boundaries and by distance D leads however to supplementary computations.

3.2. Simulation results

The LVQ1 and LVC algorithms have been compared on the same sets of data, and the computations have been carried out during the same number of epochs (see last remark of section 3.1). Figures 2 a) and 2 b) respectively show the results of the LVQ1 and LVC algorithms after 20 epochs on uniform distributions. 1000 stimuli were considered, and 50 prototypes; stimuli above the sine curve are deemed to belong to class 1, and those below the curve to class 2.

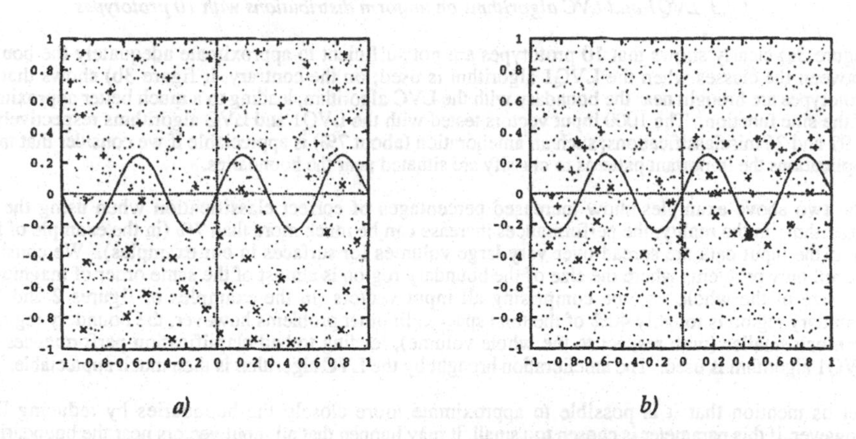

fig.2 LVQ1 and LVC algorithms on uniform distributions with 50 prototypes

Figure 2b) shows a sensible amelioration in the classification of input vectors. Indeed, it can be seen in figure 2a) that less than 20 prototypes are used to determine the approximation of the sine curve; the other prototypes are further from the boundary, and do not contribute to this approximation. In other words, if the Voronoï tessellation of the prototypes in figure 2a) is drawn, and if only the lines separating different classes are kept, more than 30 prototypes are useless. On the other hand, in figure 2b), almost all prototypes contribute to the boundary; as a result, a better approximation of the sine curve is achieved.

The two algorithms have been tested for the classification of the 1000 input vectors; 28 misclassifications came out of algorithm LVQ1, while only 17 were found with the LVC. This is however a weak result, since the approximation of the boundary between classes was already quite good with the LVQ1 algorithm; it can indeed be intuitively viewed that less than 20 prototypes are necessary to obtain such a pretty good approximation, which leaves the remaining prototypes free to scan the whole input space.

However, if the number of prototypes is not sufficient to correctly approximate the boundaries between classes in the case of the LVQ1 algorithm, i.e. when too much of them are used to scan the input space and do not contribute to the useful part of the Voronoï tessellation, the LVC algorithm can greatly improve the performances of the classification, with regards to the LVQ1 algorithm. The example of figures 3a) and 3b) shows respectively the LVQ1 and LVC algorithms applied to the same input distribution as in figure 2, but now with only 10 prototypes instead of 50.

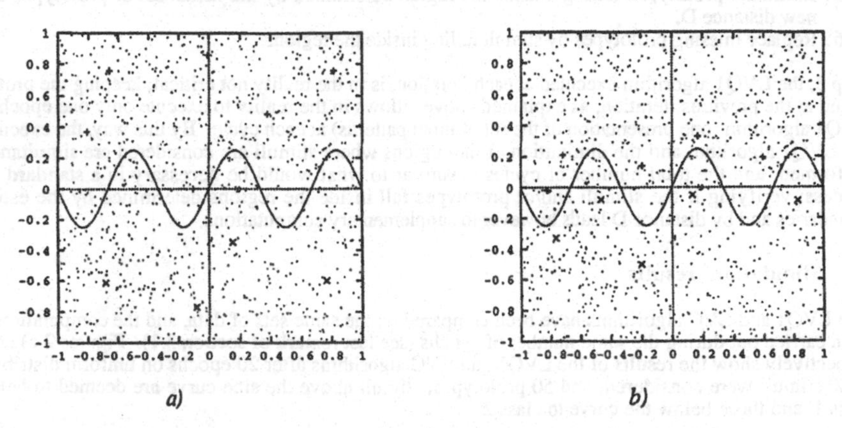

fig.3 LVQ1 and LVC algorithms on uniform distributions with 10 prototypes

Figure 3a) clearly shows that 10 prototypes are not sufficient to approximate adequately the boundary between the classes when the LVQ1 algorithm is used; on the contrary, it figure 3b) shows that most prototypes are brought near the boundary with the LVC algorithm, leading to a much better approximation of the sine function. The 1000 input vectors tested with the LVQ1 and LVC algorithms respectively lead to 97 and 26 misclassifications; such an amelioration (about 7%) is appreciable if we consider that in most applications the important patterns to classify are situated near the boundaries.

The two above examples show increased percentages of correct classification when using the LVC algorithm. In the reality, the performances increase can be much more than 7% (in the example of figure 3), if the input data are spread over very large volumes (or surfaces in our examples). We considered indeed only problems where the size of the boundary region is almost of the same order of magnitude as the size of the whole volume comprising all input vectors (in the examples of figures 2 and 3, the boundary region is roughly 40% of the total space). In other problems however, the boundary region can be much smaller (with respect to the whole volume), leading to bad classification performances if the LVQ1 algorithm is used. The amelioration brought by the LVC algorithm is then much appreciable.

Let us mention that it is possible to approximate more closely the boundaries by reducing D_{final}. However, if this parameter is chosen too small, it may happen that all input vectors near the boundaries are no more used during the last iterations, leading to a concentration of prototypes in small regions, and thus to a strong decrease of the performances. One way to avoid this phenomenon is to compute the percentage of correct classifications at each iteration, and to stop when this percentage decreases too firmly. With this method, the algorithm becomes independent from parameter D_{final}. This does not increase the number of computations, since all input patterns are already compared with all prototypes at each iteration.

A last example uses overlapping distributions of input patterns. In figure 4a) and 4b), if y is the y-coordinate of the vectors, the class of the stimuli is given by equation (8).

Again, maximizing the number of prototypes in the boundary regions leads to increased percentages of correct classifications: with 25 prototypes and 1000 input vectors, the number of misclassifications is respectively 67 and 45 for the LVQ1 and LVC algorithms.

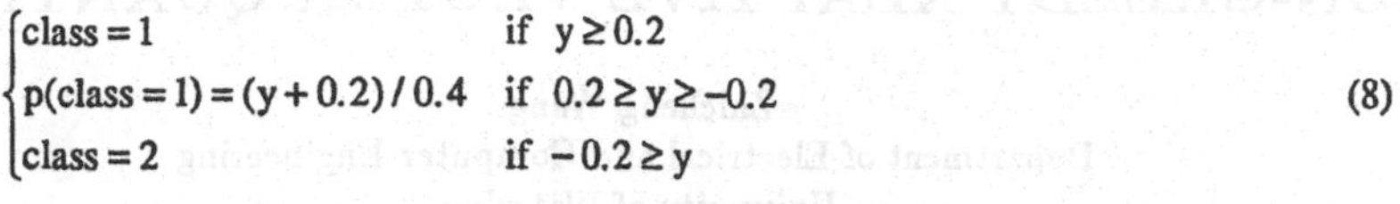

$$\begin{cases} \text{class} = 1 & \text{if } y \geq 0.2 \\ p(\text{class} = 1) = (y + 0.2) / 0.4 & \text{if } 0.2 \geq y \geq -0.2 \\ \text{class} = 2 & \text{if } -0.2 \geq y \end{cases} \qquad (8)$$

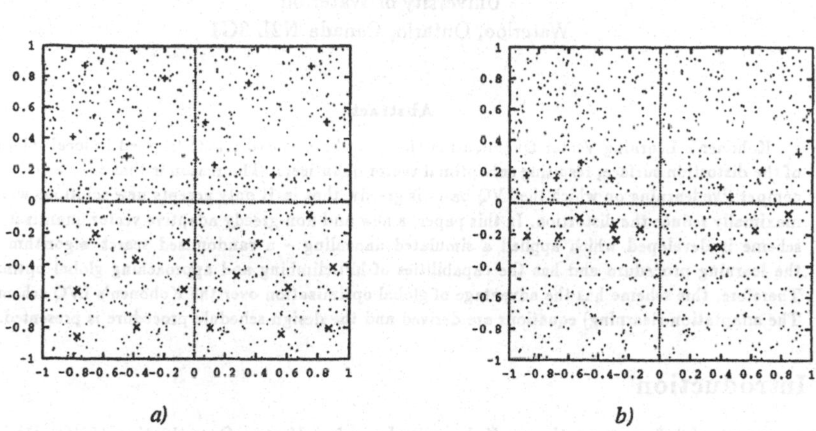

fig.4 LVQ1 and LVC algorithms on overlapping distributions (see text)

4. Conclusion

We described here an algorithm which can be used for the classification of patterns. It is based on Kohonen's LVQ1 algorithm, but has the particularity to bring all prototypes near the boundaries between classes, leading to a better approximation of these limits, and thus to increased classification performances. The prototypes found with the LVC algorithm however no more quantify adequately the input space; applications where such quantization is important should then avoid the use of this method.

We showed on several examples the well-founded of this algorithm. The increase of performances, with respect to the standard LVQ1 algorithm, is particularly sensible when the volume comprising all input vectors is very large with respect to the regions around boundaries between classes.

The way of how prototypes are generated at each iteration, while efficient, is however somewhat arbitrary. Other methods to bring the prototypes near the boundaries, like moving too distant ones in the direction of the nearest prototype from another class, could be considered. Such possibility is currently being evaluated.

5. References

[1] Kohonen, T. (1988), "Self-organization and associative memory", 2nd edition, Springer-Verlag, Berlin.

[2] Jutten, C., Guerin, A., Nguyen Thi, H.L. (1991), "Adaptive optimization of neural algorithms", in: A. Prieto ed., Artificial Neural Networks, Springer-Verlag Lecture Notes in Computer Sciences n°540, Berlin.

[3] McDermott, E., Katagiri, S. (1991), "LVQ-based shift-tolerant phoneme recognition", IEEE Transactions on Signal Processing, vol.39, n.6, June 1991, pp.1398-1411.

Acknowledgments

Part of this work has been funded by ESPRIT-BRA project n°6891, ELENA-Nerves II, supported by the Commission of the European Communities (DG XIII).

NON-GREEDY ADAPTIVE VECTOR QUANTIZERS

Zhicheng Wang
Department of Electrical and Computer Engineering
University of Waterloo
Waterloo, Ontario, Canada N2L 3G1

Abstract

Kohonen's Learning Vector Quantization (LVQ) technique easily gets trapped in local minima of the distortion surface, resulting suboptimal vector quantizers. The reason is that the behavior of competitive learning on which the LVQ bases is greedy, that is, it only accepts new solutions which maximally reduce the distortion. In this paper, a new and non-greedy adaptive vector quantization scheme is developed which applied a simulated annealing – a randomized search algorithm to the learning procedure and has the capabilities of hill-climbing and approaching global optima. Therefore, this scheme has the advantage of global optimization over the Kohonen's LVQ scheme. The adaptation (learning) equations are derived and the design schedule procedure is presented.

1 Introduction

In the context of information theory Kohonen's Learning Vector Quantization (LVQ) [1] is an alternative of the Generalized Lloyd Algorithm (GLA), better known as the Linde-Buzo-Gray (LBG) [2] algorithm: a clustering approach. In the domain of neural networks, LVQ and its extension, the topological feature map (TFM)[3], are among the best known algorithms for classification. However, these often end at local minima of the distortion surface because their behavior is essentially greedy, that is, they only accept new solutions which maximally reduce the distortion, resulting in suboptimal networks whose performance is inferior to globally optimal networks.

Consider a Euclidean space R^k in which a stochastic input variable $X \in R^k$ has the probability density function $p(X)$. Let $C_n, n = 1, 2, \ldots, N$ represent the centroids of the codebook. In the LBG algorithm – the most popular VQ scheme, there are two iterated steps. First, a training set is clustered using nearest neighbor search,

$$d(X, C_n) = ||X - C_n||^r = min_i ||X - C_i||^r, \tag{1}$$

where the rth order norm is used, and the expected reconstruction distortion is defined by:

$$E = \int ||X - C_n||^r p(X) dV_X \tag{2}$$

where dV_X is a volume element in the input space. Secondly, the codebook is updated by calculating centroids according to the clustering. Let $X_{nk}, 1 \le k \le L_n, 1 \le n \le N$ represent the inputs belonging to the nth cell, where L_n is the number of inputs that fall in the nth cell, so that the total distortion for that cell is given by

$$D_n = \sum_{k=1}^{L_n} d(X_{nk}, C_n) \tag{3}$$

the centroid is obtained by the arithmetic mean of the inputs that belong to the nth cell:

$$C_n = \frac{1}{L_n} \sum_{k=1}^{L_n} X_{nk} \tag{4}$$

In the unsupervised LVQ (called competitive learning), the only difference is at the centroid updating, There the steepest descent gradient-step optimization of E in the C_n space yields the following sequence if the mean square error (MSE) criterion ($r = 2$) is used

$$\begin{aligned} C_n(t+1) &= C_n(t) + \alpha(t)[X(t) - C_n(t)] \\ C_i(t+1) &= C_i(t) \quad for\ i \neq n \end{aligned} \tag{5}$$

where $0 < \alpha(t) < 1$, and α is decreasing monotonically with time. Convergence for both algorithms is assured in the sense that the distortion of the sequence of trial centroids is guaranteed to reduce monotonically with each iteration. However, since the multidimensional distortion function is nonconvex and the initial codebook is ad hoc, it is likely that the result will yield a local instead of global minimum of the distortion.

In this paper a simulated annealing is used as the adaptive procedure for VQ, which is one of a class of randomized search algorithms that are able to climb hills and approach global minima. Hence, a new and non-greedy adaptive vector quantization scheme is presented.

2 Annealing Processes

When the classical cost function has a single minimum, the conventional method can provide the unique ground state, and any method of gradient descent can approach the minimum. However, when the cost function has multiple extrema, a nonconvex optimization technique that allows tunnelling and variable sampling and accepting hill-climbing for escaping from local minima is required.

Simulated annealing is the artificial modeling of an annealing process which heats and then gradually cools a system to the freezing point to let the system achieve states with globally minimal energy. By appropriately defining an effective temperature for the multivariable system, simulated annealing can solve a wide collection of optimization problems. Kirkpatrick et al. [4] were the first to use simulated annealing to solve such optimization problems. B.Hajek [5] surveyed the basic theory of simulated annealing in discrete space and in continuous space and their applications. A necessary and sufficient condition for the convergence to the global minimum has been proven in 1984 by Geman and Geman [6] for the classical simulated annealing based on a strictly local sampling. It is required that the time schedule of changing the fluctuation variance, described in terms of the artificial cooling temperature $T_g(t)$ is inversely proportional to a logarithmic function of time given a sufficiently high initial temperature T_o:

$$T_g(t) = \frac{T_o}{\log(1+t)} \tag{6}$$

The simulated annealing may characterized by: (i) bounded generating probability density (thermal diffusion), e.g., Gaussian probability density

$$G(X) = \frac{1}{2\pi\sqrt{T}} \exp(-\frac{X^2}{T}), \tag{7}$$

(ii) an inversely logarithmic update cooling schedule, and (iii) the canonical hill-climbing acceptance probability

$$p(\Delta E) = \exp(-\Delta E/T), \tag{8}$$

where $\Delta E = E_{t+1} - E_t$ is the increase of cost incurred by a transition. The following probability could also be used,

$$p(\Delta E) = \frac{1}{1 + \exp(\frac{\Delta E}{T})}. \tag{9}$$

In a continuous space, assume a function V on R^n to be minimized. Simulated annealing is implemented as a diffusion with the following equation

$$X(t+1) = X(t) - \nabla V(X(t))\Delta t + \sqrt{2T}W(t) \tag{10}$$

where W(t) is white noise. The equation is a gradient descent algorithm combined with normal distribution process.

3 Non-Greedy Adaptation

The vector quantizer (codebook) design can be viewed as an optimization problem where the objective is to find the centroids (codevectors) of a codebook that best represent the training set in terms of the MSE. Since it is generally very difficult and impractical to find an analytically tractable model of a vector source and equally difficult to utilize an analytical model for codebook design, a training set of empirical data that statistically represents the source is an essential starting point for any vector quantizer design algorithm.

The main features of the LVQ algorithm are: (a) it uses a training set of typical data to define an implicit source probability density function, and (b) it performs an iterative operation where each iteration improves a winning centroid. The LVQ method is appealing because the iteration is based on the well known necessary conditions for optimality and results in a monotonically decreasing distortion, which, however, actually makes the method stick easily in poor local minima and hence is greedy competitive learning.

To obtain global minima, we developed the non-greedy adaptive vector quantization scheme in which a simulated annealing is applied to competitive learning and use the following equation for the winning centroid C_n updating which is derived from eq.(10)

$$\begin{aligned} C_n(t+1) &= C_n(t) + \alpha(t)[X - C_n(t)] + \sqrt{2T}W(t), \\ C_i(t+1) &= C_i(t) \quad for\ i \neq n, \end{aligned} \tag{11}$$

where W(t) is white noise. The non-greedy adaptation procedure is described as follows. Begin by setting the effective value of T to an initial temperature T_0, and randomly set an initial codebook.

Alter the centroids based on equation(11), and compute the resulting change in the distortion function, $\Delta E = E(t+1) - E(t)$. If $\Delta E \leq 0$, the perturbed codebook (centroids) is accepted as the new codebook. If $\Delta E > 0$, the perturbation is accepted with the probability $p(\Delta E) = \frac{1}{1+\exp(\frac{\Delta E}{T})}$. The perturbation continues until an equilibrium condition has been reached. The temperature T is then reduced to the next lower temperature in a predetermined sequence of temperatures, and perturbations are again carried out on the codebook. The annealing terminates when the convergence criterion is met and then the optimization algorithm becomes a greedy optimization algorithm.

In the procedure mentioned above, the conditions determining an annealing schedule are: (1)the initial temperature, (2)the temperature decrement, (3)the equilibrium conditions, and (4)the stopping, or convergence criterion. For an annealing schedule to be problem independent, the parameters in the four conditions should be determined by the algorithm itself and should not have any predefined values. (1) In simulated annealing, the initial temperature can be determined based on the condition proposed by White [8], $T_0 \gg \sigma$, where σ is the standard deviation of the distortion distribution. Actually, we can also take the variance of the coded training set, V_s, as a reference as σ to choose T_0. I propose $T_0 \simeq V_s$ for simulated annealing. (2) Huang,et.al. [9], proposed that the temperature should decrease by:

$$T_{t+1} = T_t \exp(-\frac{\lambda T_t}{\sigma}). \tag{12}$$

where $\lambda \leq 1$. Kirkpatrick [4] invented the intuitive temperature reducing method: $T_n = (0.9)^n T_0$, which a lot of researchers have used without observing any difficulty in convergence of the simulated annealing for most of normal problems. (3) The equilibrium condition is met intuitively when a fixed number of attempts are made or a minimum number of attempts are accepted. (4) The stopping criterion is that all the accessible codebooks at that temperature have comparable distortions. The temperature is then set to zero. At this point we can consider that we have obtained a starting codebook near a globally minimal distortion after the annealing procedure and then use a greedy algorithm like the LVQ algorithm to obtain the globally optimal codebook. This VQ scheme is named Non-Greedy Adaptive Vector Quantization (N-GAVQ).

The algorithm presented above has been applied to a simple but nontrivial example demonstrated in [2]. The global minimum is always reached no matter at which initial point the non-greedy adaptation started. However, in the LVQ scheme the local or fortunately and hardly global minima being reached are determined by their initial conditions of vector quantizers.

4 Conclusions

A new learning vector quantization scheme has been developed based upon a simulated annealing technique. The algorithm is a reliable method to improve the quality of LVQ. The main feature of this technique is that a global optimal codebook on a training set is consistently achieved regardless of the initial set of centroids, which is an improvement over the LVQ whose solutions inevitably vary with the initial centroids. Although the design (learning) complexity of the N-GAVQ is higher than that of the LVQ, the complexity of the actual encoding operation is unaffected since the design is a one-time off-line operation, similar to the optimal design of our neural vector quantizers [10][11][12]. This algorithm can also be applied to the supervied LVQ's [13] such as LVQ1 and LVQ2, presently under investigation. A fast learning vector quantization with Cauchy annealing has been developed[14].

References

[1] T.Kohonen, "An Introduction to Neural Computing," *Neural Networks* vol.1, pp.3-16, 1988.

[2] Y.Linde, A.Buzo, and R.M.Gray, "An Algorithm for Vector Quantizer Design," *IEEE Trans. Commun.*, vol.COM-28, pp.84-95, Jan. 1980.

[3] T.Kohonen, *Self-Organization and Associative Memory*, 3rd ed., Springer-Verlag, Berlin, Heidelberg, Germany, 1989.

[4] S.Kirkpatrick, C.D.Galatt, and M.P.Vecchi, "Optimization by Simulated Annealing", *Science*, vol.220, pp.671-680, May 13, 1983.

[5] B.Hajek, "A Tutorial Survey of Theory and Application of Simulated Annealing", *Proceedings of 24th Conference on Decision and Control*, pp.775-760, December 1985.

[6] S.Geman and D.Geman, "Stochastic Relaxation, Gibbs Distributions, and the Bayesian Restoration of Images", *IEEE Transactions on Pattern Analysis and Machine Intelligence*, PAMI-6, pp.721-741, 1984.

[7] H.Szu and R.Hartley, "Fast Simulated Annealing", *Physics Letters*, vol.122, no.3,4, pp.157-162, June 1987.

[8] S.White, "Concepts of Scale in Simulated Annealing", *Proceedings of the International Conference on Computer Design*, pp.646, 1984.

[9] M.F.Huang, F.Romeo, and A.Sangiovanni-Vincentelli, "An Efficient General Cooling Schedule for Simulated Annealing", pp.381, 1986.

[10] Z.Wang and J.V.Hanson, "A Neural Vector Quantizer", *Proceedings of 1992 IEEE International Symposium on Circuits and Systems*, vol.1, pp.351-354, San Diego, CA, May 1992.

[11] Z.Wang and J.V.Hanson, "Code-Excited Neural Vector Quantization", submitted to *IEEE 1993 International Conference on Acoustics, Speech and Signal Processing*, Minneapolis, Minnesota, U.S.A., April, 1993.

[12] Z. Wang and J.V.Hanson, "Design Optimization of Code-Excited Neural Vector Quantizers", submitted to *The 1993 IEEE International Conference on Neural Networks*, San Francisco, CA, March 28-April 1, 1993.

[13] T.Kohonen, "Improved Versions of Learning Vector Quantization", *Proceedings of 1990 IEEE International Joint Conference on Neural Networks*, vol.I, pp.545-550, 1990.

[14] Z. Wang and J.V. Hanson, "Cauchy Learning Vector Quantization", submitted to *The 1993 IEEE Symposium on Circuits and Systems*, Chicago, IL, May 3-6, 1993.

Hybrid Programming Environments

Philip C. Treleaven & Paulo V. Rocha

Department of Computer Science
University College London
Gower Street
London WC1E 6BT
United Kingdom

Abstract

Following the emergence of each new "Intelligent" computing technology – *expert systems, fuzzy logic, neural networks* and *genetic algorithms* – sophisticated programming environments have been developed to assist application building. Each environment being dedicated to a single intelligent technology. These environments can be classified as: *application-oriented* – dedicated to a specific application domain; *algorithm-oriented* – supporting a single algorithm; or *general-purpose* – comprehensive tool kits for any domain.

However, *hybrid* applications are now being build combining a number of intelligent techniques. Thus *hybrid programming environments* are required that allow combination of neural networks and genetic algorithms. This paper briefly reviews programming environments for neural networks, genetic algorithms and hybrid systems, using three of UCL's ESPRIT projects: GALATEA, PAPAGENA and HANSA as case studies.

1. Taxonomy

Before reviewing programming environments for neural networks and genetic algorithms, we introduce a simple taxonomy for classifying environments [13, 14], based on the target user expertise and on facilities. The taxonomy presented here divide the programming environments into three main categories: *application-oriented*, *algorithm-oriented*, and *general-purpose*.

Application-oriented

Application-oriented environments are aimed at particular business domains such as finance, engineering, marketing, and medicine. They are designed for professionals who have very little expertise in the technology being used, but want to apply it to their business domains without having to deal with the programming details.

Programming environments in this class trade-off generality for efficiency in the implementation of applications. They are usually tied to a particular hardware and have a *closed* implementation, tied to the particular application and difficult to be re-used if specifications change. This class of environments can be seen as *black-boxes*, hiding all implementation details, and offering a number of facilities for the programming of a specific application.

Algorithm-oriented

Algorithm-oriented environments support specific characteristics of the technology being considered. For instance, algorithm-oriented Neural Network programming environments support a limited number of algorithms such as Back-propagation, and Self-organising Maps. Algorithm-oriented programming environments are supplied in a form that allows easy integration with user applications such as a parametrised algorithm library

This class of programming environments can be further sub-divided in: *algorithm-specific*, supporting a single algorithm that can be applied to a broad range of applications; and *algorithm libraries*, which offer a number of parametrised algorithms or models in a standard language such as C.

The major strength of algorithm-oriented programming environments is their flexibility. Algorithm-specific environments offer an easy to use path for the quick development of applications because of the usually easy format of user interfaces and algorithm definitions. On the other hand, algorithm libraries are invaluable for its integrating capabilities with other applications written in the same language.

General-purpose

General-purpose programming environments are *tool kits* comprising not only algorithms but also other programming tools. They are usually designed for more experienced programmers and offer a great deal of flexibility in their interfaces, algorithms and execution control mechanisms.

This class of environments is the most flexible one. However, this flexibility and generality are, most of the time, achieved in detriment of the best attainable performance. To eliminate this drawback, some general-purpose programming environments are ***hardware-oriented***, offering different paths for the execution of applications on a number of sequential or parallel hardware platforms.

2. Neural Network Environments

In this section we review some important neural network programming environments and, as a case study, examine the GALATEA environment.

Application-oriented

A typical application-oriented neural network programming environment is the *Decision Learning System* (DLS) market by Nestor [9]. According to Nestor, the DLS is particularly suitable for problems that involve many interdependent variables that are difficult to process using conventional statistical methods.

The DLS is a menu driven package that includes a tutorial and examples. Nestor have developed applications covering mortgage underwriting, automobile insurance, and credit card transaction risk assessment. DLS is built on top of the *Decision Learning System*, a hierarchy of parallel neural networks. DLS is based on a proprietary model – the *Restricted Coulomb Energy* (RCE) [3].

Algorithm-oriented

The great advantage of algorithm-oriented environments is the possibility of easy integration with user applications. Examples of these environments are *BrainMaker*, an algorithm-specific environment based on the Back-propagation model and marketed by California Scientific Software[2], and *OWL* an algorithm library offered by Omlsted & Watkins [11].

OWL is a successful algorithm library written in C and targeted to the experienced computer professional. OWL supports over 19 different models, all seen by the programmer as a data structure and a set of service calls that operate on the data structures. The data structures specify the properties of the networks and the service calls invoke entry points within the network objects. To use the network, the user specifies the object and library modules, and an *include file* corresponding to the required model. The user specifies parameters needed during creation time (instantiating the network), and during execution time. The execution time parameters are defined on a per-network basis and can be altered dynamically.

General-purpose

General-purpose programming environments span many applications and algorithms. Good examples of such environments are SAIC *ANSpec* [17], NeuralWare *NeuralWorks* [10], HNC *ExploreNet* [6], and the ESPRIT II *GALATEA* [19].

ANSpec, NeuralWorks and *ExploreNet* provide similar comprehensive components and facilities. These systems comprise a graphic interface, algorithm library, specialised programming language, a simulator, and possibly a hardware accelerator. The algorithm libraries are a key component, largely responsible for the flexibility and applicability of the environments. The libraries usually incorporates popular models such as Back-propagation, in various *flavours*, and Self-organising Maps, etc.

GALATEA extends the key features of the PYGMALION Programming Environment, a research product developed as part of a ESPRIT II project. GALATEA and PYGMALION are built around industry-standard such as X-Windows, and the C and C++ programming languages to make them easily extendible. GALATEA has significantly extended the features of PYGMALION and, therefore, will be described in more detail.

2.1. Case Study: The GALATEA System

The GALATEA System comprises two interdependent domains (Figure 1): an *execution environment* and a *programming environment*. The execution environment offers hardware platforms for the simulation of any neural network model and for the integration of application specific chips. To integrate these heterogeneous platforms, GALATEA introduces the concept of a Virtual Machine (VM): a unique hardware abstraction that isolates any special characteristic of the actual execution platforms. The programming environment integrates with the execution environments through the native code of a virtual machine: the VM Language (VML), an optimised low-level neural network representation language.

VML constitutes a set of arithmetic operations that reflect the inherently parallel nature of neural network algorithm calculations. VML is completely machine-independent and addresses parallelism by implementing operators over a matrix representation of network data.

The programming environment of GALATEA supports tasks such as: the definition, programming, customisation and development of neural network algorithms; the control of the simulation of a model or the execution of an application; and debugging. The environment comprises a *high-level programming language* based on C++, algorithm and pre-programmed application *libraries*, language *translators* to VML, and four key graphical tools:

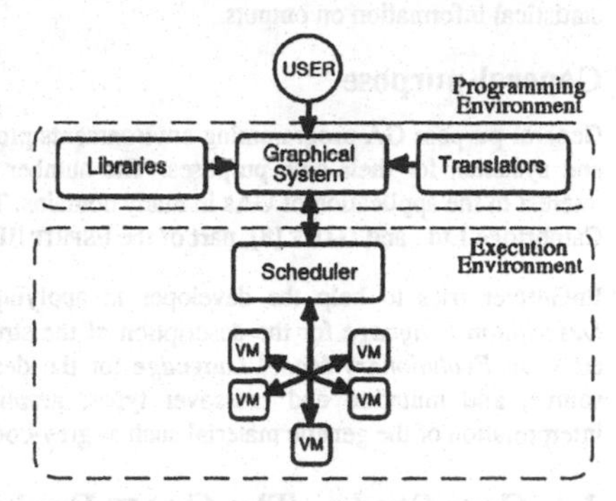

Figure 1 – The GALATEA System

- a *systems application builder*, a tool for graphically constructing neural network algorithms that are then automatically converted into executable code;
- an *execution monitor*, a tool for neurocomputer developers that measures and displays the load distribution and the traffic between the components of the neurocomputer;
- an *algorithm debugger*, perhaps the most common and desired component of a programming environment, that lets the user change parameters and fire neural network rules, at any level of the network hierarchy; and
- an *application monitor*, a windowing system that can be customised to display versions of the key parameters, inputs and outputs that are of interest for the end-user.

3. Genetic Algorithms Environments

Interest in Genetic Algorithms (GAs) is expanding rapidly and their application domain spans areas as diverse as Chemistry (e.g. protein folding), Finance (e.g. credit management), and Manufacturing (e.g. production scheduling). We next review some of the best known programming environments, focusing on the GAME Environment in detail.

Application-Oriented

Professionals that want to use Genetic Algorithms without having to acquire detailed knowledge of the workings of the technique have at their disposal a growing number of environments to choose from [14]. Many are environments analogous to a spreadsheet or word processing utility, and most follow innovative strategies. PC/BEAGLE [12] and XpertRule GenAsys [1], for example, are expert systems using GAs to generate new rules to expand their knowledge base of an application domain.

PC/BEAGLE, in particular, is a rule-finder program that turns data into knowledge by examining a database of examples and using machine-learning techniques to create a set of decision rules for classifying those examples. The system is composed by six executable blocks that are run in sequence. The GA component is a *Heuristic Evolutionary Rule Breeder* that generates the decision rules by *Naturalistic Selection.* The system runs on IBM/PC compatibles and accepts data in ASCII format. Rules are produced as logical expressions.

Algorithm-oriented

A well known programming environment in this category is the *GENEtic Search Implementation System* – GENESIS [5, 18], used to implement and test a variety of new genetic operators. Genesis is now in its Version 5.0 running on SUN Workstations and PCs.

GENESIS provides the basic procedures for *natural selection*, *crossover*, and *mutation.* Since GAs are task independent optimisers, the user must provide only an evaluation function that returns a value when given a particular point in the search space. GENESIS provides a high modifiable environment and a large amount of statistical information on outputs.

General-purpose

General-purpose GA programming environments provide users with the opportunity to customise the algorithms and systems, for their own purposes. The number of these systems is growing, stimulated by the increasing interest in the application of GAs in many domains. Two major systems are *EnGENEer* [15], developed by Logica Cambridge Ltd., and *GAME* [8], part of the ESPRIT III PAPAGENA project.

EnGENEer tries to help the developer in applying GAs to new problem domains by offering: a *Genetic Description Language* for the description of the structure of the *genetic material* (i.e. its genes, chromosomes, etc.); an *Evolutionary Model Language* for the description of options such as population size, structure and source, and mutation and crossover types; graphical monitoring tools; and a library of routines for the interpretation of the genetic material such as grey-coding, permutations, etc.

3.1. Case Study - The GAME Environment

GAME is an object-oriented environment for programming sequential or parallel GA applications and algorithms. The programming environment comprises 5 major parts (Figure 2):

- a *Virtual Machine*, the machine independent low-level abstraction representing the code responsible for the management and execution of a Genetic Algorithm application;
- the *High-level Language GA-HLL*, an object-oriented programming language for defining new GA models and applications;
- *Genetic Algorithm Libraries*, where parametrised algorithms, applications, and operator libraries, constituting validated models written in the high-level language, are stored;
- a *Graphical Monitor*, the software environment for controlling and monitoring the execution of a GA application simulation; and
- *Compilers*, to the various target machines, ranging from parallel hardware to PCs.

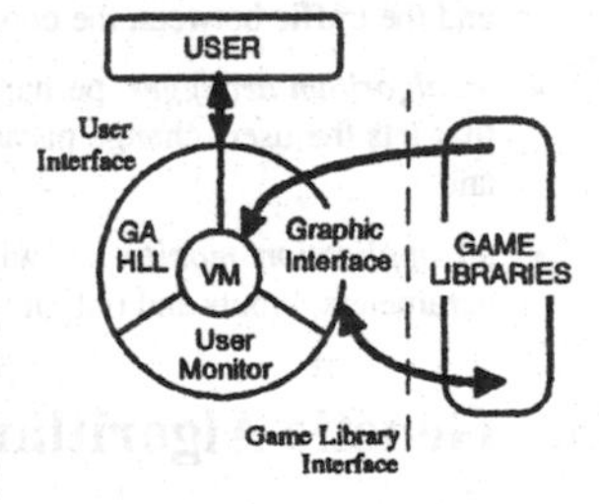

Figure 2 - The GAME Programming Environment

GAME is designed to allow a single application to use many (and possibly different) algorithms running in parallel, or a single parametrised algorithm to be configured for many applications.

To develop a Genetic Algorithm application, the user firstly loads the required modules from the library. With the high-level language it is then possible to configure, modify and extend the GA software. To execute the application, the user can invoke the graphical monitor command language for interactive monitoring and control of the simulation.

The environment is being offered in plain C++, with the ultimate goal being its development in a parallel language derived from C++, for execution in parallel machines. A preliminary version of GAME - in source code form - will be available soon free to researchers.

4. Hybrid Systems

Researchers and applications builders are beginning to realise that a combination of two or more intelligent technologies is the best strategy for tackling real-world complex problems such as those in Finance, Language Processing, and Automated Manufacturing [4]. For instance, combining a neural network with an expert system or genetic algorithm. Hybrid environments are at a very early stage of development, and therefore a review based on our taxonomy is premature.

An effective programming environment suitable for helping in the development of Hybrid Systems is POPLOG [7]. POPLOG is a powerful integrated environment comprising: POP11, a LISP-like high-level programming language; a program editor; and facilities for the implementation of easily integrable modules. One of such modules is POP-NEURAL, a set of routines for the development of neural network applications or new algorithms. Adding to the facilities of POPLOG, the POP11 language has *hooks* to other languages like PROLOG, C, and FORTRAN allowing easy integration of user-designed applications and modules. However, despite the flexibility, POPLOG is still tied to built-in facilities or specially designed modules, and cannot benefit from the advanced programming and monitoring utilities easily available for Neural Networks and Genetic Algorithms.

In line with the advances in programming environments, and with the necessity to offer a path for the re-usability of application code (mainly algorithms, user interfaces and input/output), the HANSA Project is introducing a framework for the integration of *industry-standard* tools [16].

4.1 Case Study: The HANSA Framework

The HANSA Project aims to develop an object-oriented cross-platform framework to allow developers to rapidly configure industry standard utilities and artificial intelligence shells, and is targeted at business applications for marketing, banking and insurance.

The HANSA Framework (Figure 3) utilises the object-oriented paradigm through C++ and a communication protocol to allow the exchange of information between modules within an application. The framework hides platform-related implementation details, promoting software re-use and rapid prototyping. On each of the target platforms (PCs running MS-Windows 3.1 or Windows NT, and UNIX/X-Windows workstations) the framework implements a services interface inspired by, and compatible with, Microsoft's OLE protocol. For UNIX/X-Windows systems, HANSA is developing an OLE-like interface following closely the philosophy of the *Object Management Architecture* (from the Object Management Group Consortium), and builds the framework on top of the services supplied by the *Object Request Broker*. The common OLE-like interface allows an easy port of the application code, from one platform to the other.

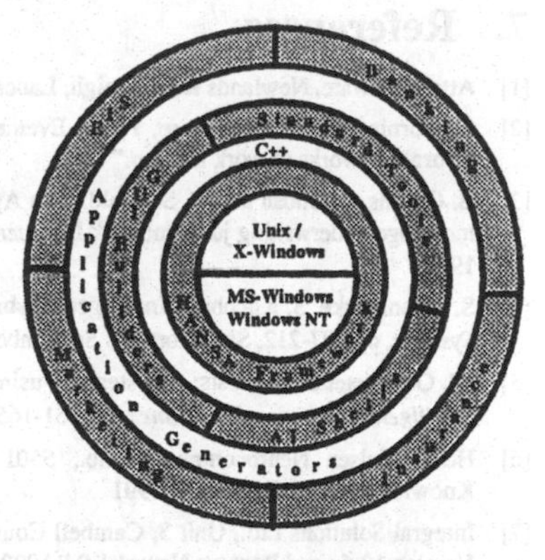

Figure 3 - The HANSA Framework

For each of the application domains targeted by the project (Executive Information Systems, Direct Marketing, Banking, and Insurance) there is at least one *application generator* in charge of controlling the configuration and execution of cooperating HANSA Tools selected from the HANSA Toolkit. This Toolkit includes: Expert Systems such as PAMELA; neural network development environments such as MIMENICE, derived from the PYGMALION and GALATEA Projects and commercialised by Mimetics; the GA programming environment GAME; as well as any industry-standard utility that complies with the HANSA-OLE protocol on PCs or Unix Workstations.

For allowing the integration of such a diverse range of Tools, HANSA constitutes a natural programming environment for the implementation of *hybrid* applications. Moreover, the possibility of utilising *off-the-shelf* Tools, putting them together in an intuitive environment, allows the user to configure and tune the application by utilising the most appropriate utilities for each task involved in the application execution.

5. Conclusions

The advances in new intelligent, adaptive technologies have created a great interest in their application in a large range of domains. However, as complex real-world problems are tackled, the limitations and strengths of these new technologies become clear. To avoid the weak points, hybrid systems are being developed putting together the adaptive, symbolic and even numeric approaches. These new systems are growing in number and acceptance as the most effective solution.

However, there is no comprehensive programming environment for Hybrid Systems, despite the great number of programming systems for each individual technology. For Hybrid Systems to reach the state-of-the-art of Neural Network and GA programming environments in terms of graphic interfaces, libraries, and high-level languages, there is still a long path to follow. The HANSA Framework aims at offering the facilities for putting together the best programming environments for particular technologies, allowing the rapid configuration and prototyping of hybrid applications.

6. Acknowledgements

We would like to acknowledge the support of our ESPRIT Partners whose work we described above.

GALATEA: *Thomson-CSF, Philips, Siemens AG, Mimetics, SGS-Thomson, INPG, IS, CRAM, INESC,* and *CTI.*

PAPAGENA: *Brainware, IFP, Telmat Informatique, CAP Gemini Innovation,, GMD, KiQ, University of Grenoble, GWI,* and *Institut für Kybernetic & Systemtheorie.*

HANSA: *CRL Ltd., O.Group, Promind, J&J Financial Consultants, IFP, Brainware,* and *Mimetics.*

7. References

[1] Attar Software, Newlands Road, Leigh, Lancashire, UK. "XpertRule GenAsys." 1992

[2] California Scientific Software, 10141 Evening Star Drive 6, Grass Valley, CA 95945, USA. "BrainMaker: Putting Neural Networks to work for you." 1991

[3] E. Collins, S. Ghosh and C. Scofield. "An Application of a multiple neural network learning system to emulation of mortgage underwriting judgements." In *International Conference on Neural Networks*, volume 2, pp 459-466. IEEE, 1988.

[4] S. Goonatilake, S. Khebbal. "Intelligent Hybrid Systems." In *First Singapore International Conference on Intelligent Systems*, pp 207-212, Singapore, 28 September - 1 October 1992.

[5] J.J. Grefenstette. "GENESIS: A system for using genetic search procedures." In *Proceedings of the 1984 Conference on Intelligent Systems and Machines*, pp 161-165. 1984.

[6] Hecht-Nielsen Neurocomputers Inc., 5501 Oberlin Drive, San Diego, CA 92121, USA. "ExploreNet 3000, KnowledgeNet, Balboa 860." 1991

[7] Integral Solutions Ltd., Unit 3, Cambell Court, Bramley, Basingstoke, Hampshire RG26 5EG, UK. "POPLOG System Version 13.6 , and POPLOG-Neural 1.0." 1992

[8] J. Kingdon, J. Ribeiro Filho, and P. Treleaven. "The GAME Programming Environment Architecture." *Technical Report TR/92/28*, Department of Computer Science, University College London. 1992.

[9] Nestor Inc., One Richmond Square, Providence, RI 02906, USA. "Learning Systems Based on Multiple Neural Networks." 1988.

[10] NeuralWare Inc., Penn Center West, Building IV, Suite 227, Pittsburg, Pennsylvania 15276, USA. "NeuralWorks Explorer." 1991.

[11] Olmsted & Watkins, 2411 East Valley Parkway, Suite 294, PO Box 3751, Escondido, CA 92025. "OWL Neural Network Library." 1988.

[12] Pathway Research Ltd., 59 Cronbrook Rd., Bristol BS6 7BS, UK. "PC-BEAGLE." 1992.

[13] M.L. Recce, P.V. Rocha and P.C. Treleaven. "Neural Network Programming Environments." In *International Conference on Neural Networks*, Brighton, UK, 4-9 September 1992.

[14] J. Ribeiro Filho, C. Alippi, P. Treleaven. "Genetic Algorithm Programming Environments." *RN/92/77*, Department of Computer Science, University College London, 1992.

[15] G. Robbins. "EnGENEer – The Evolution of Solutions." In *Proceedings of 5th Annual Seminar on Neural Networks and Genetic Algorithms*. 1992.

[16] P.V. Rocha and S. Khebbal. "HANSA Framework Specification." *TR/92/31*, Department of Computer Science, University College London, 1992.

[17] Science Applications International Corp., 10260 Campus Point Drive, Mail Stop 71, San Diego, CA 92121, USA. "SAIC Product Information -- Delta Processors, ANSpec, ANSim, CARL/BP." 1989.

[18] The Software Partnership, PO Box 991, Melrose, MA 02176, USA. "GENESIS." 1992.

[19] J. Taylor. "Enhancements for domain-specific boards - Virtual Machine concept." Technical Report (UCL GALATEA internal distribution), Department of Computer Science, University College London. July 1991.

[20] P. Treleaven and S. Goonatilake. "Intelligent Financial Technologies." In *Workshop on Parallel Solving from Nature: Applications in Statistics and Economics*. 1992

Automatic Generation of C++ Code for Neural Network Simulation

Stephan Dreiseitl (sdreisei@risc.uni-linz.ac.at)
Dongming Wang (wang@risc.uni-linz.ac.at)
Research Institute for Symbolic Computation
Johannes Kepler University, A-4040 Linz, Austria

Abstract

Coding neural network simulators by hand is often a tedious and error-prone task. In this paper, we seek to remedy this situation by presenting a code generator that produces efficient C++ simulation code for a wide variety of backpropagation networks. We define a high-level, Maple-like language that allows the specification of such networks. This language is compiled to C++ code segments that in turn are executable in link with an already given generic code for backpropagation networks. Our generator allows the specification of arbitrary network topologies (with the restriction of full connections between layers) and weightchange formulae, while the activation rule and error propagation rule remain fixed. With this tool, future research on learning rules for backpropagation networks can be made more efficient by eliminating routine work and producing code that is guaranteed to be error-free.

1 Introduction

Automatizing the process of writing computer programs — *automatic programming* — is desirable whenever the programmers need a machine to aid the construction of correct programs in an easy and efficient way. Research in this field has contributed the formal methodologies of program transformation and synthesis, which make use of the techniques of logic programming and automatic theorem proving, and the practical developments on code generation, which are intimately connected with computer algebra (CA) systems. Recent work on the latter subject has been particularly focused on the generation of efficient numerical code from symbolic expressions preprocessed by CA systems and the optimization and analysis of the generated code. The ongoing work reported in this paper has a slightly different objective. Our intention was to generate C++ code for the simulation of artificial neural networks, where the difficulties are not in coding complex symbolic expressions. We have seen from practical research that there is a need for some tools for the generation of simulation code from the informal, fragmentary specification of the network architecture and dynamics in a symbolic mathematical language.

Although neural networks are one of the most active research areas (cf. [AR88, APR90]), it is still a young developing area and its theoretical foundations have not yet been sufficiently well laid down. The behavior and performance of diverse neural models are still under observation that is largely based on computer simulation. In virtue of the diversity of neural models and their variants and the experimental nature of neural network research, a great amount of tedious, error-prone program writing has been and will continue to be involved. The researchers and programmers need often to go through the process of editing source code, correcting syntax errors and producing executable code for only a slight modification of a neural network specification. Our project started with these considerations in mind and has as its objective the development of flexible software tools as an assistant for simulating known neural models, their variants and new models in a simple, efficient way. For this end, we characterize a neural network by its topology (layers and connections) and its algorithms (activation rule, error propagation rule and learning rule), following the general framework of neural models provided in [RHW86] and the unified description of neural algorithms given in [RS90]. On this basis, we take a programming language developed for CA systems as the specification language for the generation of simulation code for neural networks.

1.1 Specification of Neural Networks for Code Generation

The mathematical nature of neural networks allows us to specify them using a mathematical language in a formal setting, suitable for automatic code generation. Proper choices for such a mathematical language are the languages developed for CA systems that are capable of specifying the majority of neural networks for the purpose of code generation. In our case, we have decided to follow the grammar of the Maple language [CGG+88] mainly because we have started some theoretical investigations of neural networks in this system. Because the Maple language is designed for the general purpose of programming symbolic algorithms, it is well structured, easy to read and to understand. We feel that this language is of sufficiently high level to be taken as our specification language. We shall consider the specification of a neural network as a Maple procedure which is not intended for execution but only for a formal description. Our choice of the Maple language and the Maple procedure construct for the specification stems also partially from the consideration of integrating this system, the neural simulator and our developed tools into a single environment for neural network research.

In our specification we make use of some basic Maple constructs, including procedure definition, selection statements, repetition statements and statement sequence. With help of these constructs it is then possible to specify the network topology and neural algorithms by using symbolic expressions in Maple syntax. As in mathematics where some notations used in the symbolic expressions should be (informally) explained, we introduce a set of key words to describe the meaning of various variable symbols. In order to describe the various aspects of neural networks, we also define some commands to declare the topology and weightchange function of a network (see Section 2.2). The network parameters may be either assigned values or passed along as run time parameters.

2 A Code Generator for ANN Simulation

In this section we present various aspects of a code generator developed for neural networks simulation. This generator is directed to the existing implementations (in C++) of elementary data types for neural network simulation with an example target code supplied by the neural networks group of Siemens Munich. Our objective was to design a code generator that could use the existing data structures and generalize the example target code by replacing pieces of code. In particular, it should be possible to specify any network topology and weight update function (with restrictions on the mathematical operations known by the system). The two other dynamical aspects of the neural network, the error propagation and activation function, remain fixed as those of a general backpropagation network (see [RHW86]). As a specification language for this task, we chose the CA system Maple as indicated before.

Given the objectives listed above, the first stage of designing the code generator was to write a compiler that accepts a subset of the Maple language and transforms it into C++ . For our purposes, the two Unix tools `flex` and `yacc` were suited very well to implement the compiler. `flex` is the faster, compatible GNU-version of the well-known lexical analyzer `lex`.

2.1 Implementation of the Maple Constructs

The syntax of our specification language is a combination of Maple control statements and neural-network specific declarations. Except for the inclusion of explicit type-information for run-time parameters (which is necessary for C++), the grammar is a subset of the one described in the Maple reference manual ([CGG+88]). The semantics are straightforward, except for the case of identifiers. This special case will be considered in more detail in Section 2.4.

With the help of `yacc`, it was possible to parse and directly transfer the source statements to C++ code, without building up a parse tree. At this point, the only temporary structure needed was a symbol table to keep track of declared variables and their types. The type information comes from the type of the first expression that the variable is assigned to, and type checking is performed on subsequent assignments.

2.2 Declaration Commands

The Maple constructs form a sufficient framework for the specification of neural networks. They need to be extended by specific commands that allow the description of various aspects of the network. We give the syntax of each command, followed by its semantics.

- Declaration of layers in network.

```
Net(⟨layerdeclaration-sequence⟩)
    ⟨layerdeclaration-sequence⟩ ::= ⟨layerdeclaration-sequence⟩ &==> ⟨layerdeclaration⟩
    ⟨layerdeclaration⟩          ::= ⟨identifier⟩[⟨number⟩]
```

This command declares the names and number of neurons of the layers that make up the net. A layer can be either an input, hidden or output layer, with its name determining its type. There can be an arbitrary number of each type of layer in the net.

- Declaration of connections between layers.

  Connection(⟨identifier⟩ ⟨connectiontype⟩ ⟨identifier⟩ ⟨symmetry-information⟩)
  ⟨connectiontype⟩ ::= &==> | &<==>
  ⟨symmetry-information⟩ ::= ε
  | (Sym)

  With this command, the connections between layers of the net are declared. The combination of connectiontype and presence (or absence) of symmetry-information determines the kind of connection between the two layers. The symbol &==> denotes a single forward connection between different layers. Declaring the connection to be symmetric is meaningless in this case, and such a connection is ignored.

  The double connection symbol &<==> declares the the connection between the respective layers to be recurrent. Using this connection symbol, it is possible to specify whether the connection should be symmetric or not by either including or omitting the symmetry-information (Sym).

- Defining the weightchange formula.

  D(⟨identifier⟩[⟨identifier⟩,⟨identifier⟩]) := ⟨expression⟩

  The weightchange formula is defined by assigning an expression to the derivative of the weight symbol. Above, the three identifiers stand for the weight symbol and two indexing symbols that can then be used on the right side of the assignment. The first index is the index of the target neuron, the second index that of the source neuron of a connection. The expression on the right side must not contain local variables. It can contain the symbols for the special variables weight, output and error. See Section 2.4 on how to declare the syntax for these predefined variables. In this expression, relative indexing of layer variables is possible by using an indexing symbol (introduced on the left side) in brackets after the variable name.

2.3 Meta-level commands

To simulate the network, it also has to be known which patterns should be learned, and for how long. Also, some information about the network, such as parameter settings and the mapping that the network currently implements, may be desirable. The following commands are such simulator-level commands.

The ReadPatterns(⟨identifier⟩,⟨number⟩) command reads in a set of patterns from a file. The filename has to be supplied as the first argument to the command, the number of patterns to be read in as the second.

The Learn() command executes one cycle through the training set, and prints the average error for this cycle. With the Check() command, a more detailed information listing about the learning process is obtained. The Info() command prints information about the network topology and default variable values and is generally invoked before learning starts.

Finally, it is possible to duplicate the starting point of simulation runs by using the same seed for the random number generator that determines the initial weight setting. This is done with the RandomSeed(⟨expression⟩) command.

2.4 Variable Kinds and Predefined Variables

In the specification of a network, we allow three kinds of variables and a set of predefined variables, some of which are updated by the system and can only be used as r-values.

- Net variables

  These variables represent the parameters of the network, such as the learning rate η. The user can define float or integer variables by the command

  NetVar(⟨identifier⟩=⟨number⟩)

The type of the variable is determined from the number it is assigned to.

- **Layer variables**

  Layer variables also denote parameters of the network. In contrast to net variables, they can have a different value for each layer in the network.

  Defining a new layer variable is done with the command

  `LayerVar(`⟨identifier⟩=⟨number⟩`)`

  After this initialization, different values can be assigned to the variables in different layers using the syntax ⟨layer-name⟩_⟨variable-name⟩.

  In the learning rule, layer variables may be indexed to use the different values of these variables in different layers.

- **Local variables**

  This third kind of variables are the meta-level variables that are part of the control-structure framework of the simulator. They should be used for loop and other auxiliary variables, and due to scoping constraints (code being produced in different C++ target files) cannot be used in the learning rule specification of the network.

- **Special variables**

  There are some variables which can only be accessed as r-values in the simulator. One of these is the net variable `error`. The other special variables can *only* be used in the learning rule. Those variables are the components of the weight vector, the error terms, the outputs and the components of the momentum vector. Unless declared otherwise, these variables can be accessed by their default syntax. This syntax is `w` for the weight, `d` for the error, `o` for the output and `m` for the momentum.

  It may sometimes be desirable to use a different syntax for these variables. This could be the case when a learning rule has been calculated with the help of a computer algebra system, in which a different syntax has been used. The learning rule can be incorporated directly after telling the simulator which syntax to use. The commands `WeightVar(`⟨identifier⟩`)`, `OutputVar(`⟨identifier⟩`)`, `ErrorVar(`⟨identifier⟩`)` and `MomentumVar(`⟨identifier⟩`)` change the syntax of the the respective variables.

2.5 Using accelerating techniques

Standard backpropagation networks can be specified and simulated using only the above commands. Various acceleration techniques can be incorporated in the simulator by the `Use(`⟨identifier⟩`)` command, where ⟨identifier⟩ is one of the following: `Momentum`, `TanHyp`, `LogisticPlus`, `TanHypPlus`, `Batchmode`, `NoScaling`, `LineSearch`, `DeltaBarDelta`, `AdaptiveGains`, `Noise`.

For a comparison of the methods using the simulator, and more bibliographical information, see [Dre92].

2.6 Inside the Code Generator

As was mentioned above, we use `flex` and yacc for the coarse structure of our compiler. The transition from source to target code is done in two interlocking steps: parsing the input and generating code fragments to link with the existing (not-changing) code. Here, these two steps are explained in more detail.

A Parsing Phase

During this phase, the input is scanned for syntactical and semantical errors, temporary structures are built up and some parts of the target code are produced. yacc automatically checks for syntax errors, and the code generator recognizes several semantical mistakes. Among them are: variable used before set, wrong type for variable, layer used before defined, no input or output layers in net, no weightchange formula given, and many more. The dynamic structures built up during this phase are the *symbol table* (for variables) and the *layer table*. No parse tree is constructed, since yacc automatically structures the input. The statements of the input are translated into target syntax immediately after they are parsed.

B Generating Phase

After the input is completely parsed, the remaining pieces of code can be generated. All in all, three code files are generated and some header files altered. As a first step, a stepping through the symbol table is performed and the declarations for the local variables are generated in a separate file. The intermediate code file is then appended to this declarations file to form the main code file of the simulator.

The remaining identifiers in the symbol table are net and layer variables, and are appended to the corresponding header files to make them known to the simulator. The layer table is used to determine the number of input, hidden and output layers in the network and their dimensions. Finally, the information that is obtained from the weightchange function is used to implement the backpropagation rule in the network. These pieces are combined to form a second file.

For the third file, it suffices to lexically substitute certain values in a pre-defined template file. This template contains, for example, the implementation of the `Learn()` command which does not depend on topology or dynamics of the network, but on some other parameters. These are substituted in with the help of the Unix tool **sed**, a stream editor.

Last but not least, a makefile is also produced to facilitate the compiling of the generated files.

3 An Example

In this section, we present a brief example input file and segments of the code that is produced by the generator.

The input specifies a network with 10 neurons in both the input and output layer, and 5 neurons in the hidden layer. The learning rule used is standard backpropagation with a momentum term. The parameter `alpha` is changed to 0.6, whereas for the learning rate a new parameter `myeta` is introduced. This parameter has a different value in the output layer than in the hidden layer. The syntax declarations for weight, output, error and momentum are omitted since the default values are used. The maximum number of learning epochs, as well as a seed for the random number generator are input at run-time.

The termination condition for learning is to learn until either all patterns have been presented `epochs` times, or until the average error `error` (a net variable) is less than 0.001, whichever comes first.

```
encoder := proc(Integer epochs, Integer seed)
     Net(Input[10] &==> Hidden[5] &==> Output[10]);
     Connection(Input &==> Hidden);
     Connection(Hidden &==> Output);
     Use(Momentum);
     RandomSeed(seed);
     ReadPatterns(encoderpattern,10);
     alpha := 0.6;
     LayerVar(myeta = 1.0);
     Hidden.myeta = 6 * myeta;
     Info();
     D(w[j,i]) := myeta[j] * d[j] * o[i] + alpha * m[j,i];
     for i from 1 to epochs while (error > 0.001) do
          Learn()
     od;
     Check()
end;
```

4 Related Work and Concluding Remarks

A great amount of work is done on numerical code generation in the field of symbolic computation. The success of generating efficient numerical code has been an important inspiration on developing integrated high-level environments for scientific computing in order to bridge the gap between symbolic computation and numerical computation and to increase the power of computing systems for scientific problem solving. Our work on code generation has followed the previous work, also with the idea of integration in mind, for the area of neural network simulation. A project with a similar idea has also been pursued in the MITRE Washington Neural Network Group [LW91]. A declarative language called Aspirin was developed to describe some standard neural network architecture. Some code generators were implemented to generate C code for simulating backpropagation networks form their declarative description. However, the neural algorithms, which are the most essential part for code generation, and CA systems were not taken into account there. In connection with CA, the use and extension of CA systems to deal with neural dynamics have been investigated in [Wan91, WS91a, WS91b]. It was indicated in [Gat90] that the symbolic solution of algebraic equations is useful for locating the critical points of neural systems. Up to this point, the inter-applications of symbolic computation and neural networks are lacking and remain open for future research. Some potential applications were proposed in [WS91c] and we are convinced that a certain amount of research efforts will make those applications possible. It is our expectation to see some successful inter-applications of symbolic computation and neural networks in the future, which will then result significant contributions to the development of both areas.

Acknowledgements. This work has been supported by a grant from Siemens AG, Munich (Dept. ZFE IS INF 2). The authors wish to thank Bernd Schürmann and his group for their cooperation and stimulating conversations in developing the project. Thanks to Wei Shen for her valuable contributions to the development of the code generator.

References

[APR90] J. A. Anderson, A. Pellionisz and E. Rosenfeld (eds.). *Neurocomputing 2: Directions for Research*, The MIT Press, Cambridge, 1990.

[AR88] J. A. Anderson and E. Rosenfeld (eds.). *Neurocomputing: Foundations of Research*, The MIT Press, Cambridge, 1988.

[Bie85] A. W. Biermann. Automatic Programming: A Tutorial on Formal Methodologies, *J. Symbolic Computation*, 1 (1985), 119–142.

[CGG+88] B. W. Char, K. O. Geddes, G. H. Gonnet, B. L. Leong, M. B. Monagan and S. P. Watt. *Maple: Reference Manual*, 5th Ed., WATCOM Publications Limited, Waterloo, 1988.

[Dre92] S. Dreiseitl. *Accelerating the Backpropagation Algorithm by Local Methods*, diploma thesis preprint, RISC-Linz, Johannes Kepler University, Austria, 1992.

[Gat90] K. Gatermann. Symbolic Solution of Polynomial Equation Systems with Symmetry, *Proc. ISSAC'90 (Tokyo, August 20–24, 1990)*, 112–119.

[LW91] R. Leighton and A. Wieland. *The Aspirin/MIGRAINES Software Tools: User's Manual*, Release V4.0, The MITRE Corporation, McLean, USA, 1991.

[RS90] U. Ramacher and B. Schürmann. Unified Description of Neural Algorithms for Time Independent Pattern Recognition, *VLSI Design of Neural Networks* (U. Ramacher and U. Rückert, eds.), Kluwer Academic Publishers, 1990, 255–270.

[RHW86] D. E. Rumelhart, G. E. Hinton, and R. J. Williams. Learning internal representations by error propagation. In *Parallel Distributed Processing*, volume 1, chapter 8. MIT Press, Cambridge, 1986.

[Wan91] D. M. Wang. A Toolkit for Manipulating Indefinite Summations with Application to Neural Networks, *Proc. ISSAC'91 (Bonn, July 15–17, 1991)*, 462-463; *ACM SIGSAM Bulletin*, 25(3)(1991), 18–27.

[WS91a] D. M. Wang and B. Schürmann. Computer Aided Investigations of Artificial Neural Systems, *Proc. IJCNN'91 (Singapore, November 18–21, 1991)*, 2325–2330.

[WS91b] D. M. Wang and B. Schürmann. Computer Aided Analysis and Derivation for Artificial Neural Systems, *IEEE Trans. Software Eng.*, to appear.

[WS91c] D. M. Wang and B. Schürmann. Computer Algebra and Neurodynamics, *Proc. Arbeitsgespräch Physik und Informatik - Informatik und Physik (Munich, November 21–22, 1991)*.

URANO:
AN OBJECT-ORIENTED ARTIFICIAL NEURAL NETWORK SIMULATION TOOL

L. Fuentes, J.F. Aldana, J.M. Troya

Dpto. de Lenguajes y Ciencias de la Computación
Facultad de Informática
Plaza El Ejido s/n
29013 Málaga.SPAIN.
aldana@ctima.uma.es
troya@ctima.uma.es

Abstract: A highly flexible environment has been developed based on the object-oriented paradigm for modelling artificial neural networks (ANNs). This paper propose a hierarchy of classes that models ANN. The design of the hierarchy is characterized by a high degree of modularity, based on parametrizable data structures and autonomous modules. Composition rules of structures and methods enable to build, step by step, more complex structures from simple ones previously defined. One of the most relevant benefits obtained from using an object-oriented approach is related with the definition of multi-networks. URANO (Universe of ANN Object oriented) is a powerful software tool developed using the C++ programming language for building and using prototypes of ANNs in which inherited mechanism can be used for building new ANNs from already existing ones. It is easy to work with and its effectiveness and efficiency have been proven by applications.

1. Introduction.

Neural computing is now in progress and the line of future development is uncertain [1].The ANN research is largely dependent upon the use of computer simulation. However, the appropiate tools for the development and simulation of ANN models and learning algorithms are currently lacking. Many neural networks researchers are not expert at computers. Therefore, they need versatil and flexible tools to express new ANN ideas in a straight forward manner.

In order to improve neural network software simulation, URANO is designed to provide a better philosophy of design and use. Neural network have many things in common with object-oriented methodology, such as construction, concurrency and message passing between the units of the system. Therefore, we find that object-oriented paradigm suits neural network design [2]. Another reason for adopting this methodology to design neural network simulation software is that we want to search for a suitable neurocomputer architecture [3].

Our first approach was the use of a concurrent logic language named Polka, which combines the logic and object-oriented paradigms. The aim of this work was to provide an environment for a flexible and rapid prototyping of new ANN models and its simulation. It should express the concurrent activity and high connectivity of the neural networks in a natural and easy way [4,5]. Owing to these attributes we have been working away at this approach, but concluding that current implementations of these languages lacks of reliability [6].

As a result, we have lastly developed a sequential and procedural ANN simulator preserving object-oriented paradigm advantages by using C++ [7]. The base of the system is a set of abstract classes that allows the design of specialized neural networks models to proceed from a higher level of abstraction than is possible in simulation systems using a procedure oriented methodology.

All of neural network models are similar in that each contains a collection of neurons connected via weighed links. An object-oriented design allows the software simulation system to capture neural models similarities and the capability to easily add distinguishing models properties. The design of URANO is object-oriented itself and allows the description of a ANN model and its dynamics in an object-oriented form too. Most neural networks simulation environments allows an "object-oriented definition". This characteristic is only present in the statements of the specification language, but not in the software system core [8,9].

It should also be mentioned that the advantages provided by an object-oriented language allow the development of a flexible simulation environment in which one can easily extend or modify the functionality of a neural network model. Furthermore, this extension may be accomplished without knowledge of the implementation details of the original neural network software product.

A significant contribution of URANO is that affords the opportunity to create neural network macro-structures. A neural network macro-structure is a set of interactives networks each one specialized in specific task. Heterogeneous macro-structures may be connected by sending a message and after that, networks communicates each other through message passing.

2. Modelling Artificial Neural Networks using an Object-Oriented Software Paradigm.

2.1. Introduction.

As mentioned previously, neural network research depend upon the use of computer simulation. Due to this fact, many ANN simulation languages and environments have been developed. These simulation systems require the user to specify the neural network arquitecture, the description of information flowing and information processing through the use of neural network language [1]. This kind of systems implies learning a new and rigid language.

The specification language normally allows one to define a specific neural model in an object-oriented form. However, if the user wishes to simulate a novel learning algorithm, he must provide the system with a procedure written in the appropiate implementation language. Polymorphism and inheriting object-oriented methods are not present in the definition of the software simulation tool.

On the other hand, if an object-oriented approach is used in the design of the system it supplies the user with generic classes each one models a different component of an ANN. This set of abstract classes provides the root upon which more complex neural networks models will be built. Therefore, the inheritance mechanism now allows system modification to be accomplished in a more abstract fashion.

2.2. Neural Networks Hierarchy.

The resurgence in neural network research has resulted in the development of a wide variety of neural network models. The issues that differentiate the various models are: network topology, computation inside neurons, learning and recall algorithms. All this characteristics are defined in an abstract class called Network. This abstract class form the framework for simulations of specific neural network models. The details of the particular architecture or learning algorithm can be incorporated into the system through the use of inheritance. The software system can be easily extended to improve its functionality, or reused in other systems that require its service.

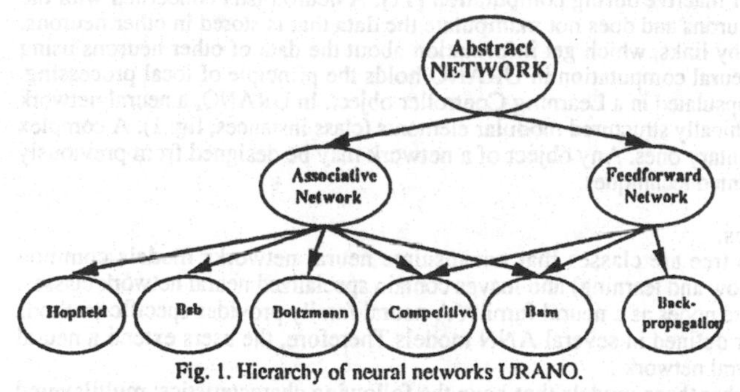

Fig. 1. Hierarchy of neural networks URANO.

The hierarchy of classes created through the use of inheritance is shown in figure 1. This object-oriented approach allows the design of specialized neural network models in an easy manner. Many of the most important aspects of neural network models such as, transfer function, or learning rule are incorporated into the system as external parameters. Users could change the function of the internal objects varying those external parameters, therefore there is no need to write new code.

3. URANO.

3.1. Neural network main components.

The design of an ANN software simulation involves a mapping of the objects and actions occurring in the problem domain to a corresponding set of data in the computer domain of the software system [10].

Basically, the objects of an ANN are the following: neuron, link, port and learning controller. A general purpose neural network abstract class, Set, was developed, which define the common characteristics of ANNs models. Two related classes, Network and Layer, were specified by inheriting the previously developed class Set. The classes Set, Network and Layer represents a group of proccesing elements such as units, layers and neurons.

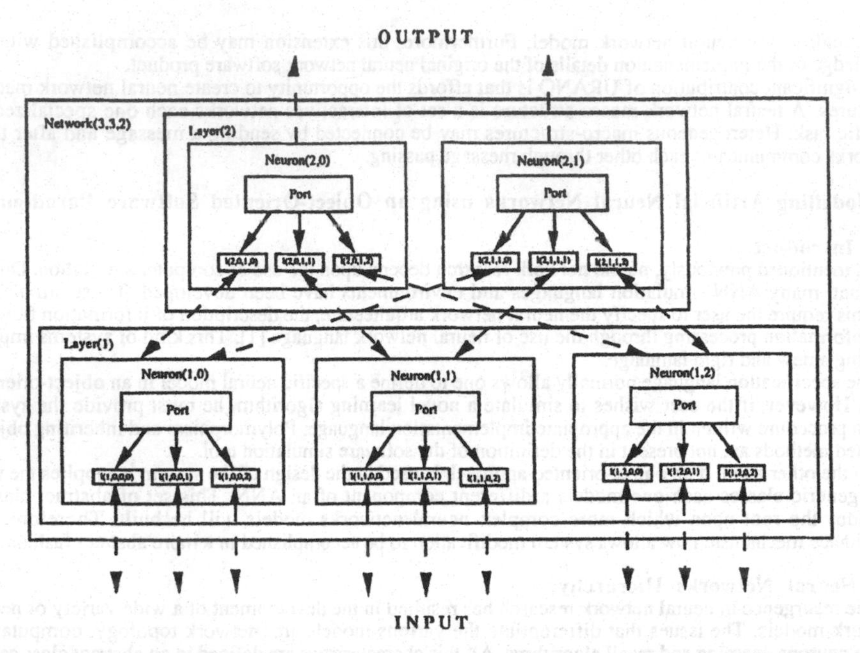

Fig. 2. Representation of a multi-layered feedforward neural network based on its description in URANO.

The neuron constitute the basic processing element of a neural network and is mapped into the Neuron class. This class defines the principal operations of the neural computing. Neurons are viewed as autonomous objects which are linked through ports for communications. Ports and synapsis are similar in the sense that both can be active or inactive during computation [11]. A neuron isn't concerned with the information processing of other neurons and does not manipulate the data that is stored in other neurons. These neurons contact each other by links, which get information about the data of other neurons using message passing. That is to say, neural computation in URANO holds the principle of local processing. Learning strategy and rule are encapsulated in a Learning Controller object. In URANO, a neural network is built from connected and hierarchically structured modular elements (class instances, fig. 1). A complex object may be composed by elementary ones. Any object of a network may be designed from previously defined models through object-oriented techniques.

3.2. Neural Network Families.

The intermediate nodes of the tree are classes that encapsulate neural networks models common characteristics (architecture, data flow and learning) and leaves contain specialized neural network classes. We refer to each of the intermediate nodes as a neural family. A neural family provides specific methods that model the common behaviour defined in several ANN models.Therefore, the users extend a neural family class instead of a whole neural network .

One of the defined families, gather those models that have the following characteristics: multilayered topology, feedforward information flow and supervised learning. URANO implements the corresponding class NetFf that is used to model neural networks like Perceptron, Adaline and Madaline. The model Backpropagation has been constructed by first inheriting the NetFf class, and then adding specialized methods that implement the learning strategy required by this specific model. In addition, a network with self-organized behaviour was simulated by inheriting the class NetOsc which represents all those models. For example Hopfield and Brain-State-In-A-Box networks only differs in the recall algorithm, therefore URANO implement two classes (NetHop, NetBsb) in order to incorporate this specific types of data retrieving.

Networks with competitive learning can be viewed as feedforward networks. Apart from that, each layer presents self organized behaviour. Therefore, the multiple inheritance mechanism allows us to implement competitive network models by inheriting the previously developed classes NetFf and NetOsc.

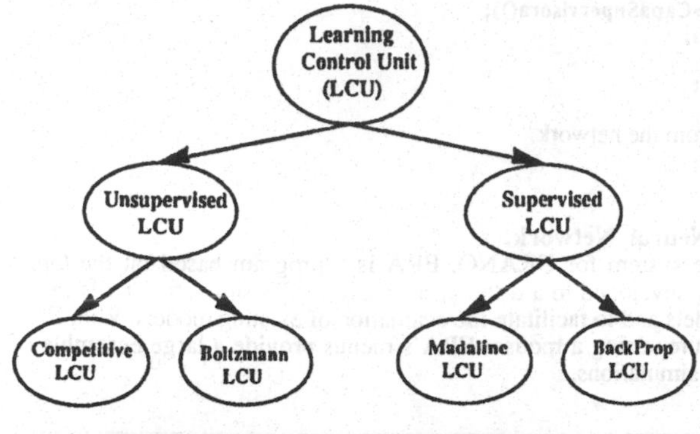

Fig. 3. Hierarchy of Learning Controllers URANO.

ANNs incorporates human learning through strategies and rules that state the adaptation process of a model. The learning methods can be classified in two main groups: supervised and unsupervised. Also, competitive learning is a certain type of unsupervised learning. In order to model different kind of neural learning, URANO defines a hierarchy of classes that is shown in fig.3. The data of the abstract class include: rule, constants and learning rate. The details of the particular learning algorithm was incorporated into the system also through the use of inheritance.

3.3. Neural Network Models implementation.

URANO as a neural network language can describe the neural network model and the process of neurocomputing. Describing the neural network model is mainly to specify the processing elements and their internal functions, and describe the arquitecture of the network. Describing the process of neurocomputing is to specify information flowing among neurons and the information processing. The structure of an ANN program unit which is written in URANO is showed in figure 4.

```
<Network_Program_URANO>::= <Create_Network>
                           <Conect_Network>
                           <Parameters_Network>
                           <Create_Learning_Controller>
                           <Parameters_Learning_Controller>
                           <Train_Network>
                           <Recall_Network>
```

Fig. 4. BNF specification of a neural network simulation model in URANO.

The classic design approach is to define a new procedural and general purpose language. URANO allows users to implement models in an object-oriented manner using the C++ programming language. Designing the neural network topology means to specify a Network object and their components (e.g., Layers, Neurons) sending the corresponding messages.

URANO supply a hierarchy of classes for the simulation of ANNs. A model has an associated class which is instantiated in a simulation model For example to build a backpropagation network we create the object netbp :

```
TipoId layers   = 3;
ListaGE lnl     = <5,3,1>
NetBp *netbp    = new NetBp(layers,lnl,FASigmoid,FPSuma);
```

The object netbp has an input, an output and various hidden layers. Neurons are connected in a feedforward manner.

```
String *port = new String("Bp");
netbp  ->  ConectFull(port,FPSumaPond,1,0.5);
```

The object netbp contains nine instances of the Link class. Neurons communicates each other through objects (Links) that encapsulate a weight. The communication is bidirectional in the sense that in one direction the neuron recieve information and in the opposite one it propagates information. Now we have to initialize neuron parameters (transfer function, threshold, bias, etc.). Neurons of a neural network has always the same attributes which can be change dynamically by sending the corresponding message.

During the search of a suitable network some kinds of units and connections always are created or lost dynamically during the process of neurocomputing. Thus methods to create or delete unit and connection dynamically must be given. Also, URANO defines statements to create, change or delete any object.

Layers in a neural network have learning controllers. In the example we create two controllers for hidden and output layers.

```
netbp -> CrearCa(RADelta);
```

After the creation of netbp, the connection of the neurons, the initialization of the parameters and associate learning controllers to the layers, the network is ready to be train.

```
CapaFf *cs = (CapaFf *)(netbp->CapaSupervisora());
ListaGE Output = AcceptOutput();
cs->BufferEnt(Output);
ListaGE Input_1 = AcceptInput();
netbp->Entrenar(Input_1);
```

At last, we can retrieve information from the network.

```
ListaGE Input_2 = AcceptInput();
netbp->Ejecutar(Input_2);
```

3.4. Dynamic Construction of a Neural Network.

ERA is the graphical user interface system for URANO. ERA is a program based on the Open Windows graphics environment. It was developed to aid the user of URANO´s simulation tools in the testing and evaluation of new ANN models and to facilitate the evaluation of existing models. With ERA the user can create new network structures using a mouse. ERA´s menus provide a large ensemble of functions to store, load, modify and run simulations.

Several common network algorithms along with several processing element activation functions, ports and links synapsis functions are provided. These algorithms and functions can be modified to create completely new networks. Communication between URANO and ERA is performed using a pipe and a simple communication protocol. The entire object-oriented facilities are available to the user. An example screen from ERA is shown in fig.5.

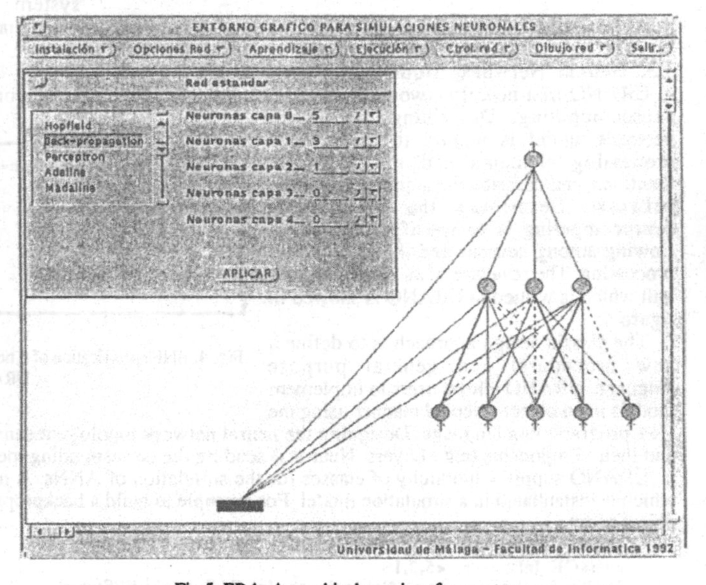

Fig.5. ERA. A graphical user interface system.

The figure illustrates the graphic nature of the program and the pull-down menu controls. Using the menu, the researcher can develop network models with new activation functions quickly and easily. ERA also allows the user to modify code that controls the simulation environment.

We could modify the backpropagation network described in 3.1. with the following methods:

```
netbp -> NuevaNeurona("",2,2,Salida,FAStep,1,FPSuma,0);
```

we add two neurons to the third layer,

```
NetFf *netff = ((NetFf *) netbp);
```

we change the type of the network from backpropagation to feedforward.

```
netff-> SetCa(1,CaSup);
```

and lastly, this statement change the learning controller of the third layer from backpropagation to a supervised one. After the user has made any ANN model modifications, the simulation program must be rebuilt in order for the changes to take effect. The process of rebuilding involves executing a makefile that recompiles models that have been modified and builds a new executable program.

3.5. Neural Networks Macro-structures.

The partition of a problem is some times useful and necessary to achieve the solution. The networks are arranged in macro-layers with the outer layers corresponding to input and output, and they communicates each other by passing real-valued activation across the inter-network links.

These networks could cooperate in the resolution of the

<Program_URANO> :: = <Network_Program_URANO> |
<Program_URANO>
<Network_Program_URANO>

Fig. 6. BNF specification of a neural network program in URANO.

problem. One of the most common fields that applies this kind of neural system is the image recognition [12]. We enlarge the specification of URANO programs (fig. 4.).

In URANO, a neural network model is built from connected and hierarchically structured modular elements. This approach allows the construction of multi-networks which have independent processing and interacts across links. It is possible to generate arbitrarily connected multi-neural networks (e.g. cascade, full-connected, with feedback). Hybrid networks can be built using multiple inheritance or by means of the specification of a macro-structure. For example we may build a two-layer structure composed of a Hopfield network (nethop) and a backpropagation (netbp). The autoassociative network cleans the input patterns, and the backpropagation classifies the M output patterns in N sets. The final neural network model implementation is shown in figure 7.

```
ListaGE In = AcceptNL();
RedHop *nethop = new RedHop(4,FAStep);
RedBp *netbp = new RedBp(3,In,FASigmoid);
String *port = new String("union");
// Conexion of the Backpropagation input layer with Hopfield
// output layer
for(int i=0;i<4;i++)
  ((RedBp *)ConjRedes[1]) -> CrearEnlace(1,0,i,0,0,i,port,1);
CapaFf *cs = (CapaFf *)(ConjRedes[1] -> CapaSupervisora());
ListaGE Output = AcceptOutput();
cs -> BufferEnt(Output);
ListaGE *Ip = AcceptInput();
((RedHop *)ConjRedes[0]) -> Entrenar(Ip);
((RedBp *)ConjRedes[1]) -> Entrenar();
```

Fig. 7. Multi-network = Hopfield network + Backpropagation network. Implementation in URANO.

4. Conclusion.

The object-oriented methodology allows the encapsulation in classes of neural network models common characteristics. We implement a set of abstract classes that may be extended by the use of inheritance. The information in URANO is distributed among objects which cooperate in neural processing. An ANN in URANO is an object that receives messages from the outside, distributes them to its sub-objects, and returns the evaluation result of an internal function. The response may go to the outside or to another network. The object-oriented design of URANO, is suitable for joining networks by links of communication. This allows an easy implementation of macro-structures whose elements cooperate in the resolution of a partitioned problem. All of this gives neural network researchers a designed software development environment for programming and testing standard ANNs models, as well as new ones. The final software product is implemented using Open Windows libraries (ERA) and the object-oriented language C++ (URANO) at a SUN4 workstation. This software is available on a large number of workstations.

5. References.

[1] Gael de la Croix, Catherine Molinoux, Benol Derot. "The N programming Language", NATO ASI Series, Vol. F68 Neurocomputing. pp. 89-92, Springer Verlag Berlin Heidelberg 1990.

[2] Mark Mullin. "Object Oriented Program Design. ", Ed. Addisson Wesley, 1990.

[3] Gregory L. Heilemen, Michael Georgiopoulos, y William D. Roome. "A General Framework for Concurrent Simulation of Neural Network Models". IEEE Transactions on Sotfware Engineering. Vol 18. No. 7, July 1992. pp. 551-562.

[4] J.F. Aldana, E. Pimentel, J.M. Troya. "Modelado de Redes neuronales mediante un lenguaje Lógico concurrente orientado a objeto". Actas de las Jornadas sobre programación declarativa. pp. 561-570. Torremolinos, Málaga, 1991.

[5] Troya, Aldana. 91 "Extending an Object-Oriented Concurrent Logic Language for Neural Networks Simulation". IWANN'91. LNCS no. 540, pp. 235-243, Springer Verlag 1991.

[6] L. Fuentes, J.F. Aldana, J.M. Troya. "Modelado Redes Neuronales mediante un lenguaje lógico concurrente orientado a objeto" Informe Técnico. Dpto de Lenguajes y Ciencias de la Computación. Universidad de Málaga. 1992.

[7] Scott Robert Ladd. "C++ Techniques & Applications", Prentice Hall.

[8] Darkui Shouren Hu. "An Object Oriented Neural Network Language", CH 3065 IEEE pp.1606-1611, 1991.

[9] Leslie S. Smith. "A Framework for Neural Net Specification". IEEE Transactions on Sotfware Engineering. Vol 18. No. 7, July 1992. pp.601-612.

[10] Gregory L. Heileman, Harold K. Brown. "Simulation of Artificial Neural Networks Models Using an Object-Oriented Software Paradigm", pp. 133-136.

[11] L. Fuentes, J.F. Aldana, J.M. Troya. "Modelado de Redes Neuronales con URANO". Informe Técnico. Dpto. de Lenguajes y Ciencias de la Computación. Universidad de Málaga. 1992.

[12] V. Cruz, G. Cristobal, T. Michaux, S. Barquin. "Distortion Invariant Image Recognition by Madaline and Back-Propagation Learning Multi-Networks". NATO ASI Series, Vol. F68 Neurocomputing. pp. 337-342, Springer Verlag Berlin Heidelberg 1990.

Realistic Simulation Tool for Early Visual Processing including Space, Time and Colour Data

William Beaudot, Patricia Palagi & Jeanny Hérault

Laboratoire de Traitement d'Images et Reconnaissance de Formes, Institut National Polytechnique de Grenoble, 46 Avenue Félix Viallet, 38031 Grenoble Cedex, France

Abstract. *This paper presents an efficient implementation on a conventional computer architecture of some early visual processing that occurs in the eye. A retinal spatiotemporal recursive filter, based on an extended version of Mead's model, has been implemented by means of an original algorithm, much faster than FFT (18 times for a 256 x 256 image). Various realistic aspects have been added, particularly non-homogeneous filtering by the crystalline lens, photoreceptor coupling, chromatic sampling and digital filtering with an irregular spatial sampling. The integration of these unconventional components of early visual organisation into a single coherent system leads to the development of clever digital realisations: among the many possibilities of this simulation tool, three of them are illustrated, concerning space-variant and colour processing. The program, written in the C language, allows synthesis, filtering and the display of colour images with moving objects.*

1. Introduction

When looking at a scene, it is hardly conceivable that our eye is not a perfect sensor like the ideal one we can imagine: first, our crystalline lens induces strong chromatic and geometric aberrations, the retina itself is an non-homogeneous sensor, blood vessels circulate on its surface, a blind spot in our visual field as large as an orange held at arm's length, visual acuity is high only in a small area called the fovea, strongly decreasing with the eccentricity of incoming light. Moreover, unremitting and unconscious micro-movements of the eye mean that images are never stable on the retina.
The question is, why, due to these many imperfections, we perceive stable visual fields without any of the above-mentioned defects. To answer this question requires an important theoretical work supported by a user-friendly simulation tool.
In this paper, we stress the role of the crystalline lens and the non-homogeneous spatiotemporal filtering of the retina for visual perception, and we show that some of their uneven aspects can be simulated in an efficient way on a conventional computer architecture. First, we present a digital model of retinal processing that takes into account the coupling between light sensors, as well as the neural structure of the foveal area in the first functional layer of the vertebrate retina. Then, a non-homogeneous, low-pass filter and an irregular sampling grid on which our retinal filter can be easily applied, are provided to simulate optical filtering of the crystalline lens and the variable sampling of the retinal area by photoreceptors. Next, a chromatic sampling of this grid combined with the same retinal filter shows a basic characteristic of colour vision: colour constancy. Finally a combination of all these processings forms an efficient system for early vision that can be at the root of the smart sensor concept (Fig. 1).

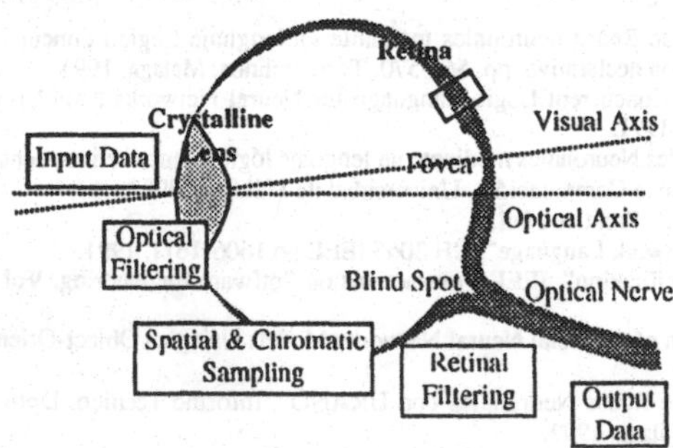

Fig. 1. Simplified Eye. Two elements are taken into account: the crystalline lens for optical processing and the retina for neural processing. Non homogeneous features are obvious: non uniform aspects of the retina (foveal depression, blind spot and variation with eccentricity), imperfect lens (geometric aberration) in retinal periphery. (Fig. 2 shows an enlargement of the framed part of the retina.) A Global Architecture of the proposed system for early visual processing is superimposed on the eye. It takes trichromatic images as input, lens filtering is applied to them, then the result is sampled by a variable spatial sampling and by a chromatic sampling which reduces data to one intensity value per pixel. Afterwards the retinal filter is applied. The output of the retina can then be used to detect edges or motion in input images.

2. Retinal Processing

Our modelling of retinal processing is based on a structural and functional approach. A structural model of the vertebrate retina is then presented that has not only a functional validity but also a predictive validity. Before presenting it, some biological considerations are worth mentioning about two aspects: neural architecture and neural processing in the retina.

2.1 Retinal Architecture

The retina is the first neural structure involved in vision. This means that it is the primary neural network for early vision processing. The vertebrate retina, from reptilian to mammalian species, is made up mainly of five neural layers (Fig. 2a): the photoreceptor layer where light is transduced into membrane potential, the bipolar cell layer that transmits information from photoreceptors to ganglion cells, while the horizontal and amacrin cell layers are involved in lateral processing.

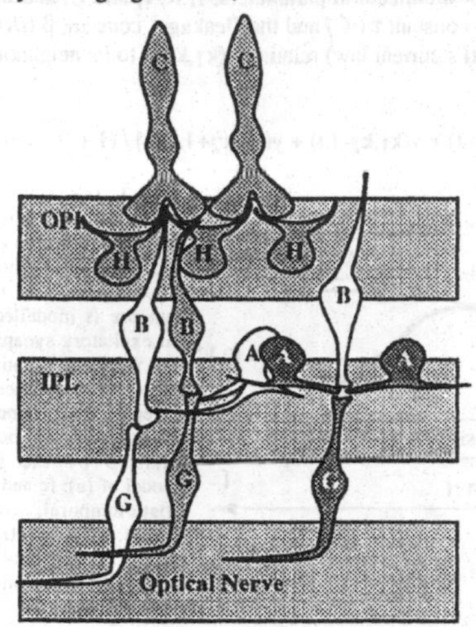

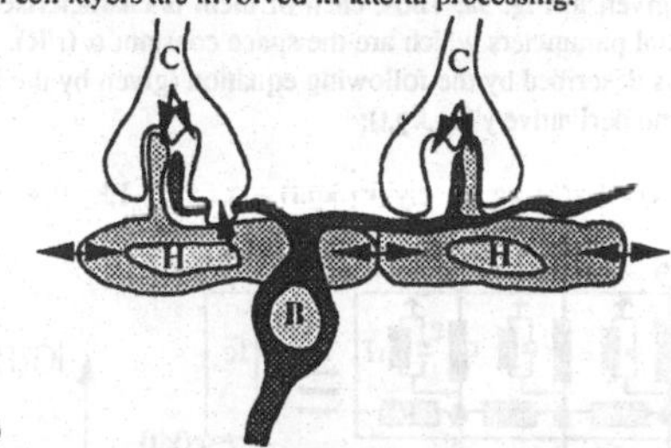

Fig. 2. (a) General architecture of vertebrate retina. Two functional layers are presented: the Outer Plexiform Layer (OPL) where synaptic communication between photoreceptors (C), horizontal (H) and bipolar (B) cells takes place, and the Inner Plexiform Layer (IPL) where synaptic communication between bipolar, amacrin (A) and ganglion (G) cells takes place. (b) The Triadic Synapse is a complex synaptic structure where interactions between three types of neurons take place. It is characterised by a ribbon-type synapse in the photoreceptor terminal and by an invagination where three dendritic processes extend: the bipolar process is central between processes of two different H-cells (Dowling 1992a). The arrows indicate synaptic communication between cells.

Basic retinal processing seems to be divided into two fine layers where synaptic interactions between retinal cells can be found: the first involved in visual processing is the outer plexiform layer (OPL) and the second is the inner plexiform layer (IPL). The division into two structural layers is accompanied by a difference in the type of information transmission in neurons: the cells in OPL present only graded potentials while the cells in IPL can generate action potentials. At present, our model only takes OPL into account. Fig. 2b shows a typical synaptic structure in OPL: the triadic synapse.

2.2 Neural Processing

The processing carried out by an artificial or biological neural network depends on its topology, on the types of synapses between neurons and on the processing of each neuron. Neurophysiological data about synaptic communication among retinal neurons help us to understand neural processing in the retina. The triadic synapse seems to be the more interesting structure from this point of view (Fig. 2b). Two types of synapse are found in it: cone-to-horizontal synapses that are excitatory chemical synapses and cone-to-bipolar inhibitory chemical synapses. H-cells are linked together through particular synapses called electric synapses or gap-junctions which make fast bi-directional electrical coupling between neighbouring H-cells.
From a functional point of view, it has been well-known for a long time that a lateral inhibition occurs in OPL. It is often modelled by a Laplacian operator or a DOG operator (Difference Of Gaussians) (Marr and Hildreth 1980, Marr 1981) that is suitable for edge detection. A more realistic understanding of the role of the retina must take into account the temporal processing. If we consider the lateral inhibition as a process of redundancy reduction, a temporal inhibition is also required (Srinivasan and al. 1982). In order to overcome these difficulties in establishing the real function (if there is one) of the retina in vision, we have developed an architectural model of the retina and to study our model according to two aspects, functional and predictive validity.

2.3 An Analog Model Involving Coupling Between Receptors

The model presented in this paper was inspired by Mead's silicon model of vertebrate retina (Mead and Mahowald 1988) to which we have added the following extensions in order to develop a more realistic model (Fig. 3a): an electrical coupling between photoreceptors (f_c layer), a leaky integrator (r_f and C components) for the membrane characteristics of each neuron, unidirectional communication via operational amplifiers and even an inhibitory feedback from H-cells to photoreceptors (not shown in the figure) as emphasised by (Siminoff 1984).
The processing that occurs in this analog retina is given by the application of the Kirchhoff's current law at each node of this circuit. Then, a spatiotemporal frequency analysis of this model is made with signal processing tools (Fourier and Z transforms) for time and space functions respectively, from which a digital filter can be deduced (Beaudot and Hérault 1992, Beaudot 1992).

2.4 Digital Model with an Efficient Cascade of Causal and Anticausal Spatial Recursive Filters

The processing of this model can be formulated by the following expression, which is schematised in Fig. 3b:

$$b(k_1,k_2,t) = f_c * [\delta - f_h] * e(k_1,k_2,t)$$

where $e(k_1,k_2,t)$ is the input data for k_1 and k_2 space samples at continuous time t, and $b(k_1,k_2,t)$ is the output data of the model, that is the bipolar cells' input. In this expression, f_c and f_h are spatiotemporal, low-pass filters and δ is the spatiotemporal Dirac function, * indicates the convolution. Both spatiotemporal low-pass filters f_c and f_h are realised by the electronic circuit given in Fig. 3a. Thus, each of them is characterised by four architectural parameters, r, R, r_f and C, and by three functional parameters which are the space constant α (r/R), the time constant τ (rC) and the "leakage" constant β (r/r_f). This circuit is described by the following equation (given by the Kirchhoff's current law) relating $y(k_1,k_2,t)$ to its neighbours and to its time derivative $y'(k_1,k_2,t)$:

$$y(k_1,k_2,t) = [\, x(k_1,k_2,t) - \tau.y'(k_1,k_2,t) + \alpha.(\, y(k_1-1,k_2,t) + y(k_1+1,k_2,t) + y(k_1,k_2-1,t) + y(k_1,k_2+1,t)\,)\,] \,/\, (1 + \beta + 4\alpha) \quad (1)$$

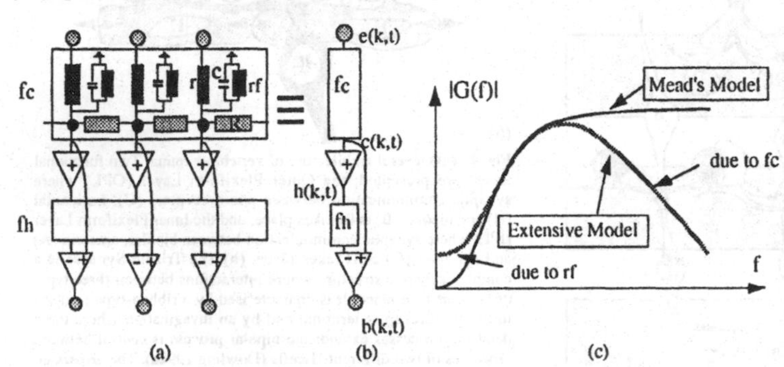

Fig. 3. (a) 1D Analog model of the vertebrate retina: it consists of two resistive lattices with leaky integrators, the triadic synapse is modelled with an excitatory synaptic (+) and with a differential amplifier between these two layers. (b) Functional description of the processing realised by the analog model of (a): fc and fh are spatiotemporal, low-pass filters. (c) Spectral response of the spatial retinal filter (Mead's model and the new one).

Thus, a recursive, non-causal spatial term appears in this expression. Moreover, a derivative term in time also appears. Then, a discrete approximation of temporal derivative is necessary to build a digital model. With the classical approximation, $y'(t) \cong [y(t) - y(t - \Delta t)] / \Delta t$, the previous expression becomes:

$$y(k_1,k_2,t) = [\, x(k_1,k_2,t) + \tau.y(k_1,k_2,t-\Delta t) / \Delta t + \alpha.(\, y(k_1-1,k_2,t) + y(k_1+1,k_2,t) + y(k_1,k_2-1,t) + y(k_1,k_2+1,t)\,)\,] \,/\, (\,1 + \beta + 4.\alpha + \tau/\Delta t\,) \quad (2)$$

This expression can be rewritten as:

$$y(k_1,k_2,t) = u(k_1,k_2,t) + \alpha\,[\, y(k_1-1,k_2,t) + y(k_1+1,k_2,t) + y(k_1,k_2-1,t) + y(k_1,k_2+1,t)\,] / (\,1 + \beta + 4.\alpha + \tau/\Delta t) \quad (3)$$

$$\text{with } u(k_1,k_2,t) = [x(k_1,k_2,t) + \tau.y(k_1,k_2,t-\Delta t) / \Delta t] \,/\, (\,1 + \beta + 4.\alpha + \tau/\Delta t\,)$$

Equation (3) leads to a purely recursive, non-causal spatial low-pass filter that can be approximated by a recursive filter separable in k_1 and k_2 and separable in spatial causal (c) and anticausal (a) filters (Beaudot 1992). That is:

$$y(k_1,k_2,t) = g^c{}_1(k_1) * g^a{}_1(-k_1) * g^c{}_2(k_2) * g^a{}_2(-k_2) * u(k_1,k_2,t) \quad (4)$$

where $g_i{}^c(k_i)$ et $g_i{}^a(-k_i)$ indicate the impulse responses of the causal filter and the anticausal filter respectively on the k_i axis. Convolutions by $g_i{}^c$ and $g_i{}^a$ are given by: $y^c(k_i) = g_i{}^c(k_i) * x(k_i)$ and $y^a(k_i) = g_i{}^a(k_i) * x(k_i)$ and can be computed respectively by the procedures:

$$y^c(k_i) = x(k_i) + \lambda\,.\, y^c(k_i-1) \qquad (k_i = 1 \text{ to } N) \quad (5a)$$

and

$$y^a(k_i) = x(k_i) + \lambda\,.\, y^a(k_i+1) \qquad (k_i = N \text{ to } 1) \quad (5b)$$

$$\text{with } \lambda = 1 - (\,1 - 4.\sigma)^{1/4} \text{ and } \sigma = \alpha / (1 + \beta + 4.\alpha + \tau/\Delta t)$$

Both filters f_c and f_h can be realised by such an algorithm. This realisation is a rather good approximation of the real processing that occurs in the analog circuit, and this algorithm is simple and fast (it is computed in $4N^2$ time steps for a NxN image). Therefore, a good efficacy is achieved without needing of high-parallelism. Moreover, we can stress the fact that the processing of each layer can be controlled by three parameters: α, β and τ, that are not independent, they can be locally altered by factors such as signal-to-noise ratio or by the introduction of a non-linear, gap-junction conductance. The previous computation can take into account this possibility, that is plausible from a neurophysiological point of view (Srinivasan and al. 1982, Piccolino 1988, Usui and al. 1988, Witkovsky and al. 1989, Dowling 1992b). In this way, space-variant and non-linear processing can be easily realised and simulated on a classical computing architecture: one 128 x 128 image is computed for the whole retinal processing every second on a 486 Personal Computer.

3. How to take into account Crystalline Lens and Irregular Sampling

Notions of attention, foveal sensing or gaze control are some of the present research directions in active vision (NSF Active Vision Workshop 1991). Two aspects of the structure of the eye are involved in these concepts: the crystalline lens and the retinal mosaic.

3.1 Space-Variant Filtering by the Crystalline Lens

The crystalline lens is the primary component of the eye's optics and is strongly involved in the accommodation mechanism. It is not a perfect lens and has some physical limitation in transfer of spatial frequencies over 60 cycles/degree, this might lead to the minimisation of moiré patterns in spatial sampling (Wässle and Boycott 1991).
The point-spread function due to the crystalline lens is narrow in the vicinity of the optical axis and widens according to the eccentricity of input light. This results in a retinal image which is more blurred (i. e. low-pass filtered) as it moves away from the centre of the image (Fig. 5a).

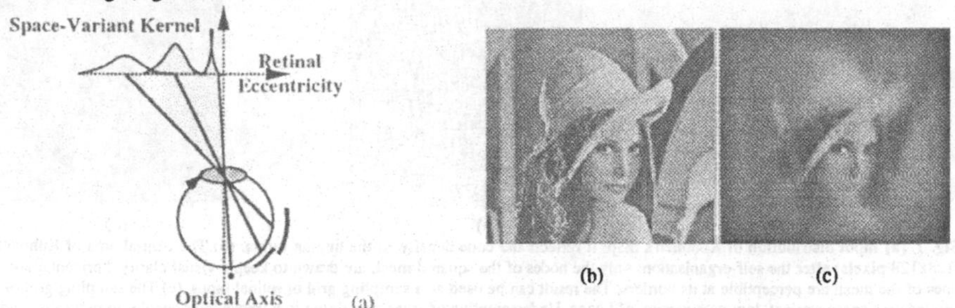

Fig. 5. (a) Crystalline lens processing is typically a space-variant filtering: the point-spread function of the eye's optics depends on the angle between the optical axis and the light beam, that is on the retinal eccentricity. (b) Original image of Lenna. (c) Result of the application of the space-variant, low-pass filter on Lenna (In expressions (5a) and (5b), the parameter λ is chosen according to the value of ki relatively to the central point of the image).

Computing this space-variant processing can be very expensive in time and in space if a non-stationary, finite-impulse-response filter is used. The previous section has shown that a spatial, low-pass filter in the retina can be computed with a recursive filter (IIR) that can be factorised into causal and anticausal filters in the two spatial dimensions (see expression 4). Moreover, the low-pass effect is controlled by a single parameter. Thus, we use the same type of filter for the computation of the space-variant processing of the lens. If $I(k_1,k_2)$ is the input image and $F_o(k_1,k_2)$ the impulse response of the optical filter, then the result of the lens filtering is written:

$$O(k_1,k_2) = \sum_{u,v} I(k_1-u,k_2-v) \,.\, F_o(k_1,k_2,u,v) = F(k_1,k_2) \circ I(k_1,k_2) \text{ with } F = g^c_1 \circ g^a_1 \circ g^c_2 \circ g^a_2$$

with g_i^c and g_i^a given by: $y^c(k_i) = x(k_i) + \lambda(k_1,k_2) \,.\, y^c(k_i-1)$ and $y^a(k_i) = x(k_i) + \lambda(k_1,k_2) \,.\, y^a(k_i+1)$ and where λ is the space-variant parameter which must be given by a function of the retinal eccentricity: $\lambda(k_1,k_2) = \xi(k_1^2+k_2^2)$.
Fig. 5b and 5c show the result of such a filter.

3.2. Model of Irregular Sampling by means of a Kohonen's Map

The organisation of the retina is not homogeneous over its whole surface. The retinal eccentricity is the main factor in the irregularity of the receptor lattice. However, even in the foveal area where the cone density is the highest, the receptor lattice is not as perfectly regular as we might think: indeed random variations of the distance between photoreceptors are always present and can serve to reduce aliasing caused by sampling (Jimenez and Agüi 1987, Bilinsky and Mikelson 1990, Ruderman and Bialek 1992). Outside the fovea, the receptor lattice becomes less regular as the low-pass optical filtering increases. These apparent imperfections in the optical quality of the eye and the retinal mosaic seem to have a biological importance (Wässle and Boycott 1991).

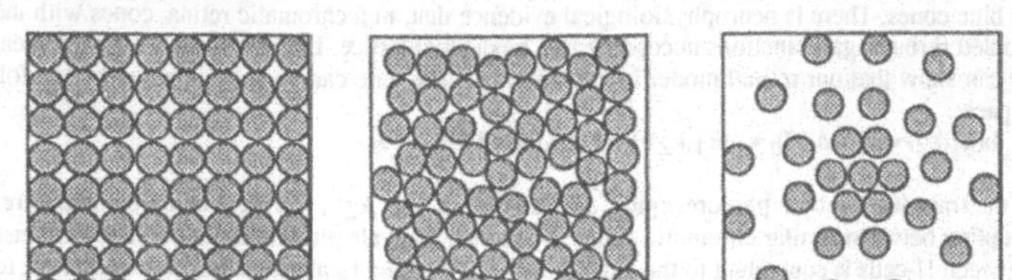

Fig. 6. (a) The regular square sampling adapted to the model of retinal filtering. (b) A more realistic sampling that occurs in the fovea. (c) A space-variant sampling with random variations in a vertebrate retina.

The simulation of the retinal filtering preceded by a variable sampling is not obvious: the retinal computation presented in section 2 was adapted to input data on a squared mesh (Fig. 6a). The introduction of a random disorder is not easy either (Fig. 6b), and furthermore, what about the addition of an eccentricity-dependent sampling (Fig. 6c)?

Thus, the main problem is to find an adequate representation of an irregular lattice while keeping the advantages of our retinal filtering. Self-organisation principles using a Kohonen's map can be very suitable to realise such a variable sampling: let us consider an imaginary neural network characterised by 4-connectivity neurons on a 2D grid that is associated with the sampling lattice. If the photoreceptor density is applied as input data to this network, and if this network self-organises according to the Kohonen's algorithm (Kohonen 1986), we obtain a receptor lattice matched with the input distribution that reflects the photoreceptors location in the retina. So the topology of this Kohonen's map is convenient for our retinal filtering. Of course, due to the space-variant sampling, the digital filter should also have space-variant constants. Fig. 7 shows the result of this realisation.

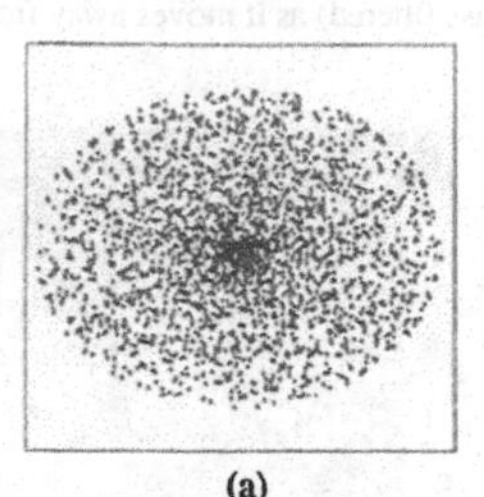
(a)

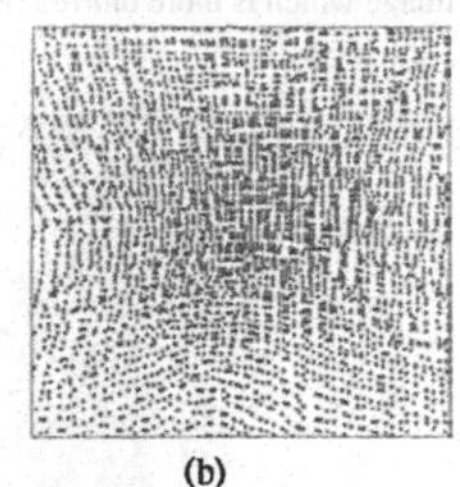
(b)

(c)

Fig. 7. (a) Input distribution of Kohonen's map: it reflects the cone density in the human retina. (b) The central area of Kohonen's map (128x128 pixels) after the self-organisation: only the nodes of the squared mesh are drawn to keep a visual clarity, horizontal and vertical lines of the mesh are perceptible at its borders. The result can be used as a sampling grid of retinal input. (c) The sampling grid of (b) was applied on a space-variant, low-pass version of Lenna. Undersampling of areas containing low frequencies enables to reduce redundancy.

3.3 Link between space-variant filter and variable sampling

In signal processing, low-pass filtering and sampling rate are linked. An undersampling leads to the addition of aliasing in the signal, so Shannon's sampling theorem is no longer respected. Indeed, this theorem states that recovering the original signal from its samples requires that the original did not contain frequencies above half the sampling rate (the so-called Nyquist frequency). To remove frequencies above the Nyquist frequency, a low-pass filtering is necessary. If the sampling is variable, like in the retina, the low-pass filtering must be space-variant, like that due to the crystalline lens. This idea is very attractive from a signal processing viewpoint. Fig. 7c shows the result of the variable sampling (Fig. 7b) on the low-pass version of Lenna (see Fig. 5c).

4. Retinal Chromatic Sampling: a process which eliminates incident light colour

The chromatic sampling is another aspect of the retina that is often left aside and its functional role is not clear yet (Siminoff 1984, 1991). The combination of the retinal filtering and the chromatic sampling can explain how colour is coded in early visual stages, and why we perceive coloured objects in a stable way in spite of the great variations of illumination (colour constancy).
Let us consider the Stockham model of images (Stockham 1972):

$$I(\mathbf{k}) = \rho(\mathbf{k}) \,.\, L(\mathbf{k}) \qquad (6)$$

where $I(\mathbf{k})$ is the light intensity received by a photoreceptor at the location $\mathbf{k}$, that is the product of the reflectance ρ by the illuminance L. This model is valid if the illuminance is slowly variable and the reflectance consists of homogeneous areas with strong transitions. It is clear that the reflectance is then associated with high frequencies while the illuminance is associated with low frequencies.
The problem of colour constancy is to determine spectral reflectance (intrinsic property of objects) from the light intensity. In signal processing, a non-linear system can separate $\rho(\mathbf{k})$ and $L(\mathbf{k})$ in (6): $\log I(\mathbf{k}) = \log \rho(\mathbf{k}) + \log L(\mathbf{k})$. Applying a high-pass filtering to this expression can then help to recover the reflectance $\log \rho(\mathbf{k})$.

Does the retina realise such a processing to code colour ? It is well known that three types of cones are present in a normal chromatic retina: red, green and blue cones. There is neurophysiological evidence that, in a chromatic retina, cones with the same spectral properties are coupled through gap-junctions according to a hexagonal lattice. Let us consider now the ideal chromatic mosaic of Fig. 8. We can show that our retinal model in a temporal steady state can be approximated by the following expression for each 1D space:

$$b(k_1,k_2) = f_c * \Delta\,[\, f_h * e(k_1,k_2)\,] = \text{Low-Pass} * \text{High-Pass}$$

where $e(k_1,k_2)$ is the result of the transduction by a photoreceptor, $e(k_1,k_2) = \log\ i(k_1,k_2)$, $i(k_1,k_2)$ being the signal received by the photoreceptor. Coupling between similar chromatic cones (f_c) is then equivalent to the finite-impulse-response filter [1 0 0 2 0 0 1], coupling between H-cells is equivalent to the FIR filter [1 2 3 2 1] and Laplacian filter Δ is equivalent to the FIR filter [-1 2 -1]. Fig. 8b shows an example of this processing on an input data with two adjacent coloured areas. For simplicity, the f_c filtering is not applied but the result is not fundamentally modified.

First, it is clear from this simple example that the colour characteristic is only coded at the boundary between coloured objects by a relative value (logarithm of intensity ratio). The second important point concerns the colour constancy mechanism, if we use (6) to model the light intensity, we obtain the new forms of the intensity received by the photoreceptors:

$$R_1 = \rho_{r1} \cdot L_r \qquad G_1 = \rho_{g1} \cdot L_g \qquad B_1 = \rho_{b1} \cdot L_b$$
$$R_2 = \rho_{r2} \cdot L_r \qquad G_2 = \rho_{g2} \cdot L_g \qquad B_2 = \rho_{b2} \cdot L_b$$

where $(\rho_{r1}, \rho_{g1}, \rho_{b1})$ stands for the reflectance property of area 1 and $(\rho_{r2}, \rho_{g2}, \rho_{b2})$ the reflectance property of area 2, while (L_r, L_g, L_b) refers to the spectral properties of the illuminance that is considered as slowly variable.

1)	Colour 1				Colour 2			
2)	B1	R1	G1	B1	R2	G2	B2	R2
3)	b1	r1	g1	b1	r2	g2	b2	r2
4)	1	1	$\frac{R1}{R2}$	$\frac{G1.R2}{R1.G2}$	$\frac{B1.G2}{G1.B2}$	$\frac{B2}{B1}$	1	1
5)	1	$\frac{R1}{R2}$	$\frac{G1}{G2}$	$\frac{B1}{B2}$	$\frac{R2}{R1}$	$\frac{G2}{G1}$	$\frac{B2}{B1}$	1

Log — LP — HP

Fig. 8. Two adjacent coloured areas (Colour 1 and Colour 2) are presented as retinal input (line 1). Each receptor receives a light intensity (upper case letters in line 2) that differs for the two areas. This signal is then transduced via a logarithmic operator (lower case in line 3). Filter [-1 2 -1] is next applied (line 4), as filter [1 2 3 2 1] (line 5). In lines 4 and 5, the logarithmic operator is removed. (Nota: r = log R)

The last line of Fig. 8 shows only monochromatic ratios, thus the monochromatic illuminance component is cancelled in each term and only monochromatic reflectance ratios remain in the retinal output. Colour constancy is then achieved through an efficient coding of the difference of *colour* between adjacent objects. However, the chromatic mosaic is not as perfect as it was presented even in the fovea (Siminoff 1991). Other mosaics have also been investigated (Palagi 1992) to take into account the non-homogeneity of the chromatic sampling in the retina (foveal and parafoveal areas) and have shown the same properties with respect to the colour coding.

5. Discussion

Several aspects of early visual processing have been presented in this paper: first an architectural model of the retina from which a computational and efficient model of the retinal processing was deduced, a space-variant filtering due to the combination of the crystalline lens, the variable sampling and finally a chromatic sampling in combination with the retinal filtering which leads to some interesting properties for the colour coding. All of these aspects have been integrated into a single software package used as a simulator of the early visual processing on a computer architecture based on a 80486 microprocessor. The adaptation of the used algorithms on a DSP (Digital Signal Processor) architecture is being considered in order to realise a real-time system. Theoretical studies of all these aspects are also investigated to always improve our understanding of the vertebrate early vision. Thus, analysing optical and retinal processing can be highly helpful in understanding biological vision. Moreover, as designers of artificial visual systems, we are interested in applying such a model to computer vision and the coding of movement for HDTV pictures which will be reported elsewhere.

References

Beaudot W (1992) Traitement Spatio-Temporel d'Images par Modèle de la Rétine. Rapport de DEA, Laboratoire TIRF, Grenoble (France)

Bilinsky I, Mikelson A (1990) Application of randomized or irregular sampling as an anti-aliasing technique. Signal Processing V: Theories and Applications, pp 505-508

Dowling JE (1992) Synapses in the Retina. In: Neurons and Networks: An Introduction to Neuroscience. The Belknap Press of Harvard University Press, Cambridge, Massachusetts, pp 318-321

Jimenez J, Agüi JC (1987) Approximate reconstruction of randomly sampled signals. Signal Processing 12, pp 153-168

Kohonen T (1986) Self-Organization and Associative Memory. Springer Verlag, Berlin

Marr D, Hildreth E (1980) Theory of edge detection. Proceedings of The Royal Society of London, B 207, pp 187-217

Marr D (1982) Vision. W-H Freeman and Company

Mead CA, Mahowald MA (1988) A silicon model for early visual processing. Neural Networks, vol. 1, n° 1, pp 91-97

NSF Active Vision Workshop (1991) Promising Directions in Active Vision. University of Chicago Technical Report CS91-27

Palagi P (1992) Vision des Couleurs pour un Modèle de la Rétine. Rapport de DEA, Laboratoire TIRF, Grenoble (France)

Piccolino M (1988) La vision et la dopamine. La Recherche n°205, vol. 19, pp 1456-1464

Ruderman DL, Bialek W (1992) Seeing beyond the Nyquist Limit. Neural Computation, Vol. 4, n°5, pp 682-690

Siminoff R (1984) Electronic simulation of cones, horizontal cells and bipolar cells of generalized vertebrate cone retina. Biological Cybernetics, Vol. 50, n° 3, pp 173-192

Siminoff R (1991) Simulated bipolar cells in fovea of human retina: I. Computer Simulation. Biological Cybernetics, vol. 64, n° 6, pp 497-510

Srinivasan MV, Laughlin SB, Dubs A (1982) Predictive coding: a fresh view of inhibition in the retina. Proceedings of The Royal Society of London, B 216, pp 427-459

Stockham T (1972) Image processing in the context of a visual model. IEEE (special issue on picture processing), 60:828-842

Usui S, Kamiyama Y, Sakakibara M (1988) Physiological engineering model of the retinal horizontal cell layer. ICNN IEEE, vol. II, pp 87-93, San Diego 1988

Wässle H, Boycott BB (1991) Functional architecture of the mammalian retina. Physiological Reviews, Vol. 71, n°2, pp 447-480

Witkovsky P, Stone S, Tranchina D (1989) Photoreceptor to Horizontal Cell Synaptic Transfer in the Xenopus Retina: Modulation by Dopamine Ligands and a Circuit Model for Interactions of Rod and Cone Inputs. Journal of Neurophysiology, vol. 62, n°4, pp 864-881

Language Supported Storage and Reuse of Persistent Neural Network Objects

CHRISTOPHER BURDORF (cb@maths.bath.ac.uk)
School of Mathematical Sciences, University of Bath, Bath, Avon, United Kingdom, BA2 7AY

Keywords: neural software, object-oriented systems, neural networks, simulation, database systems

Abstract. This paper describes a language facility which supports storage and reuse of neuron objects. Neuron objects are made persistent as a part of the POCONS language. Subsequently, these objects with all the accumulated attributes can be reused by other applications in a transparent manner. Performance improvements result from reusing objects rather than using new ones.

1 Introduction

Due to the recent fervor in the use of connectionist and neural systems, there has been active development in tools that support their representation and execution (eg. P3[15], Mirrors/II[9], RCS[10], Neula[11], Slonn [16], and NSL [17]). However, even though neural networks can require a large amount of CPU time to train, none of them provide an elegant mechanism for their permanent storage and reuse. The problem is that neural networks take large amounts of CPU time to train from which point they are used over and over again. Therefore the problem which this paper solves is that of providing an elegant mechanism for storage and retrieval of trained neural networks. The mechanism is now a built-in feature to POCONS [3] (Persistent Object-based Connectionist Simulator). It supports permanent storage and reuse of neural networks through the use of persistent objects. Persistent objects are defined as objects that outlive program execution. Their life-term is extended by residing on secondary storage. Persistent objects can be referenced in a program in the same manner as conventional objects.

The maintenance of persistency is transparent to the programmer and is handled by the underlying persistent object system (POS) which is a component of the POCONS system. Previously, POCONS only supported the storage of objects used to describe the neural network, but this new feature allows the state of the network to be checkpointed, stored as persistent objects, and later reused from it's checkpointed state.

POCONS is a component of the Persistent Simulation Environment (PSE) [5, 6, 7] which has facilities to support event, process, Petri net [2], and connectionist or neural network simulation models.

2 The POCONS Connectionist Model

The connectionist model used in POCONS consists of a set of neurons $\{n_1, n_2, ..., n_k\}$ with the weighted link between n_i and n_j being w_{ij}. It's inferencing mechanism is based on Neula's. Inferences are made by propagating neuron values using the activation propagation algorithm which is based on Hebbian learning [12].

$$a_i^{k+1} = a_i^k + sigm(\sum_j w_{ij} a_j^k) \quad (1)$$

a_i^k represents the current activity level of node a. The summation means to sum the multiplication of the outgoing arcweights with the activity levels of the nodes they link to. The *sigm* applies a threshold function to the summation to keep it within the boundaries of $[1, -1]$.

Neural networks also make use of other training algorithms such as backpropagation [15] which modifies the weights to make the network recognize different patterns. Successful training of a neural network using backpropagation can require many hours of CPU time. Backpropagation is not included as a utility in POCONS, because there are many different types of training algorithms which developers may use. For this reason, POCONS leaves it to the user to add their own training algorithms as needed.

3 Object-Oriented Connectionist Model

Amongst the concepts of object-oriented languages, the one that is of the most fundamental importance is the concept of object classes. In object-oriented languages, abstract data types called object classes, organize both data structures and functions. Each class specifies its relation to existing classes by supplying a reference to the base class, or class from which it is derived. The object class is used similar to a conventional type allowing instances of its class to be instantiated at the runtime of the program.

The notion of a object class hierarchy dates back to Simula [8]. An object class hierarchy tree grows as new classes are added to the system. Since each new class is a superset of its base class, it inherits selected data and functions, meaning that the derived class, when given access by the inheritance rules can call any of its base class functions, or modify any of the data.

POCONS is based on the object-oriented connectionist model where the user does not specify any procedural information about the network's execution. The model only requires that the user specify the neurons which represent the components of the network, their attributes, and relationships between them. POCONS can then be instructed to generate a neural network. Queries can be made on the network which initiate connectionist simulations.

4 Building an Object-Oriented Connectionist Model

The following sentences were originally presented by Anderson [1]: 1. *John is a tall lawyer.* 2. *The lawyer kicked a dog.* 3. *John kicked a model.* 4. *The model's name is Jane.* 5. *The model John kicked owns a car.*

The knowledge in these sentences can be represented in POCONS in the following way:

2pt12pt

```
(defdbneuron person (newron)  ;NEWRON is the
  ((owned initform nil)       ;top-level object.
   (height initform nil)))
(defdbneuron lawyer (person)
  ((owned  initform 'dog)))
(defdbneuron john (lawyer)
  ((kicked initform 'model)
   (height initform 'tall)))
(defdbneuron model (person) ())
(defdbneuron jane (model)
  ((owned  initform 'car)))
(defdbopposites lawyer model)
```

For a description of the primitives used in the above example see [3].

5 Conversion of Objects to a Neural Network

Once the neurons have been defined, the user must instruct the system to build the internal representation of the network by executing the function **build-neural-network**. The underlying system then converts POCONS objects into a connectionist network. The conversion algorithm examines each object and creates forward links from each subclass object to each superclass name. Links are also created from class names to their attributes. As the connectionist simulation executes, the inheritance mechanism propagates values down from class to subclass (Figure 1). Also, there are back links from subclass to class with a fraction of the weight as the links from class to subclass. The back links exist to enable, for example, the activity value for *person* if the one for *lawyer* is enabled. However, the links from class to subclass have larger weights to make the model consistent with the inheritance mechanism in object-oriented systems.

The structure of the executable neural network does not use persistent objects, so that it can contain hard pointers which are faster but cannot be stored to the database. Soft pointers can be stored to the database, but require greater overhead due to a table lookup for dereferencing. Also, the attributes of the neuron get updated frequently as it executes and there would be too much overhead if the slots were persistent. Thus, by avoiding the storage of frequently updated slots at execution time, performance is greatly improved.

6 Persistent Object Systems

Persistent object systems provide a seamless integration between the database and programming language. Persistent objects reside in secondary memory and are transparently copied into primary memory when accessed

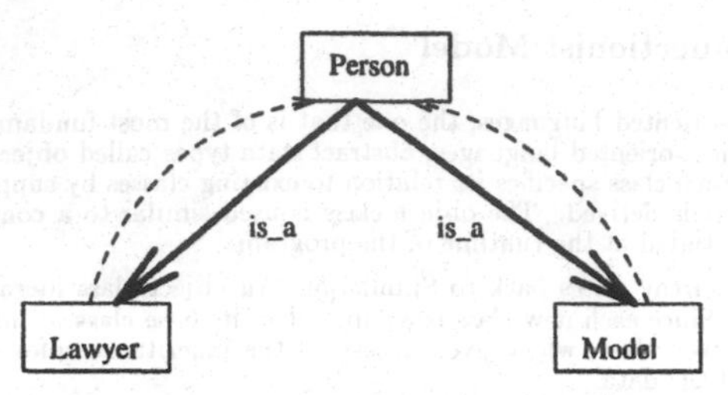

Figure 1: Inheritance in POCONS

by the user program. Depending on the update mode chosen by the user, modifications can also be done transparently to the object's secondary storage representation. Persistent objects have the advantage of being available for perusal after the application program has ceased execution.

Some of the advantages of persistent systems listed by Morrison and Atkinson [13] are *reduced complexity, reduced code size and time to execute*, and *data outlives the program*. Firstly, complexity is reduced for the application builders, because with persistent systems, there is no distraction for the programmer in dealing with the complexity of managing the database. The user need only consider the complexities involved in the mapping between the programming language and the problem to be solved. Secondly, persistent systems reduce code size, because the application program need not contain code concerned with the explicit movement of data between primary and secondary memory. Also, the time to execute is reduced, because only objects required by the system get loaded into primary memory. thus eliminating any execution time spent loading objects into primary memory that are not accessed by the application. Finally, the data outlives the program, because it resides in a database. Persistent objects are used to describe and store neural networks in the POCONS language.

7 Checkpointing and Storage

While executing a neural network simulation, it can be desirable to save the state of the network at various intermediate stages to examine the evolutionary development of the network. While the training algorithm executes, the state of the neurons andor weights get altered. The checkpointing feature saves the state of the neurons to a persistent object store but allows the simulation of the network to continue execution. Objects in the persistent object store can later be examined through queries to an object-oriented database system.

Likewise, when execution of the neural network has ceased, the state of the network can be saved using the persistent object checkpointing system. Then the network can be loaded and reused at a later date in the same state it was in when it was stored saving possibly hours of CPU time. Certainly, a trained network could be stored in files manually, then loaded and converted without the use of persistent objects. However, POCONS's built-in facility for persistent neural network objects, *transparently* stores and retrieves them. Storage and retrieval is carried out by the underlying system on a demand driven basis. It therefore requires little extra application code (opening and closing of databases) to save and load objects as will be shown in a later example.

Execution of the newly built-in function **convert-neurons-to-pos** will cause the checkpointing mechanism to convert the neuron structures into persistent objects by storing neuron data items in persistent slots and converting hard pointers to other neurons to soft pointers that can then be stored in the database as integers and dereferenced through a table lookup. The system creates one persistent object for each neuron and stores all data associated with that neuron in it. Note that these checkpointing objects are of a different class than the ones used to describe the structure of the network created by **defdbneuron** which gets represented by a single persistent object, but can be compiled into several neuron structures.

8 Reuse

POCONS supports the loading of the state of a previously trained or simulated network into primary memory, and the conversion of it into POCONS's internal neural network representation, so that it can then be reused. Once reloaded and converted training can be resumed from where it left off. In the conversion process, data items (activity values and semantic information) from persistent objects are stored in the representative neurons. Soft pointers stored in the persistent objects are converted into hard links and stored in the respective neurons.

After opening the correct databases, execution of the newly built-in function **convert-pos-to-neurons** will convert the persistent objects representing the previously checkpointed neural network into the internal neural network representation so it can be executed. The structure of the executable neural network does not use persistent objects, so it can contain hard pointers. Also, the attributes of the neuron get updated frequently as it executes and there would be too much overhead if the neuron slots were persistent. Thus, by avoiding the storage of frequently updated slots at execution time, performance is greatly improved.

9 Implication

The application program for testing this system represents a flip-flop used by a circuit analysis program written in POCONS which analizes the basic units of the circuit (eg. transistor, resistor) and determines based on the configuration, which groups of these units make up different types of gates (eg. nand, and, nor, etc.). The implementation of this program uses a POCONS construct used for the defining rules called *defimplies* [4]. POCONS resolves the rules at load or compile time and produces new links between neurons to represent the rules. Then, at execution time, the links associated with rules are given precedence over links associated with the network definition. This technique eliminates any conflict between rule links and network definition links.

The resolution of rules at load time makes loading the neural network more expensive, but causes the execution to be faster, because there is no matching of rules at runtime only propagation of values around the net. This propagation of values with the new links added for the rules produces the same inferencing results in a production rule-based system. Another advantage of resolving rules at load time rather than execution time is that the rule links can be stored with the persistent object, and reloaded with it for future program execution. This ability to store rule links greatly improves the performance of the reuse of a neural network as will be shown in the next section.

(defdbneuron name (superclasses) (slots))	The *name* argument corresponds to a neuron class which will be represented as the neuron *(is-a, name)*. The *superclasses* are a list of neuron classes which must be defined using *defdbneuron*. The neuron *name* will be linked to its superclass neurons. The contain a list of slot-attribute pairs. Each slot and it's associated *slots* value specifies a neuron which is linked to it's neuron class to support inheritance.
(build-neural-network)	Takes the persistent objects created using *defdbneuron* and converts them into a linked neural network of structures.
(convert-neurons-to-pos)	Takes the neuron structures and converts each one into a neuron object saving it's links to other neurons, the weights on those links, and the neurons activity values.
(convert-pos-to-neurons)	The *inverse* of *convert-pos-to-neurons*. Takes persistent neuron objects created by *convert-neurons-to-pos* and builds a neural net of structures linked together with weights and activity values as specified in the persistent objects.
(defimplies (condition1 [conditions]) => (result1 [results]))	A *condition* contains a neuron class name followed by a list of variables corresponding to the slot names associated with it. Each result must consist of variables or constants which will be used to match against values stored in neurons.

10 Example

The output below was generated from a session using POCONS to load and create the neural network code shown previously. After the network is created, the function **query-net** is used to generate 3 iterations of the activation propagation algorithm. The network is then converted to persistent objects and EuLisp is exited. Then EuLisp is started up again, and POCONS is loaded. After opening the database, the persistent objects are converted into a neural network which has the same state as it did when it was stored as shown. Further activation propagation iterations are then executed on the network.

```
EuLisp:0:root!0> (!> pse-net)
Loading module 'pse-net'
[ LOADING OF PSE MODULES ]
;;;OPENING OF DATABASE
eulisp:0:pse-net!1> (open-dbclass-cache "../neuron/cl.dir")
eulisp:0:pse-net!1< ()
eulisp:0:pse-net!2> (open-classes "../neuron/pmppos")
eulisp:0:pse-net!2< #<stream: 268483904 'r'>
eulisp:0:pse-net!3> (open-objects "../neuron/objects.pos")
eulisp:0:pse-net!3< #<stream: 268483920 'r'>
eulisp:0:pse-net!4> (open-object-cache "../neuron/obj.dir")
eulisp:0:pse-net!4< ()
;;;LOADING OF DEFDBNEURON DESCRIPTIONS
eulisp:0:pse-net!9> (include-forms "../neuron/act-np.em")
including '../neuron/act-np.em'
```

```
;;; CONVERSION OF DEFDBNEURON DESCRIPTIONS INTO NEURAL NETWORK
eulisp:0:pse-net!10>  (build-neural-network)
;;;EXECUTION OF 3 ITERATIONS OF ACTIVATION PROPAGATION.
;;;PRINT-OUTS SHOW STATUS OF NETWORK FOR NEURONS <> 0.0
eulisp:0:pse-net!11> (query-net '((is-a . jane)) nil 3)
(is-a model 0.330000)
(owned car 1.000000)
(is-a jane 1.000000)
(is-a person 0.108900)
(is-a model 0.551100)
(ownod car 1.000000)
(is-a jane 1.000000)
(is-a person 0.270958)
(is-a lawyer 0.108900)
(is-a model 0.748122)
(owned car 1.000000)
(is-a jane 1.000000)
eulisp:0:pse-net!11< 4
;;;CONVERSION OF NEURONS TO PERSISTENT OBJECTS FOR STORAGE.
eulisp:0:pse-net!12> (convert-neurons-to-pos)
;;;CLOSING OF THE DATABASE.
eulisp:0:pse-net!13> (close-system)
eulisp:0:pse-net!13< ()
eulisp-handler:0:pse-net!15> (exit)
Exiting EuLisp
;;;NOW A NEW EULISP IS STARTED UP TO REUSE THE SAME NETWORK.
nufeel
;;;LOAD POCONS AGAIN.
eulisp:0:root!0> (!> pse-net)
Loading module 'pse-net'
;;; DECLARE NAMES OF PERSISTENT CLASSES AND OPEN DATABASE
eulisp:0:pse-net!1> (persistent-classes (person lawyer john model jane nnode
                    db-opposites db-converse db-negconverse) "../neuron/cl.dir")
;;; OPEN DATABASES (deleted due to space limitations).
;;;CONVERT THE PERSISTENT OBJECTS TO A NEURAL NET.
eulisp:0:pse-net!7> (convert-pos-to-neurons)
;;;PRINT THE STATE OF THE NETWORK.  NOTICE IT IS THE SAME
;;;STATE AS IT WAS SAVED.
eulisp:0:pse-net!8>  (print-vals *neuron-list*)
(is-a person 0.270958)
(is-a lawyer 0.108900)
(is-a model 0.748122)
(owned car 1.000000)
(is-a jane 1.000000)
;;;COMPUTE A FURTHER ITERATION OF THE ACTIVATION PROPAGATION
;;;ALGORITHM.
eulisp:0:pse-net!9> (compute-activation-aux *neuron-list*)
(is-a person 0.477144)
(owned dog 0.108900)
(is-a lawyer 0.350351)
(is-a john 0.108900)
(is-a model 0.899490)
(owned car 1.000000)
(is-a jane 1.000000)
```

The above example was kept small for brevity, but it fully illustrates the small amount of application code required for the storage and reuse of neural networks in POCONS. The larger circuit analysis application works as well.

11 Performance Improvements

On a Stardent Titan executing EuLisp [14] it takes 238.94 seconds to load 437 neuron objects, build the neural network, and resolve three rules. It then takes 37.88 CPU seconds to convert the neural network to persistent objects converting the hard links to soft links and storing them and the other neuron values in newly created persistent objects (which the persistent object system transparently stores to secondary memory). To reload and build the same network takes 102.42 CPU seconds. Thus, by reusing the objects you get more than double speedup on the amount of time required to intialize the network.

In the case of a backpropagation trained network, more speedup would definitely be attained, because the time to save and reuse the network remains constant, but the time to execute a backpropagation training algorithm requires much greater time than creating a new network from a POCONS language description.

12 Conclusion

A built-in feature for storage and checkpointing using persistent objects has been presented. Not only is it an elegant solution to the problem, but results in improved performance as well. The implementation of this feature has been described, and an example has been presented. Performance improvements resulting in double speedup have been reported using this facility for networks trained using activation propagation, and greater performance improvements are will result from the storage and reuse of networks trained using more costly algorithms such as backpropagation.

13 Acknowledgements

Special thanks to Professor John Fitch and Kay Marie Sutcliffe for encouragement and support.

References

1. Anderson, J. R. *Language, Memory, and Thought.* Lawrence Erlbaum Associates, Publishers, 1976.

2. Burdorf, C. Per-Trans: A Persistent Stochastic Petri Net Representation Language. In *Proceedings of the 22nd Annual Pittsburgh Conference on Modeling and Simulation*, 1991.

3. Burdorf, C. POCONS: A Persistent Object-based Connectionist Simulator. In *Proceedings of the 1992 SCS Western Multiconference: Object-Oriented Simulation.* Society for Computer Simulation, 1992.

4. Burdorf, C. Representing Implication in an Object-Oriented Neural Network System Using Partitioned Connections. In *Submitted for publication*, 1992.

5. Burdorf, C. and Cammarata, S. PSE: A CLOS-Based Persistent Simulation Environment with Prefetching Capabilities. In *Proceedings of the CLOS Workshop*, 1989.

6. Burdorf, C. and Cammarata, S. PSE User's Manual. Technical Report WD-5103-DARPA, The RAND Corporation, August 1990.

7. Cammarata, S. and Burdorf, C. PSE: An Object-Oriented Simulation Environment Supporting Persistence. *The Journal of Object-Oriented Programming*, October 1991.

8. Dahl, J. and Nygaard, K. Simula: A language for programming and description of discrete event systems. User's manual, Norwegian Computing Center, 1967.

9. D'Autrechy, C., Reggia, J. A., Sutton, G. G., and Goodall, S.M. A General-Purpose Simulation Environment for Developing Connectionist Models. *Simulation*, 1988.

10. Feldman, J. A., Fanty, M. A., and Goddard, N. H. Computing with Structured Neural Networks. *IEEE Computer*, 1988.

11. Floreen, P., Myllymaki, P, Orponen, P., and Tirri, H. Compiling Object Declarations into Connectionist Networks. *AICOM*, 1990.

12. D. O. Hebb. *The Organization of Behavior.* Wiley and Sons, 1949.

13. Morrison, R. and Atkinson, M. P. Persistent Languages and Architectures. In *International Workshop on Computer Architectures to Support Security and Persistence of Information.* Springer-Verlag, 1990.

14. Padget, J. and Nuyens, G. (Eds.). The EuLisp Definition.

15. Rummelhart, D. E. and McClelland, J. L. *Parallel Distributed Processing - Volume 1: Foundations.* MIT Press, 1986.

16. Wang, D. and Hsu, C. SLONN: A Simulation Language for modeling of Neural Networks. *Simulation*, 1990.

17. A. Weitzenfeld. Neural simulation language version 2.1. Technical Report 91-05, Center for Neural Engineering, University of Southern California, August 1991.

FLEXIBLE OPERATING ENVIRONMENT FOR MATRIX BASED NEUROCOMPUTERS

John C. Taylor, Michael L. Recce, Anoop S. Mangat

Department of Computer Science
University College London
Gower Street
London WC1E 6BT, UK

This paper describes the design and implementation of a development environment for matrix based neurocomputers. A new virtual machine language provides a wide range of matrix operations and device-related input/output communications. Virtual machines may be implemented entirely on conventional workstations or may use matrix-based neurocomputer hardware. To assist in algorithm development and debugging, the virtual machine is able to generate monitoring messages. A graphical interface is used to view the workings of one or more virtual machines. The user interface allows a range of display techniques to be associated with VML scalar and matrix variables. Virtual machines and monitoring processes run under the control of a central scheduler. All communications are implemented using a message based protocol. This environment is currently being used to develop a wide range of applications.

1 Introduction

Following the growth in neural network algorithms and applications, there has been a corresponding development of new computer hardware designed to exploit the inherent parallelism of this technology. Although there is considerable variation in these machines, called neurocomputers, they can be grouped into two broad classes [6]. One class, called special purpose neurocomputers, consists of hardware accelerators for particular algorithms, while the other, general purpose neurocomputers, are designed to support the full spectrum of neural network algorithms. These two types of neurocomputers have very different software requirements. In this paper we discuss only the requirements for general purpose neurocomputers, since the software for special purpose machines is necessarily more application and machine specific.

The basic requirement for the software environment of a general purpose neurocomputer is that the algorithm is compiled to a form that can be run efficiently. However, at the same time it is important to avoid the classic long-development cycle usually associated with parallel computers. Ideally, the environment should allow development of the application in a powerful interactive programming environment, followed by seamless transfer onto a general purpose neurocomputer. Moreover, the operating system needs to contain powerful monitoring and debugging features to evaluate the algorithm as it runs on the neurocomputer.

Neural networks algorithms can, in general, have very complex architectures, without regularity or symmetry. However, most frequently, architectures are chosen which exhibit considerable symmetry, comprising homogeneous layers of identical processing elements connected in a regular way. While the first, most general, view requires models of networks to represent them as many, separate linked objects, the second view is formulated most economically as a set of arrays, and the arithmetic operations used to manipulate these arrays are expressed in matrix form.

This distinction applies also to different computer hardwares used to implement neural networks. Neurons are, of course, processing elements, meaning that each has specific local operations to perform and each, therefore, could be implemented on a separate computer processor. In practice, neuron-based networks described in this way are simulated on a restricted number of processors; either one (conventional sequential host), or else a number which is typically less than the number of neurons to be instantiated, so that one processor is shared amongst many neurons. Taking the second, matrix-based view of networks, one or more processors are used to implement matrix operations (such as matrix product). These operations are essentially atomic. Naturally, hardware which is used to implement matrix operations may use multiple processors, and so *internal parallelism* may be exploited. However, this does not necessarily correspond to the original implicit parallelism of the network. Matrix-based machines often employ systolic array techniques [8]. The software tools discussed in this paper are designed primarily to support matrix-based hardware, but can be extended to neuron-based hardware.

A large number of neural network programming environments have been developed (for a review, see [5]), and the system we describe extends these features and provides a means for efficient execution with a wide range of matrix-based general purpose neurocomputers. In this paper we first describe the software system architecture, followed by the neural network programming environment, and key components of the operating environment designed to fulfil the requirements discussed above. We discuss key features of the graphical programming environment, the implementation of the software on SUN workstations and neurocomputers developed by Siemens [4] and Philips [3], [7].

2 System architecture

In order to achieve the goals outlined above, the system architecture is designed to provide extensive tools for developing applications. The code is then compiled into a machine independent language called the virtual machine language (VML) and can run on a workstation or can be mapped onto a matrix-based neurocomputer. Figure 1 contains the basic system architecture. The development of neural network applications proceeds from left to right in the diagram and the thick arrows show the interaction between the components in the run-time environment.

The user can develop neural network applications directly in VML, which follows the structure of the 'C' programming language, or a high level object orientated language similar to 'C++'. This high level language called 'N' compiles into VML and was previously presented [1]. The graphical tools for code development, debugging and performance monitoring follow the Motif standard in X-windows.

At run time the overall control is centralised and provided by the scheduler. However, the system is not monolithic, as each of the components is a stand alone program with independent control structure. Communications between the graphical modules and the scheduler, the scheduler and the virtual machines and also with other data handling processes use a common protocol. This protocol uses a message queue, and the amount of message traffic is determined by the number of monitoring requests together with the data input and output requirements of the algorithm. Communications are currently implemented using Unix sockets. VML is executed by a virtual machine which is composed of two separate parts called 'comms' and 'exec'. The exec part of the virtual machine is treated as a co-processor, responsible for the parallel arithmetic operations, and is controlled by the 'comms' part.

Our system provides an easy development path for the implementation of neural network systems. Each application must be formulated, wherever possible, in terms of matrix operations. VML supports a range of data types, including, in addition to the usual integer and floating point types, a special fixed point type, in which a common exponent is shared between all the matrix elements, each of which stores a different mantissa. Input and output formats and data transformation operations are used to standardise data presentation format.

3 VML

VML is a general purpose programming language providing scalar and matrix arithmetic. All control flow constructs are based on the C language. The wide range of arithmetic operations, together with the inherent economy of expression provided by matrix description makes for compact and efficient code, both for neural and non-neural data processing operations.

A virtual machine, in our terms, is a conceptual stored program computer, presenting a single, well defined interface to the rest of the world. VML programs consist of one or more rules, which may be invoked on receipt of an execution directive message from the scheduler, or by calls from other rules. All explicit (coded) data transfers are performed using device-related channels; the correspondence between device numbers and actual devices is determined by an external configuration table. Each VM is able to generate additional messages containing data or other progress information for monitoring purposes in response to request messages from the scheduler.

A VML program must start with a special initialisation rule which defines the scalar and matrix variables, their types and assigns initial values. This rule may also open devices for data transfer and for look up table loading. The remainder of the program consists of one or more rules which comprise a mix of control flow, arithmetic and I/O statements.

To illustrate the compact nature of a VML description, the following rule could be used to implement the forward pass in a three layer perceptron using just four arithmetic statements :

```
DECLARE_RULE      ( recall )            /* perform forward pass       */
  flt32_mat_mul   ( A1, S0, W1 )        /* W1 : weight matrix, layer 1 */
   flt32_lut_am   ( tanh, S1, A1 )      /* activation function        */
  flt32_mat_mul   ( A2, S1, W2 )        /* repeat for layer 2...      */
   flt32_lut_am   ( tanh, S2, A2 )
ENDRULE
```

Here the activation function is calculated by applying a look up table (lut) to the accumulator matrices.

4 Communications protocol

Within the development environment, many separate processes must communicate. These include multiple instances of virtual machines including those implementing neural networks, central scheduling and control, monitoring processes, pre- and post-processing modules and peripheral device handlers. For reasons of consistency, efficiency and portability, all communicating processes use an identical message protocol. In some instances, this necessitates wrapping conventionally coded modules in a communications harness. The messages are used for control (for example to invoke rules and monitoring operations on virtual machines), to perform data transfer or initiate transfer (such as DMA) and for exception reporting.

Each communicating process creates an incoming message queue. The contents of this queue may be examined, and messages extracted according to message type or priority or receipt sequence. This permits the receiving process to handle messages in the most appropriate order, and also to discard them, if necessary.

A consistent data representation format is used for disk storage and also for data communication, for example in monitoring messages. The format is encoded in XDR [External Data Representation, Sun Network Interfaces Programmer's Guide] to ensure portability across machine architectures, and also because efficient XDR routines are available for stream and memory conversions. Various data compression options permit run length encoding and optimisation of 8 and 16 bit data representation while retaining cross-architecture XDR portability. All data items may include annotation, and filters provide conversion to and from an equivalent ASCII form.

5 Monitoring

Monitoring of running processes is required for several purposes. During algorithm development, it is necessary to be able to extract and visualise or otherwise process the data being manipulated during calculations, to debug and establish the correct functionality of code. The variables and functions under scrutiny should be presented in the context of the source language used to encode the algorithm. After initial development, the emphasis changes to tuning and performance evaluation. Monitoring of the internal data in executing processes remains necessary, but so too is measurement of interprocess communications in terms of data throughput, contention for shared resources i.e. bus bandwidth, cpu time etc. Finally, a well developed system must support error and other exception handling. This last may not necessitate graphical presentation, but in all cases, data exchanges will take place using the same protocol.

The majority of Neural Network Programming Environments include a graphical user interface (GUI) in order to configure, control and monitor the neural network. They often share common features such as menus, command lines, and other graphics such as sliders and bar charts to transfer information to the user.

The layout of neural network GUIs are constrained by the original functionality of the system, which becomes hard-coded into the environment, leading to a closed system where the user has no choice but to accept the initial design, whether it be for debugging, controlling, or monitoring. An example of this is the Graphic Monitor used in the Pygmalion Neural Network Programming Environment [2]. The user has control over which rules are fired, but cannot isolate particular parameters to monitor. This GUI displays all the parameters which are available to the system, often presenting the user with much unwanted information.

The design philosophy which we have used in developing our GUI has been influenced by that of GUI builders, such as Devguide [OpenWindows Developer's Guide, Sun Microsystems]. These builders allow the user to design an interface within a graphical environment, and then the builder automatically generates the source code, without the need for the user to be an interface programming expert.

6 GDL

Our Graphical Display Language (GDL) allows the user to design the interface independently of the neural network, and specifically for the task required. With the GDL, the user can build an interface for debugging and then design one for monitoring rather than being tied down to a fixed interface with tries to combine the two functions.

GDL contains six high-level graphical primitives which cover a wide range of common graphical display methods; graph, chart, image, tile, form and group, which are each described by lower level graphical objects such as menus, button, canvas, lines, etc. These high-level primitives are equivalent to the basic building blocks in GUI builders, such as base window, control panel, etc. Below is a brief explanation of each of the six primitives:

graph A graph primitive displays two-dimensional graphs with points and/or lines, with automatic scrolling and resizing.

chart This allows the display of two- or three-dimensional bar charts.

image An image is used to display image or matrix data.

tile This is another form of displaying matrices, but each element alters their size depending upon the data values. It is a general version of a Hinton diagram.

form The form object is used to display numeric and textual data, and it can provide the user with an input area if needed

group A group has a different purpose than the other five primitives. It is used to group together objects which share common attributes in order to reduce replication within the GDL file. For example, a graph and a chart which monitor the same variables can be grouped together to minimise redundancy.

In order to display a parameter, the user binds that parameter to one of the above primitives in the GDL code depending upon the display required. The relationship of parameters to primitives is many-to-many. Matrix parameters are limited to images and tiles, and scalars can be bound to graphs, charts and forms. Other textual and alphanumeric variables are bound to forms. It is a simple matter to change parameter to GDL primitive bindings, so the user can easily change from a debugging interface displaying extra information to one which concentrates the display on the functionality of the neural network.

The GDL obtains parameter data from the virtual machine via the communications protocol described above, therefore the GDL is highly modular, completely removed from the virtual machines or VML code. Because of this transparent communications layer, a single GUI designed with GDL can monitor data from one or more independent virtual machines. The updating of parameter data can be linked to either the rate of change of the parameter, i.e. every tenth change, or in relation to time, i.e. every 100 milliseconds. The update method for parameters are defined in GDL, but it is the responsibility of the running virtual machine to communicate the data to the interface.

An Execution Monitor (EM) is used to generate the GUI which is defined by GDL, and handles the communication with the virtual machine. The EM behaves like a GUI builder, with options for graphs, charts, images, etc., and each graphical primitive contains its own property sheet which is used to configure the primitive.

The property sheet sets which parameters to monitor, the update method, colours, title and other primitive dependent information for that particular primitive, so allowing the user to build up a number of primitives and configure them into the desired interface. The resulting configuration can then be saved to a file in the corresponding GDL format.

The EM is fully functional, and has been used to design an interface to aid in the debugging of virtual machines. This GUI is currently being used to debug several different applications written in VML. A GDL built interface is portable across a wide range of different platforms and applications. The graphical primitives can be defined in any graphical programming system, and the interface can monitor parameters from different programming languages, given support for the communications protocol.

7 Summary and conclusions

Our software is designed to permit rapid development of neural network systems and associated data processing, and to implement algorithms in a new matrix based language (VML), which permits arithmetic operations to be executed efficiently on a number of different hardwares, especially matrix-based parallel machines such as those developed by Siemens and Philips. It provides flexible and powerful monitoring tools to help with design, debugging and subsequent tuning of algorithms The graphical interface uses a description language (GDL) to customise presentation and to bind various display methods to network parameters.

One or more virtual machines, graphical monitors and other processes run together under the control of a central scheduler. All components of the environment communicate using a new protocol and prioritised message queues.

VML programs are parsed, converted to an internal form and then individual rules are executed. All external data representations i.e. file formats, data transfer and monitoring messages use a common, portable format. The standard format includes internal optimisation mechanisms for data compression which provides a flexible way to implement a trade between transmission cost and data formatting cpu time, according to context.

We are able to execute VML programs using special purpose hardware from Siemens and Philips. When accelerator hardware is available, exactly the same VML programs are used, but matrix data is instantiated in the memory space of the hardware, rather than the workstation.

The environment, which is the subject of continuing work, has already proved useful in rapid implementation of applications. VML has been use to code many algorithms, for such diverse purposes as optical character recognition, robot motion control, image processing and time series forecasting.

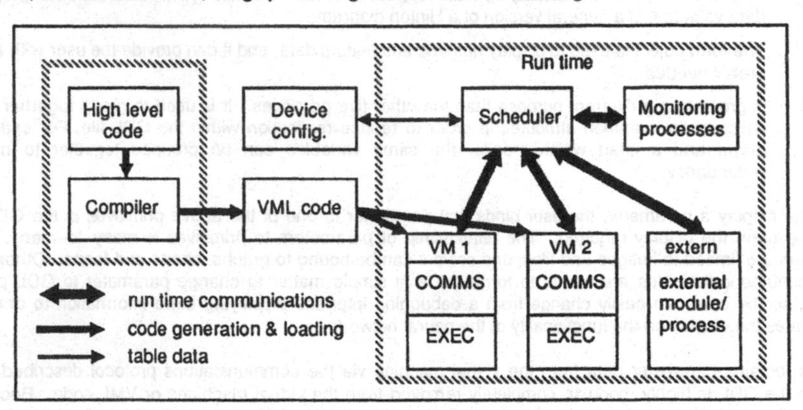

Figure 1 : System and software architecture

Acknowledgment

The work described in this paper was funded by the European Commission as part of the Galatea project (Esprit II project number 5293).

References

[1] Angeniol, B., Fogelman, F., Marcadé, E., Pimont, J.M., Simon, C., Skoda, F., Archas, Y., Bru, B., Caillaud, C., Makram, S., Zorer, J.L., Ramacher, U., Anlauf, J., Recce, M., Taylor, J. (1992) The Galatea Project, proceedings NeuroNimes '92

[2] Azema-Barac, M., Hewetson, M., Recce, M., Taylor, J.C., Treleaven, P.C., and Vellasco, M.M.B.R. (1991) PYGMALION neural network programming environment, Int. Neural Network Conference, Paris, France.

[3] Duranton, M. and Sirat, J. (1990) Learning on VLSI: A General-Purpose Digital Neurochip, Philips Journal of Research, vol. 45, no. 1, pp. 1-17, 1990

[4] Ramacher, U. and Raab, W. (1990) Fine-Grain System Architectures for Systolic Emulation of Neural Algorithms, Int. Conf. on Application Specific Array Processors, Princeton (USA), September, 1990.

[5] Recce,M.L., Rocha,P.V. and Treleaven,P.C. (1992) Neural network programming environments, 1237-1244, Artificial Neural Networks 2 (Eds. Aleksander,I. and Taylor,J.) Elsevier Science Publishers B.V.

[6] Recce,M. and Treleaven, P.C. (1988) Parallel architectures for neural computers, 487- 495, Neural Computers (Eds. Eckmiller,R. and v.d.Malsburg,C.), Springer-Verlag Berlin Heidelberg.

[7] Theeten, J.B., Duranton, M., Mauduit, N., Sirat, J.A. (1990) The L-Neuro-Chip: A Digital VLSI With On-Chip Learning Mechanism, Proc. Int. Neural Network Conf., pp 593-596, Kluwer Academic Publishers, July 1990.

[8] Vellasco, M.M.B.R. (1991) Phd thesis, University College London, London

A PARALLEL IMPLEMENTATION OF KOHONEN'S SELF-ORGANIZING MAPS ON THE SMART NEUROCOMPUTER

E. Filippi[†], J.C. Lawson[‡]

[‡] INPG, Labo TIRF, 46 Avenue F. Viallet F38031 Grenoble
[†] Dip. Elettronica, Politecnico di Torino, C.so Duca degli Abruzzi, 24 - I10129 Torino
Tel: 39 11 5644038 - Fax: 39 11 5644099

Abstract

This paper describes the implementation of a neural paradigm, the Kohonen's self organizing Map, on the SMART neurocomputer. This algorithm has been chosen as a test for the capabilities of the architecture. A high level interactive graphic monitor and a basic neural library have been developed, and the performance of the whole system analyzed.

1 Introduction

During the last years, self-organizing maps have been applied to a variety of tasks (such as signal and image processing, robot learning, optimization, data compression and restoration) with encouraging preliminary results [1, 2], but most of these applications require both fast and low cost hardware platforms to be really competitive. Also research and theoretical works require more and more powerful machines to investigate complex problems keeping the simulations computationally feasible [3].

Two main approaches have been investigated to reach the necessary computational power, memory size and input-output bandwidth: the use of general purpose supercomputers, and the development of dedicated architectures. Because of the inherent parallelism of ANNs, the interest has been focused on parallel systems: several implementations on "coarse-grained" or "fine-grained" general purpose machines can be found in literature [3, 4, 5]. They addressed for the first time issues previously never investigated and aimed at the development of specific "neurocomputers", although they may not provide cost-effective solutions. For this reason many research efforts on the design of neurocomputers are currently under way. Several neural architectures have been proposed and built [6, 7], but very few of them have ever been tested in a real application: their performance are usually analyzed only at the chip or board level and not at the system level.

This paper describes a first experimental contribution to the development of a suitable environment for the SMART machine, a novel neurocomputer under development at INPG (Grenoble): the aim is to analyze the behaviour of the architecture when a complete stand-alone system is realized.

2 Overview of the architecture

The SMART architecture [8, 9] is a distributed-memory system composed of a SPARC microprocessor as control unit and a linear processor array as processing unit.

The processor array is a pipe-line assembly of identical processing cells (one to sixteen) based on a floating-point unit served by a local RAM memory. FIFO registers are placed between the cells. These cells can work independently (in parallel) or serially (in a systolic fashion). A programming model with two basic instruction types is adopted: *move* and *arithmetic*. Arithmetic instructions proceed on the cells while move ones concurrently transfer data between the cells and a dual-port memory interface. Sparse matrices are supported using an original distributed data storage scheme handled by a dedicated parallel address generator. Zero coefficients are no longer stored. Due to the use of sequential I/O via a standard VME interface board, the amount of data exchanges required between the SMART and the remote system is a critical point, and to obtain good performances the number of arithmetical operations must be larger than I/O operations.

Figure 1 shows the SMART environment.

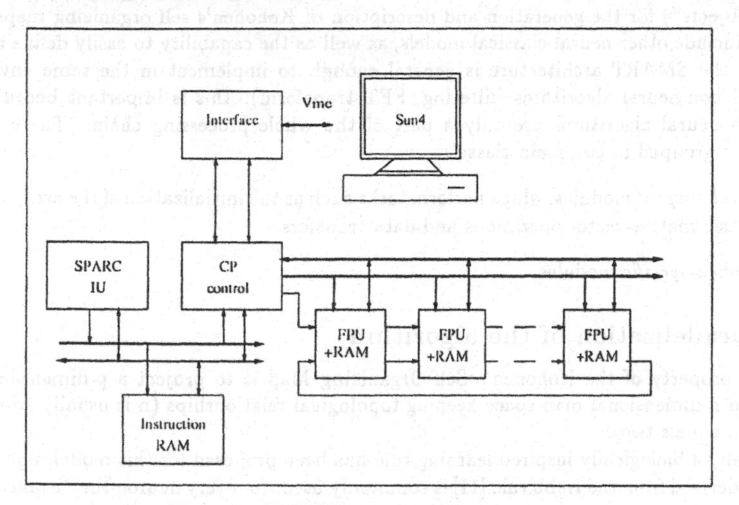

Figure 1: SMART processor environment

3 Implementation of the algorithm on the SMART machine

The software environment for a neurocomputer must provide suitable tools to define any network and to monitor and visualize its evolution.

The work here described is a contribution to the development of a SMART complete neurocomputing environment. It favours flexibility and easy monitoring, making the system suitable for demonstrations and debugging. In the following, a short description of the whole system is made.

3.1 The Sun front-end

An interactive monitor has been realized starting from a graphic user-friendly ANN simulator developed at EPFL (Lausanne), running on Unix Sun systems [10]. It displays the network state (weight space and/or neurons activations) and allows "on-line" modification of many parameters, such as network size and input distribution, distance metric, learning parameters. Learning can proceed in either automatic or manual mode. In the automatic mode, the learning parameters are controlled by a suitable heuristic. In the manual mode, the user himself sets the values of these parameters.

The initial network configuration, learning parameters and any change of them are sent to the SMART processor, which computes the required forward steps (distances and minimum calculation) and weight updates and returns the network state.

3.2 Programming the SMART machine

SMART's programming language was chosen to be a vectorial extension of the C language. This kind of languages are often used to deal with ANNs because a matrix formalism is commonly adopted to represent the network topology.

A set of routines has been defined which can be used from high-level languages, providing basic blocks ("objects") for the generation and description of Kohonen's self organizing maps. Future work will include other neural classical models, as well as the capability to easily define new ones. Note that the SMART architecture is general enough to implement in the same environment neural and non-neural algorithms (filtering, FFT transform): this is important because in any application neural algorithms are only a part of the whole processing chain. The developped modules are grouped in two main classes:

1. *general purpose* modules, which perform tasks such as the initialization of the array processor, all basic matrix-vector operations and data transfers

2. *Kohonen-specific* modules.

3.3 Parallelization of the algorithm

The main property of the Kohonen's Self-Organizing Map is to project a p-dimensional input space to an n-dimensional map space keeping topological relationships (n is usually smaller than p and often equals two).

Originally a biologically inspired learning rule has been proposed for this model, but a simpler algorithm derived from the Hebb rule [11] is commonly used: for every neuron the distance between the weight and input vectors is computed, the most active (i.e. less distant) neuron is selected, the weight vectors of all neurons in the neighbourhood of the most active one updated. The extension of the neighbourhood and the adaptation gain must decrease as functions of time to ensure network convergence. The distances and minimum computation (feed-forward phase) is the most expensive part of the algorithm.

Two main task and data distribution strategies are possible on a linear architecture such as SMART :

- to distribute weight and input vectors among the available processing cells according to the following law: the i-th weight of each neuron and the i-th input are assigned to the k-th cell if:

$$k = (i \bmod N) \tag{1}$$

- to assign a different cluster of neurons (with all the relative weights) to each cell. This solution requires more memory because a copy of the input vector must be broadcasted to every cell.

Distance The first strategy is adequate when the number of inputs is large compared to the number of cells. Each cell independently computes a 'partial' distance with the fraction of weights and inputs stored in it, then the total distance is obtained using a single systolic addition through the line of cells. Finally the distances must be distributed upon the cells for the computation of the minimum. For a N_i-input and N_o-output network, with a K cells processor-array provided with a MAC (multiply-accumulate) instruction and with the hypotesis $N_i \gg K$, the number of computing cycles is:

$$n_d \approx \frac{2N_iN_o}{K} \tag{2}$$

Otherwise the second strategy is better (the copy of the input vector components being not too damaging) and if N_o is large enough compared to the number of cells:

$$n_d \approx \frac{(3N_i - 1)N_o}{K} \tag{3}$$

Minimum The minimum selection is made in the following way: each cell locally compute independent minima (using the distances stored in it). Then one systolic operation gives the global minimum which is broadcasted to all the cells. It is compared with the distance values leading to a corresponding vector whose components are set to one where the distance matches the minimum. An inner product of a distributed vector of indexes with this binary vector selects the winner neuron index. If $N_o \gg K$ the number of computing cycles is then:

$$n_m \approx \frac{3N_iN_o}{K} \tag{4}$$

However, for sometimes more than one component matches the minimum, the SPARC control unit gives a random winner if the result of the SMART coprocessor is greater than the number of neurons. We observed the same results using this scheme as when a random neuron is picked amongst the winners.

Update First of all, the set of neurons to be updated (which depends on the actual winner) must be identified. The weight correction procedure is depending on the weight and input vectors distribution.

If they are distributed according to (1) then, for every neuron in the neighbourhood, each cell independently updates the fraction of weights stored in it. The number of computing cycles required is:

$$n_u \approx \frac{4N_iN_n}{K} \tag{5}$$

were N_n is the size of the neighbourhood.

Otherwise, sequentially for every neuron in the neighbourhood the corresponding cell must be identified and then the weights updated. In this case the number of computing cycles is:

$$n_u \approx 4N_iN_n \tag{6}$$

4 Performance evaluation

All the SMART modules and programs have been tested with C-code emulation at the SMART register level, because the basic software and hardware SMART tools were not yet completed.

Unix "pipe" connections between the Sun process implementing the graphic monitor and the one emulating the SMART machine simulate the channels of the VME interface board. Read and write instructions on a pipe do not access to disk, then this emulation correctly estimates the I/O time between SMART and the remote system. Thus the time spent on the front-end and the time needed for the I/O can be obtained from the execution profile of the Sun process. The time spent for computation in the SMART processor is difficult to estimate at this level. Anyway, knowing that a new SMART arithmetic instruction is executed every 120 ns, it can be approximated starting from the number of instructions required to implement each task (see section 3.3).

Considering that Sun and SMART processes can overlap, a lower bound and an upper bound for the total execution time can be found assuming respectively complete or null overlapping.

Table 1 and 2 compare the performance of the Sun+SMART implementation (lower and upper bound) with the performance of a Sun simulation when solving two different networks: a network with 2 inputs and 900 outputs (tab. 1), and one with 100 inputs and 100 outputs (tab. 2). They have been implemented according to respectively the second and the first strategy described in section 3.3. In all the cases 10,000 learning steps with 10 intermediate visualizations of the network state were required. A SMART array of 4 cells was considered.

The results show that the original simulator running on the Sun spends most of the time in the computation of the feed-forward phase (distances and minimum), while the 'bottleneck' of the Sun-SMART system is the data I/O. The data I/O can be strongly reduced limiting the learning monitoring (i.e. reducing the visualization of the network state), but this is obviously a problem in most applications. The total speed-up is apparently poor, but it is close to the Amdhal's low limit: in fact the parallel part of the process (feed-forward + update procedures) is about 85% of the time in the Sun simulation. This percentage increases as the network size becomes larger: then the best performance can be obtained with large networks (compatibly with the coprocessor distributed memory size). Anyway, a 20% of non-neural processing is unfortunately common in real applications, and must be taken into account if a cost-effective hardware platform is required.

30x30, 2in	Sun simul.		Sun+SMART(k=4) simul.	
task	time (sec)	time %	time (lower,sec)	time (upper,sec)
all	314	100	58.3	61.2
send inputs	-	-	3.8	3.8
read net stat	-	-	19.9	19.9
feed-forward proc.	263.7	83.7	2.45	5.05
update proc.	3.0	0.6	2.0	2.2

Table 1: Simulation results for 2 input, 30x30 neuron network

10x10, 100in	Sun simul.		Sun+SMART(k=4) simul.	
task	time (sec)	time %	time (lower,sec)	time (upper,sec)
all	503	100	101	115
send inputs	-	-	5.5	5.5
read net stat	-	-	39.4	39.4
feed-forward proc.	359	71.4	6.6	19
update proc.	48.7	9.6	2.8	3.8

Table 2: Simulation results for 100 input, 10x10 neuron network

5 Conclusions

An implementation of Kohonen's self-organizing feature map has been presented in this work as a contribution to the development of a suitable software environment for the neurocomputer

SMART. A preliminary performance evaluation for this system has been done, which points out how some compromises are needed between easiness and flexibility of use and performance. Further work is under way to implement large and sparse Kohonen-like networks on the SMART.

References

[1] M. McCord Nelson, W.T. Illingworth, 'A Practical Guide to Neural Nets', Addison-Wesley, 1990

[2] K. Goser 'Kohonen's Maps - Their Application and Implementation in Microelectronics', in *Artificial Neural Networks*, Vol 1, North-Holland pp. 703-708, 1991.

[3] K.Obermaier, H.Ritter, K. Schulten, 'Large-scale simulation of self-organizing neural network on parallel computers: application to biological modeling', *Parallel Computing*, Vol 14, pp. 381-404, 1990.

[4] A. Singer, 'Implementation of artificial neural networks on the Connection Machine', *Parallel Computing*, Vol 14, pp. 305-315, 1990.

[5] D.A. Pomerleau, G.L.Gusciora, D.S. Touretzky and H.T. Kung, 'Neural network simulation at warp speed: How we got 17 million connections per second', *Proc. IEEE Int.Conf. Neural Networks*, San Diego, 1988.

[6] C. Jutten, A. Guerin and J. Herault, 'Simulation Machine and Integrated Implementation of Neural Networks', in *Lecture Notes in Computer Sciences*, Vol. 412, Springer-Verlag pp. 244-266, 1990.

[7] D. Hammerstrom, N. Nguyen, 'An Implementation of Kohonen's Self-Organizing Map on the Adaptive Solutions Neurocomputer', in *Artificial Neural Networks*, Vol 1, North-Holland pp. 715-720, 1991.

[8] J.C. Lawson, N. Maria, C. Siegelin, 'SMART', *DHS1-D1 Final Report of NERVES EEC Project E-BRA 3049*, July 1991.

[9] J.C. Lawson, N. Maria, J. Herault, 'SMART: a Neurocomputer using sparse matrices', *Euromicro Workshop on Parallel and Distributed Processing* , January 27-29, 1993, Gran Canaria.

[10] P. Demartines and L. Tettoni, 'SOMA: Simulateur Logiciel de Modeles Neuronaux Adaptifs', internal report, LAMI-EPFL, Lausanne, 1991.

[11] T. Kohonen, 'Self-Organization and Associative Memory', *Springer Series in Information sciences*, Vol 8, Springer-Verlag, Berlin, 1987.

SIMULATION OF NEURAL NETWORKS IN A DISTRIBUTED COMPUTING ENVIRONMENT USING *NEUROGRAPH*

PETER WILKE

UNIVERSITAET ERLANGEN-NUERNBERG
LEHRSTUHL FUER PROGRAMMIERSPRACHEN
MARTENSSTR. 3 • D-8520 ERLANGEN • GERMANY
E-MAIL WILKE@INFORMATIK.UNI-ERLANGEN.DE

1 ABSTRACT

NeuroGraph is a simulator for design, construction and execution of neural networks in a distributed computing environment. The simulator either runs on single computers or as a distributed application on UNIX/X-based networks consisting of personal computers, workstations or multi-processors. The parallelization component offers the possibility to divide neural nets into concurrently executable modules, according to restrictions due to the neural net topology and computer net capabilities, i.e. *NeuroGraph* selects the best configuration out of the available distributed hardware environment to fit performance requirements.

2 THE *NEUROGRAPH* NEURAL NETWORK SIMULATOR FOR DISTRIBUTED COMPUTING ENVIRONMENTS

NeuroGraph is a neural network simulation environment for UNIX workstations. It is a software tool based on the X-Windows programming libraries and allows interactive design, training, testing and visualization of artificial neural networks (see fig. 1). The simulator consists of four major components: a simulator kernel that operates on the internal representation of the neural network models, a graphical user interface to interactively design net topologies and functionality and control network dynamics, an analyzation component to get and evaluate performance information during net execution, and a real world interface component for easy integration of the simulated networks into practical application environments.

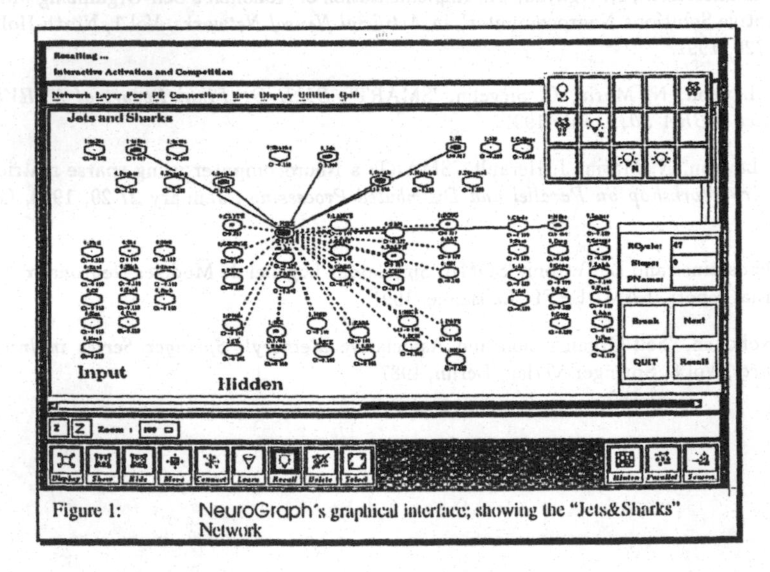

Figure 1: NeuroGraph's graphical interface; showing the "Jets&Sharks" Network

Furthermore, there are two special modules for advanced experimentation with neural nets: an easy to use toolbox for developers of neural nets offering interactive definition of specialized neuron functionality as well as net control strategies, and a parallelization component which is able to divide neural nets into parallely executable modules, and then map these modules to parallel hardware. With the parallelization module, time-consuming net control algorithms can be efficiently performed on multi-processor platforms or workstation clusters, whereas online visualization of the network dynamics is handled by graphic workstations.

3 THE SIMULATOR KERNEL

The simulator kernel (see fig. 2) provides a library of functions which are responsible for the management of the internal representation of the neural network topology and functionality. The kernel routines are subdivided into two major parts: routines which offer model-independent manipulation of data structures (installing or deleting layers, adding neurons, defining pools of neurons, installing connections between groups of neurons etc.), and routines which essentially depend on different neural network models, their connectivity and control strategy, i.e. modules containing model-specific learn and recall algorithms and interface routines (e.g., feed-forward backpropagation nets, constraint satisfaction, interactive activation and competition, pattern-associator, competitive learning, for a description of these models see [Rumelhart, McClelland 88]). The whole kernel is written in C for efficiency and portability reasons.

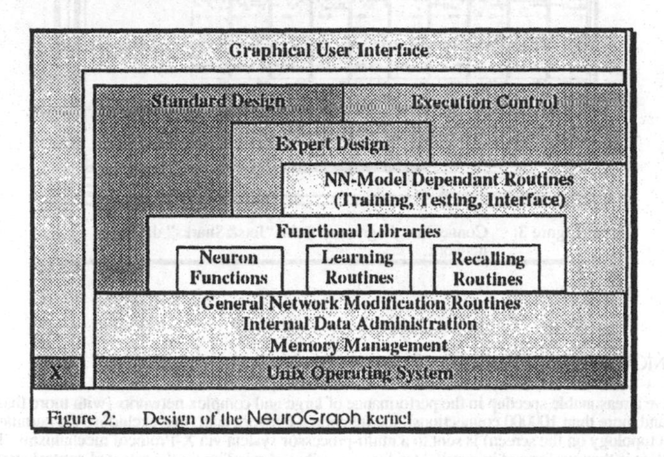

Figure 2: Design of the NeuroGraph kernel

4 THE GRAPHICAL USER INTERFACE

The graphical user interface (GUI) is based on the X-Windows system together with the OSF/Motif Widget Libraries. The GUI is (see fig. 1) an easy-to-use tool to interactively design different standard net topologies and control strategies; however, it is also possible to define new net architectures as well as neuronal and net-global functional behaviour. There are a number of standard functions (scaled summation/product, linear threshold, sigmoid, stochastic sigmoid, step, linear) available to characterize the learn and recall functionality of each processing element. But there is also a module to define user-specific signal processing and learning functions for the neurons; description of the functions is done via entering mathematical expressions for these functions.

In order to get different views on a simulated network there are several modes to visualize its topology and/or connection structures, to show only parts of a network, display neuron parameters in graphical and/or textual form etc. One special module presents connection structures in the form of a diagram (see fig. 3), as proposed by Hinton and Sejnowsky ([Rumelhart, McClelland 86], page 24).

5 SYSTEM COMPONENTS COMMUNICATION

As mentioned before, *NeuroGraph* is devided into separate modules. Common to all modules is the internal network representation which is a flexible data structure.

Using *NeuroGraph* on a single workstation this data structure is accessed via shared memory management. In a distributed computing environment *NeuroGraph* uses the X-Protocol [Young90] for communication between the software modules running on different computer network nodes [Jacob, Wilke 91][Wilke, Jacob 93]. Therefore *NeuroGraph* is able to visualize a neural net on a personal computer running an X-Server, using network parameters stored on a fileserver while the net is being trained on a multi-processor-system using back-propagation or any other learning rule.

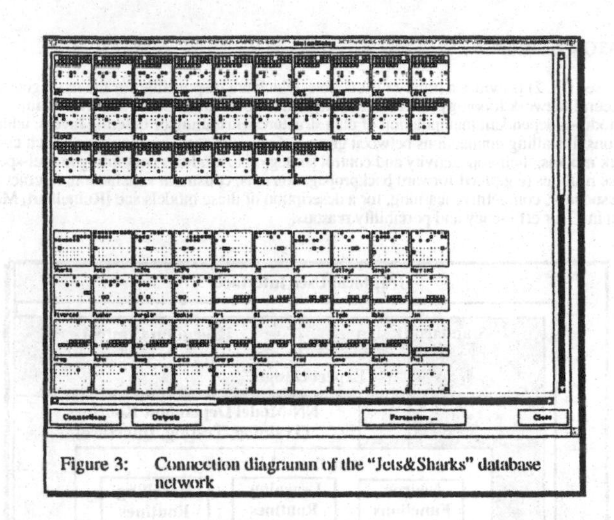

Figure 3: Connection diagramm of the "Jets&Sharks" database network

6 COMMUNICATION WITH PARALLEL HARDWARE

In order to achieve a reasonable speedup in the performance of large and complex networks (with more than 10000 processing elements and more than 100000 connections), the functional network data (i.e., excluding the parameters needed for visualizing the net topology on the screen) is sent to a multi-processor system via X-Protocol mechanisms. Time-consuming learning and recalling is then performed in parallel (as far as possible, depending on the selected control strategy and connectivity) with specially designed algorithms that efficiently use the underlying hardware. Any network data changes of which the user wants to be informed are signaled to the supervising kernel or GUI (Graphical User Interface) module, which then decides whether to reflect the changes on the screen immediately (e.g., to see a refined step by step evaluation of the network) or to wait for a general update signal (e.g., at the end of a training epoch). Alternatively the described data interchange can take place via files if the learning process is to be performed independently of the visualization GUI module (e.g., when learning may last some hours).

7 PARALLEL COMPONENTS ANALYSIS

A simple example should explain our major idea about how neural nets are subdivided into parallel executable stes of processing elements.

Given the constraint satisfaction network shown in figure 4 mapping the network to distributed hardware is a difficult task due to an irregular net topology. The problem is that directly connected neurons should not be updated concurrently in order to avoid oscillation effects.

Parallelization is done in four steps:

1. Constructing sets of neurons which can be updated concurrently, i.e. no directly connected neurons are in the same set.
2. Reducing these sets according to
   (a) the restrictions due to the available hardware, i.e. the number of processors, and
   (b) the net topology
3. Constructing a covering set with minimal cardinality whose elements are the resulting sets of step 2

4. Completing the resulting sets of step 3 to achieve total coverage of the network's set of neurons

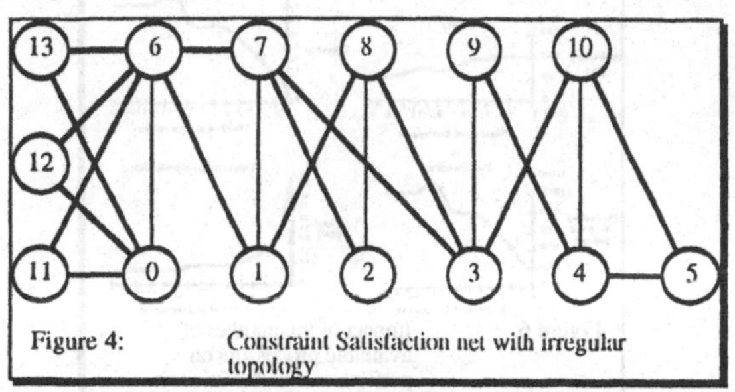

Figure 4: Constraint Satisfaction net with irregular topology

For our example net the algorithm will produce the output shown in figure 5.

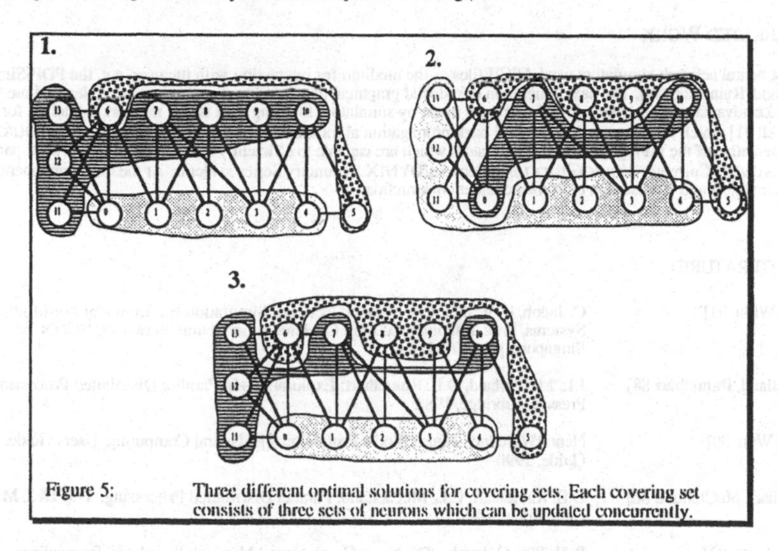

Figure 5: Three different optimal solutions for covering sets. Each covering set consists of three sets of neurons which can be updated concurrently.

Using these covering sets as a schedule the net can be executed in 3 steps because all neurons in a set can be updated concurrently. The example net consists of 14 neurons which should be processed in 3 steps; so at least 5 processors are needed (see fig. 5, third solution) to achieve fastest processing. An example schedule would be:

1. cycle: Neurons 0 to 4 are processed.
2. cycle: Neurons 5, 6, 8 and 9 are processed.
3. cycle: Neurons 7, 10 to 13 are processed.

Using the above schedule - which is automatically selected by *NeuroGraph* - a speedup of 14/3 = 4.6 is achieved; the efficiency will then be 4.6/5 = 0.93 or 93%.

Performance has been investigated using our algorithm for scheduling a Sequent S81 multiprocessor system, running DYNIX, a UNIX clone. The hardware consists of 16 processors one of which is reserved for administrational purposes. We recorded 1000 recall steps of the interactive activation & competition net "Jets&Sharks" [Rumelhart, McClelland 86, McClelland, Rumelhart 88] (see fig. 1) which consists of 68 Neurons and 1394 connections. Peak performance was 10967 neurons/sec, maximum speedup was achieved using 10 processors.

Figure 6 shows the impact of the number of available processors on the performance.

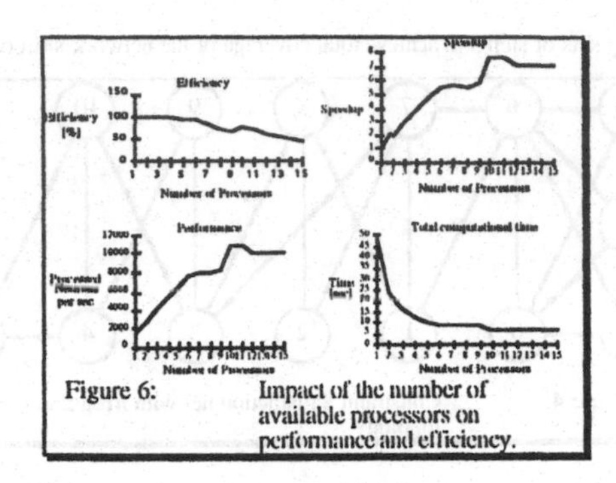
Figure 6: Impact of the number of available processors on performance and efficiency.

8 RELATED WORK

The first neural network simulators used ASCII files as the medium for interaction with the user, e.g. the PDP-Simulator [McClelland, Rumelhart 88]. The next generation introduced graphical user interfaces, as for instance [NeuralWare 90]. First steps to take advantage of distributed hardware were made by simulators focusing on a single network model, as for instance SNNS [Zell 91] which implements only parallel back propagation algorithms running on MasPar hardware. *NeuroGraph* is a representative of the third generation of simulators which are capable to efficiently execute different network paradigms on parallel hardware. Currently *NeuroGraph* uses Sequent/DYNIX Symmetry Series systems. At the time implementations for MasPar computers and workstation clusters are under construction.

9 LITERATURE

[Jacob, Wilke 91] C. Jacob, P. Wilke: A Distributed Network Simulation Environment For Multi-Processing Systems, Proc. International Joint Conference on Neural Networks, IJCNN 1991, Singapore, pp. 1178-1183.

[McClelland, Rumelhart 88] J.L. McClelland, D.E. Rumelhart: Explorations in Parallel Distributed Processing, MIT Press, Cambridge, 1988.

[Neural Ware 90] NeuralWare Inc.: NeuralWorks Professional II: Neural Computing, Users Guide, Reference Guide, 1990

[Rumelhart, McClelland 86] D.E. Rumelhart, C.L. McClelland: Parallel Distributed Processing, Vol. 1&2, MIT Press, Cambridge, 1986

[Wilke, Jacob 93] P. Wilke, C. Jacob,: The NeuroGraph Neural Network Simulator, Proceedings MASCOTS'93, San Diego, 1993.

[Young 90] Douglas A.Young: The X Window System - Programming and Applications with Xt OSF/Motif Edition, Prentice Hall, Englewood Cliffs, New Jersey, 1990

[Zell 91] A. Zell et. al.: Recent Developments of the SNNS Neural Network Simulator, Proc. Applications of Neural Networks Conference, SPIE Vol. 1294, pp. 535-544.

FULL AUTOMATIC ANN DESIGN: A GENETIC APPROACH

E. Alba, J. F. Aldana, J. M. Troya

Dpto. de Lenguajes y Ciencias de la Computación
Facultad de Informática
Universidad de Málaga
Pl. El Ejido s/n
29013 Málaga
Spain
aldana@ctima.uma.es
troya@ctima.uma.es

Abstract. ANN design is usually thought as a training problem to be solved for some predefined ANN structure and connectivity. Training methods are very problem and ANN dependent. They are sometimes very accurate procedures but they work in narrow and restrictive domains. Thus the designer is faced to a wide diversity of multimodal and different training mechanisms. We have selected Genetic Algorithms as training procedures because of their robustness and their potential application to any ANN type training. Furthermore we have addressed the connectivity and structure definition problems in order to accomplish a full genetic ANN design. These three levels of design can work in parallel, thus achieving multilevel relationships to yield better ANNs. GRIAL is the tool used to test several new and known genetic techniques and operators. PARLOG is the Concurrent Logic Language used for the implementation in order to introduce new models for the genetic work and attain an intralevel distributed search as well as to parallelize any ANN evaluation.

1.- Introduction

Artificial Neural Networks (ANNs) represent an important paradigm in AI dealing with massively parallel information processing proposed as biological models for the human brain. ANNs are widely used to offer human-like skills where they are needed, so we can find them in pattern recognition, signal processing, intelligent control and many other applications that can be faced by introducing a network as the heart of the solution system (see [Kim et al. 91]).

Wherever an ANN is to be used it must first be designed. At present, any ANN design drags along an intuitive and arbitrary path in order to reach "the better" structure and connectivity to be trained. Only the training methods are truly applied, but every ANN type seems to need its own and different training mechanism. Usually, the training mechanism is some kind of hillclimbing prosecution which is very closely related to (and so, dependent on) the problem being solved, the ANN type and/or the pattern set for it. This results in a vast landscape of different multiparameter tuning procedures that must be carried out for any individual problem and with no warranties of optimum results.

This lack of methodology and the hope of a general multifaceted quasioptimum training made Genetic Algorithms (GAs) attractive for our purposes. Defined by J. Holland in 1975 GAs simulate natural evolution. In [Goldberg 89], [Whitley & Hanson 89] and [Whitley et al. 90] can be found the basis of our GA and ANN approaches. For our purposes chromosomes will encode ANNs and genes will encode the items being optimized (weights, links or hidden layers). Alleles will be the gene components (regarding upon the coding used one or more alleles are included to compose a single gene value). Initial strings will be genetically evolved over generations of newly created offsprings searching for an optimum.

Since any ANN must be coded as a chromosome string ANN independence is achieved, just some *evaluation* procedure must be defined to recognize relative fitness of individuals to the problem. GA techniques have a stochastic behavior and so we only can expect **quasioptimum** (very frequent good or optimum) trainings. Besides local minima avoiding, generality, multiple points parallel search and robustness we can get further in using GAs to complete the ANN design. Since GAs work on some coding of the solution and not on the solution itself, we can code "any" problem as being a string of parameters and submit it to genetic optimization. Thus, it is only necessary to properly code the ANN *connectivity and structure* as strings and define an evaluation procedure to get two new levels of ANN design. In this way we can bring optimization methodology to these two design stages. This **full three levels design** is thought to help designer's work from the problem specification (patterns) up to a quasioptimum ANN to solve it (that is called a **Genetic ANN**). These three levels of design will be fully accomplished by using Genetic Algorithms. An introduction to genetic ANNs can be found in [Mühlenbein 90] and in [Mühlenbein & Kindermann 89].

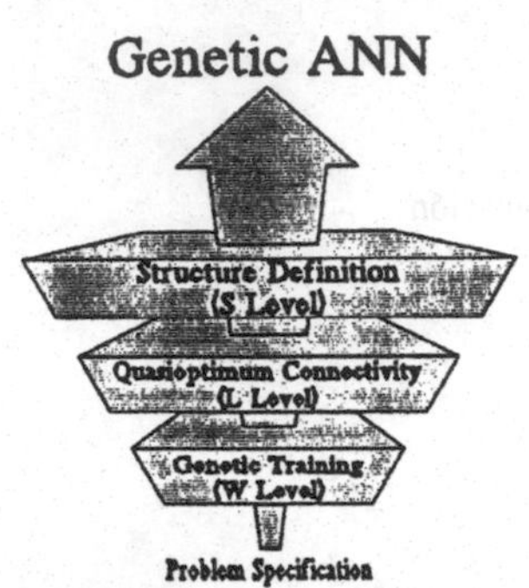

Fig 1.- *Three Levels GANN Design.*

We have designed and built up a genetic tool called **G.R.I.A.L.** (Genetic Research In Artificial Learning) [Alba, Aldana & Troya 92] to implement several known and new GA techniques and the three levels of genetic ANN design in order to test their efficacy and properties. We are furthermore concerned with scalability and computational efficiency, thus we have used a Concurrent Logic Language called **PARLOG** to implement GA and ANN behavior in GRIAL. This is a new kind of computational approach which is expected to profit from the emergent parallelism advantages.

The used test suite is made up of four problems. The XOR and the Two-bits Binary Adder (TBA) problem (but with 4 patterns) as presented in [Whitley et al. 90]. A feedforward network to make Spanish Plurals (three classes' classification) and a Hopfield network to behave as a 9-characters recognitor. The XOR and TBA problems are fully tested (training, connectivity and layered structure, either separately and together). The Spanish Plurals ANN has been trained and structure-optimized and the Hopfield network has been trained. While trying to get the optimum ANN for every problem we have tested the relative influence of the multiple GRIAL techniques.

In this work we have tested in GRIAL the effects of the *traditional selection* versus a *one-at-a-time selection* procedure. We explore the influence of coding on the ANN design by using *binary, real and diploid genotypes* (this last never tested before for ANN design). A *migration* scheme and an *adaptive mutation* similar to those used in [Whitley et al. 90] are tested against sequential single GAs and constant mutation rate. A smooth bit climber-like operator [Davis 91] called *GBit* is tried and a *Mating Restriction* similar to [Deb & Goldberg 89] is implemented. Partial genetically defined ANN jobs (as [Whitley et al. 90] where genetic training and connectivity are addressed as separate works and [Harp, Samad & Guha 89] where structure is genetically defined but backpropagation is used for training) have been considered as subproblems of the wider problem of designing a full genetic ANN.

2.- PARLOG

PARLOG [Clark & Gregory 86] is a Concurrent Logic Language which has been developed at the Imperial College. Operationally, the computational model of this kind of language consists in a concurrent processes' set which communicate by means of binding logic variables and which synchronize by waiting for unbounded logic variables. The possible behaviors of a process are defined by means of guarded horn clauses: **Head <- Guard : Body.** Head and Guard define the conditions under which a reduction can be made. Body specifies the resulting state of the processes after the reduction.

Parlog is one of these languages which exploit two kinds of parallelism: stream and-parallelism and or-parallelism. The first type of parallelism occurs when a goal is reduced to a conjunction of subgoals and they all are tested in parallel. The second type of parallelism appears when a predicate can be solved by more than one clause. In this case, all of them are tested at the same time and, if more than one fulfils its guard, one of them will be selected in an indeterministic way. Parlog also has some primitives which can be used to avoid both types of parallelism (see [Crammond et al. 89]).

This kind of language fits very well within the parallel programming paradigm. In opposition to sequential logic languages which present a transformational behavior, concurrent logic languages are well fitted to the specification of reactive system, that is, of open systems which have a high level of interaction with their environment. As stated in [Troya & Aldana 91], this is what a neural network does: it tries to reach a statistical optimum that is environment- dependent.

GRIAL is oriented to ease the changes among the GA strategies used to solve a given problem. *Unix parallelism* is achieved at *interlevel* communications while *Parlog parallelism* appears at *intralevel* searches. Real LAN distribution of intralevel Parlog parallelism can be faced by using Parlog mail-boxes. Parlog allows entering parallelism from string managing genetic operations till ANN neurons' activations. Fine and coarse grained approaches are of a straightforward implementation with Parlog. These advantages along with its lists and symbols processing make Parlog a better language than imperative ones to get good and reliable implementations.

3.- Fully Genetic Design of ANNs

In this section the three levels and full genetic ANN design is presented and analyzed by means of GRIAL. New and existing GA techniques have been tested to get a qualitative taste of the properties of this kind of design. We envisage the following exposition from bottom (genetic training) to up (genetic structure definition) passing through an intermediate genetic connectivity definition level.

3.1.- Genetic Training

To submit any ANN type to genetic training we must define some proper coding for weights to appear as strings. GA's work needs some local logic meaning to be present in strings, i.e., a chromosome must be coded to include *logic building blocks* with some meaning for the underlying problem [Goldberg 89]. Then we code any ANN string as being a sequence of its input link weights to every neuron in the ANN, from the input layer to the output layer.

The genetic crossover of two different strings (coded ANNs) profits from their best slices to create better trained offsprings. Through natural selection bad *schemata* (bad structures of the solution) are exponentially discarded and good schemata are exponentially reproduced as evolution occurs. To evaluate the fitness of a string to the *environment* (problem) strings are *expressed* (decoded) as ANNs. We use *SQE* (squared error between desired and obtained outputs extended to any output neuron and any pattern in the pattern set) as the fitness measurement to help the natural selection's work: stochastic selection picks up the best strings to be crossed (reproduced) to yield offsprings in next generation.

Weights can be coded in strings attending to several codes. Binary code (signed magnitude) is very extended in genetics. Real, Reordering and Diploid [Goldberg 89] codings are another known ones we have tried for ANN training. Binary code is very good for the GA job, but, for ANN training, it needs too large populations and evolutions even for very small problems (we want to keep population size on hundreds of individuals). Reordering schemes (genes, PMX and/or Inversion) do not seem to improve binary results, and we think this is because we are using a correct genotype representation of the problem that does not need additional genetic help. Real coding (one-weight/one-real-gene) seems to be the truly useful genotype because it allows small and quick GAs to solve adequately the trainings, despite it presents an undesired low diversity maintenance during evolution that provokes local minima appearances.

```
[1 0 1 0 0 1 1]
```

(a) Binary

```
[fd(1,1),fd(2,0),fd(3,1)...,fd(7,1)]
```

(b) Reordering

```
d( [1,-1,-1, 0, 0, 1, 0],
   [1, 0,-1, 0,-1,-1, 1])
```

(c) Diploidy

```
[-0.3]
```

(d) Real

Fig 2.- *Four chromosome structures encoding the same gene -weight- in GRIAL. If we apply Dominance on (c) we will get (a). In (b) a Parlog functor (fd) acts as a feature detector storing a logic term, in order to implement the locus allele union.*

All these codings are *Haploid* codings (one string encodes one ANN), but *Diploid* chromosomes with triallelic values [Goldberg 89] (and maybe with real values...) have much to say in allowing sophisticated behavior by helping diversity and natural adaptation to traumatic changes in the environmental conditions (they outperform the other codings using half the number of strings).

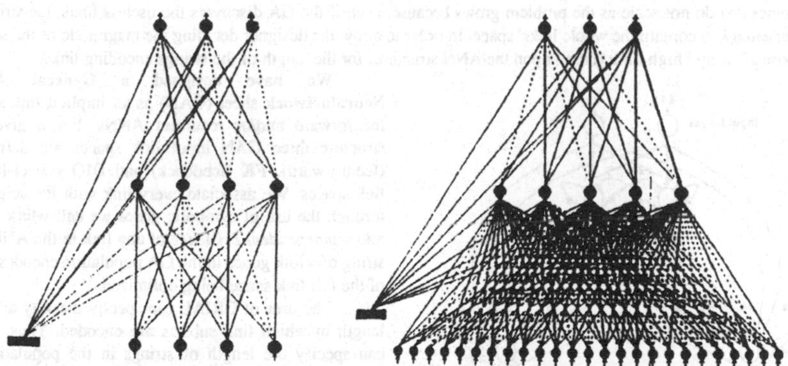

Fig 3.- *The Two-bits Binary Adder problem and the Spanish Plurals problem (24 patterns to determine one of the three Spanish plural types for any word presented to the ANN) solved by using feedforward ANNs with genetic training. In order to solve increasingly complex problems we need a better balance between exploration and exploitation and that's why we state the need of more sophisticated GA strategies and not the use of huge populations (thousands of individuals) and huge evolutions (millions of crossovers). Non-trivial training problems are actually being solved by using GAs.*

Any feedforward, recurrent, dynamic, static or any other ANN type can be trained by GA means. We have tried *constant* and *adaptive mutation* operators (this last based on Hamming Distance --hd--) to maintain diversity in population. We have stated that mutation is not a secondary operator but an essential technique, very useful in keeping population size at a relatively low value. Probability for adaptive mutation is computed as 1/hd (linear) and we have detected this as a somewhat high value (high mutation has allowed good searches with our one-at-a-time selection), but in GRIAL, a control technique called *begin-end-frequency* allows a better pursuit of GA strategies' effects by specifying how and when to apply them. In order to speed up the search of a quasioptimum weights' set we have designed the **GBit** operator, a hybrid genetic-hillclimbing procedure that makes smooth changes in the best-to-now solution string to explore its neighborhood.

Crossing two strings is not a trivial operation because these two strings may represent different functionality-neurons distributions for the problem solution and crossover will often yield two new strings (ANNs) that behave much worst than their parents and that are unable of future improvement (called *Lethal* strings). To solve this problem we have designed a **Restrictive Mating** operator based on hamming distance (similar to the Genotypic Mating Restriction described in [Deb & Goldberg 89] to impose a minimum likeness between mates to ensure that interspecies (very different couples) crossovers do not take place. This problem is very frequent in big populations where many good (but different) partial solutions often appear. GBit and RMating have shown to turn GAs quick tools of high efficacy.

Since ANN coding/decoding and evaluation operations are very complex and expensive we look for improved selection mechanisms that minimize wanderings along the search space while maintaining the GA properties. The **Generations Evolutive Scheme** (a full new generation replaces the old one) using the *Stochastic Remainder Without Replacement* seems to be very expensive in our tests despite its advantages in preserving diversity and genotype exploration. That's why we have designed the **Immediate Effects Evolutive Scheme** to keep strings ranked (from best to worst) and using the *Roulette Wheel* selection to pick up two individuals to be genetically processed and produce offsprings to be inserted in the ranking. This later selection operator is the best choice for ANN design, but population size must be kept large enough to avoid premature convergence due to its more minimum-directed search. A complete-ranked selection as that in [Whitley 89] could lessen genetic drift because in GRIAL we allow duplicates in the population and the IEES produces a high selection pressure.

We have finally tested and stated the superiority of any **Distributed search** using N GAs over any equivalent single GA search for the whole test suite. We have used a **Migration** scheme similar to [Whitley et al. 90] to define N GAs in parallel evolution with occasional exchanges of their respective best strings through Parlog streams. On every swap the Nth subpopulation sends its best string to the (X+S) mod N subpopulation, where S is the swap number. This parallelism improves time, accuracy, diversity, lethals avoiding and is more likely to yield a quasioptimum result. Real speed-up for quick GA evolutions can be brought from real machine distribution of the N GAs in a LAN or multiprocessor computer.

3.2.- Connectivity Optimization

In traditional ANN design connectivity is determined by the ANN type to be used or by experience. We have made a *genetic connectivity definition* to accomplish three goals: *(a)* GAs are used to define connectivity and weights (training), *(b)* we want our method to be scalable with the problem size and *(c)* the predefined structure to be connected must be completely used, because we expect this structure to have been built as quasioptimum by any means.

By coding links in strings we want to obtain the GAs advantages not only in that links pruning is tried, but in that links appear and disappear in parallel in strings as evolution occurs. A binary coding as the used in [Whitley et al. 90] fills every string position to encode the presence (1) or absence (0) of a link, but this coding yields excessively long chromosomes that do not scale as the problem grows because, even if the GA discovers the useless links, the strings *must hold room enough* to contain the whole links' space. In order to allow the designer deciding the magnitude of the search we only impose a low and high limit (based upon the ANN structure) for the length of the strings encoding links.

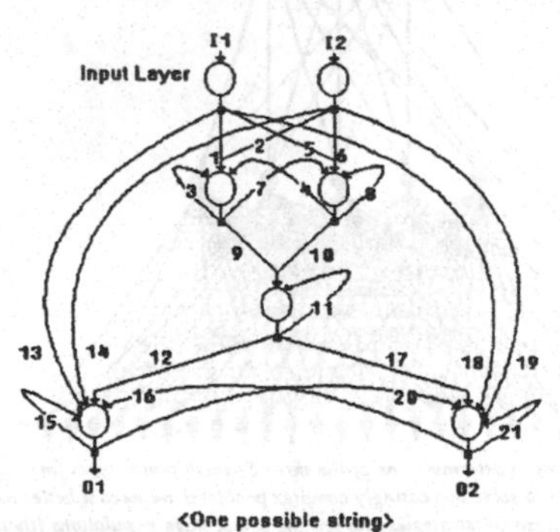

<One possible string>

```
[wlink(1,0.9),wlink(6,1),wlink(9,-0.7),wlink(11,-1),
wlink(12,-0.1),wlink(14,1),wlink(17,0),wlink(18,1)]
```

Fig 4.- *One example of a GAN specification and a possible wlink string in the connectivity level.*

We have designed a **General Artificial Neuralnetwork** sheet (GAN) as an implicit link space for feedforward and/or recurrent ANNs. For a given ANN structure three GAN input link spaces are defined: **FF** (feedforward), **FK** (feedback) and **DIO** (direct-IO) input link spaces. We associate every link with its weight value through the use of symbolic genes we call **wlink genes** to allow *independent existence* of any link in the ANN. Every string of wlink genes in the GA population encodes a *subset* of the full link space being optimized.

The user of GRIAL can specify the way and strings length in which link subsets are encoded. Thus, designer can specify the length of strings in the population, then *pruning* as desired the full link space. Excessive initial pruning has shown to be undesirable because networks are unable to learn the full pattern set. The rational use of the ANN structure brings from the results one can get by pruning links beyond a low limit: many neurons can be present in the final ANN whose input and/or output are not being considered when the ANN works.

So we have defined **Archetypes** as distinguished string positions respected by the crossover during evolution. Archetypes assure that the ANN structure is being profited by. We have got between 25% and 60% of reductions of the links' space with this kind of genetic pruning.

This linking (L) level uses the lower training (W) level to determine weights and fitness for its initial population and evolution is responsible for crossing the subsets of links (strings). Links' pruning and adding are achieved by a *link duplication effect*. We have tried penalty schemes attending to links' number (SQE*Links_Number is computed as the new fitness value instead of using the SQE value) in order to modify the fitness values. However, the results indicate that any other penalty schemes (as giving more --W-- learning facilities to strings) should be of greater success. But the real power of this technique becomes from the initial designer-driven pruning and the subsequent GA optimization. Again the best results are these obtained with N distributed GAs working in parallel. At the L level we try to make a more natural ANN design by interacting with the training lower level and then making it easy as well as more accurate and cheaper to implement (as software or hardware ANNs).

3.3.- Structure Definition

Defining the best ANN structure for a given problem is not a deterministic nor even a well-known process. This is because the complexity of recognizing every neuron's job in the whole ANN and the relationships among neurons. There exist many rules based on experience about the number and disposition of neurons, but, as for connectivity definition, designers have not a methodology to guide the design of the structure out of the *proof-and-error* mechanics.

The real problem for a genetic structure definition is *to select a good coding* to contain structure information. This coding must be general enough to be combined by crossover and specific enough to determine one unique ANN structure when decoded (expressed).

We have designed a **binary genotype** to do this job (a *strong* coding of the structure). We want to optimize the number of hidden layers and neurons per hidden layer. GRIAL puts in a chromosome one gene for every one of the potentially existing hidden layers.

[0,1,0,1, 0,0,1,0, 0,0,0,0]

(a) String encoding the structure

GN=3 GL=4

(b) Genes Number and Length

[binary,linear,binary]

(c) Transfer Functions List

Fig 5.- ***Structure Encoding. The chromosome (a) represents the existence of two hidden layers (the third gene is 0 when decoded). This search allows the generation of ANNs of between 0 and 3 hidden layers, each of them with between 0 and 15 neurons. (c) contains the transfer functions list to be used for any hidden layer in the case of existence (in this example the underlined binary transfer function will not be used because the third hidden layer doesn't exist).***

The strings at the structuring level are GN*GL bits length, where GN is the maximum number of hidden layers to be considered and GL is a value such that $2^{GL}-1$ is the maximum number of neurons to be included in any hidden layer.

We have got very good results for feedforward and GAN networks regarding useless hidden neurons pruning (in order to reduce the cost per ANN) and quasioptimum structures' generation (e.g., we have genetically duplicated the human proposed structure for the XOR problem).

Furthermore a **penalty scheme** based on changes in the fitness values according to the neurons' number of the network (multiplying the SQE value by the number of neurons in the ANN) has been successfully tried with some interesting side effects in helping natural selection decisions while searching for smaller and cheaper ANN structures.

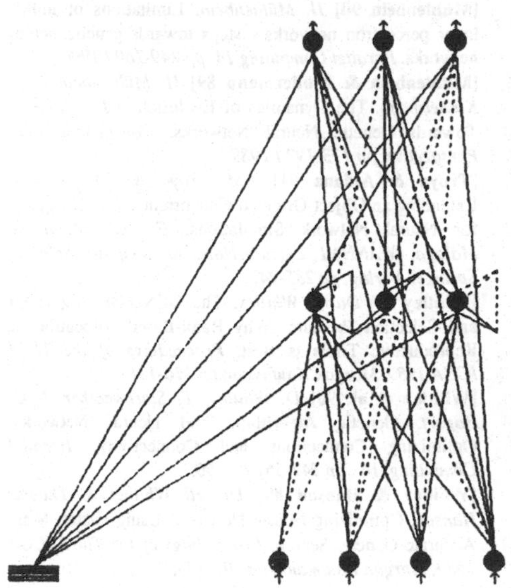

Fig 6.- ***Fully automatic genetic TBA-ANN design.***

Usually the kind of desired structure is best known by the designer than the best connectivity pattern or even the weights set to be used, and that's why S level search is accomplished by smaller GAs than those GAs needed at the L or W levels. On the other hand S strings are very complex and expensive to evaluate for fitness because they require multiple L and/or W GAs executions. Genetic operators have been extended to this level (e.g., GBit or mutation) but the simple traditional GA has shown to be a good procedure for S level. We encourage the use of the *Immediate Effects Evolutive Scheme* in order to prevent useless wanderings. For structure definition, hillclimbing-like methods may reduce the overall cost.

The final best ANN will have been designed to yield a full suited ANN to the proposed environment (the problem). Individual GAs will gain by being allowed to present a higher rate of error due to the multiple level relationships (GA parameters' tuning does not need to be of high accuracy to work well). As we can see the parallelism at many computational levels is the best way to bring efficiency and efficacy to this complex Genetic ANN design: three levels run in parallel, several GAs can work in parallel at any level and every genetic operation can be parallelized due to PARLOG software enhancement. PARLOG can be used to define an Object Oriented library of ANN simulators [Troya & Aldana 91] and to implement any kind of GA parallelism as those mechanisms outlined in [Macfarlane & East]. GRIAL includes some of these propositions.

4.- Conclusions

There exist many different non GA ways to define the target ANN to solve a problem and many other Evolutive mechanisms but the **Full Three Levels ANN Design** is thought to be the best and more natural way to take designer to quasioptimum ANNs by controlling only some GA strategies. GAs provide a good set of tools to do this work. The resulting ANNs can be later refined by any of the applicable existing mechanisms.

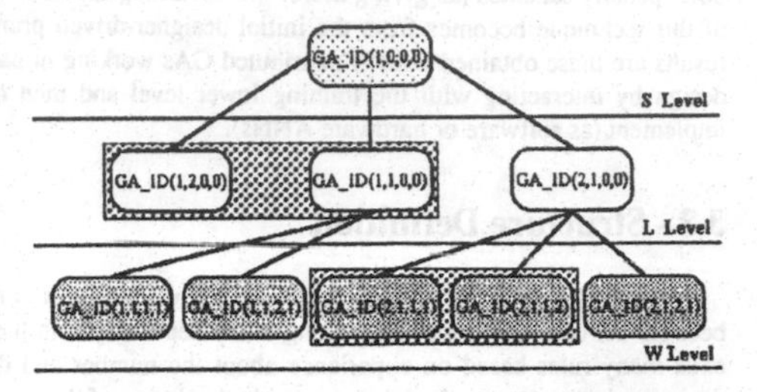

Fig 7.- *Unix and Parlog Processes Tree. Vertically disposed GA identifiers represent different parallel unix processes. Horizontally disposed GA identifiers enclosed in shadow boxes represent distributed parlog GAs.*

We think that a smart connectivity for initial pruning to be used at L strings (initial link strings' composition) can improve the connectivity definition, because, when full automatic design is being performed, the S level randomly generates this pruning and only a high number of strings at S level can overcome excessive initial pruning (because numerous initial prunings are to be tried).

The "best" GA to be used for ANN design consists in a distributed search with IEES or a similar ranked selection, using a somewhat high mutation rate, two points crossover and some help to avoid lethals and to speed up the search (GBit or another bit climber and some niche & species formation technique like RMating if no migration is to be used).

ANNs and GAs are thought to be open and reactive systems which fit very well within the logic and the concurrent programming paradigms. Thus, many advantages regarding parallel implementations and software improvements can be brought from the use of Parlog as the base language.

With the correct combination of some GA techniques we can define a general methodology to enter automatic ANN design in a unified fashion while still maintaining diversity of approaches to solve a problem. Computational requirements of this kind of design (through genetics) are to be enhanced by parallelism in order to reduce the time and memory consuming GAs that we are still forced to use in solving large problems.

References

[Alba, Aldana & Troya 92] *E. Alba Torres, J.F. Aldana Montes & J.M. Troya Linero.* Genetic Algorithms as Heuristics for Optimizing ANN Design. *Technical Report, Dpto Lenguajes y Ciencias de la Computación 1992.*

[Clark & Gregory 86] *Keith Clark & Steve Gregory.* PARLOG: Parallel Programming in Logic. *ACM Trn. on PL & S 1986, pp 1-49.*

[Crammond et al. 89] *Crammond, Davison, Burt, Huntbach & Lam.* The Parallel Parlog User Manual. *Imperial College, London, pp 1-40 / 1989.*

[Davis 91] *Lawrence Davis.* Bit-Climbing, Representational Bias and Test Suite Design. *Proceedings of the Fourth ICGA 1991, Morgan Kaufmann, pp18-23.*

[Deb & Goldberg 89] *Kalyanmoy Deb & David E. Goldberg.* An Investigation of Niche and Species Formation in Genetic Function Optimization. *Proceedings of the Third ICGA 1989, Morgan Kaufmann, pp 42-50.*

[Goldberg 89] *David E. Goldberg.* Genetic Algorithms in Search, Optimization & Machine Learning. *Addison-Wesley, 1989.*

[Harp, Samad & Guha 89] *Steven Alex Harp, Tariq Samad & Aloke Guha.* Towards the Genetic Synthesis of Neural Networks. *Proceedings of the Third ICGA 1989, Morgan Kaufmann, pp 360-369.*

[Kim et al. 91] *K.H. Kim, C.H. Lee, B.Y. Kim & H.Y. Hwang.* Neural Optimization network for minimum-via layer assignment. *Neurocomputing 3, pp 15-27 / 1991.*

[Macfarlane & East] *D. Macfarlane & Ian East.* An investigation of several Parallel genetic algorithms. *Univ.of Buckingham, MK 18 IEG, pp 60-67.*

[Mühlenbein 90] *H. Mühlenbein.* Limitations of multi-layer perceptron networks - steps towards genetic neural networks. *Parallel Computing 14, pp 249-260 / 1990.*

[Mühlenbein & Kindermann 89] *H. Mühlenbein & J. Kindermann.* The Dynamics of Evolution and Learning - Towards Genetic Neural Networks. *Connectionism in Perspective, pp 173-197 / 1989.*

[Troya & Aldana 91] *J.M. Troya & J.F. Aldana.* Extending an Object Oriented Concurrent Logic Language for Neural Network Simulations. *F. Informática de Málaga, IWANN'91, Lecture Notes in Computer Science, Springer-Verlag, pp 235-242.*

[Whitley 89] *Darrel Whitley.* The GENITOR Algorithm and Selection Pressure: Why Rank-Based Allocation of Reproductive Trials is Best. *Proceedings of the Third ICGA 1989, Morgan Kaufmann, pp 116-121.*

[Whitley et al. 90] *D. Whitley, T. Starkweather & C. Bogart.* Genetic Algorithms and Neural Networks: Optimizing Connections and Connectivity. *Parallel Computing, 14 , pp 347-361 / 1990.*

[Whitley & Hanson 89] *Darrell Whitley & Thomas Hanson.* Optimizing Neural Networks Using Faster, More Accurate Genetic Search. *Proceedings of the Third ICGA 1989, Morgan Kaufmann, pp 391-396.*

HARDWARE IMPLEMENTATIONS OF ARTIFICIAL NEURAL NETWORKS

D. Del Corso

Politecnico di Torino - Dip. Elettronica
C.so Duca Abruzzi, 24 - 10129 TORINO - ITALY
tel. 39-11-564 4044 - fax. 39-11-564 4099
Email: delcorso@polito.it

Abstract

The availability of effective and low-cost neural hardware opens the gate towards applications of neural networks which can compete with other techniques. This paper describes the design problems of silicon neural architectures, and analyzes the various implementations choices. Emphasys is on methodology and principles. Pointers to descriptions of specific neural circuits are provided.

1 INTRODUCTION

Many researches on Artificial Neural Networks (ANN) are carried out on neural systems built as software environments on workstations or PCs. Several such products are commercially available, and in some cases hardware accelerators can be added to enhance performance. On the other hand, very few neural systems are built using dedicated hardware, and there is no commercial widely diffused general purpose hardware neurocomputer.

1.1 Why neural hardware ?

Therefore before starting to deal with Hardware implementations of Artificial Neural Networks we must understand which is the motivation for developing dedicated neural hardware. The actual key question is: Do we really need specialized and expensive custom ICs (the Neural Hardware) when *almost* the same task can be done by appropriate SW on a general-purpose computer ? The answer must consider that the technology for commodity circuits (microprocessors, memories) evolves much faster than technology for custom ICs, and the design efforts dedicated to commodity VLSI allow complexity (and performance) which make questionable the benefit of custom ICs.

To put in evidence the role of custom hardware we can have a look at another class of subsystems: filters. To build a filter there are several approaches:

- stay in the *analog* domain, and build the filter with R, C, switches, other components like standard Operational Amplifiers;
- move to the *digital* domain and use A/D/A converters, DSPs, and suitable software;
- describe one of the above implementations and *run* it on an appropriate circuit *simulator* (e.g. SPICE for the analog one) on a general purpose computer.

Today, many engineers will start the design of a filter from the last approach, but none of them will choose it for the actual application. Even if a simulation environment could allow to solve some real-time tasks (e.g. in the audio frequency range with current technology), it cannot be considered the most efficient way to put a filter in a systems. However, many neural applications are handled exactly in this way !

Simulators, in any application areas, are design tools which allows to make experiments, evaluate architectures, verify performance, compare choices. As the engineer has to go towards cost effective solutions for a well defined application, the primary choice is dedicated hardware. This term includes both **custom** hardware, where the functions comes from the structure itself (type and interconnection of devices, such as the R, C, Op-Amp filter in the above example), and **programmable** hardware, where several generic function are made available by a *low level machine*, and the application function is defined at a higher abstraction level (the program for the DSP solution). The difference from this last case and the generic *CPU + simulator + description* is in the number of layers or virtual machines between the hardware and the application. In the following we will consider **simulators** those systems where the user sees an intermediate virtual neural machine, which usually has been defined on a general purpose processing system by another designer. A **programmable neurocomputer**, which belongs to the *neural hardware* family, has a low level structure optimized for neural operations, and the application designer has direct access to the lower machine levels. However, it is not easy to make a hard cut; what if the user buys a neurocomputer where lower layers are fully hidden ?

1.2 How to build neural hardware

As for any other electronic subsystem, the benefits of dedicated hardware for ANN are:

- **faster operation,** which comes from two sources:
  - *reduced number of layers* (virtual machines) between the application and the hardware.
  - lower circuit complexity which allows, for a given technology, *higher clock rate* (or faster operation for asynchronous circuits).
- **lower cost,** both in term of silicon (area), and power consumption.

In the case of ANN, the developement of dedicated hardware involves joint efforts in several fields:

- **Technology:** design and fabrication of smaller, denser ICs to increase performance.
- **Design aids,** such as simulators, libraries of cells, synthesis tools.
- **Algorithms:** this is where ANN theory and IC design experience must come to a compromise. Actual hardware has intrinsic limits: very few neural circuits use floating point computation. Complete interconnection, as required by several NN architectures, is seldom possible. Complex weight update rules are not allowed for devices with on-chip learning.

Good results can be achieved only by exploiting the peculiar features of ANN architectures:

- **Parallelism:** the required speed of computation is obtained through the cooperation of *several slow* units, instead of by a *single high speed* one.
- **Closed loop operation:** the precision of computation is obtained by correcting some parameters (e.g. weights) using *negative feedback,* instead of by high precision at each processing step.

The following sections identify the problems related with hardware implementation of neural network systems, and show how they can be solved with various design styles. Since in a single paper it is impossible to describe implementation details of actual circuits, emphasys is on principles and methodology; specific solutions are described only as examples, with pointers to appropriate references.

The ANN reference model here used is the Multi-Layer Perceptron, which embeds all the functions needed by neural systems; other architectures can be derived by changing the interconnection topology or the weight update algorithm.

2 FUNCTIONAL PARTITIONING

As any processing system, an ANN consists of three functional parts:

- **Computation Units (CU)**, which execute operations from simple (e.g. summation) to complex ones (e.g. compute weigth corrections).
- **Memory Units (MU)**, which store information used for internal computation (tipically weights) and other intermediate variables.
- **Interconnection Structure**, which supports the exchange of information CU-CU, CU-MU, and with the external world.

The operations performed by CUs are usually *multiplication* for synaptic weigthing, *addition* to compute total input activation, and *non-linear transfer function* to compute output activation. Other more complex operations may be required for systems with learning capability, to compute weight correction.

CUs can be multiplexed among different parts of the network (for instance the same synapse computes the product for several input/weight pairs).

Weights are seldom predefined at fabrication; in most cases they are loaded in the network before the recall operation. To perform the learning process weights must be changed; this requires the so-called *plastic memory*, with read-write and long term store capability.

As for conventional processing systems, the interconnection structure is frequently the bottleneck. A proposed solution (not described here) is the use of optical interconnects [KRIS92]. The interconnection problem does not exist for simulators, where all variables are in the memory and all operations are performed in the CPU. Moving an activation value from one neuron to another one is simply a matter of memory access, independent from the *distance* among neurons. As the architectures becomes parallel, with multiple processors and local memories, communication cost become less uniform (information exchange is faster and easier within the same processor or cluster).

3 INFORMATION ENCODING

The basic functions of computation, storage, communication can use different representation of information, as shown in figure 1:

- **time and amplitude continuous**: information is carried by voltage (or current) levels; there is no time sampling or quantization, both for processing and communication. These are the *fully analog* circuits.

- **discrete time, continuous amplitude**: it brings to the so-called *mixed* circuits, or more precisely *sampled data* circuits. Information is carried by voltages (currents) which become valid or are processed only at defined times.

- **discrete time, discrete amplitude**: information is carried, stored and processed in binary form, using gates, registers, and other logic devices. These are the *fully digital* circuits.

Since each subsystem (CUs, MUs, interconnections) can use different information representation, a variety of different ANN hardware families can be defined. We shall focus in the following on the benefit and drawbacks of each technique, and provide examples of specific implementations. Special attention is given to pulse-stream circuits, which belong to mixed solutions.

Both time-sampling and amplitude-quantization involve loss of information or errors, respectively in the time/frequency and in the amplitude domain:

- When the value of a variable is known or computed only at intervals T_s (sampled at a rate $F_s = 1/T_s$), the sampling theorem says that the variable is unambiguosly defined only if it has no component higher than $F_s/2$ (this avoids *aliasing* errors). Therefore time sampling puts an upper bound on the bandwidth of signals which can be processed by the system.

- When the value of a variable is discretized in the amplitude domain using N bits, the variable is affected by an error (or quantization noise) $E_q = S/2^N$, where S is the full-scale value. Therefore the amplitude quantization puts an upper bound on the dinamic range of the signal (max/min = 2^N).

Fully digital circuits are affected by both aliasing and quantization errors, but these are the only errors in such system.

Fully analog circuits do not have aliasing and quantization errors but have bandwidth and range limits too. Signals cannot change with infinite speed, nor be cleaned after injection of noise; this in turn again limits the speed and the dinamic range of the system.

4 IMPLEMENTATION APPROACH 1: DIGITAL ANN

Digital ANN use logic circuits to process information discretized in the amplitude and time domain. In principle, a neural system should use one functional unit (e.g. multiplier) for each signal, and point-to-point connections only. From this point of view, NN can be seen as the ultimate fine-grane parallel architecture.

Using each unit for a single data is not the most convenient approach for a digital implementation, where multipliers, accumulators, nonlinear transfer functions, use a considerable amount of silicon, and are therefore multiplexed among different data or different parts of the network (synapses, neurons, layers).

Digital ANN are actually, from an architectural point of view, special-purpose computer, with added parallelism in some cases. CUs are based on ALUs specialized to compute sum-of-products, MUs consist of memory banks, which store weigth as well as intermediate results and activation values, and buses make the interconnection strucures to move information between CUs and MUs. A controller handles the sequence of operatione (and takes the role of the program in a general purpose computer). Depending on speed requirements, CUs and communications can be realized in serial, parallel or intermediate forms, with different area occupation. The regularity of neural operations makes also possible to use pipelined architectures.

MUs are registers or memory banks with appropriate width; plasticity (that is the capability to store permanently changeable data) is achieved using RAMs with backup, EEPROMs, or Flash-EPROMS.

When the ANN architecture has been defined using appropriate *neural developement tools* (e.g. a simulator), the next step involves methodologies and tools for *digital design* at the architecture, register and layout levels. A variety of such tools, oriented to several design styles, is now available.

To summarize, we can put among the benefits of fully digital ANNs:

- design techniques are automated and well known, at least for synchronous circuits;
- noise immunity is high;
- resolution and precision can be high;
- plastic memory uses standard devices;
- communication and computation can be easily multiplexed;
- the digital ANN can be directly integrated in another digital processing environment (e.g. a workstation).

On the other side, if we can accept low precision, digital circuitry requires larger silicon area than the analog one with equivalent functionality. A detailed overview of several digital ANN architecture is in [TREL89]. Examples of fully digital Neural Systems with different approaches are in [HIRA89], [MAND92], [MELT92], [LAWS93].

5 IMPLEMENTATION APPROACH 2: ANALOG ANN

The term **analog** ANN is used for those circuits which work on signals continuous in time and amplitude, that is neither sampled nor quantized.

5.1 Handling the error budget

Analog ANN machines are affected by the same problems as any analog subsystem:

- **Noise** puts a lower bound on the minimum value of signals which still carries useful information, which in turn defines the dynamic range of signals (in a digital circuit this same parameter depends from the number of bits). Signal variations below the noise platform cannot be sensed or processed; this is the system *resolution* (1 LSB for digital).
- **Offset, gain, and nonlinearity** errors changes the behaviour of the system with respect to the one defined by the designer, and affect the *precision* of the circuit (this effect is not present in digital systems).
- **Bandwidth and slew rate** limit the processing speed and the rate of change of signals.

All these types of errors cumulate at each processing step, with no possibility of recovery. It is reasonable to wonder how can complex systems like ANN be built with such unprecise circuits. The answer comes from the combination of several factors:

- **Amplitude and time averaging**: The results of computation in NN come from the sum of several variables; if errors have zero mean value and are randomly distributed, their total contribution stays close to zero. The same happens in the time domain; if the analog variable is evaluated for some time (e.g. integrated, or low pass filtered), the effect of noise at frequency higher than the LPF cutoff is strongly reduced.
- **Closed loop operation**: In open-chain systems the precision of results depends on the precision of each single element in the chain; with closed-loop feedback systems and enough loop gain, the precision depends from the feedback network. Any ANN to which some kind of learning rule is applied can be considered a closed loop system, where *in-the-loop* errors are automatically compensated. The benefit applies for the machine on which learning is actually performed; if weights are computed on a simulator, and then loaded to a target machine, we must again rely on precision of computation to achieve the desired results. This is a good reason to develope neural ICs with on-board learning (at least for small weight corrections).

5.2 Weight storage

Some analog circuits use digital weight memories; in this case the quantized weigth is converted to an analog variable (current, voltage), suitable for the synaptic weight multiplier or directly used in mixed circuits (multiplying D/A converters). A variety of such mixed Analog/Digital multipliers is described in the literature, but only those with small area occupation are used for neural circuits. Some examples are in [VERL89].

For true analog plastic weight memory the two basic techniques shown in figure 2 are currently used:

- **Capacitive storage**: it consists of an array of Track/Hold circuits or *analog DRAM*. Since the capacitor discharges through the leakage of the write switch, periodic refresh from an external stable memory is required. With current technology refresh must occur every 2-100 ms. A large neural IC, where the refresh process is combined with learning (both imply the ability to change the weights), is described in [ARIM92].

- **analog EEPROM**: even this rather new technology uses the charge accumulated in a capacitor connected to the gate of a MOS transistor, but the electrode is fully isolated by Si oxide layers (*floating gate*). Voltage is adjusted by pumping charges in and out through the oxide exploiting the Tunnel effect. Special circuitry for write and erase is mandatory and this technology is not yet stabilized. The analog EEPROM principle is described in [SHIM88] and [VITT90]. A commercial general purpose neural circuit with EEPROM cells for weight storage is the Intel ETANN [ETAN91], [BRAU92].

5.3 Analog arithmetic

A variety of circuits for summation, multiplication, and other operations required in neural systems is already available from conventional analog systems.

Analog multiplication circuits use the variable resistance or transconductance of MOS transistors. Summation is accomplished by feeding current pulses to the summing node of a transresistance amplifier. The same amplifier, operated near saturation, can provide sigmoidal trasfer function. Some simple circuits are described in [PAUL88] and [VERL89].

Other specific functions, such as Manhattan distance and winner-take-all, can exploit directly some characteristiscs of MOS transistors, as described in [VITT93].

Switched-capacitor ANN circuits use combination of MOS switches and capacitor to build variable resistors, with the value controlled by the stream of pulses ([TSIV87]). An example is in figure 3. These circuits are derived from SC filters and A/D converters, operate on sampled signals, and can be classified at an intermediate place between analog and mixed circuits.

6 IMPLEMENTATION APPROACH 3: MIXED ANN

The term mixed ANN is used when the neural system has internal information represented in analog and/or numeric form, The system then contains a mix of analog and digital circuits, with appropriate A/D/A convertes. This solution allows to implement each specific function in the best way. A simple example of mixed synapses is in [ROSS89]; a complete mixed neural IC is the ANNA character recognizer [SACK93], [BOSE92].

6.1 Pulse modulations

Another class of ANN hardware which is referred to as *mixed* are pulse stream circuits. A pulse waveform is defined by the three parameters: amplitude, width, rate. Several forms of pulse modulations, act on different parameters, as shown in figure 4:

- **Pulse Amplitude (PAM)**, where information corresponds to the amplitude of the pulse;
- **Pulse Rate (PRM)**, where information corresponds to the repetition rate of the pulse (inverse of the period),
- **Pulse Width (PWM)**, where information corresponds to the width of the pulse;
- **Pulse Edge (PEM)**, where information corresponds to the absolute position of a single edge.

PAM keeps signal in the analog domain, and uses analog circuits. It is often mixed with other modulations to carry out local computations (e.g. synaptic weighting), but seldom used to move information among CUs.

PRM, PWM, and PEM use binary signals, and encode information in the time domain. Such signals can be transmitted, routed, received, and processed by digital circuits, without the loss of resolution caused by quantization. Moreover, most of these circuits can be developed using digital CAD tools. PRM is the most robust against noise and circuit variations, but the slowest one. Pulse-stream circuits have been introduced by several authors ([MURR87], [HIRA89], [MEAD91]), and extensively described and classified in [MURR91]. Their performance are analyzed in [REYN92].

A convenient variation is Coherent PWM (CPWM) [REYN92]. With coherent modulations no edge is constrained to a fixed position, but there is a global time reference which defines a fixed frame of evaluation cycles.

PEM, discussed in [DELC93], exploits the same principles of PWM, but uses the position of a single edge, or, more precisely, the delay between a fixed time reference and a single edge.

PEM halves the number of state transitions with respect to PWM (or doubles the information transmitted). If power is dissipated only on state transitions, PEM cuts the *power · delay* figure in half with respect to PWM.

Analog variables can be converted in PWM and PEM signals (or viceversa) by means of various types of converters, as described in [CHIA93].

6.2 Pulse Modulation arithmetic

Pulse modulations can use rather simple circuits for multiplication, summation, and other operations required in neural systems. The power associated to a stream of pulses is proportional to the product of three parameters: rate, width and hight; usually two parameters are constant, and CUs operate on the remaining one. In most cases, the circuits used to evaluate PRM, PWM, CPWM and PEM are very similar. It is also interesting to point out that some of these functions can be realized directly in the time domain, using standard logic gates.

- **Multiplication circuits** (for synaptic weigthing), operate on a pair of pulse parameters. *Rate · Width* circuits usually encode input activation in the pulse repetition rate, and stretch the width according to the weigth value; several examples are presented in [MURR91]. *Width · Heigth* circuits encode input activation in the width of single pulse, and modify the amplitude according to the weigth; they can use any type of analog multiplier, were one input is restricted to binary variables. This technique provide higher speed than *Rate · Width* circuits.
- **Digital summation** of pulses can use OR gates, but overlapping causes a saturation effect (which can be exploited as non-linear transfer function). If information is carried by the *rate · width · height* product, *analog* summation is accomplished by feeding current pulses to the summing node of a transresistance amplifier.
- **Sigmoidal trasfer function** can be obtained from the nonlinearity of the summing transresistance amplifier, or from synchronization effects in the output pulse generator [REYN92].
- **Manhattan distance** can be computed as absolute value of a difference. With any coherent pulse modulation this is achieved simply by X-ORing the two inputs, as in figure 5a. The output is still a coherent pulse stream, with information associated with the total pulse width within the evaluation period.
- **Winner-take-all** also requires only digital circuits. CPWM and PEM pulses can be rigth or left-aligned to a known time reference [DELC93]. With rigth-aligned pulses (shown in

figure 5b) the first edge in the cycle comes from the most-active signal, which can be selected by a digital FIFO arbiter.

In summary, coherent pulse modulations (CPWM, PEM) keep the main benefit of synchronous digital circuits: new information can be transmitted or processed on each cycle, and therefore they are well suited for time-division multiplexing, both on communication channels and on evaluation circuitry. A benefit with respect to synchronous circuits is that ground and power supply current spikes caused by state switching are spread over the evaluation cycle, thus reducing crosstalk noise towards the analog parts. Multiplexing makes possible to create large *virtual* synaptic arrays with purely digital circuits. This makes pulse-stream modulations a promising technique for the developement of complex ANN using discrete-time continuos-amplitude representations.

7 BIBLIOGRAPHY

General information and examples of implementation techniques for ANN can be found in several special issues of IEEE magazines:

- **IEEE COMPUTER**, Special issue "Artificial Neural Systems", Vol. 21, No. 3, March 1988. (introduction on NN, implementation examples mainly at the architectural level, details of an analog retina for motion detection).
- **IEEE COMMUNICATIONS**, Special issue "Neural Networks in Communications", Vol. 27, No. 11, November 1989. (mainly on algorithms and architectures, details on optical implementations)
- **IEEE MICRO**, Special issue "Silicon Neural Networks" Vol 9, no 6, December 1989. (reports from research groups on NN in Europe; details of digital, analog and pulse stream implementations).
- The **IEEE Transactions on Neural Networks** published in 1991 (March), '92 (May), and '93 Special Issues on Neural Network Hardware.

A comprehensive collection of papers on ANN implementations is: "Artificial Neural Networks: Electronic Implementations", N.Morgan Editor, IEEE-CS Press, 1990.

Papers on advanced research on Neural Network implementations appear frequently in the **IEEE Journal of Solid State Circuits**, and in the **IEEE Transactions on Circuits and Systems**, part II (Analog and Digital Signal Processing).

8 REFERENCES

[ARIM92] Y.Arima, et al.: "A Refreshable Analog VLSI NN Chip with 400 Neurons and 40K Synapses", IEEE

[BOSE92] B.E. Boser et al.: "Hardware Requirements for Neural Network Pattern Classifiers", IEEE MICRO, vol. 12, num.1, February 1992, pp. 32-40.

[BRAU92] J.Brauch et al,: "Analog VLSI Neural Networks for Impact Signal Processing", IEEE MICRO, Vol. 12, No. 6, December 1992, pp. 34-45.

[CHIA93] M.Chiaberge, et al.: "Interfacing Sensors and Actuators to CPWM and CPEM Neural Networks", Micro Neuro 1993, Edimburgh.

[DELC93] D.Del Corso, F.Gregoretti, L.Reyneri: An Artificial Neural System Using Coherent Pulse Width and Edge Modulations", Micro Neuro 1993, Edimburgh.

[ETAN91] "80170NX Electrically Trainable Analog Neural Network Data Sheet", Intel Corp., Santa Clara, CA, 1991.

[HIRA89] Y.Hirai et al.: "A Digital Neuro-Chip with Unlimited Connectability for Large Scale Neural Networks", Proc. IJCNN89, Washington, June 1989, pp. II/163-169.

[KRIS92] A.V.Krishnamoorthy, G.Yayla, S.C.Esener: "A Scalable Optoelectronic Neural System Using Free-Space Optical Interconnects", IEEE Trans. on Neural Networks, Vol. 3, No. 3, May 1992, pp. 404-413.

[LAWS93] J.C.Lawson et al.: "SMART: a Neurocomputer Using Sparse Matrices", Proc. Euromicro Workshop on Parallel and Distributed Processing, Gran Canaria, January 1993.

[MAND92] N.Manduit, et al.: "Lneuro 1.0: A Piece of Hardware LEGO for Building Neural Network Systems", IEEE Trans. on Neural Networks, Vol. 3, No. 3, May 1992, pp. 414-422.

[MEAD91] J.Meador et al.: "Programmable Impulse Neural Circuits", IEEE Trans. on NN, No. 2, 1991, pp. 101-109.

[MELT92] M.S.Melton, et al: "The TInMANN VLSI Chip" IEEE Trans. on Neural Networks, Vol. 3, No. 3, May 1992, pp. 375-384.

[MURR87] A.F. Murray, A.V.W. Smith, "Asyncronous arithmetic for VLSI neural systems", Electronics Letters, Vol 23, no 12, pp 642-643, June 1987.

[MURR91] A.F.Murray, D.Del Corso, L.Tarassenko: "Pulse stream VLSI neural networks mixing analog and digital techniques", IEEE Trans. on Neural Networks, Vol 2, no 2, March 1991, pp. 193-204.

[PAUL88] J.J. Paulos, P.W. Hollis: "Neural Networks Using Analog Multipliers", ISCAS 88, pp. 499-502.

[REYN92] L.Reyneri: "A Performance Analysis of Pulse Stream Neural Networks", Politecnico di Torino - DE, Internal Report LR9312 - December 1992.

[ROSS89] O. Rossetto et al.: "Analog VLSI Synaptic Matrices as Building Blocks for Neural Networks", IEEE MICRO, Vol 9, no 6, pp. 56-63, December 1989.

[SACK93] E. Sackinger et al., "Application of the ANNA NN Chip to High-Speed Character Recognition", IEEE Trans. on NN, vol.3, num. 3, May 1992, pp. 498-505.

[SHIM88] R.L. Shimabukuro et al.: "Dual-polarity nonvolatile MOS Analog Memory (MAM) cell for neural-type circuitry", Electronics letters, 15th September 1988, Vol 24, No. 19. pp 1231-1232.

[TREL89] P. Trelcaven et al.: "VLSI Architecures for Neural Networks", IEEE MICRO, Vol 9, no 6, pp 8-27, December 1989.

[TSIV87] Y.P. Tsividis, D. Anastassiou: "Switched-capacitor neural networks", Electronics Letters, Vol 23, no 18, pp-958959, August 1987.

[VERL89] M. Verleysen, P.G.A. Jespers: "An Analog VLSI Implementation of Hopfield's Neural Network", IEEE MICRO, Vol 9, no 6, pp 46-55, December 1989.

[VITT90] E.Vittoz et al.: "Analog Storage of Adjustable Synaptic Weights", Proc. ITG-IEEE Workshop on Microelectronics for NN, Dortmund, June 1990.

[VITT93] E.Vittoz: "Analog VLSI Signal Processing: why, where and how", to appear in the Journal of VLSI Processing, Kluwer, 1993.

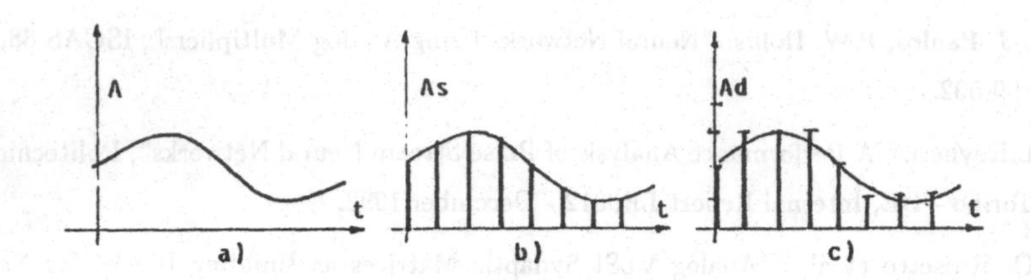

Fig. 1 - Representation of information:

a) Analog (continuous in time and amplitude);

b) Sampled (continuous in amplitude, discrete in time);

c) Sampled and quantized (discrete in time and amplitude).

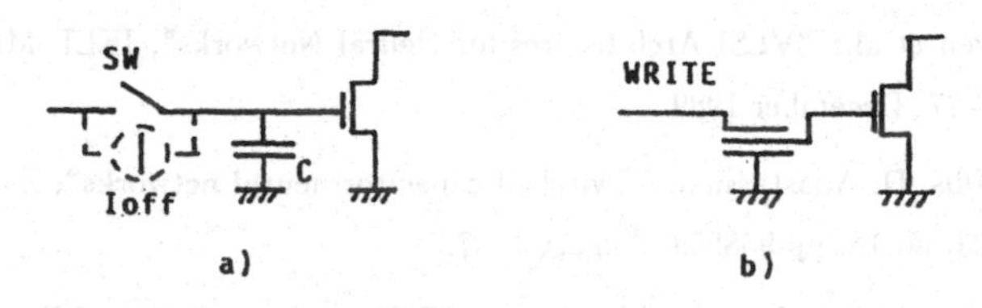

Fig. 2 - Analog plastic memories:

a) Capacitive: C is charged through SW, and discharges through the leakage current of the switch I_{off};

b) Analog EEPROM: the memory capacitor is charged through an insulating layer.

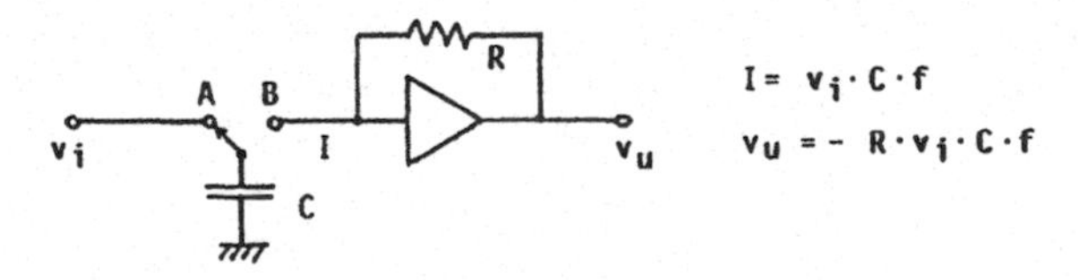

Fig. 3 - Principle of a SC multiplier. The capacitor C is switched from A to B f times per second; the current I is proportional to v_i and f. This is actually an $analog \cdot pulse - rate$ multiplier.

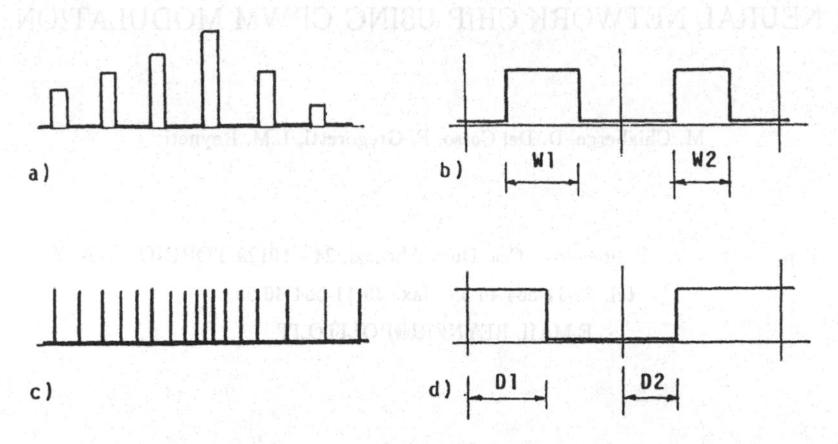

Fig. 4 - Pulse modulations: Pulse Width and Pulse Edge are coherent within the marked evaluation frames.

a) Pulse Amplitude;

b) Pulse Rate;

c) Pulse Width;

d) Pulse Edge.

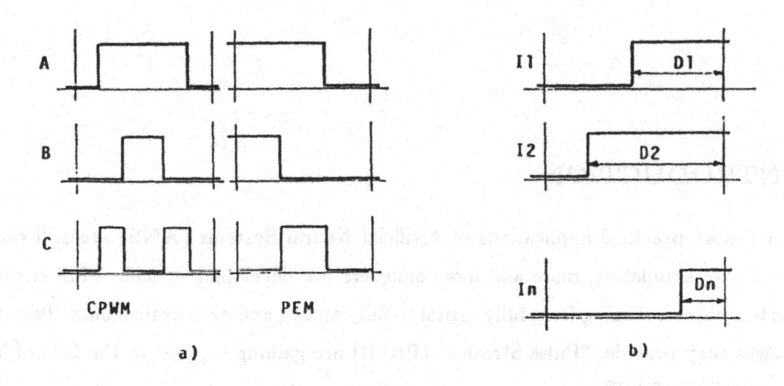

Fig. 5 - CPWM/PEM operations:

a) Manhattan distance: the energy of the output pulses C is proportional to the absolute value of A - B;

b) Winner take all: pulses (or edges) are aligned to the rigth frame boundary; the first transition defines the winner.

A NEURAL NETWORK CHIP USING CPWM MODULATION

M. Chiaberge, D. Del Corso, F. Gregoretti, L.M. Reyneri

Dip. Elettronica, Politecnico - C.so Duca Abruzzi, 24 - 10129 TORINO - ITALY
tel. 39-11-564 40 38 - fax. 39-11-564 40 99
E.MAIL REYNERI@POLITO.IT

Abstract

This paper describes a silicon implementation of an Artificial Neural Networks based on Coherent Pulse Width modulation techniques. Synapses use current generators controlled by an input Pulse Stream. Net charge generated is the product of synaptic current by pulse width. Neurons accumulate synaptic contributions and convert internal activation into an output Pulse Stream. A system optimized for lowest computation energy and highest reconfigurability has been designed, manufactured and tested.

1 INTRODUCTION

Although most practical applications of Artificial Neural Systems (**ANS**) are still carried out using software simulators, more and more designers are developing specific VLSI circuits using various techniques, ranging from fully digital to fully analog and even optical ones. Because of the advantages they provide, "Pulse Streams" (**PS**) [1] are gaining support in the field of hardware implementations of ANS.

PS are a class of modulations using "almost periodic" binary signals; information is contained in waveform timing and not in the amplitude. In Neural Systems, PS are primarily used to encode input (i_i) and output (y^j) activation signals (generically identified with α_i) and synaptic weights (w_i^j). Several examples can be found in [1, 4, 6].

Applying PS to ANS provides: high noise immunity; ease of multiplexing; low energy requirements; straightforward interface with external world; simplifies the design of synapses and neurons. A PS technique in particular has been used, namely *Coherent Pulse Width Modulation* (CPWM) which presented very good performance.

1.1 Coherent Pulse Width Modulation

As shown in fig. 1, activation pulses have constant frequency $f_o = \frac{1}{T_O}$, while their width is proportional to the activation value:

$$T_i = T_{max}\alpha_i, \tag{1}$$

where $\alpha_i \in [0 \dots 1]$ and $T_{max} \leq T_O$. All incoming streams have a known phase relationship among each other [4, 6]. As shown in fig. 1.a, every waveform is allowed to be "1" only during the *active phase* of a reference clock CCK, whereas it must always be "0" during the *idle phase*.

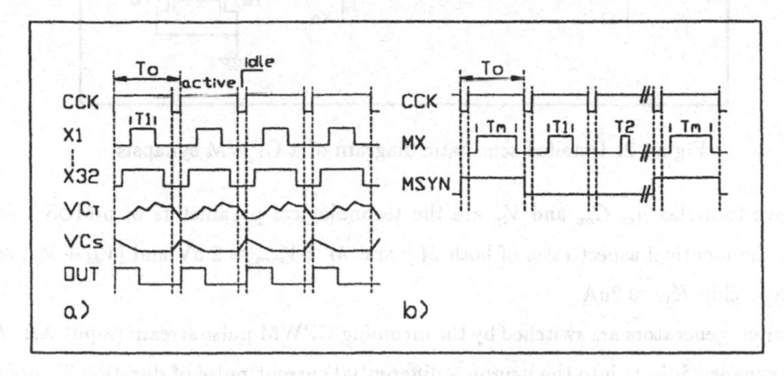

Figure 1: Timing diagram of a CPWM neuron.

2 WORKING PRINCIPLE OF CPWM NEURONS

The circuit proposed here is based on a CPWM modulation with an idle and active phase durations of 6.3μs and 0.9μs, respectively, corresponding to a stream frequency of 139kHz.

The circuit of a CPWM synapsis is shown in fig. 2. Two pMOS transistors (M^+ and M^-) implement a pair of current generators controlled by the voltage on a pair of identical capacitors C^+ and C^- ($\approx$ 1pF). When turned on by the *input buffer B*, they generate a differential current:

$$I_S = I_{M^+} - I_{M^-} \approx K_W w_i^j, \tag{2}$$

where:

$$w_i^j = \frac{(V_{C^+} - V_{C^-})(V_{C^+} + V_{C^-} - 2(V_{dd} + V_{tp}))}{(V_{min} - (V_{dd} + V_{tp}))^2} \tag{3}$$

and

$$K_W = \frac{\mu_p C_{ox} W}{2L}(V_{min} - (V_{dd} + V_{tp}))^2. \tag{4}$$

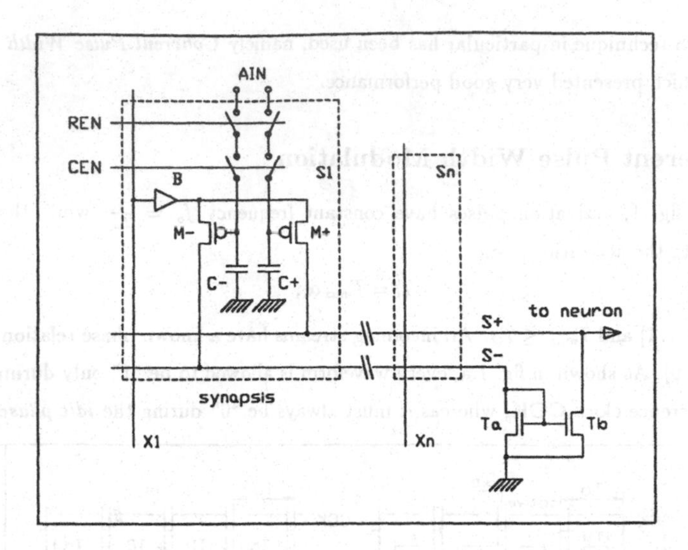

Figure 2: Detailed schematic diagram of a CPWM synapsis.

In the above formulas μ_p, C_{ox} and V_{tp} are the technological parameters of p-MOS transistors, while $\frac{W}{L}$ is the identical aspect ratio of both M^+ and M^-; $V_{min} \approx 2.5$V and $(V_{dd} + V_{tp}) \approx 4$V. In the prototype chip $K_W \approx 2\mu$A.

The current generators are switched by the incoming CPWM pulse stream (input X_i). At every cycle each synapsis injects into the neuron a differential current pulse of duration T_i proportional to the input activation i_i. From (1) and (2), the net charge contribution from each synapsis, during one cycle, is:

$$Q_i^j = K_W T_{max}(w_i^j i_i). \tag{5}$$

Several synapses are connected to a pair of summing nodes S^+ and S^- (see fig. 3), respectively for *excitatory* (I_{S^+}) and *inhibitory* currents (I_{S^-}). The neuron senses only the differential current $I_T = (I_{S^+} - I_{S^-})$. Charges coming from all the synapses are summed up together and divided by a factor K_R. The net charge either charges or discharges the integration capacitor C_T so that, at the end of the CPWM cycle, the voltage on the capacitor is (from 5):

$$V_{C_T}|_{t=T_{max}} = K_V \sum_{i=1}^{N}(w_i^j i_i), \tag{6}$$

where $K_V = \frac{K_W T_{max}}{K_R C_T}$ is an internal scaling factor.

The saturated amplifier which follows (A_1, A_2, R_1 and R_2) has a non-linear transfer function $G(x)$. The *output pulse generator* connected at the output of A_2 (made of S/H, C_S, I_Z and A_3)

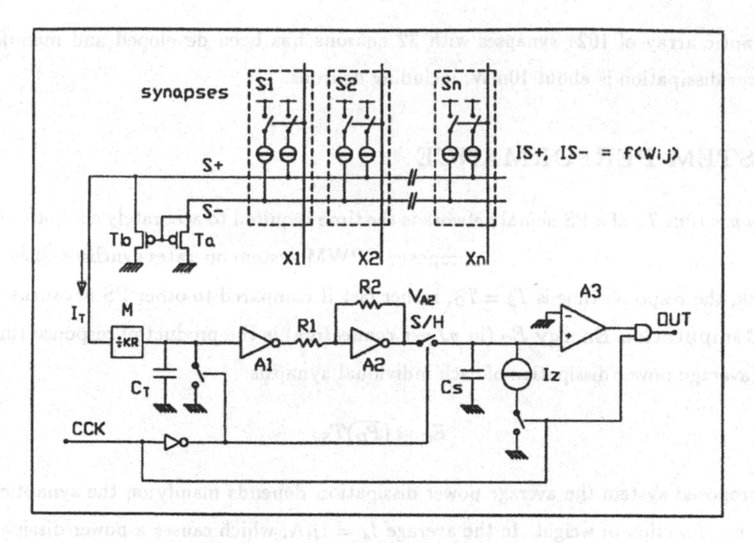

Figure 3: Simplified schematic diagram of a CPWM neuron.

generates an output pulse of duration T_Y proportional to V_{A_2} (from (6)):

$$T_Y = \frac{G\left(K_V \sum_i (w_i^j i_i)\right) C_S}{I_Z} = F^*\left(\sum_{i=1}^{N} w_i^j i_i\right), \tag{7}$$

where the output activation function

$$F^*(x) = \frac{G\left(\frac{K_W T_{max}}{K_R C_T} \cdot x\right) C_S}{I_Z} \tag{8}$$

approximates an ideal sigmoid [2], with steepness controlled by the factor K_R.

The voltage on capacitors C^+ and C^- must be periodically refreshed from an external digital memory through a D/A converter. Worst case refresh period is higher than 10ms.

From (3) there are several choices for V_{C^+} and V_{C^-}, for a given weight value. A possibility is to have always one transistor fully pinched off, that is, for $w_i^j > 0$:

$$V_{C^-} = (V_{dd} + V_{tp}) \tag{9}$$

and

$$V_{C^+} = (V_{dd} + V_{tp}) + \left(\sqrt{w_i^j} \cdot (V_{min} - (V_{dd} + V_{tp}))\right), \tag{10}$$

while, for $w_i^j < 0$, V_{C^+} and V_{C^-} shall be exchanged. This choice reduces the power consumption to a minimum, since M^+ and M^- are never conducting together.

Both M^+ and M^- are switched from their source terminal. This configuration provides a lower switching time ($T_C \approx 15$ns).

A synaptic array of 1024 synapses with 32 neurons has been developed and manufactured. Total power dissipation is about 10mW, including neurons.

3 SYSTEM PERFORMANCE

The *Response time* T_S of a PS neural network is the time required to accurately compute the whole matrix-vector multiplication. Since the proposed CPWM system operates synchronously with the CCK clock, the response time is $T_S = T_O$, rather fast if compared to other PS networks [5].

The **Computation Energy** E_C (in *pJ per connection*) is the product of response time T_S by the total average power dissipation of each individual synapsis:

$$E_C \approx (\overline{P_D})T_S \tag{11}$$

For the proposed system the average power dissipation depends mainly on the synaptic current I_d, which is a function of weight. In the average $I_d = 1\mu$A, which causes a power dissipation less than 5μW, while $T_O = 7.2\mu$s. A computation energy of less than 40pJ results.

The chip computes about 140MCPS, with an accuracy better than 1.5%. Synapsis and neuron size are about $70.000\mu m^2$ (see fig. 4) and $200.000\mu m^2$, respectively.

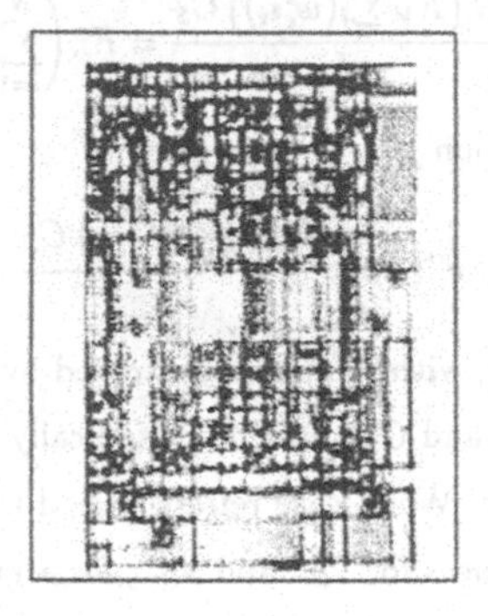

Figure 4: Microphotograph of a CPWM synapsis (size is about $65 \times 110\mu m^2$).

4 CONCLUSION

This paper has presented a complete CPWM chip set which has been developed, manufactured and tested. The interface with analog world is handled by on-chip A/PS and PS/A converters. Reconfiguration and multiplexing are simple and weight virtualization techniques allow to build large and efficient silicon ANS.

References

[1] A. F. Murray, D. Del Corso, and L. Tarassenko, "Pulse-Stream VLSI Neural Networks Mixing Analog and Digital Techniques", *IEEE Trans. on Neural Networks*, vol. 2, no. 2, March 1991, pp. 193-204.

[2] P.D. Wasserman, "Neural Computing: Theory and Practice", New York, *Van Nostrand Reinhold*, 1989.

[3] A. Hamilton, A.F. Murray, D.J. Baxter, S. Churcher, H.M. Reekie, and L. Tarassenko, "Integrated Pulse Stream Neural Networks: Results, Issues and Pointers", *IEEE Trans. on Neural Networks*, vol. 3, May 1992, pp. 404-413.

[4] L.M. Reyneri, "Silicon Implementations of Artificial Neural Systems Based on Pulse Stream Modulations", *Final Report of NERVES EEC Project*, E-BRA 3049, July 1991.

[5] H.P. Graf, E. Sackinger, B. Boser, and L.D. Jackell, "Recent Developments of Electronic Neural Nets in the USA and Canada", *Proc. Int'l Conf. on Microelectronics for Neural Networks*, Munich, October 1991, pp. 471-490.

[6] L.M. Reyneri, "A Performance Analysis of Pulse Stream Neural Networks", submitted to *IEEE Trans. on Neural Networks*, Oct. 1992.

HARDWARE IMPLEMENTATION OF A NEURAL NETWORK FOR HIGH ENERGY PHYSICS APPLICATION

J. Carrabina, F. Lisa, V. Gaitan, L. Garrido, E. Valderrama.

Centre Nacional de Microelectrònica
Universitat Autònoma de Barcelona
08193 Bellaterra Barcelona

Tel. (93) 580 26 25
FAX. (93) 580 14 96
e.mail: neures@cnmvax.uab.es

Abstract

The high speed and parallelism of VLSI Analog Neural Networks make them specially attractive for the treatment of data coming from elementary particle accelerators, which are used in high energy physics. In this paper we show the implementation of an analog neural network with low precision weights, devoted to the reconstitution of tracks: capability of handling 600 pixels/chip at about 2 10^{12} connections/second, in 40 mm^2 (1.5 μm ES2) at 100 Mhz.

1. Introduction

The high amount of information generated in a particle accelerator requires filtering processes in order to extract the good tracks from the total amount of events, those corresponding to the particles we are studying, and thus the complexity of the observation is reduced. These processes can be done at different levels [1]. There is a primary decision (level 1) at about 16ns, another one (level 2) at 10μs, and finally at 1ms (level 1), the speed of classical hardware.

By now, neural networks, including VLSI implementations [1], are becoming more and more used as an alternative to classical methods due to their good balance between quality and speed.

This problem has been studied by V. Gaitan, Ll. Garrido and G. Stimpfl-Abele [2] in the environment of the ALEPH TPC detector at LEP.

The ALEPH Time Projection Chamber (TPC) is a large three-dimensional imaging drift chamber. It is a cylinder with axial parallel magnetic and electric fields. The axis of the cylinder coincides with the beam axis which defines the Z direction of the Aleph coordinate system. The electric drift field points form each end-plate towards the central membrane that divides the chamber into two halves. The electrons produced by the ionization of traversing charged particles drift towards one end-plate, where they induce ionization avalanches in a plane of wire chambers. These avalanches are detected and give the impact point and the arrival time of the drifted electrons. This means that TPC provides three-dimensional coordinates. The X and Y coordinate are given by the impact point and the Z coordinate is obtained from the measured drift time.

The trajectory of a charged particle inside the TPC is a helix, and its projection onto the end-plate is an arc of a circle.

The purpose of the system we present is to process this projections coming form the TPC detector in order to filter the existing noise and reconstructing missing pixels.

2. A neural approach

In order to do this filtering, V. Gaitan et al. proposed a methodology based in Hopfield type neural networks. With their approach, neurons do not directly correspond to pixels in the detector, but to track segments, in such a way that with rectangular symmetry, for every pixel there are three neurons corresponding to the segments going to the pixels in the next top row.

Synaptic matrixes among these neurons are related with the curve which is to be detected. Determination of the connection values is done through the following expression

$$T_{kln} = \frac{\cos \psi_{kln}}{d_{kl} + d_{ln}}$$

that gives the weight of the synapse between the neurons kl and ln, segments of the track which correspond to the activation of the pixels k,l and n, function of the angle ψ , and d_{kl} the distance between coordinates of the pixels in the XY plane of the detector.

Ideally, the feedback of this network could reach the whole network, but dealing with specific tracks, a high degree of dilution could be imposed, reducing the total number of weights to those included in a certain range of angle and distance, and thus allowing a higher number of neurons per chip and a higher speed.

Connections between two neurons are by definition symmetrical in a locally connected feedback network.

The difference between this method and the classical methods of image filtering using convolution with fixed window templates or cellular neural networks [3], concerns to the higher attention on the slow variation of the track direction. By applying the previous expression on neurons corresponding to track segments with different direction, it can be seen that the set of synaptic weights is not the same for all the neurons.

In previous papers [4] we show the influence of dynamics in the recall phase of feedback networks. In this case dynamic effects are specially dramatic due to the different directionality represented by neurons. If, for instance, we choose neurons following a horizontal direction, tracks in that direction will be reinforced and tracks in the opposite direction won't be reconstructed.

3. IC Design strategy

In our particular application, we studied the effects of reducing weights to only unitary values, excitatory and inhibitory, having in mind the next generation of detectors. These studies show that it is possible to do an efficient filter [2] that allows the reconstruction of missing pixels of the track if their neighbours at the top and bottom are active, according to the connectivity pattern shown in Figure 1. This connectivity implies that the influence on a neuron can be reduced to five neurons in the horizontal direction, and three clusters in the vertical one, being a cluster the set of three neurons corresponding to tracks emerging from a pixel to the three above. It is also shown that there are only three different patterns for the connectivity of the neurons.

As a consequence, we conceived our chip with fixed weights corresponding to this pattern of connection.

The two strategies for the implementation of analog neural networks come from the classification of computational styles: analog and digital. These styles could be mapped in terms of VLSI implementation with higher degree of integration without the requirement of a very specialized manufacturing process. Both have a wide range of application based on their balance between parallelism versus precision.

In the implementation we propose, a direct study shows that analog implementations, based on current mode computations together with digital logic for data I/O, have higher performance than digital ones (composed of 19 full-adders, 4 basic combinational gates and 1 memory element per neuron) in terms of both area (considering full-custom design for both implementations), and speed; the recall speed will be slightly faster in analog implementations.

Analog implementations have also better performance in terms of redesign of the network, due to the fact that synaptic weights are directly mapped onto transistors, that are placed in specific layout blocks, which are easy to modify according to new requirements, whereas digital implementations require resynthesis and redesign of the logic structure.

Figure 2 shows a basic schematic of the cluster that is the set of neurons corresponding to the three segments going-up of a track pixel. In this figure it can be seen the excitatory and inhibitory weights mapped as transistors connected to a two rail bus that is sensed by a transconductance amplifier. The management of data is carried out by a dynamic shift register.

Although fixed weights mean a lose of programmability, it is the only way to obtain an IC able to deal with higher amounts of data.

The symmetry of the whole IC allows the use of most of the tools common in low level design, including design for portability among different technologies, what means design with simple structures for an easy recompaction, and structure compiler to easily build ICs with different number of rows and columns according to the structure of the detector.

Figure 3 shows the physical implementation of synaptic connections. The high density achieved by a semi-automatic compaction gives the small dimensions of 90x126 microns^2, for 30 transistors in a 1.5μm CMOS technology.

4. System architecture

The non-regular structure of the IC is due to the special characteristics of the detector, in which the radial precision is limited to twelve values, and the angular precision should be as high as possible.

In order to increase angular precision we designed the IC taking into account modularity, in such a way that chips would be connected only through their left and right pixel array sides, leaving top and bottom sides for data input and output (twelve cycles). The structure of the cylinder will then be reconstructed by a ring architecture of this neuroIC.

In addition to the neural filter, some special circuitry was designed in order to detect filtered tracks. This circuitry is also designed taking into account modularity.

5. Conclusions

In this paper we present an IC that implements and artificial neural network designed for the reconstruction of tracks coming from detectors of fundamental particles.

This task requires a very high speed (1.8 10^{12} cps.) , that could be reached through a non-programmable analog neural network, with low-connectivity of discrete weights.

Main computational parameters of the circuit are:

clusters (or pixels) : 600 (50x12)
neurons : 1800
weights per cluster: 30

area: 40 mm^2. (1.5μ ES2)
package: 120 PGA
clock speed : 100 MHz.
neural speed : 1.8 10^{12} cps.

References

[1] B. Denby. "Pattern Recognition for High Energy Physics with Neural Networks". Proc. of Neural Networks: from Biology to High Energy Physics. pp. 353-381. 1991.

[2] G. Stimpf-Abele, Ll. Grarrido, V. Gaitan. "Track Finding with Neural Networks vs. Standard Methods". First Itnl. Elba Workshop on Neural Networks: From Biology to High Energy Physics. 1991.

[3] a. L.O. Chua, L. Yang. "Cellular Neural Networks: Theory". IEEE Trans on Circuits and Systems. Vol.35, pp.1257-1272,1988.

b. L.O. Chua, L. Yang. "Cellular Neural Networks: Applications". Id, pp.1273-1290,1988.

[4] J. Carrabina. "High Speed/capacity VLSI Neural Networks". PhD. Dissertation. October 1991. Universitat Autònoma de Barcelona.

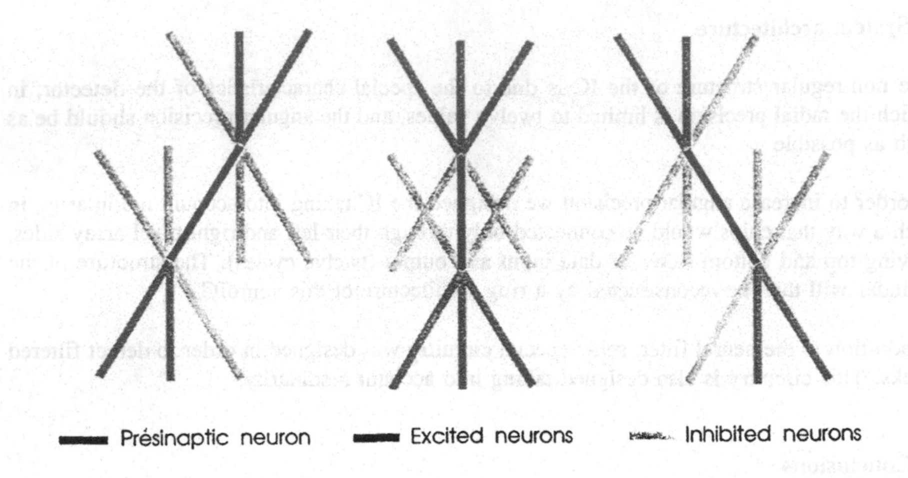

Figure 1. Three pattern types of local interconnection among neurons.

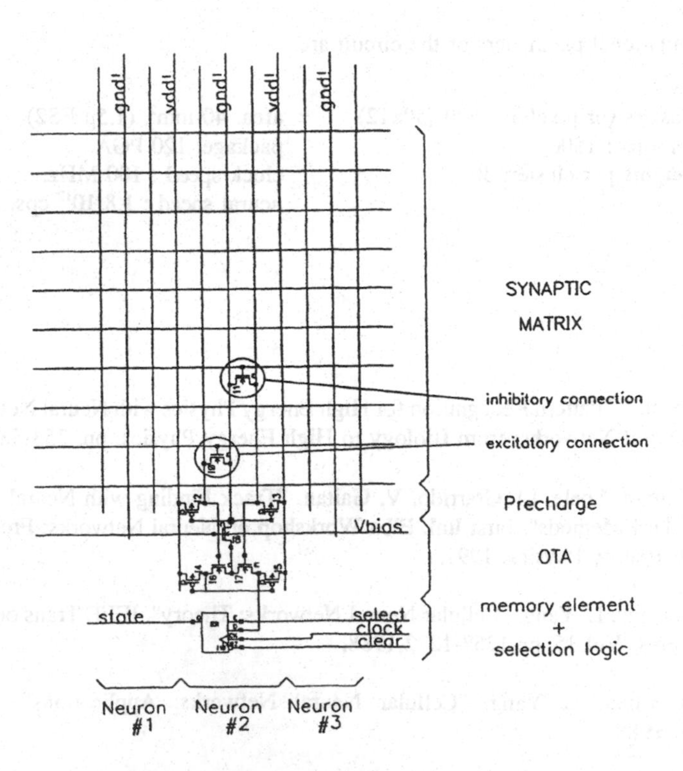

Figure 2. Basic schema of the cluster.

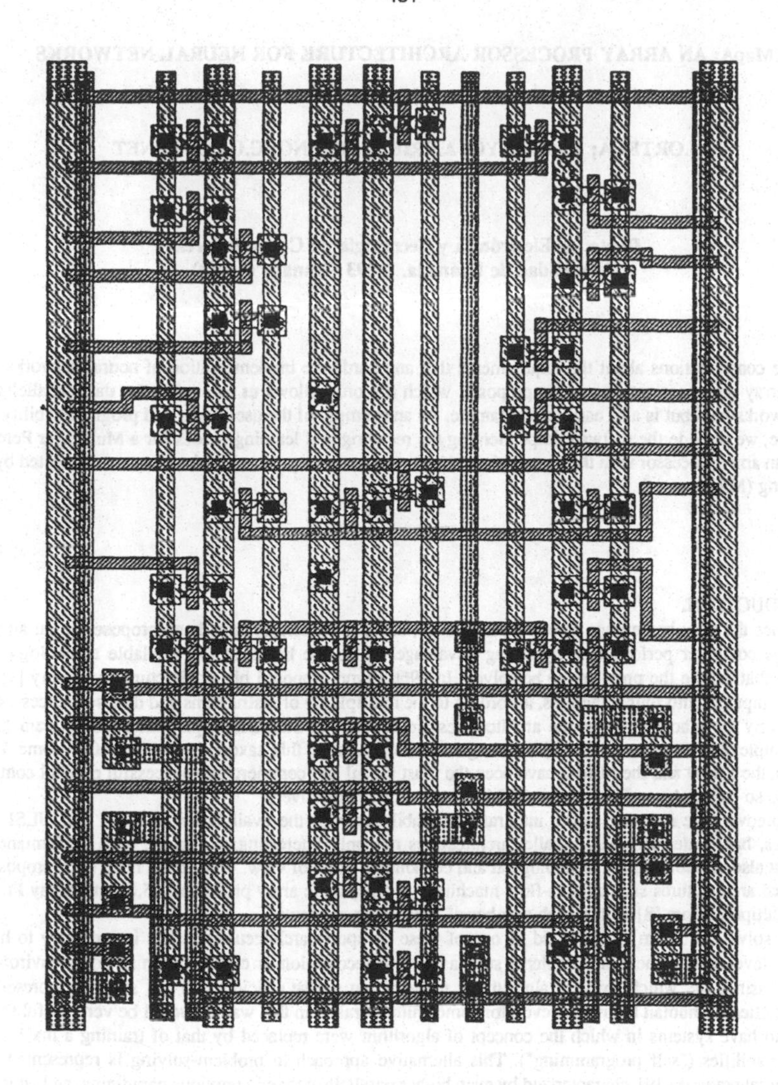

Figure 3. Layout of the synaptic connections

MapA: AN ARRAY PROCESSOR ARCHITECTURE FOR NEURAL NETWORKS

J.ORTEGA; F.J.PELAYO; A.PRIETO; B.PINO; C.G.PUNTONET

Depto. de Electrónica y Tecnología de Computadores.
Universidad de Granada. 18093 Granada (Spain)

After some considerations about the requirements that any hardware implementation of neural networks should meet, an array processor architecture is proposed which not only allows us to exploit the data parallelism that neural networks hold but is also easy to programme. As an example of the usefulness and programmability of this architecture, we provide the software implementing the recalling and learning modes for a Multilayer Perceptron (MLP) in an array processor with the proposed architecture and whose processing elements are connected by using a Multi-Ring (MR) network.

I. INTRODUCTION.

Since the very beginnings of computer science, some architectures have been proposed in an attempt to improve the computer performance by taking advantage, up to the limits of the available technology, of the parallelism that exhibe the problems to be solved. In 1966 Flynn proposed his architectural taxonomy [1] which classifies computers into four categories, according to the multiplicity of instructions and data sequences executed or operated by the machine. Thus, the architectures are denoted as Single/Multiple Instruction stream (SI/MI), Single/Multiple Data stream (SD/MD). Among the computers in this taxonomy which offer some kind of parallelism, the SIMD and the MIMD have been the most useful and commercially successful parallel computers: they are the so-called Array Processors and Multiprocessors, respectively.

Moreover, the increase in the integration capabilities with the availability of VLSI and ULSI silicon technologies, has made massively parallel architectures not only interesting due to the high performance they provide, but also possible from a technological and economical point of view. Thus, there have been proposed and implemented architectures such as data-flow machines [2-4], systolic array processors [5,6] and Array Processor [7], and Multiprocessors [8] with a high number of processing elements.

To solve a problem with the aid of one of these computer architectures makes it necessary to have an algorithm. Nevertheless, there are problems such as pattern recognition or classification in noisy environments, image processing, etc. which are very algorithmic resistant but whose efficient and fast solution represents few or no difficulties for human brains and even for some animal brains. In this way, it would be very useful for these problems, to have systems in which the concept of algorithm were replaced by that of training a machine with learning capabilities ("self programming"). This alternative approach to problem-solving is represented by the artificial neural networks [9], characterized by even higher parallelism than in previous paradigms, and which have inspired some specific architectures, using either analog or digital circuitry, in order to take efficient advantege of the parallelism that neural networks allow.

Although a neural network does not need an algorithm to solve a problem, this does not mean that the elements of a neural network could not be described by an algorithm, or that the interactions between neurons and the learning laws could not be programmed. So, it is not senseless to use a "conventional" parallel architecture to exploit the parallelism of a neural network. This paper presents an SIMD architecture called **MapA**, for **M**ultipurpose **a**rray **p**rocessors **A**rchitecture, which besides enabling an efficient exploitation of the kind of parallelism that neural networks show, is easier to programme than multiprocessor systems. With respect to the organization of the paper, Sect. II is a review of the characteristics of the neural network hardware implementations, while Sect. III describes the MapA architecture. In order to show how MapA can be used to implement neural networks, the software corresponding to the recalling and learning operation modes of a Multilayer Perceptron is provided in Sect. IV. Finally, Sect.V and VI are, respectively, the conclusion and the references.

II. HARDWARE IMPLEMENTATIONS OF NEURAL NETWORKS.

Though in the last few years many analog, digital or mixed digital/analog implementations have been described in the literature [10], the most successful and realistic applications of neural networks have been implemented by simulation in computers which does not allow the use of all the parallelism that a neural network permits.

The problems that make it difficult to have an efficient neural system can be understood if the requirements of any hardware implementation of a neural network are considered:

1. It must be possible to have a very high number of neurons simultaneously active.
2. It is necessary to provide a high grade of connectivity among neurons.
3. The architecture must be flexible enough in order to implement different neural net models and learning algorithms.
4. It must be possible to implement nonlinear functions associated with the neuron activation function and with the learning laws.
5. It is necessary to have enough local memory to store the nonlinear activation function, the weights and the local information used by the learning law.

Moreover as in any other architecture, in the design of a neural system we must take into consideration the questions related to cost, operation speed, and the possibility of either integrating the system into a general purpose computer in order to improve the accessibility to the neural resources that it provides, or designing an autonomous neural network that must agree with the corresponding limits in size, power consumption, etc. From the point of view of these considerations, in what follows in this section, we are going to analyse the analog and digital approaches to the hardware implementation of neural networks.

Analog implementations of neural networks try to materialize each neural component as a dedicated piece of semiconductor in order to obtain a full parallel emulation of the network. With these realizations, the number of processing elements per area unit are higher than the ones provided by digital approaches although they have, as a principal drawback, a limited accuracy. Despite the computing accuracy problem, if the above presented requirements of a hardware implementation of a neural network are considered, there are other difficulties to obtain efficient analog neural systems using the present VLSI technologies:

1. The in-chip storage of weights with acceptable resolution.
2. The sensibility to component mismatches.
3. The difficulty of implementing in-chip learning.
4. The impossibility of changing the neuron models and the learning laws.

Moreover, the poor programmability of these analog networks limits its use as a tool for research in the field of neural networks. Nevertheless, despite the indicated problems, the analog implementations of neural networks would provide better performances than digital ones for specific applications that require an autonomous system.

With respect to the digital hardware implementations of neural networks [11-14], they may be classified into two categories: **specific purpose processors** (with or without massive parallelism), called **neurocomputers**, and **general purpose parallel computers**. In order to select the most suitable architecture, the requirements of processing speed, learning capabilities, flexibility, autonomy and cost must be taken into account. In general, if the digital architecture implements a specific neural network model, with neurons that operate according to a given function and with a given learning law, then the system could become obsolete if the characteristics of the neural model and the learning laws changed. This is frequent in a research field as dynamic as neural networks.

With this in mind, the next sections describe the architecture we propose. It corresponds to a **general purpose** architecture which intends to provide a machine that allows not only a high speed execution of neural applications using one among several neural models, but also the acceleration if the net learning, or self-programming, process. So, to reach those goals, the neural system allows the programming of several models of neural networks with different learning laws to provide the solutions to some cognitive-like problems with an efficient use of the parallelism that a neural network shows.

III. THE MapA ARRAY PROCESSOR.

This section deals with the description of the modules and the instruction set that defines the architecture of the MapA machine. This system can be personalized by the corresponding software in order to implement any neural network model, while the neural network parallelism that it allows to use remains acceptable.

The greater or lesser effectiveness of a parallel architecture in solving a given application depends on the implicit parallelism that the algorithm used shows and the parallelism that the architecture provides. Thus, to justify the characteristics of a digital architecture for neural networks, the kind of processing that requires a neural net must be considered. A neural network comprises a lot of computational elements (artificial neurons) operating in parallel, linked by weights (synapses) that can be modified, by learning, to improve the network performance in the solution of a given problem. In any case, the computing procedure and the way in which the weights change according to learning laws is the same for all the neurons.

As all the neurons operate in the same way, it would be possible to use only one Control Unit to command the operation of all the neurons. Nevertheless, although the neurons implement the same operations, as each neuron uses a different set of weights and inputs, it would be necessary to have a different data path and a local memory for each neuron in order to compute and store the specific data used by each neuron and, so to take advantage of the spatial parallelism that a neural net offers. This is precisely the type of parallelism that an SIMD computer, according to the Flynn classification, efficiently implements. The MapA processor, which is presented in this paper, belongs to this class of machines and is described in the next two subsections.

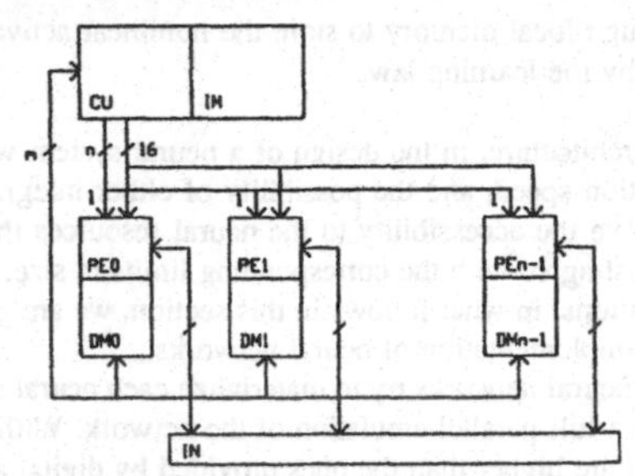

Fig.1 MapA Array Processor Structure.

III.1 MapA System Structure.

Fig.1 shows the modules of the array processor MapA. As can be seen, the MapA configuration is structured by three main elements: the Control Unit (CU), the n Processing Elements (PE_0, PE_1,.., P_{n-1}), and the Interconnection Network (IN). The CU fetches the instructions from the instruction memory (IM) where the programs are loaded, decodes the instructions, and determines where the decoded instructions should be executed. As we will see later in Sect. III.2, there are instructions directly executed in the CU (the scalar and control instructions) and instructions that the CU broadcasts to the PE's (vector instructions) in order to achieve spatial parallelism through a distributed execution in the PE's. The CU selects the PE's which participate in the execution of a vector instruction and sends them the corresponding control bits, allowing all the selected PE's perform the same operation synchronously in a lock-step fashion under the command of the CU.

To select the PE's, the CU uses its **mask register**, in which each bit corresponds to a PE; if this bit is 1 or 0, it represents, respectively, that the corresponding PE participates (is active) or not (is disabled) in the following instructions. If the PE_j is active, it will execute the corresponding operation, synchronously with the other active PE's, by using its own functional units and the data stored in its registers and in its local memory module, DM_j.

The IN allows the exchange of data between the PE's. It consists of a number of switching elements and interconnecting links that determine the possible interconnections, called **routing functions**, between PE's. These routing functions are carried out by a proper setting control of the switching elements which is managed, synchronously, by the CU.

As the routing functions establish which PE's are directly connected thus allowing data transference requiring only one clock cycle, the IN takes an important part in the processing time, and it must be carefully considered in any program executed in the machine, and particularly, in the implementation of neural networks. In this way, an IN is required which implements the connectivity of the neural model programmed, although the limits involved in its hardware implementation, and the flexibility needed to implement different neural models must be taken into account.

The memory modules DM_0, DM_1,.., DM_{n-1} are addressed by the CU, and each DM_j by each PE_j. From the point of view of the CU, all these modules build an address space of interleaved memory for storing data. This way, the CU can easily access to the memory modules of the PE's in order to load them in the initialization phase, and to communicate data with each PE's. To implement any neural model, the CU starts by storing, in the memory module DM_j associated with the PE_j where the corresponding neuron is going to be processed, the neuron weights and the rest of the parameters that define the network topology. Moreover, the CU needs to have access to the memory modules DM_j to store the input patterns and the desired outputs if supervised learning must be implemented, and to load the results of the network processing.

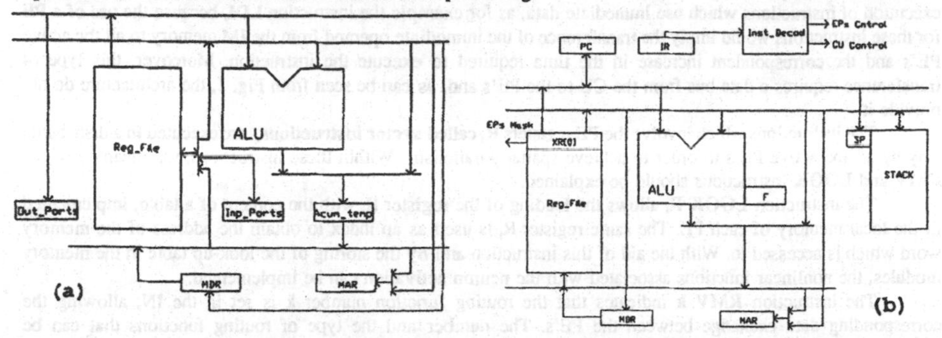

Figure 2. Internal Structure of PE's (a) and CU (b)

Fig.2 describes the structure of the MapA's PE and CU by using the registers and functional units which are visible from the instruction-set level. Table 1 gives this instruction set. As Fig.2.a shows, each PE is essentially a data path with its own ALU and does not include any control unit. This is, precisely, one of the main advantages of the MapA architecture in comparation with a multiprocessor system : because it is not necessary to implement a CU in each PE, the silicon area saved can be used to increment the amount of local memory, the speed of the functional units, and the throughput in the execution of instructions by a pipeline-based design of the system.

III.2 MapA Instruction set.

Table 1 provides the set and the meaning of the MapA array processor instructions. Basically, these instructions define a LOAD/STORE architecture, which operates with data and results stored in the CU or PE registers while the information transference between these registers and memory is carried out by using the LOAD and STORE type instructions.

Table 1. MapA Instruction Set

Instruction	Operands	Meaning
LDI	XR, i	; XR <-- i
LD	RXd, RXm	; RXd <-- m[RXm]
MV	RXd, RXs	; RXd <-- RXs
RMV	j	; R[0]$_{r_j(i)}$ <-- R[0]$_i$ (i=0, . . ,n-1) ; j-th routing function
ST	RXs, RXm	; m[RXm] <-- RXs
CLR	RX	; RX <-- 0
SET	RX	; RX <-- 1
INC	RX	; RX <-- RX+1
DEC	RX	; RX <-- RX-1
SHL	XR, i	; Shift left XR, i times
SHL	RX, RXc	; Shift left RX, [RXc] times
SHR	XR, i	; Shift right XR, i times
SHR	RX, RXc	; Shift right RX, [RXc] times
LOOK	R	; R <-- f(R) (Read from table f)
ADD	RXd, RXs1, RXs2	; RXd <-- RXs1+RXs2
SUB	RXd, RXs1, RXs2	; RXd <-- RXs1-RXs2
MUL	RXd, RXs1, RXs2	; RXd <-- RXs1*RXs2
AND	RXd, RXs1, RXs2	; RXd <-- RXs1^RXs2
OR	RXd, RXs1, RXs2	; RXd <-- RXs1vRXs2
BZ	i, XRb	; Branch to m[PC+XRb] if [F]$_i$=0
BNZ	i, XRb	; Branch to m[PC+XRb] if [F]$_i$=1
JMP	XRb	; Branch to m[PC+XRb]
BZ	i, OFFSET	; Branch to m[PC+OFFSET] if [F]$_i$=0
BNZ	i, OFFSET	; Branch to m[PC+OFFSET] if [F]$_i$=1
JMP	OFFSET	; Branch to m[PC+OFFSET]
CALL	XRb	; Subroutine at m[PC+XRb]
CALL	OFFSET	; Subroutine at m[PC+OFFSET]
RET		; Return from subroutine
NOP		; No operate

NOTES:

XR:	CU register
R:	PE register
RX:	PE or CU register
XR0:	Mask register
F:	CU state
[F]$_i$:	Bit i of register F.
m[x]:	Reference to address x in data memory DM
OFFSET:	Inmediate offset

In the instruction set there are instructions directly executed inside the CU without the participation of any PE. These are the instructions coding operations with the XR registers of the CU, called **scalar instructions**, and the control instructions which modify, or not, the CU program counter according to the branch conditions and the value of the corresponding state flag, stored in the CU register F. Moreover, the PE's can not take part in the execution of instructions which use immediate data, as for example the instruction LDI, because the use of a PE for these instructions would imply the transference of the immediate operand from the IM memory to all the active PE's and the correspondent increase in the time required to execute the instruction. Moreover, this type of transference requires a data bus from the CU to the PE's and, as can be seen from Fig. 1, the architecture do not provide it.

The instructions which involve the PE registers R, called **vector instructions**, are executed in a distributed way by all the active PE's in order to achieve spatial parallelism. Within these instructions, the meaning of the **RMV** and **LOOK** instructions should be explained.

The instruction **LOOK** R_i allows the loading of the register R_i with the content of a table, implemented in the local memory of each PE. The same register R_i is used as an index to obtain the address of the memory word which is accessed to. With the aid of this instruction and by the storing of the look-up table in the memory modules, the nonlinear functions associated with the neuron activation can be implemented.

The instruction **RMV k** indicates that the routing function number k is set in the IN, allowing the corresponding data exchange between the PE's. The number and the type of routing functions that can be established depend on the IN connecting the PE's. As examples, a hypercube interconnection network of dimension r provides r different routing functions, and so k=0,1,..,r-1 enabling the communication between those PE's whose binary coded index is different only in one bit, respectively 0,1,..,r-1; a mesh interconnection network provides four routing algorithms, k=0,1,2,3, indicating that each PE is connected with, respectively, its neighbour to the north, south, east and west.

Of the 29 instructions, the PE's can only participate in 16. This way, to control the PE operation the CU needs to generate the four lines which codify the 16 instructions plus 12 (3x4) lines that select the PE registers used in the instruction as operands and result. In addition to these 16 lines, the PE_j (j=0,1,..,n-1) receives the bit j from the CU **mask register**, XR0, which controls whether the PE_j is active or disabled, and so it interprets or not the 16 control lines.

Since the number of instructions in the set is reduced, the addressing modes are simple and few, and the instructions operate with registers, thus each instruction can be codified by using only one 32-bit word, thus facilitating the instruction decoding process, that can be done in parallel with the operand address calculation. All these characteristics correspond to a RISC style architecture. Under these circumstances, the design of the processor in order to allow a piplined execution of the instructions would be easier.

IV. MapA ARRAY PROCESSOR PROGRAMMING.

The simulation of any neural network in the MapA array processor is performed by the appropriate program. The aim of this section is to consider the characteristics of such programs by using, for example, the programs to implement the Multilayer Perceptron (MLP) [15], a well known neural network, and the backpropagation learning algorithm.

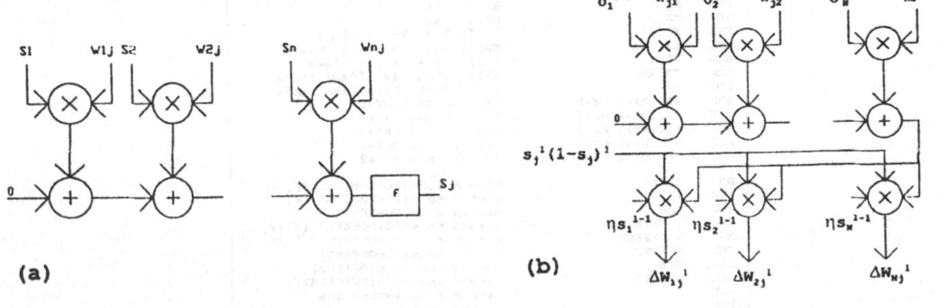

Fig.3. DDGs for a neuron in a MLP

Fig.3.a shows the Data Dependence Graph (DDG) corresponding to the expression (1) below, while Fig.3.b shows the DDG associated with the expressions (2) and (3). Expression (1), and expressions (2) and (3) give,

respectively, the activation law and the backpropagation learning algorithm for the neuron j in the layer l of an MLP with L layers and learning rate η, which has as inputs, the outputs s_i^{l-1} ($i=1,..,N_j^l$) of the N_j^l neurons, in the layer *l*-1, multiplied by the weights W_{ij}^l ($i=1,..,N_j^l$).

$$s_j^l(t+1) = f\left(\sum_{i=1}^{N_j^l} W_{ij}^l s_i^{l-1}(t)\right) \; ; \; f(x) = \frac{1}{1+e^{-x}} \qquad \textbf{(1)}$$

$$W_{ij}^l(t+1) = W_{ij}^l(t) + \eta s_i^{l-1} \delta_j^l \qquad \textbf{(2)}$$

$$\begin{aligned} \delta_j^l &= s_j^l(1-s_j^l)\left(\sum_{k=1}^{N_j^{l+1}} W_{jk}^{l+1}\delta_k^{l+1}\right) \quad (l \neq L) \\ &= y_j(1-y_j)(d_j-y_j) \quad (l=L) \end{aligned} \qquad \textbf{(3)}$$

Figs. 4 and 5 show the programs which correspond to the MLP recalling and learning modes, respectively. These programs have been written using a high level programming language similar to Pascal. To elaborate them, we have been considered a MapA system with 32 PE's, PE_j ($j=0,1,..,31$), connected by a Multiple-Ring (MR) network, as shown in Fig.6. The memory module DM_j ($j=0,1,..,31$) associated with each PE_j stores the weights of the neurons which are computed by this PE, and other additional parameters required to complete the network processing. For example, if, as it is given in Fig.7, each PE_j processes one neuron in each MLP layer, the corresponding module DM_j must have enough capacity to store:

- The learning rate, η.
- The number of weights of each neuron processed by PE_j, N_j^l ($l=0,1,..,L$).
- The corresponding component of the desired output for each input pattern, d_j.
- The corresponding component of the input pattern, s_j^0.
- The outputs of the neurons processed by PE_j, s_j^l ($l=1,..,L$).
- The weights of the neurons processed by PE_j, W_{ij}^l ($i=1,2,..,N_j^l$, $l=1,..,L$).
- The nonlinear activation function f, M_f.

Taking all this into account, the amount of local memory required by each PE_j is in the order of $L(N_{max}+1)+M_f$, where N_{max} is the maximum of N_j^l ($l=1,..,L$). So, the capacity of the local memory modules determines the maximum complexity, in terms of weights and layers, for the neural network emulated in the machine.

```
1    (* MLP in MaPA_MR *)
2    while not(STOP) do
3         for l:=1 to L do
4            begin
5               for all PE_i do
6                  begin
7                     D_i:=0;
8                     A_i:=s_i^(l-1)
9                  end;
                  (* Computation of (1) begins *)
10                for k:=1 to n do (* n = N_max *)
11                  for all PE_i do
12                     begin
13                        A_((i+1)mod n)<=A_i;
14                        D_i:=D_i+A_i W^l_(i,((i-k) mod n));
15                     end;
16                for all PE_i do s_i^l:=f(D_i);
                  (* Computation of (1) ends *)
17           end;
```

Figure 4. MLP in recalling mode

```
1    (* MLP Learning in MaPA_MR *)
2    for all PE_i do δ_i^L:=y_i(1-y_i)(d_i-y_i);
     (* Computation of (3) for l=L ends *)
3    while (niter≠0) do
4       begin
5          for l:=(L-1) to 1 do
6             begin
7                for all PE_i do
8                   begin
9                      D_i:=0; B_i:=δ_i^(l+1);
10                  end;
                 (* Computation of (3) for l≠L starts *)
11               for k:=0 to n-1 do (* n = N_max *)
12                  for all PE_i do
13                     begin
14                        A_i:=B_i*W^(l+1)_(i,((i+k) mod n));
15                        for j:=1 to k do A_((i+1)mod n)<=A_i;
16                        D_i:=D_i+A_i;
17                     end;
18               for all PE_i do
19                  begin
20                     δ_i^l:=s_i^l(1-s_i^l)D_i;
21                     A_i:=s_i^(l-1);
22                  end;
                 (* Computation of (3) for l≠L ends *)
                 (* Computation of (2) starts *)
23               for k:=1 to n do
24                  for all PE_i do
25                     begin
26                        A_((i+1)mod n)<=A_i;
27                        W^l_(i,((i-k) mod n)):=W^l_(i,((i-k) mod n))+δ_i^l A_i η;
28                     end;
                  (* Computation of (2) ends *)
29            end;
30         iter:=iter-1;
31      end;
```

Figure 5. MLP in learning mode

In the programs of Figs.4 and 5 there are some sentences which require some explanation. These are the sentences used to express communication between PE's through the interconnection network, and the sentences which indicate simultaneous execution of an operation in all the active PE's, thus achieving spatial parallelism.

The sentence B_i<=A_j means a transference from the register A in the PE_j to the register B of the PE_i. In order to compute the temporal complexity of this sentence within a program, only if there exists a direct link from the PE_j to the PE_i, will this transference take the same time as an information transfer between registers in the same PE. The translation of these transference sentences into the machine language is carried out by the instruction RMV k, where k denotes the routing function that establishes the connection between the PE_j and PE_i. In this way, it is clear that the type of IN used, determines the program that allows an efficient neural network implementation. As in the example we are considering the IN is the Multiple-Ring, then each PE has only one input port and one output port and the IN only allows one routing function. In this network, each PE_j can only make a direct transference to the $PE_{(j+1 \bmod r)}$, being r the number of active PE's.

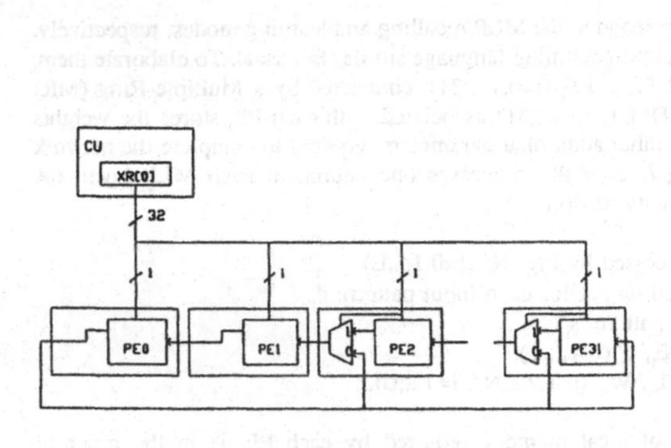

Figure 6. PE's connected with a Mult-Ring Network

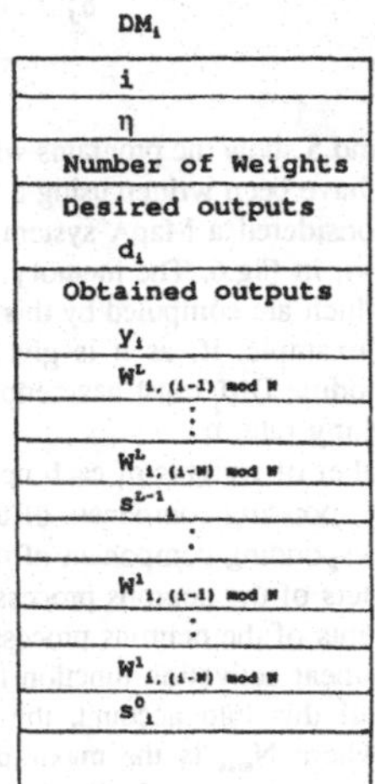

Figure 7. MLP stored in the DM modules.

```
        (* Numbers in brackets corresponds to sentences in fig. 4 *)
        LDI  XR0,FFFFh
        SHL  XR0,10h
        LDI  XR0,FFFFh      ; XR0 = FFFFFFFFh -> all PE's active
loop3   LDI  XR2,3h         ; XR2 is the number of layers
        CLR  R3
        LD   R1,R3          ; R1 <- m[0] : R1 is the EP index
                            ; [begin of 3-17]
loop2   LDI  XR1,20h        ; XR1: neurons/layer (32)
        CLR  R2             ; R2 is the accumulator : R2=0
        INC  R3
        LD   R0,R3          ; R0 <- m[R3] : R0 is the port
                            ; [Begin of 10-15]
loop1   RMV                 ; Transference through the network
        INC  R3
        LD   R4,R3          ; R4 <- m[R3] : R4 stores the weight
        MUL  R5,R4,R0
        ADD  R2,R2,R5       ; Mult. and accum.
        DEC  XR1
        BNZ  0,loop1        ; Branch if not all weights are proc.
                            ; [end of 10-15]
        LOOK R2             ; Activation function [16]
        INC  R3
        ST   R2,R3          ; m[R3] <- R2 : Store the neuron state
        DEC  XR2
        BNZ  0,loop2        ; Branch if not all layers are processed
                            ; [end of 3-17]
        JMP  loop3
```

Figure 8. MLP recalling with MapA inst.

To indicate the parallel execution of instructions in all the active PE's, we use the sentence for all EP_j where COND(j) do SENTENCE, in which COND(j) represents the condition used to select the PE's that execute SENTENCE in parallel. If all the PE's are active, then where is omitted. The compiler will generate LDI instructions which load the mask register XR[0] with 1's in the bits corresponding to the PE's whose indices verify COND(j).

In order to give a more detailed idea of the implementation of the MLP in the MapA architecture, the programs provided in Figs.4 and 5 are shown in Figs.8 and 9, respectively, written by using the machine instruction set of Table 1.

```
      {* Numbers in brackets corresponds to sent. in Fig.5 *}
      LDI   XR0,FFFFh        ; XR0 is the MASK
      SHL   XR0,10h          ;
      LDI   XR0,FFFFh        ; XR0=FFFFFFFFh -> all PE's active
      CLR   R2               ;
      LD    R1,R2            ; R1 <- m[0] (R1 = PE_index)
      LDI   XR3,#iter        ; XR3 <- iterations
loop1 SET   R2               ; [Begin of 3-30]
      LD    R10,R2           ; R10 <- Learning rate
      INC   R2
      LD    R11,R2           ; R11 <- Number of Weights
      INC   R2               ; [Begin of 2]
      LD    R3,R2            ; R3 <- Desired_Output, d_i
      INC   R2
      LD    R4,R2            ; R4 <- Obtained_Output, y_i
      SET   R5
      SUB   R5,R5,R4
      MUL   R5,R5,R4
      SUB   R3,R3,R4
      MUL   R3,R5,R3         ; R3 <- δ_j^L [End of 2]
      INC   R2
      ADD   R9,R2,R11
      LDI   XR2,#layers      ; XR2 <- layers
      JMP   loop6
loop2 LDI   XR1,#neurons     ; XR1 <- neurons/layer [Begin of 5-29]
      CLR   R5               ; R5  <- 0 (Acumulator)
loop3 LD    R6,R2            ; R6  <- Weight [Begin of 12-17]
      MUL   R6,R6,R3         ; Weight*δ
      MV    R0,R6
      MV    XR4,XR1
loop4 DEC   XR4              ; [Begin of 15]
      BZ    0,loop5
      RMV                    ; Transference
      JMP   loop4            ; [End of 15]
loop5 ADD   R5,R5,R0
      INC   R2
      DEC   XR1
      BNZ   0,loop3          ; [End of 12-17]
      SET   R7               ; [Begin of 18-22]
      SUB   R7,R7,R8
      MUL   R7,R7,R8
      MUL   R3,R7,R5         ; δ_i^l for l≠L  [End of 18-22]
loop6 MUL   R3,R3,R10
      LDI   XR1,#neurons
      LD    R0,R9
loop7 RMV                    ; [Begin of 23-28]
      INC   R9
      LD    R6,R9
      MUL   R7,R3,R0         ; Increment in weights
      ADD   R6,R6,R7         ; New weights are obtained
      ST    R6,R9            ; New weights are stored
      DEC   XR1
      BNZ   0,loop7          ; [End of 23-28]
      DEC   XR2
      BNZ   0,loop2          ; [End of 5-29]
      DEC   XR3
      BNZ   0,loop1          ; [End of 3-30]
```

Figure 9. MLP learning with MapA inst.

V. CONCLUSION.

In this paper an SIMD architecture has been described, which allows the programming of several neural network models with a satisfactory level of performance. This architecture represents an approach to the problem of hardware implementation of neural networks that combines both the flexibility to accommodate different neural models, and the ability to exploit the parallelism that neural networks provide. Moreover, if compared with other digital architectures for neural networks which are based on the multiprocessor paradigm, our array processor architecture is easier to program as has been shown with the program examples here provided.

The MapA system with 32 PE's and a Multiple-Ring Interconnection network, described in Sect.IV, has been simulated at the architectural level, by using the hardware description language Verilog, in order to visualize the execution of the instructions and to check the correctness of the programs that implement the MLP recalling and learning modes (Figs.7 and 8). The modules included in a PE with 16-bit words and the ones in a CU with 32-bit words have been generated by using the Cadence/ES^2 software. Considering the 1.6 µm CMOS technology of ES^2, and a design based on the libraries and the module generator provided by the manufacturer, the CU has a size of 29.09 mm^2 while each PE with 1024 16-bit words of memory occupies 25.28 mm^2.

Once the architecture has been designed, its utility to emulate neural networks analyzed, and the viability of implementing the architecture due to the acceptable sizes of PE's and CU checked, our purpose is to develop a MapA prototype with 32 PE's in order to attain a more detailed evaluation of the system speed performance. The next step will be to tackle a pipelining design of the system to improve the performances.

VI. REFERENCES.

[1] Flynn, M.J.:"Very High-Speed Computing Systems". Proc.IEEE, vol.54, pp.1901-1909. 1966.

[2] Treleaven, P.C.; Brownbridge, D.R.; Hopkings, R.P.:"Data Driven and Demand Driven Computer Architecture". ACM Computing Surveys, pp.93-144. March, 1982.

[3] Koren, I.; Mendelson, B.:"A Data-Driven VLSI Array for Arbitrary Algorithms". IEEE Computer, Vol.21, No.10, pp.30-43. October, 1988.

[4] Ramakrishna, B.; Yen, D.W.L. et al.:"The Cydra 5 Departamental Supercomputer". IEEE Computer, Vol.22, No.1, pp.12-35. January, 1989.

[5] Foulster, D.E.; Schreiber:"The Saxpy Matrix-1: A General-Purpose Systolic Computer". IEEE ,Computer, Vol20, No.7, pp35-43. July, 1987.

[6] Moreno, J.H.; Lang, T.:"Matrix Computations on Systolic-Type Meshes: An Introduction to the Multimesh Graph Method". IEEE Computer, Vol.23, No.4, pp.32-52. April, 1990.

[7] Potter, J.L.; Meilander, W.C.:"Array Processor Supercomputers". Proc. IEEE, vol.77, No.12, pp.1829-1841. December, 1989.

[8] Hayes, J.P.; Mudge, T.:"Hypercube Supercomputers". Proc. IEEE, vol.77, No.12, pp.1829-1841. December, 1989.

[9] Prieto, A. et al.:"Simulation and Harware Implementation of Competitive Learning Neural Networks", in "Statistical Mechanics of Neural Networks", edited by L. Garrido, Springer-Verlag, pp.189-204. 1990.

[10] Pelayo, F.J.; Pino, B.; Prieto, A.; Ortega, J.; Fernández, F.J.:"CMOS Implementation of Synapse Matrices with Programmable Analog Weights", in "Artificial Neural Networks", edited by A. Prieto, Springer-Verlag, pp.260-267. 1991.

[11] Atlas, L.E.; Suzuki, Y.:"Digital Systems for Artificial Neural Networks". IEEE Circuits and Devices Mag., pp.20-24. November, 1989.

[12] Treleaven, P.; Pacheco, M.; Vellasco, M.:"VLSI Architectures for Neural Networks". IEEE Micro, Vol.9, No.6, pp.8-27. December, 1989.

[13] Hammerstrom, D.:"A VLSI Architecture for High-Performance, Low-Cost, On-Chip Learning". Int. Joint Conf. on Neural Networks, Vol.II, pp.537-544. June, 1990.

[14] Pacheco, M.:"A Neural-RISC Processor and Parallel Architecture". PhD. Thesis, Dept. Computer Science, University College of London, University London, 1991.

[15] Widrow, B.; Lehr, M.:"30 Years of Adaptive Neural Networks: Perceptron, Madaline, and Backpropagation". Proc.IEEE, Vol.78, No.9, pp.1415-1442. September, 1990.

LIMITATION OF CONNECTIONISM IN MLP

C.V. Regueiro*, S. Barro* and A. Yáñez**

*** Departamento de Electrónica y Computación**
Facultad de Física. Universidad de Santiago de Compostela
**** Departamento de Computación**
Facultad de Informática. Universidad de La Coruña

ABSTRACT

In this work we study the behavior of restricted connectionism schemes aimed at solving one of the problems found in the implementation of Artificial Neural Networks (ANNs) in VLSI technology. We limit our study to the classical backpropagation trained MLPs and we discuss the limitations of restricted connectionism by means of a simulation of two tasks that are useful in the practical application of ANNs. From the results of these simulations it can be deduced that the perspectives given by other authors are, perhaps, excessively optimistic. The needs with respect to the number of neurons in each hidden layer increase with restricted connectionism. We have also been able to see that, independently from the type of connection structure, MLPs with a large number of layers cannot be correctly trained. Both of these problems negatively influence restrictive connectionism approaches to this type of networks.

INTRODUCTION

In the last few years the interest in Artificial Neural Networks (ANNs) has made a strong comeback. The causes of this interest can be found in the fact that the main problems of the first models have been solved and in the presentation of new more "powerful" models with new applications (Hopfield, Kohonen, ART, BAM networks, etc.). Some of these applications, although modest in size, have allowed a glimpse of the enormous possibilities of ANNs. Tasks such as character, sound or shape recognition, for instance, can be solved in a simpler and more efficient manner than by means of the application of other types of techniques.

Currently, four primary approaches exist for implementing neural networks [17][18]: their simulation in conventional computers (necessary for the evaluation of new models and variations of existing ones), their simulation in machines presenting a parallel architecture, the use of neurocomputers (architectures based on microprocessors which are specifically designed for the emulation of ANN) and, finally, the approach of producing neural chips designed in VLSI and WSI technology. This last approach is the most recent and promising. It is developed in different fronts depending on the devices used: analog, digital, resistor networks and the less developed electrooptical networks, each with its advantages and drawbacks [6]. We will not comment on them as they are beyond the scope of this work. Furthermore, before attempting commercial solutions, where the ANNs can compete with more conventional methods, current strategies have to be improved and potentiated, specially in problems such as their scaleability, the establishment of broader mathematical bases or their connectivity. In electronic devices built using VLSI technology, the connectivity is around ten and its increase negatively influences the global performance of the chip with respect to speed, power requirements and working reliability. This work is focused on this last aspect, studying one of the possible solutions for the elimination of this problem, at least partially. We want to find models which, without reducing performance, are structured using restricted connectionism and thus facilitate VLSI implementation.

SOME PROBLEMS OF VLSI IMPLEMENTATION

ANNs present some features which greatly facilitate their VLSI implementation. Their high spatial regularity and modularity permit the reduction of the length and complexity of the design process. Another very interesting feature is their resistance to structural failures, unavoidable in the chip fabrication process. Finally, the high integration density of VLSI technology nowadays permits the implementation of neural networks with a moderate number of neurons in a single chip, reasonable power requirements and a speed which is significantly higher than any simulation in a conventional computer.

There are, however, a set of important drawbacks which prevent the implementation of ANNs from reaching the performances initially expected. Some problems are not exclusive to the networks, and they have already been widely studied and can be solved without too many complications. Among these we find the implementation of non linear activation functions (for instance the sigmoidal function) and the need for a large number of connections external to the chip.

Some of the problems are new. The main one is the large number of connections assumed in the models proposed. In some of them, for example in Hopfield networks, the needs in the number of weights grow with the square of the number of neurons, $O(N^2)$. This problem has three elements to it. On one hand is the storage of that amount of information (a totally connected 1,000 neuron network has 1,000,000 connections and if we store each weight with 8 bit precision we will need 1 Mbyte). The importance of this problem depends on the type of device we use for storage: resistors, capacitors, digital memory cells, CCD, etc. Another element is routing or the physical connection between neurons, the importance of which strongly depends on whether in the implementation we use each device simulates one or several neurons at the same time. The lower the number of neurons they simulate, the higher the number of physical devices and the more important the routing problems as compared to storage [4]. The last element in the implementation of the connections is that of connectivity, that is, the maximum number of signals which can be sent or received by each neuron. Connectivity determines, in some way, the size of the neuron, as it indicates the number of weights it must implement, and is the main limitation in the realization of the networks due to the big difference between the connectivity required by current ANNs and the possibilities in VLSI technology. Do not forget that in some of the network models the connectivity coincides with the total number of neurons in the network.

In order to solve the problem of implementing the connections it might be a solution to reduce the number of weights in ANN models, either in a direct or in an indirect way. Indirect methods consist not so much in the reduction of the number of connections, but of the number of neurons. This reduction is achieved by using neurons with more complex combinational functions than the simple weighed sum, for example quadratic functions [14]. The advantage of this substitution consists in that quadratic functions present second order decision regions and can perform the same function as several classical neurones. With these structures the network needs less neurons and consequently less connections.

On the other hand, the direct reduction of connections can be considered from two points of view: to develop new models which adapt better to VLSI implementation, which would be the down-up strategy followed by Mead and his collaborators [5][10], or to adapt current models for VLSI implementation. This last solution seems to be the most promising one in the short term and is currently being studied. The literature presents two approaches which are to some extent in conflict: pruning and restricted connectionism.

The first one of these [7][13][16] is based on an 'a posteriori' pruning of the network once it has been trained and is capable of satisfactorily performing the task it has been given. The pruning process consists in the elimination of all those weights, and even neurons, which have no influence or do not perform any important action during the operation of the network, so that their elimination will not increase errors significantly. The main problem of pruning is that the resulting network is extremely irregular and consequently, its VLSI implementation might be more costly in terms of design time and silicon area used. Another problem is that the resulting network cannot solve generic problems, and if we want it to perform a different task from the one that has been specified we must start the whole process from scratch. Finally, the training starts from a totally connected network, implying that the training process must take place externally to the chip.

The other solution [1][2][3][11] is less developed than the pervious one, perhaps because it is more ambitious. It implies an 'a priori' pruning of the network, that is, the number of weights is restricted before training and independently from the task for which it is going to be trained. The objective is to be able to have new restricted connection ANN topologies which can work in an equivalent manner to that of total connection networks. The main advantages of this method is that by using regular pruning schemes the spatial regularity of the ANN can be preserved and the training process can be incorporated to the implementation of the network, increasing its flexibility and adaptability. Consequently, the use of restricted connection schemes allows a reduction of the connectivity to levels which are more acceptable for VLSI technology, thus facilitating the implementation of the resulting network.

LIMITED INTERCONNECTION MODEL (LI)

It is precisely the 'a priori' pruning or LI approach the one we are going to consider in this work, restricting the study to the multilayer perceptron network (MLP) with the backpropagation training algorithm [15], one of the best known and studied models which, initially, does not require total connection among all the neurons. In a MLP the neurons are grouped into several layers and the connections are established between neurons in consecutive layers, linking each neuron of a layer to all the neurons in the next layer. We will call this type of connection total connection, as opposed to the ones we will study here and which we will call restricted. The restriction consists in not connecting each neuron to all the neurons of the following layer, just to a smaller number (connectivity) of them according to a preset and regular connection scheme or algorithm. An example of this type of network is shown in figure 1.

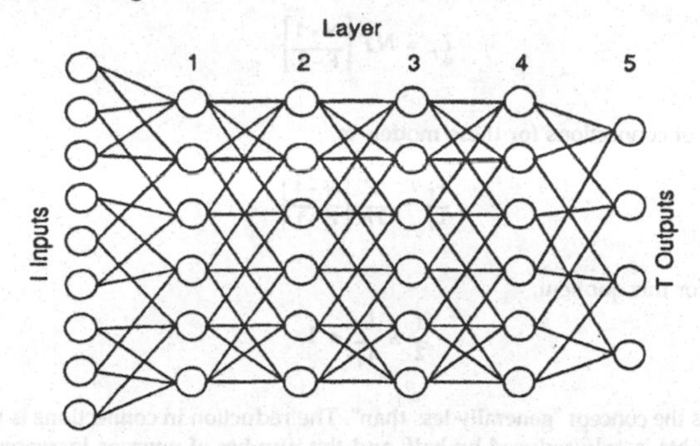

Figure 1: Limited interconnection multi-layer perceptron (LIMLP).

The limited interconnection multi-layer perceptron (LIMLP) model we consider consists of I inputs, T outputs and a group of hidden layers, each with N neurons. This restriction is needed so as not to complicate the LI schemes unnecessarily. As we have already pointed out, the neurons are not connected to all the neurons of the following layer and this generally imposes an increase in the number of intermediate layers in order to make the influence of all and each one of the inputs felt by each output. This is a basic norm we must respect if we want the resulting network to be applicable to every type of problem. The final result consists in reducing the connectivity by increasing the number of hidden layers. Our LI schemes are basically generalizations of those proposed by Akers et al., who point out the possibility of an unlimited reduction of connectionism at the cost of the corresponding increase in the number of layers of the network. Following this approach, we need a number of layers

$$L = \left\lceil \frac{N-1}{F-1} \right\rceil \tag{1}$$

where N indicates the number of neurons in each hidden layer of the LI network. This parameter is taken as equal to the number of neurons in the single hidden layer we consider if the network presents total connectionism. F is the prefixed connectivity.

LI PARAMETERS

Before considering the results of the simulations we have performed a small analysis of the possibilities of LI with respect to total connection. The advantages of LI in the direct translation of an ANN to VLSI is obviously due to the limitation of the necessary number of connections between neurons of the network to F. However, with respect to the total number of neurons it is at a disadvantage, as LI is achieved at the cost of increasing the number of hidden layers (all with the same number of neurons in our case). With regard to the total number of connections, we will calculate the needs of each type of connection structure. In general, the number of connections of a MLP is

$$C = \sum_{l=1}^{L} \sum_{i=i}^{N^l} F_i^l \tag{2}$$

where L is the number of layers, N^l is the number of neurons of layer l and F_i^l is the number of weights of the i-th neuron in layer l.

For an estimation we have used a very simple network, where all the neurons present the same fan-in (fan-out) and all the layers have the same number of neurons, that is, $F_i^l = F, \forall l \in \{1, ..., L\}, i \in \{1, ..., N^l\}$ and $N^l = N, \forall l \in \{1, ..., L\}$.

In the case of total connection, we assume that F = N and L = 2, whereas in the case of LI L is given by equation (1), and thus

$$L_t = 2N^2 \tag{3}$$

$$L_r = NF\left\lceil \frac{N-1}{F-1} \right\rceil \tag{4}$$

The ratio of the number of connections for these models is

$$\frac{L_r}{L_t} = \frac{F}{2N}\left\lceil \frac{N-1}{F-1} \right\rceil \tag{5}$$

We can establish limits for this quotient:

$$\frac{1}{2} \le \frac{L_r}{L_t} \prec 1 \tag{6}$$

the symbol "$\prec$" represents the concept "generally less than". The reduction in connections is very small. In the best case, the number of weights is only reduced by half, and the number of neurons increases, as we have already pointed out. A reduction of connections might, nevertheless, be interesting for certain types of ANN implementations such as digital ring architectures [6][9][19], where a one dimensional array of Elementary Processors (PEs) simulates the ANN in time layer by layer. In this case, a regular connection scheme (which produces, for example, a matrix of band connections around the principal diagonal) does not introduce waiting cycles in the PEs of the architecture. Thus, in this case, the reduction of the total number of connections of the network implies the reduction of its execution time by the array.

In any case, the central objective of LI is the reduction of the connectivity factor. In this sense some authors indicate that with certain LI schemes the connectivity factor can be arbitrarily set without any loss in the performance of the network. We will focus the rest of this work on this point, presenting results of simulations, whose objective is the study of performance in LIMLP.

SIMULATION RESULTS

We have used simulation as a mechanism for testing the efficiency of LI schemes. The same strategy was employed by Akers et al., who simulated very simple functions (inversion, shift and addition) with very few inputs (only 8). In our opinion the results obtained by these authors cannot be extrapolated in an immediate manner as they say. The first point in which we disagree is the selection of the tasks utilized as tests for the network. We think they must be more related to practical applications of ANNs. We have chosen two very similar tasks based on character recognition. The first one of them is a self associative memory, that is, the input patterns coincide with the output ones. The second one is a classifier where each input pattern is assigned an output neuron in the output layer. The input patterns coincide in both tasks and are obtained from the digitation of numbers and letters with a total of 35 patterns. These characters are represented as 5 column by 7 row matrices. The size of the matrix, and hence that of the network, is a compromise between the reliability of the results (in the sense of their use for extrapolations) and the complexity of the network (measured as the number of calculations it must perform and consequently the computation time).

The simulation and training of the LIMLP was carried out in a sequential computer (SPARC2) with the help of a software package called RCS (Rochester Connectionist Simulator).

The first step was to determine the minimum architectures for both types of networks, taking as minimum those architectures with the smallest set of free parameters and which can converge and learn in at least 10% of the attempts with different initial weights. In the case of networks without any connection limitation there is a single free parameter: N (number of neurons in the hidden layer), whereas in LIMLPs there are two: F (connectivity) and N' (number of neurons in each hidden layer). For the self associative memory, the minimum values found were N = 10, and F = 13 and N' = 25. For the classifier N = 5, and F = 18 and N' = 35.

In order to observe the real behavior of LI schemes we have simulated the learning process for a series of tasks and we have compared the results to those obtained using total connection two layer networks (total networks). The comparison is centered on observing how the error evolves during the training process of each task in each type of network (figures 2 and 3). Looking at the figures we can say that during the learning process the error generated by a LIMLP is quite similar to that of a total network, and this error depends more on the task to be implemented than on the type of connections.

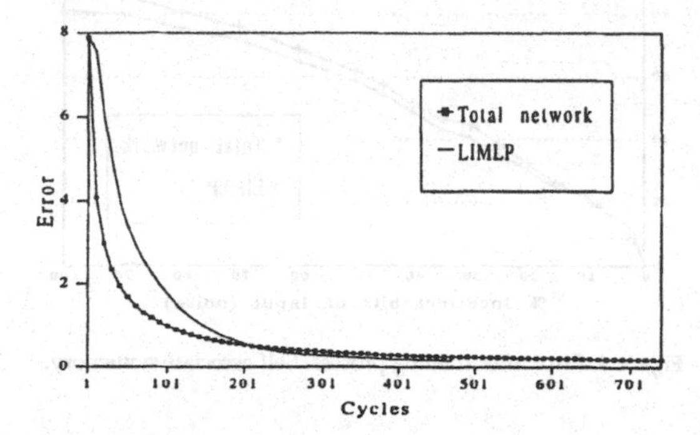

Figure 2: Error during training process. Self associative memory.

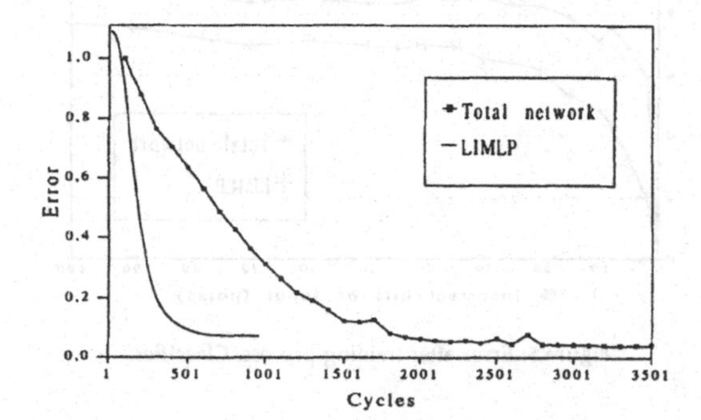

Figure 3: Error during training process. Classifier.

Another aspect we are interested in is the degree of generalization achieved by each network once it has been correctly trained (figure 4 and 5). The calculation of the generalization is carried out by introducing noise in the input patterns, randomly altering their bits. The number of incorrect outputs are counted in the whole cycle and it is expressed as a percentage of the total number of outputs. It can be seen that the curves do not differ by more than 4-6% (except with much noise), which means that generalization is not affected by the use of LI.

The simulations also reflect two adverse aspects we had not considered:

i) The first drawback is due to the fact that the needs regarding the number of neurons per layer are very different depending on the connection scheme and model of the network (total or LI). This unexpected fact invalidates all previous predictions, as this effect is not quantifiable and makes it impossible to state that a total connection network and a LIMLP network need the same number of neurons in each hidden layer. In the end we have seen that in the case of LIMLPs the number of weights increases by 140% (self associative memory) and 340% (classifier) with respect to total connection networks. Also, the increase in the number of neurons is 190% and 170% respectively.

ii) The second drawback is the limitation in the number of layers a MLP network can have if we want to successfully train it. This is a problem which directly affects the principle on which the whole LI strategy is based: to increase the number of layers in order to compensate the connectivity loss and allow all the outputs to receive information about each one of the inputs. The causes of the limited number of layers a LIMLP can handle are not due to the structure of the network, but to the training algorithm: the Backpropagation algorithm, which during training always falls into singularities or local minima from which it is not capable of coming out.

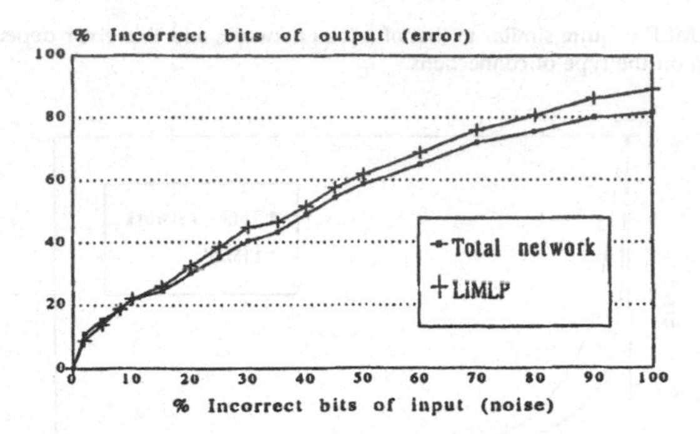

Figure 4: Error after training process. Self associative memory.

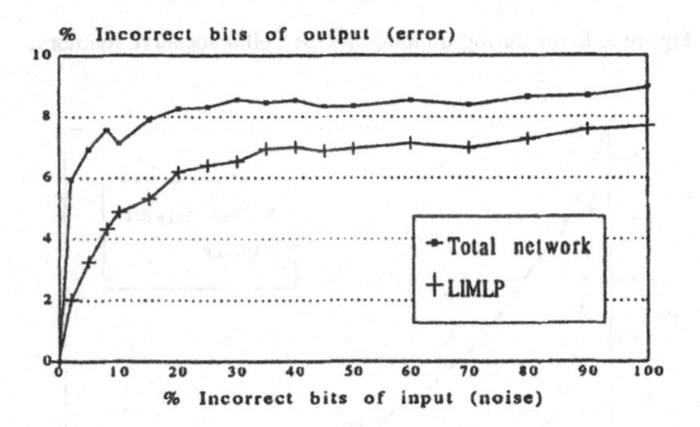

Figure 5: Error after training process. Classifier.

For analyzing this dependence of the training process on the number of layers we have simulated the XOR function. This function was chosen because the minimum network which implements it (following the minimization criteria we mentioned) has the same weight topology both in restricted and in total connection strategies, with a small number of weights, even in networks with many layers. From the simulations of the XOR function we have deduced that the difficulties in training are due, on one hand to "defects" in the Backpropagation algorithm: the bias weights are modified faster than their neighbors (bias always presents maximum output) and the algorithm is not able to correct totally erroneous outputs (the difference between the value calculated by the network and the correct value is a maximum). However, the main problem is the sensitivity of a MLP to the initial conditions [8]. If we analyze the weights in the interval [+0,5, -0,5], as Kolen [8] proposes, the maximum number of layers a MLP can have is usually 3, whereas if the initialization interval is [-5, +5] the number of layers can be larger (up to 6), although the convergence percentage is drastically reduced (10%), even in networks with few layers.

DISCUSSION

The main result of the limitation in the number of layers for MLP networks is that, unlike what some authors say, we will not be able to set the connectivity to a small enough value for VLSI implementation beforehand. The main reason for this is that it implies an increase in the number of layers and this hinders, as we have seen, the adequate convergence of the network. Therefore, the connectivity is not O(1) but, according to equation (1), it depends on the number of neurons in each layer, that is, it has a complexity of O(N), which is the same as for the total network. It is true that the connectivity can be reduced to ½ or ⅓, but it cannot be guaranteed that these reductions can be maintained for networks larger than the ones used here. Also, this reduction is always accompanied by an increase in the number of elements of the network (neurones and connections), unlike what was expected, due to the fact that the LIMLP needs more neurons in the hidden layers than the total network.

For all these reasons, restricted connectionism does not appear to be a strategy which permits the solution or elimination of the problems presented by the implementation of MLPs. Greater progress can only be expected from technological developments or the improvement of systolic ring architectures [9][19], specially if, as some authors [12] point out, for implementing gradient decrease algorithms such as Backpropagation a minimum of 8-12 bit precision is needed, as this is something that cannot be achieved with current analog technologies.

REFERENCES

[1] L.A. Akers, D.K. Ferry and R.O. Grondin, "Synthetic Neural Systems en VLSI", in *A Introduccion to Neural and Electronic Networks*, S.F. Zornetzer, J.C. Davis and C. Lau, Eds., Academic Press, pp. 317-337, 1990.

[2] L.A. Akers and M.R. Walker, "A Limited-Interconnect Synthetic Neural IC", IEEE Procedings ICNN, San Diego, CA, July 1988.

[3] L.A. Akers, M. Walker, D.K. Ferry and R. O. Grondin, "A limited-interconnect, highly layered synthetic neural architecture", in *VLSI for AI*, J.G. Delgado-Frias and W.R. Moore, Eds., Kluwer Academic, 1989.

[4] J. Bailey and D. Hammerstrom, "Why VLSI Implementations of Associative VLCNs Require Connection Multiplexing", IEEE Proceedings ICNN, San Diego, CA, July 1988.

[5] F. Faggin and C. Mead, "VLSI Implementation of Neural Networks", in *An Introduction to Neural and Electronic Networks*, S.F. Zornetzer, J.L. Davis and C. Lau, Eds., Academic Press, pp. 275-292, 1990.

[6] K. Goser, U. Hilleringmann, U. Rueckert and K. Schumacher, "VSLI Technologies for Artificial Neural Networks", IEEE Micro, pp.28-44, 1989.

[7] E.D. Karnin, "A Simple Procedure for Prunning Back-Propagation Trained Neural Networks", IEEE Transactions on Neural Networks, June 1990.

[8] J.F. Kolen and J.B. Pollack, "BackPropagation is Sensitive to Inicial Conditions", Complex Systems, Vol. 4 No. 3, June, pp. 269-280, 1990.

[9] S.Y. Kung and J.N. Hwang, "A Unifying Algorithm/Architecture for Artificial Neural Networks", International Conference on Acoustics, Speech and Signal Processing, Glasgow, Vol. 4, pp. 2505-2508, 1989.

[10] M.A.C. Maher, S.P. DeWeerth, M.A. Mahowald and C.A. Mead, "Implementing Neural Architectures Using Analog VLSI Circuits", IEEE Transactions on Circuits and Systems, May 1989.

[11] T. Markussen, "A New Architectural Approach to Flexible Digital Neural Network Chip Systems", in *VLSI for Artificial Intelligence and Neural Networks*, J.G. Delgado-Frias and W.R. Moore, Eds., pp. 315-324, Plenum Press, 1991.

[12] N. Morgan, editor. "Artificial Neural Networks: Electronic Implementations", Computer Society Press Technology Series and Computer Society Press of the IEEE, 1990.

[13] M.C. Mozer and P. Smolensky, "Skeletonization: A technique for trimming the fat from a network via relevance assessment", in *Advances in Neural Information Processing 1*, D.S. Touretzky, Ed., Morgan Kaufmann, pp. 107-115, 1989.

[14] D. Röckmann and C. Moraga, "Using quadratic perceptrons to reduce interconnection density in multilayer neural networks", in *Lecture Notes in Computer Science 540*, Springer Verlag, pp. 86-92, 1991.

[15] D.E. Rumelhart, G.E. Hinton and R.J. Williams, "Learning internal representations by error propagation", in *Parallel Distributed Processing: Explorations in the Microstructures of Cognition*, D. E. Rumelhart and J. L. McClelland, Eds., Vol. 1, Ch. 8, Cambridge MA: MIT Press, 1986.

[16] J. Sietsma and R.J.F. Dow, "Neural net prunning-- Why and how?", in Proceedings IEEE ICNN, Vol. 1, San Diego CA, pp. 325-332, 1988.

[17] P. Treleaven, "Neurocomputers",International Journal of Neurocomputing, Vol. 1, 1989.

[18] P. Treleaven, M. Pacheco and M. Vellasco, "VLSI architecturas for Neural Networks", IEEE Micro, pp. 8-27, 1989.

[19] A. Yáñez, S. Barro and A. Bugarín, "Backpropagation multilayer perceptron: a modular implementation", in *Lecture Notes in Computer Science 540*, Springer Verlag, pp. 285-295, 1991.

High Level Synthesis of Neural Network Chips

Meyer E. Nigri[1] *and Philip C. Treleaven*

Department of Computer Science
University College London
Gower Street, London WC1E 6BT
United Kingdom

ABSTRACT

In this paper we present a Neural Silicon Compiler (*NSC*) which is dedicated to the generation of Application-Specific Neural Network Chips (ASNNCs) from a high level C-based behavioural language. The integration of this tool into a neural network programming environment permits the translation of a neural application specified in the C-based input language into either binary (for simulation) or silicon (for execution in hardware). The development of the *NSC* focuses on the high level synthesis part of the silicon compilation process, where the output is a Register Transfer Level of a circuit specified in VHDL. This is accomplished through a heuristic approach, which targets the generated hardware structure of the ASNNCs in an optimised digital VLSI architecture employing both phases of neural computing on-chip: recall and learning.

1. Introduction

Implementations of artificial neural networks are divided in two basic avenues: the design of sophisticated neural programming environments (software) [9], and the design of new architectures and technologies for neurocomputers (hardware) [10]. These two approaches have been tackled independently, with no integration of software and hardware tools. While general-purpose neurocomputers have been incorporated into these software environments [3], research in special-purpose neurocomputers has been carried out separately.

In this paper, we focus on the issue of integrating software and hardware tools for neurocomputing, and describe the design of an automatic tool capable of generating ASNNCs from the same neural network description language used in the software environment [8]. This automatic tool, namely Neural Silicon Compiler (*NSC*), has been developed using the *Pygmalion* programming environment, a project funded by the European Community ESPRIT II Programme [4], which has been vastly expanded to provide a hardware route for the neural applications. The approach has taken a general framework, which allows different neural network programming systems to be used, and a generic hardware tool that can synthesise hardware structures according to the specified target architecture and adopted technology.

Figure 1 shows our proposed system, which consists of three basic components: a multi-environmental neural network programming system; an intermediate, hardware-specific, representation for the neural network; and a potential general hardware synthesis tool for generating circuits at several different target architectures.

The following sections describe the development of the *NSC*. It begins with the definition of a neural application using the *Pygmalion* environment, and concludes with the synthesis of neural chips described in VHDL, the IEEE standard hardware description language [1], as denoted by the solid lines in Figure 1.

[1]Sponsored by the Brazilian Research Agency CNPq — Conselho Nacional de Desenvolvimento Científico e Technológico.

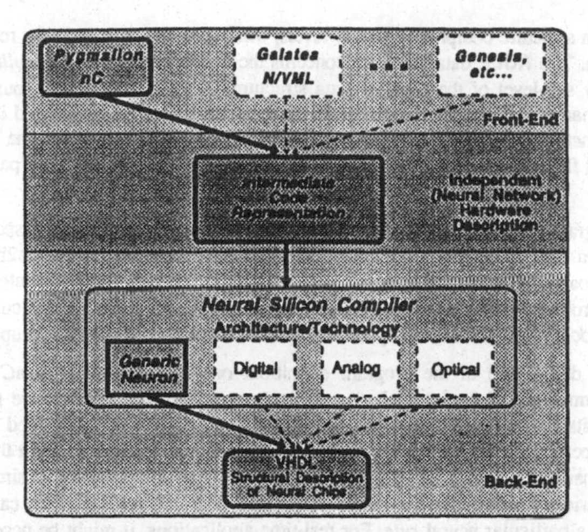

Figure 1 — A General Integrated Tool for Neural Networks Hardware

2. Extensions of the *Pygmalion* System

The link between silicon compilation and neural computing is motivated by analyses on the typical design cycle of neural network applications. Firstly, the application is defined and mapped onto a neural network, which is specified using a neural network programming language. This involves the specification of the neural model (held in the software algorithm library) and its configuration. Secondly, the network is trained using the simulation tool provided by the environment. Finally, after the network has been tuned to the particular application, the user may wish to execute the trained network in hardware, which is particularly useful in industrial and military systems [7], where typical applications encompass pattern matching, adaptive control, and signal analysis [6, 14].

Pursuing this design cycle, the *Pygmalion* system has been extended to provide a hardware path through silicon compilation [8]. This comprises four major tasks: the expansion of the *nC* programming language to include the specification of hardware-related parameters; the introduction of a hardware library for the neural models, written according to the target architectures; a hardware simulator, which provides a way for assessing in advance the performance of hardware execution, as well as representing a tool for analysing the effect of hardware constraints in digital implementations [2]; and the design of the *NSC*.

3. Neural Network Language Description

The *nC* language has proved effective for specifying neural network models and applications. The most important strengths include its ability to describe completely neural networks, in terms of topology, connectivity, and functionality, and its explicit separation between algorithm and application [12]. The language centres on the definition of some basic concepts that are closely related to neural network algorithms. These are: the **System** hierarchical data structure (Figure 2a), which groups all neural network information into one structure; and the **RULE** concept (Figure 2b), which embodies the functionality of neural network models.

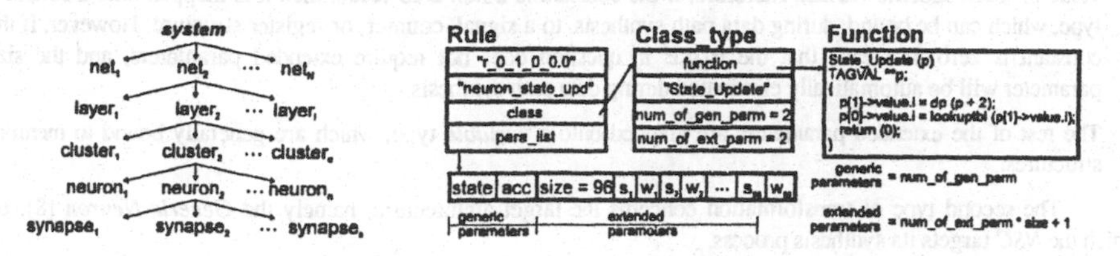

Figure 2 — (a) **System** Hierarchical Data Structure; (b) General **RULE** Data Structure

The **System** data structure comprises five sub-levels, which provide the generality required to describe any topological configuration. The **RULE** data structure concerns the *functionality* and *controllability* of the network and is specified in every sub-level of the **System** data structure. A **RULE** comprises four fields: name, class, para_list, and tag. The name field simply describes the name of the rule. The class field is a pointer that holds information about the functionality, which is defined by specific functions, and specifies the number of parameters described in the para_list field. The tag field is a special mechanism for identifying every parameter in the generic and parameter lists.

The task of programming a neural network application in *nC* basically involves specifying the *rules* that define a neural algorithm and initialising the parameters handled by each *rule*. Figure 2b illustrates the recall procedure of a Back Propagation neural network implemented by the function State_Update, which takes pairs of state and weight values from preceding layers to calculate the neuron's own state. This calculation is performed by the built-in function dp (*dot product*), followed by the activation function realised by a lookup table.

From this brief description of the program definition, one can observe that *nC* is software-oriented. Therefore, some transformations of the internal structures are necessary if the hardware route is to be able to produce optimised circuits. In addition, hardware-related parameters must be introduced in the language. The implemented extensions permit the user to specify high-level commands that directly affect the hardware synthesis, but without giving any hardware-specific details. These commands are driven by the desired performance of the final application, and basically include time and area constraints. For example, the user can specify a maximum time allowed to perform a particular neural *rule*. For real-time applications, it might be necessary that a complete recall or learning phase should be realised within a specific time. Similarly, the application might limit the number of chips due to space requirements. This would require a certain minimum number of processing elements per chip. Other, more specific hardware parameters can also be specified by the user, such as data precision, or the mechanism for implementing the activation function, such as lookup table commonly employed in digital designs [2]. These hardware extensions are implemented in the **System** and **RULE** data structures, thus preserving the original language.

4. High Level Transformations

The hardware synthesis starts by reading the **System** hierarchical structure and its associated **RULE** mechanism to extract all relevant information regarding the network's interconnection topology and neurons' functionality. During this phase several components of the *nC* specification are discarded, since they are not useful to the synthesis of a hardware structure. For example, the functions build_rules and connect, are necessary during simulation to initialise the **System** data structure by allocating memory and establishing correct pointers for the network connectivity and functionality.

The specified *rules* are then compiled into a syntax tree following conventional techniques for software compilers. Since the majority of the problems in high level synthesis can be determined and solved by graph algorithms, the syntax tree is transformed into a graph-based structure. During this process, several high level transformations are performed upon the *nC* program. The transformations are implemented through heuristic rules which are divided into *nC*-specific and hardware-specific transformations. *nC-specific* rules are defined as follows:

- Any generic parameter, associated with either the neuron's states (such as the output, error, accumulation, and target states), synapses, or any model-dependent parameter (such as learning rate) is mapped onto a *variable* of sub-type *register* type. During data path synthesis, these parameters are generally bound to register structures.
- The first item in the extended parameter field refers to its size (see Figure 2b), and thus is always a constant value for each specific **RULE**. Therefore, if the constant is a non-zero value, then it is mapped onto a *constant* type, which can be bound, during data path synthesis, to a signal, counter, or register structure. However, if the constant is zero, it means that the **RULE** in question does not require extended parameters, and the size parameter will be automatically eliminated during data path synthesis.
- The rest of the extended parameters are mapped onto a *variable* type, which are generally bound to memory structures.

The second type of transformation concerns the target architecture, namely the *Generic Neuron* [8], on which the *NSC* targets its synthesis process.

- If a RAM type is added to a constant (which results from a pointer reference, indexed by a constant in *nC*, such as p[1]), then the whole forest is transformed simply in a register node as shown in Figure 3a.

- If a ROM type is added to a register (which results from a pointer reference, indexed by a variable in *nC*, such as lookuptbl[p[i]], then the whole forest is transformed simply in the ROM node (used as operand) indexed by the register, as shown in Figure 3b. In this case, the register is simply used to address the ROM memory, and the add operation is eliminated.

- If a RAM type operation is added to a register (which results from an array indexed by a variable in *nC*, such as p[i]), then the whole forest is transformed simply in the RAM node indexed by the register, similar to the previous transformation rule and as shown in Figure 3d.

- Another transformation strategy looks for the *nC* construct such as p[i+1], which is parsed as a partial tree as shown in Figure 3c, and replaces it by a simple RAM node addressed by the variable i. This is an immediate consequence of the way in which extended parameters in *nC* are arranged, that is, as pairs or group of data. The exact RAM name is obtained again through the tag mechanism employed in all *nC* data structure.

- A more complex transformation strategy looks for the *nC* statement **for** and checks whether a control variable has been defined. If so, it checks what kind of statement is defined and carries out an analysis on the whole loop, so that a loop like:

```
for (i = 3; i < (2 * size) + 3; i += 3)
```

is transformed into an equivalent loop such as:

```
for (i = 0; i < size; i += 1)
```

which is performed in accordance with the transformations described above.

In addition, in-line subroutine expansion is performed, which unlike software compilers, gives more opportunities for hardware optimisations, since operations and storage elements can be merged. Without all transformations described above, the hardware synthesis tools would tend to create a complex hardware structure, due to the way *rules* are defined, which manipulate data as array elements.

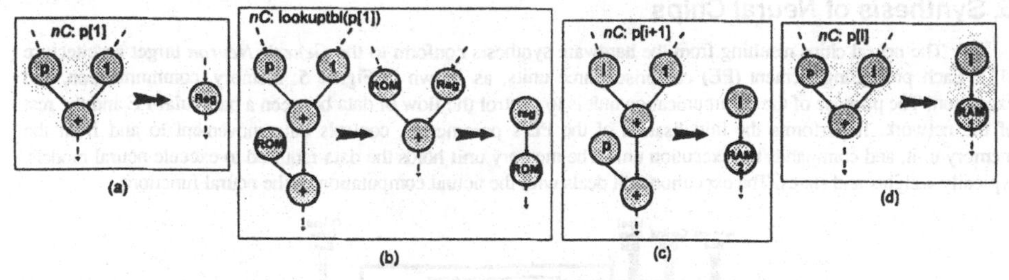

Figure 3 — Transformations Performed upon the Syntax Tree

The next step is to transform the tree based structure in a control and data flow graph (CDFG), which is used as the input for data path and control synthesis. At this stage, the *ICR* format is generated. It comprises two basic sections. The first section describes the data types defined in the *ICR*, while the second part gives the control and data flow of the neural application. This graph is then partitioned into several simple graphs (forests) according to the control information specified, which greatly simplifies its manipulation. Although this step introduces new nodes, required for temporary variables, most of them are later eliminated by the data path synthesis. Figure 4 shows the CDFG for a Back Propagation network, which has been compiled from an *nC* specification.

Data path synthesis starts by employing specialised algorithms that visit every node in the graph. The introduction of an *Activity List* for each storage element defined in the *ICR* is extremely important, during this phase, for the generation of compact structures. The *activity list* is defined as a five-value list, where each element represents the status of a storage element in each particular control state. The five possibilities are: *dead*, where no activity is performed in the variable; *idle*, in which the variable is idle but still alive, meaning that its data previously written is needed in a future state; and *rd*, *wr*, or *rdwr*, meaning that the storage element is being read, written, or read followed by a write operation in the same state, respectively.

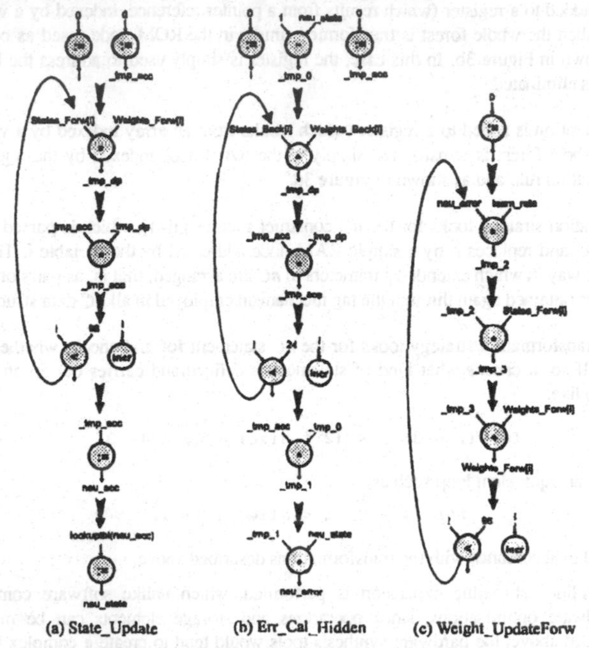

Figure 4 — CDFG for the Back Propagation *Rules*

5. Synthesis of Neural Chips

The neural chips resulting from the hardware synthesis conform to the *Generic Neuron* target architecture [11]. Each processing element (PE) comprises three units, as shown in Figure 5: memory, communication, and execution. The purpose of the communication unit is to control the flow of data between a particular PE and the rest of the network. It performs the initialisation of the PE's parameters, controls data movement to and from the memory unit, and commands the execution unit. The memory unit holds the data required to execute neural models, typically weights and states. The execution unit deals with the actual computation of the neural functions.

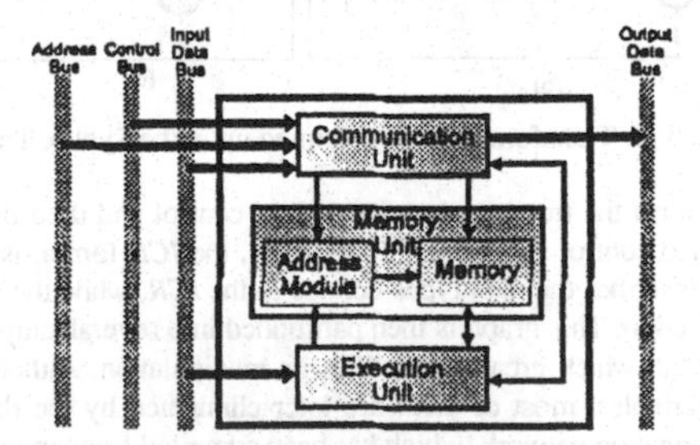

Figure 5 — The Processing Element's Internal Structure

The synthesis of these three units is performed according to the information provided by the graphs. Initially, during data path synthesis, the execution unit's data path module and the memory unit are fully synthesised from the *ICR*. This includes performing hardware allocation and scheduling, which are interdependent operations, and therefore are employed in conjunction with each other. However, depending on the user's hardware-related specification, the algorithms focus primarily on allocation or scheduling. If area is the predominant factor, than straightforward sequential scheduling is employed with no duplication of hardware resources. Conversely, if timing constraints dominate, an *ASAP* (*As Soon As Possible*) scheduling algorithm [13] is enforced, which schedules operations at the first possible state, thus providing highly parallel implementations.

Hardware allocation is performed by assigning storage elements to variables, functional units to operators, and busses to the interconnection of these modules. Storage elements are synthesised as multiport structures, according to their interconnection needs. Operators are generated from a limited library of functional units. The allocation of busses uses heuristic procedures to avoid the proliferation of redundant busses. The use of multiple busses also allows the schedule of several operations in the same clock cycle, greatly improving performance.

The synthesis of the execution unit's control module, performed by control synthesis, follows the data path synthesis, in which a finite state machine is generated for the adopted scheduling strategy. Two-phase clock mechanism is adopted, which permits that data from memory be accessed in parallel by the communication and execution units. The communication unit is then generated, which consists basically in creating a controller (using another finite state machine), and using its fixed data path, consisting of the logic to control internal and external busses.

6. Conclusions

The resulting synthesised structure for the processing elements is generated in VHDL, according to the hardware modules synthesised in the previous phase. While data path synthesis produces a highly compact structure, based on VHDL cells specially designed for the *Generic Neuron* target architecture, the generation of finite state machine follows standard practices [5], so that logic synthesis can be applied. The results obtained through the *NSC* are virtually identical to the ones obtained through manual designs. This has been the main goal of this research, and its success is highly dependent upon the high level transformation steps undertaken, and the use of heuristic rules during the synthesis of *nC* variables and operations into storage and elements and functional units, respectively.

References

[1] "IEEE Standard VHDL Language Reference Manual", *IEEE Std 1076-1987*, March 31, 1988

[2] Alippi, C. and Nigri, M.E., "Hardware Requirements for Digital VLSI Implementation of Neural Networks", in *Proc. IJCNN '91 Singapore, November 18-21 1991*, IEEE, pp. 1873-1878, New York, N.Y., 1991

[3] Angeniol, B. et al., "The Galatea Project", *Proceedings of the NeuroNimes'92*, 1992

[4] Azema-Barac, M., Hewetson, M., Recce, M., Taylor, J., Treleaven, P., and Vellasco, M., "PYGMALION Neural Network Programming Environment", *International Neural Network Conference*, Paris, France, July 9-13, 1990

[5] Carlson, S., *Introduction to HDL-Based design using VHDL*, Synopsis, Inc., 1991

[6] Josin, G., Charney, D., and White, D., "Robot Control Using Neural Networks", *IEEE International Joint Conference on Neural Networks*, vol. II, pp. 625-631, 1988

[7] Maren, A.J., Harston, C.T., and Pap, R.M., *Handbook of Neural Computing Applications*, Academic Press, Inc., 1990

[8] Nigri, M., Treleaven, P., and Vellasco, M., "Silicon Compilation of Neural Networks", *Proceedings of the IEEE CompEuro'91*, pp. 541-546, Bologna, Italy, May 13-16, 1991

[9] Recce, M., Rocha, P.V., and Treleaven, P.C., "Neural Network Programming Environments", in *Artificial Neural Networks*, ed. I. Aleksander and J. Taylor, Elsevier Science Publishers, vol. 2, pp. 1237-1244

[10] Treleaven, P., Pacheco, M., and Vellasco, M., "VLSI Architectures for Neural Networks", *IEEE Micro*, pp. 8-27, December 1989

[11] Vellasco, M. and Treleaven, P., "A Neurocomputer Exploiting Silicon Compilation", *Proc. Neural Computing Meeting, The Institute of Physics*, London, April 1989

[12] Vellasco, M., "The nC Neural Network Programming Language - Manual (Version 1.02)", University College London, *Pygmalion Project 2059*, 1990

[13] Walker, R.A. and Camposano, R., *A Survey of High-level Synthesis Systems*, Kluwer Academic Publishers, 1991

[14] Webb, A.R., "Applications of Neural Networks in Military Systems", *Military Microwaves'90*, pp. 356-361, London, July 11-13, 1990

NEURAL NETWORK SIMULATIONS ON MASSIVELY PARALLEL COMPUTERS: APPLICATIONS IN CHEMICAL PHYSICS

Bobby G. Sumpter[a], Raymond E. Guenther[b], Christian Halloy[c], Coral Getino[a], Donald W. Noid[a]

a) Chemistry Division, Oak Ridge National Laboratory
Oak Ridge, TN 37831-6182 USA
b) Dept. of Physics, University of Nebraska at Omaha,
Omaha, NE 68102 USA
c) Joint Institute of Computational Sciences
University of Tennessee, Knoxville, TN 37996 USA

A fully connected feedforward neural network is simulated on a number of parallel computers (MasPar-1, Connection Machine CM5, Intel iPSC-2 and iPSC-860) and the performance is compared to that obtained on sequential vector computers (Cray YMP, Cray C90, IBM-3090) and to a scaler workstation (IBM RISC-6000). Peak performances of up to 342 million connections per second (MCPS) could be obtained on the Cray C90 using a single processor while the optimum performance obtained on the parallel computers was 90 MCPS using 4096 processors. Efficiency such as these has enabled neural network computations to be carried out for a number of chemical physics problems. Several examples are discussed: multi-dimensional function/surface fitting, coordinate transformations, and predictions of physical properties from chemical structure.

Introduction

Recently advances in VLSI technology has led to the development of a series of massively parallel computers. With this new avenue to large computational power, algorithms or models that can exploit parallelism have a distinct advantage in obtaining peak efficiencies. In this regard, the distributed computations in neural networks are commonly believed to represent an inherent parallelism. However, the majority of simulations are still performed on sequential computers. On the other hand, continued explorations in neural network simulations has and will lead to an increasing number of applications that require huge amounts of computing resources. Such types of problems may be prime candidates for great benefits from the use of massively parallel computers.

In this paper, we present benchmarks of a simple feedforward (fully connected) neural network on 4 different massively parallel computers: MasPar-1, Connection Machine CM5, Intel iPSC-2 and iPSC-860 (beta type for the new Paragon). The peak performances obtained are compared to that of other computer architectures and some results of neural network applications in chemical physics are given.

Benchmarks

A fairly simple feedforward neural network (1 hidden layer) was designed to test on various computer architectures. The network considered consists of a certain number of hidden nodes (N)

fully connected to a certain number of outputs nodes (M). The weight matrix that determines the interconnections between nodes is thus of size NxM. The size of the matrix was varied and computations carried out on several different computers. The results of the peak performance(the maximum number of connections per second that was executed) obtained is given in Table I.

TABLE I. Peak performances for double precision computations given in million connections per second (MCPS).

number of processors	Computer	Architecture	MCPS
4096	MasPar-1	Parallel	90
128	iPSC-860	Parallel	44
64	iPSC-860	Parallel	31
64	iPSC-2	Parallel	20
32	CM5*	Parallel	59
1	Cray YMP	Serial	128
1	Cray C90	Serial	342
1	IBM-3090	Serial	89
1	RISC-6000	Serial	13

* Although the CM5 that we used was equipped with vector processing units, we were unable to take full advantage of them due to the unavailability of vector software (CMSSL).

The performance of the parallel computers was reasonably high, however, a single vector processor on the Cray C90 appears to be substantially more efficient for the present simulations. Obviously some optimization using parallel fortran [1,2] could be employed to operate more than one of the processors on the Cray and might potentially give speeds into the 1000 MCPS. Nevertheless, our simple benchmark runs demonstrate what can be typically expected from the above computers while simulating a feedforward neural network (of the backpropagation type).

It is clear that the peak performances obtained on the parallel computers depends on the number of processors that was used. As is show in Table I, there is an increase in the MCPS obtained for the iPSC-860 when going from 64 to 128 processors. This is a typical increase and is observable for all the parallel computers that we benchmarked. However, an interesting limitation that is commonly referred to as Amdahl Rule [speedup $< 1/(f+(1-f)/p)$] was clearly observed in our computations. The speedup that is obtainable by increasing the number of processors is optimally linear. However, for a given matrix size, the behavior of the speedup for any given parallel computer doesn't ever reach this goal. The problem lies in the communication times between the processors which becomes the limiting or rate controlling step. The more communication that is needed for the given computation the worse the performance as a function of the number of processors.

Finally, a notable feature of all parallel computers is the performance dependence on the size of the problem, that is the size of the connection weight matrix. The larger the matrix, the more connections per second obtained. Also of interest is that the speedup discussed above appears to increase as the matrix size decreases (note this is not the overall performance but the speedup over a single processor) and is independent of matrix size for large matrices.

Massively parallel computers potentially can open the way for a large number of computationally intensive applications. These should and do include neural network simulations. The computers discussed in this paper have demonstrated that the present state of many parallel computers is at least competitive with that of the highly developed vector computers. Furthermore, the new state-of-the art computers that are due to be released (Cray's MPP series or the Trident) or have become available this year (Intel's Paragon, for example) promise to increase performances from the typical megaflop to the teraflop range. This could translate in to billions of connections per second for neural network simulations. Future explorations in this area are essential.

Applications to Problems in Chemical Physics

The simulation of polymer properties and dynamics has always represented a grand challenge for current computers. The reasons for this are twofold: (1) to have the unique properties of polymers involves the molecular bonding of thousands of atoms, and (2) the time scale of interest for most polymer processes is quite long relative to the vibrational periods. Current work on polymer dynamics in our laboratory consists of molecular dynamics simulations of systems with >10,000 atoms which requires extensive calculations on supercomputers. Enormous efforts are needed to make realistic simulations. One approach is to develop methods which can be mapped on new parallel computers such as Kendal Square, Hypercube, Connection machine, etc. Another approach is to develop new methodologies which can supplement current molecular dynamics simulations. For the past few years, we have been formulating new techniques for polymer dynamics. Some of this work has involved utilizing new spectral methods[3], geometric statement functions[4], and more recently neural networks[5-8]. In this paper, we discuss results and simulations obtained by combining neural networks with molecular dynamics simulations.

The motion of the atoms of a molecular system are governed by the forces of interaction between each atom and its neighbors. For a nonlinear molecule, consisting of N atoms, the potential energy surface (PES) depends on 3N-6 independent coordinates. The potential energy changes as a function of the relative coordinates of the atomic nuclei. Understanding the relationship between properties of the potential energy surfaces and the dynamics and structure have recently been addressed using neural networks. For example, in a melting study of polymers, where the chains form random coils in their final state, a neural network was used to determine the relationship(s) between the coiling and potential energy parameters, temperature, number of atoms, and elapsed time[5]. The trained network was able to make accurate predictions for other simulations with 50% more atoms and extrapolations to longer times.

A backprop neural network has also been used to predict the parameters for a given anharmonic potential energy surface[6]. A single chain of polyethylene will be analyzed using normal mode calculations to obtain the $g(\omega)$ vibrational spectra. Since a fairly accurate and anharmonic potential energy surface for polyethylene exists, we used this potential energy surface in the calculations of the vibrational spectra and then, by changing the parameters of the potential functions, we obtained a large range of spectra. After the various $g(\omega)$ spectra were calculated; these data were used as inputs to a suitable neural network. After the neural network had adequately learned the relationship between vibrational spectra and the potential energy parameters, it was tested to determine the accuracy and ability to generalize this knowledge. The relative error ranged from 0 to about 1.4%, with the majority being less than 0.5%. In addition, the neural network was able to predict parameters that were <u>outside the range</u> used in training to within 4% error. These predictions showed a relatively accurate representation of the unknown data and demonstrated the ability of the neural network to determine potential energy parameters from g(ω) vibrational spectra.

Once a PES has been obtained, the dynamics of the system described by that PES can be studied by using classical trajectories. A typical problem is the discription of molecular vibrations. Although the most common descriptions of the vibrational motion were traditionally given in normal coordinates, other sets of coordinates such as local modes are more appropriate to describe vibrationally excited systems. These types of coordinates have been frequently studied in classical trajectory calculations in order to obtain mechanistic information relevant to reactive energy flow. Most of these studies obtained mode energies from a non-Cartesian coordinate reference frame, usually internal or normal coordinates. However, both the total kinetic energy of the system and the mode energies are only approximate while working in those types of coordinates, since the expansions are truncated to the first few terms. On the other hand, the total kinetic energy in a system is exactly represented in Cartesian coordinates, but only bond energies from Cartesian coordinate simulations have been extensively studied in the past. This is primarily due to fundamental difficulties in determining the time derivatives for appropriate curvilinear coordinates.

A method for calculating local mode energies in polyatomics from Cartesian coordinates has been proposed[8,9]. Basically, the mode energy is calculated as the sum of the potential plus the kinetic mode energies. The kinetic energy is computed in Cartesian coordinates; thus, a transformation to the coordinates of the potential is necessary to obtain the kinetic mode energies. This can be achieved by using a method based on Wilson's treatment of molecular vibrations [10]. Overall the evaluation of mode energies involves the calculation of the B and G matrices and the inversion of G, along the trajectory (in general, several thousand times per picosecond (ps) of dynamics). While that is more or less trivial for a small molecular system (2 - 6 atoms), it can become a problem for macromolecular systems (systems with over 1000 atoms) for which the relatively simple algebra of mode energy evaluation can greatly multiply the computer time needed for the trajectory calculation.

The advantage of the neural network approach to study internal mode energies is that once the network has been properly trained, it can be used to analyze mode energies for virtually any trajectory from hydrogen peroxide to polyethylene using stored Cartesian coordinates and momenta. These advantages have far-reaching implications and could mean substantial savings in future molecular dynamics applications, i.e., energy flow in macromolecules.

An extension of this method was used to give the kinetic mode energies for more complex systems such as macromolecules or proteins[7]. Encouraging results using the neural network/molecular dynamics techniques for studying energy flow were obtained from an overtone excited CH stretching mode in a polyethylene molecule. It was found that, using this approach, mode energy predictions can be made at any time, temperature, and excitation level without the need to carry out any additional molecular dynamics calculations or to perform additional training of the neural network.

Another type of computation that has benifited from neural network computing is the predition of physical properties from chemical structure. Among other global goals of chemistry, synthesis of new molecules, and determination of their properties are among the most important. Fulfillment of the first goal still depends more on the creativity of chemists than on the tools used, while the second goal can now be effectively realized using artificial intelligence.

A very promising and valuable approach for making predictions of properties of compounds based on the structural formulas of compounds can be achieved by combining some ideas from graph theory with neural network computing. We have used neural networks to make predictions on a set of properties for a series of saturated hydrocarbons. As the basic compounds, we have used hydrocarbons because of the availability of a large set of their properties and because they are industrially important. These compounds contain only two different atoms and have a very simple stoichiometric formula (C_nH_{2n+2}), which simplifies calculations of Wiener-like numbers. The structures of all possible isomers of aliphatic hydrocarbons up to n = 10 were "numeralized" according to Wiener, but instead of the original method only, "local sums" were calculated and used as numerical input. The intrinsic and the most important properties of hydrocarbons (boiling points, densities, heat capacity, standard enthalpy of formation, etc.) were used as the outputs. We have used the most reliable and recent sources of information to avoid experimental errors [7].

The results that were obtained are very promising, with an average % error of only 1 on both the training and test sets. The overall scheme for computation is efficient, both cost- and computer CPU-wise. Our results indicate a vast regime of productive use for such a technique, and future studies will be focused on these issues.

CONCLUSIONS

Neural networks are computational tools[11,12] which can make large contributions to current methodologies in chemical physics. Our work has focused on backpropagation feedforward networks with excellent success. Other types of networks also offer much promise in data analyses (such as the clustering algorithms of ART and SOM's). Overall, these networks have now a large proven history of success for a wide range of applications and with the development of massively parallel computers, future applications should continue to grow. In the present paper we have performed benchmarks for a feedforward neural network on a variety of computer architectures (both parallel and serial). The results have shown that relatively high efficiencies can be obtained. Futhermore, future developments in hardware and software should provide continued growth in computationally intensive simulations.

REFERENCES

[1] S. Brawer, *Introduction to Parallel Programming*, Academic Press, Inc. New York, 1989.
[2] C.-J. Wang and C.-H. Wu, Simulation 56, 223 (1992).
[3] R. Roy, B. G. Sumpter, G. A. Pfeffer, S. K. Gray, and D. W. Noid, Comp. Phys. Rep. 205, 109 (1991).
[4] D. W. Noid, B. G. Sumpter, B. Wunderlich, and G. A. Pfeffer, J. Comp. Chem. 11, 236 (1990).
[5] D. W. Noid and J. A. Darsey, Comp. Poly. Sci. 1, 157-160, (1991).
[6] B. G. Sumpter and D. W. Noid, Chem. Phys. Lett. 192, 455-462, (1992).
[7] B. G. Sumpter, C. Getino, and D. W. Noid, J. Phys. Chem. 96, 2761-2767, (1992).
[8] B. G. Sumpter, C. Getino, and D. W. Noid, J. Chem. Phys. 97, 293-306 (1992).
[9] C. Getino, B. G. Sumpter, and J. Santamaria, Chem. Phys. 145, 1 (1990).
[10] E. B. Wilson, J. C. Decias, and P. C. Cross, Molecular Vibrations (reprinted by Dover, New York, 1955).
[11] Philip D. Wasserman, Neural Computing Theory and Practice. Van Nostrand Reinhold, New York, 1989.
[12] J. M. Zurada, Introduction to Artificial Neural Systems, West Publishing Co., St. Paul, MN, 1992.

Acknowledgments

This work was supported by the Division of Material Sciences, Office ofBasic Energy Sciences, U. S. Department of Energy, under Contract No. DE-AC05-840R21400 with Martin Marietta Energy Systems, Inc. We would like to thank Drs. Karen Bennet, Sidharthan Ramachandramurthi, and Sue Smith for useful discussion and assistance on using the parallel computers.

A MODEL BASED APPROACH TO THE PERFORMANCE ANALYSIS OF MULTI-LAYER NETWORKS REALISED IN LINEAR SYSTOLIC ARRAYS

David Naylor and Simon Jones

Department of Electronic and Electrical Engineering
Loughborough University of Technology
Loughborough
Leicestershire
LE11 3TU

Abstract

An analytical model is presented for assessing the hardware performance of multi-layer neural networks realised in linearly connected systolic arrays. Metrics to assess latency, throughput, and computational and I/O bandwidth during the recall stage are derived and applied in the analysis of a variety of multi-layer structures. The effects of the performance metrics on networks with one and two hidden layers are compared in the paper. It is found that a single hidden layer is beneficial to the computational bandwidth across a wide range of hidden layer dimensions, whereas the throughput rate of networks with two hidden layers is higher than for a single layer, even when more hidden neurons are present.

INTRODUCTION

Linear arrays offer an attractive solution to the interconnect problem posed by hardware implementations of neural networks [1,2,3]. Assessing the performance of the hardware for just one of the many neural problems that exist, is often a complex task. Therefore, a mathematical model has been developed which allows the recall performance metrics of latency, throughput rate, and computational and I/O bandwidth to be calculated. It may be applied to a network with any number of hidden layers during recall.

While many neural problems may be solved with only a single hidden layer network, often a multiple hidden layer network, implementing either the same or a different learning algorithm, may provide a more accurate solution [4,5,6]. This paper uses the model to compare networks containing one and two hidden layers, to determine if the hardware performance characteristics of networks implemented in a linear systolic array may also benefit from multi-layer solutions.

MULTI-LAYER PERFORMANCE MODEL

The principal calculation performed in systolic architectures for neural networks is the Matrix Vector Multiplication (MVM) [7], which requires a very high degree of nearest neighbour communication between Processing Elements (PE). Figure 1 shows a process-time graph for a small three layer network. In this example, the array is operating sequentially on input vectors and only starts to process another after the previous result has been produced. However, the structure of the architecture is such that two pipes can be used to channel data in and out of each layer in parallel. The figure shows that the output data from one layer is the input data to another. Therefore, while each layer is performing the MVM for the incoming data, it may also be unloading the result of the previous calculation. However, computation and I/O bandwidth differences between layers in the network can cause bottlenecks which reduce the efficiency of this dual pipe technique.

The process-time graph in Figure 2 shows that the processes being performed in each layer at any particular moment in time may be classified into one of four types.

- COMPUTATION & COMMUNICATION : The MVM is being performed on the data as it is shifted through the input (upper) pipe.
- COMMUNICATION : Only the results are being shifted through the output (lower) pipe..
- COMPUTATION & COMMUNICATION / COMMUNICATION OVERLAP : Both of the above occuring simultaneously, and so making the most efficient use of the pipe and processing power.
- IDLE : No useful operation.

The figure illustrates the three states that may occur for a layer of X PEs performing an MVM a Y element data vector. For layer m of an N layer network, K_m is a non-linear function given by,

$$K_m = f(X_0, X_1, \dots X_N, Y_1, \dots Y_N) \quad (1)$$

which may be simplified to,

$$K_m = f(X_0, X_1, X_2, \dots X_{N-1}, X_N) \quad (2)$$

since,

$$X_m = Y_{m+1} \quad (3)$$

in adjacent layers. X_0 is the input layer dimension which does not physically exist in the array, but provides the input data for the first hidden layer.

The processor efficiency and array performance deteriorate as K_m increases, but within each of the three regions,

(a) $K_m = 0$,
(b) $0 < K_m <= X$,
or (c) $K_m > 0$,

the response in a particular layer is linear. Therefore, performance prediction becomes more difficult as the number of layers in the network increases.

To simplify the model, a continuous representation of process change with time has been used, rather than a cycle by cycle breakdown. Therefore, each of the above classes can be simply analysed as areas on the process-time graph, Figure 2.

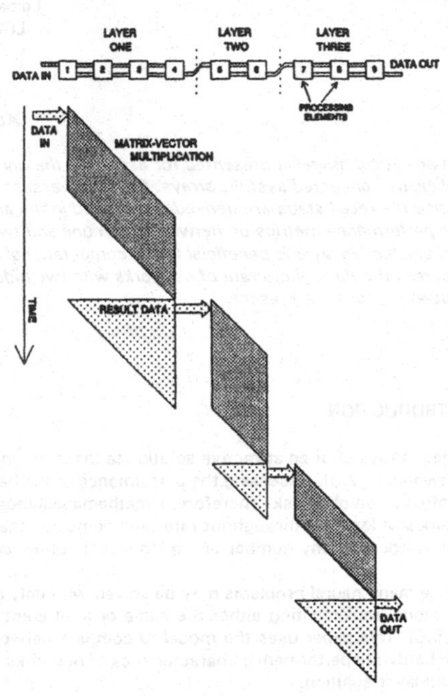

Figure 1 - Dataflow and processing breakdown in a dual piped linear systolic array, implementing an MVM calculation in three layers.

The time from the first data vector entering a layer to the first result vector emerging from that layer is the layer's latency. The network latency, L, is therefore given by the summation,

$$\textit{Array latency} \quad L = (X_1+Y_1)+(X_2+Y_2) \dots +(X_N+Y_N) \quad (4)$$

which can be simplified to,

$$L = X_0 + 2\sum_{m=1}^{N-1} X_m + X_N \quad \textit{cycles} \quad (5)$$

by reference to (3).

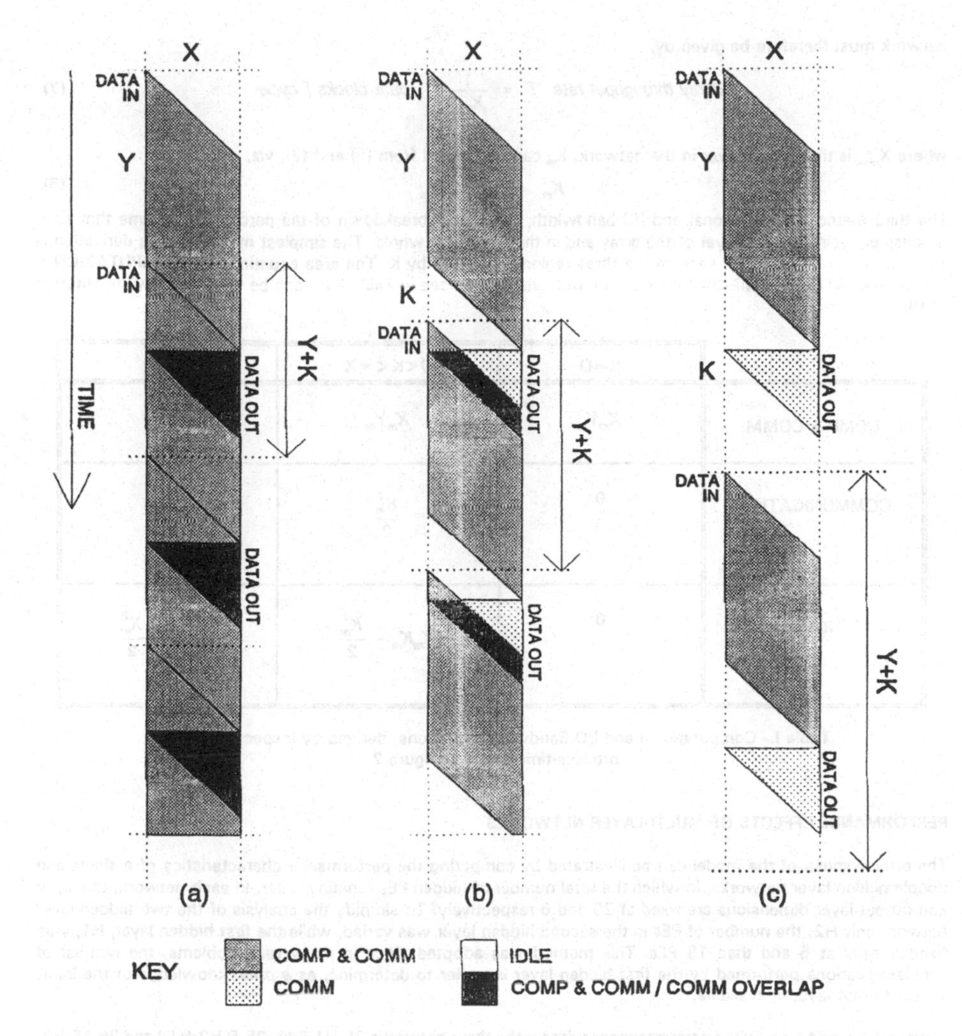

Figure 2 - Process-time graphs for (a) K=0, (b) 0<K<=X and (c) K>X. The ideal case is illustrated in (a) and as K increases the performance deteriorates.

The throughput rate for a particular layer m, is be given by,

$$\textit{Layer throughput } t(m) = \frac{1}{Y_m + K_m} \textit{ data blocks / cycle} \qquad (6)$$

which must be the same for all layers if data bottlenecks are to be minimised. The throughput rate for the whole

network must therefore be given by,

$$\textit{Array throughput rate } T = \frac{1}{X_{max}} \textit{ data blocks / cycle} \tag{7}$$

where X_{max} is the largest layer in the network. K_m can be derived from (6) and (7), viz,

$$K_m = X_{max} - Y_m \tag{8}$$

The third metric, computational and I/O bandwidth, provides a breakdown of the percentage of time that each process occupies in each layer of the array and in the array as a whole. The simplest method for its derivation is by inspection of Figure 2 in each of the three regions delimited by K. The area equations for COMPUTATION & COMMUNICATION, COMMUNICATION and IDLE are summarised in Table I. K_m can be eliminated by substitution of (8).

	K=0	0<K<=X	K>X
COMP & COMM	$X_m Y_m$	$X_m Y_m$	$X_m Y_m$
COMMUNICATION	0	$\frac{K_m^2}{2}$	$\frac{X_m^2}{2}$
IDLE	0	$X_m K_m - \frac{K_m^2}{2}$	$X_m K_m - \frac{X_m^2}{2}$

Table I - Computational and I/O Bandwidth equations, derived by inspection of the process-time graph in Figure 2.

PERFORMANCE EFFECTS OF MULTI-LAYER NETWORKS

The effectiveness of the model can be illustrated by comparing the performance characteristics of a single and double hidden layer networks, in which the total number of hidden PEs remain similar. In each network, the input and output layer dimensions are fixed at 25 and 6 respectively. To simplify the analysis of the two hidden layer network only H2, the number of PEs in the second hidden layer was varied, while the first hidden layer, H1, was fixed initially at 5 and then 15 PEs. This method was adopted since in many real problems, the number of subclassifications performed by the first hidden layer is easier to determine, as *a priori* knowledge of the input vector to that layer is available.

Figures 3,4,5 and 6 show the performance results for the three networks 25-H1-6 (i), 25-5-H2-6 (ii) and 25-15-H2-6 (iii). Figure 3 shows how the array latency and throughput rates compare. As predicted by (5), the absolute PE count and not the structure of the network affects the latency incurred in the linear array. However, this is not true for the throughput rate. All three nets show a similar performance up to the point where the total number of hidden PEs matches the dimension of the largest (input) layer. When H1 in the single hidden layer net, or H2 in the double layer net, exceed this dimension, the throughput must start to fall. However, the presence of the extra layer in (ii) and (iii), helps to dampen this effect. Therefore, larger networks may be implemented using two hidden layers before a deterioration in array throughput is observed.

Figure 4 shows the process breakdown for network (i). If this is compared with the similar graphs in Figures 5 and 6 for networks (ii) and (iii), respectively, it can be seen that in (i) the variation in computation bandwidth is much greater and in I/O bandwidth is less. At the point, H1 = 25, in Figure 4, the array processing efficiency reaches

100% due to the layer dimension mismatch being minimised. Note that as the output layer does not feed data into another layer, it does not affect the processing efficiency. This can be verified from Table I. The effect of layer mismatch is also demonstrated in the two hidden layer networks, where the low number of PEs in the first hidden layer of network (ii), causes a significant reduction in the percentage of COMP & COMM operations and a much greater percentage of pure COMMUNICATION operations.

Points of inflection are shown on each of the three graphs. These represent the points where the processing performance characteristics of the network change. In the three graphs, at the point where the hidden layer becomes the largest (marked by the line) a sharp downturn in processing efficiency is observed. The smoother response of networks (ii) and (iii), makes this change less dramatic than in (i). In fact, before this point is reached, the percentage of COMP&COMM operations are approximately constant.

Not all the points of inflection are visible on the graphs for (ii) and (iii), but it can still be seen that the characteristics are more complex for two hidden layers. This makes the task of finding the structure for optimum performance much more difficult.

CONCLUSIONS

A more in-depth discussion and analysis of the affects of multiple hidden layers is warranted, and will be addressed during the presentation of this paper. Multi-layer networks benefit from better throughput rates but suffer in terms of processor utilisation, reducing computational bandwidth. As more hidden layers are used, the complexity of the response characteristics increases, making it more difficult to establish the advantages of trading off one metric against another, when attempting to optimised the array performance.

By deriving a detailed model for multi-layer neural networks implemented in a systolic linear array, many avenues are opened up for the exploration of the architecture and its applications.

REFERENCES

1. Kung S Y and Hwang J N, 'A Unifying Algorithm/Architecture for Artificial Neural Networks', *International Conference on Application Specific Signal Processors*, pp2505-2508, Edinburgh, Scotland, 23-26 May 1989.

2. S R Jones, K Sammut and J Hunter, 'Toroidal Neural Network Processor: Architecture, Operation, Performance' *2nd International Conference on Microelectronics for Neural Networks*, pp163-169, Munich, Germany, 16-18 Oct 1991.

3. D J Myers, 'Digital Implementation of Neural Networks', *British Telecom Technology Journal*, Vol 10, No 1, pp141-148, January 1992.

4. Kung S Y and Hwang J N, 'An Algebraic Projection for Optimal Hidden Units Size and Learning Rates in Back Propagation Learning', *Proceedings of the IEEE International Conference on Neural Networks*, Vol 1, pp363-370, San Diego, 1988.

5. Sietsma J and Dow R J F, 'Creating Artificial Neural Networks that Generalise', *Neural Networks*, Vol 4, pp67-79, 1991.

6. Li-Min Fu, 'Analysis of the Dimensionality of Neural Networks for Pattern Recognition', *Pattern Recognition*, Vol 23, No 10, pp1131-1140, 1990.

7. Kung S Y and Hwang J N, 'Digital VLSI Architectures for Neural Networks', *International Symposium on Circuits and Systems*, ISCAS 89, Vol 1, pp445-448, 1989.

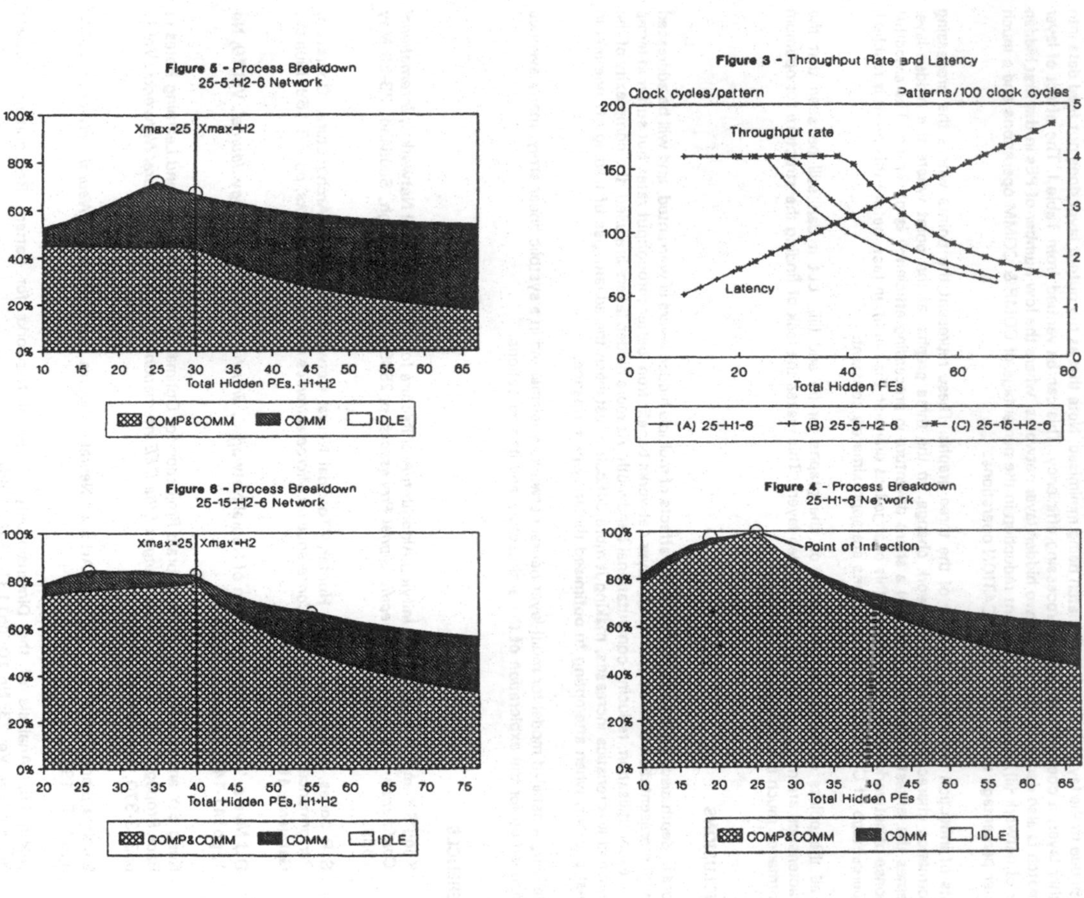

Figure 3 - Throughput Rate and Latency

Figure 4 - Process Breakdown 25-H1-6 Network

Figure 5 - Process Breakdown 25-5-H2-6 Network

Figure 6 - Process Breakdown 25-15-H2-6 Network

THE TEMPORAL NOISY-LEAKY INTEGRATOR NEURON WITH ADDITIONAL INHIBITORY INPUTS

*Chris Christodoulou**, *Guido Bugmann*+, *Trevor G Clarkson** *and John G Taylor*+

*Department of Electronic and Electrical Engineering,
+Department of Mathematics,
King's College London, Strand, London WC2R 2LS, UK

ABSTRACT

The Temporal Noisy-Leaky Integrator (TNLI) neuron model with additional inhibitory inputs is presented together with its theoretical mathematical basis. The TNLI is a biologically inspired hardware neuron which models temporal features of real neurons like the temporal summation of the dendritic postsynaptic response currents of controlled delay and duration and the decay of the somatic potential due to its membrane leak. In addition, it models the stochastic neurotransmitter release by the synapses of real neurons, as pRAMs are used at each input. Using the TNLI, we investigated the effect of synaptic integration between excitatory and inhibitory inputs on the transfer function of the neuron. We observed that inhibitory inputs increase the fluctuations of the input current and reduce the slope of the sigmoidal transfer function of the neuron, which highlights one of the differences between biological neurons and formal neurons.

1. Introduction

The concept of time is an extremely important feature of the human brain because it allows handling of dynamic environment issues. It is therefore essential for artificial neurons to exhibit temporal behaviour which will also allow effective modelling of the important features of synchronisation of neuronal activity and to preserve temporal acuity by working with output spikes.

Temporal properties are taken to be those arising from the leaky-membrane characteristics of the neuron's cell surface. Nearly all modelling approaches of these properties of real neurons today, use the Hodgkin & Huxley Leaky Integrator model [1] and are made by simulations on serial computers which makes them slow and impracticable. A hardware model of a Temporal Noisy Leaky Integrator (TNLI) using a network of pRAMs (probabilistic RAM, [2], [3]), is presented together with its mathematical theory, which can overcome the above drawbacks. The model described here includes, in addition to the model described elsewhere ([4], [5]), inhibitory inputs which induce the production of inhibitory (negative) postsynaptic current responses. The TNLI is used to investigate the effect of concurrent inhibition and excitation and the results are presented and explained.

2. Architecture and features of the TNLI model

The biological neuron consists of the cell body, the dendrites and the axon which makes synaptic contacts with dendrites of other neurons. Along the axon, propagation of nerve impulses, which correspond to digital information, is made. At the synapse, the input spikes can generate either a positive or a negative potential, depending on whether the synapse is an excitatory or an inhibitory one. The amplitude of the potential depends on the synaptic efficiency and the stochastic nature of neurotransmitter release by vesicles. These potentials propagate through the dendrites where they are delayed and their duration changes. In fact, the postsynaptic responses rise rapidly to a peak and then decline to a base line with roughly exponential trajectories which depend on the dendritic membrane time constant. If a second spike arrives at the same synapse before the first postsynaptic response has decayed completely, then the postsynaptic response of the second spike adds to the remaining tail of the first. If after a period of inactivity a long burst of spikes is delivered, then each postsynaptic response adds to the remaining tail of the preceding one building up an average effect whose magnitude reflects the rate of firing of the presynaptic neuron (Fig. 1). In other words, the essence of temporal summation that occurs in the real neuron and which we model in the TNLI neuron, is the translation of the frequency of incoming spikes into the magnitude of a net postsynaptic response. Finally, the postsynaptic responses are summed in a spatio-temporal way (on the cell body) and when the result exceeds a given threshold, then nerve impulses are generated and transmitted to other neurons.

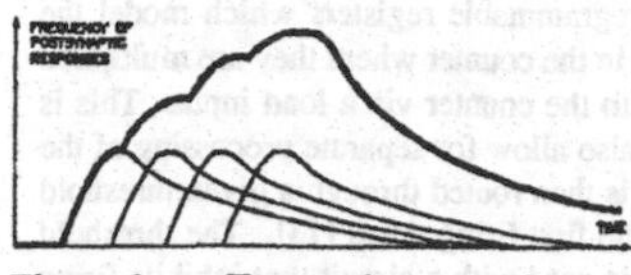

Figure 1: Temporal summation of postsynaptic responses

The hardware approach of the TNLI neuron, is based on the above biologically realistic features. An analogue hardware outline of the TNLI using a pRAM at each input and a Hodgkin and Huxley equivalent circuit for a leaky cell membrane, is shown in Fig. 2. In the TNLI, the pRAMs model the stochastic neurotransmitter release [6] by the synapses of real neurons. Neurophysiological evidence [7,8] indicates that there is a large number of synapses with a small probability to release quanta of neurotransmitter. The probability distribution of quantal release depends on the number of nerve impulses arriving at the synapse and also on the spontaneous activity of the cell body where there is no nerve impulse. This noisy nature of the synapses is perfectly matched with the stochastic pRAM behaviour. In other words, the use of the pRAMs enables generation of noise in the synaptic level of the TNLI which complies with the biological neuron, as opposed to other models in which noise is generated at the threshold level. Generalisation is also improved by this noise injection [9]. Early results described elsewhere [10], show that this intrinsic stochasticity at the input level of the TNLI increases the irregularity of the neuron's output spike train and allows us to approximate the highly irregular firing of real cortical neurons.

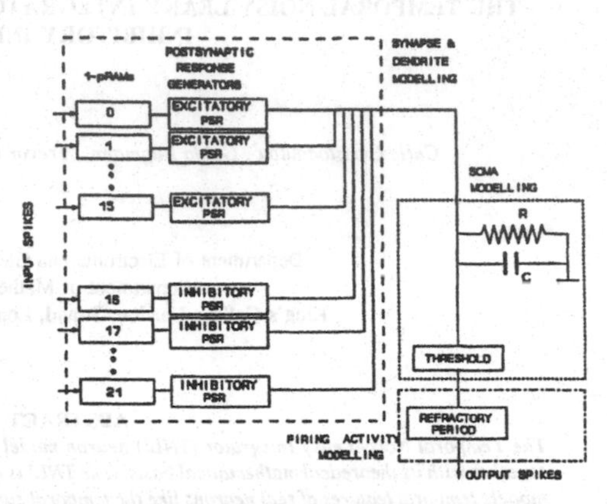

Figure 2: Block diagram of the TNLI with excitatory and inhibitory Postsynaptic Response Generators

The postsynaptic temporal response current generators (PSR) shown in the diagram of Fig. 2, model the dendritic propagation of the postsynaptic potential. For every spike generated by the pRAMs, the PSR generators produce postsynaptic current responses ($PSR_{ij}(t)$) of controlled shapes, shown in Fig. 3, which can either be excitatory or inhibitory. These particular ramp shapes were chosen for the postsynaptic responses (instead of smooth exponential ones) due to the fact that they can easily be implemented because of their defined parameters which can be trained. In addition, these shapes result in smoother responses after passing through the leaky integrator circuit, if long rise and fall times (d_r and d_f) are selected, compared to responses produced by square shapes commonly used as inputs to neurons. This enables us to reproduce the smooth postsynaptic potentials produced in distal dendrites of real neurons [11]. The postsynaptic current responses are summed temporally and the total postsynaptic current response is fed into the RC circuit (Fig. 2). The capacitance C and the resistance R represent the soma and the leaky membrane of real neurons respectively and therefore this circuit models the decay that occurs in the somatic potential of the biological neuron due to its membrane leak. The capacitance C and the resistance R are fixed at a suitable value to give the leaky membrane time constant. This intrinsic leakage of R is used to give additional temporality of biologically realistic form. Finally, if the potential of the capacitor exceeds a constant threshold (V_{th}), the TNLI neuron fires. It then waits for a refractory period (t_R) and fires again if the potential is above the threshold. Therefore, the maximum firing rate of the TNLI is given by $1/t_R$. In this model the capacitor is not reset after each output spike as in other leaky integrator neuron models [12]. Therefore the TNLI without reset will exhibit a step transfer function if fed with a continuous input current since as soon as the membrane potential passes the threshold level it will fire continuously at maximum frequency.

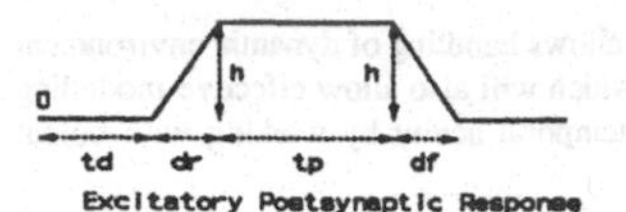

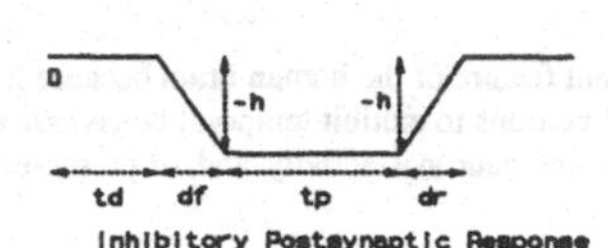

Figure 3: Shapes of the Postsynaptic Responses used in the TNLI

In the digital hardware structure of the TNLI, the pRAMS at each of the TNLI inputs will take the form of a probabilistic RAM controller with a serial update digital VLSI structure [3]. An iterative procedure is used to fetch each pRAM from the external memory and a postsynaptic response generator attached to each pRAM, produces the required postsynaptic shape. The parameter values governing this shape will be determined by the values in programmable registers which model the postsynaptic current response. These postsynaptic current responses are accumulated in the counter where they are multiplied at regular intervals by a decay rate. The decayed synaptic potential is routed back to the counter via a load input. This is a digital approach to the exponential RC-decay (see Fig. 2). The final circuitry will also allow for separate processing of the cases of pRAM spikes with high and low noise. The decayed postsynaptic response is then routed through a given threshold and if it exceeds that, the TNLI neuron fires according to the refractory period and the first firing times [13]. The threshold is designed by using a comparator and for the firing times, a shift register approach is used with a circuit that inhibits firing while in the refractory period [5].

3. Theoretical Background for the TNLI

Starting from the Hodgkin and Huxley [1] leaky integrator equation which describes the generation of an action potential in the squid giant axon, Bressloff and Taylor [13] derived a simplified version for a single-compartment leaky integrator neuron model with synaptic noise described by the shunting differential equation:

$$\underbrace{C_i \frac{dV_i}{dt}}_{(I)} = \underbrace{-\frac{V_i(t)}{R_i}}_{(II)} + \underbrace{\sum_{j\neq i} \Delta g_{ij}(t) \times [S_{ij} - V_i(t)]}_{(III)}$$

where: term (I) is the variation of accumulated charge in compartment i,
term (II) is the membrane leakage current in compartment i (negative term) and
term (III) is the synaptic input current which is excitatory for $S_{ij} > 0$ and inhibitory for $S_{ij} < 0$.

V_i (t) is the membrane potential of the ith neuron at time t, C_i is the somatic capacitance and R_i is the leakage resistance. Δg_{ij} is the increase in conductance at the synaptic connection between neuron *j* and neuron *i*, with membrane reversal potential S_{ij}, due to the release of chemical neurotransmitters.

The above equation is further simplified for our hardware TNLI neuron model. In the synaptic cleft of the real neuron, the conductance increases by the opening of ion channels (Na^+ or Na^+/K^+ channels open due to depolarising EPSP -excitatory postsynaptic potential-, or Cl^- or K^+ channels open due to a hyperpolarising IPSP -inhibitory postsynaptic potential). The increase of the conductance raises or decreases the membrane potential and if this exceeds the threshold potential (V_{th}), then an action potential (or spike) is initiated. The reverse membrane potential S_{ij}, which is a chemical constant, lies between 0 and -55mV which is approximately the threshold level in the real neuron. By using the approximation $S_{ij} >> V_i(t) \Rightarrow [S_{ij} - V_i(t)]$ in term (III) of the above equation is approximated to S_{ij} in the TNLI model, so that synaptic current flow is independent of the membrane potential V_i (t). Term (III) as a whole, which represents the current flow into the soma, corresponds in our model to the total postsynaptic response current produced by the temporal summation of the postsynaptic current responses (Fig. 3) each of which is initiated by an input spike. Thus, after the above approximation, the leaky integrator equation for the TNLI becomes:

$$C_i \frac{dV_i}{dt} = -\frac{V_i(t)}{R_i} + \sum_j \sum_{k=0}^{T} PSR_{ij}(t - t_k)$$

where $PSR_{ij}(t-t_k)$ is the postsynaptic current response caused by an input spike having arrived at time t_k on input synapse *j* and T is the total number of time steps that the system is left to operate. If the time step Δt is used as dt, then the equation can be rewritten to:

$$C_i \Delta V_i = -\frac{V_i(t)}{R_i} \Delta t + \sum_j \sum_{k=0}^{T} PSR_{ij}(t - t_k) \times \Delta t$$

We call the double summation term I(t) since it represents current, so {I(t) . Δt} is in units of charge (Coulombs). The term $C\Delta V_i$ which represents the counter contents in the TNLI can be written as $CV(t+\Delta t)-CV(t)$. So the above equation becomes:

$$CV(t+\Delta t) = CV(t) + I(t) \times \Delta t - \frac{V(t)}{R_i} \Delta t$$

In the hardware TNLI model this equation can be realised in two steps:

First step:

$$CV^*(t) = CV(t)_{before} + I(t) \times \Delta t$$

and the second step:

$$CV(t+1) = \alpha \times CV^*(t), \quad \alpha < 1$$

where α is the decay rate with which the counter contents are multiplied before they are routed back to the counter via the load input. In other words this decay rate replaces the term $\{-V(t)/R_i . \Delta t\}$ due to the hardware structure. The relationship between the decay rate α and the time constant τ = RC, can be deduced as follows:

$$\begin{aligned}
CV(t+\Delta t) &= \alpha \times CV^*(t) \\
&= CV(t) + I(t) \times \Delta t - \frac{V(t)}{R} \Delta t \\
\Rightarrow \alpha \times CV^*(t) &= CV^*(t) - \frac{CV(t)}{CR} \Delta t \\
&= CV^*(t) - \frac{CV^*(t) - I(t) \times \Delta t}{RC} \Delta t \\
&= CV^*(t)\,(1-\frac{1}{RC} \Delta t) + \frac{I(t)}{RC} \Delta t^2
\end{aligned}$$

$$\rightarrow lim_{(\Delta t \rightarrow 0)} \{ \alpha \times CV'(t) \} = CV'(t)\,(1-\frac{1}{RC}\Delta t)$$

$$So: \quad \alpha = 1-\frac{1}{RC}\Delta t$$

This relation is necessary for the software simulations of the model where the time constant is used.

4. Computational Role of inhibition in the TNLI

4.1 Biological importance of inhibition and how is it incorporated in the TNLI

Depending on the relation of the equilibrium potential for the conductance to the membrane potential level, there exist two types of effect of the inhibitory synapses: the shunting and the hyperpolarising inhibition [11]. Shunting inhibition is observed when Cl^- mediated synaptic responses increase the conductance of the membrane but not the membrane potential, since the membrane is already at the equilibrium potential of the ions involved. Hyperpolarising inhibition occurs when K^+ mediated synaptic responses hyperpolarise the membrane potential towards the K^+ ion potential. What is most interesting however, is the integration of excitation and inhibition. This depends on the relation of the membrane potential to the excitatory and inhibitory equilibrium potentials. Experimental evidence [11] shows that synaptic inhibition on concurrent synaptic excitation has a non-linear character which becomes more non-linear the more hyperpolarising the inhibition.

In the TNLI we incorporated hyperpolarising inhibition with the negative current pulses of controlled shape shown in Fig. 3. Such responses are produced by certain Postsynaptic Response generators which are assigned as inhibitory ones. The number of these generators is variable. In order to investigate the function of inhibition, the number of the inhibitory postsynaptic response generators was varied and the variation in the relationship between the Mean Input Current (I_M) and the output frequency of the TNLI was observed. I_M in the TNLI neuron i is given by:

$$I_M = \sum_{j=0}^{N} f_j \times PSR_{ij}^{*} \qquad (1)$$

where f_j is the mean input spike frequency which in our simulations is the same for each input j and PSR_{ij}^{*} is the time integral of the postsynaptic current (PSR_{ij}) produced by a spike arriving on input line j. N is the total number of input lines (or pRAMs).

4.2 Simulation data, results and discussion

The parameters used for the postsynaptic responses (Fig. 3) are: t_d (delay time) = 10ms, $d_r = d_f$ = 10ms, t_p (peak period time) = 50ms, h (postsynaptic peak current) = 5pA. The other TNLI parameters are: t_R = 4ms, R (Membrane Leakage Resistance) = 120MΩ, V_{th} = 15mV, C (Membrane Capacitance) = 25pF. The simulation time step used was Δt=1ms and the system was left to operate for T=1000ms. It must be noted that the value of the membrane time constant τ=RC=3ms might seem to be small compared to realistic values of 13.2±4.0ms [14], but due to the fall time (d_f=10ms) and the peak period time (t_p=50ms) of the postsynaptic response shape, there is a slower decay which increases the effective value of the membrane time constant. At the TNLI inputs, random spike trains of controlled mean frequency (f_j) were utilised which were unaffected by the pRAM action since the pRAM memory contents were set to '1' for an input spike and '0' for no spike and thus they fired for each input spike.

Results were taken with 16 excitatory PSR generators and 0, 2, 4 and 6 inhibitory ones. In order to obtain the same Mean Input Current we had to increase the mean input frequency (f_j), while the number of inhibitory inputs was increased. For instance, the maximum f_j required at the inputs of the TNLI to give the same I_M = 400pA, increased from 86Hz to 94Hz, 106Hz and 119Hz as the inhibitory PSR generators increased from 0 to 2, 4 and 6 respectively. The output characteristic of the TNLI for the four configurations above is shown in Fig 4. First it was observed that the TNLI gives a sigmoidal non-linear transfer function instead of a

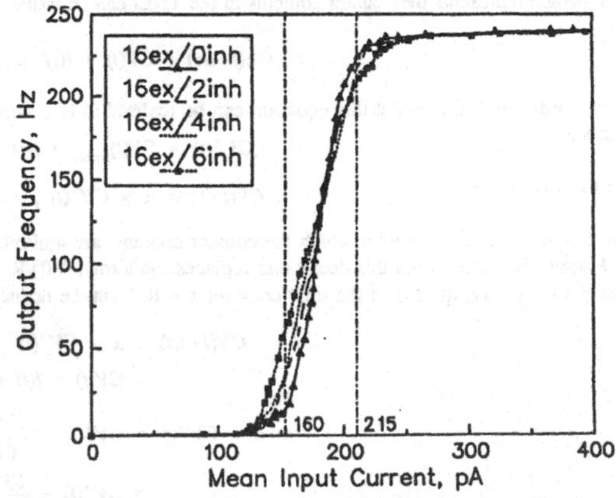

Figure 4: Effect of inhibition on the TNLI output characteristic

step function. This behaviour seems to be similar to that of the formal neuron which has a sigmoid transfer function given by: $y = 1/(1 + \exp(-\alpha A_i))$ where α is a constant that determines the slope of the sigmoid and A_i is given by: $\sum x_j w_{ij}$ where x_j is the jth input to neuron i and w_{ij} is the connection weight value from neuron j to neuron i. A_i is equivalent to I_M in the TNLI (eqn. 1). Fig. 4 shows that the introduction of the inhibition has the same effect as decreasing the value of α in the sigmoid whereas in formal neurons inhibition only affects A_i.

From Fig. 4 it can be observed that the output frequency with inhibition is higher for low Mean Input Current (I_M) values and lower for high I_M values compared to the no inhibition case. In order to explain this, two snapshots of the Membrane Postsynaptic Potential were taken for I_M = 160pA and I_M = 215pA (vertical lines on Fig. 4), for the two extreme cases of 16ex/0inh and 16ex/6inh. These are shown in Fig. 5. As can be seen, inhibition produces more fluctuations on the membrane potential though it does not change its measured mean saturation level. Therefore, in the case of low I_M (e.g., 160pA, Fig. 5a), where the mean saturation level of the membrane potential is below the threshold, the membrane potential of the 16ex/6inh case is able to exceed the threshold more frequently than in the 16ex/0inh case due to the fluctuations and thus give a higher output frequency. However, in the case of high I_M (e.g., 215pA, Fig. 5b) the mean saturation level of the membrane potential is above the threshold and so in the 16ex/6inh case, due to the high fluctuations again, the membrane potential is able to go below the threshold more frequently than in the 16ex/0inh case and thus give a lower output frequency. This explains the reduced slope of the sigmoidal characteristic curves of Fig. 4 in the presence of inhibition. The fluctuations have a similar effect in models where the capacitor is reset [12].

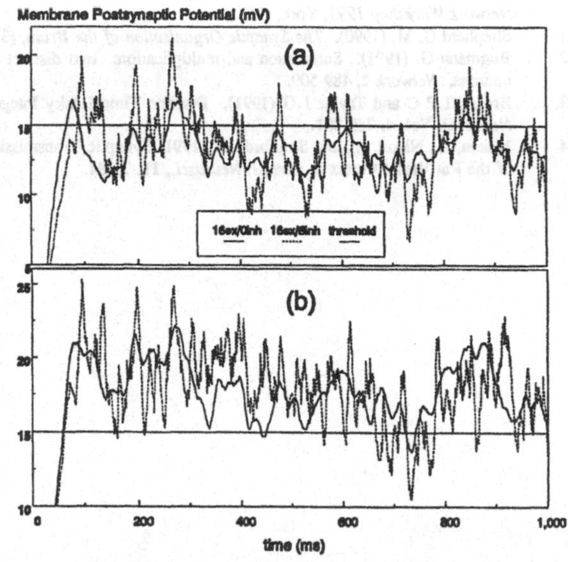

Figure 5: Membrane Potential with and without inhibition with the same I_M
(a) I_M = 160pA, (b) I_M = 215pA

5. Conclusions

We have presented in this paper the TNLI neuron model with inhibitory inputs, together with its mathematical basis and also the effect of synaptic integration between excitatory and inhibitory inputs on the transfer function of the neuron. From the results we conclude that the resulting sigmoidal transfer function of the TNLI (instead of a step function) is entirely an effect of the fluctuations. In addition, inhibition not only reduces the mean input current for the same mean input frequency, but it also modifies the transfer function of the neuron, by increasing the fluctuations of the input current around its mean saturation value. This goes beyond the assumption underlying the formal neuron used in Artificial Neural Networks where it is assumed that positive and negative inputs add linearly and then pass through a fixed sigmoidal transfer function, whereas in the TNLI the sigmoidal transfer function is modified by the signals passing through it. The effect of the fluctuations cannot be established in experimental neurobiological observations [11], since the inputs to the biological neuron cannot be controlled. Therefore, the TNLI, despite its simplicity, can be useful for modelling and understanding real neuron behaviour.

References

1. Hodgkin A L and Huxley A F (1952). A quantitative description of membrane current and its application to conduction and excitation in a nerve. *Journal of Physiology* (London) **117**, 500-544.
2. Clarkson T G, Ng C K, Gorse D & Taylor J G (1992). Learning Probabilistic RAM Nets using VLSI structures. *IEEE Transactions on Computers*, Special issue on Artificial Neural Networks, **Vol. 41**, 12, 1552-1561.
3. Clarkson T G, Ng C K, Guan Y (1993). The pRAM: an Adaptive VLSI chip. *IEEE Transactions on Neural Networks*, Special issue on Neural Network Hardware (to appear in May 1993).
4. Christodoulou C, Taylor J G, Clarkson T G & Gorse D (1992). The Noisy-Leaky Integrator model implemented using pRAMs. *Proceedings of the Int. Joint Conf. in Neural Networks 1992*, Baltimore, **Vol. I**, 178-183.

5. Christodoulou C, Bugmann G, Taylor J G and Clarkson T G (1992). An extension of the Temporal Noisy-Leaky Integrator neuron and its potential applications. *Proceedings of the IJCNN '92*, Beijing, Vol. III, 165-170.
6. Gorse D, Taylor J G (1991). A continuous Input RAM-Based Stochastic Neural Model. *Neural Networks*, Vol. 4, 657-665.
7. Katz, B. (1969). *The release of Neural Transmitter substance.* Liverpool University Press, Liverpool.
8. Raymund Y. K. Pun, Elaine A. Neale, Peter B. Cuthrie, and Philip G. Nelson (1986). Active and Inactive Central Synapses in the Cell Culture. *Journal of Neurophysiology*, Vol. 56, No. 5, 1242-1256, USA.
9. Guan Y, Clarkson T G, Taylor J G & Gorse D (1992). The application of noisy reward/penalty learning to pyramidal pRAM structures. *Proceedings of the IJCNN'92*, Baltimore, Vol. III, 660-665.
10. Christodoulou C. and Bugmann G (1993). The use of pRAMs for modelling the quantal neurotransmitter release process in the Temporal Noisy-Leaky Integrator neuron model. (to appear in the *Proc. of the Weightless Neural Network Workshop 1993*, York, UK).
11. Shepherd G. M. (1990). *The Synaptic Organisation of the Brain*, (3rd edition), Oxford University Press.
12. Bugmann G. (1991). Summation and multiplication: two distinct operation domains of leaky integrate-and-fire neurons. *Network* 2, 489-509.
13. Bressloff P C and Taylor J G (1991). Discrete Time Leaky Integrator Network With Synaptic Noise. *Neural Networks*, Vol. 4, 789-801.
14. Mason, A., Nicoll, A., and Stratford, K.. (1991) Synaptic Transmission between Individual Pyramidal Neurons of the Rat Visual Cortex *in vitro*, *J. Neurosci.*, 11, 72-84.

ARCHITECTURES FOR SELF-LEARNING NEURAL NETWORK MODULES

T G Clarkson and C K Ng

Communications Research Group
Department of Electronic and Electrical Engineering
King's College London
Strand, London WC2R 2LS, UK

ABSTRACT

The pRAM (probabilistic RAM) models the non-linear and stochastic features found in biological neurons. The pRAM is realisable in hardware and the fourth generation VLSI pRAM chip is described here. This chip contains 256 pRAM neurons and learning algorithms are built into the hardware. Several such chips can be connected together to form larger nets.

1 Introduction

The pRAM [1][2] is a stochastic and non-linear model of an artificial neuron. The pRAM generates an output in the form of a spike train. Synaptic weights are realised as multiple stored firing probabilities in the pRAM. These probability values are held in RAM and are therefore readily modified. This probability of firing corresponds to the quantal release of neurotransmitter at each synapse. The intrinsic noise is present at the synaptic level and not merely superimposed on the output as with some other models. Biologically-realistic features are expected to become increasingly advantageous in future applications of neural networks. The pRAM can implement non-linear functions and can generalise after training [3].

2 Hardware Design

Digital pRAMs were first developed because of the limited accuracy of analogue weights (approximately 6-bits). However, the functional component parts of a pRAM have been designed in VLSI analogue form and their characteristics are being measured with a view to building and evaluating analogue pRAMs in the future. The third generation of digital pRAMs incorporated on-chip learning and were fabricated in 1992 [4].

A minimum of 256 pRAM neurons per package was considered necessary for efficient use of VLSI chip area and in order to meet requirements for nets of over 1000 neurons. A pseudo-random number generator, a comparator and a learning block are shared between the 256 pRAMs in one pRAM module (Figure 1). In a module, each of the 256 pRAMs are processed serially, and each pRAM is called a virtual pRAM, since it exists only when its output is being processed, or the learning operation takes place. In this modular pRAM design, one custom integrated circuit and two external RAM devices comprise a module. Two 8-bit memory devices are required since 16-bit memory is used to store the pRAM weights.

Since nets in excess of 256 neurons are also being considered, a means of expansion is required. This has been achieved through the use of four serial links which enable one pRAM module to communicate with up to four neighbours. The connectivity of a pRAM net is reconfigurable through a lookup table. On-chip

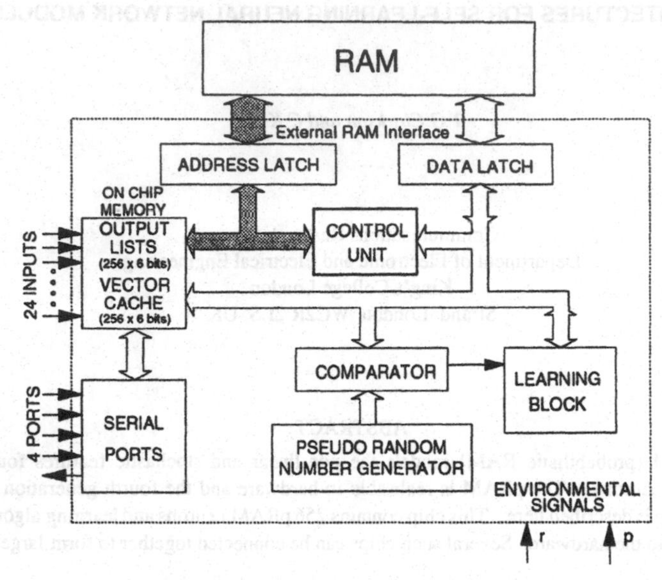

Figure 1 The pRAM modular architecture

reinforcement learning has been included in the learning block of Figure 1. The RAM array which holds the neuron weights and the connectivity data, has not been fabricated on-chip but is contained in an external RAM device to reduce costs.

2.1 Reconfigurable connectivity

This architecture employs 'virtual connections'. This means that a full set of physical connections between neurons does not exist in order to transfer activity states from one layer to the next. For each pRAM input line the address of the pRAM output to which it is connected is stored in a table, the Connectivity Table. The state of all pRAM outputs is stored in another table, the Output List. In order to determine the state of each input, its entry in the Connectivity Table, called a Connection Pointer, is used as an address, or vector, to the Output List (Figure 2). In this way, the desired activity state is copied onto the pRAM input.

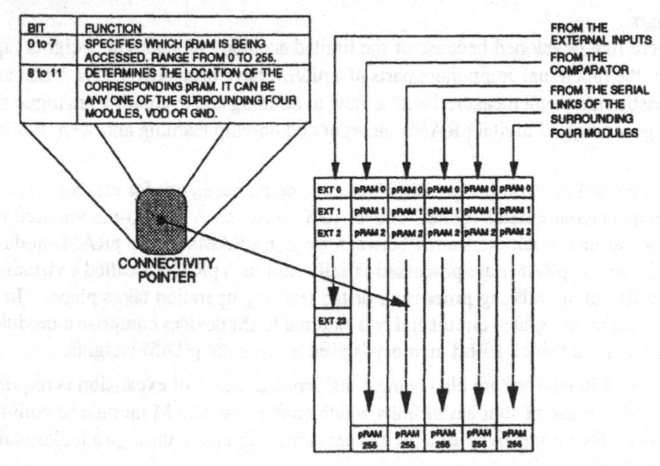

Figure 2 The operation of the Connection Pointers

Unused pRAM inputs, or inputs which are required to be fixed, may be connected to Vcc or Ground for logic '1' or '0' respectively. External logic inputs on the chip are also routed into the Output List so that the state of external hardware can form part of a pRAM's input.

In order to expand the network, connections must be made to pRAMs outside the current module. This is achieved by the serial links above, whereby the output lists from adjacent modules are exchanged. Copies of up to four external output lists are contained on the local chip so that virtual connections may be made to pRAMs in adjacent modules.

The latest design, the fourth generation pRAM module, the number of pRAM inputs has been increased from 4 to 6 and the design additionally allows connection pointers to be used in the learning circuit described below, to define the source of the reward and penalty signals (Figure 3).

2.2 On-chip learning

A number of learning algorithms have been proposed for the pRAM but the easiest to implement is a reward/penalty (or reinforcement) algorithm [5]. This also has the benefit of operating on a single spike output, rather than requiring a long pulse train before calculating the weight update. The third generation of pRAMs used only local reinforcement learning. The fourth generation device allows for local, global or competitive learning to be used. However, on-chip learning may be disabled so that other learning algorithms can be implemented off-chip. Learning is implemented as a two-pass process on the latest design since the total state of the network must be determined before reward or penalty signals are computed. The first pass is non-learning and the second pass applies the learning algorithm. When learning is disabled, the second pass is not executed so that a faster cycle time results.

When local reward/penalty learning is used, the reward and penalty signals are generated by two auxiliary pRAMs in addition to the main pRAM. Therefore 3 pRAMs are required per node for local learning. The auxiliary pRAMs are standard pRAMs taken from the pRAM modules, but their outputs are routed to the reward and penalty inputs of a main pRAM neuron using the Connection Pointers. In order to correctly implement the local reward/penalty algorithm, only 5 inputs of the main pRAM neuron can be used. The inputs to the two auxiliary pRAMs are the 5 inputs of the main pRAM and the state of the main pRAM output.

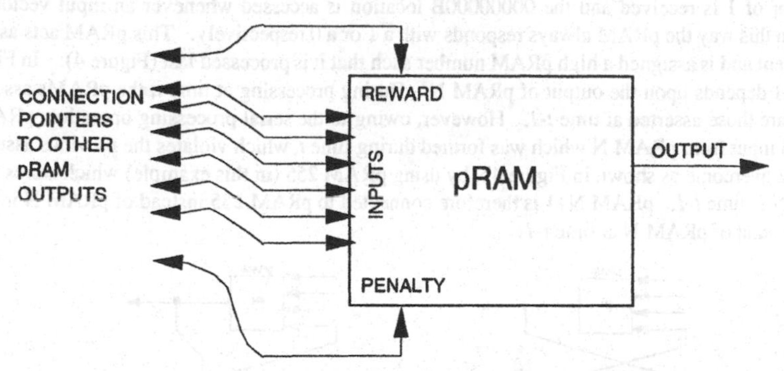

Figure 3 Reward and Penalty signals are provided with Connection Pointers

To implement global reward/penalty learning, the reward and penalty inputs of each pRAM are routed to external pins on the device where the global reward and penalty signals are applied. This is achieved by the use of Connection Pointers. For competitive learning, the outputs of competing pRAMs form the reward or penalty inputs of neighbouring pRAMs.

A variety of learning mechanisms can be implemented using the same hardware. More than one learning mechanism can be used concurrently on the same chip.

2.3 Conditions governing the sequence of operations

The pRAM controller serially processes 256 'virtual' 6-input pRAMs. When these pRAMs are arranged in layers, it is important that the input layer is processed before subsequent layers as described below. This implies that pRAMs in the input layer must all be assigned lower pRAM numbers than pRAMs in later layers. This condition must hold across all modules, if multiple modules are used.

In the earlier generation of learning pRAMs, on-chip learning took place as the virtual pRAM was being processed. This was acceptable since local learning was used and this did not place any restrictions on the order of processing pRAMs.

For global learning, the total state of the network must be known before the appropriate reward or penalty signal is determined and applied. This gives rise to a two-pass process. The first pass calculates the new output state for each of the 256 pRAMs, from the input layer first through to the output pRAMs. After each layer is processed, the outputs are handed immediately to the following layer for processing within the first pass. The reward and penalty signals are determined by an external environment, and these signals are applied to each pRAM during the second pass. The previous state of the network, which caused the present output, must be used in our reinforcement learning rule. This state could be re-created by fetching all Connection Pointers once more and forming the input vector for each neuron from the Output List as in the first pass; however, this would halve the potential performance of the module. Instead, the input vectors determined on the first pass are held in a cache and are used on the second, learning pass. In this way, extra on-chip RAM is used in order to maintain the performance of the device. This occupies previously unused space on the die.

If connections are made between pRAMs in the same layer, it can normally be arranged that the pRAMs whose outputs are required to be connected within the same layer, are processed before those pRAMs who receive such outputs on their input lines. However, if a form of cross-coupled connection is used, this condition can never be satisfied. The problem can be overcome by the use of a 'dummy' pRAM which stores the output state of one of the pRAMs and thus allows all the conditions concerning the order of processing to be met. The 'dummy' pRAM is one of the 256 pRAMs from the module which has its memory contents set to either 11111111B or 00000000B. The 11111111B memory location is accessed when an input vector of 1 is received and the 00000000B location is accessed whenever an input vector of 0 is received, in this way the pRAM always responds with a 1 or a 0 respectively. This pRAM acts as a single delay element and is assigned a high pRAM number such that it is processed last (Figure 4). In Figure 4a, pRAM N+1 depends upon the output of pRAM N. During processing at time t, the pRAMs assume that the inputs are those asserted at time t-1. However, owing to the serial processing operation, pRAM N+1 receives an input from pRAM N which was formed during time t, which violates the previous assumption. This can be overcome as shown in Figure 4b, by using pRAM 255 (in this example) which stores the state of pRAM N at time t-1. pRAM N+1 is therefore connected to pRAM 255 instead of pRAM N in order to receive the state of pRAM N at time t-1.

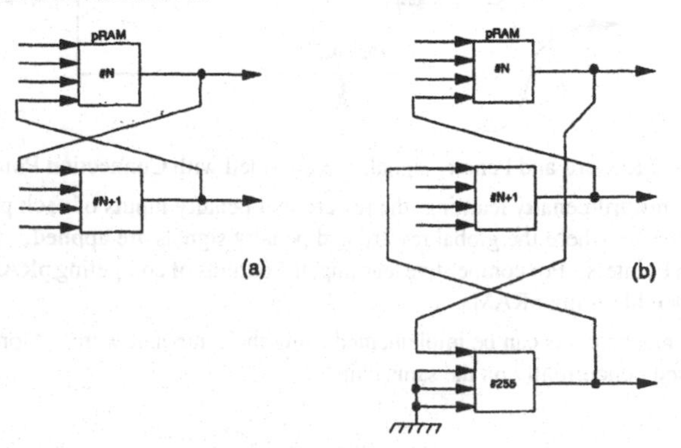

Figure 4 Use of a 'dummy' pRAM to overcome connection restrictions.

3 Fabrication of the learning pRAM

Earlier pRAM devices were fabricated in 2µm semi-custom silicon. The learning pRAM uses a 51,000 gate, 1µm CMOS gate-array with around 70% gate utilisation, packaged in an 84pin PGA. Such a high percentage utilisation is possible since most of the parts are generated as paracells, such as the RAM, or are large parts taken from a library. The amount of random logic required is small.

4 Interface to host computer

A dual-port interface to the RAM devices allows the pRAM module to be monitored by a host computer. Although the host computer is not essential to the operation of the pRAM module, it is used to set up the connection table and initialise the weights. The same interface may be used to monitor the state of the network at any time.

5 Conclusion

The pRAM has been shown to be a biologically-realistic model of a neuron. This model is capable of extension without losing its hardware-realisable features. Uses of both the non-linearity and stochasticity have been demonstrated [3][5]. Because of the high functionality in each pRAM neuron, larger or more economical networks can be built using pRAMs in comparison to other solutions. Since learning is in hardware, an autonomous pRAM neural controller can be envisaged which can act as an embedded controller in a larger system.

The features of the modular pRAM may be summarised as follows, learning is incorporated in hardware, 16-bit digital accuracy is available for learning, 256 neurons are contained in each module, the net may be expanded by using the 4 serial links and an economical VLSI implementation is achieved by the use of external RAM.

6 Acknowledgements

This project has been funded by the Science and Engineering Research Council under Grant number GR/H24495, the University of London Central Research Fund and the King's College London Research Strategy Fund. Support for fabrication has been given by GEC Plessey Semiconductors, UK.

7 References

[1] "Hardware realisable models of neural processing", T G Clarkson, D Gorse and J G Taylor, Proc. 1st IEE Int. Conf. on Artificial Neural Networks, London, 242-246, 1989.

[2] "From Wetware to Hardware: Reverse Engineering using Probabilistic RAMs", Clarkson T G, Gorse D and Taylor J G, Special Issue: "Advances in Digital Neural Networks", Journal of Intelligent Systems, 4, 11-30, Freund, London, 1992.

[3] "Generalisation in Probabilistic RAM Nets", Clarkson T G, Gorse D and Taylor J G, IEEE Transactions on Neural Networks (in print).

[4] "Learning Probabilistic RAM Nets Using VLSI Structures", Clarkson T G, Gorse D, Taylor J G, Ng C K, IEEE Transactions on Computers, Vol. 41, 12, 1992.

[5] "Biologically plausible learning in hardware realisable nets", T G Clarkson, D Gorse and J G Taylor, Proc. ICANN91 Conf., Helsinki, 195-199, 1991.

THE *GENERIC NEURON* ARCHITECTURAL FRAMEWORK FOR THE AUTOMATIC GENERATION OF ASICs

Marley M.B.R. Vellasco

Departamento de Engenharia Elétrica
Pontifícia Universidade Católica do Rio de Janeiro
Rua Marquês de São Vicente, 225
Rio de Janeiro - RJ - Brazil - 22453

Philip C. Treleaven

Department of Computer Science
University College London
Gower Street
London - WC1E 6BT - UK

ABSTRACT

Artificial neural networks have been mainly implemented as simulations on sequential machines. More recently, the implementation of neurocomputers is being recognised as the way to achieve the real potential of artificial neural networks[2]. However, current hardware implementations lean either to the optimisation of the network performance, as happens in the case of special-purpose neurocomputers, or to provide more flexibility for the execution of a large range of neural network models, as occurs with the general-purpose neurocomputers. Hence, it is desired to achieve a compromise between these two trends in order to provide high-performance application-specific neurocomputers and, at the same time, allow the user to cost-effectively execute different neural algorithms. This paper reports the results of the VLSI implementation of the so-called generic neuron architecture. This architecture serves as an architectural framework for the automatic generation of application-specific integrated circuits (ASICs), granting the necessary flexibility and high performance execution.

1. INTRODUCTION

The neural computing research area requires two specialised tools for executing artificial neural models: a flexible software tool that permits experimentation with different aspects of neural networks; and a massively parallel application-specific neurocomputer, which properly explores the intrinsic parallelism of neural networks. Various software simulators as well as ASIC chips for neural networks have been designed, with many commercial products already available. However, these products have been designed independently, with no integration of the software and hardware tools. With the incompatibility between these two tools, a neural network application tested in the software environment has to be fully designed from scratch if a high performance VLSI neuro-chip is required for the computation of the particular application. Therefore, a neural network programming environment that integrates these two tools, with the capacity of automatically generating ASIC chips from a high level specification of the neural network application, is demanded. With this complete programming environment a neural network application can be described and tested using the software tool and, after all optimal parameters have been found, an ASIC can be automatically produced without requiring the user to have any knowledge of VLSI design.

The main goal of this research has been to define a design framework that would permit the automatic generation of application-specific integrated circuits from a high-level description of a neural network application. The central point is the design of a general target VLSI architecture that encompasses the main features of neural network execution.

Some basic requirements were imposed on the processing element's internal structure and the system interconnection strategy to achieve this goal:

Flexibility

The VLSI architectural framework should provide a flexible skeleton able to execute a wide range of neural network models. The processing element's final functionality (including the learning capability) is defined by the user, which makes use of a high level specification language (the Pygmalion nC^{10} language). A silicon compiler receives this information and, based on the target architectural framework (the *generic neuron* architecture), produces customised ASICs with the required functionality. This approach offers the desired flexibility at the neural model specification phase, without affecting the final performance.

High Performance & Parallelism

In order to achieve the necessary high performance, very primitive, parallel processing elements (PEs) must be used. The resultant PEs are optimised for the computation of a specific neural network model and are automatically generated by the silicon compiler.

Modularity

To constrain communication and, consequently, reduce interconnection cost and communication delays, the ***generic neuron*** architecture should be formed by replicative, self-contained processing elements, each comprising processor, communication and memory functions.

Flexible & Regular Communication

To reduce wiring problems between PEs and also to cope with the considerable number of different interconnection patterns existent among neural models, the architectural framework should use a flexible and regular communication strategy.

Expansibility

A regular interconnection structure and the use of replicative processing elements permit an easy and fast expansion of the number of PEs, a necessary feature to conform with different network sizes.

Design Scalability

With the fast improvements in VLSI design technology, the architectural framework should provide for a scalable design as feature size decreases, allowing the addition of more PEs within the same integrated circuit without affecting the total number of pins.

Minimum Silicon Area

Because artificial neural networks make use of a large number of PEs, optimising silicon area is an important issue for the development of a successful neurocomputer. The approach taken of producing application-specific integrated circuits considerably cuts down the area, as the final chip is tunned to a specific functionality. In addition, simple and regular structures should be used to build the cell library in order to reduce wiring inside the chip.

Digital Design

In spite of the smaller area that can be achieved using analogue devices, analogue technology leads to circuits with high power consumption and low noise immunity[5]. Moreover, certain neural network models require an accuracy that is not attainable using analogue systems[2]. On the other hand, digital systems can be faster and their design techniques are more advanced. Furthermore, the learning phase is more naturally implemented using digital devices as it is difficult, using analogue technology, to devise a general learning mechanism which is suitable for a large variety of neural network algorithms[6]. For all these reasons, digital design has been utilised throughout the architecture implementation.

Based on the above requirements, the ***generic neuron*** architectural framework was designed. This paper focuses on the VLSI implementation results of a Back Propagation[8] prototype. However, before presenting the VLSI prototype, a brief description of the ***generic neuron*** architectural framework is given.

2. NEUROCOMPUTER FRAMEWORK

The ***generic neuron*** architectural framework has derived from the ***generic-neuron*** model[9, 10], which was devised to incorporate into one single structure the diverging aspects of neural network algorithms, both in terms of topology and functionality. The aim was to provide a general architectural framework that combines the high performance of special purpose hardware with the flexibility offered by general purpose neurocomputers. Based on the architectural requirements discussed above, the ***interconnection strategy*** and ***processing element structure*** were defined.

Interconnection Strategy

To achieve the above objectives, the ***generic neuron*** architecture has adopted the broadcast bus strategy as the network data communication medium. Communication occurs through a common broadcast bus, and each processing element gains access to the bus sequentially via central controller commands (Figure 1a). The interface between central controller and PEs is composed of three busses: ***data, address*** and ***control*** busses[10, 11]. The address bus specifies which PE is the owner of the data bus and the control bus contains all the necessary signals to direct the network behaviour.

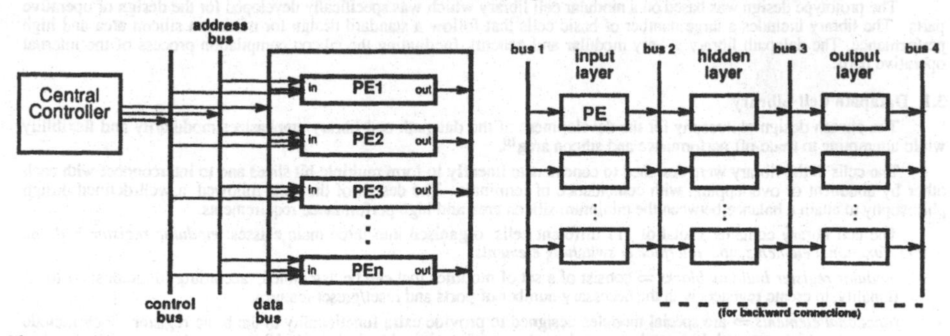

Figure 1: a) The ***generic neuron***'s bus interconnection strategy. b) The multi-bus interconnection strategy.

The bus topology provides the required ***flexibility*** to implement a vast range of different neural models, since the bus interconnection scheme can virtually realise any complex topology. Additionally, due to the regularity and simplicity of the network topology, this strategy also contributes to minimise the required silicon area as well as to achieve a ***scalable*** design, which is an essential property in massively parallel VLSI architectures. Indeed, the number of PEs produced in one chip can be unconditionally increased, with no effect on the pin count of the integrated circuit.

Another important benefit of the interconnection broadcast mode is its ability to transfer a PE 's output to all its destinations in one single cycle. This intrinsic feature is quite useful for neural network applications, enhancing the resultant network's performance. Because of these fundamental features provided by the broadcast bus, some recent hardware implementations of neural networks have also opted for this specific communication strategy[4, 12].

In the *generic neuron* architecture, the broadcast mode is extended even further, being also applied to the backward data distribution that occurs in algorithms such as Back Propagation. In such algorithms, the error values of the PEs in the output layer must be transmitted to the connected PEs of the hidden layer, in order to allow them to compute their own error values. The error value must be multiplied by the connection weight, likewise the state calculation computation on the forward phase. In current neural hardware designs, a separate *weight×error* value is transmitted for each existent connection of a particular output PE, thus reducing substantially the performance of the learning procedure. The *generic neuron* architecture avoids this problem by treating backward connections similarly to the forward connections, storing in each PE of the hidden layer, the backward synaptic weights as well. This approach increases the demand on the internal memory required to store data but the increase in silicon area (which occurs only for the hidden layer PEs) is compensated by the considerable gain in performance during the learning process execution.

In order to reduce the degradation effect when large number of PEs must be interconnected, the *generic neuron* architecture supports the implementation of multiple busses (Figure 1b). In this scheme, PEs that receive the same input are clustered onto the same data bus. This multiple bus configuration is only possible due to the PEs' separated input and output data busses. This approach is quite useful in cases that allow multiple patterns to be treated simultaneously, where the data is pipelined through the layers.

Processing Element Structure

The physical processor has been designed as a self-contained element, consisting of three basic units[11]: the *memory unit*, for storing the synaptic weights as well as the state and error values received from other neural processors; the *communication unit*, to implement all functions related to data transfer between a specific PE and the rest of the network; and finally the *execution unit* to execute the necessary operations for learning and recall.

The communication unit basically accomplishes the external address bus analysis, verifying the relevance of its contents during input and output phases. It is essentially composed of comparators and their associated registers to perform two basic tasks: determine when an output value (state or error) can be broadcasted into the data bus; and verify when the data bus value must be stored into the processing element's internal memory.

The execution unit is responsible for the computation of the neural functions. It executes the three basic functions of the *generic neuron* model (state calculation, error calculation and weight update functions)[9,10], in accordance with the high level description provided by the application designer. The mathematical operations are performed in fixed-point, 2's complement representation, which has been proved to be adequate by many simulation studies[1,10]. The control of the mathematical operations is performed by programmable logic arrays (PLAs) and the threshold function is implemented by a look-up table (ROM memory). A more detailed description of the *generic neuron* architecture can be found in other articles[10,11].

3. VLSI IMPLEMENTATION

To analyse the hardware complexity of the *generic neuron* target architecture, a prototype VLSI chip was implemented using the Back Propagation algorithm as the target neural model. The Back Propagation model has been chosen due to the strict requirements it imposes on data communication, data precision and processing capacity.

The prototype design was based on a modular cell library which was specifically developed for the design of operative parts. The library includes a large number of basic cells that follow a standard design for minimum silicon area and high performance. The datapath library is very modular and general, facilitating the silicon compilation process of the internal operative parts.

3.1. Datapath Cell Library

The chosen design philosophy for the development of the datapath cell library emphasizes modularity and flexibility while attempting to trade-off performance and silicon area[10].

The cells in the library were designed to concatenate linearly to form multiple-bit slices and to interconnect with each other by abutment or overlapping, with coincidence of terminals. The design of the cells followed a well-defined design philosophy to attain a balance between the minimum silicon area and high performance requirements.

The cell library contains a total of 117 different cells, organised into three main classes: *modular register building blocks*, *functional elements*, and *interface & switching elements*.

- *modular register building blocks* ⇒ consist of a set of modules that can be assembled, according to the desired functionality, to create registers with the necessary number of ports and reset/preset features.
- *functional elements* ⇒ are special modules designed to provide extra functionality to the basic register. They include comparators, counters, shifters, and a general-purpose arithmetic and logic unit (ALU).
- *interface and switching elements* ⇒ are cells designed to improve flexibility and compatibility among operative blocks. They include: bus drivers, data-alignment converters, and padding cells. Switching elements are simple arrays of pass transistors for interconnecting bus lines.

The resulting cell library is general, comprising a number of functional modules that can be appropriately assembled to provide the required functionality. The implemented cell modules can be stacked by abutment, substantially reducing the silicon waste in routing signals between cells. The generality and modularity characteristics of the cell modules provide an adequate cell library for the silicon compilation process, where different application-specific datapath modules can be generated without compromising the silicon area.

3.2. Chip Organization

Based on the cell library described above, the Back Propagation prototype chip was implemented. In terms of the VLSI design, the chip organisation can be separated into two units: the *operative* part, that embodies the necessary blocks to perform the neural functions; and the *control* part, that comprises the control blocks to command the neural model execution in the operative part.

3.2.1. Operative Part Organization

The processing element's operative part determines the internal hardware implementation and the topological structure of the modules defined at the functional level. The design of the operative part is based on the principle of information locality to increase performance, comprising on-chip memory and registers with multiple accesses.

The ***generic neuron*** operative part is organised in three units: a three-segment datapath, the look-up table for the threshold function, and the local memory for storing input data and synaptic weights (Figure 2).

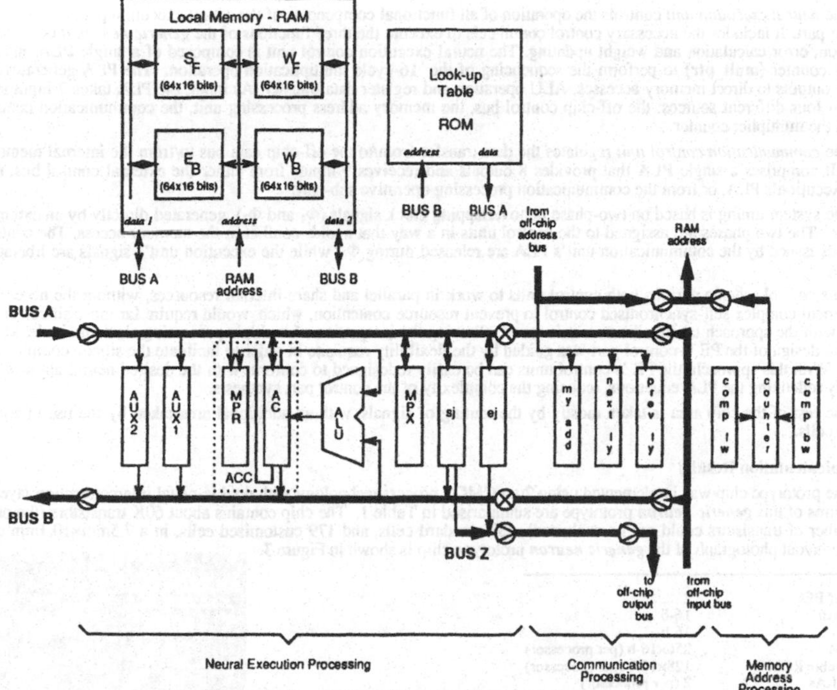

Figure 2: Organisation of the *generic neuron* Operative Part.

The local memory, which covers almost 36% of the chip area, is divided into four blocks that stores respectively: input states (S_F) and their associated weight values (W_F); as well as backward errors (E_B) with their related synaptic weights (W_B). All memory blocks share the same address bus, while the data busses are separated into two groups, one for the input values (states and errors) and another for the synaptic weights (forward and backward). The look-up table is implemented by a ROM memory with 128 words, occupying merely 6% of the silicon area.

The datapath occupies approximately 9% of the chip area. The whole datapath is 16-bits wide, and has a structure which consists of three perfect rectangles, with busses in metal-1 and control lines in metal-2. This rectangular datapath is assembled from a large number of small cells on a multiple bus structure. The bus system is interrupted by sets of tri-state drivers which allow data transfer or independent processing by the operative sub-parts. These tri-state drivers are made sufficiently large to drive the capacitive loads of internal and external busses lines.

Functionally, the datapath is composed of three operative sub-parts (Figure 2): *Neural Execution Processing*; *Communication Processing*; and *Memory Address Processing*. The *neural execution processing* unit includes all essential modules to compute the necessary mathematical operations, such as: the ALU; accumulator/shifter (ACC); multiplicand register (MPX); state and error registers (s_j and e_j); and two auxiliary registers (Aux1 and Aux2) used to store intermediate results of the Back Propagation execution. The *communication processing* unit contains the three necessary comparators: one that verifies the processing element's own address (**my_add**) and two that determine the layer addresses from which the processing element should receive inputs (**prev_ly** and **next_ly**). Finally, the *memory addressing processing* unit encompasses a counter to sequentially access the memory blocks, as well as its associated comparators to determine when the whole memory block has been searched, either in the forward (**comp_fw**) or backward (**comp_bw**) calculation.

3.2.2. Control Part Organization

The control part of a sequential machine commands the operative part by activating its control lines at the right time according to the system timing. The control part of the ***generic neuron*** prototype takes less than 4% of the chip area and it is implemented with PLAs. Each PLA synthesizes a "nondeterministic" finite state machine[3], in the sense that a machine can be, simultaneously, in more than one state. The PLA implementation of nondeterministic finite automata provides a considerable reduction of silicon area in contrast with conventional implementations.

The organisation of the control part is extremely simple and includes two control units: *Neural Execution Control Unit* and *Communication Control Unit.*

The *neural execution unit* controls the operation of all functional components of the neural execution processing in the operative part. It includes the necessary control commands to execute the three functions of the ***generic neuron*** model: state calculation, error calculation and weight updating. The neural execution control unit is composed of a single **PLA** and an auxiliary counter (**mult_ptr**) to perform the sequencing of the 16-cycle multiplication operation. The PLA generates 29 different outputs to direct memory accesses, ALU operation and register data transfer. As input, the PLA takes 7 input signals from four different sources: the off-chip control bus, the memory address processing unit, the communication control unit, and the multiplier counter.

The *communication control unit* regulates the data transfer from/to the off-chip data bus to/from the internal memory blocks. It comprises a single **PLA** that provides 8 outputs and receives 9 inputs from either the external control bus, the neural execution's PLA, or from the communication processing operative sub-part.

The system timing is based on two-phase nonoverlapping clock signals (Φ_1 and Φ_2), generated directly by an external oscillator. The two phases are assigned to the control units in a way that avoids conflict in the memory access. The control commands issued by the communication unit's PLA are released during Φ_1, while the execution unit's signals are liberated during Φ_2.

This control scheme permits both control units to work in parallel and share internal resources, without the necessity to implement complex self-synchronised control to prevent resource contention, which would require far more silicon area. Instead, with the approach taken, silicon area is saved, allowing the integration of multiple processing elements in the same chip. The design of the PE 's control part was guided by the flexibility requisite in order to facilitate the silicon compilation process. With this approach, the PE 's control units can be easily redesigned to conform with the desired neural application by simply redefining the PLA equations, reducing the complexity of the control part synthesis.

The rest of the chip area is taken mostly by the routing of signals, with an additional area taken by the use of some standard cells.

3.3. Implementation Results

The prototype chip was implemented using 2μm CMOS process technology with double-metal interconnection layers. The features of this ***generic neuron*** prototype are summarised in Table 1. The chip contains about 60K transistors (the precise number of transistors could not be evaluated), 433 standard cells, and 179 customised cells, in a 7.5mm×10.1mm die area. The layout photograph of the ***generic neuron*** prototype chip is shown in Figure 3.

Number of PEs	2
Data Length	16-b
ALU	16-b
Data RAM	256x16-b (per processor)
Look-up table ROM	128x16-b (per processor)
Control PLAs	2 (per processor)
Package	68 pin CLCC
Number of standard cells	433
Number of full-custom cells	179
Device technology	2μm CMOS
Die Size	7.5x10.1mm ($76mm^2$)

Table 1: The prototype basic features.

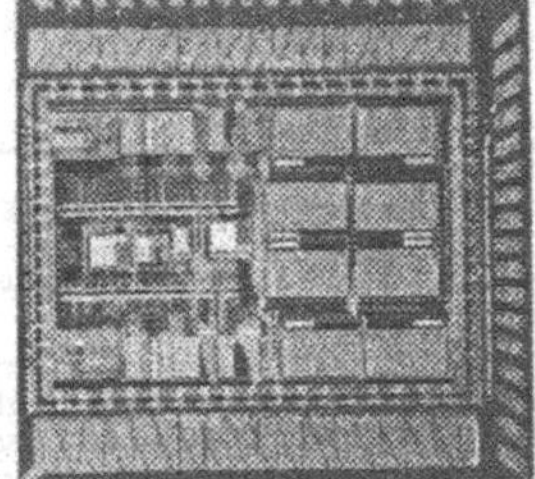

Figure 3: Photograph of the B.P. prototype.

This first prototype implementation has been quite important in providing feedback about the constraints of the hardware implementation and in verifying the critical modules in terms of silicon area. Table 2 summarises the results of the ***generic neuron*** packing density investigation using five different CMOS fabrication technologies, ranging from 2μm to 0.8μm. For each processing technology, seven different PE configurations (in terms of the number of input connections provided) have been experimented, establishing for each memory configuration, the total number of PEs that can be produced per chip.

As can be seen from these results, the ***generic neuron*** architecture yields good packing density in most of the fabrication technologies, except for the 2μm that affords multiple PEs per chip only with 1K or less number of connections. It must

Number of Processing Elements Integrated per Chip								
PE Internal Configuration				*Fabrication Technology*				
				2.0μ	*1.5μ*	*1.2μ*	*1.0μ*	*0.8μ*
64	16-bit connections	(256	bytes)	6	11	18	27	42
128	16-bit connections	(512	bytes)	5	9	15	22	35
256	16-bit connections	(1K	bytes)	4	7	11	16	26
512	16-bit connections	(2K	bytes)	3	5	8	12	18
1024	16-bit connections	(4K	bytes)	2	3	5	8	12
2048	16-bit connections	(8K	bytes)	1	2	3	4	7
4096	16-bit connections	(16K	bytes)	-	1	1	2	4

Table 2: Packing Density Investigation of the ***generic neuron*** architecture.

be noted that the above figures have been originated considering the RAM cells generated by the software package used. If state-of-the-art RAM cells had been utilised, the number of PEs per chip would have increased substantially.

Additionally, these evaluations have been carried out using the Back Propagation prototype example, with learning procedure on-chip. For simpler models or recall phase only applications, the maximum number of integrated PEs can be increased even further, since the Back Propagation learning algorithm is quite demanding in terms of silicon area.

4. CONCLUSIONS

Artificial neural networks are progressing to the point where, for some specific applications, sequential computers will no longer be adequate for their computation. It is believed that massively parallel computers, composed of very simple, replicative processing elements are the answer to reach the natural capability of artificial neural networks[2]. Furthermore, neural network algorithms are continuously evolving, originating various neural programming environments that allow researchers to experiment with existent neural algorithms as well as develop new ones. There is, therefore, a need to integrate these two systems - neurocomputer and neural programming environment - to create an automatic route from the software environment to the manufacture of neuro-chips dedicated to the specified and tested application.

The ultimate goal of this research was to define a design framework for the automatic generation of dedicated neural network chips from a high level specification. This objective has been achieved by the definition of the ***generic neuron*** architectural framework. This configurable architecture can be tuned according to the user's definition, and then be automatically translated into ASICs by the silicon compiler under development[7]. The final application-specific neuro-chip is correct-by-construction and grants the required high performance for executing neural network models.

References

1. Alippi, C. and Nigri, M.E., "Hardware Requirements for Digital VLSI Implementation of Neural Networks," *Int. Joint Conf. on Neural Networks*, Singapore, November 18-21, 1991.
2. Atlas, L. and Suzuki, Y., "Digital Systems for Artificial Neural Networks," *IEEE Circuits and Dev. Magazine*, Nov. 1989.
3. Floyd, R.W. and Ullman, J.D., "The Compilation of Regular Expressions into Integrated Circuits," Report STAN-CS-80-798, Computer Science Department, Stanford University, April 1980.
4. Hammerstrom, D., "A VLSI Architecture for High-Performance, Low-Cost, On-Chip Learning," *Int. Joint Conf. on Neural Networks - IJCNN 90*, vol. II , pp. 537-544, San Diego, California, June 17-21, 1990.
5. Murray, A.F., Smith, A.V.W., and Butler, Z.F., "Bit-Serial Neural Networks," *Neural Information Processing Systems (Proc. 1987 NIPS Conf.)*, p. 573, Denver, November 1987.
6. Myers, D.J. and Brebner, G.E., "The Implementation of Hardware Neural Net Systems," *The First IEE Int. Conf. on Artificial Neural Networks* , pp. 57-61, October 16-18, 1989.
7. Nigri, M.E., Treleaven, P.C., and Vellasco, M.M.B.R., "Silicon Compilation of Neural Networks," *Proceedings of the IEEE CompEuro'91*, pp. 541-546, Bologna, Italy, May 13-16, 1991.
8. Rumelhart, D.E. and McClelland, J.L., "Parallel Distributed Processing: Explorations in the Microstructure of Cognition," in *MIT Press, Cambridge, Mass.*, vol. 1 & 2, 1986.
9. Vellasco, M.M.B.R. and Treleaven, P.C., "A Neurocomputer Exploiting Silicon Compilation," *Proc. Neural Computing Meeting, The Institute of Physics*, pp. 163-170, London, April 1989.
10. Vellasco, M.M.B.R., "A VLSI Architecture for Neural Network Chips," PhD Thesis, Dept. Computer Science, University College London, University of London, February 1992.
11. Vellasco, M.M.B.R., "A VLSI Architecture for the Automatic Generation of Neuro-Chips," *Int. Joint Conf. on Neural Networks - IJCNN'92*, Beijing, China, November 3-6, 1992.
12. Yasunaga, M.et al, "Design, Fabricatioon and Evaluation of a 5-inch Wafer Scale Neural Network LSI Composed of 576 Digital Neurons," *Int. Joint Conf. on Neural Networks - IJCNN 90*, vol. II, San Diego, June 17-21, 1990.

A RISC ARCHITECTURE TO SUPPORT NEURAL NET SIMULATION†

Marco Pacheco

Departamento de Engenharia Elétrica
Pontifícia Universidade Católica do Rio de Janeiro
Rua Marques S. Vicente 225, Gávea
22453 - RJ - Rio de Janeiro, Brazil

Philip Treleaven

Department of Computer Science
University College London
Gower St.
London WC1E 6BT, UK

***Abstract*—The Neural-RISC architecture consists of a primitive microprocessor and a parallel architecture, designed to optimise the computation of neural network models. The Neural-RISC system architecture consists of linear arrays of microprocessors connected in rings. Rings end up in an interconnecting module forming a cluster. Clusters of rings are arranged in different point-to-point topologies and are controlled by a host computer. The Neural-RISC node architecture comprises a 16-bit reduced instruction-set processor, a communication unit, and local memory—all integrated into the same silicon die. A VLSI prototype chip was implemented to demonstrate the system and node architecture. Using the standard 2μ CMOS technology, the chip integrates an array of two Neural-RISC microprocessors. This paper discusses the Neural-RISC design issues, presents a system overview and describes the VLSI implementation.**

1. Introduction

Many neurocomputer designs have already been implemented, some of them are available as commercial products[9]. The design approaches are depicted in Figure 1. They include: simulators using conventional computers, accelerator boards, processor arrays, and dedicated hardware implementations of a specific neural network model. The choice of available neurocomputers varies significantly in performance and flexibility[4]. Neurocomputers range from flexible, general-purpose systems to high performance, special-purpose systems. An optimal neurocomputer, if such an architecture exists, lies at some point between the parameters of flexibility and performance.

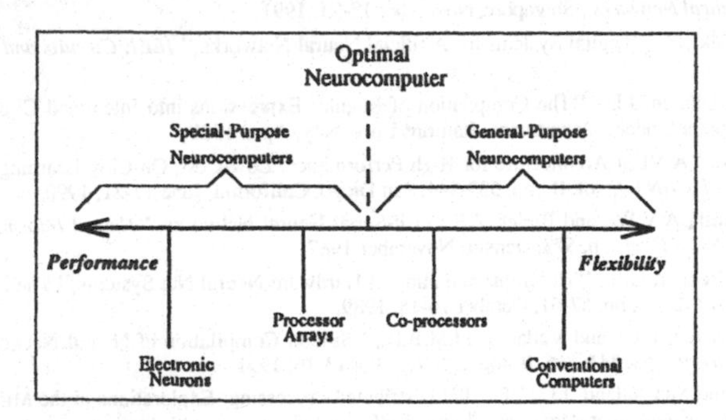

Figure 1: Spectrum of Neurocomputer Architectures

Our research is directed toward the development of such optimal neurocomputer architecture. To achieve this objective, we have designed a highly parallel architecture which combines the following issues:

†This work was supported by: CAPES and CNPq, Research Funding Agencies, Brazil; the British Council; LABO Eletrônica; and SID Informática, Brazil.

• ***Parallelism*:**
The neurocomputer architecture must employ real parallelism to exploit the distributed nature of artificial neural networks and to attempt the high performance required by real-world applications.

• ***Performance*:**
Overall performance must also be increased by improving the speed of individual processing elements and the bandwidth of data communication structures.

• ***Flexibility*:**
To support a range of neural network models and to provide application portability, the neurocomputer must be programmable, both in terms of processing units and node connections.

• ***Cost*:**
The neurocomputer must achieve practical and cost-effective implementation using current VLSI packing density and interconnection technologies.

• ***Scalability*:**
The neurocomputer architecture must scale properly in terms of VLSI design, performance and cost, so that it can be expanded into larger systems, according to the application requirements.

2. Neural-RISC Architecture

The Neural-RISC architecture comprises two basic building structures: Neural-RISC rings and clusters of rings. A Neural-RISC ring is a linear array of identical microprocessors (Neural-RISC nodes) connected in a ring topology. Variable-size rings end up in a multi-ring interconnect module, forming a cluster. The interconnect module acts as a communications server supporting inter-ring and inter-cluster message routing. Clusters can be arranged in diverse point-to-point topologies to achieve the optimal system size and configuration. A system composed of a number of clusters includes a host computer as the master controller (Figure 2). The host, which typically consists of a workstation, supports network initialisation and provides programming facilities. Each ring is crossed by two communication channels, in opposite directions to reduce the distance covered by message packets. During operation, messages in the form of packets, are transferred between nodes that are logically connected. Messages are of variable length, prefixed with the address of the destination node. The system can address messages from any node to any other particular node, to a group of nodes (corresponding to a layer or a cluster, as in neural network models), to all nodes, or to the host. Downloading simple programs into each processor node configures the neurocomputer. The architecture can be scaled up to 65,536 nodes to meet the application requirements and to achieve an optimal balance between performance and implementation cost.

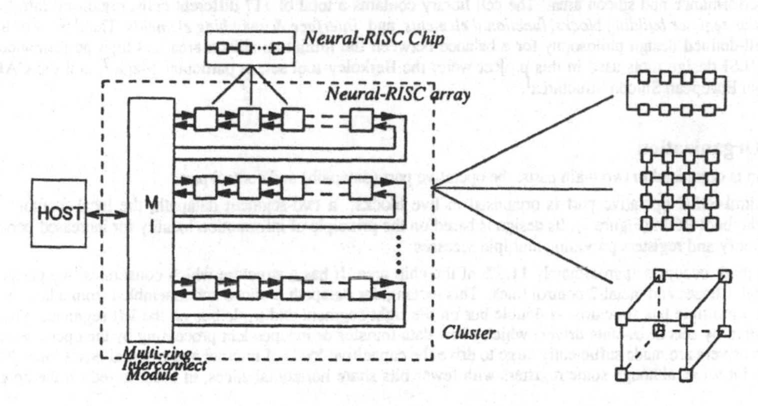

Figure 2: Neural-RISC Architecture

The Neural-RISC architecture exhibits a compromise between wirability and performance, given the current constraints of the VLSI technology. As a homogeneous parallel system, the architecture can be expanded to match application requirements. Larger systems can be implemented by growing the architecture in the number of chips (or nodes per chip), number of rings and in the number of clusters. The method by which to scale the architecture can be chosen to reduce implementation cost, to optimise communication throughput, or to increase its capacity. An important characteristic of the Neural-RISC architecture is its ability to expand into larger systems, virtually, with no loss of performance. In the Neural-RISC architecture, the communication delay decreases as the number of rings increases, as it can be seem by its network average distance (number of processors or links a message must cross, to reach the destination node)[2]. For a Neural-RISC cluster, composed of *N* processors, evenly distributed across *R* rings ($R>1$), the average distance (Ad_{NR}), is found to be:

$Ad_{NR}=(N+R)/2R$. For instance, using clusters with 128 rings, the average distance is under 33 links for systems with up 4096 processors[6].

The Neural-RISC node architecture combines performance and flexibility, by integrating computation, communication and storage capabilities into a single node design, which is clustered in many per chip. The node architecture consists of a self-contained microprocessor composed of a 16-bit RISC processor, a communication unit, and a local memory. The processor adopts a reduced instruction set approach (16 instructions); an expanding opcode branches into 14 (no-operand) memory-mapped instructions. Single-cycle execution is achieved for most of the instructions by employing a simple arithmetic pipeline structure and a two-phase, nonoverlapping clock. The processor also includes a programmable timer and an interrupt controller: the timer is a 16-bit resolution count register to support timed applications; the interrupt system supports five (maskable) interrupt requests, connected in daisy chain, to attend the communication unit, the timer and the initial synchronisation of the network[6].

The communication unit implements a simple network, organised as a linear-array with effectively point-to-point connections. Each processor in the array is connected to each neighbour by two 16-bit links: an input and an output link. A link employs two standard handshaking signals to clock data. Separated controllers allow bidirectional communication to occur simultaneously with neighbour processors. Packets are automatically forwarded to their destination using a very simple protocol. Two (variable-length) FIFO buffers provide asynchronous transfer of packets between the communication unit and the processor unit. The processor can access the buffers at any cycle, using any of the memory access instructions. In the Neural-RISC node architecture, processing and communication completely overlap thus relieving the I/O bottleneck frequently seen in parallel systems.

The local memory is composed of four blocks: a RAM memory for instructions and data, two data RAM blocks to implement the I/O buffers of the communication unit, and a boot-strapping ROM.

3. VLSI Implementation

For the VLSI implementation, the primary objective was to investigate the hardware complexity of the Neural-RISC node architecture. The node complexity is the chief measurement to determine the degree of real parallelism a neurocomputer can achieve in practical implementations. In this respect, *node area* and *packing density* form the essential criteria for assessment.

In this project, we have considered several design methodologies and have carefully evaluated their influence in all design attributes. Design effort has been concentrated on the requirements of minimum silicon area and high performance. To this end, we have designed a datapath cell library, which emphasizes modularity and flexibility, and attempts a rational balance for performance and silicon area. The cell library contains a total of 117 different cells, organised into three main classes: *modular register building blocks, functional elements*, and *interface & switching elements*. The design of these cells follows a well-defined design philosophy for a balance between the minimum silicon area and high performance requirements. The VLSI design tools used in this project were: the Berkeley tool set, in particular Magic[7], and the CAD system Solo 2000 from European Silicon Structures[1].

3.1. Chip Organization

The chip is organized in two main parts: the operative part (datapath) and control part.

The Neural-RISC operative part is organised in five blocks: a two-segment datapath, the local memory, two I/O buffers, and the boot ROM (Figure 3). Its design is based on the principle of information locality for increased performance, with local memory and registers providing multiple accesses.

The datapath occupies approximately 11.3% of the chip area. It has a structure which constitutes two perfect rectangles, with metal-1 buses and metal-2 control lines. This rectangular datapath is sliced and assembled from a large number of small cells on a multiple bus structure—a double bus on the right segment and triple bus on the left segment. The bus system is interrupted by sets of tri-state drivers which allow data transfer or independent processing by the operative sub-parts. These tri-state drivers are made sufficiently large to drive the capacitive loads of internal and external buses lines. The whole datapath is 16-bit wide, although some registers with fewer bits share horizontal slices, in order to reduce the height of the datapath.

Functionally, the datapath is composed of three operative sub-parts:

- Data processing
- I/O buffer management
- Communication processing

The data processing sub-part includes the execution unit and the timer (Figure 3). At the the right end of the first datapath segment are the registers and counters used for I/O buffer management. The second datapath segment (right side) contains the hardware (registers, latches, comparators, etc.) that implements the data ports and the input arbiter of the communication unit.

The control part of the Neural-RISC takes less than 5% of the prototype chip area and it is implemented with three PLAs: the processor control PLA and two communication channel PLAs. Each PLA synthesizes a "nondeterministic" finite state machine[3], in the sense that a machine can be in more than one state at the same time. The PLA implementation of

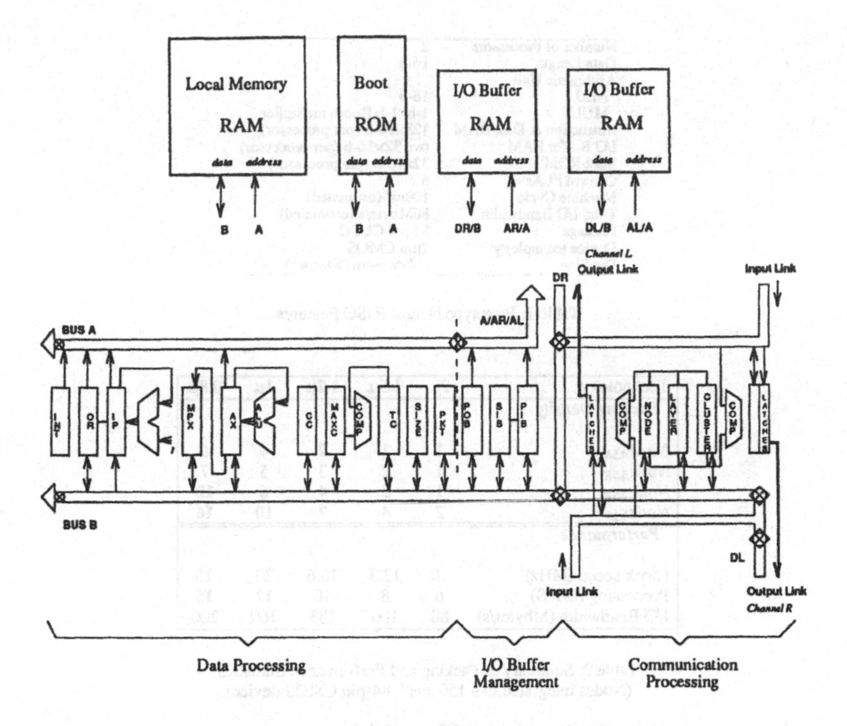

Figure 3: Organisation of the Neural-RISC Operative Part

nondeterministic finite automata[8] provides a considerable reduction of silicon area, in contrast with conventional implementations[5].

The system timing is based on two-phase nonoverlapping clock signals (Φ_1 and Φ_2), generated directly by an external oscillator. Control operations are executed in a scheme where each clock phase corresponds to a conventional machine cycle. This strategy accelerates the instruction flow, enabling instructions to execute in fewer clock cycles.

4. Implementation Results

The preliminary results from the Neural-RISC VLSI implementation were very encouraging. Despite the use of 2μ CMOS processing technology, the Neural-RISC prototype chip integrates two complete nodes in an 8.7×9.6mm die size. The chip is expected to operate at about 6 MIPS, with an I/O bandwidth of 80Mbytes/s, driven by a 10MHz clock. In this chip, local memory was reduced to allow for a multiple-node prototype, which could be used for statistical purposes. The Neural-RISC features for this implementation are summarised in Table 1. The chip contains about 60K transistors, 891 standard cells, and 331 customised cells. The total number of 84 pins includes 64 I/O data pins.

Prototyping was essential for the architecture development. The VLSI prototype provided us realistic feedback about the constraints of the hardware implementation and allowed us to evaluate reimplementations of the Neural-RISC chip with denser technologies.

To assess the VLSI implementation, we have performed a rigorous statistical analysis in order to determine the packing and performance capabilities of the Neural-RISC architecture at *chip* level. In this analysis, we took node examples with balanced configurations for local memory: from the minimum workable local memory ($Node_{1.25K}$ = 1.25Kx16-b), to the maximum implementable instruction&data memory, ($Node_{4.5K}$ = 4.5Kx16-b). Table 2 summarises the packing and performance statistics of the Neural-RISC chip.

As shown by these results, $Node_{1.25K}$ yields good packing density in most of the technologies, except 2μ. On the other hand, $Node_{4.5K}$ only achieves good density (6 nodes) with the 0.8μ process. In this case, a single chip can integrate six 16-bit processors, each with its own 4-link parallel communication unit and 9K bytes of local memory. Overall, low packing density is obtained for processing technologies from 2μ to 1.2μ, while a satisfactory number of integrated nodes is

Number of Processors	2
Data Length	16-b
Arithmetic Unit	
ALU	16-b
MULT	1-bx1-b Booth multiplier
Instruction & Data RAM	128x16-b (per processor)
I/O Buffer RAM	two 32x16-b (per processor)
Boot ROM	32x16-b (per processor)
Control PLAs	6
Machine Cycle	100ns (estimated)
Total I/O Bandwidth	80Mbyte/s (estimated)
Package	84 pin CLCC
Device technology	2μm CMOS
Die Size	8.7x9.6mm ($84mm^2$)

Table 1: Prototype Neural-RISC Features

Technology	2μ	1.5μ	1.2μ	1μ	0.8μ
Packing Density					
$Node_{4.5K}$	1	1	2	4	6
$Node_{3.5K}$	1	2	3	5	7
$Node_{2.5K}$	1	2	4	6	10
$Node_{1.25K}$	2	4	7	10	16
Performance					
Clock speed (MHz)	10	13.3	16.6	20	25
Processing (MIPS)	6	8	10	12	15
I/O Bandwidth (Mbytes/s)	80	106	133	160	200

Table 2: Summary of Packing and Performance Statistics
(Nodes integrated in a 150 mm^2, 84-pin CMOS device)

achieved for the current most advanced technologies (i.e., 1.0μ and 0.8μ).

It should be noted that, if a state-of-art RAM had been used in this analysis, the maximum number of processors per chip would have increased significantly. The reason is that local memory occupies between 62% ($Node_{1.25K}$) and 85% ($Node_{4.5K}$) of the total silicon area.

The performance of each Neural-RISC chip is presented in terms of its processing and communication efficiency in Table 2. The clock rate (expressed in mega hertz) indicates the chip operating speed, while processing performance (in MIPS) is the estimated performance of each integrated processor node. The I/O bandwidth represents the total bidirectional bandwidth of a Neural-RISC processor through its four 16-bit links.

5. Conclusions

The primary goal of the Neural-RISC project was to investigate a RISC microprocessor and a parallel architecture for building a general-purpose neurocomputer. The motivation for the Neural-RISC node and system architectures was to develop a cost-effective hardware, suitable for supporting a range of neural network algorithms and to work cooperatively with existing neural network programming systems. The architecture objectives were parallelism, flexibility, performance, low cost, scalability, and the ease of system integration. Characteristically, some of these objectives often conflict. The Neural-RISC architecture is the result of a set of rational compromises that were made in attaining these objectives.

The Neural-RISC microprocessor approaches the processing and communication speeds necessary for solving large applications, while retaining the flexibility required for supporting neural network environments. This flexibility yields benefits in the execution efficiency of neural models and in the density achievable in large systems. The parsimony of the Neural-RISC design allows, for instance, 1024 processor nodes, each provided with 9K bytes of local memory, to be packed onto two Sun VME cards. Such a system can deliver 15 billion instruction per second peak performance, competitive with many current neurocomputers on a broad variety of neural network applications.

The VLSI implementation demonstrated the system and node architectures, and allowed us to assess the hardware requirements and performance of neurocomputer systems based on the Neural-RISC architecture. Considering the limited experience and manpower available for the VLSI design, much of the design effort was directed towards attaining a deliverable prototype in due course. We therefore believe that the next chip will be much closer to our goals.

References

1. ES2-European Silicon Structures, "SOLO 2000," Product Information, ES2, Germering, West Germany, 1987.
2. Agrawal, D.P., Janakiram, V.K., and Pathak, G.C., "Evaluating the performance of multi-computer configurations," *IEEE Computer*, no. 19, pp. 23-37, May 1986.
3. Floyd, R.W. and Ullman, J.D., "The Compilation of Regular Expressions into Integrated Circuits," Report STAN-CS-80-798, Computer Science Department, Stanford University, April 1980.
4. Hecht-Nielsen, R., "Performance Limits of Optical, Electro-Optical, and Electronic Neurocomputers," *Optical and Hybrid Computing SPIE*, vol. 634, pp. 277-306, 1986.
5. Obrebska, M., "Efficiency and Performance Comparison of Different Design Methodologies for Control Parts of Microprocessors," *Microprocessing and Microprogramming*, vol. 10, pp. 163-178, North-Holland, 1982.
6. Pacheco, M. and Treleaven, P., "Neural-RISC: A Processor and Parallel Architecture for Neural Networks," *IJCNN'92 - Int. Joint Conf. on Neural Networks*, Beijing, 3-6 November 1992.
7. Scott, W.S., Mayo, R.N., Hamachi, G., and Ousterhout, J.K., *1986 VLSI Tools: Still More by the Original Artists*, Computer Science Division, EECS Dep., University of California at Berkeley, 1986.
8. Silva, H.T., "Estruturas Regulares para Controle em Circuitos Integrados," Msc Thesis, COPPE - Universidade Federal do Rio de Janeiro, 1985.
9. Treleaven, P., Pacheco, M., and Vellasco, M., "VLSI Architectures for Neural Networks," *IEEE Micro*, vol. 9, no. 6, pp. 8-27, December 1989.

HARDWARE DESIGN FOR SELF ORGANIZING FEATURE MAPS WITH BINARY INPUT VECTORS

S. Rüping, U. Rückert and K. Goser

University of Dortmund
Dept. of Electrical Engineering
P.O. Box 500500
W-4600 Dortmund 50
Germany

Abstract:

A number of applications of self organizing feature maps require a powerful hardware. The algorithm of SOFMs contains multiplications, which need a large chip area for fast implementation in hardware. In this paper a resticted class of self organizing feature maps is investigated. Hardware aspects are the fundamental ideas for the restictions, so that the necessary chip area for each processor element in the map can be much smaller then before and more elements per chip can work in parallel. Binary input vectors, Manhatten Distance and a special treatment of the adaptation factor allow an efficient implementation. A hardware design using this algorithm is presented. VHDL simulations show a performance of 25600 MCPS (Million Connections Per Second) during the recall phase and 1500 MCUPS (Million Connections Updates Per Second) during the learning phase for a 50 by 50 map. A first standard cell layout containing 16 processor elements and full custom designs for the most important parts are presented.

1. Introduction

Self organizing feature maps as described in [1] have applications in the fields of data analysis and pattern recognition. Especially for real time application a fast and efficient hardware is needed. Well known "von Neumann computers" work serially on each element of a self organizing feature map, so that they have to be very fast and can work only on rather small maps when time is the critical aspect.
A speedup is possible when a calculation unit is available for each element of the map or a special coprocessor is used. There are a number of hardware designs known in literature [e.g. 2, 3, 4]. Due to the mathematical operations which are necessary for the algorithm of self organizing feature maps, calculation units are rather complex and need large chip area. For increasing map size multiplexing of the available units is required to restrict the number of chips.
The basic idea of this paper is to look at a special class of self organizing feature maps. The aim is to get a most simplified hardware by making restrictions on the algorithm and the used operations. For example binary input vectors and Manhatten distance make it possible to calculate the distance between two vectors without multiplication. Chapter 2 will describe all the simplifications according to hardware aspects which are made for the presented circuits.
The investigation leeds to a hardware design that is described in chapter 3 and 4. Simulations using the hardware description language VHDL result in numbers for the performance of the hardware. This is discussed in chapter 5. Different layouts are already designed. First there is a standard cell layout containing 16 processor units for self organizing feature maps and second there are full custom layouts for the most important parts of the unit. Chapter 7 will close the paper dicussing the results.

2. Simplifications according to hardware

The algorithm of self organizing feature maps is summarized in equation 1.1. The calculation of the weight vector for the time t+1 is made with the old weight vector w(t), the distance between the old weight vector and the input vector and an adaptation factor alpha.

$$\vec{w}(t+1) = \vec{w}(t) + \alpha(t,x,y)\cdot(\vec{x}(t) - \vec{w}(t))$$

$\vec{w}(t+1);\ \vec{w}(t+1)$	:	new and old weight vector
$\alpha(t,x,y)$	:	adaptation factor
t	:	time
x,y	:	coordinates of the element

(1.1)

The distance between the weight vector and the input vector is very often defined by the euclidean distance which is calculated by

$$(\vec{x} - \vec{w}) = \sqrt{\sum_i (x_i - w_i)^2} \tag{1.2}$$

Due to the difficulties to realize the square root in hardware and to the fact that not the absolute value is important but the relation between all the distances on the map, most implementations use the distance

$$(\vec{x} - \vec{w}) = \sum_i (x_i - w_i)^2 \tag{1.3}$$

where the square root is not used. But nevertheless multiplications are needed. A distance which is very simple to implement in hardware is shown in equation 1.4. It is called the Manhattan distance and can be realized by a modified adder and a register.

$$(\vec{x} - \vec{w}) = \sum_i |x_i - w_i| \tag{1.4}$$

Further multiplications are necessary for the product of the adaptation factor alpha and the calculated distance. A restriction to discrete values of alpha can simplify this multiplication. Alpha has a value between 0 and 1. If only the values shown in 1.5 are allowed, the operation can be handled by a shifter.

$$\text{alpha} \in \left\{1, \frac{1}{2}, \frac{1}{4}, \frac{1}{8}, \frac{1}{16}, \frac{1}{32}, \ldots\right\} \tag{1.5}$$

A multiplication with 1/2 is the same as shifting one bit to the right. Which means the restricted class of self organizing feature maps using the Manhattan distance and discrete values for alpha can be designed in hardware without using multipiers.
Another simplification that decreases the number of necessary I/O ports and the chip area can be done by restricting the components of the input vector to be binary.

$$\vec{x} = (x_1, x_2, \ldots, x_n)^T \qquad x_i \in \{0,1\} \quad \text{input vector}$$
$$\vec{w} = (w_1, w_2, \ldots, w_n)^T \qquad w_i \in \{0,1,2,3,\ldots,2^4 - 1\} \quad \text{weight vector} \tag{1.6}$$

The precision of the weights is set to 4 bits. First simulations show that this seems to be a reasonable value but further investigations have to be done in this field. For the definition of the distance equation 1.7 is given. The bit of the inputvector codes the minimum and the maximum of the possible weight values.

$$|\vec{x} - \vec{w}| = \sum_{i=1}^{n} |x_i \cdot (2^4 - 1) - w_i| \tag{1.7}$$

3. Hardware design

The restrictions described in chapter 2 allow a simple hardware design with very few ports between the units. Figure 1 shows an example of a 2 by 2 map, which can easily be extended to larger map sizes. Each unit, in Figure 1 called PE (Processor Element), has a port for the row, the column, the clock and data. A three bit command bus sends the instructions to the units.

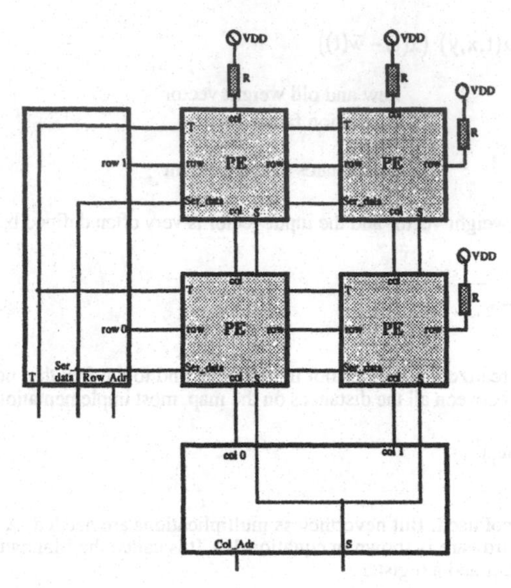

Fig. 1: Example for a 2 by 2 map

Each column and row line is connected to a seperate pull up resistor. The unit's outputs can be 'low' or can have a high impedance. So it is possible to realize a 'wired or' and use this for the minimum search. After calculation of the distance all units start to count down the distance register (see chapter 4). The unit, which is the first to be zero, put a low voltage on it's row and column line. At this time the control logic knows the minimum unit and it's position. If there are more than one unit that are the minimum at the same time, a request to the elements can solve the problem.
All units are designed in the same way. Therefore the system can easily be extended by adding new rows and columns to the map, when the control logic has the capability to handel the number of rows and columns.

4. Internal structure

The internal structure of the units is shown in Figure 2. The connections of the unit described in chapter 3 can be found above the controller. The main blocks are the calculation block, the weights, sum and alpha register and the controller. The weight register stores the weight vector and presents a component per clock cycle to the calculation block. The calculation block produces the distance of the corresponding input and weight components and add it to the sum register. During the learning phase it calculates the new weight value using the bits in the alpha register. All this is done in one clock cycle.
The input vector is send to all units via the data line. One component per clock cycle is handled and all units work in parallel. In order to decrease the chip area the input vector components are not stored in the units. They must be sent again during the learning phase. Investigations have shown that this is no disadvantage. Due to the binary type of the input vectors it is possible to send these bits as fast as the unit could read them from an internal register.
To keep the design most flexible, the adaptation factor is send before each adaptation step. That makes it possible to control the adaptation function and neighborhood size without restrictions. On the other hand all units with the same alpha can work in parallel during the adaptation step. This will be explained in more detail in chapter 5.

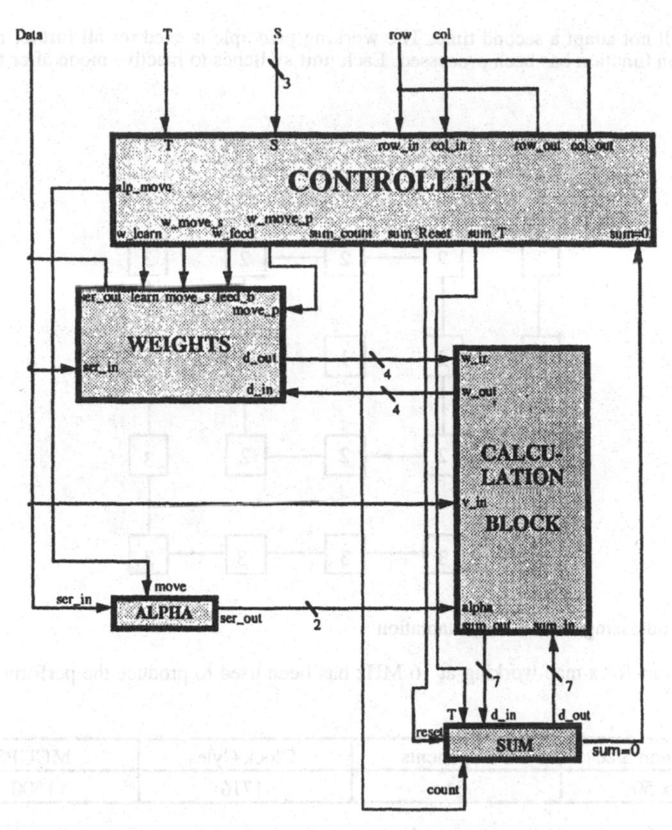

Fig. 2: The internal structure of the unit

5. Performance of the system

As mentioned before the unit is capable of processing one component per clock cycle, which means the system needs 64 clock cycles for calculating all distances on a map for an input vector with 64 components. The number of clock cycles to find the minimum depents on the minimum distance. The distance might be small for an already learned map, so for this example an average value of 36 is taken, which leads to a result of 100 clock cycles for each input vector finding the position of the most similar element. Figure 3 lists the performance for a 50 by 50 map working at 16 MHz.

Mapsize	Components	Clock Cycles	MCPS
50 x 50	64	~ 100	25600

Fig. 3: Performance during recall phase

During the recall phase all units can work in parallel. This is not possible during learning phase, when only units with the same value of alpha can calculate the new weight vectors at the same time. Figure 4 explains the process. First the minimum element adapts it's weights and switches into inactive mode. The element was addressed by the row and the column lines. Then a field of 3 by 3 units is addressed with 3 row and 3 column lines. This field can now adapt in parallel while the central unit (the minimum) is

inactive and will not adapt a second time. The working principle is used for all further rings until the whole adaptation function has been processed. Each unit switches to inactive mode after the adaptation step.

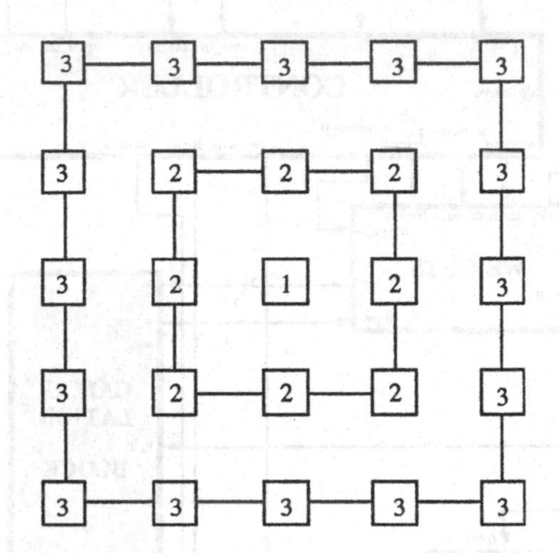

Fig. 4: Addressing the units for adaptation

VHDL simulations for a map working at 16 MHz has been used to produce the performance numbers listed in Figure 5.

Adaptation Size	Components	Clock Cyles	MCUPS
50 x 50	64	1716	1500

Fig. 5: Performance during learning phase

6. Layout of the unit

A 1.5 um CMOS technology has been used to design the different parts of the proposed unit in full custom layout. The calculation block has a size of about 0.15 mm^2. The weight register with a capacity of 256 bits (64x4) needs about 1.85 mm^2. Both are manufactured at the moment and will be tested soon. A 10 by 10 mm chip could contain about 25 units, where the majority of the area is used by the on chip memory. Further versions of the units will be designed in 0.8 um CMOS technology, so that the number of units per chip will increase.
On the other hand a standard cell layout containing 16 processing units (each with 64 bits (8x4) memory) has been developed and is shown in Figure 6. The cells are based on a 2 um CMOS process. The chip has a size of 10 by 10 mm and works with 16 MHz clock rate.

7. Discussion

A hardware design for self organizing feature maps is presented in this paper. Only a restricted class of feature maps can be handle by the hardware, but with this restrictions it is possible to simplify the design and decrease the necessary chip area. It seems, that there are a number of applications where binary data is processed. On the other hand it might be possible to find an efficient coding for continous data. This would increase the number of applications.
The hardware is very simple to extend. A chip containing full custom designed units is expected to contain a much higher number of units than the standard cell layout. Therefore large map sizes with all elements working in parallel can be built.

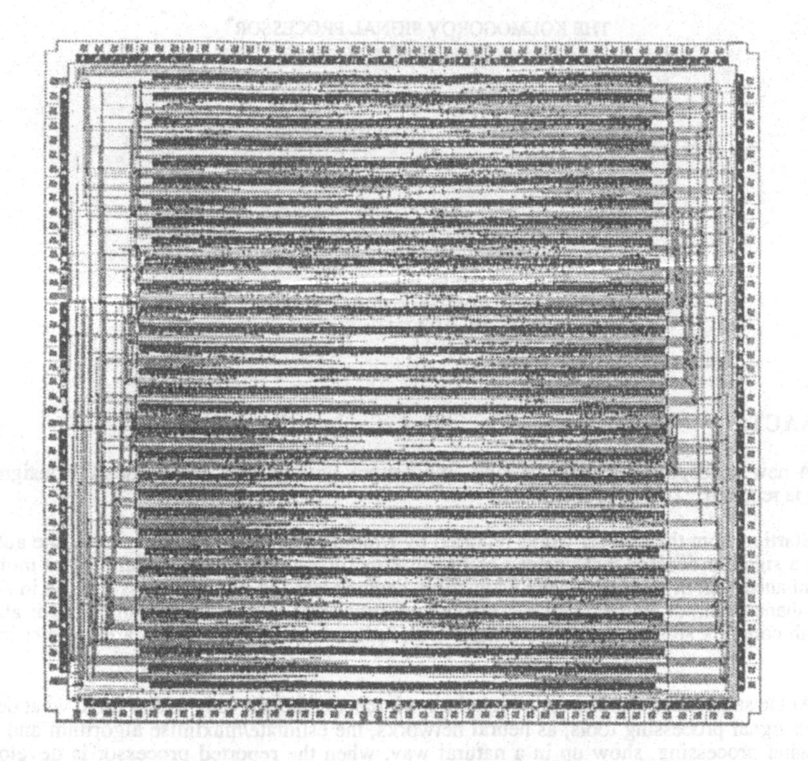

Fig. 6: Standard cell layout of 16 units (10x10 mm, 2 um CMOS)

A software for simulating the restricted class of self organizing feature maps is tested at the moment and will be used to investigate the necessary weight precision and other specifications of the maps.

8. Acknowledgement

This work has been partly supported by the german ministry of research and technology BMFT, contract number 01-IN 103 B/O.

9. References

[1] T. Kohonen. *Self-Organization and Associative Memory*. Springer Verlag Heidelberg New York Tokio, 1984

[2] D. Hammerstrom, N. Nguyen. *An Implementation of Kohonen's Self-Organizing Map on the Adaptive Solutions Neurocomputer*. Proceedings of Artificial Neural Networks 1991, pp. 715 - 720. North-Holland, 1991

[3] V. Tryba: *Selbstorganisierende Karten: Theorie, Anwendung und VLSI-Implementierung*. (in German). Dissertation an der Universität Dortmund, Abteilung Elektrotechnik. 1992

[4] U. Ramacher, U. Rückert, J.A. Nossek. Proceedings of the 2nd International Conference on Microelectronics for Neural Networks. Kyrill &Method Verlag, München, 1991

THE KOLMOGOROV SIGNAL PROCESSOR*

Miguel A. Lagunas, Ana Pérez-Neira, Montse Nájar, Alba Pagés

TSC Department, Mod. D5
ETSI Telecomunicacion
Apdo. 30002
08080 Barcelona
SPAIN

ABSTRACT:

A new concept of signal processing architecture, with general guide-lines for design and learning is reported.

Starting from the pioneer works reported by Kolmorogov [2] and J. Herault [4], the authors develop a signal processor that, instead of facing directly the problem of simultaneous multiple signal enhancement or multiparameter estimation, decomposes the signal processing task in single signal enhancement, or single parameter estimation, problems. The resulting framework is able to cope with complex signal processing designs with reduced complexity and in a more distributed manner.

At the same time, the reader, through the work reported herein, may realise up to what degree attractive signal processing tools, as neural networks, the estimate/maximise algorithm and high order signal processing, show up in a natural way, when the reported processor is developed, designed and non-supervised learning is considered.

I. INTRODUCTION

The problem of multiple component signal scenarios or multiparameter estimation is the most important issue of statistical signal processing research. Simultaneous waveform decomposition and/or parameter estimation still remain as an unsolved problem in a general and well supported manner.

Several approaches, apparently disconnected among them, have been reported. To be more specific, the EM (estimate and maximise) algorithm, the multicomponent signal analysis; the high order signal processing and a vast literature on neural networks for signal processing attempt, and sometimes arise, to spectacular solutions to many problems, which are difficult to face from a classical signal processing approach.

We will try to show that all of them are strongly related trough the fundamental work of one of the best mathematicians of our century. In 1957 Andrei Kolmogorov proved the existence of a solution for the problem of finding a continuous function which produces a mapping from n inputs to m outputs using only functions of a single variable. The theorem describes the basic architecture to implement the mentioned function, which basically consist on two processing stages.

Regardless the Kolmogorov's theorem had been emerged in signal processing from the neural network literature, it will be shown that somehow the two stages of the EM algorithm are

---

* This work was done under Basic Esprit ATHOS and has been supported by the National Research Plan of Spain, CICYT, Grant number TIC92-0800-C05-05.

closely related with the two stages mentioned above. Also, the so-called independent component analysis can be viewed as a particular case of the Kolmogorov's architecture. Finally, the high order processing reveals to be the basic tool for the learning (adaptive version) of the reported processor.

The scope of applications of the resulting signal processor framework is quite vast; nevertheless, and due to our limited experience, this paper will focuss two basic applications. The problems selected were the multiple line enhancement and tracking, which is of great interest in many spectral estimation problems and of capital importance in co-channel interference processing in communication systems. The second application selected is the source separation in narrowband array processing. This second application will attempt to solve the problem of simultaneous DOA estimation and source beamforming in a multiple signal scenario. Further work is currently devoted to the case of broadband processing and non linear filtering

II THE EXISTENCE THEOREM

The theorem we use to start with this presentation was due to Andrei Kolmogorov [2], solving the problem 13 given by Hilbert, and can be summarised as follows: Given any continuous function g(x)=y, where x and y are vectors with n and m components respectively, the function g(.) can be implemented exactly by two stages. The first stage produces 2n+1 outputs with the same number of processors, and the second stage produces the desired m components of the output vector.

Being x_q the components of vector x, the first stage provides 2n+1 components x'_q with the following transfer functions:

$$x'_q = \sum_{s=1}^{n} \lambda^q . \Phi(x_s+q\,\varepsilon)+ q \; ; q=1,2n+1 \qquad (1)$$

where λ and $\Phi(.)$ do not depend on g(.), but they do on the number of components m of the output vector y; and ε is an arbitrary chosen positive value.

The second stage contains the m functions $\varphi_q(.)$ that provide the y-components.

$$y_q = \sum_{s=1}^{2n+1} \varphi_q (x'_s) \; ; q=1,m \qquad (2)$$

This description has been obtained from R. Hecht-Nielsen [1], and will be used as reference on this paper. A recommended reference on the proof of the theorem can be found in the work of Lorentz [7].

An illustration of the above described architecture can be viewed in figure 1. Note that the first stage has been represented as a single function for all the inputs, being inherent the dependence of their output on the branch index

One detail of capital importance in exploiting the theorem resides in the stated independence of function $\Phi(.)$ on the specific g(.) to be implemented.

Even assuming such independence, some details deserve our attention at this moment. First, note that ε plays the role of a perturbation over the corresponding input component; and, second note that, basically, this architecture converts the multivariate problem in an univariate one since all the functions depend on a single variable. This second point is the most important feature to retain in our work, which reveals that finding an optimum over many variables can be reduced to a single variable problem.

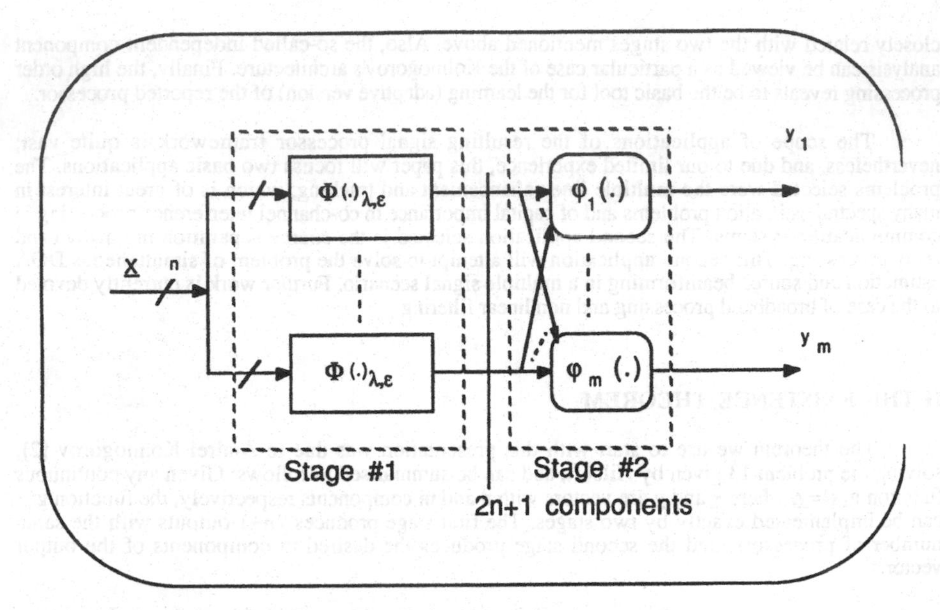

Figure 1. The Kolmogorov signal processor.

With respect the perturbation qε, we will see that, allowing a less restrictive formulation, the perturbation is highly dependent on the number of outputs. Furthermore, the design of this perturbation will be done under the so-called inhibition stage.

Again the problem of designing $\Phi(.)$,λ and ε independent of the specific function to be implemented remains as the big unknown and astonishing part of the theorem. Looking at the first stage as a mapping of the input vector $\underline{x}$, by a set of generic functions $\Phi_i(\underline{x})$, into a set of 2n+1 variables, will enlarge our opportunity to arise to a more specific and practical contribution of the mathematical theorem. We will see, in this presentation that, the task we have assigned to the family of functions $\Phi_i(\underline{x})$ is almost the same and may be considered almost independent of the application under analysis; only parameters or arguments will be responsible of the differences that may occur between these functions. In summary, the basic transformation, involved in the theorem, which transforms input vector components x_q in components x'_q, will be as (3),

$$\Phi_i(\underline{x}) = \Phi(\underline{x}, \alpha_i) \qquad (3)$$

where α_i is a parameter to be found. A more detailed description of these functions will be provided in the next section.

Other important point concerning the first stage is that, from the input vector, the functions defined in (3) will provide, in general, 2n outputs that can be grouped in vectors of two components, as it is shown in (4).

$$\Phi(\underline{x}, \alpha_i) = \underline{x}'_i \quad ; \; i=1,n \qquad (4)$$

This formulation provides 2n outputs out of the 2n+1 stated in the theorem.

Concerning the second stage, and considering that the outputs of the functions $\varphi_m(x'_s)$ will provide the m components of vector $\underline{y}$, the second set of functions can be described as a new generalised function of vector $\underline{x}'$,

$$y_q = \phi_q(\underline{x}') = \sum_{s=1}^{2n} \phi_q(x'_s) \quad (5)$$

note that $\underline{x}'$ is a vector containing the 2n components of the n $\underline{x}'_i$ vectors which result from the first stage.

$$\underline{x}' = (\underline{x}'^H_1, \underline{x}'^H_2, \ldots\ldots\ldots, \underline{x}'^H_n) \quad (6)$$

Furthermore, assuming some involved constraints to the processor, y_q can be viewed as a function only of the corresponding $\underline{x}'_q$. This reduces the design complexity of the system, whenever this approach is possible.

$$y_q = \phi_q(\underline{x}'_q) \;;\; q=1,m \quad (7)$$

Note that, with this formulation, up to m equal n outputs can be implemented.

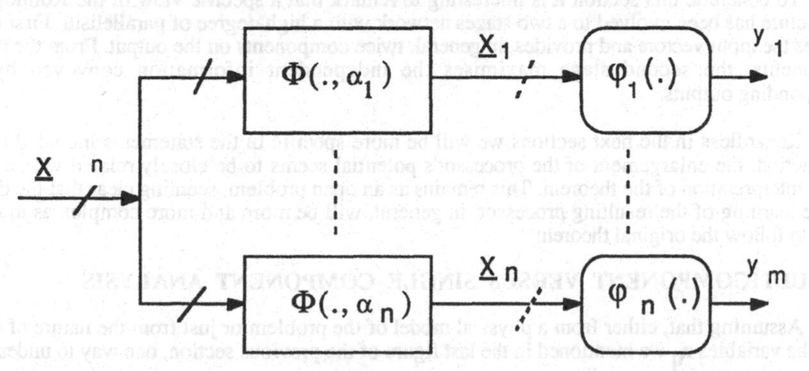

Figure 2. A specific implementation of the Kolmogorov signal processor

Looking carefully to the resulting scheme, depicted in figure 2, we assign to the first stage the task of isolating in the output vector,or scalar, the information required to obtain the output component y_q.

This architecture does not differ significantly, in the second stage, since it can be assumed that the final m-components of vector $\underline{y}$ can be derived from a weighted sum of the resulting outputs in figure 2. This new scheme is shown in figure 3, being the output components as in (8).

$$y_q = \sum_{s=1}^{n} \varphi_q^{new}(\underline{x}'_s) \quad ; \quad \varphi_q^{new}(\underline{x}'_s) = a_{qs} \cdot \varphi_s(\underline{x}'_s) \quad (8)$$

At this point, it is important to remark the role of the z_i which are like the set of independent variables able to describe all the information required to obtain the vector output $\underline{y}$. This is the reason why we will focuss our effort in the processor that from the input data $\underline{x}$ provides these independent variables $\underline{z}$. Whenever $\underline{z}$ is available, finding weights a_{qs} is an easy to solve problem, provided we know the desired $\underline{y}$.

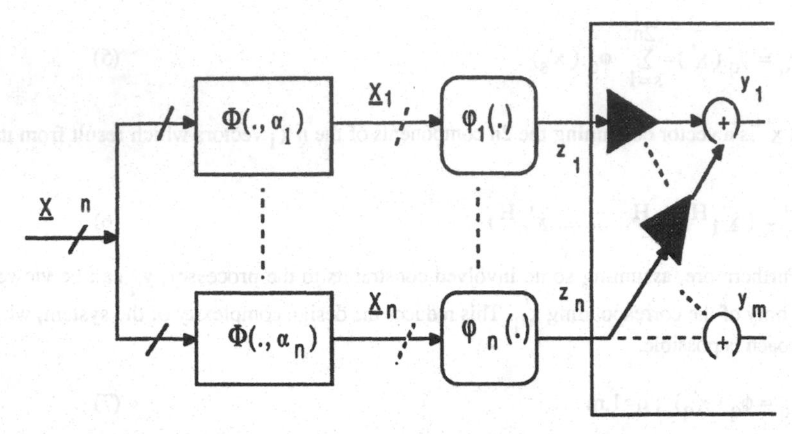

Figure 3. Further extensions of the processor shown in figure 2.

To conclude this section it is interesting to remark that a specific view of the Kolmogorov architecture has been evolved to a two stages network with a high degree of parallelism. First stage, receives the input vectors and provides, in general, twice components on the output. From the output components, the second stage maximises the independent information conveyed by the corresponding outputs.

Regardless in the next sections we will be more specific in the statements included in this introduction, the enlargement of the processor's potential seems to be closely related with a more formal interpretation of the theorem. This remains as an open problem, seeming clear that the design and the learning of the resulting processor, in general, will be more and more complex as much as we try to follow the original theorem.

III MULTICOMPONENT VERSUS SINGLE COMPONENT ANALYSIS

Assuming that, either from a physical model of the problem or just from the nature of it, we know the variables z_q we mentioned in the last figure of the previous section, one way to understand the role of the first stage is to convert the initial problem, of searching from the n variables the z_q's simultaneously, into one problem where each z_q is isolated. In other words, given $\underline{x}$ as a function of all the variables z_q q=1,m (m<=n), find the processor that produces at every branch the set of variables z_q by means of an adequate design of functions $\Phi(.,\alpha_q)$.

$$\underline{x} = g_c(\underline{z}) \;\text{-----}\!> x'_q = \rho(z_q) \qquad (9)$$

Let us take one example from the linear filtering case just to illustrate the previous statements. Being $\underline{x}$ a given data record of a given signal, and $\underline{y}$ the corresponding record of output signal data samples; looking for the independent set z_q, we will arise to the DFT of the original data record and the weights a_{is} will be derived directly from the DFT of the desired output signal data record.

$$\begin{aligned} x'_1 &= \text{DFT}(\underline{x})_{\text{beam } l} \\ z'_1 &= \varphi_1(x'_1) = \text{DFT}(\underline{y})_{\text{beam } l} \,/\, x'_1 \qquad (10) \\ a_{iq} &= \exp(-j\,2\,\pi\, i\, q\,/N) \end{aligned}$$

In summary, the independent components will be the frequency components of the input data record. Note that the first stage is independent of the output record, and the feasibility of the network. Also, the most important issue is that the first stage is just producing and independent variable z_q, without leakage from the other independent components z_i for $i \neq q$. When x is just a windowed function of a given signal (i.e. finite data length record), the problem of leakage removal motivates signal dependent processing associated with the $\Phi\ (\,.\,,\alpha_q)$, this makes them different but with the same role in the processor (i.e. to remove as much as possible the leakage between frequency bands).

A basic example of the power of the Kolmorov's architecture can be found in the problem of finding multiple local maxima of a function g(x). Being x_o and x_1 the local maxima and having an initial guess of them as x_{on}, x_{1n}; an iterative procedure will be to search for a function $\varphi(x,x_{1n})$ such that when x_{1n} coincides with x_1, the resulting function have a single maxima in x_o. In other words, the function $g_o(x)$ has a single maxima in x_o.

$$g(x).\ \varphi(x,x_1) = g_o(x) \tag{11}$$

This condition implies (12) for the desired function.

$$\frac{g(x)}{dx}.\ \varphi\ (x,x_1)\ + g(x)\frac{\varphi(x,x_1)}{dx} = \begin{cases} \neq 0 \text{ for } x \neq x_o \\ 0 \text{ for } x = x_o \end{cases} \tag{12}$$

The above application involves the basic procedure of the EM algorithm. In the maximise steep, an approximation from x_{on} and x_{1n} are found from functions $g_o(x)$ and $g_1(x)$. The resulting values are passed to functions $\varphi\ (x,x_{on})$ and $\varphi\ (x,x_{1n})$ to produce a new set of $g_i(x)$. Note the similarity of this procedure with the previous architectures in figure 4.

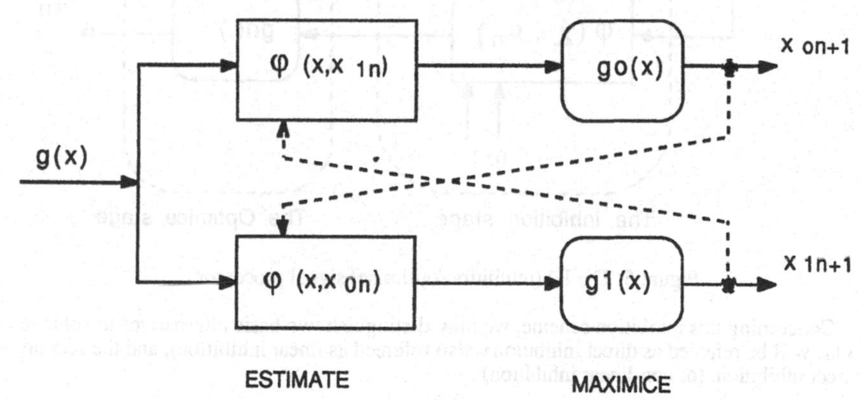

Figure 4. Structure to find multiple maxima of a function g(x). Note the similarity with previous figures and the feedback, inherent with the iterative character of the procedure.

We have shown in this section two examples of the potential of the two layers network involved in the Kolmogorov's theorem. Both examples will be used later on in the applications reported hereafter. It is important to remark that most of the problems envisaged for the design of this network resides in the first stage, since the second one is done under a classical procedure of maximising a function of a single variable. Due to the mentioned reasons, next section will be devoted to the design of the first processing layer. Also, it will be very important that the processing in the first stage, by means variable inhibition, converts the multivariate problem in a single variable one as it was indicated in the Kolmogorov's theorem.

IV THE INHIBITION LAYER

The basic feature assigned to the first layer, of our approach to the Kolmogorov's theorem, has been to convert the multivariate problem in a single variable one. To gain more insight in our objective, let us imagine that a set of narrowband array snapshots are the input to the processor; any array processing technique will end up with the direction of arrival of the impinging sources as the most valuable information about the signal scenario. At the first glance, the problem seems to be like a multiple source separation where Esprit, Music etc, could be used in order to find the sources' DOAs. Assuming only that two sources are present in a given scenario, our claim is that the first stage of the processor has to convert the problem into two problems of single source location, where both DOAs will be determined in an easier context that the initial one. Because this will be the objective in designing the first layer, we will refer to it as the inhibition layer; in the sense that every processor on this stage inhibits the effects of all variables except one, which has to be estimated on the second layer after it has been isolated from the others. In summary, the architecture we arise is still remembering the initial one, but much more defined than before.

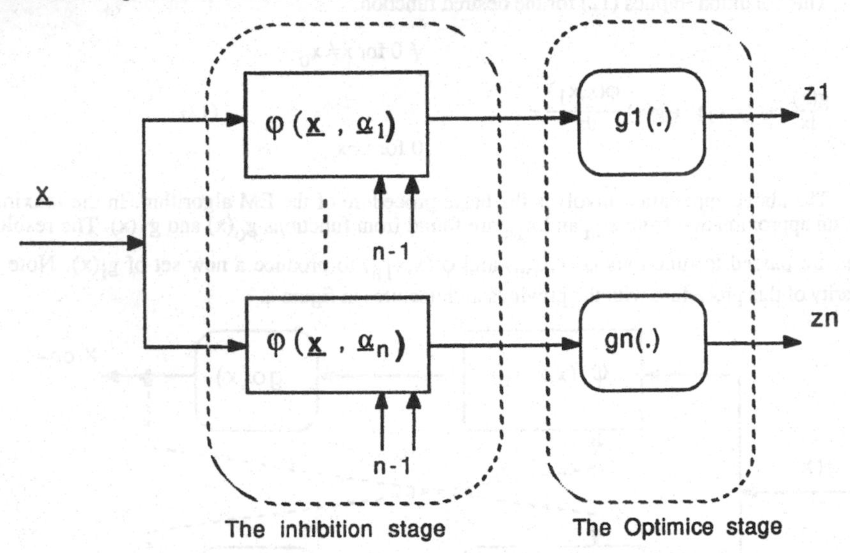

Figure 5. The IO (inhibition/optimise) signal processor.

Concerning this inhibition scheme, we may distinguish two basic alternatives to achieve our goal. One will be refereed as direct inhibition (also refereed as linear inhibition), and the second one as indirect inhibition. (or non-linear inhibition)

Direct inhibition could be named as linear inhibition in the sense that it is necessary that, either the z_i or a linear processing of them (y_i in figure 3) can be used to subtract their influence in other branches. To be more specific, under direct or linear inhibition, it is assumed that a linear combination of the z_i (i=1,n) i≠j allows the inhibition of their effects on branch j. In other words, by direct subtraction of the z_i's we obtain an input to the second layer processor of branch j that only depends on the specific z_j. The objective of the inhibition layer is given in (13);

$$\varphi(\underline{x},\alpha_j) = \underline{x} - L\,[\,z_1,\ldots,z_{j-1},z_{j+1},\ldots\ldots,z_n,\,\alpha_j] = \Omega\,(z_j) \qquad (13)$$

being $\Omega(.)$ a function of a single variable, which will be just the variable assigned to this branch.

Just to provide an easy example of direct inhibition, let us take a look of the scheme reported by J. Herault, C. Jutten and P. Comon [3] and [4]. Being the input of a function of time x(t), and, provided that it is a linear mix of two independent functions s_1 (t) and s_2(t),

$$E(s_1(t).\ s_2(\tau+t))=0 \text{ for any } \tau \quad (14)$$

the network of figure 6, with adequate learning is an example of the use of linear inhibition for signal separation. The learning to ensure convergence of the outputs to the actual components of x(t) will be briefly described hereafter, being strongly related with the independent character of the mixture.

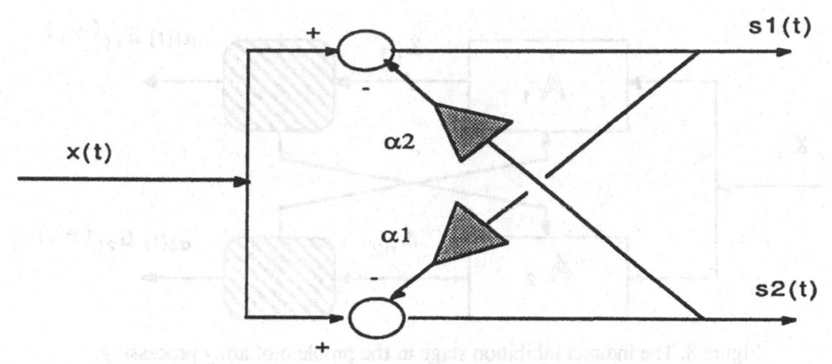

Figure 6. The network reported by Herault et al. for source separation as an example of direct inhibition.

Note that in this case,

$$\varphi(x, \alpha_1) = x - \alpha_1 s_2(t)$$
$$\varphi(x, \alpha_2) = x - \alpha_2 s_1(t) \quad (15)$$

and, in general, for the case of source separation,

$$\varphi(x, \alpha_i) = x - \underline{\alpha}_i^H \underline{S}_i$$
$$\underline{S}_i = (s_1, \ldots\ldots, s_{i-1}, s_{i+1}, \ldots\ldots, s_n) \quad (16)$$

The case of indirect inhibition needs to model the input data, being in consequence very dependent on the application. Unfortunately, regardless it could be more useful than direct inhibition, it is difficult to provide a general rule to asses the design of this kind of inhibition. We will use two examples we have experienced in the Kolmogorov processor with unexpected success. The first case concerns with line enhancement, and the second one is related with the narrowband array processing problem.

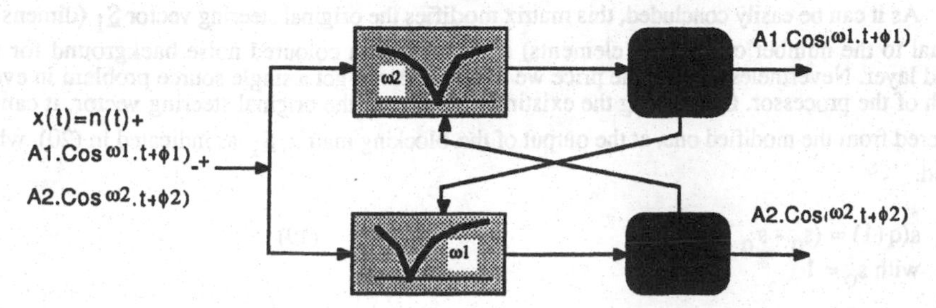

Figure 7. The indirect inhibition stage in the problem of multiple line separation and tracking.

The first case involves the problem of co-channel interference or collision in multiple access system by single carrier demand. In this case, the input signal is the addition of two sinusoids with frequencies w_o and w_1, under severe doppler and doppler rate conditions. Assuming that the estimates are accurate enough, despite the necessary learning rule, the indirect inhibition does not require the full waveform of the two lines, as it was the case when using linear inhibition. Note that just two notch filters will act as desired by removing the undesired line in every branch.

A second example, used by the authors in array processing problems, is the inhibition of a given source by a blocking matrix which removes the steering vectors produced at the output of the remaining branches. This is depicted in figure 8.

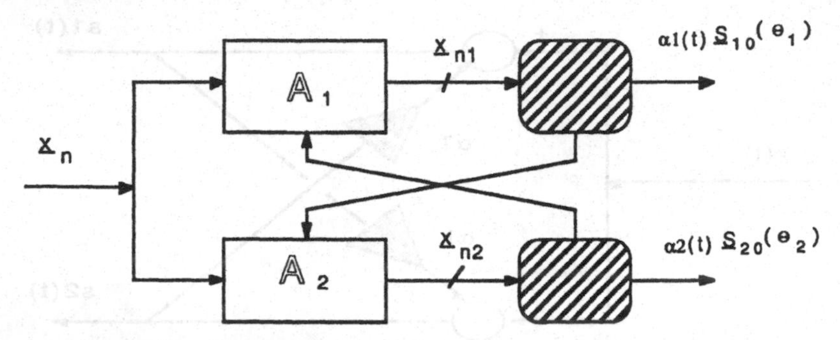

Figure 8. The indirect inhibition stage in the problem of array processing.

For the narrowband case, given the direction of arrival (DOA) θ_2, the blocking matrix $\underline{\underline{A}}_1$ will be formed as it is shown in (17).

$$\underline{\underline{A}}_1 = \begin{bmatrix} 1 & -\alpha_1 & 0 & 0 & \dots & 0 \\ 0 & 1 & -\alpha_2 & 0 & \dots & 0 \\ \dots & \dots & \dots & \dots & \dots & \dots \\ 0 & 0 & 0 & & 1 & -\alpha_{nq-1} \end{bmatrix} \quad (17)$$

being $\underline{\underline{A}}_1$ a (q-1)xq matrix.

The value of each α_q will be designed so that $\underline{\underline{A}}_1.\underline{S}2 = 0$; in this way, $\underline{\underline{A}}_1$ blocks the DOA corresponding to source 2.

$$\alpha_q = \exp(j\frac{2\pi f_c}{c}(d_q \cos\varphi_q - d_{q+1}\sin\varphi_{q+1}).\sin(\theta_2)) \quad (18)$$

As it can be easily concluded, this matrix modifies the original steering vector $\underline{S}_1$ (dimension Q equal to the number of aperture elements) and produces a coloured noise background for the second layer. Nevertheless, this is the price we pay in order to get a single source problem in every branch of the processor. Concerning the existing mapping of the original steering vector, it can be recovered from the modified one, at the output of the blocking matrix, $\underline{S}_1'$ as indicated in (20), when desired.

$$s(q+1) = (s_q - s'_q) / \alpha_q \quad (19)$$
with $s_o = 1$

For the case of a full-filled linear array the expression (18) will have the simple form showed in (20)

$$\alpha = \exp(-j\frac{2\pi f_c}{c}.d. \sin(\theta_2)) \quad (20)$$

where "d" is the distance between sensors.

In this latter case, the matrix written in (17) could be viewed as one whose rows are spatial FIR filters of first order. Each of this rows acts on the spatial data vector, $\underline{x}n$, as a notch filter that rejects the spatial frequency (or DOA) corresponding to the source 2 (θ_2)

In some situations, it may be more convenient to increase the order of the filter so as to reduce the notch-band. This is the case, for instance, of having very closely located sources, because then the nulling introduced by the inhibitory stage not only eliminates the direction to block, θ_2, but also reduces the power of source 1, that is, the one which should be preserved in the branch; the narrower the notch-band, the less will be affected θ_1.

The aforementioned idea gives rise to other possibilities for the blocking matrix $\underline{\underline{A}}_1$ depending on the order of the nulling desired to introduce. The proposed formulation is the one showed in (21)

$$\underline{\underline{A}}_1 = \begin{bmatrix} 1 & \ldots 0 \ldots & -\alpha_j & \ldots 0 \ldots & \\ 0 & 1 & \ldots 0 \ldots & -\alpha_{j+1} & \ldots 0 \\ \ldots & \ldots & \ldots & \ldots & \ldots \\ 0 & \ldots 0 \ldots & 1 & \ldots 0 \ldots & -\alpha_{nq-j} \end{bmatrix} \quad (21)$$

where

$$\alpha_j = \alpha^j$$

that is, the subindex "j" stands for both, the column number and the order of the filter.

As in the first example of multiple line separation,the indirect inhibition stage requires less information than the linear inhibition, where the full waveform is needed in order to inhibit the other variables. This represents the big advantage of indirect inhibition.

The main drawback of the indirect is that it requires of a better knowledge of the input data structure (data modelling); and, it may uses some mapping of the selected variables that may have some impact in the design of the second layer.

V THE OPTIMISATION STAGE

This stage is dedicated to find a single variable or waveform in the output of the processor allocated in the inhibition layer. The degree of complexity of this stage depends very much on the inhibition layer selected.

For indirect inhibition or blocking just a parameter may be enough to obtain a satisfactory result in the overall performance of the processor. In this case, both the multi-line tracking or the array processing examples need a spectral estimate analysis to find or to estimate either frequencies or DOAs. Ii is important to remark that the maximise stage, under indirect inhibition, requires the estimation of a single frequency or DOA in a coloured background. A further refinement on the blocking matrix (17) could be to establish orthogonality, such that the resulting noise remains white.

$$\underline{\underline{A}}_1 \cdot \underline{S}_1 = 0$$
$$\underline{\underline{A}}_1 \cdot \underline{\underline{A}}_1^H = \underline{\underline{I}} \quad (22)$$

Nevertheless, the refinement involved in (22) increases, above reasonable limits, the complexity in the learning or adaptive version of the processor. In other words, the above choice for $\underline{\underline{A}}$ has to be traded with the computational burden in constructing these matrices from the corresponding steering vectors. The reduction on the complexity in the inhibition stage, claimed before, forces to increase the performance on the second stage. Regardless that Music or MLM can be envisaged from the second stage, it has to be keep in mind that both of them require stabilised data covariance matrix and this will relent the overall convergence of the network. Due to the expected single line nature of the problem to be faced in this second stage, it mat be better to use just the magnitude of the spatial Fourier transform of the snapshot resulting from the first stage; doing this, the convergence of the processor is faster than with other DOA estimation procedures.

When direct inhibition is used, no coloured noise will be observed. This is at the expense of a more complex second stage, which evolves to a signal enhancer. The output of the second stage has to be the full waveform, including the original steering vector together with the source waveform.

Let us concentrate in the case of the multitone application. It is clear that the best signal enhancer for this problem will be an Extended Kalman Filter, which can be viewed as a steep ahead from the classical PLL [9]. The authors have been reported the potential of the Kolmogorov´s processor for this case [5]. The network is depicted in figure 9.

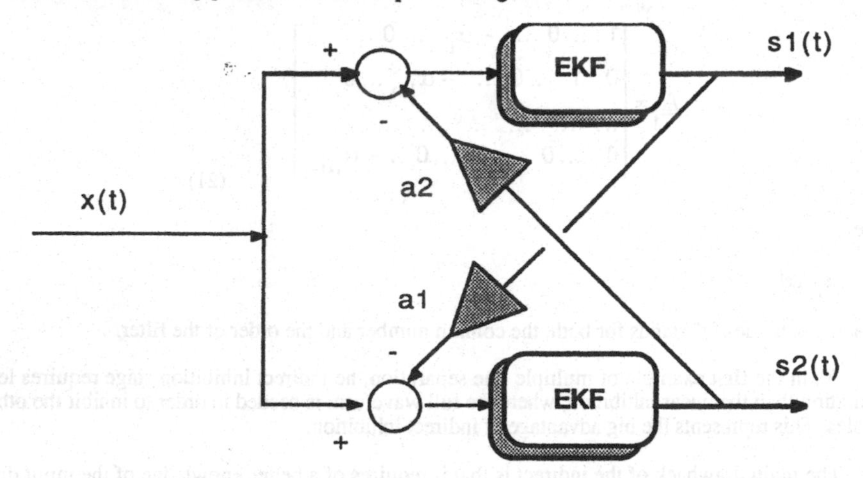

Figure 9. The network for the multitone application with direct inhibition.

As mentioned before, the issue of learning will be reported in a section hereafter. This learning is strongly related with the design of the inhibition weights shown in figure 9. In any case, any choice unsymmetrical, which avoids the unstable case of the product $a_o \cdot a_1$ equal to one [3], shows evidence of the quality associated to the scheme of figure 9.

Concerning the second application on array processing, and aiming the same results that in the multitone case, the authors develop an EKF for the error associated to the covariance columns. To be more explicit, given the cross-correlation vector between the received snapshot and a reference sensor (labeled as zero in the formulation), the measured and modelled vector are:

$$\underline{XS}_n = \underline{X}_n \cdot x^*_n(0)$$
$$\underline{XS}_i = \bar{\alpha}^2_i \; \underline{S}_i + \bar{\sigma}^2 \cdot (1,0,....,0)^T \qquad (23)$$

and being $\bar{\alpha}_i$, $\bar{\sigma}$ and $\underline{S}_i$ the estimates , an EKF driven by the error $\underline{\varepsilon}_i$, defined as in (24), updates the estimates of the source power and the steering vector (see [6] for more details on the objective and the dedicated EKF).

$$\varepsilon_i = (\underline{XS}_n - \underline{XS}_i) \qquad (24)$$

In this case, the processor changes, since the input is the first column of the data estimated covariance from the received snapshots.

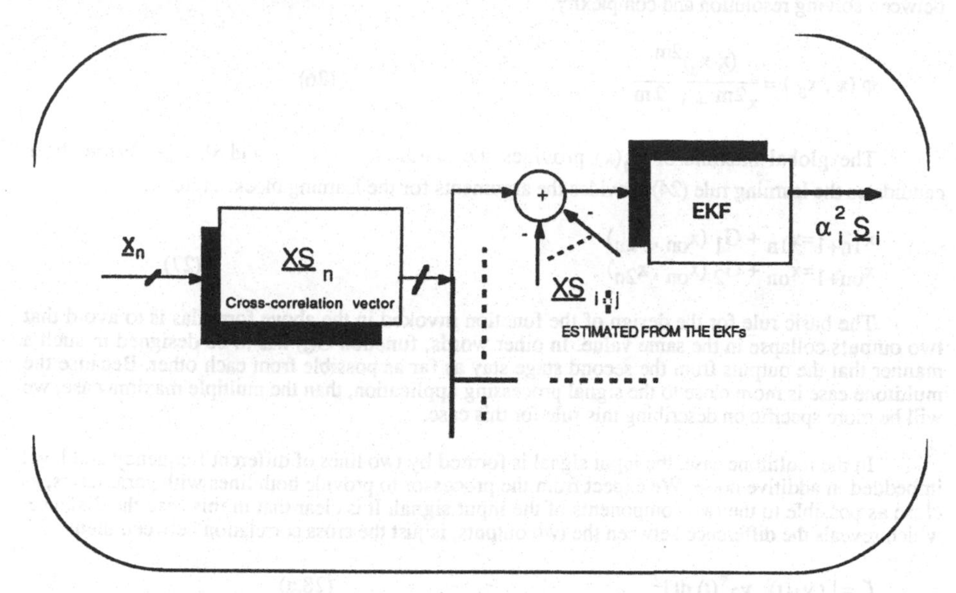

Figure 10. The direct inhibition processor for the estimation of the components of an array data covariance.

The above network estimates accurately spatial stationary scenarios, even in the case of full coherent sources present on it [10]. The computation of the covariance matrix avoids the requirement of estimating every source waveform, as it would be the case when using direct inhibition over the received snapshots.

Note that the flexibility of the processor architecture remains quite open. The two examples reported herein illustrate its contribution to alternative schemes based in either classical filtering or spectral estimation.

As a general rule it can be said that indirect inhibition together with a signal or parameter enhancer, in the second stage, constitutes the most innovative contribution of the Kolmogorov processor. The examples reported in section VII prove this claiming.

6 LEARNING RULES

Since we started this presentation on the Kolmogorov processor, it was assumed that the output functions or variables, in any suggested configuration, were independent one of each others. This independence becomes crucial for the unsupervised learning of the processor, which is the most interesting contribution that this architecture provides to signal processing. We will examine hereafter two different cases introduced before namely: the multiple maxima of a function and the multitone cases.

Taking the multiple maxima of a function, as described in section III, and assuming the inhibition layer is designed properly, the second layer reduces to find the single maxima of the function shown in (25).

$$\varphi(x, x_{1n}) \cdot g(x) = g_o(x)$$

$$\varphi\,(x\,,x_{on}\,)\,.\,g(x) = g_1(x) \qquad (25)$$

Being the two actual maxima of g(x) x_o and x_1, an inhibition function adequate for the purposes assigned to the first layer will be as it is shown in (26), where m is an integer trading between solving resolution and complexity.

$$\varphi\,(x\,,x_a\,) = \frac{(x-x_a)^{2m}}{x^{2m} + x_a{}^{2m}} \qquad (26)$$

The global maxima of $g_o(x)$ provides the candidates x'_{1n+1} and x'_{on+1}. From these candidates the learning rule (24) provides the arguments for the learning blocks at iteration n+1.

$$\begin{aligned} x_{1n+1} &= x_{1n} + G_1\,(x_{on}\,, x_{2n}) \\ x_{on+1} &= x_{on} + G_2\,(x_{on}\,, x_{2n}) \end{aligned} \qquad (27)$$

The basic rule for the design of the function invoked in the above formulas is to avoid that two outputs collapse in the same value. In other words, function G(,) has to be designed in such a manner that the outputs from the second stage stay as far as possible from each other. Because the multitone case is more close to the signal processing application, than the multiple maxima case, we will be more specific on describing this rule for this case.

In the multitone case, the input signal is formed by two lines of different frequency and level imbedded in additive noise. We expect from the processor to provide both lines with parameters, as close as possible to the two components of the input signal. It is clear that in this case the distance, which reveals the difference between the two outputs, is just the cross correlation between them.

$$\zeta = |\int y_1(t)\,.\,y_2{}^*(t)\,dt\,|^2 \qquad (28.a)$$

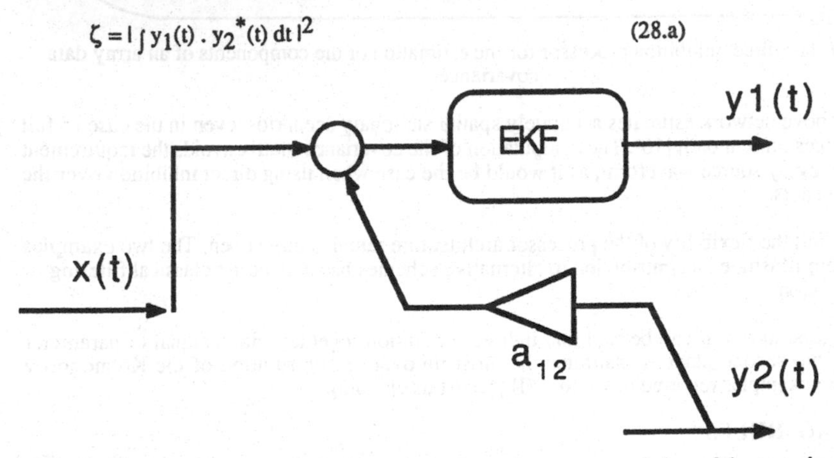

Figure 11. The learning of the weights depends on the cross-correlation of the two signal outputs.

As it was reported by the authors in [5] , when y1() and y2() at two independent random processes, (28) should be extended to more high order moments of both outputs. In fact, in the reference mentioned, it is suggested to use two instantaneous non linear functions $\rho_1(y_1(t))$ and $\rho_2(y_2(t))$ as in (25b).

$$\zeta = |\int \rho_1(y_1(t))\,.\,\rho_2(y_2(t))\,dt\,|^2 \qquad (28.b)$$

With this learning objective, some combination of high order cross moments, which depend on the polynomial approximation for both non-linearities, of the two output processes is minimised. Of course, the choice for these functions is crucial depending on up to what degree they are relevant concerning the statistical independence of the two signals involved. A formal description on the selection of these non-linear functions does not exist and it remains open in the field of high order signal processing. Once the objective is representative of the degree of the independence between the two outputs, the learning proceeds in the same manner that with the order four objective of (28.a). The reader may check in the next section that the simplified objective provides adequate results and that the use of non-linear functions results in a refinement of the learning algorithm.

Taking the instantaneous value of the objective, the gradient to be used in the learning of the weights (see figure 11) will be (29).

$$\frac{d\zeta}{da_{12}^*} = |y_2(t)|^2 \cdot y_1(t) \cdot \frac{dy_1^*}{da_{12}} \qquad (29)$$

Regardless $y_1(t)$ is a non-linear function of the weight, under the small error approximation valid for the second stage processor (the EKF), we can assume a linear behaviour of the second layer with respect their corresponding inputs (i.e. $y_1(t)$ is proportional to $x(t)-a_{12} \cdot y_2(t)$ in figure 11). Doing this, the suggested learning rule will be (30).

$$a_{12}(n+1) = a_{12}(n) + \mu_1 \cdot |y_2(t)|^2 \cdot y_1(t) \cdot y_2^*(t) \qquad (30)$$

Note that high order character of the correction term in front of the second order associated to adaptive algorithms in classical signal processing. This kind of learning has been used successfully by the authors of references [11], [12] and [13], in the blind equalisation problem. Independence between symbol samples was used to remove intersymbol interference in digital communications.

One important feature of the above learning, which results in a fundamental guide-line in selecting non-linear functions mentioned above, is the unsymmetrical nature of the learning. This means that any choice of $\rho_i(y_i(t))$ has to preserve the unsymmetrical learning, and it can be done with odd functions. This choice will avoid the undesired state of the processor, when it provides the same signal at every output.

It is important to remark that, no matter the analysis has been restricted to the case of two outputs, it could be easily extended to the multiple outputs processor.

A similar approach can be used for non-linear inhibition. Taking figure 8 as reference, and assuming that only elevation is needed, figure 12 depicts the architecture which includes the unsupervised learning.

Being two parameters, instead of two time functions, we search for independent perturbations $\tilde{\Theta}_1$ and $\tilde{\Theta}_2$, defined as the difference between the current values and the corresponding running average mean;

$$\tilde{\Theta}_{1n} = \Theta_{1n} - \bar{\Theta}_1$$

$$\tilde{\Theta}_{2n} = \Theta_{2n} - \bar{\Theta}_2 \qquad (31)$$

the objective will be to reduce as much as possible (32).

$$\zeta = E[\, \tilde{\Theta}_{1n} \cdot \tilde{\Theta}_{2n} \,]^2 \qquad (32)$$

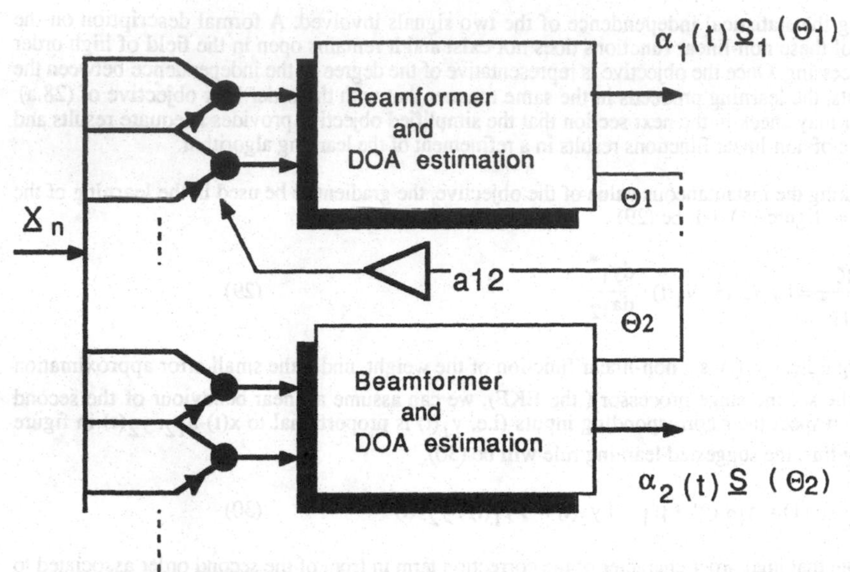

Figure 12. Inhibition block (non-linear case) where the learning of a_{12} is related with the separation between estimates Θ_1 and Θ_2.

Again, the derivative of the instantaneous objective, with respect one of the two weights will be:

$$\frac{d\zeta}{da_{12}} = |\tilde{\Theta}_{2n}|^2 \cdot \tilde{\Theta}_{1n} \cdot \frac{d\tilde{\Theta}_{1n}}{da_{12}} \qquad (33)$$

As before, no matter the elevation angle is obtained from the maxima of the scanning with a phased array beamformer, we assume a linear dependence between the observed perturbations around the mean $\tilde{\Theta}_{1n}$ and the driving variable in the inhibition layer. This results in the same rule as it was in the multitone case described before.

$$a_{12}(n+1) = a_{12}(n) + \mu \, (\tilde{\Theta}_{2n})^3 \cdot \tilde{\Theta}_{1n} \qquad (34)$$

Even the hard approximations involved in deriving this rule, it is easy to compute and reveals, see section VII and [8], the excellent performance of the processor. It is worthwhile that apparently the use of the source waveform will take advantage of the previous learning derived for the multitone case (i.e. independence in the source waveforms). Nevertheless, it requires to estimate not only the source DOA's but also the source waveforms; at the same time, this choice will be very sensitive to coherent sources or multipath propagation. The rule reported in (34) works well even under coherent sources scenarios.

To summarise what is reported in this section, we recall that any unsupervised learning of the Kolmogorov processor is based in independent outputs or perturbations. The issue of suitable metrics for measuring independence, trading between complexity and performance remains open. In any case, the unsymmetrical character of the learning for every inhibition block remains crucial to avoid that different second layer processors collapse in the same output.

VII SOME EXAMPLES.

In this section some specific examples revealing the potential of the processor herein described will be reported.

The first case considered is the line enhancement example. Following the structure of figure 9, two lines, at 0 dBs of signal to noise ratio, evolving in the same frequency of 0.25 during 500 samples split in two lines during samples 500 up to 1500; both lines collapse again between samples 1500 and 2000. The learning is set as described in section VI and the system uses direct inhibition. Figure 13 shows the frequency tracking performed by the two EKFs. Figure 14 shows the corresponding envelope of the two lines enhanced by the system.

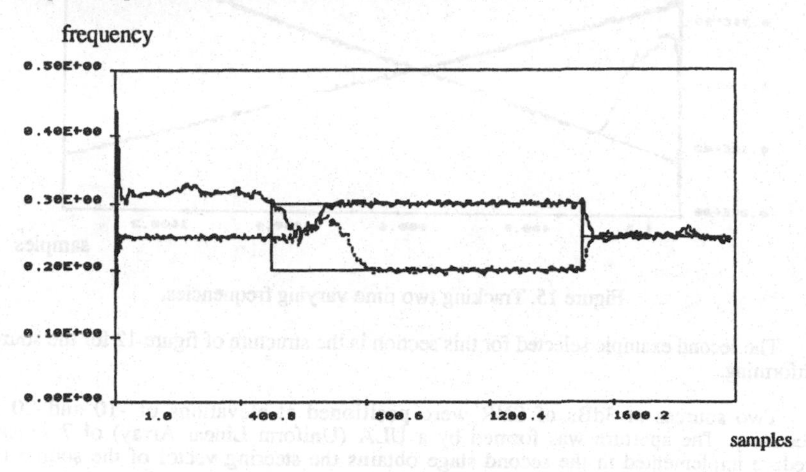

Figure 13. Line enhancement example. Estimated instantaneous frequencies versus samples of the input data signal. Both lines in a SNR of 0 dBs.

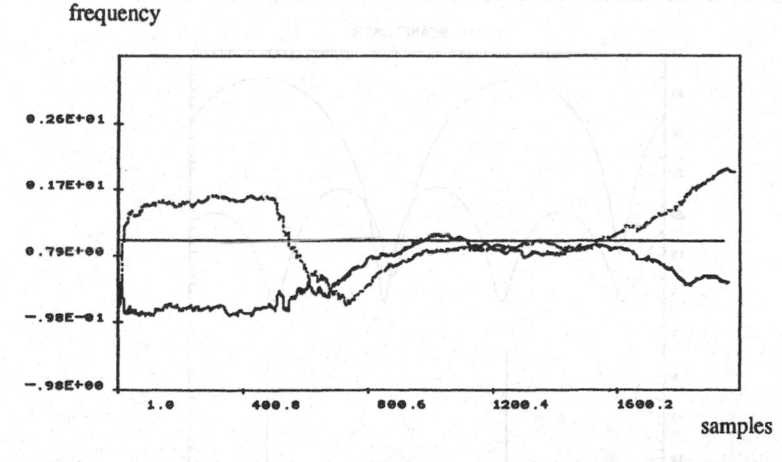

Figure 14. Estimated envelope of the lines enhanced corresponding to the signal scenario of figure 13.

Note from these plots one of the EKFs tracks the adequate line with a good envelope estimate; at the same time the second EKFs evolves erratic with very low envelope estimate. During the second zone, after a convergence time both EKFs track the corresponding lines. Finally, in the last zone the behaviour of zone one show up again.

A second example of the line-enhancement problem is shown in figure 15, also with the same SNR for two crossing instantaneous frequencies.

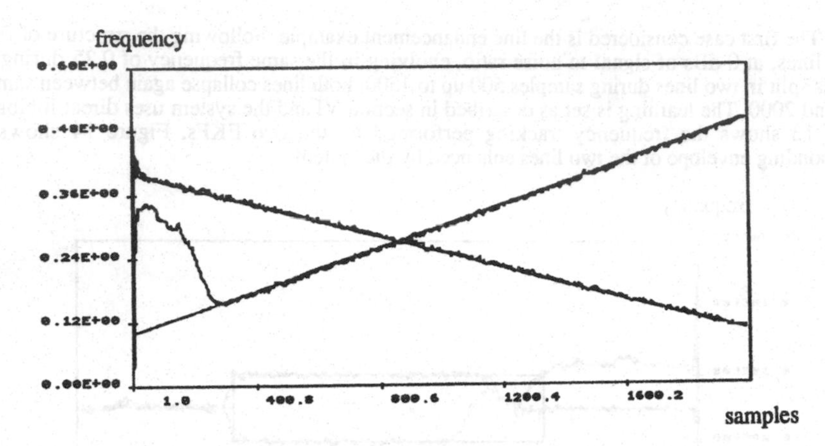

Figure 15. Tracking two time varying frequencies.

The second example selected for this section is the structure of figure 12 for the source array beamforming.

Two sources at 3dBs of SNR were positioned at elevations of -10 and 20 degrees respectively. The aperture was formed by a ULA (Uniform Linear Array) of 7 sensors. The procedure implemented in the second stage obtains the steering vector of the source from the maximum eigenvector of the input covariance matrix at every second stage processor. This procedure take into account the coloured character of the spatial noise at the output of the non-linear inhibition stage implemented. Figure 16 shows the response of the estimated steering vectors after 10 iterations over the stabilised covariance matrix with 500 snapshops.

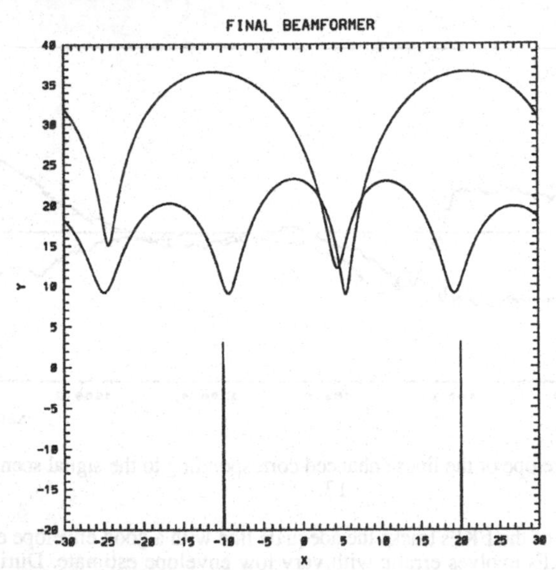

Figure 16. Steering vectors response corresponding to the array processor with non-linear inhibition and the signal scenario described in the text.

It is important to remark that the processor is not sensitive to coherent sources due to its architecture which includes the spatial inhibition of sources.

Another example, for the same aperture, is shown in figure 17. In this case, sources are located at 0 and 5 degrees with the same SNR; this source separation is far below the bandwidth of the aperture and the figure proves the superresolution behaviour of the processor.

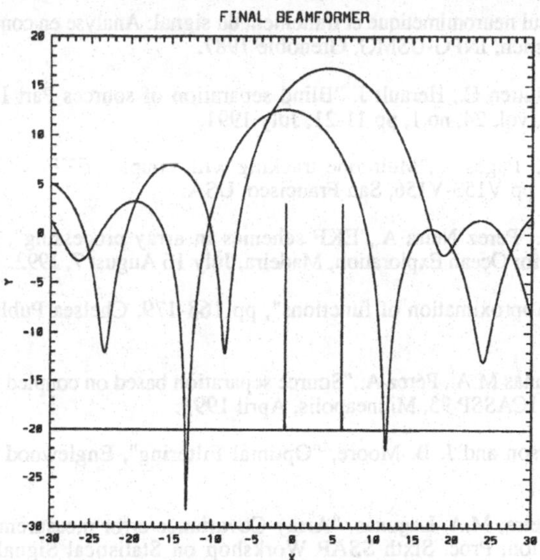

Figure 17. Superresolution performance

From the examples included in this section it can be envisaged the extraordinary potential of the Kolmogorov processor in signal processing. Note that the cases reported include both direct and non-linear inhibition. Note also that the second stages used in the examples (i.e. EKF and GSVD) only affect the global convergence and accuracy of the structure; other alternatives as analog PLL or spatial Fourier transform could be used when trading performance with complexity.

VIII CONCLUSIONS

The so-called Kolmogorov's processor has been described as a new architecture for signal processing. The new architecture can accommodate on it well supported algorithms for signal processing, improving their performance due to the reduction to the single source or signal problem, performed by the inhibition stage.

It is very important to point out that the best of the processor comes from the combination of the following guide-lines:

- It uses two stages of processing;
- The first stage has to be devoted to inhibition or to isolate a single component from the multiple components present on the input data;
- The so-called non-linear inhibitions highly recommended, when it is possible, since this inhibition becomes more robust to the estimation noise or misadjusment;
- Use high order learning directly, as described in the text, or by using non-linear mixtures, in order to maximise the distance between second stage outputs.

With these basic rules the processor may show a performance far above from classical signal processing architectures. Finally, it is worthwhile to mention that sophisticated second stage processors will enable fast convergence at the expense of hardware and tuning complexity.

IX REFERENCES

[1] R. Hecht-Nielsen. "Neurocomputing". Addison Wesley, 1990, ISBN 0-201-09355-3, 1990, pp.122-124.

[2] A.N. Kolmogorov. "On the representation of continuous functions of many variables by superposition of continuous functions of one variable and addition". (in Russian), Dokl. Akad. Nauk USSR, 114, pp 953-956, 1957.

[3] Jutten C. "Calcul neuromimetique et traitement du signal: Analyse en composants independents". Ph.D. Tesis, In french, INPG-USMG, Grenoble 1987.

[4] Common P., Jutten C., Herault J. "Blind separation of sources Part II: Problem statement", Signal Processing, vol. 24, no.1, pp 11-21, July 1991.

[5] Lagunas M.A., Pagés A.,"Multitone tracking with coupled EKFs and high order learning". Proc. ICASSP 92, pp V153-V156, San Francisco, USA.

[6] Lagunas M.A., Pérez Neira A.,"EKF schemes in array processing", NATO-ASI Acoustics Signal Processing for Ocean Exploration, Madeira, July 16 August 7, 1992.

[7] Lorentz G. "Approximation of functions", pp 168-179, Chelsea Publishing Co., New York 1986.

[8] Nájar M., Lagunas M.A., Pérez A.,"Source separation based on coupled single DOA estimation processors", Proc. ICASSP 93, Minneapolis, April 1993.

[9] B.D.O. Anderson and J. B. Moore, "Optimal Filtering", Englewood Cliffs, N. J. Prentice-Halls, 1979.

[10] Ana Pérez-Neira, M.A. Lagunas, "Array Covariance error measurement in adaptive source parameter estimation, Proc. Sixth SSAP Workshop on Statistical Signal & Array Processing. October 7-9, 1992.

[11] Y. Chen, C.L. Nikias, ,J.G. Proakis, "CRIMNA: Criterion with Memory Nonlinearity for blind equalization", HOS, Proc. Signal Processing Workshop on High Order Statistics, Chanrrouse, Francia, July, 1991, Ed. J.l. Lacoume.

[12] M.Gaeta, J.L. Lacoume, "Source Separation versus Hypothesis", HOS, Proc. Signal Processing Workshop on High Order Statistics, Chanrrouse, Francia, July, 1991, Ed. J.l. Lacoume.

[13] C. Jutten, L. Nguyen Thi, E. Dijkstra, E. Vittoz, J. Caelen, "Blind Separation of Sources: an Algorithm for Separation of Convolutive Mixtures", HOS, Proc. Signal Processing Workshop on High Order Statistics, Chanrrouse, Francia, July, 1991, Ed. J.l. Lacoume.

PROJECTIVITY INVARIANT PATTERN RECOGNITION WITH HIGH-ORDER NEURAL NETWORKS

G. Joya(*) and F. Sandoval(**)

(*)Dept. Arquitectura y Tecnología de Computadores y Electrónica
(**)Dept. Tecnología Electrónica
Universidad de Málaga, Pza. El Ejido S/N, 29013-Málaga, SPAIN

Abstract

The need to provide a neural network for pattern recognition with invariance to a new transformation, projectivity, is considered. This invariance is justified when working with object images that can appear rotated in relation to an axis contained in its own plane. An invariable relation to the transformation is found, the double ratio of four points, and incorporated to the network as a restriction to the weights. A projectivity invariant pattern classifier has been simulated. Besides, some considerations about high order neural networks are expounded.

1.- INTRODUCTION.

A high order neuron is one whose potential is the weighted sum not only of its inputs but also of products of the latter ones. Thus, its output will be given, in general, by the expression:

$$y_i = F\left(\sum_j T_j(i)\right) \qquad (1)$$

being the nth order term:

$$T_n(i) = \sum_j \ldots \sum_k W_n(i,j\ldots k)\,x(j)\ldots x(k) \qquad (2)$$

In pattern recognition tasks, high order terms can be used to incorporate previus information about the system to the network structure. Thus, by means of a set of appropriate restrictions, the network can be invariant to a certain transformation on the patterns to be recognized. Besides, this information incorporated does not need now to be learned, which simplifies the learning algorithm and reduces the training time.

In what follows, we will refer to a high order neural network without hidden layers, with just one bipolar neuron in the output layer and a set of input neurons with 0 and 1 values corresponding to the pixels of an NxN bidimensional input field.

In this sense, C. L. Giles and T. Maxwell [1] provide the restriction that must be applied to the weights of a second order neural network to classify two patterns with invariance to translation: the value of the weight associated with the product of two inputs depends only on the distance between their corresponding pixels in the input field. Thus, if the output neuron activation function is

$$y_i = sgn(\sum_j \sum_k W_2(i,jk)x(j)x(k)) \qquad (3)$$

the weights must verify:

$$W_2(i,jk) = W_2(i,j-k) \qquad (4)$$

The invariances to scale and rotation are studied by M.B Reid, L. Spirkovska y E. Ochoa [2] and S. Kollias et al. [3]. A second order network distinguishes between two patterns with scale invariance if the weight associated with the product of two inputs depends only on the slope of the line drawn between their respective pixels:

$$W_2(i,jk) = W_2(i,(y_k - y_j)/(x_k - x_j)) \qquad (5)$$

being (x_k, y_k) the coordinates of pixel k in the input field.

On the other hand, a third order network with a topology similar to the former ones will be invariant to rotation if the weight associated to the product of the three inputs depends only on the angles of the triangle which form their respective pixels:

$$W_3(i,jkl) = W_3(i,\alpha\beta\gamma) = W_3(i,\gamma\alpha\beta) = W_3(i,\beta\gamma\alpha) \qquad (6)$$

being alpha, beta and gamma the angles of the triangle formed by the pixels j, k, l.

In this paper we justify the need to introduce invariance to a new transformation: projectivity. This invariance allows us to solve a problem without solution by using the classical transformations (translation, scale, rotation): the recognition of a pattern rotated in relation to an axis contained in its own plane (section 2). We find a relationship that remains invariant to this transformation and will provide us with the restriction that weights must fulfil to incorporate it (section 3). We reproduce the simulated network topology as well as the experimental results achieved (section 4). We carry out some considerations on the high order neural network use (section 5) and eventually expound our conclusions and future works (section 6).

2.- WHY INCORPORATE PROJECTIVITY INVARIANCE.

The incorporation in a network of invariance to the three transformations above described: translation, scale, rotation, allows us to recognize objects regardless of their position on the plane (translation), their withdrawal and approaching on parallel planes (scale) or their turn around an axis perpendicular to the pattern plane (rotation).

In a number of cases it is necessary to recognize a rotated turned around an axis contained in its plane, (the objects in a transporting band can appear in different slopes, as well as a plane flying). This transformation, which cannot be achieved as a composition of the former ones, would be perceived by a viewfinder as different projections of the object. In these cases the construction of a network invariant to this new transformation, projectivity, will be necessary.

3.- THE DOUBLE RATIO OF FOUR POINTS AS INVARIANT TO PROJECTIVITY.

To provide the network with projectivity invariance, we must find a relationship between the points in the pattern that will remain constant when applying this transformation.

If we use Porcelet's projectivity concept, taken from [4], we will say that two forms of the first and second category are projective to each other when one can be obtained from the other one by means of a series of projections and sections.

Thus, in figure 1, the series formed by the points ABCD is transformed by the projection from point P on the one formed by A'B'C'D', and this one, by its projection with respect to axis g on the series A''B''C''D''.

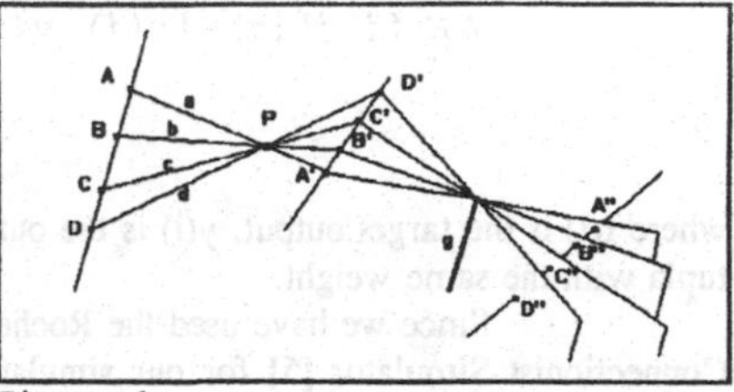

Figure 1

Let us study the two projective series shown in figure 2. The simple ratio of three points in them is given by:

$$(BCD) = \overline{BC}/\overline{BD} = (\sin(bc)/\sin(bd))\cdot(VC/VD) \qquad (7)$$

Let us remark that in our series, (BCD) ≠ (B'C'D'), since VC/VD ≠ VC'/VD'.

In order to find a relationship that remains invariant, we will build the Double Ratio of four points, defined as:

$$(ABCD) = (ACD)/(BCD) = (\sin(ac)/\sin(ad))\cdot(\sin(bd)/\sin(bc)) \qquad (8)$$

Which, as we can see, remains constant for both series, since it only depends on the angles that the beam of straight lines abcd form.

This relation will be the one we will use to incorporate projectivity invariance to the network.

4.-SIMULATION AND RESULTS.

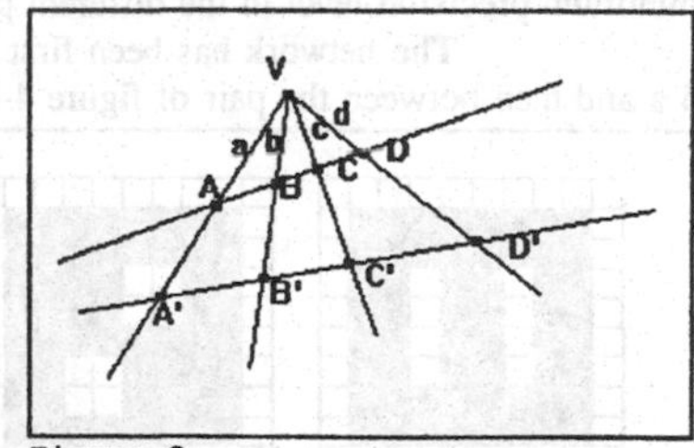

Figure 2

Since our invariant, the double ratio, involves four points, a network distinguishing two patterns invariant to projectivity will have the topology described at the beginning, with a fourth order output neuron.

Let $W_4(i,jklm)$ be the weight associated with the input $x_j x_k x_l x_m$ and $W_4(i,pqrs)$ the weight associated to the input $x_p x_q x_r x_s$. Then $W_4(i,jklm)$ will equal $W_4(i,pqrs)$ if the double ratio (jklm) equals the double rate (pqrs).

Let us remember that here p represents the position on the input plane of the pixel associated to x_p.

For pixels arranged vertically and horizontally we will have:

$$W_4(i,jklm)=W_4(i,(x_l-x_j)*(x_m-x_k)/(x_m-x_j)*(x_l-x_k)) \quad (9)$$

$$W_4(i,jklm)=W_4(i,(y_l-y_j)*(y_m-y_k)/(y_m-y_j)*(y_l-y_k)) \quad (10)$$

being (x_j, y_j) the coordinates of pixel j.

Since we only have one weight layer, the learning algorithm used will be a high order generalization of the perceptron rule:

$$\Delta W_4(i,jklm)=[t(i)-y(i)]*\sum_{N(jklm)} x_j \cdot x_k \cdot x_l \cdot x_m \quad (11)$$

where t(i) is the target output, y(i) is the output obtained and the summation will spread to all the 4-tupla with the same weight.

Since we have used the Rochester Connectionist Simulator [5] for our simulation, which is not a simulator that facilitates high order use, the simulated network topology is that of figure 3, where intermediate layers π and Σ are limited to carry out the input 4-tupla products and the sums of those products with the same weight. Thus, the neurons a and b in the layer π make the product of four inputs whose positions correspond to the form (n,n+1,n+2,n+3), therefore they have the same double ratio. This is because their outputs are added in the same neuron e in the layer Σ. In the same way, the outputs of the neurons c and d in π, which make the product of pixels (n,n+1,n+4,n+5), are added in the same neuron f of the layer Σ. Actually it is a network without hidden layers since we only have one weights layer.

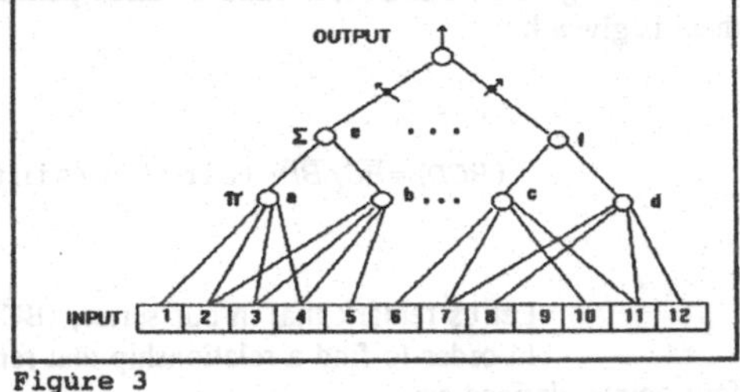

Figure 3

The input layer consists of one hundred neurons with activation values 1 and 0 corresponding to the black and white ones in the pixel field that represents the pattern image. The input field 10x10 dimension responds to a commitment between our memory capacity available and a minimum precision level in the different projected patterns.

The network has been first trained to distinguish between the pair of patterns of figure 4-a and then between the pair of figure 4-b.

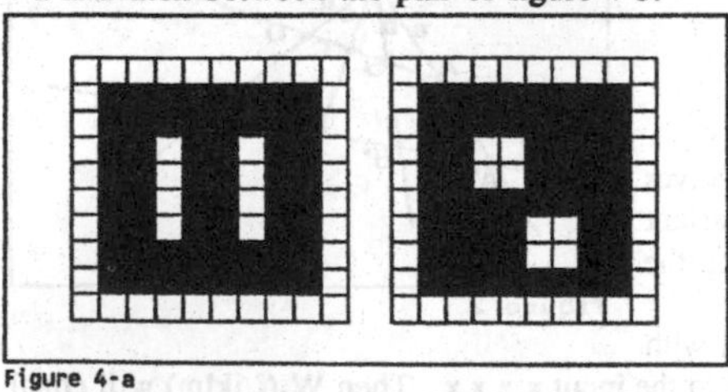
Figure 4-a

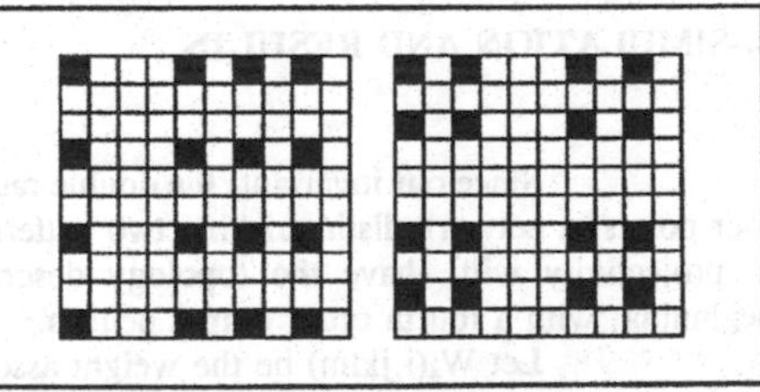
Figure 4-b

The transformed patterns are their projections turned around a vertical or horizontal axis contained in the image plane. The turns carried out are the ones whose sine is 1/3, 1/2, 2/3. In the same way, we have used translated and rotated patterns to verify that our restriction includes invariance to classical transformation. Given the limited number of pixels some of these transformations produce patterns which cannot be represented accurately (the end of a line falls in the middle of a pixel), in these cases defect and excess patterns are used. In the first case, although the number of pixels activated, and consequently the output neuron potential varies with each pattern, the classification is right in all the cases presented. In the second case we have worked with dot figures in order to check more directly the efficiency of the restrictions. Here, the number of pixels activated remains constant, thus, any transformation regarding the Double Ratio will have the same effects on the neuron. If our hypothesis is right the group of set neurons in the layer Σ,- which represent all possible combinations of four black pixels in the pattern-, will alway be the same for any rotation angle.This condition is satisfied in our case and for these patterns recognition has also been complete.

5.- CONSIDERATIONS ON HIGH ORDER NEURAL NETWORKS APPLIED TO INVARIANT PATTERN RECOGNITION.

As it has already been said, the mission of high order in a network for invariant pattern recognition is to incorporate a previous knowledge to the network structure, i.e., to find a relationship invariant to our transformation, and incorporate it as a restriction for the weights, in such a way that the network does not find any difference between the two patterns that are transformed in each other.

In the case of translation, in a network like the one described at the beginning [1], with the restriction that a weight depends only on the distance of the pair of the inputs it is associated with, this mission is completely fulfilled. The output neuron activation, as well as the potential, will remain the same for any position the pattern is in. As a consequence, to teach the network how to distinguish between the two patterns, we will only need one example of each pattern during the training, and not several transformed examples of each one, as it would be the case in a first order network trained with Backpropagation. This simplification of the number of patterns is shown as one of the great advantages of high order neural networks.

However, if we want to incorporate scale and rotation or projectivity invariance, the former assertion is not altogether true, since the output neuron potential depends on the number of pixels activated, which will increase as the pattern size increases.In the same way, the fact that a segment of infinite points is not represented by an infinite number of pixels makes appear new relationships between new pairs of pixels when applying a pattern, which would not be taken into account if we use the smallest one as learning pattern.

On the other hand, the big problem of high order neural networks, the combinatorial increase of connectionism with the network order, is not satisfactorily solved. Thus, the use of coarse-coding, as introduced in [6], to reduce the number of inputs, produces two coarse fields which, in fact, have less neurons, but neither do they produce a univocal image of the pattern, nor do they keep its proportions when the pattern is subject to different transformations. In the same way, the solution brought forward by X. Zhou, M.W. Koch and m.W. Roberts in [7], [8], requires a selective attention selection modulus which controls the part of the pattern to be focused at each time, by means of a focusing pre-defined strategy. This combinatorial increase restrains the input field dimensions that we can use, which aggravates even more the former problem, since the less number of pixels we have, the more coarse the approximations will be.

6.CONCLUSIONS AND FUTURE WORKS.

The need to incorporate projectivity invariance in a pattern recognition network has been justified. An invariant relation, the Double Ratio of four points, has been described, as well as its incorporation to the network structure by a restriction in fourth order weights. A fourth order network invariant to projectivity has been described and tried. The limitation in pattern dimension has prevailed us from carrying out a test field wide enough. It is therefore necessary to find a new method that can decrease connectionism in high order neural networks, in such a way that we can incorporate a bigger number of pixels in our input field and thus decrease the representation error.

On the other hand, our network must be improved by introducing relationships between any four pixels in any direction, which will allow to distinguish patterns that at the moment cannot be distinguished. The automatized search of a series of key points that will remain for any of its transformations, seems to us an interesting research way to achieve that the output neuron potential not be affected by the variation of the number of pixels in every pattern.

ACKNOWLEDGEMENT.

This work has been partially supported by the Spanish Comisión Interministerial de Ciencia y Tecnología (CICYT), Project No. TIC91-0965.

REFERENCES.

[1].- C.L. Giles and T. Maxwell,"Learning, Invariance, and Generalization in High-Order Neural Networks".Applied Optics,vol.26,No 23,pp. 4972-4978,(1987).

[2].- M.B.Reid,L.Spirkovska, and E.Ochoa,"Rapid Trainig of Higher-Order Neural Networks for Invariant Pattern Recognition".Proc.Joint Int. Cnf. on Neural Networks,Washington D.C.,(1989), pp. 689-692.

[3].- S.Kollias,A. Tirakis and T. Milos,"An Efficient Approach To Invariant Recognition of Images Using Higher-Order Neural Networks", Artificial Neural Networks.Elsevier Science Publishers B.V. (North-Holland)(1991),pp. 87-92.

[4].- P.Puig Adam,"Curso de Geometría Métrica II".Ed. R. Puig,(1.965).

[5].- N.H. Goddard et Al,"Rochester Connectionist Simulator", Technical Report 233, The University of Rochester (USA),(1989).

[6].- L.Spirkovska, M.B. Reid,"Coarse-Coding Applied to HONNs for PSRI Object Recognition".Int.Joint Conf. on Neural Networks, Vol. II, p. A-931, Seattle, WA(1991).

[7].-X. Zhou, M.W. Koch, and M. W. Roberts,"A Selective Attention Neural Network for Invariant Recognition of Distorted Objects", Int. Joint Conf. on Neural Networks, Vol. II, p. A-942, Seattle, WA 1991.

[8].- X. Zhou, N.W. Koch, and M.W. Roberts,"Selective Attention in High-Order Neural Networks for Invariant Object Recognition", Int. Joint Conf. on Neural Networks, Vol. II, p. A-937, Seattle (WA), 1991.

REJECTION OF INCORRECT ANSWERS FROM A NEURAL NET CLASSIFIER

F. J. Śmieja

German National Research Centre for Computer Science (GMD),
Schloß Birlinghoven, 5205 St. Augustin 1,
Germany.
e-mail: (Internet) smieja@zi.gmd.de, (Bitnet) smieja@gmdzi.uucp

Abstract

The notion of approximator rejection is described, and applied to a neural network. For a real world classification problem the residual error is shown to decrease with the inverse exponential of the fraction of patterns rejected. The trade-off of "good" patterns rejected and "bad" patterns rejected is shown to increase approximately linearly with rejection rate. A compromise is therefore necessary between trade-off/rejection rate and residual error. A meta-level solution is proposed for removal of the residual error, through use of a modular system of parallel approximators.

Introduction

The problem of function approximation can be summarized as:

$$y = f(x) \tag{1}$$

where x is a pattern selected from the input space and y is the approximation (the output space). f represents the approximation function used. All function approximators are more or less involved with estimating their own function f. Many have additional properties, but the necessary qualification is equation (1). Neural networks, in particular, possess only this characteristic of function approximators. However, an additional very important component of an approximator possessing any degree of "intelligence" is that of **reflection**. Reflection can appear in many forms, all of which involve a degree of self-assessement and interpretability from the approximator. This subject is too extensive to be discussed in full here (see [9, 1, 2, 10]). It suffices to mention that one aspect of reflection is the ability to reject an x because the y produced by f is considered (by the approximator) to be unsatisfactory.

Classification problems are a special form of approximation, and the expected form of the output y (it represents one of N classes) may be profitably used to execute the process of reflection for any type of approximator. Even neural networks can perform reflection about their y for such problems.

In this paper this form of reflection is used to limit the set of patterns approximated positively by a 3-layer feed-forward network. It works by rejecting a pattern on the basis of inappropriate y form and/or when the hidden representations indicate border cases. Relations are described between residual error and rejection rate, and approximation trade-off and rejection rate, for an optical character recognition (OCR) problem. This problem is shown to have the characteristic that the residual error can only be removed with an exponentially increasing rejection rate. Trade-off is shown to increase linearly with rejection rate. A compromise must therefore be made for the reliability of the approximator.

We conclude that a more promising way of increasing performance is to use a modular system, containing specialist approximators working in parallel with different forms of f.

Rejection and unavoidable trade-offs

An approximator possessing a rejection capacity answers in the following to an input x: (a) "accept", approximation is y (b) "reject", tentative approximation is y. Whether the y offered is believed by the encompassing system is a matter for it, and no concern of the approximator. It provides the additional information as to the estimated correctness of the approximation.

Rejection can proceed either through the positive affirmation of outputs (and rejection of all others), or through the positive rejection of outputs (and acceptance of all others). In fact, rejection is a discrete form of a more general "confidence" measure, estimating the degree of believability of y [9, 1].

Rejection rate, R, is the (absolute) percentage of patterns rejected by the classifier.

Approximation trade-off τ, is defined:

$$\tau := \frac{\text{rejected} \wedge \text{correct}}{\text{rejected} \wedge \text{wrong}}$$

where rejection is measured relative to the entire test set. τ measures the cost of rejection. In rejecting a set of patterns, it is to be expected that not only those patterns that are incorrectly classified are rejected, but also those that are correctly classified. We suggest later that a trade-off of around 1 is reasonable for real-world classifiers.

Residual error of the classifier, E, is the absolute percentage of patterns that are not rejected, but nevertheless incorrectly classified.

Reliability, r, is defined:

$$r := 1 - \frac{E}{1-R} \quad (2)$$

r is a measure of the "trustworthyness", or believability, of the classifier. A high reliability implies a very trustworthy approximator. Such trustworthy systems may, however, turn out to have a very restricted area of expertise. Such a compromise must necessarily be accepted, at least when single systems are used. A major benefit expected from modular systems is to remove the necessity for such compromises.

The nature of OCR problems

Real-world problems are in general characterized by large dimensional input spaces. Furthermore, they are in general a form of classification problem, i.e. the number of possible different output vectors (classes) N is highly constrained. OCR is a typical real-world problem, which possesses another property we believe also to be characteristic of large real-world problems [1]. This property is that the classes are approximately (linearly) separable up to a limit (generally around a 70% success rate [4]). After that the classes overlap and envelop one another so thickly that the decision planes that have to be drawn become ever more numerous and particular in order to increase success rate further. This all leads to exponentially increasing costs of improvement during the adaptation phase.

We show empirically below that this error cannot be removed any more cheaply through rejection techniques during the testing phase. Indeed, the residual error E likewise has an exponential relationship with rejection rate.

Methods of rejection in a neural net

When the approximator is a feed-forward neural network with a single hidden layer, there exist the following possibilities for filtering patterns:

Acceptance based on cleanness of output vector

A y is clean if
$$\{[y_k = \max_i y_i] > clean\_top\} \wedge \{y_i < clean\_bottom \, \forall i \neq k\}$$
where *clean_top* and *clean_bottom* are set so as to force the output to have the form of a target class (the target classes in our implementation are binary with only one output element of value 1).

Rejection based on dirtiness of output vector

A y is dirty if
$$[y_k = \max_i y_i] < dirty\_bottom$$
Thus if there is no sign of the output looking like a class target, the x is rejected.

Hyperplane class separation method

The nodes of the hidden layer can be viewed as hyperplanes [8, 7, 3], performing the separations between classes in the input space. Each pair of points in the input space required to be separated must be separated in this space with a hyperplane.

Patterns that are liable to cause error are most likely to be border patterns. These are patterns near the overlap areas between classes. The hyperplane rejection method thus works by rejecting the patterns lying too near to important hyperplanes. An **important** hyperplane is one which, were it to have its polarity reversed, would produce a different class as approximation for x to that which was formally given. Reversing polarity of a hyperplane means changing the sign of the accumulated input to the hidden node associated with this hyperplane. The distance from a hyperplane is the perpendicular distance, and for the cut-off value we use the symbol ε below.

Experiments on rejection for an OCR problem

For the OCR problem we used the NIST (National Institute of Standards and Technology (USA)) Special Database 3, a database of handwritten upper-case letters (26 classes), with input dimension 256. For the training of the network we selected a set of 4616 patterns, and for testing a separate set of 4363 patterns. The network used was trained with back-propagation [5], had 100 hidden nodes, and the classes were coded as one–class–one–position in the output vector.

Rejection based on the output vector (scheme 1)

Here the cleanness of the output vector only was taken into account. The parameters (*clean_top*, *clean_bottom*) were varied from $(0.95, 0.05)$ (high rejection) to $(0.55, 0.45)$ (low rejection) in steps of 0.05.

Rejection based on the output vector and hidden layer (scheme 2)

Here dirtiness of output vector was used, as well as a series of hyperplane rejections, each producing extra rejected patterns. This method begins with low rejection: *dirty_bottom* $= 0.3$, then uses the hyperplane rejection, increasing the value of ε as: $0.05, 0.1, 0.5$. Fear of increasing the value of τ led to the use of more complicated rejection, to increase the value of R. Now, not only were patterns with $\varepsilon < 0.1$ from an important hyperplane rejected, but also those with 2 important hyperplanes < 0.5 away. Thus, if the pattern is (reasonably) near to two important hyperplanes, it will be considered quite uncertain. The next addition rejected those with 3 important hyperplanes < 0.75 away, and the next those with 4 important hyperplanes < 1.0 away.

Results

Scheme 1 produced values of R in $18.9 \leq R \leq 70.2$, and scheme 2 values in $5.92 \leq R \leq 20.02$. Additionally the point $R = 0$ was used (the performance without rejection).

Figure 1 shows the trade-off τ as a function of R for the two schemes, and also the residual error E. Remarkable is the way the two lines nearly fit onto one another. It is surprising because it is evident from this that the hyperplane rejection method, despite its sophistication, turns out to produce practically the same R–behaviour as the more heavy-handed clean-class method. The form of the curves are similar, indicating an approximate linear dependence for $\tau(R)$ and a convincingly exponential one for $E(R)$.

E was plotted against the logarithm of R, yielding fairly accurately the following relationship:

$$E(R) = A\exp(-\frac{R}{\lambda}) \quad A = 11.48,\ \lambda = 15.1 \pm 0.2.$$

Thus the residual error is approximately halved with about 15% rejection. For this approximator it also seems to be around $R = 15\%$ where the value of $\tau \approx 1$.

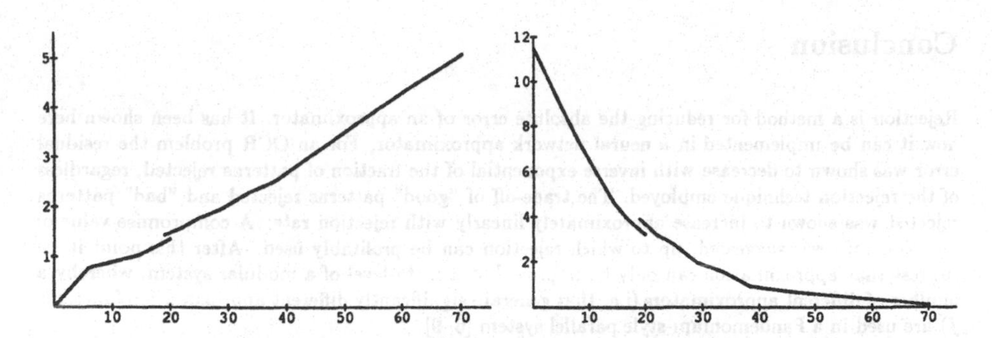

Figure 1: (left) Trade-off vs. rejection, (right) residual error vs. rejection. The two segments in each graph indicate rejection scheme 1 and rejection scheme 2. The abscissa is the rejection rate R.

Dealing with the residual error

It seems therefore that the residual error is not to be reduced without enormous cost. In order to bring down E, R must be increased ever higher the smaller E is, and from the first curve in figure 1 this means τ will correspondingly increase. 100% reliability is clearly not practicable for single neural network approximators. What should be done?

This problem is in fact not confined to the neural network: it is a feature of the OCR problem itself that the misclassified patterns are simply not well-behaved and remain around the decision areas. This fact becomes clear when one observes that the rejection method was largely unimportant so far as E and τ were concerned. The classes are simply too intermingled to make such a "lopping-off" of difficult areas profitable at low E.

Our solution is to employ a number of approximators, each with their own (quite different) f, and each with a reasonable stand-alone performance factor. We suggest $\tau \approx 1, R \approx 15\%$ (or, if known, $R \approx \lambda$). They work independently and when any one of the approximators confidently accepts a pattern it is believed (unless two approximators both accept but produce different y's). This alone will reduce E considerably for the whole system, providing the functions f are sufficiently different. Current work indicates this indeed to be the case, and at least 5% improvement has been observed, even with quite similar f's [1]. Additionally, when none of the approximators wants to accept the pattern, a consensus can be taken over their y's. If a consensus exists, this answer should be accepted.

It is necessary that the approximators can produce a reasonably good performance when working alone. Otherwise the performance of the joint system deteriorates, for the following reasons:

1. If the f's are sufficiently different, they will presumably indicate quite different areas of specialization, and it is important that the approximator itself can approximate sufficiently well in these areas. The consensus can only be profitable in the overlapping regions.

2. The approximators must be good enough in the overlapping regions to make a consensus possible!

3. (Related to 1.) it is desired to keep the individual speicalists in a modular structure as autonomous as possible, so that the cost of computing is broken down. This is more favourable when the specialists compute as much as possible within themselves, and limit the set of cases that require cooperation with other approximators.

Conclusion

Rejection is a method for reducing the absolute error of an approximator. It has been shown here how it can be implemented in a neural network approximator. For an OCR problem the residual error was shown to decrease with inverse exponential of the fraction of patterns rejected, regardless of the rejection technique employed. The trade-off of "good" patterns rejected and "bad" patterns rejected was shown to increase approximately linearly with rejection rate. A compromise value of rejection rate was suggested, up to which rejection can be profitably used. After this point it we suggest that approximation can only be improved at a meta-level of a modular system, whereby a number of different approximators (i.e. that generate significantly different approximation functions f) are used in a Pandemonium-style parallel system [6, 9].

We further hold that most real-world, large-scale problems are of the OCR type, and that steps can most usefully be made in the direction of meta-levels of approximation and reflective systems, rather than that of the details of the approximator itself.

Acknowledgements

I thank Uwe Beyer and Heinz Mühlenbein for regular fruitful and inspiring discussions. This work was funded by the German Ministry of Research and Technology, grant number 01 IN 111 A/4.

References

[1] U. Beyer and F. J. Śmieja. Quantitative aspects of data-driven information processing. Technical Report #812, Gesellschaft für Mathematik und Datenverarbeitung, St Augustin, Germany, March 1993.

[2] S. Hubrig-Schaumburg. Handwritten character recognition using a reflective modular neural network system. Master's thesis, Bonn University, Germany, 1992.

[3] R. P. Lippmann. An introduction to computing with neural nets. *IEEE ASSP Magazine*, April 1987.

[4] H. Mühlenbein. Editorial. *Parallel Computing*, 14(3):247–248, August 1990. special edition on neural networks.

[5] D. E. Rumelhart, G. E. Hinton, and R. J. Williams. Learning internal representations by error propagation. *Nature*, 323(533), 1986.

[6] O. G. Selfridge. Pandemonium: a paradigm for learning. In *The Mechanisation of Thought Processes: Proceedings of a Symposium Held at the National Physical Laboratory, November 1958*, pages 511–527, London: HMSO, 1958.

[7] F. J. Śmieja. Neural network constructive algorithms: Trading generalization for learning efficiency? *Circuits, Systems and Signal Processing*, 12(2):331–374, 1993.

[8] F. J. Śmieja and H. Mühlenbein. The geometry of multilayer perceptron solutions. *Parallel Computing*, 14:261–275, 1990.

[9] F. J. Śmieja and H. Mühlenbein. Reflective modular neural network systems. Technical Report #633, GMD, Sankt Augustin, Germany, February 1992.

[10] F. Weber. Self-reflective exploration of the kinematics of a two-joint robot arm. Diplomarbeit, University of Bonn, Germany, 1992. in German.

NONLINEAR TIME SERIES MODELING BY COMPETITIVE SEGMENTATION OF STATE SPACE

Carlos J. Pantaleón-Prieto*, Aníbal R. Figueiras-Vidal**

* Dpto. Electrónica, ETSI Telecom. Univ. Cantabria, Av los Castros s.n., 39005 Santander, Spain.
* *DSSR, ETSI Telecom. UPM, Ciudad Universitaria, 28040 Madrid, Spain.

ABSTRACT

In this paper we propose a general approach for modeling nonlinear time series. We define a finite memory dynamical system and use a state-space representation. The given time series represents a trayectory through the state space of the dynamical system, and we consider each state it passes through as a pattern with associated output the predicted value. We attempt to model the (probably) nonlinear relation between the state and the output as a combination of several simple models, each of them aproximating the function in a part of the state space. The models are obtained iteratively. Starting with random coefficients models, each pattern is presented to all the models, and the one that better fits the predicted value is trained with that pattern. This process is continued till stability is reached. We finish with a group of patterns associated to each model (i.e., a segmentation of the state space) and a model for each group.

A preliminary application to nonstationary signal segmentation is presented, and future lines of research are suggested.

1. INTRODUCTION

Nonlinear modeling has become a field of very intense research and a lot of literature has been produced presenting different approaches to the problem [1,2].

Given a time series y[k], we want to estimate a non-anticipative finite memory model of the form:

$$y[k+1]=f(y[k],\ y[k-1]...y[k-l+1])+e[k] \tag{1}$$

where l is the memory of the model and e[k] the prediction error. We will determine the best model when e[k] is a strictly white process.

The vector {y[k],y[k-1]...,y[k-l+1]} can be interpreted as a sample of the state space of f with associated output y[k+1]. We can view the time series as a collection of pairs {state vector, signal value}, so our problem becomes the approximation of a non-linear function from samples. This approach is called the codebook paradigm [3].

This functional approximation can be accomplished with different methods. We can classify them as global or local , considering if we use one or several models to cover the state space.

The normal approach when a global model for f is pursued is to use some set of parametric functions (kernels) and to express f as a linear combination of them. Polynomials can be used, as in [4], where NARMAX models are introduced. A NARMAX model takes the form of a functional similar to (1), but considering as parameters not only the past outputs, but the past inputs and past prediction errors [4]. Bilinear models are based on a polynomical approximation too, but only adding cross term products between the input and the output (bilinear terms) to an ARMA model [2]. Other types of functions can be used, as for example splines or radial basis functions [5]. The general problem of this kind of approach is the strong dependence on the chosen kernel and the difficulty of analysing issues like stability or robustness.

It is also possible to introduce the nonlinear behaviour by considering local approximations of f(x) over its state space.The so called State-dependent Models [1] follow this approach and consider a general ARMA model whose parameters at time k depend on the state at that time.

Codebook Prediction is also based on fitting local state-space models to the data [3]. Under the codebook paradigm one can view the signal as a collection of patterns. We consider that f is a smooth function so patterns that lie near in the state space will lead to similar outputs. Patterns are grouped by nearest neighbour techniques. A local AR model is fit to every group of patterns [3]. Several issues remain to be analyzed, as how to select the order of the model, as well as finding some performance measure, interpreted as the compromise between complexity and precision.

Our approach uses the same framework as in [3], but under a different philosophy. We also approximate **f** by a collection of local models (see Fig. 1), each of them covering part of the state space; but we do not follow a nearest neighbour technique. We let the local models compete for every pattern in the state space. Using a neural paradigm, we present each training vector to all the models we consider, and the one that gives less error is 'trained' with that pattern. At the end , we get the models, as well as a segmentation of the state space. This competitive segmentation of the training data has been successfully applied in paralell MLP architectures [6].

Note that following a local approach is looking for two types of models: a *segmentation* model that selects what local model a particular state belongs to, and several local *prediction* models, that, given a particular state, predict the next value of the series.

For example, we can deal with locally stationary signals, characterized by abrupt changes of their statistical properties at unknown instants and a stationary behaviour in between those instants [7]. If we consider the direct (nearest neighbour) method we can find that two patterns that lay very close in state space can have very different outputs due to the fact that they belong to different time intervals with different models. On the other hand, an indirect (competitive) approach will take this fact into account.

To this extent, the competitive approach can be considered more general, in part because we expect that, if a nearest neighbour clustering of the state space makes sense, the competition will make each model concentrate on groups of patterns that are near one to another in the state space.

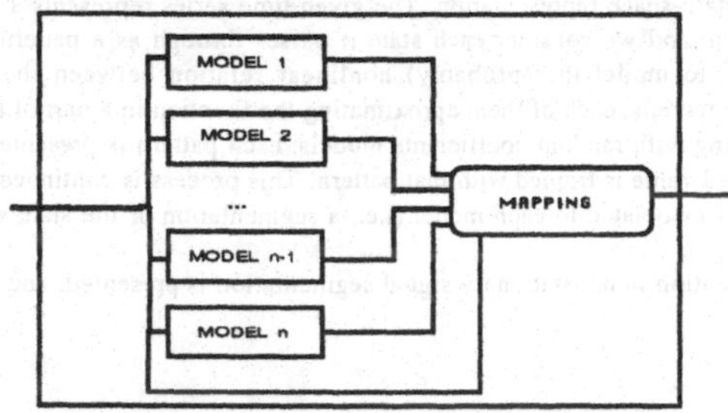

Fig. 1 Graphical description of the method as a mapping of models to parts of the state space.

2. PROBLEM FORMULATION

A broad class of models, for example (1), can be represented in state space notation [3]:

$$\mathbf{x}[k+1] = G(\mathbf{x}[k], \mathbf{u}[k], k) \tag{2a}$$

$$\mathbf{y}[k] = H(\mathbf{x}[k], \mathbf{u}[k], k) \tag{2b}$$

The vector $\mathbf{x}[k]$ is the state, $\mathbf{u}[k]$ the input and $\mathbf{y}[k]$ the ouptut. If we consider unavailable the input we have to rely on a functional only of the past outputs, leading to a particular case:

$$\mathbf{y}[k+1] = F(\mathbf{y}[k]) + \mathbf{u}[k] \tag{3}$$

The class of processes that can be expressed in the form (3) are called Nonlinear Autorregresive Processes (NLAR) [3]. Note that now $\mathbf{y}=\mathbf{x}$ (the output vector and the state vector are the same). $\mathbf{u}[k]$ is the driving error term, which we assume white.

If we want to predict the future state of our dynamical system, we need to estimate the functional F all over the state space, or, at least, over the part of the state space we are interested in. From the above expression it is obvious that the best estimate of the output vector $\mathbf{y}[k+1]$ given $\mathbf{y}[k]$ is [3]

$$\hat{\mathbf{y}}[k+1] = E\{F(\mathbf{y}[k]) + \mathbf{u}[k] \mid \mathbf{y}[k]\} = F(\mathbf{y}[k]) \tag{4}$$

This expression is equivalent to (1). So, the problem can be viewed as an approximation of **f** from some noisy samples nonuniformly distributed over its state space [3].

To interpolate this functional we will use a set of local models, each of them covering a part of the state space. The competition among models as they are trained generates the segmentation of the state space.

3. COMPETITIVE SEGMENTATION

Competitive learning is a well known neurocomputing paradigm [8]. The method can be stated as follows. We have a collection of vectorial observables $\{\mathbf{x}(t)\}$ and a set of reference vectors $\{\mathbf{m}_i(t)\}$, initialized at random. We choose iteratively one of the $\{\mathbf{x}(t)\}$ and compare it with all the reference vectors using some metric. The winner of this competition gets reduced its distance (in the reference metric) to the training vector. When stability is reached, every reference vector represents a group of the training data [9].

It is well known that a simple competitive training does not, in general, assure the activation of all the reference vectors [8]. The *conscience* method of DeSieno [8] and the FSCL (Frequency Sensitive Competitive Learning) [10] have been proposed to overcome this difficulty. We will apply the FSCL approach which basically uses a distance measure that takes into account how many times a particular vector (in our case a particular model) is the winner. More detailed information about the subject can be obtained in [11].

Competitive learning makes each reference vector concentrate on a particular group of patterns. It does not seem too risky therefore to think of extending this idea to models, and thinking that *if several models compete for training patters each will concentrate on some group of them that share some kind of similarity*. So we can expect that a competition among different models will produce a meaningful segmentation of the state space.

Now, as in [3], we will concentrate in AR models, but all the facts we will consider can be extended to any model that can be built in an adaptive fashion. An equivalent architecture with MLP has been studied [6], and any parametric model can be used in the same way.

Consider an AR model of the form:

$$y[k+1]=\sum_{i=0}^{l-1} a_i y[k-i]+u[k] \tag{5}$$

where $\mathbf{a}=\{a_0,a_1..,a_{l-1}\}$ is the vector of AR coefficients and y[k] is the kth sample in the serie; u[k] is the error term. We define the vector $\mathbf{y}_k$=(y[k],y[k-1],...,y[k-n+1]) as the pattern number k with desired response d_k=y[k+1]. We train our models taking these patterns at random from our codebook.

To estimate **a** in an adaptive fashion, the LMS algorithm can be used [8]. This gradient-based algorithm updates the coefficients in every iteration making a stochastic estimation of the gradient:

$$\mathbf{a}^{j+1}=\mathbf{a}^j+\alpha e^j \mathbf{y}_k \tag{6}$$

the error in iteration j being defined as:

$$e^j=d_k-(\mathbf{a}^j)^T\mathbf{y}_k \tag{7}$$

Using the above, we propose the following competitive segmentation algorithm:

1) Form a codebook of pairs ({y[k],y[k-1]...y[k-l]}; y[k+1]) from the signal history;
2) Select the number of models you wish to consider and initialize them at a random value;
3) Take a data vector and predict with every model. The one that better matches the desired response is trained with that data by the LMS algorithm;
4) go to 3) till stability has been reached.

At this point we have a set of models as well as a segmentation of the state space; different approaches can be taken at this time depending on the kind of problem we are dealing with:

5a) We have a model for the segmentation of the state-space. The classification we have obtained can be used to confirm the validity of the model as well as to estimate the parameters of such a model. We can switch back and forth between the segmentation model and the prediction model until we obtain a reasonable solution (i.e., once we have adjusted the data to the segmentation model we estimate again the prediction models in the segmented state space we have obtained).

5b) We can have several candidate models for the segmentation. The obtained classification can serve to validate any of these models.

5c) We have no idea about how to model the state space segmentation. In this case we can consider a general purpose classificator as the multilayer perceptron [8].

4. A SIMULATION EXAMPLE

4.1 Example

We apply our model to the segmentation of a particular kind of nonstationary signals referred to as quasi-stationary or locally stationary. This type of signals are composed of segments of stationary behaviour combined with abrupt changes of their statistical properties in the transitions between different segments (see Fig. 2). They appear frequently in fields like speech analysis, seismology, vibration analysis and econometrics [7].

We assume that we have observed a quasi-stationary signal over a period of time, and we want to determine: a) the number of segments; b) the samples where the transitions take place; and c) the model for every segment [7]. Classical approaches have been taken to solve this problem [11]. We will only consider the approach followed in [7].

In this reference, a Bayesian solution is presented. It is based on maximizing the a posteriori probability of all the desired parameters given the signal. This approach ends with the minimizaton over all the possible values of the parameters of a determinate functional: the numerical problem is formidable, but it can be tackled by dynamic programming. The assumptions made in this technique are: a) The maximum number of possible segments is known; b) the segments can be modeled parametrically from a known set of models with a maximum order; and c) the data can be associated with parametric density functions.

The problem can also be formulated under our model. The corresponding algorithm is as follows:

1) Choose a number of models n.
2) For i=1 to n, define i models with the maximum order, and let them compete for the data until stability is reached.
3) Given the final models, fit a segmentation model with i-1 transitions to the resulting clouds of points.
4) Reestimate the parameters of all the models once the time segmentation is defined.
5) Repeat 2) for every allowable order.
6) The model that reaches a better performance is elected.

In the next paragraph we will consider the segmentation modeling of step 3).

4.2 Segmentation model

Once we have obtained the clouds of points, as shown for example on Fig. 3, we have to estimate where the transitions take place in the time index, i.e. the segmentation model.

The most obvious possibility is to try all the possible distributions of the boundaries and the one that gests less error when matched to the cloud of points is chosen. Given n models and a sequence of length N the number of possible ditributions of the transitions is (N-1)...*(N-n). In the simplest case of two models we only have N-1 possible solutions. With a reasonable number of models the complexity is not too high.

In our simulations, we use a different approach to reduce even further the complexity of the method. In first place we have to realize that a given signal segment has to have a minimum length to be detected. This *minimum length of a time segment* wil be called L.

Now suppose there is a signal like the one shown in Fig. 3, and we need to decide how many models are there. The first step will be to decide if a particular model has won enough times in a given time segment to assign that time interval to the model. If a model has really concentrated in a particular time segment it will have a lot of winning points in a window of length L centered at that particular time segment. How many? It seems reasonable to say that more than L/n (n number of models) and less that L+1 (The threshold can be set in between these two values, a higher one making it more difficult to find winning models).

We will now calculate the number of winning points of all the models in every possible window of length L. Once we applied this comparison we will have the group of the good models (those that, at least in one window of length L, have a winning number of points greater than the threshold). These "good models" have a window where they get the greatest number of winning points (and, of course, this window will very rarely be the same for two models). We will center the time segment for each model in that "best" window.

So, we have found p "good models" ordered in the time index. How do we estimate the transitions? We take the first and the second model and we estimate the transition between them; then, the second and the third; and so on. Given two models, the transition instant is located where the second model starts to have more winning elements than the first.

Of course, we do not claim that this approach is optimum in any sense: but it is simple, intuitive, fast, and it works quite nicely.

4.3 Simulation results

We have carried the same simulation example (correcting some obvious mistakes) as in [10] and compared the results of both approaches. The aim is not to *improve* the results of the Bayesian technique, but to test how our *general* model behaves against the other. The scope of our model is much wider, and its complexity much more reduced.

We generate 100 realizations of a quasi-stationary process, each 300 samples long. Each realization has three segments of AR processes. The segment boundaries are at points 81 and 211. The first process is of second order with parameters a_1= 1.37 and a_2=-0.56. The second is fourth order with parameters a_1=1.6, a_2=-1.73, a_3=0.924, and a_4=0.3816. The third one is first order with a parameter a_1=-0.8. This parameters follow the notation in (6). The variance of the excitation noise is 1. The maximum number of assumed segments is 5, and the maximum order of the AR models is set to 5. The minimum length of a time segment is L= 25 samples. The threshold for validating models is set to L*(n-1)/n winning points in a window of length L. To estimate the order of the AR models we use the MDL (Minimum Descriptive Length) criterion [12].

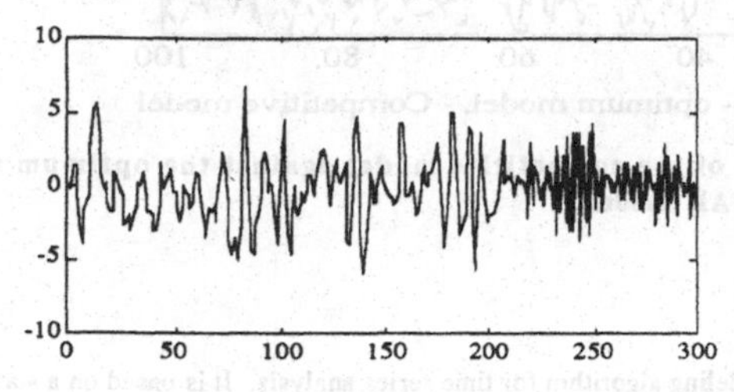

Fig. 2 A typical realization of the locally stationary process

border 1 (81)		border 2 (211)	
range	cases	range	cases
< 65	9	< 195	3
65-75	13	195-205	4
75-85	29	205-215	74
85-95	15	215-225	8
95-105	7	> 225	5
> 105	15		

Table 1 Ranges of estimated boundaries (When they were detected).

The results are shown in Tables 1 and 2. A typical cloud of points obtained in a three model competition can ben seen in Fig. 3.

k=1	k=2	k=3	k=4	k=5
0	23	73	4	0

Table 2 Number of times that the estimated number of segments was equal to k.

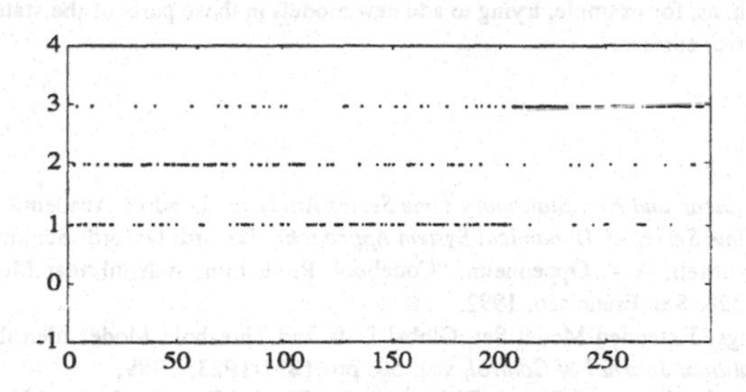

Fig. 3 Graph of the time instants each of the three competitive models have won.

We can see that our algorithm correctly estimates the number of segments in 73% of the cases, while the Bayesian approach gets about a 91%. But, as you can see in Figure 4, the MDL obtained in our competitive modeling (even when the algorithm gets a wrong number of segments) is reasonably close to the optimum.

We conclude that our model shows, in this specific problem, a reasonable performance at a low computational cost.

This kind of modeling can be applied, for example, to speech coding, if you realize that when the coding window is centered in a transition between two phonemes the AR modeling becomes unreliable: it is possible to think of a speech coder that chooses if a transition has taken place in the actual window by means of a competition among models. We can use one bit to determine if the AR coefficents we transmit stand for one single AR model or not. In the case we have two models we reduce their orders to fit in the same data flow using some additional bits to code the transition point. Our method will be easily implementable; on the other side, it does not seem very realistic to use the dynamic programming involved in the Bayesian approach in this kind of application where on line processing is compulsory.

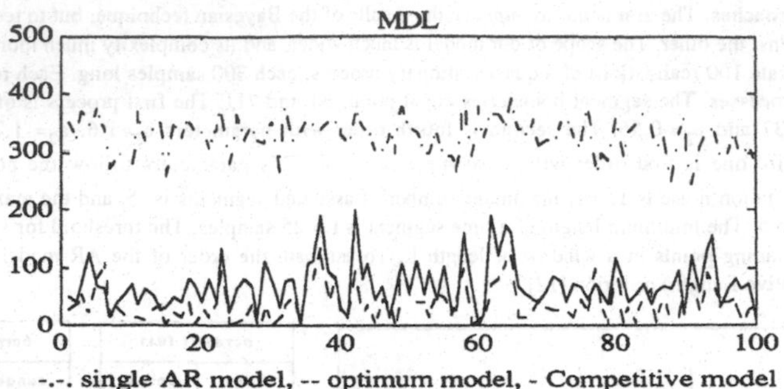

Fig 4. MDL obtained in 100 simulations of the competitive model against the optimum model (three AR segments) and a unique AR model.

5. CONCLUSIONS

We have presented a nonlinear signal modeling algorithm for time series analysis. It is based on a state space representation for finite memory dynamical systems. We view the time series as a collection of patterns each with it associated output (the sample to be predicted). We consider a local model approach and try to fit parametric models to groups of patterns. The election of the model for each pattern is made by competition among models while they are trained. We start with a number models with random parameters and we train them in an adaptive fashion. We present every pattern to each model and the one that gets the best prediction is trained with that particular pattern. When stability is reached we end with a group of patterns associated to each model, i.e., a segmentation of the state space, and a model for each group. We have discussed the range of applicability of this technique, and, in a particular case, we demonstrate how our model shows a reasonable behaviour with a reduced complexity.

We can think of improving the performance of our competitive method of segmentation by controlling the evolution of the competition, so it will lead to a distribution following the a priori model for the segmentation. A sort of weigth function can be used. The same approach as the one shown here is currently being applied to threshold models [2], which have a structure similar to the locally stationary signals we have considered in this work.

The development of other approaches for nonlinear extension of linear models represents an important direction of future research, as, for example, trying to add new models in those parts of the state-space where the error is greatest given the linear error surface.

REFERENCES

[1] M.B. Priestley: *Non -linear and Non-Stationary Time Series Analysis*; London: Academic Press; 1989.

[2] H. Tong: *Non-linear Time Series -A Dynamical System Approach-;* Oxford: Oxford Science Publications; 1990.

[3] A.C. Singer, G.W. Wornell, A.V. Oppenheim, "Codebook Prediction: A Nonlinear Modeling Paradigm"; Proc. ICASSP, vol. 5, pp. 325-328; San Francisco, 1992.

[4] S. Chen, S. A. Billings,"Extended Model Set, Global Data and Threshold Model Identification of Severely Non-Linear Systems"; *International Journal of Control*, vol. 50, pp. 1897-1923; 1989.

[5] M. Casdagli, "Nonlinear Prediction of Chaotic Time Series" ; *Physica D*, vol. 35, pp. 335-356; 1989.

[6] M. Jordan, "Recent Developments in Supervised Learning"; Tutorial #6, IJCNN; Baltimore, 1992.

[7] P.M. Djuric, S.M. Kay, G.F. Boudreaux-Bartels, "Segmentation of Nonstationary Signals"; Proc. ICASPP vol. 5, pp. 161-164; San Francisco, 1992.

[8] R. Hecht-Nielsen: *Neurocomputing;* Reading (Mass):Addison Wesley; 1990.

[9] T. Kohonen, "The Self-Organizing Map"; *Proc.IEEE*,vol. 78, no. 9, pp. 1464-1480; 1990.

[10] S.C. Ahalt, "Vector Quantization Using Artificial Neural Network Models"; Proc. COST #229, pp. 111-130; Bayona (Spain), 1991.

[11] M. Basseville, A. Beneviste (editors), *Detection of Abrupt Changes in Signals and Dynamical Systems*; Berlin:Springer-Verlag; 1980.

[12] J.A. Rissanen, "Universal Prior for the Integers and Estimation by Minimum Description Length", *Ann. Stat.*, vol. 11, pp. 417-431; 1983.

IDENTIFICATION AND PREDICTION OF NON-LINEAR MODELS WITH RECURRENT NEURAL NETWORK

ADAM Olivier ZARADER Jean-Luc MILGRAM Maurice

Laboratoire de Robotique de Paris
URA 1305, email : adam@robo.jussieu.fr
Boîte 164 - T 66 - 2e étage - 66-56
4, place Jussieu, 75252 Paris Cedex 05, FRANCE

ABSTRACT

Using a neural network to identify models and predict signals allows to go beyond the linear domain. In this paper, we show the advantage of using neural network for these signal processing applications. Thus, the function charactering the cell (sigmoïd or others) allows the study of non-linear models. Using feedback links specific to a recurrent network, the time is taken into account. Two different goals are assigned to the two phases in using this type of network : 1) the neural network training method uses a gradient backward propagation method. During the learning phase, the weights of the network are modified to identify the parameters of the given model. 2) during the test phase, the network predicts the output for each time step. Results are presented in the case of a Non-Linear AutoRegressive filters and they confirm the good responses of neural networks both for identification of paramaters and for prediction of output for these non-linear models.

INTRODUCTION

A lot of papers about filtering problems in Signal Processing have been published in the specialized litterature. An interesting approach for signal analysis is to make a model of the signal generating process in order to extract a few pertinent parameters best characterizing (according to a specific criteria) the signal. Identifing a model defined by its C_i coefficients is equivalent to find the set $\hat{}C_i$ characterizing a same class filter, i.e. with no error between C_i and $\hat{}C_i$. And for the prediction of a signal, the difference between the output y(k) of the model and $\hat{}y(k)$ of filter built with $\hat{}C_i$ coefficients must be minimized. Up to now, the various methods are using linear systems. Thus, we took a two steps approach by identifying a non-linear model and then predicting the signal generator by the model. We use neural networks which have a built-in non-linear function in each cell. Backpropagation algorithm is already well known. Identification of paramaters of the model is associated to the training phase of the network and the prediction is done during the test phase (with the usage of previous signal values). Time is very important : it is used in the internal structure of the networks which are recurrent i.e. with links for the output layers to input layers (look-back).

In the first part, we consider the signal as a non-linear filter output, excited by a white noise. There is a close analogy with this non-linear filter and the already known linear filters.
Afterwards, we move to the study of dynamical neural networks and a specific use of the gradient backward propagation algorithm.
Then, different tests allows us to evaluate the networks performances for model's identification, and non-linear models signal prediction.

1. CREATION OF NON LINEAR MODEL

We must define a non-linear filter containing known paramaters. The filter structure has to be identified by a specific neural network. We add a non-linear function to a classical linear filter.
Infinite Impulsionnel Response filters are of particular interest : 1) their specific structure include an output-input reacting loop, analog to a recurrent network. 2) They are well known and several methods of computation exist.

1.1. AutoRegressive Moving Average Model.

In the linear domain, the filter chosen to modelise a process has a rational transfert function. It is described by a differential equation (continuous representation) or by a recurrent equation with constant coefficients (discret representation). We will use the second form for numerical process.
The filter is represented by the equation :

$$y(k) = -\sum_{i=1}^{p} a_i y(k-i) + \sum_{j=0}^{q} b_j u(k-j)$$

Using the z-transform, the transfert function is :

$$H(z) = \frac{Y(z)}{U(z)} = \frac{b_0 + b_1 z^{-1} + \ldots + b_q z^{-q}}{1 + a_1 z^{-1} + \ldots + a_p z^{-p}} = \frac{B(z)}{A(z)}$$

where U(z) is the input and Y(z) the output.

1.2 Definition of a non-linear model.

Using the algorithms builted previously, it is not possible to identify a non-linear model and predict its output. The linear domain constrains the use of the above methods to a few specific applications. Thus, we introduce a non-linear model with a closed ARMA filters structure with the addition of a non-linear function.

$$y(k) = f\left(-\sum_{i=1}^{p} a_i y(k-i) + \sum_{j=0}^{q} b_j u(k-j) + c\right)$$

f is a derivable non-linear function; u is a gaussian, centered, unit standard deviation, white noise; c is a real constant. We use the sigmoid function for our study:

$$f(x) = m\,\frac{e^{kx} - 1}{e^{kx} + 1}$$

k is a stiffness factor of the sigmoid function. The function f is derivable at each point and vary between -m to +m for x values included between $-\infty$ and $+\infty$.

This type of filter will be called NLARMA for Non Linear AutoRegressive Moving Average filter. The same methodology could be used to define NLMA and NLAR by analogy with MA filter (Moving Average) and AR filter (AutoRegressive).

2. SPECIFIC RECURRENT NEURAL NETWORKS

2.1. Recurrent neural networks.

During the training phase, it is possible to present at feedforward networks the examples set in random instead of chronological order. This characteristic is widely used for classification and recognition of patterns but prevents any usage in prediction of time series. Rumelhart and al. (1986)[1] describes recurrent network as feedforward multi-layers network which increases by one layer at each time step. E. Levin (1990)[2] define a recurrent network as equivalent to N inditical perceptrons. Jordan (1986), Stanelta (1987) and Elman (1988)[3] use a particular structure of the network (fig. 1), with a few recurrent carefully chosen connections. The originality of the structure resides in a specific layer, called context, dedicated to the reception of ouput value from the previous time step. But, Servan-Schreiber (1988) has shown the limit of this method in learning and detecting long time sequences. Mozer (1988) tries to overcome the difficulty by using neurons with an internal waiting loop allowing for t-n input values.

For our study, we define the structure of network (fig.2). A network is completely defined by the number of the external input, the number of the internal input and the number of the output. An internal input is the loop-back of the output. The context layer (Jordan model) is replaced by delayed cells. The depth of the historical past depends of the number of internal input cells.

2.2. The backpropagation.

The training phase consists in minimizing the difference between computed output and desired output. Weights are modified according to the gradient backward propagation method[4]. This algorithm is summarized below :

1. presentation of a (external and internal inputs) at step k.
2. propagation in the network.
3. computation of criteria (error).
4. backpropagation in the network.
5. modification of network's weights.
6. presentation of (external and internal inputs) at step k+1.

remarks:

* initial weights are at random.

* backpropagation algorithm allows for several possibilities of evaluation of the error criteria going from the global modification of the weights after evaluation of the whole set of examples (gradient) to the modification after process of each example (stochastic gradient). The latest method is applied because the studied models are stationary relative to the parameters.

* for our work, we used two options : 1) recurrent method : in this case, we present the computed output values at the internal inputs. For example, a component of training set is {x(k-M), x(k-M+1), ..., x(k), y(k-N), y(k-N+1), ..., y(k-1),

$y_{des}(k)\}$ for a network with M+1 external inputs, N internal inputs and one output. **2) pseudo-recurrent method** : now, we present, at the internal input, the desired values at proper time. So, a component of training set is $\{x(k-M), x(k-M+1), ..., x(k), y_{des}(k-N), y_{des}(k-N+1), ..., y_{des}(k-1), y_{des}(k)\}$ for the same network.
In the first case, with a stochastic gradient learning method, we are in the adaptative case : each new example has a direct impact on the evolution of the network. In the second case, we don't use computed values as internal inputs and therefore the error betwenn desired output and computed output is not propagated.

3. RESULTS : IDENTIFICATION AND PREDICTION.

3.1. Identification and prediction of simple models.

We started with simple models to validate our method by using NLAR filters equivalent in the linear domain to purely recursive second order cells. A gaussian centered, unit standard deviation, white noise (u(k)) is used as external input and the filter provide the expected output y(k). These values are presented to network's inputs and y(k) will be the desired ouputs. And it is possible to know when sigmoïd function of the output cell is used in its linear part, non-linear part (rounded part), or saturated part.

3.1.1. same dimension between model and network.

We got the first results by using a network of the same dimension as the filter to be identified (tab.1).

same size network/model	recurrent learning identification	no identif.	pseudo-recurrent learning identification	no identif.	total
linear part	101	0	101	0	101
non linear part	92	12	101	3	104
saturation	70	9	79	0	79
total	263	21	281	3	284

Tab.1 : Identification of non-linear models using networks of the same size of models.

For the networks giving good identification of models, the precision of the weights is in the order of 10^{-6} (computational noise) after about 500 iterations, and the relative error between the output of network and the desired output is always less than 10^{-4} after 500 time steps.
We observe that identification using pseudo-recurrent learning method is better than identification using recurrent learning method. During the training phase, the pattern is constant for the first method and evolutive for the second method.

3.1.2. Network oversized.

If we do not have any information on the model to be identified, we must start with a neural network of arbitrary size. It is reasonable to use a rather large network and watch for its behaviour to a filter with less parameters. For example, a 5 inputs - 1 output network has been used to study the same NLAR filter (tab.2).

oversized network/model	recurrent learning identification	no identif.	pseudo-recurrent learning identification	no identif.	total
linear part	49	2	49	2	51
non linear part	62	39	101	0	101
saturation	41	34	73	2	75
total	152	75	223	4	227

Tab.2 : Identification of non-linear models using oversized networks.

1 case : linear part of the sigmoïd function.
This is the case of small amplitude signals and the output stays in the linear part of the sigmoïd function ($y(k)<0.7$). After the trainig phase, the most of networks converge to their model. For smallest magnitude of input signals, the weights are a linear combination of the filter's coefficients. The network could converge to any of these combinations, according to the initials weights. Anyway, the error of prediction stays less than 10^{-4}. After about 700 000 change of weights, the network converge to the coefficients of the filter, zeroing the weights of cells in excess and the error between weights and coefficients of filter is about 10^{-5}.
Using neural network in this case is not justified because we can use linear methods.

2 case : non-linear part of the sigmoïd function.
Close to the saturation part of sigmoïd, the network converge after 50 000 change of weights, to the coefficients of the filter, zeroing the weights of cells in excess. During the prediction phase, the error is less than 10^{-4} : the weights are equal to the filter's coefficients. We will see reasons of no convergence for some networks in the case 3.

3 case : saturated part of the sigmoïd function.

Some networks do not converge because of their noise sensitivity (different noises are not studied here) :

*** In the ideal case, (the network is in the exact configuration of the filter), there is no error and :**

$$y_{des}(k) = f(a_0 u(k) + b_2 y_{des}(k-2) + b_1 y_{des}(k-1)) = f(p_1(k))$$

where $p1(k)$ **is the potential at step k.**

*** In the pseudo-recurrent method :**

$$y(k) = f(v_0(k)u(k) + w_2(k)y_{des}(k-2) + w_1(k)y_{des}(k-1))$$

$$y(k) = f(p_1(k) + e_{v0}(k)u(k) + e_{w2}(k)y_{des}(k-2) + e_{w1}(k)y_{des}(k-1))$$

where

$$v_0(k) = a_0 + e_{v0}(k)$$
$$w_i(k) = b_i + e_{wi}(k)$$

*** In the recurrent method :**

$$y(k) = f(m_0(k)u(k) + n_2(k)y(k-2) + n_1(k)y(k-1))$$

$$y(k) = f(p_1(k) + e_{m0}(k)u(k) + e_{n2}(k)y_{des}(k-2) + e_{n1}(k)y_{des}(k-1) + b_2 e_y(k-2) + b_1 e_y(k-1) + e_{n2}(k)e_y(k-2) + e_{n1}(k)e_y(k-1))$$

where

$$m_0(k) = a_0 + e_{m0}(k)$$
$$n_i(k) = b_i + e_{ni}(k)$$
$$y(k) = y_{des} + e_{ydes}(k)$$

Thus, the number of the error terms is 3 for the pseudo-recurrent method and 7 for the recurrent one. This penalizes the success and the speed of weights convergence.

3.2. Evolving models.

The Filter considered above uses constant parameters. Using the stochastic gradient learnig method leads to the fastest convergence (1 iteration = 1 weight change). Now, let's study models with time dependent parameters, namely coefficients variations.

a. linear. Coefficients of the model describe straight line equation . The weights of the network vary according to the variation of the coefficients of the filter (fig.3). After 200 iterations, the error between weights and coefficients are of the same order of magnitude as in the previous case.

b. non-linear. The filter uses sinusoidal varying parameters (fig.4). After the learning period, the error between the weights of the network and the coefficients of filter is less than 10^{-3}**.**

CONCLUSION

Starting from the classical methods for parameters identification and signals prediction, used in the linear domain, we defined ARMA filters with non-linear function. Recurrent neural networks have shown their advantage for studying such filters. We are currently investigating larger filters (up to 10 parameters) and larger networks (up to 30 weights). The first results are consistent with above results.

REFERENCES

[1] D. Rumelhart, G.E. Hintonand R.J. Williams, "Learning Internal Representations by Error Propagation", *Parallel Distributed Processing : Explorations in the Microstructure of Cognition*. MIT Press.

[2] E. Levin, "A Recurrent Neural Network: limitations and training", *Neural Networks*, vol.3, pp.641-650,1990.

[3] J.L. Elman, "Finding Structure in Time", *Cognitive Science*, n°14, pp.179-211, 1990.

[4] Y. Le Cunn, "Modèles Connexionnistes de l'Apprentissage", *Thèse de l'Université Paris 6*, 1987

[5] Ronald J. Williams and David Zisper, "Experimental Analysis of Real-Time Recurrent Learning Algorithm", *Connection Science*, Vol.1, n°1, 1989.

[6] F.J. Pineda, "Generalization of back propagation to recurrent networks", *Phys. Rev. Lett.*, vol.59, n°19, pp.2229-2232, Nov.87.

[7] B. Pearlmutter, "Dynamic Recurrent Neural Networks", *School of Computer Science*, Carnegie Melln University, Pittsburgh, PA 15 213, Déc.90.

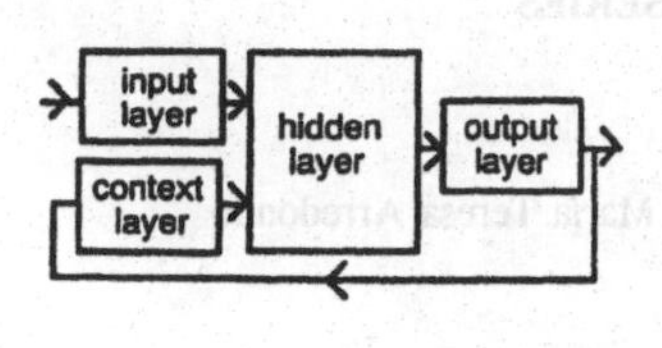

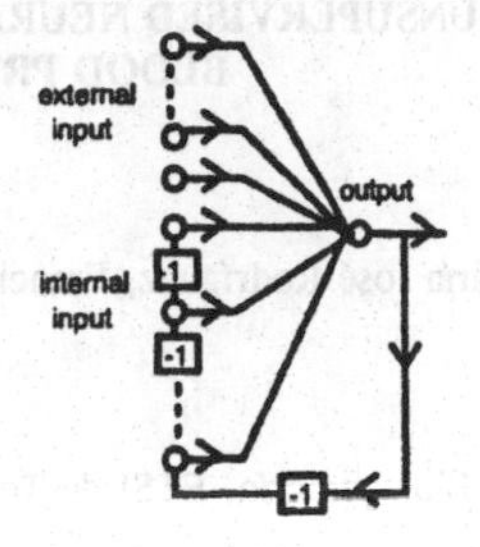

Fig.1 : recurrent network type JORDAN

Fig.2 : type of recurrent networks used

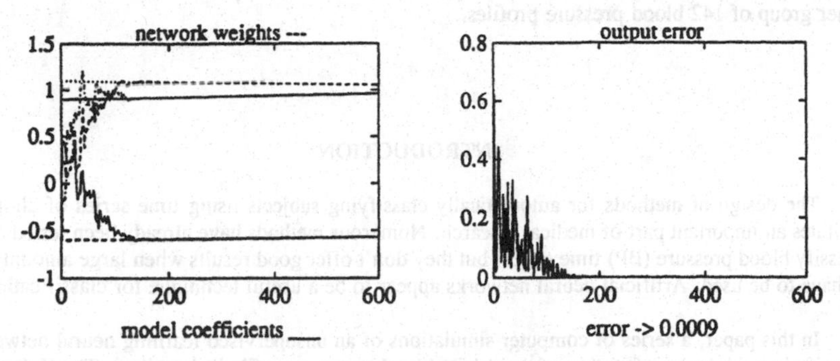

$m_0(k) = 0.9 + 0.0001*k \; ; \; n_1(k) = 1.1 - 0.00008*k \; ; \; n_2(k) = -0.6 + 0.00005*k$

Fig.3 : linear variation of model's coefficients

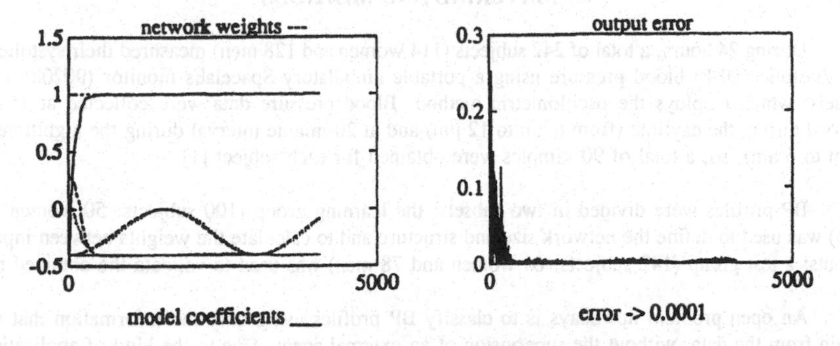

$m_0(k) = 1 \; ; \; n_1(k) = -0.4 \; ; \; n_2(k) = -0.2 * \cos(2*\pi*k/3000) - 0.2$

Fig. 4 : non-linear variation of model's coefficients

USE OF UNSUPERVISED NEURAL NETWORKS FOR CLASSIFICATION OF BLOOD PRESSURE TIME SERIES

María José Rodríguez, Francisco del Pozo and María Teresa Arredondo

Dep. de TE y Bioingeniería, ETSI de Telecomunicación, Univ. Politécnica de Madrid, España

Abstract: This paper describes a method to classify blood pressure time profiles using artificial neural network with unsupervised learning. Kohonen's Topology Preserving Maps were used to identify similar characteristics in 100 profiles from different subjects. Afterwards, obtained results were validated using another group of 142 blood pressure profiles.

INTRODUCTION

The design of methods for automatically classifying subjects using time series of clinical data constitutes an important part of medical research. Numerous methods have already been tested and used to classify blood pressure (BP) time series, but they don't offer good results when large amount of input data have to be used. Artificial neural networks appear to be a useful technique for classification tasks.

In this paper, a series of computer simulations of an unsupervised learning neural network used to classify individuals as a function of their daily blood pressure profile is described. The Kohonen self-organizing neural network was chosen for this purpose. Our study aims to test its behaviour and usefulness for blood pressure time series classification tasks.

MATERIAL AND METHODS

During 24 hours, a total of 242 subjects (114 women and 128 men) measured their systolic (SBP) and diastolic (DBP) blood pressure using a portable ambulatory Spacelabs monitor (90202 or 90207 models) which employs the oscillometric method. Blood pressure data were collected at 15-minute interval during the daytime (from 6 am to 12 pm) and at 20-minute interval during the nighttime (from 0 am to 6 am), so, a total of 90 samples were obtained for each subject [1].

BP profiles were divided in two subsets: the learning group (100 subjects: 50 women and 50 men) was used to define the network size and structure and to calculate the weights between inputs and outputs; a test group (142 subjects: 64 women and 78 men) was used to validate the obtained results.

An open problem nowadays is to classify BP profiles using only the information that we can obtain from the data, without the supervision of an external agent. Due to the kind of application, we have to use a neural network that allows continued-value inputs. There are some neural networks that are useful for classification tasks and allow unsupervised learning and analogue inputs: Kohonen's Topology Preserving Maps, Counterpropagation network, Adaptive Resonance Theory 2, etc ([2], [3]). After an extensive study, Kohonen's network was chosen because of the simplicity of its implementation, its capacity to classify similar patterns in clusters and the capacity of the weights to approximate the distribution of the input vectors, that is, the weight values of an output are equal to the mean value of all BP profiles classified in that output.

Kohonen's algorithm is applied to an architecture were there are N continuous-valued inputs (x_i), and M outputs arranged in an array (one- or two-dimensional). The outputs are fully connected to the inputs via the weights (w_{ij}). A competitive learning rule is used, choosing the winner (i^*) as the output unit with a weight vector closest to the current input. The learning rule is:

$$w_{ij}(t+1) = w_{ij}(t) + \eta\, \lambda(i,i^*)(x_j - w_{ij}) \quad \forall i, j$$

The neighbourhood function $\lambda(i,i^*)$ is 1 for $i=i^*$ and falls with the distance between units i and i^* in the output array. Thus, units close to the winner, as well as the winner i^* itself have their weights changed appreciably, while those further away, where $\lambda(i,i^*)$ is small, experience little effect.

The gain term, η, should start with a value close to the unit and decrease monotonically thereafter. The learning phase stops when the gain term is zero.

The learning phase of Kohonen's algorithm consists of repeatedly presenting numerous pattern to the network without specifying to each class the pattern belongs to (unsupervised learning). During this phase, the network self-organizes his weights, in such a way that patterns that resemble each other are mapped into clusters that are neighbours in the output array.

A software was developed that allows to assess the optimal size and structure (one- or two-dimensional array) of the output array and also the gain, neighbourhood and number of iterations during the learning phase.

Once the optimal conditions were chosen, the test data set was used to validate the results obtained during the learning phase.

RESULTS

As inputs to the neural network we used the time specified SBP and DBP values. To make the neural network learning phase practical is mandatory to reduce the number of inputs. Then, in a previous step, we calculate the average value within each hour interval of the SBP and DBP samples, in this way the input dimension is reduced from 180 to 48 (24 SBP and 24 DBP values). Normalized values (subtracting the individual mean value and dividing by the individual standard deviation) were used. Also, the sex, age and SBP and DBP daily time average for each subject can be used as inputs to the network.

The outputs are limited by practical reasons to a number between six and twelve; they can be distributed in a linear (1x6, 1x7, 1x8, 1x9, 1x10, 1x11, 1x12 outputs) or planar array (2x3, 2x4, 3x3, 2x5, 2x6, 3x4 outputs). All of these options were tested to obtain the one that allows a better classification.

After some trials, a Gaussian curve was chosen as the neighbourhood function; this curve has its maximum value in the winner output and decrease as the distance between the winner and the other outputs increases. During the learning phase, the statistical deviation of the curve is decreased, so the number of neurons that change its weight diminish in each epoch. At the end of this phase, only the weight of the winner neuron should change.

In the learning phase, there are two steps that have a slightly different nature: initial formation of the correct order and final convergence of the map into an asymptotic form. For good results, the latter phase may take around 10 times as many steps as the former. These two steps can be distinguished by the gain term used in each one. The time dependence of the gain factor that we have chosen take the form of a-bt. In the ordering phase, it starts with a value near to the unit (0.9) and ends with a value near to 0.01. In the convergence phase, the gain starts with this last value and ends with a value of 0. At this moment, the learning process automatically stops.

Results obtained using different sizes in the output array show that maximum classification power is obtained when using a 1x8 linear array. Figures 1a,b to 8a,b represent the different BP profiles classified in each output (each one of them representing a different cluster) when using as inputs the 24 SBP and 24 DBP values for each subject. Figure 1a to 8a represent the SBP values of each subject in mmHg in an 24-h interval and figure 1b to 8b represent the DBP values in the same interval. In each plot, the wider line represent the mean value of all profiles belonging to this cluster (or weights of the corresponding output).

It can be seen that profiles with different morphologies belong to different clusters; SBP and DBP mean values have not a big influence in the classification, that is, normotensive and hypertensive people can be classified in the same cluster if they have the same kind of BP profile.

If the SBP and DBP mean values are used as inputs, another different classification is obtained. In this case, BP profiles are classified in different groups, more as a function of their SBP and DBP mean values than by their similar morphologies. In figures 9a,b, the SBP and DBP profiles for each cluster obtained using the mean values are shown. In this way, it is possible to distinguish between normotensive and hypertensive people defined according to the current criteria worldwide in use..

In both cases, results were validated using the test data set. All BP profiles of the test set were classify in the cluster with the most similar characteristics.

CONCLUSIONS

The main conclusion is that it is possible to use artificial neural network to classify BP time series without external supervision. Depending on the kind of inputs that are used, a different classification is obtained.

If only the normalized SBP and DBP values are used, Kohonen's Topology Preserving Maps can be used to detect different morphologies in the BP time series. If SBP and DBP mean values are added, the network is able to separate hypertensive and normotensive people in different levels.

REFERENCES

[1] J.L. Palma: "Control ambulatorio continuo de la presión arterial". In: *Avances en Electrocardiología*. A. Bayés de Luna Ed. Barcelona: Doyma 1981, 177-182.

[2] T. Kohonen: "The Self-Organizing Map", *Proceedings of the IEEE* 1990; 78: 1464-1480.

[3] R.P. Lippmann: "An Introduction to Computing with Neural Nets", *IEEE ASSP Magazine*; April 1987: 4-22.

ACKNOWLEDGMENT

This work has been supported by the Spanish CICYT Grant No. TIC-271/92, and also by the European Program AIM: EPIC, project number A.2007 and IREP, project number A.2018.

Blood pressure data have been provided by Dr. José Luis Palma Gámiz from the "Hospital Ramón y Cajal" of

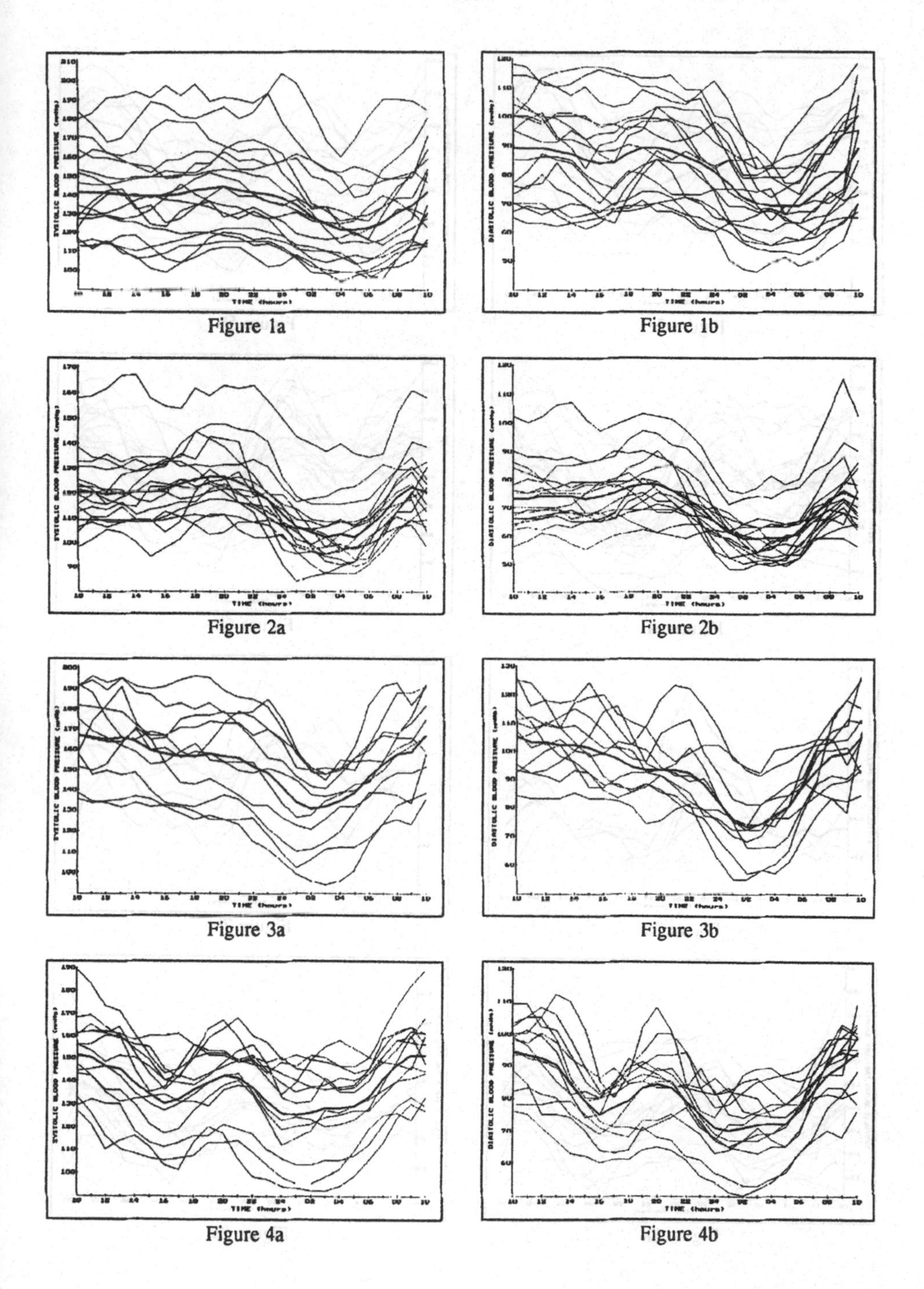

Figure 1a

Figure 1b

Figure 2a

Figure 2b

Figure 3a

Figure 3b

Figure 4a

Figure 4b

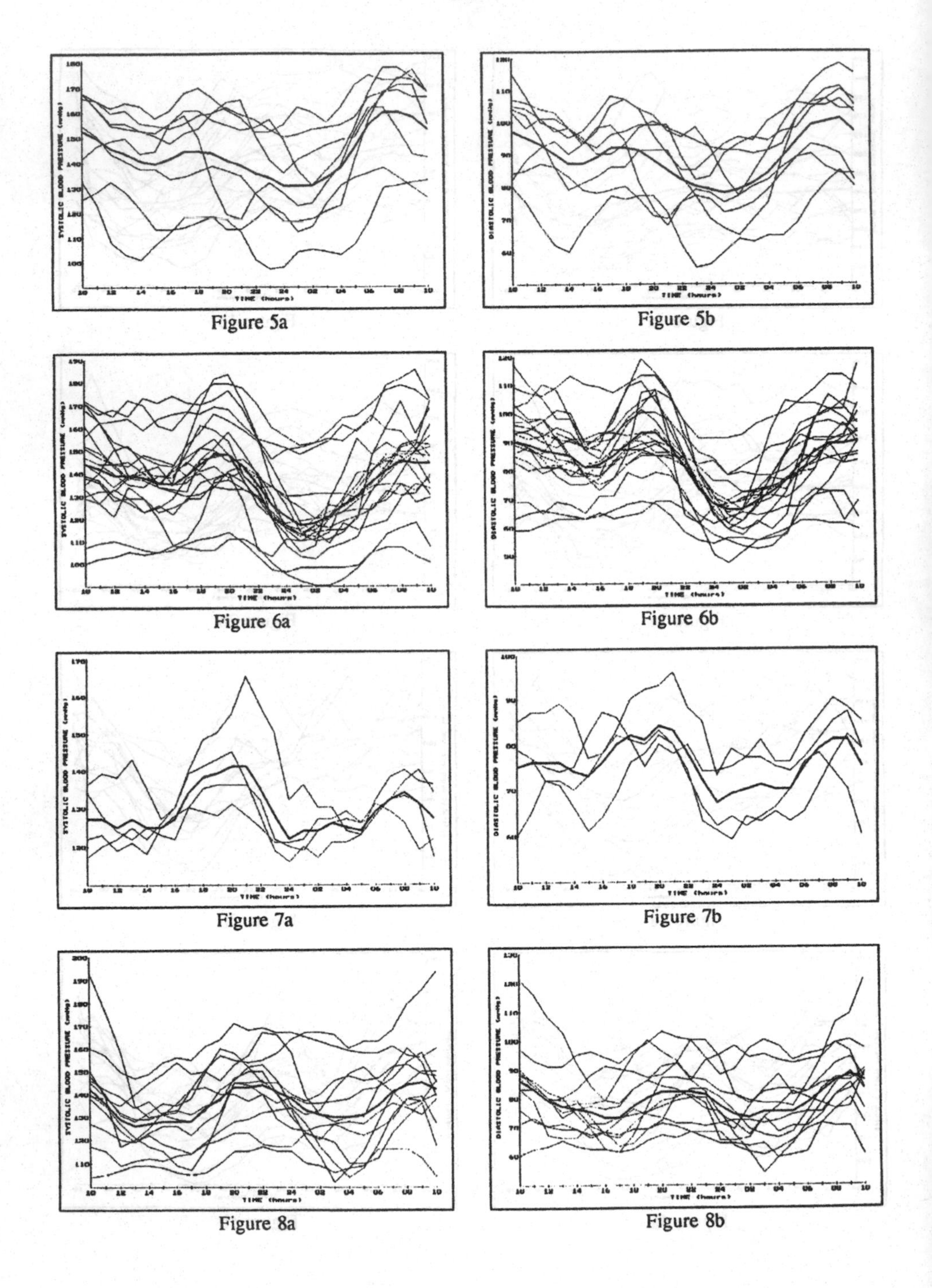

Figure 5a

Figure 5b

Figure 6a

Figure 6b

Figure 7a

Figure 7b

Figure 8a

Figure 8b

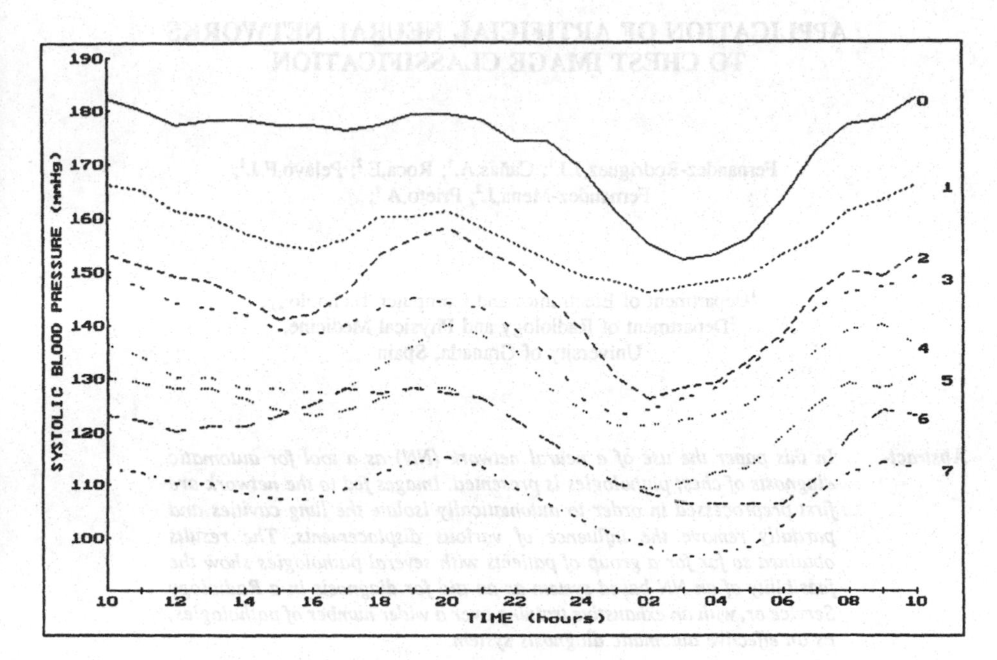

Figure 9a

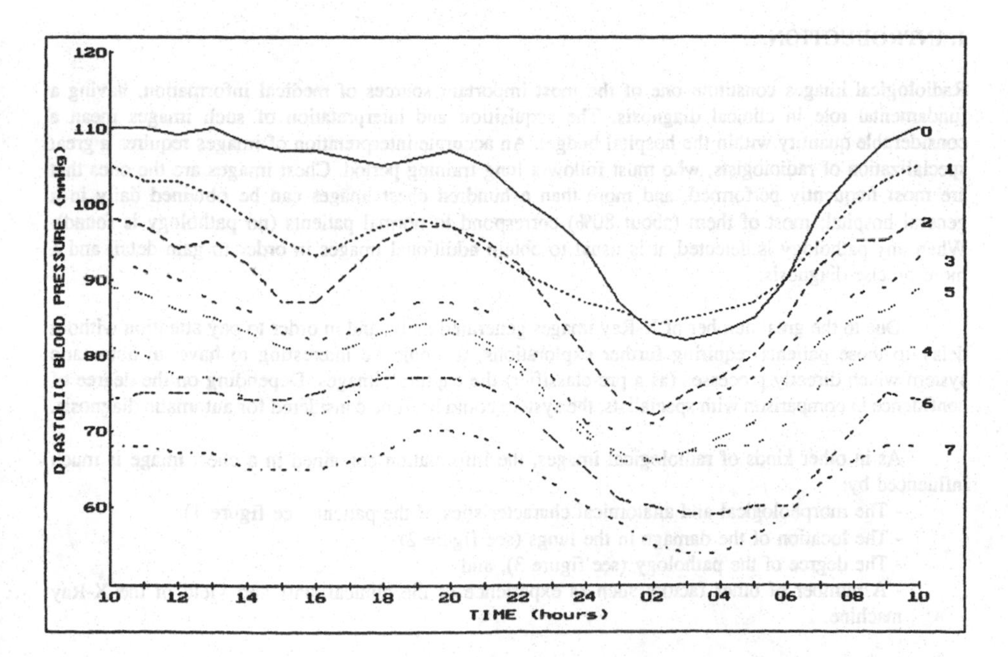

Figure 9b

APPLICATION OF ARTIFICIAL NEURAL NETWORKS TO CHEST IMAGE CLASSIFICATION

Fernandez-Rodriguez,J.J.[1]; Cañas,A.[1]; Roca,E.[2]; Pelayo,F.J.[1]; Fernandez-Mena,J.[2]; Prieto,A.[1];

[1]Department of Electronics and Computer Technology,
[2]Department of Radiology and Physical Medicine,
University of Granada, Spain

Abstract- *In this paper the use of a neural network (NN) as a tool for automatic diagnosis of chest pathologies is presented. Images fed to the network are first preprocessed in order to automatically isolate the lung cavities and partially remove the influence of various displacements. The results obtained so far for a group of patients with several pathologies show the feasibility of an NN-based system as an aid for diagnosis in a Radiology Service or, with an exhaustive training over a wider number of pathologies, as an effective automatic diagnosis system.*

1. INTRODUCTION.

Radiological images constitute one of the most important sources of medical information, having a fundamental role in clinical diagnosis. The acquisition and interpretation of such images mean a considerable quantity within the hospital budget. An accurate interpretation of images requires a great specialization of radiologists, who must follow a long training period. Chest images are the ones that are most frequently performed, and more than a hundred chest images can be obtained daily in a general hospital; most of them (about 80%) correspond to normal patients (no pathology is found). When any pathology is detected, it is usual to obtain additional images in order to gain detail and a more precise diagnosis.

Due to the great number of X-Ray images generated daily, and in order to pay attention without delay to those patients requiring further explorations, it would be interesting to have an automatic system which directly processes (as a pre-classifier) the captured images. Depending on the degree of confidence in comparison with specialists, the system could itself be considered for automatic diagnosis.

As in other kinds of radiological images, the information contained in a chest image is much influenced by:

- The morphological and anatomical characteristics of the patient (see figure 1).
- The location of the damage in the lungs (see figure 2).
- The degree of the pathology (see figure 3), and
- A number of other factors such as experience of the clinical staff and yield of the X-Ray machine.

Taking these problems into account it is difficult to imagine a system not based on learning which could work well as an automatic classifier of chest images. On the other hand, most contributions in this field (using classical image processing techniques) try fundamentally to enhance the quality of images only to have a better diagnosis by the radiologist and to reveal the details related to certain pathologies.

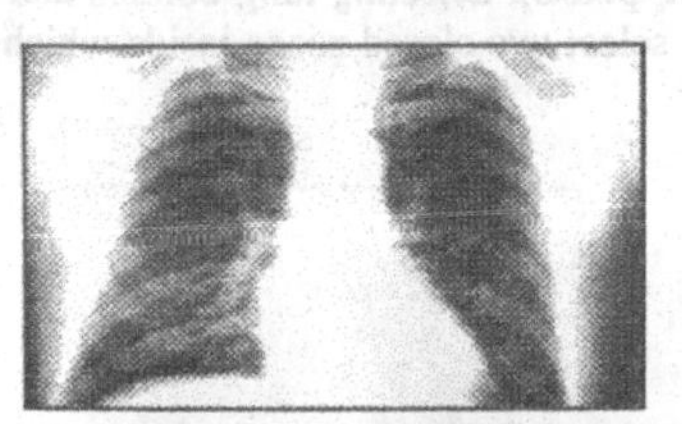
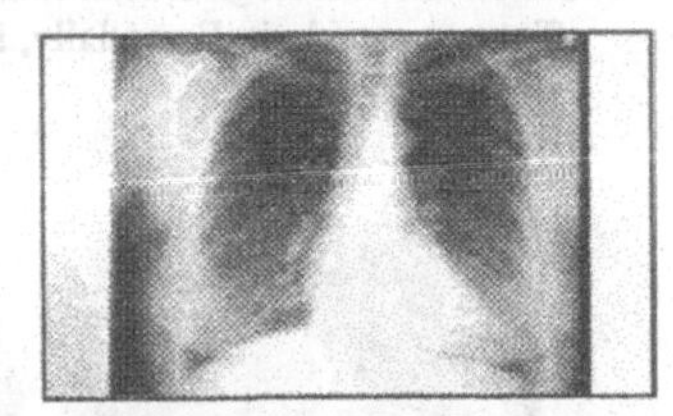

Fig. 1: Digitized chest images of male and female normal patients.

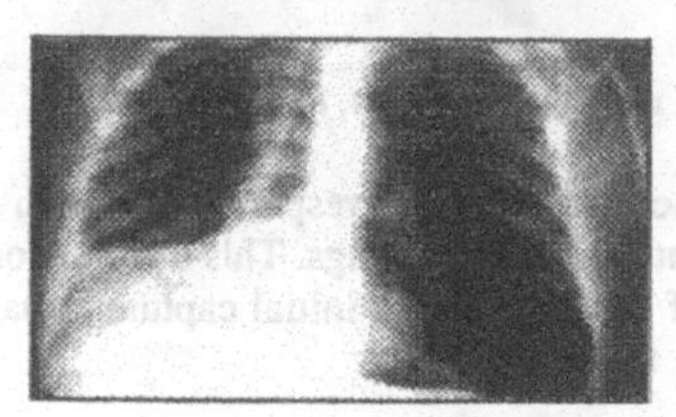
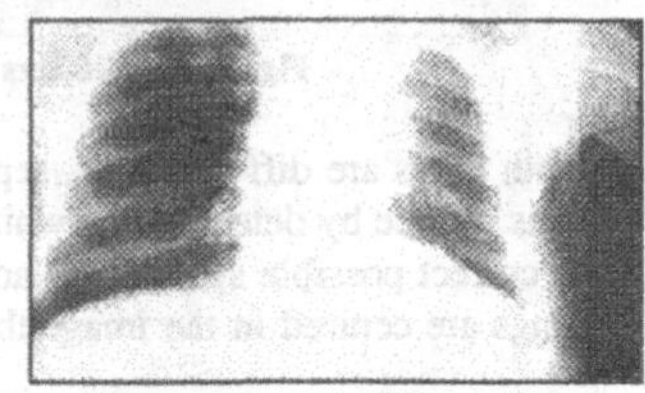

Fig. 2: Chest images of two patients with epidermoid carcinoma located in different lungs.

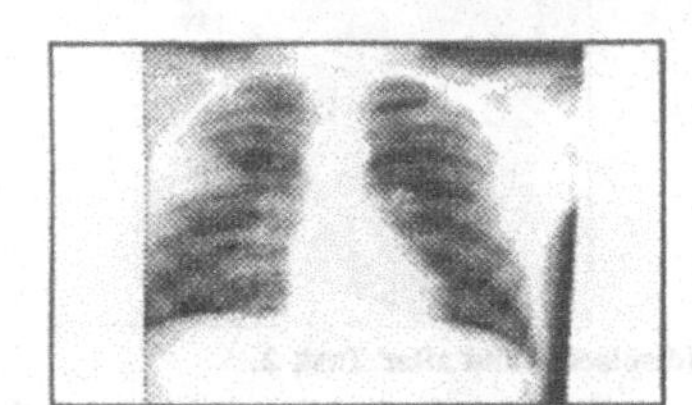
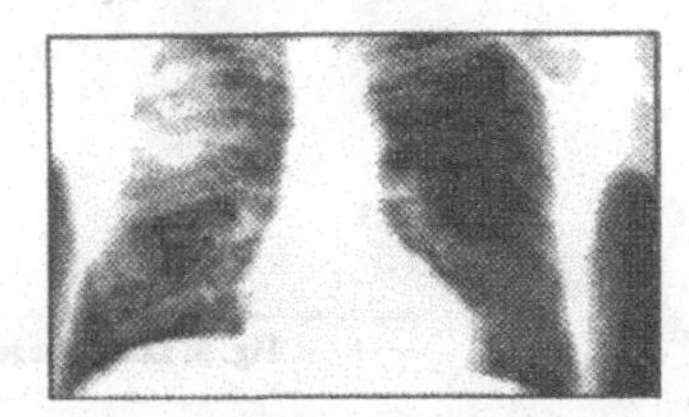

Fig. 3: Chest images of two patients with pneumonia in different degrees.

Neural networks, due to their capability of extracting features by learning, can play a significant role in the field of radiological image classification. In fact, the results we have obtained so far with Multi-Layer Perceptrons (MLPs) for a limited number of training samples are very promising. When a statistically significant number is completed, for several pathologies, the expected accuracy in classification will be similar to that obtained by an expert radiologist. Due to the limited number of training samples we have available, the generalization capabilities of the NN have not yet been fully exploited. We present simulation results illustrating this point, showing the improvement in classification accuracy when the number of samples increases for a fixed net size.

The samples used so far correspond to normal patients and four different pathologies (31 samples of each group, i.e. 155 samples in all). Each sample is first captured and digitized from the conventional screen film image, and is also preprocessed to remove irrelevant information, intensifying the interesting parts (the lungs).

The rest of this paper is organized as follows: Section 2, the preprocessing algorithm; in Section 3, the neural network characteristics are described; Section 4 is devoted to presenting the results of the classifications currently performed, and Section 5 shows some conclusions and possible improvements that we intend to introduce in our system.

2. PREPROCESSING OF RADIOLOGICAL IMAGES.

A preprocessing program has been developed, which performs the following tasks for each digitized image:

Task 1 **Cut off the radiological film image ($N_i \times M_i$ pixels), detecting lung borders and shaving off zones outside it. Essentially, it tries to select two closed zones inside which there is valid information.**

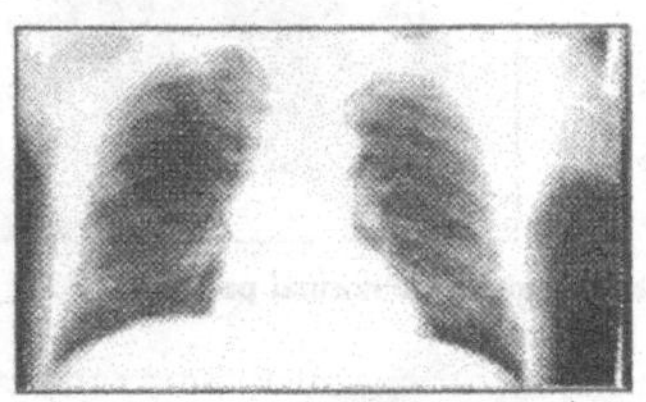
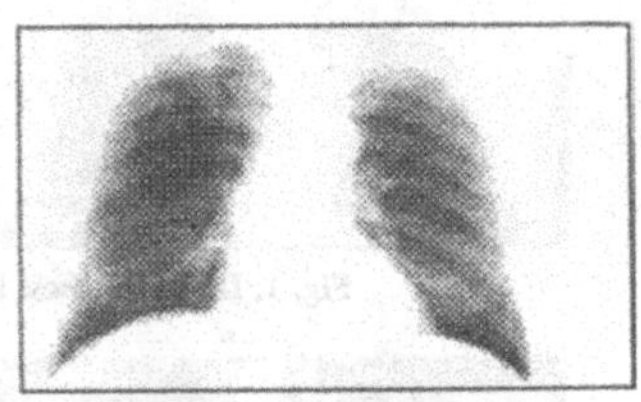

Fig. 4: Chest images before and after Task 1 (N_i=451, M_i=301).

Task 2 **Both lungs are differentiated, separating the information corresponding to each of them. This is done by detecting the white strip that separates the lungs. This information is used to correct possible spatial displacements of the lungs in the initial captured image. Both lungs are centred in the image that will be fed to the NN.**

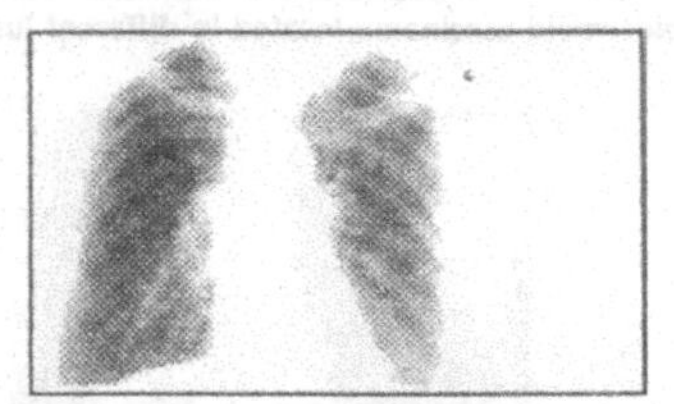
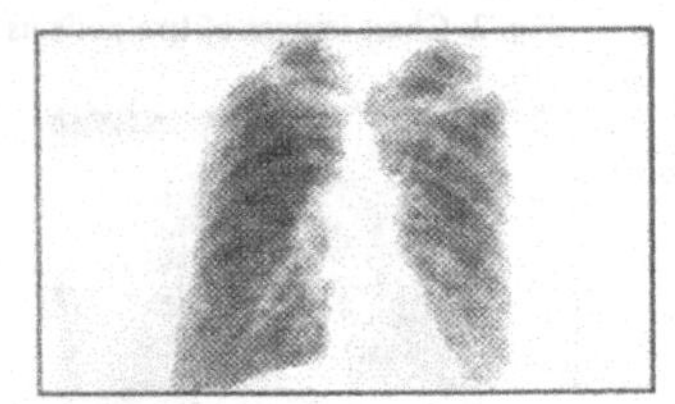

Fig. 5: Lung images before (displaced) and after Task 2.

Task 3 **A spacial alignment is imposed, in such a way that each lung is scaled to a box of preset dimensions. It is a linear alignment; and in order to do this the longest vertical and horizontal lines limited by the lung contour are detected, and compared to the dimensions of the desired rectangular grid ($N_a \times M_a$). Then, a redistribution of image pixels by interpolation or compression is performed, in such a way that it is completely adapted to the box $N_a \times M_a$.**

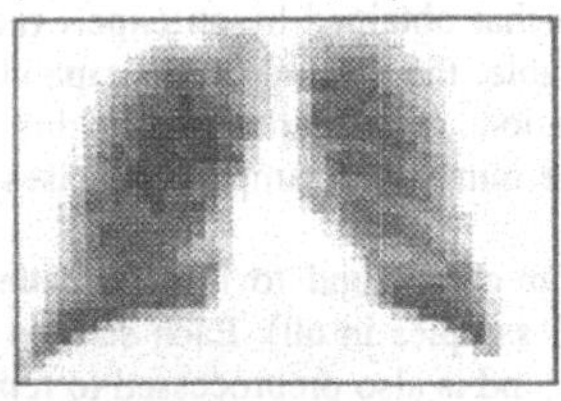

Fig. 8: Image obtained after Task 3 for the lungs in figure 4 ($N_a \times M_a$=40×30).

Task 4 **The resulting information is compressed, in such a way that the data quantity in the picture is reduced to that of the neural network input. The image is divided to as many boxes as pixels are wanted (n=$N_o \times M_o$). Each box (pixel) is assigned a grey value equal to the mean of the initial pixels inside the box.**

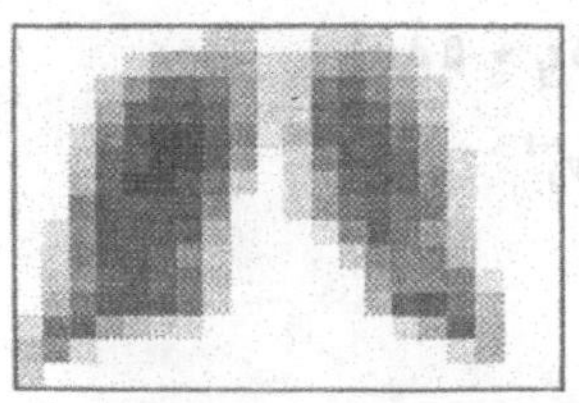

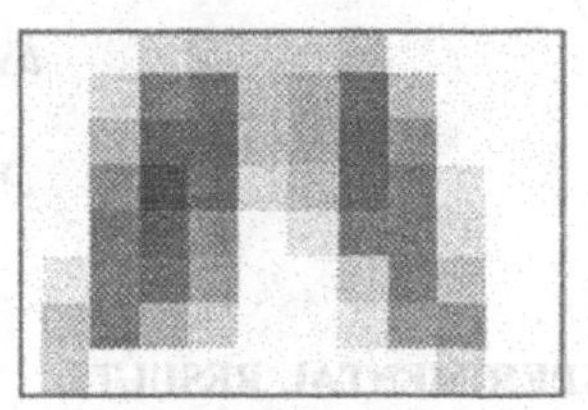

Fig. 7: Images produced by Task 4 from figure 6, for n=20×15 and n=10×8.

An additional preprocessing task was carried out in an intent to improve the information contained in a fixed size image. It consists of a sophisticated algorithm which rotates each lung and puts it in a "vertical" position. However, this task, in the end, was not used since it does not improve the recognition rate of the neural network.

3. CHARACTERISTICS OF THE NEURAL NETWORK.

As a result of the preprocessing, each initial $N_i \times M_i$ picture is spatially normalized and represented by a matrix with $n = N_o \times M_o$ components, which is fed to an n-input MLP [WID]. It consists of a fully-connected feed-forward net with n inputs, one or two layers of hidden units, and one output layer of 5 units which correspond to *normal* patients and four different pathologies: *epidermoid carcinoma, pneumonia, bronchiectasis,* and *lung fibrosis.* The output unit with highest activation indicates which group the input belongs to. Specifically, the configuration which has provided the best results (with one hidden layer) is shown in figure 8.

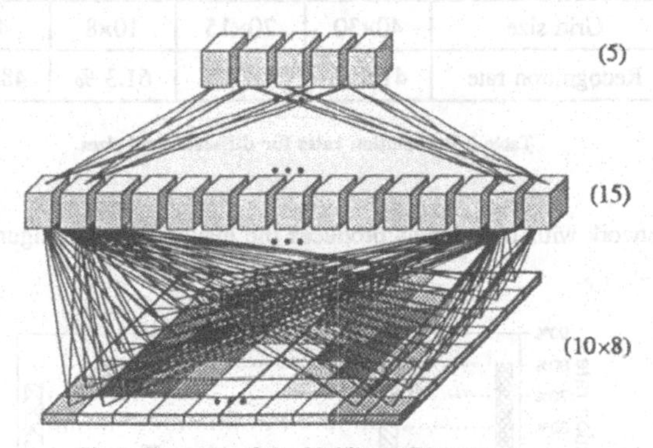

Fig. 8: Structure of the Multilayer Perceptron.

To perform the training of the NN we have used the well-known Backpropagation algorithm [RUM,WID], including a *smoothing* technique as is shown in equation 1, where α depicts the learning rate and β is a smoothing factor which ranges from 0 to 1 [HNC,WAS]. The weight adjustment is smoothed in a quantity proportional to β. If the term $(1-\beta)$ is ignored, this method coincides with the *Momentum* technique. *Smoothing* is preferred to *Momentum* because it decouples the learning rate from the smoothing effect [HNC]. For our application, the best results are obtained when learning parameters $\alpha=0.1$ and $\beta=0.9$ are used.

$$\Delta w_{ij}^{t+1} = (1-\beta)\alpha\,\delta z_j + \beta\,\Delta w_{ij}^{t} \qquad (1)$$
$$w_{ij}^{t+1} = w_{ij}^{t} + \Delta w_{ij}^{t+1}$$

4. EXPERIMENTAL RESULTS.

Using the above described preprocessing technique and the MLP, several simulations were carried out for different sizes of the grid that defines the input layer. To evaluate the classification accuracy, 31 training and testing runs were performed. In each run, we took 30 samples per group for training (i.e. 150 samples in all), taking the remaining one from each group for testing (5 test samples in all). The sets of test samples are disjoint for all runs, i.e. in each run we took a sample from each group for testing that we did not take in a previous run. The classification percentages are obtained as mean values of all runs. Due to the limited number of training samples available, the classification accuracy is better for coarse grids (i.e. for less weights in the first layer). Thus, although less detail is maintained in the input image (see Fig.7), a 10×8-input matrix produces better results than a 20×15 one. This is due to the higher generalization capabilities of the perceptron in the first case. In fact, for the 20×15 matrix the network "memorizes" all the training samples, giving a 100% classification accuracy for those samples. Using test samples different from those used for training, the results in Table I have been obtained for differents grid sizes. When the size of the grid is reduced below 10x8 pixels the loss of information in the input image (particularly in those images corresponding to weak lesions) causes a decrease of recognition rate.

Grid size	40×30	20×15	10×8	4×3
Recognition rate	41.5 %	55.7 %	61.3 %	48.6 %

Table I: Recognition rates for different grid sizes.

In particular, the network with 10×8 inputs produces the results shown in figure 9.

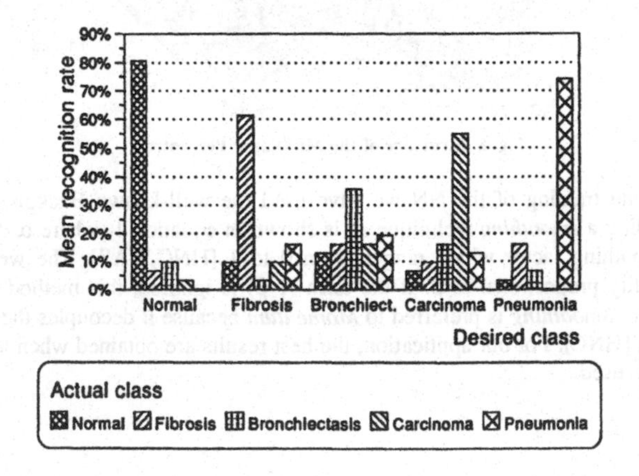

Fig. 9: Classification and Confusion rates obtained for the 10×8-input network.

These results must be interpreted taking into account that, with the number of input images used, the network capabilities are not fully exploited. This point may be easily corroborated by performing various classification tests for different numbers of training samples, as is illustrated in figure 10. In this figure we show how classification accuracy increases with the number of training samples. The number of training samples (of each group) per run is represented in the horizontal axis (X). Each point in the chart represents the mean classification accuracy obtained from 31 runs in which we take as many training samples per group as its X axis value (x_i), and the remaining ones in the group ($31-x_i$) to test. In the figure, there are four traces. From the bottom to upwards, different traces correspond to the mean classification accuracy when the desired output unit is the one with highest activation, or is among the two, three, or four units with highest activation, respectively. For 30 training samples, the percentage obtained in the bottom trace is just the average of those appearing in figure 9 (i.e. that of the third colum in Table I).

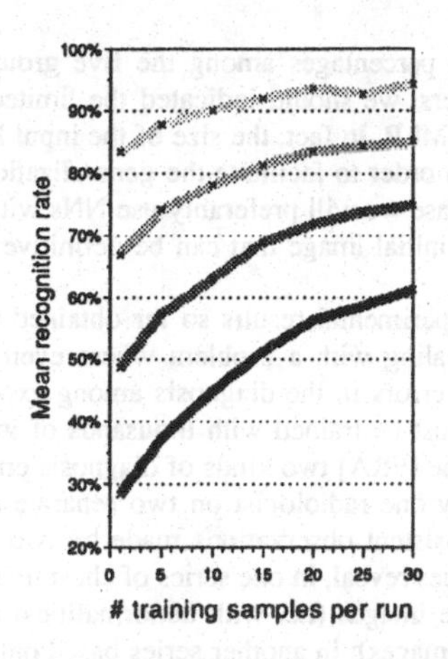

Fig. 10: Increase of classification accuracy when the number of training samples grows.

On the other hand, if we consider the output neuron with the lowest activation and we use this information to state that the supplied image can be discarded as belonging to the corresponding group, a similar average percentage to that shown in the top trace (for 30 training samples) is obtained. Specifically, the input images can be discarded from the groups: normal, fibrosis, bronchiectasis, carcinoma, and neumonia, with confidence degrees of 93.9%, 100%, 96.8%, 87.9%, and 93.5%, respectively.

5. CONCLUSIONS AND FUTURE TRENDS.

In this paper an approach to radiological image classification based on neural networks has been presented. Chest images have been considered since they are the most frequently performed in a general hospital, and in most cases they correspond to normal patients. Thus, an efficient classifier which detects abnormalities in the images (although it does not give a precise diagnosis) would be very useful to improve the eficiency of a Radiology Department.

The classification system we have used consists of an MLP and a preprocessor that removes irrelevant information from the initial image and adapts it to the size of the NN input layer. A set of 155 radiological images (proceeding from conventional screen film images performed in several X-Ray Departments), belonging to normal patients and four different pathologies, have been used in the application.

Using test samples different from those used for training, an average classification (diagnosis) accuracy of about 61% is obtained. In particular, normal chest images are correctly classified in 80% of es. An interesting percentage results if we consider the output neuron with the lowest activity, which indicates the input image category (pathology or normal chest) that can be discarded. An average confidence degree of 94.4% is obtained to discard a chest image from a group when the image presented makes the output neuron of such group to have the lowest activity. In particular, if the image makes the neuron of the normal group to have the lowest activity, we can affirm with a 93.9% of confidence that some pathologie exists.

Although the classification percentages among the five groups may seem low in the general context of NNs as pattern classifiers, we should indicated the limited number of training samples we have had available for training the MLP. In fact, the size of the input layer of the NN had to be reduced (at the expense of loss of detail) in order to facilitate the generalization capabilities of the MLP. When we have an extensive image data base we will preferably use NNs with more inputs (see Sections 3 and 4) to maintain more details of the initial image that can be definitive for diagnosis.

The expectations of the experimental results so far obtained may be better appreciated if it is taken into account that we are dealing with a problem where even specialists frequently fail. Many studies reveal a high incidence of errors in the diagnosis among experienced radiologists, and to gain such experience, the radiologist must be trained with thousands of image readings. As an example, in the studies of Garland and Cochrane [FRA] two kinds of diagnosis errors are considered: intra-observer (inconsistent observations made by one radiologist on two separate readings of the same radiological images); and inter-observer (inconsistent observations made by two or more radiologists of the same radiological images). As these studies reveal, in one series of chest images the interpreters missed almost one third of radiologically positive images (i.e. with abnormalities) and overread about 1 per cent of negative images (i.e normal chest images); in another series based only on positive radiological images, inter-observer errors ranged from 9 to 24 per cent and intra-observer errors ranged from 3 to 31 per cent. Even more surprising results are reported by Felson et al. [FEL] for inter-observer errors in the detection and evaluation of radiological abnormalities in coal workers' pneumoconiosis. Three groups of experienced "readers" were employed: one group of radiologists and other doctors residing in the mining areas being studied, a second group of 24 radiologists with considerable experience with pneumoconiosis, and a third group of 7 radiologists with extensive experience in pneumoconiosis radiological interpretation. More than ten thousand radiological images were used in the study. The first and second group agreed on only 75% of their readings. However, excluding the chest images interpreted as normal by both groups of readers, one reader disagreed with the other in 82.2% of cases. Agreement between the first and third group was even worse: of 14,369 readings, there was disagreement in 75.3% of cases; in fact, in only 10.1% did both group readers agree that the chest was normal. The second and third groups showed lower levels of disagreement in their readings (but as much as 30%) due to their higher experience of the pathology being studied.

Another aspect to take into account is that the NN diagnoses only with regard to radiological images. However, a radiologist has, besides radiological images, other laboratory evidence (e.g. blood analysis), and previous studies of the patient. In fact, most errors in readings and diagnosis of radiological images are made due to not analysing previous studies of the patient [PED]. This is a kind of knowledge that could also be included in an NN to implement a more precise diagnosis system.

The utility of a diagnosis tool based on NNs would be better exploited by using a digital imaging system, by means of which some of the preprocessing tasks we now perform on the (indirectly captured and digitized) images could be simplified or replaced by others already included in the system, and the access to big digital image data bases would greatly facilitate an extensive training of the NN.

Acknowledgements: We would like to thank J.J. Merelo and F. Rios for their collaboration.

REFERENCES

[FEL] Felson, B. et al: "Observations on the results of multiple readings of chest films in coal's miners pneumoconiosis". Radiology, 109:19-, 1973.

[FRA] Fraser; Paré; Paré; Fraser; Genereux: "Diagnosis of Diseases of the Chest". Tomo I. Ed Saunders. 1988.

[HNC] "HNC Neurosoftware". Release 2.3. March, 1990.

[PED] Pedrosa, C.S.: "Diagnóstico por Imagen. Tratado de Radiología Clinica". Tomo I. Ed Interamericana. 1986.

[RUM] Rumelhart, D.E.; Hinton, G.E.; Williams, R.J. (1986): "Learning internal representations by error propagation" in Rumelhart, D.E.; McClelland, J.L. (Eds.): "Parallel distributed processing: Explorations in the microstructure of cognition: Vol. I. Foundations" (MIT Press).

[WAS] Wasserman, P.D.: "Neural computing: theory and practice". ANZA Research, Inc. Van Nostrand Reinhold. New York, 1989.

[WID] Widrow, B.; Lehr, M.:"30 Years of Adaptive Neural Networks: Perceptron, Madaline, and Backpropagation". Proceedings of the IEEE, Vol. 78, No. 9, pp. 1415-1442, September 1990.

COMBINATION OF SELF-ORGANIZING MAPS AND MULTILAYER PERCEPTRONS FOR SPEAKER INDEPENDENT ISOLATED WORD RECOGNITION *

J. Tuya, E. Arias, L. Sánchez, J. A. Corrales

Universidad de Oviedo, Campus de Viesques,
Escuela Técnica Superior de Ingenieros Industriales
e Ingenieros Informáticos.
Área de Lenguajes y Sistemas Informáticos.
Carretera de Castiello, s/n. E-33394 GIJON/SPAIN

Abstract:

A new Neural Network architecture that combines the Kohonen Self-Organizing Maps and Multilayer Perceptrons for a speech recognition task is presented. This architecture overcomes the problem of time-alignement of the succesive frames obtained from one utterance of one word: the succesive frames of a word generate a trajectory in a two-dimensional space using the Self-Organizing Map. These are classified using the Perceptron. Comparation with other techniques are made, and results are better than the obtained wich Trace Segmentation. The vocabulary used in the experiments is a highly difficult subset from the Spanish alphabet: the Spanish E-Set.

1 Introduction

Artificial Neural Networks are used in a high variety of pattern classification fields, including in speech recognition. A high amount of Neural Network paradigms and combinations between Neural Networks and classical methods have been tryed [Lippmann, 89].

Multilayer perceptrons (MLP) [Rummelhart, 86] [Widrow, 90] have been used to clasify static speech patterns [Huang, 88]. But because of speech is inherently dynamic, it is essential to represent adecuately the temporal component and time-align the patterns of speech. Multilayer perceptrons (MLP) combined with Trace Segmentation (TS) [Castro, 91] [Casacuberta, 92] attain better results than single MLP. There exist another architectures derived from MLP, as, for example, the Time Delay Neural Networks (TDNN) [Waibel, 89] [Lang, 90], and modular TDNN architectures [Waibel, 89b]. Results are usually compared with clasical techniques [Bottou, 89]. Also, recurrent neural networks [Robinson, 89] and Dynamic Programming Networks [Sakoe, 89] have been studied.

Self-Organizing Feature Maps (SOM) have been introduced and used in classifiyng static patterns of speech (phonemes) [Kohonen, 88] [Kohonen, 88b] Combinations of SOM with MLP variants have been tried, some of them with success: combination of SOM and TDNN for isolated words [McDermott, 90], combination of SOM with MLP to make phonetic transcriptions [Kokonen, 90], etc.

In this paper we present a new architecture that combines the SOM and MLP. This architecture overcomes the problem of time-alignement of the succesive frames obtained from one utterance of one word. Results are compared against a MLP with TS, using the same train and test set. The vocabulary used in the experiments is the most confusable subset from the Spanish alphabet: the Spanish E-Set (SES).

* This work has been supported by the CICYT (Spain) under project PRONTIC-326/90, and by the FICYT (Spain) under project 91/8.

2 Network Description

As is stated previously, this network is a combination of SOM acting as an acoustic-fonetic decoder, a number of intermediate layers that convert the temporal information of succesive frames of speech into a bi-dimensional mapping (for each utterance) and a MLP as classifier. This architecture can be shown in Figure 1 and is explained below:

Figure 1: Achitecture of the network

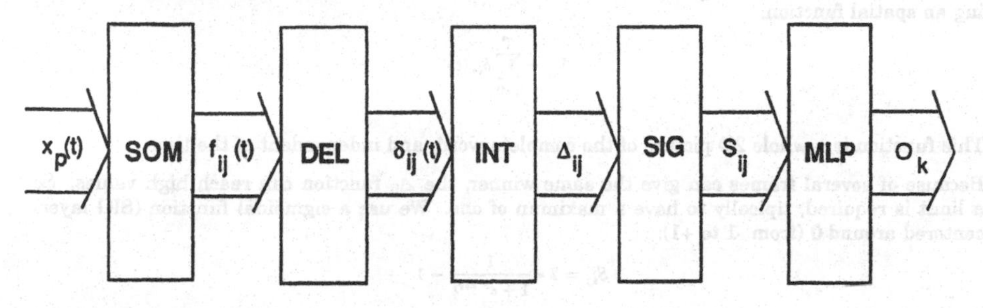

2.1 The Kohonen Self-Organizing Map

The Self-Organizing Map (SOM) is a "nonlinear projection" of the probability density function from an high-dimensional space into a low-dimensional space (tipically bi-dimensional). His training and properties have been exhaustively studied [Kohonen, 90].

The network consists on one two-dimensional layer, having $N_i * N_j$ neurons. Each neuron is connected with all inputs (a vector x of dimension N_p) via the adaptative weigths $m_{ij,p}$. Each time that an input vector x is presented, the output neurons compute their activation:

$$f_{ij} = \sum_{p=1}^{N_p} m_{ijp} x_p$$

The winner is the neuron having the vector m_{ij} that is closest to the input x.

In a speech recognition task the input vector will be some parametric representation of speech for an instant of time (frame). A word will be a sucesion of vectors $x_p(t)$. For different phonemes the winner will be different. In this architecture we take advantage of an interesting property of SOM and spech: Because of the speech is essentially continuous, the winner obtained for a frame at the time t will be the same or the neighbour than the obtained at the time $t+\Delta t$. If $\Delta t \longrightarrow 0$ the succesive frames of speech parameters will generate a continuous trajectory.

Figure 2 is a sample of the winners obtained with the succesive frames for the utterance of the words $|efe|$, $|ele|$, $|e\lambda e|$ $|eme|$, $|e\underline{r}e|$ in a 12x12 map. The lines connect succesive points obtaining the trajectory.

2.2 The Time Integration Layers

Each frame of speech, the input vectors $x_p(t)$ are transformed by the SOM layer in an activation value for each neuron: $f_{ij}(t)$, $1 \leq i \leq N_i$, $1 \leq j \leq N_j$. The delta layer (DEL) computes the winner, resulting the function $\delta_{ij}(t)$:

$$\delta_{ij}(t) = \begin{cases} 1 & if\ ij\ is\ the\ winner \\ 0 & elsewhere \end{cases}$$

Because of excessive tiny trayectories are not desired, the DEL layer provides a value between 0

and 1 (tipically. 0.25) for the nearest neighbours of the winner. So:

$$\delta_{ij}(t) = \begin{cases} 1 & \textit{if ij is the winner} \\ 0.25 & \textit{for the neighbours of the winner} \\ 0 & \textit{elsewhere} \end{cases}$$

The spatio-temporal values $\delta_{ij}(t)$ are integrated over the time by the integration layer (INT), resulting an spatial function:

$$\Delta_{ij} = \sum_{t=0}^{T} \delta_{ij}(t)$$

This function is a whole 2D picture of the complete word, and independent of the time.

Because of several frames can give the same winner, the Δ_{ij} function can reach high values. So a limit is required, tipically to have a maximun of one. We use a sigmoidal function (SIG layer) centered around 0 (from -1 to +1):

$$S_{ij} = 2 * \frac{1}{1 + e^{-\Delta_{ij}}} - 1$$

Figure 3 represents the activation values S_{ij} obtained in this layer, using the same words than the trajectories shown at the Figure 2.

Figure 2: Trajectories generated for the utterance of several words

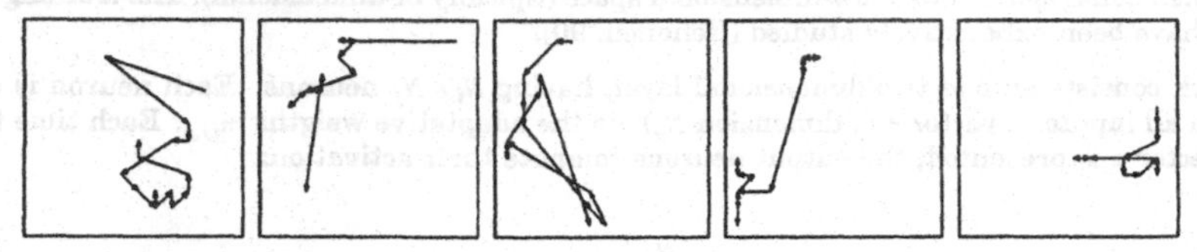

Figure 3: Values of the activations generated for the INT layer

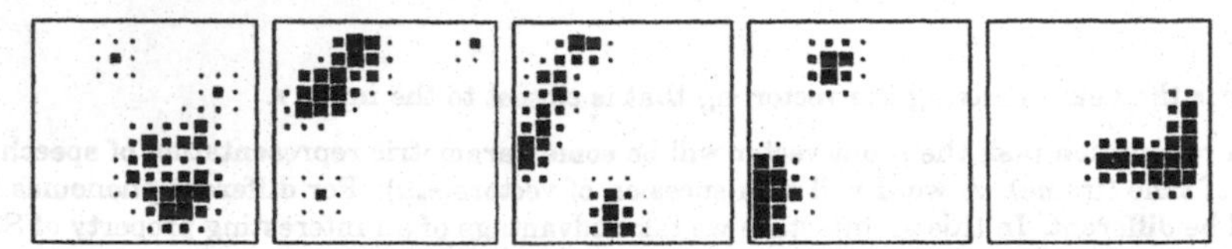

2.3 The MLP classifier

Finally, S_{ij} is treated as the input for a multilayer perceptron (MLP layer) with N_h hidden nodes. The architecture and learning for it has been exhaustively studied: [Rummelhart, 86]. This MLP receives 2D images S_{ij}, $1 \leq i \leq N_i$, $1 \leq j \leq N_j$. for each word, and classifies it into the target categories O_k, $1 \leq k \leq N_k$, where N_k is the number of target words.

2.4 Training and Testing

The training of the whole network consists on three stages:

1. **Unsupervised training of the SOM layer using all speech frames.**
2. **Generation of the activation maps (layer INT) for each word using the previously trained SOM layer.**
3. **Supervised training of the MLP layer for each word using the previously generated maps.**

The process process is similar for testing: First the S_{ij} activations of the SIG layer are computed, and then they are put as input for the MLP.

3 Experiments

This chapter explains the speech database used, their preprocessing and the experiments performed on it.

3.1 Corpora

As is stated previously, the experiments with this architecture have been made using the Spanish E-Set. This consists on nine words: $|efe|$, $|ele|$, $|e\lambda e|$ $|eme|$, $|e\eta e|$, $|ene|$, $|ere|$, $|e\underline{r}e|$, $|ese|$. All these nine words start and finish by the same vowel $|e|$, and are bi-sylabic.

The speech database consists of 10 repetitions of the nine words, wich where uttered by 10 speakers, 8 of them are used for training, and 2 for testing. The test set has been selected in order to have two different types of speakers:

1. Standard speaker (speak the same Spanish dialect as the other in the train set).
2. Extremely difficult speaker. He belongs to another country in Spain (speaks a different dialect than the other in the train set).

3.2 Preprocessing

For the experiments shown below we have selected one of the most common parametric representations for speech [Davis, 80].

Each word utterance is sampled at 10KHz and low-pass filtered at 5KHz. The resulting sequence is transformed into frames by a 256 points FFT using a hamming window. The resulting 128 points are grouped into 19 spectral bands using a mel-spaced filter bank [Zwicker, 61]. The logarithm of the square of each band is used to produce a total of 8 cepstral coefficients [Schafer, 75]. Succesive frames are separed by 64 points (6.4 ms at 10 KHz). An additional parameter is added to the 8 cepstral coefficients: it is the energy, computed from the original time-discrete signal using the same hamming window.

The resulting vector x has nine components for each frame of time t ($N_p = 9$).

3.3 Results

Two series of experiments are sumarized: one using TS combined with MLP, and the second with the architecture proposed. In both experiments the entire word is presented for train and test. None information exists about the phonemes that integrate the word.

There exists previous experiments made with the SES using TS and MLP [Castro, 91b], but using different speakers. In this communication, the experiments are sumarized in the Table 1. The learning rate is selected at $\eta = 0.001$ and the momentum at $\alpha = 0.4$. All words are trace segmented resulting 20 frames for each, and results are obtained for 15 and 20 hidden nodes.

Table 1: Summary of experiments using Trace Segmentation and a Multilayer Perceptron (% Rec.)

Speaker	15 hidden nodes	20 hidden nodes
Difficult	46.7%	50.0%
Standard	55.6%	73.4%

Because the SES is very dificult to learn, the network can never learn all the train set, so, the stop rule can not be based on the error observed during the training phase. Several preliminar experiments have shown that the learning is complete after 3000 epochs. If the network is trained with more epochs, the overlearning problem appears.

The experiments made with the architecture proposed are shown in the Table 2. The training of SOM layer is made in two phases using a square layer of 12x12 nodes: Phase one has an initial gain $\alpha = 0.2$ that decreases linearly to 0. The initial neighborhood radius $N_c(t) = 6$ decreases linearly to one. The number of cycles is 400000. Phase two uses $\alpha = 0.05$, $N_c(t) = 3$ and 1000000 cycles.

Table 2: Summary of experiments using the proposed architecture (% Rec.)

Speaker	1st. phase 10 hidden nodes	1st. phase 15 hidden nodes	1st. phase 20 hidden nodes	2nd. phase 10 hidden nodes	2nd. phase 15 hidden nodes	2nd. phase 20 hidden nodes
Difficult	51.2%	55.6%	58.9%	54.5%	56.7%	52.3%
Standard	82.3%	80.0%	78.9%	84.5%	86.7%	87.8%

Different MLP classifiers have been tried with 10,15 and 20 hidden nodes, using $\eta = 0.05$ $\alpha = 0.6$ and 400 training epochs. It is interesting to note:

- This MLP learns faster than the used with TS. More than 400 epochs results in an overlearning of the network.
- Requires a lower number of weights.
- It can learn the entire train set, the errors measured at 400 epochs are near 0.4% and descending when overtraining.

4 Conclusions

This work shows some preliminary results about how a Self-Organizing Map can overcome the problem of time-alignement applied to the problem of isolated word recognition for an small, but very difficult dictionary. The results are significantly better than the obtained using Trace Segmentation with the same train and test sets. The architecture obtains good results for short utterances, but not for long and complex utterances, because the trajectories formed by the SOM, can overlap, loosing relevant information.

This architecture requires more training time than a single MLP because of the need for training the SOM layer, but otherwise, the number of weights of the MLP layer is reduced by approximately a half, and the number of epochs needed for trainning is reduced by a factor of five.

Additionally, the network is able to learn the entire train set. So, this network could be suitable for speaker dependend speech recognition.

5 References

- Bottou, L.; Foguelman Soulié, F.; Blanchet, P.; Liénard, J. S. (1989) *"Speaker-Independent Isolated Digit Recognition: Multilayer Perceptrons vs. Dynamic Time Wraping"*. In: Neural Networks, Vol. 3, pp. 453-465
- Casacuberta, F.; Castro, M. J.; Puchol, C. (1992) *"Isolated Word Recognition Based on Multilayer Perceptrons"*. In: Pattern Recognition and Image Analysis. Pérez de la Blanca (Ed.)
- Castro, M. J.; Casacuberta, F.; Puchol, C. (1991) *"Trace Segmentation with Artificial Neural Networks"*. Universidad Politécnica de Valencia. Technical Report DSIC-II/13/91.
- Castro, M. J.; Casacuberta, F. (1991) *"The use of Multilayer Perceptrons in Isolated Word Recognition"*. Proc. IWANN-91, pp. 348-354
- Davis, S. B.; Mermelstein, P. (1980) *"Comparison of Parametric Representations for Monosyllabic Word Recognition in Continuous Spoken Sentences"*. IEEE Transactions on Acoustics, Speech and Signal Processing. Vol, ASSP-25, no. 4, pp. 357-366
- Huang, W. M.; Lippmann, R. P.; Nguyen, T. (1988) *"Neural Nets for Speech Recognition"*. In: Conference of the Acoustical Society of America, Seattle WA.
- Kohonen, T. (1988) *"The neural Phonetic Typewriter"*. IEEE Computer, March 1988, pp. 11-22
- Kohonen, Teuvo (1988b) *"Self-Organization and Associative Memory"*. Springer-Verlag, 3rd Edition, pp. 119-157

- Kohonen, Teuvo (1990) *"The Self-Organizing Map"*. Proceedings of the IEEE, Vol. 78, no. 9, pp. 1464-1479
- Kokonen, M.; Torkkola, K. (1990) *"Using Self-Organizing Maps and Multi-Layered Feed-Forward Nets to obtain Phonemic Transcriptions of Spoken Utterances"*. Speech Communication, 9 (1990) pp. 541-549
- Lang, K.; Waibel, A. H. (1990) *"A Time-Delay Neural Network Architecture for Isolated Word Recognition"*. Neural Networks, Vol. 3, pp. 23-43
- Lippmann, R.P. (1989) *"Review of Neural Networks for Speech Recognition"*. In: Readings in Speech Recognition. Alex Waibel and and Kai-Fu Lee (Eds.), pp 374-392
- McDermott, E.; Iwamida, H.; Katagari, S.; Tohkura, Y. (1990) In: Readings in Speech Recognition. Alex Waibel and and Kai-Fu Lee (Eds.), pp 425-438
- Robinson, A. J. (1989) *"Dynamic Error Propagation Networks"*. Ph. D. Thesis. Trinity Hall and Cambridge University Engineering Department
- Rummelhart, D. E., Hinton G. E., Williams, R. J. (1986) *"Learning Internal Representations by Error Propagation"*. In: Paralell Distributed Processing: Explorations in the Microstructure of Cognition. Rumelhart, D. E. and McClelland, J. L. (Eds.). Cambridge, MA: MIT Press, pp. 318-362.
- Sakoe, H.; et Al. (1989) *"Speaker Independent Word Recognition Using Dynamic Programming Neural Networks"*. Proc. ICASSP-89, pp. 29-32
- Schafer, R. W.; Rabiner, L. R. (1975) *"Digital Representations of Speech Signals"*. In: Readings in Speech Recognition. Alex Waibel and and Kai-Fu Lee (Eds.), pp. 49-64
- Waibel, A.; Hanazawa, T.; Hinton, G.; Shikano, Kiyohiro; Lang, Kevin, J. (1989) *"Phoneme Recognition Using Time-Delay Neural Networks"*. IEEE Transactions on Acoustics, Speech and Signal Processing, March 1989.
- Waibel, A.; Sawai, H.; Shikano, K. (1989) *"Consonant Recognition by Modular Construction of Large Phonemic Time-Delay Neural Networks"*. In: Readings in Speech Recognition. Alex Waibel and and Kai-Fu Lee (Eds.), pp. 405-407
- Widrow, B.; Lehr, M. A. (1990) *"30 Years of Adaptative Neural Networks: Perceptron, Madaline, and Backpropagation"*. Proceedings of the IEEE, Vol. 78, no. 9, pp. 1415-1441
- Zwicker, E. (1961) *"Subdivision of the Audible Frecuency Range into Critical Bands (Frequenzgruppen)"*. In: The Journal of the Acoustical Society of America, Vol. 33, no. 2, pp. 248

AN INDUSTRIAL APPLICATION OF NEURAL NETWORKS TO NATURAL TEXTURES CLASSIFICATION.

Gérard YAHIAOUI (*) (%), Bertrand BOROCCO (#)

(*) : advance, 15, rue des champs, 92600 ASNIERES, FRANCE

(%) : Ecole Spéciale de Mécanique et d'Electricité, 4, rue Blaise Desgoffes, 75006 PARIS, FRANCE

(#) : PSA Peugeot Citroën, chemin de la malmaison, 91570 Bièvre, FRANCE

ABSTRACT :

In this paper, we describe an application of neural network for the classification of natural materials textures. We developped this solution in the context of leather quality control. This leather is used in car sits manufacturing (c.f. figure 1). The aim of this control is to make sure of the compatibility of every visual aspect in the whole car. This job is currently processed by human experts that cannot inspect every sits with the same attention. The automation of such a process is very complicated because it is necessary to build a model of human vision in order to take into account how a texture is interpreted as an aspect. As a matter of fact, human perception is processed with subjectivity that makes very hard to propose an efficient explicit mathematical model. We explain why neural networks can be useful in such an application, and we expose our solution. We describe technical gears of this solution (fractals preprocessing, neural architecture, ...) and we explain how we built the global solution with the help of the A.G.E.N.D.A. methodology.

Then, we show simulations results.

For correspondence : ADVANCE
Mr. Gérard YAHIAOUI
15 rue des champs
92600 ASNIERES
FRANCE

I - INTRODUCTION : THE PROBLEM OF CLASSIFICATION AND SUBJECTIVITY MODELING IN QUALITY CONTROL :

In quality control applications, if it is not possible to inspect each piece of a production, then, mathematical theories (like the so-called χ^2 test) provide efficient tools to deduce the quality of the whole production from the quality of just a few pieces.
The result of such a control is a percentage of good quality pieces in the whole production. But in certain kinds of applications (security, energy, luxury, ...) it is important to locate and qualify each default. It means that each piece must be inspected. Then, if the control does not mean only to compare a measure to a standard value, it becomes very difficult to automate the process. For instance, the quality control of a perfume cannot be processed easily without the help of human perception. Then, human experts would have to inspect themselves every piece.
The problem that we expose in this paper is the quality control automation of leather sits aspects (c.f. Fig. 1). This application needs to model how human vision does interpret a natural texture as a global aspect.

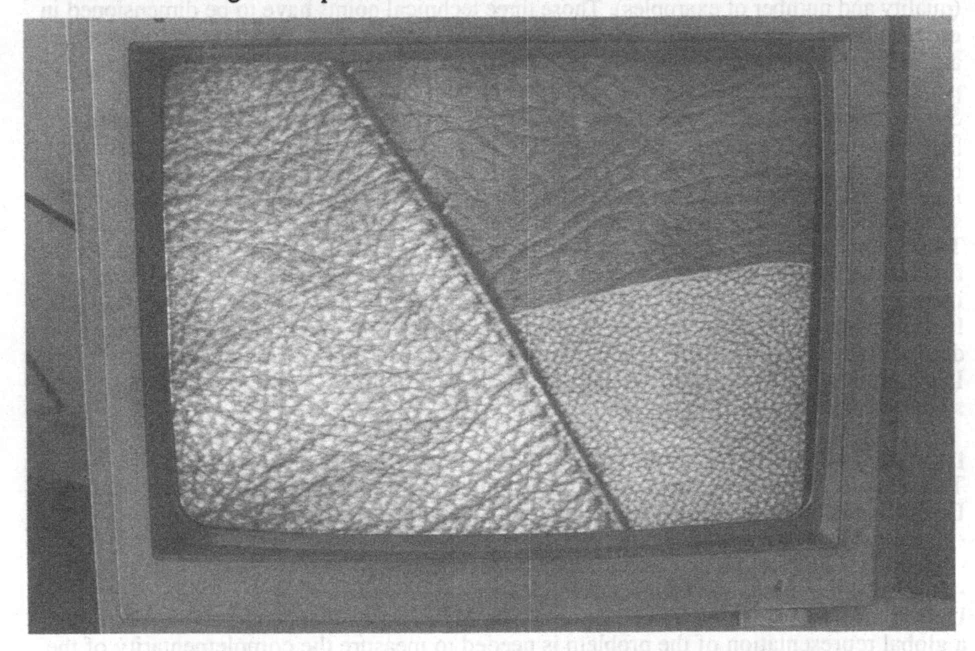

Figure 1 : leather textures (from left to right : thin, middle, and rough textures).

In this perception task, it is important to take into account the subjectivity of human interpretation. Indeed, it is not sufficient to extract from the image of leather some mathematical estimators that characterize the relations between gray levels : the more important is to know how human perception interprets these relations.
Then, in one hand, classical algorithmic solutions that consist on an explicit modeling of the decision process are not easy to design.
In the other hand, knowledge based systems do not seem to be efficient for this kind of applications because it is impossible either to define rules neither to choose a distance criterium to compare cases.
Neural networks, and especially their supervised learning ability, provide a relevant tool that permits to model human subjectivity in classification tasks. Indeed, this subjectivity is present inside the learning examples (one learning example is a couple composed of the input vector and of the decision vector that corresponds to the subjective classification the human expert would make), and the supervised learning rule teaches the network to process this subjective association.

II - THE CHOICE OF NEURAL NETWORKS TECHNOLOGY FOR TEXTURE CLASSIFICATION :

As mentioned before, neural networks provide an efficient tool in supervised learning operations that permits to design an implicit model of a subjective decision.
Furthermore, we dare say that non-linear processing abilities of neural nets make them more adapted to image analysis than classical statistical classifiers. As a matter of fact, an image is the result of a non-linear mix between two bi-dimensional functions as it is described in the Stockham model [1]. This non-linear production model gives to images a special character that other bi-dimensional signal do not have. For this reason, non-linear processing is theoreticly more adapted to image analysis than linear processing.
Then, supervised learning and non-linear processing abilities make more and more researchers apply neural networks for image understanding tasks.
However, the disadvantage of neural networks technology is that a real industrial solution is often difficult to design. In particular, in complex problems, it is hard to choose, a priori, preprocessing, parallel architecture characteristics, and relevant learning examples (quality and number of examples). Those three technical points have to be dimensioned in order to build a coherent solution. The theoretical supports of a complete neural solution are different : algorithmic notions for preprocessings, linear algebra and topology for parallel architectures, statistics for the choice of relevant examples.
So, we decided to use a new functional representation approach called A.G.E.N.D.A. [2]. This methodology has been designed to help specialists to build a coherent solution by giving a global view of the project. In particular, it is possible to propagate constraints inherent to a technical choice to the other choices, and then to measure the global cost of a solution before beginning any simulation. Each global solution is represented by a graph. This graph can be read as a trace of the conception phase. Indeed, this graph explains the functionalities that are used in each technical choice, and it is easy to understand for instance why a given algorithm or a given architecture has been used, and why the learning has been processed with a given number of examples. This functional description amplifies the efficiency [3] of an expert to design a neural networks solution. In the following paragraph, we use A.G.E.N.D.A. notions in the description of our solution conception.

III - DESCRIPTION OF OUR TEXTURE CLASSIFICATION PROBLEM :

This chapter deals with the way of designing our complete neural network solution. Designing a complete industrial solution using neural networks means :
- to choose relevant preprocessing,
- to build an adapted parallel architecture of network,
- to select a representative learning examples set.

Unfortunately, these three technical points cannot be chosen independently. It means that a global representation of the problem is needed to measure the complementarity of the technical points mentioned before.
Then, the first step of the method is to describe the problem in terms of variabilities. This is what is done for our leather texture classification problem in the figure 2.

TRANSLATION,
ROTATION,
GLOBAL LUMINANCE,
AVERAGE SIZE OF GRAINS,
HETEROGENEOUSNESS,
CONTRAST,
GRAIN PATTERNS ,
PERIODICITY OF PATTERNS,
FLUCTUATIONS OF MAIN DIRECTIONS,
FLUCTUATIONS OF THE SIZE OF GRAINS

figure 2 : description of our classification problem in terms of variabilities.

In this list of variabilities, some points induce a variation of the texture, and others can be considered as noise for our classification problem. For instance, if you change the grain size, you will see a different texture. But you do not change the texture by processing any rotation of the image. Then, the estimators built by preprocessing and neural network will

have to take into account the rotation problem in order to give the same result for any rotation (it is not so easy if the texture has strong direction dependence).
A.G.E.N.D.A. provides a representation tool for this kind of problem by linking in a graph representation the variabilities of the problem and the technical gears of the neural solution (c.f. Figure 5).The graph links the description of the variabilities with the technical gears of the solution.
Then, it is possible to see on this graph what are the functionalities that the designer wants to implement when he chooses each technical gear.
Before proposing a graph adapted to our texture classification problem, we propose to give a theoretical description of our preprocessing algorithm : the fractal signature.

IV - IMAGE TEXTURE, AND FRACTALS :
Fractals are a new chapter of mathematics that has been written by the mathematician Benoît MANDELBROT [4]. This new mathematics chapter provides an efficient tool to describe some geometric figures that the Euclidian geometry do not characterize properly. It is the case, for instance, of the Peano curves (c.f. figure 3).

figure 3 : at each iteration, each segment is replaced with a complete curve

The curve shown in figure 3 tends to a surface. However, in the Euclidian theory, it is always a curve (then its dimension is 1) and its lengh becomes infinite. The fractals geometry will give to this curve a dimension which is a non integer number between 1 (because it is a curve) and 2 (because it tends to a surface). For a long time, the mathematicians considered this kind of object as a mathematical monster without any application to the real world geometry description. Benoît MANDELBROT showed that a lot of natural objects can be considered as fractals, when the resolution of the observation process changes its space resolution. In particular, a lot of image processing applications has been built with fractals.
Indeed, an image is classicly considered as a 2 dimensions object. However, it is possible to represent it in the three dimensions Euclidian space (O, x, y, z), when x and y are the space variables, and z is the value of the luminance at each point (x, y).
As for the Peano curve described before, the more there are irregularities in this image, and the more it tends to a volume. Then fractals theory gives to such an object a dimension which is a non-integer number between 2 and 3. Now, let us give a mathematical description. In the following lines, we are going to define the entity $A(\varepsilon)$, that will be used in our application.
Let us consider an iterative non-linear filtering of the image : a new pixel is computed from the values of the initial pixel and its neighbours. The difference between the new and the old gray levels defines an elementary volume. When you apply such a process on each pixel of the image, then it leads to a global volume. If the elementary volume is 1, then this global volume can be considered as a global surface. Let us imagine a blanket that would cover the 3D representation of the image. If the thickness of the blanket is 2ε, then the surface of this blanket is its volume divided by 2ε. It is possible to consider ε as a variable. Then this variable defines the space resolution of the observation. It is possible to define with this process two surfaces : the upper surface $U\varepsilon$, and the lower surface $B\varepsilon$ as followed

Let $G(i, j)$, and $U_0(i, j) = B_0(i, j) = G(i, j)$.be the initial space functions of the iterative process.

Then, if ε is an entire positive number, and for n=4 neighbours :

$U_\varepsilon(i, j) = \max \{ U_{\varepsilon-1}(i, j) + 1, \max(U_{\varepsilon-1}(m, n)), \text{ where } | (m, n) - (i, j) | < 1 \}$

$B_\varepsilon(i, j) = \min \{ B_{\varepsilon-1}(i, j) - 1, \min(B_{\varepsilon-1}(m, n)), \text{ where } | (m, n) - (i, j) | < 1 \}$

The total volume of the blanket is :

$V_\varepsilon = \Sigma_{i,j} (U_\varepsilon(i, j) - B_\varepsilon(i, j))$

and the measure of its surface is given by : $A(\varepsilon) = (V_\varepsilon - V_{\varepsilon-1}) / 2$

This function $A(\varepsilon)$ is classicaly used to compute the fractal non integer dimension defined by HAUSDORFF and BESICOVITCH [5]. This fractal dimension is used in several image textures classification applications. However, we show in the figure 4 that this dimension does not discriminate our leather textures :

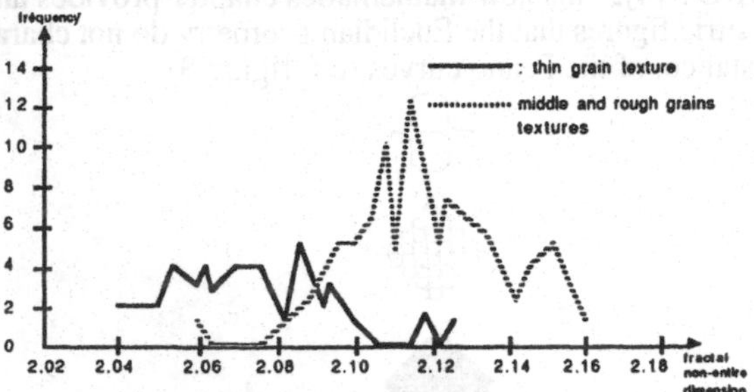

figure 4 : histograms of fractals dimensions.

In the following paragraph, we use the properties of $A(\varepsilon)$ to build our texture classifier. It will be possible to read on the A.G.E.N.D.A. graph that we propose the functionalities of $A(\varepsilon)$ that are used in this application. In the following lines, A(e) will be called "fractal signature".

V - OUR NEURAL NETWORKS SOLUTION :

At this step of our presentation, we gave the main points of our conception method and we described our preprocessing algorithm. Now, we propose a synthesis of this presentation. This synthesis is the adapted A.G.E.N.D.A. graph that we built. This graph links the problem variabilities to the three technical points : preprocessing, network architecture, and learning examples choice.

DESCRIPTION OF THE PROBLEM : TECHNICAL ELEMENTS :

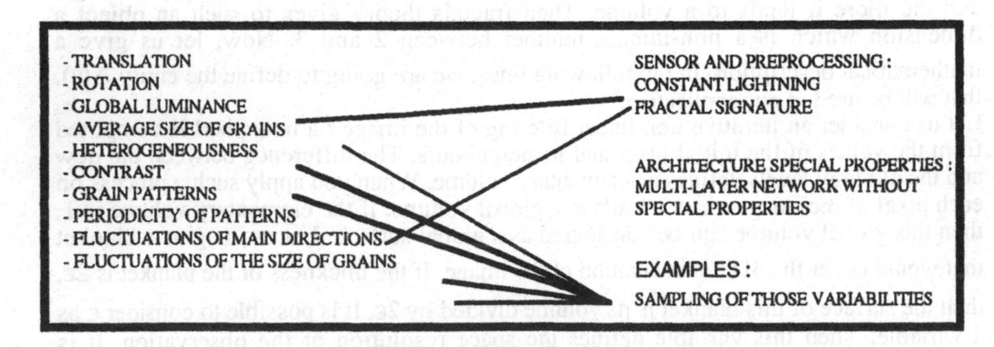

figure 5 : our.solution represented as an A.G.E.N.D.A. graph.

It is easy to see on the graph what are the properties of A(ε) that we used in this solution. For instance, it is possible to verify that this estimator gives the same result for any rotation.
Furthermore, this graph shows very clearly the role of the learning examples set. Indeed, the learning examples must make a relevant sampling of the space generated by the variabilities linked with the third technical pole (i.e. learning examples characteristics). It means that the learning phase does not consist on a relevant sampling of the input vector space, but on a relevant sampling of the space generated by the variabilities mentioned before. This new result permits to calculate very easily the number of examples that is necessary for a given graph. As a matter of fact, it is easy to apply the sampling theory to these variabilities, and then to obtain the number of examples that must be used to train the network. It is important to notice that this number of examples depends on the preprocessing and on the network architecture characteristics. The A.G.E.N.D.A. graph permits to see the dependance of the learning examples set. Our adapted graph leaded to an set of 120 learning examples.

VI - THE SIMULATION PHASE :
As shown in figure 6, the best solution was obtained with a network of 6 cells. This number of neurons has been determined iteratively. Indeed, the best solution is the solution that is precise enough, but it is also the solution that permit to generalize enough. Theoretical works [6] show that the best generalization results are obtained for the lowest entropy neural network (i.e. for the smallest network, in the sense of the number of connections). This is why A.G.E.N.D.A. proposes to make the network grow from the minimum number of cell (here it is 1 neuron !) to the optimal solution. This method permits to get the smallest network that is precise enough for the application.

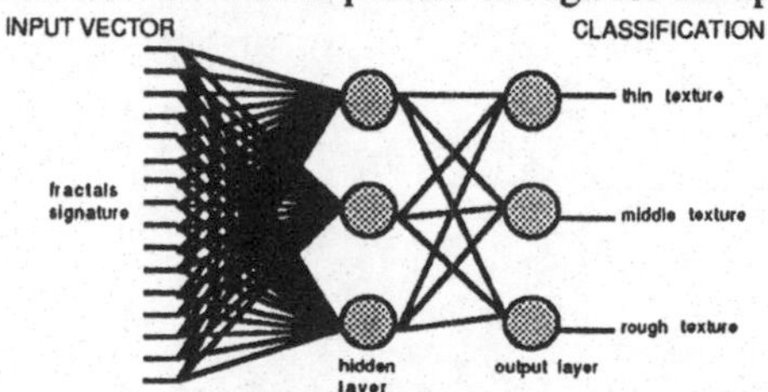

figure 6 : our neural network texture classifier.

It is interesting to notice that some new learning rules for neural networks provide an efficient tool to automate the growth of the network [7], even if these learning rules cannot support yet every kind of multi-layered networks and every kind of data.

VII - SIMULATION RESULTS AND CONCLUSION :
We tested our neural solution on a big amount of data. The results seem to prove that this solution is robust (c.f. Figure 7).

	Class 1 (big grain)	Class 2 (middle grain)	Class 3 (thin grain)
success :	96%	91%	99%
fail :	4%	9%	1%

Figure 7 : simulation results.

This solution is currently tested in the industrial research center of the French Company "PSA Peugeot Citroën". We think that our method could be used in the classification of visual aspects for a lot of materials : paper, wood, glass, steel, ...
This solution is very easy to compute (fast preprocessing and little number of connections), and the A.G.E.N.D.A. graph permits to modify our solution to apply it to another problem. Indeed, if you change the problem, then you change the variabilities list, and it is easy to see what are the modifications involved in the technical solution.

VIII - REFERENCES :

[1] : "Image Processing in the context of a Visual Model", T.G. STOCKHAM, Proceedings of IEEE, Vol60, N°7, Jul. 1972.

[2] : "Un cadre méthodologique dédié à la conception de solutions neuronales : la méthode AGENDA", Gérard YAHIAOUI, Proceedings of NeuroNîmes 92, Nov. 1992, France.

[3] : " Guide d'Intégration des Techniques Emergentes de l'Informatique ", Jean-Louis AMAT & Gérard YAHIAOUI, Editions Eyrolles, 1993, France.(to be published).

[4] : " Les objets fractals", Benoît MANDELBROT, Editions Flammarion, 2ème éd. 1984,

[5] : " Multiple Resolution Texture Analysis and Classification", S. PELEG, J. NAOR, R. HARTLEY, D. AVNIR, PAMI-6, N°4, IEEE 1984,

[6] : " Learning and Generalization in Neural Networks : the Contiguity Problem", S. SOLLA, in "Neural Networks, from models to applications", L. PERSONNAZ & G. DREYFUS réd., IDSET 1988.

[7] : Knerr, Personnaz, Dreyfus, 1990, "Single Layer Learning Revisited : A Stepwise Procedure for Building and Training a Neural Network" : Neurocomputing : Algorithms, Architectures and Applications, F. Fogelman, J Herault eds, NATO ASI Series, Springer

USE OF A LAYERED NEURAL NETS AS A DISPLAY METHOD FOR N-DIMENSIONAL DISTRIBUTIONS

Lluís Garrido[a,b], Vicens Gaitan[b] Miquel Serra-Ricart[c], Xavier Calbet[c]
(a) Departament d'Estructura i Constituens de la Materia
Universitat de Barcelona
Diagonal 647, E-08028 Barcelona, Spain
(b) Institut de Física d'Altes Energies
Universitat Autònoma de Barcelona
E-08193 Bellaterra (Barcelona), Spain
(c) Instituto de Astrofísica de Canarias
E-38200 La Laguna (Tenerife), Spain

November 30, 1992

Abstract

In this paper we present some examples of a method based on layered neural nets, trained with multi-seed backpropagation, to display a n-dimensional distribution in a projected space of 1, 2 or 3 dimensions. The method can be used as encoder.

1 Introduction

A very common problem in many areas of science is the classification of objects in clusters. It has been said many times that the human eye is the best pattern recognizer when the objects are represented by only one, two or three variables. If this is not the case, one of the strategies is to reduce the number of original variables by some transformation to 1, 2 or 3, giving a complementary view of the input data set.

To avoid confusion we would like to write a few words in order to explain the difference between clustering, encoding and dimension reduction of multidimensional data. In the first two we try to "compress" the original data in such a way that it will be possible to recover the maximum information from the compressed data afterwards, losing some of the original information on the final results. On the other hand the dimension reduction works with redundant data and the problem is to define a computational mapping between locations on an n-dimensional input space, and locations on an m-dimensional constrain surface embedded in the n-dimensional input space without any loss of information , in this way, abstraction and simplification in the description of data is reached [5].

The difference between encoding and clustering can be found in the requirements over the functional form of the transformation from the input data into the compressed one. For clustering we wish to maintain the general structure of the input data, that is, the aim is to preserve the original topology of the data set, i.e., points that are close/far away in the input have to be close/far away in the compressed data. These requirements necessary if one want to extract visual conclusions from the compressed data, are not necessary in the encoder case.

In fact the best encoder is such that the compressed data has a uniform distribution (max. information).

One method used most frequently in clustering and dimension reduction is the "Principal component analysis (PCA)" [1] where the reduced dimensions are linear combinations of the original variables. Here we want to present an alternative, based on Neural Nets, which is not restricted to work with linear combinations, and therefore, permit more complicated transformations between the input distribution and the projected one.

2 The method

The method is based on the technique of Self-Supervised Backpropagation (SSBP) also known as the "encoder" problem [4]. This algorithm is implemented on a Neural Net that has n units in the input layer activated with the quantities of the n-dimensional distribution under study, n units in the output layer which are forced to match the activation, one by one, of the input layer units; and one hidden layer, that contain only 1,2 or 3 units whose activation is going to be interpreted as the final projection (from here on we will call this units the "neck units").

This method based on neural nets can be explained with the help of two functions, f1 and f2, such that

$$INPUT \in R^n \xrightarrow{f1} COMPRESS \in R^m \xrightarrow{f2} OUTPUT \in R^n$$

where $m < n$.

The first function, f1, transforms the input n-dimensional data into the compressed data ,the neck units, with a lower dimension ($m < n$). The second function, f2, transforms the compressed data into the output data, which has the same dimension as the input (n). These functions are required to minimize the Euclidean distance between an input point and its output ($\chi^2 = \sum_{examples}(INPUT - OUTPUT)^2$).

We can see f2 as a m-dimensional surface embedded in the n dimensional input/output space. As a result of the minimization this surface f2 will try to be as close as possible to the input points, depending on the flexibility given to the function. The other function, f1, will project an input point to another one on the f2 surface as close as possible to it, depending again on the flexibility given to f1. Is this freedom given to functions f1 and f2 which will be related to the preservation of the input topology and by it, with the possibility of using the method for clustering or encoding.

Sanger [4] showed that the SSBP with only one hidden layer and a linear response function for the neurons, is equivalent to PCA (in this case f1 and f2 are linear and the preservation of the input topology is assured in some directions). Saund [5] describe an approach for performing the dimension reduction task using also a Neural Net with only one hidden layer, but with the sigmoidal function for the neuron response. Saund shows that the assumption of linearity need not be made about the underlying constraint surface (now f1 and f2 are not restricted to be linear), though the method fails when the constraint surface doubles back on itself sharply.

The method used here was proposed by Oja [3] and is a generalization of the Saund approach. The net contains an additional hidden layer before and after the layer containing the neck units and a sigmoid response function is used for each neuron. This 5 layer network is more powerful because it will allow a more complicated functional form of f1 and f2, but still it preserves the topology of the input data.

Backpropagation (BP) was used to self-train the network. It is known that BP algorithm has two major problems. The first one is that the energy surface may have local minima, therefore finding the optimal solution is not guaranteed; the second one is slowness on the convergence process. To minimize these problems we applied the Multi-Seed Backpropagation (MSBP) method. It consists of using a wide range of random number generator seeds to find the initial weights of the network and calculating, for every seed, the final minimum energy with classical BP, making use of the power given by parallel processing using a Sparc IPX workstation net. After the MSBP was applied, an energy versus seed sample was obtained in order to easily distinguish local minima points from global minima one.

2.1 Dimensional reduction application

First a strongly nonlinear example is considered. A two-dimensional input space is taken, and the input points are concentrated on the circle

$$(x = 0.5 + (0.5 - n)cos(\theta),\ y = 0.5 + (0.5 - n)sin(\theta)) \tag{1}$$

where n is an added random factor in order to simulate real noise (Fig. 1, up). The input data can then be parametrized with only one parameter, the angle θ. It is fairly evident that a linear reduction process, such as the PCA, is not able to solve the presented circle problem. A 3 layer neural network with one neck unit has the problem of forming the mapping $f1(x, y) = \theta$ from the input layer to the neck unit, and the mapping f2 gives by Ec. 1, from the neck unit to the output layer. With general activation functions for the neuron response this is not possible [3]. However, as both functions f1 and f2 are continuous, a 5 layer neural network with three hidden layers is theoretically able to do the job. Fig. 1 shows the results yield by a 2:5:1:5:2 neural network using a sigmoid function for the neuron response after the convergence is reached using the MSBP. On the top we can see the learned mapping f2 (a one-dimensional surface, the continuous line) superposed on the two-dimensional input points, whereas on the bottom the activation of the neck unit versus the angle θ is plotted, showing that this angle has been approximated by the neural network.

2.2 Clustering application

Here we illustrate an example of a cluster task using real data. Galaxies collected in the ESO-LV catalog (Lauberts & Valentijn 1989) are taken as input data. Galaxies with ESO visual diameter $\geq$ 1 arcmin and high galactic latitude (b> 30°) were selected. Eleven catalog parameters (Table 1), were used to describe each input galaxy. These 11 parameters are very similar to those used by Lauberts & Valentijn [2] to perform the automated classification presented in the ESO-LV catalog. Only galaxies with all 11 parameters available were chosen in order to avoid problems of missing data. The final working sample has 4997 galaxies. The mentioned 11 parameters were normalized between 0 and 1, so the PCA and the NNA (using a 11:23:1:23:11 neural network) were applied in order to find cluster structures. Results are shown in Fig. 2. The PCA projection onto the first principal component do not show data clustering, whereas the NNA projection is able to separate different groups of objects. A morphological classification performed by visual examination of the galaxy image are also available. The main visual morphological types of galaxies (Sa, Sb, Sc, Sd, SO, E) were grouped into three major classes: Sa+Sb, Sc+Sd, and SO+E. The PCA and NNA results corresponding to the E+SO, Sa+Sb and Sc+Sd classes can be seen in Fig. 2, and we deduce that a nonlinear process, as the NNA, is needed to separate the three classes of galaxies.

3 Conclusions

A method based on artificial neural networks to display a n-dimensional distribution data set has been described. As principal component analysis (PCA), neural network analysis (NNA)

Table 1: The galaxy parameters.

Symbol	Description	Tutorial
$(B-R)_e$	B-R integrated in ellipse enclosing half total B light	Colour
$(B-R)_{10''}$	B-R integrated in central 10″ diameter circle	"
$(B-R)_T$	B-R total colour	"
N^B_{oct}	exponent of the fit of a generalized de Vaucouleurs law to B octants	Profile gradients
N^R_{oct}	exponent of the fit of a generalized de Vaucouleurs law to R octants	"
$log(D^B_{80}/D^B_e)$	D^B_{80} and D^B_e are the major diameter of the ellipses at 80% and half total B light, respectively	Dimensions
$log(D^B_{26}/D^B_e)$	D^B_{26} is the major diameter of the ellipse at 26 B mag arcsec^{-2}	"
$log(b/a)$	b/a is the galaxy axial ratio	Orientation
μ^B_0	average B surface brightness within 10 arcsec diameter circular aperture	Surface Brightness
μ^B_e	B surface brightness at half total B light	"
μ^R_e	R surface brightness at half total R light	"

offers powerful ways of extracting information of the data structure useful to, a) reduce the number of input variable to its inherent dimensionality, and b) identify different groups of objects. However the NNA can improve the PCA method due to the fact that the projected variables are not restricted to be linear combinations of original ones, as it has been shown with an artificial example and a real astronomical application.

At present we are working on an approach that combines the minimization of the Euclidean distance between the input and output points, and a desired functional form for the f1 transformation.

References

[1] Kendall M.G. (1957) "A course in Multivariate Analysis", Griffin & Co., London.

[2] Lauberts A., Valentijn E.A. (1989) "The Surface Photometry Catalogue of the ESO-Uppsala Galaxies". ESO

[3] Oja E. (1991) "Artificial Neural Networks", in: Kohonen T., Mäkisara K., Simula O., Kangas J. (eds.) Proceedings of the 1991 International Conference on Artificial Neural Networks. North-Holland, Amsterdam, p. 737.

[4] Sanger T.D. (1989) "Optimal Unsupervised Learning in a Single-Layer Linear Feedforward Neural Network", Neural Networks Vol.2, 459-473.

[5] Saund E. (1989) "Dimensionality-Reduction Using Connectionist Network", IEEE Transactions on Pattern Analysis and Machine Intelligence Vol. 11, 304-314.

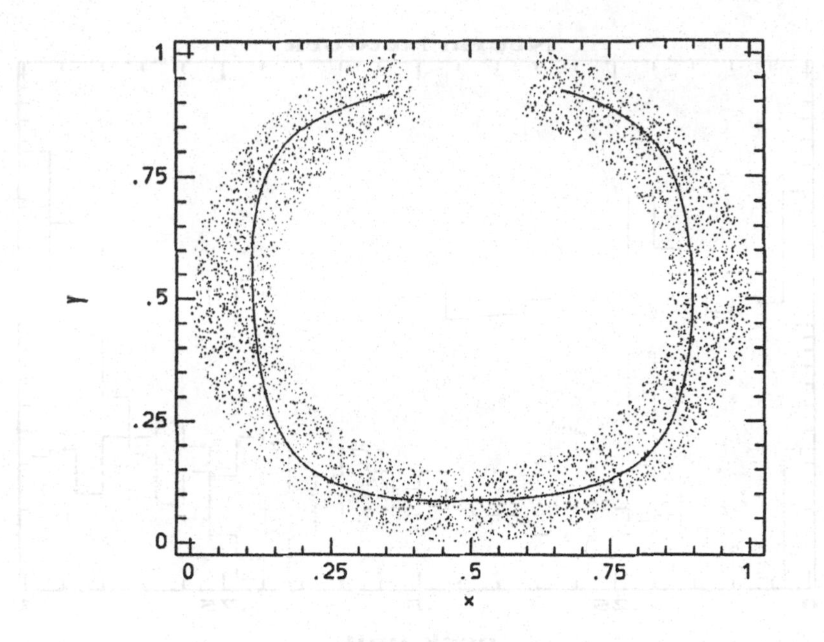

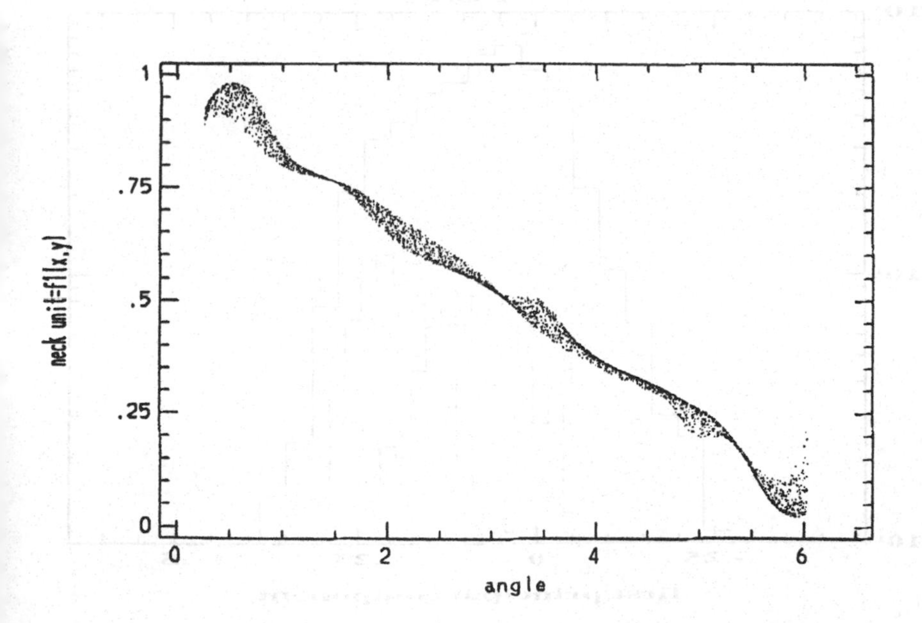

Figure 1: On the top we can see the learned mapping f2 (a one-dimensional surface, the continuous line) superposed to input two-dimensional points yields by a 2:5:1:5:2 Neural Net using a sigmoid function for the neuron response, after the MSBP was applied, whereas down we plot the activation of the neck unit versus the angle θ.

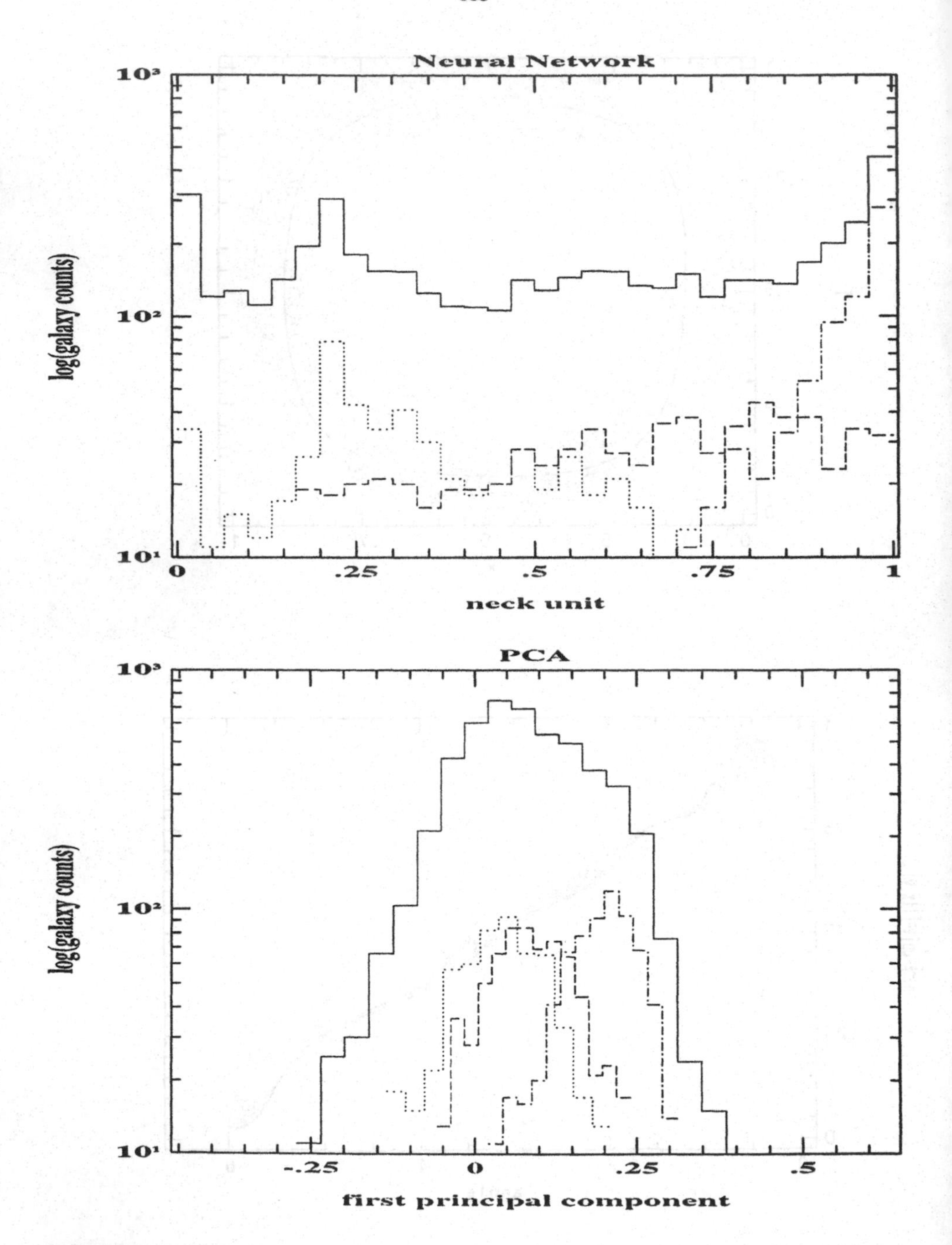

Figure 2: Projection results made by the PCA (down) and the neural network (up) for the total galaxy sample (solid lines) and the three galaxy classes Sc+Sd (dotted line), Sa+Sb (dashed line) and E+SO (dotdash line).

MLP MODULAR VERSUS YPREL CLASSIFIERS

Yves LECOURTIER*, Bernadette DORIZZI**, Philippe SEBIRE**, Abdel ENNAJI*

* La3i, Université de Rouen
UFR des Sciences et Techniques, BP 118
F-76134 Mont Saint Aignan Cedex, France

** Institut National des Télécommunications
9, rue Charles Fourier
F-91011 Evry Cedex, France
nad@etna.int-evry.fr

Abstract

We present two connectionist modular approaches which are potentially able to deal with real applications as their size does not increase drastically with the size of the problem. The first model relies on a very simple cooperation of modular MLP networks specially designed for some sub-tasks. The second is based on a new methodology using a particular processing element ("neuron") called yprel. The main characteristics of the approach are: (i) An yprel classifier is a set of yprel nets, each net being associated to a particular class; (ii) the learning is supervised and conducted class by class; (iii) the structure of the net is not a priori chosen, but is determined step by step during the learning process. Both approaches are compared on a well-known classification task (recognition of typographic characters) in terms of performance rates.

1. Introduction

Connectionist learning algorithms have been studied for many years now and the large amount of applications in which neural techniques have been used tends to demonstrate that they are very useful tools for different classification tasks such as character recognition, speech processing etc.... Moreover, in order to give more precise and qualitative results, some studies concentrated on the comparison between well known classical classification algorithms (K Nearest Neighbors, Parzen Windows etc...) and well known connectionist ones [IDA92, LEE91, MAR91]. They all assert that the backpropagation learning algorithm provides classification performance rates equivalent to those of the best classical algorithms on the same task. However this is only true if the training data base is big enough (namely of order 10 000 for digit recognition) and statistically representative of the problem at hand.

Now an important problem emerges: if we want to use a standard MLP (Multi Layer Perceptron) to deal with a huge data base and with a relatively big number of classes, we have to design an adequate network architecture and its size will increase with the size of the problem. Consequently, the time which will be necessary for the simulations will be very cumbersome. Moreover, a realistic application is submitted to changes in its specifications (for example, a new class is added, some new examples are considered etc...). So, as no incremental learning is possible with the standard MLP due to its fixed architecture and its mode of functioning, it does not seem to be a very practical tool in real size applications.

A general idea to tackle this problem could be to split the task into several smaller ones, for which the processing will be faster. The corresponding networks will thus cooperate to produce a result for the global task. This idea has been previously explored by several authors, in different ways [WAI89, JAC91, BOL91]. Once this general frame has been set, we are led to the following problems : on what criteria does one split the task into subparts, what kind of network does one use to solve the subtasks, and how does one perform the cooperation of the different sub-networks ?

In this paper, we present different answers, and we compare performances on the problem of recognition of typographic characters composed of digits, upper- and lower-case letters (49 classes). The first division scheme is to have a network dedicated to each class. We compare two approaches, the first with MLP networks and the second with networks based on a particular processing element called "yprels"[LEC92] which will be described in the following. With MLP, the size of each net remains quite large. So, others division schemes have been tested. These divisions which are based on common sense, reduce the complexity of the problem, but lead to a low number of nets.

Before giving details about these two models, let us describe the data base we have used. It is made of a set of multifont digits, upper- and lower-case letters, digitized by scanning. This leads to a 49 classes problem (some letters are not differentiated between upper- and lower-case). Both methods have been tested on a test set of 3888 characters. In the MLP approach, the classification is made directly from the pixel image that we normalize to a 16x32 matrix with values in {-1,0,1}. On the contrary, the inputs to the yprel classifier are primitives (44 in the total) which have been extracted from the images. They correspond to classical primitives used in OCR, namely the dimensions of the surrounding rectangle, the number of intersections with lines, the distances between the surrounding rectangle and the first pixel of the character, the densities of pixels in some regions etc...

When trying to compare several methods, we are faced to the problem of the choice of comparison criterion. Various points of view can be adopted. We can either compare methodology capacities, using qualitative criteria only, or compare performances on a test set. In this last case, we must decide on a well-suited performance criterion. The answer influences the decision rule to be used with the classifier. With MLP, this decision rule is not part of the classifier. So several different choices can be adopted. The simpler is to take as answer of the classifier the output with maximal magnitude. However, two problems are not taken into account with such a rule: first, if no output has "significant" magnitude, is the best answer (i) to select the output with maximal magnitude or (ii) to reject with no classifying decision? Secondly, what should we do if two (or more) outputs have close significant magnitude? With the MLP classifier, the answer to these two questions leads to the problem of choosing two thresholds: the threshold of "significant" magnitude and the threshold of difference between the two maximal magnitudes. Depending on these choices, the obtained result can be analyzed in term of good recognition rate, rejection rate and ambiguity rate, the complement to one of these three rates being the confusion rate. With Yprels classifier, the only possible answer (without choice of particular threshold) is in term of the previous rates. So, the comparisons of the two methods will be based on performances of MLP with equivalent either confusion or rejection rates.

2. The MLP Modules

2.1 A single architecture for the global task

Before presenting a modular architecture, and in order to have some elements of comparison, we have built a general network to solve this task. The structure is inspired from the one introduced by LeCun [LEC90] for handwritten digit recognition.

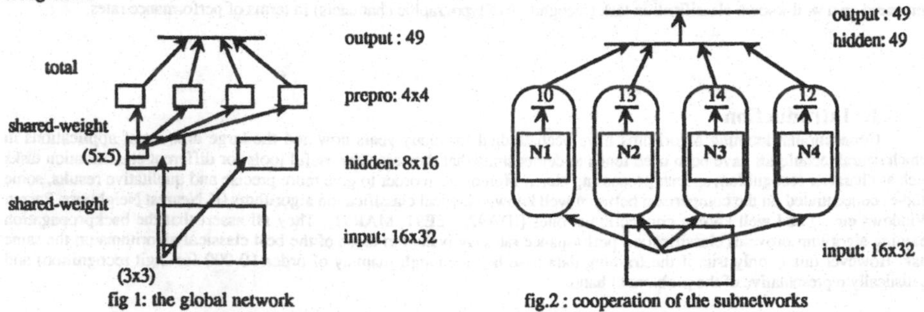

fig 1: the global network

fig.2 : cooperation of the subnetworks

The particularity of that type of network lies in the local and shared-weight connections. This kind of architecture seems particularly well suited to raw image processing because, through the implementation of filters which are known to be essential in image processing, it produces a reduction of the number of the free parameters of the system. This point is very important as it ensures a better generalization.

We give the results obtained by our best simulation. Changes in the architecture of the network (number of hidden and prepro maps, number of hidden layers) could slightly improve the performances of the network. No exhaustive study has been performed for the moment, as we are not interested in finding the network which gives the best performances for this problem but to compare this type of architecture with a modular one.

2.2 Modular Networks

We have tried two kinds of decomposition :

2.2.1. One network per class

One MLP network has been built for each class in order to discriminate between the elements which are in this class and the others. Each network has a very simple structure similar to that of the global network (see fig 1) except that the output is composed of only two neurons. Each net is learned through the presentation of the whole learning base, divided in two parts. One is composed of the samples of the corresponding character and the other is made of all the others characters. The frequency of presentation of the examples is balanced between both classes. Considered independently, all these networks present a good generalization rate (around 99,8-100%). When we consider a character that must be recognized, we present it to the different networks and we select the net for which the class output response is the highest (up to a certain threshold). This simple cooperation scheme is the one which gives the best results.

2.2.2 Separation in four modules

The preceding decomposition leads to a great number of networks. Wishing to design less networks, our first idea was to split the data base into 3 "natural" parts corresponding to digits, upper-case and lower-case letters. However, even if the results in learning and generalization are excellent for each sub-network, we have not been able to obtain more than 90% of generalization for the global problem, and that, with several design of cooperation modules.

We thus tried another division of the base in 4 groups ("big circle": 0,O,Q,8,G,B,C,2,R,P, "little circle": 9,6,3,4,A,e,5,a,b,d,g,h,q", "lines": ,I,J,t,1,L,7,E,F,T,H,f,i,j,r "cross": M,X,Y,W,N,K,S,Z,U,V,m,n). The idea is to group together the characters which look alike or which may confuse by the global network. Each network N_i has a structure similar to that of fig 1 and learns on its corresponding part of the training base. In consequence, the network dedicated to "big circle" has never been trained with a "g" by example.

To make these networks cooperate, we have trained an MLP network with one hidden layer of 49 cells and complete connections, on the whole training set. (see fig 2)

2.2.3 Results

	global net		one net per class		4 nets	
recognition	98.0	95.5	97,4	95,93	97,4	96,35
confusion	2.0	0.7	2,59	1,46	2,6	1,21
ambiguity		0.1		0,28		0,15
rejection		3.7		2,31		2,28

The results are the best that we have obtained from 3 simulations. In fact for the global net, the recognition rate without reject goes from 96,8 to 98%. This shows that the performances of both modular methods are similar to those of the global network. The main advantages of modularity seem thus to be i) first, to reduce slightly the calculation time because there are less outputs and ii) second, the ability to add new classes without a complete relearning. Moreover, as the performances are highly dependent on the way the base is split into parts, the "one network per class" approach appears more attractive for cooperation, even if it demands the design of more networks.

3. Yprel Classifier

As mentioned in the introduction, several problems are not well solved with classical neural methods. So we have developed a new approach of the classifying problem using a classifier based on particular processing elements called yprels. Several features exhibit the key ideas of the approach: (i) An yprel classifier is a set of yprel nets, each net being associated to a particular class; (ii) the nets are non recurrent; (iii) the learning is a competitive supervised learning conducted class by class; (iv) each yprel has either one or two inputs; (v) the structure of the net is not a priori chosen, but is determined step by step during the learning process.

Among these characteristics, the last one exhibits one of the main differences with methods like MLP. With these methods, the choice of a well suited structure of the network is always a key problem and is based on empirical knowledge of the user. With yprel nets, the structure determination is part of the learning process.

Another key point of the proposed methodology lies on the choice of a particular processing element with only one or two inputs. Conducted experiments show that this restriction on the number of inputs does not limit practically learning capacities. On another viewpoint such a restriction has several advantages: (i) it leads to simple calculations for each processing element during learning phase, (ii) it allows an easy display of the behavior of each processing element and so an easy understanding of the internal behavior of the net on a set of examples, (iii) eventually it would facilitate an implementation on VLSI chip.

Finally, another important characteristic is the learning methodology: the goal of the learning for each yprel, and so for the whole net, is to build a function, its answer being among the three following:
- the prototype belongs to the class;
- the prototype does not belong to it;
- the yprel does not conclude for this prototype.

Before explaining the way to build such a function, we will give some basic ideas on the learning methodology. Then we will discuss the structure of the processing element and the way to determine its parameters (similar to weights or synaptic coefficients on classical neural networks). Finally we will give preliminary results obtained on the OCR problem presented in section 1.

3.1 Learning process for an yprel net.

The learning process for a net associated with a given class k starts with an initialization stage. During this stage an yprel with only one input is associated with each input feature. This set of yprels constitutes an input layer of the net. After this stage is performed, learning is based on the following algorithm:

1- among a list L of previously built yprels choose two of them as inputs of a new yprel;

2 - calculate the yprel parameters such that, among the set of prototypes non previously classified by the input yprels, the output of the new yprel realizes the best split between prototypes of class k and prototypes of the others classes ;

3- evaluate the residual number of non classified prototypes;

4- if this number is lower than the number of non classified prototypes of both inputs insert the new yprel on the list L else discard it;

5- if the number of non classified prototypes is non zero and the maximum number of trial is non reached go to step 1;

6- select the best yprel of L (the yprel with minimal number of non classified prototypes) and all the yprels useful to build its inputs as the resulting network for the class k.

This algorithm shows that calculation for a new network reduces to calculation of the parameters of only one yprel, and so is a fast procedure. However, the amount of possible choices for the two inputs of a new yprel is huge. As an example, it is easy to show that for a problem with 20 features, it is possible to generate more than 4.10^{16} combinations of yprels to form nets with less than 25 yprels. So, it is necessary to have powerful rules to guide the choice of combinations to be tested. Two natural rules have proven their efficiency: first select with a higher probability the best yprels of L, that is those with lower number of non classified prototypes, and second select with a higher probability the yprels of the input layer, that is those related on the basic information lying on features.

Note that during the learning process, each selected yprel has a number of unclassified prototypes lower than those of its inputs. Two useful consequences can be deduced of this fact: (i) as already mentioned on step 5 of the algorithm, if the number of unclassified prototypes becomes zero, learning stops; (ii) an upper bound of the size of a network is known: the number of yprels of a network will be lower than the number of prototypes of the base.

Two main differences can be noted with learning rules for MLP. First, parameters of each yprel are calculated once when it is used as terminal element of a new net and then frozen and never modified later. Such an approach is close to the GMDH (Group Method of Data Handling) developed mainly by Ivakhnenko for function approximation purpose [HEC90]. The second difference lies on the belonging/rejection decision. The MLP nets do not take any decision: they calculate likelihoods. Decision is an extra mechanism, the simpler being the determination of the output with maximal likelihood. Conversely, every yprel tries to come to a decision for as many prototypes as possible: the decision process is part of the net processing. Moreover, after a decision is coming for an yprel, the only action of yprels of higher level is to transmit it.

3.2 The yprel.

Two kind of yprels must be distinguished: the yprels of the input layer with only one input, and the general yprel with two outputs.

3.2.1) Input layer

The goal of these yprels is to normalize data. As mentioned, such an yprel is associated to each feature. The algorithm used to calculate parameters of yprel associated to feature x is:

1- If the net is built for class k, for all prototypes of the learning base belonging to class k, determine minimal and maximal values of the feature. Let $[f_{min}, f_{max}]$ be the resulting interval.

2- In order to improve generalization increase slightly this interval. Let $[b_{min}, b_{max}]$ be the resulting interval.

3- Let y the output of the yprel, and $y=k_1x+k_2$ the linear part of the transfer function of the yprel; determine the parameters k_1 and k_2, such that interval $[b_{min}, b_{max}]$ is mapped on interval [0 , 1].

After learning, the decision rule of an yprel of the input layer will be

- compute $y=k_1x+k_2$;
- if y does not belong to [0 1], then y=-1.

The value -1 is arbitrary. It means only "this prototype is rejected: it does not belong to class k". A value of y remaining into [0,1] indicates "no decision for this prototype".

Therefore, all outputs of yprels of the input layer will be always either within the interval [0 , 1] or equal to -1. Note that another particular value, arbitrarily chosen equal to 2 will also be possible for general yprels: it will means "the prototype belongs to class k".

3.2.2) General yprel

Let $c_{n,1}$ and $c_{n,2}$ the inputs of yprel n, these inputs being the outputs of previous yprels selected as mentioned in section 3.1. Because of the possible values of these outputs defined in the previous section, the set of prototypes of the learning base can be displayed on an $(c_{n,1}, c_{n,2})$ coordinate system as illustrated on figure 3.

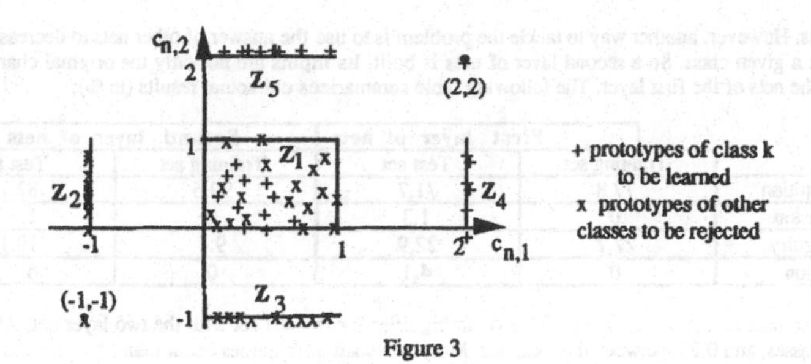

Figure 3

This figure exhibits seven different zones:
-Z1 holds prototypes of class k and prototypes of other classes that neither $c_{n,1}$ nor $c_{n,2}$ are able to classify;
-Z2 holds prototypes that $c_{n,1}$ reject, but that $c_{n,2}$ is not able to classify;
-Z3, Z4 and Z5 are similar zones where only one input come to a decision;
-(-1,-1)point corresponds to all the prototypes rejected both by $c_{n,1}$ and $c_{n,2}$;
-(2,2)point contains all the prototypes that both $c_{n,1}$ and $c_{n,2}$ classify as belonging to class k.

Learning parameters of a general yprel is performed with the following algorithm:
1- Determine the list of prototypes remaining in Z1;
2- Compute a linear combination of the inputs $a_1 c_{n,1} + a_2 c_{n,2}$ such that
- elements of class k are grouped on the shortest possible interval
- elements of other classes are rejected as well as possible outside this interval
under the constraint $a_1^2 + a_2^2 = 1$;
3- Compute intervals $[f_{min}\ f_{max}]$ that holds all prototypes belonging to the class k and interval $[f^0_{min}, f^0_{max}]$ that holds all primitives of other classes. From the 4 boundaries of these intervals, deduce two parameters k_1 and k_2 such that if $y=k_1(a_1 c_{n,1} + a_2 c_{n,2})+k_2$ is outside the interval [0 , 1], then a decision of rejection or belonging to the class can be reached without error.

Step2 uses a combination linear with respect to parameters a_1 and a_2. So, choosing a quadratic criterion to evaluate distance between prototypes, the problem of calculating the parameters a_1 and a_2 is a least-square problem.

3.3 Preliminary results

Some preliminary results have been obtained with the previous methodology. As an illustrative example, a simple problem with two input features and two classes (class 0 being composed of three separated areas in the input coordinate system) leads to the following nets:

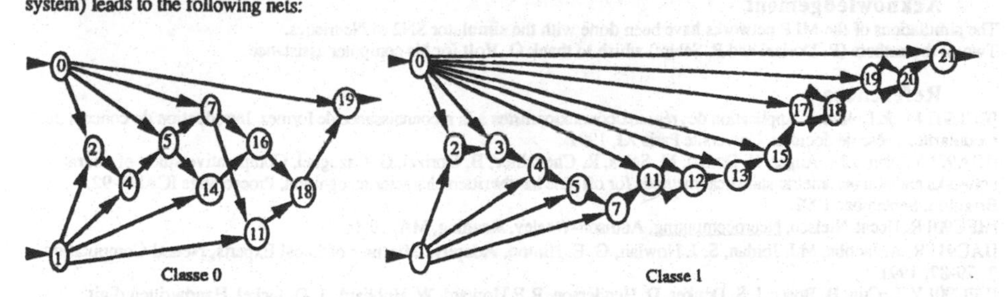

This figure exhibits the general aspect of any net, composed of interconnected yprels with only two inputs, (but possib multiple outputs), and for which the structure is not organized in layers (as was the case of MLP).

Tests on the OCR problem mentioned in the introduction have been conducted with a preliminary version of the presented methodology. The main difference with the actual version relies on the following point : yprels of the net of class k do not take any decision on elements belonging to class k, they only reject elements not belonging to it. It has been proven that, with such a decision rule, an yprel net defines a convex domain containing all the elements of class k in the characteristic space. So a class associated with a non convex domain cannot be well isolated from the others : the corresponding net gives a positive answer for some remaining elements of other classes, generating thus ambiguities. The improved methodology presented in this paper is presently under implementation. It will give better results because it has been shown that, with the new decision rule, an yprel net defines the domain associated with a class as the union of several

convex domains. However, another way to tackle the problem is to use the answer of other nets to decrease the percentage of ambiguities for a given class. So a second layer of nets is built. Its inputs are not only the original characteristics but also the outputs of the nets of the first layer. The following table summarizes our actual results (in %):

	First layer of nets		Second layer of nets	
	Training set	Test set	Training set	Test set
recognition	77,8	71,7	90.6	82.7
confusion	0	1,3	0	1.2
ambiguity	22,2	22,9	9.4	10.1
rejection	0	4,1	0	6

It must be noticed that amoung the 10.1% of ambiguities for the test set with the two layer net, 9.9% are ambiguities between two classes, and 0.2% between three classes. Moreover such ambiguities occur mainly for classical confusing classes on OCR systems like (I 1) or (S 5). So, as the classifier is able to detect such situations, the solution can come from a specific treatment (either using new information from the image or using context knowledge like figure/letter). Finally, the only remaining problem with the yprel approach lies on the level of rejected elements in the test set. A possible solution, due to the "one class-one net" point of view, is to perform an incremental learning by adding rejected elements to the training set of the faulty net. Such an approach can also be used for confusions and ambiguities and is currently under implementation. The most promising result of these preliminary tests is the low level of confusion which can be compared with the one obtained with the MLP classifiers.

Conclusion

This paper shows that, for supervised classification, good performances can be obtained with a particular network associated to each apriori class and a cooperation rule using the outputs of all the nets. This approach has three main advantages: (i) it is possible to make incremental learning on one class without changing the nets corresponding to the other classes; (ii) each particular net can be simpler than a global net, so learning can be faster; (iii) it is possible to add new classes, even partially overlapping without retraining the whole system.

Two kinds of nets have been considered. With MLP, the performance rates are as good as the ones obtained with a global network. Moreover, some change in the choice of a rejection level does not affect significantly the results. Yprel nets illustrate another way to tackle the problem. One of the main advantages of the proposed methodology is the automatic building of the structure of each net.

At this point it is not possible to make a fair comparison of the two approaches. The preliminary results obtained with the yprels are promising but some complementary studies remain necessary in order to validate the approach. However, it is true that, as far as different improvements are brought in the yprel methodology, the performance rates increase significantly.

Let us just stress at this moment one major difference between MLP and Yprels nets, namely the decision mechanism. The MLP nets do not take any decision: they calculate likelihoods. Conversely, every yprel tries to come to a decision for as many prototypes as possible: the decision process is part of the net processing. Comparing advantages of each of these two points of view is an interesting open problem and would be the object of forthcoming theoretical studies.

Acknowledgement

The simulations of the MLP networks have been done with the simulator SN2 of Neuristics.
Two of the authors (B. Dorizzi and P. Sebire) whish to thank O. Volt for his computer assistance.

References

[BOL91] M. de Bollivier, Application des réseaux connexionnistes à la reconnaissance de formes. Introduction du concept de modularité. Thèse de doctorat, Université Paris XI, 1991

[IDA92] Y. Idan, J.M. Auger, N. Darbel, M. Sales, R. Chevallier, B. Dorizzi, G. Cazuguel, Comparative study of neural networks and non parametric statistical methods for off-line handwritten character recognition, Proceedings ICANN 92, Brighton, September 1992.

[HEC90] R. Hecht-Nielsen, Neurocomputing, Addison-Wesley, Reading, MA, 1990.

[JAC91] R. A. Jacobs, M.I. Jordan, S. J. Nowlan, G. E. Hinton, Adaptive Mixtures of Local Experts, Neural Computation 3, 79-87, 1991.

[LEC90] Y. LeCun, B. Boser, J. S. Denker, D. Henderson, R.E.Howard, W. Hubbard , L.D. Jackel, Handwritten digit recognition with a backpropagation network. In D.Touretzky (ed.) Neural Information Processing Systems, 2, Morgan Kaufmann.

[LEC92] Y. Lecourtier, A. Ennaji, J. Le Bas, Réseau d'yprel et classification. BIGRE, 80, 286-294, 1992.

[LEE91] Y. Lee, Handwritten digit recognition using K nearest neighbor, radial basis function and backpropagation neural networks, Neural Computation 3, 440-449, 1991.

[MAR91] G. L. Martin, J. A. Pittman, Recognizing hand printed letters and digits using backpropagation learning, Neural Computation 3, 258-267, 1991.

[WAI89] A. Waibel, Modular Construction of Time-Delay Neural Networks for Speech Recognition, Neural Computation 1, 39-46, 1989.

HOW MANY HIDDEN NEURONS ARE NEEDED TO RECOGNIZE A SYMMETRICAL PATTERN ?

PATINEL Jocelyn[1], LEONE Gilles[2], MILGRAM Maurice[3]

[1]Laboratoire d'Intelligence Artificielle, CEMAGREF
Parc de Tourvole, BP 121, 92185 Antony Cedex FRANCE

[2]Laboratoire de Physiologie Neurosensorielle, CNRS, Université Paris VI,
15 rue de l'Ecole de Médecine, 75270 Paris Cedex 06 FRANCE

[3]Laboratoire de Robotique de Paris, Université Paris VI, CNRS
4 place Jussieu, 75252 Paris Cedex 05 FRANCE

ABSTRACT

We study the detection of axial symetries in binary images with artificial layered neural networks, trained with the backpropagation rule. The number of hidden neurons necessary to classify correctly images is almost independant from the images' size. We show experimental results obtained with different networks, and we note that only a few hidden neurons are really usefull. These neurons are caracterised by the regular spatial structure of their input weights. We then propose a new and more efficient training algorithm that yields networks with the minimum number of neurons necessary to perform the classification. In the last part, we take up the theoretical analysis of this problem that leads to an interesting superior limit for the number of hidden neurons needed for its solution.

1. A NETWORK FOR THE DETECTION OF SYMMETRY

1.1 The input patterns

An input pattern is a square image I consisting in nxn pixels of value +1 or -1 (white or black). I is said to be symmetrical if it is globally invariant under application of one of the four following symmetries : H: symmetry relative to the horizontal axis ; V: symmetry relative to the vertical axis ; D1: symmetry relative to the first diagonal and D2: symmetry relative to the second diagonal. Each axis goes through the central point if n is odd, through the center of the square if n is even. Each image is associated with a single desired output from the network (+1 if symmetrical, -1 if not symmetrical).

1.2. Neural networks and learning

1.2.1. Structure of the networks

In our simulations we used layered neural networks, totally connected and without loops. The neurons have sigmoïdal activation functions, yielding output values within the interval [-1 ; 1]. The output of a neuron is written as follows :

$$S_i = f(\textstyle\sum_j W_{ij} S_j + t_i) \qquad (1)$$

where S_i is the output of neuron i ; S_j the output of neuron j ; W_{ij} the weight of the connection from neuron j to neuron i ; t_i the threshold of neuron i ; and f the sigmoïd activation function : $f(x) = (e^{kx} -1)/(e^{kx} +1)$ where k is the steepness and was always equal to 1 (when $k \rightarrow \infty$, we obtain a threshold activation function).

The learning procedure was the Backpropagation of the error (Le Cun, 1987 and Hinton, 1988). The number of hidden layers varied from 1 to 3, the number of neurons of those layers varied also from one network to another. The number of input neurons was determined by the size of the input images. (See § 1.1). The network had only one output neuron, that was supposed to give the value +1 or -1 when the input pattern was symmetrical or not symmetrical respectively. Yet the output values varied continuously within [-1 ; 1], therefore we had to choose an external decision rule to classify the pattern. Thus :

$-1 < S_i < -T$ meant "non-symmetrical pattern" ;

$-T < S_i < +T$ meant "undetermined pattern" ;

$+T < S_i < 1$ meant "symmetrical pattern".

T was the detection threshold, equal to 0.75 in all our simulations. The network performance was measured as the percentage of well classified patterns relative to this threshold T. It would be illusory to use the difference between the desired output and the obtained output to measure the performance since this difference is largely dependent on the scale of value of the weights multiplied by the steepness k.

1.2.2. Learning algorithm

Our simulations were based on supervised learning : a series of symmetrical and non-symmetrical patterns was presented to the network, each time with the desired ouput S_d (resp. 1 or -1). The initialisation of the weights was done arbitrarilly, following a uniform distribution between -0.5 and 0.5 , the thresholds were initially set to 0.

We used the backpropagation algorithm with a momentum factor, meant to accelerate the convergence (Vogl et al.,1988). The modification of weight W_{ij} (connecting neuron j with neuron i) is written as follows :

$$\Delta W_{ij}^{(t)} = -\lambda.\ \partial E/\partial W_{ij} + \mu.\ \Delta W_{ij}^{(t-1)} \quad \text{where } \lambda = 0.7 \text{ and } \mu = 0.2 \qquad (2)$$

The weight update was applied only after the entire repertoire of patterns to be learned had been presented, which made the learning indifferent to the presentation order. The learning was continued until the mean square error was inferior to 0.001.

2. STUDY OF THE STRUCTURATION OF THE FIRST HIDDEN LAYER NEURONS

2.1. The structuration index

A neuron i of the first hidden layer is connected to n^2 neurons (j) of the retina (input layer) through connection weights W_{ij}. For each neuron i of the hidden layer, we can write the matrix of connection weights to the retina W(i) with elements W_{ij} for j=1 to n^2, where n^2 is the number of pixels in the retina. Defined in this manner, W(i) is a matrix of $M_n(IR)$. It is then possible to use the usual notions of distance and norm belonging to this metric space, in particular the norm $L^2(IR^n)$ defined by the relation :

$$N(W(i)) = \sqrt{(\Sigma_j ((W(i)_j)^2))} \qquad j= 1..n^2 \qquad (3)$$

We found in our simulations that the matrix W(i) for some neurons (later described as "structured" : see § 2.2.3) seems to be antisymmetrical relative to each of the symmetry axes (H, V, D1, D2). The notion of antisymmetry for a matrix A of $M_n(IR)$ is generally defined relative to the first diagonal and is expressed by the following relation between its elements :

$a_{i,j} = - a_{j,i}$ for i= 1 to n, j = 1 to n, where i represents the ligne et j the column.

From now on, we will call the "transpose of A along D1", noted ${}^tA_{D1}$, the matrix of elements $a_{j,i}$, the index D1 refering to the first diagonal. Similarly, by extension, the transpose of A along the second diagonal, noted ${}^tA_{D2}$, is the matrix of elements $a_{n+1-j,n+1-i}$; the transpose of A along the vertical axis, noted ${}^tA_V$, is the matrix of elements $a_{n+1-i,j}$; finally the transpose of A along the horizontal axis, noted ${}^tA_H$, is the matrix of elements $a_{i,n+1-j}$. Consequently, the distance from W(i) to $-{}^tW(i)_{D1}$, noted $d(W(i) , -{}^tW(i)_{D1})$ and equivalent to $N(W(i) + {}^tW(i)_{D1})^2$, will be null if and only if $W(i) = -{}^tW(i)_{D1}$. In addition, the triangle inequality gives us the relation :

$$0 \leq d(W(i),-{}^tW(i)_{D1}) \leq d(W(i),M0) + d(M0,-{}^tW(i)_{D1}) \qquad (4a)$$

where M0 is the matrix of $M_n(IR)$ of elements 0.

As $d(W(i),M0) = N^2(W(i)) = d(M0,-{}^tW(i)_{D1})$, we can define our structuration index STR(i) :

$$STR(i) = (d(W(i),-{}^tW(i)_{D1}) + d(W(i),-{}^tW(i)_{D2}) + d(W(i),-{}^tW(i)_H) + d(W(i),-{}^tW(i)_V)) / 4xN^2(W(i)) \qquad (4b)$$

(4a) and (4b) => $0 \leq STR(i) \leq 2$ (4c)

The equality STR(i) = 0 can only be true when W(i) is antisymmetric relative to each axis of symmetry. However, the closer this index is to 0, the more W(i) tends, in the mathematical sense, towards being totally antisymmetrical and, consequently, the more the pattern of connection weights seems structured. It is the value of this index that will now account for what we call "structuration of the hidden layer neurons". In what follows, we will consider that a neuron of the first hidden layer is structured if its STR is inferior to 0.7. This index can be used to find the correlation between the performance of a network and the structuration of some hidden neurons, that is the construction of a relevant internal representation for the problem to be solved, in the hidden layer. Finally the evolution of this index during the learning of the network may allow us to understand how the network builds this internal representation.

2.2. Learning

2.2.1. The input patterns

In a first series of simulations we limited ourselves to 6x6 pixels images, giving us 2^{36} (~6.87 E+10) possibles images. Among these possibilities are 2^{18} (~2.62 E+5) invariant images relative to the symmetry H, the same number for V, 2^{21} (~2.10 E+6) for D1 and for D2, and the remainder (99.3 %) non-symmetrical. Nevertheless we chose to build our learning set with 50 % non-symmetrical and 50 % symmetrical patterns for which we kept the "natural" proportions of 1 H or V symmetry for 8 D1 or D2 symmetries. The patterns were randomly generated, the total number of patterns being 2592. The generalisation capabilities of a network were tested with a generalisation set of 5184 patterns, generated with the same proportions as the learning set.

Next, using 10x10 pixels images, we obtained 2^{100} (~1.27 E+30) possible configurations of which $2x(2^{50}$ (~1.13 E+15) + 2^{60} (~1.15 E+18)) ~2.31 E+18 are symmetrical. We kept the proportions of the different types of symmetries as previously. For these simulations the number of input patterns was raised to 4320 for the learning and was equal to 2160 for the generalisation.

2.2.2. Simulation results

A total of 10 "6x6" simulations (see table 1) and 5 "10x10" simulations (see table 2) were done. The results are expressed in percentage of correctly classified patterns, according to the decision rule described earlier (§ 1.2.1).

We studied the internal structure of the networks at the end of the learning, by visualising the incoming connections to the first hidden layer (Hinton diagram), as done by Lehki et Sejnovski (1989). Some almost antisymmetrical distributions of weights relative to the four axes, are found in all the networks at the end of the learning period (see figure 1) ; the values of these connections are distributed in a very specific way : they are either inferior to -2, or superior to 2, or very close to 0. These weights can be considered to form a mask with which are convolved the inputs, thus coding these in some neurons of the first hidden layer. The number of such neurons was only 2 or 3 per network, regardless of the image sizes.

We defined the "contribution" of each neuron of the hidden layers as being the difference between the performance of the network before and after the deletion of the studied neuron (all the outgoing weights from this neuron are set to 0) ; the neurons having the highest contribution were those that we found to be "structured" ; on the other hand the deletion of certain other neurons sometimes increased the performance of the network. See figure 2.

2.2.3. Structuration and contribution

By considering the values of our index we can now classify the hidden neurons into two well separated categories : structured neurons (for which the index varies from 0.1 to 0.7, it is inferior to 0.3 in most cases) and non structured (for which the index varies from 0.9 à 1.6, it is superior to 1.4 in most cases). We found *2 or 3 structured neurons in each network at the end of the learning period*: : it is worth noticing that *the transition from 6x6 to 10x10 pixels images does not modify this number* ; this issue will be discussed in § 3.

In the same way as before, we observed the lowering of the generalisation results when certain neurons were deleted. We tried in particular to find out whether the structured neurons alone were sufficient to insure good performance for the networks. The results show that for 3 out of the 5 networks we could keep only 2 structured neurons in the first hidden layer yielding a performance even greater than that obtained with the whole network.

The detailed results for on one hand the symmetrical input patterns and on the other hand the non-symmetrical ones show that, with the few structured neurons left, the networks recognise far more symmetrical patterns than with all the neurons. On the other hand they recognise fewer non-symmetrical patterns. We can conclude therefore that the structured neurons encode the symmetry, whereas the effect of the others is to globally shift the output value towards negative values.

2.4. "Epigenetic" learning

Having established that very few hidden neurons are necessary to achieve a good performance, we tried to train directly such networks. Some simulation attempts having shown that a network with initially too few neurons in the first hidden layer cannot learn, we built up a new procedure that we call "epigenetic learning" (see also early works of Le Cun, 1989). It consists, starting from a network having a usual number of hidden neurons (10 to 25), in deleting the less structured neuron with a regular frequency (every 60th presentation of the learning set, in our simulations), if both following conditions are met :

C1 : at least one neuron has a structuration index inferior to 0.7, this is meant to let the structuration process begin ;

C2 : at least one neuron has a structuration index superior to 0.7, so that there will be remaining neurons.

Thus, as soon as the structuration process has begun, the learning goes on with fewer and fewer neurons in the first hidden layer. Eventually only highly structured neurons remain. It should be pointed out that the decrease in number of neurons is based on their structuration and not on their contribution. The use of these two possible criterions would have yielded different networks, for at the beginning of the learning period, the least structured neuron is not bound to have the smallest contribution.

A "decremental" learning algorithm has two main points of interest.
First the duration of the learning is significantly reduced as the backpropagation algorithm has a duration of order o(n), where n is the number of neurons.

Second this kind of learning may produce better performing networks through the selection of neurons particularly suited to solving the problem. In order to test this hypothesis we compared two networks (SYM4_13E et SYM4_13C), initially identical (same number of neurons and same initial values of weights) and trained with the same learning parameters (number of presentation of the learning set, learning rate...). The networks trained by the epigenetic algorithm, from now on called epigenetic networks, display much better results than the networks previously obtained. This can be seen on table 3 : an epigenetic network gives generalisation performances of at least 9% higher.

The epigenetic network SYM4_13E, which possess only 4 neurons in the first hidden layer (12 initially), also shows better performance than the normally trained network with only these 4 neurons active : SYM4_13D (more than 30%). Figure 3 shows simultaneously the evolution of the structuration index in the two networks, as a function of the number of iterations of the learning set, for the 4 neurons remaining in the epigenetic network. This graph shows clearly that the epigenetic learning induces a better structuration of the hidden neurons with a ratio of about 2. We can see that the neurons kept by the epigenetic algorithm are those that are most structured in SYM4_13C except neuron 10. We can also observe that the future structured neurons are not always those which have a low initial value of STR. The number of 4 neurons found in the first hidden layer is superior to the ones previously found (2 or 3) ; however a modification of the epigenetic learning parameters (C1=0.7 and C2=0.3), gives us a new network with 3 hidden neurons (SYM3_11E) and displaying excellent performances (table 3).

These results show that a limited number (3) of hidden neurons can solve this symmetry detection problem with a good performance and that this performance is correlated to the values of the structuration index of the neurons. In addition, the epigenetic algorithm introduced above has demonstrated better efficiency than the classical backpropagation algorithm.

3. PROBLEM OF SYMMETRICAL ORBIT CONFIGURATIONS, THEORETICAL ANALYSIS

Our simulations have shown that a network exhibiting good performance in symmetry detection has few structured neurons (about 2 or 3). Moreover, our results show that the performance rests mainly on these neurons. We were not able to determine precisely any bound for the number of these hidden neurons but it seems that this number does not depend on the size of the retina. Simulations with 6x6 and 10x10 retinas lead to approximatively the same amount of structured hidden neurons. In conclusion, we expect that the number of hidden neurons needed to do the discrimination between symmetrical and non-symmetrical patterns is constant and independant of the retina size. We give now a proof of this assertion.

Let I be a square image of MxM points ; black points take the value -1 and white points the value +1. We recall that I is symmetrical if it is globaly invariant for at least one of the four axial symmetries (H, V, D1 and D2), all axes passing by the center of the retina. Let us consider the partition of I by the orbits O_j : such an orbit is built from any point p of the retina when we apply the four symmetries to p, to images of p and so on. One can easily check that an orbit has 8 points except for a few degenerate cases, when p belongs to some symmetry axis. The process of deciding whether I is symmetrical or not can be divided into K decisions (K is the number of orbits), one for each orbit, with the evident restriction concerning the fact that the same symmetry has to work for all orbits. For a typical orbit with 8 points, there is 256 configurations, 54 of which are symmetrical. Let $x_0, \ldots, x_7$ denote the 8 values of orbit points, numbered clockwise. If we consider the image as a vector (M^2 coordinates), the symmetry condition becomes a conjunction of conditions :

$x_i = x_{K-i}$ for some $K \in \{1,3,5,7\}$ and for $i \in \{0,...,7\}$ with 8 or 4 such conditions for each orbit.

The first question is : how many hidden neurons do we need in a network (with only one hidden layer) to ensure that it will give an output +1 if the symmetry for the horizontal axis is verified and -1 if it is not ? We shall prove that this number is two, whatever is the value of M. The second question is : how many hidden neurons do we need in a network (with only one hidden layer) to ensure that it will give an output +1 if the symmetry for any axis is true and -1 if it is false?

Let $E=\{-1,+1\}^{MxM}$ be the set of all possible images and H_M the subset of symetrical images for the H axis. If we find an hyperplane L of IR^{MxM} giving : $E \cap L = H_M$ we have also found a linear function u, the kernel of u being L. With u, we get easily a positive real number ε such that :

$X \in E$ and $|u(X)| < \varepsilon \Rightarrow X \in H_M$

We take : $\varepsilon = 0.5$ Min$\{ |u(X)| \mid X \in E\text{-}H_M \}$ and the fact that $E\text{-}H_M$ is a finite set implies that $\varepsilon > 0$.

If we have some linear function u as above, we can write : $u(X) = \Sigma_i u_i X_i$

Let us consider a network with 2 hidden units, with weights from the neuron number i of the input layer set to u_i for the first hidden neuron and $-u_i$ for the second hidden neuron. By tuning thresholds of these two neurons, we can produce for each neuron an output very close to +1 when u(X)=0. We have only to make an AND function between these two outputs to get what we wish, a global output very close to +1 if and only if X belongs to H_M. In fact, if X is not in H_M , we will get $u(X) > \varepsilon$ or $u(X) < -\varepsilon$ and so one of the 2 hidden units will produce an output far enough from +1, thus ensuring that the global output is close to -1. To make the AND function, we can use two very large equal weights w and a threshold T such as : f(2w-T) close to +1 and f(w-T) close to -1

For a sigmoïde f, we know that $f_n(x)=f(n.x)$ can be as close as we want to the "Sign" function. Then if 2w-T>0 and w-T<0, the AND function is achieved ; this can be done with : T=3w/2.

This result is somehow trivial but this link between the size of the hidden layer and a linear function is the main point for what follows.

In conclusion, we need only 2 hidden neurons to realize a slice of a vector space, without considering the dimension of this space.

Now, we just have to prove that such an hyperplane (or the linear function of which it is the kernel) exists for our problem.

Avoidance lemma

Consider p vectors x_i of an euclidian space S of dimension greater than 2 ; if every x_i is different from the null vector, then there exists a positive real ε and at least one linear function u from S to IR so that : for i=1...p $|u(x_i)| > \varepsilon$

We can also formulate the lemma as follows : for any set of p points of an euclidian space of dimension greater than 2, and for any point O not belonging to this set, there is an hyperplane that contains O and passes away from any of the p points.

Proof is quite simple by reccurence on p.

To use the "avoidance lemma", let us consider a linear function D from the space IR^{MxM} to IR^n that associates to an image I a vector D(I) :

$D(I) = (x_{i1}-x_{i2},...,x_{in}-x_{in+1})$ where each pair of subscripts meets some elementary condition implied by the global symmetry condition (for the axis H for instance). Notice that n depends on the size of the image.

So, the symmetry condition (for H) is equivalent to : D(I) = 0 =(0,0,...,0).

Image of $E=\{-1,+1\}^{MxM}$ by D is a finite set D(E). We see that H_M is the kernel of D and : $(0,0,........,0) \notin D(E\text{-}H_M)$.

We can apply the "avoidance lemma" to $D(E\text{-}H_M)$ and find a linear function u.

Consider now the new linear function : $U = u_o.D$; U is a linear function from IR^{MxM} to IR that verifies :

$X \in E$ & $|U(X)| < \varepsilon \Rightarrow X \in H_M$

using that linear function U as above, we can build a network with 2 hidden units that solves the problem of the symmetry (for one symmetry only).

Experimentally, we were able to train a neural network with four hidden neurons to correctly classify the 256 configurations.

CONCLUSIONS

In conclusion, we have presented here a set of theoretical and experimental results. From a theoretical standpoint, we have been able to prove the existence of a pessimistic bound for the number of hidden neurons (8 neurons) to solve the 4 axes symmetry problem. Nevertheless, the independance of this number from the size of the images seems to be a very interesting result. Empiricaly, it appears that a network with 3 hidden units is able to recognize almost any symmetrical pattern.

It should be stressed that the backpropagation algorithm yielded networks that use only a small number of hidden neurons without the help of any competition or selection mecanism. These few neurons have a specifically structured connection pattern with the input layer. By defining a numerical index that accounts for this structuration, we were able to correlate the performance of the networks with the emergence of these structures. We also defined a selective learning algorithm which allows us to train networks faster and more efficiently. These points seem to be more general than the symmetry problem. In fact, we have succeeded, in a particular case, to synthesize a complex boolean function (if one measures the complexity by, for instance, the number of terms of the normal disjunctive form) with a much simpler network. Computation of the output of such a network would be more efficient than a direct evaluation of the boolean function involving 1 bit operations. There remains to examine how to apply this kind of synthesis to other boolean functions.

REFERENCES

Hinton GE (1986)Learning distributed representations of Concepts. *in Proc of the 8th ann conf of the Cognitive Science Society.* Hillsdale, Erbaum.

Le Cun Y and alii (1989) Optimal brain damage. *in Advances in Neural Information Processing Systems , Denver, Colorado* San Mateo, C.A.Kaufmann 1989.

Lehky SR, Sejnowski TJ (1989) Sensory Processing in the Mammalian Brain : Neural Substrates and Experimental Strategies. Edited by JS Lund. New York: Oxford University Press, pp. 331-344.

Vogl TP, Mangis JK, Rigler AK, Zink WT, Alkon DL (1988) Accelerating the Convergence of the Back-Propagation Method. *Biological Cybernetics,* 59, 257-264.

Neural networks			% correct in learning			% correct in generalisation		
Designation	Architecture	ld	sym	ns	Total	sym	ns	Total
SYM3_2	36/25/1	1	100	99	99.5	97	98	97.5
SYM3_3	36/19/1	1	97	96.5	97	92	96	94
SYM3_1	36/12/1	1	94.5	97.5	96	88	95	91
SYM3_S1	36/12/1	2	100	100	100	97	97	97
SYM3_4	36/6/1	2	100	98.5	99	96	97	96.5
SYM3_S4	36/6/1	2	99	98.5	99	92	96	94
SYM3_51	36/3/1	1	?	?	?	86	92	89
SYM3_52	36/3/1	2	97.5	97	97	94	98	96
SYM4_1	36/10/7/1	1	100	99	99.5	98	97	97.5
SYM5_1	36/10/5/2/1	1	99.5	99.5	99.5	96	98	97

Table 1 : Simulation results for 6x6 pixels images

One ligne represents one simulation.

The architecture of each network is given by the number of neurons of each layer. The "ld" column gives the relative learning duration. Results are given in % of the number of well classified input patterns relative to the total number of input patterns (here 2592 for the learning set and 5184 for the generalisation set) ; the "sym" column gives the results for only the symmetrical input patterns, the "ns" column for the non-symmetrical ones, the "Total" column for all the input patterns of the considered set.

The networks SYM3_S1 et S4 have been trained without neuron thresholds. The network SYM3_52 is the network SYM3_51 after a new series of presentation of the learning set.

Neural networks			% correct in learning			% correct in generalisation		
Designation	Architecture	nb iter	sym	ns	Total	sym	ns	Total
SYM3_11	100/10/1	9114	100	99.5	99.5	86.2	80.2	83.2
SYM3_20	100/19/1	7009	100	99.5	100	69.3	69.1	69.2
SYM4_10	100/9/3/1	19693	100	99.5	100	90.9	80.6	85.7
SYM4_20	100/19/5/1	6695	100	100	100	78.9	82.0	80.5
SYM5_15	100/14/4/2/1	8582	100	100	100	94.8	89.1	92.0

Table 2 : Simulation results for 10x10 pixels images

Refer to the legend of table 1.

The "nb iter" column gives the number of iterations of the learning period. Here the results are given for 4320 input patterns in the learning set and 2160 in the generalisation set.

Network	Architecture	sym	non sym	Total
SYM3_11 E	100/10->3/1	99	95	97
SYM4_13 E	100/12->4/4/1	100	95.5	98
SYM4_13 C	100/12/4/1	82	86.5	84
SYM4_13 D	100/4/4/1	35.5	64.5	50

Table 3 : epigenetic simulation results

The columns have the same meaning as in the previous tables. The results are given for 2880 generalisation input patterns. Two networks have undergone epigenetic learning : their designation ends with the letter E. The network SYM4_13 C initially identical to SYM4_13 E, was trained classically. The network SYM4_13 D is obtained from SYM4_13 C by preservation of the only hidden neurons present in SYM4_13 E (see text).

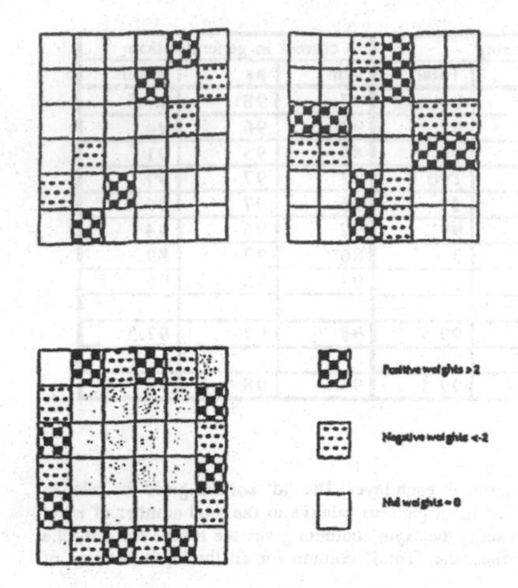

Figure 1 : Examples of input weight patterns to some structured neurons of the first hidden layer. These patterns are almost antisymmetrical and at least some of them are found in all the networks trained for the detection of symmetry in 6x6 pixels images. This figure shows the spatial structure of the patterns and the relative values of the connections.

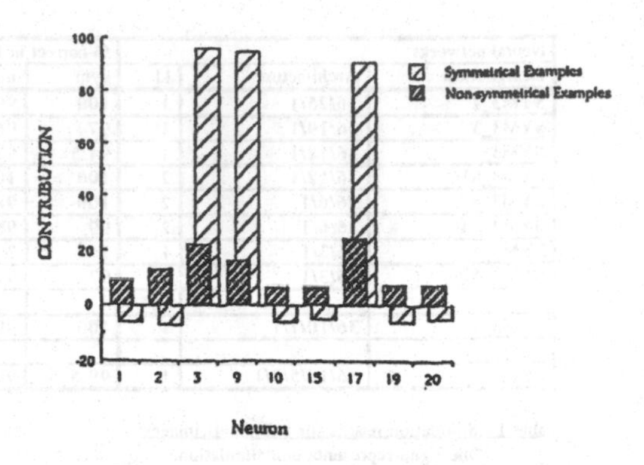

Figure 2 : Contributions of neurons of the first hidden layer of network SYM3_2 calculated on a generalisation set, for symmetrical and non-symmetrical input patterns. Neurons having a contribution inferior to 1% in absolute value have not been presented. A negative contribution means that the performance of the network increases when the neuron is deleted. Neurons 3, 9 and 17 are structured (see text).

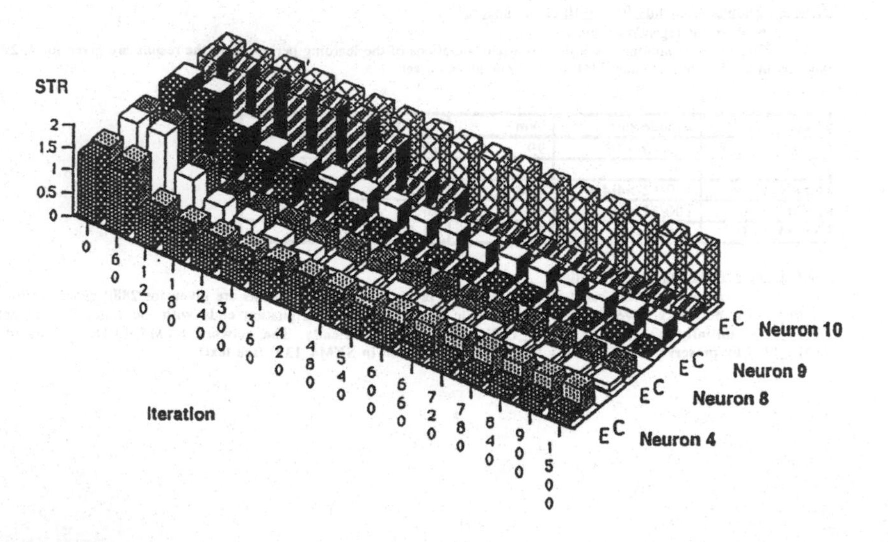

Figure 3 : Various evolutions of the structuration index during the classical learning (C) and the epigenetic learning (E) of two initially identical networks. The evolutions are compared for the 4 neurons that are preserved in the epigenetic network. Note that the index evolve no more after the 900th iteration.

Hopfield Neural Network for Routing

S. Cavalieri, A. Di Stefano, O. Mirabella

Istituto di Informatica e Telecomunicazioni
Facolta' di Ingegneria, Universita' di Catania
V.le A. Doria 6, 95125 Catania, Italy
tel +3995 339449 fax +3995 338887

Abstract.

In the context of the Packet Switching Networks, the main need is to optimize the paths connecting the source and the destination. In this paper the authors deal with the problem of routing by means of a neural approach. This is based on the assumption that each switching node is made up by a Hopfield network which is capable to determine the optimal path which connects the local node with a generic node in the packet switching network. On this basis, the suitable output link on which each incoming packet must be routed is chosen.

1. Introduction.

From a topological viewpoint computer networks can be seen as multipath systems in the sense that there is often no single path between the source node to which the computer generating the information is connected and the destination node. The communication structure is particularly complex in the case of Packet Switching Networks in which each switching node has different possible outputs for the routing of data packets (elementary fragments into which each message is divided), or the Internetworking of Local, Metropolitan or Wide Area networks in which a packet may cross different networks before reaching its destination. Although the problems related to Packet Switching Networks and Internetworking are in some respects similar, they have peculiar characteristics which suggest adopting different approaches in order to reach a solution. In this paper the authors deal with the problem of routing with reference to a Packet Switching Network, even though the results obtained can easily be transferred to the problems of Internetworking.

In the context of Packet Switching Networks, the main need is to optimize the paths connecting the source and the destination, which can be achieved by using suitable routing algorithms. This optimization takes account of various parameters such as the length of the paths, the traffic in the various links, the length of the queues in the switching nodes, which are summarized in a cost parameter associated to each communication link. To route a packet properly, each switching node holds a routing table which indicates the outputs to be used for each destination. As network conditions may change (on account of variations in traffic, the number of nodes, the single interconnections, etc.), the tables have to be updated periodically to allow dynamic adaptive routing. This activity can be performed both at a centralized and at a distributed level [1]. In the first case a Routing Control Center (RCC) collects information from all the nodes in the network and then, by means of a shortest path algorithm, processes the optimal paths for each destination and periodically distributes a new routing table to the nodes. In the case of distributed routing, on the other hand, each node updates its own routing table, usually on the basis of local information from adjacent nodes.

In this paper the authors deal with the problem of routing by means of a neural approach. More specifically, by entrusting the routing mechanism in each node to a neural network, they attempt to obtain distributed routing, at the same time keeping a global vision of the network. This allows the main advantage of centralized routing to be maintained, avoiding problems linked to the transfer of routing tables from the RCC to all the nodes (which mainly leads to communication overloading and a loss of consistency between the various tables, due to communication delays).

The lecterature offers different solution to be adopted in order to reach this goal. Using classical single source shortest path algorithms [2][3] (generally adopted in centralized routing) allows each switching node to prepare its routing table, however it needs powerful computational resources in order to obtain low processing time.

In this paper, the routing problem has been approached considering the Hopfield associative neural network used as optimizer. The Hopfield network would seem, by nature, to be suitable for solving the optimization of routing, although, as will be shown below, it is remarkably critical in the phase devoted to mapping the conditions surrounding the problem to be optimized with the internal parameters of the network itself. This approach is particularly powerful if we refer to a hardware implementation which guarantees low implementation cost and short processing time.

The paper is made up by the following sections. Section 2 shows the neural network model which solves the optimal path searching adopted by the authors for the distributed routing. Section 3 characterizes the behaviour of the NN by the performance evaluation. Finally in Section 4, the authors will discuss the problėms relevant with the real use of the NN approach in a distributed routing scenario, suggesting possible solutions.

2. The Neural Network Model for the Distributed Routing.

The neural approach that here is presented (and that is similar to the one used in [4]), it is based on the assumption that each switching node is made up by a Hopfield Network (the continuos and deterministic model described in [5] was adopted), which is capable to determine, for each incoming packet characterized by a destination address, the optimal path which connect the local node with the destination node. On this basis the suitable output link on which the packet must be routed is chosen. This solution consists of obtaining from the stabilized output of the neural network information about the optimal path between two nodes in the graph, which models the computer network. Alternative approaches in which the network simultaneously provides all the optimal paths between the source node and all the other nodes, or all the optimal paths between each possible pair of nodes in the graph, were immediately discarded as they would have required a much more complex neural network in terms of the number of neurons.

Indicating the number of nodes in the graph as n, a Hopfield neural network with n^2 neurons was considered. The neurons were logically subdivided into n groups of n neurons each. Henceforward we will identify each neuron with a double index, xi, (where the index x=1..n relates to the group, whereas the index i=1..n refers to the neurons in each group), its output with OUT_{xi}, the weight for neurons xi and yj with $W_{xi,yj}$, and the external bias current for neuron xi with I_{xi}.

If the output of the generic neuron xi, OUT_{xi}, assumes a value of 1, it indicates whether the optimal path determined is such that node x in the graph is connected with node i. Thus if the optimal path between the source node (s) and the destination node (d) is made by the following node sequence: $n_0, n_1, \ldots, n_h$ with n_i being the node index (i=0..h), $n_0 = s$ and $n_h = d$, the outputs are:

$$\begin{cases} OUT_{n_j n_{j+1}} = 1 & j = 0..h-1 \\ OUT_{xi} = 0 & \text{othewise} \end{cases}$$

Solving two-node optimal path problem by Hopfield networks requires the definition of a suitable energy function, based on the conditions surrounding the problem. Below we will show energy function terms, describing the inherent surrounding conditions. The results given will relate to determination of the optimal path connecting the generic node s to the node d. The expression of the energy function is:

$$E = \frac{A}{2} \bullet \sum_x \sum_i \sum_{j \neq i} OUT_{xi} \bullet OUT_{xj} + \qquad (1)$$

$$\frac{B}{2} \bullet \sum_i \sum_x \sum_{y \neq x} OUT_{xi} \bullet OUT_{yi} + \qquad (2)$$

$$\frac{C}{2} \bullet \left(\sum_i OUT_{si} - 1 \right)^2 + \qquad (3)$$

$$\frac{D}{2} \bullet \left(\sum_i OUT_{di} \right)^2 + \qquad (4)$$

$$\frac{E}{2} \bullet \sum_i \sum_{j \neq i} OUT_{ii} \bullet OUT_{jj} + \qquad (5)$$

$$\frac{G}{2} \bullet \left(\sum_{x} OUT_{xs} \right)^2 + \qquad (6)$$

$$\frac{F}{2} \bullet \sum_{x} \sum_{\substack{i \neq s \\ i \neq x}} \sum_{\substack{j \neq i \\ j \neq x}} d_{xi} \bullet OUT_{xi} \bullet OUT_{ij} \quad (7)$$

Each function term respectively refers to the following surrounding conditions:
1) each node can only be connected to one other, that is there can be at most one output in each switching node.
2) no node can be crossed twice.
3) As the source node s must be connected to another one, the output OUT_{si} will have to be a 1.
4) It is necessary for the destination node to be number d. For this purpose it is sufficient for node d to have no outputs ($OUT_{di} = 0$, $i=1..n$).
5) The generic i-th node cannot be connected to itself ($OUT_{ii} = 0$, $i=1..n$).
6) The optimal path can only cross the source node s once ($OUT_{xs} = 0$, $x=1..n$).
7) Finally, it is necessary to introduce a term which will take account of the distance to be covered. This term represents the overall length of each valid path connecting the source and destination nodes. Validity consists of discarding all the closed paths (i.e. all those connecting the source nodes with two different nodes), all those in which a node is crossed more than once, and those in which each node reconnects with itself (i.e. $OUT_{ii} = 1$ $(i=1..n)$). The expression of this condition (((7) in the energy function) presents the term $d_{xi} \bullet OUT_{xi} \bullet OUT_{ij}$, that represents the generic component of each valid path between node s and node d.
From a comparison between the Liapunov function expressed by :

$$E = -\frac{1}{2} \bullet \sum_{x} \sum_{i} \sum_{y} \sum_{j} W_{xi,yj} \bullet OUT_{xi} \bullet OUT_{yj} - \sum_{x} \sum_{i} I_{xi} \bullet OUT_{xi} \qquad (8)$$

and the energy function shown before, the weights of the neural network were determined according to the optimization problem being considered. In this way reaching a local minimum by the Liapunov equation corresponds to the solution to the problem. The weights matrix thus obtained was asymmetric and so the stability conditions established in [5] were not satisfied. By adding the following term:

$$\frac{F}{2} \bullet \sum_{y} \sum_{\substack{j \neq s \\ j \neq y}} \sum_{\substack{i \neq j \\ i \neq y}} d_{yj} \bullet OUT_{ji} \bullet OUT_{yj} \qquad (9)$$

the weights matrix is made symmetrical without altering the surrounding condition relating to distance. The weights conditions determined on the basis of the comparison between the energy function obtained by summing the symmetrization term (9) to the terms seen previously and the Liapunov function (8), are expressed by:

$$
W_{xi,yj} = \begin{cases}
-A & \text{if } x = y \text{ and } i \neq j \\
-B & \text{if } x \neq y \text{ and } i = j \\
-C & \text{if } x = y \text{ and } x = s \text{ and } i = j \\
-2 \cdot C & \text{if } x = y \text{ and } x = s \text{ and } i \neq j \\
-D & \text{if } x = y \text{ and } x = d \text{ and } i = j \\
-2 \cdot D & \text{if } x = y \text{ and } x = d \text{ and } i \neq j \\
-E & \text{if } x \neq y \text{ and } x = i \text{ and } y = j \\
-F \cdot d_{xy} & \text{if } x \neq i \text{ and } i \neq s \text{ and } j \neq i \text{ and } y = i \text{ and } j \neq x \\
-F \cdot d_{xy} & \text{if } x = j \text{ and } j \neq s \text{ and } j \neq i \text{ and } y \neq j \text{ and } i \neq y \\
-G & \text{if } i = j \text{ and } i = s \text{ and } x = y \\
-2 \cdot G & \text{if } i = j \text{ and } i = s \text{ and } x \neq y
\end{cases}
$$

As seen before, the weights strongly depend on the energy function coefficients, which must be tuned in each node for any instance of source, destination and costs distribution. It is evident that the source is fixed, since the NN is resident in each node. In order to test the real applicability of the proposed neural approach it is necessary to verify that the coefficients are not strongly influenced by varying the destination and cost distribution. The following section aims to verify this two requirements.

3. Some Considerations on the Results Obtained.

In the tests performed, the authors made use of the Anza Plus neural simulator [6], made up of a neural accelerator plugged into an IBM PC and software management of the board (User Interface Subroutine Library - UISL). The tests were carried on considering a communication network made up by 10 nodes. The graph which models the network is shown in Fig.1. The numbers near each link represent the cost of the link.

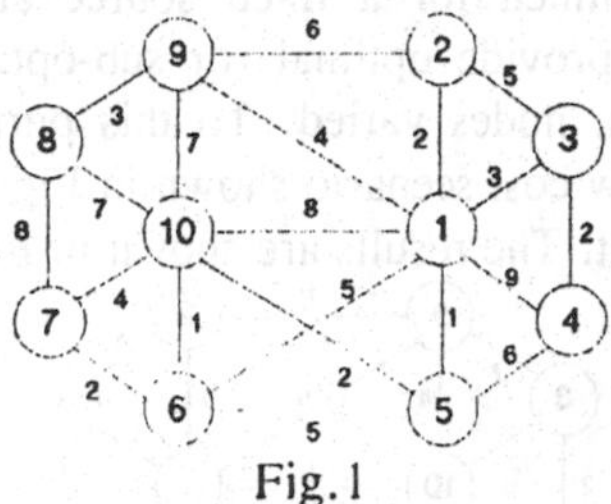

Fig.1

From the very first tests carried out it was also seen that when the destination node varied with respect to the same source, it was necessary to vary the coefficients even further in order to determine an optimal solution. However, through continuous adjustments to the coefficients it was possible to fix values for each source, in such a way as to ensure a high number of valid solutions. The values found for each source also made it possible to make the neural network converge on optimal or almost optimal paths for all destinations. Fig.2 shows the results obtained for source 7, considering ten iteration for each destination node. As can be seen, for each destination the number of wrong paths is very low. In addition, for most destinations (nodes 1,2,4,5,6,8,10) the percentages of optimal paths are very high. Only for some nodes (nodes 3 and 9) optimal paths were not obtained, but the paths mainly present a cost near to optimal, as can be seen in Table 1, which specifies all the paths indicating their overall cost as compared with the optimal cost.

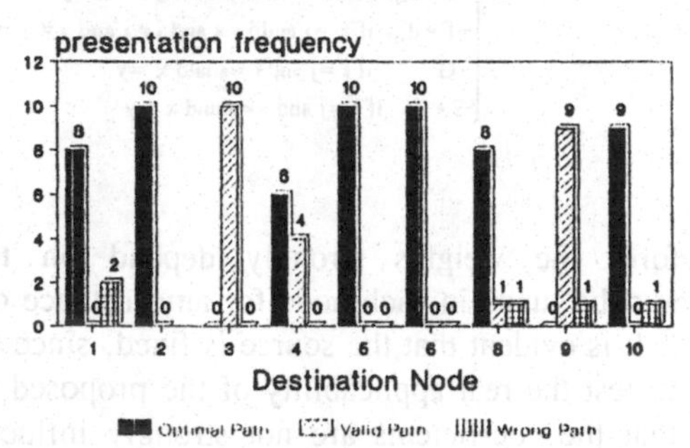

Fig2

Node	Valid Paths	Cost	Optimal Cost	Number of Valid Paths
Node 3	7-6-10-5-4-3	11	9	10
Node 4	7-6-10-5-1-4	15	11	1
	7-6-10-5-1-2-3-4	15	11	3
Node 8	7-6-10-8	10	8	1
Node 9	7-8-9	11	10	9

Table 1

The subsequent investigation aimed at ascertaining whether, once the optimal coefficients had been determined for a fixed source and graph configuration, the network would continue to provide optimal (or sub-optimal) results even when the costs of connections between nodes varied. To this purpose, we used node 7 as a benchmark, considering a new cost scenario shown in Fig.3, and the same sequence of tests as before was carried out. The results are shown in Fig.4 and table 2.

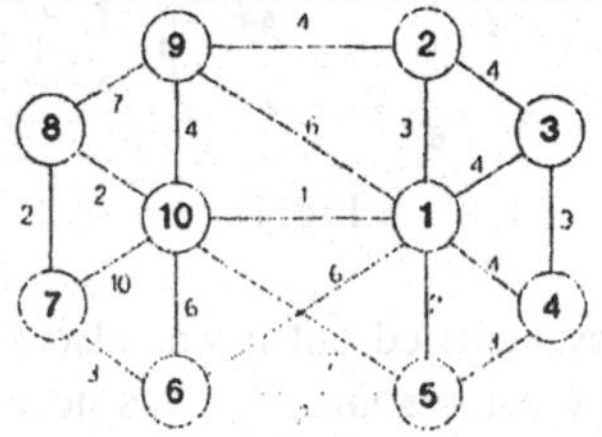

Fig.3

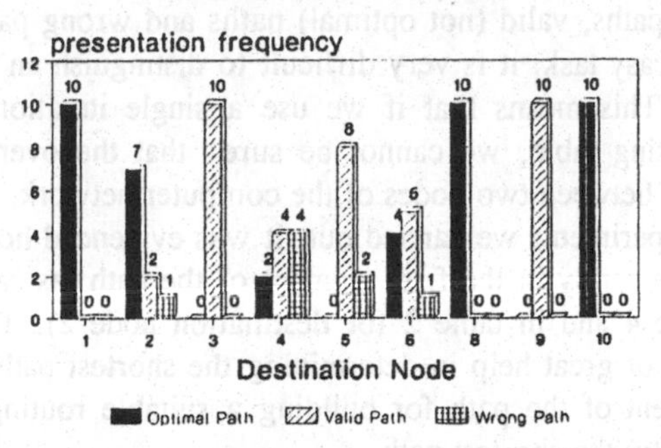

Fig.4

Node	Valid Paths	Cost	Optimal Cost	Number of Valid Paths
Node 2	7-8-9-2	13	8	1
	7-8-10-1-9-2	15	8	1
Node 3	7-8-10-1-4-3	12	9	10
Node 4	7-8-10-5-4	9	8	4
Node 5	7-8-10-1-5	7	5	8
Node 6	7-8-10-1-5-6	9	3	5
Node 9	7-8-10-1-2-9	12	8	10

Table 2

From comparison with Fig.2 it emerges that although no further adjustment has been made to neural network coefficients, the NN continues to converge on a high number of optimal paths. This shows that the awareness acquired by the NN is only linked to the framework of the graph and is relatively independent of the values of the costs of the arcs, thus greatly increasing the applicability of the approach proposed.

4.Distributed Routing Performance by Hopfield Neural Approach.

From the results we have shown in the previous section, the ability of the Hopfield neural network in finding a valid path (often the shortest one) in a packet switching network is clearly assessed. Moreover, for each source node, the coefficients of the energy function only depend on the computer network topology and are valid for all the destination nodes set, and costs configuration. These are undoubtedly encouraging results, but they are not enough good if we have a real application in mind.

The basic idea concerning the applicability of the Hopfield-based approach to the distributed routing, consists in the use of the outputs provided by the NN to build the routing table for the local node. For each final destination only the first segment of the total path is taken into account: it represents the adjacent node to which route the incoming packet for a final destination node.

As seen before, the crucial point is that once we determine the energy function, the convergence of the network is not univocally determined, because sometimes we can find both local minima or wrong solutions. As we have shown in

Figs.2,4, if we run several times the NN, for each destination we can find a certain number of optimal paths, valid (not optimal) paths and wrong paths. While the error detection is a very easy task, it is very difficult to distinguish an optimal path from a non-optimal one. This means that if we use a single iteration to determine each element of the routing table, we cannot be sure that the overall routing obtained optimizes the paths between two nodes of the computer network.

From the experiments we carried out, it was evidenced how the errors in non-optimal paths occurs rarely in the first segment of the path (as can be seen in table 1 for destination node 4 and in table 2 for destination node 2). This is an interesting result which can be of great help in determining the shortest path. In fact, if we use only the first segment of the path for building a suitable routing table, this will be (with high probability) the shortest path.

At this point it is necessary to verify the correctness of the approach we have proposed. In order to evaluate the NN performance, we considered two different scenario according the result provided by Hopfield outputs:

- in the first approach, (we call it "minimum cost") we have chosen the best path among those obtained on a large spectrum of iterations.
- in the second approach, (we call it "maximum cost") we select the highest cost path obtained in the same iteration set.

On the basis of to above mentioned scenarios, we have computed two different routing tables for each node considering the cost scenario shown in Fig.1. These tables have been used to obtain the complete path followed by a packet from a generic source node to a destination. In particular we selected, as done before, the source node 7 and considered all the other nodes as destinations.

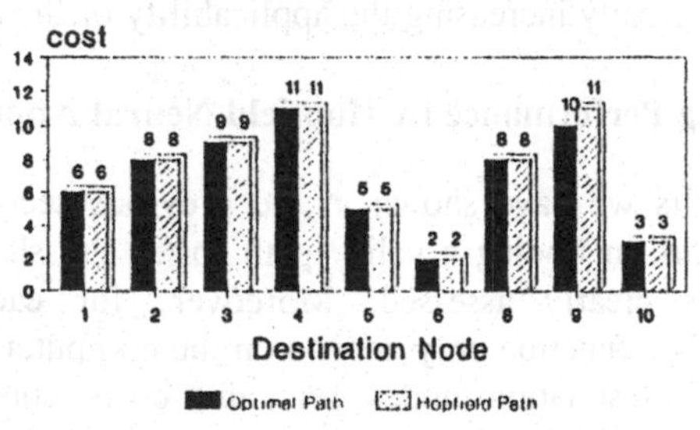

Fig.5.a

Figs.5.a and 5.b compare the total cost of the paths from node 7 to all the others using the "minimum cost approach" and "maximum cost approach" respectively. We can see that the two approaches offer similar performances with the only exception of destination 8.

Costs from Node 7
Maximum Cost Scenario

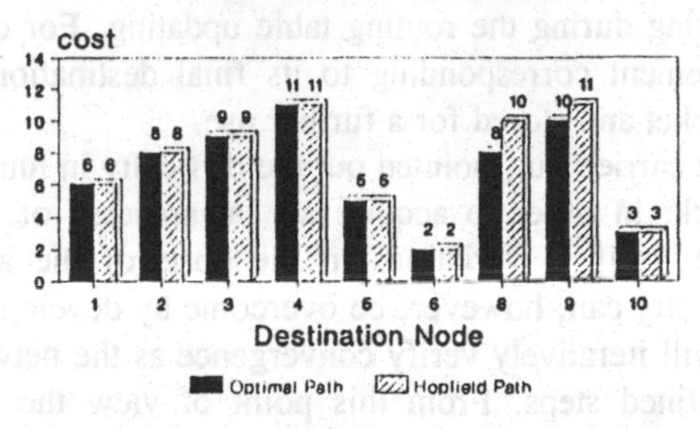

Fig.5.b

Moreover, the performances we have obtained are close to the optimal one. Summarizing, two main aspects may be pointed out from the results shown since now. Firstly, if we compute the complete path summing all the first segment of the paths provided by the NN each node, we strongly reduce errors made by the NN. Infact the results obtained in this way and shown in Fig.5.a and b are much better than those shown in Fig.2 where each complete path was provided by an unique NN. Secondly, since the two scenarios above presented provide similar performances, we infer that the first valid path found by each NN is suitable to built the routing path, and we do not need to adopt an additional strategy to select the best result in an iteration set.

Final Remarks.

In this paper the authors have presented a solution to the routing problem in a Packet Switching Computer Network based on a Hopfield network. Convergence on optimal solutions is guaranteed with an acceptable percentage of error. This is inevitable in approaches of this kind and is due to the high number of local minima the energy function presents. However, this problem has been mostly overcome through a suitable strategy for preparing the routing tables: the responsibility of preparing the shortest path from source to destination is distributed among several NN, each one relevant with a different source node. From an analysis of the approach considered some considerations must be carried out.

Each NN provides the whole path from its source and destination, but only one segment of this path is used. From a pure computational point of view this is a waste of processing power, but, on the other hand, in this way we overcome the problem of errors related to local minima.

An open problem, that requires an appropriate solution, is the overload in the computer network caused by message required to exchange the information inherent modifications in the costs of the network interconnections. This is required since each NN must have a global vision of the network. This is a common problem offered by all the routing algorithms and can be made less critical by a suitable cost communication strategy (its discussion is out of the scope of this paper).

The NN approach proposed seems suitable to an off-line utilization, since the iterative process to build the routing table is time-consuming. Nevertheless we could have an on-line routing during the routing table updating. For each incoming packet, the routing table element corresponding to its final destination, may be processed, used to route the packet and stored for a further use.

The tests we carried out, pointed out the difficulty in tuning the coefficients of the Hopfield network, in order to acquire an "awareness" of the network topology alone, and to adapt itself to variations in the costs of the arcs in the Computer Network. This difficulty can, however, be overcome by developing suitable automatic instruments which will iteratively verify convergence as the network parameters vary, according to pre-defined steps. From this point of view the existence of classical algorithms to determine the shortest path may provide valid support in controlling the validity of the paths found, step by step.

References.

[1] A.S.Tanenbaum, "Computer Networks", Second Edition,Prentice-Hall International Editors, pp.284-309.

[2] E.W.Dijkstra, "A Note on Two Problems in Connexion with Graphs", Numer.Math. vol.1, pp.269-271, October 1959.

[3] R.W.Floyd, "Algorithm 97:Shortest Path", Communication ACM, vol.7:12; pp.701.

[4] S.Cavalieri, A.Di Stefano, O. Mirabella, "A Neural-Network-Based Approach for Routing in a Packet Switching Network", IJCNN92, Baltimore, Maryland, USA, June 7-11, 1992, pp.913-918, Vol.II.

[5] J.J.Hopfield, "Neurons with Graded Response Have Collective Computational Properties Like those of two-state Neurons", proceedings National Academy of Sciences 81:3088-3092, May 1984.

[6] Anza Plus User's Guide and Neurosoftware Documents Release 2.2 15 May, 1989.

NEURAL NETWORK ROUTING CONTROLLER FOR COMMUNICATION PARALLEL MULTISTAGE INTERCONNECTION NETWORKS

A. García-Lopera, A. Díaz Estrella, F. García Oller, and F. Sandoval

E.T.S.I. Telecomunicación, Dpto. de Tecnología Electrónica
Universidad de Malága. Plaza El Ejido, s/n, 29013 Málaga, SPAIN

ABSTRACT

In this paper we propose a centralized neural network contention controller for Asynchronous Transfer Mode (ATM) input buffered switches fabrics. These switches are self-routing type, and suffer of head of line (input), internal and output blocking possibilities. Messages (cells) through the switch can collide at these three points. The neural controller will be able to choose within each cell slot, the maximum possible number of connection-requesting cells from the inputs queues with legal routes. The switch will be able of connecting these selected cells to desired output addresses without internal or output blocking possibility. Increased throughput performance will result. We propose and analyze new high performance architectures of parallel connected switches and associated controllers, which, in addition, are fault-tolerant. Simulation results show low cell loss ratio for small size of the input queues.

1.- INTRODUCTION

Recent technological advances, particularly in fibre optics and microelectronics have made possible to consider Broadband Integrated Services Data Network (B-ISDN) as the most promising technique for the development step after the narrowband integrated environment (Integrated Services Data Network, ISDN) [1]. Fast Packet Switching has emerged as a prime candidate to support the high data rates and the wide variety of traffic requirements existing in B-ISDN. The Asynchronous Transfer Mode (ATM) in multiplexing and switching has been proposed as a flexible mechanism for handling the different traffic types in broadband networks. Information in ATM systems is sent in small fixed length packets, that are called cells, and routing is based on the information contained in the header, that accompanies each cell. Latest standards define the ATM transport format with a cell size of a 48-byte information payload plus a 5 byte header, requiring a mere total of 3 μs for transmission on a 150 Mb/s ATM link. Actually, the bottleneck in the system is not in the transmission (fiber optic with several Gigabits Bandwith) but in the interconnection network (or switching fabric) that is able to work at rates up to just a few Mb/s. Switching thousands of ATM channels requires a broadband switch with a capacity ranging from hundreds of Gb/s to several Terabits. We assume that the ATM links are slot synchronized at the switch and the pathway within a switch have to be reconfigured per cell time.

Conventional circuit switching methods are too slow, with respect to the cell time, to set up that transfers paths for the ATM links. Several architectural designs of fast packet switches offering high performance have been proposed in recent years [2]. The challenge today is to realize these architectures in hardware at the required speeds, and integrate means in these switching systems to achieve fault tolerance. The most common architectures of interconnection networks proposed are based in space division multistage network. This structure has been extensively studied and used in circuit switching and considered for ATM switching [3]. Specially interesting is a class of them whose main property is to be self-routing, e.g. Omega, Flip, Baseline, Shuffle...All of them are included into the generic Banyan network. An interconnection network is self-routing if there is only one predeterminated path through the switch to get each output outlet from each input inlet. These networks are capable of simultaneously switching several cells and in parallel to obtain a high throughput. Unfortunately they are blocking networks in the sense that cells can collide with each other at the internal switching nodes and get lost. To overcome this drawback different approaches has been applied. One approach is to implement parallelized control algorithms [4]. It suffer limitations of software speed at ATM frequencies (higher than 155 Mbit/s). Also hardware architectures based on sorting networks [5] or buffered input-output like the Buffer-Space-Buffer (BSB) [6] has been proposed.

The routing of a set of cell through a multistage interconnection network can be modelled as a constraint satisfaction or assignment problem. Recently, the neural networks has been used to solve a variety of this type of problems [7] [8]. The Neural Networks are architectures of massive parallel processing and they are also topologically similar to the switching fabrics. So, they suggest its application to the control of these systems. It is expected to increase the switching time and routing capability. A. Marrakchi and T. Troudet [9] propose a Hopfield neural network to control a crossbar switch in real time with maximum throughput. Ghosh and Varma [10] use the Hopfield model based neural controller to reduce the switching noise in crossbar switches. Brown studys the problem of finding routes using neural networks in rearrangeable [11] and Banyan [12] multistage switching networks. In this context we propose a new architecture of neural controlled multistage interconnection fabric based in the parallel array of banyan switching networks (PBSF). The neural controller model proposed is based in the Winner-Take-All (WTA) circuit extended to allow multiple overlapping WTA [7]. The paper is organized as follows. We first describe the multistage interconnection network object of our neural application and the traffic control problem. Then we describe the neural controller proposed . Then we consider the new architecture proposed which provides better throughput and fault tolerance. Finally we show the simulation results for several switching fabric configurations and traffic loads, and some concluding remarks are given.

2. SWITCHING FABRICS AND THE BLOCKING PROBLEM.

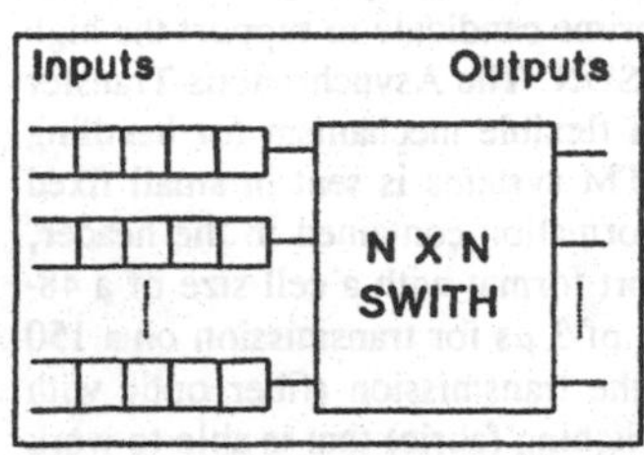

Figure 1: NxN Input Buffered Switch

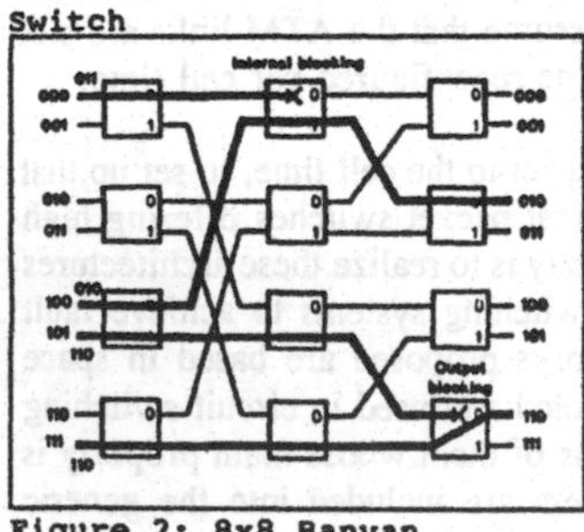

Figure 2: 8x8 Banyan Switch

A NxN switch is defined as a device that takes a set of N inputs, and reproduces them at its N outputs in some desired permutation. The switch is blocking if not all the possible permutation of the inputs are allowed, otherwise is non-blocking. In Figure 1 is represented this switch with input buffers. In the input buffers are the cells waiting to be transmitted through the switch. Output blocking situation is generated when two or more inputs of the switch have cells with the same destination address: their cells will be routed to the same output at the same slot and will collide. Internal blocking is produced when two or more cells with distinct inputs and outputs have overlapped paths through the switch. The traffic control problem is to find the maximum number of conflict-free cells to be transmitted from the input to the output at each slot. The maximum number of cells that the switch can transfer from the inputs to the outputs during a slot is referred to as the switching throughput capacity, while the switching capacity per input is assumed to be one; i.e., at most one cell per input can be transferred to an output in one slot. In Figure 2 an 8x8 (N=8) regular square Banyan switch is represented showing internal and output blocking. The network considered consists of $n = \log_2 N$ (n=3) stages composed of 2x2 non blocking switching elements. A cell self-routes through the banyan network using its destination address. The **kth** most significant bit in the output address of a cell controls the state (cross or bar) of the switching elements of the stage **k**, sending the cell to the top outlet (bit=1) or to the bottom outlet (bit=0).

2.1 Banyan Blocking Constraints.

Let a NxN ($N=2^n$) regular square Banyan switch (Figure 2). There is only one link between each input and output of these nodes. Each cell at the input is defined by a pair (I/O) where I is the input address and O is the output address. The blocking is produced if two cells try to share the same link between two switches. Therefore, none of the cells are blocked if and only if there is no link used by more than one cell. Let L_k(I/O) the label of the link that the cell (I/O) uses in the kth stage to go to the (k+1)th stage. Stage 0 is the input stage, and stage n+1 is the output stage. S^k_j is the set of all input/output pairs that use link j after stage k. For each stage k it exist N different links, so there are N different S^k_j. T. Brown [12] has defined the blocking constraints for a Banyan switch in terms of "class of equivalence". Each L_k, $0 \leq k \leq n$, partitions the N^2 input-output pairs into N equivalence classes. For N=8 there are n+1=4 different sets of N=8 equivalence classes. A set C of cell is non blocking if and only if each equivalence class S^k_j contains at most one pair (I/O) ϵ C.

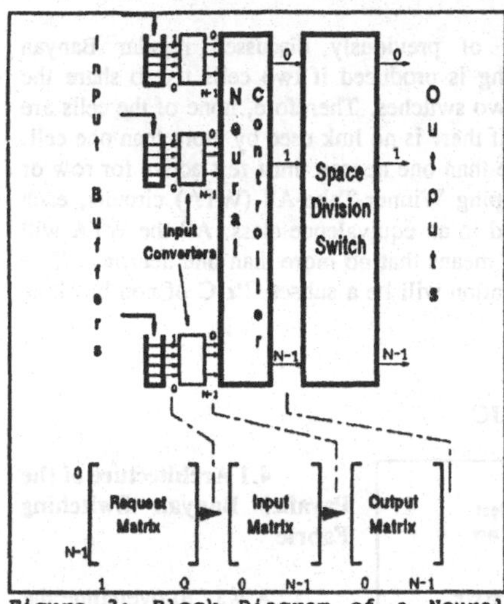

Figure 3: Block Diagram of a Neural Controller

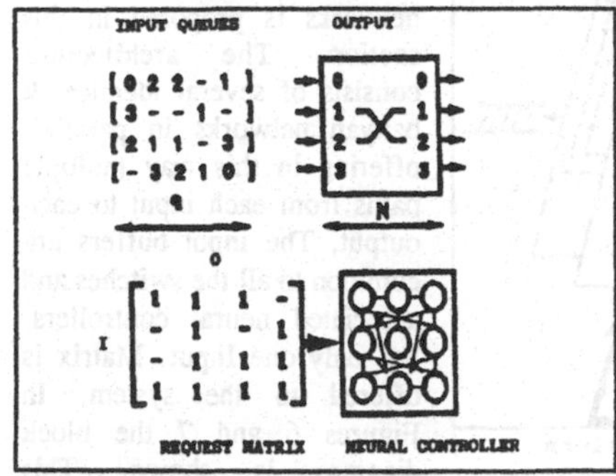

Figure 4: Matrix assignment example

3. NEURAL CONTROLLER MODEL

3.1 General Structure.

In Figure 3 a general structure of a switch and associated neural controller is shown. The switch has N x N input/output lines and the buffer size is Q. A cell arriving to the input buffer is defined by its input line arriving address, I_j, and its output line destination address, O_j. A label (I/O) is assigned to each cell. Given a set C of cells queuing at the inputs of the switch, the elements of the Input Matrix N x Q are the pair (I_j,O_j) waiting at the input queues for a given slot time. The Request Matrix N x N (Input x Output) to the controller is a binary matrix obtained mapping the Input Matrix into the Request Matrix, with a "1" in the element (I/O) if and only if there exist a cell (I/O) owing to C. The controller applies the specific blocking constraints for the space switch to choose a subset C' that is non blocking and has maximum overlap with C. If there are several cell with the same label (I/O), i.e. in an input buffer several cells going to the same output address, they will be represented in the Request Matrix by only one "1".

The neural controller matchs the topology of the switch network and its design is derived from the specific set of constraint for the controlled interconnection network. It consists of an array of N x N neurons that corresponds to the input matrix of Figure 1. The input to the neural networks is the input matrix. The neuron (I/O) receives a +1 if the corresponding element in the input matrix is +1. Otherwise the input is -1. The constraint conditions will define the way of connecting the neurons. After computation, the set of neurons that are on will map a subset C'ϵ C of non blocking cells. An example of this situation is shown in Figure 4.

3.2 The Winner-Take-All Circuit

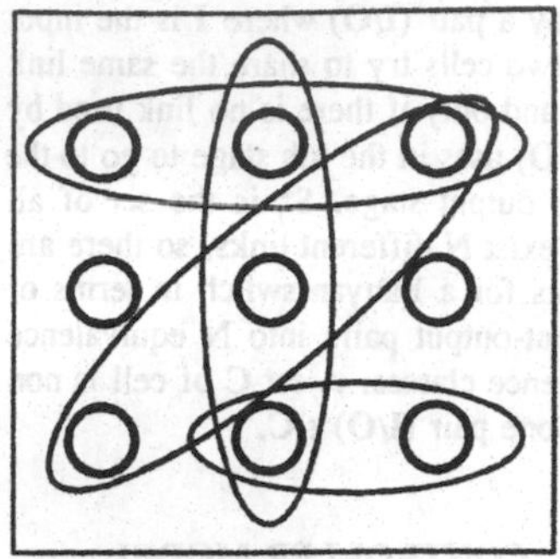

Figure 5: Multiple overlapped WTA

Inhibition plays an important control function in our selection task. One form of inhibitory system is a subnetwork with a WTA mechanism. The WTA subnetwork consists of mutually inhibitory links among all the neurons. Starting at the same initial internal state and with different inputs for each neuron, a competitive dynamic collective mechanism takes place and as a result of this competition only the neuron with the largest input is selected and all other units with weaker inputs are inhibited. The WTA may be implemented with O(N) connectivity [13]. Brown [12] found that WTA and k-WTA circuits still work properly when they are multiple overlapped in a Multiple Overlapping Winner-Take-All-Circuit as it is represented in Figure 5.

In the case of previously discussed regular Banyan network, the blocking is produced if two cells try to share the same link between two switches. Therefore, none of the cells are blocked if and only if there is no link used by more than one cell. Translated to the neural controller, it means that no more than one neuron must rest active for row or column. The neurons are connected in multiple overlapping Winner-Take-All (WTA) circuits, each circuits corresponding to an S_j^k. So, each WTA is related to an equivalence class. All the WTA will compete and only one neuron will win for each WTA; it means that no more than one neuron will be on per constraint set. Therefore, the output after computation will be a subset $C' \epsilon C$ of non blocking cells.

4. PARALLEL BANYAN SWITCHING FABRIC

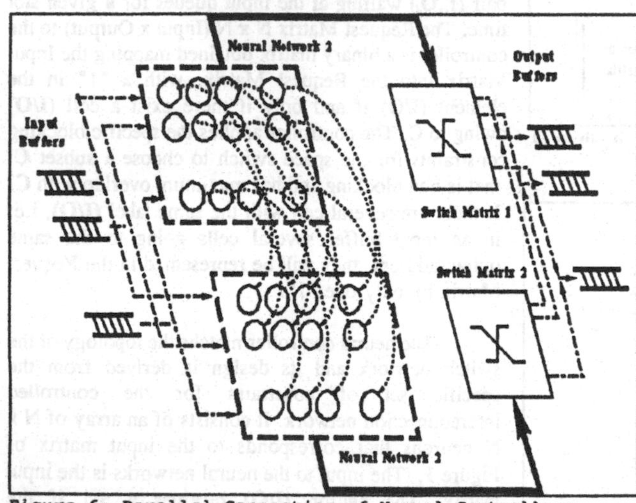

Figure 6: Parallel Connection of Neural Controllers

4.1 Architecture of the Parallel Banyan Switching Fabric

A way to overcome the undesirable blocking effect of the banyan interconnection networks is proposed in this section. The architecture consists of several identical k banyan networks in parallel, offering in this way multiple paths from each input to each output. The input buffers are common to all the switches and associated neural controllers. So, only one Input Matrix is offered to the system. In Figures 6 and 7 the block diagram is shown. This architecture achieve output buffering, and each output of every banyan network is connected to the output buffer belonging to the corresponding output port. At the output of a banyan network, all those cells selected from the input queues to be transmitted are addressed to the output port's buffers. At this output buffers the cells must be pulled out at a rate of k times the input frequency cell in the input buffers.

4.2 Connection of the neural controllers

In Figure 7, it is represented the request binary matrix corresponding to each Banyan switch (k=1....k=n). The elements a_{ij1}, a_{ij2},... a_{ijn}, corresponding to an I/O pair in the input matrix, common to all the Banyan switches. The I/O cell must be addressed through only one of the parallel matrix. So, we connect all the corresponding a_{ij1}....a_{ijn} of each neural controller in a WTA. The output of each neuron inhibits all the another. The winner neuron avoids that the rest of them win in their respectives overlapped WTA. In this way the request matrix is subdivided in n maximum disjoint sets overlapping with the set C that forms the input matrix. Each parallel neural controller sends an output matrix to its corresponding banyan switch. Therefore, the output after computation will be a maximum of k different subsets of non blocking cell. With this architecture we reduce the internal blocking, offering different paths across the parallel switches.

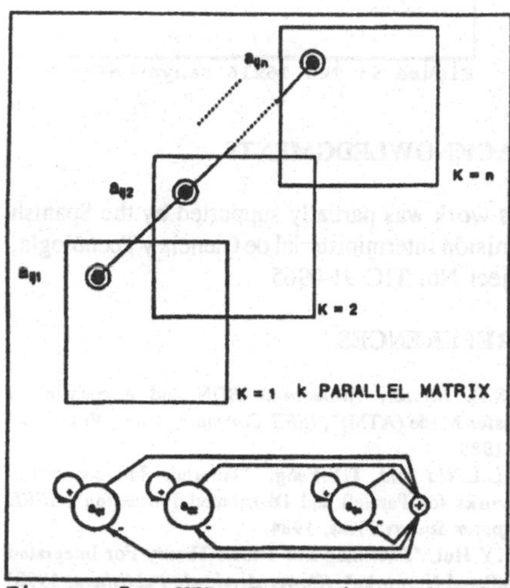

Figure 7: WTA Connection of the Parallel Neural Controller

4.3 Fault tolerance

The parallel configuration provide a fault tolerance scheme. If an internal switch or a link is faulty the corresponding WTA belonging to its class of equivalence S^k_j, as defined in 3.2, will receive an inhibit input in all his neurons. Alternatives routes through another Banyan switch will be selected. Neither the weights nor the WTA connections need to be changed.

5. SIMULATION AND RESULTS

For simulation purposes we have defined a Constant Bit Rate model of traffic. This is a simple Bernouilli process, where one cell arrives at the beginning of each time slot with probability "p". The input cell are stored in the input buffer of variable buffer lenght or "queue depth". The destination addresses of the cell are uniformly distributed with probability 1/N that a given output is chosen. The Neural controller has been simulated by the Rochester Connectionist Simulator [14]. Banyan switching networks with differents capacity (NxN) and parallel configurations was simulated. For three 16x16 Banyan networks in parallel the neural controller contained 16x16x3 = 768 neurons. Each simulation consisted in three cicles of 10.000 cells for each input. They were repetead for differents traffics loads "p" (probability of cell arrival per slot). The size of the input queue (queue depth) were increased in stpes of "1" until to get zero cell lost. In the Figures 8, 9, and 10 results for different configurations are shown. The cell loss in the routing fabric is function of the number N of inputs and output ports in the switching banyan network, the input load "p" and the number of banyan networks "n". Simulation of PBSF architecture has shown that it achieves very low cell loss with only three parallel banyan switching and small size of the input buffers for a high traffic load (high "p").

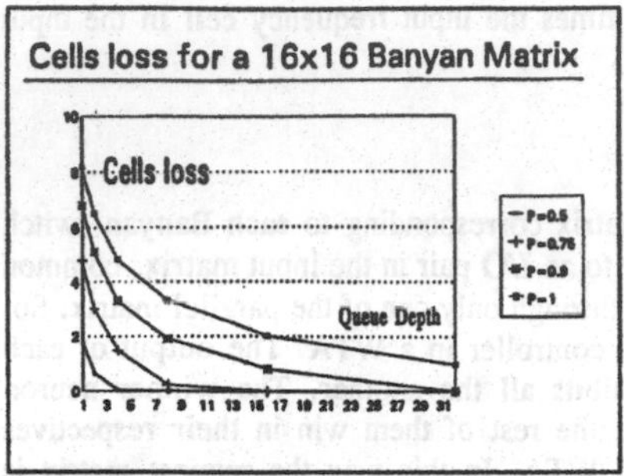

Figure 8: 16x16 Banyan Matrix

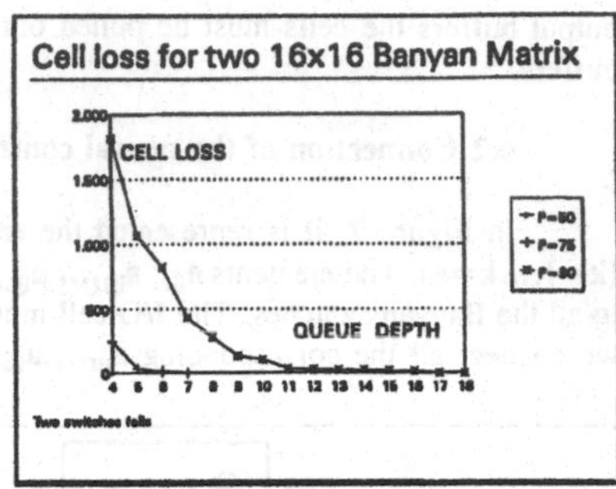

Figure 9: Two 16x16 Banyan Matrix

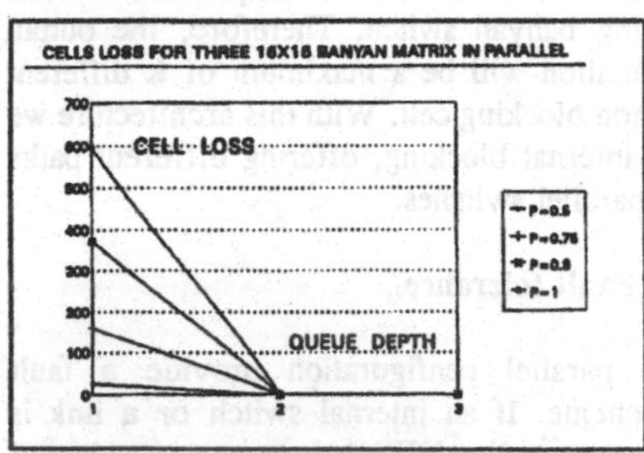

Figure 10: Three 16x16 Banyan Matrix

7. ACKNOWLEDGMENTS

This work was partially supported by the Spanish Comisión Interministerial de Ciencia y Tecnología, Project No. TIC 91-0965.

8. REFERENCES

[1] S.E. Minzer, "Broadband ISDN and Asynchronous Transfer Mode (ATM)", *IEEE Communi. Mag.*, Vol.27 Nº 9. , 1989.

[2] C.L.Wu and T.Y.Feng, "Tutorial: Interconnection Networks for Paralell and Distributed Processing", *IEEE Computer Society Press*, 1984.

[3] J.Y.Hui, "Switching and Traffic Theory For Integrated Broadband Networks", *Kluwer Academic Publishers*, 1990.

[4] J. Munkrees, "Algorithms for the Assignment and Transportion Problems," *Journal of the Society for Industrial Applications of Mathematics*, Vol. 5, pp 32-38, 1957.

[5] K. E. Batcher, , "Sorting Networks and their Applications", *Proceedings Coference AFIPS*, 1968, SJCC, pp 307-314.

[6] H.Obara, S. Okamoto y Y. Hamazumi, "Input and Output queueing ATM switching architecture with spatial and temporal slot reservation control", *Electronics Letters*, Vol. 28, No. 1, Jan.uary 1992.

[7] S.P. Eberhardt, Daud T., D.A. Kernst, T.X. Brown and A.P. Thakoor., "Competitive Neural Architecture for Hardware Solution to The Assignment Problem", *eural Networks*, Vol. 4, pp. 431-442, 1991.

[8] Hakim N.Z. and Meadows H.E., "A neural network approach to the setup of the Benes switch",in *Infocom90*, pp397-402

[9] A. M. Marrakchi y T. Troudet, "A Neural Net Arbitrator for Larger Crossbar Packet Switches," *IEEE Trans. on Circ. and Sys.*, Vol. 36, pp. 1039-1041, Jul.1989.

[10] J.Ghosh y A. Varma, "Neural Networks for Fast Arbitration and Switching Noise Reduction in Large Crossbar", *IEEE Trans. On Circuits and Systems*, Vol. 38, No.1, January. 1991.

[11] T. X. Brown, "Neural Networks for Switching,"*IEEE Commun. Mag.*, Vol. 27, pp.72-81, Nov 1989.

[12] T. X. Brown y K.H. Liu, "Neural Network Design of a Banyan Network Controller,"*IEEE J. on Selected Areas of Comm.*, pp 1428-1438, Oct. 1990.

[13] J. Lazzaro., S. Ryckebush. M.A. Mahowald and C.A. Mead., "Winner-Take-All of O(N) Complexity" in D.Touretzky (Ed.), *Advances in Neural Information Processing Systems*, pp 703-711, Palo Alto CA: Morgan Kauffman Publishers, Inc.

[14] N.H. Goddard, K.J. Lynne, T. Mintz, y L.Bukys, "Rochester Connectionist Simulator", *University of Rochester*, Computer Science Department, Rochester , New York, October 1989.

ADAPTIVE ROUTING USING CELLULAR AUTOMATA

Joël MINOT*
Laboratoires d'Electronique Philips
22, av. Descartes
94453 LIMEIL-BREVANNES
FRANCE

Abstract

This paper presents the use of connectionnist techniques for optimum routing in packet-switched telecommunication networks. The aim is to minimize connection costs and optimize resource utilization. The core of the presented routing algorithm is based on a cellular automaton. Simulations have been run based on realistic X.25 network topologies and show that this connectionnist routing algorithm performs better than classical ones. Thanks to these simulations, this algorithm has been selected for the realization of the adaptive routing task in Philips X.25 switches.

1 Introduction

Routing is one of the most important task in a telecommunication network. The purpose of routing is to determine or select a path for the data flow between two nodes or switches in a network. The use of adaptive routing mechanisms in data network corresponds to pre-defined strategies for increased resource utilization, congestion avoidance, increased throughput rate, or financial cost reduction of the communication.
Adaptive routing procedures exist in many public and private data networks such as ARPANET, Local Area Networks (Ethernet LANs for example) or packet switched network (X.25 networks like TRANSPAC in France). These networks become more and more complex: the number of switches (several thousands) and the communication throughput rates (several Gbits/s) are increasing and the services provided to the users are increasingly real-time oriented (multi-media applications). Protocols have to manage real time dynamic routing, topology control and the dynamic load of the links and the switches. As the delays introduced by the two latter problems are dependent on the network "hardware", routing procedures have to be strongly optimized. As conventional algorithms often fail to solve the computationally intensive problems posed by routing and by the real time constraints, new techniques have to be introduced.
Neural networks and connectionnist models have been successfully used in optimization [6] and in the field of telecommunications [1, 2, 7, 9]. In our laboratories, we also investigate the applicability of connectionnist concepts to communication networks. Although connectionnist techniques (such as neural networks) do not always compute optimal solutions, they rapidly compute near optimal solutions. Optimization is then a trade-off between the quality of the solution and the time to reach that solution. Moreover, their learning mechanisms avoid some parts of the formal definition and of the heuristic definition steps generally encountered in problem solving. In this paper, we study some particular connectionnist models called cellular automata and present the use of such techniques in X.25 routing procedure. Experiments based on real X.25 network topology are then shown.

2 Shortest path determination

In general, the core of the routing task consists in a shortest path tree building, in a backtracking algorithm that converts the tree into paths and in a module that converts paths into network logical addresses [14].
According to the informations describing the status of the links between the switches, every switch $\{s\}$ computes the tree of the shortest path between $\{s\}$ and all the other switches. This tree is a topological representation of the network where tree nodes are the switches and tree branches correspond to inter-switch links. From that representation, one is able to extract the entire minimal cost route between $\{s\}$ and all the other switches.
This shortest path tree determination has been studied for many years in the field of system optimization, and particularly in the fields of transportation and telecommunications [3].

*Work supported by TRT / Philips Communication Systems

2.1 The Dijkstra algorithm

The most famous algorithm has been invented by E.W. Dijkstra. It is a local minimization technique and can be described in its simplest form as follows [3]:
At any step of the iterative algorithm, one holds a list of candidate switches initialized to the root switch $\{s\}$. The best path length between the root switch $\{s\}$ and network switches are all initialized to infinity.
A candidate switch is selected and removed from the candidate list. Then, the path length between one neighbour of this switch and the root switch $\{s\}$ is computed by descending one level of the shortest path tree. If the path length decreases, this neighbour is elected for addition to the candidate list and the best path length between itself and the root switch $\{s\}$ is updated. The algorithm then iterates on the neighbours and on the switches of the candidate list. It stops when the candidate list becomes empty.
Other algorithms [4, 5, 11, 12] are inspired by the algorithm described above. The variations mainly consist in speeding up the management of the candidate list.

2.2 Shortest path tree building using cellular automaton

2.2.1 Principles

The telecommunication network is represented by a *cellular automaton*. A cellular automaton is a structure composed by non-linear *units*, that are in general all identical and in which every unit communicates with its nearest neighbours according to the wiring topology or a given strategy [15]. Such an unit is called a *cell*. Every switch $\{s\}$ has its own cellular automaton in order to compute its own shortest path tree.
In our case, a telecommunication switch named $\{i\}$[1] is represented by a cell (i) of the cellular automaton. The communications between two cells (i) and (j) of the cellular automaton are authorized if both switches $\{i\}$ and $\{j\}$ are physically connected in the real telecommunication network.
We set the problem as the building of a shortest path tree building from a base switch $\{s\}$. We suppose that the matrix $C(t) = [C_{ij}(t)]$ indicating the cost of the link between switches $\{i\}$ and $\{j\}$ is known at time t (t is function of the updating frequency of $C_{ij}(t)$ for the whole network and is generally set by the network operator). According to ISO recommendations concerning the resource optimization [8], $C_{ij}(t)$ can be a function of the number of available virtual paths in switch $\{i\}$, of the number of available buffers in the process queue of $\{i\}$, of the processing load of $\{i\}$, and of the available throughput of the link $\{i\}$-$\{j\}$. It could also be the financial cost of the $\{i\}$-$\{j\}$ link (in case of leased lines). Recommendations for the management and the diffusion of $C(t)$ (dynamical behaviour) over the whole network are provided by [8].

2.2.2 Dynamics of a cell (i)

Let L_{sj} be the distance between cells (s) and (j) which may not be neighbour. Cell (k) is a neighbour of cell (i) if both cells are directly connected with link values set to $C_{ik}(t)$ and $C_{ki}(t)$. The distance L_{sj} is computed by the sum of the costs $C_{ik}(t)$ corresponding to the crossed links $\{i\}$-$\{k\}$ along a given path between cells (s) and (j) or switches $\{s\}$ and $\{j\}$. Every cell (i) has to minimize the distance between (s) and itself, as a function of the informations delivered by its neighbours.

1. Initialization:
   (a) of the step counter: $k(i) = 0$,
   (b) of the temporary path between (s) and (i) at the kth step: $L_{si}^{k(i)}(t) = \infty$.
2. Relaxation:
   For every neighbour (j) of the cell (i):
   (a) The cell (i) reads the temporary distance between cells (s) and (j): $L_{sj}^{k(j)}(t)$.
   (b) The cell (i) adds the communication cost $C_{ji}(t)$ between cells (j) and (i) to $L_{sj}^{k(j)}(t)$ and gives the temporary path cost of (s) towards (i) via (j).
   (c) If the neighbouring cell (j) has induced the local minimization of $L_{si}^{k}(t)$, the cell (i) stores the neighbouring cell (j) and the temporary path cost of (s) towards (i) $L_{si}^{k(i)}(t)$ given by

$$min(L_{sj}^{k(j)}(t) + C_{ji}(t), L_{si}^{k(i)}(t))$$

[1]NOTATIONS: switch names are written between braces and cell names between parentheses

3. Then, when all the neighbour cells (j) have been read once, the cell (i) broadcasts its minimum value $L_{si}^{k(i)}(t)$ to its neighbours (j).

4. The step counter is incremented $k(i) = k(i) + 1$ and the cell (i) loops on steps 2 to 4 without waiting for any synchronization signal between cells (asynchronous behavior).

2.2.3 Cellular automaton behavior

Every cell (i) works in an asynchronous way and converges towards a minimal distance $L_{si}^{\infty}(t)$. The automaton reaches a stable state when the minimum value $L_{si}^{k(i)}(t)$ of every cell (i) remains unchanged between two iterations (k(i)) and (k(i)+1):

$$\forall i, L_{si}^{k(i)}(t) = L_{si}^{k(i)+1}(t)$$

The convergence of the algorithm towards a global minimum for all the cells can be proved under the assumption that all the $C_{ij}(t)$ are positive [10], that is verified in our application.

2.2.4 Implementation in a sequential process

The problem to solve is to implement a parallel process on a sequential machine.
In order to build the shortest path tree based on switch $\{s\}$, every cell is activated sequentially as the true parallelism is not feasible (see section 2.2.3). However, in order to speed up the computations, we will impose an order for activating the cells. The activation order of cell (i) is determined by the minimal hop (or crossed link) number between switches $\{s\}$ and $\{i\}$.
This *a priori* order can be computed during the initialization phase of the switch or redefined on-line. The overhead of this heuristic is very low because it is equivalent to run the shortest path tree builder with a random activation order and with all link costs set to 1.0. In that respect, the resulting L_{si}^{∞} will directly give the number of crossed links between $\{s\}$ and $\{i\}$. We have verified that this heuristic increases convergence speed in a ratio of two to three.
Moreover, while running with a random $C(t)$, this heuristic provides the paths that uses the minimal number of cells in the case of several possible paths between (s) and (i) with the same L_{si}^{∞}.

2.3 Tests

Tests have been performed on three network topologies:

- A small network, called X25-SWITCH, inspired by the existing internal network of a Philips X.25 switch. The aim here is to route between modules of the switch (see map on figure 7): N = 3 x 3 modules and K = 4. K is the connectivity, defined as $K = \frac{\text{number of links}}{\text{number of nodes}}$: it estimates the network compacity;
- An "improbable" network (not realistic in X.25 telecommunication area because of its high connectivity), called BIG, that is a 9x9 meshed array (see map on figure 8): N = 9 x 9 = 81 switches and K = 3.55;
- The backbone network, called BACKBONE, inspired by a national X.25 network (see map on figure 9): N = 28 switches and k = 3.14;

This algorithm has been compared to two of the best algorithms that compute the Dijkstra method in a classical way [13]: Ford-Moore [4, 11] and Pape [12].
The test methodology is the following:

- For every base node $\{s\}$ and for T=20 different network states, we will compute the shortest path tree building. Those T states correspond to T random matrices $C(t) = [C_{ij}(t)]$ with $0 < C_{ij}(t) \leq 1.0$ for the existing $\{i\}$-$\{j\}$ links and for $0 \leq t \leq T$.
- Then, we plot the measures performed on the algorithms as functions of the node $\{s\}$ $(0 \leq s < N)$:
  - the average time for the building of T shortest path trees (see charts 2, 4, 6).
  - the maximum value of allocated memory during the building of T shortest path trees (see charts 1, 3, 5).
- Besides, those charts plot for the virtual $(N+1)^{th}$ node:
  - the average time for the building of T shortest path trees and for the N base nodes.
  - the maximum allocated memory for the building of T shortest path trees and for the N base nodes.

3 Conclusions

The tests proved that our cellular automaton algorithm is faster than the classical algorithm in realistic cases. Code complexity is smaller and allocated memory is low and constant. We shall notice that, as network complexity increases, the convergence speed of our algorithm decreases faster than the convergence speed of the classical ones. However, as the trends in network architecture are the design of hierarchical networks split in domains and sub-domains with less than a hundred switches (with less links than BIG network) in every routing domain or sub-domain [8], we are convinced that our algorithm is still suitable for that kind of organization.
At present, the extension of this work is the implementation on a real Philips X.25 switch in order to check the interaction with the network with respect to the real-time aspects, the overhead in communication between switches (mainly the load informations of all the links), the load oscillations on the links and the global performances of the network in terms of transmitted packet throughput.
Another application under study is the use of this algorithm for multi-service networks like B-ISDN/ATM based networks. In these cases, the complexity increases because of different service requirements (throughput, error rates, priority, ...), but, thanks to the generality of the algorithm, these requirements can be incorporated in link costs.

References

[1] M. Aicardi, F. Davoli, R. Minciardi, and R. Zoppoli. Decentralized routing, teams and neural networks in communications. In *Proc. of GLOBECOM'91, Phoenix*, pages 2386–2390, December 1991.

[2] N. Ansari and D. Liu. The performance evaluation of a new neural network based traffic management scheme for a satellite communication network. In *Proc. of GLOBECOM'91, Phoenix*, pages 0110–0114, December 1991.

[3] E.W. Dijkstra. A note on two problems in connection with graphs. *Numeric Math.*, 1:269–271, 1959.

[4] L.R. Ford and D.R. Fulkerson. *Flows in network*, pages 130–134. Princeton university, Princeton, Cambridge, 1962.

[5] F. Glover, R. Glover, and D. Klingman. Computational study of an improved shortest path. *Networks*, 14:25–36, 1984.

[6] J. J. Hopfield and D. Tank. "Neural" computation of decisions in optimization problems. *Biological Cybernetics*, 5:141–152, 1985.

[7] I. Iida, A. Chugo, and R. Yatsuboshi. Autonomous routing scheme for large scale network based on neural processing. In *1989 IEEE International Conference on Systems, Man and Cybernetics*, pages 194–199, November 1989.

[8] ISO/IEC DIS 10589:1990 Information technology - Telecommunication and information exchange between systems. *Intermediate system to intermediate system intra-domain routeing exchange protocol for use in conjunction with the protocol for providing the connectionless-mode network service (ISO 8473)*, October 1990. (Draft International Standard).

[9] F. Kamoun and M.K. Mehmet Ali. A neural network shortest path algorithm for optimum routing in packet switched communication networks. In *Proc. of GLOBECOM'91, Phoenix*, pages 0120–0124, December 1991.

[10] J. Minot. Recherche du plus court chemin par automate cellulaire. technical report, Laboratoires d'Electronique Philips, April 1992.

[11] E.F. Moore. The shortest path through a maze. In *International Symposium on the Theory of Switching, Cambridge*, volume II, pages 285–292, 1957.

[12] U. Pape. Algorithm 562: Shortest path length. *ACM Trans. Math. Software*, 6:450–455, 1980.

[13] G. Scheys. Description et comparaison des algorithmes de calcul de l'arbre des plus courtes routes. Technical note, Philips Research Lab. Brussels, May 1991.

[14] A. Tanenbaum. *Computer networks.* Prentice Hall, Englewood Cliffs, NJ, 1989.

[15] G.Y. Vichniac. Cellular automata models of disorder and organization. In G. Weisbuch E. Bienenstock, F. Fogelman-Soulié, editor, *Disordered systems and biological organization*, chapter 1, pages 3–20. Springer-Verlag Berlin, 1986.

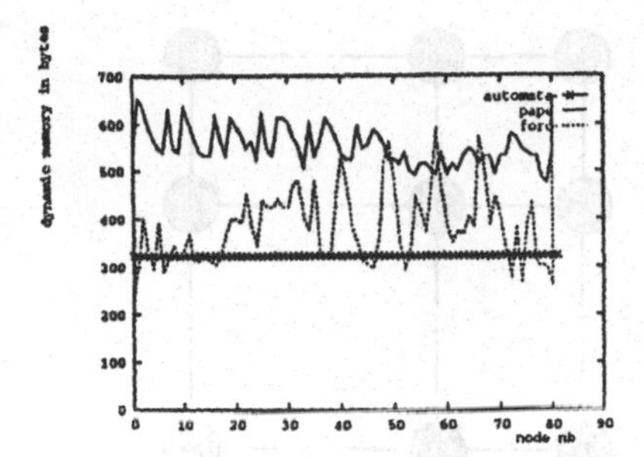

Fig.1 : BIG: Allocated memory vs base node s

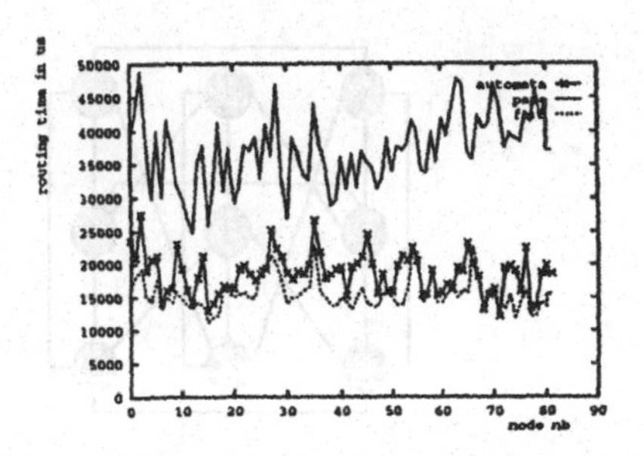

Fig. 2: BIG: Computation time vs base node s

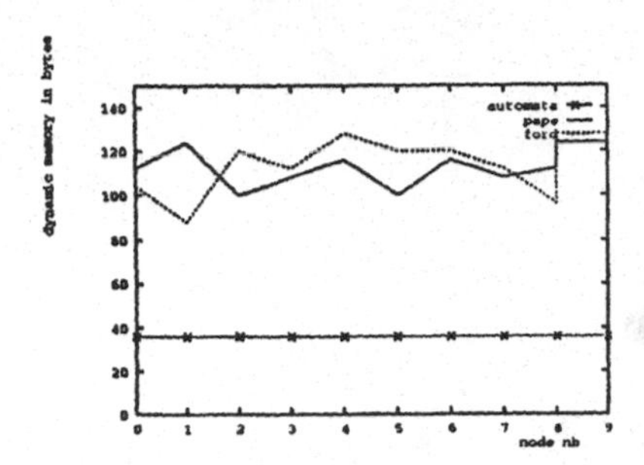

Fig.3 : X.25 SWITCH: Allocated memory vs base node s

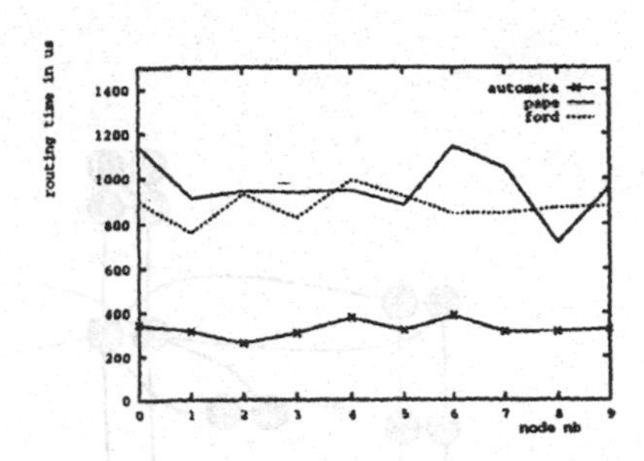

Fig. 4: X.25 SWITCH: Computation time vs base node s

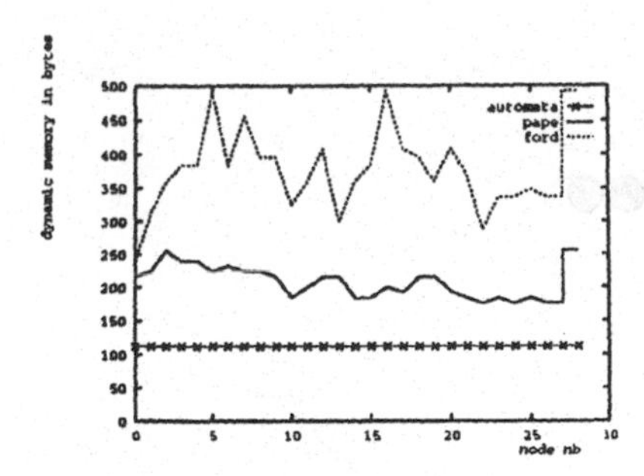

Fig.5 : BACKBONE: Allocated memory vs base node s

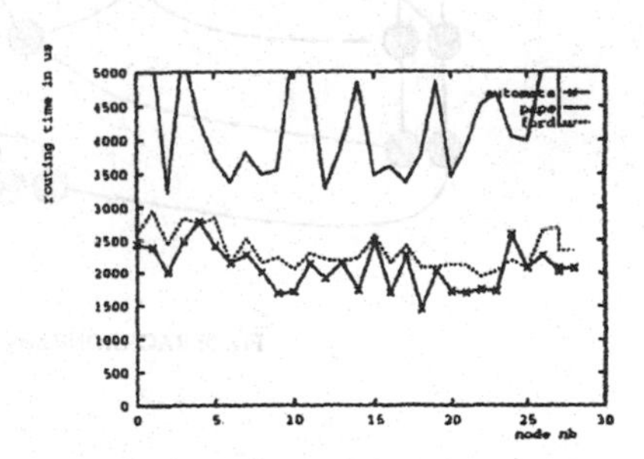

Fig. 6: BACKBONE: Computation time vs base node s

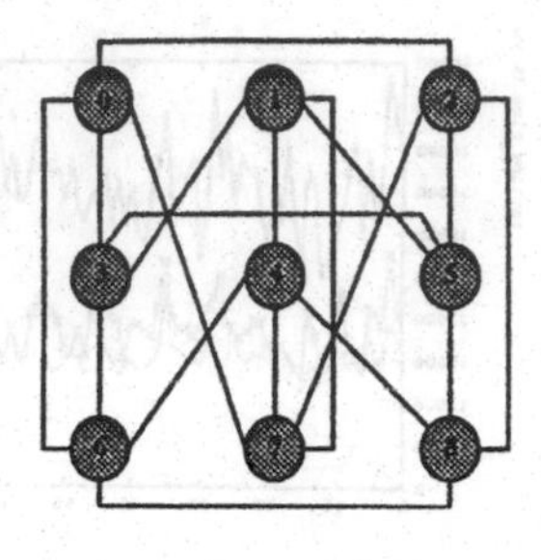

Fig. 7: X.25 SWITCH network

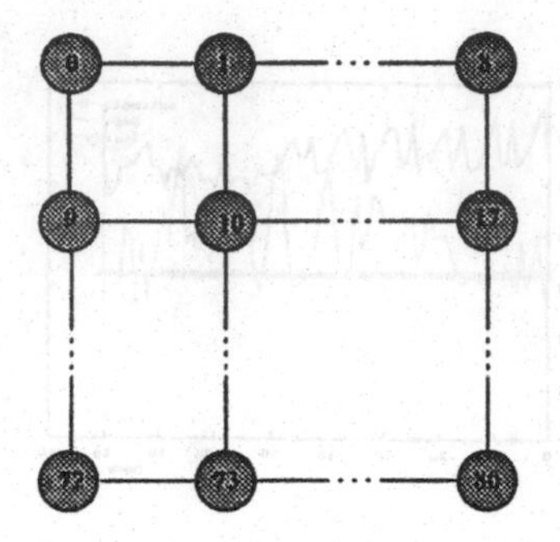

Fig. 8: BIG network

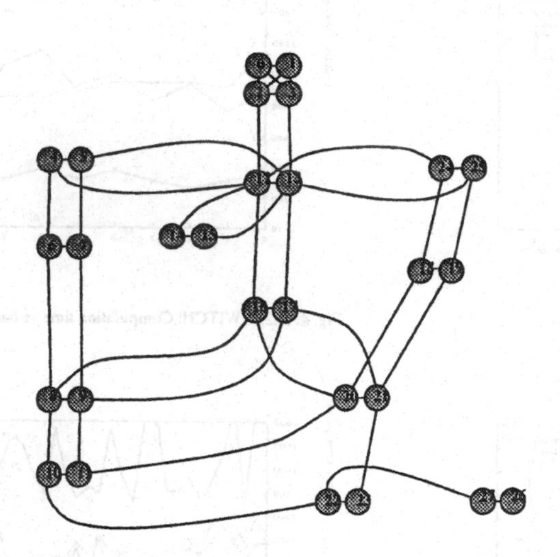

Fig. 9: BACKBONE network

OPTIMAL BLIND EQUALIZATION OF GAUSSIAN CHANNELS

Jesús Cid-Sueiro*, Luis Weruaga-Prieto, Aníbal R. Figueiras-Vidal

* ETSI Telecomunicación-UV, 47011, C/ Real de Burgos s/n, Valladolid, Spain
DSSR, ETSI Telecomunicación-UPM, Ciudad Universitaria, 28040 Madrid, Spain[1]

ABSTRACT

In this paper we show that a recurrent version of a Radial Basis Functions network can compute optimal symbol-by-symbol decissions for equalizing Gaussian channels in digital communication systems, when the (linear or not) channel response and the noise variance are known. In order to do this, the recurrent RBF (RRBF) computes several statistics about the possible states of the channel, which can be used to estimate the channel response during operation. Taking advantage of this fact, a novel technique for learning the channel parameters in a non-supervised way is theoretically derived, resulting a simple and fast algorithm that can be used for tracking in time variant environments, or for blind equalization purposes.

Finally, we show several simulation results which support the theoretically derived results.

1. INTRODUCTION

Communication channel equalizers based on neural networks have been recently proposed in order to solve the low detection capabilities or the excessive complexity problems of classical schemes: linear equalizers, DFE or Viterbi detectors. Several non-linear structures, based on Multilayer Perceptrons [1], Radial Basis Functions [2][3], Pao networks [4] or Self-Organizing Maps [5] have demonstrated a good performance in many situations.

One important problem in most neural equalizers is that network parameters are not related to channel parameters in a direct way. Thus, it is difficult to take into account the probably known characteristics of the transmission: additive Gaussian noise, linearity, etc. Moreover, a direct estimation of the optimal boundaries requires large systems, increasing the usually long training times associated with neural schemes.

It has been pointed out [3][5] that, when the channel is corrupted by an additive Gaussian noise, a Gaussian RBF network can compute optimal symbol-by-symbol decisions, and there is a direct relationship between optimal network parameters and channel statistics. In this case, equalization of linear or quasi-linear channels can be easily done by learning the channel response [6][7][8], and, after this computing networks parameters. In [9], we have also shown that a recurrent version of the RBF network can compute the optimal-symbol-by-symbol decisions based on all the received samples till the detection instant.

Mulgrew et al. [10] have also demonstrated that combined RBF-DFE equalizers outperform Viterbi detectors in time invariant environments. However, a usual decision-directed training algorithm is used, and then performance is bad when the signal-to-noise ratio is low.

An important feature of optimal (recurrent or not) RBF equalizers is that they use the received samples to compute statistics of the transmitted symbols. In this paper, we take advantage of this fact, deriving a theoretically optimal adaptive algorithm for learning the channel response in a non-supervised, non-decision directed way.

2. NOISY CHANNEL MODEL

Let us consider a transmission of equiprobable symbols belonging to a D-dimensional alphabet $X=\{x_i, i=1,...,L\}$ through some finite-memory communication channel corrupted by additive white

[1] This work has been partially supported by CAICYT grant TIC 92 # 0800-C05-01

Gaussian noise. If $\mathbf{r}(k)$ represents the received vector at time k, we can write

$$\mathbf{r}(k) = \mathbf{y}(k) + \mathbf{n}(k) \tag{2.1}$$

where $\mathbf{n}(k)$ is the white noise vector and

$$\mathbf{y}(k)=\mathbf{h}\{\mathbf{x}(k),...,\mathbf{x}(k-m)\} \tag{2.2}$$

where m is the channel memory and $\mathbf{h}\{.\}$ is a possibly non-linear function of the last m transmitted symbols, characterizing the channel response. The transmitted alphabet is finite, so $\mathbf{y}(k)$ can only take a finite number of values, which define some alphabet in reception $Y=\{\mathbf{y}_i,\ i=1,...,L^{m+1}\}$.

The noise distribution is Gaussian, so the received samples form Gaussian clusters centered at the different points of Y. For this reason, the equalizer can be interpreted as a cluster classifier, which assignes each cluster to some point of Y, and cluster sets to some point of X. An RBF network with Gaussian nodes is perfectly suited to do it [5], and the adaptive learning algorithm we propose in [11] can be used to train the network withouth supervision.

However, this approach does not take advantage of the very frequent linearity or quasi-linearity of $\mathbf{h}\{.\}$. For simplicity, consider a one dimensional transmitted alphabet (D=1) of possibly complex symbols; if the channel is linear, we can define vector $\mathbf{h}=(h_0,...,h_m)^T$ of the coefficients of the linear response, expressing the received samples as

$$\mathbf{y}(k) = \mathbf{h}^T \mathbf{x}(k) \tag{2.3}$$

where vector $\mathbf{x}(k)$ is now equal to $(x(k),...,x(k-m))^T$. Instead of learning L^{m+1} possible values of $\mathbf{y}(k)$, learning $\mathbf{h}$ (i.e, m+1 coeficients) is, in fact, faster and easier.

Anyway, an RBF equalizer does not use the information provided by all the past samples. Only the present sample (or a few amount of past samples) are used to compute decisions with RBF networks. In the next section, we show that recurrent RBF networks can compute optimal symbol-by-symbol decisions based in all the received sequence.

3. RECURRENT RBF NETWORKS FOR OPTIMAL DETECTION

The optimal symbol-by-symbol equalizer has been studied in [12]: a recursive algorithm is proposed, making at each time instant k the decision about transmitted symbol x(k) minimizing the expected number of symbol errors, with the information provided by all the received sequence till k (delayed decissions are also possible) (see Tab. I). An important fact is that this algorithm can be implemented in a recurrent version of the RBF architecture, as shown in Fig. 2, in which the weights of the linear combiner are previous results computed by the network ($g_j(k)$ in Tab.I), and the centers of the activation functions are all possible values of y. The noise probability density function f_n is computed by the activation nodes. In the following, we call RRBF to this network.

1. Initialization:

$$g_j(0)=\begin{cases}1 & j=00...0\\ 0 & \text{resto}\end{cases}$$

2. For i = 0,1, and k > 0;

$$g_j(k)=f_n(r_k-y_{ix0})g_{x0}(k-1)+f_n(r_k-y_{ix1})g_{x1}(k-1)$$

$$G(k,i)=\sum_{j=i0...0}^{i1...1} g_j(k)$$

3. Decide $\hat{s}k = i$ maximizing G(k,i)
4. k= k+1
5. Return to 2.

Table I. Optimum symbol-by-symbol detection algorithm with zero delay for a binary transmission. f_n is the noise probability density function. The subindexes of scalar centroids 'y' must be interpreted as a concatenation of binary values. For example, y_{ix0} denotes the centroid with a subindex resulting from the concatenation of binary values i, x, and "0".

One advantage of optimal RBF schemes is that they provide statistics about the state of the channel: the output of the multipliers are proportional to the probabilities of $\mathbf{x}(k)$ being equal to each possible combination of (m+1) transmited symbols, $\mathbf{x}_i$, $i=0,\dots,2^{m+1}-1$. That is, if, at time k, the multiplier outputs are $g_i(k)$, $i=0,\dots,2^{m+1}-1$, then

$$p\{x(k)=x_i \mid r(0),r(1),\dots,r(k)\} = \frac{g_i(k)}{\sum_{j=0}^{N} g_j(k)} \tag{3.1}$$

This information is essential in the new adaptive algorithm that we propose in the following, which takes into account the linear relationship among observations. It works in a non-supervised way, so it is perfectly suited to solve blind equalization and time-varying channel equalization problems.

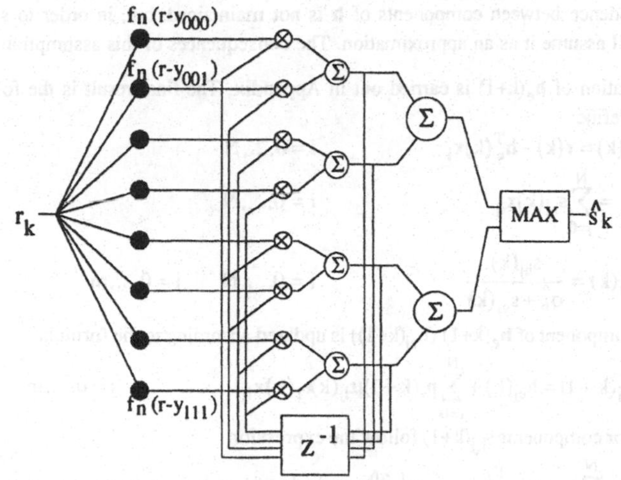

Fig. 2. Recurrent RBF equalizer, for a binary transmission in a channel with memory m=2.

4. NON-SUPERVISED CHANNEL ESTIMATION ALGORITHM

Our data generation model for the symbol detection problem is the following: the observed sample at each time instant k follows the expression

$$r(k) = \mathbf{h}^T \mathbf{x}(k) + n(k) \tag{3.2}$$

where

$$\mathbf{x}(k) = \begin{cases} \mathbf{x}_0 & \text{with probability } p_0(k) \\ \dots & \\ \mathbf{x}_N & \text{with probability } p_N(k) \end{cases} \tag{3.3}$$

where $N=2^{m+1}-1$ and $p_i(k)$ is the 'a priori' probability that $\mathbf{x}(k)=\mathbf{x}_i$. i.e, p_i represents the statistics of $\mathbf{x}(k)$ before receiving $r(k)$, so

$$p_i(k) = p\{\mathbf{x}(k)=\mathbf{x}_i \mid r(0),r(1),\dots,r(k-1)\} \tag{3.4}$$

and, as we have seen, is computed by the RRBF network. $n(k)$ is a Gaussian white noise sequence $N(0,\sigma_n^2)$. Let us assume that everything is known, except the channel response (classical methods can be used for learning the noise variance).

A model for our lack of knowledge about the desired parameters is also necessary. If at the k-th step of the adaptive algorithm, we get channel estimate $\mathbf{h}_e(k)$, then we will assume that $\mathbf{h}$ (i=0,...,N) is a Gaussian random vector with independent components, mean $\mathbf{h}_e(k)$ and variance vector $\mathbf{s}_h(k)$, (i.e, the j-th component of $\mathbf{s}_h(k)$ is the variance of j-th component of $\mathbf{h}$, and it will be denoted by $s_{hj}(k)$).

At each iteration of the clustering algorithm, we will select an estimate

$\mathbf{h}_e(k+1)$ (i=0,...,N), minimizing the cost function:

$$C_{k+1} = E\left\{\|\mathbf{h} - \mathbf{h}_e(k+1)\|^2 \mid r(k)\right\} \tag{3.5}$$

and it is easy to show that this minimum is got when

$$\mathbf{h}_e(k+1) = E\{\mathbf{h} \mid r(k)\} \tag{3.6}$$

In that case, the cost function becomes the conditional variance of $\mathbf{h}$

$$C_{k+1} = \mathrm{Var}\{\mathbf{h} \mid \mathbf{r}_k\} = \|\mathbf{s}_h(k+1)\|_1 \tag{3.7}$$

where $\mathbf{s}_h$ (k+1) is the vector with its j-th component equal to the conditional variance of h_j.

The information provided by r(k) modifies our lack of knowledge about $\mathbf{h}$. In the next iteration of the adaptive algorithm, $\mathbf{h}$ is considered a Gaussian random vector with mean $\mathbf{h}(k+1)$ and variance vector $\mathbf{s}_h(k+1)$. Independence between components of $\mathbf{h}$ is not maintained, but, in order to simplify the final algorithm, we will assume it as an approximation. The consequences of this assumption are discussed in Sections 5 and 6.

The derivation of $\mathbf{h}_e(k+1)$ is carried out in Appendix. The final result is the following adaptive algorithm: let us define

$$\varepsilon_i(k) = r(k) - \mathbf{h}_e^T(k)\mathbf{x}_i \qquad i = 0,\ldots,N \tag{3.8}$$

$$s_{yi} = \sum_{j=0}^{N} s_{hj}(k)x_{ij} \qquad i = 0,\ldots,N \tag{3.9}$$

$$\mu_{ij}(k) = \frac{s_{hj}(k)}{\sigma_n^2 + s_{yi}(k)} \qquad i = 0,\ldots,N \quad j = 0,\ldots,m \tag{3.10}$$

The j-th component of $\mathbf{h}_e(k+1)$ ($h_{ej}(k+1)$) is updated according to the formula

$$h_{ej}(k+1) = h_{ej}(k) + \sum_{i=0}^{N} p_i(k+1)\mu_{ij}(k)\varepsilon_i(k)x_{ij} \qquad j = 0,\ldots,m \tag{3.11}$$

and variance vector components $s_{hj}(k+1)$ follow the expression

$$s_{hj}(k+1) = \sum_{i=0}^{N} p_i(k+1)\left[\left(1-\mu_{ij}(k)\left|x_{ij}^2\right|\right)s_{hj}(k) + \left|h_{ej}(k+1) - h_{ej}(k) - \mu_{ij}(k)\varepsilon_i(k)x_{ij}\right|^2\right] \tag{3.12}$$

Finally, in the usual constant modulus transmission case, $|\mathbf{x}_i|^2=\sigma_x^2$, $s_{yi}(k)=s_y(k)$, $\mu_{ij}(k+1)=\mu_j(k)$ (i=0,...N), and

$$h_{ej}(k+1) = h_{ej}(k) + \mu_j(k)\sum_{i=0}^{N} p_i(k+1)\varepsilon_i(k)x_{ij} \qquad j = 0,\ldots,m \tag{3.13}$$

$$s_{hj}(k+1) = \left(1-\mu_j(k)\sigma_x^2\right)s_{hj}(k) + \sum_{i=0}^{N} p_i(k+1)\left|h_{ej}(k+1) - h_{ej}(k) - \mu_j(k)\varepsilon_i(k)x_{ij}\right|^2 \tag{3.14}$$

5. SIMULATION RESULTS

In order to show the differences between the proposed method and a decision directed learning rule, we train an RRBF network in a supervised way, for equalizing the linear channel $\mathbf{h}$=(0.5,0.6,0.3) (noise variance 0.2) in a binary transmission case. After that, at each iteration, we change $\mathbf{h}$ by adding a random perturbation $\mathbf{d}(k)$ with variance 0.00001. Both methods are then used for tracking the channel variations. In 50 simulations 300 samples long each, the final symbol error probability was 0.10 using the proposed method, and 0.17 with a decission directed procedure.

In a second experiment, non-supervised training is carried out from the beginning (noise variance 0.1). No initial reference sequence is used. In that case, it is known that a phase ambiguity exists, and and there are two global minima of the cost function. For experimental purposes, we consider succesfully training when one of both minima is got.

In 75% out of 40 simulations, results are similar to that shown in Fig. 3.: training is succesfull, and much faster than that of a decision directed method (with successful training in just 55% simulations).

In 20% simulations, however, a global minimum is not found along the first 150 training samples, and there is a strong misadjustement between predicted and observed errors. The explanation for it is the simplification done during the derivation of formulas (3.11) and (3.12) (incorrelation between the components of **h** was assumed). The consequence is a slower learning than that predicted by $\mathbf{s}_h(k)$, and the adaptation steps $\mu_i(k)$ decrease faster than necessary.

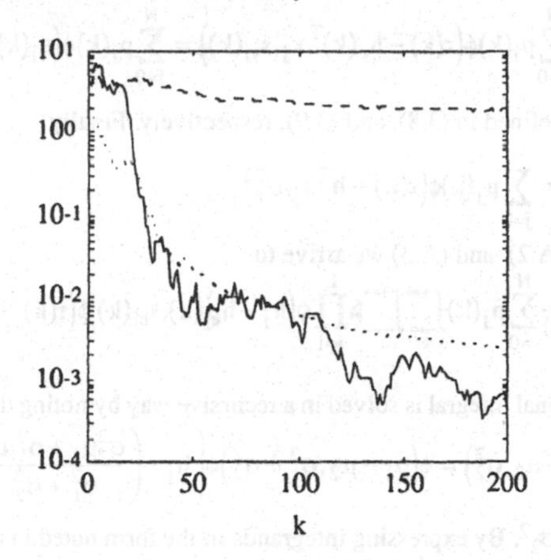

Fig. 3: RRBF blind equalizer. Evolution of the norm of the difference between the channel response and the estimations with decision directed (dashed line) and non decision-directed (continuous line) rules. The dotted line is the theoretical prediction.

6. CONCLUSIONS

Recurrent RBF networks, trained with the proposed non-supervised algorithm, have demonstrated to work well in solving tracking problems in time varying environments, improving the capabilities of a decision directed learning rule. The key of this difference is that we train the network as a function of the probabilities of the symbols being transmitted. In a decision directed way, training is independent of the certainty in the decissions, and bad training results when wrong decisions are made. With the method derived here, if a doubtfull decisions is made, weights have a small variation. These are preliminary results, and a further study must be done in order to test the capabilities of the novel method in realistic selective fading channel models, in both theoretical and practical parts.

We have also shown that RRBF networks can also be used for blind equalization purposes. The resulting model was succesfully compared with a decision directed model. However, local minima problems have been found.The problem is partially solved in an indirect way by adapting the noise variance with a classical estimator. But two other lines can be explored:

1. Re-starting the algorithm if training faults
2. Reducing exponencially the adaptation steps (μ_{ij}) which, in addition, simplifies (3.11) avoiding the computation of (3.12).

APPENDIX

The adaptive rule for $\mathbf{h}(k+1)$ is derived as follows:

$$\mathbf{h}(k+1) = E\{\mathbf{h} \mid r(k)\} = \int_{-\infty}^{+\infty}\!\!\cdots\!\int_{-\infty}^{+\infty} \mathbf{h} f_{h/r}(\mathbf{h} \mid r(k))\, dh_0 \ldots dh_L \tag{A.1}$$

where the conditional density can be written as

$$f(\mathbf{h} \mid r(k)) = \frac{f_h(\mathbf{h}) f_{r/h}(r(k) \mid \mathbf{h})}{f_r(r(k))} \tag{A.2}$$

If **h** is a random vector with mean $\mathbf{h}_e(k)$ and variance vector $\mathbf{s}_h(k)$, then

$$f_h(\mathbf{h}) = \phi(\mathbf{h} - \mathbf{h}_e(k), \mathbf{s}_h(k)) = \prod_{j=1}^{L} \phi(h_j - h_{ej}(k), s_{hj}(k)) \tag{A.3}$$

where $\phi(\mathbf{x}, \mathbf{s})$ is the Gaussian probability density function with zero mean and variance vector **s**. Also, it is inmediate to show that

$$f_r(r(k)) = \sum_{j=0}^{N} p_j(k)\phi(r(k) - \mathbf{h}_e(k)^T \mathbf{x}_j, s_{yj}(k)) = \sum_{j=0}^{N} p_j(k)\phi(\varepsilon_j(k), s_{yj}(k)) \tag{A.4}$$

where $\varepsilon_j(k)$ and $s_{yj}(k)$ are defined in (3.8), and (3.9), respectively. Finally

$$f_r(r(k) \mid \mathbf{h}) = \sum_{j=0}^{N} p_j(k)\phi(r(k) - \mathbf{h}^T \mathbf{x}_j, \sigma_n^2) \tag{A.5}$$

Combining (A.1), (A.2) and (A.3) we arrive to

$$E\{\mathbf{h} \mid r(k)\} = \frac{1}{f_r(r(k))} \sum_{j=0}^{N} p_j(k) \int_{-\infty}^{+\infty} \cdots \int_{-\infty}^{+\infty} \mathbf{h} \prod_{j=1}^{L} \phi(h_j - h_{ej}(k), s_{hj}(k)) \phi(r(k) - \mathbf{h}^T \mathbf{x}_j, \sigma_n^2)\, dh_0 \ldots dh_L \tag{A.6}$$

The (L+1)-dimensional integral is solved in a recursive way by noting that

$$\phi(h_j - \mu_1, \sigma_1^2)\phi(h_j - \mu_2, \sigma_2^2) = \phi(\mu_1 - \mu_2, \sigma_1^2 + \sigma_2^2)\phi\left(h_j - \left(\frac{\sigma_2^2\mu_1 + \sigma_1^2\mu_2}{\sigma_1^2 + \sigma_2^2}\right), \frac{\sigma_1^2\sigma_2^2}{\sigma_1^2 + \sigma_2^2}\right) \tag{A.7}$$

for every μ_1, μ_2, $\sigma_1{}^2$, and $\sigma_2{}^2$. By expressing integrands in the form noted in (A.7), integral solutions are inmediate and the (L+1) integral is solved, with final result (3.11).

Identical method can be used for computing variance vector $\mathbf{s}_h(k+1)$.

8. REFERENCES

[1] S. Chen et al, "Adaptive Equalization of Finite Non-linear Channels Using Multilayer Perceptrons", *Signal Processing*, vol. 20, no. 2, pp. 107-119; 1990.

[2] S. Chen et al, "Reconstruction of Binary Signals Using an Adaptive Radial Basis Function Equalizer"; *Signal Processing*, vol. 22, no. 2, pp. 77-93; 1991.

[3] G. J. Gibson et al, "The Application of Nonlinear Structures to the Reconstruction of Binary sSignals", *IEEE Trans. Signal Processing*, vol. 39, no. 8, pp. 1887-1884; 1991.

[4] S. Arcens, J. Cid-Sueiro, A. R. Figueiras-Vidal, "Pao Networks for Data Transmission Equalization"; Proc. of the International Joint Conference on Neural Networks, vol. 2, pp. 963-967; Baltimore, MA, Jun. 1992.

[5] T. Kohonen, K. Raivio, O. Simula, O. Venta, J. Henriksson, "Combining Linear Equalization and Self-Organizing Adaptation in Dinamic Discrete-Signal Detection", Proc. of the International Joint Conference on Neural Networks, vol. 1, pp. 223-228; San Diego, CA, Jun 1990.

[6] J. Cid-Sueiro, A. R. Figueiras-Vidal, "Optimal Symbol-by Symbol and Sequential Nonlinear Channel Equalizers Using Neural Network Structures"; First Cost #229 WG. 4 Workshop; Leysin (Switzerland), March 1992.

[7] S.Haykin: *Adaptive Filter Theory (2nd ed.)*. Englewood Cliffs, NJ: Prentice-Hall; 1991.

[8] L. Weruaga-Prieto, J. Cid-Sueiro, A. R. Figueiras-Vidal, "Optimal Variable-Step LMS Look-Up-Table Plus Transversal Filter Nonlinear Echo Cancellers"; Proc. of the IEEE ICASSPP'92; vol. 4, pp.229-232; San Francisco (CA), March 1992.

[9] L. Weruaga-Prieto, J. Cid-Sueiro, A. R. Figueiras-Vidal, "A Fast RLS Look-Up-Table Plus FIR Adaptive Echo Cancellers"; Proc. of the European Signal Processing Conference (EUSIPCO'92), vol. 3, pp. 1615-1618; Brussels (Belgium), Aug. 1992.

[10] J. Cid-Sueiro, A. R. Figueiras-Vidal, "Igualación no Lineal Optima para Comunicaciones Digitales por Satélite", Actas del II Congreso INTA, pp. 485-490; Madrid (Spain), Oct. 1992.

[11] S. Chen, B. Mulgrew, S. Mc Laughlin, P.M. Grant, "Adaptive Bayesian Equaliser with Feedback for Mobile Radio Channels"; 2nd Cost #229 WG. 4 Workshop on Adaptive Algorithms in Communications; Bordeaux (France), Oct. 1992.

[12] J. F. Hayes, T.M. Cover, J.B. Riera, "Optimal Sequence Detection and Optimal Symbol-by Symbol Detection: Similar Algorithms", *IEEE Trans. on Communications*, vol. COM-30, pp. 152-157, Jan. 1982.

Noise Prediction in Urban Traffic by a Neural Approach

G.Cammarata [1],S.Cavalieri [2],A.Fichera [1],L.Marletta [1]

(1) Istituto di Macchine
(2) Istituto di Informatica e Telecomunicazioni
Universita' di Catania
V.le A.Doria, 6 - 95125 Catania, Italy
fax +39 95 338887 - e-mail ad@iit.unict.it

Abstract.

The aim of this paper is to determine functional relationships between the road traffic noise and some physical parameters. Should this goal achieved, it is possible to modify the causes of traffic noise, in order to sensibly reduce it. Correlations are usually derived trough multiple regression analysis. In this paper an alternative solution based on the use of a neural approach is proposed. Its advantage is due to the capability of the neural networks to model non-linear systems such as the one treated in the paper. After an overview about the neural approach, the learning and production phase results are shown and discussed. They point out how good is the approach proposed to model noise pollution in urban areas.

1.Introduction.

The noise pollution in urban areas compromises the quality of life. One of the most important noise source is the road traffic. In recent years a strong need is felt to determine functional relationships linking the noise to some physical parameters, aimed at providing means to reduce the noise pollution.

The meaningful parameters of the traffic noise refer to the number of vehicles and flow velocity (traffic parameters), type of pavement and slope of the road (road parameters), road width and building height (urban parameters).

Many researchers have looked for correlations, linking the noise emissions to the above mentioned parameters. Problems are mainly due to the large number of variables and the suitable choice of the search approach. Whereas correlations are usually derived trough multiple regression analysis, in this paper a neural network will be used. The main advantage of this approach is the possibility to model non-linear systems such as the problem dealt with in this paper.

The model identification is based on acoustic survey of medium and small towns of Sicily [1]. Data were collected in roads with typical features of commercial, residential and industrial areas, and three times for every location point and for every one of the subsequent four time intervals :(7-7:30 am; 10:30-11 am;0:45-1:15 pm; 8-8:30 pm). All the measurements were done in working days excluding all atypic conditions. The surveys consist of the following parameters :

- number of cars
- number of trucks
- height of buildings in both sides of the road
- width of the road.
- sound pressure level L_{eq} expressed by :

$$L_{eq} = 10 \log \frac{1}{T} \sum 10^{\frac{L_i}{10}} \Delta t$$

being T the time of observation and L_i the sound pressure level measured in the time Δt (in particular Δt was 10 min.). In order to get a satisfactory identification of the model, data were classified according to the width of the roads. Following this approach the model obtained strictly holds for roads with similar features.

2.The Neural Network Approach.

The particular problem to solve has led to an approach based on a mapping neural network. More specifically, a Back Propagation Network (BPN) [2] was considered, because of the efficiency of the training algorithm in a broad spectrum of applications, as acknowledged in literature.

This network was trained using a set of data mentioned in the previous section, concerning the city of Messina. Then, in the production phase, the correctness of the approach was tested using data measured in some other cities in Sicily.

The neural approach included a preliminary analysis in order to verify if the data were qualitatively significant for the extrapolation of a functional relationship between them. This preliminary analysis was quite difficult because the data concerning Messina were characterized by a certain number of poorly significant values. For example, some measurements were taken at cross-roads, where the noise level was related to the stop at the traffic lights and not to the vehicle flow rate. Discarding the measurements which were not significant to characterize the problem of noise prediction, we obtained a set of examples to be used during the learning phase of the neural network.

2.1. Description of the Neural Network Input/Output Vectors.

The aim of this section is to describe the I/O interfaces used in the neural approach. The training set, obtained from the process of analysis and filtering described in the previous section, is made up of the number of cars, the number of trucks, the height of the buildings on the sides of each road, the width of the latter and the equivalent level of noise pressure. In accordance with the aim, it was necessary to train the neural network so that it was able to determine the level of noise produced by a particular number of cars and trucks in a road of a certain length flanked by buildings of a certain height. The architecture of the neural network was determined on this basis. More specifically, at least in an initial phase it was characterized by 5 inputs and only 1 output. The NN output supplies the value of the equivalent level of noise for a particular value of the four inputs - the number of cars, the number of trucks, the height of the buildings on the sides of the road and the width of the road. The number of neurons in the hidden layer, on the other hand, was varied (together with the set of parameters characterizing the network, such as bias or learning rate) in order to reach full network convergence. Convergence was never reached in any of the tests carried out. This was presumably caused by the excessive number of inputs as compared with the limited number of training examples (the number of measurements available was quite low, on account of the analysis and filtering phase performed on them). It was therefore indispensable to reduce these inputs drastically.

Several examples have been given in leterature [3][4][5] of attempts to model the problem dealt with here, by means of linear regressions in which the level of sound pressure is made to depend, among the other, on equivalent number of vehicles. In order to obtain a neural model directly comparable with most classical approaches, and in order to reduce the number of inputs as far as possible, it was decided to combine the number of cars and trucks into a single parameter: the equivalent number of vehicles. From the results shown in [6], where a equivalence between the number of trucks and the number of cars was stated (on this basis a truck is equivalent to 6 cars as concerning its noise pollution), the total number of vehicles was calculated by summing the number of cars with the number of trucks, previously multiplied by 6. It was then considered appropriate to combine the heights of the buildings on the sides of each road into a single parameter relating to their average value. We believed, in fact, the average height of buildings to be sufficiently significant when the difference in height between the buildings on the two sides of the road is not excessive. This latter condition is plausible in that the modelling presented here was performed with reference to a particular kind of urban architecture characterized by the presence of buildings on both sides of the road.

On the basis of what was said above, it is thus clear that a BPN model characterized by 3 inputs and only 1 output (shown in Fig.1) was considered. The NN inputs were the equivalent number of vehicles, the average height of the buildings and the width of the road. The output was always the equivalent level of sound pressure. The number of neurons in the hidden layer was made to vary along with the other network parameters, until convergence was reached. The number of neurons in the hidden layer finally considered was 30.

2.2 The Learning and Production Phases.

The learning and production phases were carried out by the NN simulator "Explorenet" [7] running on a IBM-pc. After the training phase, which was successfully completed, a production phase was carried out with the with the aim of verifying whether the NN was able to extrapolate the functional relationship between the input and the output variables. With this aim in mind, different measurements taken in the city of Messina and not included in the training set were considered. Such an attempt was unsuccessful. From a subsequent analysis of the data, we ascribed the reason to the presence of a strong dependence of the functional relationship of the data on the characteristics of the road (i.e. the width). In particular we observed similar features in data distribution for roads wider than 30 meters, and for roads narrower than 30 meters. For this reason, we split the training examples into two groups. In the first group only the examples characterized by roads over 30 meters wide were considered. Obviously the second group was made up of the remaining examples.

As a consequence, we considered two different BPNs each of which was characterized by 3 inputs, 1 output and 30 hidden neurons. After the training phase, which was successful for the two NNs, the production phase was carried out.

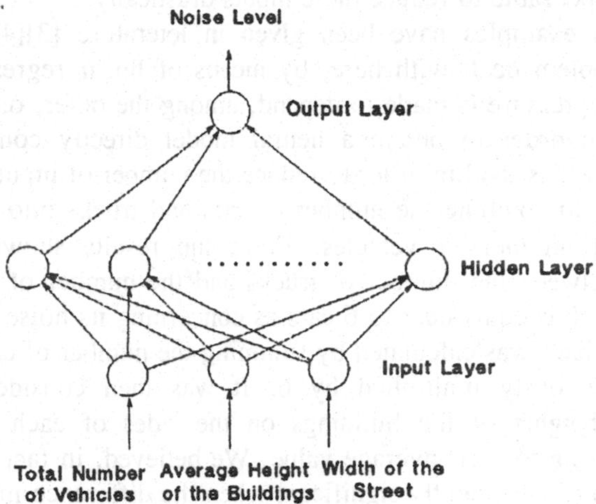

Fig.1 - BPN model.

Tabs.1 and 2 show the data used respectively for the two networks in order to verify the correctness of the generalization. The data refer to the measurements taken in Messina, and not included in the training sets.

Tab.1-Set of data concerning the production phase in the first NN

Number of Total Vehicles per hour (x 10000)	Average Building Height (x 100) meters	Road Width (x 100) meters	Measured Noise (x 100) db	NN Output (x100) db
0.4575	0.085	0.35	0.794	0.7870
0.375	0.085	0.35	0.767	0.7677
0.0525	0.125	0.30	0.714	0.7147
0.2685	0.10	0.30	0.754	0.7646
0.038	0.10	0.30	0.703	0.7025

Tab.2-Set of data concerning the production phase in the second NN

Number of Total Vehicles per hour (x 10000)	Average Building Height (x 100) meters	Road Width (x 100) meters	Measured Noise (x 100) db	NN Output (x 100) db
0.21	0.15	0.28	0.799	0.8068
0.1065	0.15	0.28	0.740	0.7256
0.007	0.10	0.26	0.632	0.6218
0.1375	0.15	0.25	0.763	0.7451
0.121	0.10	0.25	0.745	0.7287
0.008	0.10	0.20	0.599	0.6055

3. The Use of NN for the Noise Prediction.

The goal of this section is to demonstrate the validity of the NN approach shown before. This valuation will be based on the verification of the noise prediction capability offered by the neural networks considered in the previous section. With this aim in mind, a certain number of data concerning the noise pollution in two cities different from Messina were given to the NNs in the production-phase mode. In particular two set of data from the measurements taken in Palermo and Catania were considered. This choice is essentially based on the presence in Palermo and Catania of the same great variety of road width found in Messina. This allows to obtain a validation of the neural approach on a wide spectrum of real scenarios.

According to the methodology described before, the measurements concerning Palermo and Catania were divided into two sub-sets depending on the width of each road. The first group of data was characterized by road larger than 30 meters, while in the second group the road width was under 30 meters. These groups of data were prepared to be given the two NNs respectively in production-phase mode. Fig.2 shows the noise level trend measured in Palermo, in roads larger than 30 meters. It is compared with the noise level provided by the first NN. As can be seen the predicted noise is very close to the real noise. In table 3 the real scenarios concerning the noise levels shown in Fig.2 are represented. As shown in the table a very large spectrum of road width was considered, ranging from 30 meters up to 58 meters. The absolute error between the measurements and the predictions is very low for every scenario considered.

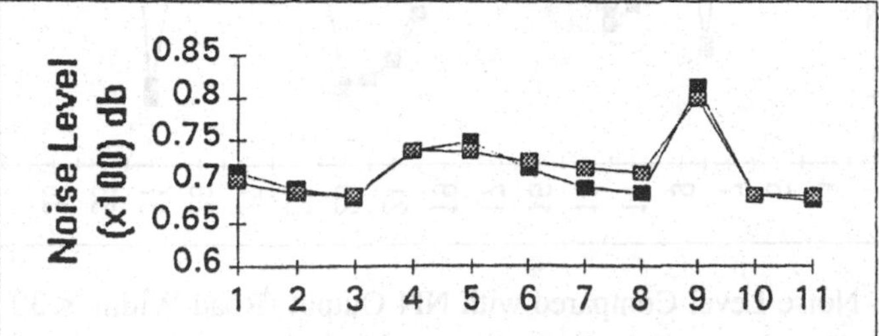

Fig.2 - Real Noise Level Compared with NN Output (Road Width >30 meters)

Table 3 -Set of data obtained by measurements in Palermo and concerning the production phase in the first NN

	Number of Total Vehicles per hour (x10000)	Average Building Height (x 100) meters	Road Width (x 100) meters	Measured Noise (x 100) db	NN Output (x 100) db	Absolute Error
1	0.429	0.125	0.58	0.712	0.699412	0.012588
2	0.3816	0.125	0.58	0.691	0.685386	0.005614
3	0.378	0.125	0.58	0.681	0.684307	0.003307
4	0.2874	0.1425	0.411	0.736	0.737698	0.001698
5	0.2538	0.1485	0.4	0.748	0.735027	0.012974
6	0.1878	0.0660	0.323	0.715	0.723965	0.008965
7	0.162	0.0660	0.323	0.693	0.716785	0.023785
8	0.141	0.0660	0.323	0.686	0.710862	0.024862
9	0.1302	0.38	0.32	0.813	0.799682	0.013318
10	0.0516	0.0495	0.314	0.683	0.683515	0.000515
11	0.0498	0.0495	0.314	0.677	0.682979	0.005979

In a similar way, a certain number of measurements concerning road width under 30 meters were prepared for the second NN in production-phase mode. Fig.3 shows the real noise levels measured in Palermo compared with the one provided by the neural network. In table 4 the set of data are represented. As can be seen, the table contains a larger number of data than the table 3. This is to be ascribed to the presence in Palermo of a much larger number of road whose width is under 30 meters.

The measurements concerning Catania were characterized by road narrower than 30 meters. For this reason, only the second NN was considered in the production phase. Fig. 4 and table 5 show the noise level and the set of data concerning the measuraments taken in Catatnia. Again, the very little differences between real and the predicted noise, point out how suitable is the neural approach considered here.

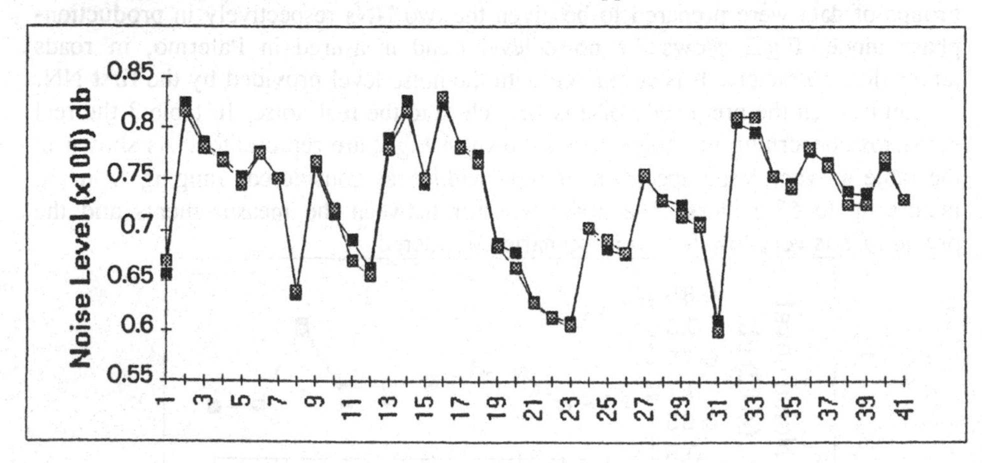

Fig.3 - Real Noise Level Compared with NN Output (Road Width <30 meters)

Table 4 -Set of data obtained by measurements in Palermo and concerning the production phase in the second NN

	Number of Total Vehicles per hour (x10000)	Average Building Height (x 100) meters	Road Width (x 100) meters	Measured Noise (x 100) db	NN Output (x 100) db	Absolute Error
1	0.0534	0.099	0.286	0.657	0.67	0.013
2	0.2172	0.198	0.28	0.824	0.815	0.009
3	0.1806	0.245	0.225	0.785	0.779	0.006
4	0.1668	0.245	0.225	0.77	0.767	0.003
5	0.1392	0.245	0.225	0.752	0.745	0.007
6	0.1764	0.23	0.22	0.775	0.774	0.001
7	0.1482	0.23	0.22	0.75	0.7506	0.0006
8	0.0384	0.205	0.22	0.637	0.64	0.003
9	0.1722	0.1485	0.22	0.76	0.766	0.006
10	0.1194	0.1485	0.22	0.714	0.72	0.006
11	0.0606	0.27	0.2	0.69	0.67	0.02
12	0.0468	0.27	0.2	0.662	0.655	0.007
13	0.1914	0.16	0.2	0.788	0.778	0.01
14	0.2364	0.1485	0.2	0.825	0.815	0.01
15	0.1566	0.1485	0.2	0.745	0.749	0.004
16	0.2808	0.231	0.196	0.823	0.83	0.007
17	0.2022	0.066	0.1915	0.781	0.78	0.001
18	0.1818	0.066	0.1915	0.772	0.765	0.007
19	0.0786	0.25	0.19	0.687	0.683	0.004
20	0.0588	0.24	0.19	0.677	0.663	0.014
21	0.0252	0.215	0.19	0.629	0.627	0.002
22	0.0156	0.2	0.177	0.614	0.613	0.001
23	0.0096	0.25	0.16	0.609	0.604	0.005
24	0.1092	0.245	0.16	0.701	0.702	0.001
25	0.099	0.245	0.16	0.681	0.69	0.009
26	0.0822	0.245	0.16	0.675	0.677	0.002
27	0.171	0.2	0.155	0.751	0.753	0.002
28	0.1428	0.16	0.151	0.73	0.727	0.003
29	0.1266	0.16	0.151	0.723	0.713	0.01
30	0.1134	0.215	0.15	0.707	0.703	0.004
31	0.0108	0.2145	0.15	0.61	0.6	0.01
32	0.261	0.132	0.15	0.804	0.81	0.006
33	0.2532	0.198	0.134	0.795	0.81	0.015
34	0.1818	0.18	0.117	0.751	0.752	0.001
35	0.1674	0.18	0.117	0.744	0.74	0.004
36	0.2178	0.132	0.117	0.776	0.779	0.003
37	0.1938	0.132	0.117	0.764	0.761	0.003
38	0.1488	0.1815	0.1145	0.736	0.724	0.012
39	0.1524	0.0825	0.111	0.734	0.723	0.011
40	0.2148	0.297	0.0722	0.762	0.77	0.008
41	0.1608	0.297	0.0722	0.73	0.729	0.001

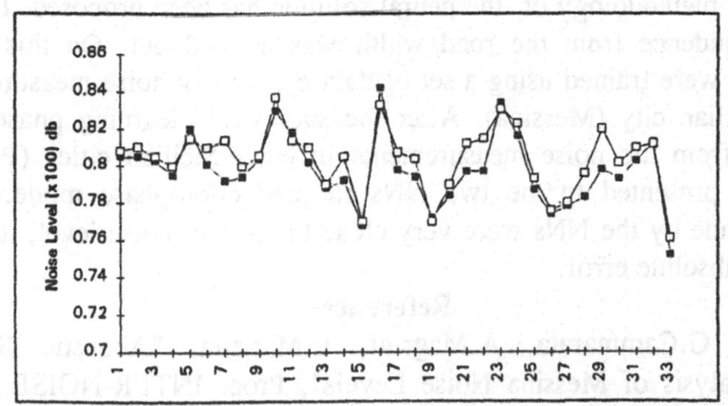

Fig.4 - Real Noise Level Compared with NN Output (Road Width <30 meters)

Table 5 - Set of data obtained by measurements in Catania and concerning the production phase in the second NN

	Number of Total Vehicles per hour (x10000)	Average Building Height (x100) meters	Road Width (x100) meters	Measured Noise (x100)db	NN Output (x100) db	Absolute Error
1	0.2574	0.14	0.2	0.805	0.80816	0.00316
2	0.2676	0.14	0.2	0.805	0.810326	0.005326
3	0.2676	0.14	0.2	0.803	0.805	0.002
4	0.246	0.14	0.2	0.795	0.8	0.005
5	0.2496	0.14	0.2	0.82	0.815	0.005
6	0.3168	0.14	0.2	0.801	0.81	0.009
7	0.285	0.14	0.2	0.792	0.813976	0.021976
8	0.3042	0.14	0.2	0.794	0.8	0.006
9	0.2856	0.105	0.2	0.804	0.805667	0.001667
10	0.2064	0.1575	0.18	0.83	0.837	0.007
11	0.2766	0.12	0.18	0.818	0.813918	0.004082
12	0.2742	0.12	0.18	0.802	0.81	0.008
13	0.2292	0.12	0.18	0.788	0.79	0.002
14	0.291	0.12	0.18	0.792	0.805	0.013
15	0.2466	5.25E-02	0.18	0.768	0.77	0.002
16	0.3222	0.11	0.14	0.842	0.833027	0.008973
17	0.2538	5.25E-02	0.135	0.798	0.80758	0.00958
18	0.237	5.25E-02	0.135	0.794	0.80397	0.00997
19	0.1896	5.25E-02	0.135	0.772	0.77	0.002
20	0.2598	5.25E-02	0.135	0.784	0.79	0.006
21	0.2262	8.75E-02	0.12	0.797	0.812	0.015
22	0.2448	7.00E-02	0.12	0.797	0.814747	0.017747
23	0.2574	0.1225	0.115	0.834	0.830767	0.003233
24	0.2586	0.1225	0.115	0.816	0.823	0.007
25	0.1098	8.75E-02	0.11	0.787	0.793188	0.006188
26	0.114	8.75E-02	0.11	0.774	0.776	0.002
27	0.141	8.75E-02	0.11	0.778	0.78	0.002
28	0.1728	8.75E-02	0.11	0.783	0.796	0.013
29	0.2454	8.75E-02	0.11	0.798	0.82	0.022
30	0.1854	5.50E-02	0.11	0.793	0.801773	0.008773
31	0.2406	5.50E-02	0.11	0.803	0.81	0.007
32	0.1698	0.105	0.105	0.813	0.812264	0.000736
33	2.04E-02	4.00E-02	0.105	0.752	0.760905	0.008905

Conclusion.

The authors have presented a neural approach to the urban traffic noise prediction, based on the use of a BackPropagation Network. The advantage of this solution versus the classic approach, is due to the neural network capability of modelling non-linear functions as the functional relationships in noise prediction seem to be. A methodology of the neural solution has been proposed. In particular a strong dependence from the road width was pointed out. On this basis, two different NNs were trained using a set of data concerning noise measurements in a particular Sicilian city (Messina). After the successful learning phases, a set of data coming from the noise measurements in other Sicilian cities (Palermo and Catania) was presented to the two NNs in production-phase mode. The noise predictions made by the NNs were very close to the real noise level, as shown by the very low absolute error.

References

[1] G.Bisio, G.Cammarata, A.Magrini, L.Marletta, "Acoustic Survey and Statistical Analysis of Messina Noise Levels", Proc. INTER-NOISE 91, Sidney 1991, Vol. II, pp.797-800.

[2] J.L.McClelland, D.E.Rumelhart, "Explorations in Parallel Distributed Processing", The MIT Press, pp.121-159.

[3] C.S.T.B., "L'outil des ètudes acoustiques sur maquettes urbaines", Centre de Maquettes de Grenoble.

[4] M.A. Burgess, "Urban Traffic noise prediction from measurements in the metropolitan area of Sidney", Appl. Acoustics, 10 [1977],1.

[5] R. Josse, "Notions d'Acoustique", Ed. Eyrolles, Paris 1972.

[6] Nelson P., "Transportation Noise", Reference Book, Butterworths, 1987.

[7] ExploreNet, HNC, Release 2.0, User Guide Manual, April 1991.

A CONNECTIONIST APPROACH TO THE CORRESPONDENCE PROBLEM IN COMPUTER VISION

Hiroshi Sako* and Hadar Itzhak Avi-Itzhak**

*Hitachi Dublin Laboratory, Hitachi Europe Ltd., O'Reilly Institute, Trinity College, IRELAND
**Stanford University, Stanford, CA 94305, USA

Abstract: A problem which often arises in computer vision is that of matching corresponding feature points within images. The correspondence is complicated by the fact that the feature sets are not only randomly ordered but have also been distorted by some transformation whose parameters are usually unknown. If these parameters are completely unknown, then all n! permutations must be compared. A neural computational method is proposed to cope with this combinatorial problem. The method is applied to 2D point correspondence and 3D-to-2D point correspondence, and is also extended to a Boltzmann machine implementation which is less sensitive to local minima, regardless of initial conditions.

1. Introduction

The correspondence problem is important in computer vision, especially in stereo vision [1], motion analysis [2],[3], and model-based image understanding [4] since they require corresponding parts of an object's images to be analysed. In stero vision images are detected from different directions in the case of stereo vision, whereas in motion analysis the tracing of the corresponding points of every image frame is needed. In addition, the correspondence between the feature points of a model and its input image is required for model-based image understanding.

A quick solution to this problem is essential to upgrade automatic machines in the production systems, such as machines for the alignment and measurement of an object's pattern using image processing and pattern recognition techniques [5],[6]. The reason is that these machines usually require matching between the image and template pattern.

Several methods [7],[8],[9] have been developed to date to cope with the correspondence problem in the computer vision field by using Hopfield type neural networks. We showed in Reference [7] that the neural computation has the potential to solve the difficult point correspondence problem between two point-groups, one of which is transformed from the other by an unknown affine transformation and has co-ordinates that are quite different from those of the other.

In this paper, a neural computation method named MatchNet1 is introduced again [7]. Then, MatchNet2 for the correspondence of points in three-dimensional space is proposed. Finally, a modified MatchNet1 for use with a Boltzmann machine is presented. It has the potential to provide a global minimum solution independent of the transformation conditions.

2. MatchNet1

MatchNet1 is used in the two-dimensional(2D) point correspondence problem. The input points are assumed to be transformed by an affine transformation with additive positional noise from the original model points. This situation occurs when a flat picture, which naturally has feature points, moves in three-dimensional(3D) space and the points are projected to the 2D image plane of an imaging device.

A. *Formulations*

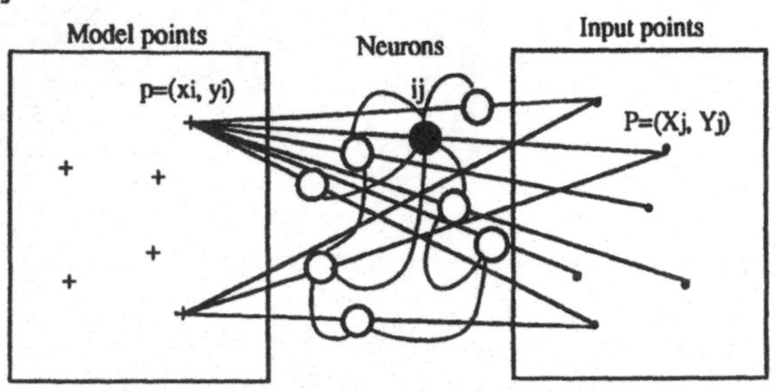

Fig. 1. Conceptual relationship between neurons and points

The conceptual relationship between neurons and the features that are to be matched is illustrated in Fig. 1. From each of the model's features, a virtual line is drawn to each of the input's features. One neuron is placed on each such line. The output level of this neuron will denote the validity of a match between this pair of model and input features. Each neuron receives data from all other neurons in the network, modifies the data, and makes its output available to the network. This processing will be done with the goal of decreasing the total network energy E. We will define E in such a way that minimising it will be equivalent to finding the best match.

The neurons can be thought of as a matrix, where neuron output V_{ij} will converge to one if there is a match between model feature p_i and input feature P_j. If these two features do not correspond to each other, it will converge to zero. For a one-to-one correspondence, this matrix should have a permutations matrix form. The first component of the network energy function will facilitate this constraint:

$$E_1=(K_1/2)(\Sigma_i\Sigma_j\Sigma_{k\neq j}V_{ij}V_{ik}+\Sigma_j\Sigma_i\Sigma_{k\neq l}V_{ij}V_{kj})+(K_2/2)(\Sigma_i\Sigma_jV_{ij}/n-1.0)^2. \quad (1)$$

The first two sums assume a value of zero when there is no more than one neuron active per row or per column. As the neuron outputs are always positive, this is the minimal value for these two sums. The third sum stimulates the activity of exactly n neurons, where n is the number of features to be matched.

In the 2D problem, the input point $P_j=(X_j, Y_j)$ is assumed to be transformed by an unknown affine transformation with additive positional noise from the original model point $p_i=(x_i, y_i)$. In this case we use the following as the second component of the network energy function:

$$E_2=(K_3/2)\Sigma_i\Sigma_jV_{ij}^2\,[G_1^2+G_2^2],\ \text{where}\ G_1=X_j-(ax_i+by_i+c)\ \text{and}\ G_2=Y_j-(Ax_i+By_i+C). \quad (2)$$

K_1, K_2, and K_3 are constants. a, b, c, A, B and C are the coefficients of the affine transformation. Initially they may be guessed; however, as the network runs, it uses the least-squared estimation which satisfies the following linear equation so that $\Sigma(G_1^2+G_2^2)$ is always at its minimum:

$$\begin{bmatrix} \Sigma_i x_i^2 & \Sigma_i x_i y_i & \Sigma_i x_i & 0 & 0 & 0 \\ \Sigma_i x_i y_i & \Sigma_i y_i^2 & \Sigma_i y_i & 0 & 0 & 0 \\ \Sigma_i x_i & \Sigma_i y_i & n & 0 & 0 & 0 \\ 0 & 0 & 0 & \Sigma_i x_i^2 & \Sigma_i x_i y_i & \Sigma_i x_i \\ 0 & 0 & 0 & \Sigma_i x_i y_i & \Sigma_i y_i^2 & \Sigma_i y_i \\ 0 & 0 & 0 & \Sigma_i x_i & \Sigma_i y_i & n \end{bmatrix} \begin{bmatrix} a \\ b \\ c \\ A \\ B \\ C \end{bmatrix} = \begin{bmatrix} \Sigma_i\Sigma_j V_{ij}X_j x_i \\ \Sigma_i\Sigma_j V_{ij}X_j y_i \\ \Sigma_i\Sigma_j V_{ij}X_j \\ \Sigma_i\Sigma_j V_{ij}Y_j x_i \\ \Sigma_i\Sigma_j V_{ij}Y_j y_i \\ \Sigma_i\Sigma_j V_{ij}Y_j \end{bmatrix}. \quad (3)$$

E_2 is proportional to the total squared error between each input point and its corresponding transformed model point and is minimised when the correct match is found. Once the network is initiated, it will run based on the neurodynamics $du_{ij}/dt=-\partial E/\partial V_{ij}$, where u_{ij} is the input to neuron ij related by $V_{ij}=1/[1+\exp(-u_{ij})]$ and $E=E_1+E_2$.

B. Experimental results

The correspondence between the points and a transformation of these points by an unknown rotation and distortion are presented in Fig.2. Each result shows one with the smallest error out of ten trials that were executed, each in a different initial state. How to choose such initial states was presented in Reference [7].

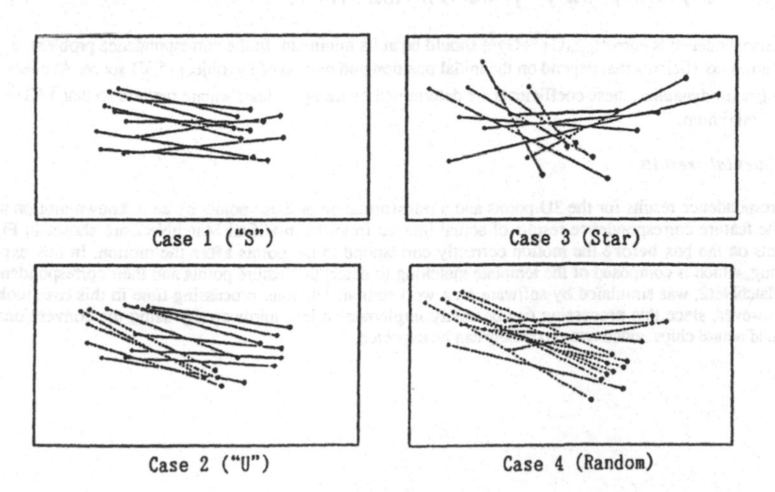

Fig.2. 2D correspondence results (·: Model points •: Input points)

3. MatchNet2

In this section, MatchNet1 is modified to solve the correspondence problem between 3D model points and 2D input points. This situation occurs when a 3D object moves in 3D space and its feature points are projected onto the 2D image plane of an imaging device. If correspondence can be achieved, 3D information will be obtained from the 2D image of the object.

A. Formulations

For the 3D-to-2D correspondence problem, the input feature $P_j=(X_j', Y_j')$ is assumed to be a projection of the three-dimensional model feature $p_i=(\zeta_i, \eta_i, \xi_i)$ onto the image plane, following some motion in 3D space. In this case the error terms G_1 and G_2 are obtained in the following manner. Consider a particular point, $q=(x, y, z)$, in a view-centred 3D co-ordinate system. The point is perspectively projected onto its corresponding point, $Q=(X, Y)$, on the image plane of a device such as a TV camera. The geometry of the perspective projection is expressed by $X=fx/z$, $Y=fy/z$, where f is the focal length. Next, consider a particular point on a rigid object with co-ordinates of (ζ_m, η_m, ξ_m). It is in an object-centered co-ordinate system whose origin is located at $(0, 0, z_0)$ in the 3D co-ordinate system. Any rigid object motion is composed of rotation around the object's centre and its translation (dx_0, dy_0, dz_0). The coordinates of the point after the motion in the 3D co-ordinate system are expressed by the following equation using the rotation matrix.

$$\begin{bmatrix} x_m' \\ y_m' \\ z_m' \end{bmatrix} = \begin{bmatrix} r_1 & r_2 & r_3 \\ r_4 & r_5 & r_6 \\ r_7 & r_8 & r_9 \end{bmatrix} \begin{bmatrix} \zeta_m \\ \eta_m \\ \xi_m \end{bmatrix} + \begin{bmatrix} dx_0 \\ dy_0 \\ z_0+dz_0 \end{bmatrix}, \tag{4}$$

The co-ordinates of the point after the motion on the image plane are

$$X_m'=x_m'/z_m'=(r_1\zeta_m+r_2\eta_m+r_3\xi_m+dx_0)/(r_7\zeta_m+r_8\eta_m+r_9\xi_m+z_0+dz_0),$$
$$Y_m'=y_m'/z_m'=(r_4\zeta_m+r_5\eta_m+r_6\xi_m+dy_0)/(r_7\zeta_m+r_8\eta_m+r_9\xi_m+z_0+dz_0), \tag{5}$$

where f can be assumed to be 1.0 without losing generality. These equations lead to the following relationship between (ζ_m, η_m, ξ_m) and (X_m', Y_m'):

$$e_1\zeta_mX_m'+e_2\eta_mX_m'+e_3\xi_mX_m'+X_m'+e_4\zeta_m+e_5\eta_m+e_6\xi_m+e_7=0,$$
$$e_1\zeta_mY_m'+e_2\eta_mY_m'+e_3\xi_mY_m'+Y_m'+e_8\zeta_m+e_9\eta_m+e_{10}\xi_m+e_{11}=0, \tag{6}$$

where $e_1=r_7/(z_0+dz_0)$, $e_2=r_8/(z_0+dz_0)$, $e_3=r_9/(z_0+dz_0)$, $e_4=-r_1/(z_0+dz_0)$, $e_5=-r_2/(z_0+dz_0)$, $e_6=-r_3/(z_0+dz_0)$, $e_7=-dx_0/(z_0+dz_0)$, $e_8=-r_4/(z_0+dz_0)$, $e_9=-r_5/(z_0+dz_0)$, $e_{10}=-r_6/(z_0+dz_0)$, $e_{11}=-dy_0/(z_0+dz_0)$.

These equations express the constraint that must be satisfied between the 3D co-ordinates of any point on the object before the motion and the 2D image plane co-ordinates of the point after the motion. Therefore, function G_1 and G_2 can be defined as:

$$G_1=e_1\zeta_iX_j'+e_2\eta_iX_j'+e_3\xi_iX_j'+X_j'+e_4\zeta_i+e_5\eta_i+e_6\xi_i+e_7,$$
$$G_2=e_1\zeta_iY_j'+e_2\eta_iY_j'+e_3\xi_iY_j'+Y_j'+e_8\zeta_i+e_9\eta_i+e_{10}\xi_i+e_{11}. \tag{7}$$

If the correspondence is correct, $\Sigma(G_1^2+G_2^2)$ should be at its minimum. In the correspondence problem, e_1 through e_{11} are unknown coefficients that depend on the initial position and motion of the object in 3D space. At every iteration when solving neurodynamics, these coefficients are determined by using the least square method so that $\Sigma(G_1^2+G_2^2)$ is always at its minimum.

B. Experimental results

The correspondence results for the 3D points and a transformation of these points by an unknown motion are given in Fig.3. The feature correspondence results of actual images from the modified MatchNet2 are shown in Fig.4. The feature points on the box before the motion correctly correspond to the points after the motion. In this experiment, the processing, which is composed of the template matching to detect the feature points and their correspondence using modified MatchNet2, was simulated by software on a workstation. The total processing time in this case took several seconds. However, since this processing can be easily implemented into hardware by using the conventional image processing and neural chips, video rate processing can be expected.

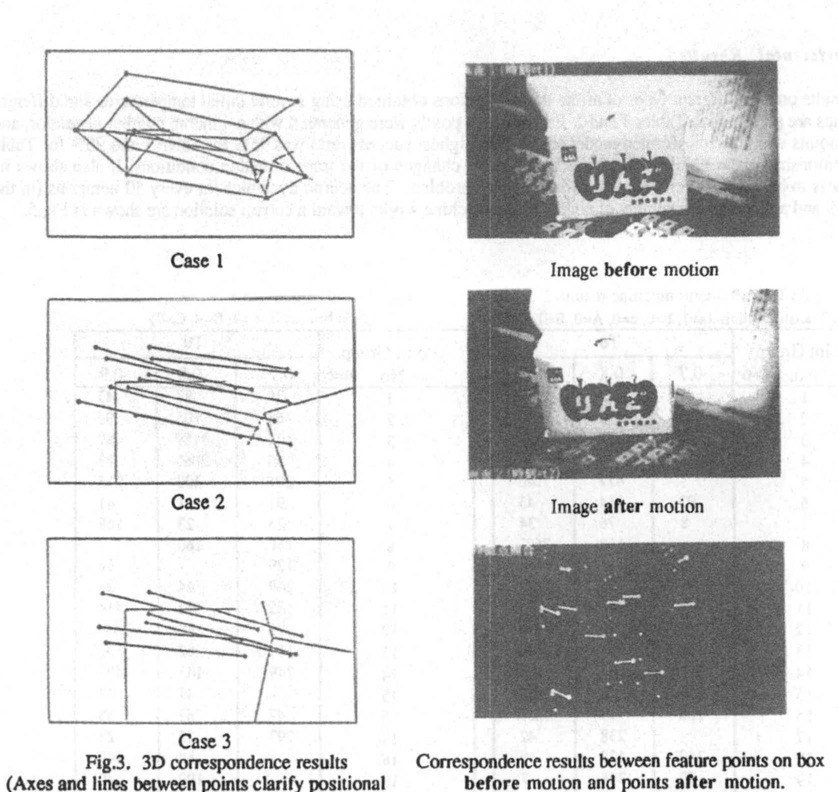

Fig.3. 3D correspondence results
(Axes and lines between points clarify positional relationship in 3D space.)

Correspondence results between feature points on box **before** motion and points **after** motion.
Fig.4. Feature correspondence results

4. MatchNet1 Boltzmann Machine

MatchNet1 can be utilised with a Boltzmann machine. Such a pairing has the potential to always provide a global minimum solution independent of the transformation conditions.

A. Calculation of weights

The E2 of MatchNet1 was changed slightly so that the total energy, E, could be translated into the well known form of $E=(1/2)\Sigma_i\Sigma_j\Sigma_k\Sigma_l T'_{ijkl}V_{ij}V_{kl}+\Sigma_i\Sigma_j I'_{ij}V_{ij}$. This translation was done by substituting in the solutions of Eq.3, which form the linear functions of V_{ij} ($a=\Sigma_i\Sigma_j a_{ij}V_{ij}$, $b=\Sigma_i\Sigma_j b_{ij}V_{ij}$, $c=\Sigma_i\Sigma_j c_{ij}V_{ij}$, $A=\Sigma_i\Sigma_j A_{ij}V_{ij}$, $B=\Sigma_i\Sigma_j B_{ij}V_{ij}$, $C=\Sigma_i\Sigma_j C_{ij}V_{ij}$) for coefficients a, b, c, A, B, and C and by using $V_{ij}{}^m=V_{ij}$ $(m\geq 1)$, which holds true because V_{ij} is always zero or one in this machine. The weight from neuron kl to neuron ij is

$$T'_{ijkl}=(T_{ijkl}+T_{klij})(1-\delta_{ik}\delta_{jl})/2, \qquad (8)$$

where $T_{ijkl}=K_1[\delta_{ik}(1-\delta_{jl})+\delta_{jl}(1-\delta_{ik})]+(K_2/n^2)+K_3\{\Sigma_p[(\delta_{pk}X_l-a_{kl}x_p-b_{kl}y_p-c_{kl})(\delta_{pk}X_j-a_{ij}x_p-b_{ij}y_p-c_{ij})+\Sigma_p[(\delta_{pk}Y_l-A_{kl}x_p-B_{kl}y_p-C_{kl})(\delta_{pk}Y_j-A_{ij}x_p-B_{ij}y_p-C_{ij})]\}$, and $\delta_{ij}=1$ when i=j. Otherwise $\delta_{ij}=0$. The other term of the E is $I'_{ij}=-(K_2/n)+(T_{ijij}/2)$.

B. Boltzmann Machine [10]

The machine is governed stochastically by probability function $Prob(V_{ij}=1)=1/(1+\exp[-\Delta E_{ij}/T])$, where ΔE_{ij} is the difference in the energy function induced by altering the state of neuron ij, i.e. $\Sigma_k\Sigma_l T'_{ijkl}V_{kl}+I'_{ij}$, and T is defined as the machine's temperature. For such a thermodynamic machine, the probability of convergence at thermal equilibrium to any two states, α and β, is related to $Prob(\alpha)/Prob(\beta)=\exp[-(E_\alpha-E_\beta)/T]$. Thus, the lower the temperature is made, the higher the probability becomes for arriving at the correct answer at thermal equilibrium. However, at low temperatures, more time is required to achieve the equilibrium. Therefore, annealing must be applied with a slowly declining temperature profile: $T(t)=T_0/[1+\ln(t)]$, where t is the number of iterations, and T_0 is the initial temperature.

C. Experimental Results

The results on two different types of affine transformations obtained using several initial temperatures and different point groups are presented in Tables 1 and 2. Each model's points were generated with a random number generator, and the input points were the transformed model points. The highest success ratio was 95% for Table 1 and 90% for Table 2. This demonstrates the stability of the machine despite changes of the transformation conditions. It also shows its effectiveness in helping to overcome the correspondence problem. The neuron dynamics for every 50 iterations (in the case of n=9 and a diagonal assignment of points) as the machine works toward a correct solution are shown in Fig.5.

Table 1. Boltzmann machine results 1
Affine Transformation (a=1, b=0, c=0, A=0, B=1, C=0)

Point Group No. n=6	T0		
	0.7	0.8	0.9
1	41	13	-
2	23	492	58
3	-	271	135
4	37	17	30
5	-	477	140
6	93	114	43
7	8	76	34
8	-	168	3627
9	77	-	176
10	76	25	157
11	252	243	44
12	282	21	49
13	66	93	115
14	103	61	166
15	236	1366	92
16	108	73	1235
17	-	238	42
18	340	125	30
19	47	346	7
20	13	179	296
Success Ratio	80%	95%	95%

Table 2. Boltzmann machine results 2
(a=1, b=2, c=0, A=3, B=4, C=0)

Point Group No. n=6	T0		
	0.7	0.8	0.9
1	30	87	41
2	64	105	92
3	143	1189	87
4	89	2065	83
5	201	200	218
6	91	41	41
7	23	23	168
8	166	160	-
9	129	-	56
10	247	64	94
11	62	152	314
12	71	49	2544
13	111	62	130
14	249	145	497
15	-	41	33
16	47	62	33
17	297	25	23
18	-	60	149
19	-	100	40
20	125	-	-
Success Ratio	85%	90%	90%

Number: Number of iterations needed to produce correct solution. Hyphen : Incorrect solution

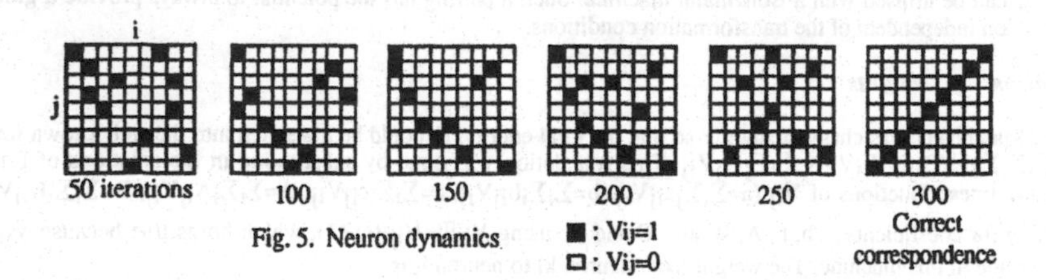

Fig. 5. Neuron dynamics

5. Conclusion

In this paper, three kinds of neural computation methods for solving the correspondence problem in the computer vision field have been presented, including one with the Boltzmann machine. These methods have the potential to solve difficult problems such as two point groups being transformed by an transformation with unknown coefficients and having quite different co-ordinates. The generality and practicality of these methods were confirmed in several experiments, including one using a real image.

Acknowledgements: **The authors would like to thank Professor Shun-ichi Amari of the University of Tokyo for his useful comments on this research and Mr. Darrin R. Uecker of the University of California at Santa Barbara for his co-operation in developing MatchNet1. Thanks are also due Dr. Masakazu Ejiri and Dr. Shigeo Nagashima of the Central Research Laboratory, Hitachi Ltd., and, Dr. Yutaka Kuwahara and Dr. Nobuo Hataoka of Hitachi Europe Ltd. for their encouragement and support throughout this research.**

References:

[1] S.T.Barnard and M.A.Fischler: "Computational stereo," *Computing Surveys*, vol.14, no.4, pp.553-572(1982).

[2] J.K.Aggarwal and N.Nandhakumar: "On the computation of motion from sequences of images -A review," *Proc. of IEEE*, vol.76, no.8, pp.917-935(1988).

[3] R.Y.Tsai and T.S.Huang: "Uniqueness and estimation of 3-D motion parameters of rigid bodies with curved surfaces," *IEEE Trans., Pattern Anal. Machine Intell.*, vol.PAMI-6, no.1, pp.13-17(1984).

[4] R.A.Brooks: "Model-based three-dimensional interpretation of two-dimensional images," *IEEE Trans., Pattern Anal. Machine Intell.*, vol.PAMI-5, no.2, pp.140-150(1983).

[5] H.Sakou, T.Miyatake, S.Kashioka, and M.Ejiri: "A position recognition algorithm for semiconductor alignment based on structural pattern matching," *IEEE Trans., Acoust., Speech, Signal Processing*, vol.ASSP-37, no.12, pp.2148-2157(1989).

[6] H.Sakou, H.Yoda, and M.Ejiri: "An algorithm for matching distorted waveforms using a scale-based description," *Proc. of IAPR Workshop on CV - Special hardware and industrial applications -*, Tokyo, pp.329-334(1988).

[7] D.R.Uecker and H.Sakou: "Point pattern matching using a Hopfield-type neural network," *Proc. of IJCNN at Washington D.C.*, pp.II-449-452(Jan. 1990).

[8] W.Lin, F.Liao, C.Tsao, and T.Lingutla: "A connectionist approach to multiple-view based 3-D object recognition," *Proc. of IJCNN at San Diego*, pp.II-835-844(Jun. 1990).

[9] P.Zhu, T.Kasvand, and A.Krzyzak: "Motion estimation based on point correspondence using neural network," *Proc. of IJCNN at San Diego*, pp.II-869-874(Jun. 1990).

[10] D.E.Rumelhart, J.L.McClelland, and the PDP Research Group: *Parallel distributed processing*, vol.1, MIT Press(1987).

self-organizing feature maps for image segmentation

René Natowicz (1), Robert Sokol (1,2)

(1) E.S.I.E.E. Laboratoire de Traitement de l'Information et des Systèmes
Cité Descartes - B.P. 99 - 93162 Noisy le Grand Cedex - France
Phone : 33 1 45926714 - Fax : 33 1 45926699 - Email : natowicr@apo.esiee.fr

(2) Université de Paris XII - Génie Biologique et Médical
bât. P2 - pièce 230 - Av. du Général de Gaulle - 94010 Créteil Cedex - France

Abstract : a connectionist method for segmenting digital images in grey level is defined. This method relies on the topology preserving property of Kohonen's self-organizing feature maps. This method is adaptive in the sense that the most present on the image an interval of grey values is, the most accurate the segmentation in this range is. Segmentation of various pictures illustrates the method.

1. Introduction

Self-organizing feature maps [1,2] have been extensively studied for vector quantization purposes and specially for image coding [3,4]. When appropriately computed, a self-organizing feature map that quantizes a vector set V preserves the topology of set V in the sense that two vectors of set V, close according to the distance defined on V, will be "coded" by two cells spatially close on the map.

We propose a new method relying on this property for segmenting digital images in grey level using self-organizing maps. This method has two stages :

1. a vector quantization of the image grey levels by a self-organizing map
2. a computation of the set of segmentation pixels.

Because of map's topology preserving property, two close grey level values will be coded by two cells spatially close. We will say that two pixels spatially close on the image and whose grey levels are coded by two distant cells are to be considered as segmentation pixels. To put it into concrete terms, one defines a spatial neighborhood for the pixels and a spatial neighborhood for the cells : segmentation pixels are thoses within the neighborhood of which the grey level topology preserving property does not hold spatially.

We will prove that the resulting segmentation is adaptive in the sense that the most represented on the picture an interval of grey level values is, the most accurate the segmentation in this range is.

2. Image coding through self-organizing feature maps

For presenting the method itself we need to keep in mind the definition of vector quantization and the topological property of vector quantizations computed by self-organizing feature maps as follows [1,2].

2.1 Vector quantization and topology preserving property

Let V be a set of vectors (including the scalar case), $M = \{m_1, ..., m_n\}$ a vector quantization of set V, $d_V(.,.)$ a distance on set V, and $d_M(.,.)$ a distance on set M. The mapping that assigns any vector $v \in V$ the closest vector $m_{c(v)} \in M$ is a coding function.

An appropriately built self-organizing feature map computes a coding function K, $K: V \to M$, where M is the map's cell vector set. This coding function has the property of preserving the topology of vector set V in the following sense [1,2] : let v, v', v'' be three vectors of set V. If vectors v, v', v'' are such that $d_V(v, v') \le d_V(v, v'')$ then their respective closest vectors $m_{c(v)}, m_{c(v')}, m_{c(v'')}$ on map's cells are such that $d_M(m_{c(v)}, m_{c(v')}) \le d_M(m_{c(v)}, m_{c(v'')})$.

3. Segmenting an image using a self-organizing feature map

For sake of clarity we only consider in this chapter the case of one-dimension maps (composed of chain connected cells) and the set of grey level values as being the vector set to be quantized. In this case map cells are scalar values.

Let I be an image in grey levels, $P(I)$ its set of pixels, $M = \{m_1, \ldots, m_n\}$ be a one-dimension self-organizing map, and $gl(p)$ the grey level of pixel p. Any pixel of the image is characterized by its spatial position and grey level, i.e. $p=((x,y), g)$. As self-organizing map M quantizes the set of grey level values G, it computes a coding function K over this set, $K: G \to M, g \mapsto K(g) = m_{c(g)}$, where $m_{c(g)}$ is the map's cell having the scalar value the closest to g. Furthermore, as any image is nothing else than a spatial arrangement of grey level values, the coding function K computed by map M can be extended to a coding function over the set of pixels $P(I)$ by assigning any pixel the cell coding for its grey level : $\forall p = ((x,y), g) \in P(I), K(p) = K(gl(p))$.

Having defined the coding function K over the set of pixels $P(I)$, we assign every pixel p a spatial neigborhood $N_I(p)$ on the image (the most classical spatial neighborhoods are 9-neighborhood and 5-neighborhood of central pixel p) and we assign any cell m_i of the map a spatial neighborhood $N_M(m_i)$ defined as $N_M(m_i) = \{m_k \in M, |i-k| \le r\}$, i.e. $N_M(m_i)$ is the set of cells whose path length to cell m_i is less than r, integer r being the radius of the ball centered on cell m_i.

We consider pixel p to be a segmentation pixel if and only if there is a pixel q in its neighborhood such that its coding cell $K(q)$ lies outside the neighborhood of coding cell $K(p)$. Then, the segmentation of image I is the set $S(I)$ of such pixels: $S(I) = \{p \in p(I), \exists q \in N_I(p), K(q) \notin N_M(K(p))\}$.

One can notice that the segmentation step of the method can be computed in a highly parallel fashion as all the pixels can be processed simultaneously.

3.1 Adaptive segmentation

The segmentation process just described has the property to be adaptive in the sense that if r is an interval of grey level values and p_ρ is the probability that the grey level value of a pixel lies in range r, then the higher the probability p_ρ is, the most accurate the segmentation is in range r. We now prove this property.

Let us remind that vector set $M = \{m_1, ..., m_n\}$ is a vector quantization of set V if the error functional $\int_V d_V(v, m_{c(v)})\, p(v)dv$ is minimal. Therefore, the quantization of set V is a sampling of V by the elements of set M such that if V_1 and V_2 are subsets of V such that $\int_{V_1} p(v)dv > \int_{V_2} p(v)dv$, then subset V_1 is more sampled than subset V_2.

Turning back to pictures, let $\rho_i = K^{-1}(m_i)$, $\rho_i \subset V_1$, and $\rho_j = K^{-1}(m_j)$, $\rho_j \subset V_2$, be the set of grey level values respectively coded by cell m_i and cell m_j. As the self-organizing map M is a vector quantization of pixel set $P(I)$, one has $\int_{V_1} p(v)dv > \int_{V_2} p(v)dv \Rightarrow |\rho_i| < |\rho_j|$ because subset V_1 is more sampled than V_2.

Furthermore, the topology preserving property of self-organizing feature maps implies that sets ρ_i and ρ_j are intervals. Therefore, cell m_j codes an interval of grey level values wider than cell m_i's one.

Now let p_i and p_j be two pixels whose values respectively belong to intervals ρ_i and ρ_j. One can see that pixel p_i is a segmentation pixel iff some pixel q_i in its neighborhood is coded by a cell $K(q_i)$ out of cell m_i's neighborhood, i.e. $K(q_i) \notin N_M(K(p_i))$, or, otherwise stated, $K(q_i) \in M - R_i$, with $R_i = \bigcup_{i-r \le k \le i+r} K^{-1}(m_k)$.

Because set R_i is an union of disjoint sets one has $|R_i| = \sum_{i-r \le k \le i+r} |K^{-1}(m_k)|$. With the same notation, pixel p_j is a segmentation pixel iff $K(q_j) \in M - R_j$.
Because map M quantizes the set of pixel values and because by hypothesis V_1 is more sampled than V_2, if $R_i \subset V_1$ and $R_j \subset V_2$ then

$$\forall k, i-r \le k \le i+r, \forall k', i-r \le k' \le i+r, |K^{-1}(m_k)| < |K^{-1}(m_{k'})|.$$

Therefore one has $|R_i| < |R_j|$, so, $|M - R_i| > |M - R_j|$, which means that the segmentation is more accurate in the neighborhood of pixel p_i than in neighborhood of pixel p_j (the set of grey level values yelding a discontinuity has more elements for pixel p_j).

3.2 Multiresolution segmentation

For a given self-organizing map, the parameters of the method are the spatial neighborhoods defined respectively on the picture and on the map. The neighborhood defined on the map is parametrized by radius r, and for the same map, one obtains segmentations at varying levels of details depending upon radius values : S_r and $S_{r'}$ being the sets of pixel segmentation computed for respective radius r and r', one has $r \le r' \Rightarrow S_{r'} \subseteq S_r$. The upper adaptive property holds at every level of details.

For sake of clarity we only considered one-dimension maps and a pixel segmentation. Generalisation to maps of higher dimensions or to block segmentation is straightforward because the definition of neighborhoods is the only notion involved in this generalisation.

4. Examples of image segmentation

A property of the method is that it can, without modification, process images of different characteristics. In appendix are given three examples of segmentation resulting from the depicted method, where pictures range from contrasted with little noise to very noisy with little contrast.
Figure 1. is a multiresolution segmentation of a 256 grey level image. A map of 50 chain connected cells was used to quantize the set of grey level values. The spatial neighborhood of any interior pixel is the 9-neighborhood centered on it. The values of radius r defining cells' spatial neighborhood range from 0 (uppermost left image) to 5 (downmost right image).
Figures 2 a. is a picture in grey levels (metallic tools) and figure 1 b. is the corresponding segmented picture. A one-dimension map of 30 chain connected cells was used for grey level quantization. The value of radius r defining cells' spatial neighborhood is 2 and the spatial neighborhood of any interior pixel is the 9-neighborhood centered on it.
Figure 3 a. is an echocardiography. The vector set quantized is the set of 3*3-blocks of the picture, therefore each cell is a 9-component vector, each component of which codes an interval of grey level values. On the picture, the neighborhood of any interior block is the 4-block neighborhood centered on it. The self-organizing map had only 2 cells, therefore radius r's value is 0. In this example, the segmentation is a classification of the 3*3-blocs of the image in only two classes. If there are n different possible grey levels for the pixels, the vector quantization is the process of seaching among the at most n^9 3*3-blocks of the image, the two seeds of the forthcoming classification.

5. Conclusion

We have set out a connectionist method of grey level image segmentation which relies on the topology preserving property of self-organizing feature maps used for vector quantization. This method has, for a given map, two parameters which are spatial neighborhood defined on the picture and spatial neighborhood defined on the map. We proved that the method is adaptive in the sense that the most represented on the picture an interval of grey level value is, the most accurate the segmentation in this range of values is.

Further researches on the outcomes of segmentation as a function of the number of map's cells, the dimension of the map and the spatial neighborhoods defined on pictures and maps are in progress. Besides, one expects to obtain an adaptive segmentation of image sequences by letting a self-organizing map continuously adjust itself to the time dependent grey level probability distribution involved by the sequence.

References

[1] **T. Kohonen,** "Self-organization and associative memory", Springer-Verlag Berlin, 1984.
[2] **T. Kohonen,** "The self-organizing feature map", proceedings of the I.E.E.E., vol. 78, n° 9, September 1990.
[3] **N.M. Nasrabadi, Y. Feng,** "Vector quantization of images based upon the Kohonen self-organizing feature map", I.E.E.E. Int. Conf. on Neural Networks, pp. 101-108, San Diego California, 1988.
[4] **E. le Bail, A. Mitchie,** "Quantification vectorielle par le réseau neuronal de Kohonen", Traitement du Signal, vol. 6, n° 6, 1989.

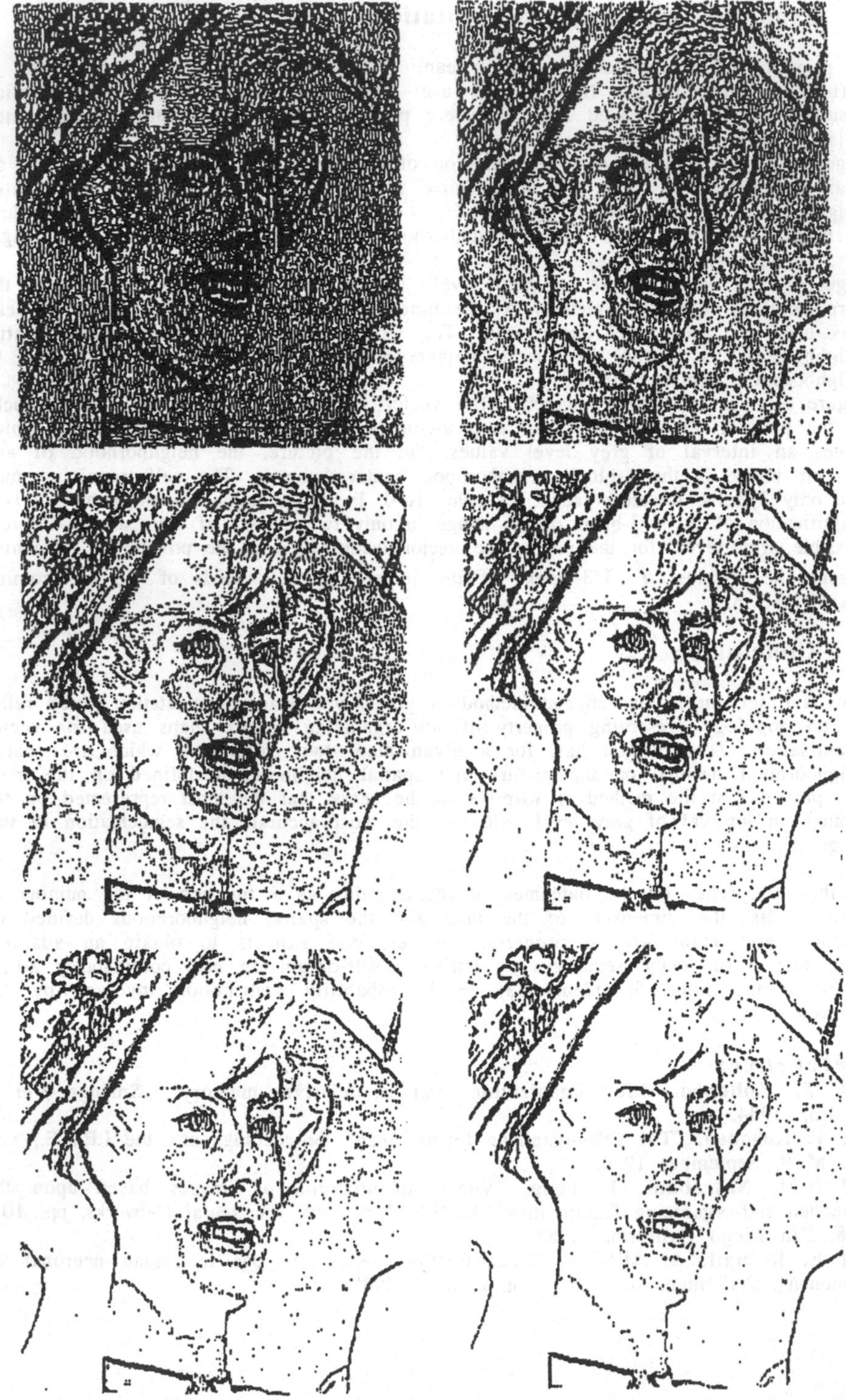

Figure 1. : multiresolution segmentation. Map neighborhood's radius : 0 to 5

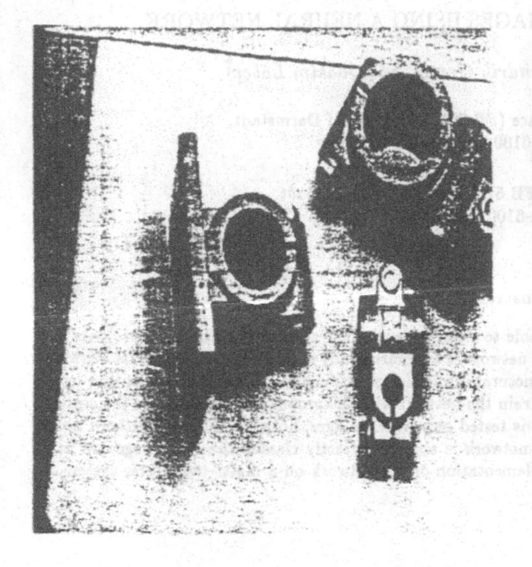

figure 2 a. (metallic tools)

figure 2 b. (segmented picture)

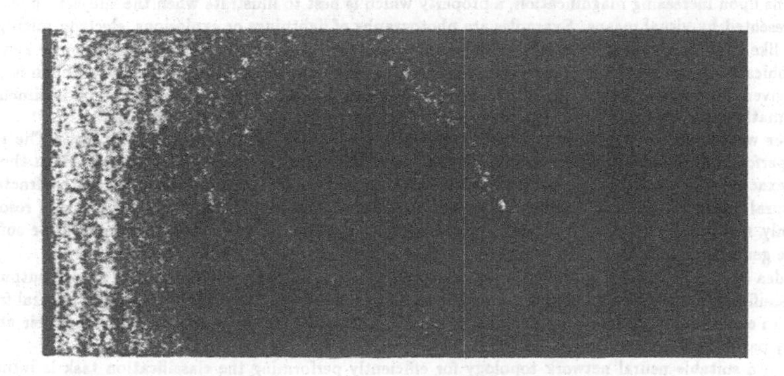

figure 3 a. (echocardiography)

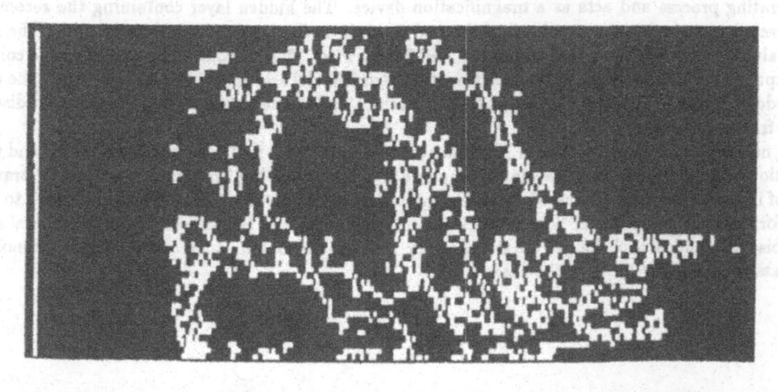

figure 3 b. (segmented picture)

RECOGNITION OF FRACTAL IMAGES USING A NEURAL NETWORK

Bernd Freisleben, John–Hendrik Greve† and Joakim Löber†*

*Department of Computer Science (FB 20), University of Darmstadt,
Alexanderstr. 10, D–6100 Darmstadt, Germany

†Department of Physics (FB 5), University of Darmstadt,
Schloßgartenstr. 7, D–6100 Darmstadt, Germany

Abstract

In this paper we present a neural network that is able to determine whether a given pixel image represents the visualization of a fractal structure or not. The proposed network is a hierarchically organized, multi–level feedforward architecture which has been designed to exploit the structural properties of artificially generated fractals. The basic idea is to extract the generator of a fractal image and train the network via backpropagation to produce the correct classification. The classification quality of the network is tested on several images, both fractal with/without noise and non–fractal, and it will be demonstrated that the network is able to correctly classify the test images up to a certain signal–to–noise ratio. An efficient parallel implementation of the network on a multi–transputer system is described.

1 Introduction

Fractal geometry provides the mathematical background for exploring the seemingly random or chaotic behaviour associated with various natural phenomena and many systems of physical, biological, chemical, electrical and artistic interest [1, 3, 5]. Loosely speaking, fractals are rough–edged, bumpy objects which continue to exhibit similar structural details upon increasing magnification, a property which is best to illustrate when the subject under investigation is represented by visual means. Examples are photographs of lightnings or explosions, electron micrographs of fineparticles like dust or sand grains, satelite pictures surveying the surface of the earth, and computer generated images for graphically displaying the dynamics of diffusion and growth processes [2]. The general problem is to find out whether a given picture manifests a fractal structure, and in case it does, its *fractal dimension* [3] is aimed to be determined to mathematically characterize the properties of the object shown in the picture.

In this paper we present an approach to decide whether a given pixel image is a fractal or not. The system developed for performing the classification task recognizes artificially created fractals which are ideal in the sense that they are exactly *self–similar* at any magnification. The restriction to mathematically generated fractals, as opposed to natural fractals which may exhibit different fractal boundaries or surfaces at various levels of resolution and are thus only *statistically* self–similar [2], reduces the complexity of the problem, but it appears to be sufficient to illustrate the general methodology developed.

The basic idea of our proposal is to design a neural network which upon presentation of an image outputs the result of the classification. The use of a neural network is motivated by the fact that many images of natural fractals are most likely to contain more or less large portions of noise; since neural networks are acclaimed for their abilities to process noisy patterns, they seem to be a good choice.

The design of a suitable neural network topology for efficiently performing the classification task is intimately related to the structural properties of artificially created fractals. Since such a fractal image may be fully reconstructed once its *generator* (the particular pixel pattern reoccuring when the image is magnified) is known, the network is set up as a multi–layer feedforward architecture in which the original image presented at the input layer is successively reduced in size by several hierarchically connected hidden layers until the size of the generator is reached. The number of hidden layers corresponds to the number of iterations executed to generate the image; the network effectively reverses the generating process and acts as a magnification device. The hidden layer containing the reconstructed generator is connected to a single output unit which is used to indicate the result of the classification. The network learns to emit a value near 1 when the image is a fractal and 0 otherwise. This is achieved by updating the connection weights of the output unit via the well known backpropagation algorithm [6] and appropriately replicating the updated weights in the hidden layers at the lower levels of the hierarchy. The trained network is then used to distinguish fractal from non–fractal images.

The proposed network architecture has been implemented in parallel on a multi–transputer system, and the parallel implementation achieves satisfactory speedups compared to a sequential implementation on a single transputer. A large number of images have been generated, both fractal with/without noise and non–fractal, in order to test the classification performance of the network. The network classified all non–noisy test images correctly; fractal patterns where random noise was added took longer for the network to converge, but up to a certain signal–to–noise ratio they were also classified correctly in all of the cases.

The paper is organized as follows. Section 2 describes the fractal images used as inputs to the neural network. In section 3 the functionality of the network is explained. Section 4 discusses the parallel implementation of the proposed approach on the multi–transputer system. The performance of the implemented neural network is evaluated in section 5. Section 6 concludes the paper and presents areas for further research.

2 Fractal Images

The images used to evaluate the classification performance of the neural network are two–dimensional arrays of (black or white) pixels, and all of them are 256×256 pixels in size. The fractal images were created by applying appropriate *affine transformations* [3] to a particular generator pattern. The size of all generator patterns is 4×4 pixels, and 4 loops of affine transformations were executed to arrive at the final 256×256 image. An example is shown in figure 1, where the 4 stages starting with the 4×4 generator pattern (top left) over the 16×16 pattern (top right) and the 64×64 pattern (bottom left) to the 256×256 image (bottom right) are graphically displayed. The final image shown is known as *Sierpinski's triangle* [3].

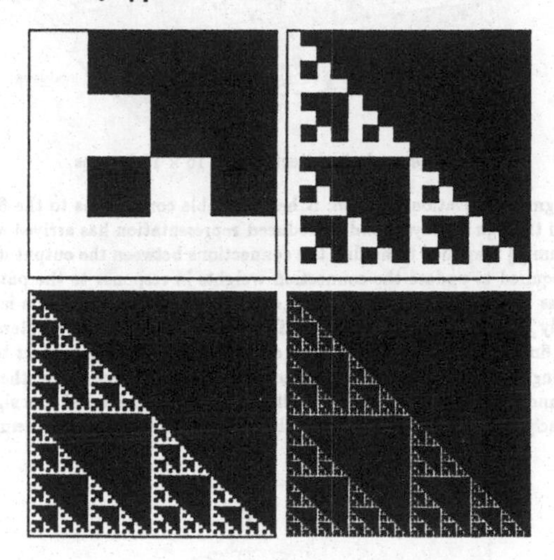

Figure 1: Generation of a Fractal Image

3 Network Architecture

The input layer of the neural network developed for classifying the images consists of N units, where N is the number of pixels in the image. Assuming that the generating pattern consists of G pixels, then sets of G input units are grouped together and each set is connected to a particular unit in a first hidden layer, resulting in a layer with N/G units. The grouping procedure is continued by introducing further hidden layers until a final hidden layer consisting of G units has evolved, and these are combined into a single output unit which serves to indicate the result of the classification. Applied to the 256×256 images used in our experiments, the network architecture consists of the input layer with 256×256 units, hidden layer 1 with 64×64 units, hidden layer 2 with 16×16 units, hidden layer 3 with 4×4 units, and the output layer with a single unit. Figure 2 illustrates the proposed architectural design; however, only a network for processing 16×16 pixel images is displayed in order to keep the illustration simple.

The weights of all connections up to the final hidden layer are initially set to a value of 1.0, and the connections of the output unit are initialized to random values between 0 and 1. All hidden units are equipped with linear activation functions. In addition, the activations of all units in the final hidden layer are multiplied by a normalization factor which is computed as the fraction between the fractal density d_f and the number n_p of pixels set (the number of units with non–zero activation values). The fractal density d_f is obtained by determining the k-th root of n_p, where k is the number of layers (including the input layer) below the output unit. This type of normalization enables the network to reconstruct the original generator in the final hidden layer, provided that the image is a fractal.

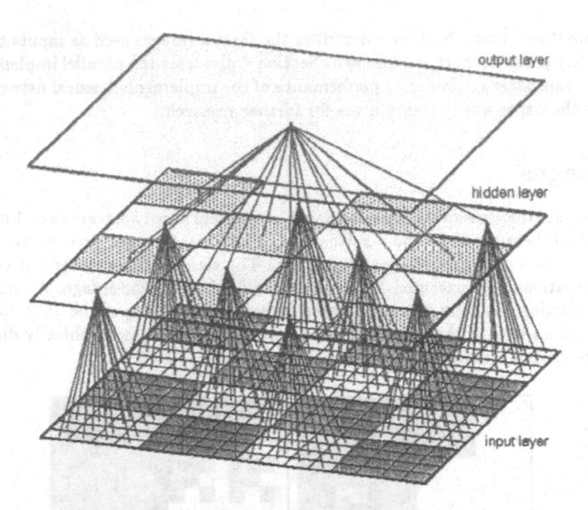

Figure 2: Network Architecture for 16 × 16 Images

The output unit has a sigmoid activation function; it has trainable connections to the final hidden layer. After an image has been forwarded through the layers and its reduced representation has arrived at the final hidden layer, the backpropagation algorithm [6] is applied to modify the connections between the output unit and the final hidden layer. The desired output required to update the connection weights in response to the pattern extracted from the image is set to 1.0, regardless of the image being a fractal or not, since this information is not known in advance. Additionally, a few randomly created patterns with a similar density are used as counterexamples and explicitly presented to the units in the final hidden layer. The desired output for these patterns is set to 0.0. Once the training phase is finished, the resulting weights are copied to the corresponding connections of the units in all preceeding layers, as shown in figure 3, and the activation function of all hidden units is switched to a sigmoid function, in order to distribute the knowledge acquired at the output unit to the hidden units and enable them to recognize the fractal generator.

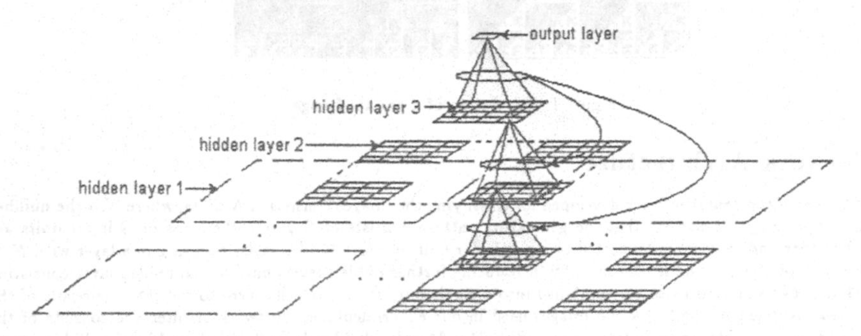

Figure 3: Weight Replication

The original image is then again presented to the network. In case the image is a fractal, all hidden units will compute an activation value near 1.0, and the fractal generator will consequently evolve at the final hidden layer. Since the network has been trained to emit a value of 1.0 when the generator is present, the image will be correctly classified as a fractal. In case the image is not a fractal, a pattern quite different from the one learned will appear at the final hidden layer, and therefore the network will output a value near 0.0 to indicate non-fractality.

The entire procedure is re-applied whenever an image is aimed to be classified. This means that the network, in contrast to the usual manner of applying backpropagation, does not need to be trained on a large number of fractal and non-fractal images before it can perform the classification task. It is able to process a completely unknown image independently of other images used as training examples and immediately classify it correctly.

In order to use the network as described above, the size of the fractal generator must be known in advance, which in most cases is not a realistic assumption. However, the proposed approach can easily be extended to handle this problem by designing an architecture for the maximal generator size desirable and starting to process the patterns by initially leaving portions of the network unused until the topology for the corresponding generator size has been revealed.

4 Implementation

The network has been implemented in parallel on a message passing multi-computer system, the MEIKO Computing Surface [4], consisting of 72 transputers T800 (25 MHz) with 1 MB, 2 MB or 4 MB RAM. A SUN 4/390, equipped with 4 transputers which are connected to the remaining 68 transputers installed in a separate cabinet, is used as the host computer.

The implementation was written in the programming language *C*, employing the MEIKO cross compiler and the communication features offered by the *MEIKO CS Tools* programming environment [4].

The hierarchical properties of the network architecture suggest an almost natural way of mapping the network onto the multi-transputer system. A master/slave organization is used as the logical interconnection topology and fundamental communication pattern. Considering that the images investigated are 256×256 pixels in size, the input and hidden layers of the network (the units and their connections) are vertically split into 16 equally sized parts. Each of these is mapped to a dedicated transputer, on which a slave task is running that is responsible for performing the processing steps required for 1/16 of each the layers mentioned. The output unit and the final hidden layer, including the connections between the two, are mapped to a further transputer and processed by the so called master task, as shown in figure 4.

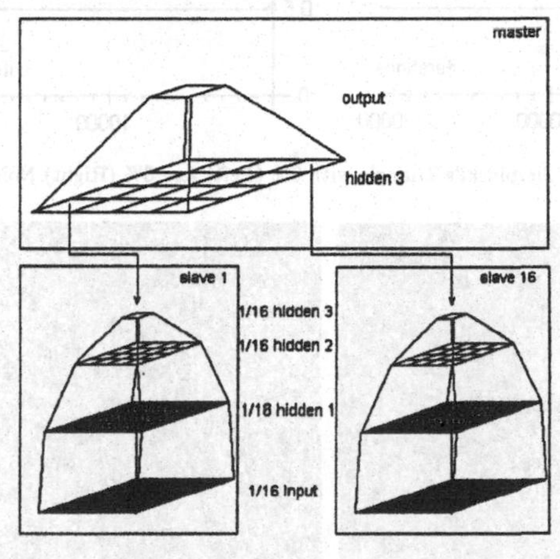

Figure 4: Mapping the Network to the Multi-Transputer System

Each of the slaves communicates the activation value computed for their final hidden layer unit to the master. The master in turn runs the backpropagation algorithm in order to train the output connections as described above.

5 Performance

In this section we present the classification quality of the proposed network. A large number of experiments have been conducted, both with fractal/non-fractal images with/without noise, to evaluate the network performance, but due to space limitations only a few examples can be shown. However, it should be said that all non-noisy images generated for the experiments, no matter if fractal or not, were classified correctly by the network.

The following examples have been selected to demonstrate the classification behaviour when noisy images are presented to the network. Figure 5 shows the results for the non-noisy image of Sierpinski's triangle (left) and a version of the same image where 3% of the bits, chosen at random, were inverted (right). The diagrams below the images indicate the output values as a function of the number of learning iterations.

As expected, the non-noisy image is correctly classified as a fractal. The noisy version takes a little longer for the network to converge to the desired output value, but the network nevertheless manages to classify the image correctly, too. Figure 6 shows the performance when 30% (left) and 40% (right) of the bits of Sierpinski's triangle were randomly selected and inverted.

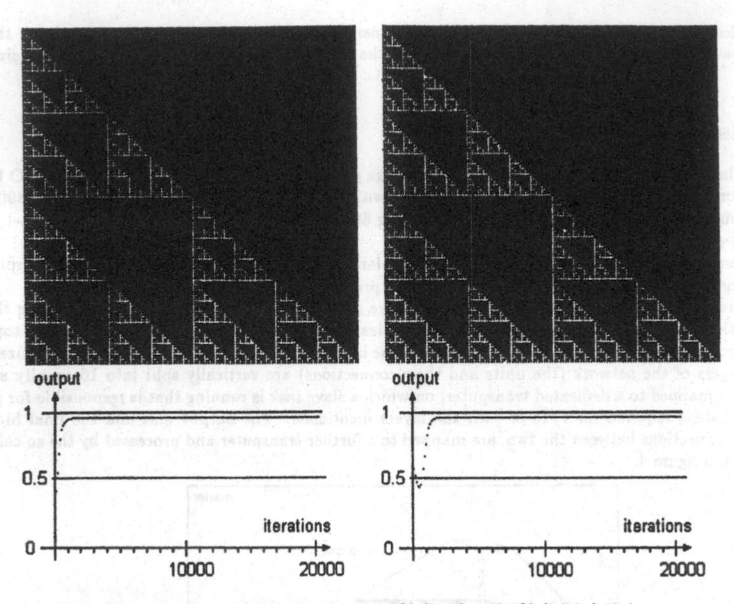

Figure 5: Sierpinski's Triangle with 0% (Left) and 3% (Right) Noise

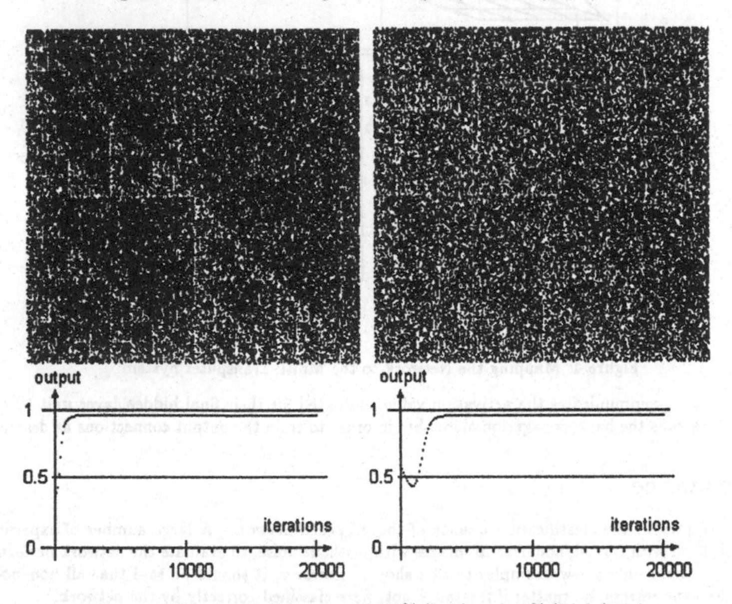

Figure 6: Sierpinski's Triangle with 30% (Left) and 40% (Right) Noise

Although the noise present in the images considerably increases the complexity of the classification task, both examples were correctly classified by the network. It approached the desired output values after less than 1500 learning iterations. Figure 7 shows two examples of Sierpinski's triangle with 50% noise (left) and 55% noise (right) added in the manner described above.

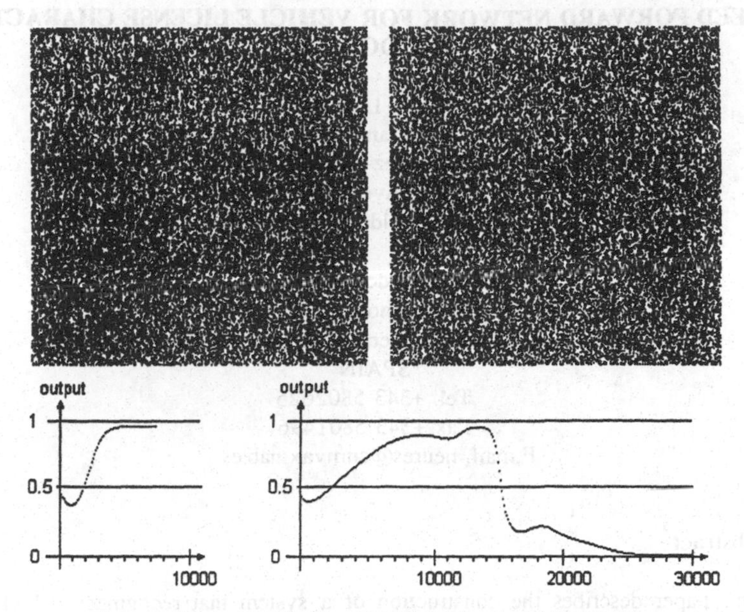

Figure 7: Sierpinski's Triangle with 50% (Left) and 55% (Right) Noise

Both images are impossible to recognize by a human viewer, but the network is able to correctly classify the image with 50% noise after approximately 4000 learning iterations. When more noise is present, the network starts to oscillate and finally approaches an output value of 0.

The results indicate that the network is well suited for achieving our aims. In particular, for noisy images the results seem to be quite impressive. The computation times required to obtain the result of a classification are very short (between 5 and 120 seconds). The parallel implementation on 17 transputers achieves a speedup factor of 6.72 as compared to a sequential implementation on a single transputer. This is quite satisfactory considering that the main portion of the computation is performed sequentially by the master when it executes the backpropagation algorithm.

6 Conclusions

In this paper we have presented a neural network approach for classifying given pixel images into fractals or non-fractals. The network was designed as a hierarchically organized, multi-layer feedforward architecture in order to make use of the structural properties of artificially generated fractals. The backpropagation algorithm was employed for training the network. The network has been implemented in parallel on a multi-transputer system. A large number of images have been generated, both fractal with/without noise and non-fractal, in order to test the classification performance of the network. The network classified all non-noisy test images correctly; fractal patterns where random noise was added took longer for the network to converge, but up to a certain signal-to-noise ratio they were also classified correctly. Among the issues for further research are: a comparative evaluation of alternative learning algorithms applied to the classification of fractal images, a study of other types of parallel implementation, and an investigation to see how the general idea behind our approach can be applied to natural fractals.

References

[1] M. Barnsley. *Fractals Everywhere*. Academic Press, 1988.

[2] D. Kaye. *A Random Walk Through Fractal Dimensions*. VCH Publishers, 1989.

[3] B. Mandelbrot. *The Fractal Geometry of Nature*. Freeman, 1983.

[4] Meiko Ltd. *Meiko Computing Surface, CS Tools Documentation*, 1989.

[5] H. Peitgen and P. Richter. *The Beauty of Fractals*. Springer-Verlag, 1986.

[6] D.E. Rumelhart, G. Hinton and R.E. Williams. Learning Internal Representations by Error Propagation. In *Parallel Distributed Processing*, Vol. 1, 318–362, MIT Press, 1986.

FEED FORWARD NETWORK FOR VEHICLE LICENSE CHARACTER RECOGNITION

F. Lisa
J. Carrabina
C. Pérez-Vicente
N. Avellana
E. Valderrama

Centre Nacional de Microelectrònica
Universitat Autònoma de Barcelona.
08193 Bellaterra. Barcelona.
SPAIN
Tel. +343 5802625
Fax +343 5801496
E.mail. neures@cnmvax.uab.es

Abstract

This paper describes the construction of a system that recognizes vehicle license numbers using feed forward neural networks, once they have been extracted using classical methods. The system has been trained and tested on real-world data. In order to reduce the total amount of required memory and increase the process speed, an additional step has been added to the learning algorithm, that produces low precision weights {+1,0,-1}. The network obtained after this training process has a similar behaviour to those networks using a floating point representation for weights. A special hardware accelerator has been developed to achieve high speed recognition.

1. Introduction

Image recognition is one of the fields of research where neural networks show a higher degree of competitiveness against the classical methods, due to the requirements of high speed processing of large amount of data. Parallelism, inherent in neural network processing, could be more efficient for these purposes.

Vehicle license recognition is usually viewed as a two-step process: segmentation of the image and recognition of each symbol. This scheme is still the same in both the classical and the neural approaches.

In the segmentation process, the starting image, usually represented as a gray scale, is processed until the isolation of several binary images representing individual symbols. The main problem is the difficulty in extracting the vehicle license number under conditions of changeable and uneven illumination, so that one part of the license plate is darker than the others, besides the problem of getting the image from the vehicle.

Several methods are used in order to obtain binary images. These methods mainly deal with the choice of the threshold between the two states. Otsu's method [1] obtains the threshold after inspecting a grey level histogram of the image, whereas later improvements uses a variable threshold to overcome uneven illumination problems. These methods require several iterations through the whole image, and this makes them slower without parallelism.

Neural network approaches to solve image segmentation use several kernels, which represent features of the image such as edges, corners or others obtained through the training process. The presence of these features can be computed in parallel for all of them in a portion of the image. Using specialized neural network chips, the AT&T system can process 32*32 pixels in parallel [2]. For an image of 512*512 pixels, they need 256 executions of the recall phase of the network. Very simple one-layer networks, with 20 input units, suffice to determine the presence of a character from the feature representation.

The symbols obtained from the segmented image are in a fixed window ranging from 5*8 to 20*20, so that compression or expansion should be necessary before the recognition phase. This is true for both the classical approach, mainly consisting in the evaluation of the Hamming distance to the different patterns, and the neural approach.

There are different neural approaches ranging from those that operate over the whole image, to the ones that use feature extraction. The methods that deal with the whole image, which are of small size, use classical feedback networks -i.e Hopfield networks or BAM networks- or feed forward networks -i.e. multilayer perceptron networks, trained with the backpropagation algorithm-. In these cases, images usually do not come from real images, since compression process highly degrades theyr quality.

For large size windows, these methods are not useful since the size of the network becomes too large for the most powerful learning algorithms. The most common way to proceed is to use a two-step approach. There is a first step for the extraction of basic features, and a second step for the recognition of the set of features.

The highest degree of parallelism can be achieved with the smallest resolution for weight and state values, which is specially useful for analog chips, but also for the digital ones since the throughput is increased. This requires a slightly different learning scheme in order to obtain this set of weights. In the case of the AT&T chip [3], weights are represented by 6 bits and states by 3, and it can process 1000 characters per second.

2. Network topology.

The system developed at our laboratories focusses on the recognition of characters of vehicle licenses. These characters came from a segmentation process that uses a non-neural method developed at the Centre de Tractament d'Imatges at the UAB [4]. This method is a knowledge-based algorithm that warranties the correct segmentation of the images independently of the position and orientation of the vehicle.

Our network recieves an image of 20x30 binary pixels for each character. We use a two-level strategy to reduce the complexity of the learning process. Our basic goal is to obtain low resolution weights for the whole network and, therefore, reduce the processing time and complexity.

The first step lies in the extraction of features from the image using a three layer network. Since car plates have specific font characters, the set of features could be reduced (42 units in the F2 layer) and it is reasonable to write them by hand, simplifying the learning process. The first layer (F1) detects variations of the basic features, and the second groups them. The corresponding weights are binary and the activation functions are hard-limiter. The neurons in layer F1 have a very high threshold and behave like multi-input "and" gates, whereas the neurons in layer F2 have the opposite behaviour and we can see them as "or" gates.

The second step is a feed forward network whose input are features and with an output for every class (34 classes are considered). There is a hidden layer with 100 units. Figure 1 shows the structure of the network. A special learning algorithm used to obtain low-resolution weights is explained further below.

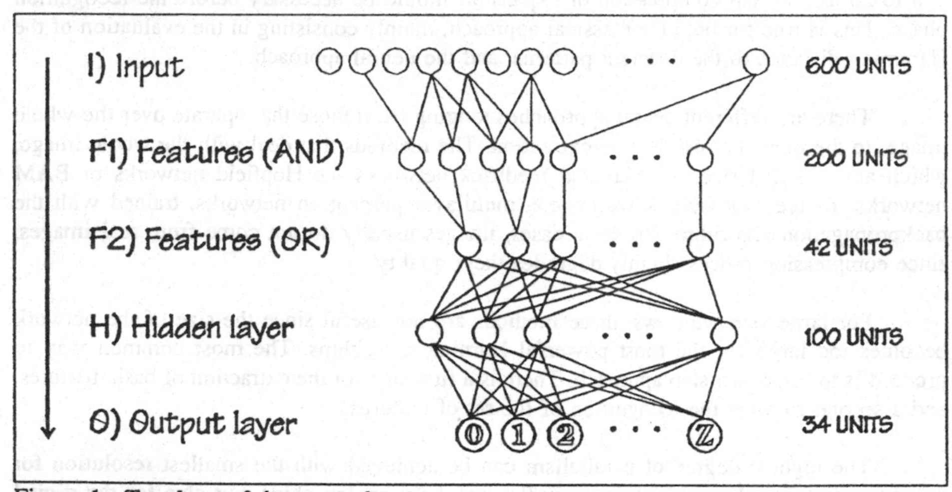

Figure 1.- Topology of the network

The connectivity degree between layers I, F1 and F2 is very reduced due to the locality of the features being extracted. That structure is not of convolutional type, so we do not need to shift a window to the whole image. Therefore the computing time will also be reduced. The last two layers are fully connected but, as we will see later, they have low-precision weights (+1, 0 or -1).

3. The learning rule.

The method used to train the network is derived from the well-known backpropagation algorithm [5]. The main addition to this algorithm is that it looks for low resolution weights (-1, 0, +1), which are optimal in terms of computational cost; they allow higher degree of parallelism, faster execution, less memory and simpler resources. The algorithm, described in [6], introduces a penalty term to the backpropagation cost function that drives the learning evolution to local minima, verifying the constraints imposed to the set of weights.

The cost function can be expressed as follows:

$$Ep = E_1 + E_2$$

where E_1 is the cost function used in the backpropagation method (the squares of the differences between the actual and the desired output value), and E_2 is a function with local minima in the desired weight values:

$$E_2 = \sum_{ij} w_{ij}^2 (w_{ij}^2 - 1)^2$$

According to the backpropagation methodology, we evaluate the gradient for this function:

$$\Omega = -\frac{\partial E_1}{\partial w_{ij}} = -6w_{ij}^5 + 8w_{ij}^3 - 2wij$$

This expression is added to the backpropagation rule, obtaining the following rule:

$$\Delta_p W_{ij}(n+1) = (\delta_{pj}\, O_{p_i} + \lambda\Omega)\eta + \alpha \Delta_p W_{ij}(n)$$

where η is the learning rate, α is the momentum factor and λ is a constant which determines the effect of the penalty. In the first iterations we set λ=0 to insure that the network learns, using high resolution weights. When the network is stable, the value of λ is increased until we obtain a set of weights verifying all the constraints.

Finally we obtain a network composed of low-resolution weights and also simpler activation functions (figure 2). After training with sigmoidal functions, discrete weights allow us to simplify the activation function to a look-up table function with two bits of resolution for the hidden layer, and to a step function for the output layer.

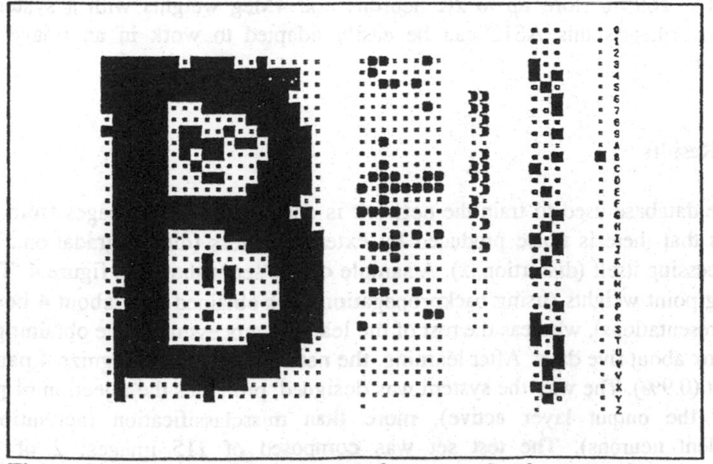

Figure 2.- Example of the process of a pattern by the network.

Training and simulation programs were developed in the Rochester Connectionist Simulator environment [7] using a general-purpose Sun Sparcstation II.

4. Hardware implementation

Neural networks with low resolution weights can be processed through very simple resources. If the number of weights is small, analog computations produce a very high speed [3]. If it is large and weights require external memory, digital processing can also be used. In this case the main limitation to the process speed is the number of input/output pins in the processor [8], since computational resources are logical gates.

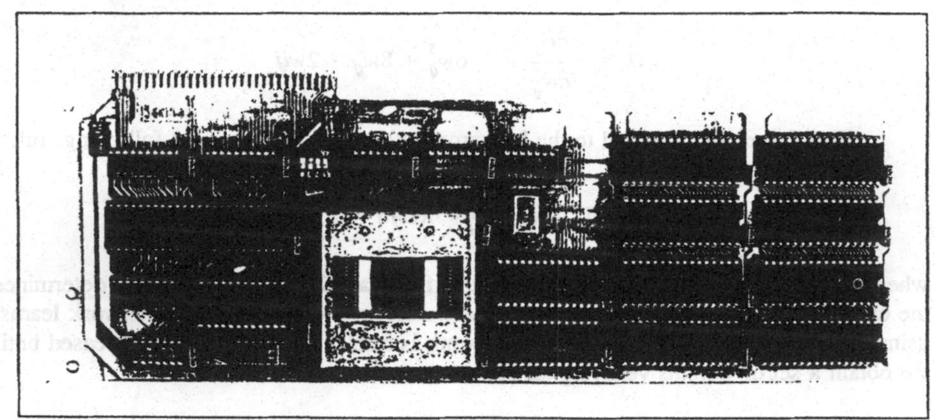

Figure 3.- The NDN4 board.

The final configuration of our network has 10000 two-bit weights and 976 neurons, so we decided to implement a digital processor, the NDN4. This processor can use an already developed PC board specialized in dealing feedback networks [8], by only changing the CPU. This board is able to store up to 2K neurons and 4Meg weights with a system clock of 48Mhz. Nevertheless this ASIC can be easily adapted to work in an image processing environment.

5. Results

The database used to train the network is composed of 438 images from real world, this means that there is noise produced by external factors (dirt, degradation,..) or by the image processing itself (distortion,...). A sample of this set is shown in figure 4. The solution for floating point weights (using backpropagation) was obtained after about 4 hours of CPU (118000 presentations), whereas the rest of the learning process up to the obtaining of two-bit weights took about five days. After learning, the network failed to recognize 4 patterns of the training set (0.9%). The way the system was designed, promotes the rejection of patterns (no neuron of the output layer active), more than missclassification (activation of non-correspondent neurons). The test set was composed of 215 images; 2 of them were misclassified and 8 rejected (4.5% total error).

Figure 4. Sample of vehicle license symbols.

Using simulation programs, the process speed is 10 characters per second, whereas with the system above described, it is possible to process 10000 characters per second. As usual in this kind of systems, the process speed is mainly restricted by the segmentation process.

6. Conclusions

In this paper we show that it is possible to solve real problems using neural networks with low resolution weights, by making more effort in the learning phase. These networks can be easily implemented in hardware through simple (and cheap) resources, and introduced in a specific image processing system.

In the vehicle license number recognition, we deal with images of 20*30 pixels, showing that large dimensions for input layers are not so important restrictions for the implementation of neural networks.

7. Acknowledgements

This work has been parcially supported by the CICYT under grant TIC 91-1049-C02-01.

8. References

[1] N. Otsu, "An automatic threshold selection method based on discriminant and least square criteria". Trans. IECE of Japan, Vol. J63-D, N.4, pp 349-355, 1980.

[2] H.P. Graf et al. "Recent developments of electronic neural nets in the US and Canada". Proc of the 2nd Intl. Conference on Microlectronics for Neural Networks. Munich, pp. 471-488. March 1991.

[3] B.E. Boser et al. "An analog neural network processor and its application to high-speed character recognition". Proc. of Int. Joint Conf. Neural Networks, pp I-415 - I-420, July 1991.

[4] J. Massa. "Detecció Automàtica de Matrícules en un entorn obert". Master Dissertation. Universitat Autonoma de Barcelona. September 1991.

[5] D.E. Rumelhart, G.E. Hinton and R.J. Williams. "Learning internal representations by error propagation", in PDP, vol 1, pp. 318-362. MIT press, 1986.

[6] C.J. Perez-Vicente, J. Carrabina and E. Valderrama."Discrete learning in Feed-Forward Neural Networks". International Journal of Neural Systems, vol.2, no.4 (1991) 323-329.

[7] N.H. Goddard et al. "Rochester connectionist simulator". Technical Report 233. October 1989.

[8] J. Carrabina. "High speed/capacity VLSI neural networks". PhD Dissertation. Universitat Autonoma de Barcelona. September 1991.

INTERPRETATION OF OPTICAL FLOW THROUGH COMPLEX NEURAL NETWORK

Minami MIYAUCHI, Masatoshi SEKI[†], Akira WATANABE and Arata MIYAUCHI

School of Management
and Informatics,
SANNO College
1573 Kami-kasuya, Isehara,
Kanagawa 259-11 Japan
miyauchi@mi.sanno.ac.jp

Department of Electronics
and Communication
Musashi Institute of Technology
1-28-1 Tamazutsumi, Setagaya-ku,
Tokyo 158 Japan
miyauchi@ec.musashi-tech.ac.jp

Abstract In computer vision, the interpretation of optical flow (motion vector field calculated from images) and estimation of motion are important tasks. This study proposes a motion interpretation network which enables optical flow (OF) interpretation and describes motions on a plane through the use of a neural network with complex back propagation learning. Furthermore, an OF normalization network for optical flow normalization is proposed for the interpretation of diverse flow patterns, such as real image optical flow. Two types of output function of a neuron unit are examined. Using test patterns and real image optical flow, the generalization capacity of proposed network is investigated. And the ability is confirmed experimentally.

1 Introduction

In computer vision, the interpretation of optical flow [1] (motion vector field calculated from images) and estimation of motion are important tasks [2]. This study proposes a motion interpretation network which enables optical flow (OF) interpretation and describes motions on a plane through the use of a neural network with complex back propagation learning. Furthermore, an OF normalization network for optical flow normalization is proposed for the interpretation of diverse flow patterns, such as real image optical flow.

Methods for estimating motion from optical flow include a method that obtains the optimum solution by using several flow vectors to solve motion equations [3], [4], [5]. However, this method is time consuming and prone to noise, and solutions are for actual images cannot easily be obtained.

Neural networks are frequently utilized in pattern translation and are far less affected by noise [6], [7]. The calculation time required after learning is short, and the network are suitable for interpretation of motion. In addition, the networks proposed in this paper utilize complex BP and thus can naturally accommodate optical flow, a two-dimensional vector, as a complex number.

[†]Now with Toshiba medical engineering Co.

2 Complex Back Propagation Learning

Complex back propagation learning has been developed by Nitta and Furuya [8],[9], who expanded the weight of connection and the threshold of each unit in conventional neural networks to complex numbers. It is shown as an effective method for graphic conversion.

3 Motion Interpretation Network

Figure 1 shows the architecture of the motion interpretation network proposed in this study. In the flow vectors to be fed to the neural network, it is assumed that motion develops all over the frame, centering around the center of the frame.

- **Input layer**

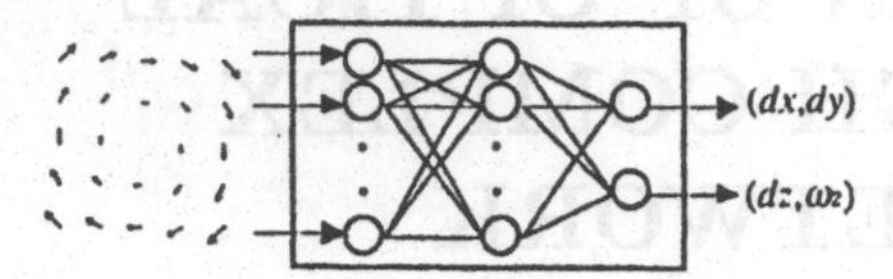

Figure 1: Motion interpretation network

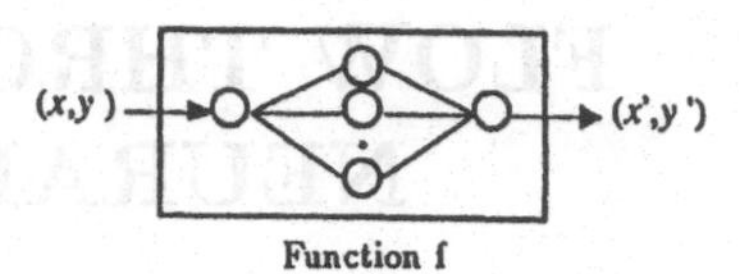

Figure 2: OF normalization network

Units corresponding to each vector of optical flow are arranged two dimensionally. Inputs to each unit are of complex numbers $u + v \cdot i$ corresponding to optical flow vectors $(u, v) = (dx/dt, dy/dt)$ at each pixel calculated from input images. Flow vectors are transformed to be in the range of [0, 1] in advance. Units of 25 (= 5×5) was used in the computer simulation.

- **Output layer**

  Two complex output units corresponding to the displacement components parallel to the frame $(dx,\ dy)$, expansion and contraction component dz and rotational component parallel to the frame ω_z $(dz,\ \omega_z)$ are available. They are denoted as $dx+dy \cdot i$ and $dz+\omega_z \cdot i$. As output values are in the range of [0, 1], they should be inverse transtormed to express actual velocity components.

- **Hidden layer**

  There is only one hidden layer, and 16 units are used.

4 Motion Interpretation Networks with Normalization Capacity

Motion interpretation networks interpret the optical flow in an entire frame. However, in reality, optical flow frequently cannot be obtained for the entire frame. Hence, OF normalization networks utilize the graphic conversion networks proposed by Nitta and Furuya to normalize sparce optical flow, partially defined optical flow and optical flow of arbitrary size and shape to $n \times n$.

In OF normalization networks, a complex function that translates a point on a two-dimensional plane to another point can be estimated by supplying a point before translation and a point after translation as learning data.

By having the network learn the starting point (x, y) of each vector (u, v) of optical flow as a point before conversion and the terminal point $(x', y') = (x + u, y + v)$ as a point after conversion, it is possible to have the neural network estimate a function f, to describe the optical flow. Here the network is expected to output a value of the complex function f, at all locations on the frame. By providing points aligned in an $n \times n$ lattice format to this network as starting points, and obtaining the final points from each starting point, a normalized $n \times n$ optical flow can be created. Figure 2 shows the architecture of OF normalization network. The cordinates of all points are normalized by considerng the size of frame is 1.0×1.0.

5 Output function of a neuron unit

In complex back propargation network complex sigmoid function has been used for nonlinearities of a neuron unit [8]. (See eq. (1).)

$$f(z) = \frac{1}{1+\exp(-x)} + \frac{1}{1+\exp(-y)} \cdot i \qquad (1)$$
$$(z = x + y \cdot i)$$

By the way, in usual network (real back propargation network) not only sigmoid function but also threthold logic is used for output nonlinear function. We apply threthold logic for complex back propargation learning. Because output values of inperpretation network and normalization network must be proportional to the norm of input optical flow vecters. Threthold logic for complex back propargation network is as eqs.(2) and (3).

$$f(z) = g(x) + g(y) \cdot i, \qquad (z = x + y \cdot i) \qquad (2)$$

$$g(x) = \begin{cases} 0 & x \leq 0 \\ ax & ax \leq 1 \\ 1 & ax > 1 \end{cases}, \quad (a : constant > 0) \qquad (3)$$

6 Experimental

6.1 Interpretation network

A test pattern consisting of 25 (= 5×5) vectors is provided to a motion interpretation network which had learned basic motions, so that the generalization capacity of the motion interpretation network can be studied. The teacher patterns and test patterns are provided motion components, and these are arranged so that motion developed all over the frame, centering around the center of the frame.

The number of types of tercher patterns used is 25, and these patterns contain a single motion component. Then 47 types of test patterns are provided.

Table 1 shows the motion vectors of teacher pattern supplied. A motion vector of test pattern and motion interpretation network output using sigmoid and threshold logic corresponding to the pattern are given

Table 1: Motion vector of teacher patterns

	dx	dy	dz	ω_z
1	1	0	0	0
2	0	1	0	0
3	-1	0	0	0
4	0	-1	0	0
5	1	1	0	0
6	-1	1	0	0
7	1	-1	0	0
8	-1	-1	0	0
9	0	0	1	0
10	0	0	-1	0
11	0	0	0	1
12	0	0	0	-1
13	0.5	0	0	0
14	0	0.5	0	0
15	-0.5	0	0	0
16	0	-0.5	0	0
17	0.5	0.5	0	0
18	-0.5	0.5	0	0
19	0.5	-0.5	0	0
20	-0.5	-0.5	0	0
21	0	0	0.5	0
22	0	0	-0.5	0
23	0	0	0	0.5
24	0	0	0	-0.5
25	0	0	0	0

in Table 2, Table 3 (a) sigmoid and Table 3 (b) threshold logic, respectively. Patterns 1 to 24 are patterns for investigating generalization capacity with respect to unknown speeds; patterns 25 to 35 are patterns for investigating the generalization capacity for a combination of multiple motion conponents.

The results indicate that ununiformity was present for interpretation using sigmoid, the mean square error is 0.0016 on the average, with a maximum of about 0.055. And interpretation using threshold logic estimates accurate value. Motion interpretation networks that have learned basic mation are thought to have the capacity to generalize unknown patterns.

6.2 Normalizaion network

Next, examples of experinental results on normalization capacity are shown. In the experiments, patterns in which optical flow is not obtained for the entire frame (test patterns 36–47) are created, then normalized by the OF normalization network and interpreted by the motion interpretation network. Figure 3 shows an input optical flow, the same optical flow after normalization, and the results of interpretation for each of them. In Figure 3, an input optical flow of test pattern is shown above, its normalized pattern is shown below, resul of interpretation and provided motion vector are indicated as (dx, dy)(dz, ω_z) at the bottom. The results shows the normalization capacity of network using threshold logic is superior to that of network using sigmoid.

Table 2: Motion vector of test patterns

No.	dx	dy	dz	ω_z
1	0.75	0	0	0
2	0	0.75	0	0
3	-0.75	0	0	0
4	0	-0.75	0	0
5	0.75	0.75	0	0
6	-0.75	0.75	0	0
7	0.75	-0.75	0	0
8	-0.75	-0.75	0	0
9	0	0	0.75	0
10	0	0	-0.75	0
11	0	0	0	0.75
12	0	0	0	-0.75
13	0.25	0	0	0
14	0	0.25	0	0
15	-0.25	0	0	0
16	0	-0.25	0	0
17	0.25	0.25	0	0
18	-0.25	0.25	0	0
19	0.25	-0.25	0	0
20	-0.25	-0.25	0	0
21	0	0	0.25	0
22	0	0	-0.25	0
23	0	0	0	0.25
24	0	0	0	-0.25
25	0.5	0	0.5	0
26	0	0.5	0.5	0
27	0.5	0.5	0.5	0
28	0.5	0	0	0.5
29	0	0.5	0	0.5
30	0.5	0.5	0	0.5
31	0.33333	0	0.33333	0.33333
32	0	0.33333	0.33333	0.33333
33	0.33333	0.33333	0.33333	0.33333
34	1	0.5	0	0
35	0.5	1	0	0

test pattern 42
(0.04, 0.20) (0.12, 0.21)

test pattern 42
(-0.00, 0.13)(0.00, 0.19)

test pattern 42
after normalization
(0.23, 0.51) (-0.20, 0.64)

test pattern 42
after normalization
(0.01, 0.04)(-0.02, 0.96)

(a) sigmoid (b) threshold logic

Figure 3: Experimental results on normalization capacity

Table 3: Output of interpretation network

(a) sigmoid

No.	dx	dy	dz	ω_z
1	0.7568	0.0046	0.0096	-0.0031
2	0.0065	0.7546	0.0106	-0.0016
3	-0.7853	-0.0037	0.0034	-0.0191
4	0.0084	-0.7851	0.0265	-0.0295
5	0.7572	0.7529	0.0077	-0.0104
6	-0.7806	0.7608	0.0046	-0.0118
7	0.7745	-0.7818	0.0108	-0.0334
8	-0.7888	-0.7964	0.0114	-0.0363
9	-0.0046	0.0039	0.7853	-0.0115
10	-0.0084	-0.0125	-0.7810	-0.0071
11	-0.0074	-0.0017	0.0114	0.7668
12	0	-0.0016	-0.0174	-0.7857
13	0.2875	0.0005	0.0156	-0.0014
14	0.0016	0.2868	0.0123	-0.0008
15	-0.2884	-0.0023	0.0114	-0.0085
16	-0.0008	-0.2894	0.0194	-0.0106
17	0.2892	0.2875	0.0130	0.0010
18	-0.2867	0.2865	0.0085	-0.0048
19	0.2875	-0.2874	0.0193	-0.0077
20	-0.2891	-0.2915	0.0158	-0.0155
21	-0.0013	0.0013	0.3169	-0.0054
22	-0.0007	-0.0038	-0.2904	-0.0036
23	-0.0009	-0.0012	0.0138	0.2947
24	-0.0009	-0.0008	-0.0166	-0.3041
25	0.5261	-0.0005	0.5519	0.0144
26	-0.0011	0.5508	0.5695	-0.0200
27	0.5279	0.5450	0.5414	-0.0053
28	0.5230	0.0076	-0.0291	0.5452
29	0.0012	0.5188	-0.0029	0.5363
30	0.5247	0.5195	0.0115	0.5204
31	0.3668	-0.0053	0.3881	0.3943
32	0.0083	0.3697	0.3876	0.3762
33	0.3696	0.3634	0.3779	0.3771
34	0.9119	0.5442	0.0043	-0.0108
35	0.5507	0.9094	0.0110	-0.0096

(b) threshold logic

No.	dx	dy	dz	ω_z
1	0.75	0	-0.001	0
2	0.0001	0.75	-0.0016	-0.0003
3	-0.7499	0.0001	-0.0013	-0.0009
4	0	-0.7499	-0.0007	-0.0006
5	0.7501	0.75	-0.0015	0.0002
6	-0.7499	0.75	-0.0017	-0.0007
7	0.75	-0.75	-0.0006	-0.0001
8	-0.7499	-0.7499	-0.0008	-0.001
9	0.0005	0.0001	0.7417	-0.0016
10	-0.0003	-0.0001	-0.744	0.0007
11	0	0.0004	0	0.7424
12	0.0002	-0.0003	-0.0023	-0.7433
13	0.2501	0	-0.0011	-0.0003
14	0.0001	0.25	-0.0013	-0.0004
15	-0.2499	0	-0.0012	-0.0006
16	0.0001	-0.2499	-0.001	-0.0005
17	0.2501	0.25	-0.0013	-0.0002
18	-0.2499	0.25	-0.0013	-0.0005
19	0.2501	-0.25	-0.001	-0.0003
20	-0.2499	-0.2499	-0.001	-0.0006
21	0.0002	0.0001	0.2465	-0.0008
22	0	0	-0.2488	0
23	0	0.0002	-0.0008	0.2472
24	0.0001	-0.0001	-0.0015	-0.2481
25	0.5003	0.0001	0.4942	-0.0009
26	0.0003	0.5001	0.4938	-0.0011
27	0.5003	0.5001	0.4939	-0.0008
28	0.5	0.0003	-0.0003	0.4951
29	0	0.5003	-0.0007	0.4949
30	0.5	0.5002	-0.0006	0.4952
31	0.3333	0	0.3297	0.3295
32	0	0.3335	0.3295	0.3294
33	0.3334	0.3334	0.3295	0.3296
34	1	0.5	-0.0013	0.0003
35	0.5001	1	-0.0017	0.0001

6.3 Noisy optical flow

Figure 4 shows an input optical flow with aditive noise($\sigma^2 = 0.00067, 0.0053$ and burst noise), the same optical flow after normalization, and the results of interpretation for each of them. This result indicates the interpretation network and the normalization network works well even under the additive noise.

6.4 Real imamge

Figure 5(a) shows the first frame of real image sequence(128 pixels $\times$ 128 pixels) used in this experiment and the object moves about 2.5 pixels per a time interval to the left. Figure 5(b) shows a part of optical flow (10 pixels $\times$ 10 pixels) caluculated from the real image sequence. Figure 6 shows the pattern after normarization. The result of this experiment shows the effectiveness of our networks.

7 Conclusions

Through the use of the method proposed in this paper, it was possible to interpret the optical flow and obtain motion parameters on a plane from optical flow. Experimental results show the motion interpretation network lerned basic motions have the capacity to generalize unknown patterns including patterns consists of multiple motion components. Equipping the network with normalization capacity further enabled us to obtain motion components from various optical flows, such as, sparse flow fields, partially defined flow patterns and optical flow of varions size and shape. By using threshold logic as output function of a neuron unit, accurate value of motion companents are obtained. Test patterns with noise and real image optical flow are also tested and motion components are estimated correctly. In future studies, we shall attempt to develop the present method to enable extraction of 3D motion.

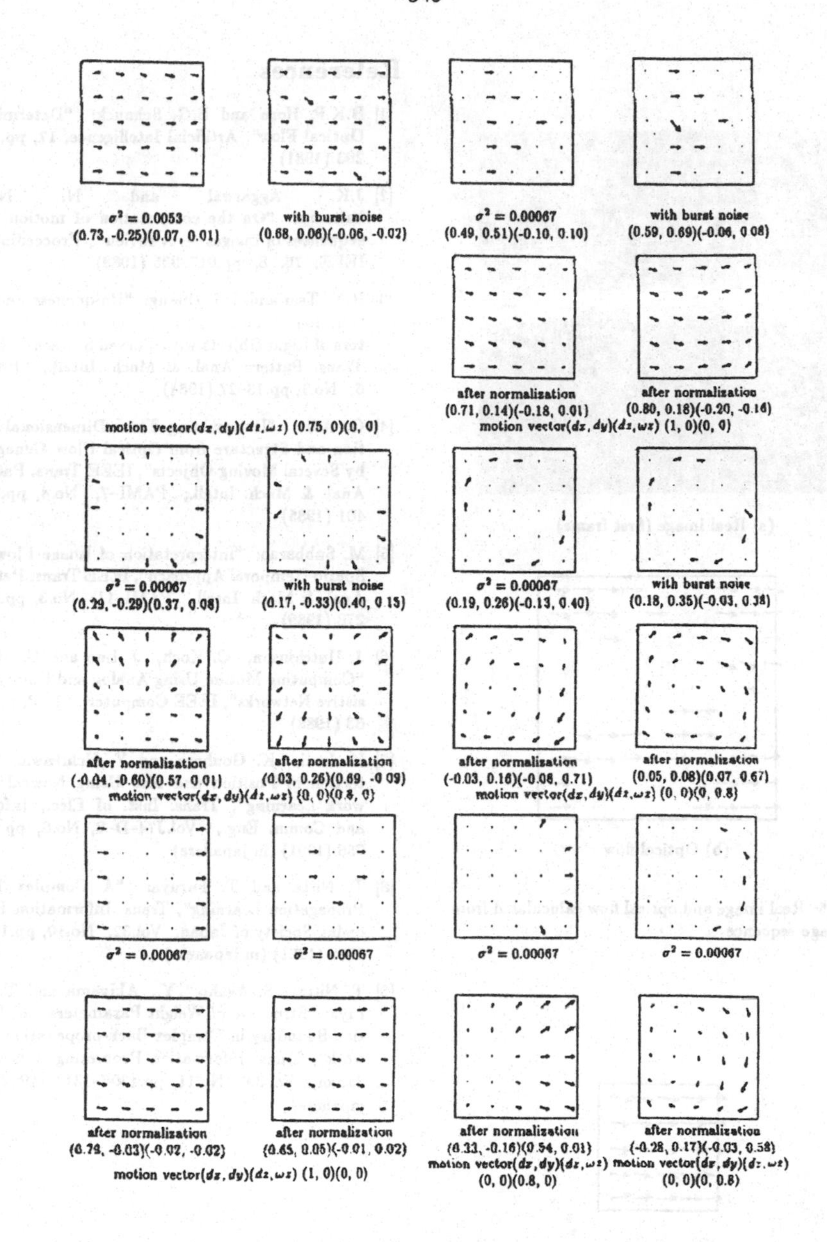

Figure 4: Experimental results with noise added inputs

(a) Real image (first frame)

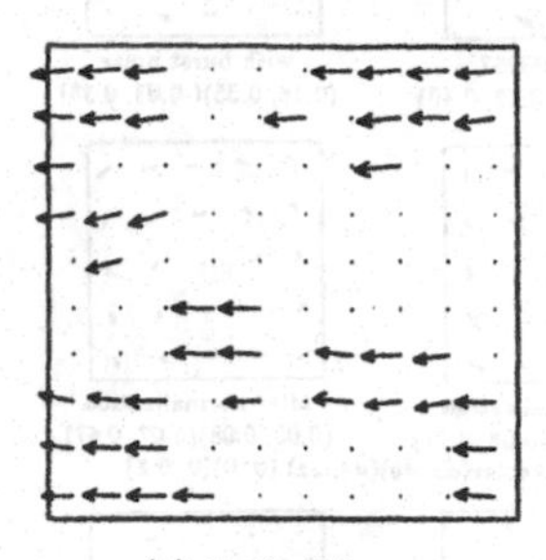

(b) Optical flow

Figure 5: Real image and optical flow caluculated from real image sequence

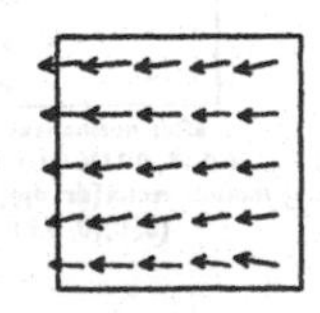

(-2.44, -0.15)(-0.07, -0.20)
motion vector$(dx, dy)(dz, \omega_z) \approx (-2.5, 0)(0, 0)$

Figure 6: Experimental results with real image sequence

References

[1] B.K.P. Horn and B.G. Schunck: "Determining Optical Flow", Artificial Intelligence, 17, pp.185–203 (1981)

[2] J.K. Aggarwal and N. Nandhakumar: "On the computation of motion from sequences of images — A review", Proceedings of IEEE, 76, 8, pp.917–935 (1988)

[3] R.Y. Tsai and T.S. Huang: "Uniqueness and Estimation of Theree-Dimensional Motion Parameters of Rigid Objects with Curved Surfaces", IEEE Trans. Pattern Anal. & Mach. Intell., PAMI-6, No.1, pp.13–27 (1984)

[4] G. Adiv: "Determining Three-Dimensional Motion and Structure from Optical Flow Generated by Several Moving Objects", IEEE Trans. Pattern Anal. & Mach. Intell., PAMI-7, No.4, pp.384–401 (1985)

[5] M. Subbarao: "Interpretation of Image Flow: A Spatio-Temporal Approach", IEEE Trans. Pattern Anal. & Mach. Intell., PAMI-11, No.3, pp.266–278 (1989)

[6] J. Hutchinson, C. Koch, J. Luo and C. Mead: "Computing Motion Using Analog and Binary Resistive Networks", IEEE Computer, 21, 3, pp.52–63 (1988)

[7] K. Imai, K. Gouhara and Y. Hchikawa: "Detection of Position and Size Using Newral Network Learning", Trans. Inst. of Elec., Inform., and Comm. Eng., Vol.J74-D-2, No.6, pp.748–756 (1991) (in japanese)

[8] T. Nitta and T. Furuya: "A Complex Back Propagation Learning", Trans. Information Proccesing Society of Japan, Vol.32, No.10, pp.1319–1329 (1991) (in japanese)

[9] T. Nitta, S. Akaho, Y. Akiyama and T. Furuya: "Structure of Weight Parameters and Decision Boundary in Complex Back-propagation Network", Trans. Information Proccesing Society of Japan, Vol.33, No.11, pp.1306–1313 (1992) (in japanese)

CT IMAGE SEGMENTATION BY SELF-ORGANIZING LEARNING

D. Cabello*, M.G. Penedo*, S. Barro*, J.M. Pardo* and J.Heras**

*: Departamento de Electrónica y Computación. Facultad de Física.
**: Servicio de Cirugía Ortopédica. Hospital General de Galicia
Universidad de Santiago de Compostela. Spain.

Abstract

In this paper we approach the segmentation of tibia CT images using a self-organizing feature map. This type of Artificial Neural Network carries out a competitive learning process which permits the discrimination of different structures found in the images with sensitivity to changes in the distribution and value of the gray levels of the pixels. The results obtained show that this technique is adequate for the segmentation of images with complex structures and a low signal/noise ratio.

Introduction

The segmentation of an image is a basic problem in automatic image analysis systems. Segmentation consists in the division of an image into a set of homogeneous elementary regions. In this work we approach the segmentation of tibia computational tomography (CT) images aimed at the three dimensional reconstruction of the structure. In order to do this we analyze a sequence of slices which correspond to transverse sections of the bone. The final objective is the automatic construction of a patient specific 3D model of the tibia using finite elements. Given the fact that the configuration of these elements is of paramount importance for a correct dynamic simulation of the structure when it has to withstand stresses, these elements must respond to the elementary region distribution. Their mechanical properties are related to the densitometric properties of the regions. It is therefore important to obtain a detailed and precise segmentation of each slice and consequently, of the global structure.

A large quantity of strategies for approaching the segmentation process can be found in the literature on the topic. They respond to two basic ideas: detection of local discontinuities (edge/boundary detection) or detection of local areas with homogeneous properties (amplitude and clustering segmentation, region growing) (Pratt, 1991). However, given the complexity of the structures in this type of images and their low signal to noise ratio, these classical segmentation techniques produce poor results when applied to them. In order to solve this problem, we present an alternative scheme based on the use of the self organizing topological maps proposed by Kohonen (1982). Several authors suggest employing these connectionist structures for the segmentation of medical images obtained by means of ultrasounds (Silverman, 1991) or magnetic resonance (Springub et al, 1991). On the other hand, the results would be better if instead of basing the segmentation only on the intensity of the pixels, their spatial properties were also taken into account. For this, each pixel is characterized by means of a feature vector whose components are the gray levels of the pixel and of its neighbors in the image plane, if we process a single slice, or in space if we process the whole 3D structure. The segmentation problem is thus transformed into a problem of classifying a set of n-dimensional feature vectors. Using a competitive, non supervised, learning process the neural structure generates a topological map which responds to the structures found in the data set. In our case, this map is a linear structure with J units, where each unit contains an n-dimensional vector. Each point of the feature space corresponds to the nearest point (class) in the generated map. The elementary regions resulting from the process arise when this information about classes is transferred to the image space: adjacent pixels whose feature vectors are projected onto the same unit will make up an elementary region. The preliminary results obtained for tibia CT images evidence the validity of the structure we propose.

Methodology

Material

In order to approach the global problem of the automatic construction of a patient specific three dimensional tibial model we use a sequence of 24 contiguous transverse CT scan proximal tibia slices obtained in vivo by means of a scanner operating under the following conditions: 120 KVp, 200 mA, 3.7 ms exposure and 2 mm slice thickness. These images were obtained with 256 gray levels, a 180*180 pixel spatial resolution and a pixel size of 0.71 mm^2.

Our initial objective was to determine both the perimeter (external cortical, inner cortical and bone thickness) and bone morphology. For this, we have processed these images slice by slice using the Artificial Neural Network (ANN) and the procedure we describe below.

Self-organizing feature map

We consider a self-organizing feature map made up of J neurons. Each neuron N_j, receives the pixels of a window or partial area of a tibia CT image as inputs. We represent this window by means of a vector $P=\{p_1,...p_I\}$, where each element p_i represents the gray level of the i-th pixel of the window. Each neuron N_j receives the values p_i as inputs modulated by a weight vector $W_j=\{w_{1j},...w_{Ij}\}$, where w_{ij} is the weight associated with the i-th input to neuron j. This way, when a input P is applied to the network, an output vector $O=\{o_1,...o_J\}$ is obtained. The component o_j, output of the j-th neuron, is:

$$o_j = (P - W_j)^T * (P - W_j) \quad (1)$$

Therefore, the neuron with the lowest output value will be the one which will better "correspond" to the pattern or window used as input to the network. Figure 1 shows the structure of the network.

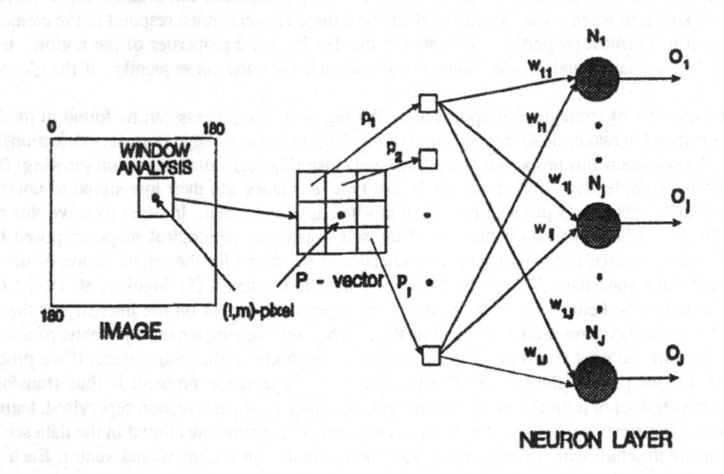

Figure1. Structure of the network

The training process in this type of AANs produces an adjustment of the weight vectors associated with the different neurons of the network. This training process is aimed at making the neurons sensitive to the different underlying structures in the patterns chosen for training. We are going to describe in detail how this training of the network is carried out.

Step 1: Initialize all the weights to random values in the value range of the inputs. Initialize the neighborhood factor to NF=J/2, where J is the number of neurons in the network. Initialize ΔW_j, $1 \leq j \leq J$, to zero. Initialize the learning factor, L, to a value between 0 and 1.

Step 2: We consider a training set made up of patterns $\{P_1,...,P_K\}$. For each input pattern to the network, P_k, a neuron j, whose output is minimum, is selected. The weights W_s of neurons N_s, $\forall$ $(j\text{-}NF) \leq s \leq (j+NF)$ / $s \in \{1,...,J\}$ (one dimensional treatment of the neurons in the network) are modified according to the following expressions:

$$W_s = W_s + L*\Delta W_s \quad (2)$$

$$\Delta W_s = P_k - W_s \quad (3)$$

Step 2 is repeated until for every P_k, $\Delta W_s < \varepsilon$.

Step 3: If NF=0 then convergence and end
Else NF = NF -1 and return to step 2.

Network training process

Obviously, the learning of the neural network will be strongly conditioned by the training set chosen. Other crucial parameters in our case are the number of neurons making up the network and the size of the input image window. In this section we are going to study these aspects from the framework of the results obtained in our application.

In our case we are interested in a training process which accommodates the response of the network to an appropriate segmentation of tibia CT images. We are interested in a segmentation which is sensitive not only to distribution changes, but also to changes in the values of the gray levels of the pixels. For this reason we have not considered normalizing the input patterns and the weight vectors of the neurons.

In order to choose the appropriate number of neurons for the network and input window size, we have used the visual information provided by the segmentation of the image achieved by the network after its convergence over a 180*180 pixel training image. For this, after the training process, we associate the gray level j*255/J to neuron j. We then apply the network, already trained, over the training image, reassigning to each pixel the gray level associated to the resulting winner neuron after providing the network with the window centered in that pixel. The visual analysis of this image, derived from the training image, provides information on the performance of the segmentation process after the training of the network. Following this process, and for the images we consider, we have decided that 12 is an adequate number of neurons.

With respect to the size of the input window, we must point out that it conditions the segmentation process as it must be in accordance with the size of the structures of interest found in the image. In our case, we have obtained the best segmentation results using small sized windows, more specifically, 3x3 pixel windows.

A drawback of the learning by competition methods is that some neurons might not be winners during training, staying out of the learning process. Kohonen's self-organizing feature map prevents this problem, at least partially, using a process for updating the neurons which involves not only the winning neuron but also the neurons which are a part of an adaptive neighborhood factor. Nevertheless, some cases in which this learning process can cause instabilities and slowness in the network convergence process have been described. These aspects strongly depend on the neighborhood factor chosen. In our case we have managed to avoid non convergence problems by means of an adequate selection and ordering of the input patterns or windows provided to the network during the training process. We have managed to obtain an appropriate convergence of the network by scanning the training image using a process in which the temporal proximity in the introduction of windows to the network does not correspond to their spatial proximity.

Experimental results.

The examples we now include demonstrate the suitability of the structure we propose for the segmentation of single slices using a twelve unit map and a 3*3 information window for each pixel.

Referencing the images from left to right and from top to bottom, figure 2.a shows an original gray-scale image responding to a transverse slice of the tibia in its higher shaft zone. The network was trained with 10 iterations. Figure 2.b shows the segmented image; in order to obtain it we have assigned to the pixels the gray levels corresponding to the winning neurons. Consequently, the elementary regions are represented by means of masks. As can be observed, the segmentation obtained responds to the underlying anatomic structure. The units of the map have extracted the class structure found in the data set and the resulting classification represents the distribution of the densities of the different tissues appearing in the image, from the skin (units 2 and 3, gray level masks 21 and 42) to the cortical bone (unit 12, mask with a value of 255). Level 0 corresponds to the background. A later analysis permits uniting the responses of adjacent neurons in the map which correspond to similar densities in the same tissue. Figure 2.c shows the result of this process considering 7 different responses. The contours of these regions are shown superposed on the original image in figure 2.c. As can be observed, they delimit the anatomic structures found in the image. We have correctly extracted both the geometry of the bone and a map with the distribution of densities inside it which corresponds to the different trabecular structures of the bone tissue in the area covered by the slice. Even though it is of less importance for our particular application we have also delimited the muscle and other soft tissues.

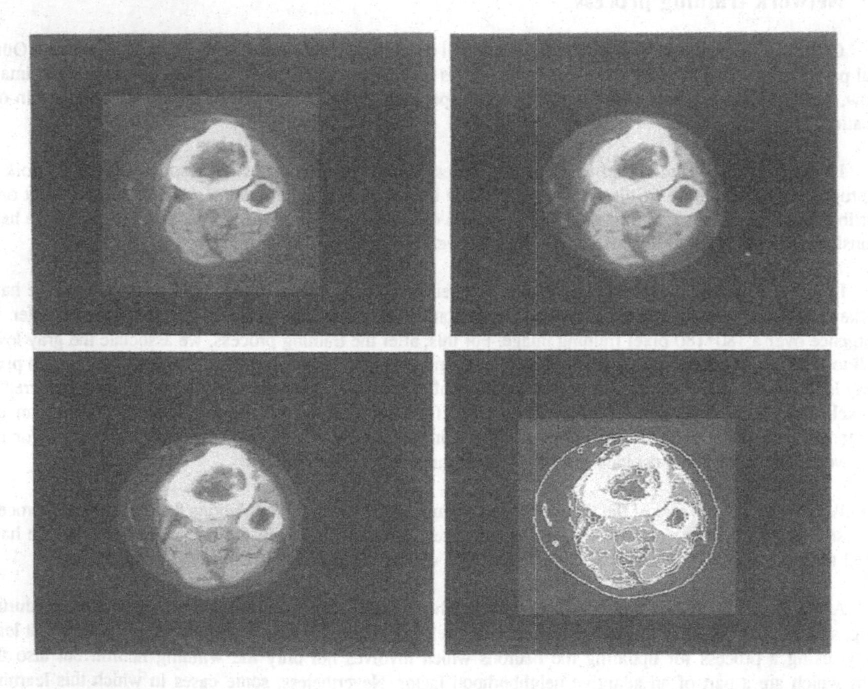

Figure 2: (a):original gray-scale tibia image. (b): Results from the analysis using the neuronal network (12 masks). (c): Elementary regions after the process of uniting responses (7 masks). (d): Contours of (c) superposed on (a).

Figure 3.a shows another original image, now corresponding to a slice of the tibia in its plateau area. The results of the segmentation process using the network trained with the image of figure 2.a is shown in figure 3.b. In this figure we have united the same neurons as in figure 2.c. We again correctly extract the geometry of the bone, segmenting its interior into bone tissue areas of homogeneous densities.

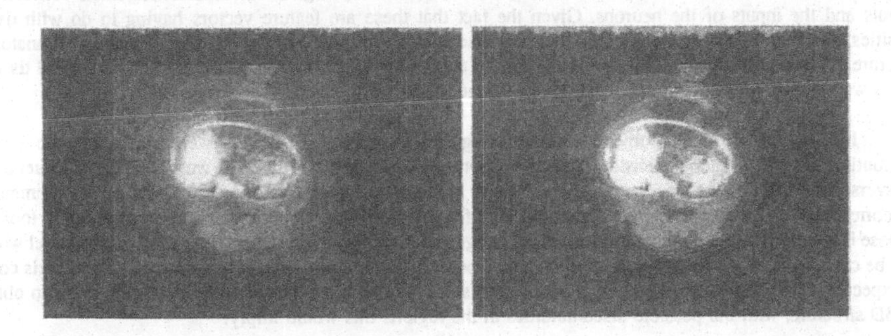

Figure 3. (a): original tibia gray-scale image in the plateau area. (b): Results from the analysis (7 masks).

In order to compare results, figure 4.a shows the segmentation of the image of figure 3.a by means of thresholding the histogram and figure 4.b the result of applying the Canny (1986) edge detector with parameter $\sigma = 0.75$. With respect to the image resulting from the thresholding process, we can point out that even when the valleys of the histogram are well defined, the resulting segmentation is quite coarse, it is not able to extract, for example, the geometry of the bone. A finer segmentation would imply more elaborate algorithms, which would require a higher computational effort than that of the structure we propose. On the other hand, as can be inferred from figure 4.b, we would find an even more complicated situation if we tried to segment the image by means of edge detection techniques. We would then have to answer questions such as: Which edgess are due to the noise of the image and which to the anatomic structures found there?, What is the exact position of the boundary, given the fact that the result of the operator is a function of σ?, How can we close edges?, etc... The technique we propose yields a fine and complete segmentation of the image with a lower computational cost than those of any of the other techniques found in the literature.

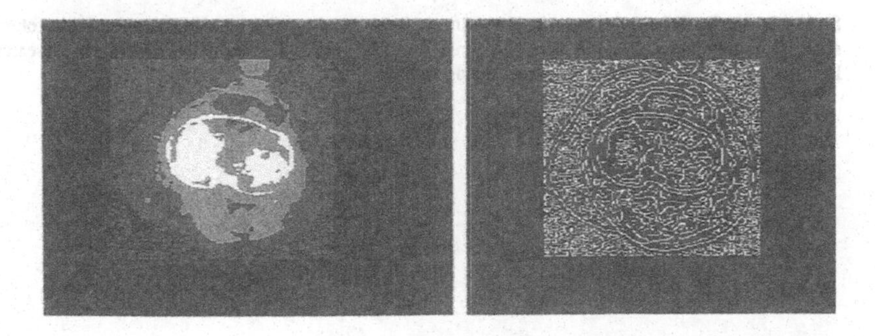

Figure 4. (a): Result of segmenting the image of figure 3.a by means of thresholding the histogram. (b): Result of applying a Canny operator ($\sigma = 0.75$).

Conclusions

In this work we present the capabilities of a self-organizing feature map for solving segmentation problems in proximal tibia computational tomography images. The training process converges rapidly. For the image sizes we work with it only requires a few seconds in a IBM RISC6000 workstation. After the training process, the weight vectors associated to the neurons are organized so that a topological correspondence exists between the outputs and the inputs of the neurons. Given the fact that these are feature vectors having to do with tissue densities, that correspondence permits achieving a segmentation which responds to the underlying anatomic structure, and thus differentiating the distinct types of tissues found and even discriminating pathological tissues, which would permit the identification of possible tumors.

In this work we had the objective of extracting the geometry of the tibia as well as a bone tissue density distribution map for a later dimensional reconstruction and modeling. To this end we processed a sequence of transverse slices, slice by slice, classifying each pixel based on the spatial information about its environment in the corresponding image plane. An alternative idea for the future development of the computational tool we propose is to approach the segmentation of the 3D image in a direct manner. The classification of each pixel would now be carried out based on its local information in space. A better definition of the structure in the Z axis could be expected as the results from processing each slice separately would not have to be packed in order to obtain the 3D structure, with the possible discontinuities in the regions this would imply.

Acknowledgements.

This work was supported by Spanish CICYT under Grant No. TIC91/0816.

References.

[1]. Canny, J.F.: A computational approach to edge detection. IEEE Trans. Patt. Anal. Machine Intell., vol. PAMI-8, 679-698, Nov., 1986.

[2]. Kohonen, T.: Self-Organized Formation of Topologically Correct Feature Maps. Biological Cybernetics, 43, 59-62, 1982.

[3]. Pratt, W.K.: Digital Image Processing. Jhon Wiley & Sons, Inc. New York, 1991.

[4]. Silverman, R.H.: Segmentation of Ultrasonic Images with neural networks. International Journal of Pattern Recognition and Artificial Intelligence, vol. 5, n.4, 619-628, 1991.

[5]. Springub, A.; D. Scheppelmann and H. Meinzer: Segmentation of multisignal images with kohonens selflearning topological map. In Computer Analysis of Images and Patterns (R. Klette, ed.) Research in Informatics, Vol. 5, 148-152. Akademie Verlag, Berlín, 1991.

TEXTURE IMAGE SEGMENTATION USING A MODIFIED HOPFIELD NETWORK

A. Mosquera, D. Cabello, M. J. Carreira and M. G. Penedo.

Departamento de Electrónica y Computación. Facultad de Física.
Universidad de Santiago de Compostela.
15706 Santiago de Compostela. SPAIN.

Abstract. In this work we describe the implementation of an artificial neural network, an extension of Hopfield's model, for the supervised segmentation of textured images. We use a Markov random field in order to model the textures in the image. The problem is approached in terms of the minimization of a objective function which integrates statistical and spatial information and which is projected onto the network. It provides a locally optimal solution to the problem of the classification of M*M pixels into K classes (textures). The experimental results obtained on artificial and natural images show the validity of the architecture we propose.

1. INTRODUCTION.

Image segmentation is an important problem in artificial vision, image analysis, etc. It consists in the partition of an image into a set of elementary regions characterized by the fact that some property is constant. Many objects in the real world present textured surfaces; the information texture provides is determinant in order to obtain a correct segmentation of the image. Several authors have developed statistical model-based approaches to texture segmentation/classification. These methods imply a first step of parameter estimation and a later classification, supervised or not, of the feature vector thus obtained (Vickers and Modestino, 1982; Kashyap and Khotanzad, 1984; Chellappa and Chatterjee, 1985). In this work we present a model-based approach to supervised texture image segmentation adopting Markov Random Field (MRF) as image model. For the classification of each pixel we consider statistical information (its feature vector) as well as spatial information in its neighborhood (label distribution). The segmentation problem is approached in terms of the minimization of a objective function that integrates both informations.

The inherent parallelism that both the problems of image processing and neural network architectures present have increased the interest in their use for the solution of difficult computational problems, such as the segmentation problem, (Manjunath et al. 1990; Chung-Lin Huang, 1992). In most cases, these neural networks are designed so as to minimize an energy function given by the architecture of the network itself, (Hopfield and Tank, 1986). Its parameters are thus obtained from the objective function we want to minimize for the solution of the problem. In practical cases, the networks must have the lowest possible number of connections. In this context, Markov's random field plays an important role as it limits the dependencies of each pixel to a small set of neighbors. In this work we describe the implementation of an artificial neural network for the segmentation of textured images which is an extension of Hopfield's model. The objective function is projected onto the network. This function provides a locally optimal solution for the problem of the classification of M*M pixels into K classes (textures). The results obtained, both on artificial images, a visualization of Markov's field, and on natural images extracted from Brodatz's album (Brodatz, 1966) prove the validity of the architecture we propose.

2. IMAGE MODEL.

The use of Markov's random field model in image processing applications has been investigated by many authors, and in the literature there is a detailed discussion of its use in images where textures are present, (Hassner and Sklansky, 1980; Cross and Jain, 1983; Chellappa, 1985; Cohen and Cooper, 1987; Mosquera et al., 1991). In this work we will use a second order Markov random field in order to model the probability density of the image intensities matrix.

In a Markov field model, the intensity of the pixels of an image is a linear combination of the intensities of neighboring pixels. We denote as Ω the set of points in the image of dimension M*M, $\Omega = \{ (i,j) ; 0 \leq i,j \leq M-1 \}$, and s is a two dimensional vector such that $s \in \Omega$. If we assume that the observations y(s) of a texture are Gaussian and have a zero mean value, Markov's random field model can be described by means of the following equation (Woods, 1972):

$$y(s) = \sum_{r \in N_s} \theta_r \, y(s+r) + e(s) \tag{1}$$

where e(s) is a sequence of Gaussian noise with a null mean value and with a value of υ for the variance, N_s is the neighbor vector considered and θ_r is the parameter vector of the model. This implies that:

$$p(y(s) \,/\, y(r), r \neq s) = p(y(s) \,/\, y(s+r), r \in N_s) \tag{2}$$

That is, $\{y(\cdot)\}$ is strictly a Markov field with respect to its neighbors. For a second order field $N_s = \{ (0,1), (1,0), (0,-1), (-1,0), (-1,1), (1,1), (1,-1), (-1,-1)\}$.

The unknown parameters $\{\theta,\upsilon\}$ can be estimated for each pixel in a given image by means of the equations

$$\begin{aligned} \theta^* &= \left[\sum_{\Omega'} q(s)q'(s)\right]^{-1}\left[\sum_{\Omega'} q(s)y(s)\right] \\ \upsilon^* &= \frac{1}{M^2}\sum_{\Omega'}\left[y(s)-\theta^{*\prime}q(s)\right]^2 \end{aligned} \tag{3}$$

where q(s) is the neighbor vector of y(s) and the sums extend to parameter estimation windows of size Ω'. These equations constitute a maximum probability estimation for the parameters of the model, (Kashyap and Chellappa, 1983).

Once the image model has been established and its parameters estimated in each position using (3), each pixel is characterized by a feature vector. The segmentation problem become then a problem of classifying in the feature space. Nevertheless, as the elementary regions resulting from the process arise when this information about classes is transfered to the image space, segmentation will improve considerably if we also consider spatial information for pixel classification. In order to do this, we lay out the segmentation problem in terms of the minimization of an objective function which integrates statistical (feature vectors) and spatial information (label distribution). This objective function is the sum of two terms. The first term corresponds to a quadratic error of the assignment of an image point to a given texture and the second favors the assignment of neighboring points to the same texture. This error function is given by

$$\varepsilon \,\alpha \sum_{s}^{M*M} \sum_{l=1}^{K} \left[\theta_s - \hat{\theta}_l\right]^2 l_{sl} - \lambda \sum_{s}^{M*M} \sum_{r \in N_s} L_s' \, L_r \tag{4}$$

where K is the number of textures in the image, $\hat{\theta}_l$ is the characteristic parameter vector of the l-th texture, θ_s is the parameter vector estimated using (3) for the pixel s, N_s is the neighbor vector considered and λ is a weighting factor. L_s is the label vector of the pixel s, $L_s = col(l_{s1},...,l_{sl},...,l_{sk})$, $l_{sl} \in \{0,1\}$ and $\sum_l l_l = 1$. That is, we adopt a hard classification with only one label for each pixel.

3. NEURAL NETWORK FOR TEXTURE CLASSIFICATION.

We will approach the segmentation problem by means of the implementation of an artificial neural network which minimizes the function of the expresion (4). The neural network we use is made up of K layers, each one of them consists in a matrix of M*M neurons, where K is the number of textures existing in the image and M the dimensionality of the image. We assume that the elements (neurons) of the network are binary and are indexed as (i,j,l), where (i,j) indicates the position in the image and l the layer. The (i,j,l)-th neuron is ON if its output V_{ijl} is 1, and this will indicate that pixel (i,j) in the image belongs to texture l. Let $T_{ijl:i'j'l'}$ be the weight associated with the interconnection between neurons (i,j,l) and (i',j',l') and I_{ijl} the input of neuron (i,j,l) and let us use a network with symmetric connections, $T_{ijl:i'j'l'}=T_{i'j'l':ijl}$, and without feedback, that is $T_{ijl:ijl}=0$. Then, the general expression of the energy of the network is:

$$E=-\frac{1}{2}\sum_{i=1}^{M}\sum_{j=1}^{M}\sum_{l=1}^{K}\sum_{i'=1}^{M}\sum_{j'=1}^{M}\sum_{l'=1}^{K}T_{ijl:i'j'l'}V_{ijl}V_{i'j'l'}-\frac{1}{2}\sum_{i=1}^{M}\sum_{j=1}^{M}\sum_{l=1}^{K}I_{ijl}V_{ijl} \tag{5}$$

Figure 1 shows the general structure of the network.

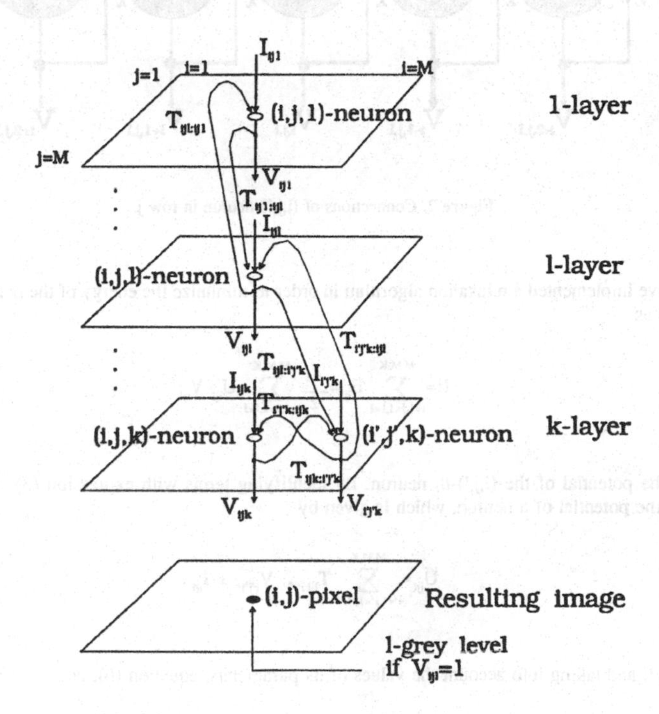

Figure 1. The general structure of the network.

The use of this network implies to obtain a set of states of the neurons so that its energy is a minimum. In order to employ the spontaneous process of energy minimization of the neural network to do the segmentation, we must project over the network the objective function which solves the problem. By inserting the multiplying factor ½ in (4) and identifying with (5) we can obtain the parameters of the network. Considering s=(i,j), $L_{sl} = V_{ijl}$ and r=(i',j'), these parameters will be

$$I_{ijl} = -\,[\theta_{ij} - \hat{\theta}_l]^2$$

$$T_{ijl:i'j'l'} = \begin{cases} \lambda & \text{if } (i',j') \in N_s \text{ and } l'=1 \\ 0 & \text{other case} \end{cases} \tag{6}$$

That is, there are no connections betwen neurons belonging to different layers and, within a layer, the connections of a neuron only reach its neighbors in the field we are considering. Figure 2 illustrates this. In this figure we show the connections between neurons in row j of layer l. This distribution presents a radial symmetry within the layer.

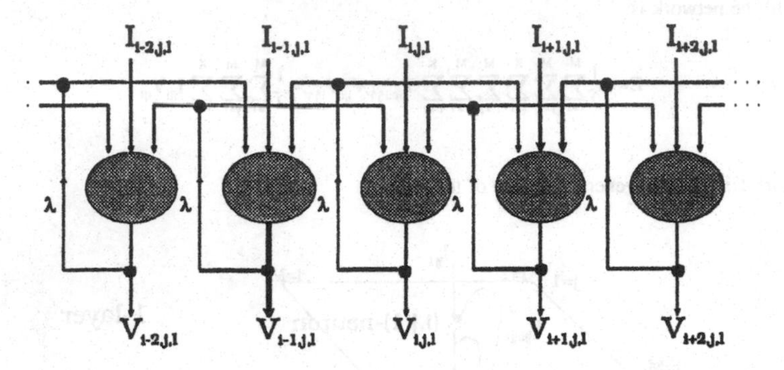

Figure 2. Connections of (i,j,l)-neuron in row j.

We have implemented a relaxation algorithm in order to minimize the energy of the network. This energy can be written as

$$E=\sum_{i=1,j=1,l=1}^{M,M,K} E_{ijl}=\frac{-1}{2}\sum_{i=1,j=1,l=1}^{M,M,K} U_{ijl}\, V_{ijl} \tag{7}$$

where U_{ijl} is the potential of the (i,j,l)-th neuron. By identifying terms with expression (5), we can obtain the expression of the potential of a neuron, which is given by

$$U_{ijl}=\sum_{i'=1,j'=1,l'=1}^{M,M,K} T_{ijl;i'j'l'}\, V_{i'j'l'} + I_{ijl} \tag{8}$$

For our network and taking into account the values of its parameters, equation (6), is

$$U_{ijl}=\lambda \sum_{(i',j')\in N_s} V_{i'j'l}-(\theta_{ijl}-\hat{\theta}_l)^2 \tag{9}$$

In the relaxation algorithm, in order to generate a new state of the neural network, we assume that each pixel has a single label, which means that only one neuron is active in each column. The new state of the network is calculated by means of the "Winner-takes-all" method over each column so that the neuron with the highest potential is turned ON and the other are OFF. This process is assumed as simultaneous for all the columns of the network. This scheme makes the system converge to a stable state, generally a local minimum of the function we want to minimize.

Before applying the network to a general image, it is necessary to know what textures it may find, that is, it is necessary to undergo a learning process. As we're making a supervised segmentation, this process consists in the calculation of the Markov field parameters characterizing the texture. These parameters are going to be the ones characterizing each layer of the neural network.

4. RESULTS.

In order to test the validity of the artificial neural network and the relaxation algorithm we propose, we have generated images which are a composition of different textures and attempted their segmentation. Figure 3.a shows a composition of textures which were artificially generated as a visual representation of Markov random fields with different values of the parameters. Figure 3.b, on the other hand, shows a composition of natural textures (beach sand, water and wood grain) extracted from Brodatz's album (Brodatz,1966). Both images have 256*256 pixels, 256 levels of gray and present a combination of three different textures; the network necessary in order to segment them will have three layers. The parameters $\hat{\theta}_l$ characterizing each texture/layer of the network are estimated from 25*25 pixel samples for the textures in figure 3.a and from 49*49 pixel samples for those in figure 3.b.

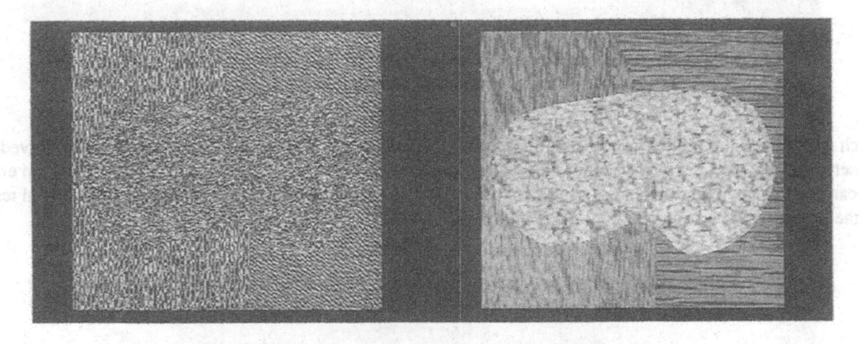

Figure 3. Texture composition images. (a): artificially generated textures, (b): natural textures (beach sand, water and wood grain).

An important aspect in the segmentation process is the determination of the optimum window size Ω' for the estimation, using (3), of the parameter vector θ_{ij} (local properties). This window must be large enough for the estimation of the parameters to be correct and small enough for us to be able to detect small regions in the image. Its size will be determined automatically by means of the analysis of the codifference matrices obtained as a generalization of the operation of binary images erosion to multiple gray level images using a structural element. The distance between the minima of the codifference matrices sum function provides information about the size of the 'motif' of the texture, (Mosquera et al., 1991). This method fixed a window size Ω' of 9*9 pixels for figure 3.a and 17*25 for figure 3.b. Also, in this last case, we have generalized the concept of neighborhood, taking as closest neighbors to a given pixel those located at a distance of half the size of the motif (Mosquera et al., 1993). This implies that in the second term of equation (4), N_s includes the neighbors we consider and all the pixels inside them.

The value of parameter λ is not critical in the segmentation process. Within a small interval, higher values lead to more compact segmentations; outside this interval, its value does not affect the results obtained very much. In the examples considered we have taken values in the interval (1,3).

The results obtained when applying the network to the image of figure 3.a considering $\lambda = 1$, starting from a random distribution of neuron activation and after 40 iterations are shown in figure 4; in figure 4.a the regions obtained appear as masks with different gray levels and in figure 4.b, we show their contours superimposed on the original image. As we can see, the boundaries obtained agree with those noticed in the original image.

Figure 5 shows the segmentation obtained from the image in figure 3.b. From left to right and from top to bottom, figures 5.a and 5.b show the results provided by the network and the deterministic relaxation algorithm considered, $\lambda = 2$, starting again from a random distribution of the activation of the neurons and after 40 iterations. In order to compare, figures 5.c and 5.d show the segmentation provided by a minimum distance classifier, taking into account three classes whose characteristic patterns are parameters $\hat{\theta}_l$, assigning to each pixel the class (texture)

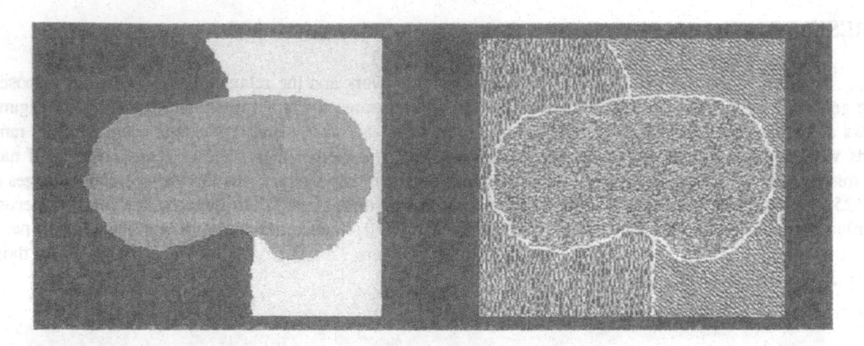

Figure 4. Results of the segmentation of the artificially generated texture composition image. (a): Mask image. (b): contours superimposed on original image.

which is closest in the property space and transferring this information to the image space. It can be observed that the network, as it also integrates spatial information, presents a better performance, reducing segmentation errors. We can conclude by pointing out the validity of the architecture we propose, which provides almost optimal results for the problem of the segmentation of textured images.

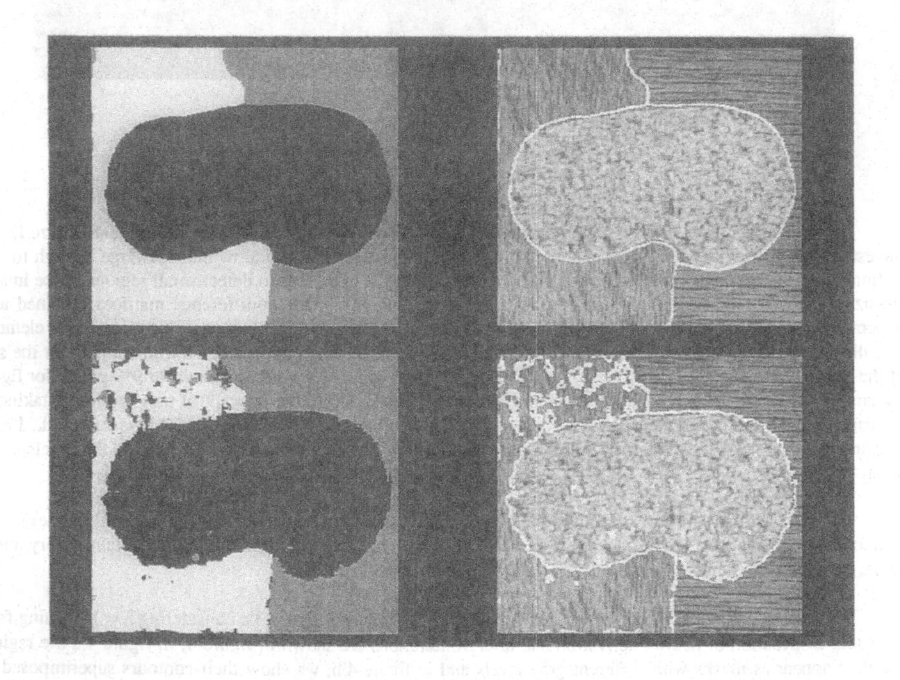

Figure 5. Segmentation of natural textures. (a),(b): results obtained using the neural network, (c),(d): results obtained using a minimum distance classifier.

ACKNOWLEDGEMENTS.

This work was supported by Spanish CICYT, TIC91-0816, and Xunta de Galicia, XUGA20602A90.

REFERENCES.

[1]. P. Brodatz. "Textures: A photographic album for artistis & designers". Dover Publications, New York, 1966.

[2]. R. Chellappa. "Two-dimensional discrete Gaussian Markov random field models for image processing", in *Progress in Pattern Recognition 2* (L.N Kanal and A. Rosenfeld, Eds.), pp. 79-112, Elsevier, New York, 1985.

[3]. R. Chellappa and S. Chatterjee. Classification of textures using Gaussian Markov Random fields. IEEE Trans. Acoustic, Speech and Signal Processing, vol. ASSP-33, n. 4, pp. 959-963. 1985.

[4]. S.F. Cohen and D.V. Cooper. "Simple Parallel Hierarchical and Relaxation Algorithms for Segmenting Noncausal Markovian Random Fields". IEEE Trans. Pattern Anal. Machine Intell., vol. PAMI-9, pp. 195-219, 1987.

[5]. G.R. Cross and A.K. Jain. "Markov random field texture models". IEEE Trans. Pattern. Anal. Machine Intell., vol. PAMI-5, pp. 25-39, 1983.

[6]. M. Hassner and J. Sklansky. "The use of Markov random fields as model of texture". Comput. Graphics Image Processing, vol. 12, pp. 357-370, 1980.

[7]. Chung-Lin Huang. "Parallel image segmentation using modofied Hopfield model". Pattern Recognition Letters vol. 13, pp.345-353, 1992.

[8]. J.J. Hopfield and D. W. Tank. "Computing with Neural Circuits: A model". Science, vol. 233, pp.625-633, 1986.

[9]. R.L. Kashyap and R. Chellappa. "Estimation and Choice of Neighbors in Spatial-Interaction Model of Images". IEEE Trans. Information Theory, vol. IT-29, pp. 60-72, 1983.

[10]. R.L. Kashyap and A. Khotanzad. A sthocastic model based technique for texture segmentation. Seventh International Conference on Pattern Recognition, Montreal, 1984, pp. 1202-1205.

[11]. B.S. Manjunath, Tal Simchony and R. Chellappa. "Stochastic and deterministic networks for texture segmentation". IEEE Trans. Acoust., Speech, Signal Process. vol. 38, pp.1039-1049, 1990.

[12]. A. Mosquera, D. Cabello, M.J. Carreira and M.G. Penedo. "Unsupervised textured image segmentation using Markov random field and clustering algorithms". In *Computer Analysis of Images and Patterns* (R. Klette, Ed.). Research in Informatics, vol. 5, pp.139-147, Akademie Verlag, Berlín, 1991.

[13]. A. Mosquera, D. Cabello, J.M. Mallo, M.J. Carreira and M.G. Penedo. "Functional Neighborhood in Markov Random Fields: Generalized Texture Models". The 8th Sacandinavian Conference on Image Analysis. Tromsø, Norway, may, 1993.

[14]. A.L. Vickers and J.W. Modestino. A maximum likelihood approach to texture classification. IEEE Trans. Pattern Anal. Machine Intell. vol. PAMI-4, n.1, pp. 61-68, 1982.

[15]. J.W. Woods. Two-dimensional discrete Markivian Fields. IEEE Trans. Information Theory, vol. IT-18, n.2, pp. 232-240, 1972.

IMAGE COMPRESSION WITH SELF-ORGANIZING NETWORKS

Bernd Freisleben and Maximilian Mengel

**Department of Computer Science (FB 20), University of Darmstadt,
Alexanderstr. 10, D-6100 Darmstadt, Germany**

Abstract

In this paper we evaluate and compare the performance of self-organizing neural networks applied to the task of image compression. The networks investigated are two-layered architectures with linear neurons, and variants of Hebbian learning rules are used to reduce the dimensionality of the inputs while preserving a maximum of information in the output units. Although in theory all networks considered are effectively equivalent to performing the Karhunen-Loève transform, which is the optimal image compression method in the sense that it allows linear reconstruction of the input information with minimal squared error, the results obtained in practice reveal significant differences between the networks. An experimental study has been conducted to demonstrate these differences and thus some light is shed on the suitability of self-organizing neural networks for image compression, particularly in comparison to more conventional methods.

1 Introduction

The proliferation of multi-media tools in computer communication networks has increased the demand for techniques to improve the efficiency of transmission and storage of images. A large variety of algorithms for image compression has been proposed [4]. Their performance, measured by their data compressing ability, the reconstruction error to the original image and their implementation complexity, depends on the different types of images and the requirements of the applications considered.

The general idea behind a popular class of image compression algorithms is to exploit the fact that nearby pixels in images are often highly correlated. A given image is therefore divided into several blocks of pixels, and each block, treated as vector, is linearly transformed into a vector whose components are mutually uncorrelated. These components, called *transform coefficients*, are then independently quantized for transmission or storage. The reconstruction of the original image is obtained by using an inverse linear transform operation on the quantized coefficient vector. Compression is achieved by transmitting or storing only those coefficients that account most for the image data's variance and among those, additionally quantizing the more important ones more finely than others. It has been shown [4] that the overall average mean-squared reconstruction error is minimized when the rows of the transformation matrix used for decorrelation are the orthonormalized eigenvectors of the covariance matrix of the input vectors. The corresponding optimal transform under this error citerion is called *Karhunen-Loève Transform (KLT)* [4], also known as *Principal Component Analysis (PCA)* [5] in the field of statistics. Since the KLT is quite time-consuming to compute, it is in practice substituted by a suboptimal but fast transform. An example is the *Discrete Cosine Transform (DCT)* [13] which performs very closely to the KLT and has therefore been selected as the basis of the *JPEG* image compression standard [2]. In contrast to the KLT, the design of the DCT is input-independent, since the same transformation matrix is used for any image that needs to be compressed.

Beyond the classical transformations, neural network techniques have quite successfully been applied to compressing images, in particular multilayer architectures employing the well known backpropagation algorithm [1, 8]. These approaches are based on the idea that in a three-layer network with a hidden layer smaller than the input and the (equally sized) output layer, the input data must somehow be squeezed down to a compact representation in the hidden layer which is then again expanded at the output layer. It has been shown that the type of compact representation emerging in the hidden layer is essentially similar to performing a PCA [8], with the variances spread uniformly across the hidden units such that they are equally important. Therefore, coding the more important features with higher resolution is not easily possible. Furthermore, the long computation times usually associated with backpropagation training are certainly disadvantageous for suggesting a neural approach in favour of the classical methods.

In this paper we consider self-organizing neural networks applied to the image compression problem. Oja [9] was the first to observe that a modified Hebbian learning rule applied to a simple linear network model with N input units and a single output unit converges to a weight vector which is identical to the first principal component of the input distribution. Thus, the network adaptively extracts the first principal component from the given data set without explicity computing the data's covariance matrix. Several proposals have been made to extend Oja's work in developing networks that perform a complete PCA [3, 6, 10, 11, 12]. We present the architectures on which these approaches are based and the learning algorithms associated with them. An experimental study is carried out to evaluate their performance in an image compression task. The results indicate that in practice not all of them fulfill the theoretical expectations.

The paper is organized as follows. Section 2 presents the network architectures considered for achieving image compression. In section 3 the implementation of the networks and the performance results obtained in the experiments are described. Section 4 concludes the paper and discusses areas for further research.

2 Network Architectures

All networks considered in this paper consist of an input and an output layer, with full connections between the two layers. To achieve compression, the output layer contains fewer units than the input layer. In some of the architectures, there are either hierarchical or full lateral connections among the output units, as shown in figure 1.

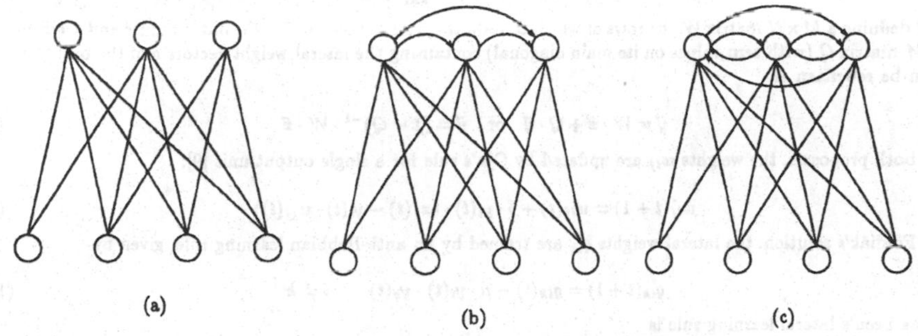

Figure 1: Architectures of the Networks Investigated

The networks proposed by Oja [10] and Sanger [12] are based on the architecture depicted in figure 1 (a). The output units have linear activation functions, with the output $y_i(t)$ of unit i at time t given by

$$y_i(t) = \sum_{j=1}^{N} x_j(t) \cdot w_{ij}(t) \tag{1}$$

where N is the number of input units, $x_j(t)$ is the input presented to input unit j at time t and $w_{ij}(t)$ is the weight of the connection between output unit i and input unit j at time t. In Oja's proposal, the w_{ij}'s are modified according to

$$w_{ij}(t+1) = w_{ij}(t) + \delta \cdot y_i(t) \cdot (x_j(t) - \sum_{k=1}^{M} y_k(t) \cdot w_{kj}(t)) \tag{2}$$

while Sanger has suggested a learning rule of the form

$$w_{ij}(t+1) = w_{ij}(t) + \delta \cdot y_i(t) \cdot (x_j(t) - \sum_{k=1}^{i} y_k(t) \cdot w_{kj}(t)) \tag{3}$$

where M is the number of output units, and $\delta > 0$ is a learning parameter.

The only difference is in the upper limit of summation. The weight vectors in Oja's network converge to span the same subspace as the eigenvectors associated with the M largest eigenvalues of the input covariance matrix. In Sanger's network, the eigenvector directions are extracted themselves, individually in decreasing order of the eigenvalues.

Rubner and Tavan [11] have introduced trainable lateral connections among the output units to design a network that also performs PCA. In their architecture, a lateral connection q_{ik} between output unit i and output unit k is present only if $i < k$ (see figure 1 (b)). The activation $y_i(t)$ of output unit i is computed by

$$y_i(t) = \sum_{j=1}^{N} x_j(t) \cdot w_{ij}(t) + \sum_{k=1}^{i-1} y_k(t) \cdot q_{ik}(t) \tag{4}$$

The weights w_{ij} between the input and output layer are trained with a plain Hebbian learning rule, followed by normalization of the resulting weight vector $\vec{w}_i'(t+1)$ to unit length:

$$w_{ij}(t+1) = w_{ij}(t) + \delta \cdot y_i(t) \cdot x_j(t), \qquad \vec{w}_i(t+1) = \vec{w}_i'(t+1)/|\vec{w}_i'(t+1)| \tag{5}$$

The lateral weights q_{ik} are modified according to an *anti-Hebbian* rule, equivalent to a Hebb rule with a negative learning parameter μ:

$$q_{ik}(t+1) = q_{ik}(t) - \mu \cdot y_i(t) \cdot y_k(t) \qquad i < k \tag{6}$$

The lateral connections are used to decorrelate the outputs; the principal components are extracted in order, just like in Sanger's proposal.

Földiák's [3] and Leen's [6] proposals are based on the architecture displayed in figure 1 (c), where full lateral connections q_{ik} in the output layer are employed. In both approaches, the activation $y_i(t)$ of output unit i at time t is given by

$$y_i(t) = \sum_{j=1}^{N} x_j(t) \cdot w_{ij}(t) + \sum_{\substack{k=1 \\ k \neq i}}^{M} y_k(t) \cdot q_{ik}(t) \tag{7}$$

By defining a $M \times N$ matrix W the rows of which contain the weight vectors $\vec{w}$ of the output units and a symmetric $M \times M$ matrix Q (with zero values on its main diagonal) containing the lateral weight vectors $\vec{q}$ of the output units, (7) can be rewritten as

$$\vec{y} = W \cdot \vec{x} + Q \cdot \vec{y} \quad \text{or} \quad \vec{y} = (1 - Q)^{-1} \cdot W \cdot \vec{x} \tag{8}$$

In both proposals, the weights w_{ij} are updated by Oja's rule for a single output unit [9]:

$$w_{ij}(t+1) = w_{ij}(t) + \delta \cdot y_i(t) \cdot (x_j(t) - y_i(t) \cdot w_{ij}(t)) \tag{9}$$

In Földiák's solution, the lateral weights q_{ik} are trained by an anti–Hebbian learning rule, given by

$$q_{ik}(t+1) = q_{ik}(t) - \mu \cdot y_i(t) \cdot y_k(t) \qquad i \neq k \tag{10}$$

whereas Leen's lateral learning rule is

$$q_{ik}(t+1) = q_{ik}(t) - \mu \cdot (d \cdot q_{ik}(t) - C \cdot y_i(t) \cdot y_k(t)) \tag{11}$$

where d is a rate constant and C is a coupling constant which depends on the eigenvalue spectrum. Similar to Oja's network, Földiák's proposal does not find the eigenvectors themselves, but converges to the subspace spanned by the eigenvectors, while Leen's network performs a complete PCA. Both proposals are quite sensitive to the parameters required to be specified for the learning algorithms, particularly in Leen's approach decreasing learning parameters (proportional to $1/t$) are necessary to achieve convergence.

In all networks presented, image compression is achieved as follows. The image to be compressed is divided into equally sized blocks of pixels; each of these blocks is represented by an N–dimensional vector. The image is scanned from left to right and top to bottom, and these vectors are consecutively presented to the input units of the network, as shown in figure 2. The input/output connections are initialized to small random values (the random values used in our experiments were drawn from a uniform distribution in the interval [-0.5, 0.5]), whereas the lateral connections, if applicable, are initially set to zero.

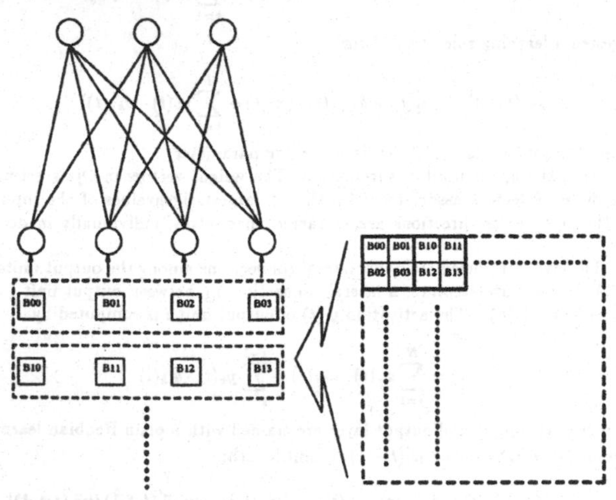

Figure 2: Applying Self–Organizing Networks to Image Compression

In the learning mode, the input vectors representing an image are propagated through the network by computing the output activities and performing the corresponding weight updates. The whole image is presented to the network several times to give the network time to converge. After the training phase is finished, the image is again processed by the network, however without any connection updates. The image is compressed by multiplying each block represented as an N-dimensional vector by each of the M weight vectors obtained after training to generate M coefficients for coding the block. Assuming that each block consists of $p \times p$ pixels, the coefficients for the block $b^{n,m}$ starting at position $(n \cdot p + 1, m \cdot p + 1)$ in the image I are given by

$$y_i^{n,m} = \sum_{q=1}^{p} \sum_{r=1}^{p} w_{i\,(q-1)\cdot p+r} \cdot I_{n\cdot p+q, m\cdot p+r} \tag{12}$$

To decompress the image, each block of the image is reconstructed by adding together all the weight vectors multiplied by their coefficients:

$$\hat{I}_{n\cdot p+q, m\cdot p+r} = \sum_{i=1}^{M} w_{i\,(q-1)\cdot p+r} \cdot y_i^{n,m} \tag{13}$$

3 Implementation and Performance

All the networks described in the previous section have been implemented in *C* on a SUN Sparcstation, together with appropriate programs for image processing/visualization, numerical computation of the eigenvectors of the input covariance matrix, and code for the discrete cosine transform in order to be able to compare the network results with a classical method. In fact, the code written for, say Oja's network, can be used to produce compression results for the DCT by simply initializing the weight vectors with the transform coefficients computed for the DCT and doing a recall. The KLT is obtained in a similar manner by using the numerically computed eigenvectors instead.

In order to compare the performance of the different approaches, a 256 × 256 8-bit-pixel image with 256 grey levels is used. The 256 grey level values are normalized to values between −1 and 1. The image is split into 8 × 8 pixel blocks, leading to a network with 64 input units. In all experiments conducted, this 64-dimensional input is compressed to a 3-dimensional output, i.e. the networks consist of 3 units in the output layer. If no adaptive quantization scheme is used, this leads to coding each 8 × 8 block of pixels (512 bits) with a total of 24 bits, so the data rate is 0.375 bits per pixel and the compression factor is 21.33. The blocks do not overlap, and the complete image is presented 8 times, resulting in 8192 input vectors presented to the networks during the training phase.

In order to evaluate the quality of the reconstructed images, the mean squared error (MSE)

$$MSE = \sum_{s,t} (I_{s,t} - \hat{I}_{s,t})^2 \tag{14}$$

between the original and the decompressed image is computed, where s, t denote the number of pixels in the x-direction and the y-direction, respectively.

Figure 3 shows the original image used for the comparison and the same image after compression/decompression with the KLT. The MSE for the KLT under the experimental conditions described above is $MSE = 29.63$.

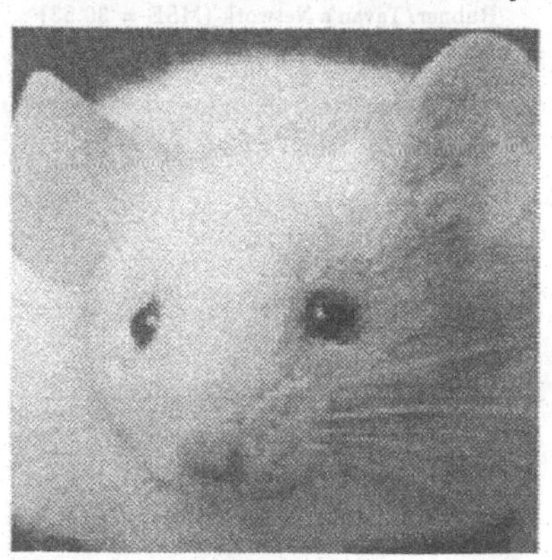

Original Image

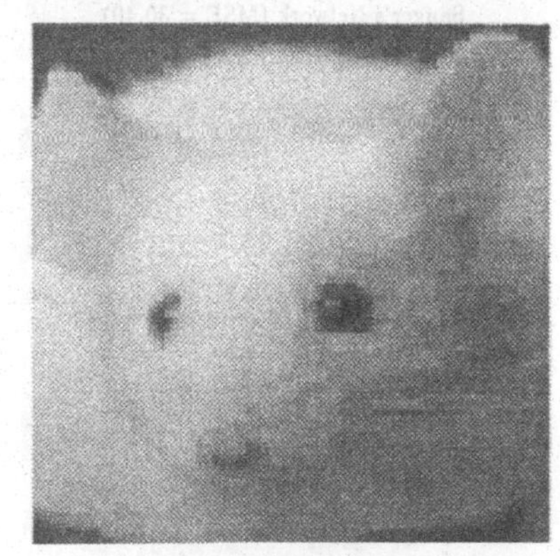

KLT (MSE = 29.63)

Figure 3: Original Image and Image Compressed/Decompressed with the KLT

The results of the DCT and all the networks considered in this paper are summarized in figure 4. The networks used for compression/decompression are listed below the individual images. The MSE is also given for each of the experiments conducted.

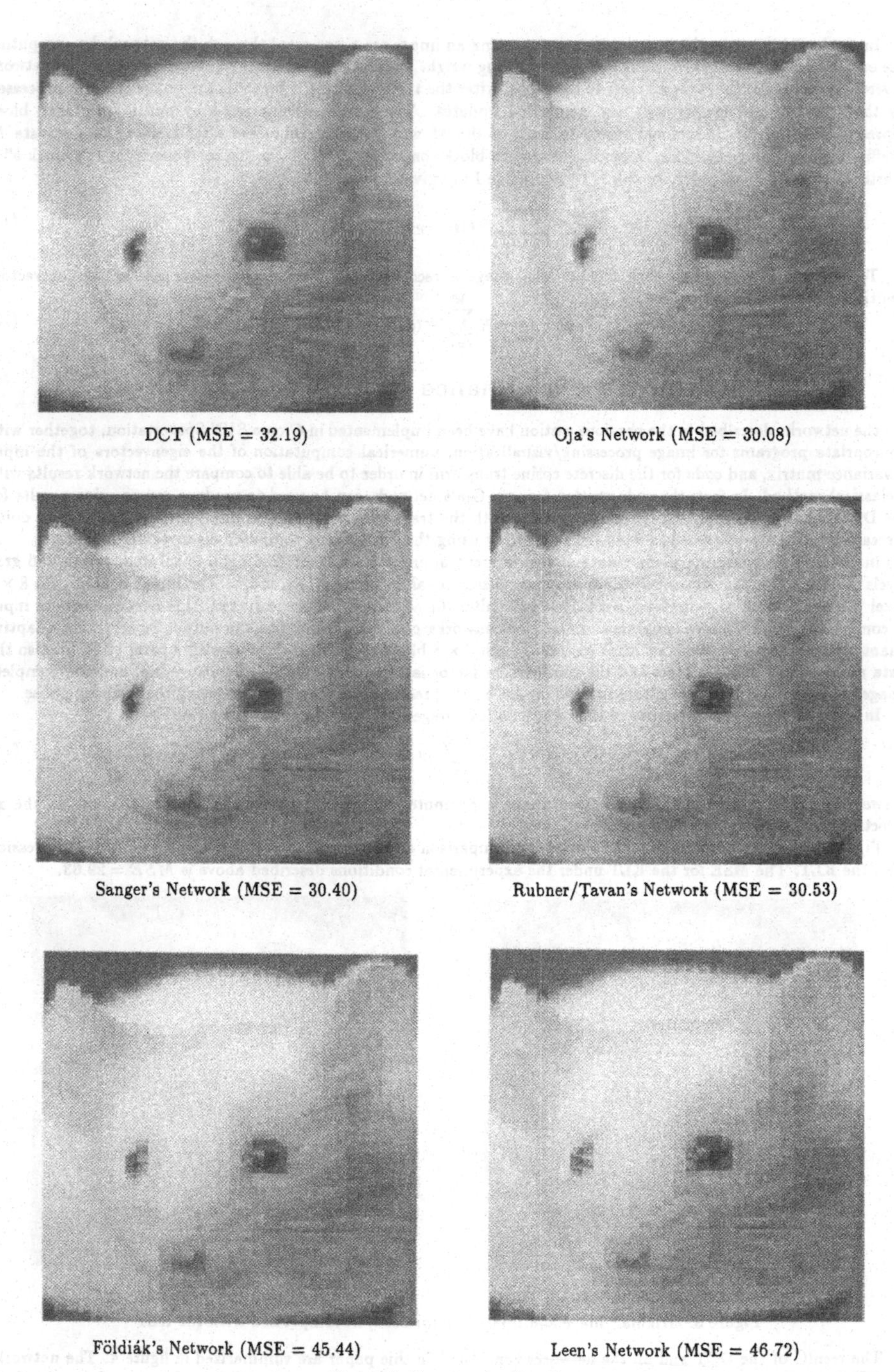
DCT (MSE = 32.19)

Oja's Network (MSE = 30.08)

Sanger's Network (MSE = 30.40)

Rubner/Tavan's Network (MSE = 30.53)

Földiák's Network (MSE = 45.44)

Leen's Network (MSE = 46.72)

Figure 4: Compression Results

The MSE criterion separates the networks into two categories, those that perform better than the DCT (Oja, Sanger, Rubner/Tavan) and those that are worse (Földiák, Leen). Similar results have been obtained in experiments with other images, but due to space limitations we cannot present them here. None of the networks produces the optimal results obtained by the KLT, most probably because of numerical inaccuracies, but the ones in the first category come very close to it. The block structure is quite evident in the images resulting from Földiák's and Leen's network, but this effect might be avoided by letting the blocks overlap as in [8]. The convergence behaviour, and consequently the resulting compression quality, clearly depends on the proper settings of the learning parameters, and a large number of simulations have been carried out to determine the most appropriate learning parameters for the networks considered. One way to achieve a good compression quality and also speed up convergence is to start the training phase with the weight vectors initialized to the DCT coefficients. This method, however, cannot be recommended for the networks that extract the eigenvectors in order, since the DCT coefficients do not easily convey information about the ranking of variances.

4 Conclusions

In this paper we have evaluated the performance of linear neural networks with modified Hebbian learning rules applied to the task of image compression. Although all of the networks considered are theoretically capable to perform as good as the optimal Karhunen–Loeve transform, they exhibit significant differences when they are applied in practice. An experimental investigation has been made to illustrate these differences. It has been shown that some of the networks outperform the discrete cosine transform, the most commonly used method for image compression, while other networks do not produce results of comparable quality. Among the issues for future research are a comparison to further self-organizing networks used for image compression [7], an integration of a suitable neural network into a image compression software package like JPEG [2] and an analysis of the generalization ability of the networks by exposing them to unknown images.

References

[1] G. Cottrell, P. Munro and D. Zipser. Image Compression by Backpropagation: An Example of Extensional Programming. In: *Advances in Cognitive Science*, Vol. 3, Norwood, 1987.

[2] Digital Equipment Corporation. JPEG Still Picture Compression Standard, *Communications of the ACM*, 34:31-44, 1991.

[3] P. Földiák. Adaptive Network for Optimal Linear Feature Extraction. In: *Proceedings of the International Joint Conference on Neural Networks.* San Diego, SOS Printing, pp. 401-405, 1989.

[4] A.K. Jain. Image Data Compression: A Review. *Proceedings of the IEEE*, 69(3):349-389, 1981.

[5] I.T. Jolliffe. *Principal Component Analysis.* Springer–Verlag, 1986.

[6] T.K. Leen. Dynamics of Learning in Feature–Discovery Networks. *Networks*, 2, pp. 85-105, 1991.

[7] L.K. Liu and P.A. Limogenides. Unsupervised Orthogonalization Neural Network for Image Compression. *Proceedings of the SPIE'92 Conference on Intelligent Robots and Computer Vision XI*, Vol. 1826, pp. 215–225, 1992.

[8] M. Mougeot, R. Azencott and B. Angeniol. Image Compression With Back Propagation: Improvement of the Visual Restoration Using Different Cost Functions. *Neural Networks* 4, pp. 467-476, 1991.

[9] E. Oja. A Simplified Neuron Model as a Principal Component Analyzer. *Journal of Mathematical Biology*, 15, pp. 267–273, 1982.

[10] E. Oja. Neural Networks, Principal Components, and Subspaces. *International Journal of Neural Systems* 1, pp. 61–68, 1989.

[11] J. Rubner and P. Tavan. A Self-Organizing Network for Principal–Component Analysis. *Europhysics Letters* 10, pp. 693–698, 1989.

[12] T.D. Sanger. Optimal Unsupervised Learning in a Single–Layer Linear Feedforward Neural Network. *Neural Networks* 2, pp. 459-473, 1989.

[13] L.P. Yaroslavsky. *Digital Picture Processing.* Springer–Verlag, 1985.

Neural Networks as Direct Adaptive Controllers

Mohammad Bahrami

School of Electrical Engineering, University of New South Wales, P. O. Box 1, Kensington, NSW 2033 AUSTRALIA.
e–mail : bahrami@syscon.ee.unsw.edu.au

Abstract

A learning scheme for multilayer feedforward neural networks used as direct adaptive controllers of nonlinear plants is suggested. This scheme is a supervised steepest descent one that does not require backpropagation of the error. Using a neural network controller trained with this method does not require the identification stage and this makes it superior to the other methodologies. Methods of using neural networks for plant control suggested in the literature are discussed and compared with the proposed system. Simulations based on model reference control of nonlinear plants show satisfactory performance.

1. Introduction

Artificial neural networks have been suggested for identification and control of nonlinear plants. They have certain characteristics that make them particularly good candidates for this task. Among these characteristics are the inherent parallel and distributed nature that makes them fast when implemented in hardware, learning from examples, ability to approximate any mapping function, and fault tolerance. A neural network can be used as a feed forward controller as depicted in Figure 1.

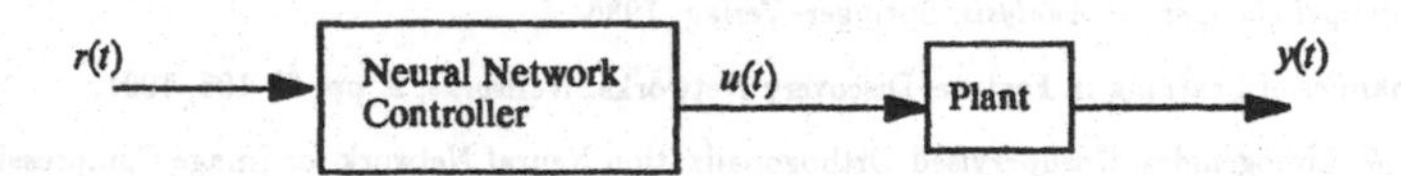

Figure 1. A neural network as a feedforward controller.

The function of the controller in this structure is to provide the $u^*(t)$ that can generate the desired $y^*(t)$. Naranrdra and Parthasurathy in their award winning paper [1] state : " At present methods for directly adjusting the control parameters based on the output error (between the plant and the reference model outputs) are not available. This is because the unknown nonlinear plant lies between the controller and the output error. Hence, until such methods are developed, adaptive control of nonlinear plants has to be carried out using indirect methods ". In this paper a *direct* controller for nonlinear systems using a neural network is proposed.

2. Neural Networks for Control

Using backpropagation for training a neural network controller as in Figure 1 to generate the required $u^*(t)$ for the plant, requires knowledge of the plant's Jacobian matrix, the partial derivative of the plant's output with respect to its input. If Jacobian matrix of the plant is known, we can back propagate the error $| y(t) - y^*(t) |$ through the plant and adjust the weights of the neural network controller accordingly. If the Jacobian matrix is not known, we can use another neural

network as shown in Figure 2 to identify or approximate the plant.

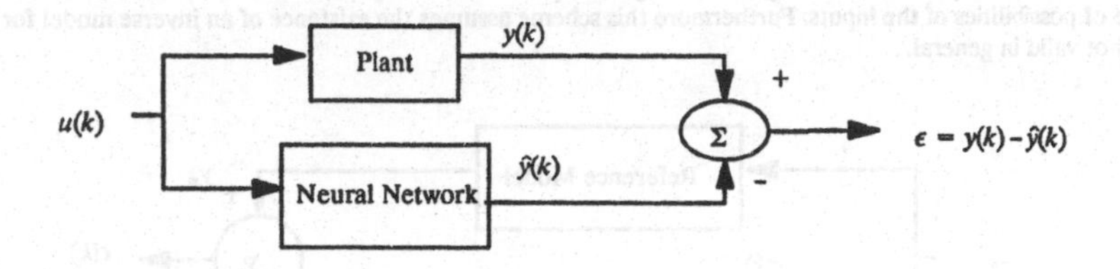

Figure 2. Identification of a plant using a neural network.

After learning to provide the same input–output mapping as the plant, the neural network indicated by NN2 in Figure 3, can be used to replace the plant.

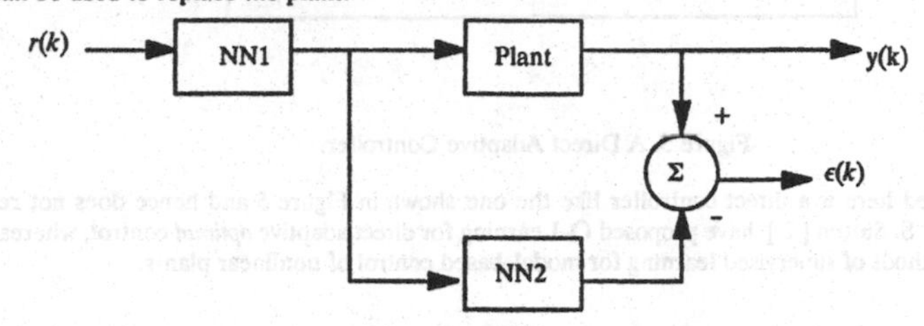

Figure 3. Indirect control of a nonlinear system.

The system composed of NN1 and NN2 can be regarded as a single neural network for which the desired output $y^*(t)$ is known. We can back propagate the error through the weights of NN2 and NN1 and adjust the weights of NN1 accordingly. In this structure the weights of NN2 remain unchanged.

This scheme requires training of NN2 as a system identifier. For an identifier to capture the input–output characteristics of a plant for its entire range of inputs, different possible values of the inputs should be introduced to the system. Exact identification of a system is not an easy task especially when one considers nonlinearities and dynamics of the system. Furthermore such scheme can not be used in real time because in that case we do not have any control on the form of inputs to the plant.

Another method for indirect control of nonlinear plants makes use of a neural network trained on the inverse model of the plant as shown in Figure 4.

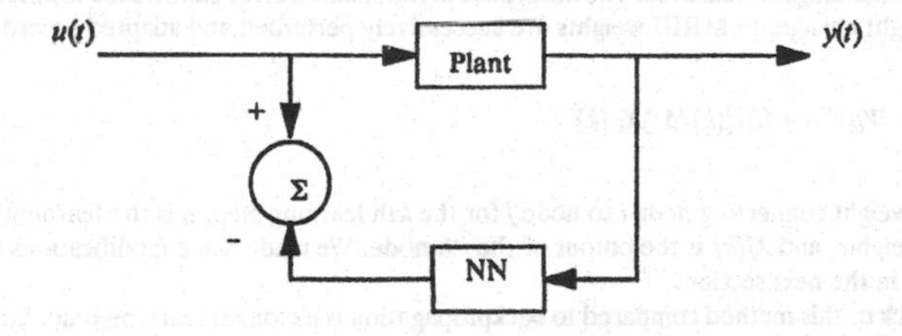

Figure 4. Training a neural network on the inverse model of a plant.

In this method the input to the neural network NN is the output of the plant. The output of the network is compared with the input of the plant $u(t)$, and weights of the network are adjusted until it can provide the desired mapping between $y(t)$ and $u(t)$. A network trained according to this scheme can be used in a system as shown in Figure 1. Since this network has acquired the inverse model of the plant, giving the desired output as the input to the system ($r(t) = y^*(t)$) will generate

the $u(t)$ necessary for the plant to produce $y^*(t)$. Identification of the inverse model of a plant is not necessarily easier than identification of its normal form. Again the range of inputs need to vary to enable the neural network to capture the range of possibilities of the inputs. Furthermore this scheme assumes the existence of an inverse model for a plant which is not valid in general.

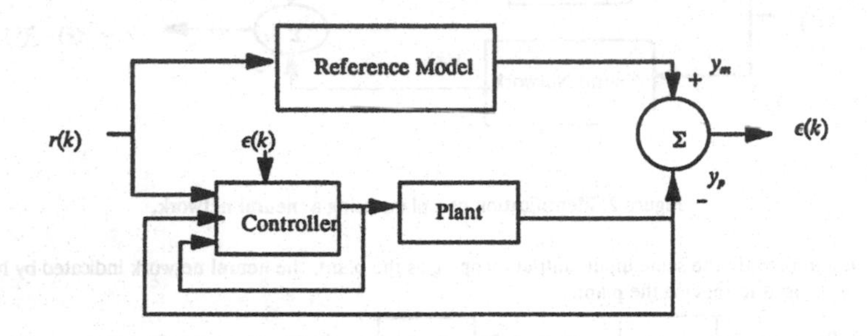

Figure 5. A Direct Adaptive Controller.

The controller proposed here is a direct controller like the one shown in Figure 5 and hence does not require the identification phase. R. S. Sutton [2] have proposed Q-Learning for direct adaptive *optimal* control, whereas here we worked on specific methods of supervised learning for model-based control of nonlinear plants.

3. Method of Training

The learning method used here is a supervised steepest descent training method that also uses some ideas of reinforcement learning. It is different from backpropagation training in that it does not require backpropagation of the error. A similar learning method was used in Madaline Rule III (MRIII) [3] which was originally proposed for analog implementation of neural networks. Since this method does not need prior knowledge about the transfer characteristics of the computing devices, it is relatively immune to the effects of neuron to neuron variations. The difference between backpropagation and this method lies primarily in the method used to determine gradient estimate. Backpropagation uses a priori knowledge about the characteristics of the network's computing elements to estimate the gradient, but here we determine gradients by explicit measurements.
Gradient estimation in this method involves perturbing the values of network weights and measuring the change in the network's instantaneous squared output error. Like backpropagation the appropriate weight adjustments are determined from these gradient estimates. A training pattern is presented as input of the network and squared error $\epsilon^2 = | D - Y|^2$ is measured, where Y is the neuron output and D is the desired output. Then a small perturbation is added to a weight and the squared response is again measured. The difference in the squared error can be used to determine the appropriate direction of the weight change. In MRIII weights are successively perturbed and adapted according to:

$$W_{ij}[k+1] = W_{ij}[k] - \eta\,(\Delta\epsilon_{ij}^2[k]/\delta\,)X_i\,[k] \qquad (1)$$

where $Wij[k]$ is the weight connecting node i to node j for the kth learning step, η is the learning rate, δ is the amount of perturbation in weights, and $Xi[k]$ is the output of the ith node. We made some modifications to the MRIII method which are explained in the next section.
The primary drawback of this method compared to backpropagation is its longer learning time. Unlike backpropagation in which one presentation of input and the associated error measurement is used for adjustment of all weights, this method requires introduction of input and measurement of the error for adjustment of each weight.

4. The Proposed Control Scheme

The structure of the proposed system for direct adaptive control is illustrated in Figure 6 in which Δ refers to one unit time delay. The output of the plant $y(t)$, for the current reference input $r(t)$ is compared with the desired output $y^*(t)$, taken from the reference model. The weights of the controller are temporarily changed and the difference in the error is used for estimation of gradient that is used in the actual weight updates.

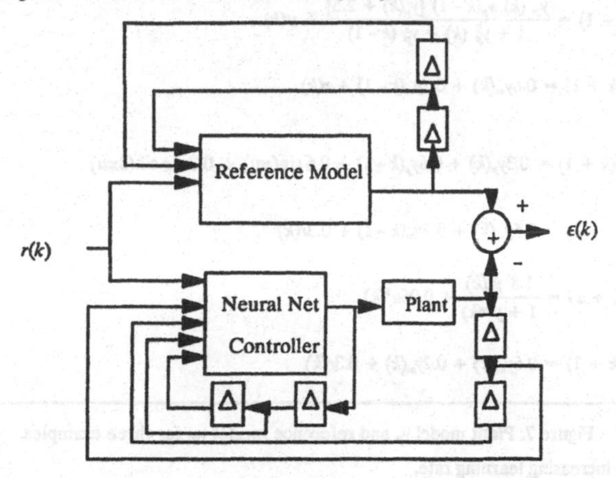

Figure 6. Structure of the proposed system.

In the original MRIII training, errors for all samples in the training set are accumulated and one adjustment is applied for all of the samples. Since adaptive control problems need immediate response, we used per sample training and did the adjustments after introduction of each sample. Gradient descent in the one-dimensional space of one weight is only determined by the sign of the derivative, so it is not necessary to use the $X_i[k]$ term in the weight updates. Another modification involves perturbation of the weights with small random values when change of weights determined by this rule does not actually reduce the error. Since MRIII is basically a steepest descent search, it is always possible that network may get stuck in a local minima. Perturbation of weights with small random values serves to get the network out of this situation. Finally for purpose of better stability we made use of a momentum factor to take into account the previous change of weights in any weight update.

Weight updates were carried out according to :

$$\Delta W_{ij}(k+1) = \eta \Delta \epsilon^2 + \alpha \Delta W_{ij}(k) \qquad (2)$$

Where ΔWij is the amount of adjustment in weight Wij and α is the momentum factor.

5. Simulation Results

We worked on three examples involving nonlinear plants. The plant model y_p and reference model y_m for these examples are shown in Figure 7. In these equations r denotes the reference and u is the input to the plant. Figures 8 to 10 show the response of the reference model (dotted line), plant without controller (solid line) and plant with controller (dashed line but mostly coincident with reference model) for sinusoidal and step inputs.

6. Discussion

In this paper a neural network based direct adaptive controller for nonlinear plants is proposed. Using this method of control does not require assumption of a model for the plant and it makes it different from classical direct adaptive control methodologies. Furthermore, since training of this network does not require backpropagation of error, it makes *direct* adaptive control possible, a scheme beyond the capabilities of backpropagation-based neural networks

The structure we used for the controller was an MLP neural network with 6 input units, 21 hidden units and one output unit. Learning rate and momentum were chosen 0.1 and 0.8 respectively. The best value for these parameters in the case of no other information, is normally obtained through trial and error.

Like any other feedback controller of a nonlinear system, the system proposed is prone to instability. Providing an assured stable controller for any arbitrary nonlinear system is a difficult task and is not attempted here. Using the momentum factor as described in Section 4 was one precaution for damping high frequency oscillations. It effectively prohibits the output of the network having large fluctuations. Another precaution is to start the system with a small learning rate and increase it gradually while monitoring the output of the system. As soon as the output starts to show unstable

$$\begin{cases} y_p(k+1) = \dfrac{y_p(k)\, y_p(k-1)\, [y_p(k) + 2.5]}{1 + y_p^2(k) + y_p^2(k-1)} + u(k) & (3) \\ y_m(k+1) = 0.6y_m(k) + 0.2y_m(k-1) + r(k) & (4) \end{cases}$$

$$\begin{cases} y_p(k+1) = 0.3y_p(k) + 0.6y_p(k-1) + 0.6\sin(\pi u) + 0.3\sin\sin(3\pi u) & (5) \\ y_m(k+1) = 0.6y_m(k) + 0.2y_m(k-1) + 0.3r(k) & (6) \end{cases}$$

$$\begin{cases} y_p(k+1) = \dfrac{1.8\, y_p(k)}{1 + y_p^2(k)} + 0.06u^3(k) & (7) \\ y_m(k+1) = 0.6y_m(k) + 0.2y_m(k) + 0.3r(k) & (8) \end{cases}$$

Figure 7. Plant model y_p and reference model y_m for three examples.

performance, we stop increasing learning rate.

We can consider a learning stage for the controller during which the weights of the network can be adjusted according to the rules described here and when the weights have settled to their stable value, we can stop learning and use the network as a feedforward controller. In that case since there would be no feedback of the plant for further adjustment of weights, the controller would not introduce instability to the system. We did not use this scheme in our experiments because this structure would not be an adaptive control.

References

[1] K. S. Narendra and K. Parthasarathy," Identification and Control of Dynamical Systems Using Neural Networks " IEEE Transactions on Neural Networks Vol. 1, No. 1, pp 4–27, 1990, the winner of the 1991 IEEE Transactions on Neural Networks Outstanding Paper Award.

[2] R. S. Sutton *et al* " Reinforcement Learning is Direct Adaptive Optimal Control " IEEE Control Systems Magazine Vol 12, No. 2, pp 19–22, 1992.

[3] D. Andes, B. Widrow, M. Lehr and E. Wan,"MRIII : A Robust Algorithm for Training Analog Neural Networks " International Joint Conference on Neural Networks, pp I–533–536, January 15–19, 1990.

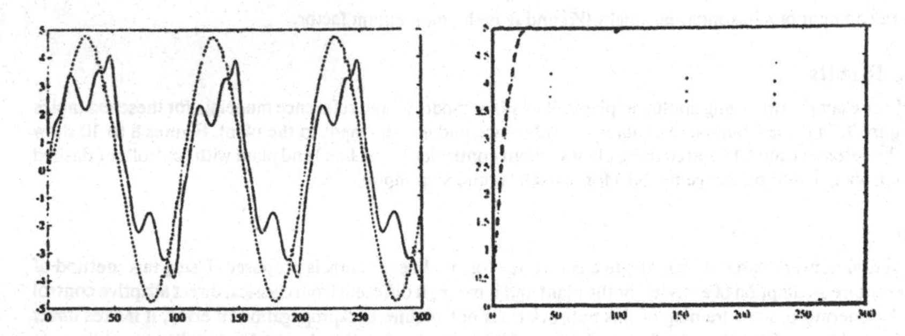

Figure 8. Plant and the reference model sinusoidal and step response in the first example.

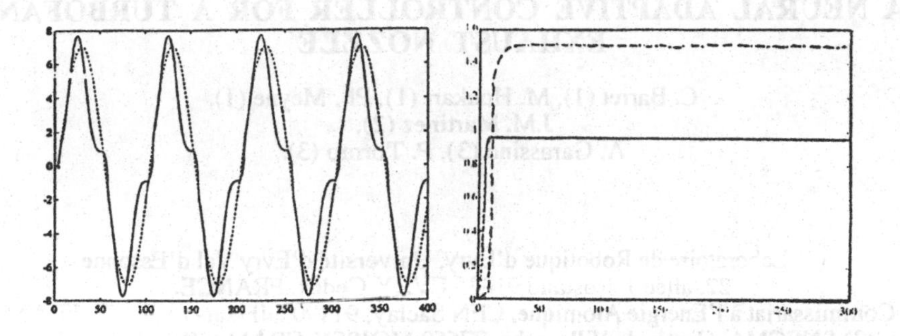

Figure 9. Plant and the reference model sinusoidal and step response in the second example.

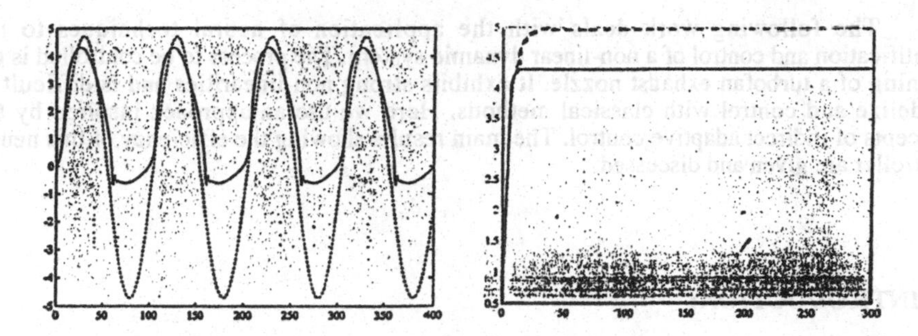

Figure 10. Plant and the reference model sinusoidal and step response in the third example.

A NEURAL ADAPTIVE CONTROLLER FOR A TURBOFAN EXHAUST NOZZLE

C. Barret (1), M. Houkari (1), Ph. Meyne (1),
J.M. Martinez (2),
A. Garassino (3), P. Tormo (3).

(1) Laboratoire de Robotique d'Evry, Université d'Evry Val d'Essonne
22, allée J. Rostand 91025 EVRY Cedex, FRANCE.
(2) Commissariat à l'Energie Atomique, CEN Saclay, 91190 GIF-sur-YVETTE, FRANCE
(3) SNECMA, Centre de Villaroche, 77550 MOISSY-CRAMAYEL, FRANCE.

ABSTRACT:

The following work deals with the application of neural techniques to the identification and control of a non-linear dynamic system. The process to be controlled is the opening of a turbofan exhaust nozzle. It exhibits strong non-linearities and is difficult to modelize and control with classical methods. Here we use an approach inspired by the concepts of indirect adaptive control. The main results showing the efficiency of this neural controller are given and discussed.

1- INTRODUCTION:

The theory of control provides tools perfectly suited for the analysis and synthesis of linear systems. However the treatment of non-linear processes is still confrontedwith theoretical and practical difficulties.

It has recently been shown, by Hornik, Stinchcombe and White (1), using the Stone-Weierstrass theorem, that a multilayered neural network with one hidden layer can approximate any continuous function **f** belonging to $\mathbf{C(R^n xRP)}$ over a compact subset of $\mathbf{R^n}$ with the desired accuracy. On these theoretical grounds, neural networks are candidates for identification and control of non-linear systems. Their use is particularly attractive if a mathematical model of the process is not available, but only empirical laws or experimental data exist, as is frequent in practical cases.

Interesting results have been already obtained in the case of static systems, in particular the problem of the inversion of coordinates in robotics (2-4). Further works have been generalized to dynamic systems for which various types of architectures have been proposed (5-9). A particularly interesting structure of neural controller is based on the principle of indirect adaptive control (10) and uses two distinct neural nets. The first identifies the process and the second controls it (5), (6), (11).

The work presented here follows this line and uses such architecture to control the opening of a turbojet nozzle. This problem is at present inadequately handled by classical methods due to the presence of nonlinearities which cannot be easily modelized and are very sensitive to the working conditions of the nozzle.

In the second part of the paper, we describe the architecture of the controller and the main steps of its conception. In part 3 the results on the identification of the process are presented and discussed. In part 4, the results significant for the control are given. These results which are markedly better than those obtained with the classical controller used until now.

2- STRUCTURE OF THE NEURAL ADAPTIVE CONTROLLER:

The general architecture used for the identification and control of the process is given in Fig. 1. It is inspired by the techniques of indirect adaptive control (5), and "distal learning" (6), but it takes into account the actual constraints of industrial control, particularly in the field of security during initializing and setting phases. It would be clumsy indeed to ignore the presence of a preexisting classical controller which, although imperfect, partially solves nevertheless the problem raised.

The main steps of the conception of the neural adaptive controller are the following.The first one concerns the identification of the process. This is done by adjusting the synaptic weights of the "model" neural network by backpropagation of the identification error:

$$\varepsilon = \Sigma\, (y-y^*)^2$$

where y and y^* are the outputs of the plant to be identified and of the model respectively. During this phase, the values of input u of the plant and the neural network model can be directly chosen inside the safe domain of the process or supplied by the classical controller. In that case, the neural network controller, until now unconnected to the process (a=0), can be initialized by identifying the behavior of the classical controller. The control of the process can then be transferred from the classical to the neural controller (a=1).

At that time the both controllers behave exactly the same, and the second phase aims at improving the performances of the neural controller. This is done as follows. The residual error ε_c between the reference input y_d and the output y of the plant is backpropagated through the neural model whose synaptic weights are fixed. This operation displays an error δU on the input control which appears as the product of ε_c by the transposed neural Jacobian matrix of the the neural model (12). This neural Jacobian matrix is itself an approximate form of the actual Jacobian matrix of the plant around its operating point. The error δU is used as a cost function to be minimized by adjusting the synaptic weights of the "control" neural network. So the neural controller will improve its performance and will tend to give to the plant the correct input values u nullifying the output error ε_c .

If the process to be controlled is not time variable, the neural controller can now work "freely" with fixed synaptic weights, both in the neural model and controller. It is however very easy and inexpensive to fully keep the adaptive properties of the controller. Periodic updates, with T_m and T_c periods, of the synaptic weights of both model and control neural nets will be enough.

3- IDENTIFICATION RESULTS:

The simplified architecture indicated in Fig. 2 is used to identify the plant. We have to choose several factors: the size of the neural net, the parameters of the learning algorithm, the kind of applied signals and the organization of the learning data base.

The most critical point for the structure of the neural net is to choose the number of input neurons and their associated variables. According to the specialists of the process, it can be described by a state function using the last four samples of the output: y_k, y_{k-1}, y_{k-2} and y_{k-3}, and the last three samples of the inputs u_k, $u_{k-1}$and u_{k-2}. This information is retained in the conception of the neural model which will have seven inputs. The inputs of the neural model associated with the y values can be connected, through a tapped delay line, either to the output of the plant, or to the output of the neural net itself (5). In the first case, known as "parallel-series", the multilayered neural net is a standard feedforward one and the classical backpropagation algorithm.can be used In the second case, known as "parallel", the neural network becomes a recurrent one and requires the use of a dynamic backpropagation method. In our case, the y signal is supposed noiseless during the simulations and will be weakly noisy in the actual benchmark. So the first solution can be retained.

Between this input layer, made of neurons with a linear output function, and the output neuron, the structure of the hidden layer(s) is empirically adjusted. After a few trials, we have chosen two intermediate layers with three and four neurons.These neurons and the output neuron have a sigmoidal output function.

The architecture of the net is determined, we must now choose the parameters, (learning rate and momentum) of the backpropagation algorithm. A systematic study of the residual error after a given learning time and in generalization led us to retain a learning rate of 0.7 and a momentum of 0.2. The choice of the signals applied to the plant and the neural model and the organization of the learning base are the most important points of the identification procedure. In order to carry out a further and easier on-line identification of the actual process, we must choose only input signals corresponding to acceptable functioning states of the plant.

If we use a squared input signal of randomly distributed heigth between 0 and Umax, the neural net reproduces the behavior of the system correctly except for input signal values between 0,7Umax and Umax. The result is markedly improved by sequentially presenting to the system signals with randomly distributed amplitude fist between 0,7Umax and Umax, then between 0,4Umax and Umax and finally between 0 and Umax. So we obtain results as shown in Fig. 3, corresponding to a quadratic error of an average 3% during a generalization experiment. This result is enough in order to obtain an efficient control of the process by means of the architecture in Fig. 1. Nevertheless, we can observe in Fig. 4, how useful it is to maintain periodically a re-learning step of the neural model whenever a sequence of input is sent to the plant. In the case of Fig. 4, an updating of the synaptic weights is done every 10 sampling times. This allows a reduction of the mean identification error below 1% (the outputs of the plant and of the neural model are nearly indistinguishible on Fig. 4). Moreover this procedure allows us to avoid problems due to a drifting of the process or memory loss of the neural model.

4- CONTROL RESULTS:

The structure of the control network is based upon the fact that it does represent an inverse model of the plant. Then it is logical that its inputs should be comparable to those of the model network, except that the error $\varepsilon_c = y_d - y$ is substituted to the output value y and an additionnal input is directly linked to the desired output y_d. So the neural network used in the following consists of an input layer of eight linear neurons, two hidden layers with three and four neurons and an output layer with only one neuron. The latter and those in the hidden layers have a sigmoidal output.

As explained in part 2, a first step is the learning by the control network of the behavior of the classical controller. This is done in the same way as the learning of the model network and during or after the identification phase. When the behavior of the control network is satisfactory and the plant is not likely to be provided with dangerous inputs, the control network is connected to the process (a=0 in Fig. 1). Then the second step can begin, namely the optimization of the neural controller. We can recall briefly the procedure. The difference ε_c between the desired output y_d and the output of the plant y is backpropagated through the model network with fixed synaptics weights. This operation provides a value δU of the error on the output of the control network which is backpropagated through it in order to adjust its weights. In fact we have observed experimentally that it is useful to place a multiplicative coefficient A_j on δU before the backpropagation through the control network. The coefficient A_j appears to the user as an adjustable parameter, whose role seams to be similar to the one of the gain of an integrator : if A_j is too low, a static error remains; if A_j is too high, the step response tends to oscillate.

Some examples of results obtained by simulation are presented in Fig. 5 to 7. Fig. 5 shows the input and output of the plant with the neural controller and with the classical one in use at the moment, in the case of a squared periodic desired output y_d. One can observe that the neural controller allows us to obtain a rise time about half as much and no static error with a very low overshoot ($<3\%$) . One can note in Fig. 6 that these features are maintained if the desired output is a sequence of steps of various heights and widths. It is worth noting that the results given in Fig. 6 are obtained with fixed synaptic weights. If updates of the weights are carried out every ten sampling times for example, the results are even better (Fig. 7). If the desired output follows variations as in a real use of the turbojet, the value of the output error ε_c is about three times smaller. This will allow significant improvements in the performance of the turbofan.

5- CONCLUSION:

We have presented here an application of the concept of adaptive neural control to the real case of the control of the opening of a turbofan nozzle which is a non-linear, imperfectly modelized dynamic system. The proposed architecture and methodology allow us to proceed with maximum safety as far as the process is concerned. The results obtained by simulation are very encouraging and clearly better than those obtained by means of a classical control method. Nevertheless, it is now necessary to validate these results on an actual benchmark.

Acknowledgements:

This work is supported by the French Ministry of Research and Space under grant N° 91 P0821.

REFERENCES:

1- K.M. Hornik, M. Stinchcombe, H.White: "Multilayer feedforward networks are universal approximators", Neural Networks, 2, 359-366, 1989.
2- D. Psaltis, A. Sideris, A. Yamamura: "A multilayered neural net controller", IEEE Control System Mag., 8, 17-21, 1988.
3- H. Miyamoto, M. Kawato, T. Setoyama: "Feedback error learning neural network for trajectoty control of a robotic maniplulator" , Neural Networks, 1, 251-265, 1988.
4- M. Kuperstein: "Adaptive visual-motor coordination in multijoint robot using parallel architecture" , Proc. IEEE Conf. on Robotics and Automation, 1592-1602, 1987.
5- K.S. Narendra, K. Parthasarathy: "Identification and control of dynamical systems using neural networks", IEEE Tr. on Neural Networks, 1, 4-27, 1990.
6- M.J. Jordan, D.E. Rumelhart : "Forward model: supervised learning with a distal teacher", MIT Internal Report n° 40, 1991.
7- V.C. Chen , Y.H. Pao: "Learning control with neural networks" Proc. IEEE Conf. on Robotics and Automation, 1448-1453, 1989.
8- C.G. Atkeson, D.J. Reinkensmeyer: "Using associative content-adressable memories to control robots", Proc. IEEE Conf. on Decision and Control, 792-797, 1988.
9- M. Kawato, K. Furukawa, R. Suzuki : "A hierarchical neural network model for control and learning of voluntary movements", Biological Cybernetics, 57, 169-185, 1987.
10- K.S. Narendra, A.M. Annaswamy: "Stable adaptive systems", Englewood Cliffs, NJ, Prentice Hall, 1989.
11- J.M. Martinez, C. Parey, M. Houkari, C. Barret, P. Grizzo : "Backpropagation under the point of view of the theory of control" 4th. Int. Conf. on Neural Networks and Applications , 279-292, 1991 (in french).
12- M. Houkari, C. Parey, J.M. Martinez: "Process control using artificial neural networks" , Int. Conf. on Industrial Automation, Montreal, June 1992.

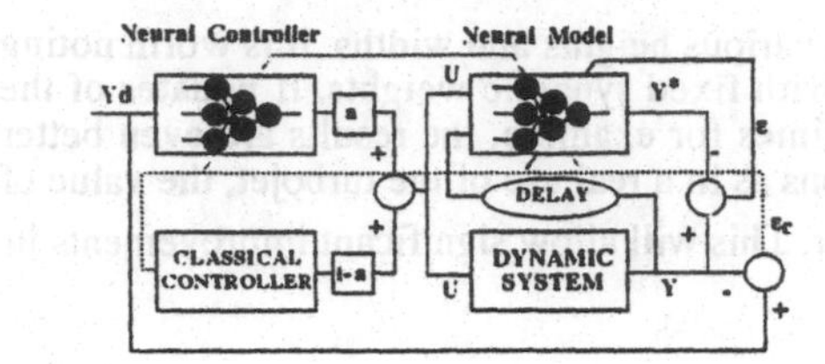

Fig.1 : General Adaptive Learning Architecture.

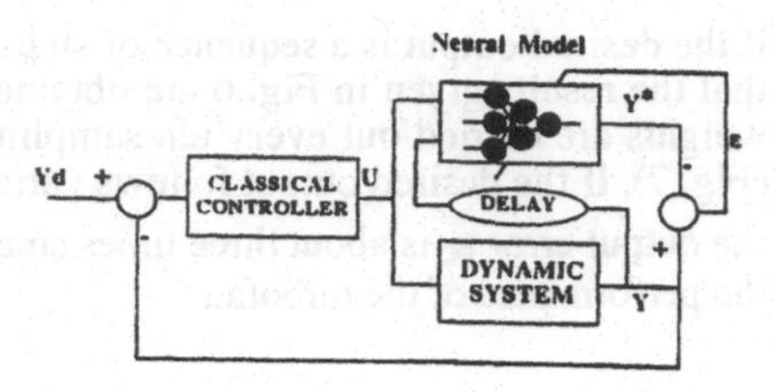

Fig.2 : Scheme for Identification.

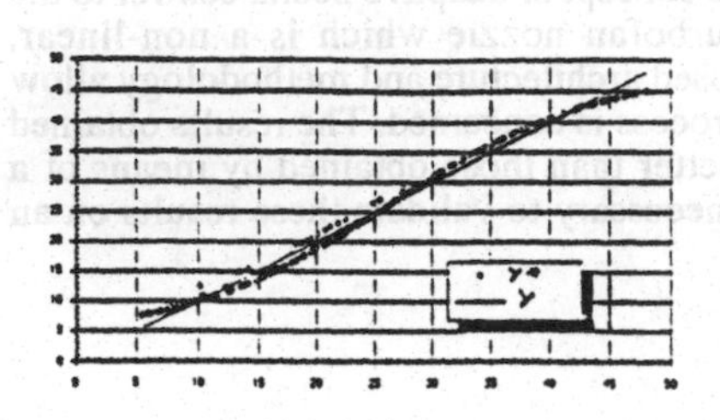

Fig.3 : Generalization with Reinforced Learning.

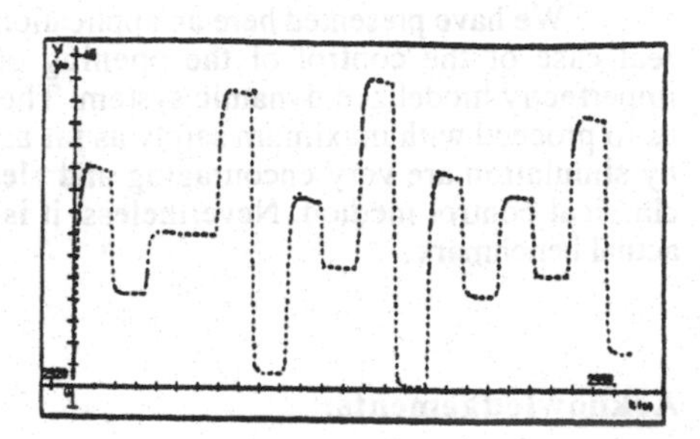

Fig.4 : Y and Y* with Re-learning every ten sampling times.

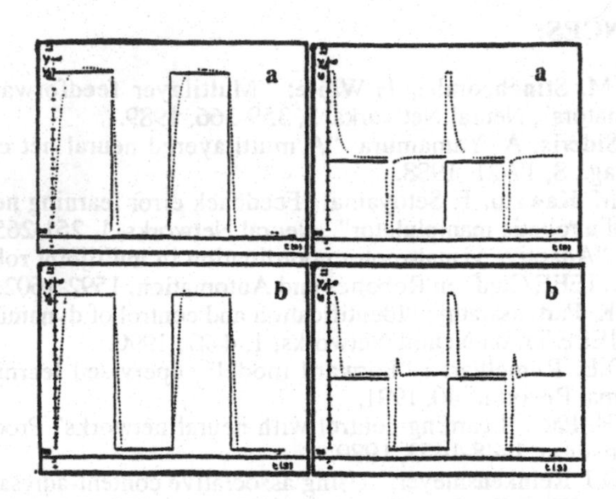

Fig.5 : Yd, Output and Input of the Process with the Classical (a) and Neural(b) Controller.

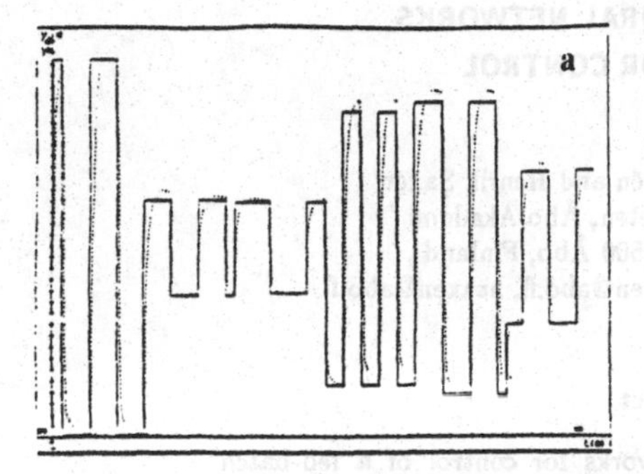

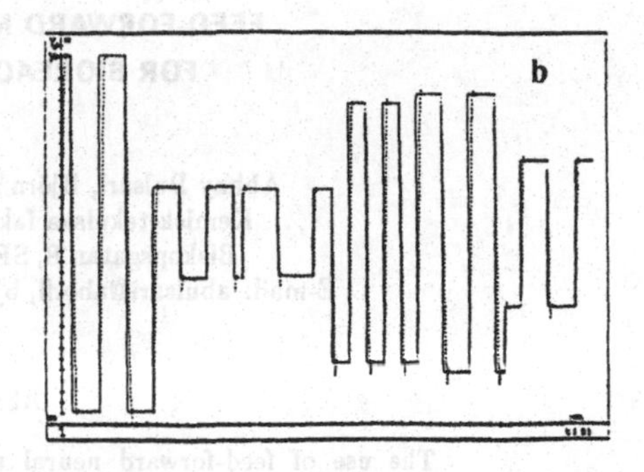

Fig.6 : Yd and Y with Classical (a) and Neural Controller in Generalization (b).

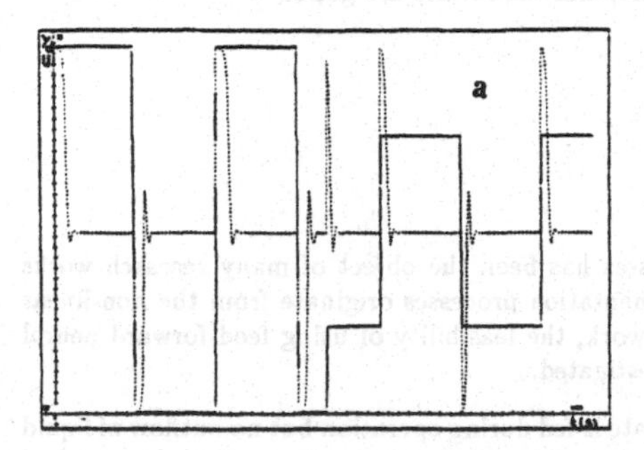

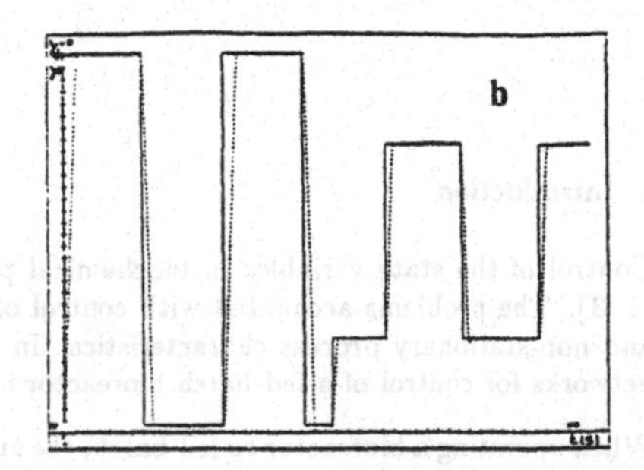

Fig.7 : Input (a) and Output (b) of the Process with Re-learned Neural Controller.

FEED-FORWARD NEURAL NETWORKS FOR BIOREACTOR CONTROL

Abhay Bulsari, Björn Saxén and Henrik Saxén
Kemisk-tekniska fakulteten, Åbo Akademi
Biskopsgatan 8, SF-20500 Åbo, Finland
E-mail: abulsari@abo.fi, bjsaxen@abo.fi, hsaxen@abo.fi

Abstract

The use of feed-forward neural networks for control of a fed-batch bioprocess is investigated. The process is simulated using kinetic expressions for growth of *Saccharomyces.* In addition to the non-linear properties common for most bioprocesses, the process studied in this work has non-steady state characteristics due to the fed-batch operation. A model of the inverse plant is identified by training a neural network, after which the network is used for control. Several network configurations are considered, and the control results are compared. The approach is straightforward and simple, and the results are good.

1 Introduction

Control of the state variables in biochemical processes has been the object of many research works [1–3]. The problems accounted with control of fermentation processes originate from the non-linear and non-stationary process characteristics. In this work, the feasibility of using feed-forward neural networks for control of a fed-batch bioreactor is investigated.

When operating a bioreactor as fed-batch, the substrate is fed during operation but no outflow of liquid occurs. The fed-batch operation enables a longer period of appropriate growth conditions compared to batch operation. On the other hand, certain problems associated with continuous processes can be avoided in a fed-batch process; the risk of contamination is reduced and equipment requirements are lower.

Different strategies are used for the substrate feed. Traditionally, the feed rate is programmed to follow some function in time, *e.g.* a ramp or an exponential function, designed based on experience of the process. An obvious limitation in this approach is the absence of feedback from the process. A more refined method is to adjust the feed rate using feedback control, where the aim is to follow some given trajectories for the state variables. This requires the possibility to measure or estimate state variables in the process. Certain other problems also exist; bioprocesses are normally highly non-linear, and the non-steady state of the fed-batch reactor leads to difficulties in tuning standard controllers, like PID.

In this work, the approach is to let the ethanol concentration in a fed-batch yeast fermentation follow a given trajectory by manipulating the feed rate. A model of the inverse process is identified by training a neural network with process data, and the trained network is used as a controller for the process. Earlier work has demonstrated the feasibility of using a plant inverse neural model for control of a continuous biochemical process [4]. However the results are not as good as the ones obtained in this work. Similar approaches for neural control have been applied on other processes [5,6].

2 The Process

The process studied is a yeast growth process in a fed-batch reactor, where the yeast exhibits diauxic behaviour with respect to glucose and ethanol. Thus, ethanol can both be formed as a result of fermentation, but also consumed as a substrate for the biomass. The equations used for simulation of the process have been proposed by Niklasson, presented in [7].

The balance equations for the reactor are given by

$$\frac{dV}{dt} = Q; \quad \frac{dS}{dt} = -(R_1 + R_2) + \frac{Q(S_0 - S)}{V} \tag{1a,b}$$

$$\frac{dX}{dt} = Y_1R_1 + Y_2R_2 + Y_4R_3 - \frac{XQ}{V}; \quad \frac{dP}{dt} = Y_3R_2 - R_3 - \frac{PQ}{V}; \quad \frac{dE_0}{dt} = -R_4 - \frac{E_0Q}{V} \tag{1c,d,e}$$

$$\frac{dE_1}{dt} = R_4 - R_5 - \frac{E_1Q}{V}; \quad \frac{dE_2}{dt} = R_5 - \frac{E_2Q}{V} \tag{1f,g}$$

where V is liquid the volume, Q, the inflow, S, the substrate (glucose) concentration, X, the biomass (yeast) concentration and P, the product (ethanol) concentration. E_0, E_1 and E_2 are concentrations of enzymes catalyzing the transformation of ethanol to biomass. The enzyme E_0 is initially present in the medium. Y denotes the yield factors for the conversions. The reaction rates are given by

$$R_1 = \frac{S}{S + K_{S1}}K_1\alpha X; \quad R_2 = \frac{S}{S + K_{S2}}K_2(1-\alpha)X \tag{2a,b}$$

$$R_3 = \frac{P}{P + K_{S3}}K_3E_2X; \quad R_4 = \frac{K_4}{S^3 + K_{S4}}E_0X; \quad R_5 = K_5XE_1 \tag{2c,d,e}$$

where R_1 gives the rate for the respirative oxidation on glucose, R_2, the fermentation to ethanol, R_3, the conversion of ethanol to biomass and, R_4 and R_5 give the formation rates for the enzymes E_1 and E_2. K stands for constants in the rate expressions. The enzyme formation is strongly inhibited by the substrate, as can be seen from (2d).

Simulations were carried out using equations (1) and (2). Since the purpose was to create data for identification of the dynamics in different states of the process, the feed rate was randomly changed with a certain time interval. The process behaviour is shown in Figure 1. Only data in the area between the vertical lines in the figure ($25 \leq t \leq 52$) is used for the identification since the feed is not started until $t > 25$ and the process is insensitive to control actions in the area where $S \approx 0$.

3 The Neural Network

Feed-forward neural networks [8] of different sizes were used for the identification of the inverse process model. Both networks with linear activation functions and networks with non-linear functions in one hidden layer were used. Training a network with linear activation functions is equivalent to linear regression analysis, and the linear network controller performs similar to proportional control. For the non-linear networks, the symmetric logarithmoid [9] was used as the activation function, given by

$$x_i = \frac{\beta a_i}{|\beta a_i|} \ln(1 + |\beta a_i|) \tag{3}$$

where a_i is the net input of a node and x_i the output. β is a gain term and is usually set to 1. Similar results were obtained with the well established sigmoidal activation function, but convergence could not be achieved as fast and as easily. The Levenberg-Marquardt method [10,11] was used to train the networks, *i.e.* minimise the sum of squares of errors between calculated and desired output by adjusting the weights in the network.

The aim was to train the network for the relation between the deviation in the controlled variable and the required action in the manipulated variable. Various sets of network input signals were tested. The difference in the ethanol concentration at one sampling interval, $P_{t+\Delta t} - P_t$, was always fed to the network. Other input signals are shown in Figure 2. For all network configurations the dilution rate, $D_t (= Q_t/V_t)$, was the only network output. The reason for using dilution rate instead of feed rate was the influence of the liquid volume on the dynamics. A measurement or estimation of the volume was assumed to be available.

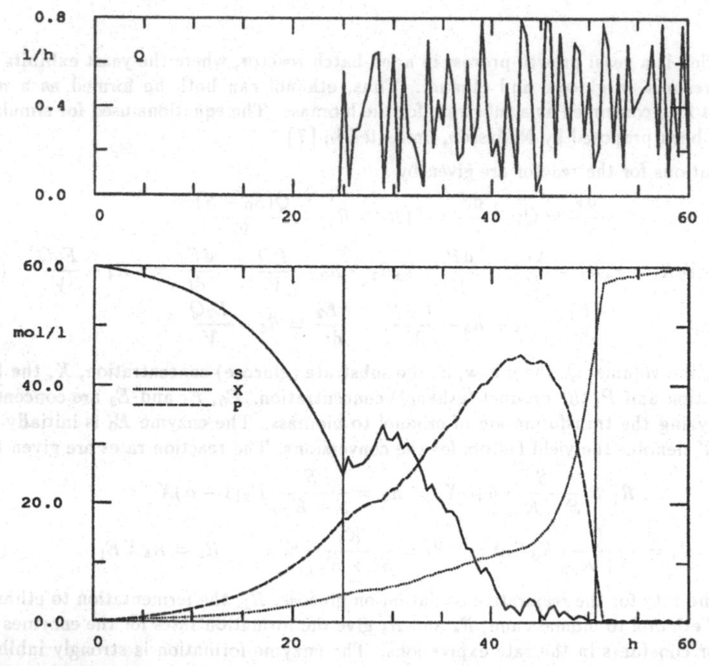

Fig. 1 The state variables (lower graph) and the manipulated variable (upper graph) for a simulated process run. Data in the range between the vertical lines has been used for training the controllers.

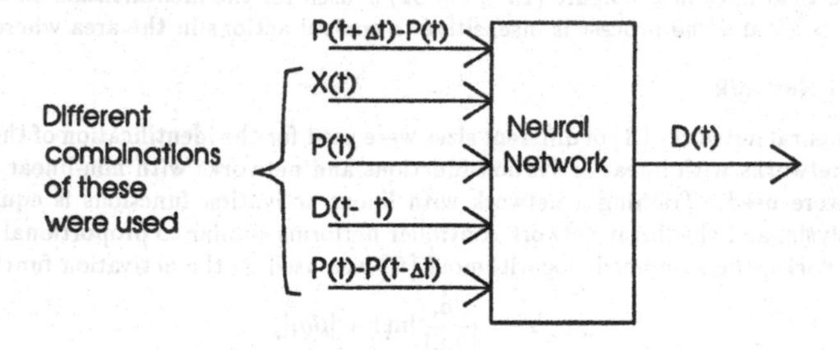

Fig. 2 Considered network input and output signals.

4 Data Used for Training

Data from five simulations, like the one shown in Figure 1, was used for training. The simulations were identical except for the randomly generated sequences in the feed rate. During the simulations, the state variables and the feed rate were recorded with a time interval of 0.5 h. This was also the time between the changes in the feed rate. This resulted in a data set of 255 patterns, enough to train the network sizes considered. Some stochastic behaviour was introduced by adding white noise to the dilution rate before training. The noise was in the range $[-0.1D,\ 0.1D]$. For numerical reasons, the network input signals were scaled to the same order of magnitude. The desired network output, D, was multiplied by a factor of 5.

The reason for using a rather sparse sampling is mainly the measurement delay in a real process, since on-line measurements of the biomass and product concentrations are normally not carried out. A possible approach is to estimate the state variables from other measurement signals, *e.g.* using a model based on elemental balances. Even in this case, the state variables may not be reliably estimated very frequently due to the sensitivity of the method.

5 Training the Networks

First, the networks were trained for a simple case, where only $P_{t+\Delta t} - P_t$ was used as input. A linear network resulted in a rms error of 0.137 for the output signal and introducing non-linearities only reduced the error to a limited extent.

The result was improved by giving the network information about one state variable, X or P. The errors for the linear networks were not reduced very much, but introducing non-linearities had a remarkable effect, and the rms error dropped to about 0.05. For all network configurations tested, X seemed to give more information about the state of the process than P. Feeding both X and P to the net did not improve the result remarkably.

The result was further improved by feeding D_{t-1} and $P_t - P_{t-\Delta t}$ to the net. This is reasonable, since these signals describe the dynamic behaviour at the previous sampling interval.

Table 1 lists the results from the training of different network configurations. In all networks with hidden nodes, the symmetric logarithmoid was used as activation function. Both rms error and maximum error, *i.e.* the biggest error in the trained patterns, are given in the table. The results do not necessarily represent optimal solutions, *i.e.*, the global minima for the objective function. However, the trends are probably accurate.

Table 1 Results from training various network configurations. In the table, both rms error (upper) and the maximum error (lower) for the network output ($5D$) are given.

Network inputs	Linear network	One hidden layer with 3 nodes	6 nodes
$(P_{t+\Delta t} - P_t)$	0.137 0.366	0.102 0.328	
$P_t, (P_{t+\Delta t} - P_t)$	0.134 0.329	0.0629 0.199	0.0576 0.211
$X_t, (P_{t+\Delta t} - P_t)$	0.114 0.378	0.0537 0.172	0.0488 0.196
$P_t, D_{t-\Delta t}, (P_t - P_{t-\Delta t}), (P_{t+\Delta t} - P_t)$	0.102 0.378	0.0500 0.243	0.0409 0.194
$X_t, D_{t-\Delta t}, (P_t - P_{t-\Delta t}), (P_{t+\Delta t} - P_t)$	0.105 0.475	0.0464 0.193	0.0333 0.163

6 Control Performance

The trained network controllers were implemented in the routine for simulation of the process. The network input and output signals were scaled and the network output was limited to positive values only. During control, the difference between the desired product concentration at next sampling time and the actual concentration, $P_{t+\Delta t,des} - P_t$, was used as input instead of $P_{t+\Delta t} - P_t$, which was used during training. A ramp-shaped trajectory for the desired product concentration set point was chosen for evaluation of the control performances.

The networks where only $P_{t+\Delta t,des} - P_t$ was used as input did not show good results. One reason is probably the abnormal process behaviour occurring at the end of the training data, which cannot be described by $P_{t+\Delta t} - P_t$ alone.

Better results were obtained with networks where also P_t was used as input. The linear network did not perform well, but the nonlinear ones showed better capabilities. The networks with X as input instead of P showed better control performance, and achieved reasonably well also in the linear case.

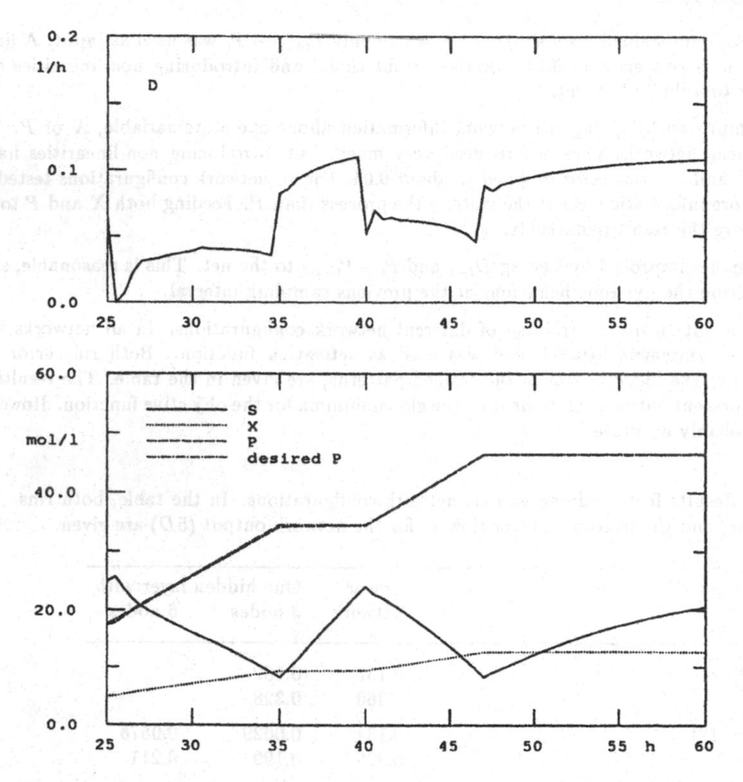

Fig. 3 Control using a network with six hidden nodes and $P_{t+\Delta t,des} - P_t$, X_t, $D_{t-\Delta t}$ and $P_t - P_{t-\Delta t}$ as input signals.

As expected, the networks fed with D_{t-1} and $P_t - P_{t-\Delta t}$ in addition to $P_{t+\Delta t,des} - P_t$ and one state variable showed the best control performance. Even in these cases, the networks with X as input performed better. The controlled state variable follows the desired ramp function without significant errors, as can be seen in Figure 3. The errors in the controlled variable were of the same size for both alternatives of input state variables, but oscillations occurred in the control signal obtained from the networks fed with P.

7 Conclusion and Discussion

Feed-forward neural networks have been shown to be capable of controlling a state variable in a fed-batch bioprocess. The non-linearities in the process require non-linear properties in the controller, realised by hidden nodes with non-linear activation functions in the networks. Non-stationary characteristics of the process is handled by feeding information of the state variables, and not only the control error, to the network. These are the major advantages of the neural controller compared to well established control algorithms, *e.g.* PID control.

Compared to control based on a mathematical model of the process, the network has certain advantages and drawbacks. In a neural controller, the structure of the model is formed from training data, and little a priori knowledge is required. This can save time-consuming modelling of the process. On the other hand, enough representative data to learn the process behaviour is required, and available knowledge of the process behaviour cannot be included into the network directly.

The neural network controllers performed well for the process studied. The results are decidedly better than those obtained in earlier work on neural control of a bioprocess [4], where some instability in the control was noticed. The use of a plant inverse model as controller is a very straightforward approach, with certain limitations. One major shortcoming is the choice of objective function for the network training. The target for process control is normally to achieve minimal error in the controlled variable. However, the network is trained to produce a control signal as near the required as possible, which may not lead to a minimum in the control error. The use of a plant inverse controller is also limited to processes with small time delays, whereas a controller based on a non-inverse model of the process does not have this limitation. Other architectures for neural network controllers have been developed to overcome these problems [12]. Another weakness in the approach for neural control used in this work is the lack of adaptation to changes in the process, caused by *e.g.* disturbances during operation. The desired change in the controlled variable is fed to the network, but previous control errors are not. A combination of a plant inverse neural controller and a standard feedback controller to compensate for stationary control errors [6] seems a reasonable approach.

References

[1] Fish, N.M., R.I. Fox and N.F. Thornhill (eds), *Computer Applications in Fermentation Technology: Modelling and Control of Biotechnological Processes*, Elsevier 1989.

[2] Dochain, D. *On-Line Parameter Estimation, Adaptive State Estimation and Adaptive Control of Fermetation Process*, Doctoral diss., Université Catholique de Louvain, 1986.

[3] Axelsson, J. P., *Modelling and control of fermentation process*, Doctoral diss., Lunds tekniska högskola, Lund, Sweden 1989.

[4] Bulsari, A. B., "Biokemiallisen prosessin säätö neuraaliverkoilla", Report 91-9, Institutionen för värmeteknik, Åbo Akademi, 1991.

[5] Werbos, P. J., "Backpropagation and Neurocontrol: A Review and Prospectus", IJCNN Washington 1989, 209–216.

[6] Psaltis, D., A. Sideris, and A. A. Yamamura, "A Multilayered Neural Network Controller", IEEE Control Systems Magazine, April 1988, 17–21.

[7] Dunn, I. J., J. Ingham, E. Heinzle and J. E. Prenosil, *Biological Reaction Engineering, Part III*, Lecture notes, Braunwald, Switzerland 1991.

[8] Lippmann, R. P., "An Introduction to Computing with Neural Nets", *IEEE ASSP Magazine*, April 1987, 4–22.

[9] Bulsari, A., B. Saxén and H. Saxén, "Application of the Symmetric Logarithmoid as an Activation Function for Neurons in a Feed-Forward Neural Network", *Scandinavian Conference on Artificial Intelligence*, Roskilde, Denmark, May 1991, 62–70.

[10] Levenberg, K. "A Method for the Solution of Certain Nonlinear Problems in Least Squares", *Quart. Appl. Math.*, 2 (1944), 164–168.

[11] Marquardt, D. W., "An Algorithm for Least-Squares Estimation of Non-Linear Parameters", *J. Soc. Indust. Appl. Math.*, 11 (1963), 431–441.

[12] Sørensen, O., "Optimal Control with Neural Networks", IASTED International Symposium on Applied Informatics, Zürich, July 1991, 90–95.

LEARNING NETWORKS FOR PROCESS IDENTIFICATION AND ASSOCIATIVE ACTION

Lisa Borland and Hermann Haken
Institute for Theoretical Physics and Synergetics
University of Stuttgart
Pfaffenwaldring 57/4, W-7000 Stuttgart 80

November 26, 1992

Abstract

Using short-time correlation function measurements of an observed process as input, we show that it is possible to train a network to learn the non-linear stochastic dynamics underlying the process. Alternatively this can be formulated as a neural network with non-linear stochastic synapses, which can, after training, be used to associate actions.

1 Basic Ideas

In many processes in physics, biology and other fields, we observe time series of one or several variables, and we are interested in determining the underlying dynamics, or forces, driving the process. In this paper we describe a method [1-4] which enables us to "learn" these underlying deterministic and stochastic forces, starting solely from observations. We will assume that the processes in question are Markovian and stationary. Due

to fluctuations, the observations that we begin with are certain short-time correlation function measurements of the type

$$< \mathbf{q}_{i+\tau}\mathbf{q}_i > \tag{1}$$

where $\mathbf{q}_i$ represents the values of the observed variables at time $t = i$ while their values at a slightly later time $t = i + \tau$ are denoted by $\mathbf{q}_{i+\tau}$. The correlation functions can either be measured directly or calculated from time series of the process. These are offered to our system, which for now can be considered as a black box which learns the non-linear stochastic transfer function $\mathbf{h}(\mathbf{q}_i)$ that brings the state from $\mathbf{q}_i$ to $\mathbf{q}_{i+\tau}$ in that

$$\mathbf{q}_{i+\tau} = \mathbf{h}(\mathbf{q}_i). \tag{2}$$

How the system works will be discussed in the following section. We can show [2-4] that $\mathbf{h}(\mathbf{q}_i)$ consists of a deterministic part and a stochastic part according to

$$h_l = q_{l,i} + K_l(\mathbf{q}_i) + \sum_{n=1}^{N} g_{nl}(\mathbf{q}_i)w_n \tag{3}$$

where h_l, $l = 1 \cdots N$ - with N corresponding to the number of variables whose time series we observe - are the components of $\mathbf{h}$ and $q_{l,i}$,$l = 1 \cdots N$ are the components of $\mathbf{q}_i$. The w_m are random variables satisfiying

$$< w_m >= 0 \qquad \text{and} \qquad < w_m(t)w_n(t') >= \delta_{tt'}\delta_{mn} \tag{4}$$

Our system is capable of simultaneously learning both of the non-linear functions $K_l(\mathbf{q}_i)$ and $g_{nl}(\mathbf{q}_i)$. This is the step that we call process identification, because, having determined $\mathbf{h}(\mathbf{q}_i)$, we have determined the dynamics driving the observed process, in the sense that we can now construct the discrete Ito-Langevin equation

$$\frac{\mathbf{q}_{i+\tau} - \mathbf{q}_i}{\tau} = \frac{1}{\tau}(\mathbf{h} - \mathbf{q}_i) \tag{5}$$

which becomes, for the components $q_{l,i}$ of $\mathbf{q}_i$:

$$\frac{q_{l,i+\tau} - q_{l,i}}{\tau} = \frac{1}{\tau}(K_l(\mathbf{q}_i) + \sum_{n=1}^{N} g_{nl}w_n) \tag{6}$$

Here, the $\frac{K_l}{\tau}$ correspond to the deterministic force, while the $\frac{g_{nl}}{\tau}$ are connected with the stochastic forces underlying the process. In the following section, we shall also describe how the system can be used to perform associative actions once the transfer function $\mathbf{h}(\mathbf{q}_i)$ has been learnt.

2 Network Implementation

Consider the state vector $\mathbf{q}_i$ with components $q_{k,i}$, $k = 1 \cdots N$, where N is the number of variables of the observed time series. We can treat the components $q_{k,i}$ as a set of unconnected neurons. Similarly, $\mathbf{q}_{i+\tau}$ is an N-dimensional vector with components $q_{k,i+\tau}$, each of which can also be identified as a neuron. We now let the neurons $\mathbf{q}_i$ be connected to the neurons $\mathbf{q}_{i+\tau}$ via a set of nonlinear stochastic synapses described by $\mathbf{h}(\mathbf{q}_i)$. The idea now is to use certain correlation function measurements between $\mathbf{q}_i$ and $\mathbf{q}_{i+\tau}$ as in eq(1) to train the network to learn the synapses, or transfer functions, $\mathbf{h}(\mathbf{q}_i)$. To this end we make the following ansatz for the $K_l(\mathbf{q}_i)$:

$$K_l(\mathbf{q}_i) = \sum_{\kappa=0}^{K} a_{l,\kappa} q_{l,i}^{\kappa} \tag{7}$$

and for the $g_{ln}(\mathbf{q}_i)$:

$$g_{ln}(\mathbf{q}_i) = \sum_{\mu=0}^{M} b_{l,\kappa} q_{l,i}^{\mu} \tag{8}$$

where we choose K and M depending on the order up to which we wish to characterize the nonlinearities of the process. (Reasonable choices for processes with additive or multiplicative noise are $K = 3$ and $M = 1$, if we assume that it is sufficient to characterize the deterministic force up to the third order). All in all, the network must therefore learn $L = K + M + 2$ parameters.

The learning procedure now consists in determining the parameters $a_{l,\kappa}$ and $b_{l,\kappa}$. We do this as presented in our earlier papers [2-4], the method of which we will merely sketch here. A set of appropriate short-time correlation function measurements $< \Omega_v(\mathbf{q}_{i+\tau}, \mathbf{q}_i) >$, $v = 1, \cdots L$ are chosen as training

set, or input, to our system. Note that these can be seen as averages over a joint probability distribution in the sense that

$$< \Omega_v(q_{i+\tau}, q_i) >= \int \Omega_v(q_{i+\tau} q_i) P(q_{i+\tau}, q_i) dq_{i+\tau} dq_i. \quad (9)$$

The first step of our procedure is to make an unbiased guess on $P(q_{i+\tau}, q_i)$ which we do by applying the maximum information principle of Jaynes [5]. In this formalism, the input measurements form a set of constraints under which we vary the probability distribution $P(q_i, q_{i+\tau})$, using the method of Lagrange multipliers, so that the information

$$i = -\int P(q_{i+\tau}, q_i) \ln P(q_{i+\tau}, q_i) q_i) dq_{i+\tau} dq_i \quad (10)$$

is maximized. This provides us with a formal expression for $P(q_{i+\tau}, q_i)$, containing a set of Lagrange multipliers λ_v, $v = 1, \cdots L$. The main point here is, as we have shown elsewhere [2-4], that these are directly connected to the parameters $a_{l,\kappa}$ and $b_{l,\kappa}$ occurring in the transfer functions of eqs(5) and (6). The problem therefore boils down to determining the Lagrange multipliers. To this end, we have developed a method based upon minimizing the information gain as defined by

$$k = \int\int \tilde{P}(q_{i+\tau}, q_i) \ln \frac{\tilde{P}(q_{i+\tau}, q_i)}{P(q_{i+\tau}, q_i)} dq_{i+\tau} dq_i \quad (11)$$

Here, $\tilde{P}$ represents the actual probability function according to which the input measurements $< \Omega_v(q_{i+\tau}, q_i) >$ are distributed, which we do not know but are trying to guess in the form of P. k can be interpreted as the distance in probability space between $\tilde{P}$ and P. Obviously, our best guess of $\tilde{P}$ is that P which gives $k = 0$, providing us with a tool to find the Lagrange multipliers; we must vary k with respect to the λ_v's until k is minimal. This can be done by means of a gradient strategy technique which results in a kind of learning dynamics for the λ_v's, namely

$$\dot{\lambda}_v = < \Omega_v(q_{i+\tau}, q_i) >_P - < \Omega_v(q_{i+\tau}, q_i) >_{\tilde{P}} \quad (12)$$

where the term with subscript $\tilde{P}$ represents the input correlation function measurements while those with subscript P are the corresponding correlation functions calculated using the probability distribution $P(q_{i+\tau}, q_i)$

which is a function of the Lagrange multipliers λ_v. It is necessary that our learning system calculates these values for each time-step of the learning dynamics. The dynamics run until the internally calculated correlations reproduce the input ones.

Once the λ_v's are learnt, we immediately obtain values for the parameters $a_{l,\kappa}$ and $b_{l,\kappa}$ in eqs(7) and (8) which allows us in turn to completely formulate the transfer function $\mathbf{h}(\mathbf{q}_i)$ as defined in eq(3). The network therefore succeeds in learning the non-linear stochastic synapses, or transfer functions, which bring the state from $\mathbf{q}_i$ to $\mathbf{q}_{i+\tau}$. By inserting the numerical value of the sampling time τ into the expression eq(6), we are also able to determine the deterministic and stochastic forces underlying the observed process.

Another aspect of our method is that, having learnt the stochastic transfer function $\mathbf{h}(\mathbf{q}_i)$, we can let $\mathbf{h}$ operate on some initial state $\mathbf{q}_t$. In this way, we can make a stochastic prediction of the state at a slightly later time point $t+\tau$ to be $\mathbf{q}_{t+\tau}$. Iterating this procedure by letting $\mathbf{h}$ now operate on $\mathbf{q}_{t+\tau}$, we can make a stochastic prediction of $\mathbf{q}_{t+2\tau}$. Continuing in this fashion, we can extrapolate the temporal evolution of the state vector up to time $t+N_t\tau$. This idea can easily be visualised as a kind of neural net consisting of neurons $\mathbf{q}_{t+\mu\tau}$, $\mu = 0, \cdots N_t$ with the synapses represented by the transfer function $\mathbf{h}(\mathbf{q}_{t+\mu\tau})$. Within this framework we can interprete the stochastic temporal extrapolation as a kind of "associative action" performed by a synergetic network.

3 An Example

We present a test of our method for process identification on a simulated one-dimensional process with multiplicative noise according to

$$dq_i = (\alpha q_i + \beta q_i^3)dt + \tilde{w}\sqrt{Q}q_i\sqrt{dt} \tag{13}$$

with

$$\alpha = 0.5 \qquad \beta = -1 \qquad \text{and} \qquad Q = 1 \tag{14}$$

We quote results which were obtained in a recent publication of ours [3]. As input, our system was offered the following short-time correlation function

measurements $< \Omega_v >_{\hat{p}}$:

$$\begin{aligned} &< q_i^\kappa q_{i+\tau} > \qquad && \kappa = -1, 0, 1 \\ &< q_i^\mu q_{i+\tau}^2 > \qquad && \mu = -2 \end{aligned} \tag{15}$$

The correlation functions above were calculated for the parameter-values shown in eq(14) with a sampling-time of $\tau = 0.01$. The learning dynamics of eq(9) were then started, numerically implemented as described earlier [2]. To simplify the numerical calculations, we assumed apriori that $a_0 = b_0 = 0$. The system soon converged to give the following values of a_κ and b_κ

$$a_1 = 0.49, \qquad a_2 = -0.05, \qquad a_3 = -0.97 \quad \text{and} \quad b_1 = 0.99 \tag{16}$$

and delivered internally calculated values of the joint moments $< \Omega_v >_P$ in very good agreement with the actually measured values $< \Omega_v >_{\hat{p}}$. The values of a_κ and b_κ together with that of τ were inserted into formulas (7) and (8) to give

$$K(q_i) \approx 0.5 q_i + 1.0 q_i^3 \tag{17}$$

and

$$g(q_i) \approx 1.0 q_i \tag{18}$$

which do indeed describe the forces underlying the test-process.

References

[1] H.Haken, Information and Self-Organization, Springer-Verlag, Berlin-Heidelberg-New York 1988

[2] L.Borland, H. Haken, Unbiased Determination of Forces Causing observed Processes, Z. Phys. B - Condensed Matter 81 (1992) 95

[3] L.Borland, H. Haken, Unbiased Estimate of Forces from Measured Correlation Functions, including the Case of Strong Multiplicative Noise, Ann. Physik 1 (1992) 452

[4] L.Borland, H. Haken, Learning the Dynamics of Two Dimensional Stochastic Markov Processes, to be published in Open Systems and Information Dynamics

[5] E.T. Jaynes, Phs. Rev 106 (1957) 4,620

On-line Performance Enhancement of a Behavioral Neural Network Controller

J.R. Pimentel[1], D. Gachet[2], L. Moreno, and M.A. Salichs[3]

Dpto. Ingeniería de Sistemas y Automática
Universidad Politécnica de Madrid (UPM)
e-mail : salichs@disam.upm.es

Abstract

We present a method to enhance the performance of a neural nework controller which learns to coordinate the behaviors of a real-time, autonomous mobile robot. The enhancement is achieved by defining several measures to evaluate the performance of the neural network controller. If the performance is adequate, we use the outputs of the neural network controller, otherwise we use a heuristic controller which is based on a rule-based methodology. Both, the neural network and heuristic controllers have been described elsewhere. The system is designed so that it exhibits the advantages of both controllers and minimizes their drawbacks. The implementation has been tested with OPMOR, a simulation environment for mobile robots and several results are presented. Portions of this work have been performed under the EEC ESPRIT 2483 PANORAMA Project.

1 Introduction

We have presented in [8] and [21] two methods for generating the appropriate simultaneous activation of primitive behaviors of mobile robots to execute more complex behaviors. Whereas the first method [21] uses a neural network which learns to coordinate the behaviors, the second method [8] uses a rule-based, heuristic method to generate the appropriate behaviors to execute a given task.

Although the performance of each method is adequate for relatively simple robot environments, each has a number of difficulties when executing in more complex environments. The disadvantages of the neural network based controller when executing in complex environments are two: generalization capability and poor data association. The disadvantage of the heuristic based controller is that in order to have a robust system a large number of rules is required. A heuristic system with many rules is difficult to manage, test, expand, and the exact nature of the rules depend on the detailed characteristics of the system (e.g., the number of ultrasonic sensors used, maximum velocity of the vehicle, etc).

Thus the idea to combine both methods appears interesting. The neural network could handle situations for which it can generalize appropriately and the heuristic rules could handle the remaining situations. In this way the problem with the generalization capability of neural networks is corrected and the complexity of the set of heuristic rules is kept manageable. This paper presents a method which combines features of a neural network controller and a rule-based heuristic system. We assume that the robot is non-holonomic and is equipped with a belt of 24 ultrasonic sensors which are fired every T secs. The operating environment is unstructured, unknown, and dynamic. In the context of Fig. 2, the method in this paper is classified as reactive, behavior based, with simultaneous activation of behaviors.

2 Behavioral Control

Several researchers have already argued the importance of looking at a mobile robot as a set of primitive behaviors . Primitive behaviors are also important components of reactive control which is a recently emerged paradigm for guiding robots in unstructured and dynamic environments [1-5]. Mobile robots must permanently interact with its environment and this is the essential characteristic of reactive programs. By reactive, we mean that all decisions are based on the currently perceived sensory information. As depicted in Fig. 1, the fundamental idea of behavioral control is to view a robot task (also referred to as an emergent behavior) as the temporal execution of a set of primitive behaviors. We define an emergent behavior as a

[1] On leave from the GMI Engineering & Management Institute, Flint, Michigan, USA
[2] On leave from Army Polytechnic School, Quito, Ecuador
[3] Dpto. Ingeniería, Universidad Carlos III, Madrid, Spain

simple task or a task which is made up of more elementary (primitive) behaviors.

We have developed an architecture for experimenting with a wide range of reactive control methodologies [8]. The architecture is depicted in Fig. 3 and basically consists of the following modules: behavioral analysis, fusion supervisor, behavioral training, and executor.

Let $c_1, c_2, \ldots, c_N$ be the output of each primitive behavior. Then the output of an emergent behavior is:

$$c_0 = \sum_{i=1}^{N} a_i c_i$$

where the a_i coefficients, with $0 \leq a_i \leq 1$ are found by an appropriate combination of measurement information provided by the perception system. The main function of the fusion supervisor is the calculation of the weights a_i so that the performance of the robot for the execution of the tasks is adequate.

2.1 Behavioral Primitives

We define a set of primitive behaviors whose outputs are fused in a linear combination fashion. The set of primitive reflexive behaviors active at any given time depends upon the occurrence of specific events in the environment (e.g., the detection of an obstacle, time for the robot to recharge itself, the issuance of an interactive command, etc). The primitive behaviors being considered are:

c_1 : goal attraction

c_2 : perimeter following (cw, clockwise)

c_3 : perimeter following (ccw, counterclockwise)

c_4 : free space

c_5 : keep away (from objects)

The goal attraction primitive produces an output which directs the robot towards a specific goal. The perimeter following behavior follows the perimeter of a fix (i.e., static) obstacle maintaining a prespecified distance away from it. This behavior take into account the minimun radius of curvature of the robot in order to avoid uneven corners. The radius of curvature is defined as $R = v/w$ where v is the linear velocity and w the angular velocity of the robot. The movement can be performed clockwise (CW) or counterclockwise (CCW) [12].

The free space behavior will cause the movement of the robot in a direction so that it is free from objects. This behavior takes the information from measurements provided by the front sensors and chooses the direction of the longer adjacentes measures in order to ensure a safe course in that direction. The output of this behavior is the direction of motion at a constant speed.

Finally the keep away behavior basically avoids obstacles in the proximity of the robot, chosen the direction that ensures no collision with the nearby obstacles.

3 Performance Enhancement

The primary goal of this section is to briefly describe the neural network and heuristic controllers and the enhancement rules. We assume that the task to be performed is to go from an initial position to a final position (i.e., the goal) while avoiding all objects in the environment.

3.1 The Neural Network Controller

The fusion supervisor basically identifies a condition (e.g., too close to an object, goal in sight) and based on the task to be performed determines the relative contribution of each primitive behavior to carry out the task, a natural application for a neural network (NN).

By taking advantage of some features of OPMOR we have deviced a method for training the NN where a human acts as the trainer. The idea is to allow a human to train the NN by means of an OPMOR control window providing control for each behavior independently as depicted in Fig. 4. When all behavior coefficients are 0 the robot simply stops. The values are changed using a slider panel of X Windows in an on-line fashion when the simulator is running. The slider values are read by the OPMOR simulator and are used to control the robot. It takes some practice on the part of a human trainer to learn how to control the robot just as it takes some effort for humans to learn a new behavior (e.g., driving a car). After some experience (taking several hours) a human trainer is ready to teach the NN.

During the learning phase, the human trainer leads the robot using the control panel in terms of the five behaviors in a set of five rich robot environments containing many objects and hopefully most of the situations that the robot will encounter in actual cases. The control values set by the human trainer and the additional boolean variable are the training patterns for the NN. It is important to point out that during the training phase, the human trainer uses his own perception system to set appropriate values for the control coefficients a_i and the boolean variable *in sight*. An advantage of using a human trainer is the elimination of conventional programming or the definition of analytical functions to specify how learning should be accomplished.

A back-propagation network has been used with an input layer with 13 nodes, two hidden layers with 30 nodes each, and an output layer with 5 nodes. The 13 inputs correspond to 11 ultrasonic range measurements (out of the 24 available on the Robuter), the current heading of the robot, and its current velocity. The 5 outputs are the a_i coefficients of the five primitives robot motions listed above. The equations describing the network are derived and described in detail in [13].

The neural network is trained based on the set of samples provided by the human trainer and is found to converge after about 500 iterations. The neural network internal weights w_{ij} are stored for the execution phase. During the execution phase, the neural network uses the stored w_{ij} coefficients and the 13 inputs to directly generate the a_i coefficients which can be used to guide the robot in any environment.

3.2 The Heuristic Controller

The heuristic controller uses the following rules:

1. If there is no immediate obstacles in the direction of the goal, and the goal is close to the robot then: $a_1 =$ *high*, $a_2 =$ *verylow*, $a_3 =$ *verylow*, $a_4 =$ *verylow*, and $a_5 =$ *low*.

2.(a) If there are obstacles blocking a direct path to the goal and to the right of the robot then: $a_1 =$ *verylow*, $a_2 =$ *verylow*, $a_3 =$ *high*, $a_4 =$ *low*, and $a_5 =$ *verylow*.

2.(b) If there are obstacles blocking a direct path to the goal and to the left of the robot then: $a_1 =$ *verylow*, $a_2 =$ *high*, $a_3 =$ *verylow*, $a_4 =$ *low*, and $a_5 =$ *low*.

3. If the goal is far away from the robot and there is no immediate obstacles in the direction of the goal then: $a_1 =$ *medium*, $a_2 =$ *verylow*, $a_3 =$ *verylow*, $a_4 =$ *medium*, and $a_5 =$ *verylow*.

Where the terms *very low, low, medium, high*, and *very high* range between 0 and 1 and depend on the desired motion control characteristics (e.g., not too close to obstacles).

The first rule gives preponderance to the *goal attraction* behavior by directing the robot toward the goal. The contribution of the *free space* and *keep-away* behaviors are small but necessary because of unexpected situation (e.g., bad measurements of the ultrasonic sensors) and they contribute to a improved performance of the task.

The second rule (a) and (b) generates the coefficients a_i in order to avoid a static obstacle by the use of *contour following clock-wise* and *contour following counter-clock-wise* behaviors respectively, with a small contribution of the *free space* and *keep-away* behaviors for the same reason as that in the first rule, our experience has demonstrated that the inclusion of this small contributions is necessary in order to improve the performance of navigational rule.

The last rule ensures the motion of the robot across the empty space if the goal position is far away form the robot, this rule is important because the navigation across the empty space is better for security reasons. These rules obviously do not contain all the possibilities in the mix of the primitive behaviors, but considering the complexity of the environments used for testing the implementation, they provide a good set for the completion of the task. The previous rules satisfying a collision free motion between the start and goal positions of the robot and they were chosen by extensive experience and works well in all environments considered.

3.3 Enhancement Rules

The performance evaluator evaluates the performance of the neural network based controller periodically every *H_steps* cycles of the control algorithm. We use the following performance measures which are similar to those defined by Clark, Arkin, and Ram [6]:

a_step_size: The average step size.

a_o_count: The average number of measurements less than a threshold.

a_progress: The average of the progress towards the goal. The progress is defined as the slope of the function given by the remaining euclidean distance from the current position of the robot to the goal.

In all the previous measures, the average is performed over the last *Hsteps* cycles.

Based on the previous performance measures, the enhancement rules are performed under the following situations:

1. No-Movement.

if *a_step_size* < *Tmotion* then the robot is not really moving and it is driven by the heuristic controller.

2. Movement-Toward-Goal.

if *a_step_size* > *Tmotion* and *a_progress* > *Tprogress* then the robot is moving towards the goal and driven by the neural controller.

3. No-Progress-With-Obstacles

If $a\_step\_size > Tmotion$ and $a\_o\_count > Tobstacles$ and $a\_progress < Tprogress$ then the robot is moving with obstacles detected but no progress is being made towards the goal. If the *contour_following* behavior is active, the neural controller still drives the robot, otherwise the robot is driven by the heuristic controller.

4. No-Progress-No-Obstacles

If $a\_step\_size > Tmotion$ and $a\_o\_count < Tobstacles$ and $a\_progress < Tprogress$ then the robot is moving without obstacles detected and no progress is made towards the goal. In this case the heuristic controller drives the robot.

For our implementation we have used the following numerical values: $Hsteps = 10, Tmotion = 80, Tobstacles = 8, Tprogress = 0$.

4 Results

Fig. 5(a) depicts the performance of the neural network in a relatively complex environment which is different from the ones used for training, where the robot moves to a goal while exhibiting the following behaviors: contour following behavior, keep away and goal attraction behaviors. Likewise, Fig. 5(b) shows de performance of the heuristic controler in the same environment. Finally, Fig. 5(c) shows the performance of the enhanced controller.

It is noted that the performance of the neural network is adequate for most of the trajectory except that it makes the robot bump into one of the obstacles. This problem is attributed to the generalization ability problem exhibited by the neural network. The performance of the heuristic controller is adequate but the path taken is rather long as the set of heuristic rules used does not attempt to optimize the path taken. As depicted in Fig. 5(c) the enhanced controller performs better than any of its constituent components acting alone.

Tables 1 through 3 show a numerical comparison of the enhanced controller relative to the neural network based controller and the rule-based, heuristic controller in terms of the current cycle, total distance travelled , average step size, the average progress, the average number range measurements below a threshold, and number of times the robot bumped into obstacles. Again, the performance of the enhanced system is better than the other controllers.

5 Discussion

The learning experiments are not uniform and incomplete because they do not have complete coverage of all possible situations in an environment, and it is necessary the use of some form of selective training patterns in order to ensure that the neural network learn certain difficults aspects like turning in an uneven corner or when the robot finds a critical situation (local minima).

The selection of the training set for the neural network is the most critical decision that affects its final performance. The training set must attempt to represent the total range of inputs covering most environments of interest to produce the desired final results.

The advantage of this NN-based scheme as a local navigation controller for the mobile robot is the reduction of programming effort and that there is no need to define special functions to guide learning when compared to other methods.

The inclusion of performance measures has resulted in an enhancement of the previous controllers. It remains to be seen if these performance measures will be enough to have a truly robust motion control system.

5.1 Limitations of our Work

Although we have improved the performance of two previous methods for controlling the activation of primitive behaviors in a mobile robot simultaneously, our method suffer from the following limitations:

1. Incomplete coverage of performance measures. The performance measures defined do not guarantee to cover all situations of interest. To have a more robust system we need to define additional performance functions and hence increase the number of enhancement rules.

2. Poor data association. One of the main difficulties of perception-based mobile robot motion control is data association, the problem of determining the exact nature of the environment which corresponds to a particular measurement. For our case, the main reason for poor data association is the poor resolution fo the ultrasonic sensors (18 degrees).

5.2 Related Work

Maes and Brooks [5] developed an on-line learning algorithm for teaching a legged robot to walk. The application domain used by Clark, Arkin, and Ram [6] is the same as ours but the method is not based on a connectionist approach. Pearce, Arkin, and Ram [7] use a genetic algorithm to develop an unsupervised learning method for navigating in a crowded environment. The work of Tuijnman and Krose [10] is similar to ours in that a backpropagation network is trained to perform a task: that of avoiding a collission with another robot but the control is non-behavioral.

Our method uses the same methodology for training the system as that used by Shepanski and Macy [9] which have developed a supervised system for teaching a vehicle to stay on course or change lanes. The system is non-behavioral and uses the experience of a human trainer to train a neural network. Bourbakis [17] has developed a non behavior-based, heuristic algorithm to guide a robot to a goal but suffers from a number of limitations: the robot is a point, holonomic, there are only two robots in the environment, the perception system is ideal with the respect to the number of sensors and their accuracy.

Mahadevan and Connell [18] have developed a reinforcement learning algorithm for a box pushing task where three pre- specified behaviors are learned assuming a priority ordering in their execution. Another result with an application domain similar to that of Mahadevan and Connell but not ours is that of Berns, Dillmann, and Zachmann [19]. They present a learning algorithm for following another mobile robot. In the context of Fig. 2, this correspond to learning one individual behavior. Millan and Torras [20] also used reinforcement learning to guide a robot to a goal but their approach is non-behavioral. Lozano-Perez [15] has developed the configuration space (C-space) method which can be used for mobile robots [16]. However these methods are non-reactive and has not been pursued further.

6 Conclusions and Future Work

A method has been presented to combine the best features of a neural network controller and that of a heuristic controller. The combined system uses effectively the available generalization capability of the neural network controller and reduces the complexity of an heuristic system with similar performance. Results show that the enhanced system performs better than any of its constituent components working separately. Our system suffer from the following limitations: incomplete coverage of the performance measures and poor data association. Future work will concentrate in removing these limitations.

Acknowledgments

The first two authors wish to thank the Dirección General de Investigación Científica y Técnica (DGICYT) of the Spanish Ministry of Education and Science for research funding. This research has been founded by the Commission of the European Communities (Project ESPRIT 2483 PANORAMA) and the Comisión Interministerial de Ciencia y Tecnología CICYT (Projects ROB90-159 and ROB91-64).

References

[1] T.L. Anderson, and M. Donath, "Animal Behavior as a Paradigm for Developing Robot Autonomy," in Designing Autonomous Agents, P. Maes, Editor, pp. 145-168, MIT Press, 1990.

[2] R.C. Arkin, "Motor Schema-Based Mobile Robot Navigation," The International Journal of Robotics Research, Vol. 8, No. 4, pp. 92-112, Aug. 1989.

[3] R. A. Brooks, "A Robust Layered Control System for a Mobile Robot", IEEE J. Robotics and Automation, RA-2, April, pp 14-24, 1986.

[4] L. P. Kaelbling and S. J. Rochestein, "Action and Planning in Embedded Agents", Robotics and Autonomous Systems 6,pp., 35-48, 1990.

[5] P. Maes and R. A. Brooks, " Learning to Coordinate Behaviors", Autonomous Mobile Robots: Control, Planning and Architecture, pp. 224-230, IEEE Computer Society Press, 1991.

[6] R.J. Clark, R.C. Arkin, and A. Ram, "Learning Momentum: On-line Performance Enhancement for Reactive Systems", Proc. IEEE Int. Conf. on Robotics and Automation,pp. 111-116, Nice, France, 1992.

[7] Pearce M., R.C. Arkin, and A. Ram, "The Learning of Reactive Control Parameters Trough Genetic Algorithms", Proc. IROS 92., pp. 130-137, Raleigh, NC., 1992.

[8] D. Gachet, M.A. Salichs, J.R. Pimentel, L. Moreno, A. de la Escalera, "A Software Architecture for Behavioral Control Strategies of Autonomous Systems", Proc. IECON'92, pp. 1002-1007, San Diego CA, Nov. 1992.

[9] J.F. Shepanski and S.A. Macy, "Teaching Artificial Neural Systems to Drive: Manual Training Techniques for Autonomous Systems," Proc. AIP Conference, pp. 693-700, 1988.

[10] F. Tuijnman and B.J.A. Krose, "Neural Networks for Collision Avoidance Between Autonomous Mobile Robots," Proc. AIP Conference, pp. 407-416, 1990.

[10] F. Tuijnman and B.J.A. Krose, "Neural Networks for Collision Avoidance Between Autonomous Mobile Robots," Proc. AIP Conference, pp. 407-416, 1990.

[11] Robosoft, "Robuter User's Manual" Robosoft, Paris, France 1992.

[12] M.A Salichs, D. Gachet, E.A Puente and L. Moreno, "Trajectory Tracking of a Mobile Robot. An Application to Contour Following", Proc. IECON'91, pp., 1067-1070 Kobe, Japan 1991.

[13] D.E. Rumelhart, G.E. Hinton, and R.J. Williams, "Learning Internal Representation by Error Propagation," in Parallel Distributed Processing: Explorations in the Microstructure of Cognition, Vol 1,D.E. Rumelhart and J.L. McClelland(Eds), chap. 8, (1986), Bradford Books/MIT Press, Cambridge.

[14] J.R. Pimentel, E.A. Puente, D. Gachet, and J.M Pelaez, "OPMOR: Optimization of Motion Control Algorithms for Mobile Robots," Proc. IECON'92, pp. 853-861, San Diego CA., Nov. 1992.

[15] T. Lozano-Perez, "Spatial Planning: A Configuration Space Approach," IEEE Trans. on Computers, Vol. 32, No. 2, pp. 108-120, 1983.

[16] J. Ilari, and C. Torras, "2D Path Planning: A Configuration Space Heuristic Approach," The International Journal of Robotics Research, Vol.9, No 1, pp. 75-91, Feb. 1990.

[17] N.G. Bourbakis, "Real-Time Path Planning of Autonomous Robots in a Two-Dimensional Unknown Dynamic Navigation Environment," Journal of Intelligent and Robotic Systems, Vol. 4, pp. 333-362, 1991.

[18] S. Mahadevan and J. Connell, "Automatic Programming of Behavior- Based Robots Using Reinforcement Learning," Artificial Intelligence, Vol. 55, pp. 311-365, 1992.

[19] K. Berns, R. Dillmann, and U. Zachmann, "Reinforcement-Learning for the Control of an Autonomous Mobile Robot," Proc. IROS'92, Raleigh, N.C., pp. 1808-1814, July 1992.

[20] J. del R. Millan, and C. Torras, "Learning to Avoid Obstacles Through Reinforcement: Noise-tolerance, Generalization, and Dynamic Capabilities," Proc. IROS'92, Raleigh, N.C., pp. 1801-1807, July 1992.

[21] E.A. Puente, D. Gachet, J. R Pimentel, L.E. Moreno, and M.A. Salichs, "A Neural Network Supervisor for Behavioral Primitives of Autonomous Systems," Proc. IECON'92, pp. 1105-1110, San Diego CA., Nov. 1992.

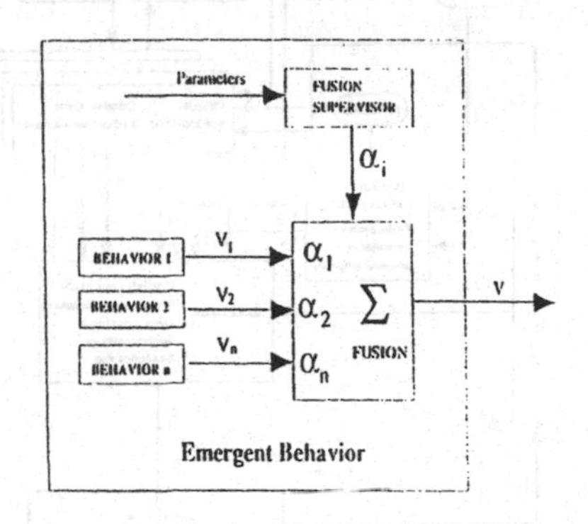

Figure 1: Schema for Behavioral Control

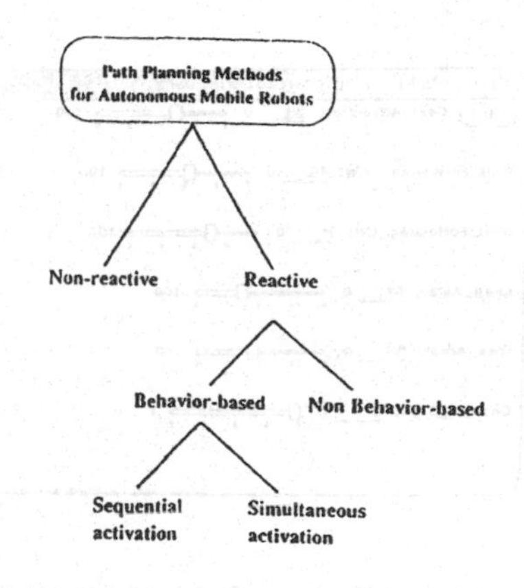

Figure 2: Classification of Path Planning Methods for Autonomous Mobile Robots

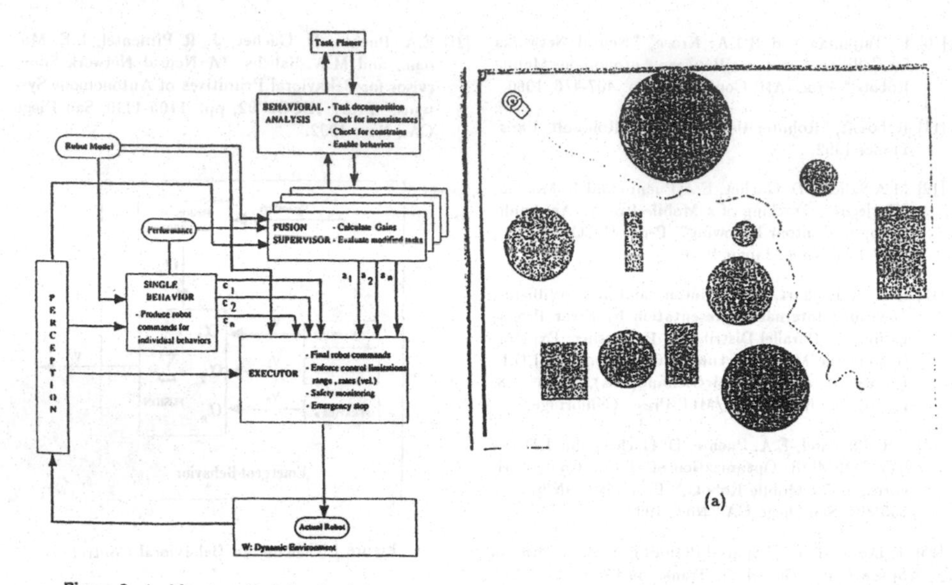

Figure 3: Architecture for Behavioral Control

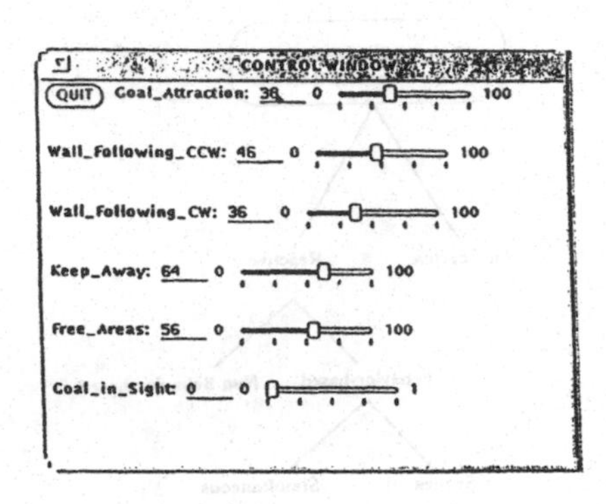

Figure 4: Control Window for Training the Neural Network

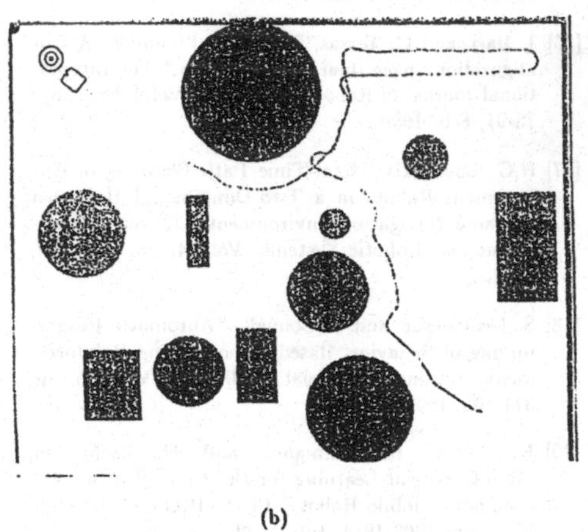

Figure 5: Performance of (a) Neural Network Controller (b) Rule- based Heuristic Controller (c) Enhanced Controller

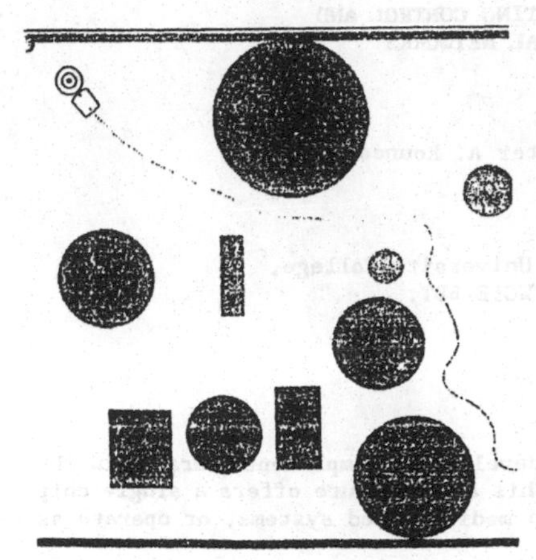

(c)

Table 1. Numerical Results for the Neural Network Controller

step	d_traveled	a_step_size	a_progress	a_o_count	hits
105	9734	101	24	5	0
210	21366	104	96	8	3
225	25757	145	142	3	13

Table 2. Numerical Results for the Heuristic Controller

step	d_traveled	a_step_size	a_progress	a_o_count	hits
105	8316	89	38	4	0
210	18987	89	27	4	0
225	20851	134	12	4	0
300	28296	97	2	6	0
420	41612	93	43	4	0
600	60599	186	146	1	0

Table 3. Numerical Results for the Enhancement Controller

step	d_traveled	a_step_size	a_progress	a_o_count	hits
105	9348	101	26	3	0
210	19193	104	96	6	0
225	20667	88	80	7	0
300	30425	147	141	2	0

AN ARCHITECTURE FOR IMPLEMENTING CONTROL AND SIGNAL PROCESSING NEURAL NETWORKS

Richard P. Palmer and Peter A. Rounce

Department of Computer Science, University College,
Gower Street, London WC1E 6BT.

Abstract

This paper presents an architecture that has been developed to implement neural networks for control and signal processing applications. This architecture offers a single chip solution that can be used standalone in small and medium sized systems, or operate as a preprocessor in larger applications.

The constraints imposed on this architecture are discussed, and the methods used to satisfy these are presented. The design of a fully integrated neural processor using this architecture is proposed, and the application development tools are discussed.

1. Introduction.

This paper first considers the special requirements for an architecture to implement control and signal processing neural networks, and attempts to match these to existing solutions. This study highlights the lack of suitable hardware at present, leading towards the development of the architecture presented in this paper. This work is at the stage where a test chip has recently been fabricated, and full system level simulations have been completed[1,2].

It is now hoped that we can continue with this research using external funding. The developments that we propose involve the fabrication of a fully integrated neural processor using this architecture, and the creation of a suite of application development tools.

2. Control and Signal Processing Networks.

This section considers some recent research into control and signal processing using neural networks. From this research the need, and requirements for specialised hardware, is apparent[3].

One common feature found in both control and signal processing systems is the need to process continuously varying signals; temperature readings for control applications or speech signals in signal processing. To enable the representation of these signals requires special techniques to be adopted. Diagram 1 shows three techniques that have been reported in recent research[4].

- Spatial representation uses the input layer of the neural network to hold a sequence of incoming data items. This sequence is continuously updated, with new data being entered at one end, being shifted through the input layer, and removed from the other end. This action does not necessarily have to be synchronised with the network, allowing data to be entered faster, or slower, than the network iteration time.

- Time-delay networks use a similar technique to spatial representations, except the shifting of input data items is now synchronised with each network iteration. This

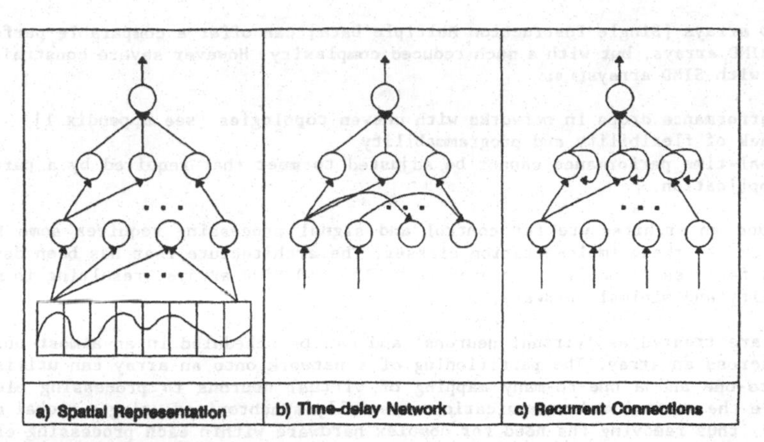

a) Goldberg, K. Y., Pearlmutter, B. A. Using Backpropagation with Temporal Windows to Learn the Dynamics of the CMU Direct-Drive Arm II. in Touretzky, D. S., (ed), Advances in Neural Information Processing Systems 1,Morgan Kaufmamm, San Mateo, CA. pp 356-363.
Gorman, R. P., Sejnowski, T. J. Learned Classification of Sonar Targets using a Massively Parallel Network. IEEE Trans. Acoustics, Speech and Signal Processing. pp. 1135-1140.

b) Lang, K. J., Waibel, A. H., Hilton, G. E. A Time-delay Neural Network Architecture for Isolated Word Recognition. Neural Networks, 3. pp. 23-43.
Smyth, E. J. Temporal Representations in a Connectionist Speech System. in Touretzky, D. S., (ed), Advances in Neural Information Processing Systems 1,Morgan Kaufmamm, San Mateo, CA. pp 240-247.

c) Jordan, M. I. Generic Constraints on Underspecified Target Trajectories Proc. 1st Int. Joint Conf. Neural Networks. Washington, D. C. 1989. pp 335-340.
Brown, G. D. A. Short Term Memory Capacity Limitations in Recurrent Speech Production and Perception Networks. Proc. 2nd Int. Joint. Conf. Neural Networks. Washington, D. C. 1990. pp 43-46.

Diagram 1 Time Representation in Neural Networks

method has been used to recognise temporal periodic data and out of phase patterns.

- Recurrent connections provide contextual representations within a network's hidden layers. This has been used to reproduce stored patterns when stimulated, and in timing cyclic operations.

For control and signal processing in real-time applications, additional constraints are imposed; sample period[1] and response time[2] being the two most important. Synchronisation to external events also needs considering; *START* and *FINISH* signals provide the most basic form of real-time synchronisation.

3. Available Solutions

The solutions currently being used to implement neural networks can be broken down into distinct classes:-

- Analogue implementations use a hardwired approach to provide very fast asynchronous operation[5]. However their lack of programmability and poor predictability, make them unsuitable for general purpose use, and for the production of large numbers of similarly programmed devices.

- Digital pulse implementations provide very similar properties to analogue device, but with increased predictability and stability[6].

- MIMD arrays [Multiple Instruction Multiple Data] offer the flexibility and performance required by most neural network models, however they tend to be very large and expensive, putting them beyond the reach for many applications[7].

1 Sample period determines the speed that data is input to a system.
2 Response time represents the time that occurs between a change on the inputs affecting the output.

- SIMD arrays [Single Instruction Multiple Data] can offer a comparable performance to MIMD arrays, but with a much reduced complexity. However severe constraints are met with SIMD arrays[8,9]:

  - performance drops in networks with uneven topologies [see appendix 1]
  - lack of flexibility and programmability
  - real-time performance cannot be adjusted to meet that required by a particular application.

To produce an architecture for control and signal processing requires some form of hybrid between these implementation classes. The architecture that has been developed combines features normally associated with MIMD and SIMD arrays, resulting in maximum flexibility and minimal hardware.

Neurons are treated as 'virtual neurons' and can be scheduled in an almost unlimited manner across an array. The partitioning of a network onto an array can utilise both a many-to-one and a one-to-many mapping of virtual neurons to processing elements. Meanwhile the control and communications are fully synchronised using a global control strategy, thus removing the need for complex hardware within each processing element.

4. Architecture Developed

Diagram 2 presents an overview of a proposed fully integrated neural processor. This consists of a number of processing elements (PE), each with a small amount of local memory in EEPROM (Weights Table). These memory blocks hold the weights for each virtual neuron, and the code required to schedule these weights. The contents of each processor's weighs table is determined at compile time, and is programmed into the EEPROM via the Bus Interface Unit.

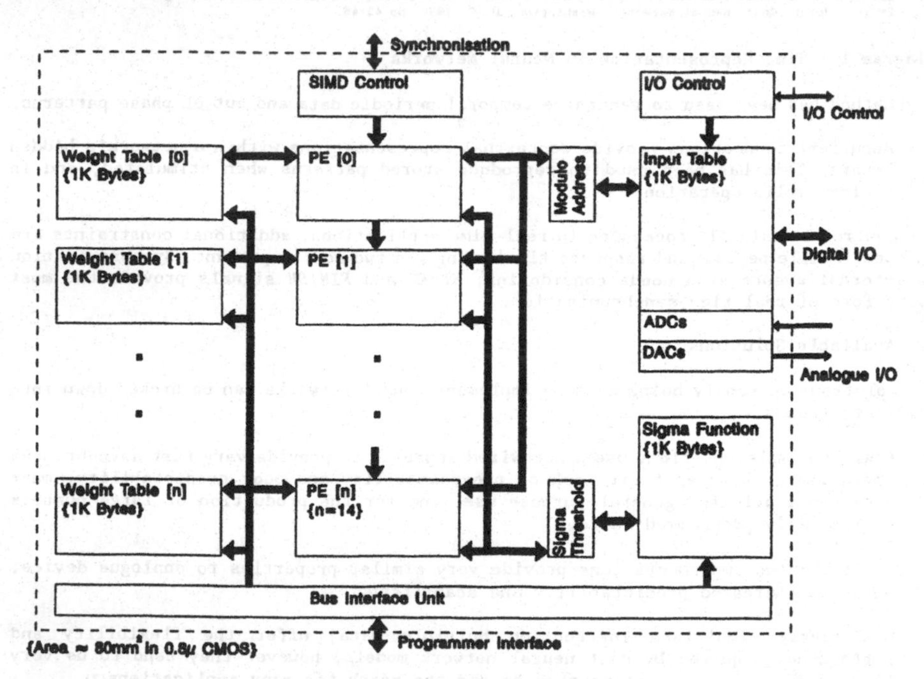

Diagram 2 Overview of Chip Layout

Each processor is linked via a shared bus to the Input Table. This is a common section of memory providing all the communications between processing elements and input/output devices. Input and output can be performed via the Digital I/O Bus, or on-chip ADCs and DACs. These converters can be memory mapped into the input table, to provide direct access to real-time events, or accessed via modulo buffers under control of the I/O Control Unit and the Modulo Address function.

These modulo buffers are used to store sequences of data to create spatial representations of continuously varying signals; Diagram 3 shows such an arrangement. In this diagram an ADC writes data into the modulo buffer, and the network uses this modulo buffer as its input layer. The use of this modulo buffer enables the input data to be effectively shifted through memory, without the need for complex and time consuming block move instructions, that would have been required otherwise.

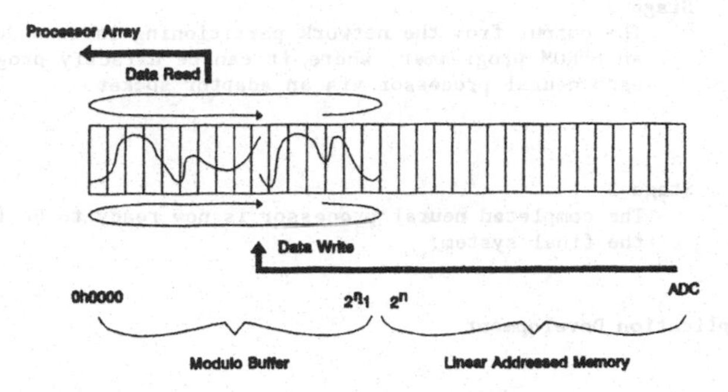

Diagram 3 Spatial Representation using a Modulo Buffer

The Sigma Function and Sigma Threshold units are used to compute the neural output from each sum-of-products; these units are again shared between the processors to minimise on the hardware requirements. The sigma function is held in an EEPROM lookup table, allowing this function to be specified during programming.

The control for this neural processor is considered the most crucial feature in this architecture. The following section considers the control mode used, and shows how it is implemented.

5. Control of Neural Processor

The control for this processor uses features normally associated with both MIMD and SIMD control modes. This provides a high level of flexibility and programmability, while ensuring that the complexity is kept minimal. The employment of both these control modes, and the selection between them is considered below:-

- SIMD control is provided by the SIMD control unit, which continuously broadcasts a simple sequence of instructions. This sequence can update a single synapse in every processor upon each iteration. Individual processing elements can select between this instruction stream, and their own MIMD instruction stream.

- MIMD control is used to configure each processing element, and to schedule each virtual neuron onto a specific processor. This is achieved by inserting short sequences of MIMD instructions between the weights of each virtual neuron. Special logic switches between the two control modes, enabling MIMD control to schedule each virtual neuron, and for them to be updated under SIMD control.

This strategy results in a fully deterministic architecture, enabling compiler techniques to be used in the optimisation and management of each processing element. Meanwhile synchronisation to real-time events is performed via SIMD control, thus minimising the control and synchronisation hardware required by each processing element.

6. Application Development

The application development environment proposed uses a combination of off-the-shelf software packages, and custom tools; Diagram 4 outlines the stages involved.

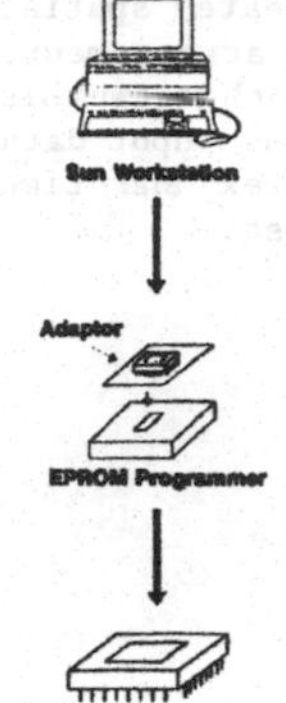

Stage 1
Network development and training is performed using an off-the-shelf neural simulation package. Partitioning is then performed on the network using custom software tools.

Stage 2
The output from the network partitioning is then downloaded to an EPROM programmer, where it can be directly programmed into each neural processor via an adaptor socket.

Stage 3
The completed neural processor is now ready to be installed in the final system.

Diagram 4 Application Development

To ensure a fluent transition between the stages in the application development requires careful design of the network partitioning tool. This must hide the details of the architecture from the application developer, while ensuring that the maximum performance is obtained from the processor. Work at present is looking into VLSI and ASIC cell placement algorithms, which is hoped can offer a suitable solution to this task.

Diagram 5 shows a very simple partitioning problem. A three layer network is partitioned onto three processing elements; a neuron from each layer is partitioned onto a processor, with the output layer neuron being 'cut' into three, and scheduled across all three processors.

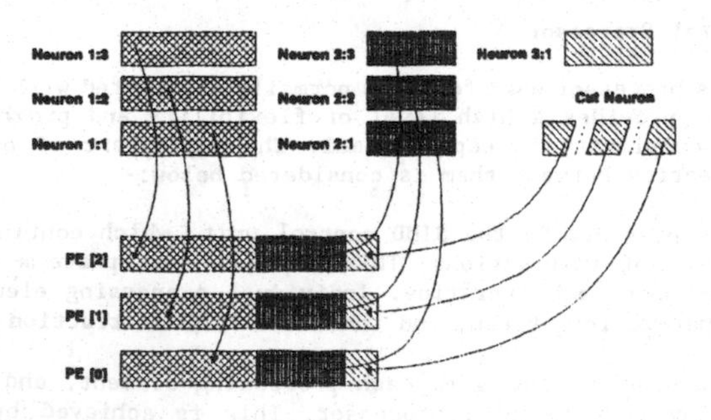

Diagram 5 Network Partitioning

7. Conclusion

This paper has described an architecture developed at University College London for the implementation of control and signal processing neural networks. This architecture has been developed over the last three years, and a test chip incorporating a single processing element has been fabricated. It is now hoped that this work may continue with external funding, to produce a fully integrated neural processor as described above.

8. Acknowledgements

The authors would like to thank their colleagues at University College London, and to SERC for the initial funding via a PhD student grant.

Appendix 1

The analysis below is used to calculate the performance efficiency for a SIMD array implementing a feedforward network, using a 1:1 mapping of neurons to PEs.

This performance efficiency is calculated as the average efficiency of all the processing elements. Each processing element's efficiency is the ratio of the number of inputs to that neuron, to the largest number of inputs in the network.

$$\frac{\sum_{i=1}^{k} M_i \dfrac{M_{i-1}}{\max_{i=0}^{k} M_i}}{\sum_{i=1}^{k} M_i} \qquad \text{[Eq. 1]}$$

Where:

K	Number of layers in the network (Input layer K=0)
M_i	Number of neurons in layer i
M_{i-1}	Number of inputs to layer i (Feedforward networks, 100% connectivity)
$\max_{i=0}^{k} M_i$	The largest number of inputs in the network (Largest layer in network)

This analysis was used to show the low efficiency of SIMD arrays implementing many control and signal processing networks. This low efficiency is due to the uneven layer lengths found in these networks, this being used represent continuously varying signals. In a preliminary study many signal processing networks were found to have a performance efficiency of less that 50%, when implemented on a SIMD array.

References

1. Palmer, R. P. High Performance Digital Neural Network Implementation for Small-Scale Portable Applications. *Proc. 2nd Int. Conf. on Microelectronics for Neural Networks*. IEEE German Section, Oct 1991, Munich. pp. 207-216.
2. Palmer, R. P. A Novel Architecture for a High Performance Low Complexity Neural Device. Phd. Thesis, Department of Computer Science, University College, London. February 1992.
3. Soucek, B. Neural and Concurrent Real-Time Systems Wiley Interscience 1989.
4. Maren, A., Harston, C., Pap, R. Handbook of Neural Computing Applications.
5. Holler, M., Tam, S., Castro, H., Benson, R. An Electrically Trainable Artificial Neural Network (ETANN) with 10240 "Floating Gate" Synapses *IEEE Int. Conf. on Neural Networks*. 1989.
6. Murray, A. F., Smith, A. V. W., Butler Z. F. Bit-Serial Neural Networks. *IEEE Conf. On Neural Information Processing - Natural and Synthetic* Denver 1987.
7. Garth, S. C. J. A Chipset for High Speed Simulation of Neural Network Systems. *Proc. IEEE First Int. Conf. on Neural Networks*. San Diego 1987.
8. Myers, D. J., and Brebner, G. E. The Implementation of Hardware Neural Net Systems. *Proc. First IEE Int. Conf. on Artificial Neural Networks*. London 1989 IEE Conference Publication 313, pp. 57-61.
9. Kashai, Y., and Be'ery, Y. Comparing Digital Neural Networks. *IFIP Workshop on Silicon Architectures for Neural Networks*. St. Paul-De-Vance, France. 1990.

PLANLITE: ADAPTIVE PLANNING USING WEIGHTLESS SYSTEMS

Janko Mrsic-Flögel

Neural Systems Engineering
Department of Electrical Engineering
Imperial College, London SW7 2BT
e-mail: jmf@doc.lc.ac.uk (JANET)

Abstract
This paper proposes an adaptive planning system design based on the weightless net paradigm[1]. The system learns local state transitions through 'exploration' of the application environment and then generalizes on the learnt data - enabling the system to produce optimal plans based on action cost. The generalization also allows data to be trained into the planner during the generalization process, allowing the planner to adapt relevant plans according to the new data. A simulation of the proposed system has been implemented and successfully applied to several problem domains. The amount of computation during operation of this system is less than that of a conventional rule-based planner.

1.Introduction

The task of a planning system is to generate a plan, which is a solution to a given problem. The plan is usually composed of operators (actions), which are provided to the system for a given application environment. A typical problem is described to the planner by an initial state of the system and a goal state description. Given the current/initial, S(i), and goal, S(g), states the planner should be able to produce the next operator. Applying this operator to the current state should move the planning system closer to the goal state in state space.
The motivation for this work was to investigate how weightless neural systems can be applied to planning problems. This paper is focusing on:
a) **Representation** of planning problems in a neural system
b) **Learning** the application environment
c) **Generalization** of the trained neural system for the given representation
d) **Performance** of neural planning system
e) **Integration** strategy for the Planlite system

2.Weightless Systems

Weightless neural systems are distinguished from those classically used in neural computing by the fact that they store the required responses to their input patterns in addressable locations rather than as combination of connections strengths.One type of element used for the design of weightless systems is the G-RAM[1].
The G-RAM is a device from the logical nodes by Aleksander which include the RAM neuron, as used by the WISARD, and the PLN (probabilistic logic node) [2]. As well as the usual "read" and "write" phases in other logic nodes, a "spread" phase is introduced in this node. During the "spread" phase, the trained information is generalized by means of a "spreading" algorithm. The G-RAM node has K binary inputs addressing 2^K memory locations. Each memory location stores a B-bit word which is output when addressed. Therefore the RAM can output 2^B messages. Each message may be regarded as a "firing probability" of a neuron.

3.Planning System Design

A planning problem can be specified by the initial and goal state descriptions.These states are in effect the total information that the system uses as input. Thus, the input vector to the weightless system has to be segmented into two parts - the encoded initial state and the encoded goal state. One can imagine the state space of such a neural system being represented as a two dimensional state transition matrix with rows representing current states and columns the goal states. Each element in such a matrix would represent the next action/ state leading to the goal state. Such a representation can be seen in Fig.2. The output of the weightless system has to include the next action for the given plan as it is required for the plan. Other information can be stored in the output of the system such as the next state. In effect the system could also serve as a next state predictor/indicator.

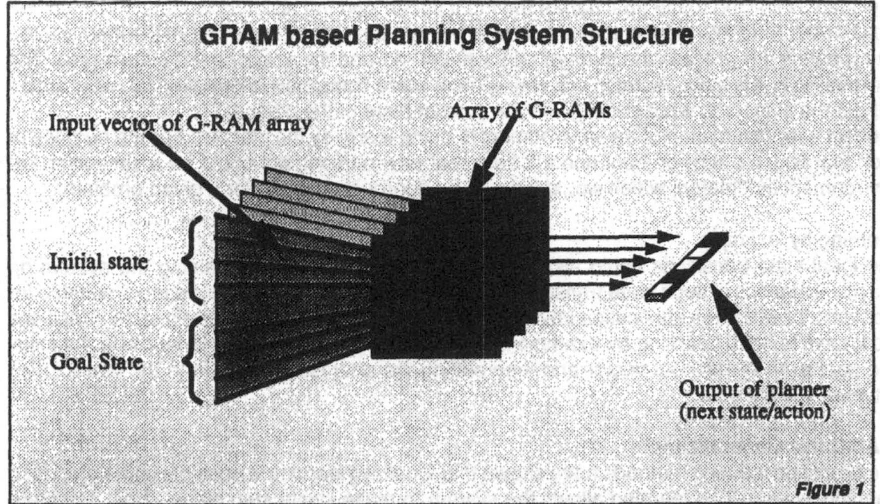

Figure 1

A weightless system for planning can be constructed by using only a single G-RAM with 2*s bits input vector, where s is the number of bits used to represent a single state of the application environment. Such a RAM would therefore be able to address 2^(2*s) locations. The output and storage locations have to have enough bits to encode the next state/next action information as well as the cost of the action = s + a + c bits, where a is the number of bits to encode any action and c is the action cost (fig. 2). The next state information is used as feedback to the system as will be described later in the paper. Fig.1 shows such a system.

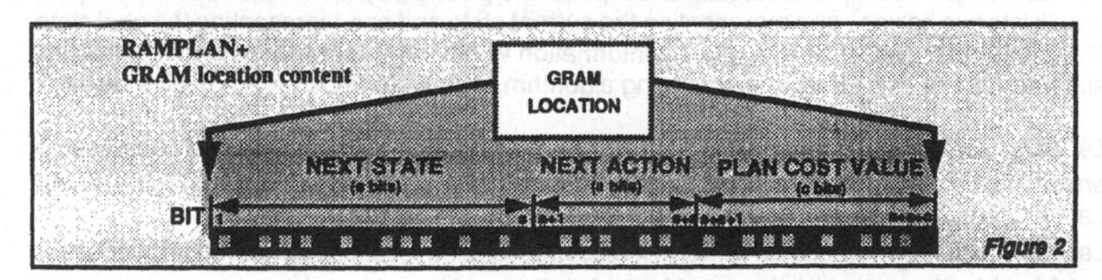

Figure 2

4.Learning

Having designed a system which can takes as input the current and goal states, one has to show what and how the system can learn the information supplied by the application environment.

Learning state transitions

In the proposed weightless system we have a unique location for each combination of current state and goal state. We can therefore store a different next action/next state information in each such location.

Learning by 'exploration' (Localized learning)

One of the qualities associated with neural systems is that they adapt to their environment. In tho same manner the planner has to learn information about its environment. The system can learn by exploring its environment. The system starts with the encoded goal state equaling the encoded current state or position of the system in the environment. Given the set of actions, the planner outputs a random action, which leads the system into a new current state. This new state is first fed to the goal part of the input vector and the action/new state is trained into the G-RAM. After this training the new state is fed into the current state part of the input vector. By repeating this process the planner can 'wander' around the environment state space. The information that is being learnt in this way is **local state transition** information i.e. what action will get the planner to what state, given the current state of the planner.

In effect a representation of the environment is being learnt by the planner. For the purpose of such learning, the environment needn't be a real environment, but could be a simulation i.e. environment automaton, which would have to be rigorously defined for a given problem.

Learning by 'example'

The local state transitions could also be trained into the system by training the system on specific instances of such state transitions. This learning approach might be more suitable for problems for which the application environment is not well defined, but certain state transitions are known to hold. Learning by example is achieved by presenting the planner with the initial and goal state in the input vector and then setting the addressed location to the correct next action/next state data.

5. Adaptive Generalization

The proposed learning methods have shown that localized learning is possible in the proposed planning system. This learning allows the system to make single step plans - from one state to one of its neighbouring states. By generalizing on this local state transition information within the G-RAM, it is possible to form plans from any state to any state, provided that the environment allows it. This happens during the 'spreading' phase of the G-RAM.

Spreading Algorithm

To understand how the spreading algorithm should generalize on the learnt data we have to look at the nature of the internal data and input of the planning system. The input vector space is split into two halves - one representing the current state and one representing the goal state. Given this input vector, we can use the information within the G-RAM location to find the next state/action.The definitions and spreading algorithm, name RAMPLAN+, are outline below:

Let $Adr_{ig}= S(i) <> S(g)$, where Adr_{ig} is the input vector (address) to the G-RAM composed of encoded bit patterns representing the system states - S(i) (initial state) and S(g) (goal state).

Let $LOC(Adr_{ig})$ = the total contents of the location.

Let $DEF(Loc(Adr_{ig}))$ = boolean function indicating wether a location has stored information.

Let C(adr) be the cost value encoded in the location addressed by adr.

Let c_{undef} be the cost value for undefined locations - location in 'u' state.

$$\forall S_i:$$
$$\forall S_g:$$
$$if$$
$$\exists (S_x)$$
$$\text{such that } DEF\,(|S_i|\diamond|S_x|)\ \&\ \mathrm{DEF}\,(|S_x|\diamond|S_g|)$$
$$then$$
$$\mathrm{Let}\ l = C\,(|S_i|\diamond|S_x|) + C\,(|S_x|\diamond|S_g|)$$
$$if$$
$$l < C\,(|S_i|\diamond|S_g|)$$
$$then$$
$$\mathrm{LOC}\,(|S_i|\diamond|S_g|) = \mathrm{LOC}\,(|S_i|\diamond|S_x|),$$
$$C\,(|S_i|\diamond|S_g|) = l..$$

if no generalization for LOC(|Si|<>|Sg|) in single generalization cycle then set to location 'u' ($C = c_{undef}$), except for one-action plan location (where goal state = next state).

Fig.3 shows how the data trained using the bar stacking example is spread during the generalization.The initial matrix shows the G-RAM contents after localized learning had occurred. Only three subsequent 'spread' cycles are necessary to completely spread the learnt contents. It can be shown that if the number of 'spread' cycles necessary to spread the information completely, from ungeneralized information, depends on the length of the longest plan: Number of 'spread' cycles to cover state space = **ciel($\log_2$ (length of longest path)).** The logarithmic relation to the longest path length is due to the nature of the spreading algorithm. The generalization effectively spreads the goal information backwards through the states of a plan, forming an optimal path. This learning has certain analogies with the experiments concerning animal learning performed by Tollman and Hull [3], where they tried to show that mice learn the path backwards from the reward (goal) to the initial state.

6.Planner Operation / Simulation

Plan formation: Following the learning and spread phases the system is able to produce plans for the learnt application state space. By setting the input vector to the desired initial and goal state the G-RAM will produce the next state / next action at the outputs. By feeding back this next state representation that was output by the planner into the initial state segment of the input vector, the planner will produce the following state and by repeating this feedback process will eventually reach the goal state. By noting the produced actions at the feedback system output, the plan is created. The time cost of producing a single plan is n memory read cycles, where n is the length of the plan. The system basically gives real time solutions to problems, as the plans have been learnt previously by generalization or training.

Simulation: A simulation of the weightless planning system was implemented on Sun Sparc workstation using the NEMESYS simulator[4]. An X-windows graphical interface and environment state representation language were incorporated to aid in the training of the planning system. The example used to describe the generalization of memory content is based on the Blocksworld problem of stacking blocks. The application environment contains three

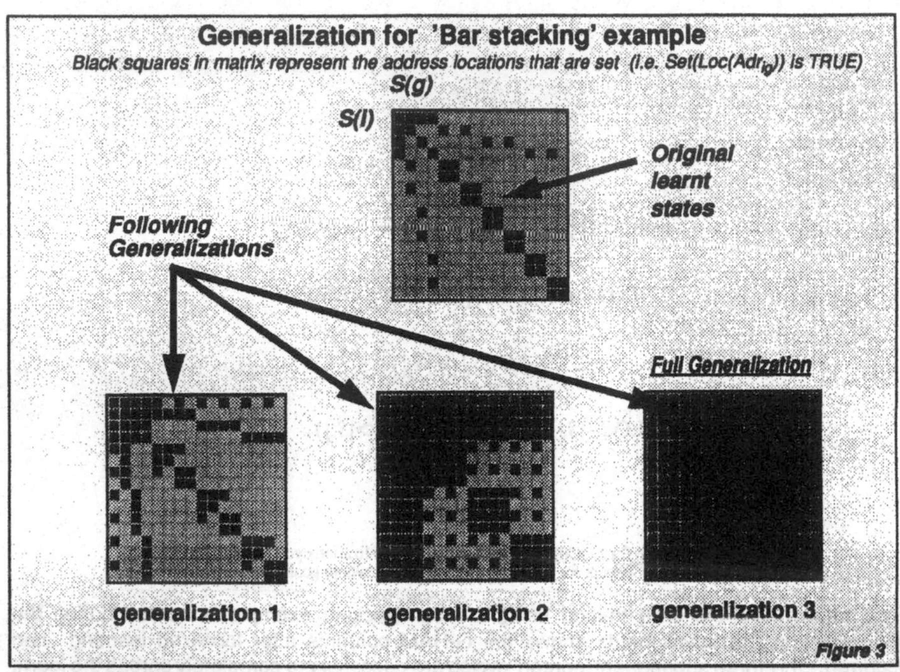

Figure 3

bars of different lengths: a short bar of one pixel length, a medium bar of two pixels and a long bar of three pixels. These bars can be stacked on top of each other by the following set of actions: PUT BAR 1 (places bar of length 1 on top of stack, if bar is available), PUT BAR 2 (places bar of length 2 on top of stack, if bar is available), PUT BAR 3 (places bar of length 3 on top of stack, if bar is available), TAKE OFF BAR (removes bar from top of the stack), REST (performs no action). The environment state which is presented to the system includes a 3x3 binary retina showing the state of the stacked bars, as can be seen in figure 4.

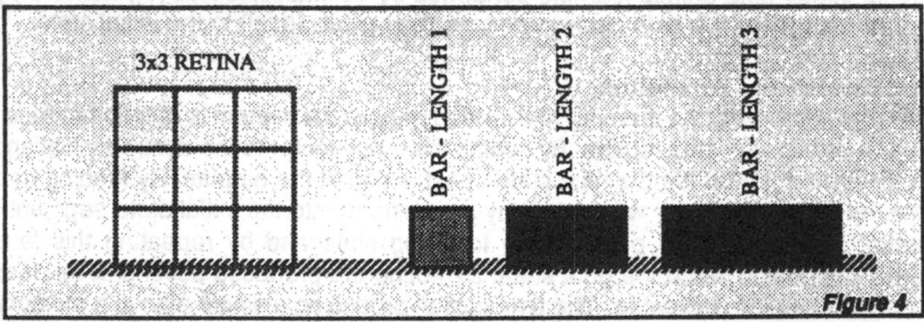

Figure 4

A Blocksworld environment automaton has been designed, for the purpose of system training. The automaton provides the information of the new state after execution of any action. In this case, the training of the GRAM is caused by a random action generator executing actions and thus effecting state changes. The state change information and the causal action are stored in the GRAM by environmental learning. There are 46 local state changes (depicted by black squares in the state diagram in figure 3) that can be learnt in this application domain. Figure 5 shows a sample of learning trials using the random action generator. The figure also shows that the state transition for this problem are usually learnt within 600 generated actions.

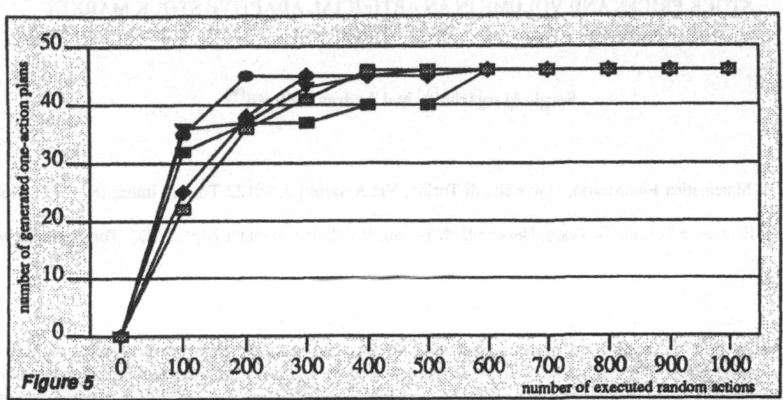

Figure 5

The system has been applied to several problems including two-dimensional path finding and underground railway passenger route planning[7] based on the London Underground network.

7. Hybrid Integration Strategy - Further Work

Recently there has been a rise in interest about hybrid systems: systems which are composed of several different technologies. The integration of the Planlite unit and an expert system into a hybrid system is to be considered, for planning applications where the application domain is dynamic. A dynamic application domain requires the Planlite unit to adapt and learn changes as they occur. In the hybrid system the task of creating the new training data for the planner, based on environment changes, is handled by the expert system. The expert system receives sensory data from the environment. Using a knowledge base it then extracts the necessary changes and forms a new training set for the planning unit. The training set consists of one-action plan relations, which when trained into the Planlite unit will then generalize over the entire location content, with newly trained information. In such a system the Planlite unit does not learn from its own 'experience', but the learning is based on expert knowledge relating the sensory data top the planning application domain.Applications that could benefit from such solutions may include aircraft object avoidance systems.

7.Conclusions

A weightless system for adaptive planning problems was proposed, based on the G-RAM node. The proposed planner can learn about the environment state space by 'exploration' or example and thus learns localized state transitions in state space. The spreading phase of the G-RAM allows for the localized state transition to 'spread' throughout the state space, using the outlined generalization algorithm. The number of 'spread' cycles required to reach a state of full generalization of the learnt data is logarithmically related to the length of the longest plan in the application environment. The system was implemented and tested on the NEMESYS simulator. The weightless system shows fast reaction compared to conventional rule-based techniques, as only direct memory reading is involved.

References

[1] I.Aleksander, Ideal Neurons for Neural Computers, Parallel Processing on Neural Systems and Computers, Ed. R. Eckmiller/G.Hartmann/G.Hauske,pp.225-228, North-Holland, 1990

[2] I.Aleksander, The logic of Connectionist Systems, Neural Computing Architectures, MIT Press, 1988

[3] Thomas J. Tighe, Modern Learning Theory, Oxford University Press, 1982

[4] J.Mrsic-Flogel, NEMESYS - modelling system for weightless nets,ICANN91,Elsevier,1991

[5] M.Drummond,1989, AI Planning-A Tutorial and Review, NASA Report,AIAI-TR-30

[6] J.Mrsic-Flogel,Cognitive Planning with Weightless Systems,Proc. of ICANN92 conference, Ed. I.Aleksander and J. Taylor,Elsevier,1992

[7] J.Mrsic-Flogel,Aspects of Planning with Neural Systems, PhD Thesis, Imperial College, November 1992

STOCK PRICES AND VOLUME IN AN ARTIFICIAL ADAPTIVE STOCK MARKET

Sergio Margarita(+) **and Andrea Beltratti**(*)

(+) Istituto di Matematica Finanziaria, Università di Torino, Via Assarotti 3, 10122 Torino, Italia; tel. (011) 546805; fax (011) 544004
(*) Istituto di Economia Politica G. Prato, Università di Torino, Via della Cittadella 10/E, 10122 Torino, Italia; tel. (011) 540900; fax (011) 541497

Abstract: we present an application of neural networks to financial markets, experimenting with various learning mechanisms that may describe reasonable behavioral rules followed by agents acting under incomplete information about the environment. Each agent is described by a neural network who decides the price she is willing to pay for an asset, and the quantity she wants to buy or sell. Agents differ as to a number of dimensions, and in particular the trading strategy that may be used to divide the traders in two different categories. The interactions among the different agents determine every day the market price and the volume of transactions. We analyze the behavior of the market as a function of the proportions of traders in the two categories, showing that increasing heterogeneity positively affects the market volume, without visibly increasing the volatility of prices. We also experiment with different learning mechanisms, that can be interpreted in economic terms as forcing on the agents differing degrees of risk aversion.

1. INTRODUCTION

Neural networks are increasingly used in economics and finance. Their powerful computational abilities make them ideal to understand the evolution of complex systems of which agents have an imperfect knowledge, and therefore have to learn over time some of the basic features of the environment. In this paper we focus on the learning behavior of heterogeneous agents interacting in a stock market. Some basic economic principles are used to determine the appropriate targets for traders who evaluate the best action by means of neural networks.

Available evidence of the way agents behave in actual markets show that agents differ as to a number of factors (for example the information set, the interpretation of the information, wealth, risk aversion, trading strategies). Our paper considers agents who (i) have different models of the world and heterogeneous information sets (ii) over time learn some features of the other agents' models of the world and affect the environment with their own actions (iii) evaluate assets according to a specific methodology based either on past dividends or on past prices. Even more importantly, our agents do not behave in order to maximize a well-specified time-invariant utility function, but learn to decide the valuation of the share, the quantity and the direction (purchase or sale) of the trade with the purpose of increasing their wealth over time. This is not done according to some demand function coming from utility maximization, but by learning to optimize with respect to the relationship between signals and wealth. In this sense our model has some behavioristic features of the bounded rationality hypothesis of Simon (1979), Nelson and Winter (1982) and others. Arthur (1991) has recently shown the relevance for economic behavior of learning methods similar in spirit to the one we use in this paper, while Holland and Miller (1991) argue for the importance of modelling artificial adaptive markets. Nottola et al. (1991) also have a model that describes the working of a financial market, even if their methodology is completely different.

We pursue these different goals by using artificial neural networks (from now on ANNs). We use ANNs both as algorithms that may simulate human behavior as they learn from the environment and as a black-box device that allows the building of models that do not assume homogeneous utility-maximizing agents. We consider ANNs that create their own procedures for deciding the action that is appropriate for any given information set; the networks then interact in a market and exchange at prices that depend on the characteristics of the pair of agents performing the trade. The model determines endogenously both the price and the volume of transactions.

The organization of the paper is the following: section 2 describes the agents from the point of view of their economic behavior, while section 3 considers their description as ANNs. Section 4 describes the marketplace and section 5 the learning behavior. Section 6 presents the results and section 7 concludes. An Appendix at the end of the paper reports all the relevant notation.

2. DESCRIPTION OF THE AGENTS

Agent i enters the early morning of day t with a given amount of money $M_{i,t}$, a stock of shares $S_{i,t}$ and an information set $I_{i,t}$. She has two choices for carrying her wealth to the following morning: money and shares, about which she will have to make a decision at noon of day t. Each day t shares can be bought or sold against money in a stock market opening at noon at a price $p_{ij,t}$ depending on the value assigned to shares by the agent j who will be (randomly) met in the marketplace (see section 4); in fact we consider a decentralized marketplace where all the exchanges take place at different values. Money is a riskless asset; investing $1 in money at t means having $1 of money at t+1. Buying shares is risky because there is uncertainty about the price which will prevail at t+1, and about the dividend that will be paid at the end of period t, denoted with D_t. Each transaction involves buying or selling a number of shares which depends on the choices of the two agents who meet in the market (see section 4). The actions of the traders depend on the specific structure, described in the next sections.

3. THE STRUCTURE OF THE ARTIFICIAL TRADERS

3.1 Technical traders

Technicals compare at each moment of time the expected rate of return on the two assets. In order to form an expectation about the rate of return on stocks (the rate of return on money is 0) technical trader i is modelled as a network that forms an expectation about the price at which the share may be sold in period t+1 and about the dividend that will be paid by stocks at the end of period t, D_t.

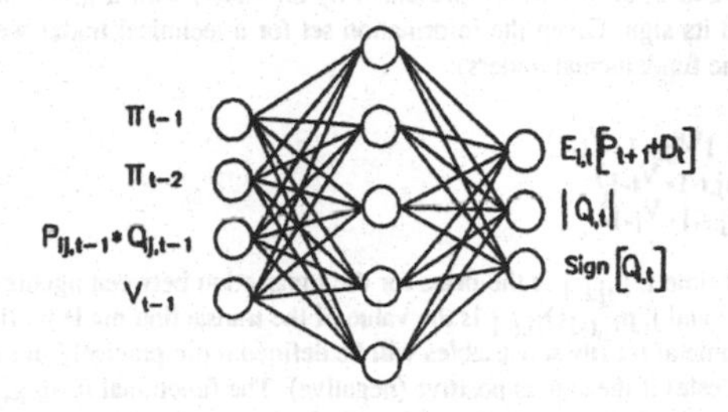

Figure 1. The structure of technical traders

We denote the expectation of agent i for the price with $E_{i,t}p_{t+1}$, a general symbol to underscore that agent i does not know the identity of agent j with whom she will transact; the expectation of the dividend is $E_{i,t}D_t$. These expectations are formed on the basis of an information set containing the previous history of market prices, the value of the trade made in the previous period, and the market volume in the previous period (Figure 1).

3.2 Fundamental traders

Fundamental traders adopt a long-run perspective of the market, basing their action on the estimated fundamental value of the asset. The main differences between the technical and fundamental traders lie in the information sets and in the interpretation of the first output (Figure 2).

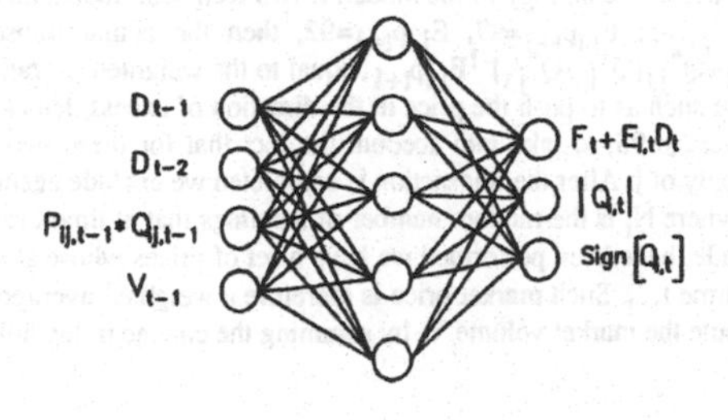

Figure 2. The structure of fundamental traders

As figure 2 shows, the information set contains two past values of dividends rather than two past prices. This is done in order to estimate the fundamental value of the asset at time t, that we define $F_{i,t}$. The sum of such value and the expectation of the dividend is the first output of the agent, and also in this case this sum is an important reference point. We assume that such fundamental value differs from agent to agent; we let agent j learn over time a fundamental value equal to:

$$F_{jt} = g_j\, m + 0.5[(1+\exp(-\sum_{i=0}^{4} D_{t-i})^{-1} - 0.5] \tag{1}$$

where g_j is extracted from a uniform distribution $U(g_{min},g_{max})$. Therefore each trader recognizes an estimate of the fundamental value of the asset that is a smooth function of the realization of the past values of dividends. This is the reason why two values of dividends are incorporated into the information set: they can help the trader to learn more efficiently the relationship between dividends and fundamental values.

3. THE NEURAL NETWORK REPRESENTATION OF AN AGENT

As we have seen in figures 1 and 2, an agent is represented by an ANN, with a given information set. The net has 3 outputs, a price, a quantity and its sign. Given the information set for a technical trader we can therefore write (with an obvious similar extension for the fundamental traders):

$$E_{i,t}p_{t+1} = g_{i1t}(\Pi_{t-1}, \Pi_{t-2}, p_{ij,t-1}Q_{ij,t-1}, V_{t-1}) \tag{2a}$$
$$Q_{i,t} = g_{i2t}(\Pi_{t-1}, \Pi_{t-2}, p_{ij,t-1}Q_{ij,t-1}, V_{t-1}) \tag{2b}$$
$$T_{it} = g_{i3t}(\Pi_{t-1}, \Pi_{t-2}, p_{ij,t-1}Q_{ij,t-1}, V_{t-1}) \tag{2c}$$

where Π_t is the market price at time t, $p_{ij,t-1}$ is the price for the transaction between agents i and j, $Q_{ij,t}$ is the amount of the transaction between agents i and j, $p_{ij,t-1}Q_{ij,t-1}$ is the value of the transaction made by the agent during the last period of time, V_{t-1} is the market volume at t-1 (these variables will be defined more precisely in section 4), T_{it} is the sign of the transaction chosen, a purchase (sale) if the sign is positive (negative). The functional form g_{ijt} in (2) describes the function mapping the inputs of the i-th network at time t to the j-th output at time t. Such form is equal for each network, but the weights vary from one to the other and from one period to the other.

For example $T_{it}=1$ ($T_{it}=-1$), $Q_{i,t}=3$ and $E_{i,t}p_{t+1}=100$ means that agent i wants to buy (sell) 3 shares at a price equal to or lower than (equal to or greater than) 100. Of course agents face a budget constraint; they can go to the market only after making sure that their desired position (selling or buying at a given price) is not incompatible with available wealth. We impose a cash-in-advance constraint and a no short selling constraint. To take into account the budget constraint we check whether the planned purchase is affordable on the basis of available money, that is whether $Q_{i,t}<(M_{i,t}/E_{i,t}p_{t+1})$. The actual planned quantity is therefore equal to $Q^*_{i,t}=\min(Q_{i,t}, (M_{i,t}/E_{i,t}p_{t+1}))$. In case of selling we check that the sale is compatible with previous holdings, so that the actual quantity is $Q^*_{i,t}=\min(-Q_{i,t}, S_{i,t})$

4 THE MARKETPLACE

After forming their reservation prices agents meet randomly in a decentralized marketplace with no auctioneer. From the sample of J agents we randomly select two agents and match them. Given the assumptions we have made about the demand schedule, a transaction may or may not occur in our model. There is no transaction for example if the two agents are both seller or both buyers; in this case we just put them again in the sample of potential traders and extract other two. If a trading possibility exists we organize a transaction which is dominated by the short side of the market. An example is the best way to describe the transaction technology of the model. If two technical traders meet in the marketplace with the following outputs $Q^*_{i,t}=10$, $Q^*_{j,t}=-5$; $E_{i,t}p_{t+1}=97$, $E_{j,t}p_{t+1}=92$, then the actual transaction is $Q_{ij,t}=5$ at a price $p_{ij,t}=Q^*_{i,t}[Q^*_{i,t}-Q^*_{j,t}]^{-1}E_{i,t}p_{t+1}+Q^*_{j,t}[Q^*_{i,t}-Q^*_{j,t}]^{-1}E_{j,t}p_{t+1}$, equal to the weighted average between the expected prices of the two agents; the weights are such as to push the price in the direction of excess demand. In our example the price is 95.33, larger than the simple average 95, to take into account the fact that for these two particular traders i and j the demand of i is larger than the supply of j. After the transaction is completed we exclude agents i and j from the sample.

When N_t transactions, where N_t is the random number of meetings that at time t satisfy the necessary conditions for a successful conclusion of trade, have been performed we have a set of prices whose simple average is considered for determining the market price at time t, Π_t. Such market price is therefore a weighted average of the expected prices of the single traders. We can also compute the market volume V_t by summing the change in the holdings of shares of each single agent.

5 LEARNING

At the end of the day agents know the market price Π_t, and can therefore spend the night calculating the value of their wealth and updating their view of the world. As to wealth dynamics, for an agent i who bought during the day wealth (evaluated at market values) is equal to $W_{i,t+1}=(M_{i,t}-Q_{ij,t}p_{ij,t})+\Pi_t(S_{i,t}+Q_{ij,t})$; if the agent was instead a seller we have $W_{i,t+1}=(M_{i,t}+Q_{ij,t}p_{ij,t})+\Pi_t(S_{i,t} - Q_{ij,t})$. Agents can also update their "view of the world" (the weights of the network); this gives rise to two sets of problems that will be discussed separately.

As to quantity, in order to understand how the agent learns we need to know what she would like to have done at t-1 if she had known the actual price at t. From the point of view of agent i we can say that if $p_{ij,t} > p_{il,t-1}$ then i would have liked to buy more shares at t-1. The target we compute is equal to the quantity she could have bought had all the money been spent in shares, that is $M_{i,t-1}/p_{ij,t-1}$. As a consequence we measure the error by comparing the quantity output at t-1, $Q_{i,t-1}$ with $M_{i,t-1}/p_{ij,t-1}$. If one wanted to interpret the decision rules that are learnt from the agent in this market as demand functions coming out from utility maximization then one might say that our training procedure amounts to teaching the agents to become risk neutral. We also try in the simulations to decrease the importance of the comparison of the actual quantity with the risk-neutrality benchmark. In a second set of simulations we compare the output of the network with the simple average between the quantity that was transacted last period and the "optimal" quantity $M_{i,t-1}/p_{ij,t-1}$, in this way trying to downplay the importance of the "optimal" benchmark. This therefore is not teaching the agent to go towards corner solutions where all the amount of wealth is allocated to either money or shares. In a third set of simulations we use a still different teaching method, that modifies only the sign of the transaction (and the price, as will be explained later on) and not the amount. In this case the trader learns to trade in a way to increase her wealth over time, without learning to allocate all her wealth in either money or shares. Wealth maximization is pursued with a more balanced portfolio.

On the contrary, if $p_{ij,t} < p_{il,t-1}$ then i would have liked to sell more shares at t-1, regretting not keeping all wealth in the form of money. She compares $Q_{i,t-1}$ with $-S_{i,t-1}$, the maximum quantity she could have sold at t, give the no-short-sale constraint. The same general considerations that were advanced before apply to this case, with the same alternative learning methods.

As to the price, note that the forecast of the price and of the dividends made at the beginning of day t by traders cannot be compared to anything known at the end of period t, since such forecast is about variables which will be known during period t+1. What the network can do at the end of day t is to compare the forecast which was made at time t-1 with the new information available on day t. We compare the forecast of the net at time t-1 with the actual transaction price at t plus the dividend paid at t. As to fundamental traders, we compare the price and dividend output of the network with the sum of the fundamental value we have defined in (1) and the actual dividend.

6. RESULTS

Figures 3 to 7 contain the market price at each time t and the volume that is exchanged on the market. In figure 3 there 50 technicals and 50 fundamentals, the dividend is stochastic and uniformly distributed between 0.01 and 0.04, while the parameter g_j is uniformly distributed between 0.9 and 1.1. The price path is reasonably stable, even in the face of much variability in transactions. Transactions do not tend to disappear in 300 periods, contrary to what happens in figure 4, where all the agents are fundamental traders. The importance of divergences of opinions for the volume has been noted among the others by Tauchen and Pitts (1983). The divergence in the opinions about the fundamental value of the shares is not enough to generate trading in the long period. The same is true of figure 5, where everybody is a technical, and where a new feature appears, that is the increased volatility in price. It seems therefore that technicals reach a consensus opinion by trading and by destabilizing the price. The extrapolative nature of their behavior may in fact increase volatility with respect to what is done by long-term traders looking at dividends.

Figures 6, 7 and 8 analyze the consequences of the different hypotheses about learning behavior. For a given set of parameters about dividends and heterogeneity of fundamental traders they compare the case where the comparison is with the quantities corresponding to risk-neutrality (figure 6), to the case where the comparison is softened by a target being equal to the simple average between the previous quantity and the one transacted in the previous period (figure 7) and finally to the case of what we called qualitative learning, where there is no learning about volume of transaction. The results do not show significant differences, either in terms of prices or in terms of volume. We deduce that the results of our model are not very sensitive to the exact learning procedure that we impose.

8. CONCLUSIONS

We have presented a model of a financial market where agents learn to determine their own behavioral rules in order to maximize their wealth. The rules consist of a choice of price and quantity that meets a certain budget constraint, and that can or cannot be satisfied completely in the trade with another agent. We have shown the robustness of the results to various possible learning rules, and the relationship between market prices and volumes. We have provided some evidence that increasing the heterogeneity of the trading strategies looked at by the agents affect most of all the volume of trades, while increasing the proportion of agent following a chartist strategy increases the volatility of prices.

There are many possible extensions of this model, that we hope to pursue in our future work. First, we plan to experiment more with the current version, in order to analyze the consequences of learning for accumulation of wealth of the various agents. Also, we want to study the volume and price outputs of the simulations of the model to see whether they present some of the statistical features (e.g. autocorrelations, cross-correlations, and so on) that is usually possible to find in real data. Second, we plan to elaborate on this version to create others that are more heterogeneous as to behavior of the agents. One possibility is to force the agents to follow some predetermined trading strategies, in order to see whether one strategy takes over the others. A second possibility is to consider a market populated by both risk-neutral and risk-averse agents, in order to see whether any type dominates in the long run.

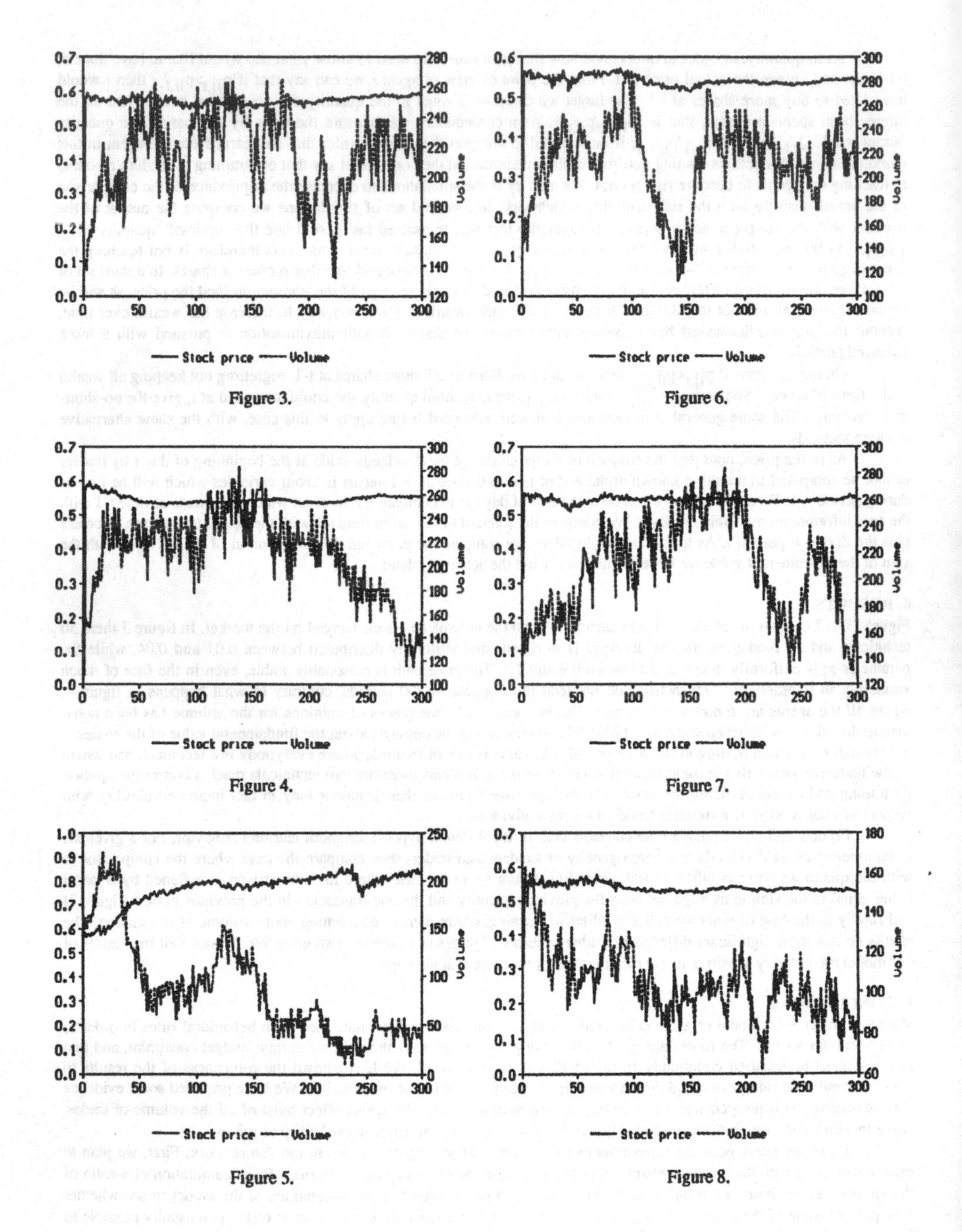

Figure 3.

Figure 6.

Figure 4.

Figure 7.

Figure 5.

Figure 8.

References

Arthur W.B., 1991, Designing economic agents that act like human agents: a behavioral approach to bounded rationality, American Economic Review, vol. 81, n.2, pp. 353-359

Beltratti A. and S. Margarita, 1992a, Simulating an artificial adaptive stock market, Quaderni di ricerca n. 22, LUISS University, Rome

Beltratti, A. and S. Margarita, 1992b, Evolution of trading strategies among heterogeneous artificial economic agents, Nota di Lavoro 17.92, Fondazione ENI Enrico Mattei, Milano, to be presented at SAB 92, From animals to animats, Honolulu 1992

Holland, J.H. and J.H. Miller, 1991, Artificial adaptive agents in economic theory, American Economic Review, vol. 81, n.2, pp. 365-370

Nelson R.N. and S. Winter, 1982, An evolutionary theory of economic change, Harvard University Press, CA

Nottola, C., F. Leroy and F. Davalo, 1991, Dynamics of artificial markets, in Bourgine P., F. Varela (eds.), Towards a practice of autonomous systems, MIT Press, Cambridge, MA

Simon, H.A., 1979, Rational decision making in business organizations, American Economic Review, 69

Tauchen G.E. and M. Pitts, 1979, The price variability-volume relationship on speculative markets, Econometrica, 51, pp. 485-505

Appendix

The notation is the following:

Π_t stock market price in terms of money;
$E_{it}p_{t+1}$ i-th agent's forecast of the price at which a transaction involving shares against money will be executed at t+1;
I_{it} information set of agent i at the beginning of day t;
$p_{ij,t}$ price at which the actual transaction between agents i and j is executed during period t;
$p_{i,t}$ price of the ith transaction at t;
$Q_{i,t}$ quantity which agent i is willing to sell (if $Q_{i,t}<0$) or buy (if $Q_{i,t}>0$) before looking at her budget constraint;
$Q^{*}_{i,t}$ quantity which agent i is willing to sell or buy after looking at her budget constraint;
$Q_{ij,t}$ actual quantity of shares which is transacted between agents i and j at t;
$T_{i,t}$ sign output of net i at time t;
D_t dividend paid by the stock at the end of period t;
$M_{i,t}$ available stock of money at the beginning of day t;
$S_{i,t}$ number of shares available by the agent at the beginning of day t;
N_t number of transactions performed during day t.
V_t market volume at time t,

APPLICATION OF THE FUZZY ARTMAP NEURAL NETWORK ARCHITECTURE TO BANK FAILURE PREDICTIONS

Luis J. de Miguel Elena Revilla J. Miguel Rodríguez
J. Manuel Cano *

Department of Systems Engineering and Control
and
Department of Business Administration,
University of Valladolid
Paseo del Cauce, s/n, Valladolid, 47011, Spain
Phone: +34 83 423358
Fax: +34 83 423310
e-mail: epi@cpd.uva.es

March 1993

Keywords: bank failure predictions, Fuzzy Sets Theory, Adaptive Resonance Theory, supervised neural network

Abstract

This paper describes an application of a neural nertwork architecture for bank failure predictions. The experiences of financial instability has motivated a great interest in the question of bank solvency. Although several prediction models have been developed, their reliability is not enough. The paper presents a supervised neural networt model, based on the Adaptive Resonance Theory (ART), which was introduced by Stephen Grossberg in 1976. Also Fuzzy Sets Theory is used as processing scheme. The data sample have been taken from the Spanish bank crisis during 1978-1983. The model obtained is able to predict the complete learning data set in two different cases, data one year prior the failure and data two years prior the failure. The obtained results, using a set of test data, reach a high level of performance, with a low number of training cycles.

*We would like to thank the members of the Neural Networks Group of D.I.S.A. (F. J. Díaz, Y. A. Dimitriadis, C. García, I. García, J. López Coronado, A. Muñoz, E. Zalama) for their contributions to the preparation of this paper.

1 Introduction

A good deal of modern banking legislation owes its existence to the preoccupation of public officials with guaranteeing the soundness of the system of payments and safeguarding the accounts of depositors. Experience gained during periods of financial instability has motivated a great number of efforts to perfect control mechanisms. It is not strange, therefore, the question of bank solvency has once again become a matter of great interest. For example, at the end of 1977, there existed in Spain 110 private banks. In the six following years, fifty-one of these banks —around 46%— found themselves enmeshed in solvency crises to a greater or lesser degree. It is therefore desirable and warranted to explore new techniques to predict these crisis situations in order to take the proper measures beforehand. Several U.S. institutions, such as the Federal Deposit Insurance Corporation, the Controller of the Currency and the New York Federal Reserve Bank, have been working on the desing of a fundamentally empirical-based, early warning system. Certain European supervising authorities have been working on similar systems. Different academic specialists have also concerned themselves with the matter, especially now that the banking sector is evolving so quickly.

Within the indicated lines of work, this paper tries to prove the efficiency of the Fuzzy ARPMAP architecture to predict bank crisis. The intention is to provide an early warnings system which would be capable of identifying, before the fact, those institutions whose economic and financial characteristics make them good candidates for joining, at some time in the near future, the group of establishments in difficulty. These systems can also serve as an alarm for shareholders, money market institutions, uninsured depositors, etc.

In Section 2, an overview of the bank failure predictions is presented. In the next section an introduction to the ART models is made. Section 4 is devoted to the Fuzzy ARTMAP architecture, which has been used as predictive model. Sections 5 and 6 report the data sample and the experimental results, while conclusions are exposed in Section 7.

2 Bank Failure Prediction Models

The bank failure prediction models proposed in the literature on the subject may be grouped into three categories:

- Theoretically-derived EWS *(Early Warning Systems)* are designed to estimate the probability of a difficult financial position by using models defined a priori by an explicit theory on suspension of payments or of bankruptcy, that formalizes the causes and interactions of the factors bringing about these events. [1, 10]
- Simulation-based EWS are designed not only to detect problems but, in addition to analyze alternative lines of action for solving them. The model is supposed to be an analog of the economic-financial process carried out in a depository institution, and reproduces its essencial characteristics and relations. [9]
- Ex post empirical EWS. In order to construct these models, the usual way is to gather sample information on institutions that are functioning normally and on establishments that have gone into crisis over a certain period. By comparing the characteristics of the latter at a given time prior to distress with those of the former group, it is possible to identify where the main systematic differences lie. From there, it should be possible to design a prediction procedure. The characteristics thus studied are usually economic and financial ratios derived from accounting statements or observed on the stock market. These indicators are implicity considered as potentially relevant for the purpose of the study. While, in some cases, the analysis is limited to the individual examination of each indicator, in other studies an integrated -or multivariate- perspective is introduced, in order to obtain a sample function that offers a more or less wide combination of predictive variables. There is quite a range of techniques nowadays used: from simple graphic and tabular comparison, through discriminant analysis, the linear probability model and the arctangent regression,as logit and probit to neural network models. [3]

We will focus our interest on the Fuzzy ARTMAP model as an ex-post empirical EWS for predictiving banking crisis.

3 Adaptive Resonance Theory

The Adaptive Resonance Theory (ART) was introduced by Stephen Grossberg and Gail Carpenter, of Boston University, in 1976 [4, 5, 6]. The ART models are based on a few basic neural modules with serious neurobiological foundations, but they have not yet gained an important influence in commercial applications. ART architectures are mainly unsupervised neural networks that self-organize stable recognition codes in real time in response to sequences of input patterns. The code self-stabilization property is obtained by using two adaptive filters of weights. The bottom-up adaptive filter is contained in pathways leading from a feature representation field to a category representation field. The top-down adaptive weights filter is used in the feedback pathway. This structure lets us solve the dilemma between plasticity,—potential for rapid change in the weights—and stability,—not forgetting the past learnt patterns—. The real time response means that there is no border between training and working time. That means that the neural system works in both learning and operating modes simultaneously. Different specific architectures have been proposed, depending on the type of the input pattern: ART1 for binary vectors, ART2 for analog ones and ART3, which incorcoporates the possibility of distributed pattern recognition codes. Also, some supervised architectures have been proposed such as ARTMAP or Fuzzy ARTMAP. [7, 8] These supervised learning systems are built up from a pair of unsupervised ART modules, (ART_a and ART_b), which are linked by an associative learning network. The already mentioned ART properties imply large differences between these supervised architectures and the Back-Propagation networks.[7] For example, Fuzzy ARTMAP learns in five training epochs a benchmark that requires twenty thousands epochs for Back Propagation to learn. [6]

4 Fuzzy ARTMAP

The Fuzzy ARTMAP system is a neural network architecture based on the Fuzzy Sets Theory and the ART architectures. The complete Fuzzy ARTMAP architecture includes two ART modules called ART_a and ART_b. Each of them creates stable recognition categories in response to arbitrary sequences of input patterns. During the supervised learning period, the ART_a module recives a vector of input patterns I_a^r, while ART_b recives the vector of input patterns I_b^s, which is the known prediction given I_a^r. These modules are linked by an associative learning network and an internal controller, which is designed to create the minimal number of recognition categories, at F_2 field, needed to get accuracy criteria. It is done by using a Minimax Learning Rule that enables the system to learn quickly, looking for minimizing the predictive error and maximizing the predictive generalization.

The main parameters of the system are:

- Vigilance parameters ρ_a and ρ_b are dimensionless parameters used to establish a matching criterion. Vigilance parameter ρ_a of ART_a is increased by the minimal amount needed to correct a predictive error at ART_b. However, lower values of ρ_a lead to form larger categories, broader generalization and higher code compression.

- Choice parameter, α, which tunes the choice function at the category field. Small values of α tend to minimize recoding during learning.

- Learning rate parameter, β, enable to get fast learning, with $\beta = 1$, and slow recoding learning, with $\beta < 1$.

The convergence of the learning algorithm is ensured because all adaptive weights, in the Fuzzy ART systems, are monotone increasing. This leads to a problem of categories proliferation, which is solved by a pre-processing scheme called complement coding. It consists of representing each feature by two input variables, one to encode the presence of a feature and the other to encode its absence. Therefore, we should obtain two post-processing input variables for every input one. The first one, y_j^p, represents the presence of the characteristic, and the second one, y_j^a its absence.

$$y_j^p = \frac{I_a^j - min_j(I_a^j)}{max_j(I_a^j) - min_j(I_a^j)}$$

$$y_j^a = 1 - \frac{I_a^j - min_j(I_a^j)}{max_j(I_a^j) - min_j(I_a^j)}$$

This post-processing format have been chosen in order to be suitable for the input of the neural network.

5 Data sample

We have focused our attention on the 1978-1983 peak of the spanish banking crisis. On December 31, 1981 there existed in Spain 128 private banks, including Extebank, which is usually considered as a private bank in spite of its mixed ownership. However, 28 banks which were considered as *foreing*, 17 banks that were at the time the object of reorganization processes and 3 banks absolutely atypical within the panorama of spanish banking have been taken off the study cases set. If, therefore, we exclude these 48 establishments, we are left with a sample of 80 Spanish private banks which, at the end of 1981, were apparently functioning in normal way. Between December 31, 1981, and February 23, 1983, a crisis situation arose in 31 out of these 80. The series of 15 ratios have been taken from a previous study[12], and have been calculated for each of the 80 members of this sample, from data and accounts corresponding to 1981. The list of ratios are:

R1: Total loan portfolio/total assets
R2: Private securities/total assets
R3: Fixed assets/total assets
R4: Cash assets/total liabilities
R5: Cash assets/borrowings
R6: Net positions in the credit system/customers' total resources and other liabilities
R7: Ordinary investment/ordinary resources
R8: Customers' term resources/customer' resources
R9: Free equity capital/total loan portfolio
R10: Equity capital/contingent liabilities
R11: Interest expenses/total operating revenues
R12: Gross margin/average customers' resources
R13: Operating expenses/gross margin
R14: Depreciation and provisions/gross operating earnings
R15: Earnings after tax/average total assets

Then, we may do an analysis or prediction one year prior to failure. The data have been taken from the *Anuario Estadistico de la Banca Privada*, published by the Higher Banking Council. For the purpose of the model, we must define what we mean by insolvent bank. In early empirical studies[1], we found a variety of definitions. We shall here apply the term insolvent to institutions in crisis—that means, institutions that have been required by the Bank of Spain to restore their impaired stated capital—.

6 Experimental results

Let N be the total number of observations (banks), M the number of explicative variables, x_j^i the *jth* explicative variable corresponding to the *ith* observation (accounting ratio j of the i bank), and y^i the binary dependent variable

where,

$$y^i = \begin{cases} 0 & \text{if bank } i \text{ were solvent one or two years later} \\ 1 & \text{if bank } i \text{ were insolvent one or two years later} \end{cases}$$

During the supervised learning period the inputs of the Fuzzy ARTMAP are:

$$I_a^i = \{x_0^i, x_1^i, ..., x_j^i, ..., x_M^i\}$$

and

$$I_b^i = \{y^i\}$$

where,

$$i \in (1, N)$$

Let us define:

- Type I error: an actually insolvent bank is classified as solvent one.
- Type II error: an actually solvent bank is classified as insolvent one.

Trying to predict crisis from data of one year prior, we choose the complete data set of 80 banks ($N = 80$), to make the net learn (case A). After 4 training epochs, the system is able to predict correctly the solvency of all banks in this set. Therefore, 100% of success was obtained using different vigilance parameters, $\rho_a = \rho_b = 0.95, \rho_a = \rho_b = 0.85, \rho_a = \rho_b = 0.75$. The vigilance parameter in test epoch was in all the cases 0.05 units lower than the one used in the learning mode. In a second experiment (case B), we choose a random sample of 41 banks (25 of them were solvent and 16 insolvent), to train the neural network, and the rest of the banks (39) as test data set. The parameters used were, $\rho_a = \rho_b = 0.88, \alpha = 0.001, \beta = 1.0$ in the training epochs, and $\rho_a = \rho_b = 0.82. \alpha = 0.001, \beta = 1.0$ in the test epoch. Finaly, in the third experiment (case C), we take the same set of 80 banks ($N = 80$), with data of two years prior to the bank crisis, to make the neural network learn. After 4 training epochs, the system is also able to predict correctly the solvency of all banks in this set, two years prior to the crisis. Therefore, 100% of success was obtained using vigilance parameters, $\rho_a = \rho_b = 0.85$. The vigilance parameter in test epoch was 0.05 units lower than the one used in the learning mode. The comparative results are shown in the table.

Experimental	*LOGIT*		*L.P.M.*		*NORMIT*		*FuzzyARTMAP*	
Errrors (%)	I	II	I	II	I	II	I	II
A	3.2	2	3.2	0	3.2	2	0	0
B	18.8	0	6.2	8.16	47	45	6.2	6.5
C	6.5	0	9.6	6.3	6.4	0	0	0

7 Conclusions

We have presented a new approach to the bank failure prediction using a Fuzzy ARTMAP architecture. The supervised neural network studied in this paper offers an alternative to the existing failure bank prediction models. Empirical results show that this model has a good behaviour. The Fuzzy ARTMAP system incorporates the well-known advantages of the ART models, such as the reduced number of training cycles and the convergence of the weights to the supervised neural network models. The future research lines lead to analyse the efficiency of the different models in a comparative study.

References

[1] E.I. Altman, R.B. Avery, R.A. Eisenbeis and J.F. Jr. Sinkey, Application of classification techniques in bussiness, banking and finance, JAI Press, Greenwich, CT, 1981.

[2] E.I. Altman, R.G. Haldeman and P. Narayanan, "ZETA analysis: a new model to identify bankruptey risk of corporations", Journal of Banking and Finance, 1 (1), pp. 29-54, 1977.

[3] J. F. Bovenzi, J. A. Marino and F. E. McFadeen., "Commercial bank failure prediction models", Economic Review (Federal Reserve Bank of Atlanta), 68 (11) , pp. 14-26, 1983.

[4] G.A. Carpenter and S. Grossberg, "ART 2: Self-Organization of Stable Category Recognition Codes for Analog Input Patterns", Applied Optics, 26 (23), pp. 4919-4930, December 1987.

[5] G.A. Carpenter and S. Grossberg, "ART3: Hierarchical Search Using Chemical Transmitters in Self-Organizing Pattern Recognition Architectures", Neural Networks, 3(1), pp. 129-152, January/February 1990.

[6] G.A. Carpenter, S. Grossberg and D.B. Rossen, "Fuzzy ART: Fast Stable Learning and Categorization of Analog Patterns by an Adaptive Resonance System", Neural Networks, 4(1), pp. 759-771, January/February 1990.

[7] G.A. Carpenter, S. Grossberg and J. H. Reynolds, "ARTMAP:Supervised real-time learning and classification of nonstationary data by a self-organizing neural-network.", Neural Network, 4 (5), pp. 565-588, 1991.

[8] G.A. Carpenter, S. Grossberg, N. Markuzon, J.H. Reynolds and D.B. Rosen, "FUZZY ARTMAP: A neural network architecture for incremental supervised learning of analog multidimensional maps.", IEEE Transactions on Neural Networks, 3, pp. 698-713, 1992.

[9] G.A. Hanweck, "Using a Simulation Model Approach for the Identification and Monitoring of Problem Banks.", Research Papers in Banking and Financial Economics, Washington DC, 1977.

[10] R. W. Nelson, "Bank capital and banking risk", Research Paper, 7902, New York, Federal Reserve Bank of New York, 1979.

[11] A.N. Refenes, M. Azema-Barac and P.C. Treleaven, "Financial Modeling using Neural Networks", Technical Report UCL-CS-92-94, University College London, 1992.

[12] J.M. Rodríguez, "The Crisis in Spanish Private Banks: A Logit Analysis", Finance, 10(1), pp. 69-88, 1989.

[13] K.Y. Tam and M.Y. Kiang, "Managerial Applications of the Neural Networks: the case of bank failure predictions", Management Science, 38(7), pp. 926-947, July 1992.

[14] E. Zalama, C. García and J. López. "Adaptive Resonance Architecture for Authorization of Deficits", Proc. of the International Conference Avignon'92, Banking and Insurance Conference June 1992.

[15] Fondo de Garantía de Depósitos en Establecimientos Bancarios,Memoria correspondiente al ejercicio 1983. Madrid, 1983.

COMBINATION OF NEURAL NETWORK AND STATISTICAL METHODS FOR SENSORY EVALUTION OF BIOLOGICAL PRODUCTS: ON-LINE BEAUTY SELECTION OF FLOWERS.

F. ROS*, A. BRONS*, F. SEVILA*, G. RABATEL*, C. TOUZET**
* CEMAGREF, BP 5095, 34033 Montpellier CEDEX 1, France
** LERI-ERIEE, Parc Scientific George Besse, 30000 Nîmes, France

Abstract
In order to automize on-line selection of biological products, it is necessary to determine relationships between human sensory evaluation (like the beauty of flowerplants), and physical measurements on objects (like machine vision images). Classical methods of image processing and statistics, are combined with neural network techniques. The research deals with methods for the selection of significant parameters for the judgement, and methods for decision learning and generation: for both types of methods, classical statistics and neural network technics are either compared or combined. Interest of the various combinations are discussed, through the application on beauty selection of flowerplants.

KEYWORDS : Multilayer neural networks, Principal Component Analysis, Backward stepwise selection, Sensory evaluation, Image processing.

1. INTRODUCTION

Many complex automation tasks in agriculture or the food industry are described in the literature and are showing the important role of machine vision (Tillett 1991). Images grabbed from various types of cameras into a computer memory are processed in order to extract some appropriate features used as decision elements. In most of the cases, the decision making procedure based on image processing parameters is quite easily solved. For instance, in the case of appreciation or detection of natural variable objects, Bayesian statistical classifiers are now in common use (Slaughter and Harrel 1989).
These decision methods are however limited : indeed, for an easy implementation the number of classifying parameters cannot be numerous, 2 to 3 being the common situation.
The problems left for automation in agriculture or the food processing industry correspond to the more complex operations, demanding human experience. For these tasks, a learning phase based on multiple perception is needed to afford rapid decisions and actions.
Existing statistical methods can help to solve these kind of problems: the easiest methods to implement are mainly dealing with linear situations, where the decision variables could significantly be computed through a linear combination of the descriptive parameters. However, most of the decisions on biological objects happen to be non-linear.
Another difficulty is found when human experts are not making coherent decisions on these products: building an artificial decision making system means also to overcome this problem.
Neural networks have recently been employed as classifying tools for situations of non-linearity: their "black box" type functioning (descriptive variables at the input of the box generate decision values at the output) makes them very attractive for engineers who have to implement effective solutions for automatic classification.
For such complex situations (numerous parameters, non-coherent experts, non-linear decision), the purpose of the research is to look for methods of implementing on-line automatic decision-making based on image features of biological objects.

2. DESCRIPTION OF THE SCIENTIFIC APPROACH

The areas which are adressed in this research are:

*<u>how to handle informations and data produced by subjective evaluations of groups of non-coherent human experts</u>: using state of the art technics, a method to generate a VIRTUAL EXPERT to be used as a reference has been built.

*<u>how to reduce the number of significant parameters used for the artificial decision system</u>: the number of initial variables is important and it favorises the presence of noise and redundancy. The literature proposes methods for variables reduction, which have been adapted and applied.

*<u>how to modelise the human sensory and subjective evaluation</u>: models using only statistical methods, or only neural networks have been built in order to compare their performances. The criteria of comparaison between models is the correlation with the virtual expert judgements (see above).

Endly, the above steps for data processing and decision modelisation are combined in PROTOTYPE CLASSIFIERS which use statistics (to condense information) and neural network (to make the decisions). Their design and application are discussed.

Evaluation of the various methods and prototypes in this research have been made on the case of flowerplants classification, usually based on beauty evaluation. Descriptive variables are measurements of image features. Decisions are based on the gradings of these plants made by groups of experts.

3. STATE OF THE ART

3.1 Statistical methods:

Multiple linear regression (MLR)

In food quality evaluation, the most common use of MLR is to predict the magnitude of some sensory attribute on the basis of a series of objective measures of food (Resurreccion 1988). This method is used to approximate the output decision variables, using a linear combination of the descriptive variables.

Principal Component Analysis (PCA)

PCA constructs linear combinations of the original data to construct other variables or components, which are independent, orthogonal and of known significance.

PCA can be used:

*to reduce the number of variables down to a smaller number of significant components with as little loss of information as practicable. The first component will account for the greatest portion of variance, the second for the second largest portion, and so on (Resurreccion 1988).

*in sensory analysis, to determine if a panellist is so singular in his or her responses as to be an outlier, or to subdivide the panel into groups more homogeneous in their responses than the panel as a whole (Powers 1984, 1988).

3.2 Multilayer neural networks:

Artificial neural networks or connectionist models are artificial intelligence systems which attempt to achieve good performance via dense interconnection of simple computational elements. The behaviour of an artificial network is given by its structure and the strength of the connections. The learning algorithm consists in modifying the connection strengths to improve performance.

A multilayer neural network is a feed-forward model with one or more layers of computing elements between the input and output layer. The elements of each layer are connected with elements of the previous layer(s). In a layer, there are no connections between the elements. A multilayer network functions as follows : the inputs are copied in the input layer. Each neurone computes its output value by an activation function f(I), where I is given by :

$$I = \sum_{i=1}^{N} wi * xi \quad + \quad q$$

where N is the number of inputs, wi a weight connection, xi the output of the connected element and q a bias. The activation function is often a sigmoid function. The most widely used learning algorithm for this type of network is the backpropagation rule (Rumelhart and McClelland 1986). It is an iterative gradient algorithm designed to minimise the mean square error between the output signal of the network and the desired output. For detailed descriptions of this algorithm, the reader is referred to (Rumelhart and McClelland 1986, Lippmann 1987 and Pao 1989).

4 MATERIALS AND METHODS

4.1. objets for sensory evaluation: the flowerplants data

Red cyclamen is, like most horticultural products, a pot plant for wich no quality standard is used in Europe. Therefore it has been chosen for our research because they are among the main marketed flowered-pot-plants in France. A large number of cyclamen plants were sampled in order to gather a large range of quality levels. Each expert has given three grades: one for the quality of the leaves, one for the quality of the flowers and one for the general quality of the plant.

We organised two experiments. In the first one with 13 experts and 100 plants (EXP1). In the second one we had 450 plants and 7 experts (EXP2).

4.2 Image features Measurements

The machine vision equipment used consisted of: a 512*512 colour CCD camera, a BYTECH colour frame grabber board, a 386 PC computer, CEMAGREF software for colour images features extraction.

By colour segmentation of each of these images, those pixels representing the plant are selected , then the leaves are separated from the flowers. Two segmented binary images are produced for each image of the plant (see figure 1).

On each of these binary image, the following 10 features are computed: total surface area, total perimeter, co-ordinates x and y of the centre of gravity, moments of inertia along Ox and Oy, perimeter and surface area of the convex hull, co-ordinates of the convex hull centre of gravity.

So, for each of the cyclamen potplants of EXP1 or EXP2, we have descriptive features (V_I) out of this images processing, and sets of 3 grading numerical values (O_K) given by experts.

4.3 Statistics and Neural Network packages

For the statistical calculations we have used the statistical software library STATGRAPHICS from STSC Inc. The neural networks were simulated with the software from NeuralWorks Professional II/Plus from NeuralWare Inc.

4.4 Learning and testing procedure for the Artificial Classifiers
In the experiments calibration groups are randomly chosen among the plants to train the models (50 among 100 in the first set, and 300 among 450 in the second one). We apply this procedure 4 times on the same set of plants, making 4 different calibration samples. Each prototype classifier is thus built 4 times, being calibrated and tested each time on different samples of our data base.
We want the model to react as an expert, so its behaviour has to be similar to the experts ones. Which imply to evaluate its performance compare to these human experts or panellists. To do so, we first have to analyse the results of the panellists: McDaniel et al. (1987) and Malek et al. (1986) chose correlation analysis to judge the performance of the panellists vis-à-vis the others.
In our case a virtual expert is produced out of the evaluations of the panel of experts (see method in next chapter). We decide that the model is evaluated by its correlation with this virtual expert. In the figures, this correlation is also compared with the correlations of the various real experts, with the virtual one.

5. RESULTS OF THE RESEARCH
5.1. Method for the production of a virtual human expert
Since sensory evaluation is made by a panel of human experts, it is intended to extract from their decisions, a representative global behaviour which can be used as a reference for the training of the artificial decision system.
Powers (1988) proposes various methods to analyse the decisions of a group of N experts, in order to determine those who have a special behaviour compared to the group. Among these methods, a PCA on the grades given by the M experts is described: the purpose is to replace the non-coherent values from the M experts, by the values computed using the first m most significant components given by the PCA (with $m < M$).
If the first component of this PCA is taking more than 75% of the total variance of the experts judgements, we propose to chose it for the design of a Virtual Human Expert who will replace the panel for futur comparison and training.
If the first component of this PCA is not taking enough variance, it means that the panel is not coherent enough to be used as a reference for futur training of the artificial model. In such case Powers method is applied to detect and eliminate the expert (s) whose behaviour (s) is (are) the less coherent with the group. Then the above procedure is run another time, until obtaining a virtual expert.

5.2 Preparation of descriptive variables
The descriptive variables from physical measurements are generally too numerous for the system to be efficiently implemented: it is thus necessary to reduce them. Furthermore they are generally not independant and noisy.
Two methods are described in the literature to reduce and improve the input variables in a statistical model: through PCA based methods and a stepwise selection by multiple linear regression, see Tomassone 1983.
According to the type of data (to their linearity, to their redundancy, to the noise levels) each of them, or the combination of the 2 can be applied.

5.3 Artificial classifiers design
Descriptive variables (V_i) on the biologic objects have been selected and improved: a new set of descriptive components (I_j) is thus produced to be used for artificial classification. A classifier will use (I_j) as inputs, and should produce decision parameters (O_K) as close as possible to the human expert ones. The classifier is trained with real data and decisions coming from a panel of experts.
Two kinds of model are considered for such a task:

- a multiple linear regression model, called MLR-model
- a neural network model, called NN-model

Both can easily be implemented and computed or trained with existing software (see section 2.6).

- In the MLR-model, each decision parameters (O_K) is computed through a linear combination of the descriptive variables (I_j)

.* for the NN-model, a 3-layer neural network type is empirically chosen. The values of the selected variables (I_j) for the object are fed into the input layer. The readings of the output layer give the decision numerical values (O_K) for this object.
In both cases, classification is directly obtained from the object image features measurements, allowing their use on-line for automatic classification.

5.4. Test of the methods on a flowerplant Data Base
To evaluate the methods described in the last 3 chapters, they have been implemented for the first flowerplant data base EXP1.

virtual human expert

By application of the method in chapter 5.1. the virtual human expert for EXP1 is obtained by a standardised mean of the panel of the 13 experts. Figure 2 shows how the 13 human experts of the panel correlate with this virtual human expert.

descriptive variables preparation

The original set of 40 descriptive variables for EXP1 has been considered as the first possible input set of variable $(I_j)_1$.
Then the 3 possible methods of data set improvement described in chapter 5.2. have been applied:

1 A PCA is run and the first 15 components $(I_j)_2$ are selected.

2 A selection through linear regression is run and produces 16 most significant data $(I_j)_3$.

3 A linear regression is first run to select 16 among the 40 original variables. A PCA is then run on the 16 selected variables leading to 8 input components $(I_j)_4$.

prototype artificial classifiers

Both classifiers described in chapter 5.3. are applied successively to the 4 set of input data $(I_j)_p$. This led us to 8 different prototypes of artificial classifier for the evaluation of flowerplants according to data set EXP1. For the NN classifier, one hidden layer with 5 neurones has been found as the best solution for this modelisation, according to various experimental attempts made with various other number of layers and number of neurones. The learning rate was 0.15 with a momentum value of 0.4.

experimental results

The 8 prototypes of classifiers have been evaluated according to procedures described in chapter 3: see figures 3 and 4.

In figure 3, we see that a linear model has never obtained the minimum performance of the human experts. This confirms the results of Journot (1991).

The NN-models (figure 4) do obtain the minimum performance of the human experts. In the case of the flower quality all the models are satisfactory, and we can note a slight amelioration when the image features are pre-processed with the statistical methods. For the other gradings, the pre-processing is necessary to obtain satisfactory results.

Although results shown in figure 4 appear to be encouraging, there are still some difficulties: Since each performance of tthe prototypes are computed with an average of 4 different calibrations, as indicated in chapter 3.4., it appears that the results for each model vary a lot depending on the patterns used for training. This might be due to the limited size of the pattern sets (50). The pattern set for validation could be out of range of the training set for both the input and output values, since the choice has been made randomly with no quality balancing constraints.

This is why further developments have been necessary.

6. COMPLEMENTARY RESEARCH DEVELOPMENTS

methodological discussion

To limit variability of the results due to unsufficient number of training examples, as obtained in the experiments in chapter 5.4., first idea is to produce set of data with increased number of patterns.

With more training patterns, the network can analyse more details of the grading procedure, but the capacity of the network to treat the non-linearities appears by experience to stay limited.

We can increase the number of hidden layers, but the backpropagation algorithm becomes less efficient (Pao 1989).

An other method to treat non-linearity has been proposed by Pao (1989). A pre-processing of the input variables of a multilayer neural network uses functional links. Initial input variables are transformed with a set of independent functions, such as:

$$f_k(I_i) = a_k.\sin(k*\pi*I_i) + b_k.\cos(k*\pi*I_i)$$

$$f_k(I_i) = a_k.(I_i)^k \quad pour \quad k = 1,2,3,\ldots$$

These functions are a kind of non-linear pre-processing that is otherwise done by the network itself. This method can handle those problems with high non-linearity without increasing the number of hidden layers. It increases however the number of inputs and so the number of connections.

complementary experiments

A second experiment with flowerplants has been made using data set EXP2 described in chapter 3.1. It includes 450 plants instead of 100.

Since in the previous step of the research, best results were obtained with a NN classifier using a $(I_j)_4$ type of selected descriptive data, this method will be considered as a reference one for this second experiment.

An other classifier has been designed by applying the functional links method. It is called FL-model. The links are computed on the 8 initial $(I_j)_4$ data to produce different inputs in the network. The chosen links are:

$$f_k(I_i) = \sin(k*\pi*I_i)$$

$$f_k(I_i) = I_i; \quad I_i * I_j \quad pour \quad i = 1,2,\ldots,8; \quad j = i+1,\ldots,8$$

This gives us 68 inputs for the neural network which gives us a more detailed description without increasing the redundancy or the noise. The model generated by this method gives better results then the NN-model on the EXP2 data set of flowerplants (see figure 5).

7. CONCLUSION

The purpose of this research has been to look for methods of implementations of on-line automatic decision based on image features of biological objects, in the case of complex situations (numerous parameters, non-coherent experts, non-linear decisions).
Various aspects have been looked at:
* non-coherent expert judgements: a method to generate a virtual human expert which significantky can represent the a panel of human expert has been designed.
* improvement of descriptive variables: 2 state-of-art reduction and improvement procedures have been studied, and a combination of the 2 has been proposed .
* design of artificial classifiers: 2 types of classifiers have been considered. One is based on the computation of linear regression between selected descriptive variables and the virtual expert decision variables. The other is based on a 3-layer neural network trained with the same variables.
Evaluation of the various methods and classifying prototypes proposed in this research has been made on the case of cyclamen potplant classification. Descriptive parameters are features extracted from images of these plants. Decisions are based on the gradings of these plants made by a group of experts.
On a first potplant database, the neural network classifier using statiscally improved variables was the lone to give acceptable result compare to human expertise. Non stability of the results have brought however some difficulties. They were related to unsufficient size of the training base.
A complementary methodological approach has been made to overcome such difficulties: apart from increasing the size of the data base for better training, a functional link method has been applied to unable the classifier to take non-linearities into better account.
Compare to the first neural network classifier developped in this research, the functional link one gave better performances when implemented on a second experiment on flowerplants.
Combination of statistical data pre-processing, with functional link neural network modelisation appears to be the best design for an artificial classifier for complex sensory evalutions of biological products.

REFERENCES

1. Journot V. 1991. Détermination d'un indice de la qualité des produits agricoles en frais. Thesis at the Institut National Agronomique, Paris-Grignon (in French).
2. Malek D.M., Munroe J.H., Schmitt D.J. and Korth B.. 1986. Statistical evaluation of sensory judges. Am. Soc. Brew. Chem. J.. 49(1):23.
3. Martens M. 1985. Sensory and chemical quality criteria for white cabbage studied by multivariate data analysis. Lebensmittel Wissenschaft und Techn. 18:100-104.
4. McDaniel M., Henderson L.A., Watson B.T.Jr. and Heatherball D.. 1987. Sensory panel training and screening for descriptive analysis of Pinot Noir wine fermented by several strains of malolactic bacteria. Journal Sensory Studies. 2:149.
5. Lippmann R.L.. 1987. An introduction to computing with neural nets. IEEE ASSP Magazine. Vol.4. p4-22.
6. Pao Y.. 1989. Adaptive Pattern Recognition and Neural Networks. Addison Wesley Pub. Comp.
7. Papelier S. 1990. Applications de la vision assistée par ordinateur à l'étude de la qualité du cyclamen. Report CNIH (in French).
8. Powers J.J. 1984. Using general statistical programs to evaluate sensory data. Food Technology 38(6):74-84.
9. Powers J.J. 1988. Uses of multivariate methods in screeening and training sensory panelists. Food Technology 42(11):123-127.
10. Resurreccion A.V.A. 1988. Applications of multivariate methods in food quality evaluation. 42(11):128-136.
11. Rumelhart, D.E., J.L.McClelland. 1986. Parallel Distributed Processing. MIT Press.
12. Slaughter, D.C., R.C. Harrell 1989. Discriminating fruit for robotic harvest using colour in natural outdoor scenes. TRANS. of the ASAE March-April 32(2):757-763.
13. Tillett R.D.. 1991. Image Analysis for Agricultural Processes : A Review of Potential Opportunities. J. Agricultural Engineering Research. Vol.50(4): 247-258.
14. Tomassonne R. 1983. La régression linéaire : Nouveaux regards sur une ancienne méthode statistique. MASSON (in French).

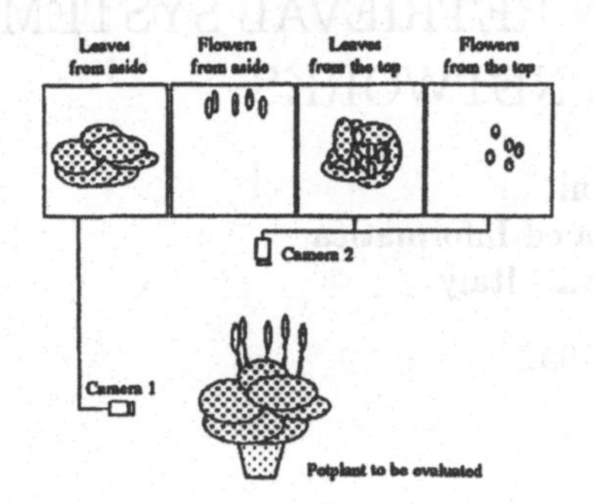

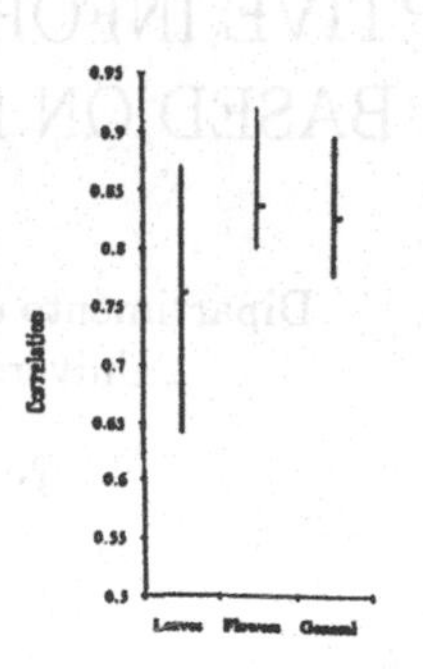

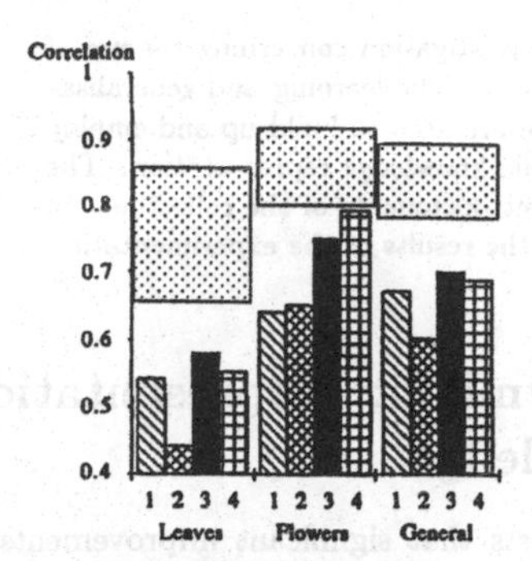

figure 3:The correlation of the 4 types of MLR-models with the virtual expert:
1 : with 40 input variables 2: with the 15 PCA-components
3 : with the 16 LR-variables 4: with the 8 C-components
The box in the background is representing the correlations of the panellist with the virtual expert.

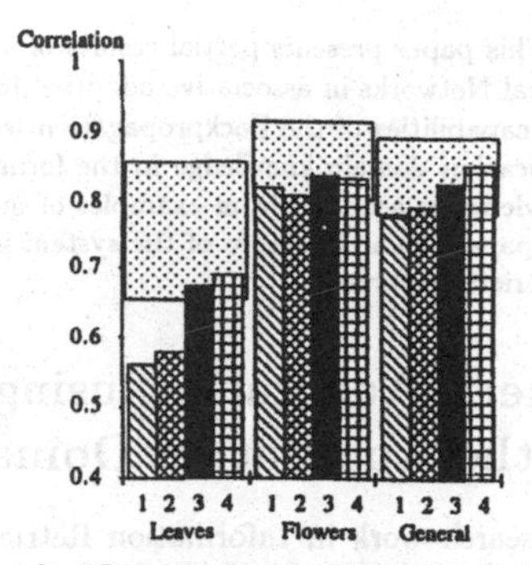

figure 4: The correlation of the 4 types of NN-models with the virtual expett :
1: with 40 inpt variables 2 : with the 15 PCA-components
3: with the 16 LR-variables 4 : with the 8 C-components
The box in the background is representing the correlations of the panellist with the virtual expert.

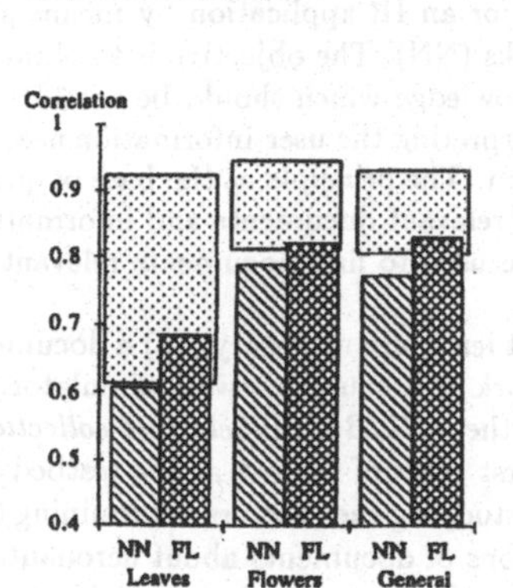

figure 5 : The correlation of the NN and FL-model with the vitual expert. The results can be compared with the correlation of the expert, represented the box in the background.

AN ADAPTIVE INFORMATION RETRIEVAL SYSTEM BASED ON NEURAL NETWORKS

Fabio Crestani*
Dipartimento di Elettronica ed Informatica
Universita' di Padova - Italy

November 22, 1992

Abstract

This paper presents partial results of an experimental investigation concerning the use of Neural Networks in associative adaptive Information Retrieval. The learning and generalisation capabilities of the Backpropagation learning procedure are used to build up and employ application domain knowledge in the form of a sub-symbolic knowledge representation. The knowledge is acquired from examples of queries and relevant documents of the collection. In this paper the architecture of the system is presented and the results of the experimentation are briefly reported.

1 Query Adaptation using a Subsymbolic Representation of the Application Domain Knowledge

Recent research work in Information Retrieval (IR) suggests that significant improvements in retrieval performance requires techniques that, in some sense, "understand" the content of documents and queries. Recently IR researcher have tried to use application domain knowledge to determine relevant relationships between documents and queries. The aim of my research is to show the possibility of learning and use domain knowledge for an IR application by means of a the learning and generalisation capabilities of Neural Networks (NN). The objective is to obtain a sub-symbolic representation of an IR application domain knowledge which should be used by an IR system to adapt an original user formulated query, so interpreting the user information need in the light of particular characteristics of the application domain. The principle at the base of query adaptation is that similar queries should have similar sets of relevant documents and information acquired about documents relevant to other queries could be used to find documents relevant to a new query.

In order to perform an experimental analysis, two tools, at least, are necessary: (a) a document collection with relevance judgements, and (b) a neural network or a neural network simulator.

The document collection chosen for the investigation is the *ASLIB Cranfield test collection*. This collection was built up with considerable effort in the first years of the 60s as the testbed for the ASLIB-Cranfield research project. This project aimed at studying the "factors determining the performance of indexing systems" and produced two collections of documents about aeronautics. They were comprehensive of documents and relative requests and relevance judgements developed in order to perform various retrieval experiments. The two main collections have different sizes. The largest is made up of 1400 documents with 279 requests and relevance judgements. The second one, which is the most used, is made of 200 documents with 42 requests and relevance judgements. For a full description of the collection see [1].

---

*The author is currently a visiting Research Fellow at the Department of Computing Science, University of Glasgow (UK)

Because of operative limits, at present only the 200 document collection has been used in the investigation. It is of course understood that this limits the generality of the results obtained. However, the main purpose of this investigation is to demonstrate the feasibility of the proposed approach. There are still many open issues in the application of NN to IR, and the problem of scaling the results is one of the major ones. Nevertheless, it must be noticed that the dimension of the data set used in the these experiments is larger than that used in any other application of NN techniques to IR, to the present author's knowledge.

For the investigations reported in this paper a NN simulator, *PlaNet 5.6* ([2]), running on a fast conventional computer has been used. The details about the particular learning algorithm used and the experimental setting of its parameters are fully reported in [3].

2 An Adaptive IR System

A simple Adaptive IR System (henceforth simply called AIRS) was developed. It is made up of the following components:

Query Processor : transforms a query expressed using index terms into a binary vector (1 indicates the presence of the term in the query and 0 its absence). The dimension of the vector must enable the representation of all the possible queries a user might formulate.

Neural Network Simulator : a 3 layers feedforward NN using the Back Propagation (BP) learning rule simulated on PlaNet.

Matcher : evaluates the similarity between two binary vectors using the Dice's coefficient (see [4]) and produces a value indicating that similarity.

Document Processor : transforms documents, which are usually represented using index terms, into a binary vector representation.

Using AIRS implies performing two phases: a training phase and a retrieval phase.

Before using the AIRS for retrieval purposes, it must be trained. The structure of the system during the **training phase** is depicted in Figure 1 on the left. During the training phase there is no use of the Matcher. On one side, the Query Processor gets the query in the form of a set of index terms. The Query Processor transforms it into a binary vector whose dimension is that of the input layer of the NN model. The input and the output layer of the NN model used by the NN simulator are set to represent a query and a relevant document, and the BP algorithms is used for the learning. This is monitored by the NN simulator control structure and when some predetermined conditions are met the learning phase is halted. Link matrices are produced, representing the application domain knowledge acquired. They are stored for their further use in the retrieval phase. The AIRS is fed according to various teaching strategies as it will be explained in Section 4.

During a **retrieval phase** the components of the AIRS interact with each other in the following way (see Figure 1, on the right). After the query processor has transformed the query into a binary representation, the NN is activated. The activation spreads from the input layer to the output layer using the weight matrices produced during the training phase. The vector representing the query is therefore modified or, better, adapted according to the application domain knowledge and a new query representation vector is produced on the NN simulator's output layer. On the other side, the entire collection of documents is transformed into a large representation matrix by the Document Processor. This big matrix is then fed, together with the result of the query adaptation into the Matcher. The Matcher produces a ranked list of document identification numbers. The ranking reflects the evaluated relevance of the documents to the query. The interface to the AIRS display the documents to the user according to their evaluated relevance to the query. The user can the assess the actual relevance of the documents presented and mark them according to his personal perceived relevance. This can be used for further training also inside the retrieval phase. The results of this training, however, will be discarded at the end of the retrieval phase in order to avoid AIRS to be influenced by personal, and not supervised, relevance relationships.

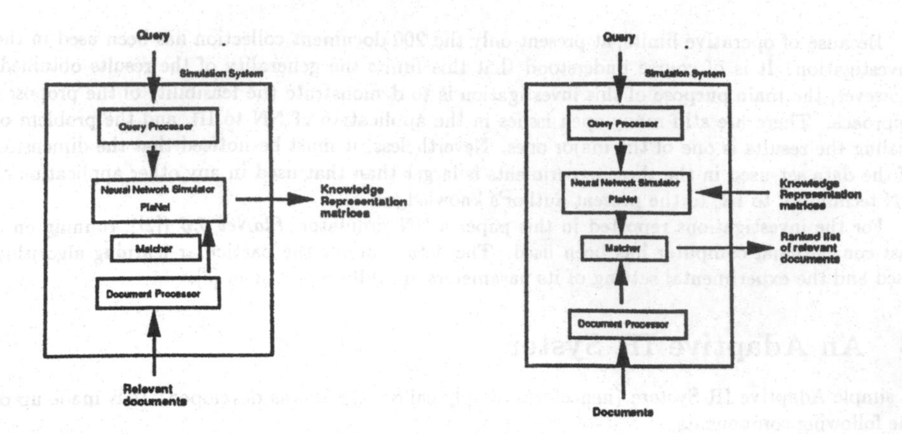

Figure 1: Schematic view of the simulation system during the training and the retrieval phases

3 The evaluation criteria

The main features which were considered for evaluation were the following.

First the learning performance is evaluated, that is the ability of the system to acquire domain knowledge. This is evaluated using an approach which is classical in NN research. It consists of the evaluation of the "mean error" between the the training (target) results and the obtained results.

Then the generalisation performance, that is the ability of the system to generalise domain knowledge in the retrieval phase is evaluated. The recall and precision performance of the AIRS are determined. These are two well known effectiveness measures in IR. They are respectively, the proportion of all documents in the collections that are relevant to a query and that are actually retrieved, and the proportion of the retrieved set of documents that is relevant to the query. They have been evaluated at different stages of learning and with different sets of training examples. If some learning of the domain knowledge has taken place, and if the representation structure can generalise it, then an improvement in the performance obtained in the retrieval of new queries has to be expected.

Finally, the retrieval performance is evaluated. This is done comparing the retrieval performance of the AIRS with that of a classical IR systems. This comparison enables the evaluation of the query adaptation strategy versus the use of the original query. The results of this comparison are again presented using recall and precision graphs.

4 Experimental results

The typical form of the following experiments consists, first, in training the system using a subset of the examples provided by the relevance assessment. The knowledge acquired by means of the training phase, is then used to adapt the original user query to take into account the experience acquired by solving similar queries in the training phase. So, after the training has been performed and its effectiveness evaluated, the AIRS is tested to see if it is able to generalise the associations learned and to respond correctly for the remaining part of the examples. The solutions provided by the AIRS are compared to the solutions provided by domain experts. After this has been done, the effectiveness of the AIRS is tested against the effectiveness of classical system, which is based on the evaluation of similarity between the original user query and the documents. This in order to assure that the solutions provided by the AIRS are also better than those provided by methods which make no use of application domain knowledge.

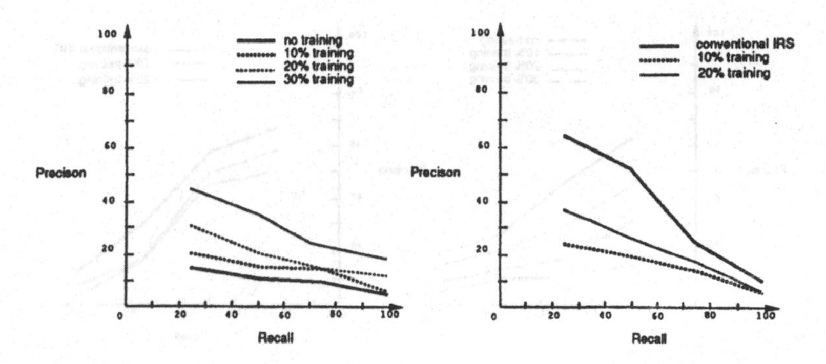

Figure 2: Generalization and retrieval performance of Total Learning

Three different learning strategies were employed in the experimentation, resulting in three different ways of training the AIRS. The first and second kinds of learning are called *Total Learning* (TL) and *Horizontal Learning* (HL). This is because the application domain knowledge is learned by training the AIRS using examples spanning over the entire application domain at the same time. The third kind of learning is called *Vertical Learning* (VL), because it tends to build up knowledge going straight on a specialisation dimension, without bothering to build a large base.

In the following a brief summary of the characteristics and performance of these three learning strategies is reported.

Total learning

In this learning strategy a single training example was made up of a query and a document known, from the relevance assessment, to be relevant to that query. The set of training examples is made of all documents relevant to a set of queries. Each training example is considered by the NN module of the AIRS as a pattern to be learned. Experimental results concerning generalization and retrieval obtained for various dimensions of the training set are reported in Figure 2.

Horizontal learning

Here a single training example was made up of a query and the cluster representative of the set of documents known to be relevant to that query.

The motivation for performing such an experimentation came from an analysis of the results obtained in the TL. The use of different training examples where the input is constant while the output varies sometimes quite considerably, causes the NN to be subject to too much noise to be able to generalise what it has learned. A possible way of avoiding this problem could be in using some kind of synthesis of the characteristics of the set of documents relevant to a query. This is equivalent to using a single document representation for each query in the training phase, thus having a single different output for each input. This unique document representation should characterise all the relevant documents for a query. The most common way of obtaining such a representation is by clustering the set of documents in order to produce a cluster representative, which summarises and represents the objects in the cluster. A "centroid" cluster representative was determined for each query, according to the intuition that terms occurring more than once in the cluster should be taken into consideration as representative of the cluster. Results obtained evaluating the entire set of queries for various dimensions of the training set are reported in Figure 3.

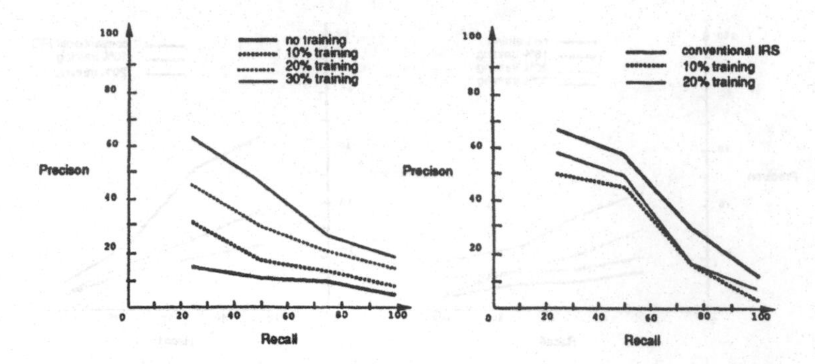

Figure 3: Generalization and retrieval performance of Horizontal Learning

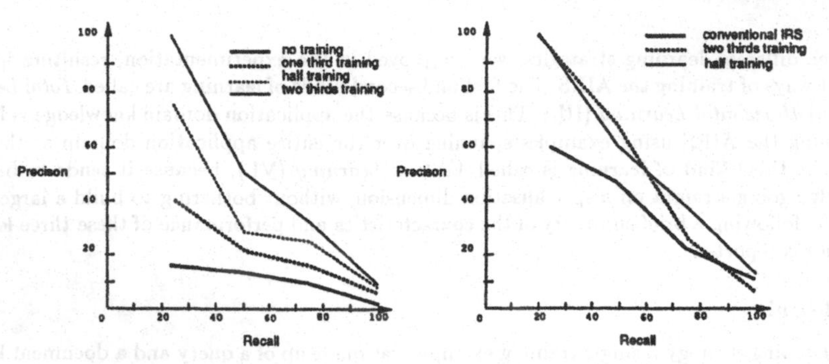

Figure 4: Generalization and retrieval performance of Vertical Learning

Vertical learning

In VL a training example is made of a single query representation and a single relevant document representation. The set is made of only a subset of all the documents known to be relevant to that particular query. Different dimensions of the learning set were used. Experiments are identified as the ratio of the cardinality of the learning set over the entire set of documents known to be relevant to the query. Three illustrative values of the ratio were used: 1/3, 1/2, and 2/3. Results are depicted in Figure 4.

There are various differences between VL and TL. The main one is that VL concerns a single query and uses information about some documents known to be relevant to finding other relevant documents to the same query. The heuristic rule the AIRS uses is: "if these documents are relevant to this query, then these other documents must be relevant too". In this way it is similar to classical IR relevance feedback. In TL the AIRS has to generalise the information about several queries and their relative relevant documents acquired during the training phase to find the proper query adaptation of a new user formulated query. In this case the AIRS uses this heuristic rule: "if these sets of documents were relevant to these queries, than this set of document must be relevant to this query". These two tasks are not mutually exclusive, but they can be thought to work together at different stages of a query session, as previously explained.

5 Conclusions

The query adaptation produced by TL does not give good results. The NN is not able to learn and generalise the characteristics of the application domain knowledge. The amount of information submitted to the system seems to be too much and the system shows a form of "confusion". It is necessary the filter and prearrange the information to be leaned.

Query adaptation produced by IIL gives performance which are similar to those provided by the use of the original query. The interesting thing is that the adapted query is most of the time quite different from its original formulation. Accordingly, the two sets of documents resulting from the use of the original query and the adapted one are sometimes quite different. The adapted query is often able to retrieve relevant documents which the original query is not able to retrieve. This is because the adaptation process determines useful terms which are not specified in the original query, but that are useful for the determination of the set of the document relevant to the information need. Why then there is more or less the same level of performance? What happens is that the AIRS gives too much importance to the domain knowledge acquired in the training phase modifying the query accordingly, but doing so it looses some of the information contained in the original query, that is why some terms specified in the original formulation of the query were not used in the adapted version.

The results produced by VL show that it is easier to learn about a very narrow topic when there is no other knowledge which can interfere with the learning. The advantage of such a result is that it is possible to distinguish two different kind of query: (a) a generic query, in which the user expresses his not well defined information need; (b) a very specific query in which the user is able to point out some document he knows to be relevant. In the first case the query adaptation resulting from HL or a combination of traditional and adaptive retrieval can be used. The result provided by that retrieval should be good enough to enable the user to point out some relevant documents among those retrieved. In the second case the user, either in a process of relevance feedback or in a formulation of the query by giving example of relevant documents (known in IR as "query by example"), can provide a more specific formulation of his information need. In this case the system can use this information to retrieve other relevant documents without taking into account its knowledge of the entire application domain, but focussing only on that particular topic. The two situations are typical of an interactive query session and they can be combined together as it is done in many systems using relevance feedback devices. A major point is that the result of a good query session, ie. a query specification and a set of relevant documents, can be stored and used in training sessions. This process should keep improving the performance of the system's response to the first kind of query. This has not been proved yet and some other experimental investigations are going to be devoted to the analysis of the improvement of the domain knowledge base.

References

[1] C. Claverdon, J. Mills, and M. Keen. *ASLIB Cranfield Research Project: factors determining the performance of indexing systems.* ASLIB, 1966.

[2] Y. Miyata. *A user's guide to PlaNet version 5.6: a tool for constructing, running and looking into a PDP network.* Computer Science Department,, University of Colorado, Boulder, USA, December 1990.

[3] F. Crestani. A network model for Adaptive Information Retrieval. Departmental Research Report 1992/R6, Department of Computing Science, University of Glasgow, Glasgow, UK, April 1992.

[4] C.J. van Rijsbergen. *Information Retrieval.* Second edition, Butterworths, London, 1979.

SOFTWARE PATTERN EEG RECOGNITION AFTER A WAVELET TRANSFORM BY A NEURAL NETWORK

CLOCHON P. , CLARENCON D.*, CATERINI R.* (CA), ROMAN V.*

* INSERM U 320 CHRU Côte de Nâcre,
14033 CAEN Cedex, France

** CRSSA U 18, BP 87
38702 GRENOBLE-LA TRONCHE Cedex, FRANCE

Abstract. The recent development of micro-computers in association with the improvement of data acquisition techniques and signal treatment has made easier the analysis of cerebral electrical activity.

But the methods based on classical harmonic analysis reveal ineffective to detect some activities as epileptiform spike-and-waves of paroxystic origin.

In order to detect those spike-and-waves, we developed a signal treatment based on Morlet's wavelets. This treatment generates a 2-D representation including the time/frequency componants of the EEG signal splitted into 5 seconds spans. In these figures, the spike-and-waves are detected by a neuronal network. The result is then stored into a file, for a delayed use.

I - Introduction

The most commonly used treatment methods of biological signals are based on frequency analysis and power density spectrum (FFT), but unfortunately this frequency representation is not suited to the study of transient events. For example, in EEG analysis, many methods have been developed for detection and quantification of paroxystic activities, using spike and seizure recognition programs, but the analyses based on morphological aspect of these paroxystic events are strictly localized in the time domain. So it is tempting to speculate whether time-frequency analysis would be useful for detection and quantification of paroxystic EEG activities.

The wavelet transform analysis deserves some attention for it allows such time-frequency representation of the signal. It has been used in applied signal processing, and first in geophysics [7]. This method was recently introduced in biological domains [3,8,9] and in the case of EEG [4] we obtained characteristic shapes on analyzing experimental spikes-and-waves.

In these conditions, image analysis can be carried out by comparing expected patterns to given templates [1]. Another way would passibly consist in using neural networks.

As a matter of fact, one of the aims of wavelet transform is to provide an easily interpretable visual representation of signals. This is a prerequisite for further applications, such as pattern recognition [5].

In this way, a neural networks processes the whole time–frequency representation of the recording every 5-second epochs. The network recognizes the typical representative patterns of epileptiform phases occuring during the totality of the experiment and then records the temporal occurences of the detected events. The recording file allows a delayed reconstitution of the unfolding of the signal paroxystic phase.

2. Material and Method

Paroxystic epileptiform events are induced by intra–peritoneal injection of a GABA antagonist (picrotoxine $2 mg.kg^{-1}$) to chronically implanted rats, with cortical electrodes for EEG recording. The EEG signal is amplified and directed to a PC 386 computer. A 12 bits ADC card samples and numerizes the signal at 200 Hz.

We may now formally define the problem of wavelet detection. This operation is carried out in delayed time, from the stored signal.

The chosen wavelet was the same as that written by Morlet. The Morlet's wavelets are complex functions, concentrated in time and frequency, each presenting the same shape and are mainly function of two parameters namely *a* and *b*, but unlike Gabor's wavelet, the *a* parameter quantifies the dilatation (or compression) of the time scale rather than an actual frequency change.

The basic wavelet expression is:

$$w_k(t) = e^{ikt} \left(e^{-t^2/2} - \sqrt{2}\; e^{-k^2/4} . e^{-t^2}\right) \quad (1)$$

k is an arbitrary constant.

According to the value of the constant *k*, we can obtain several shapes of basic wavelets within the same gaussian envelope.

We can generate other analyzing wavelets by dilatation (or compression) in frequency domain using the parameter *a*. Thus, we obtain a class of wavelets (namely a wavelet family) the elements of which all have the same shape and are spread in frequency domain:

$$w_k(t/a) = e^{ikt/a} \left(e^{-t^2/2a^2} - \sqrt{2}\; e^{-k^2/4} . e^{-t^2/a^2}\right) \quad (2)$$

Moreover, the energy of these wavelets ($\int [wk(t/a)]^2\, dt$) has to be constant whatever the value of *a*. This is obtained by multiplying $w_k(t/a)$ by $1/\sqrt{a}$. Thus:

$$w_{k,a}(t) = 1/\sqrt{a}\; [w_k(t/a)] \quad (3)$$

For application of the wavelet transform method to EEG time–frequency analysis, we have to choose an adequate frequency of the basic analysing wavelet of the family, which is called the "mother"; in practice, we fixed this frequency at 10 Hz. This 10 Hz mother wavelet was obtained using the formula (3) where the compression parameter *a* is the ratio (0.08) of the period of the mother wavelet (0.1 sec.) by the period of the basic wavelet $w_k(t)$ (1.25 sec.).

Let us condider a wavelet S(n). The (n) index denotes that the wavelet may appear according to many variations. We denote by S the wavelet which has the general shape of a paroxystic spike and by S(i) the particular, exact, shape of the *ith* paroxystic spike in the given EEG signal. The wavelet S(i) is a function of time. Since we work with digital computers, we have to use the sampled wavelet, namely the string of D_i equally spaced samples of the wavelet.

These will be denoted by :

$$S_k\ (i):\ (k=k_i,k_i+1,\ldots,k_i+D_i-1) \tag{4}$$

Equation (4) has to be understood as follows : the ith sampled wavelet starts at sample number k_i and ends at sample number k_i+D_i-1 (i.e: its duration is D_i samples). We assume that outside the D_i samples the wavelet is zero namely $S_k(i) = 0$ for all $k>k_i+D_i-1$. The signal we are monitoring consists of many wavelets, each of them having a different "gain", G(i). We may therefore write the signal, S_i as the summation of all wavelets:

$$S_k = \Sigma_i\ G(i)S_k(i);\quad (i=1,2\ldots) \tag{5}$$

The signal S_k (represented by its *kth* sample) is therefore the collection of the wavelets as they appear along the time axis.

The root wavelet is determined at 10Hz. The six root-derived wavelets are spread over 3 octaves distributed on both sides of this central frequency (i.e.: 1.25Hz, 2.5Hz, 5Hz, 20Hz, 40Hz and 80Hz). Ten bands per octave give an adequate frequency resolution to the form recognition executed by the neural networks.

3. Image Preprossessing

The wavelet transform enables a time/ frequency image (Fig. 1) of the signal which is analysed in this way. A region of interest is defined between 30 Hz to 80 Hz. Indeed, the variations in frequency caused by wave points create an increase of the high frequencies in comparison to a normal recording.

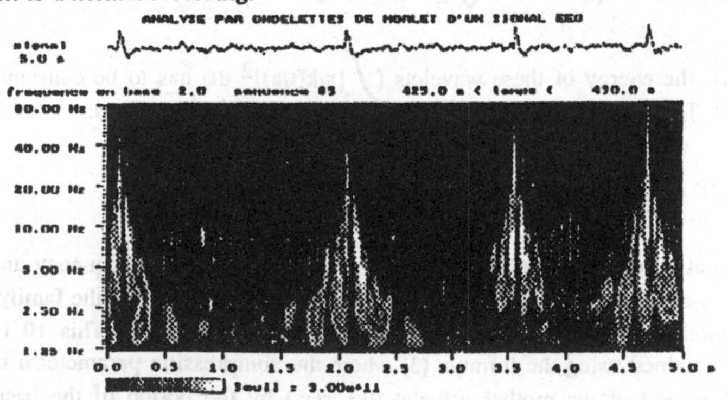

Fig. 1. Time/frequency wavelet transform

3.1 Threshold Image

We carry out a threshold of the image in the region of interest (Fig. 2), in .order to eliminate any possible low amplitude frequency which would not be generated by spikes.

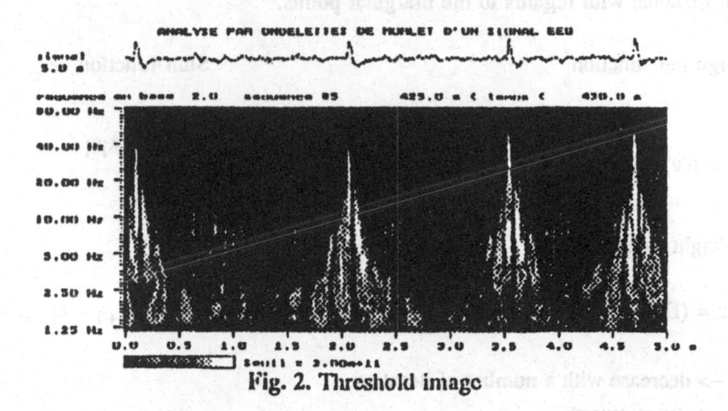

Fig. 2. Threshold image

3.2 Extractions of Parameters

A mean pixel number of 30–80 Hz is calculated using the threshold image. This provides a representation of the time/frequency information between 0 and 1, as well as a lower output of data.

$$X_t = \frac{1}{N} \sum_{i=30Hz}^{80 Hz} \text{Pixel}(t,i) . \qquad N = \text{number of pixels.} \qquad (6)$$

4. Neural Networks

4.1 The task

The construction of neural networks is done automatically by localizing the learning for each neuron. This method consists in taking a neuron, carrying out the learning process on the whole data base containing 2 forms to be classified (form *A* and form *B*). If there is no convergence after N iterations (N=50), the 2 forms are not geometrically separable. The unfiled data is sorted so as to keep the 'well filed' data only. A new learning session is carried out from this data which enables us to obtain a pre-classification for a neuron. The system then creates a new neuron, and ensures the learning process of the data form *A* which was badly filed in comparison to the data of form *B* as a whole. It carries out the same operation as previously (sorting of badly filed data and relearning) to obtain a convergence. The system repeats this once more, recreates a neuron until all the vectors of the form *A* are filed with regards to the form *B*.

The input layer is thus obtained in the same way. This implies that the data which characterizes the 2 classes are separable in an N-dimensional space. For hidden and output layers, it is synthetized by means of logical functions [2, 6]. This method enables a very easy hardware implementation of the system.

4.2 Learning rules

The rule used is the generalized delta rule. Its tranfer function is a sigmoid which guarantees a convergence if data are geometrically separable and places the hyperplane at maximum distance with regards to the marginal points.

Sigmoid function

$$S = f(v) = \frac{e^{\beta v} - 1}{e^{\beta v} + 1}$$

Sum function

$$v = \sum_{i=1}^{N} x_i c_i \qquad (7)$$

Weight adaptation

$$\Delta c = (D - S)\, \delta f(v) / \delta(v) \qquad c_{i+1} = c_i + \mu \Delta c_i$$

(8)

v -> decrease with a number of iteration
D -> want output

During the learning period, we used the crossed-validation rule. This rule consists in dividing the data into n sections. The learning is carried out in *n-1* set and is tried out on the last one. This operation is carried out *n* times.
During the test, the transfer function is replaced by a threshold function, and 100% of the learning examples have been correctly classified.

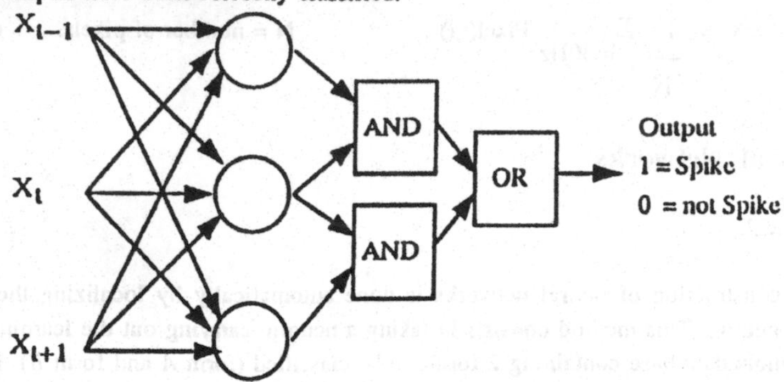

Fig. 3. Neural Networks

5. Results and Conclusion

When only very few data remain to be filled, it is better to duplicate so as to ensure a convergence if the form B contains a large quantity of data.
The classifier is undergoing tests on a larger data base for it to be validated. With this type of network construction we are no longer obliged to resolve the problem of the definition of the number of neurons in a hidden layer.

The results which are obtained in this way are stored in file and the expert analyses the sequences of spikes to discover when they first occurred.

Acknowledgements

This research was supported by grant provided by DRET 91/046.
The authors are grateful to Mrs M. Galonnier for her helpful technical assistance.
We are grateful to Mrs Anita Gomez and Miss Sally Turner for their clerical assistance.

References

1. R. Caterini, J.L. Vernet, G. Delhomme, A. Dittmar : A sofware for pattern recognition of skin potentiel responses. Innov. Techn. Med., 13, 3, 256–268, (1992).

2. P. Clochon, G. Perchey, C. Couque, H. Rebeyrolle, D. Bloyet, P. Etevenon: Automatized classification by neural networks of EEG signals with artifact rejection. 14th annual international conference of the IEEE Engineering in Medecine and Biology Society, Satellite symposium on neuroscience and technology, 51–55 (1992).

3. J.A. Crowe, N.M. Gibson, M.S. Woolfson, M.G. Somek : Wavelet transform as a potential tool for ECG analysis and compression. J Biomed. Eng., 14, 268–272 (1992).

4. P. Gourmelon, D. Clarençon, H. Vignal, J.M. Brun, E. Macioszczyk et L. Fontenil : Intérêt de la transformée en ondelettes dans l'analyse des activités paroxystiques épileptiformes de l'activité électrique cérébrale. S.S.A. Trav. Scient., 11, 295–296 (1990).

5. A. Grossmann, J. Morlet, T. Paul : Transforms associated to square integrable group representations. J. Math. Phys. 27, 2473–2479 (1985).

6. S. Knerr, L. Personnaz, G. Dreyfus, Senior Member, IEEE: Handwritten digit recognition by neural networks with single-layer training. IEEE Transactions on Neural Networks,1992 , in press.

7. J. Morlet, G. Arens, I. Fourgeau, D. Giard : Wave propagation and sampling theory, Geophysics, 47, 203–206 (1982).

8. A.W. Przybyszewski : An analysis of the oscillatory patterns in the central nervous system with the wavelet method. J. Neurosci. Methods, 38, 247–257 (1991).

9. C. Tismer, M. Jobert : The application of wavelet transformation for sleep EEG analysis, Sleep Research, 20A, 518 (1991).

Authors Index

Springer-Verlag and the Environment

Lecture Notes in Computer Science

For information about Vols. 1–610
please contact your bookseller or Springer-Verlag

Vol. 611: M. P. Papazoglou, J. Zeleznikow (Eds.), The Next Generation of Information Systems: From Data to Knowledge. VIII, 310 pages. 1992. (Subseries LNAI).

Vol. 612: M. Tokoro, O. Nierstrasz, P. Wegner (Eds.), Object-Based Concurrent Computing. Proceedings, 1991. X, 265 pages. 1992.

Vol. 613: J. P. Myers, Jr., M. J. O'Donnell (Eds.), Constructivity in Computer Science. Proceedings, 1991. X, 247 pages. 1992.

Vol. 614: R. G. Herrtwich (Ed.), Network and Operating System Support for Digital Audio and Video. Proceedings, 1991. XII, 403 pages. 1992.

Vol. 615: O. Lehrmann Madsen (Ed.), ECOOP '92. European Conference on Object Oriented Programming. Proceedings. X, 426 pages. 1992.

Vol. 616: K. Jensen (Ed.), Application and Theory of Petri Nets 1992. Proceedings, 1992. VIII, 398 pages. 1992.

Vol. 617: V. Mařík, O. Štěpánková, R. Trappl (Eds.), Advanced Topics in Artificial Intelligence. Proceedings, 1992. IX, 484 pages. 1992. (Subseries LNAI).

Vol. 618: P. M. D. Gray, R. J. Lucas (Eds.), Advanced Database Systems. Proceedings, 1992. X, 260 pages. 1992.

Vol. 619: D. Pearce, H. Wansing (Eds.), Nonclassical Logics and Information Proceedings. Proceedings, 1990. VII, 171 pages. 1992. (Subseries LNAI).

Vol. 620: A. Nerode, M. Taitslin (Eds.), Logical Foundations of Computer Science – Tver '92. Proceedings. IX, 514 pages. 1992.

Vol. 621: O. Nurmi, E. Ukkonen (Eds.), Algorithm Theory – SWAT '92. Proceedings. VIII, 434 pages. 1992.

Vol. 622: F. Schmalhofer, G. Strube, Th. Wetter (Eds.), Contemporary Knowledge Engineering and Cognition. Proceedings, 1991. XII, 258 pages. 1992. (Subseries LNAI).

Vol. 623: W. Kuich (Ed.), Automata, Languages and Programming. Proceedings, 1992. XII, 721 pages. 1992.

Vol. 624: A. Voronkov (Ed.), Logic Programming and Automated Reasoning. Proceedings, 1992. XIV, 509 pages. 1992. (Subseries LNAI).

Vol. 625: W. Vogler, Modular Construction and Partial Order Semantics of Petri Nets. IX, 252 pages. 1992.

Vol. 626: E. Börger, G. Jäger, H. Kleine Büning, M. M . Richter (Eds.), Computer Science Logic. Proceedings, 1991. VIII, 428 pages. 1992.

Vol. 628: G. Vosselman, Relational Matching. IX, 190 pages. 1992.

Vol. 629: I. M. Havel, V. Koubek (Eds.), Mathematical Foundations of Computer Science 1992. Proceedings. IX, 521 pages. 1992.

Vol. 630: W. R. Cleaveland (Ed.), CONCUR '92. Proceedings. X, 580 pages. 1992.

Vol. 631: M. Bruynooghe, M. Wirsing (Eds.), Programming Language Implementation and Logic Programming. Proceedings, 1992. XI, 492 pages. 1992.

Vol. 632: H. Kirchner, G. Levi (Eds.), Algebraic and Logic Programming. Proceedings, 1992. IX, 457 pages. 1992.

Vol. 633: D. Pearce, G. Wagner (Eds.), Logics in AI. Proceedings. VIII, 410 pages. 1992. (Subseries LNAI).

Vol. 634: L. Bougé, M. Cosnard, Y. Robert, D. Trystram (Eds.), Parallel Processing: CONPAR 92 – VAPP V. Proceedings. XVII, 853 pages. 1992.

Vol. 635: J. C. Derniame (Ed.), Software Process Technology. Proceedings, 1992. VIII, 253 pages. 1992.

Vol. 636: G. Comyn, N. E. Fuchs, M. J. Ratcliffe (Eds.), Logic Programming in Action. Proceedings, 1992. X, 324 pages. 1992. (Subseries LNAI).

Vol. 637: Y. Bekkers, J. Cohen (Eds.), Memory Management. Proceedings, 1992. XI, 525 pages. 1992.

Vol. 639: A. U. Frank, I. Campari, U. Formentini (Eds.), Theories and Methods of Spatio-Temporal Reasoning in Geographic Space. Proceedings, 1992. XI, 431 pages. 1992.

Vol. 640: C. Sledge (Ed.), Software Engineering Education. Proceedings, 1992. X, 451 pages. 1992.

Vol. 641: U. Kastens, P. Pfahler (Eds.), Compiler Construction. Proceedings, 1992. VIII, 320 pages. 1992.

Vol. 642: K. P. Jantke (Ed.), Analogical and Inductive Inference. Proceedings, 1992. VIII, 319 pages. 1992. (Subseries LNAI).

Vol. 643: A. Habel, Hyperedge Replacement: Grammars and Languages. X, 214 pages. 1992.

Vol. 644: A. Apostolico, M. Crochemore, Z. Galil, U. Manber (Eds.), Combinatorial Pattern Matching. Proceedings, 1992. X, 287 pages. 1992.

Vol. 645: G. Pernul, A M. Tjoa (Eds.), Entity-Relationship Approach – ER '92. Proceedings, 1992. XI, 439 pages, 1992.

Vol. 646: J. Biskup, R. Hull (Eds.), Database Theory – ICDT '92. Proceedings, 1992. IX, 449 pages. 1992.

Vol. 647: A. Segall, S. Zaks (Eds.), Distributed Algorithms. X, 380 pages. 1992.

Vol. 648: Y. Deswarte, G. Eizenberg, J.-J. Quisquater (Eds.), Computer Security – *ESORICS* 92. Proceedings. XI, 451 pages. 1992.

Vol. 649: A. Pettorossi (Ed.), Meta-Programming in Logic. Proceedings, 1992. XII, 535 pages. 1992.